Birkhäuser

Advances in Mathematical Fluid Mechanics

Series editors

Advances in Mathematical Fluid Mechanics is a forum for the publication of high quality monographs, or collections of works, on the mathematical theory of fluid mechanics, with special regards to the Navier-Stokes equations. Its mathematical aims and scope are similar to those of the *Journal of Mathematical Fluid Mechanics*. In particular, mathematical aspects of computational methods and of applications to science and engineering are welcome as an important part of the theory. So also are works in related areas of mathematics that have a direct bearing on fluid mechanics.

More information about this series at http://www.springer.com/series/5032

Tomáš Bodnár • Giovanni P. Galdi • Šárka Nečasová
Editors

Fluids Under Pressure

Birkhäuser

Editors
Tomáš Bodnár
Department of Technical Mathematics
Czech Technical University
Prague, Czech Republic

Giovanni P. Galdi
Department of Mechanical Engineering
University of Pittsburgh
Pittsburgh, PA, USA

Šárka Nečasová
Academy of Sciences of the Czech Republic
Prague, Czech Republic

ISSN 2297-0320 ISSN 2297-0339 (electronic)
Advances in Mathematical Fluid Mechanics
ISBN 978-3-030-39638-1 ISBN 978-3-030-39639-8 (eBook)
https://doi.org/10.1007/978-3-030-39639-8

Mathematics Subject Classification: 35Q30, 35Q35, 65M08, 76A02, 76D03, 76D05, 80M10

This book is published under the imprint Birkhäuser, www.birkhauser-science.com by the registered company Springer Nature Switzerland AG.
The registered company address is: Gewerbestrasse 11, 6330 Cham, Switzerland

Preface

This book is based on lectures presented at a thematic summer school named "Fluids under Pressure," which was held in Prague in August 2016 in the series "Prague-Sum," started back in 2011. The aim of this monograph is to cover various roles of pressure in physics as well as in mathematical modeling and analysis of fluids flow problems. The pressure is a common denominator in all the chapters of the book. Besides several theoretical problems concerning the well-posedness of the Navier-Stokes equations, some chapters are devoted to the role of pressure in finite element and finite volume methods and their Computational Fluid Dynamics (CFD) applications. All the chapters are written by world-renowned experts in the corresponding fields, which makes this volume an excellent summary of state-of-the-art knowledge in this area.

The summer school as well as the present book was made possible due to generous direct support from Czech Academy of Sciences (special project to support research and educational activities for young people), Institute of Mathematics (institutional support Research Plan RVO 67985840), European Research Center for Flow Turbulence and Combustion (ERCOFTAC), and Czech Science Foundation under the project No. 16-03230S.

This book is dedicated to the loving memory of Christian Simader, our loving friend and excellent mathematician.

Prague, Czech Republic — Tomáš Bodnár
Pittsburgh, PA, USA — Giovanni P. Galdi
Prague, Czech Republic — Šárka Nečasová

Chapter 1
An Approach to the Primitive Equations for Oceanic and Atmospheric Dynamics by Evolution Equations

Matthias Hieber and Amru Hussein

Abstract The primitive equations for oceanic and atmospheric dynamics are a fundamental model for many geophysical flows. In this chapter we present a summary of an approach to these equations based on the theory of evolution equations. In particular, we discuss the hydrostatic Stokes operator, well-posedness results in critical spaces within the $L^p(L^q)$-scale, within the scaling invariant space $L^\infty(L^1)$ for Neumann boundary conditions, and within the $L^\infty(L^p)$ space for mixed boundary conditions and $p > 3$. In addition, we investigate real analyticity of solutions, convergence of the scaled Navier-Stokes equations to the primitive equations, and the existence of periodic solutions for large forces.

1.1 Introduction

The primitive equations for oceanic and atmospheric dynamics are a fundamental model for many geophysical flows. This set of equations is derived from the Navier-Stokes or Boussinesq equations for viscous flows by assuming that the vertical motion can be approximated by the hydrostatic balance. This means in the isothermal setting, roughly speaking, that the pressure depends linearly on the height.

The aim of this chapter is to discuss various analytical aspects of these equations and to give an overview over recent results obtained by the authors and their co-authors, Yoshikazu Giga, Mathis Gries, Takahito Kashiwabara and Marc Wrona, see [24, 28, 29, 34, 35]. Here, we present only an overview about the results obtained and the methods being used; complete proofs can be found in the articles cited above. In this article we consider the primitive equations from the point of view of evolution equations and regard it as a semi-linear parabolic evolution equation in certain solenoidal subspaces of L^p-spaces. We thus use mainly techniques from the

M. Hieber (✉) · A. Hussein
Technische Universität Darmstadt, Darmstadt, Germany
e-mail: hieber@mathematik.tu-darmstadt.de

T. Bodnár et al. (eds.), *Fluids Under Pressure*, Advances in Mathematical Fluid Mechanics, https://doi.org/10.1007/978-3-030-39639-8_1

theory of evolution equations as well as maximal L^q-L^p-regularity estimates for results within the L^p-L^q-setting for $1 < p, q < \infty$. Additionally, we are interested in the situation of bounded initial data and consider hence the primitive equations, depending on the boundary conditions chosen, then in the scaling invariant space $L^\infty(L^1)$ as well as in $L^\infty(L^p)$ for $p > 3$, respectively.

Our main focus is to prove global strong well-posedness results for these equations for various types of arbitrary large initial values. Note that the uniqueness of weak solutions for initial data in L^2 constitutes an open problem until today, while global well-posedness for initial data in H^1 is well-established. So, taking into account the weak–strong uniqueness principle, well-posedness results for "rough" initial data and strong solutions enlarge the spaces of initial values for which uniqueness holds towards the L^2-setting.

In addition to questions of well-posedness, regularity properties of solutions are investigated in some detail. Furthermore, the existence of periodic solutions for arbitrary large periodic forces is being discussed. Finally, we analyze the convergence of the solutions of scaled Navier-Stokes equations to solutions of the primitive equations, hereby giving a rigorous justification of the hydrostatic approximation in the L^p-L^q-setting.

This chapter is organized as follows: first we state the primitive equations subject to various boundary conditions and introduce our functional analytic setting. After investigating the linear hydrostatic Stokes equation and the associated hydrostatic Stokes semigroup we present in the following sections local and global well-posedness results. Results on periodic solutions as well as a rigorous justification of the hydrostatic approximation finish this chapter.

1.1.1 The Primitive Equations

To begin with, we state the three-dimensional primitive equations in the isothermal setting, neglecting gravity and Coriolis force, modeling the velocity of the fluid

$$u\colon \Omega \to \mathbb{R}^3, \quad u = (v, w), \quad \text{where } v = (v_1, v_2)$$

denotes the horizontal component and w the vertical one, and the pressure

$$\pi\colon \Omega \to \mathbb{R},$$

where $\Omega \subset \mathbb{R}^3$. The primitive equations are given by

$$\begin{cases} \partial_t v + u \cdot \nabla v - \Delta v + \nabla_H \pi = 0, & \text{in } \Omega \times (0, T), \\ \partial_z \pi = 0, & \text{in } \Omega \times (0, T), \\ \operatorname{div} u = 0, & \text{in } \Omega \times (0, T), \\ v(0) = a. & \end{cases} \tag{PE}$$

Here we consider the model situation of a cylindrical domain

$$\Omega := G \times J,$$

where $J = (-h, 0)$ or $J = (z_0, z_1)$ is a general bounded interval, $G = (0, 1) \times (0, 1)$ and in some situations $G = \mathbb{R}^2$, and $T \in (0, \infty]$. Denoting the horizontal coordinates by $x, y \in G$ and the vertical one by $z \in J$, we use the notation $\nabla_H = (\partial_x, \partial_y)^T$, whereas Δ denotes the three-dimensional Laplacian, ∇ and div the three-dimensional gradient and divergence operators, respectively. The equations (PE) are supplemented by boundary conditions on

$$\Gamma_u = G \times \{0\}, \quad \Gamma_b = G \times \{-h\} \quad \text{and} \quad \Gamma_l = \partial G \times (-h, 0),$$

i.e. the upper, bottom, and lateral parts of the boundary $\partial\Omega$, respectively, given by

$$\begin{aligned} &v, w, \pi \text{ are periodic on } \Gamma_l \times (0, \infty), \\ &w = 0 \text{ on } \Gamma_b \cup \Gamma_u \times (0, \infty), \\ &v = 0 \text{ on } \Gamma_D \times (0, \infty) \quad \text{and} \quad \partial_z v = 0 \text{ on } \Gamma_N \times (0, \infty), \end{aligned}$$

where Dirichlet, Neumann, and mixed boundary conditions are comprised by the notation

$$\Gamma_D \in \{\emptyset, \Gamma_u, \Gamma_b, \Gamma_u \cup \Gamma_b\} \quad \text{and} \quad \Gamma_N = (\Gamma_u \cup \Gamma_b) \setminus \Gamma_D.$$

Note that the divergence free condition comprises not only $\nabla \cdot u = 0$ but also $n \cdot v = 0$, where n is the outer normal vector on the boundary. This boundary condition translates in this cylindrical setting to $w = 0$ on top and bottom. The boundary conditions on v have different physical interpretations.

1.1.2 Primitive Equations Compared to Navier-Stokes Equations

The primitive equations are derived from the Navier-Stokes equations, where the evolution equation for the vertical velocity w is reduced to $\partial_z \pi = 0$. This assumption has strong implications on the analysis of these equations and before going into the details, it should be recalled that the three-dimensional primitive equations admit a unique, global, strong solution for arbitrary large data in H^1. This breakthrough result was proved by Cao and Titi [14] in 2007 using energy methods to prove a suitable *a priori* bound in

$$L^\infty(0, T; H^1(\Omega)) \cap L^2(0, T; H^2(\Omega)) \quad \text{for any } T > 0.$$

The question whether such results for Navier-Stokes equations hold or not is open. Nevertheless, the primitive equations are closely related to the Navier-Stokes equations. First by the derivation of the primitive equations from the Navier-Stokes equations by the hydrostatic approximation, second by its structure as a semi-linear parabolic evolution equation with quadratic non-linearity, and eventually the role of the pressure is similar in both sets of equations.

Comparing both equations, one observes that one major difference is that the primitive equations have a particularly anisotropic structure, that is, they behave differently with respect to horizontal and vertical coordinates, while Navier-Stokes equations have an isotropic structure where all directions can be interchanged. Furthermore, by the hydrostatic approximation, the evolution equation for the horizontal velocity w cancels out and is replaced by $\partial_z \pi = 0$, so that the vertical velocity $w = w(v)$ is completely determined by the horizontal velocity by the condition $\nabla \cdot u = 0$ and its boundary conditions to be

$$w(v)(x, y, z) = -\int_{-h}^{z} \mathrm{div}_H v(x, y, \xi) d\xi, \quad \text{where} \quad \mathrm{div}_H v = \partial_x v_1 + \partial_y v_2.$$

Therefore, in the non-linearity the term $w(v)\partial_z v$ has derivatives in both arguments. Note that using the divergence free condition, the whole non-linearity can be written as $\nabla \cdot (u(v) \otimes v)$, while for Navier-Stokes equations, one has the representation $\nabla \cdot (u \otimes u)$. This is relevant for the respective weak formulations.

Moreover, using the zero boundary conditions of w, the divergence free condition can be replaced by a condition on the vertical average of the horizontal velocity

$$\mathrm{div}_H \overline{v} = 0, \quad \text{where} \quad \overline{v}(x, y) = \frac{1}{h} \int_{-h}^{0} v(x, y, \xi) d\xi.$$

Another important feature of this anisotropic structure is that the pressure in the primitive equations is actually a function of only two variables, this means that the full pressure is determined by the surface pressure. It has turned out that this is the entrance point for the proof of suitable *a priori* bounds. More precisely, setting

$$v = \overline{v} + \tilde{v}, \quad \text{where} \quad \tilde{v} = v - \overline{v},$$

allows one to decompose the primitive equations with respect to the projection defined by taking the vertical average $\overline{v}$ and its complementary projection $\tilde{v}$, and one obtains the equations

$$\begin{aligned}
\partial_t \bar{v} - \Delta_H \bar{v} + \nabla_H \pi &= -\bar{v} \cdot \nabla_H \bar{v} - \frac{1}{h} \int_{-h}^{0} (\tilde{v} \cdot \nabla_H \tilde{v} + \tilde{v} \mathrm{div}_H \tilde{v})\, dz, \\
\mathrm{div}_H \bar{v} &= 0, \\
\bar{v}(0) &= \bar{a},
\end{aligned}$$

and

$$\partial_t \tilde{v} - \Delta \tilde{v} + \tilde{v} \cdot \nabla_H \tilde{v} + v_3 v_z + \bar{v} \cdot \nabla_H \tilde{v} = -\tilde{v} \cdot \nabla_H \bar{v} + \frac{1}{h} \int_{-h}^{0} (\tilde{v} \cdot \nabla_H \tilde{v} + \tilde{v} \operatorname{div}_H \tilde{v})\, dz,$$
$$\tilde{v}(0) = \tilde{a}.$$

This decomposition into a system of coupled 2-D Navier-Stokes equations involving the pressure term and 3-D nonlinear heat equations without pressure suggests that the 3-D primitive equations can be interpreted as an intermediate model between 2-D and 3-D Navier-Stokes equations.

The evolution equation approach in L^p-spaces for the Navier-Stokes equations has given many relevant results identifying initial conditions which guarantee local strong well-posedness or global strong well-posedness for small data. In this context critical spaces and invariance under the parabolic scaling

$$v_\lambda(t, x_1, x_2, x_3) = \lambda v(\lambda^2 t, \lambda(x_1, x_2, x_3)), \quad \lambda > 0, \tag{1.1}$$

play an important role. For the three-dimensional Navier-Stokes equations, Giga [26] and Kato [46] proved well-posedness for initial values in the scaling invariant space L^3_σ using semigroup methods. Another critical function space—the Besov space $B^{3/p-1}_{pq}$—has been introduced by Cannone [9] considering the full space $\mathbb{R}^3$, and Prüss and Wilke considered similar spaces for general semi-linear evolution equations, see [68], which applies in particular to Navier-Stokes equations on bounded domains. For these results one has to take into account general L^p-spaces rather than the classical L^2-setting, where due to the classical work by Fujita and Kato, see [23], initial values have been obtained which are roughly speaking in $H^{1/2}$.

These achievements inspired our co-authors and us to analyze the primitive equations from the viewpoint of evolution equation in L^p-spaces with the aim to obtain well-posedness results for initial conditions with little regularity, that is, for "rough data." In addition, it turns out that also many known results can be re-obtained in a very transparent and straightforward way using evolution equation methods.

1.1.3 Main Results

Most of the results concerning the analysis of the primitive equations are based on energy methods. In particular, the seminal result by Cao and Titi [14] on strong global well-posedness of these equations relies on these methods. In this article we present a different approach to the primitive equations based on the theory of evolution equations. In a certain sense the approach presented here parallels the one initiated by Fujita, Kato and Giga for the Navier-Stokes equations. In order to do so, we introduce first the hydrostatic Helmholtz projection allowing to write the primitive equation as an evolution equation in $L^p_{\overline{\sigma}}(\Omega)$, the hydrostatic solenoidal subspace of $L^p(\Omega)$, see [35]. Of central importance is then the hydrostatic Stokes operator A introduced in Sect. 1.5, which is shown to have the property of maximal L^p-L^q-regularity, see Sect. 1.5. In particular, A generates a bounded holomorphic semigroup. We then regard the primitive equations as a semi-linear evolution equation in the space $L^p_{\overline{\sigma}}(\Omega)$ and present several approaches for local and global strong well-posedness of the primitive equations, see [27, 35]. In particular, we obtain in Sect. 1.9 well-posedness results in critical spaces within the L^p-L^q-scale for $1 < p, q < \infty$.

We then turn over attention to the situation of the scaling invariant space $L^\infty(\mathbb{R}^2; L^1(J))$ and prove that the primitive equations subject to Neumann boundary conditions are strongly globally well-posed for horizontally periodic initial data of the form $a = a_1 + a_2$ with $\nabla \cdot_H \overline{a_1} = \nabla \cdot_H \overline{a_2} = 0$ and

$$a_1 \in BUC(\mathbb{R}^2, L^1(J)) \text{ and } a_2 \text{ being a small perturbation in } L^\infty(\mathbb{R}^2; L^1(J)).$$

Note that this space is invariant under the parabolic scaling (1.1). The results in Sects. 1.10, 1.9 can be hence seen as the counterpart of the classical results by Giga [26] and Kato [46] for the Navier-Stokes equations with initial values in L^3_σ. This result depends on L^∞-estimates for

$$e^{t\Delta_H} R_i R_j (-\Delta_H)^{-\alpha/2} \quad \text{and} \quad e^{t\Delta_H} R_i R_j \partial_k, \quad i, j \in \{1, 2\},$$

where $R_i = \partial_i \Delta_H^{-1/2}$ with $\Delta_H = \partial_x^2 + \partial_y^2$ are the two-dimensional Riesz transforms.

For the case of Dirichlet boundary conditions the analysis is more involved and described in Sect. 1.11. Here we obtain global strong well-posedness for data of the same type, but $L^1(J)$ is being replaced by $L^p(J)$ for $p \in (3, \infty)$. The main difficulty when dealing with the primitive equations on spaces of bounded functions is that the hydrostatic Helmholtz projection $\mathbb{P}$ *fails to be bounded* with respect to the L^∞-norm. We show that nevertheless the combination of the three ingredients $\nabla, \mathbb{P}, e^{tA}$ give rise to bounded operators on $L^\infty_H L^p_z(\Omega)$ for $p > 3$, which satisfy for $t > 0$ typical second order parabolic estimates of the form

$$t^{1/2} \|\partial_i e^{tA} \mathbb{P} f\|_{L^\infty_H L^p_z(\Omega)} \leq C e^{t\beta} \|f\|_{L^\infty_H L^p_z(\Omega)},$$

$$t^{1/2} \|e^{tA} \mathbb{P} \partial_j f\|_{L^\infty_H L^p_z(\Omega)} \leq C e^{t\beta} \|f\|_{L^\infty_H L^p_z(\Omega)},$$

$$t \|\partial_i e^{tA} \mathbb{P} \partial_j f\|_{L^\infty_H L^p_z(\Omega)} \leq C e^{\beta t} \|f\|_{L^\infty_H L^p_z(\Omega)},$$

where $\partial_i, \partial_j \in \{\partial_x, \partial_y, \partial_z\}$.

Second, concerning the regularity of solution, we notice that the primitive equations are subject to parabolic smoothing and thus they regularize to become even C^∞ for $t > 0$. Such a result is—more or less—part of the folklore. In contrast, real-analyticity in time and space is usually rather difficult to prove. Using methods from maximal L^q-regularity, we prove analyticity of solutions provided the linear part of the primitive equations has maximal L^q-regularity. The latter property and more generally, the existence of a bounded $\mathcal{H}^\infty$-calculus for the hydrostatic Stokes operator can be proved as follows: since the linearized problem is given by

$$\begin{cases} \partial_t v - \Delta v + \nabla_H \pi_s = f, \\ \nabla \cdot_H \overline{v} = 0, \\ v(0) = v_0. \end{cases}$$

we may solve first for the surface pressure to obtain a representation of the hydrostatic Stokes operator as

$$Av = \Delta v + \nabla_H \Delta_H^{-1} \mathrm{div}_H \tfrac{1}{h} \left(\partial_z v \mid_{\Gamma_u} - \partial_z v \mid_{\Gamma_b} \right).$$

Third, the convergence of the solutions of scaled Navier-Stokes equations to solutions of the primitive equations can be obtained in an elegant way by methods from evolution equation. Furthermore, these methods allow to prove convergence results in the L^q_t-L^p_x-norm and give a rigorous justification of the hydrostatic balance assumption.

Finally, we prove the existence of periodic solutions for *large* forces. This can be seen as a special feature of the primitive equations since for many semi-linear evolution equations period solutions are known to exist only under a smallness condition. Our approach is based on Brouwer's fixed point theorem for the construction of a—possibly non-unique—weak periodic solution. the global *a priori* bound and a weak–strong uniqueness argument. The uniqueness of these periodic solutions is related to the uniqueness of weak solutions for initial data in L^2.

1.1.4 Related Literature

The mathematical analysis of the primitive equations was pioneered by Lions, Teman, and Wang in 1992 by a series of articles [56–58]. There, modeling of the primitive equations of the ocean, the atmosphere and the coupled model for atmosphere and ocean was elaborated in detail and the existence of a global weak solutions for initial data in L^2 was proven. Let us emphasize that the uniqueness problem for weak solutions, as in the case of the Navier-Stokes equations, remains open until today.

The analysis of the linearized problem goes back to the work of Ziane [76, 77]. He proved H^2-regularity of the corresponding resolvent problem. Taking advantage of this result, the existence of a local, strong solution with data $a \in H^1$ was proved first by Guillén-González, Masmoudi, and Rodiguez-Bellido in [32]. They used a Galerkin scheme for the construction of this solution.

A landmark in the mathematical analysis of these equations has been reached in 2007 by Cao and Titi [14]. They proved a breakthrough result for this set of equation, which, roughly speaking, says that there exists a unique, global strong solution to the primitive equations for *arbitrarily large* initial data $a \in H^1$. Cao and Titi considered in [14] the case of Neumann conditions on the bottom and top of the underlying domain. Kukavica and Ziane also considered in [50, 51] the primitive equations subject to Dirichlet boundary conditions on the bottom. They proved global strong well-posedness of the primitive equations with respect to arbitrarily large H^1-data also for this case. For a different approach which needs initial data in H^2, see also Kobelkov [47]. The existence of a global attractor for the primitive equations was proved by Ju [44] and its properties were investigated by Chueshov [17].

Two major mathematical questions related to the primitive equations are open until today: they are the uniqueness question for weak solutions for initial data in L^2 and the well-posedness of the inviscid Euler-type primitive equations. Concerning the uniqueness of weak solutions, we mention here the work of Bresch, Kazhikhov, and Lemoine [8], who proved the uniqueness of weak solutions in the two-dimensional setting for initial data a with $\partial_z a \in L^2$. The existence of a global, strong solution in the two-dimensional setting was proved by the same authors as well as by Petcu, Teman, and Ziane in [65, Section 3.4]. The problem was revisited recently by Kukavica, Pei, Rusin, and Ziane in [49]. The authors show uniqueness of weak solutions under the assumption that the initial data are only continuous in the space variables. Recent regularity results for weak solutions by Li and Titi [53], Ju [45] and Kukavica, Pei, Rusin, and Ziane [49] are also pointing in this direction. More specifically, starting from a weak solution to the primitive equations, these authors investigated regularity criteria for weak solutions for the primitive equations, following hereby in a certain sense the spirit of Serrin's condition in the theory of the Navier-Stokes equations and methods of weak–strong uniqueness. Li and Titi proved in [53] that weak solutions are unique for initial values in C^0 or in $\{v \in L^6 : \partial_z v \in L^2\}$ including a small perturbation belonging to L^∞. By the weak–strong uniqueness property, it follows that these weak solutions regularize and become strong solutions. Ju included the case $\{v \in L^2 : \partial_z v \in L^2\}$, cf. [45].

For the inviscid Euler-type primitive equations, finite time blow-up for some initial data has been proven in the three-dimensional setting by Cao, Ibrahim, Nakanishi, and Titi [10] and by Wong [75] in two-dimensional case. Local existence results have been proven earlier by Masmoudi and Wong in [60] and Kukavica, Masmoudi, Vicol, and Wong [48]. Ill-posedness results for Sobolev spaces have been obtained by Han-Kwan and Nguyen [33].

Also, important modifications of the primitive equations dealing with either only horizontal viscosity and diffusion or with horizontal or vertical eddy diffusivity were recently investigated in a series of articles by Cao and Titi in [15] and by Cao, Li, and Titi in [11–13]. Here, global well-posedness results are established for initial data belonging to H^2. For related results using different methods, we refer to Saal and [70] and Saal et al. [42].

The question of the convergence of the scaled Navier-Stokes equations to the primitive equations has been investigated recently by Li and Titi [55]. They proved convergence within the L^2_t-L^2_x-setting by energy methods.

Recently, the coupling to moisture and its analysis has come into focus, see [18, 38] and the references given therein. The model studied in [18] involves some Heaviside functions and it is treated using variational methods.

The free boundary value problem for the primitive equations, either for the ocean and atmosphere separately or for the coupling of both, is of high practical relevance. Starting with the work by Lions, Temam, and Wang [56] in the weak setting there have been recent results by Honda and Tani [39] and Ignatova, Kukavica, and Ziane [43].

The primitive equations with linearly growing data arising naturally from a rotating coordinate frame have been considered by Saal, Sawada, and the second author in [41]. Full and partial hyper-viscosity has been considered by the second author in [40].

For a survey of known results and further references to the literature, we refer to the recent article by Li and Titi [54]. For introductory works for the primitive equations, we refer to the works of Majda [59], Pedlosky [64], Vallis [73], Washington-Parkinson [74], and Petcu, Temam, and Ziane [65].

1.1.5 Structure of the Chapter

This chapter is structured as follows: Sect. 1.2 gives a short introduction on how the primitive equations are derived. Section 1.3 discusses the cylindrical model situation and introduces basic function spaces.

In Sect. 1.4 we derive a fundamental $L^2(0, T; H^2(\Omega)) \cap H^1(0, T; L^2(\Omega))$ *a priori* estimate. The *a priori* bound is discussed already at the beginning of this chapter, since this estimate is particular for the primitive equations and it furnishes furthermore a good understanding of the structure of these equations. The *a priori* bound leads to the key insight that solutions to the primitive equations cannot blow up in finite time in the maximal L^2-L^2-regularity norm. Moreover, most of the estimates need little preparations and therefore this estimate is a very good starting point for our analysis of the primitive equations.

In Sect. 1.5 a wide range of problems related to the linearized primitive equations in isotropic L^p-spaces are addressed. For the evolution equation approach to semi-linear equations a profound understanding of the linearization and its mapping properties is crucial. Estimates for the linearized problem in anisotropic $L^\infty((0, 1)^2; L^p(-h, 0))$ spaces are being discussed in Sects. 1.6 and 1.7, for Neumann and Dirichlet boundary conditions, respectively.

The information on the linearized problem allows us to construct unique, local strong solutions in Sects. 1.8 and 1.10. The Fujita-Kato approach as well as the maximal L^q-regularity approach is presented in Sect. 1.8. Iteration schemes inspired by the classical iteration schemes for the Navier-Stokes equations due to Kato and Giga are given for the case of anisotropic spaces and for the case of mixed Dirichlet-Neumann boundary conditions, respectively. These local solutions are extended to global solutions using the previously derived *a priori* bound in Sects. 1.9 and 1.11. Additional regularity properties are given as well.

The existence of time-periodic solutions for large forces is addressed in Sect. 1.12. In Sect. 1.13 the full primitive equations including temperature and salinity are discussed. The rigorous justification of the hydrostatic balance assumption is investigated in Sect. 1.14.

1.2 Modeling: Conservation Laws and Approximations

The primitive equations for oceanic and atmospheric dynamics are considered to be a fundamental model for many geophysical flows. The primitive equations of the ocean are described by a system of equations which are derived from the Navier-Stokes or Boussinesq equations for incompressible viscous flows by assuming that the vertical motion is modeled by the hydrostatic balance. This assumption is considered to be valid since for large scale geophysical flows the vertical scale is much smaller compared to the horizontal one.

Here we consider for simplicity and brevity only the modeling of the primitive equations of the ocean. The primitive equations of the atmosphere are derived differently, since the atmosphere consists of compressible gas. However, surprisingly after a coordinate change in vertical direction the resulting primitive equations can be described just as a compressible fluid. To this end, taking for granted that by the hydrostatic approximation the pressure is strictly decreasing with the altitude, the pressure can be taken as new vertical coordinate.

This very brief introduction is borrowed from the much more detailed outline by Petcu, Temam, and Ziane, cf. [65], which refers also to the original works by Lions, Temam, and Wang, see [56–58].

1.2.1 Conservation Laws

So, to start with, consider the following conservation laws. Assume that we are in the three-dimensional space, where we denote by x, y the horizontal coordinates and z for the vertical coordinate. First, conservation of momentum gives

$$\rho \frac{du}{dt} + 2\rho\omega \times u + \nabla\pi + \rho\mathbf{g} = D.$$

Here $u = (v, w)\colon \mathbb{R}^3 \to \mathbb{R}^3$ denotes the velocity field with

$$v = (v_1, v_2) \quad \text{the horizontal velocity and} \quad w \quad \text{the vertical velocity,}$$

$\rho\colon \mathbb{R}^3 \to \mathbb{R}$ the (mass) density and $p\colon \mathbb{R}^3 \to \mathbb{R}$ the pressure. The term $2\rho\omega \times u$ represents the Coriolis force of the rotating earth with ω as angular velocity, the term $\rho\mathbf{g}$, where $\mathbf{g} = (0, 0, g)$ is assumed to be constant, represents the gravitational force, and D stands for the viscosity term.

Conservation of mass reads as

$$\frac{d\rho}{dt} + \rho \nabla \cdot u = 0.$$

Temperature $T : \mathbb{R}^3 \to \mathbb{R}$ and salinity $S : \mathbb{R}^3 \to \mathbb{R}$, i.e., the salt concentration in the ocean, will be modeled by diffusion-transport equations represented at first by

$$\frac{dT}{dt} = Q_T \quad \text{and} \quad \frac{dS}{dt} = S_T,$$

for some Q_T and S_T.

The equation of state is of the general form

$$\rho = f(T, S, p)$$

for some function f.

1.2.2 Boussinesq Approximation

Now, applying the Boussinesq approximation one replaces $\rho = \rho_0$, where $\rho_0 > 0$ is a constant, in each term except in the gravity term $\rho\mathbf{g}$, where one has the equation of state

$$\rho = \rho_0(1 - \beta_T(T - T_r) + \beta_S(S - S_r)),$$

where T_r and S_r are (constant) reference values of T and S, and $\beta_T, \beta_S \geq 0$. This means that the non-constant part of the density has only a notable impact on the gravitational force. Inserting for the material derivative

$$\frac{d}{dt} = \partial_t + u \cdot \nabla, \quad \text{where } u \cdot \nabla u' = u_1 \partial_x u' + u_2 \partial_y u' + u_3 \partial_z u',$$

and modeling the divergence of the viscosity part of the stress tensor by the anisotropic Laplacian

$$D = Du = \mu_v \Delta_H + \nu_v \partial_z^2, \quad \Delta_H = \partial_x^2 + \partial_y^2, \quad \mu_v, \nu_v > 0,$$

one obtains the Boussinesq equations of the ocean

$$\begin{cases} \partial_t v + v \cdot \nabla_H v + w \cdot \partial_z v - \mu_v \Delta_H v - \nu_v \partial_z^2 v + \frac{1}{\rho_0} \nabla_H \pi + 2\omega \times v = 0, \\ \partial_t w + v \cdot \nabla_H w + w \cdot \partial_z w - \mu_v \Delta_H w - \nu_v \partial_z^2 w + \frac{1}{\rho_0} \partial_z \pi + \rho g = 0, \\ \nabla \cdot_H v + \partial_z w = 0, \\ \partial_t T + v \cdot \nabla_H T + w \cdot \partial_z T - \mu_T \Delta_H T - \nu_T \partial_z^2 T = g_T, \\ \partial_t S + v \cdot \nabla_H S + w \cdot \partial_z S - \mu_S \Delta_H S - \nu_S \partial_z^2 S = g_S, \\ \rho = \rho_0(1 - \beta_T(T - T_r) + \beta_S(S - S_r)) = 0, \end{cases}$$

where g_T and g_S represent sources for temperature and salinity. These are supplemented by initial conditions $v(0) = v_0$, $w(0) = w_0$, $T(0) = T_0$, $S(0) = S_0$ and suitable boundary conditions. Here heat and salinity have been modeled by diffusion transport equations.

1.2.3 Hydrostatic Approximation

The hydrostatic approximation is based on the fact that the vertical scale, which for the ocean is of order $10^3 - 10^4$km, is much larger than the horizontal scale, which is smaller or equal to 11km. Consider a situation with small aspect ratio $\varepsilon > 0$, i.e.,

$$G = (0, 1) \times (0, 1) \quad \text{and} \quad \Omega_\varepsilon = G \times (-\varepsilon, \varepsilon).$$

Related to this small aspect ratio of $\varepsilon > 0$ is the anisotropic behavior of the viscosity which is in accordance with physical observations. Namely focusing for simplicity on the velocity equation, assume that

$$\mu_v = O(1) \quad \text{while} \quad \nu_v = O(\varepsilon^2).$$

Note that this choice is a "good guess" since it leads—as will come out *a posteriori*—to the primitive equations with full viscosity while lower orders would lead to only partial viscosity. Now, omitting temperature and salinity, neglecting the Coriolis force, one ends up with the anisotropic Navier-Stokes equations on Ω_ε for the velocity $u = (v, w)$ and the pressure π:

$$\begin{aligned} \partial_t v - \Delta_H v - \varepsilon^2 \partial_z v + v \cdot \nabla_H v + w \partial_z v + \nabla_H \pi &= 0, \\ \partial_t w - \Delta_H w - \varepsilon^2 \partial_z^2 w + v \cdot \nabla_H w + w \partial_z^2 w + \partial_z \pi &= -\rho g, \\ \nabla \cdot_H v + \partial_z w &= 0, \\ v(0) &= v_0, \\ w(0) &= w_0. \end{aligned}$$

Transforming Ω_ε to a fixed domain $\Omega = \Omega_1$, one can define scaled quantities on Ω by

$$\begin{aligned} v_\varepsilon(x, y, z) &= v(x, y, \varepsilon z), \\ w_\varepsilon(x, y, z) &= \frac{1}{\varepsilon} w(x, y, \varepsilon z), \\ p_\varepsilon(x, y, z) &= p(x, y, \varepsilon z), \\ \rho_\varepsilon(x, y, z) &= \rho(x, y, \varepsilon z). \end{aligned}$$

These solve the scaled Navier-Stokes equations

$$\begin{aligned}
\partial_t v_\varepsilon - \Delta_H v_\varepsilon - \partial_z v_\varepsilon + v_\varepsilon \cdot \nabla_H v_\varepsilon + w_\varepsilon \partial_z v_\varepsilon + \nabla_H \pi_\varepsilon &= 0,\\
\varepsilon^2(\partial_t w_\varepsilon - \Delta_H w_\varepsilon - \partial_z w_\varepsilon + v_\varepsilon \cdot \nabla_H w_\varepsilon + w_\varepsilon \partial_z w_\varepsilon) + \partial_z \pi_\varepsilon &= -\rho_\varepsilon g,\\
\nabla \cdot_H v_\varepsilon + \partial_z w_\varepsilon &= 0,\\
v_\varepsilon(0) &= v_{\varepsilon,0},\\
w_\varepsilon(0) &= w_{\varepsilon,0}.
\end{aligned}$$

Taking now formally the limit $\varepsilon \to 0+$ gives the hydrostatic approximation by replacing the evolution equation for w by

$$\frac{\partial p}{\partial z} = -\rho g.$$

This means that the full pressure $p\colon \Omega \to \mathbb{R}$ is determined by the surface pressure $p_s\colon G \times \{-1\} \to \mathbb{R}$. For constant temperature and salinity this means that the pressure grows linearly with respect to the z-direction. Moreover, there is no evolution equation left for w which is only determined by the condition $\nabla \cdot u = 0$, i.e. $\partial_z w = -\partial_x v_1 - \partial_y v_2$. Integrating with respect to z gives that

$$w(x, y, z, t) = w(x, y, -1, t) - \int_{-1}^{z} \partial_x v_1(x, y, \xi, t) + \partial_y v_2(x, y, \xi, t) d\xi.$$

Hence w is determined by v and the boundary conditions imposed on w. So, the primitive equations of the ocean are

$$\begin{cases}
\partial_t v + v \cdot \nabla_H v + w \cdot \partial_z v - \mu_v \Delta_H v - \nu_v \partial_z^2 v + \frac{1}{\rho_0} \nabla_H \pi + 2\omega \times v = 0,\\
\frac{\partial p}{\partial z} = -\rho g,\\
\nabla \cdot_H v + \partial_z w = 0,\\
\partial_t T + v \cdot \nabla_H T + w \cdot \partial_z T - \mu_T \Delta_H T - \nu_T \partial_z^2 T = g_T,\\
\partial_t S + v \cdot \nabla_H S + w \cdot \partial_z S - \mu_S \Delta_H S - \nu_S \partial_z^2 S = g_S,\\
\rho = \rho_0(1 - \beta_T(T - T_r) + \beta_S(S - S_r)) = 0,
\end{cases}$$

where we have included into this formulation again the temperature and the salinity.

To focus on the mathematical difficulties, in the following we normalize all constants to one, assume for v Neumann boundary conditions and for w Dirichlet boundary conditions on bottom and top and restrict ourselves to the velocity equations setting $T, S \equiv 0$. For simplicity the hydrostatic approximation is simplified to

$$\frac{\partial p}{\partial z} = 0.$$

1.3 Preliminaries

In this section we introduce several model situations and describe relevant function spaces.

1.3.1 Model Situation

Here, a cylindrical setting is assumed, i.e., for $h > 0$ consider

$$\Omega = G \times (-h, 0), \quad \text{where} \quad G = (0, 1) \times (0, 1),$$

where the bottom, the upper part and the lateral part of the boundary are denoted by

$$\Gamma_b = G \times \{-h\}, \quad \Gamma_u = G \times \{0\} \quad \text{and} \quad \Gamma_l = \partial G \times (-h, 0).$$

Assuming lateral periodicity we actually have $G = \mathbb{T}^2$, where $\mathbb{T}^2$ denotes the 2-D flat torus. Sometimes we will consider also the case $G = \mathbb{R}^2$. Concerning domains we follow the guideline to consider complicated equations on simple domains focusing first on the mathematical difficulties of the equation rather than on general assumption on the domain.

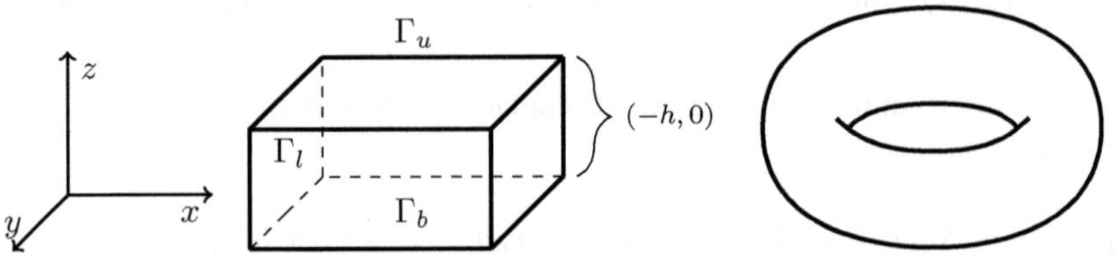

(a) Coordinates (b) Cylindrical domain (c) Cylindrical domain with periodicity

The original primitive equations for the velocity in the isothermal setting, i.e. omitting temperature and salinity, can be equivalently reformulated as

$$\begin{cases} \partial_t v + v \cdot \nabla_H v + w(v) \cdot \partial_z v - \Delta v + \nabla_H \pi_s = f, & \text{in } \Omega \times (0, T), \\ \operatorname{div}_H \overline{v} = 0, & \text{in } \Omega \times (0, T), \\ v(0) = v_0, & \text{in } \Omega, \end{cases} \tag{1.2}$$

Here $v = (v_1, v_2)$ denotes the horizontal velocity of the fluid and π_s the surface pressure. Moreover, denoting the horizontal coordinates by $(x, y) \in G$ and the vertical one by $z \in (-h, 0)$, we use the notations

$$\nabla_H = (\partial_x, \partial_y)^T, \quad \nabla \cdot_H v = \partial_x v_1 + \partial_y v_2 \quad \text{and} \quad \overline{v} := \frac{1}{h} \int_{-h}^{0} v(\cdot, \cdot, \xi) d\xi,$$

whereas Δ denotes the three-dimensional Laplacian. The vertical component of the velocity $w = w(v)$ is determined by

$$w(x, y, z) = -\int_a^z \operatorname{div}_H v(x, y, \xi) d\xi. \tag{1.3}$$

Furthermore, the full pressure π is determined by the surface pressure $\pi_s : G \to \mathbb{R}$, since in the original model one has $\partial_z \pi = 0$.

The equations (1.2) are supplemented by the boundary conditions

$$\begin{aligned} &v, \pi_s \text{ are periodic on } \Gamma_l \times (0, \infty), \\ &v = 0 \text{ on } \Gamma_D \times (0, \infty) \quad \text{and} \quad \partial_z v = 0 \text{ on } \Gamma_N \times (0, \infty), \end{aligned} \tag{1.4}$$

where Dirichlet, Neumann, and mixed boundary conditions are comprised by the notation

$$\Gamma_D \in \{\emptyset, \Gamma_b, \Gamma_u, \Gamma_b \cup \Gamma_u\} \quad \text{and} \quad \Gamma_N = (\Gamma_b \cup \Gamma_u) \setminus \Gamma_D.$$

Note that in the literature several sets of boundary conditions are considered. In [57, equation (1.37) and (1.37)'] Dirichlet and mixed Dirichlet Neumann boundary conditions are considered, respectively, while in [14] Neumann boundary conditions are assumed.

1.3.2 *Isotropic Function Spaces*

In the following, horizontal periodicity is modeled by the function spaces where smooth functions are periodic only with respect to x, y coordinates but not in z direction. To this end, we will need a terminology to describe periodic boundary conditions. Let $m \in \mathbb{N}_0$, we then say that a smooth function $f : \overline{\Omega} \to \mathbb{R}$ is *periodic of order m on Γ_l* if

$$\frac{\partial^\alpha f}{\partial x^\alpha}(0, y, z) = \frac{\partial^\alpha f}{\partial x^\alpha}(1, y, z) \text{ and } \frac{\partial^\alpha f}{\partial y^\alpha}(x, 0, z) = \frac{\partial^\alpha f}{\partial y^\alpha}(x, 1, z),$$

for all $\alpha = 0, \dots, m$. If the two quantities above are anti-symmetric, then v is said to be *anti-periodic*. In the same way we define the periodicity of order m on ∂G for a function defined on $\overline{G}$. Set

$$C^\infty_{\text{per}}(\overline{\Omega}) := \{f \in C^\infty(\overline{\Omega}) \mid f \text{ is periodic of arbitrary order on } \Gamma_l\},$$

$$C^\infty_{\text{per}}(\overline{G}) := \{f \in C^\infty(\overline{G}) \mid f \text{ is periodic of arbitrary order on } \partial G\}$$

We define for $p, q \in (1, \infty)$ and $s \in [0, \infty)$ the Besov spaces

$$B^s_{pq,per}(\Omega) := \overline{C^\infty_{per}(\overline{\Omega})}^{\|\cdot\|_{B^s_{pq}(\Omega)}} \quad \text{and} \quad B^s_{pq,per}(G) := \overline{C^\infty_{per}(\overline{G})}^{\|\cdot\|_{B^s_{pq}(G)}},$$

where B^s_{pq} denotes Besov spaces, which are defined as restrictions of Besov spaces on the whole space $B^s_{p,q}(\mathbb{R}^3)$, compare, e.g., [72, Definitions 3.2.2]. Note that for the Bessel potential spaces $H^{s,p}_{per} = B^s_{p2}$ holds, and in particular

$$H^{s,p}_{per}(\Omega) = \overline{C^\infty_{per}(\overline{\Omega})}^{\|\cdot\|_{H^{s,p}(\Omega)}} \quad \text{and} \quad H^{s,p}_{per}(G) = \overline{C^\infty_{per}(\overline{G})}^{\|\cdot\|_{H^{s,p}(G)}}.$$

Here $H^{s,p}(\Omega)$ and $H^{s,p}(G)$ denote the Bessel potential spaces, which are defined as restrictions of Bessel potential spaces on the whole space, compare, e.g., [72, Definition 3.2.2.]. It is well known that the space $H^{s,p}(\Omega)$ coincides with the classical Sobolev space $W^{m,p}(\Omega)$ provided $s = m \in \mathbb{N}$. Of course, we interpret $H^{0,p}_{\text{per}}$ as L^p. Equivalently, these Sobolev spaces equipped with periodic boundary conditions in the horizontal directions are given by

$$H^{m,p}_{\text{per}}(\Omega) = \{f \in H^{m,p}(\Omega) \mid f \text{ is periodic of order } m-1 \text{ on } \Gamma_l\},$$

$$H^{m,p}_{\text{per}}(G) = \{f \in H^{m,p}(G) \mid f \text{ is periodic of order } m-1 \text{ on } \partial G\}.$$

For an open set $M \subset \mathbb{R}^n$ $(n = 2, 3)$, $L^p_0(M)$ is defined by

$$L^p_0(M) := \{v \in L^p(M) \mid \int_M v = 0\}.$$

For a Banach space X, and an interval $I \subset \mathbb{R}$ we denote by

$$H^{m,p}(I; X), \quad m \in \mathbb{N}_0, \quad p \in [1, \infty],$$

vector valued Sobolev spaces with respect to time, where $H^{0,p} = L^p$. For $X = H^{n,q}(\Omega)$, sometimes the short-hand notation

$$H^{m,p}_t H^{n,q}_x := H^{m,p}(J; H^{n,q}(\Omega))$$

is used.

1.3.3 Anisotropic Function Spaces

In order to take into account the *anisotropic* nature of the primitive equations, we will be also dealing with anisotropic L^p-spaces on cylindrical sets. More precisely, if $\Omega = \Omega' \times \Omega_3 \subset \mathbb{R}^2 \times \mathbb{R}$ is a product of measurable sets and $q, p \in [1, \infty]$, we define

$$L^q_H L^p_z(\Omega) := L^q(\Omega'; L^p(\Omega_3)) = \{f : v \to \mathbb{K} \text{ measurable}, \|f\|_{L^q_H L^p_z(\Omega)} < \infty\},$$

for $\mathbb{K} \in \{\mathbb{R}, \mathbb{C}\}$ with norm

$$\|f\|_{L^q_H L^p_z(\Omega)} := \begin{cases} \left(\int_{\Omega'} \|f(x', \cdot)\|^q_{L^p(\Omega_3)} \, dx'\right)^{1/q}, & q \in [1, \infty), \\ \operatorname{ess\,sup}_{x' \in \Omega'} \|f(x', \cdot)\|_{L^p(\Omega_3)}, & q = \infty. \end{cases}$$

Endowed with this norm, $L^q_H L^p_z(v)$ is a Banach space for all $p, q \in [1, \infty]$.

Moreover, for $\Omega = G \times J$, where $G = (0, 1)^2 \subset \mathbb{R}^2$ and $J \subset \mathbb{R}$ is an interval, we define for $s, r \geq 0$ and $1 \leq p, q \leq \infty$ the function spaces

$$H^{r,q}_z H^{s,p}_{xy} := H^{r,q}(J; H^{s,p}(G)).$$

Equipped with the norms $\|v\|_{H^{r,q}_z H^{s,p}_{xy}} = \big\| \|v(\cdot, z)\|_{H^{s,p}(G)} \big\|_{H^{r,q}(J)}$, these become Banach spaces.

Taking Hölder's inequality independently with respect to z and (x, y) we obtain

$$\|fg\|_{L^q_z L^p_{xy}} \leq \|f\|_{L^{q_1}_z L^{p_1}_{xy}} \|g\|_{L^{q_2}_z L^{p_2}_{xy}} \quad \text{for} \quad \tfrac{1}{p} = \tfrac{1}{p_1} + \tfrac{1}{p_2}, \ \tfrac{1}{q} = \tfrac{1}{q_1} + \tfrac{1}{q_2}. \tag{1.5}$$

Embedding relations will also be performed separately in z and x, y; in fact, we have

$$H^{r,q}_z H^{s,p}_{xy} \hookrightarrow H^{r',q'}_z H^{s,p}_{xy} \quad \text{provided} \quad H^{r,q}(J) \hookrightarrow H^{r',q'}(J),$$

$$H^{r,q}_z H^{s,p}_{xy} \hookrightarrow H^{r,q}_z H^{s',p'}_{xy} \quad \text{provided} \quad H^{s,p}(G) \hookrightarrow H^{s',p'}(G).$$

Let us also remark that

$$H^{r,p}_z H^{s,p}_{xy} \subset H^{r+s,p}(\Omega).$$

The above relations hold also replacing H by W.

1.4 A Priori Bounds

A priori bounds for the primitive equations have been found first by Cao and Titi. Their strategy is nowadays the standard procedure to obtain such bounds for the primitive equations. More precisely, Cao and Titi proved L^∞_t-H^1_x and L^2_t-H^2_x bounds with respect to time and space, and bounds of this quality are sufficient to prevent finite time blow-up of strong solutions. In their original work, Cao and Titi considered Neumann-boundary conditions; Kukavica and Ziane gave a proof for the case of Dirichlet boundary conditions, see [50, 51].

In this section we present a proof of the *a priori* bounds only for the case of mixed Dirichlet Neumann boundary conditions, i.e.,

$$v|_{\Gamma_b} = 0 \quad \text{and} \quad v_z|_{\Gamma_u} = 0$$

with lateral periodicity. This case has been considered in [35] for the case of $f \equiv 0$ and has been adapted in [25] for $f \neq 0$. The case of pure Dirichlet boundary conditions $v|_{\Gamma_b} = 0$ and $v|_{\Gamma_u} = 0$ is analogous while the case of Neumann boundary conditions $v_z|_{\Gamma_b} = 0$ and $v_z|_{\Gamma_u} = 0$ needs minor adjustments, and this case is discussed, for instance, in [27] and of course in the original work [14].

Theorem 1 (*A Priori Bound* in the Maximal L^2-Regularity Space) *Let $T \in (0, \infty)$,*

$$v_0 \in \{H^1_{per}(\Omega)^2 \colon \nabla \cdot_H \overline{v} = 0 \text{ with } v_0|_{\Gamma_D} = 0\} \quad \text{and} \quad f \in L^2(0, T; L^2(\Omega)^2).$$

Then there exists a continuous function $B_{(\|v_0\|, \|f\|)} \colon [0, T] \to [0, \infty)$ such that if there is a solution

$$v \in L^2(0, T'; H^2(\Omega)) \cap H^1(0, T'; L^2(\Omega)), \quad T' \leq T,$$

to the primitive equations, then one has

$$\|v\|_{L^2(0,T';H^2(\Omega)) \cap H^1(0,T';L^2(\Omega))} \leq B_{(\|v_0\|, \|f\|)}(T').$$

Remark 1 By the embedding

$$L^2(0, T; H^2(\Omega)) \cap H^1(0, T; L^2(\Omega)) \hookrightarrow C^0([0, T]; H^1(\Omega))$$

the statement of an $L^\infty(0, T; H^1(\Omega))$ is included in the above statement. Nevertheless proving the $L^\infty(0, T; H^1(\Omega))$ *a priori* bound is the crucial part of the proof. In [14] the authors prove $L^\infty(0, T; H^1(\Omega))$ and $L^2(0, T; H^2(\Omega))$ bounds.

One can prove even higher order *a priori* bounds.

Theorem 2 (L^∞_t-H^2_x-*A Priori Bound*) *Let $T \in (0, \infty)$,*

$$v_0 \in \{H^2_{per}(\Omega)^2 \colon \nabla \cdot_H \overline{v} = 0 \text{ with } (1.4)\} \quad \text{and} \quad f \in H^1(0, T; L^2(\Omega)^2).$$

Then there exists a continuous function $B'_{(\|v_0\|_{H^2}, \|f\|_{H^1_t L^2_x})} \colon [0, T] \to [0, \infty)$ such that if there is a strong solution

$$v \in C^0([0, T']; H^2(\Omega)) \cap C^1([0, T']; L^2(\Omega)), \quad T' \leq T,$$

to the primitive equations, then one has

$$\|v\|_{L^\infty(0,T';H^2(\Omega))} \leq B'_{(\|v_0\|_{H^2},\|f\|_{H^1_t L^2_x})}(T').$$

1.4.1 Energy Inequality and Splitting of the System

Assume that v is a solution to (1.2) for

$$v_0 \in L^2(\Omega) \quad \text{and} \quad f \in L^2(0,T;L^2(\Omega)), \quad T \in (0,\infty].$$

Multiplying (1.2) by v, integrating over Ω, and making use of Lemma 2 below, i.e., the cancellation property for the non-linearity, yields for $t \in (0,T)$

$$\|v(t)\|^2_{L^2(\Omega)} + 2\int_0^t \|\nabla v(s)\|^2_{L^2(\Omega)}\,ds = \|v_0\|^2_{L^2(\Omega)} + 2\int_0^t f(s)\cdot v(s)\,ds. \tag{1.6}$$

In particular, by a compensation argument which uses the Poincaré inequality for functions on $(-h,0)$,

$$\|v\|_{L^2(\Omega)} \leq \frac{1}{h}\|\partial_z v\|_{L^2(\Omega)}, \qquad v \in \{v \in H^1_{per}(\Omega)\colon v|_{\Gamma_u} = 0 \text{ or } v|_{\Gamma_b} = 0\},$$

one obtains for $t \in (0,T)$

$$\|v(t)\|^2_{L^2(\Omega)} + \int_0^t \|\nabla v(s)\|^2_{L^2(\Omega)}\,ds \leq \|v_0\|^2_{L^2(\Omega)} + h^2\int_0^t \|f(s)\|^2_{L^2(\Omega)}\,ds. \tag{1.7}$$

Therefrom it follows that

$$\|v\|^2_{L^\infty(0,T;L^2(\Omega))} + \|v\|^2_{L^2(0,T;H^1(\Omega))} \leq C(\|v\|^2_{L^2(\Omega)} + \|f\|^2_{L^2(0,T;L^2(\Omega))}).$$

Note that $\bar{v}$ and $\tilde{v}$ also admit $L^\infty(L^2)\cap L^2(H^1)$-type bounds as above. The energy inequality (1.6) will be the starting point of our proof. Recall that

$$\overline{v} := \frac{1}{h}\int_{-h}^0 v(\cdot,\cdot,\xi)d\xi \quad \text{and} \quad \tilde{v} := v - \overline{v},$$

define complementary orthonormal projections which define a splitting of $L^2(\Omega)$. Following the strategy of Cao and Titi in [14], taking the vertical average and its complement of (1.2) induces a splitting into the following set of equations for $\bar{v}$ and $\tilde{v}$:

$$\begin{aligned}\partial_t \bar{v} - \Delta_H \bar{v} + \nabla_H \pi &= \overline{f} - \bar{v} \cdot \nabla_H \bar{v} \\ &\quad - \tfrac{1}{h} \int_{-h}^{0} (\tilde{v} \cdot \nabla_H \tilde{v} + \operatorname{div}_H \tilde{v}\, \tilde{v})\, dz - \tfrac{1}{h} v_z|_{\Gamma_b} \quad \text{in } G, \\ \operatorname{div}_H \bar{v} &= 0 \qquad\qquad\qquad\qquad\qquad\qquad\qquad\qquad \text{in } G,\end{aligned} \tag{1.8}$$

and

$$\begin{aligned}\partial_t \tilde{v} - \Delta \tilde{v} + \tilde{v} \cdot \nabla_H \tilde{v} + v_3 v_z + \bar{v} \cdot \nabla_H \tilde{v} &= \tilde{f} - \tilde{v} \cdot \nabla_H \bar{v} \\ + \tfrac{1}{h} \int_{-h}^{0} (\tilde{v} \cdot \nabla_H \tilde{v} + \operatorname{div}_H \tilde{v}\, \tilde{v})\, dz + \tfrac{1}{h} v_z|_{\Gamma_b} & \qquad \text{in } \Omega.\end{aligned} \tag{1.9}$$

This is a coupled system of a 2-D Stokes and a 3-D heat equation, where the main difficulty lies in the coupling terms. In particular, this splitting allows us to take advantage of the fact that the pressure appears only in the equation for $\overline{v}$.

Remark 2 For Neumann boundary conditions on top and bottom, the term $\frac{1}{h} v_z|_{\Gamma_b}$ disappears. In particular for initial conditions v_0 with $v_0 = \overline{v}_0$, i.e., $\tilde{v}_0 = 0$, the dynamics is completely given by the 2-D Navier-Stokes equations. In this sense, the primitive equations can be seen as an intermediate model between the 2-D and the 3-D Navier-Stokes equations.

1.4.2 Some Inequalities

Before proceeding with the proof of the *a priori* bound some inequalities are needed. In particular, the interpolation inequality

$$\|f\|_{L^4(G)} \leq C \|f\|_{L^2(G)}^{1/2} \|f\|_{H^1(G)}^{1/2}, \qquad f \in H^1(G), \tag{1.10}$$

for a constant $C > 0$ independent of f, plays an important role as well as the classical Hölder, Poincaré, Young, and Grönwall inequality, Sobolev embeddings, and trace theorem for Sobolev spaces amongst others. Moreover, the following lemmas give estimates for the two-dimensional Stokes system and the three-dimensional heat equation.

Lemma 1 (Estimate for the 2-D Stokes and the 3-D Heat Equation)

(a) *Let*
$f \in L^2(G)^2$ and $(\overline{v}, \pi)$ be a solution of the equation

$$\partial_t \overline{v} - \Delta_H \overline{v} + \nabla_H \pi = f,$$

satisfying $\operatorname{div}_H \overline{v} = 0$ *in G and such that $\overline{v}$ and π are periodic on ∂G. Then there exists a constant $C > 0$ such that for all $f \in L^2(G)^2$*

$$8\partial_t \|\nabla_H \overline{v}\|_{L^2(G)}^2 + \|\Delta_H \overline{v}\|_{L^2(G)}^2 + \|\nabla_H \pi\|_{L^2(G)}^2 \leq C \|f\|_{L^2(G)}^2.$$

(b) *Let $f \in L^2(\Omega)$ and v be a solution of $\partial_t u - \Delta v = f$ in Ω such that $v_z = 0$ on Γ_u, $v = 0$ on Γ_b and v is periodic on Γ_l. Then there exists a constant $C > 0$ such that for all $f \in L^2(\Omega)$*

$$\partial_t \|\nabla v\|^2_{L^2(\Omega)} + \|\Delta v\|^2_{L^2(\Omega)} \le C\|f\|^2_{L^2(\Omega)}.$$

Proof

(a) Multiplying the equation by $\partial_t u$ or $-\Delta_h u$, integrating by parts over G, and adding the resulting equations gives

$$\partial_t \|\nabla_h u\|^2_{L^2(G)} + \|\partial_t u\|^2_{L^2(G)} + \|\Delta_h u\|^2_{L^2(G)} = (f, \partial_t u - \Delta_h u).$$

Note that the pressure terms give no contributions, thanks to the periodic boundary conditions. Then, evaluating the pressure term by

$$\|\nabla_H \pi\|^2_{L^2(G)} \le 3(\|f\|^2_{L^2(G)} + \|\partial_t u\|^2_{L^2(G)} + \|\Delta_h u\|^2_{L^2(G)}),$$

one obtains

$$4\partial_t \|\nabla_h u\|^2_{L^2(G)} + \|\partial_t u\|^2_{L^2(G)} + \|\Delta_h u\|^2_{L^2(G)} + \|\nabla_H \pi\|^2_{L^2(G)} \le 3\|f\|^2_{L^2(G)} + 4(f, \partial_t u - \Delta_h u),$$

and eventually an absorbing argument gives the desired result.

(b) This can be proved by multiplying the equation with $-\Delta v$ and integrating by parts over Ω.

□

Remark 3 We note that there is a constant $C > 0$ such that

$$\|\nabla_H^2 \overline{v}\|_{L^2(G)} \le C\|\Delta_H \overline{v}\|_{L^2(G)} \quad \text{and} \quad \|v\|_{H^2(\Omega)} \le C\|\Delta v\|_{L^2(\Omega)},$$

respectively, if lateral periodicity and Dirichlet boundary conditions on top or bottom are assumed.

Lemma 2 (Cancellation Properties and Anisotropic Estimates) *Let $p, q, r \in (1, \infty)$ and $g : \Omega \to \mathbb{R}^2$. Then*

(a) $\int_\Omega (\tilde{v} \cdot \nabla_H g + w g_z) \cdot |g|^{q-2} g = 0$ *and* $\int_\Omega (\bar{v} \cdot \nabla_H g) \cdot |g|^{q-2} g = 0$, *whenever the integrals are well defined.*

(b) *Let $\frac{1}{p}(\frac{1}{2} + \frac{1}{q}) \ge \frac{1}{r}$, and $z \in (-h, 0)$. Then there exists a constant $C > 0$ such that for $g : \Omega \to \mathbb{R}^2$*

$$\||g(\cdot, z)|^p\|_{L^q(G)} \le C(\|g(\cdot, z)\|^p_{L^r(G)} + \|g(\cdot, z)\|^{p-r/2}_{L^r(G)} \|\nabla_H |g(\cdot, z)|^{r/2}\|_{L^2(G)}).$$

Proof

(a) The assertion follows from integration by parts. In fact, the volume integrals disappear since $\operatorname{div} u = 0$ or equivalently $\operatorname{div}_H \bar{v} = 0$. The same is true for the surface integrals since $w = 0$ on $\Gamma_u \cup \Gamma_b$ and $\nu_{\partial\Omega} = 0$ on $\Gamma_u \cup \Gamma_b$ and due to the periodic boundary conditions on Γ_l.

(b) For simplicity of notation, write g instead of $g(\cdot, z)$. Observe that

$$\||g|^p\|_{L^q(G)} = \||g|^{r/2}\|^{\beta}_{L^\alpha(G)}, \text{ where } \alpha = 2pq/r \text{ and } \beta = 2p/r.$$

On the other hand, the embedding $H^\gamma(G) \hookrightarrow L^\alpha(G)$ for $\gamma = 1 - 2/\alpha$ together with interpolation gives

$$\|f\|_{L^\alpha(G)} \leq C\|f\|^{1-\delta}_{L^2(G)}\|f\|^{\delta}_{H^1(G)} \text{ for } \delta \in [\gamma, 1]$$

for smooth functions f. Hence,

$$\||g|^p\|_{L^q(G)} \leq C\||g|^{r/2}\|^{\beta(1-\delta)}_{L^2(G)} \||g|^{r/2}\|^{\beta\delta}_{H^1(G)}.$$

Next, choosing δ such that $\beta\delta = 1$, which is possible since $\frac{1}{p}(\frac{1}{2} + \frac{1}{q}) \geq \frac{1}{r}$, and noting that $\beta(1 - \delta) = 2p/r - 1$ and $\|\cot\|_{H^1(G)} = \|\cdot\|_{L^2(G)} + \|\nabla_H \cdot\|_{L^2(G)}$, the desired estimate follows.

□

1.4.3 *Proof of the A Priori Bound*

In the sequel, we will derive bounds for $\bar{v}$, v_z, and $\tilde{v}$. Note that each of the three bounds is depending on the other two and hence, each of these bounds is not closed by itself alone. However, adding these estimates yields a closed estimate, see (1.15) below, to which the classical Grönwall inequality is applicable. Comparing this to other derivations of *a priori* bounds, the novelty of our approach lies in the fact that we are dealing with L^2_t-L^2_x-estimates for $\nabla_H \pi$ and L^∞_t-L^4_x-estimates for $\tilde{v}$, whereas the authors in [18, 22, 50, 51] performed $L^2(L^{3/2})$-estimates for $\nabla_H \pi$ and $L^\infty(L^6)$-estimates for v. We subdivide our proof into six steps.

Step 1 (L^∞_t-H^1_x and L^2_t-H^2_x-Estimates for $\overline{v}$)
Equation (1.8) implies using Lemma 1 (a) that

$$8\partial_t \|\nabla_H \bar{v}(t)\|^2_{L^2(G)} + \|\Delta_H \bar{v}\|^2_{L^2(G)} + \|\nabla_H \pi\|^2_{L^2(G)} \leq$$
$$C\left(\big\||\bar{v}||\nabla_H \bar{v}|\big\|^2_{L^2(G)} + \big\||\tilde{v}||\nabla_H \tilde{v}|\big\|^2_{L^2(\Omega)} + \|v_z\|^2_{L^2(\Gamma_b)}) + \|\bar{f}\|^2_{L^2(G)}\right)$$
$$=: I_1 + I_2 + I_3 + \|\bar{f}\|_{L^2(G)}.$$

Now, one estimates each term of the right-hand side above separately. Note that $\|\bar{f}\|_{L^2(G)} \le C\|f\|_{L^2(\Omega)}$. Then, Hölder's inequality and the interpolation inequality (1.10) yield

$$\begin{aligned}
I_1 &\le C\|\bar{v}\|^2_{L^4(G)}\|\nabla_H \bar{v}\|^2_{L^4(G)} \\
&\le C(\|\bar{v}\|^2_{L^2(G)} + \|\bar{v}\|_{L^2(G)}\|\nabla_H \bar{v}\|_{L^2(G)}) \\
&\quad \cdot (\|\nabla_H \bar{v}\|^2_{L^2(G)} + \|\nabla_H \bar{v}\|_{L^2(G)}\|\Delta_H \bar{v}\|_{L^2(G)}) \\
&= C(\|\bar{v}\|^2_{L^2(G)}\|\nabla_H \bar{v}\|^2_{L^2(G)} + \|\bar{v}\|^2_{L^2(G)}\|\nabla_H \bar{v}\|_{L^2(G)}\|\Delta_H \bar{v}\|_{L^2(G)} \\
&\quad + \|\bar{v}\|_{L^2(G)}\|\nabla_H \bar{v}\|^3_{L^2(G)} + \|\bar{v}\|_{L^2(G)}\|\nabla_H \bar{v}\|^2_{L^2(G)}\|\Delta_H \bar{v}\|_{L^2(G)}) \\
&\le C[(\|v\|^2_{L^2(\Omega)} + \|v\|^4_{L^2(\Omega)})\|v\|^2_{H^1(\Omega)} + \|\Delta_H \bar{v}\|^2_{L^2(G)} \\
&\quad + (\|v\|_{L^2(\Omega)} + \|v\|^2_{L^2(\Omega)})(\|v\|_{H^1(\Omega)} + \|v\|^2_{H^1(\Omega)})\|\nabla_H \bar{v}\|^2_{L^2(G)}],
\end{aligned}$$

where we have used the estimates $\|\nabla_H^2 \bar{v}\|_{L^2(G)} \le C\|\Delta_H \bar{v}\|_{L^2(G)}$, cf. Remark 3 for the second inequality and $\|\nabla_H \bar{v}\|_{L^2(G)} \le C\|v\|_{H^1(\Omega)}$ for the last inequality.

Since $|\nabla_H \tilde{v}| \le |\nabla \tilde{v}|$, we obtain for the second term I_2 on the right-hand side above that

$$I_2 \le C_1\||\tilde{v}||\nabla \tilde{v}|\|^2_{L^2(\Omega)}.$$

In view of the trace theorem, one has for $\delta > 0$

$$\begin{aligned}
I_3 = \|v_z\|^2_{L^2(\Gamma_b)} &\le \|v_z\|^2_{H^{1/2+\delta}} C\|v_z\|^{2(1/2-\delta)}_{L^2(\Omega)}\|\nabla v_z\|^{2(1/2+\delta)}_{L^2(\Omega)} \\
&\le C\|v_z\|^2_{L^2(\Omega)} + 1/4\|\nabla v_z\|^2_{L^2(\Omega)}.
\end{aligned}$$

Consequently, there exists a constant $C > 0$ such that

$$\begin{aligned}
&8\partial_t\|\nabla_H \bar{v}\|^2_{H^1(G)} + \|\nabla_H \pi\|^2_{L^2(G)} \le \\
&\quad C(\big\||\tilde{v}||\nabla \tilde{v}|\big\|^2_{L^2(G)} + (1 + \|v\|^2_{L^2(\Omega)} + \|v\|^4_{L^2(\Omega)})\|v\|^2_{H^1(\Omega)} \\
&+(\|v\|_{L^2(\Omega)} + \|v\|^2_{L^2(\Omega)})(\|v\|_{H^1(\Omega)} + \|v\|^2_{H^1(\Omega)})\|\nabla_H \bar{v}\|^2_{L^2(G)} \\
&+\|\bar{f}\|^2_{L^2(G)}) + \tfrac{1}{4}\|\nabla v_z\|^2_{L^2(G)},
\end{aligned} \tag{1.11}$$

where we made use of the fact that $\|v_z\|_{L^2(\Omega)} \le \|v\|_{H^1(\Omega)}$. Therefore,

$$\begin{aligned}
&\partial_t\|\nabla_H \bar{v}\|^2_{L^2(G)} + \|\nabla_H \pi\|^2_{L^2(G)} + \|\Delta_H \bar{v}\|^2_{L^2(G)} \le \\
&\quad C(\|v\|_2 + \|v\|_2^2)(\|v\|_{H^1} + \|v\|^2_{H^1})\|\nabla_H \bar{v}\|^2_{L^2(G)} + C\big\||\tilde{v}||\nabla \tilde{v}|\big\|^2_{L^2(G)} \\
&\qquad + C(1 + \|v\|_2^2 + \|v\|_2^4)\|v\|_{H^1}{}^2 + C\|f\|_2^2 + \frac{1}{4}\|\nabla v_z\|^2_{L^2(G)}.
\end{aligned}$$

Step 2 (Estimates for $v_z \in L_t^\infty L_x^2$)

We multiply (1.2) by $-\partial_z v_z$, integrate over Ω, and use Lemma 2. Then all the boundary integrals vanish except the one involving $\nabla_H \pi$. The resulting equation reads as

$$\frac{1}{2}\partial_t \|v_z\|_{L^2(\Omega)}^2 + \|\nabla v_z\|_{L_2(\Omega)}^2 = -\int_G \nabla_H \pi \cdot v_z|_{\Gamma_b} - \int_\Omega (v_z \cdot \nabla_h u) \cdot v_z \tag{1.12}$$

$$+ \int_\Omega \operatorname{div}_H v\, v_z \cdot v_z - \int_\Omega f \cdot \partial_z v_z$$

$$=: I_4 + I_5 + I_6 + I_f.$$

Recalling that $v = \bar{v} + \tilde{v}$, some further integration by parts yields

$$I_5 = -\int_\Omega (v_z \cdot \nabla_H \bar{v}) \cdot v_z - \int_\Omega (v_z \cdot \nabla_H \tilde{v}) \cdot v_z$$

$$= -\int_\Omega (v_z \cdot \nabla_H \bar{v}) \cdot v_z + \int_\Omega \operatorname{div}_H v_z\, \tilde{v} \cdot v_z + \int_\Omega (v_z \cdot \nabla_h u_z) \cdot \tilde{v}$$

$$=: I_{51} + I_{52} + I_{53},$$

and that

$$I_6 = \int_\Omega \operatorname{div}_H \tilde{v}\, v_z \cdot v_z = -2 \int_\Omega (\tilde{v} \cdot \nabla_h u_z) \cdot v_z.$$

We now estimate each of the above terms. Using first the Cauchy-Schwartz inequality, we obtain analogously to the case of I_3 that

$$|I_4| \le \frac{1}{4}\|\nabla_H \pi\|_{L^2(G)}^2 + C\|v_z\|_{L^2(\Omega)}^2 + \frac{1}{6}\|\nabla v_z\|_{L^2(\Omega)}^2.$$

Furthermore, by Fubini's theorem, Hölder's and Minkowski's inequality, and Lemma 2

$$|I_{51}| \le \int_G |\nabla_H \bar{v}| \int_{-h}^0 |v_z|^2\, dz$$

$$\le C\|\nabla_H \bar{v}\|_{L_2(G)} \left\| \int_{-h}^0 |v_z|^2\, dz \right\|_{L_2(G)}$$

$$\le C\|\nabla_H \bar{v}\|_{L_2(G)} \int_{-h}^0 \left\| |v_z|^2 \right\|_{L_2(G)} dz$$

$$\le C\|\nabla_H \bar{v}\|_{L_2(G)} \int_{-h}^0 \left(\|v_z\|_{L^2(G)}^2 + \|v_z\|_{L^2(G)} \|\nabla_H v\|_{L^2(G)} \right) dz$$

$$\le C\|\nabla_H \bar{v}\|_{L_2(G)} \|v_z\|_{L^2(\Omega)}^2 + C\|\nabla_H \bar{v}\|_{L^2(G)}^2 \|v_z\|_{L^2(\Omega)}^2 + \frac{1}{6}\left\|\nabla_H v_z\right\|_{L_2(\Omega)}^2.$$

For the remaining terms we have

$$|I_{52}| + |I_{53}| + |I_6| \le C \int_\Omega |\tilde{v}||\tilde{v}_z||\nabla_h u_z| \le C \big\| |\tilde{v}||\nabla \tilde{v}| \big\|^2_{L^2(\Omega)} + \frac{1}{6} \|\nabla v_z\|^2_{L^2(\Omega)},$$

where we have used $v_z = \tilde{v}_z$, $|v_z| \le |\nabla v|$, and $|\nabla_h u_z| \le |\nabla v_z|$. Moreover,

$$I_f \le C\|f\|^2 + \frac{1}{6}\|\nabla v_z\|^2.$$

Therefore, by using Cauchy-Schwartz inequality and a compensation argument

$$\begin{aligned}\partial_t \|v_z\|_2^2 + \|\nabla v_z\|_2^2 \le C \left(\|v\|_{H^1} + \|v\|^2_{H^1}\right)\|v_z\|_2^2 + \\ \frac{1}{2}\|\nabla_H \pi\|^2_{L^2(G)} + 2C_2 \big\| |\tilde{v}||\nabla \tilde{v}| \big\|_2^2 + C\left(\|v\|^2_{H^1} + \|f\|_2^2\right), \qquad (1.13)\end{aligned}$$

where we have used $\|v_z\|_{L^2(\Omega)} \le \|v\|_{H^1(\Omega)}$ and $\|\nabla_H \bar{v}\|_{L^2(G)} \le C\|v\|_{H^1(\Omega)}$.

Step 3 (Estimates for $\tilde{v} \in L_t^\infty L_x^4$)

Multiplying (1.9) by $|\tilde{v}|^2\, \tilde{v}$, integrating by parts in Ω, and using Lemma 2, we obtain

$$\begin{aligned}&\frac{1}{4}\partial_t \|\tilde{v}\|^4_{L^4(\Omega)} + \frac{1}{2}\big\|\nabla |\tilde{v}|^2\big\|^2_{L^2(\Omega)} + \big\| |\tilde{v}||\nabla \tilde{v}| \big\|^2_{L^2(\Omega)} \\ &= -\int_\Omega (\tilde{v} \cdot \nabla_H \bar{v}) \cdot |\tilde{v}|^2\, \tilde{v} + \frac{1}{h}\int_\Omega \int_{-h}^0 (\tilde{v} \cdot \nabla_H \tilde{v} + \operatorname{div}_H v\, \tilde{v})\, dz \cdot |\tilde{v}|^2\, \tilde{v} \\ &\quad + \frac{1}{h}\int_\Omega v_z|_{\Gamma_b} \cdot |\tilde{v}|^2\, \tilde{v} + \int_\Omega \tilde{f} \cdot |\tilde{v}|^2\, \tilde{v}. =: I_7 + I_8 + I_9 + I_f'. \qquad (1.14)\end{aligned}$$

The last term I_f' is bounded by $\|\tilde{f}\|_2 \big\| |\tilde{v}|^3 \big\|_2$. We see moreover that

$$\begin{aligned}\big\| |\tilde{v}|^3 \big\|_2 &= \big\| |\tilde{v}|^2 \big\|_3^{3/2} \le C \big\| |\tilde{v}|^2 \big\|^{3/2}_{H^{1/2}(\Omega)} \le C \big\| |\tilde{v}|^2 \big\|_2^{3/4} \left(\big\| |\tilde{v}|^2 \big\|_2 + \big\| \nabla |\tilde{v}|^2 \big\|_2 \right)^{3/4} \\ &= C\|\tilde{v}\|_4^{3/2} \left(\big\| |\tilde{v}|^2 \big\|_2 + \big\| \nabla |\tilde{v}|^2 \big\|_2 \right)^{3/4} \le C\|\tilde{v}\|_4^3 + C\|\tilde{v}\|_4^{3/2} \big\| \nabla |\tilde{v}|^2 \big\|_2^{3/4},\end{aligned}$$

where we have used the embedding $H^{1/2}(\Omega) \hookrightarrow L^3(\Omega)$ and the interpolation inequality together with Poincaré's inequality to estimate

$$\|v\|^2_{H^{1/2}} \le \|v\|_{L^2}\|v\|_{H^1} \le C\|v\|_{L^2}\|\nabla v\|_{L^2}.$$

It then follows that

$$
\begin{aligned}
I_f' &\le C\|\tilde f\|_2\|\tilde v\|_4^3 + C\|\tilde f\|_2\|\tilde v\|_4^{3/2}\|\nabla|\tilde v|^2\|_2^{3/4}\\
&= C\big(\|\tilde f\|_2^2\|\tilde v\|_4^4\big)^{1/2}\big(\|\tilde v\|_4^2\big)^{1/2} + C\|\tilde f\|_2^{2/8}\big(\|\tilde f\|_2^2\|\tilde v\|_4^4\big)^{3/8}\big(\|\nabla|\tilde v|^2\|_2^2\big)^{3/8}\\
&\le \frac{1}{6}\|\nabla|\tilde v|^2\|_2^2 + C\|\tilde f\|_2^2\|\tilde v\|_4^4 + C\|\tilde v\|_4^2 + C\|\tilde f\|_2\\
&\le \frac{1}{6}\|\nabla|\tilde v|^2\|_2^2 + C\|f\|_2^2\|\tilde v\|_4^4 + C\|v\|_{H^1}{}^2 + C\|f\|_2.
\end{aligned}
$$

By Fubini's theorem, Hölder's and Minkowski's inequalities as well as by Lemma 2, the term I_7 can be bounded as

$$
\begin{aligned}
I_7 &\le C\int_G |\nabla_H \bar v| \int_{-h}^0 |\tilde v|^4\,dz\\
&\le C\|\nabla_H\bar v\|_{L_2(G)}\left\|\int_{-h}^0 |\tilde v|^4\,dz\right\|_{L_2(G)}\\
&\le C\|\nabla_H\bar v\|_{L_2(G)}\int_{-h}^0 \big\||\tilde v|^4\big\|_{L_2(G)}\,dz\\
&\le C\|\nabla_H\bar v\|_{L_2(G)}\int_{-h}^0 \Big(\|\tilde v\|_{L^4(G)}^4 + \|\tilde v\|_{L^4(G)}^2\|\nabla_H|\tilde v|^2\|_{L^2(G)}\Big)\,dz\\
&\le C\|\nabla_H\bar v\|_{L_2(G)}\|\tilde v\|_{L^4(\Omega)}^4 + C\|\nabla_H\bar v\|_{L^2(G)}^2\|\tilde v\|_{L^4(\Omega)}^4 + \frac{1}{6}\|\nabla_H|\tilde v|^2\|_{L_2(\Omega)}^2.
\end{aligned}
$$

In a similar manner, the term I_8 may be estimated as

$$
\begin{aligned}
I_8 &\le C\int_G \left(\int_{-h}^0 |\tilde v||\nabla_H\tilde v|\,dz\right)\left(\int_{-h}^0 |\tilde v|^3\,dz\right)\\
&\le C\left\|\int_{-h}^0 |\tilde v||\nabla_H\tilde v|\,dz\right\|_{L^{4/3}(G)}\left\|\int_{-h}^0 |\tilde v|^3\,dz\right\|_{L^4(G)}\\
&\le C\int_{-h}^0 \big\||\tilde v||\nabla_H\tilde v|\big\|_{L^{4/3}(G)}\,dz\int_{-h}^0 \big\||\tilde v|^3\big\|_{L^4(G)}\,dz\\
&\le C\left(\int_{-h}^0 \|\tilde v\|_{L^4(G)}\|\nabla_H\tilde v\|_{L^2(G)}\,dz\right)\\
&\quad\cdot\int_{-h}^0 \Big(\|\tilde v\|_{L^4(G)}^3 + \|\tilde v\|_{L^4(G)}\big\|\nabla_H|\tilde v|^2\big\|_{L^2(G)}\Big)\,dz\\
&\le C\|\tilde v\|_{L^4(\Omega)}\|\nabla_H\tilde v\|_{L_2(\Omega)}(\|\tilde v\|_{L^4(\Omega)}^3 + \|\tilde v\|_{L^4(\Omega)}\big\|\nabla_H|\tilde v|^2\big\|_{L^2(\Omega)})\\
&\le C\|\nabla_H\bar v\|_{L_2(G)}\|\tilde v\|_{L^4(\Omega)}^4 + C\|\nabla_H\bar v\|_{L^2(G)}^2\|\tilde v\|_{L^4(\Omega)}^4 + \frac{1}{6}\|\nabla_H|\tilde v|^2\|_{L_2(\Omega)}^2.
\end{aligned}
$$

Finally, by the trace theorem and by Poincaré's inequality as well as Lemma 2, we estimate the term I_9 as

$$\begin{aligned}
I_9 &\le C \int_G |v_{z|\Gamma_b}| \int_{-h}^{0} |\tilde{v}|^3 \, dz \\
&\le C \|v_z\|_{L^2(\Gamma_b)} \int_{-h}^{0} \big\| |\tilde{v}|^3 \big\|_{L^2(G)} \, dz \\
&\le C \|v_z\|_{L^2(\Gamma_b)} \int_{-h}^{0} (\|\tilde{v}\|_{L^4(G)}^3 + \|\tilde{v}\|_{L^4(G)} \big\| \nabla_H |\tilde{v}|^2 \big\|_{L^2(G)}) \, dz \\
&\le C \|v_z\|_{L^2(\Gamma_b)} \|\tilde{v}\|_{L^4(\Omega)}^3 + C \|v_z\|_{L^2(\Gamma_b)} \|\tilde{v}\|_{L^4(\Omega)} \big\| \nabla_H |\tilde{v}|^2 \big\|_{L^2(\Omega)} \\
&\le C \|v_z\|_{L^2(\Omega)}^{1/2} \|\nabla v_z\|_{L^2(\Omega)}^{1/2} \|\tilde{v}\|_{L^4(\Omega)}^3 \\
&\quad + C \|v_z\|_{L^2(\Omega)}^{1/2} \|\nabla v_z\|_{L^2(\Omega)}^{1/2} \|\tilde{v}\|_{L^4(\Omega)} \big\| \nabla_H |\tilde{v}|^2 \big\|_{L^2(\Omega)} \\
&\le C \|v_z\|_{L^2(\Omega)}^{2/3} \|\tilde{v}\|_{L^4(\Omega)}^4 + C \|v_z\|_{L^2(\Omega)}^2 \|\tilde{v}\|_{L^4(\Omega)}^4 \\
&\quad + \frac{1}{4C_3} \|\nabla v_z\|_{L^2(\Omega)}^2 + \frac{1}{6} \big\| \nabla_H |\tilde{v}|^2 \big\|_{L_2(\Omega)}^2 .
\end{aligned}$$

Combining these estimates with (1.14) we conclude that

$$\begin{aligned}
\partial_t \|\tilde{v}\|_4^4 + \big\| |\tilde{v}| |\nabla \tilde{v}| \big\|_2^2 \le\ & C\big(\|v\|_{H^1}^{2/3} + \|v\|_{H^1} + \|v\|_{H^1}^2 + \|f\|_2^2\big) \|\tilde{v}\|_4^4 \\
&+ \frac{1}{4} \|\nabla v_z\|_2^2 + C \|v\|_{H^1}^2 + C \|f\|_2,
\end{aligned}$$

where we made use of the estimates $\|\nabla_H \bar{v}\|_{L^2(G)} \le C \|v\|_{H^1(\Omega)}$ and $\|v_z\|_{L^2(\Omega)} \le \|v\|_{H^1(\Omega)}$.

Step 4 (Adding the Above Estimates)

Addition of (1.11), (1.13), and (1.14) enables us to absorb the terms $\big\| |\tilde{v}| |\nabla \tilde{v}| \big\|_{L^2(\Omega)}^2$, $\|\nabla v_z\|_{L^2(\Omega)}^2$ and $\|\nabla_H \pi\|_{L^2(G)}^2$ into the left-hand side. This leads us to

$$\begin{aligned}
\partial_t \Big(8 \|\nabla_H \bar{v}\|_{L^2(G)}^2 + \|v_z\|_2^2 + \frac{C_3}{4} \|\tilde{v}\|_4^4\Big) + \frac{1}{2} \Big(\|\nabla_H p\|_{L^2(G)}^2 + \|\nabla v_z\|_2^2 + C_3 \big\| |\tilde{v}| |\nabla_H \tilde{v}| \big\|_2^2\Big) \\
\le K_1(t) \big(8 \|\nabla_H \bar{v}\|_{L^2(G)}^2 + \|v_z\|_2^2 + (C_3/4) \|\tilde{v}\|_4^4\big) + K_2(t), \qquad (1.15)
\end{aligned}$$

where, thanks to (1.7), the functions K_1 and K_2 given by

$$\begin{aligned}
K_1(t) &:= C(1 + \|v\|_2 + \|v\|_2^2)(\|v\|_{H^1}^{2/3} + \|v\|_{H^1} + \|v\|_{H^1}^2 + \|f\|_2^2), \\
K_2(t) &:= C(1 + \|v\|_2^2 + \|v\|_2^4) \|v\|_{H^1}^2 + C(\|f\|_2 + \|f\|_2^2),
\end{aligned}$$

are integrable on $[0, T]$. It follows from (1.6) and Hölder's inequality that for $t \geq 0$

$$\int_0^t K_1(s)\,ds \leq C(1 + \|a\|_{L^2(\Omega)} + \|a\|^2_{L^2(\Omega)}) \cdot (\|a\|^{2/3}_{L^2(\Omega)} t^{2/3} + \|a\|_{L^2(\Omega)} t^{1/2} + \|a\|^2_{L^2(\Omega)}) < \infty,$$

and

$$\int_0^t K_2(s)\,ds \leq C(1 + \|a\|^2_{L^2(\Omega)} + \|a\|^4_{L^2(\Omega)})\|a\|^2_{L^2(\Omega)} < \infty.$$

Applying Grönwall's inequality thus yields

$$\begin{aligned} \|\nabla_H \bar{v}(t)\|^2_{L^2(G)} &+ \int_0^t \|v_z(s)\|^2_{L^2(\Omega)} + \|\tilde{v}(s)\|^4_{L^4(\Omega)} ds \\ &\leq \left(C(\|a\|^2_{H^1(\Omega)} + \|a\|^4_{H^1(\Omega)}) + \int_0^t K_2(s)\,ds \right) e^{\int_0^t K_1(s)\,ds} \\ &=: B_1(t, \|a\|_{H^1(\Omega)}), \quad t \geq 0, \end{aligned} \tag{1.16}$$

where we estimated the left-hand side of (1.16) for $t = 0$ by $C(\|a\|^2_{H^1(\Omega)} + \|a\|^4_{H^1(\Omega)})$. By (1.6), we obtain further that $\|\bar{v}(t)\|^2_{H^1(G)} \leq C\|a\|^2_{L^2(\Omega)} + \frac{1}{8} B_1(t, \|a\|_{H^1(\Omega)})$ and that

$$\frac{1}{2}\int_0^t (\|\nabla_H \pi\|^2_{L^2(G)} + \|\nabla v_z\|^2_{L^2(\Omega)} + C_3 \big\| |\tilde{v}||\nabla \tilde{v}| \big\|^2_{L^2(\Omega)})\,ds \leq B_1(t, \|a\|_{H^1(\Omega)}), \quad t > 0. \tag{1.17}$$

Step 5 (Estimates for $v \in L^\infty_t H^1_x$)

We consider (1.2) as an inhomogeneous heat equation of the form

$$\partial_t v - \Delta v = -v \cdot \nabla_H v - w v_z - \nabla_H \pi + f$$

and apply Lemma 1b). Since

$$v \cdot \nabla_H v = \bar{v} \cdot \nabla_H \bar{v} + \bar{v} \cdot \nabla_H \tilde{v} + \tilde{v} \cdot \nabla_H \bar{v} + \tilde{v} \cdot \nabla_H \tilde{v},$$

we obtain

$$\partial_t \|\nabla v\|_{L^2(\Omega)}^2+\|\Delta v\|_{L^2(\Omega)}^2 \le C(\|\bar{v}\cdot\nabla_H\bar{v}\|_{L^2(G)}^2+\|\bar{v}\cdot\nabla_H\tilde{v}\|_{L^2(\Omega)}^2+\|\tilde{v}\cdot\nabla_H\bar{v}\|_{L^2(\Omega)}^2$$
$$+ \|wv_z\|_{L^2(\Omega)}^2 + \|\tilde{v}\cdot\nabla_H\tilde{v}\|_{L^2(\Omega)}^2 + \|\nabla_H\pi\|_{L^2(G)}^2 + \|f\|_{L^2(\Omega)}^2). \tag{1.18}$$

One estimates each term on the right-hand side of (1.18), where the last two terms were already estimated in (1.17). The interpolation inequality (1.10) yields

$$\begin{aligned}\|\bar{v}\cdot\nabla_H\bar{v}\|_{L^2(G)}^2 &\le C\|\bar{v}\|_{L^4(G)}^2\|\bar{v}\|_{H^1(G)}\|\bar{v}\|_{H^2(G)} \le C\|\bar{v}\|_{L^4(G)}^2\|v\|_{H^1(\Omega)}\|v\|_{H^2(\Omega)}\\ &\le C\|\bar{v}\|_{H^1(G)}^4\|v\|_{H^1(\Omega)}^2 + 1/8\|\Delta v\|_{L^2(\Omega)}^2,\end{aligned}$$

where we have used $\|v\|_{H^2(\Omega)} \le C\|\Delta v\|_{L^2(\Omega)}$, cf. Remark 3. The second term in (1.18) is estimated by interpolation as

$$\begin{aligned}\|\bar{v}\cdot\nabla_H\tilde{v}\|_{L^2(\Omega)}^2 &\le C\|\bar{v}\|_{L^6(G)}^2\|\tilde{v}\|_{W^{1,3}(\Omega)}^2\\ &\le C\|\bar{v}\|_{L^6(G)}^2\|v\|_{W^{1,3}(\Omega)}^2\\ &\le C\|\bar{v}\|_{H^1(G)}^2\|v\|_{H^1(\Omega)}\|v\|_{H^2(\Omega)}\\ &\le C\|\bar{v}\|_{H^1(G)}^4\|v\|_{H^1(\Omega)}^2 + 1/8\|\Delta v\|_{L^2(\Omega)}^2,\end{aligned}$$

where we used the embedding $H^{3/2}(\Omega) \hookrightarrow W^{1,3}(\Omega)$.

Similarly, since $H^{3/2}(G) \hookrightarrow W^{1,4}(G)$, the third term above is bounded by

$$\begin{aligned}\|\tilde{v}\cdot\nabla_H\bar{v}\|_{L^2(\Omega)}^2 &\le C\|\tilde{v}\|_{L^4(\Omega)}^2\|\bar{v}\|_{W^{1,4}(G)}^2\\ &\le C\|\tilde{v}\|_{L^4(\Omega)}^2\|\bar{v}\|_{H^1(G)}\|\bar{v}\|_{H^2(G)}\\ &\le C\|\tilde{v}\|_{L^4(\Omega)}^2\|v\|_{H^1(\Omega)}\|v\|_{H^2(\Omega)}\\ &\le C\|\tilde{v}\|_{L^4(\Omega)}^4\|v\|_{H^1(\Omega)}^2 + 1/8\|\Delta v\|_{L^2(\Omega)}^2.\end{aligned}$$

Finally, for the fourth term in (1.18) we use the anisotropic estimate given in the proof of Lemma 14. This combined with the interpolation inequality (1.10) and with Poincaré's inequality yields

$$\begin{aligned}\|wv_z\|_{L^2(\Omega)}^2 &\le C\|w\|_{L_z^\infty L_{xy}^4}^2\|v_z\|_{L_z^2L_{xy}^4}^2\\ &\le C\|\nabla_h u\|_{L_z^2L_{xy}^4}^2\|v_z\|_{L_z^2L_{xy}^4}^2\\ &\le C\|\nabla_h u\|_{L_z^2L_{xy}^2}\|\nabla_h u\|_{L_z^2H_{xy}^1}\|v_z\|_{L_z^2L_{xy}^2}\|v_z\|_{L_z^2H_{xy}^1}\\ &\le C\|\nabla v\|_{L^2(\Omega)}\|v\|_{H^2(\Omega)}\|v_z\|_{L^2(\Omega)}\|\nabla v_z\|_{L^2(\Omega)}\\ &\le C\|v_z\|_{L^2(\Omega)}^2\|\nabla v_z\|_{L^2(\Omega)}^2\|\nabla v\|_{L^2(\Omega)}^2 + 1/8\|\Delta v\|_{L^2(\Omega)}^2.\end{aligned}$$

Combining these estimates with (1.18) we arrive at

$$\partial_t \|\nabla v\|_2^2 + \|\Delta v\|_2^2 \le C\|v_z\|_2^2\|\nabla v_z\|_2^2\|\nabla v\|_2^2 + C(\|\bar{v}\|_{H^1(G)}^2 + \|\tilde{v}\|_4^4)\|v\|_{H^1}^2 + C\|f\|_2^2.$$

Since $\int_0^t C\|v_z\|_{L^2(\Omega)}^2\|\nabla v_z\|_{L^2(\Omega)}^2\, ds$ and since $\int_0^t C(\|\bar{v}\|_{H^1(G)}^2 + \|\tilde{v}\|_{L^4(\Omega)}^4)\|v\|_{H^1(\Omega)}^2$ ds are bounded by $CB_1(t, \|a\|_{H^1(\Omega)})^2$ and by

$$L(t) := C(B_1(t, \|a\|_{H^1(\Omega)}) + \|a\|_{L^2(\Omega)}^2)\|a\|_{L^2(\Omega)}^2,$$

respectively, it follows from Grönwall's inequality that for $t \ge 0$

$$\|\nabla v(t)\|_{L^2(\Omega)}^2 + \frac{1}{2}\int_0^t \|\Delta v\|_{L^2(\Omega)}^2\, ds \le (\|\nabla a\|_{L^2(\Omega)}^2 + L(t))e^{CB_1(t,\|a\|_{H^1(\Omega)})^2} < \infty.$$

Consequently, for a suitable function B_2

$$\|v\|_{L^\infty(0,T;H^1(\Omega))} \le B_2(t, \|a\|_{H^1(\Omega)}).$$

Step 6 (Estimates for $\partial_t v \in L_t^2 L_x^2$)
Having $L^\infty(0, T; H^1(\Omega))$ and $L^2(0, T; H^2(\Omega))$ bounds, the bound on $H^1(0, T; L^2(\Omega))$ follows considering

$$\partial_t v = +\Delta v - \nabla_H \pi - v \cdot_H v - w\partial_z v + f,$$

where each term on the right-hand side can be estimated. For instance,

$$\begin{aligned}
\|w\partial_z v\|_{L_t^2 L^2}^2 &= \int_0^t \|w(s)\partial_z v(s)ds\|_{L^2}^2 \le C\int_0^t \|v(s)\|_{H^{3/2}}^4 ds \\
&\le C\int_0^t \|v(s)\|_{H^1}^2\|v(s)\|_{H^2}^2 \le C\|v\|_{L^\infty(0,T;H^1(\Omega))}^2\|v\|_{L^2(0,T;H^2(\Omega))}^2.
\end{aligned}$$

Proof of Theorem 2 ($v \in L_t^\infty H_x^2$) For the proof of Theorem 2 one needs an $L_t^\infty L_x^2$ bound on $\partial_t v$. This can be done by means of difference quotients. Then one uses that

$$-\Delta v = -\partial_t v - \nabla_H \pi - v \cdot_H v - w\partial_z v + f,$$

and estimates the right-hand side. □

1.5 Hydrostatic Stokes Equations in Isotropic L^p-Spaces

The linearization of the primitive equations (1.2) are the *hydrostatic Stokes equations*, which are given by

$$\begin{cases} \partial_t v - \Delta v + \nabla_H \pi_s = f & \text{in } \Omega \times (0,T), \\ \nabla \cdot_H \overline{v} = 0 & \text{in } \Omega \times (0,T), \\ v(0) = v_0 & \text{in } \Omega. \end{cases} \tag{1.19}$$

The name "hydrostatic Stokes equations" is motivated by the assumption of a hydrostatic balance when deriving the full primitive equations. The equations (1.19) are supplemented by the boundary conditions (1.4).

One strategy to solve the classical Stokes equations is to define first spaces of solenoidal functions and to prove existence of a bounded projection thereon—the Helmholtz projection. Applying this to the Stokes equations the pressure cancels out, and then one can solve this equation for the velocity from which the pressure can be reconstructed. This strategy can be adapted to the hydrostatic case too. However, since the pressure is a function of two variables only, here it constitutes sometimes a simplification to solve for the pressure first.

To deal with the linear problem we use a combination of different strategies. First, form methods give basic information in the L^2-setting. Many properties such as resolvent estimates carry over to the L^p-situation for the reflexive range $p \in (1,\infty)$. Also the pressure can be obtained by solving the two-dimensional weak Poisson problem with lateral periodicity.

The second approach is by perturbation techniques. One can solve the equation for the pressure explicitly, and the linear problem becomes a perturbation of the heat equation and its restriction to solenoidal functions.

Combining these two approaches we can derive several properties of the operator governing the linear problem such as sectoriality including the angle of sectoriality, structure of the spectrum, invertibility, regularity of solutions, elliptic regularity, boundedness of the $\mathcal{H}^\infty$-calculus including the angle. In particular the boundedness of the $\mathcal{H}^\infty$-calculus has many implications. First maximal L^q_t-L^p_x-regularity follows for $p, q \in (1,\infty)$. Second the domains of the fractional powers are given by complex interpolation spaces which can be computed explicitly including the boundary conditions.

1.5.1 Solenoidal Functions and the Hydrostatic Helmholtz Projection

Similarly to the case of the Stokes equation, for $1 < p < \infty$ we consider a solenoidal subspace of $L^p(\Omega)^2$ which is defined by

$$L^p_{\overline{\sigma}}(\Omega) = \overline{\{v \in C^\infty_{per}(\Omega)^2 : \nabla \cdot_H \overline{v} = 0\}}^{L^p(\Omega)^2}. \tag{1.20}$$

This space will play the analogous role in our investigations of the primitive equations as the solenoidal space $L^p_\sigma(\Omega)$ plays in the theory of the Navier-Stokes equations. Note that $v \in L^p_{\overline{\sigma}}(\Omega)$ implies $u(v) = (v, w(v)) \in L^p_\sigma(\Omega)$, where

$$L^p_\sigma(\Omega) = \overline{\{u = (v, w) \in C^\infty_{per}(\Omega)^3 : \nabla \cdot u = 0 \text{ and } w|_{z\in\{0,-h\}} = 0\}}^{L^p(\Omega)^3}. \tag{1.21}$$

As in the case of the classical Helmholtz projection, the existence of the hydrostatic Helmholtz projection is closely related to the unique solvability of the Poisson problem in the weak sense.

Proposition 1.5.1 (Weak Solvability of the Poisson Problem) *Let $p \in (1, \infty)$ and $f \in L^p(G)^2$. Then there exists a unique $\pi \in W^{1,p}_{\mathrm{per}}(G) \cap L^p_0(G)$ satisfying*

$$\langle \nabla_H \pi, \nabla_H \phi \rangle_{L^{p'}(G)} = \langle f, \nabla_H \phi \rangle_{L^{p'}(G)}, \quad \phi \in W^{1,p'}_{\mathrm{per}}(G) \cap L^{p'}_0(G). \tag{1.22}$$

Furthermore, there exists a constant $C > 0$ such that

$$\|\pi\|_{W^{1,p}(G)} \leq C\|f\|_{L^p(G)}, \quad f \in L^p(G)^2. \tag{1.23}$$

The solution operator is given by the projector onto the horizontal gradient fields

$$\nabla_H \pi = \nabla_H \Delta_H^{-1} \nabla \cdot_H f,$$

where Δ_H denotes the two-dimensional Laplacian defined on $H^{2,p}_{per}(G)^2$. Its inverse is considered as an operator in $L^p_0(G) = \{\overline{v} \in L^p(G)^2 \mid \int_G \overline{v} = 0\}$. Here, $\nabla_H \Delta_H^{-1} \nabla \cdot_H$ denotes actually the closure of the composition which is a bounded operator in $L^p(G)^2$. This projector can also be represented in terms of Riesz transforms on the torus, which can be defined by means of Fourier series, compare [62, Chapter 9], where a slightly more general situation is considered.

The classical Helmholtz projection onto the solenoidal space

$$L^p_\sigma(G) = \overline{\{\overline{v} \in C^\infty_{per}(G)^2 : \nabla \cdot_H \overline{v} = 0\}}^{L^p(G)^2}$$

on the unit square with horizontal periodicity is given by

$$Q : L^p(G)^2 \to L^p_\sigma(G), \quad Q\overline{v} = \overline{v} - \nabla_H \pi = (\mathbb{1} - \nabla_H \Delta_H^{-1} \nabla \cdot_H)\overline{v}.$$

Hence, there exists a continuous projection $\mathbb{P}$, called the *hydrostatic Helmholtz projection*, from $L^p(\Omega)^2$ onto $L^p_{\overline{\sigma}}(\Omega)$ which is defined using the decomposition

$$v = \overline{v} + \tilde{v}, \text{ where } \tilde{v} := v - \overline{v}, \quad \tilde{v} = \overline{v} \text{ if and only if } v = 0,$$

as

$$\mathbb{P}\colon L^p(\Omega)^2 \to L^p_{\overline{\sigma}}(\Omega), \quad \mathbb{P}v = \tilde{v} + Q\bar{v}. \tag{1.24}$$

In particular $\mathbb{P}$ annihilates the pressure term $\nabla_H \pi$ and

$$\operatorname{Ran}\mathbb{P} = \left\{ v \in L^p(\Omega)^2 \colon \int_{-h}^{0} v(\cdot,\cdot,\xi)d\xi = 0 \right\} \oplus L^p_\sigma(G) = L^p_{\overline{\sigma}}(\Omega).$$

The above Proposition 1.5.1 allows us to define equivalently the hydrostatic Helmholtz projection $\mathbb{P} : L^p(\Omega)^2 \to L^p(\Omega)^2$ as follows: given $v \in L^p(\Omega)^2$, let $\pi \in W^{1,p}_{\text{per}}(G) \cap L^p_0(G)$ be the unique solution of equation (1.22) with $f = \bar{v}$. We then set

$$\mathbb{P}v = v - \nabla_H \pi. \tag{1.25}$$

Remark 4 The space $L^p_{\overline{\sigma}}(\Omega)$, $p \in (1,\infty)$, coincides with the following subsets of $L^p(\Omega)^2$, where $\nu_{\partial G}$ denotes the outer unit normal assigned to ∂G.

(a) $X^1_p := \{v \in L^p(G)^2 : \langle \bar{v}, \nabla_H \phi \rangle_{L^{p'}(G)} = 0 \text{ for all } \phi \in W^{1,p'}_{\text{per}}(G)\}$, where $\frac{1}{p} + \frac{1}{p'} = 1$;
(b) $X^2_p := \{v \in L^p(G)^2 : \operatorname{div}_H \bar{v} = 0, \bar{v} \cdot \nu_{\partial G} \text{ is anti-periodic of order 0 on } \partial G\}$,

see [35, Section 3].

1.5.2 Hydrostatic Stokes Operator

In this subsection we introduce the hydrostatic Stokes operator within the L^p-setting. It can be viewed as the analogue of the classical Stokes operator.

The hydrostatic Helmholtz projection $\mathbb{P}$ defined as in (1.25) allows us to define the hydrostatic Stokes operator as follows. In fact, let $1 < p < \infty$ and $L^p_{\overline{\sigma}}(\Omega)$ be defined as above. Then the *hydrostatic Stokes operator* A_p on $L^p_{\overline{\sigma}}(\Omega)$ is defined by

$$A_p v := \mathbb{P}\Delta v, \quad D(A_p) := \{v \in H^{2,p}_{per}(\Omega)^2 : \partial_z v|_{\Gamma_N} = 0,\ v|_{\Gamma_D} = 0\} \cap L^p_{\overline{\sigma}}(\Omega).$$

The corresponding resolvent problem is

$$\begin{aligned} \lambda v - \Delta v + \nabla_H \pi &= f \ \text{ in } \Omega, \\ \operatorname{div}_H \bar{v} &= 0 \ \text{ in } G, \end{aligned} \tag{1.26}$$

subject to the boundary conditions (1.4).

To study spectral properties of A_p we start by deriving a weak formulation of the problem (1.26). To this end, we introduce the function spaces V and W associated with the velocity and the pressure of the fluid by

$$V := \{\varphi \in H^1_{\text{per}}(\Omega)^2 : \varphi = 0 \text{ on } \Gamma_D\}, \qquad W := L^2_0(G).$$

Note that V and W are closed subspaces of $H^1(\Omega)^2$ and $L^2(G)$, respectively. If (v, π) is a classical solution of (1.26), then multiplying $(1.26)_1$ and $(1.26)_2$ by test functions $(\varphi, \phi) \in V \times W$ and integrating over Ω, we obtain

$$\begin{cases} \lambda(v,\varphi)_{L^2(\Omega)} + (\nabla v, \nabla\varphi)_{L^2(\Omega)} - (p, \operatorname{div}_H \bar{\varphi})_{L^2(G)} = (f,\varphi)_{L^2(\Omega)}, \\ \qquad\qquad\qquad\qquad\qquad\qquad -(\phi, \operatorname{div}_H \bar{v})_{L^2(G)} = 0, \end{cases} \tag{1.27}$$

where with a slight abuse of notation $(\cdot,\cdot)_{L^2(G)}$ is the $L^2(G)$ inner product for either scalars, vectors, or matrices. Conversely, if (v, π) is smooth and satisfies (1.27), then it defines a classical solution of (1.26). Assuming $f \in V^*$, we rephrase equation (1.27) as

$$\begin{cases} a_\lambda(v,\varphi) + b(\pi,\varphi) = \langle f, \varphi\rangle_V, & \varphi \in V, \\ \qquad\qquad\quad b(\phi, v) = 0, & \phi \in W, \end{cases} \tag{1.28}$$

where $a_\lambda : V \times V \to \mathbb{C}$ and $b : W \times V \to \mathbb{C}$ are bounded sesquilinear forms. In the following lemma we state the coerciveness of a and the inf-sup condition for b.

Lemma 3 *Let $\varepsilon \in (0, \pi/2)$ and $\lambda \in \Sigma_{\pi-\varepsilon} \cup \{0\}$.*

a) *There exists a constant $C > 0$ such that*

$$|a_\lambda(\varphi,\varphi)| \geq C(|\lambda|\|\varphi\|^2_{L^2(\Omega)} + \|\varphi\|^2_V), \quad \varphi \in V.$$

b) *There exists a constant $C = C_\Omega > 0$ such that*

$$C\|\phi\|_W \leq \sup_{\varphi\in V} \frac{|b(\phi,\varphi)|}{\|\varphi\|_V}, \quad \phi \in W.$$

The following result follows from the above observations and the Babuška-Brezzi theory on mixed problems (see, e.g., [31, Corollary I.4.1]).

Proposition 1.5.2 (Weak Solvability in L^2) *Let $f \in V^*$. Then there exists a unique solution $(v, \pi) \in V \times W$ of equation* (1.28) *and a constant $C > 0$ such that*

$$\|v\|_V + \|\pi\|_W \leq C\|f\|_{V^*}.$$

In particular, $-A_2 = -\mathbb{P}\Delta$ is associated to the closed symmetric form defined by

$$\mathfrak{a}[v, v'] := \langle \nabla v, \nabla v' \rangle_{L^2(\Omega)^{2\times 2}}, \quad \text{where } v, v' \in \{H^1_{per}(\Omega) \cap L^2_{\overline{\sigma}}(\Omega) : v|_{\Gamma_D} = 0\}.$$

A key observation is that by solving the equation (1.19) for the pressure and applying $(\mathbb{1} - \mathbb{P})$ to (1.19) one can compute the pressure as

$$\nabla_H \pi = (\mathbb{1} - \mathbb{P})\overline{f} - (\mathbb{1} - \mathbb{P}) D_z v \mid_{\Gamma_D},$$

where $(\mathbb{1} - \mathbb{P})v = \nabla_H \Delta_H^{-1} \mathrm{div}_H \overline{v}$, and for brevity we set

$$D_z v \mid_{\Gamma_D} = \tfrac{1}{h} \left(\Gamma_u \partial_z v \mid_{\Gamma_u} - \gamma(b) \partial_z v \mid_{\Gamma_b} \right),$$

and for $c \in \{b, v\}$ we set $\gamma(c) = 1$ if $\Gamma_c \subset \Gamma_D$ and $\gamma(c) = 0$, otherwise. Hence, we get

$$A_p v = \Delta v + \nabla_H \Delta_H^{-1} \mathrm{div}_H D_z v \mid_{\Gamma_D}, \quad v \in D(A_p).$$

Moreover since the perturbation term involves traces

$$D_z v \mid_{\Gamma_D} \in H^{1-1/p-\delta}(G)^2 \quad \text{for } v \in D(A) \subset H^2(\Omega)^2$$

and for some $\delta > 0$. This allows us to analyze the above linear problem by perturbation methods for the Laplacian, where the pressure is in fact a lower order perturbation. Moreover, the additional regularity for the pressure allows one to apply some "bootstrap"-argument to extend results from L^2 to L^p-spaces.

1.5.3 Spectra and Resolvent

The perturbation approach gives the analyticity of the semigroup on some sector. To specify the sector one can study the resolvent problem associated with the hydrostatic Stokes equations within the L^p-setting. More precisely, let

$$\lambda \in \Sigma_{\pi-\varepsilon} := \{\lambda \in \mathbb{C} : |\arg \lambda| < \pi - \varepsilon\}$$

for some $\varepsilon \in (0, \pi/2)$ and $f \in L^p(\Omega)^2$ for some $1 < p < \infty$.

Theorem 3 (Resolvent Estimate in L^p) *Assume that $\Gamma_D \neq \emptyset$. Let $\lambda \in \Sigma_{\pi-\varepsilon} \cup \{0\}$ for $\varepsilon \in (0, \pi/2)$. Moreover, let $p \in (1, \infty)$ and $f \in L^p(\Omega)^2$. Then equations* (1.19) *admit a unique solution $(v, \pi) \in H^{2,p}_{\mathrm{per}}(\Omega)^2 \times H^{1,p}_{\mathrm{per}}(G) \cap L^p_0(G)$ and there exists a constant $C > 0$ such that for $f \in L^p(\Omega)^2$*

$$|\lambda|\,\|v\|_{L^p(\Omega)} + \|v\|_{H^{2,p}(\Omega)} + \|\pi\|_{H^{1,p}(G)} \le C\|f\|_{L^p(\Omega)}. \tag{1.29}$$

The resolvent estimates for equation (1.19) given in Theorem 3 yield that $-A_p$ generates a bounded analytic semigroup on $L^p_{\overline{\sigma}}(\Omega)$ of angle zero.

Proposition 1.5.3 (Elliptic Regularity and Spectral Properties) *Let $p \in (1, \infty)$ and $\lambda \ge 0$ if $\Gamma_D \ne \emptyset$ and $\lambda > 0$ if $\Gamma_D = \emptyset$.*

(a) *Let $v \in D(A_p)$ and $(A_p - \lambda)v = f$. If $\mathbb{P}f \in H^{s,p}(\Omega)^2$ for $s \ge 0$, then $v \in H^{s+2,p}(\Omega)^2$ and there exists a constant $C > 0$ such that*

$$\|v\|_{H^{s+2,p}(\Omega)^2} \le C\|\mathbb{P}f\|_{H^{s,p}(\Omega)^2}.$$

(b) *The spectrum of A_p is purely discrete and all the eigenfunctions of A_p belong to $C^\infty(\overline{\Omega})^2$. In particular, the spectrum of $-A_p$ is independent of p, $\sigma(-A_p) \subset [0, \infty)$ and $\sigma(-A_p) \subset [c, \infty)$ for some $c > 0$ provided $\Gamma_D \ne \emptyset$.*

(c) *The semigroup generated by $A_p - \lambda$ is exponentially stable, i.e., there exist constants $C > 0$, $\beta \ge 0$ such that*

$$\|T_p(t)f\|_{L^p_{\overline{\sigma}}(\Omega)} \le Ce^{-\beta t}\|f\|_{L^p_{\overline{\sigma}}(\Omega)}, \quad t > 0, \tag{1.30}$$

where $\beta > 0$ if $\Gamma_D \ne \emptyset$.

Remarks 1.5.4

(a) In the quantitative analysis of the primitive model the Coriolis force, which may be incorporated by replacing f in (1.2) by $f_0 k \times v$, plays an important role. For the qualitative analysis we omit this term since it is a zero order term, which can be included easily into our analysis by setting

$$A_p v = \mathbb{P}\Delta + \mathbb{P}(f_0 k \times v).$$

(b) In [18] the hydrostatic Stokes operator with Robin boundary conditions is considered. The strategy to solve first for the surface pressure can be applied in this case as well.

1.5.4 Bounded $\mathcal{H}^\infty$-Calculus

For a given operator A a perturbation B is of lower order if for $\alpha \in (0, 1)$

$$D(A^\alpha) \subset D(B) \quad \text{and} \quad |Bv| \le C|v| + c|A^\alpha v|, \quad v \in D(A),$$

where $C > 0$ and $c > 0$ is sufficiently small. Perturbation results for the H^∞-calculus including lower order perturbations were studied by many authors, see, e.g., [2, 19, 20, 52]. In order to deal with the present situation we apply [52, Proposition 13.1] and [20, Proposition 2.7].

Since many mapping properties are known for the Laplacian on cylindrical domains, cf. [62], these now carry over to the hydrostatic Stokes operator.

Theorem 4 (Bounded H^∞-Calculus) *Let $p \in (1,\infty)$ and $\nu \geq 0$. Then the operator $-A_p+\nu$ admits a bounded H^∞-calculus on $L^p_{\overline{\sigma}}(\Omega)$ with $\phi^\infty_A = 0$ provided $\nu > 0$. If $\Gamma_D \neq \emptyset$, then the above assertion holds true even for $\nu = 0$.*

The existence of the bounded $\mathcal{H}^\infty$-calculus for $-A_p$ implies that

$$D((-A_p)^\theta) = [L^p_{\overline{\sigma}}(\Omega), D(A_p)]_\theta \quad \text{for} \quad \theta \in [0,1],$$

where $[\cdot,\cdot]_\theta$ denotes the complex interpolation functor, compare, e.g., [20, Theorem 2.5]. Since $D(A_p) \subset H^{2,p}(\Omega)^2$, we may conclude analogously to [35, Lemma 4.6 (a)] that $D(A_p^\theta) \subset H^{2\theta,p}(\Omega)^2$. A suitable retract to compute the interpolation spaces in terms of boundary conditions will be discussed in the next subsection.

Corollary 1 (Fractional Powers) *Let $1 < p < \infty$ and $\theta \in [0,1]$. Then*

$$D((-A_p)^\theta) = [L^p_{\overline{\sigma}}(\Omega), D(A_p)]_\theta.$$

Considering $(-A_p)^{1/2}$ we obtain from Corollary 1 the L^p-boundedness of the *hydrostatic Riesz transformations* associated with A_p.

Corollary 2 (Boundedness of Hydrostatic Riesz Transformations) *Let $1 < p < \infty$. Then the hydrostatic Riesz transform*

$$R_p\colon L^p_{\overline{\sigma}}(\Omega) \to L^p(\Omega)^{2\times 2} \quad \textit{given by} \quad R_p v := \nabla(-A_p)^{-1/2} v$$

is bounded provided $\Gamma_D \neq \emptyset$.

As a further consequence, we obtain maximal L^q-L^p-regularity estimates for the linearized primitive equations.

Corollary 3 (Maximal L^q_t-L^p_x-Regularity) *Let $p,q \in (1,\infty)$ and $T \in (0,\infty)$. Then*

$$-A_p \in \mathcal{M}_q((0,T); L^p_{\overline{\sigma}}(\Omega)).$$

In particular, A_p is the generator of an analytic semigroup on $L^p_{\overline{\sigma}}(\Omega)$. If $\Gamma_D \neq \emptyset$, then the above assertion also holds true for $T = \infty$.

Recall that $-A_p \in \mathcal{M}_q((0,T); L^p_{\overline{\sigma}}(\Omega))$ means that

$$\partial_t v + A_p v = f, \quad v(0) = v_0$$

admits for

$$f \in L^q(0,T; L^p_{\overline{\sigma}}(\Omega)) \quad \text{and} \quad v_0 \in (D(A_p), L^p_{\overline{\sigma}}(\Omega))_{1-1/q,q}$$

a unique solution in the maximal regularity space

$$v \in L^q(0,T;D(A_p)) \cap H^{1,q}(0,T;L^p_{\overline{\sigma}}(\Omega)).$$

Here $(\cdot,\cdot)_{\theta,q}$ denotes the real interpolation functor. This means that both $\partial_t v$ and $A_p v$ are in $L^q(0,T;L^p_{\overline{\sigma}}(\Omega))$. This property is essential for the construction of local strong solutions in Sect. 1.8. Recall that

$$A \text{ has bounded } H^\infty\text{-calculus} \Rightarrow A \text{ has maximal } L^q\text{-regularity} \Rightarrow A \text{ is sectorial.}$$

1.5.5 Interpolation and Trace Spaces

In this subsection we give an explicit characterization of the complex and real interpolation spaces

$$V_\theta := [L^p_{\overline{\sigma}}(\Omega), D(A_p)]_\theta \quad \text{and} \quad X_{\theta,q} := (L^p_{\overline{\sigma}}(\Omega), D(A_p))_{\theta,q}$$

related to the hydrostatic Stokes operator, where $[\cdot,\cdot]_\theta$ denotes the complex interpolation functor and $(\cdot,\cdot)_{\theta,q}$ the real interpolation functor for $0 \le \theta \le 1$ and $q \in (1,\infty)$. Then the above spaces are characterized as follows.

Proposition 1.5.5 (Interpolation Spaces) *Let* $\theta \in (0,1)$ *and* $p,q \in (1,\infty)$ *with* $\theta \notin \{1/2p, 1/2+1/2p\}$,

(a) *then one has*

$$V_\theta = \begin{cases} \{v \in H^{2\theta,p}_{per}(\Omega) \cap L^p_{\overline{\sigma}}(\Omega) : \partial_z v|_{\Gamma_N} = 0,\ v|_{\Gamma_D} = 0\}, & \frac{1}{2} + \frac{1}{2p} < \theta \le 1, \\ \{v \in H^{2\theta,p}_{per}(\Omega) \cap L^p_{\overline{\sigma}}(\Omega) : v|_{\Gamma_D} = 0\}, & \frac{1}{2p} < \theta < \frac{1}{2} + \frac{1}{2p}, \\ \{v \in H^{2\theta,p}_{per}(\Omega) \cap L^p_{\overline{\sigma}}(\Omega)\}, & \theta < \frac{1}{2p}, \end{cases}$$

(b) *and*

$$X_{\theta,q} = \begin{cases} \{v \in B^{2\theta}_{pq,per}(\Omega) \cap L^p_{\overline{\sigma}}(\Omega) \colon \partial_z v|_{\Gamma_N} = 0,\ v|_{\Gamma_D} = 0\}, & \frac{1}{2} + \frac{1}{2p} < \theta < 1, \\ \{v \in B^{2\theta}_{pq,per}(\Omega) \cap L^p_{\overline{\sigma}}(\Omega) \colon v|_{\Gamma_D} = 0\}, & \frac{1}{2p} < \theta < \frac{1}{2} + \frac{1}{2p}, \\ B^{2\theta}_{pq,per}(\Omega) \cap L^p_{\overline{\sigma}}(\Omega), & 0 < \theta < \frac{1}{2p}. \end{cases}$$

Proof Let us note first that, following, for instance, the work of Amann [1], results on the interpolation of boundary conditions for Sobolev spaces are known for elliptic second operators on domains with C^∞-boundaries subject to mixed boundary conditions on disjoint parts of the boundaries. This carries over to the present situation provided there are suitable retractions of interpolation couples from such a situation to the one considered here.

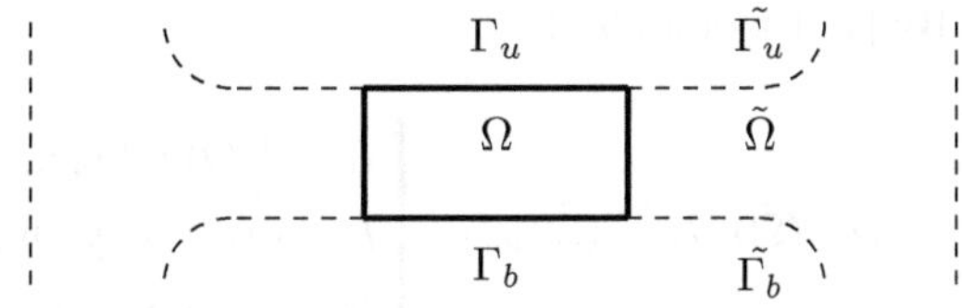

Fig. 1.1 Extension of Ω to $\tilde{\Omega}$

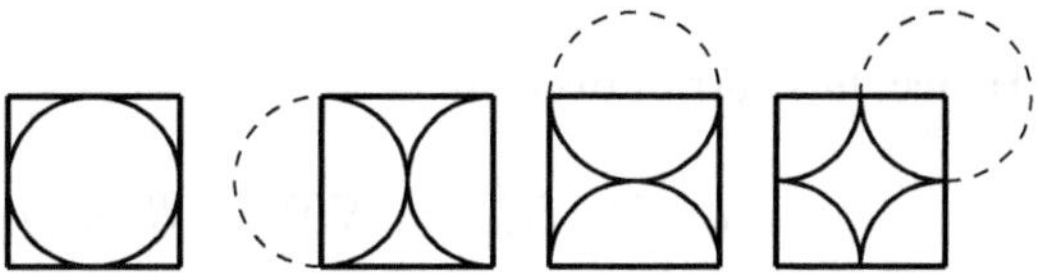

Fig. 1.2 Covering $\{G_i\}_{i=1,\dots,4}$ for $G \cong S^1 \times S^1$. (**a**) G_1. (**b**) G_2. (**c**) G_3. (**d**) G_4

Note first that there is a C^∞ domain $\tilde{\Omega}$ extending Ω such that the boundary of $\tilde{\Omega}$ extends $\Gamma_u \subset \tilde{\Gamma}_u$ and $\Gamma_b \subset \tilde{\Gamma}_b$ for $\tilde{\Gamma}_u, \tilde{\Gamma}_b \subset \partial\tilde{\Omega}$. Such an $\tilde{\Omega}$ is schematically depicted in Fig. 1.1.

Let v_i be a finite covering of $S^1 \times S^1 \times (-h, 0)$, cf. Fig. 1.2, $i \in I$, $|I| < \infty$, and consider a smooth partition of unity $\varphi_i : \Omega \to [0, 1]$ with $\operatorname{supp} \varphi_i \subset v_i$ such that

$$\sum_{i \in I} \varphi_i \equiv 1. \tag{1.31}$$

Let for each $i \in I$ be $\tilde{\Omega}_i$ a copy of $\tilde{\Omega}$. With v_i as in Fig. 1.2, each v_i is sufficiently small such that v_i can be identified given with $\tilde{v}_i = \tilde{G}_i \times (-h, 0)$, $\tilde{G}_i$ dashed. Now, we can define the co-retract S and the corresponding retract R by

$$Sv = \{\chi_i v\}_{i \in I}, \quad Rv = \sum_{i \in I} \chi_i v_i, \quad \text{where } \chi_i := \sqrt{\varphi_i},$$

respectively. This defines for $s \in [0, \infty)$ and $p \in (1, \infty)$ maps which are preserving boundary conditions imposed on Γ_u, Γ_b and $\tilde{\Gamma}_v, \tilde{\Gamma}_b$, respectively,

$$S\colon H^{s,p}_{per,b.c.}(\Omega) \to \bigoplus_{i \in I} H^{s,p}_{b.c.}(\tilde{\Omega}_i), \quad R\colon \bigoplus_{i \in I} H^{s,p}_{b.c.}(\tilde{\Omega}_i) \to H^{s,p}_{per,b.c.}(\Omega),$$

where abbreviating $H^{s,p}_{per,b.c.}(\Omega)$ and $H^{s,p}_{b.c.}(\tilde{\Omega})$ denote spaces with boundary conditions as considered here. From (1.31) we conclude that $S \circ R \equiv \mathbb{1}$ and that R is indeed a retraction.

Using now general properties of interpolation couples and their interplay with retractions and co-retractions, compare [71, Theorem 1.2.4], yields for $\theta \in [0, 1]$

$$\begin{aligned}[L^p(\Omega), H^{2,p}_{per,b.c.}(\Omega)]_\theta &= [R(\oplus_{i \in I} L^p(\tilde{\Omega}_i)), R(\oplus_{i \in I} H^{2,p}_{b.c.}(\tilde{\Omega}_i))]_\theta \\ &= R(\oplus_{i \in I}[L^p(\tilde{\Omega}_i), H^{2,p}_{b.c.}(\tilde{\Omega}_i)]_\theta).\end{aligned}$$

By [1, Theorem 5.2]

$$[L^p(\tilde{\Omega}), H^{2,p}_{b.c.}(\tilde{\Omega})]_\theta = \begin{cases} H^{2\theta,p}(\tilde{\Omega}) \text{ all b.c.}, & 1+1/p<\theta\le 2,\\ H^{2\theta,p}(\tilde{\Omega}) \text{ only Dirichlet part}, & 1/p<\theta\le 2,\\ H^{2\theta,p}(\tilde{\Omega}) \text{ without b.c.}, & 0\le\theta\le 1/p. \end{cases}$$

Hence, this carries over to

$$[L^p(\Omega), H^{2,p}_{per,b.c.}(\Omega)]_\theta = R(\oplus_{i\in I}[L^p(\tilde{\Omega}_i), H^{2,p}_{b.c.}(\tilde{\Omega}_i)]_\theta),$$

and by [71, 1.7.1 Theorem 1] we conclude for the velocity

$$\begin{aligned}[L^p_{\overline{\sigma}}(\Omega), D(A_p)]_\theta &= [L^p(\Omega)\cap L^p_{\overline{\sigma}}(\Omega), H^{2,p}_{per,b.c.}(\Omega)\cap L^p_{\overline{\sigma}}(\Omega)]_\theta\\ &= [L^p(\Omega), H^{2,p}_{per,b.c.}(\Omega)]_\theta \cap L^p_{\overline{\sigma}}(\Omega).\end{aligned}$$

For (*b*) the proof works analogously replacing complex by real interpolation. □

1.5.6 Semigroup Estimates

Note that from Proposition 1.5.3 (c), we obtain also the following interpolation inequalities for the semigroup generated by A_p. This semigroup smoothing with respect to interpolation spaces is crucial for the construction of local solutions via the Fujita-Kato approach, in Sect. 1.8.

Proposition 1.5.6 (Semigroup Estimates in Interpolation Spaces) *Let* $p,q\in(1,\infty)$, $0\le\theta_1,\theta_2\le 1$ *with* $\theta_1+\theta_2\le 1$. *Then*

(a) *There exist constants* $C>0$ *and* β *such that for* $t>0$

$$\|e^{tA_p}f\|_{V_{\theta_1+\theta_2,p}} \le Ct^{-\theta_1}e^{-\beta t}\|f\|_{V_{\theta_2}}, \qquad f\in V_{\theta_2};$$

(b) $t^{\theta_1}\|e^{tA_p}f\|_{V_{\theta_1+\theta_2,p}} \to 0$ *as* $t\to 0$.

From Corollary 1 we obtain that the hydrostatic semigroup admits global L^p-L^q-smoothing properties.

Proposition 1.5.7 (L^p-L^q-Smoothing) *Let* $\Gamma_D\neq\emptyset$ *and* $p,q\in(1,\infty)$ *such that* $p\le q$. *Then there exists a constant* $C>0$ *such that*

$$\|e^{tA_p}P_pf\|_{L^q(\Omega)^2} \le Ct^{-\frac{3}{2}\left(\frac{1}{p}-\frac{1}{q}\right)}\|f\|_{L^p(\Omega)^2}, \qquad \text{for } f\in L^p(\Omega)^2, \quad t>0,$$

$$\|\nabla e^{tA_p}P_pf\|_{L^q(\Omega)^2} \le Ct^{-\frac{3}{2}\left(\frac{1}{p}-\frac{1}{q}\right)-\frac{1}{2}}\|f\|_{L^p(\Omega)^2}, \qquad \text{for } f\in L^p(\Omega)^2, \quad t>0,$$

$$\|e^{tA_p} P_p \nabla \cdot f\|_{L^q(\Omega)^2} \leq C t^{-\frac{3}{2}\left(\frac{1}{p}-\frac{1}{q}\right)-\frac{1}{2}} \|f\|_{L^p(\Omega)^{2\times 2}}, \quad for\ f \in L^p(\Omega)^{2\times 2}, \quad t>0.$$

1.6 Hydrostatic Stokes Equations in Anisotropic $L^\infty(L^p)$-Spaces for Neumann Boundary Conditions

This section concerns the linear hydrostatic Stokes equations subject to Neumann boundary conditions. We establish estimates on the hydrostatic semigroup needed for the iteration scheme later on.

1.6.1 Interpolation Inequalities for the Caputo Fractional Derivative

In this subsection we consider the Caputo fractional derivatives in vertical direction and collect some anisotropic interpolation inequalities.

First, recall that the Riemann-Liouville operator is given by

$$I_{z_0}^\alpha f = \frac{z_+^{\alpha-1}}{\Gamma(\alpha)} * f_+, \quad f \in L^\infty(J),$$

where the zero extension of f to $\mathbb{R}$ is denoted by f_+, and the zero extension of z from $(0, h)$ to $\mathbb{R}$ is denoted by z_+. We also set $I_{z_0}^0 f = f$. Then $I_{z_0}^\alpha f$ is called the α-times integral of f from z_0 whenever $\alpha > 0$, and

$$I_{z_0}^{\alpha_1+\alpha_2} = I_{z_0}^{\alpha_1} I_{z_0}^{\alpha_2} \quad \text{for all} \quad \alpha_1, \alpha_2 > 0,$$

cf. [36, Section 23.16]. The Caputo derivative ∂_z^α for $\alpha \in (0, 1)$ is defined by

$$\left(\partial_z^\alpha f\right)(z) := \left(I_{z_0}^{1-\alpha}(\partial_z f)\right)(z), \ z \in \overline{J},$$

where $\partial_z f = \partial f/\partial z$. This formula is well-defined for $f \in W^{1,p}(J)$. In fact, by the Hausdorff-Young inequality for convolutions

$$\left\|\partial_z^\alpha f\right\|_{L^p(z_0,z_0+\mu)} = \left(\int_{z_0}^{z_0+\mu} \left|\partial_z^\alpha f(z)\right|^p \mathrm{d}z\right)^{1/p} \leq \frac{\mu^{1-\alpha}}{\Gamma(2-\alpha)} \|\partial_z f\|_{L^p(z_0,z_0+\mu)} \tag{1.32}$$

for $\mu \in (0, h)$, since $\int_0^\mu z^{-\alpha} \mathrm{d}z = \mu^{1-\alpha}/(1-\alpha)$. Here we identified $\partial_z f$ with $(\partial_z f)_+ \cdot \chi_{(z_0,z_0+\mu)}$ denoting by $\chi_{(z_0,z_0+\mu)}$ the characteristic function.

Lemma 4 (Interpolation Inequality for the Caputo Derivative) *Let* $\alpha \in (0, 1)$ *and* $p \in [1, \infty]$. *Then the estimate*

$$\left\|\partial_z^\alpha f\right\|_p \leq \frac{2}{\Gamma(1-\alpha)} \|f\|_p^{1-\alpha} \|\partial_z f\|_p^\alpha \tag{1.33}$$

holds true for all $f \in W^{1,p}(J)$ *satisfying* $f(z_0) = 0$.

1.6.2 Interpolation Inequalities for the Horizontal Derivatives

We next state an interpolation inequality for the horizontal Laplace operator in the space $L^\infty(L^p)$. Denote by G_t the 2-dimensional Gauss kernel, i.e.,

$$G_t(x) = (4\pi t)^{-1} \exp\left(-|x|^2/4t\right), \quad x \in \mathbb{R}^2, t > 0,$$

and let $e^{t\Delta_H} f = G_t *_H f$, where $*_H$ denotes convolution in the horizontal variables, only. Then the negative fractional powers of $-\Delta_H$ are defined by

$$(-\Delta_H)^{-\alpha/2} f = \frac{1}{\Gamma(\alpha/2)} \int_0^\infty s^{\frac{\alpha}{2}-1} e^{s\Delta_H} f \, \mathrm{d}s, \quad \alpha \in (0, 2). \tag{1.34}$$

Lemma 5 (Interpolation Inequality for Horizontal Derivatives) *Let* $\alpha \in (0, 1)$ *an* $p \in [1, \infty]$. *Then there exists a constant* $C > 0$ *such that*

$$\left\|\nabla_H (-\Delta_H)^{-\alpha/2} f\right\|_{\infty,p} \leq C \|f\|_{\infty,p}^\alpha \|\nabla_H f\|_{\infty,p}^{1-\alpha} \tag{1.35}$$

for all $f \in L^\infty(\mathbb{R}^2; L^p(J))$ *with* $\nabla_H f \in L^\infty(\mathbb{R}^2; L^p(J))$.

1.6.3 Pointwise and L^∞ Bounds for the Heat Semigroup, Riesz Transforms, and Fractional Powers of the Laplacian

In this section, we derive estimates in time and space for fractional derivatives and for the semigroups $e^{t\Delta_N}$ and $e^{t\Delta_H}$, where Δ_N denotes the three-dimensional Laplacian on $\mathbb{R}^2 \times J$ with Neumann boundary conditions.

Lemma 6 (Decay Estimates for the Heat Semigroup Acting on Fractional Derivatives on an Interval) *Given* $\alpha \in [0, 1]$ *and* $p \in [1, \infty]$, *there exists a constant* $C > 0$ *such that*

$$\left\|e^{t\Delta_N} \partial_z I_{z_0}^\alpha f\right\|_p \leq C t^{-(1-\alpha)/2} \|f\|_p, \quad t > 0,$$

for all $f \in L^p(J)$ *satisfying* $I^\alpha_{z0} f(z_1) = 0$.

As mentioned above, the operator I^0_{z0} is interpreted as identity.

Proof By duality

$$\left\| e^{t\Delta_N} \partial_z I^\alpha_{z0} f \right\|_p = \sup \left\{ \langle e^{t\Delta_N} \partial_z I^\alpha_{z0} f, \varphi) \rangle \mid \varphi \in C^\infty_c(J), \|\varphi\|_{p'} \le 1 \right\},$$

where $\langle \varphi, \psi \rangle = \int_J \varphi\psi \mathrm{d}z$ and $\frac{1}{p} + \frac{1}{p'} = 1$ for $p \in [1, \infty)$. Moreover,

$$\left\langle e^{t\Delta_N} \partial_z I^\alpha_{z0} f, \varphi \right\rangle = \left\langle \partial_z I^\alpha_{z0} f, e^{t\Delta_N} \varphi \right\rangle = -\left\langle I^\alpha_{z0} f, \partial_z e^{t\Delta_N} \varphi \right\rangle,$$

where in the last identity we used that $(I^\alpha_{z0} f)(z_1) = 0$ and $(I^\alpha_{z0} f)(z_0) = 0$. Since

$$\left\langle I^\alpha_{z0} f, \psi \right\rangle = \left\langle f, \overline{I}^\alpha_{z_1} \psi \right\rangle$$

with

$$\overline{I}^\alpha_{z_1} \psi(z) = \frac{1}{\Gamma(\alpha)} \int_z^{z_1} (\xi - z)^{\alpha-1} \psi(\xi) \mathrm{d}\xi,$$

one concludes that

$$\left\langle e^{t\Delta_N} \partial_z I^\alpha_{z0} f, \varphi \right\rangle = -\left\langle f, \overline{I}^\alpha_{z_1} \partial_z e^{t\Delta_N} \varphi \right\rangle.$$

Since $\overline{I}^\alpha_{z_1} \partial_z$ is similar to the Caputo fractional derivative and $\partial_z e^{t\Delta} \varphi(z_1, t) = 0$, one can adapt Lemma 4 to obtain

$$\left\| \overline{I}^\alpha_{z_1} \partial_z e^{t\Delta_N} \varphi \right\|_{p'} \le \frac{2}{\Gamma(\alpha)} \left\| e^{t\Delta_N} \varphi \right\|^\alpha_{p'} \left\| \partial_z e^{t\Delta_N} \varphi \right\|^{1-\alpha}_{p'}.$$

One can prove that $\left\| e^{t\Delta_N} \varphi \right\|_{p'} \le \|\varphi\|_{p'}$ for all $t > 0$ and there is a $C > 0$ independent of φ such that

$$\left\| \partial_z e^{t\Delta_N} \varphi \right\|_{p'} \le C t^{-1/2} \|\varphi\|_{p'} \quad \text{for all} \quad t > 0.$$

This follows by extending the problem to the whole space problem, cf. Lemma 10 below. To this end, we extend φ periodically in a suitable way to $\mathbb{R}$ to obtain $e^{t\Delta_N} \varphi = G_t * \tilde{\varphi}$, where $\tilde{\varphi}$ denotes the extension of φ. Thus

$$\left\| \overline{I}^\alpha_{z_1} \partial_z e^{t\Delta_N} \varphi \right\|_{p'} \le C t^{-(1-\alpha)/2} \|\varphi\|_{p'}, \quad t > 0,$$

with $C > 0$ depending on α, only. We therefore conclude that

$$\left|\langle e^{t\Delta_N}\varphi_z I^\alpha_{z_0} f, \varphi\rangle\right| \le \|f\|_p \left\|\overline{I}^\alpha_{z_1} \partial_z e^{t\Delta_N}\varphi\right\|_{p'} \le C t^{-(1-\alpha)/2} \|f\|_p \|\varphi\|_{p'}, \quad t > 0.$$

Note that the case $p = \infty$ follows by duality from the case $p = 1$, and for $\alpha = 1$ the assertion remains true since the $\partial_z I^\alpha_{z_0} f = f$ if $f(z_0) = 0$.

□

We proceed with pointwise estimates for the heat semigroup $e^{t\Delta}$ on $L^\infty(\mathbb{R}^d)$ combined with Riesz transforms and fractional powers of the Laplacian. We use the Bochner representation formula for the fractional powers of the Laplacian given by

$$(-\Delta)^{-\alpha/2} f = \frac{1}{\Gamma(\alpha/2)} \int_0^\infty s^{\alpha/2-1} (G_s * f)\, ds, \quad \alpha > 0, \tag{1.36}$$

where G_s is the d-dimensional Gauss kernel. Using the smoothing effect of $e^{t\Delta}$ for $t > 0$ we obtain

$$e^{t\Delta}(-\Delta)^{\alpha/2} f = (-\Delta)^{-(1-\alpha/2)}(-\Delta) e^{t\Delta} f,$$

and the representation (1.36) yields

$$e^{t\Delta}(-\Delta)^{\alpha/2} f = \frac{1}{\Gamma(1-\alpha/2)} \int_0^\infty s^{-\alpha/2} (-\Delta G_{s+t}) * f\, ds,$$

interpreting $(-\Delta)^0$ as the identity operator. Recall that the i-th Riesz transform is defined by

$$R_i = \partial_i (-\Delta)^{-1/2}, \quad \text{where } \partial_i = \partial/\partial x_i \quad \text{for all } 1 \le i \le d.$$

Lemma 7 (Pointwise Bounds for $e^{t\Delta}(-\Delta)^{\alpha/2}$ and $e^{t\Delta} R_i R_j (-\Delta)^{\alpha/2}$) *Consider $\mathbb{R}^d$, $d \in \mathbb{N}$.*

(i) *Let $\alpha \in [0, 2]$. Then there exists $H_t \in L^1(\mathbb{R}^d)$ satisfying $|H_t\|_1 \le C$ for some $C > 0$ independent of $t > 0$ such that all $f \in L^\infty(\mathbb{R}^d)$*

$$\left|e^{t\Delta}(-\Delta)^{\alpha/2} f(x)\right| \le t^{-\alpha/2} (H_t * |f|)(x), \quad x \in \mathbb{R}^d,\ t > 0.$$

In particular,

$$\left\|e^{t\Delta}(-\Delta)^{\alpha/2} f\right\|_\infty \le C t^{-\alpha/2} \|f\|_\infty, \quad t > 0.$$

(ii) *Let $\alpha \in (0, 2]$. Then there exists $\tilde{H}_t \in L^1(\mathbb{R}^d)$ satisfying $|\tilde{H}_t\|_1 \le C$ for some $C > 0$ independent of $t > 0$ such that for all $f \in L^\infty(\mathbb{R}^d)$*

$$\left|e^{t\Delta}R_iR_j(-\Delta)^{\alpha/2}f(x)\right| \le t^{-\alpha/2}(\tilde{H}_t * |f|)(x), \quad x \in \mathbb{R}^d,\ t > 0.$$

In particular,

$$\left\|e^{t\Delta}R_iR_j(-\Delta)^{\alpha/2}f\right\|_\infty \le Ct^{-\alpha/2}\|f\|_\infty, \quad t > 0.$$

(iii) *There exists* $\breve{H}_t \in L^1(\mathbb{R}^d)$ *satisfying* $|\breve{H}_t\|_1 \le C$ *for some* $C > 0$ *independent of* $t > 0$ *such that for all* $f \in L^\infty(\mathbb{R}^d)$

$$\left|e^{t\Delta}R_iR_j\partial_k f(x)\right| \le t^{-1/2}(\breve{H}_t * |f|)(x), \quad x \in \mathbb{R}^d,\ t > 0.$$

In particular,

$$\left\|e^{t\Delta}R_iR_j\partial_k f\right\|_\infty \le Ct^{-1/2}\|f\|_\infty, \quad t > 0.$$

Remark 5 The Riesz transforms are unbounded operators on $L^\infty(\mathbb{R}^d)$, nevertheless the compositions of the operators $e^{t\Delta}R_iR_j(-\Delta)^{\alpha/2}$ and $\partial_k e^{t\Delta}R_iR_j$ define nevertheless bounded operators on $L^\infty(\mathbb{R}^d)$ for all $t > 0$.

Proof of Lemma 7 Let $\beta \in \mathbb{N}^d$. Then there exists a constant $C = C_{d,\beta} > 0$ such that for all $t > 0$ one has

$$|\partial^\beta G_t| \le Ct^{-|\beta|/2}G_{2t}. \tag{1.37}$$

Therefrom, it follows that

$$\begin{aligned}|e^{t\Delta}(-\Delta)^{\alpha/2}f| &\le \frac{C}{\Gamma(1-\alpha/2)}\int_0^\infty s^{-\alpha/2}(s+t)^{-1}G_{2(s+t)} * |f|\,\mathrm{d}s \\ &= \frac{C}{\Gamma(1-\alpha/2)}t^{-\alpha/2}\int_0^\infty v^{-\alpha/2}(v+1)^{-1}G_{2t(v+1)} * |f|\,\mathrm{d}v.\end{aligned}$$

Setting

$$H_t := \frac{C}{\Gamma(1-\alpha/2)}\int_0^\infty v^{-\alpha/2}(v+1)^{-1}G_{2t(v+1)}\,\mathrm{d}v$$

and observing that $|H_t\|_1 \le C < \infty$ for all $t > 0$ provided $\alpha \in (0,2)$ yields estimate (i) for those values of α. For $\alpha = 0$ and $\alpha = 2$, we set $H_t := G_t$ and $H_t := G_{2t}$, respectively, and apply (1.37).

In order to prove estimate (ii), observe that

$$e^{t\Delta}R_iR_j(-\Delta)^{\alpha/2}f = (-\Delta)^{-(1-\alpha/2)}\partial_i\partial_j e^{t\Delta}f, \quad 1 \le i,j \le d.$$

The case $\alpha = 2$ then follows from (1.37) by setting $\tilde{H}_t := G_{2t}$, and for $\alpha \in (0, 2)$ we have

$$e^{t\Delta} R_i R_j (-\Delta)^{\alpha/2} f = \frac{1}{\Gamma(1 - \alpha/2)} \int_0^\infty s^{-\alpha/2} (\partial_i \partial_j G_{s+t}) * f \, ds,$$

and then the same argument used to derive (i) applies. For (iii) we write

$$e^{t\Delta} R_i R_j \partial_k f = (-\Delta)^{-1} \partial_i \partial_j \partial_k e^{t\Delta} f = \int_0^\infty \partial_i \partial_j \partial_k G_{s+t} * f \, ds$$

and since by (1.37) we have $|\partial_i \partial_j \partial_k G_{s+t}| \le C(s + t)^{-3/2} G_{2(s+t)}$ for $s, t > 0$ we can set

$$\breve{H}_t := \int_0^\infty (v + 1)^{-3/2} G_{2t(v+1)} \, dv.$$

Then $|\breve{H}_t\|_1 \le \int_0^\infty (v+1)^{-3/2} \, dv < \infty$, which yields estimate (iii). The corresponding norm estimates then follow from estimates (i)–(iii) and the Hausdorff-Young inequality for convolutions. □

1.6.4 Anisotropic Estimates for the Hydrostatic Stokes Semigroup

Due to the Neumann boundary conditions, the hydrostatic Stokes semigroup S on $L^\infty_\sigma(L^p(J))$ for $p \in [1, \infty]$ is given by

$$S(t) = e^{t\Delta_H} \otimes e^{t\Delta_N}, \quad t > 0,$$

and that its extension to the larger space $L^\infty(L^p)$ for $p \in [1, \infty]$ will be denoted by S_∞. The main result of this subsection is

Proposition 1.6.1 (Linear Estimates for the Hydrostatic Stokes Semigroup) *Let $p \in [1, \infty]$. Then the following assertions hold:*

(i) *There exists a constant $C > 0$ such that for all $f \in L^\infty(L^p)$*

$$\|\nabla S_\infty(t) f\|_{\infty,p} \le C t^{-1/2} \|f\|_{\infty,p},$$
$$\|S_\infty(t) \nabla_H \cdot f\|_{\infty,p} \le C t^{-1/2} \|f\|_{\infty,p}, \quad t > 0.$$

(ii) *For $\alpha \in [0, 1)$, there exists a constant $C > 0$ such that for all $f \in L^\infty(L^p)$ satisfying $I^\alpha_{z_0} f(z_1) = 0$*

$$\left\| S_\infty(t)\partial_z I^\alpha_{z_0} f \right\|_{\infty,p} \le C t^{-(1-\alpha)/2} \|f\|_{\infty,p}, \quad t > 0.$$

(iii) *For* $\alpha \in (0, 2]$, *there exists a constant* $C > 0$ *such that for all* $f \in L^\infty(L^p)$

$$\left\| S_\infty(t)\mathbb{P}(-\Delta_H)^{\alpha/2} f \right\|_{\infty,p} \le C t^{-\alpha/2} \|f\|_{\infty,p}, \quad t > 0.$$

(iv) *There exists a constant* $C > 0$ *such that for all* $f \in L^\infty(L^p)$

$$\|S_\infty(t)\mathbb{P}\nabla_H \cdot f\|_{\infty,p} \le C t^{-1/2} \|f\|_{\infty,p}, \quad t > 0.$$

(v) *There exists a constant* $C > 0$ *such that for all* $f \in L^\infty(L^p)$

$$\|S_\infty(t) f\|_{\infty,p} \le C t^{-(1-1/p)} \|f\|_{\infty,1}, \quad t > 0.$$

Remark 6 In the case where $\alpha = 0$ in assertion (ii), the operator $I^0_{z_0}$ is interpreted as the identity operator and there is no restriction for f other than $f \in L^\infty(L^p)$.

To prove this, it is helpful to investigate first the periodic heat semigroup on $L^p(J)$.

Lemma 8 (Estimate for the Periodic Heat Semigroup) *Let* $\mathbb{T} = \mathbb{R}/\omega_0\mathbb{Z}$ *for some* $\omega_0 > 0$, $p \in [1, \infty]$ *and* $f \in L^p(\mathbb{T})$. *Then*

$$(G_t * f)(z) = \int_0^{\omega_0} E_t(z - y) f(y) \mathrm{d}y, \quad z \in \mathbb{T}, \quad t > 0,$$

where $E_t(z) = \sum_{k=-\infty}^{\infty} G_t(z - k\omega_0)$ *for* $z \in \mathbb{T}$. *In particular,*

$$\|G_t * f\|_{L^p(\mathbb{T})} \le \|f\|_{L^p(\mathbb{T})}, \quad t > 0.$$

Proof of Lemma 7 The above representation for $G_t * f$ follows by noting that

$$(G_t * f)(z) = \sum_{k=-\infty}^{\infty} \int_{k\omega_0}^{(k+1)\omega_0} G_t(z - y) f(y) \mathrm{d}y, \quad z \in \mathbb{T}, \quad t > 0,$$

and

$$\int_{k\omega_0}^{(k+1)\omega_0} G_t(z - y) f(y) \mathrm{d}y = \int_0^{\omega_0} G_t(z - y - k\omega_0) f(y + k\omega_0) \mathrm{d}y, \quad t > 0,$$

where $f(y+k\omega_0) = f(y)$ for all $k \in \mathbb{Z}$ by periodicity. The estimate claimed follows now from Young's inequality since $\int_0^{\omega_0} E_t(z - y)\mathrm{d}z = \int_{-\infty}^{\infty} G_t(z - y)\mathrm{d}z = 1$ for all $t > 0$ and since $E_t \ge 0$ for all $t > 0$. □

Lemma 9 (Derivative Estimate for the Periodic Heat Semigroup) *Given the assumptions of Lemma 8, there exists a constant $C > 0$, independent of ω_0, such that*

$$|\partial_z(G_t * f)(z)| \leq Ct^{-1/2} \int_0^{\omega_0} E_{2t}(z-y)\,|f(y)|\,\mathrm{d}y, \quad z \in \mathbb{T}, \quad t > 0.$$

In particular,

$$\|\partial_z(G_t * f)\|_{L^p(\mathbb{T})} \leq Ct^{-1/2}\|f\|_{L^p(\mathbb{T})}, \quad t > 0.$$

The following lemma can be traced back to the previous lemma by periodization.

Lemma 10 (Derivative Estimate for the Heat Semigroup) *Given $p \in [1, \infty]$, then there exists a constant $C > 0$ such that for all $f \in L^p(J)$*

$$\left\|e^{t\Delta_N} f\right\|_{L^p(J)} \leq \|f\|_{L^p(J)} \quad \text{and} \quad \left\|\partial_z e^{t\Delta_N} f\right\|_{L^p(J)} \leq Ct^{-1/2}\|f\|_{L^p(J)}, \quad t > 0.$$

Proof of Proposition 1.6.1

(*i*) These assertions follow from Lemma 10, and from the pointwise estimates

$$\left|\nabla_H e^{t\Delta_H} f\right| \leq Ct^{-1/2} G_{2t} * |f|, \quad \left|e^{t\Delta_H} f\right| \leq G_t * |f|,$$

and $e^{t\Delta_H}\partial_{x_i} f = \partial_{x_i} e^{t\Delta_H} f$ for $i = 1, 2$.

We first prove that $\|\partial_z S_\infty(t) f\|_{\infty,p} \leq Ct^{-1/2}\|f\|_{\infty,p}$ for all $t > 0$. By Lemma 10

$$\left\|\partial_z S_\infty(t) f(x', \cdot)\right\|_{L^p(J)} \leq Ct^{-1/2}\|e^{t\Delta_H} f(x', \cdot)\|_{L^p(J)}$$

holds for almost all $x' \in \mathbb{R}^2$. By Minkowski's inequality and due to the positivity of $e^{t\Delta_H}$

$$\|e^{t\Delta_H} f(x', \cdot)\|_{L^p(J)} \leq e^{t\Delta_H}\|f(x', \cdot)\|_{L^p(J)},$$

and thus

$$\|\partial_z S_\infty(t) f\|_{\infty,p} \leq Ct^{-1/2} \operatorname{ess\,sup}_{x'} \left(e^{t\Delta_H}\|f(x', \cdot)\|_{L^p(J)}\right) \leq Ct^{-1/2}\|f\|_{\infty,p}, \; t>0.$$

Next, we prove that $\|\nabla_H S_\infty(t) f\|_{\infty,p} \leq Ct^{-1/2}\|f\|_{\infty,p}$ for all $t > 0$. Note that

$$\|e^{t\Delta_N}\nabla_H e^{t\Delta_H} f(x', \cdot)\|_{L^p(J)} \leq \|\nabla_H e^{t\Delta_H} f(x', \cdot)\|_{L^p(J)}.$$

We observe that

$$|\nabla_H e^{t\Delta_H} f(x',z)| \le Ct^{-1/2}\,(G_{2t} *_H |f|)\,(x',z),$$

and applying Minkowski's inequality yields

$$\|\nabla_H e^{t\Delta_H} f(x',\cdot)|_{L^p(J)} \le Ct^{-1/2}(G_{2t} *_H \|f(x',\cdot)\|_{L^p(J)}).$$

We thus conclude that

$$\|\nabla_H S_\infty(t) f\|_{\infty,p} \le Ct^{-1/2}\|f\|_{\infty,p}, \quad t>0.$$

(ii) Since

$$\left|e^{t\Delta_H} g\right|(x') \le (G_t * |g|)(x'), \quad t>0,$$

Fubini's theorem implies

$$\left\|e^{t\Delta_H} e^{t\Delta_N} \partial_z I^\alpha_{z0} f(x',\cdot)\right\|_{L^p(J)} \le G_t * \left\|e^{t\Delta_N}\partial_z I^\alpha_{z0} f(x',\cdot)\right\|_{L^p(J)}, \quad t>0,$$

for almost all $x' \in \mathbb{R}^2$. By Lemma 3

$$\left\|e^{t\Delta_N}\partial_z I^\alpha_{z0} f(x',\cdot)\right\|_{L^p(J)} \le Ct^{-(1-\alpha)/2}\|f(x',\cdot)\|_{L^p(J)}, \quad t>0,$$

which allows us to conclude that

$$\left\|S_\infty(t)\partial_z I^\alpha_{z0} f\right\|_{\infty,p} \le Ct^{-(1-\alpha)/2}\|G_t\|_1\|f\|_{\infty,p} = Ct^{-(1-\alpha)/2}\|f\|_{\infty,p}, \quad t>0.$$

The proof is also valid for the case $\alpha = 0$ yielding $\|S_\infty(t)\partial_z f\|_{\infty,p} \le Ct^{-1/2}\|f\|_{\infty,p}$ for all $t>0$.

(iii) We verify by Lemma 7 (i) and (ii) that

$$\begin{aligned}\left\|S_\infty(t)\mathbb{P}(-\Delta_H)^{\alpha/2} f(x',\cdot)\right\|_{L^p(J)} &\le \left\|e^{t\Delta_H} e^{t\Delta_N}(-\Delta_H)^{\alpha/2} f(x',\cdot)\right\|_{L^p(J)} \\ &\quad + \sum_{1\le i,j\le 2}\left\|e^{t\Delta_H} e^{t\Delta_N} R_i R_j(-\Delta_H)^{\alpha/2}\overline{f}\right\|_{L^p(J)} \\ &\le t^{-\alpha/2}\left(\left\|H_t *_H |f|(x',\cdot)\right\|_{L^p(J)} + h\,(\tilde{H}_t *_H \overline{f})(x')\right), \quad t>0,\end{aligned}$$

for almost all $x' \in \mathbb{R}^2$ since $\overline{f}$ is independent of z. By Fubini's theorem

$$\int_J |H_t *_H |f|(x',z)|\,\mathrm{d}z = \left(H_t *_H \int_J |f(\cdot,z)|\,\mathrm{d}z\right)(x'), \quad \text{for a.a. } x' \in \mathbb{R}^2,$$

which allows us to conclude that

$$\left\| S_\infty(t)\mathbb{P}(-\Delta_H)^{\alpha/2} f \right\|_{\infty,p} \le t^{-\alpha/2}\left(\|H_t\|_{L^1(\mathbb{R}^2)} + \|\tilde{H}_t\|_{L^1(\mathbb{R}^2)} \right) \|f\|_{\infty,p}.$$
$$\le 2Ct^{-\alpha/2}\|f\|_{\infty,p}, \quad t > 0.$$

(iv) As above we have

$$\|S_\infty(t)\mathbb{P}\nabla_H \cdot f\|_{L^p(J)} \le \left\|\nabla_H e^{t\Delta_H} f\right\|_{L^p(J)} + \sum_{1\le i,j\le 2} \left\|e^{t\Delta_H} R_i R_j \nabla_H \cdot \overline{f}\right\|_{L^p(J)}.$$

The first term was already estimated and the second one is treated in the same way as in *(iii)*.

□

1.7 Hydrostatic Stokes Equations in Anisotropic $L^\infty(L^p)$-Spaces: The Case of Dirichlet-Neumann Boundary Conditions

In this section we consider the linear hydrostatic Stokes equation (1.19) subject to mixed boundary conditions, i.e. subject to (1.4). Note that the choice of boundary conditions has a severe impact on the linearized primitive equations. In the setting of layer domains, i.e., $\Omega = G \times (-h, 0) \subset \mathbb{R}^3$ with $G = (0, 1)^2$ and $h > 0$, this is illustrated best by the hydrostatic Stokes operator A. The latter can be represented as

$$Av = \Delta v + \frac{1}{h}\nabla_H(-\Delta_H)^{-1}\mathrm{div}_H\left(\partial_z v\big|_{z=-h}\right), \tag{1.38}$$

where for $z = -h$ Dirichlet and for $z = 0$ Neumann boundary conditions are imposed and periodicity is assumed horizontally. In particular, in the case of pure Neumann boundary conditions, the hydrostatic Stokes operator reduces to the Laplacian, i.e., $A = \Delta$. It is the aim of this section to study properties of the hydrostatic Stokes semigroup and terms of the form $\nabla e^{tA}\mathbb{P}$ on spaces of bounded functions, which then yield global, strong well-posedness results for the case of mixed Dirichlet-Neumann boundary conditions.

The main difficulty when dealing with the primitive equations on spaces of bounded functions is that the hydrostatic Helmholtz projection $\mathbb{P}$ *fails to be bounded* with respect to the L^∞-norm. In the following we sketch that the combination of the three ingredients ∇, $\mathbb{P}$, e^{tA} nevertheless gives rise to bounded operators on $L^\infty_H L^p_z(\Omega)$, which in addition satisfy typical second order parabolic decay estimates of the form

$$t^{1/2}\|\partial_i e^{tA}\mathbb{P}f\|_{L^\infty_H L^p_z(\Omega)} \le Ce^{t\beta}\|f\|_{L^\infty_H L^p_z(\Omega)},$$
$$t^{1/2}\|e^{tA}\mathbb{P}\partial_j f\|_{L^\infty_H L^p_z(\Omega)} \le Ce^{t\beta}\|f\|_{L^\infty_H L^p_z(\Omega)},$$
$$t\|\partial_i e^{tA}\mathbb{P}\partial_j f\|_{L^\infty_H L^p_z(\Omega)} \le Ce^{\beta t}\|f\|_{L^\infty_H L^p_z(\Omega)},$$

for $t > 0$, where $\partial_i, \partial_j \in \{\partial_x, \partial_y, \partial_z\}$. Since $\mathbb{P}$ fails to be bounded on $L^\infty(\Omega)^2$ it is not evident which space is a suitable substitute for $L^p_{\overline{\sigma}}(\Omega)$ in the case $p = \infty$. Here we consider the spaces

$$X := C_{\text{per}}([0,1]^2; L^p(-h,0))^2 \quad \text{and} \quad X_{\overline{\sigma}} := X \cap L^p_{\overline{\sigma}}(\Omega), \quad p \in (1,\infty). \tag{1.39}$$

The *hydrostatic Stokes equation* is given by

$$\partial_t v - \Delta v + \nabla_H \pi = f, \quad \nabla \cdot_H \overline{v} = 0, \quad v(0) = a \tag{1.40}$$

subject to the mixed Dirichlet and Neumann boundary conditions

$$\partial_z v = 0 \text{ on } \Gamma_u \times (0,\infty), \quad \pi, v \text{ periodic on } \Gamma_l \times (0,\infty), \quad v = 0 \text{ on } \Gamma_b \times (0,\infty).$$

The dynamics of this evolution equation is governed by the hydrostatic Stokes operator, and its $X_{\overline{\sigma}}$-realization $A_{\overline{\sigma}}$

$$A_{\overline{\sigma}} v := \mathcal{A} v, \quad D(A^{\overline{\sigma}}_\infty) = \{v \in W^{2,p}_{\text{per}}(\Omega)^2 \cap X_{\overline{\sigma}} : \partial_z v|_{\Gamma_u} = 0, v|_{\Gamma_b} = 0, \mathcal{A} v \in X_{\overline{\sigma}}\},$$

where $\mathcal{A} v$ is defined by (1.38). We will show that $A_{\overline{\sigma}}$ generates a strongly continuous, analytic semigroup $e^{tA_{\overline{\sigma}}}$ on $X_{\overline{\sigma}}$. Our main result on the hydrostatic semigroup acting on $X_{\overline{\sigma}}$ reads as follows.

Theorem 5 (Semigroup Estimates for the Hydrostatic Stokes Operator with Mixed Boundary Conditions) *Let $p \in (3,\infty)$. Then the following assertions hold true:*

a) *$A_{\overline{\sigma}}$ is the generator of a strongly continuous, analytic, and exponentially stable semigroup $e^{tA_{\overline{\sigma}}}$ on $X_{\overline{\sigma}}$ of angle $\pi/2$.*
b) *There exist constants $C > 0$, $\beta > 0$ such that for $\partial_i, \partial_j \in \{\partial_x, \partial_y, \partial_z\}$*

$$t^{1/2}\|\partial_j e^{tA_{\overline{\sigma}}} f\|_{L^\infty_H L^p_z(\Omega)} \le Ce^{\beta t}\|f\|_{L^\infty_H L^p_z(\Omega)}, \quad t > 0,\ f \in X_{\overline{\sigma}}, \tag{i}$$
$$t^{1/2}\|e^{tA_{\overline{\sigma}}}\mathbb{P}\partial_j f\|_{L^\infty_H L^p_z(\Omega)} \le Ce^{\beta t}\|f\|_{L^\infty_H L^p_z(\Omega)}, \quad t > 0,\ f \in X_{\overline{\sigma}}, \tag{ii}$$
$$t\|\partial_i e^{tA_{\overline{\sigma}}}\mathbb{P}\partial_j f\|_{L^\infty_H L^p_z(\Omega)} \le Ce^{\beta t}\|f\|_{L^\infty_H L^p_z(\Omega)}, \quad t > 0,\ f \in X_{\overline{\sigma}}; \tag{iii}$$

c) *For all* $f \in X_{\overline{\sigma}}$

$$\lim_{t\to 0+} t^{1/2}\|\nabla e^{tA_{\overline{\sigma}}} f\|_{L^\infty_H L^p_z(\Omega)} = 0.$$

We start the proof of Theorem 5 with a resolvent estimates for the Laplacian subject to mixed boundary conditions. More precisely, consider

$$\lambda v - \Delta v = f \text{ on } \Omega, \tag{1.41}$$

and for $\partial_i \in \{\partial_x, \partial_y, \partial_z\}$

$$\lambda w - \Delta w = \partial_i f \text{ on } \Omega. \tag{1.42}$$

Lemma 11 (Resolvent Estimate) *Let* $\theta \in (0,\pi)$ *and* $f \in L^q_H L^p_z(\Omega)$ *for* $q \in [1,\infty]$, $p \in [1,\infty)$. *Then there exists* $\lambda_0 > 0$ *such that for* $\lambda \in \Sigma_\theta$ *with* $|\lambda| \geq \lambda_0$ *the problems* (1.41) *and* (1.42) *have unique solutions* $v \in L^q_H L^p_z(\Omega)$ *and* $w \in L^q_H L^p_z(\Omega)$, *respectively, and there exists a constant* $C_\theta > 0$ *such that*

$$|\lambda|\cdot\|v\|_{L^q_H L^p_z(\Omega)} + |\lambda|^{1/2}\|\nabla v\|_{L^q_H L^p_z(\Omega)} + \|\Delta v\|_{L^q_H L^p_z(\Omega)} \leq C_\theta \|f\|_{L^q_H L^p_z(\Omega)}, \tag{1.43}$$

$$|\lambda|^{1/2}\|w\|_{L^q_H L^p_z(\Omega)} \leq C_\theta \|f\|_{L^q_H L^p_z(\Omega)}. \tag{1.44}$$

In particular, for $q = \infty$ *and* $p \in (2,\infty)$ *one can choose* $\lambda_0 = 0$.

The proof transfers the corresponding estimates for the whole space by a subtle cut-off method to the given situation. For details, see [29].

Since $\Omega = G \times (-h, 0)$ is a cylindrical domain, the semigroup generated by the Laplacian with the above boundary conditions satisfies

$$e^{t\Delta}(f \otimes g) = e^{t\Delta_H} f \otimes e^{t\Delta_z} g, \quad f : G \to \mathbb{R}^2, \quad g : (-h, 0) \to \mathbb{R},$$

where $(f \otimes g)(x, y, z) := f(x, y)g(z)$ is an elementary tensor, $\Delta_H := \partial_x^2 + \partial_y^2$ is the Laplacian on G with periodic boundary conditions and Δ_z is defined by

$$\Delta_z v := \partial_z^2 v, \quad D(\Delta_z) = \{f \in W^{2,p}(-h, 0) : f(-h) = \partial_z f(0) = 0\}.$$

The two operators involved, Δ_H and Δ_z, have the following properties.

Lemma 12 *Let* $p \in (1,\infty)$. *Then the operator* Δ_z *generates a strongly continuous, exponentially stable, analytic semigroup on* $L^p(-h, 0)$.

Lemma 13 *Let $\theta \in (0, \pi/2)$. Then there exists a constant $C_\theta > 0$ such that for all $\tau \in \Sigma_\theta$ we have*

$$|\tau|^{1/2}\|\nabla_H e^{\tau\Delta_H} Qf\|_{L^\infty(G)} \le C_\theta \|f\|_{L^\infty(G)}, \quad f \in L^\infty(G).$$

The assertion of Lemma 12 is standard, the one of Lemma 13 follows by the theory of Fourier multipliers for $L^\infty(\mathbb{R}^n)$ given in Lemma 8.2.3 and 8.2.4 of [5].

In order to prove Theorem 5 we first collect several facts concerning the corresponding theory in $L^p_{\overline{\sigma}}(\Omega)$. To this end, let $p \in (1, \infty)$, and recall that $A_{p,\overline{\sigma}} \colon D(A_{p,\overline{\sigma}}) \to L^p_{\overline{\sigma}}(\Omega)$ was defined by

$$A_{p,\overline{\sigma}} v = \mathbb{P}\Delta v, \quad D(A_{p,\overline{\sigma}}) = \{v \in W^{2,p}_{\text{per}}(\Omega)^2 : \text{div}_H \overline{v} = 0, \partial_z v\big|_{\Gamma_u} = 0, v\big|_{\Gamma_b} = 0.\}.$$

Consider furthermore $A_p \colon D(A_p) \to L^p(\Omega)^2$ defined by

$$A_p v := \Delta_p v + Bv, \quad D(A_p) := D(\Delta_p)^2, \quad Bv := \frac{1}{h}(1 - Q)\partial_z v\big|_{\Gamma_b},$$

where Δ_p denotes the Laplacian in $L^p(\Omega)^2$. The idea is that the pressure term may be recovered by applying the vertical average and horizontal divergence to (1.40), yielding

$$\Delta_H \pi = \text{div}_H \overline{f} - \text{div}_H \frac{1}{h}\partial_z v\big|_{\Gamma_b}, \tag{1.45}$$

or equivalently since $1 - Q$ agrees with $\nabla_H(-\Delta_H)^{-1}\text{div}_H$ one has $\nabla_H \pi = (1 - Q)\overline{f} - Bv$. We also have the inclusions

$$A \subset A_p \quad \text{and} \quad A_{\overline{\sigma}} \subset A_{p,\overline{\sigma}}, \tag{1.46}$$

and note that the semigroups $e^{tA_{p,\overline{\sigma}}}$, e^{tA_p}, e^{tA} and $e^{tA_{\overline{\sigma}}}$ are consistent semigroups. Let $\lambda_0 > 0$ with $\lambda_0 \in \rho(A_p)$, $\theta \in (0, \pi/2)$ and $\lambda \in \Sigma_{\theta+\pi/2} \cap B_{\lambda_0}(0)^c \subset \varrho(A_p)$. By (1.46), $\lambda - A$ is injective for $\lambda \in \rho(A_p)$ and likewise $\lambda - A_{\overline{\sigma}}$ is injective for $\lambda \in \rho(A^{\overline{\sigma}}_p)$. Since $X \hookrightarrow L^p(\Omega)^2$, the existence of a unique $v \in D(A_p)$ for $p \in (1, \infty)$ follows from the L^p-theory for A_p, see Section 3.5. Moreover, since $W^{2,p}_{\text{per}}(\Omega)^2 \hookrightarrow X$ for $p \in (3/2, \infty)$ we obtain $v \in D(A)$. Furthermore, $(A_p - \lambda)^{-1}$ leaves $L^p_{\overline{\sigma}}(\Omega)$ invariant and thus $f \in X_{\overline{\sigma}}$ implies $v \in D(A_{\overline{\sigma}})$. Hence,

$$\rho(A_p) \subset \rho(A) \quad \text{and} \quad \rho(A_{p,\overline{\sigma}}) \subset \rho(A_{\overline{\sigma}}). \tag{1.47}$$

In particular, the resolvent sets are non-empty and thus the operators are closed. Observing $v = (\lambda - A)^{-1} f$ is equivalent to

$$v = (\lambda - \Delta_p)^{-1}(f + Bv), \tag{1.48}$$

and using the fact that Q is continuous on $C^{0,\alpha}_{per}([0,1]^2)$ for $\alpha \in (0,1)$ we conclude

$$\begin{aligned}\|Bv\|_{L^\infty_H L^p_z(\Omega)} &\le h^{1/p}\|Bv\|_{L^\infty(\Omega)} \le h^{1/p}\|Bv\|_{C^{0,\alpha}([0,1]^2)}\\ &\le C\|\partial_z v|_{\Gamma_b}\|_{C^{0,\alpha}([0,1]^2)} \le C\|v\|_{C^{1,\alpha}(\overline{\Omega})}.\end{aligned}$$

Assuming $p \in (3,\infty)$ we have $W^{2,p}(\Omega) \hookrightarrow C^{1,\alpha}(\overline{\Omega})$ for some $\alpha = \alpha_p \in (0, 1 - 3/p)$. Using the resolvent estimate for A_p in $L^p(\Omega)^2$ we obtain

$$\begin{aligned}\|v\|_{C^{1,\alpha}(\overline{\Omega})} &\le C_p\|v\|_{W^{2,p}(\Omega)}\\ &\le C_p\left(\|v\|_{L^p(\Omega)} + \|Av\|_{L^p(\Omega)}\right) \le C_p(1+|\lambda|^{-1})\|f\|_{L^p(\Omega)}.\end{aligned}$$

This and $|\lambda| > \lambda_0$ yield $\|Bv\|_{L^\infty_H L^p_z(\Omega)} \le C_p(1+\lambda_0^{-1})\|f\|_{L^\infty_H L^p_z(\Omega)}$. Hence, Lemma 11 yields

$$|\lambda|\cdot\|v\|_{L^\infty_H L^p_z(\Omega)} + |\lambda|^{1/2}\|\nabla v\|_{L^\infty_H L^p_z(\Omega)} + \|Av\|_{L^\infty_H L^p_z(\Omega)} \le C_{\theta,p,\lambda_0}\|f\|_{L^\infty_H L^p_z(\Omega)}, \tag{1.49}$$

where we used that for λ as above and $p \in (3,\infty)$ one has

$$\|Av\|_{L^\infty_H L^p_z(\Omega)} \le \|\Delta v\|_{L^\infty_H L^p_z(\Omega)} + \|Bv\|_{L^\infty_H L^p_z(\Omega)} \le C_{\theta,p,\lambda_0}\|f\|_{L^\infty_H L^p_z(\Omega)}.$$

Note that if we consider instead $f \in X_{\overline{\sigma}}$, then $\lambda_0 > 0$ can be taken to be arbitrarily small and θ arbitrarily close to $\pi/2$ by Theorem 3. Since $0 \in \rho(A_{p,\overline{\sigma}}) \subset \rho(A_{\overline{\sigma}})$, see Proposition 1.5.3, it follows that the spectral bound

$$\beta := \sup\{\mathrm{Re}(\lambda) : \lambda \in \sigma(A_{\overline{\sigma}})\}$$

is negative and estimate (1.49) is valid for all $\lambda \in \Sigma_\theta$, $\theta \in (0,\pi)$ and $f \in X_{\overline{\sigma}}$. To verify that $D(A)$ and $D(A_{\overline{\sigma}})$ are dense in X and $X_{\overline{\sigma}}$, respectively, observe that the space

$$C^\infty_{\mathrm{per}}([0,1]^2; C^\infty_c((-h,0)))^2$$

is contained in $D(A)$ and dense in X, so the semigroup generated by A is strongly continuous on X. Since it leaves $L^p_{\overline{\sigma}}(\Omega)$ invariant, the restriction of the semigroup on $X \cap L^p_{\overline{\sigma}}(\Omega) = X_{\overline{\sigma}}$ is strongly continuous as well and generated by the restriction of A onto $D(A) \cap L^p_{\overline{\sigma}}(\Omega) = D(A_{\overline{\sigma}})$, i.e. $A_{\overline{\sigma}}$, which is therefore densely defined on $X_{\overline{\sigma}}$. We thus proved assertions *a*) and estimate (*i*) in *b*). The remaining assertions b) and d) are proved in a similar manner provided $\partial_j \in \{\partial_x, \partial_y\}$.

The remaining estimates for vertical derivatives, i.e. estimates (ii) and (iii) in Theorem 5 for $\partial_j = \partial_z$ are more difficult to obtain. For a detailed proof of these estimates we refer to [29] and note here only that the strategy of our proof is based on a method arising in L^∞-type error estimates for finite elements, see [69].

1.8 Local Well-Posedness in the L^p-L^q-Setting for $1 < p, q < \infty$

In this section, the local well-posedness of the primitive equations is discussed for various spaces and using different methods. After the linear part has been studied in detail in the previous sections, now the balance between the linear part and the non-linearity is essential to construct local solutions. Local mild or strong solutions are constructed using fixed point arguments.

1.8.1 Non-linearity in Isotropic L^p-Spaces

Consider an abstract semi-linear evolution equation

$$\begin{aligned}\partial_t u - Au &= F(u,u),\\ u(0) &= u_0,\end{aligned}$$

where A is the generator of a strongly continuous analytic semigroup in X_0 with domain $D(A) = X_1$, and where F is bilinear and for some $C > 0$ and $\gamma \in (0,1)$ such that

$$F\colon D(A^\gamma) \times D(A^\gamma) \to X_0, \quad \|F(u,u)\|_{X_0} \leq C\|v\|_{D(A^\gamma)}\|u\|_{D(A^\gamma)}.$$

The Fujita-Kato method gives for this estimate on the non-linearity mild solutions for initial values

$$u_0 \in D(A^{2\gamma-1}) = [X_1, X_0]_{2\gamma-1}$$

which is heuristically two-times the difference between the smoothness of the linear part and the non-linearity, compare Fig. 1.3, where the factor two stems from the fact that the non-linearity is quadratic.

The maximal regularity approach gives a similar result for initial values with the same differentiability in real interpolation rather than complex interpolation spaces, i.e., for some $q \in (1,\infty)$

$$u_0 \in (X_1, X_0)_{2\gamma-1,q}.$$

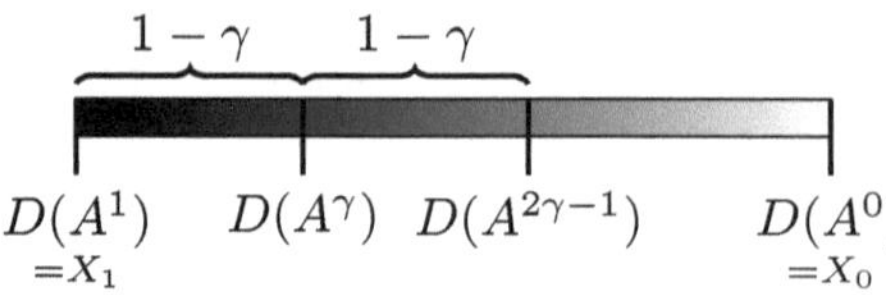

Fig. 1.3 Fujita-Kato method for semi-linear equations

Throughout this section, let $p \in (1, \infty)$. We represent the nonlinear terms by the bilinear map F_p by

$$F_p(v, v') := \mathbb{P}(v \cdot \nabla_H v' + w(v)\partial_z v'),$$

and set $F_p(v) := F_p(v, v)$. Since $u' = (v', w(v'))$ is divergence-free, we also obtain the representation

$$F_p(v, v') = \mathbb{P}\nabla \cdot (u' \otimes v).$$

Observe that $w(v)$ is less regular than v with respect to (x, y), but that w has good regularity properties with respect to z. We start by collecting various facts concerning the map F_p.

Lemma 14 (Estimate on the Non-linearity) *Let $p \in (1, \infty)$ Then*

$$F_p\colon H^{1+1/p,p}(\Omega)^2 \times H^{1+1/p,p}(\Omega)^2 \to L^p_{\overline{\sigma}}(\Omega)$$

and there exists a constant $M > 0$ such that

a) *for $v \in H^{1+1/p,p}(\Omega)^2$*

$$\|F_p(v, v)\|_{L^p_{\overline{\sigma}}(\Omega)} \leq M\|v\|^2_{H^{1+1/p,p}},$$

b) *and for $v, v' \in H^{1+1/p,p}(\Omega)^2$ one has*

$$\|F_p(v, v) - F_p(v', v')\|_{L^p_{\overline{\sigma}}(\Omega)} \leq M(\|v\|_{H^{1+1/p,p}(\Omega)} + \|v'\|_{H^{1+1/p,p}(\Omega)}) \cdot \|v - v'\|_{H^{1+1/p,p}(\Omega)}.$$

Proof In view of the bilinearity of $v \cdot \nabla_h v' + w(v)\partial_z v'$ with respect to v and v', assertion b) may be proved similarly as in a). We hence only prove a). Since $\mathbb{P}$ is bounded in $L^p_{\overline{\sigma}}(\Omega)$ it suffices to bound the $L^p(\Omega)^2$-norms of $v \cdot \nabla_h u$ and $w\partial_z v$ separately.

By Hölder's inequality,

$$\|v \cdot \nabla_h v\|_{L^p(\Omega)} \leq C\|v\|_{L^{3p}(\Omega)}\|v\|_{W^{1,3p/2}(\Omega)}$$

for some $C > 0$. The desired bound follows from the embedding of $H^{1+1/p,p}(\Omega)$ into $L^{3p}(\Omega)$ and $W^{1,3p/2}(\Omega)$. Next, thanks to (1.5) we obtain

$$\|w\partial_z v\|_{L^p(\Omega)} \le \|w\|_{L^\infty_z L^{2p}_{xy}} \|\partial_z v\|_{L^p_z L^{2p}_{xy}}.$$

It then follows that

$$\begin{aligned}\|w\|_{L^\infty_z L^{2p}_{xy}} &\le C\|w\|_{W^{1,p}_z L^{2p}_{xy}} \le C\|\partial_z w\|_{L^p_z L^{2p}_{xy}} = C\|\mathrm{div}_H\, v\|_{L^p_z L^{2p}_{xy}} \\ &\le C\|v\|_{L^p_z W^{1,2p}_{xy}} \le C\|v\|_{L^p_z H^{1+1/p,p}_{xy}} \le C\|v\|_{H^{1+1/p,p}(\Omega)},\end{aligned}$$

where we have used the embedding $W^{1,p}(-h,0) \hookrightarrow L^\infty(-h,0)$, Poincaré's inequality as well as the embedding $H^{1+1/p,p}(G) \hookrightarrow W^{1,2p}(G)$. We also have

$$\|\partial_z v\|_{L^p_z L^{2p}_{xy}} \le C\|v\|_{W^{1,p}_z L^{2p}_{xy}} = C\|v\|_{H^{1,p}_z L^{2p}_{xy}} \le C\|v\|_{H^{1,p}_z H^{1/p,p}_{xy}} \le C\|v\|_{H^{1+1/p,p}(\Omega)}.$$

Hence, $\|w\partial_z v\|_{L^p(\Omega)} \le C\|v\|^2_{H^{1+1/p,p}(\Omega)}$ for some $C > 0$. □

The estimate in Lemma 14 can be extended to higher order Sobolev spaces.

Lemma 15 (Higher Order Estimate on the Non-linearity) *There exists a constant $C > 0$, depending only on Ω, $p \in (1,\infty)$ and $s \ge 0$, such that for $v, v' \in H^{s+1+1/p,p}(\Omega)^2$*

$$\|F_p(v, v')\|_{H^{s,p}(\Omega)^2} \le C\|v\|_{H^{s+1+1/p,p}} \|v'\|_{H^{s+1+1/p,p}},$$

i.e., $F_p(\cdot,\cdot)\colon H^{s+1+1/p,p}(\Omega)^2 \times H^{s+1+1/p,p}(\Omega)^2 \to H^{s,p}(\Omega)^2$ is a continuous bilinear map.

1.8.2 Local Mild Solutions via the Fujita-Kato Scheme

Here, we prove the existence of a unique, mild solution to the system (1.2). Our approach is inspired by the so-called Fujita-Kato approach for the Navier-Stokes equations, see, e.g., [23].

Recall that (1.2) can be rewritten equivalently as

$$\begin{cases} \partial_t v(t) + A_p v(t) = \mathbb{P} f(t) + F_p v(t), & t > 0, \\ v(0) = a. \end{cases} \tag{1.50}$$

Let $T > 0$ and $\delta = \delta(p) := 1/p$. Then $v \in C([0,T]; V_\delta)$ is called a *mild solution* to equation (1.50) provided v satisfies the integral equation

$$v(t) = e^{-tA_p}a + \int_0^t e^{-(t-s)A_p}\big(\mathbb{P}f(s) + F_p v(s)\big)\,ds, \qquad t \geq 0. \tag{1.51}$$

In order to formulate the local well-posedness result, we define for $T > 0$, $\delta = 1/p$ and $\gamma := \frac{1}{2} + \frac{1}{2p}$ the space $\mathcal{S}_T$ as

$$\mathcal{S}_T := \{v \in C([0,T]; V_\delta) \cap C((0,T]; V_\gamma) : \|v(t)\|_{V_\gamma} = o(t^{\gamma-1}) \text{ as } t \to 0\}.$$

When equipped with the norm

$$\|v\|_{\mathcal{S}_T} := \sup_{0\leq s\leq T} \|v(s)\|_{V_\delta} + \sup_{0\leq s\leq T} s^{1-\gamma}\|v(s)\|_{V_\gamma},$$

the space $\mathcal{S}_T$ becomes a Banach space. Our local well-posedness result reads as follows.

Proposition 1.8.1 (Local Existence and Uniqueness of Mild Solutions) *Let $T > 0$, $\delta = 1/p$ and $\gamma = \frac{1}{2} + \frac{1}{2p}$ and assume that $a \in V_{\delta,p}$ and $\mathbb{P}f \in C((0,T]; L^p_{\overline{\sigma}}(\Omega))$ with $\|\mathbb{P}f(t)\|_{L^p_{\overline{\sigma}}(\Omega)} = o(t^{2\gamma-2})$ as $t \to 0$. Then there exist $T^* > 0$ and a unique mild solution $v \in \mathcal{S}_{T^*}$ to (1.50). If in addition $\|a\|_{V_{\delta,p}} + \sup_{0\leq s\leq T} s^{2-2\gamma}\|\mathbb{P}f(s)\|_{X_p}$ is sufficiently small, then $T^* = T$.*

Proof We subdivide our proof into several steps.

Step 1 (Approximating Sequence) Consider an approximating sequence $(v_m) \in \mathcal{S}_T$ which is defined by

$$\begin{aligned} v_0(t) &:= e^{-tA_p}a + \textstyle\int_0^t e^{-(t-s)A_p}\mathbb{P}f(s)\,ds, \\ v_{m+1}(t) &:= v_0(t) + \textstyle\int_0^t e^{-(t-s)A_p}F_p v_m(s)\,ds, \qquad t > 0. \end{aligned} \tag{1.52}$$

These v_m are well defined in $\mathcal{S}_T$ since

$$\begin{aligned} \|v_0(t)\|_{V_\gamma} &\leq \|e^{-tA_p}a\|_{V_\gamma} + \int_0^t \|e^{-(t-s)A_p}\|_{\mathcal{L}(V_0,V_\gamma)} s^{2\gamma-2}s^{2-2\gamma}\|\mathbb{P}f(s)\|_{L^p_{\overline{\sigma}}(\Omega)}\,ds \\ &\leq Ct^{\gamma-1}\|a\|_{V_\delta} + C\int_0^t (t-s)^{-\gamma}s^{2\gamma-2}\,ds \sup_{0\leq s\leq t}(s^{2-2\gamma}\|\mathbb{P}f(s)\|_{L^p_{\overline{\sigma}}(\Omega)}), \end{aligned}$$

where Lemma 1.5.6 b) is used. Hence

$$t^{1-\gamma}\|v_0(t)\|_{V_\gamma} \leq C\|a\|_{V_\delta} + CB(\gamma, 2-2\gamma)\sup_{0\leq s\leq t}(s^{2-2\gamma}\|\mathbb{P}f(s)\|_{L^p_{\overline{\sigma}}(\Omega)}), \quad t \in (0,T),$$

where $B(\cdot,\cdot)$ denotes the Beta function. The fact that

$$t^{1-\gamma}\|v_0(t)\|_{V_\gamma} \to 0 \quad \text{as} \quad t \to 0$$

follows from Lemma 1.5.6 c). A similar computation combined with Lemma 14a) gives that for $t \in (0, T)$

$$t^{1-\gamma}\|v_{m+1}(t)\|_{V_\gamma} \le t^{1-\gamma}\|v_0(t)\|_{V_\gamma} + CB(\gamma, 2-2\gamma)M \sup_{0\le s\le t}(s^{1-\gamma}\|v_m(s)\|_{V_\gamma})^2.$$

By induction, we then see that $v_m \in \mathcal{S}_T$ for all $m \ge 0$.

Step 2 (Quadratic Inequality) One defines

$$k_m(t) := \sup_{0\le s\le t} s^{1-\gamma}\|v_m(s)\|_{V_\gamma}$$

and $C_1 := CB(\gamma, 2-2\gamma)M$. Then one deduces that

$$k_{m+1}(t) \le k_0(t) + C_1 k_m(t)^2, \quad t > 0, \quad \text{with} \quad k_m(0) = 0$$

for $m \ge 0$. This quadratic inequality implies for $0 < t < T$,

$$\text{if} \quad k_0(t) < 1/(4C_1), \text{ then } k_m(t) < K(t) := (1-\sqrt{1-4C_1k_0(t)})/(2C_1) < 1/(2C_1).$$

The first inequality is satisfied provided one of the following assertions is true:

(1) T is sufficiently small (note that $k_0(t)$ is continuous and that $k_0(0) = 0$);
(2) $\|a\|_{V_\delta} + \sup_{0\le s\le T} s^{2-2\gamma}\|\mathbb{P}f(s)\|_{L^{\frac{p}{\sigma}}(\Omega)}$ is sufficiently small.

Note that the cases (1) and (2) will lead to the local and global existence, respectively.

The proofs of the convergence and the uniqueness are now straightforward. □

Remark 7 (Estimate on the Existence Time) Let us clarify the dependency of T^*, i.e. the length of the existing time of the solution constructed above, on the initial data a for the case $f \equiv 0$. In this case, T^* is chosen in such a way that

$$k_0(T^*) = \sup_{0\le s\le T^*} s^{1-\gamma}\|e^{-tA_p}a\|_{V_\gamma} < 1/(4C_1),$$

whereas $k_0(t)$ is estimated by the use of Lemma 1.5.6 b) as

$$k_0(t) \le Ct^{\min\{1-\gamma,\varepsilon\}}\|a\|_{V_{\delta+\varepsilon}}$$

for all $t > 0$, provided that $a \in V_{\delta+\varepsilon}$ with $0 \le \varepsilon \le 1-\delta$. Therefore, if $\varepsilon > 0$, then we may set

$$T^* = \frac{1}{2}\left(\frac{1}{4CC_1\|a\|_{V_{\delta+\varepsilon}}}\right)^{\max\{1/(1-\gamma), 1/\varepsilon\}},$$

which depends only on the $V_{\delta+\varepsilon}$,-norm of the initial data.

Note, however, that we cannot assume $\varepsilon = 0$ above. In fact, for $a \in V_\delta$ the dependency of T^* on a cannot be controlled merely by the V_δ-norm of a.

1.8.3 Hölder Regularity with Respect to Time for Mild Solutions

The mild solution to (1.50) constructed above is in fact a strong solution. An important step in the proof is the following assertions.

Lemma 16

a) *Let $\theta_1, \theta_2 \geq 0$ such that $\theta_1 + \theta_2 \leq 1$. Then there exists a constant $C > 0$ such that*

$$\|I - e^{-tA_p}\|_{\mathcal{L}(V_{\theta_1+\theta_2,p}, V_{\theta_1,p})} \leq Ct^{\theta_2}, \quad t \geq 0.$$

b) *For $a \in L^p_\sigma(\Omega)$ set $z(t) := e^{-tA_p}a$. Then there exists a constant $C > 0$ such that*

$$\|z(t+s) - z(t)\|_{V_{\tau,p}} \leq Ct^{-1+\tau}s^{1-\tau} \textit{ for all } \tau \in [0,1] \textit{and all } s \geq 0.$$

We now collect mapping properties of the convolution integral

$$H(t) := \int_0^t e^{-(t-s)A_p} f(s)\, ds,$$

where $f \in C((0,T]; L^p_\sigma(\Omega))$ satisfies certain assumptions as $t \to 0$ and $t \to \infty$. Using the above Lemma 16 one can show the following lemma. For related results, see, e.g., [37, Lemmas 3.4 and 3.5].

Lemma 17 *Let $\kappa \geq 0$, $\tau \in (0,1)$ and $f \in C((0,T]; L^p_\sigma(\Omega))$.*

a) *Assume that $\|f(t)\|_{L^p_\sigma(\Omega)} \leq Ct^{-\kappa}$ for all $t \in (0,T]$. Then there exist $\varepsilon > 0$ and $\tilde{C} > 0$ such that*

$$\|H(t+s) - H(t)\|_{V_\tau} \leq C\tilde{C} \max\{t^{\varepsilon-\kappa}s^{1-\tau-\varepsilon}, t^{-\kappa}s^{1-\tau}\}, \quad s \in [0, T-t].$$

b) *Assume that $f \in C^\theta((0,T]; L^p_\sigma(\Omega))$ and that $\|f(t)\|_{L^p_\sigma(\Omega)} \leq L_1 t^{-\kappa}$ for $t \in (0,T]$ as well as $\|f(t+s) - f(t)\|_{L^p_\sigma(\Omega)} \leq L_2 t^{-\tau} s^\theta$ for $t \in (0,T]$ and $s \in [0, T-t]$. Then there exists a constant $c > 0$ such that*

$$\|\partial_t H(t)\|_{L^p_\sigma(\Omega)} + \|H(t)\|_{D(A_p)} \leq ct^{-c}(L_1 + L_2 e^{-ct}), \quad t > 0.$$

Using these estimates on the convolution terms one can derive additional time regularity for the mild solutions.

Proposition 1.8.2 (Hölder Regularity in Time) *Let* $f \in C^{\theta}((0,T]; L^p_{\overline{\sigma}}(\Omega))$ *with* $\theta \in (0,1)$. *Then the mild solution* v *to* (1.51) *given in Proposition 1.8.1 is a strong solution. More precisely,*

$$v \in C^{1,\mu}((0,T]; L^p_{\overline{\sigma}}(\Omega)) \cap C^{\mu}((0,T]; D(A_p))$$

satisfies (1.50) *for all* $t \in (0,T]$ *and where* $\mu = \min\{\theta, 1-\gamma-\epsilon\}$ *and* $\epsilon > 0$ *can be chosen arbitrarily small.*

Remark 8 (Maximal Hölder Regularity) Given the situation of Lemma 17 b), we obtain maximal Hölder regularity of v, i.e.,

$$v \in C^{1,\theta}((0,T]; L^p_{\overline{\sigma}}(\Omega)) \cap C^{\theta}((0,T]; D(A_p));$$

see [63, Thm 4.3.5].

1.8.4 Local Strong Well-Posedness and Maximal L^q-Regularity in Time-Weighted Spaces

In this subsection we present the maximal L^q-regularity approach to the local strong well-posedness of the primitive equations. To this end, we collect first some results on semi-linear evolution equations, which then will be applied to the primitive equations.

Let X_0, X_1 be Banach spaces such that $X_1 \hookrightarrow X_0$ is densely embedded, and let $A\colon X_1 \to X_0$ be bounded. The aim is to solve the semi-linear problem for $0 < T \le \infty$

$$\partial_t v + Av = F(v) + f, \qquad 0 < t < T, \qquad v(0) = v_0. \tag{1.53}$$

For a Banach space X, a time weight $\mu \in (1/q, 1]$, a time interval $J \subset [0,\infty)$, and $k \in \mathbb{N}$, we set

$$\begin{aligned}
L^q_{\mu}(J;X) &= \{v \in L^1_{loc}(J;X)\colon t^{1-\mu}v \in L^q(J;X)\},\\
H^{1,q}_{\mu}(J;X) &= \{v \in L^1_{loc}(J;X) \cap H^{1,1}(J;X)\colon t^{1-\mu}v' \in L^q(J;X)\},\\
&\ \ \vdots\\
H^{k+1,q}_{\mu}(J;X) &= \{v \in L^1_{loc}(J;X) \cap H^{k+1,1}(J;X)\colon v' \in H^{k,q}_{\mu}(J;X)\},
\end{aligned}$$

Here v' denotes the time derivative of v in the distributional sense.

The problem (1.53) is considered for initial data within the real interpolation space

$$v_0 \in X_{\gamma,\mu} = (X_0, X_1)_{\mu-1/q,q} \quad \text{and for} \quad f \in \mathbb{E}_{0,\mu}(J) := L^q_\mu(J; X_0),$$

where $q \in (1, \infty)$. We aim for solutions lying in the maximal regularity space

$$\mathbb{E}_{1,\mu}(J) := H^{1,q}_\mu(J; X_0) \cap L^q_\mu(J; X_1)$$

and define for $\beta \in [0, 1]$ the complex interpolation space

$$X_\beta := [X_0, X_1]_\beta.$$

Theorem 6 ([68, Theorem 1.2]) *Assume that*

(H1) *A has maximal L^q-regularity for $q \in (1, \infty)$.*
(H2) *$F : X_\beta \to X_0$ satisfied the estimate*

$$\|F(v_1) - F(v_2)\|_{X_0} \le C(\|v_1\|_{X_\beta} + \|v_1\|_{X_\beta})(\|v_1 - v_2\|_{X_\beta})$$

for some $C > 0$ independent of v_1, v_2.
(H3) *$\beta - (\mu - 1/q) \le \frac{1}{2}(1 - (\mu - 1/q))$, that is $2\beta - 1 + 1/q \le \mu$.*
(S) *X_0 is of class UMD, and the embedding (mixed derivative theorem)*

$$H^{1,q}(\mathbb{R}; X_0) \cap L^q(\mathbb{R}; X_1) \hookrightarrow H^{1-\beta,q}(\mathbb{R}; X_\beta)$$

is valid for each $\beta \in (0, 1)$ and $q \in (1, \infty)$.

and let

$$v_0 \in X_{\gamma,\mu} \quad \text{and} \quad f \in L^q(0, T; X_0).$$

Then there exists a time $T' = T'(v_0)$ with $0 < T' \le T$ such that problem (1.53) *admits a unique solution*

$$v \in H^{1,q}_\mu(0, T'; X_0) \cap L^q_\mu(0, T'; X_1).$$

furthermore, the solution v depends continuously on the data.

Condition (S) holds true whenever X_0 is of class UMD and there is an operator $A_\# \in \mathcal{H}^\infty(X_0)$ with domain $D(A_\#) = X_1$ satisfying $\phi^\infty_{A_\#} < \pi/2$, see Remark 1.1 of [68]. To verify condition (S) in the situation considered here, i.e., $X_0 = L^p_\sigma(\Omega)$ and $X_1 = D(A_p)$, note first that $L^p_\sigma(\Omega)$ is of class UMD as closed subspace of $L^p(\Omega)^2$, and second, considering $A_\# = A_p - \lambda$ with $\lambda > 0$, it was proved in Theorem 4 that $A_\# \in \mathcal{H}^\infty(X_0)$ with $\phi^\infty_{A_\#} = 0 < \pi/2$.

Here for $1 < p < \infty$ we set

$$X_0 = L^p_{\overline{\sigma}}(\Omega) \quad \text{and} \quad X_1 = D(A_p).$$

Recalling that X_β is given by $X_\beta = [L^p_{\overline{\sigma}}(\Omega), D(A_p)]_\beta$ we obtain

$$[L^p_{\overline{\sigma}}(\Omega), D(A_p)]_{(1+1/p)/2} \subset H^{1+1/p,p}(\Omega)^2,$$

and the estimate on the non-linearity follows from Lemma 14 with $\beta = (1+1/p)/2$. Hence local well-posedness for the primitive equations follows.

Proposition 1.8.3 (Local Strong Well-Posedness in Maximal L^q-Regularity Spaces) *Let $p, q \in (1, \infty)$ with $1/p + 1/q \leq 1$, $\mu \in [1/p + 1/q, 1]$ and $T > 0$. Assume that*

$$v_0 \in X_{\mu-1/q,q} \quad \text{and} \quad \mathbb{P}f \in L^q_\mu(0, T; L^p_{\overline{\sigma}}(\Omega)).$$

Then there exists $T' = T'(v_0)$ with $0 < T' \leq T$ and a unique, strong solution v to (1.2) *on $(0, T')$ with*

$$v \in H^{1,q}_\mu(0, T'; L^p_{\overline{\sigma}}(\Omega)) \cap L^q_\mu(0, T'; D(A_p)).$$

1.9 Global Well-Posedness: The Setting $1 < p < \infty$

Starting from our local existence results we present in this section several strategies to prove global well-posedness. First, controlling the existence time in a suitable norm, which is subject to an *a priori* bound allows us to extend solutions to any finite time interval. This is the strategy applied for solutions constructed by the Fujita-Kato scheme. Second, we apply a "no blow-up"-argument and arguing by contradiction, the local solution extends to global one. This can be done for the local solution in the maximal L^2-regularity space since the maximal L^2-regularity norm remains bounded. Third, having a global solution in one space, we may use smoothing properties of solutions, where a local solution smoothens, for instance, into the maximal L^2-regularity setting and extends thus to a global solution even in a more regular space.

1.9.1 Extending Local Fujita-Kato Type Solutions to Global Ones

Note that for the solutions from Theorem 1.8.3 one has control on the existence time provided that the initial values are slightly more regular than necessary for the local existence result, compare Remark 7. More precisely, one needs control of the $V_{\delta+\varepsilon}$-

norm for $\varepsilon > 0$. This is equivalent to the $H^{2/p+\varepsilon,p}(\Omega)$-norm, and for $p = 2$, this means that one needs control on the $H^{1+\varepsilon}(\Omega)$-norm. By Theorem 2 one has even a H^2-*a priori* bound.

Theorem 7 (Existence of Unique, Global Strong Solutions) *Let $p \in (1, \infty)$ with and suppose that*

$$f \in H^{1,2}_{loc}((0,\infty); L^p(\Omega)^2 \cap L^2(\Omega)^2).$$

a) *Assume that*

$$a \in \{v \in H^{2/p,p}_{per}(\Omega)^2 \cap L^p_{\overline{\sigma}}(\Omega) \mid v\,|_{\Gamma_D} = 0\}.$$

Then there is a unique, global, strong solution to (1.2) *and* (1.4) *satisfying*

$$v \in C^1((0,\infty); L^p_{\overline{\sigma}}(\Omega)) \cap C^0((0,\infty); D(A_p)),$$
$$\pi_s \in C^0((0,\infty); H^{1,p}_{per}(G) \cap L^p_0(G)).$$

b) *If in addition $a \in D(A_p)$, then the above solution extends to* $[0, \infty)$.

The proof uses the following steps:

Step 1 (Local Existence and Regularization to L^2) According to Proposition 1.8.1(c) there exists $T^* > 0$ such that there is a strong solution to (1.2) and (1.4) on $[0, T^*]$. For $p \in [2, \infty)$ we have for $t \in (0, T^*]$

$$v(t) \in D(A_p) \subset D(A_2)$$

and hence the solution constructed in Proposition 1.8.1 is also a solution in L^2.

Consider $p \in (1, 2)$. Proposition 1.8.1 implies that v is a strong solution to (1.2) and (1.4) with

$$v(t) \in D(A_p) \subset H^{2,p}(\Omega)^2 \quad \text{for } t \in (0, T^*].$$

By Sobolev's embeddings

$$D(A_p) \hookrightarrow V_{1/p_1} \quad \text{for } p \geq \frac{3p_1}{2p_1+1}. \tag{1.54}$$

This fact has been used in [35] to prove global existence for the case $p = 6/5$ with $p_1 = 2$, see Fig. 1.4. In the following we iterate this procedure by defining the recursive sequences (p_n) by

$$p_0 := 2, \quad p_{n+1} := \frac{3p_n}{2p_n+1}, \quad n \in \mathbb{N}_0.$$

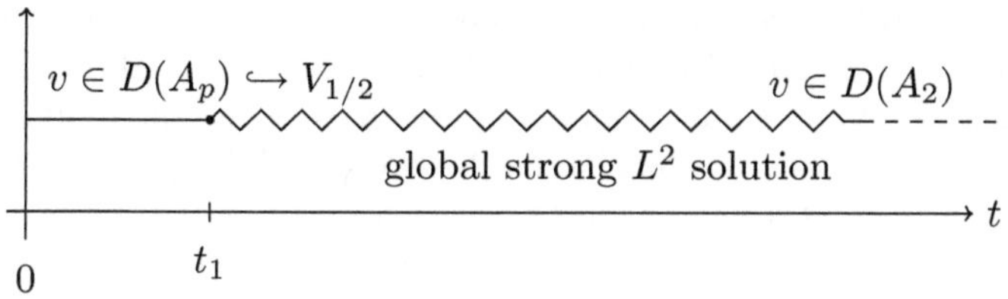

Fig. 1.4 Regularization for the velocity v and $p \geq 6/5$

By induction, (p_n) are strictly decreasing with

$$\lim_{n\to\infty} p_n = 1.$$

Hence, for any $p \in (1, 2)$, there exists $m \in \mathbb{N}_0$ such that $p_m < p \leq p_{m-1} \leq 2$. So, for $t_0 > 0$ we have

$$v(t_0) \in D(A_p) \subset V_{1/p_{m-1}}.$$

We now use $v(t_0)$ as new initial values for a solution in $L^{p_{m-1}}_{\overline{\sigma}}(\Omega)$. By uniqueness of strong solutions, both the original solution and the newly constructed solution coincide in $L^{p_{m-1}}_{\overline{\sigma}}(\Omega) \times L^{q_{n-1}}(\Omega)^2$. By Proposition 1.8.1 for $t_1 > t_0$ and by (1.54)

$$v(t_1) \in D(A_{p_{m-1}}) \subset V_{1/p_{m-2}}.$$

Again we construct solutions in $L^{p_{m-2}}_{\overline{\sigma}}(\Omega)$ by using $v(t_1)$ as new initial values. Iterating this procedure we arrive at a solution at time $0 < t_m < T^*$ satisfying

$$v(t_m) \in V_{1/2}.$$

Using these values as initial values it follows by uniqueness of strong solutions that for $t_m < t \leq T^*$, the local strong solution constructed in Proposition 1.8.1 is already an L^2 solution. We hence may assume without loss of generality initial values at $t_m < t_{m+1} < T^*$

$$v(t_{m+1}) \in D(A_{p'}), \quad \text{where } p' := \max\{p, 2\}.$$

Step 2 (Global Existence for $p \in [2, \infty)$) Consider

$$I_0 = [0, T_1], \quad I_1 = [T_1' - \varepsilon_1, T_1], \quad I_2 = [T_1', T_2], \ldots, \quad I_n = [T_n', T_{n+1}], \ldots,$$

where $T^* = T_1 < T_2 < \ldots$ with $T_n \to \infty$ and $T_n' - \varepsilon_n$ with suitable $\varepsilon_n > 0$ such that $T_n' - \varepsilon_n > T_{n-1}$. We thus obtain a sequence of finite intervals with neighbor intervals overlapping, where the union of all these intervals covers the whole interval $[0, \infty)$.

Iteratively, we may then construct unique strong solutions on each of these intervals. The induction basis is provided by the local existence in $I_0 = [0, T_1]$. Assume that there is a unique strong solution on all I_m for $m \leq n$. By the uniqueness of the local solutions, the solutions coincide on the overlaps $I_l \cap I_k$, $k, l \leq n$ and hence by assumption there is a unique strong solution on $[0, T_n] = \cup_{m\leq n} I_n$. Now, by assumption $v(T_n') \in D(A_p)$are bounded since by induction hypothesis

$$v \in C^0((0, T_n]; D(A_p)).$$

For $\varepsilon > 0$ small, one has for all $p \geq 2$

$$D(A_2) \subset V_{\delta+\varepsilon}.$$

Therefore, combining the *a priori* estimates from Theorem 2 on $\|v(s)\|_{D(A_2)}$ with the assumption

$$\sup_{s\in I_{n+1}} \|f(s)\|_{L^p(\Omega)^2} < \infty$$

and using Proposition 1.8.1 (b), the time interval length T_n^* can be chosen uniformly. This allows one to construct local solutions within the intervals

$$[T_n' + kT_n^*, T_n' + (k+1)T_n^*] \cap [T_n', T_{n+1}], \quad k = 0, 1, 2, \ldots k_n,$$

where $k_n \in \mathbb{N}$ is the smallest number such that $T_n' + (k_n)T_n^* \geq T_{n+1}$. By the uniqueness of the local strong solutions this extended solution is unique as well, and hence it is proven that there is a unique solution even on $(0, T_{n+1}]$.

Step 3 (Global Existence for $p \in (1, 2)$)

If $p \in (1, 2)$, then by step 2 there is a global strong solution in $L^{p'}$ for

$$p' := \max\{p, 2\}.$$

Since

$$D(A_2) \subset D(A_p), \quad p \in (1, 2),$$

this is already a unique, global, strong solution in L^p, respectively.

Step 4 (Recovering the Pressure) The pressure can be recovered to be

$$\nabla_H \pi_s = (\mathbb{1} - \mathbb{P}) \{f - (\Delta v + v\nabla_H v + w\partial_z v)\},$$

and it exists globally since $\|v\|_{D(A_p)}$ and exists globally.

1.9.2 Decay at Infinity

Considering the primitive equations with Dirichlet boundary conditions on top or bottom one obtains exponential decay of the solutions for large times provided that the force term has this property, in particular for $f \equiv 0$.

Theorem 8 (Decay at Infinity) *Let $\Gamma_D \neq \emptyset$ and let $f \in C[0,\infty; L^p(\Omega)^2)$ with*

$$\|f\|_{L^p(\Omega)^2} = O(e^{-\beta t}) \text{ as } t \to \infty,$$

where β are given as in Proposition 1.5.3, and let (v, π_s) be the strong solution to the primitive equations on $(0,\infty)$. Then

$$\|\partial_t v\|_{L^p} + \|\Delta v\|_{L^p} = O(e^{-\beta t}), \quad \|\nabla_H \pi\|_{L^p} = O(e^{-\beta t}) \quad \text{as} \quad t \to \infty.$$

By the regularization properties we may assume without loss of generality that $v(0) \in D(A_p)$. As in the proof of Proposition 1.8.1 we consider a fixed point iteration, this time with exponential weight. To this end, let $\tilde{\beta}$ be such that

$$0 < \tilde{\beta} < \beta.$$

Then we define

$$\tilde{k}_m^\infty := \sup_{s\in(0,\infty)} e^{\tilde{\beta}s}\|v_m(s)\|_{V_\gamma}, \quad m \in \mathbb{N}_0.$$

Similarly to the proof of Proposition 1.8.1 we show that

$$e^{\tilde{\beta}t}\|v_0(t)\|_{V_\gamma} \le Ce^{(\tilde{\beta}-\beta)t}\|a\|_{V_\delta} + C \sup_{s\in(0,t)}\{e^{\beta_f s}\|\mathbb{P}f(s)\|_{L^p_{\overline{\sigma}}(\Omega)}\},$$

and furthermore

$$e^{\tilde{\beta}t}\|v_{m+1}(t)\|_{V_\gamma} \le e^{\tilde{\beta}t}\|v_0(t)\|_{V_\gamma} + C \sup_{s\in(0,t)}\{e^{\tilde{\beta}s}\|v_m\|_{V_\gamma}\}^2 + C \sup_{s\in(0,t)}\{e^{\tilde{\beta}s}\}$$

Hence one obtains again a quadratic inequality

$$\tilde{k}_{m+1}^\infty \le \tilde{k}_0^\infty + C(\tilde{k}_m^\infty)^2$$

Now, observe that for $\tilde{k}_0^\infty$ small enough, the sequence $(\tilde{k}_m^\infty)$, $m \in \mathbb{N}_0$ is uniformly bounded. In particular, the limits of $e^{\tilde{\beta}s}v_m(s)$ exist in V_γ and define a global solution with exponential decay. Since locally solutions are unique the constructed solution coincides with the global solution constructed before. In particular, then the nonlinear remainders $F_p(v, \tau, 0)$ are exponentially decaying.

It remains to prove that there are initial values such that $\tilde{k}_0^\infty$ is indeed sufficiently small. Consider first the case $p = 2$. By assumption $f \in L^2(0, \infty; L^2(\Omega)^2)$. We conclude from the energy inequality that $\|v(t)\|^2_{H^1(\Omega)}$ is integrable on $(0, \infty)$. Hence,

$$\inf_{t\in(0,\infty)} \left(\|v(t)\|^2_{H^1(\Omega)} + \|\tau(t)\|^2_{H^1(\Omega)} \right) = 0,$$

and it follows that there is $t_0 \geq 0$ such that $\|v(t)\|^2_{H^1(\Omega)} = \|v(t)\|^2_{V_{1/2}}$ is small. Hence, it follows form the above that $\|v(t)\|^2_{D(A_2)}$ is exponentially decaying.

For $p \in (1, \infty)$ one has $D(A_2) \subset V_{1/p}$. Thus, there for all $p \in (1, \infty)$ there exist initial values for which $\tilde{k}_0^\infty$ is small enough. We conclude as in [35, Theorem 6.1, Step 9] that

$$\|\partial_t v\|_{L^p} + \|v\|_{D(A_p)} \leq Ce^{-\beta_v t} \leq Ce^{-\beta t}, \text{ for some } C > 0.$$

Reconstructing the pressure term we obtain

$$\|\nabla_H \pi\|_{L^p} \leq C \left(\|v\|^2_{D(A_p)} + \|f\| \right) \leq Ce^{-\beta t}.$$

1.9.3 Global Existence in Maximal L^2-Regularity Spaces and Higher Regularity

Before considering general maximal L^q-regularity spaces we consider the case $p = q = 2$.

Proposition 1.9.1 (Global Well-Posedness and Higher Regularity) *Let* $0 < T < \infty$ *and* $v_0 \in \{H^1 \cap L^2_{\overline{\sigma}}(\Omega) : v|_{\Gamma_D} = 0\}$.

(a) *If* $\mathbb{P}f \in L^2(0, T; L^2_{\overline{\sigma}}(\Omega))$, *then there exists a unique, strong solution* v *to the primitive equations* (1.2) *in*

$$v \in H^1(0, T; L^2_{\overline{\sigma}}(\Omega))) \cap L^2(0, T; D(A_2)).$$

(b) *If in addition* $t \mapsto t \cdot \mathbb{P}f_t(t) \in L^2(0, T; L^2_{\overline{\sigma}}(\Omega))$, *then*

$$t \cdot v_t \in H^1(0, T; L^2_{\overline{\sigma}}(\Omega))) \cap L^2(0, T; D(A_2)).$$

Note that *(b)* implies already boundedness of the $L^\infty_t H^2_x$-norm provided that f is regular enough. The proof of the higher regularity result relies on the implicit function theorem. Masuda [61] introduced first an extra parameter to prove the spatial analyticity of solutions to the Navier-Stokes equations using the implicit function theorem. Angenent [3, 4] systematically developed the nowadays called parameter trick valid for general quasilinear evolution equations. For proofs within the L^p-setting we are working in and for further refinements of this method, we refer, e.g., to [67].

The following elementary lemma is needed to extend regularity of solutions from $(0, T')$ for any $0 < T' < T$ to $(0, T)$.

Lemma 18 (No Blow-Up Norms Implies Global Existence) *Let* $v \in \mathbb{E}_{1,\mu}(0, T')$ *for any* $0 < T' < T$, *and* $\sup_{0<T'<T} \|v\|_{\mathbb{E}_{1,\mu}(0,T')} < C$ *for some constant* $C > 0$. *Then* $v \in \mathbb{E}_{1,\mu}(0, T)$.

Proof Proof of Proposition 1.9.1. To prove assertion (a) consider

$$t_+(v_0) := \sup\{T' > 0\colon \text{Equation (1.2) has a solution in } \mathbb{E}_{1,1}(0, T')\}.$$

By Proposition 1.8.3 $t_+(v_0) > 0$ and the solutions in $\mathbb{E}_{1,1}(0, T')$ are unique. Indeed, if we assume that there are two solutions $v, v' \in \mathbb{E}_{1,1}(0, T')$, then setting

$$t_1(v_0) := \sup\{s > 0\colon \|(v - v')(s)\|_{X_{\gamma,1}} = 0\},$$

we see that $t_1(v_0) > 0$ by Proposition 1.8.3. Further, by continuity, $\mathbb{E}_{1,1}(0, T') \hookrightarrow C([0, T']; X_{\gamma,1})$ and the above supremum is attained.

Assuming that $t_1(v_0) < T'$, again by Proposition 1.8.3, the solution with new initial value at $t_1(v_0)$ is unique on some time interval, thus contradicting the assumption.

Assume now that $t_+(v_0) < T$. By Theorem 1

$$\|v\|_{\mathbb{E}_{1,1}(0,T')} \leq B(\|v_0\|_{H^1(\Omega)}, \|\mathbb{P}f\|_{L^2(0,T;L^2(\Omega))}, t_+(v_0))$$

for any $0 < T' < t_+(v_0)$. Hence by Lemma 18 we have $v \in \mathbb{E}_{1,1}(0, t_+(v_0))$. Since the trace in $\mathbb{E}_{1,1}(0, t_+(v_0))$ is well-defined $v(t_+(v_0))$ can be taken as new initial value, thus extending the solution beyond $t_+(v_0)$ contradicting the assumption. Hence $t_+(v_0) = T$, and again combining Theorem 1 and Lemma 18 we have $v \in \mathbb{E}_{1,1}(0, T)$. This proves part (a).

Lemma 19 *Let* p, q, μ *and* $v_0, \mathbb{P}f$ *be as in Proposition 1.8.3. Assume that*

$$v \in H^{1,q}_{\mu}(0, T; L^p_{\overline{\sigma}}(\Omega)) \cap L^q_{\mu}(0, T; D(A_p))$$

is a solution to (1.2). *If in addition* $t \mapsto t \cdot \mathbb{P}f_t \in L^q_{\mu}(0, T; L^p_{\overline{\sigma}}(\Omega))$, *then*

$$t \cdot v_t \in H^{1,q}_{\mu}(0, T; L^p_{\overline{\sigma}}(\Omega)) \cap L^q_{\mu}(0, T; D(A_p)),$$
$$\|t \cdot v_t\|_{H^{1,q}_{\mu}(0,T;L^p_{\overline{\sigma}}(\Omega)) \cap L^q_{\mu}(0,T;D(A_p))} \leq C(v, f, f_t, T),$$

for some finite constant C.

Assertion (b) follows directly from Lemma 19. □

Proof of Lemma 19 Let v be a solution to 1.2 in $\mathbb{E}_{1,\mu}(0,T)$. Consider for $0<\varepsilon<\frac{T-T'}{T'}$, $0<T'<T$, the map

$$G\colon (-\varepsilon,\varepsilon)\times \mathbb{E}_{1,\mu}(0,T')\to \mathbb{E}_{0,\mu}(0,T')\times X_{\gamma,\mu},$$
$$(\lambda,\nu)\mapsto (\nu'+(1+\lambda)F(\nu)-(1+\lambda)f_\lambda,\nu(0)-v(0)),$$

where $f_\lambda(t,\cdot):=f((1+\lambda)t,\cdot)$. As in [16, Section 9.2] one can prove that the implicit function theorem applies and there is an implicit function $g_\lambda(-\varepsilon',\varepsilon')\to \mathbb{E}_{1,\mu}(0,T')$, $\varepsilon'\le\varepsilon$ which solves $G(\lambda,g_\lambda(\lambda))=(0,0)$. By uniqueness we conclude that $g_\lambda=v_\lambda$. The implicit derivative at $\lambda=0$ is

$$\partial_\lambda g_\lambda|_{\lambda=0}=t\cdot v_t=-(\partial_v G)(0,v)(A_pv+F_p(v)-f-t\cdot f_t,0),$$

i.e., $t\mapsto -t\cdot v_t$ is the solution to the equation

$$\partial_t h-A_ph+F_p(h,v)+F_p(v,h)=A_pv+F_p(v)-f-t\cdot f_t,\quad h(0)=0,$$

and by Lemma 20

$$\|t\cdot v_t\|_{\mathbb{E}_{1,\mu}(0,T')}\le C(v,\|A_pv+F_p(v)-f-t\cdot f_t\|_{\mathbb{E}_{0,\mu}(0,T')}).$$

Note that

$$\|A_pv+F_p(v)-f-t\cdot f_t\|_{\mathbb{E}_{0,\mu}}\le C\left(\|v\|_{\mathbb{E}_{1,\mu}}+\|v\|^2_{\mathbb{E}_{1,\mu}}+\|f\|_{\mathbb{E}_{0,\mu}}+\|t\cdot f_t\|_{\mathbb{E}_{0,\mu}}\right).$$

Now since $\|v\|_{\mathbb{E}_{1,\mu}(0,T)}+\|f\|_{\mathbb{E}_{0,\mu}(0,T)}+\|t\cdot f_t\|_{\mathbb{E}_{0,\mu}(0,T)}$ are bounded by assumption, $\sup_{T'<T}\|t\cdot v_t\|_{\mathbb{E}_{1,\mu}(0,T')}$ is bounded as well, and therefore $t\cdot v_t\in\mathbb{E}_{1,\mu}(0,T)$ by Lemma 18. □

1.9.4 Global Existence in Maximal L^q-Regularity Spaces

We state here the global, strong well-posedness of the primitive equations in maximal L^q_t-L^p_x-spaces for arbitrarily large data in the critical Besov spaces characterized in Proposition 1.5.5.

Theorem 9 (Global Well-Posedness in Maximal L^q-Regularity Spaces) *Let $p, q \in (1, \infty)$ such that $1/p + 1/q \leq 1$. For $0 < T < \infty$ let $\mu \in [1/p + 1/q, 1]$,*

$$v_0 \in X_{\mu-1/q,q} \qquad \text{and} \qquad \mathbb{P}f \in H^{1,q}_\mu(0, T; L^p_{\overline{\sigma}}(\Omega)) \cap H^{1,2}(\delta, T; L^2_{\overline{\sigma}}(\Omega))$$

for some $\delta > 0$ sufficiently small.

Then there exists a unique, strong solution v to the primitive equations (1.2) *satisfying*

$$v \in H^{1,q}_\mu(0, T; L^p_{\overline{\sigma}}(\Omega)) \cap L^q_\mu(0, T; D(A_p)).$$

This theorem can be proven using higher order *a priori* bounds. For instance, Proposition 1.9.1 already provides an $L^\infty(\delta, T; H^2(\Omega))$ bound. Having suitable embedding properties of the trace spaces one can apply [67, Theorem 5.7.1] to obtain global well-posedness.

1.9.5 Parabolic Smoothing and Higher Regularity

The following theorem deals with the parabolic smoothing effect and the real analyticity of the solution. The proof of this regularity results relies on the implicit function theorem via the so-called parameter-trick which has already been used in the proof of Lemma 19. Note that the additional regularity assumption on f is needed, together with Proposition 1.9.1, to prove the global existence of the solution in Theorem 9 above. We set $v^{(j)} := \partial_t^j v$ and denote by C^ω the space of real analytic functions.

Theorem 10 (Higher Time and Space Regularity) *Let*

$$v \in H^{1,q}_\mu(0, T; L^p_{\overline{\sigma}}(\Omega)) \cap L^q_\mu(0, T; D(A_p))$$

be the global strong solution to the primitive equations for $v_0 \in X_{\mu-1/q,q}$ and $\mathbb{P}f \in L^q_\mu(0, T; L^p_{\overline{\sigma}}(\Omega))$ for p, q, μ as in Theorem 9 below.

(a) *If $\mathbb{P}f \in H^{k,q}_\mu(0, T; L^p_{\overline{\sigma}}(\Omega))$ for $k \in \mathbb{N}_0 = \mathbb{N} \cup \{0\}$, then for any $0 < T' < T$*

$$t^j \cdot v^{(j)} \in H^{1,q}_\mu(0, T'; L^p_{\overline{\sigma}}(\Omega)) \cap L^q_\mu(0, T'; D(A_p)), \quad j = 0, \ldots, k,$$
$$v \in H^{k+1,p}_{loc}(0, T; L^p_{\overline{\sigma}}(\Omega)) \cap H^{k,q}_{loc}(0, T; D(A_p)) \cap C^k((0, T); X_{1-1/q,q});$$

(b) *If $\mathbb{P}f \in C^\infty((0, T); L^p_{\overline{\sigma}}(\Omega))$ or $\mathbb{P}f \in C^\omega((0, T); L^p_{\overline{\sigma}}(\Omega))$, then*

$$v \in C^\infty((0, T); D(A_p)) \qquad \text{or} \qquad v \in C^\omega((0, T); D(A_p)),$$

respectively;

(c) *If* $\mathbb{P}f \in C^\infty((0,T); C^\infty_{per}(\Omega)^2)$ *or* $\mathbb{P}f \in C^\omega((0,T); C^\omega_{per}(\Omega)^2)$*, then*

$$v \in C^\infty((0,T); C^\infty_{per}(\Omega)^2) \qquad or \qquad v \in C^\omega((0,T); C^\omega_{per}(\Omega)^2),$$

respectively.

Let $F: X_1 \to X_0$ be a continuously differentiable function and f an integrable function. We consider the problem

$$v' + F(v) = f. \tag{1.55}$$

Define $F(v) := A_p v + F_p(v)$.

Lemma 20 *Let* $p, q \in (1,\infty)$ *with* $1/p + 1/q \le 1$, $\mu \in [1/p + 1/q, 1]$, $T > 0$. *Then the mapping*

$$\mathcal{F}: H^{1,q}_\mu(0,T; L^p_\sigma(\Omega)) \cap L^q_\mu(0,T; D(A_p)) \to L^q_\mu(0,T; L^p_\sigma(\Omega)), \quad v \mapsto F(v)$$

is continuously differentiable and even real analytic with

$$DF(v)h = A_p h + F_p(v,h) + F_p(h,v).$$

Moreover, for any $g \in L^q_\mu(0,T; L^p_\sigma(\Omega))$ *the equation*

$$\partial_t h - DF(v)h = g, \qquad h(0) = 0,$$

admits a unique solution $h \in H^{1,q}_\mu(0,T; L^p_\sigma(\Omega)) \cap L^q_\mu(0,T; D(A_p))$ *with*

$$\|h\|_{H^{1,q}_\mu(0,T;L^p_\sigma(\Omega)) \cap L^q_\mu(0,T;D(A_p))} \le C(v,g)$$

where $C(v,g) > 0$ *remains bounded for* $v \in H^{1,q}_\mu(0,T; L^p_\sigma(\Omega)) \cap L^q_\mu(0,T; D(A_p))$ *and* $g \in L^q_\mu(0,T; L^p_\sigma(\Omega))$.

Proof of Theorem 10 The assertions (a) and (b) follow from [16, Theorem 9.1] by using Lemma 20.

Note that the abstract result [16, Theorem 9.1] is based on the Banach space version of the implicit function theorem applied to maps of the type $(v(t,\cdot), \lambda) \mapsto \lambda F(v)(\lambda t, \cdot)$. Now, in order to prove (c), the implicit function theorem needs to be applied to both space and time variables. For the most direct approach we need that

$$\Phi_{\lambda,\eta}: (0,\infty) \times \Omega \to (0,\infty) \times \Omega, \ (t,x) \mapsto (\lambda \cdot t, x + t\eta), \ \lambda \in (0,\infty), \ \eta \in \mathbb{R}^3,$$

defines an isomorphism for parameters satisfying $|\lambda - 1| < \epsilon$ and $\|\eta\| < \epsilon$, $\epsilon > 0$, see, e.g.,[66, Section 5]. This is not true for general domains, but it is true for the whole space and also for the torus taking into account periodicity.

The strategy applied here is to first prove analyticity with respect to the horizontally periodic x, y variables, thus proving analyticity of the pressure, and second, to apply a localization procedure in z direction close to $\Gamma_D \neq \emptyset$, while on Γ_N solutions are extended by even reflection onto a larger domain.

To this end, let $\Omega_{per} = \Omega \cup \Gamma_l$ be equipped with the topology of $S^1 \times S^1 \times (-h, 0)$, where $S^1 = \mathbb{R}/\mathbb{Z}$, that is, taking into account lateral periodicity which induces a group structure in the lateral direction. Then

$$\Phi_{\lambda,\eta_H} \quad \text{for} \quad \eta_H = (\mu_x, \mu_y, 0), \quad \eta_x, \eta_y \in S^1$$

defines an isomorphism on Ω_{per} and for $v_{\lambda,\eta_H} = v \circ \Phi_{\lambda,\eta_H}$ we obtain by the chain rule

$$\partial_t v_{\lambda,\eta_H} = \lambda(\partial_t v) \circ \Phi_{\lambda,\eta_H} + \eta_H \cdot (\nabla v) \circ \Phi_{\lambda,\eta_H}.$$

Moreover, for $\epsilon > 0$, we define the real analytic map

$$H : \mathbb{E}_{1,\mu} \times (1-\epsilon, 1+\epsilon) \times (-\epsilon, \epsilon)^2 \to \mathbb{E}_{0,\mu} \times X_{\gamma,\mu}$$

by

$$H(\lambda, \eta_H, v) := (\partial_t v_{\lambda,\eta_H} - \lambda(A_p v + F_p(v))_{\lambda,\eta_H} - \eta_H \cdot \nabla v_{\lambda,\eta_H}, v_0 - v),$$

where the solution v to (1.2) with initial data v_0 solves $H(1, 0, v) = (0, 0)$.

Note that $(A_p v + F_p(v))_{\lambda,\eta_H} = A_p v_{\lambda,\eta_H} + F_p(v_{\lambda,\eta_H})$. The Fréchet derivative $\partial_v H$ is then an isomorphism by arguments similar to the ones given in the proof of Lemma 20 and by using that H is polynomial in v. Therefore, the implicit function theorem yields that $v(\lambda t, x + t\eta_H)$ is real analytic around $(1, 0)$ in η_H and λ. From this we deduce real analyticity of v around (x, t) with respect to time and the horizontal directions, compare, e.g., [66, Section 5]. One can also adapt the approach in [21] for locally symmetric spaces to the situation of a symmetry in only two space directions. In particular, this proves analyticity of the surface pressure π.

Concerning the z-direction, we note first that (1.2) is compatible with even reflections along the Neumann part of the boundary. Thus for $\Gamma_D = \emptyset$ solutions v may be extended to the full torus—a feature used in the literature dealing with Neumann boundary values, see, e.g., [54]—and replacing η_H by general $\eta \in \mathbb{R}^3$ in the above arguments implies analyticity of solutions including the boundary.

If $\Gamma_D \neq \emptyset$, we need to apply a localization procedure with respect to z-variable. The details of this method are neglected here and we refer to [67, Section 9] for details.

Since the main non-locality in the primitive equation arise from the pressure term, we consider finally

$$\partial_t v - \Delta v + v \cdot \nabla_H v + w(v) \cdot \partial_z v = f_s, \quad f_s = f - \nabla_H \pi, \quad v(0) = v_0,$$

where f_s is real analytic by the considerations above and the assumption on f. The above proofs can now be adapted by using the fact that the non-linearity $v \mapsto w(v)\partial_z + v \cdot \nabla_H v$ is real analytic. □

A different strategy to prove smoothness, but not real analyticity, of solutions is to consider higher order time derivatives, which are well defined according to Theorem 10 (a) and (b). Then, by Lemma 15

$$F_p(\cdot)\colon H^{s,p}(\Omega)^2 \to H^{s-(1+1/p),p}(\Omega)^2, \quad s \geq (1/2 + 1/2p), \quad p \in (1, \infty),$$

is well-defined and bounded and we write

$$\partial_t^{(n)} v = A_p^{-1}(\partial_t^{(n)} F_p(v) - \partial_t^{(n+1)} v).$$

Since $\partial_t^{(n)} v \in D(A_p)$ for all $n \in \mathbb{N}_0$, we conclude first that

$$\partial_t^{(n)} F_p(v) \in H^{2-(1+1/p),p}(\Omega)^2.$$

Second, applying elliptic regularity from Proposition 1.5.3, we conclude that $\partial_t^{(n)} v \in H^{(3-1/p),p}(\Omega)^2$. Iterating this argument, that is, "trading time for space regularity" and using Sobolev embeddings we arrive at

$$v \in C^\infty((0,\infty); C^\infty_{per}(\overline{\Omega})^2),$$

thereby proving smoothness including the boundary.

Another strategy for smoothness of solutions, namely proving first additional space regularity and deriving therefrom additional time regularity has been developed in [30] in the case of the Navier-Stokes equations.

1.10 Local and Global Existence in the $L^\infty(L^p)$-Setting: The Case of Neumann-Neumann Boundary Conditions

We consider now the primitive equation in the anisotropic function space $L^\infty(\mathbb{R}^2; L^1(J))$, which is invariant under the scaling

$$v_\lambda(t, x_1, x_2, x_3) = \lambda v(\lambda^2 t, \lambda(x_1, x_2, x_3)), \quad \lambda > 0.$$

This means that $\|v_\lambda\|_{L^\infty(\mathbb{R}^2; L^1(\lambda^{-1}J))} = \|v\|_{L^\infty(\mathbb{R}^2; L^1(J))}$ for all $\lambda > 0$. Moreover, v_λ is a solution to the primitive equations whenever v has this property.

Based on the L^∞-type estimates for the underlying hydrostatic Stokes semigroup S on $L^\infty(L^1)$ and its gradient obtained in Sect. 1.6, we develop an iteration scheme yielding first the existence of a unique, local mild solution for initial data of the form $a = a_1 + a_2$ with

$a_1 \in BUC_\sigma(\mathbb{R}^2, L^1(J))$ and a_2 being a small perturbation in $L^\infty_\sigma(\mathbb{R}^2; L^1(J))$.

The main idea of our approach is as follows: As a first step, we extend the hydrostatic Stokes semigroup S from the $L^p(L^p)$-setting to the $L^\infty(L^1)$-setting. Duhamel's formula leads us then to consider terms of the form $S(t)\mathbb{P}\nabla \cdot (u(v) \otimes v)$. Recall that $u = (v, w)$ involves first derivatives through $w = w(v)$ and thus second order derivatives appear in the above term. This implies a singularity of order t^{-1} for $S(t)\mathbb{P}\nabla \cdot (u(v) \otimes v)$ for $t > 0$, which is non-integrable.

In order to surpass this difficulty, we smoothen the horizontal derivatives by inserting fractional powers of the horizontal Laplacian and the vertical derivative by inserting fractional vertical derivatives and obtain

$$S(t)\mathbb{P}\nabla \cdot (v \otimes v) = \underbrace{S(t)\mathbb{P}(-\Delta_H)^{(1-\alpha)/2}}_{\text{decay term}} \nabla_H \cdot (-\Delta_H)^{-(1-\alpha)/2}(v \otimes v)$$

$$+ \underbrace{S(t)\partial_z I^\alpha_{z_0}}_{\text{decay term}} \; I^{1-\alpha}_{z_0} \partial_z(wv), \qquad t > 0.$$

1.10.1 Nonlinear Estimates

We start by estimating the integral term for functions with vanishing vertical average.

Lemma 21 *For $\alpha \in [0, 1)$ there exists a constant $C > 0$ such that for $t > 0$*

$$\|S(t)\mathbb{P}\nabla \cdot (\tilde{v} \otimes v)\|_{\infty,1} \le Ct^{-(1-\alpha)/2}\Big(\|\nabla\tilde{v}\|_{\infty,1}\|v\|_{\infty,1} + \|\tilde{v}\|_{\infty,1}\|\nabla v\|_{\infty,1}\Big)^{1-\alpha} \cdot \Big(\|\nabla v\|_{\infty,1}\|\nabla\tilde{v}\|_{\infty,1}\Big)^{\alpha},$$

for all $v \in L^\infty_\sigma(L^1)$ satisfying $\overline{v} = 0$ and all $\tilde{v} = (\tilde{v}, \tilde{w})$ with $\tilde{v} \in L^\infty_\sigma(L^1)$ satisfying $\overline{\tilde{v}} = 0$ as well as $\tilde{w} = \int_z^{z_1} \operatorname{div}_H \tilde{v} \, dx_3$.

Proof We first note that

$$\nabla \cdot (\tilde{v} \otimes v) = \nabla_H \cdot (\tilde{v} \otimes v) + \partial_z(\tilde{w}v).$$

Since $\operatorname{div}_H \overline{\tilde{v}} = 0$ we obtain $\tilde{w} = 0$ at $z = z_0$ and since $\tilde{w} = 0$ at $z = z_1$ by definition, we see that $\overline{\partial_z(\tilde{w}v)} = 0$. Hence,

$$\mathbb{P}\nabla \cdot (\tilde{v} \otimes v) = \mathbb{P}\nabla_H \cdot (\tilde{v} \otimes v) + \partial_z(\tilde{w}v).$$

The case $\alpha = 0$ is now straightforward using Proposition 1.6.1 (ii),(iv). Consider now the case $\alpha \in (0,1)$. Noting that $(-\Delta_H)^{(1-\alpha)/2}$,$(-\Delta_H)^{-(1-\alpha)/2}$ and ∇_H commute, we write for $t > 0$

$$\begin{aligned} S(t)\mathbb{P}\nabla \cdot (\tilde{v} \otimes v) &= S(t)\mathbb{P}(-\Delta_H)^{(1-\alpha)/2}\nabla_H \cdot (-\Delta_H)^{-(1-\alpha)/2}(\tilde{v} \otimes v) \\ &\quad + S(t)\partial_z I_{z_0}^{\alpha} I_{z_0}^{1-\alpha} \partial_z(\tilde{w}v) \\ &=: I + II. \end{aligned}$$

Applying Proposition 1.6.1 (iii) and Lemma 5 yields

$$\begin{aligned} \|I\|_{\infty,1} &\le Ct^{-(1-\alpha)/2} \left\| \nabla_H \cdot (-\Delta_H)^{-(1-\alpha)/2} \tilde{v} \otimes v \right\|_{\infty,1} \\ &\le Ct^{-(1-\alpha)/2} \|\nabla_H(\tilde{v} \otimes v)\|_{\infty,1}^{\alpha} \|\tilde{v} \otimes v\|_{\infty,1}^{1-\alpha}, \quad t > 0. \end{aligned}$$

Since $\overline{v} = 0$ and $\overline{\tilde{v}} = 0$, we obtain the estimates

$$\begin{aligned} \|\nabla(\tilde{v} \otimes v)\|_{\infty,1} &\le \|\nabla\tilde{v}\|_{\infty,1} \|v\|_{\infty,\infty} + \|\tilde{v}\|_{\infty,\infty}\|\nabla v\|_{\infty,1}, \\ \|\tilde{v} \otimes v\|_{\infty,1} &\le \|\tilde{v}\|_{\infty,1}\|v\|_{\infty,\infty} + \|\tilde{v}\|_{\infty,\infty}\|v\|_{\infty,1}, \\ \|v\|_{\infty,\infty} &\le \|\partial_z v\|_{\infty,1}, \\ \|\tilde{v}\|_{\infty,\infty} &\le \|\partial_z \tilde{v}\|_{\infty,1} \end{aligned}$$

and the term $\|I\|_{\infty,1}$ can be thus estimated as claimed.

In order to estimate $\|II\|_{\infty,1}$ we observe that Proposition 1.6.1 (ii) and Lemma 4 yield

$$\|II\|_{\infty,1} \le Ct^{-(1-\alpha)/2} \left\|\partial_z^{\alpha}(\tilde{w}v)\right\|_{\infty,1} \le Ct^{-(1-\alpha)/2}\|\tilde{w}v\|_{\infty,1}^{1-\alpha} \|\partial_z(\tilde{w}v)\|_{\infty,1}^{\alpha}, \quad t > 0.$$

Here we invoked the fact that

$$I_{z_0}^{\alpha}\left(I_{z_0}^{1-\alpha}\partial_z(\tilde{w}v)\right)(z_1) = (\tilde{w}v)(z_1) = 0.$$

Since

$$\|\tilde{w}\|_{\infty,\infty} \le C\,\|\partial_z\tilde{w}\|_{\infty,1} \le C\,\|\nabla_H\tilde{v}\|_{\infty,1}$$

we are able to estimate $\|II\|_{\infty,1}$ in the same way as I. This completes the proof. □

The next step consists of proving a similar estimate for the above integral term, however, without assuming that the vertical average of the functions involved is vanishing. To this end, we set

$$\|v\|_{1,\infty,1} := \|v\|_{\infty,1} + \|\nabla v\|_{\infty,1}.$$

We argue similarly as in the proof of Lemma 21 but considering $v = (v - \overline{v}) + \overline{v}$.

Proposition 1.10.1 (Estimate for the Nonlinear Term) *Let $\alpha \in [0, 1)$, then there exists a constant $C > 0$ such that for $t > 0$*

$$\|S(t)\mathbb{P}\nabla \cdot (\tilde{v} \otimes v)\|_{\infty,1} \le Ct^{-(1-\alpha)/2}\Big(\|\tilde{v}\|_{1,\infty,1}\|v\|_{\infty,1} + \|v\|_{1,\infty,1}\|\tilde{v}\|_{\infty,1}\Big)^{1-\alpha} \cdot \Big(\|\tilde{v}\|_{1,\infty,1}\|v\|_{1,\infty,1}\Big)^{\alpha},$$

for all $\tilde{v} = (\tilde{v}, \tilde{w})$ with $\tilde{v} \in L^\infty_\sigma(L^1)$, $\nabla\tilde{v} \in L^\infty(L^1)$ where $\tilde{w} = \int_z^{z_1} \operatorname{div}_H \tilde{v}\, dx_3$, and $v \in L^\infty_\sigma(L^1)$ satisfying $\nabla v \in L^\infty(L^1)$.

1.10.2 Iteration Scheme for the Neumann-Neumann Case

The iteration scheme developed below yields sequences defined for $m \in \mathbb{N}$ by

$$K_m(t) := \sup_{0<\tau<t} \tau^{1/2}\|v_m(\tau)\|_{1,\infty,1},$$
$$\|v_m(\tau)\|_{1,\infty,1} := \|v_m(\tau)\|_{\infty,1} + \|\nabla v_m(\tau)\|_{\infty,1}.$$

We show that this sequence and related sequences converge provided

$$K_0(t) = \sup_{0<\tau<t} \tau^{1/2}\, \|S(\tau)a\|_{1,\infty,1}$$

is sufficiently small. This smallness can be balanced by taking the time as well as the rough part a_2 sufficiently small. The sequence (v_m) is then a Cauchy sequence and converges to the unique solution of (1.2).

Theorem 11 (Local Strong Well-Posedness)
Let $a \in L^\infty_\sigma(L^1)$.

a) *Then there exists a constant $\varepsilon_0 > 0$ such that if $\overline{\lim}_{t\to 0}\, t^{1/2}\|S(t)a\|_{1,\infty,1} \le \varepsilon_0$, there exists $T > 0$ such that (1.2) subject to Neumann boundary conditions, i.e. $\Gamma_D = \emptyset$ in (1.4), admits a unique, local mild solution*

$$v \in C\big((0,T); BUC_\sigma\big(L^1(J)\big)\big) \cap L^\infty\big((0,T); BUC_\sigma\big(L^1(J)\big)\big)$$

with

$$\overline{\lim}_{t\to 0}\, t^{1/2}\|v(t)\|_{1,\infty,1} \le \varepsilon_0.$$

b) *The above condition* $\limsup_{t\to 0} t^{1/2}\|S(t)a\|_{1,\infty,1} \le \varepsilon_0$ *is in particular satisfied for* a *of the form* $a = a_1 + a_2$ *with* $a_1 \in BUC_\sigma(L^1)$ *and* $a_2 \in L^\infty_\sigma(L^1)$ *provided* $\|a_2\|_{L^\infty_\sigma(L^1)}$ *is sufficiently small. In this case*

$$\overline{\lim_{t\to 0}}\, t^{1/2}\|\nabla v(t)\|_{L^\infty(L^1)} \le C\|a_2\|_{L^\infty_\sigma(L^1)}$$

for some $C > 0$*, independent of* a*.*

c) *If* $a_2 = 0$*, then* $v \in C\left([0,T);\, BUC_\sigma\left(L^1(J)\right)\right)$*.*

Remark 9 Since $e^{t\Delta_H}$ as well as $\mathbb{P}$ and the non-linearity leave horizontal periodicity invariant we observe that if a is in addition to the assumption of Theorem 11 periodic with respect to the horizontal variables, then this is also true for $v(t)$ for any $t > 0$.

Proposition 1.10.2 (Local Existence for $p > 1$**)** *Let* a *and* $T > 0$ *be as in Theorem 11. If in addition to the assumptions of Theorem 11 the initial data* a *satisfies*

(i) $a \in L^\infty_\sigma(L^p)$ *for some* $p \in (1,\infty]$*, then*

$$t^{1/2-1/2p}v,\, t^{1-1/2p}\nabla v \in L^\infty\left(0,T;\, L^\infty_\sigma(L^p)\right);$$

(ii) $a \in BUC_\sigma(L^p)$ *for some* $p \in (1,\infty]$*, then*

$$t^{1/2-1/2p}v,\, t^{1-1/2p}\nabla v \in C\left([0,T),\, BUC_\sigma(L^p)\right);$$

(iii) $a \in BUC_\sigma(BUC)$*, then*

$$t^{1/2}v,\, t\nabla v \in C\left([0,T),\, BUC_\sigma(BUC)\right).$$

The local mild solution constructed in Theorem 11 exists at least on some short time interval $[0,T)$, where $T > 0$ depends on a. Later we will use smoothing properties to obtain a global strong solution, instead we may also estimate the existence time $T > 0$ explicitly from below in terms of the $|||\cdot|||$-norm, defined for $a \in L^\infty_\sigma(\mathbb{R}^2; L^1(J))$ and for $\mu \in [0,1/2)$ by

$$|||a||| := [a]_\mu + \|a\|_{\infty,1}, \quad \text{where} \quad [a]_\mu := \sup_{0<t<1} t^\mu\, \|\nabla S(t)a\|_{\infty,1}\,.$$

Proposition 1.10.3 (Estimate on the Life Span)

Let a *and* $T > 0$ *be as in Theorem 11. Assume in addition that* $[a]_\mu < \infty$ *for some* $\mu \in [0,1/2)$*. Then there exists* $C > 0$*, depending on* μ *only, such that*

$$1/T \le \min\left(C|||a|||,\, 1\right)^{2/(1/2-\mu)}\,.$$

We now give a proof of Theorem 11. The proof of Proposition 1.10.3 is similar.

Proof of Theorem 11 Consider the sequence (v_m) recursively defined for $t \geq 0$ by

$$v_{m+1}(t) := S(t)a - \int_0^t S(t-s)\mathbb{P}\nabla \cdot (v_m(s) \otimes v_m(s))\mathrm{d}s, \quad m \in \mathbb{N}$$
$$v_0(t) := S(t)a.$$

Applying Proposition 1.6.1 (i), (ii) with $\alpha = 0$, there exists $C > 0$ such that

$$\begin{aligned}\|v_{m+1}(t)\|_{\infty,1} &\leq \|S(t)a\|_{\infty,1} + C\int_0^t (t-s)^{-1/2}\|v_m(s) \otimes v_m(s)\|_{\infty,1}\mathrm{d}s \\ &\leq \|S(t)a\|_{\infty,1} + C\int_0^t (t-s)^{-1/2}\|v_m(s)\|_{\infty,\infty}\|v_m(s)\|_{\infty,1}\mathrm{d}s \\ &\leq \|S(t)a\|_{\infty,1} + C\int_0^t (t-s)^{-1/2}\|v_m(s)\|_{1,\infty,1}\|v_m(s)\|_{\infty,1}\mathrm{d}s.\end{aligned} \tag{1.56}$$

Note that constants $C > 0$ here and below are independent of v_m, v_m, and t. We now estimate $\|\nabla v_{m+1}(t)\|_{\infty,1}$ by Proposition 1.10.1. Since

$$\nabla S(t-s) = \nabla S(\tfrac{t-s}{2})S(\tfrac{t-s}{2})$$

Proposition 1.6.1 (i) and Proposition 1.10.1 with $\alpha = 1/2$ yield

$$\begin{aligned}&\|\nabla v_{m+1}(t)\|_{\infty,1} \leq \|\nabla S(t)a\|_{\infty,1} \\ &+ C\int_0^t (t-s)^{-1/2}(t-s)^{-1/4}\|v_m(s)\|_{1,\infty,1}^{3/2}\|v_m(s)\|_{\infty,1}^{1/2}\mathrm{d}s, \quad t > 0.\end{aligned} \tag{1.57}$$

Note that in the above estimate we may also take any $\alpha \in (0, 1)$. For $m \in \mathbb{N} \cup \{0\}$ and $t > 0$ we now set

$$\begin{aligned}K_m(t) &:= \sup_{0<\tau<t} \tau^{1/2}\|v_m(\tau)\|_{1,\infty,1}, \\ H_m(t) &:= \sup_{0<\tau<t} \|v_m(\tau)\|_{\infty,1}, \\ M_m(t) &:= \sup_{0<\tau<t} \tau^{1/2}\|v_m(\tau)\|_{\infty,1}.\end{aligned}$$

Estimate (1.56) combined with $\|S(t)a\|_{\infty,1} \leq \|a\|_{\infty,1}$ for all $t > 0$ yields

$$H_{m+1}(t) \leq \|a\|_{\infty,1} + CK_m(t)H_m(t), \quad t > 0. \tag{1.58}$$

multiplying (1.57) by $t^{1/2}$ yields

$$\sup_{0<\tau<t} \tau^{1/2} \|\nabla v_{m+1}(\tau)\|_{\infty,1} \leq \sup_{0<\tau<t} \tau^{1/2} \|\nabla S(\tau)a\|_{\infty,1} + C K_m(t)^{3/2} H_m(t)^{1/2}, \quad t > 0, \tag{1.59}$$

and by multiplying (1.56) with $t^{1/2}$, we obtain

$$M_{m+1}(t) \leq \sup_{0<\tau<t} \tau^{1/2} \|S(\tau)a\|_{\infty,1} + Ct^{1/2} K_m(t) H_m(t), \tag{1.60}$$

provided $t \leq T$ for some $T \leq 1$. Adding (1.59) and (1.60) yields

$$K_{m+1}(t) \leq K_0(t) + C K_m(t)^{3/2} H_m(t)^{1/2} + Ct^{1/2} K_m(t) H_m(t), \quad m \geq 0 \tag{1.61}$$

with

$$K_0(t) = \sup_{0<\tau<t} \tau^{1/2} \|S(\tau)a\|_{1,\infty,1} .$$

By assumption $K_0(t) \leq \varepsilon_0$, and by the following Lemma 17 the sequences (H_m) and (K_m) are thus uniformly bounded provided ε_0 is sufficiently small.

To conclude the proof of assertion (a), one can show now that (v_m) is a Cauchy sequence in the space $L^\infty\left((0, t_0), L^\infty_\sigma(L^1)\right)$ and that also $(t^{1/2}\nabla(v_m))$ defines a Cauchy sequence in $L^\infty\left((0, t_0), L^\infty(L^1)\right)$. Hence, v as the limit of (v_m) satisfies the desired estimate. Moreover $v_m, \nabla v_m \in C\left((0, t_0); L^\infty(L^1)\right)$ and therefore, $v \in C\left((0, t_0); BUC_\sigma(L^1)\right)$.

In order to prove (b) let $a_1 \in BUC_\sigma(L^1)$, then

$$\tau^{1/2} \|\nabla S(\tau)a_1\|_{\infty,1} \to 0 \quad \text{as} \quad \tau \searrow 0, \quad \text{and} \quad \tau^{1/2} \|\nabla S(\tau)a_2\|_{\infty,1} \leq C \|a_2\|_{\infty,1}$$

by Proposition 1.6.1 (i). Thus, by (1.58) and (1.61), the sequences (H_m) and (K_m) fulfill the assumptions of the following Lemma 17 provided t is small enough, say $t \leq t_0$, since $K_0(t) \to 0$ as $t \searrow 0$, and $\|a_2\|_{\infty,1}$ is sufficiently small. The sequences (H_m) and (K_m) are thus uniformly bounded.

To prove (c), assume that $a_2 = 0$. Then one can use strong continuity of the semigroup to prove that (v_m) is a Cauchy sequence in $C\left([0, t_0], BUC_\sigma(L^1)\right)$ and $(t^{1/2}\nabla v_m)$ is a Cauchy sequence in $C\left([0, t_0], L^\infty(L^1)\right)$. Hence, v as the limit of (v_m) has the desired regularity.

The proof of the uniqueness follows in both cases by a similar line of arguments. □

It remains to prove the uniform boundedness of the sequences (H_m) and (K_m) defined in the above proof.

Lemma 22 *Let $A, \varepsilon > 0$ be constants and assume that $(H_m) \subset \mathbb{R}$ and $(K_m) \subset \mathbb{R}$ are sequences satisfying*

$$H_0 \le A, \qquad H_{m+1} \le A + C H_m K_m,$$

$$K_0 \le \varepsilon, \qquad K_{m+1} \le \varepsilon + C K_m^{3/2} H_m^{1/2} + (4A)^{-1} K_m H_m,$$

for all $m \ge 0$ *and a constant* $C > 0$ *independent of* m*. Then there exists* $\varepsilon_0 = \varepsilon_0(C, A) > 0$ *such that* (K_m) *and* (H_m) *are bounded sequences provided that* $\varepsilon \le \varepsilon_0$.

It is now not difficult anymore to show that the local solution constructed in Theorem 11 extends to a global one.

Theorem 12 *Suppose in addition to the assumptions of Theorem 11 that a is periodic with respect to the horizontal variables. Then, given* $T^* > 0$*, the unique mild solution v constructed in Theorem 11 extends a to a unique, strong solution to the primitive equations subject to Neumann conditions on* $(0, T^*)$.

In order to extend the local solution to a global one, we make use of the regularization of the solution. By Theorem 11 $v(t_0) \in BUC(W^{1,1})$ for some $t_0 > 0$ and thus $v(t_0) \in BUC(L^p)$ for all $p \in [1, \infty]$. By Proposition 1.10.2, $v(t_1), \nabla v(t_1) \in BUC(L^p)$ for $t_1 > t_0$. Choosing $p = 2$ and $v(t_1)$ as new initial data, it follows from Theorem 9 that v extends to global solution.

1.11 Local and Global Existence: The $L^\infty(L^p)$-Setting and Mixed Dirichlet-Neumann Boundary Conditions

We start this section with the following estimates needed for the iteration scheme for the nonlinear problem.

Lemma 23 (Estimate on the Non-linearity) *Let* $p > 3$*. Then there exists a constant* $C > 0$ *such that for all* $t > 0$ *and all* $v_i \in X_{\overline{\sigma}}$ *satisfying* $\nabla v_i \in X$ *and* $v_i\big|_{\Gamma_b} = 0$ *with* $v_i = (v_i, w_i)$ *as in* (1.3) *for* $i = 1, 2$

$$\|e^{tA}\mathbb{P}(v_1 \cdot \nabla)v_2\|_{L^\infty_H L^p_z} \le C t^{-1/2} \|\nabla v_1\|_{L^\infty_H L^p_z} \|v_2\|_{L^\infty_H L^p_z}, \tag{i}$$

$$\|\nabla e^{tA}\mathbb{P}(v_1 \cdot \nabla)v_2\|_{L^\infty_H L^p_z} \le C t^{-1/2} \|\nabla v_1\|_{L^\infty_H L^p_z} \|\nabla v_2\|_{L^\infty_H L^p_z}, \tag{ii}$$

$$\|\nabla e^{tA}\mathbb{P}(v_1 \cdot \nabla)v_2\|_{L^\infty_H L^p_z} \le C t^{-1} \|\nabla v_1\|_{L^\infty_H L^p_z} \|v_2\|_{L^\infty_H L^p_z}, \tag{iii}$$

as well as

$$\|e^{tA}\mathbb{P}(v_1 \cdot \nabla)v_2\|_{L^\infty_H L^p_z} \le$$

$$C\left(t^{-1/2} \|\nabla v_i\|_{L^\infty_H L^p_z} \|v_j\|_{L^\infty_H L^p_z} + \|\nabla v_1\|_{L^\infty_H L^p_z} \|\nabla v_2\|_{L^\infty_H L^p_z}\right) \tag{iv}$$

where $\{i, j\} = \{1, 2\}$.

Proof Note that

$$\|(v_1 \cdot \nabla)v_2\|_{L^\infty_H L^p_z} \le \left(\|v_1\|_{L^\infty(\Omega)} + \|w_1\|_{L^\infty(\Omega)}\right) \|\nabla v_2\|_{L^\infty_H L^p_z}.$$

By Sobolev embeddings, Poincaré's inequality and $X \hookrightarrow L^p(\Omega)^2$ we obtain

$$|v_i\|_{L^\infty(\Omega)} \le C\|v_i\|_{W^{1,p}(\Omega)} \le C\|\nabla v_i\|_{L^p(\Omega)} \le C\|\nabla v_i\|_{L^\infty_H L^p_z},$$

and similarly

$$\|w_i\|_{L^\infty(\Omega)} \le C\|\mathrm{div}_H v_i\|_{L^\infty_H L^p_z} \le C\|\nabla v_i\|_{L^\infty_H L^p_z}.$$

Theorem 5 then implies (ii) as well as

$$\begin{aligned}\|\nabla e^{tA}\mathbb{P}(v_1 \cdot \nabla)v_2\|_{L^\infty_H L^p_z} &\le Ct^{-1/2}\|(v_1 \cdot \nabla)v_2\|_{L^\infty_H L^p_z} \\ &\le Ct^{-1/2}\|\nabla v_1\|_{L^\infty_H L^p_z}\|\nabla v_2\|_{L^\infty_H L^p_z}.\end{aligned}$$

Assertion (i) is proved in a similar manner. In fact, since $\mathrm{div}\, u_i = 0$ we have

$$(u_1 \cdot \nabla)v_2 = \nabla \cdot (u_1 \otimes v_2) = \nabla_H \cdot (v_1 \otimes v_2) + \partial_z(w_1 v_2)$$

and Theorem 5 yields

$$\begin{aligned}\|e^{tA}\mathbb{P}(v_1 \cdot \nabla)v_2\|_{L^\infty_H L^p_z} &= \|e^{tA}\mathbb{P}\nabla \cdot (v_1 \otimes v_2)\|_{L^\infty_H L^p_z} \\ &\le \|e^{tA}\mathbb{P}\nabla_H \cdot (v_1 \otimes v_2)\|_{L^\infty_H L^p_z} + \|e^{tA}\mathbb{P}\partial_z(w_1 v_2)\|_{L^\infty_H L^p_z} \\ &\le Ct^{-1/2}\left(\|v_1 \otimes v_2\|_{L^\infty_H L^p_z} + \|w_1 v_2\|_{L^\infty_H L^p_z}\right) \\ &\le Ct^{-1/2}\|\nabla v_1\|_{L^\infty_H L^p_z}\|v_2\|_{L^\infty_H L^p_z}.\end{aligned}$$

The remaining estimates (iii) and (iv) are obtained analogously. □

1.11.1 Global Strong Well-Posedness: Dirichlet-Neumann Boundary Conditions

The following two theorems represent our global strong well-posedness results for the primitive equations subject to mixed Dirichlet-Neumann boundary conditions.

Theorem 13 *Let $p \in (3, \infty)$. Then for all $a \in X_{\overline{\sigma}}$ there exists a unique, global, strong solution v to the primitive equations* (1.2) *and* (1.4) *with $v(0) = a$ satisfying*

$$v \in C([0,\infty); X_{\overline{\sigma}}),\ \ t^{1/2}\nabla v \in L^\infty((0,\infty); X),\ \ \limsup_{t\to 0+} t^{1/2}\|\nabla v(t)\|_{L^\infty_H L^p_z(\Omega)}=0.$$

The corresponding pressure satisfies $\pi \in C((0,\infty); C^{1,\alpha}([0,1]^2))$ for some $\alpha \in (0, 1-3/p)$ and is unique up to an additive constant.

Theorem 14 *Let $p \in (3, \infty)$. Then there exists a constant $C_0 > 0$ such that if $a = a_1 + a_2$ with $a_1 \in X_{\overline{\sigma}}$ and $a_2 \in L^\infty_H L^p_z(\Omega)^2 \cap L^p_{\overline{\sigma}}(\Omega)$ with*

$$\|a_2\|_{L^\infty_H L^p_z(\Omega)} \le C_0,$$

then there exists a unique, global, strong solution v to the primitive equations (1.2) *and* (1.4) *with $v(0) = a$ satisfying*

$$v \in C([0,\infty); L^p_{\overline{\sigma}}(\Omega)) \cap L^\infty((0,T); L^\infty_H L^p_z(\Omega))^2$$

as well as

$$t^{1/2}\nabla v \in L^\infty((0,\infty); X),\quad \limsup_{t\to 0+} t^{1/2}\|\nabla v\|_{L^\infty_H L^p_z(\Omega)} \le C\|a_2\|_{L^\infty_H L^p_z},$$

where $C > 0$ does not depend on the data, and the pressure has the same regularity as in Theorem 13.

Remarks 1.11.1

a) We note that when in the situation of Theorem 14 the initial data do not belong to X, i.e. when $a_2 \neq 0$, the solution fails to be continuous at $t = 0$ with respect to the $L^\infty_H L^p_z$-norm.
b) The condition $p > 3$ is due to the embeddings

$$v_{\text{ref}}(t_0) \in B^{2-2/q}_{pq}(\Omega)^2 \hookrightarrow C^1(\overline{\Omega})^2 \text{ and } W^{2,p}(\Omega) \hookrightarrow C^{1,\alpha}(\overline{\Omega}) \quad \text{for } p \in (3,\infty).$$

Some words about our strategy for proving the global well-posedness results are in order: We will first construct a local, mild solution to the problem (1.2) and (1.4), i.e. a function satisfying the relation

$$v(t) = e^{tA_{\overline{\sigma}}}a + \int_0^t e^{(t-s)A_{\overline{\sigma}}}\mathbb{P}F(v(s))\,ds, \quad t \in (0,T) \tag{1.62}$$

for some $T > 0$ and where $F(v) = -(v\cdot\nabla)v$. We will then show that v regularizes for $t_0 > 0$. By Theorem 9 we may take $v(t_0)$ as a new initial value to extend the mild solution to a global, strong solution on (t_0, ∞) and then on $(0, \infty)$ by uniqueness. The additional regularity for $t \to 0+$ results forms the construction of the mild solutions. In order to construct a mild solution, we decompose $a = a_{\text{ref}} + a_0$ such that a_{ref} is sufficiently smooth and a_0 can be taken to be arbitrarily small. Using our results concerning the existence of solutions to the primitive equations for *smooth* data, we obtain a reference solution v_{ref} and construct then $V := v - v_{\text{ref}}$ by an iteration scheme using L^∞-type estimates for terms of the form $\nabla e^{tA_{\overline{\sigma}}}\mathbb{P}$ given in Theorem 5.

Let us recall from Theorem 9 that for

$$a \in X_\gamma := (L^p_{\overline{\sigma}}(\Omega), D(A_p))_{1-1/q,q} \subset B^{2-2/q}_{pq}(\Omega)^2 \cap L^p_{\overline{\sigma}}(\Omega)$$

with $p, q \in (1, \infty)$ satisfying $1/p + 1/q \leq 1$, the primitive equations subject to (1.4) admit a unique global strong solution.

The iteration scheme will make use of the following fact.

Lemma 24 *Let $(a_n)_{n\in\mathbb{N}}$ be a sequence of positive real numbers such that*

$$a_{m+1} \leq a_0 + c_1 a_m^2 + c_2 a_m \quad \textit{for all } m \in \mathbb{N}$$

and constants $c_1 > 0$ and $c_2 \in (0, 1)$ such that $4c_1 a_0 < (1 - c_2)^2$. Then $a_m < \frac{2}{1-c_2} a_0$ for all $m \in \mathbb{N}$.

Proof of Theorem 13 We subdivide the proof into several steps.

Step 1: ***Decomposition of Data***

Given $a \in X_{\overline{\sigma}}$, we will split a into a smooth part a_{ref} and a small rough part a_0, i.e. $a = a_{\text{ref}} + a_0$, as follows: Since $A_{\overline{\sigma}}$ is densely defined on $X_{\overline{\sigma}}$ we may take $a_{\text{ref}} \in D(A_{\overline{\sigma}})$ such that $a_0 := a - a_{\text{ref}}$ is arbitrarily small in $X_{\overline{\sigma}}$. Now let $q \in (1, \infty)$ be such that $1/q + 1/p \leq 1$ and $2/q + 3/p < 1$. By Sobolev embeddings $X_\gamma \hookrightarrow C^1(\overline{\Omega})^2$. Moreover, since $D(A_{\overline{\sigma}}) \subset D(A_{p,\overline{\sigma}}) \subset X_\gamma$, Theorem 9 implies that, taking a_{ref} as initial data, there exists a function $v_{\text{ref}} \in C([0, \infty); X_\gamma)$ solving the primitive equations with initial data $v_{\text{ref}}(0) = a_{\text{ref}}$.

Step 2: ***Estimates for the Construction of a Local Solution***

In this step we show there existence of a constant $C_0 > 0$ such that if $a_0 \in X_{\overline{\sigma}}$ satisfies $\|a_0\|_{L^\infty_H L^p_z} < C_0$, then there exist $T > 0$ and a unique function

$$V \in \mathcal{S}(T) := \{V \in C([0, T]; X_{\overline{\sigma}}) : |\nabla V(t)\|_{L^\infty_H L^p_z} = o(t^{-1/2})\},$$

where

$$\|V\|_{\mathcal{S}(T)} = \max\left\{\sup_{0<t<T} |V(t)\|_{L^\infty_H L^p_z}, \sup_{0<t<T} t^{1/2}|\nabla V(t)\|_{L^\infty_H L^p_z}\right\}$$

such that $v = v_{\text{ref}} + V$ solves equation (1.2) on $(0, T)$ with initial value $v(0) = a$. In order to construct V we define an iterative sequence of functions $(V_m)_{m\in\mathbb{N}}$ by

$$V_0(t) = e^{tA}a_0, \quad V_{m+1}(t) = e^{tA}a_0 + \int_0^t e^{(t-s)A} F_m(s)\, ds \tag{1.63}$$

where

$$F_m := -\mathbb{P}\left((v_m \cdot \nabla)V_m + (v_m \cdot \nabla)v_{\text{ref}} + (v_{\text{ref}} \cdot \nabla)V_m\right),$$

$v_m = (V_m, W_m)$ and $v_{\text{ref}} = (v_{\text{ref}}, w_{\text{ref}})$ where w is given by (1.3). Since $\mathbb{P}a_0 = a_0$ we have

$$\|V_0\|_{\mathcal{S}(T)} \leq C\|a_0\|_{L^\infty_H L^p_z}, \quad T \in (0, \infty)$$

by Theorem 5. For $m \geq 1$ set

$$K_m(t) := \sup_{0<s<t} s^{1/2}\|\nabla V_m(s)\|_{L^\infty_H L^p_z}, \quad H_m(t) := \sup_{0<s<t} \|V_m(s)\|_{L^\infty_H L^p_z}.$$

It can be then shown that the sequences $(K_m)(t)$ and $(H_m)(t)$ satisfy for $t > 0$ the relation

$$\begin{aligned} K_{m+1}(t) &\leq C_1\Big(\|a_0\|_{L^\infty_H L^p_z} + K_m(t)H_m(t) + K_m(t)^2 + Rt^{1/2}K_m(t)\Big), \\ H_{m+1}(t) &\leq C_1\Big(\|a_0\|_{L^\infty_H L^p_z} + K_m(t)H_m(t) + Rt^{1/2}H_m(t) + Rt^{1/2}K_m(t)\Big). \end{aligned} \tag{1.64}$$

Now let $T \in (0, T_0)$ where $T_0 > 0$ is chosen such that $8C_1RT_0^{1/2} < 1$. Then for all $0 < t \leq T < T_0$ we have

$$\|V_m\|_{\mathcal{S}(t)} \leq \frac{8}{3}C_1\|a_0\|_{L^\infty_H L^p_z}, \quad t \in (0, T]. \tag{1.65}$$

Step 3: ***Convergence***

We now show that $(V_m)_{m\in\mathbb{N}}$ is a Cauchy sequence in $\mathcal{S}(T)$ provided $\|a_0\|_{L^\infty_H L^p_z}$ is sufficiently small. To this end, consider the sequence

$$\tilde{V}_m := V_{m+1} - V_m, \quad m \geq 0.$$

It can be then shown that for $m \geq 1$ and $0 < t \leq T < T_0$ we have

$$\|\tilde{V}_m(t)\|_{\mathcal{S}(t)} < \frac{1}{2}\|\tilde{V}_{m-1}(t)\|_{\mathcal{S}(t)}.$$

Hence, $(V_m)_{m\in\mathbb{N}}$ converges in $\mathcal{S}(T)$ and denoting its limit by V we see that

$$V(t) = e^{tA}a_0 - \int_0^t e^{(t-s)A}\mathbb{P}\Big((v(s)\cdot\nabla)V(s) + (v(s)\cdot\nabla)v_{\mathrm{ref}}(s) + (v_{\mathrm{ref}}(s)\cdot\nabla)V(s)\Big)\,ds \tag{1.66}$$

$t \in (0, T)$. Thus $v := V + v_{\mathrm{ref}}$ is a solution to the primitive equations (1.2).

Step 4: ***Extending to a Global Solution***

The embedding $L^\infty_H L^p_z(\Omega) \hookrightarrow L^p(\Omega)$, as well as the estimates

$$t^\vartheta \|e^{tA}\mathbb{P}f\|_{D((-A_{p,\overline{\sigma}})^\vartheta)} \le C\|f\|_{L^p(\Omega)}, \quad t^{1/2}\|e^{tA}\mathbb{P}\nabla\cdot f\|_{L^p(\Omega)} \le C\|f\|_{L^p(\Omega)}$$

for $t > 0$ and $\vartheta \in [0, 1]$, implies $V(t_0) \in D((-A_{p,\overline{\sigma}})^\vartheta)$ for $t_0 > 0$, and thus $v(t_0) \in D((-A_{p,\overline{\sigma}})^{1/p})$ as well. Hence, v can be extended to a global strong solution on (t_0, ∞).

Step 5: ***Uniqueness and Additional Regularity***

This can be seen similarly as in the above steps. □

Proof of Theorem 14 Let $a = a_1 + a_2$ be as in Theorem 14. Set $a_0 := a_2 + (a_1 - a_{\mathrm{ref}})$ with

$$a_{\mathrm{ref}} \in D(A_{\overline{\sigma}}), \quad a_1 \in X_{\overline{\sigma}} \quad \text{and} \quad a_2 \in L^\infty_H L^p_z(\Omega)^2 \cap L^p_{\overline{\sigma}}(\Omega),$$

and where a_{ref} is such that a_0 satisfies the smallness condition of Theorem 13. The above iteration scheme allows to construct a solution V for a_0 and, in turn, v for a. The regularity property

$$v \in C([0,\infty); L^p_{\overline{\sigma}}(\Omega)) \cap L^\infty((0,T); L^\infty_H L^p_z(\Omega))^2$$

follows from the boundedness and exponential stability of the associated semigroup on $L^p_{\overline{\sigma}}(\Omega)$ and $L^\infty_H L^p_z(\Omega)^2 \cap L^p_{\overline{\sigma}}(\Omega)$, as well as from the strong continuity on $L^p_{\overline{\sigma}}(\Omega)$. By Step 4 in the previous proof, the solution regularizes at $t_0 > 0$ and we obtain $v \in C((0,\infty); X_{\overline{\sigma}})$. The condition

$$\limsup_{t\to 0+} t^{1/2}\|\nabla v\|_{L^\infty_H L^p_z(\Omega)} \le C\|a_2\|_{L^\infty_H L^p_z}, \tag{1.67}$$

is verified as follows. Theorem 5 yields

$$\limsup_{t\to 0+} t^{1/2}|\nabla e^{tA}(a_1 - a_{\mathrm{ref}})\|_{L^\infty_H L^p_z} = 0, \quad t^{1/2}|\nabla e^{tA}a_2\|_{L^\infty_H L^p_z} \le C_4|a_2\|_{L^\infty_H L^p_z}, \; t > 0,$$

and thus

$$\limsup_{t\to 0+} t^{1/2}|\nabla V_0(t)\|_{L^\infty_H L^p_z} \le C_4|a_2\|_{L^\infty_H L^p_z}.$$

The remaining assertion $\limsup_{t\to 0+} t^{1/2}\|\nabla V_m(t)\|_{L^\infty_H L^p_z} \le 2C_4\|a_2\|_{L^\infty_H L^p_z}$ is finally proved by induction. Finally, the uniqueness assertion is proved by a modification of the proof of Theorem 13. □

1.12 Time-Periodic Solutions

In this section we discuss time-periodic solutions to the primitive equations, i.e. the set of equations

$$\begin{cases} \partial_t v + v\cdot\nabla_H v + w(v)\cdot\partial_z v - \Delta v + \nabla_H \pi = f, & \text{in } \Omega\times(0,T),\\ \operatorname{div}_H \overline{v} = 0, & \text{in } \Omega\times(0,T),\\ v(0) = v(T), & \text{in } \Omega, \end{cases} \tag{1.68}$$

is discussed for some $T > 0$. We admit *arbitrary* time-periodic $f \in L^2(0,\mathcal{T}, L^2(\Omega))$, *without assuming any smallness condition on* f. As a corollary steady-state solutions are established.

Our approach is based on the following three steps: First, we construct a suitable *weak* time-periodic solution, v_w, to the primitive equations corresponding to the given f, by combining standard Galerkin's method with Brouwer's fixed point theorem. To apply Brouwer's fixed point theorem, the energy inequality is crucial. Second, we show the existence of a unique, *strong* solution v_s to the *initial-value* problem (1.2) for arbitrary $f \in L^2(0;\mathcal{T}; L^2(\Omega))$, and a in a subspace of $H^1(\Omega)$. Finally, we look at v_w as a weak solution to the *initial-value* problem and employ a weak–strong uniqueness result of the type proved by Guillén-González, Masmoudi, and Rodriguez-Bellido in [32], which then implies $v_s \equiv v_w$.

Theorem 15 (Existence of Strong Time-Periodic Solutions for Periodic Forces) *Let $T > 0$ and let $f \in L^2(0,\mathcal{T}; L^2(\Omega))$, all $\mathcal{T} > 0$, be T-periodic. Then problem* (1.68) *has at least one corresponding strong T-periodic solution. If, in particular, f is time-independent, this solution is a strong steady-state solution.*

1.12.1 Weak Time-Periodic Solutions

Given an interval $I \subset \mathbb{R}$, the space $C_w(I; L^2_{\overline{\sigma}}(\Omega))$ stands for the class of functions $v : I \to \mathbb{H}(\Omega)$ such that $t \mapsto (v(t), \psi)$ is continuous for all $\psi \in L^2_{\overline{\sigma}}(\Omega)$. Here $(\cdot,\cdot)$ stands for the L^2 scalar product.

We say that a function $f \in L^1(0,\mathcal{T}; L^1(\Omega))$, all $\mathcal{T} > 0$, is T-periodic ($T > 0$) if $f(t,x) = f(t+T,x)$ for a.a. $(t,x) \in [0,\infty)\times\Omega$.

Definition 1 (Weak Time-Periodic Solutions) Let $T > 0$ and for all $\mathcal{T} > 0$ let $f \in L^1(0, \mathcal{T}; L^2(\Omega))$. A function $v : [0, \infty) \times \Omega \to \mathbb{R}^2$ is called *weak T-periodic solution* to (1.68) if

i) $v \in C_w([0, \mathcal{T}]; L^2_{\overline{\sigma}}(\Omega)) \cap L^2(0, \mathcal{T}; V_{1/2})$ for all $\mathcal{T} > 0$,
ii) For all $\mathcal{T} > 0$ and all $\varphi \in C^1([0, \mathcal{T}]; V_{1/2}) \cap L^2(0, \mathcal{T}; D(A_2))$

$$\int_0^t \big\{(v, \partial_t \varphi) - (\nabla v, \nabla \varphi) + (v \cdot \nabla_H \varphi + w(v)\, \partial_z \varphi, v)\big\} =$$

$$-\int_0^t (f, \varphi) + (v(t), \varphi(t)) - (v(0), \varphi(0)), \quad t \in (0, \mathcal{T}),$$

iii) For all $\mathcal{T} > 0$, all $t \in (0, \mathcal{T}]$ and a.a. $s \in [0, t)$, v satisfies the *strong energy inequality*

$$\|v(t)\|_2^2 + 2 \int_s^t \|\nabla v(\tau)\|_2^2 d\tau \leq \|v(s)\|_2^2 + 2 \int_s^t (f(\tau), v(\tau)) d\tau,$$

iv) $v(t + T, x) = v(T, x)$ for all $t \geq 0$ and a.a. $x \in \Omega$.

A weak T-periodic solution v is called a *strong* if, in addition to i)–iv), it holds

$$v \in C([0, \mathcal{T}]; V_{1/2}) \cap L^2(0, \mathcal{T}; D(A_2)), \quad \partial_t u \in L^1(0, \mathcal{T}; L^2(\Omega)),$$

for all $\mathcal{T} > 0$.

If, in particular, $f \in L^2_{\overline{\sigma}}$ is independent of $t \geq 0$, we say that $v_s \in V_{1/2}$ is a *weak steady-state solution* to (1.68) if v_s satisfies condition ii) for all $\varphi \in D(A_2)$. A weak steady-state solution is called *strong* if $v_s \in D(A_2)$.

First, one shows that the class of weak T-periodic solutions to (1.68) is not empty under suitable assumptions on f.

Proposition 1.12.1 (Existence of Weak Time-Periodic Solutions) *Let $T > 0$, and let $f \in L^2(0, \mathcal{T}; L^2(\Omega))$, all $\mathcal{T} > 0$, be T-periodic. Then, there exists at least one weak T-periodic solution, v, to* (1.68). *If, in particular, f does not depend on t, then v is a weak steady-state solution.*

Proof The proof is based on the Faedo-Galerkin method, and an application of Brouwer's fixed point theorem which uses the energy inequality.

Let $\{\psi_n\} \subset D(A_2)$ be an orthonormal basis of A_2 dense in $V_{1/2}$ and $D(A_2)$. For example, we may take the eigenvectors of the hydrostatic Stokes operator in the L^2 setting. Let

$$v_m := \sum_{k=1}^{m} c_{mk}(t) \psi_k(x)$$

where, for all $r = 1, \dots, m$,

$$\frac{d}{dt}(v_m, \psi_r) = (v_m \cdot \nabla_H \psi_r, v_m) + (w(v_m)\, \partial_z \psi_r, v_m) - (\nabla v_m, \nabla \psi_r) + (f, \psi_r)\,. \tag{1.69}$$

We now show a uniform bound in time on the functions $c_{mk}(t)$ for $k = 1, \dots, m$, which implies that the system (1.69) has a solution $c(t) := (c_{m1}(t), \dots, c_{mm}(t))$ for all $t \geq 0$ and all $m \geq 1$. In fact, multiplying both sides of (1.69) by $c_{mr}(t)$, summing over r, and taking into account $(v_m \cdot \nabla_H v_m, v_m) + (w(v_m)\, \partial_z v_m, v_m) = 0$, we obtain

$$\frac{d}{dt}\|v_m\|_2^2 + 2\|\nabla v_m(t)\|_2^2 = 2(f, v_m)\,. \tag{1.70}$$

Since $v_m = 0$ at $x_3 = -h$, we have

$$\|v_m\|_2^2 \leq h^2 \|\partial_z v_m\|_2^2, \tag{1.71}$$

so that by Schwartz's inequality and (1.70) we infer

$$\frac{d}{dt}\|v_m\|_2 + \frac{2}{h^2}\|v_m\|_2 \leq 2\|f\|_2\,.$$

Integrating the latter from $t = 0$ to arbitrary $t > 0$ we get

$$e^{2t/h^2}\|v_m(t)\|_2 \leq \|v_m(0)\|_2 + 2\int_0^t e^{2\tau/h^2}\|f(\tau)\|_2 d\tau\,, \tag{1.72}$$

thus deducing the claimed uniform bound for $c(t)$, once we observe that, by the orthonormality property of $\{\psi_n\}$, $|c(t)| = \|v_m(t)\|_2$.

Next, choose $R > 0$ such that

$$R(e^{2t/h^2} - 1) \geq 2\int_0^T e^{2\tau/h^2}\|f(\tau)\|_2\, d\tau$$

and let $\mathbb{B}_R^m$ the ball in $\mathbb{R}^m$ centered at the origin with radius R. It follows from (1.72) that $|c(T)| = \|v_m(T)\|_2 \leq R$ provided $|c(0)| = \|v_m(0)\|_2 \leq R$. Thus the map

$$S : \mathbb{R}^m \ni c(0) \mapsto c(T) \in \mathbb{R}^m$$

maps $\mathbb{B}_R^m$ into itself. Since it is also continuous, we conclude by Brouwer's theorem that for each $m \geq 1$ there exists $v_m(0)$ such that $v_m(0) = v_m(T)$. From here on the proof follows the standard Galerkin scheme. □

1.12.2 Weak–Strong Uniqueness

In this section we give a proof of Theorem 4 based on a weak–strong uniqueness argument for the *initial-value* problem. More precisely, given $f \in L^2(0, T; L^2(\Omega))$, let v_w be a weak T-periodic solution corresponding to Proposition 1.12.1. We assume f to be time-dependent and note that the case where f does not depend on time is treated in entirely analogous way.

By properties i) and iii) we infer that there exists $t_0 > 0$ such that $v_w(t_0) \in V_{1/2}$ and for $t \geq t_0$

$$\|v_w(t)\|_2^2 + 2\int_{t_0}^{t} \|\nabla v_w(\tau)\|_2^2 d\tau \leq \|v_w(t_0)\|_2^2 + 2\int_{t_0}^{t} (f(\tau), v_w(\tau))d\tau. \tag{1.73}$$

We now look at our weak time-periodic solution as a weak solution v_w to the *initial-value problem* with initial data $a \equiv v(t_0)$. Since $v(t_0) \in V_{1/2}$ we may use it also as initial value for the *global* strong solution determined in Theorem 7. The assertion of Theorem 15 follows provided we are able to show that the weak solution coincides with the strong one.

Such a weak–strong uniqueness result is already known, under slightly different boundary conditions; see [32, Theorem 1.3]. The arguments used there would equally apply to the case considered here, nevertheless we will sketch a proof here. To this end, let v_s be the velocity field of the strong solution. Considering $\sigma = v_s - v_w$ we verify that

$$\|\sigma(t)\|_2^2 + 2\int_0^t \|\nabla\sigma(\tau)\|_2^2 d\tau \leq 2\int_{t_0}^t \{(\sigma(\tau)\cdot\nabla_H\,\sigma(\tau) + w(\sigma)\partial_z\sigma(\tau), v_s(\tau))\} d\tau\,.$$

Estimating the above right-hand side by

$$\begin{aligned}\|\sigma(t)\|_2^2 + 2\int_{t_0}^t \|\nabla\sigma(\tau)\|_2^2 d\tau &\leq C\int_{t_0}^t \{\|\sigma(\tau)\|_2^{\frac{1}{2}} \|\sigma(\tau)\|_{H^1}^{\frac{3}{2}}(\|\nabla_H\, v(\tau)\|_2 \\ &\quad + \|\partial_z v(\tau)\|_2^{\frac{1}{2}} \|\partial_z v_s(\tau)\|_{H^1}^{\frac{1}{2}})\} d\tau \\ &\leq C\int_{t_0}^t g(\tau)\|\sigma(\tau)\|_2^2 d\tau + \int_{t_0}^t \|\nabla\sigma(\tau)\|_2^2 d\tau,\end{aligned} \tag{1.74}$$

where

$$g(\tau) := \|\nabla_H\, v_s(\tau)\|_2^4 + \|\partial_z v_s(\tau)\|_2^2 \|\partial_z v_s(\tau)\|_{H^1}^2.$$

By the regularity properties of v_s it follows that $g \in L^1(t_0, t)$ for arbitrary $t > t_0$, so that by Grönwall's lemma we conclude that

$$\sigma \equiv v_s - v_w = 0 \text{ a.e. in } \Omega \times (t_0, t) \text{ for all } t > t_0.$$

By the time-periodicity of v_w this proves the desired regularity for v_w. The proof of the Theorem 15 is complete.

1.13 Primitive Equations with Temperature and Salinity

The full primitive equations, i.e. the equations describing the conservation of momentum and mass of the fluid coupled with the equations for temperature as well as salinity are given by

$$\begin{cases} \partial_t v + v \cdot \nabla_H v + w \cdot \partial_z v - \Delta v + \nabla_H \pi_s = f + \Pi(\tau, \sigma), \\ \operatorname{div}_H \overline{v} = 0, \\ \partial_t \tau + v \cdot \nabla_H \tau + w \cdot \partial_z \tau - \Delta \tau = g_\tau, \\ \partial_t \sigma + v \cdot \nabla_H \sigma + w \cdot \partial_z \sigma - \Delta \sigma = g_\sigma. \end{cases} \tag{1.75}$$

Furthermore, the pressure π is determined by the surface pressure $\pi_s(x, y) = \pi(x, y, -h)$, while the part of the pressure due to temperature and salinity is given by

$$\Pi(\tau, \sigma) = -\nabla_H \int_{-h}^{z} \beta_\tau \tau(\cdot, \xi) - \beta_\sigma \sigma(\cdot, \xi) d\xi, \quad \beta_\tau, \beta_\sigma > 0,$$

compare [58, Equations (2.4) and (2.60)], [50, Section 4] for the case with salinity and [14, Subsection 2.1] for the case only with temperature. One imposes initial conditions $v(0) = a$, $\tau(0) = b_\tau$, $\sigma(0) = b_\sigma$ and forcing terms f, g_τ, and g_σ. The temperature and salinity are denoted by τ and σ, respectively, and π denotes the pressure of the fluid, while v denotes as usual the horizontal velocity. The system is complemented by the boundary conditions

$$\begin{cases} \partial_z v = 0, \quad w = 0, \ \partial_z \tau + \alpha \tau = 0, \ \partial_z \sigma = 0 \text{ on } \Gamma_u \times (0, \infty), \\ v = 0, \quad w = 0, \quad \partial_z \tau = 0, \ \partial_z \sigma = 0 \text{ on } \Gamma_b \times (0, \infty), \\ v, \pi, \tau, \sigma \quad \text{are periodic} \quad \text{on } \Gamma_l \times (0, \infty), \end{cases} \tag{1.76}$$

where $\alpha > 0$ and

$$\Gamma_u = G \times \{0\}, \quad \Gamma_b = G \times \{-h\} \quad \text{and} \quad \Gamma_l = \partial G \times (-h, 0).$$

Global well-posedness can be proven using a Fujita-Kato scheme as for the isothermal setting or using maximal L^q-regularity. Note that integrability for temperature and salinity compared to the velocity must not agree.

Theorem 16 (Existence of Unique, Global Strong Solutions) *Let* $p, q_\tau, q_\sigma \in (1, \infty)$ *with* $q_\tau, q_\sigma \in [\frac{2p}{3}, p] \cap (1, p]$ *and suppose that*

$$f \in H^{1,2}_{loc}((0,\infty); L^p(\Omega)^2 \cap L^2(\Omega)^2),$$
$$g_\tau \in H^{1,2}_{loc}((0,\infty); L^{q_\tau}(\Omega) \cap L^2(\Omega)), \qquad g_\sigma \in H^{1,2}_{loc}((0,\infty); L^{q_\sigma}(\Omega) \cap L^2(\Omega)).$$

a) *Assume that*

$$a \in \{v \in H^{2/p,p}_{per}(\Omega)^2 \cap L^p_{\overline{\sigma}}(\Omega) | v \mid_{\Gamma_b} = 0\}, \quad b_\tau \in H^{2/q_\tau, q_\tau}_{per}(\Omega), \; b_\sigma \in H^{2/q_\sigma, q_\sigma}_{per}(\Omega).$$

Then there is a unique, global, strong solution to (1.75) *and* (1.76) *satisfying*

$$v \in C^1((0,\infty); L^p_{\overline{\sigma}}(\Omega)) \cap C^0((0,\infty); D(A_p)),$$
$$\pi \in C^0((0,\infty); H^{1,p}_{per}(G) \cap L^p_0(G)),$$
$$\tau \in C^1((0,\infty); L^{q_\tau}(\Omega)) \cap C^0((0,\infty); D(\Delta_\tau)),$$
$$\sigma \in C^1((0,\infty); L^{q_\sigma}(\Omega)) \cap C^0((0,\infty); D(\Delta_\sigma)).$$

b) *If in addition*

$$a \in D(A_p) \text{ and } b_\tau \in D(\Delta_\tau), \quad b_\sigma \in D(\Delta_\sigma)$$

then the above solution extends to $[0,\infty)$.

1.13.1 Linearized Temperature and Salinity Equations

Considering the diffusion equations

$$\partial_t \tau - \Delta \tau = g_\tau, \quad \tau(0) = b_\tau, \quad \partial_t \sigma - \Delta \sigma = g_\sigma, \quad \sigma(0) = b_\tau,$$

we define operators Δ_τ for $\alpha > 0$ and Δ_σ by

$$\Delta_\tau \tau = \Delta \tau, \quad D(\Delta_\tau) = \{\tau \in H^{2,q_\tau}_{per}(\Omega) \mid (\partial_z \tau + \alpha \tau) \mid_{\Gamma_u} = 0, \quad \partial_z \tau \mid_{\Gamma_b} = 0\},$$
$$\Delta_\sigma \sigma = \Delta \sigma, \quad D(\Delta_\sigma) = \{\sigma \in H^{2,q_\sigma}_{per}(\Omega) \mid \partial_z \sigma \mid_{\Gamma_u} = 0, \quad \partial_z \sigma \mid_{\Gamma_b} = 0\}.$$

Consider the interpolation spaces

$$V_{\theta,p} := [L^p_{\overline{\sigma}}(\Omega), D(A_p)]_\theta, \; \hat{V}^\tau_{\theta,q} := [L^q(\Omega), D(\Delta_\tau)]_\theta, \; \hat{V}^\sigma_{\theta,q} := [L^q(\Omega), D(\Delta_\sigma)]_\theta$$

for $0 \le \theta \le 1$ and where $[\cdot,\cdot]_\theta$ denotes the complex interpolation functor.

These Laplacians in a cylindrical setting were investigated in detail, e.g., by Nau in [62, Section 8.2.2], and therefrom we conclude the following results.

Proposition 1.13.2 (Semigroup Estimates) *Let $p, q \in (1, \infty)$, $0 \leq \theta_1, \theta_2 \leq 1$ with $\theta_1 + \theta_2 \leq 1$. Then*

(a) $V_{\theta,p} \subset H^{2\theta,p}(\Omega)^2$ *and* $\hat{V}^\tau_{\theta,q} \subset H^{2\theta,q}(\Omega)$, $\hat{V}^\sigma_{\theta,q} \subset H^{2\theta,q}(\Omega)$, $0 \leq \theta \leq 1$;
(b) *There exist constants* $C_v, C_\tau, C_\sigma > 0$ *and* $\beta_v, \beta_\tau > 0$ *such that for* $t > 0$

$$\begin{aligned}
\|e^{tA_p} f\|_{V_{\theta_1+\theta_2,p}} &\leq C_v t^{-\theta_1} e^{-\beta_v t} \|f\|_{V_{\theta_2,p}}, && f \in V_{\theta_2,p},\\
\|e^{t\Delta_\tau} g_\tau\|_{\hat{V}^\tau_{\theta_1+\theta_2,q}} &\leq C_\tau t^{-\theta_1} e^{-\beta_\tau t} \|g_\tau\|_{\hat{V}^\tau_{\theta_2,q}}, && g_\tau \in \hat{V}^\tau_{\theta_2,q},\\
\|e^{t\Delta_\sigma} g_\sigma\|_{\hat{V}^\sigma_{\theta_1+\theta_2,q}} &\leq C_\sigma t^{-\theta_1} \|g_\sigma\|_{\hat{V}^\sigma_{\theta_2,q}}, && g_\sigma \in \hat{V}^\sigma_{\theta_2,q};
\end{aligned}$$

(c) $t^{\theta_1} \|e^{tA_p} f\|_{V_{\theta_1+\theta_2,p}} \to 0$,
$t^{\theta_1} \|e^{t\Delta_\tau} g_\tau\|_{\hat{V}^\tau_{\theta_1+\theta_2,q}} \to 0$ *and* $t^{\theta_1} \|e^{t\Delta_\sigma} g_\sigma\|_{\hat{V}^\sigma_{\theta_1+\theta_2,q}} \to 0$ *as* $t \to 0$.

1.13.2 Non-linearity and Couplings

The main task compared to the isothermal setting is to control the non-linearity. Adopting the short-hand notation

$$\zeta := (\tau, \sigma), \quad g := (g_\tau, g_\sigma) \quad \text{and} \quad b := (b_\tau, b_\sigma),$$

and setting for simplicity $q_\tau = q_\sigma$ along with

$$q := q_\tau = q_\sigma, \quad \hat{V}_{q,\theta} := \hat{V}^\tau_{q,\theta} \times \hat{V}^\sigma_{q,\theta} \quad \text{and} \quad \Delta_\zeta := \Delta_\tau \oplus \Delta_\sigma,$$

where Δ_ζ is now a vector valued Laplacian writing $\Delta_{q,\zeta}$ in case of ambiguity. Also we represent the nonlinear terms by

$$F_p(v, \zeta) := -P_p\left(v \cdot \nabla_H v + w\partial_z v - \Pi(\zeta)\right), \tag{1.77}$$

$$G_q(v, \zeta) := -\left(v \cdot \nabla_H \zeta + w\partial_z \zeta\right). \tag{1.78}$$

Lemma 25 (Estimate on the Couplings and Nonlinearities) *Let $p, q \in (1, \infty)$ be as in Theorem 16, and let $\gamma(r) := \frac{1}{2} + \frac{1}{2r}$. Then (F_p, G_q) maps $V_{\gamma(p),p} \times \hat{V}_{\gamma(q),q}$ into $L^{\frac{p}{\sigma}}(\Omega) \times L^q(\Omega)^2$, and there exists a constant $C > 0$ such that*

(a) *For* $v \in V_{\gamma(p),p}$ *and* $\zeta \in \hat{V}_{\gamma(q),q}$

$$\|F_p(v, \zeta)\|_{L^{\frac{p}{\sigma}}(\Omega)} \leq C \left(\|v\|^2_{V_{\gamma(p),p}} + \|\zeta\|_{\hat{V}_{\gamma(q),q}}\right),$$

$$\|G_q(v,\zeta)\|_{L^q(\Omega)^2} \leq C\left(\|v\|^2_{V_{\gamma(q),q}} + \|\zeta\|^2_{\hat{V}_{\gamma(q),q}}\right).$$

(b) *For* $v, v' \in V_{\gamma(p),p}$ *and* $\zeta, \zeta' \in \hat{V}_{\gamma(q),q}$

$$\begin{aligned}\|F_p(v,\zeta) - F_p(v',\zeta')\|_{L^{\frac{p}{\sigma}}(\Omega)} \leq& C\left(\|v\|_{V_{\gamma(p),p}} + \|v'\|_{V_{\gamma(p),p}}\right)\|v - v'\|_{V_{\gamma(p),p}}\\ &+ C\|\zeta - \zeta'\|_{\hat{V}_{\gamma(q),q}},\\ \|G_q(v,\zeta) - G_q(v',\zeta')\|_{L^q(\Omega)^2} \leq& C\left(\|v\|_{V_{\gamma(p),p}} + \|v'\|_{V_{\gamma(p),p}}\right)\|\zeta - \zeta'\|_{\hat{V}_{\gamma(q),q}}\\ &+ C\left(\|\zeta\|_{\hat{V}_{\gamma(q),q}} + \|\zeta'\|_{\hat{V}_{\gamma(q),q}}\right)\|v - v'\|_{V_{\gamma(p),p}}.\end{aligned}$$

1.14 Hydrostatic Approximation

In this section we show that the primitive equations can be obtained as the limit of an-isotropically scaled Navier-Stokes equations. This gives a rigorous justification of the formal derivation presented in Sect. 1.2. Recall that the scaling parameter $\varepsilon > 0$ represents the ratio of the depth to the horizontal width. This approximation is motivated by the fact that for large-scale oceanic dynamics, this aspect ratio ε is rather small and implies anisotropic viscosity coefficients. For an aspect ratio ε, i.e., in the case where the spacial domain can be represented as $\Omega_\varepsilon = G \times (-\varepsilon, +\varepsilon)$ for some $G \subset \mathbb{R}^2$, and a horizontal and vertical eddy viscosity 1 and ε^2, respectively, the system can be rescaled into the form

$$\left\{\begin{aligned} \partial_t u_\varepsilon + v_\varepsilon \cdot \nabla v_\varepsilon - \Delta v_\varepsilon + \nabla_H p_\varepsilon &= 0,\\ \varepsilon(\partial_t w_\varepsilon + v_\varepsilon \cdot \nabla w_\varepsilon - \Delta w_\varepsilon) + \tfrac{1}{\varepsilon}\partial_z p_\varepsilon &= 0,\\ \operatorname{div} v_\varepsilon &= 0.\end{aligned}\right. \tag{1.79}$$

in the time-space domain $(0, T) \times \Omega_1$, which is *independent* of the aspect ratio, compare Sect. 1.2.

First convergence results for the above system in the steady state case go back to Besson and Laidy [7]. The convergence of the above system has been studied first by Azérad and Guillén in [6] in the setting of weak convergence and without a uniform convergence rate. More recently, Li and Titi [55] investigated the strong convergence of the above system within the L^2_t-L^2_x-setting for horizontal initial velocities belonging to H^1 and H^2. In addition, they showed a convergence rate of order $\mathcal{O}(\varepsilon)$.

Here, we present convergence results of the above system in the strong sense within the more general L^p_t-L^q_x-setting. Our approach is based on maximal L^p_t-

L^q_x-regularity estimates for the heat equation. This allows us to give a very short proof of the convergence result in the more general L^p_t-L^q_x-setting, which even in the L^2_t-L^2_x-setting allows for a slightly larger class of initial data compared to the one introduced by Li and Titi in [55]. In contrast the approach by Li and Titi in [55] relies on second order energy estimates.

1.14.1 Model Situation

Consider the cylindrical domain $\Omega := (0,1)^2 \times (-1,1)$. In the following we will consider primitive equations on the full torus with parity conditions which correspond to Neumann boundary conditions on top and bottom, i.e., boundary conditions (1.4) with $\Gamma_D = \emptyset$. First, let $v = (v, w)$ be the solution of the primitive equations

$$\begin{cases} \partial_t u + v \cdot \nabla v - \Delta v + \nabla_H p = 0 & \text{in } (0,T) \times \Omega, \\ \partial_z p = 0 & \text{in } (0,T) \times \Omega, \\ \operatorname{div} v = 0 & \text{in } (0,T) \times \Omega, \\ p \text{ periodic in } x, y & \\ v, w \text{ periodic in } x, y, z, \text{ even and odd} & \text{in } z, \\ v(0) = v_0 & \text{in } \Omega. \end{cases} \tag{PE}$$

Second, let $v_\varepsilon = (v_\varepsilon, w_\varepsilon)$ be the solution of the anisotropic Navier-Stokes equations

$$\begin{cases} \partial_t u_\varepsilon + v_\varepsilon \cdot \nabla v_\varepsilon - \Delta v_\varepsilon + \nabla_H p_\varepsilon = 0 & \text{in } (0,T) \times \Omega, \\ \partial_t w_\varepsilon + v_\varepsilon \cdot \nabla w_\varepsilon - \Delta w_\varepsilon + \frac{1}{\varepsilon^2} \partial_z p_\varepsilon = 0 & \text{in } (0,T) \times \Omega, \\ \operatorname{div} v_\varepsilon = 0 & \text{in } (0,T) \times \Omega, \\ p_\varepsilon \text{ periodic in } x, y, z, \text{ even} & \text{in } z, \\ v_\varepsilon, w_\varepsilon \text{ periodic in } x, y, z, \text{ even and odd} & \text{in } z, \\ v_\varepsilon(0) = v_0 & \text{in } \Omega. \end{cases} \tag{NS$_\varepsilon$}$$

Here $v = (v_1, v_2)$ and $v_\varepsilon = (v_{\varepsilon,1}, v_{\varepsilon,2})$ denote the horizontal velocities, w and w_ε the vertical velocities, and p and p_ε denote the pressure term for the primitive equations as well as the Navier-Stokes equations, respectively. As before, since w is assumed to be odd, the divergence free condition for the primitive equation translates into $\operatorname{div}_H \overline{v} = 0$, where $\overline{v}(x,y) = \frac{1}{2}\int_{-1}^{1} v(x,y,z)\, dz$, and

$$w(\cdot,\cdot,z) = -\int_{-1}^{z} \operatorname{div}_H v(\cdot,\cdot,\zeta)\, d\zeta.$$

1.14.2 Strong Solutions and Maximal Regularity Spaces

For given $p, q \in (1, \infty)$ we set

$$X_0 := L^q(\Omega), \quad X_1 := H^{2,q}_{per}(\Omega),$$

$$X_0^v := \{v \in L^q_{\overline{\sigma}}(\Omega) : v \text{ even in } z\}, \quad X_1^v := \{v \in H^{2,q}_{per}(\Omega)^2 \cap L^q_{\overline{\sigma}}(\Omega) : v \text{ even in } z\},$$

$$X_0^{\mathrm{v}} := \{(v_1, v_2, w) \in L^q_\sigma(\Omega) : v_1, v_2 \text{ even } w \text{ odd in } z\},$$

$$X_1^{\mathrm{v}} := \{(v_1, v_2, w) \in H^{2,q}_{per}(\Omega)^3 \cap L^q_\sigma(\Omega) : v_1, v_2 \text{ even } w \text{ odd in } z\}.$$

For $T \in (0, \infty]$ we also define the maximal regularity spaces

$$\mathbb{E}_0(0, T) := L^p(0, T; X_0), \quad \mathbb{E}_1(0, T) := L^p(0, T; X_1) \cap H^{1,p}(0, T; X_0),$$

and analogously $\mathbb{E}_0^v(0, T)$, $\mathbb{E}_1^v(0, T)$, and $\mathbb{E}_0^{\mathrm{v}}(0, T)$, $\mathbb{E}_1^{\mathrm{v}}(0, T)$ with respect to X_0^v, X_1^v, and X_0^{v}, X_1^{v}, respectively. In order to simplify our notation we sometimes omit the subscripts v and v and write only $\mathbb{E}_0(0, T)$ and $\mathbb{E}_1(0, T)$.

The corresponding trace spaces are given by real interpolations and X_γ is defined by

$$X_\gamma = (X_0^{\mathrm{v}}, X_1^{\mathrm{v}})_{1-1/p,p}.$$

Following the lines of Proposition 1.5.5 the trace space X_γ can be characterized as follows.

Lemma 26 (Characterization of the Trace Space) *Let $p, q \in (1, \infty)$. Then*

$$X_\gamma = \begin{cases} \{(v_1, v_2, w) \in B^{2-2/p}_{q,p,per}(\Omega)^3 \cap L^q_\sigma(\Omega) : v = (v_1, v_2) \text{ even, } w \text{ odd in } z, \\ \qquad\qquad (\partial_z v, w) = 0 \text{ at } z = -1, 0, 1\}, \\ \qquad\qquad 1 > \frac{2}{p} + \frac{1}{q}, \\ \{(v_1, v_2, w) \in B^{2-2/p}_{q,p,per}(\Omega)^3 \cap L^q_\sigma(\Omega) : v = (v_1, v_2) \text{ even, } w \text{ odd in } z, \\ \qquad\qquad w = 0 \text{ at } z = -1, 0, 1\}, \\ \qquad\qquad 1 < \frac{2}{p} + \frac{1}{q}. \end{cases}$$

Finally, we say that $\mathrm{v} = (v, w)$ is a *strong solution to the primitive equations* (in the L^p-L^q-setting), if $v \in \mathbb{E}_1^v$ and (PE) holds almost everywhere. We say that v_ε is a *strong solution to the Navier-Stokes equations*, if $\mathrm{v} \in \mathbb{E}_1^{\mathrm{v}}$ and (NS$_\varepsilon$) holds almost everywhere.

1.14.3 Convergence Result

In order to prove convergence, our strategy consists of controlling the maximal regularity norm of the differences $(v_\varepsilon - v, \varepsilon(w_\varepsilon - w))$ by the aspect ratio ε. To this end, we introduce the difference equations of $(\mathrm{NS}_\varepsilon)$ and (PE): setting

$$V_\varepsilon := v_\varepsilon - v, \quad W_\varepsilon := w_\varepsilon - w, \quad U_\varepsilon := (V_\varepsilon, W_\varepsilon), \quad P_\varepsilon = p_\varepsilon - p,$$

we obtain

$$\left\{\begin{array}{rll} \partial_t V_\varepsilon - \Delta V_\varepsilon = & F_H(V_\varepsilon, W_\varepsilon) - \nabla_H P_\varepsilon & \text{in } (0,T)\times\Omega, \\ \partial_t \varepsilon W_\varepsilon - \Delta \varepsilon W_\varepsilon = & \varepsilon F_z(V_\varepsilon, W_\varepsilon) - \frac{1}{\varepsilon}\partial_z P_\varepsilon & \text{in } (0,T)\times\Omega, \\ \operatorname{div} U_\varepsilon = & 0 & \text{in } (0,T)\times\Omega, \\ P_\varepsilon \text{ periodic in } x, y, z, & \text{even} & \text{in } z, \\ V_\varepsilon, W_\varepsilon \text{ periodic in } x, y, z, & \text{even and odd} & \text{in } z, \\ U_\varepsilon(0) = & 0 & \text{in } \Omega, \end{array}\right. \tag{1.80}$$

where the forcing terms F_H and F_z are given by

$$\begin{aligned} F_H(V_\varepsilon, W_\varepsilon) &:= -U_\varepsilon \cdot \nabla v - u \cdot \nabla V_\varepsilon - U_\varepsilon \cdot \nabla V_\varepsilon, \\ F_z(V_\varepsilon, W_\varepsilon) &:= -U_\varepsilon \cdot \nabla w - u \cdot \nabla W_\varepsilon - U_\varepsilon \cdot \nabla W_\varepsilon - \partial_t w - u \cdot \nabla w + \Delta w. \end{aligned}$$

Applying the maximal regularity estimate given in Proposition 1.14.3 to (1.80), we are able to estimate $\|U_\varepsilon\|_{\mathbb{E}_1}$ in terms of the right-hand sides. The latter will be estimated in a series of lemmas to obtain a quadratic inequality for the norm of the differences, cf. Corollary 4, and we need to ensure that the constant term and the coefficient in front of the linear term are sufficiently small. This can be achieved provided the aspect ratio ε is small enough and provided the vertical and horizontal solution of the primitive equations exist globally in the maximal regularity class, cf. Proposition 1.14.4.

Assumption (A) Let $q \in \left(\frac{4}{3}, \infty\right)$ and $p \geq \max\left\{\frac{q}{q-1}; \frac{2q}{3q-4}\right\}$, i.e.,

$$1 \geq \left\{\begin{array}{l} \frac{1}{p} + \frac{1}{q}, \text{ if } q \geq 2, \\ \frac{2}{3p} + \frac{4}{3q}, \text{ if } q \leq 2. \end{array}\right.$$

Theorem 17 (Convergence of Scaled Navier-Stokes to Primitive Equations) *Assume that p, q fulfill Assumption (A) and let $v_0 \in X_\gamma$, $T > 0$ and (v, w) and $(v_\varepsilon, w_\varepsilon)$ be solutions of (PE) and (NS_ε), respectively. Then there exists a constant $C > 0$, independent of ε, such that for ε sufficiently small it holds*

$$\|(V_\varepsilon, \varepsilon W_\varepsilon)\|_{\mathbb{E}_1(T)} \leq C\varepsilon.$$

In particular

$$(v_\varepsilon, \varepsilon w_\varepsilon) \to (v, 0) \text{ in } L^p(0,T; H^{2,q}(\Omega)) \cap H^{1,p}(0,T; L^q(\Omega))$$

as $\varepsilon \to 0$ *with convergence rate* $\mathcal{O}(\varepsilon)$.

Remarks 1.14.1

(a) The case $p = q = 2$, investigated before in [55], is covered by our result, cf. Assumption (A). More specifically, there it is assumed that $v_0 \in H^2$ whereas for our purposes $v_0, \operatorname{div}_H v_0 \in H^1$ suffices. This constitutes a slight improvement.
(b) The scaled Navier-Stokes equations are locally well-posed in the maximal regularity spaces on the torus and the parity conditions are preserved. Theorem 17 yields that for each time T there exists an $\varepsilon > 0$ such that the solution exists on $(0, T)$. However, this holds only for initial values of the particular form $u_0 = (v_0, w(v_0))$.
(c) Our method can be adjusted to the case with perturbed initial data. That is, given initial data $(u_{0,\varepsilon})_{\varepsilon>0} \subset X_\gamma$ converging to u_0 in X_γ as $\varepsilon \to 0$ of order $\mathcal{O}(\delta_\varepsilon)$ for some null-sequence $(\delta_\varepsilon)_{\varepsilon>0}$, then Theorem 17 holds with $(v_\varepsilon, w_\varepsilon)$ replaced by the solution of $(\mathrm{NS}_\varepsilon)$ with initial data $v_{0,\varepsilon}$. In that case the maximal regularity norm of the differences is bounded by $C \max\{\varepsilon, \delta_\varepsilon\}$ and consequently the convergence rate is of order $\mathcal{O}(\max\{\varepsilon, \delta_\varepsilon\})$.

The proof of Theorem 17 relies upon estimates on the terms F_H and F_z in equations (1.80) within the L^p-L^q-framework. These estimates imply eventually a quadratic inequality for the difference of the velocities, see Corollary 4. In order to establish these estimates we need to ensure that the solution of the primitive equations belongs to the maximal regularity class, see Proposition 1.14.4, and that the nonlinear terms can be estimated in $\mathbb{E}_0(0, T)$ which will be shown in the subsequent subsection. Throughout this section let $T < \infty$.

1.14.4 Nonlinear Estimates

Here, we estimate the bilinear terms and keep track of the T-dependence of the norms involved. To this end consider for $T \in (0, \infty)$ embeddings of the type

$$H^{s,p}(0,T) \overset{\eta}{\hookrightarrow} H^{s',p}(0,T)$$

where $\overset{\eta}{\hookrightarrow}$ stands for an embedding with embedding constant CT^η, $C > 0$ independent of T.

Lemma 27 *Let* $s \geq s' \geq 0$ *and* $p \in [1, \infty)$. *Then* $H^{s,p}(0,T) \overset{s-s'}{\hookrightarrow} H^{s',p}(0,T)$.

Proof Let

$$m := \lfloor p(s-s')\rfloor \quad \text{and} \quad 1/r := m + 1/p + s' - s \in [1/p, 1).$$

Sobolev embeddings and Hölder's inequality yield

$$H^{s,p} \hookrightarrow H^{m+s',r} \overset{1-\frac{1}{r}}{\hookrightarrow} H^{m+s',1} \hookrightarrow H^{m-1+s',\infty} \overset{1}{\hookrightarrow} H^{m-1+s',1} \hookrightarrow$$

$$\ldots \hookrightarrow H^{s',\infty} \overset{\frac{1}{p}}{\hookrightarrow} H^{s',p}.$$

□

We will make also use of the following classical mixed derivative theorem, see, e.g., [67, Corollary 4.5.10].

Proposition 1.14.2 (Mixed Derivative Theorem) *If* $\theta \in [0,1]$, *then*

$$\mathbb{E}_1(0,T) \hookrightarrow H^{\theta,p}(0,T; H^{2-2\theta,q}(\Omega)).$$

To estimate the non-linearities we will need estimates of the following type

Lemma 28 (Estimate on the Non-linearity in Space and Time)

(a) *Let* $p,q \in (1,\infty)$ *such that* $2/3p + 1/q \le 1$. *Then for all* $v_1, v_2 \in \mathbb{E}_1(0,T)$, $\partial \in \{\partial_x, \partial_y, \partial_z\}$ *and* $\eta \in \left[0, \frac{3}{2}\left(1 - \frac{2}{3p} - \frac{1}{q}\right)\right]$ *there exists a constant* $C > 0$ *such that*

$$\|v_1 \partial v_2\|_{\mathbb{E}_0(0,T)} \le C T^{\eta} \|v_1\|_{\mathbb{E}_1(0,T)} \|v_2\|_{\mathbb{E}_1(0,T)}, \qquad C > 0.$$

(b) *Let* $q \in (1,\infty)$, $v_1, v_2 \in H^{1+1/q,q}(\Omega)$ *and* $w_1 := \int_{-1}^{z} \operatorname{div}_h u_1$. *Then there exists a constant* $C > 0$ *such that*

$$\|w_1 \partial_z v_2\|_{L^q} \le C \|v_1\|_{H^{1+1/q,q}} \|v_2\|_{H^{1+1/q,q}}.$$

(c) *Let* $p,q \in (1,\infty)$ *such that* $1/p + 1/q \le 1$ *and* $\eta \in \left[0, 1 - \frac{1}{q} - \frac{1}{p}\right]$. *Then for all* $v_1, v_2 \in \mathbb{E}_1(T)$ *and* w_1 *given by* $w_1 := \int_{-1}^{z} \operatorname{div}_h u_1$ *there exists a constant* $C > 0$ *such that*

$$\|w_1 \partial_z v_2\|_{\mathbb{E}_0(T)} \le C T^{\eta} \|v_1\|_{\mathbb{E}_1(T)} \|v_2\|_{\mathbb{E}_1(T)}.$$

Proof

(a) Set $\theta_1 = \frac{2\eta}{3} + \frac{2}{3p}$ and $\theta_2 = \frac{1}{2}\theta_1$. The mixed derivative theorem, Lemma 27 and Sobolev's embedding yield

$$\mathbb{E}_1(T) \hookrightarrow H^{\theta_1,p}(0,T;H^{2-2\theta_1,q})$$
$$\overset{2\eta/3}{\hookrightarrow} H^{2/3p,p}(0,T;H^{2-2\theta_1,q}) \hookrightarrow L^{3p}(0,T;L^{3q}(\Omega)),$$

and

$$\mathbb{E}_1(T) \hookrightarrow H^{\theta_2,p}(0,T;H^{2-2\theta_2,q})$$
$$\overset{\eta/3}{\hookrightarrow} H^{1/3p,p}(0,T;H^{2-2\theta_2,q}) \hookrightarrow L^{3p/2}(0,T;H^{1,3q/2}(\Omega)).$$

Hölder's inequality thus implies

$$\begin{aligned}\|v_1\partial v_2\|_{L^p(L^q)} &\leq \big\| \|v_1\|_{L^{3q}} \|\partial v_2\|_{L^{3q/2}} \big\|_{L^p} \\ &\leq \|v_1\|_{L^{3p}(L^{3q})}\|v_2\|_{L^{3p/2}(H^{1,3q/2})} \leq CT^\eta \|v_1\|_{\mathbb{E}_1}\|v_2\|_{\mathbb{E}_1}.\end{aligned}$$

(b) Similarly as in Lemma 15 we obtain by anisotropic Hölder's inequality and Sobolev inequalities

$$\begin{aligned}\|w_1\partial_z v_2\|_{L^q} &\leq \|w_1\|_{L^{2q}_{xy}L^\infty_z}\|\partial_z v_2\|_{L^{2q}_{xy}L^q_z} \\ &\leq C\|\operatorname{div}_H v_1\|_{L^{2q}_{xy}L^1_z}\|\partial_z v_2\|_{L^{2q}_{xy}L^q_z} \\ &\leq C\|v_1\|_{H^{1+1/q}_{xy}L^1_z}\|v_2\|_{H^{1/q,q}_{xy}H^{1,q}_z}.\end{aligned}$$

(c) Set $\theta = \frac{\eta}{2} + \frac{1}{p}$. By Lemma 27 and the mixed derivative theorem, Proposition 1.14.2, one obtains

$$\begin{aligned}\mathbb{E}_1(0,T) \hookrightarrow H^{\theta,p}(0,T;H^{2-2\theta,q}) &\overset{\frac{\eta}{2}}{\hookrightarrow} H^{1/2p,p}(0,T;H^{2-2\theta,q}) \\ &\hookrightarrow L^{2p}(0,T;H^{1+1/q,q}(\Omega)),\end{aligned}$$

Using the short-hand notation

$$X := H^{1+1/q,q} \quad \text{and} \quad L^p(L^q) := \mathbb{E}_0(0,T),$$

part (a), (b), and the above embeddings imply

$$\begin{aligned}\|w_1\partial_z v_2\|_{L^p(L^q)} &\leq C\, \| \|v_1\|_X \|v_2\|_X \|_{L^p} \\ &\leq C\|v_1\|_{L^{2p}(X)}\|v_2\|_{L^{2p}(X)} \leq CT^\eta \|v_1\|_{\mathbb{E}_1(T)}\|v_2\|_{\mathbb{E}_1(T)}.\end{aligned}$$

□

1.14.5 Maximal Regularity Results

In this subsection we state that the vertical and horizontal solution of the primitive equations belong to the maximal regularity class $\mathbb{E}_1^v(0, T)$ as well as that the solution the linearized system associated with (1.80) fulfills a maximal regularity result estimate. We start by considering the linearization of (1.80). It corresponds to the difference equation of (NS_ε) and (PE).

Given $F \in \mathbb{E}_0(0, T)$ and initial data $U_0 \in X_\gamma$ we consider the linear problem

$$\begin{cases} \partial_t U - \Delta U = F - \nabla_\varepsilon P & \text{in } (0, T) \times \Omega, \\ \nabla_\varepsilon \cdot U = 0 & \text{in } (0, T) \times \Omega, \\ U, P \text{ periodic} & \text{in } x, y, z, \\ U(0) = U_0 & \text{in } \Omega, \end{cases} \tag{1.81}$$

where $\nabla_\varepsilon := (\partial_x, \partial_y, \varepsilon^{-1}\partial_z)^T$. The functions U and P are the unknowns and represent the velocity and pressure differences.

We now aim to prove a maximal regularity estimate for U, where the constants are *independent* of the aspect ratio and the pressure gradient.

Lemma 29 (Estimate on the Pressure) *Let $\varepsilon > 0$, $q \in (1, \infty)$ and assume that $F = (f_H, f_z) \in L^q(\Omega)$ and $P \in H^{1,q}_{per}(\Omega)$ are satisfying the equation $-(\Delta_H + \varepsilon^{-2}\partial_z^2)P = \operatorname{div}(f_H, \varepsilon^{-1} f_z)$ for $\varepsilon > 0$. Then there exists a constant $C > 0$, independent of ε, such that*

$$\|(\nabla_H P, \varepsilon^{-1}\partial_z P)\|_q \leq C\|F\|_q.$$

The proof of this lemma takes advantage of the periodicity, i.e., in particular of the Neumann boundary conditions of the primitive equations. This allows one to interchange the Helmholtz projection and the Laplacian, and to apply Mikhlin's theorem in the period setting.

Proposition 1.14.3 (Maximal Regularity Estimate) *Let $p, q \in (1, \infty)$, $T > 0$, $F \in \mathbb{E}_0(T)$, $U_0 \in X_\gamma$ and $\varepsilon > 0$. Then there is a unique solution U, P to the equation (1.81) with $U \in \mathbb{E}_1(0, T)$ and $\nabla_\varepsilon P \in \mathbb{E}_0(0, T)$, where P is unique up to a constant. Moreover, there exist constants $C > 0$ and $C_T > 0$, independent of ε, such that*

$$\|U\|_{\mathbb{E}_1(T)} \leq C\|F\|_{\mathbb{E}_0(T)} + C_T\|U_0\|_{X_\gamma}.$$

Proof First, one defines the ε-dependent Helmholtz projection

$$\mathbb{P}_\varepsilon := \mathrm{Id} - \nabla_\varepsilon \Delta_\varepsilon^{-1} \operatorname{div}_\varepsilon, \quad \text{where } \Delta_\varepsilon = \nabla_\varepsilon \cdot \nabla_\varepsilon,$$

By Lemma 29 this is a bounded projection with uniform norm bound independent of ε.

First, we apply this to (1.81). Taking into account that due to periodicity $\mathbb{P}_\varepsilon \Delta = \Delta \mathbb{P}_\varepsilon$ and $\mathbb{P}_\varepsilon \nabla_\varepsilon P = 0$ hold, the equation (1.81) reduces to the heat equation with right-hand side $\mathbb{P}_\varepsilon F$.

Maximal L^p-regularity of the three-dimensional Laplacian in the periodic setting yields

$$\|U\|_{\mathbb{E}_1(T)} \leq C\|\mathbb{P}_\varepsilon F\|_{\mathbb{E}_0(T)} + C_T\|U_0\|_{X_\gamma} \leq C\|F\|_{\mathbb{E}_0(T)} + C_T\|U_0\|_{X_\gamma}.$$

□

The primitive equations admit a unique solution $v \in \mathbb{E}_1^v(T)$, which satisfies in addition $v \in C^\infty((0,T), C^\infty(\overline{\Omega})^2)$ and hence $w \in C^\infty((0,T), C^\infty(\overline{\Omega}))$ for any $T > 0$. It remains to show that w belongs to the maximal regularity class $\mathbb{E}_1(0, T^*)$ for some $T^* > 0$. To this end , one applies $\int_{-1}^z \operatorname{div}_H(\cdot)$ to (PE) which yields an evolution equation for w

$$\partial_t w - \Delta w = f(v, w) \text{ in } (0,\infty) \times \Omega,$$

where

$$f(v, w) = -\int_{-1}^z \operatorname{div}_H (\nabla_H p + u \cdot \nabla v).$$

Using $\operatorname{div}_H \overline{v} = 0$, for $z = 1$ we obtain $2\Delta_H p = -\operatorname{div}_H \int_{-1}^1 u \cdot \nabla v$ and thus

$$f(v, w) = \frac{1}{2}\int_z \operatorname{div}_H u \cdot \nabla v = \frac{1}{2}\int_z \operatorname{div}_H \operatorname{div} u \otimes v,$$

where $\int_z := \int_z^1 - \int_{-1}^z + z\int_{-1}^1$. Observe that

$$\operatorname{div}_H \operatorname{div} u \otimes v = \partial_z(w \operatorname{div}_H v + v \cdot \nabla_H w) + (\operatorname{div}_H v)^2 + 2v \cdot \nabla_H \operatorname{div}_H v + \nabla_H v \cdot (\nabla_H v)^T.$$

Hence, $f(v, w) =: f_1(v, w) + f_2(v) + f_3(v, w)$ with

$$f_1 = (w \operatorname{div}_H v - v \cdot \nabla_H w)|_z^1, \quad f_2 = \frac{1}{2}\int_z \left(\nabla_H v \cdot (\nabla_H v)^T + (\operatorname{div}_H v)^2\right),$$

$$f_3 = \int_z \partial_z v \cdot \nabla_H w.$$

Here we used the fact that $\int_z v \cdot \nabla_H \operatorname{div}_H v = -2(v \cdot \nabla_H w)|_z^1 + \int_z \partial_z v \cdot \nabla_H w$, which follows by integration by parts. Each of these terms can now be estimated using the estimates on the non-linearity given above. Hence one obtains

Proposition 1.14.4 (Additional Regularity of Solutions to the Primitive Equations) *Let p, q fulfill Assumption (A) and let v be the strong solution of the primitive equations associated to v_0 satisfying $(v_0, w_0) \in X_\gamma$. Then*

$$u = (v, w) \in \mathbb{E}_1^u(T) \text{ for all } T > 0.$$

Corollary 4 (Quadratic Inequality) *Let $\mathcal{T} > 0$ and $p, q \in (1, \infty)$ such that $1/p + 1/q \leq 1$. Let $(V_\varepsilon, W_\varepsilon) \in \mathbb{E}_1^u(0, \mathcal{T})$ denote the solution of equation* (1.80) *for some $u = (v, w) \in \mathbb{E}_1(0, \mathcal{T})$ and initial data $U_0 \in X_\gamma$. Then for $X_\varepsilon(T) := \|(V_\varepsilon, \varepsilon W_\varepsilon)\|_{\mathbb{E}_1(0,T)}$ and for any $\eta \in [0, 1 - 1/p - 1/q]$ there exists a constant $C > 0$, independent of ε, such that*

$$X_\varepsilon(T) \leq C T^\eta \left[X_\varepsilon(T)\|u\|_{\mathbb{E}_1(0,T)} + X_\varepsilon^2(T)\right] + \varepsilon C \left[\|u\|_{\mathbb{E}_1(0,T)} + T^\eta \|u\|^2_{\mathbb{E}_1(0,T)}\right] + C\|U_0\|_{X_\gamma},$$

for all $T \in [0, \mathcal{T}]$.

Proof Since

$$F_H = -V_\varepsilon \cdot \nabla_H v - W_\varepsilon \partial_z v - v \cdot \nabla_H V_\varepsilon - w \partial_z V_\varepsilon - V_\varepsilon \cdot \nabla_H V_\varepsilon - W_\varepsilon \partial_z V_\varepsilon,$$

we obtain with the help of Lemma 28 (a), (b)

$$\|F_H\|_{\mathbb{E}_0(T)} \leq C T^\eta \|V_\varepsilon\|_{\mathbb{E}_1(T)} (\|V_\varepsilon\|_{\mathbb{E}_1(T)} + \|(v, w)\|_{\mathbb{E}_1(T)}). \tag{1.82}$$

Similarly, since

$$\varepsilon F_z = \varepsilon(-V_\varepsilon \cdot \nabla_H w - w \operatorname{div}_H V_\varepsilon) - \varepsilon W_\varepsilon \operatorname{div}_H (v + V_\varepsilon) - (v + V_\varepsilon) \cdot \nabla_H \varepsilon W_\varepsilon - \varepsilon(\partial_t w - u \cdot \nabla w + \Delta w),$$

Lemma 28 (a) yields

$$\|\varepsilon F_z\|_{\mathbb{E}_0(T)} \leq C T^\eta [\|V_\varepsilon\|_{\mathbb{E}_1(T)} \|w\|_{\mathbb{E}_1(T)} + \|\varepsilon W_\varepsilon\|_{\mathbb{E}_1(T)} (\|V_\varepsilon\|_{\mathbb{E}_1(T)} + \|v\|_{\mathbb{E}_1(T)})] + C\varepsilon[\|w\|_{\mathbb{E}_1(T)} + T^\eta \|w\|^2_{\mathbb{E}_1(T)}].$$

Combining this estimate with (1.82), Proposition 1.14.3 yields the assertion. □

1.14.6 Proof of Theorem 17

Fix $\mathcal{T} > 0$ and denote by u the solution of equation (PE). Proposition 1.14.4 yields $u \in \mathbb{E}_1(\mathcal{T})$. We now show that

$$X_\varepsilon(T) := \|V_\varepsilon, \varepsilon W_\varepsilon\|_{\mathbb{E}_1(T)} \leq \varepsilon C(\|(v, w)\|_{\mathbb{E}_1(\cdot)}, \mathcal{T}, p, q)$$

for all $T \in [0, \mathcal{T}]$ and $\varepsilon > 0$ small enough. By the uniform continuity of $T \mapsto \|u\|_{\mathbb{E}_1(T)}$ on $[0, \mathcal{T}]$ there is a $T^* \in [0, \mathcal{T}]$ such that

$$\|u\|^p_{\mathbb{E}_1(T+T^*)} - \|u\|^p_{\mathbb{E}_1(T)} \leq (2C\mathcal{T}^\eta)^{-p}$$

for all $T \in [0, \mathcal{T} - T^*]$ and where $C > 0$ denotes the constant given in Corollary 4. Moreover, Corollary 4 with $U_0 = 0$ implies

$$CX_\varepsilon^2(T) - \tfrac{1}{2}X_\varepsilon(T) + \varepsilon \geq 0, \quad T \in [0, T^*]. \tag{1.83}$$

Observe that $(X_\varepsilon)^p : t \mapsto \left(\int_0^t \ldots\right)$ is continuous in $[0, T]$ and $X_\varepsilon(0) = 0$. Thus, for $\varepsilon < (16C)^{-1}$, we may solve this quadratic inequality and obtain $X_\varepsilon \leq 2\varepsilon$ on $[0, T^*]$.

Note that inequality (1.83) holds indeed on a time interval independent of ε. More specifically, if one replaces T^* by $T_\varepsilon < T^*$, where T_ε is the maximal existence time of (NS$_\varepsilon$) with initial data u_0, then it holds similarly as above that $X_\varepsilon \leq 2\varepsilon$ on $[0, T_\varepsilon]$ which implies a contradiction to the maximality of the existence time.

By induction one can iterate this procedure for ε sufficiently small. So, assume there is some $m \in \mathbb{N}$ such that

$$mT^* < \mathcal{T} \quad \text{and} \quad X_\varepsilon \leq \varepsilon 2K_m \quad \text{in } [0, mT^*],$$

where

$$K_1 = 1 \quad \text{and} \quad K_m = 2^{1/p}\left[(2Cc_{mT^*} + 1)\, K_{m-1} + 1\right]$$

and c_T denotes the embedding constant of $\mathbb{E}_1(0, T) \hookrightarrow L^\infty(0, T; X_\gamma)$. Let

$$(\tilde{V}_\varepsilon, \varepsilon\tilde{W}_\varepsilon)(T) = (V_\varepsilon, \varepsilon W_\varepsilon)(T + mT^*)$$

be the unique solution of problem (1.80) with respect to $\tilde{u}(T) = u(T + mT^*)$ and initial data $U_0 = (V_\varepsilon, \varepsilon W_\varepsilon)(mT^*)$. Setting

$$\tilde{X}_\varepsilon^p(T) := \|(\tilde{V}_\varepsilon, \varepsilon\tilde{W}_\varepsilon)\|^p_{\mathbb{E}_1(T)} = X_\varepsilon^p(T + mT^*) - X_\varepsilon^p(mT^*),$$

Corollary 4 and the argument about the ε-independency of the time interval given above imply

$$C\tilde{X}_\varepsilon^2(T) - \tfrac{1}{2}\tilde{X}_\varepsilon(T) + \varepsilon + C\|U_0\|_{X_\gamma} \geq 0, \quad T \in [0, \min\{T^*; \mathcal{T} - mT^*\}].$$

By assumption $\|U_0\|_{X_\gamma} \leq c_{mT^*} X_\varepsilon(mT^*) \leq c_{mT^*}\varepsilon 2K_m$. Since $\tilde{X}_\varepsilon(0) = 0$ and $\tilde{X}_\varepsilon$ is continuous in $[0, \min\{T^*; \mathcal{T} - mT^*\}]$, we may solve the quadratic inequality for

$$\varepsilon < (16C(1 + 2Cc_{mT^*}K_m))^{-1}$$

and obtain

$$\tilde{X}_\varepsilon \leq \varepsilon 2(1 + 2Cc_{mT^*}K_m) \quad \text{in } [0, \min\{T^*; \mathcal{T} - mT^*\}].$$

Hence, by the assumption on m

$$\begin{aligned} X_\varepsilon^p(T) &\leq (\varepsilon 2(1 + 2Cc_{mT^*}K_m))^p + X_\varepsilon^p(mT^*) \\ &\leq 2\,[2\varepsilon(1 + 2Cc_{mT^*}K_m) + \varepsilon 2K_m]^p \\ &= (\varepsilon 2K_{m+1})^p, \end{aligned}$$

for all $T \in [mT^*, \min\{(m+1)T^*; \mathcal{T}\}]$. The assumption on m implies

$$X_\varepsilon \leq \varepsilon 2K_{m+1} \quad \text{in } [0, \min\{(m+1)T^*; \mathcal{T}\}].$$

By induction we get

$$X_\varepsilon \leq \varepsilon 2K_M \quad \text{in } [0, \mathcal{T}]$$

with $M = \left\lceil \frac{\mathcal{T}}{T^*} \right\rceil$. This completes the proof of Theorem 17.

Bibliography

1. Herbert Amann. Nonhomogeneous linear and quasilinear elliptic and parabolic boundary value problems. In *Function spaces, differential operators and nonlinear analysis (Friedrichroda, 1992)*, volume 133 of *Teubner-Texte Math.*, pages 9–126. Teubner, Stuttgart, 1993.
2. Herbert Amann, Matthias Hieber, and Gieri Simonett. Bounded H_∞-calculus for elliptic operators. *Differential Integral Equations*, 7(3-4):613–653, 1994.
3. Sigurd Angenent. Parabolic equations for curves on surfaces. I. Curves with p-integrable curvature. *Ann. of Math. (2)*, 132(3):451–483, 1990.
4. Sigurd B. Angenent. Nonlinear analytic semiflows. *Proc. Roy. Soc. Edinburgh Sect. A*, 115(1-2):91–107, 1990.

5. Wolfgang Arendt, Charles J. K. Batty, Matthias Hieber, and Frank Neubrander. *Vector-valued Laplace Transforms and Cauchy Problems*, volume 96 of *Monographs in Mathematics*. Birkhäuser/Springer Basel AG, Basel, second edition, 2011.
6. Pascal Azérad and Francisco Guillén. Mathematical justification of the hydrostatic approximation in the primitive equations of geophysical fluid dynamics. *SIAM J. Math. Anal.*, 33(4):847–859, 2001.
7. O. Besson and M. R. Laydi. Some estimates for the anisotropic Navier-Stokes equations and for the hydrostatic approximation. *RAIRO Modél. Math. Anal. Numér.*, 26(7):855–865, 1992.
8. Didier Bresch, Alexandre Kazhikhov, and Jérôme Lemoine. On the two-dimensional hydrostatic Navier-Stokes equations. *SIAM J. Math. Anal.*, 36(3):796–814, 2004/05.
9. Marco Cannone. A generalization of a theorem by Kato on Navier-Stokes equations. *Rev. Mat. Iberoamericana*, 13(3):515–541, 1997.
10. Chongsheng Cao, Slim Ibrahim, K. Nakanishi, and Edriss Titi. Finite time blow-up for inviscid primitive equations of ocean and atmospheric dynamics. *Comm. Math. Phys.*, 337:473–482, 2015.
11. Chongsheng Cao, Jinkai Li, and Edriss S. Titi. Global well-posedness of strong solutions to the 3D primitive equations with horizontal eddy diffusivity. *J. Differential Equations*, 257(11):4108–4132, 2014.
12. Chongsheng Cao, Jinkai Li, and Edriss S. Titi. Local and global well-posedness of strong solutions to the 3D primitive equations with vertical eddy diffusivity. *Arch. Ration. Mech. Anal.*, 214(1):35–76, 2014.
13. Chongsheng Cao, Jinkai Li, and Edriss S. Titi. Global well-posedness of the three-dimensional primitive equations with only horizontal viscosity and diffusion. *Comm. Pure Appl. Math.*, 69(8):1492–1531, 2016.
14. Chongsheng Cao and Edriss S. Titi. Global well-posedness of the three-dimensional viscous primitive equations of large scale ocean and atmosphere dynamics. *Ann. of Math. (2)*, 166(1):245–267, 2007.
15. Chongsheng Cao and Edriss S. Titi. Global well-posedness of the $3D$ primitive equations with partial vertical turbulence mixing heat diffusion. *Comm. Math. Phys.*, 310(2):537–568, 2012.
16. Ralph Chill and Eva Fašangová. *Gradient Systems. Lecture Notes of the 13th International Internet Seminar.* Matfyzpress, Prague, 2010.
17. Igor Chueshov. A squeezing property and its applications to a description of long-time behaviour in the three-dimensional viscous primitive equations. *Proc. Roy. Soc. Edinburgh Sect. A*, 144(4):711–729, 2014.
18. Michele Coti Zelati, Aimin Huang, Igor Kukavica, Roger Temam, and Mohammed Ziane. The primitive equations of the atmosphere in presence of vapour saturation. *Nonlinearity*, 28(3):625–668, 2015.
19. Robert Denk, Giovanni Dore, Matthias Hieber, Jan Prüss, and Alberto Venni. New thoughts on old results of R. T. Seeley. *Math. Ann.*, 328(4):545–583, 2004.
20. Robert Denk, Matthias Hieber, and Jan Prüss. R-boundedness, Fourier multipliers and problems of elliptic and parabolic type. *Mem. Amer. Math. Soc.*, 166(788):viii+114, 2003.
21. Joachim Escher and Gieri Simonett. Analyticity of solutions to fully nonlinear parabolic evolution equations on symmetric spaces. *J. Evol. Equ.*, 3(4):549–576, 2003. Dedicated to Philippe Bénilan.
22. Lawrence Christopher Evans and Robert Gastler. Some results for the primitive equations with physical boundary conditions. *Z. Angew. Math. Phys.*, 64(6):1729–1744, 2013.
23. Hiroshi Fujita and Tosio Kato. On the Navier-Stokes initial value problem. I. *Arch. Rational Mech. Anal.*, 16:269–315, 1964.
24. Ken Furukawa, Yoshikazu Giga, Matthias Hieber, Amru Hussein, Takahito Kashiwabara, and Marc Wrona. Rigorous justification of the hydrostatic approximation for the primitive equations by scaled Navier-Stokes equations. arXiv:1808.02410, 2018.
25. Giovanni P. Galdi, Matthias Hieber, and Takahito Kashiwabara. Strong time-periodic solutions to the 3D primitive equations subject to arbitrary large forces. *Nonlinearity*, 30(10):3979–3992, 2017.

26. Yoshikazu Giga. Solutions for semilinear parabolic equations in L^p and regularity of weak solutions of the Navier-Stokes system. *J. Differential Equations*, 62(2):186–212, 1986.
27. Yoshikazu Giga, Mathis Gries, Matthias Hieber, Amru Hussein, and Takahito Kashiwabara. Analyticity of solutions to the primitive equations. arXiv:1710.04860, 2017.
28. Yoshikazu Giga, Mathis Gries, Matthias Hieber, Amru Hussein, and Takahito Kashiwabara. The primitive equations in the scaling invariant space $L^\infty(L^1)$. arXiv:1710.04434, 2017.
29. Yoshikazu Giga, Mathis Gries, Matthias Hieber, Amru Hussein, and Takahito Kashiwabara. The hydrostatic stokes semigroup and well-posedness of the primitive equations on spaces of bounded functions. arXiv:1802.02383, 2018.
30. Yoshikazu Giga and Tetsuro Miyakawa. Solutions in L_r of the Navier-Stokes initial value problem. *Arch. Rational Mech. Anal.*, 89(3):267–281, 1985.
31. Vivette Girault and Pierre-Arnaud Raviart. *Finite element methods for Navier-Stokes equations*, volume 5 of *Springer Series in Computational Mathematics*. Springer-Verlag, Berlin, 1986. Theory and algorithms.
32. F. Guillén-González, N. Masmoudi, and M. A. Rodríguez-Bellido. Anisotropic estimates and strong solutions of the primitive equations. *Differential Integral Equations*, 14(11):1381–1408, 2001.
33. Daniel Han-Kwan and Toan T. Nguyen. Ill-posedness of the hydrostatic Euler and singular Vlasov equations. *Arch. Ration. Mech. Anal.*, 221(3):1317–1344, 2016.
34. Matthias Hieber, Amru Hussein, and Takahito Kashiwabara. Global strong L^p well-posedness of the 3D primitive equations with heat and salinity diffusion. *J. Differential Equations*, 261(12):6950–6981, 2016.
35. Matthias Hieber and Takahito Kashiwabara. Global strong well-posedness of the three dimensional primitive equations in L^p-spaces. *Arch. Ration. Mech. Anal.*, 221(3):1077–1115, 2016.
36. Einar Hille and Ralph S. Phillips. *Functional analysis and semi-groups*. American Mathematical Society Colloquium Publications, vol. 31. American Mathematical Society, Providence, R. I., 1957. rev. ed.
37. Toshiaki Hishida. Existence and regularizing properties of solutions for the nonstationary convection problem. *Funkcial. Ekvac.*, 34(3):449–474, 1991.
38. Sabine Hittmeir, Rupert Klein, Jinkai Li, and Edriss S. Titi. Global well-posedness for passively transported nonlinear moisture dynamics with phase changes. *Nonlinearity*, 30(10):3676–3718, 2017.
39. Hirotada Honda and Atusi Tani. Small-time existence of a strong solution of primitive equations for the ocean. *Tokyo J. Math.*, 35(1):97–138, 2012.
40. Amru Hussein. Partial and full hyper-viscosity for Navier-Stokes and primitive equations. arXiv:1809.03954, 2018.
41. Amru Hussein, Martin Saal, and Okihiro Sawada. Primitive equations with linearly growing initial data. *Ann. Sc. Norm. Super. Pisa Cl. Sci.*, (to appear) accepted 2018.
42. Amru Hussein, Martin Saal, and Marc Wrona. Primitive equations with horizontal viscosity: The initial value and the time-periodic problem for physical boundary conditions. arXiv:1902.03186, 2019.
43. Mihaela Ignatova, Igor Kukavica, and Mohammed Ziane. Local existence of solutions to the free boundary value problem for the primitive equations of the ocean. *J. Math. Phys.*, 53(10):103101, 17, 2012.
44. Ning Ju. The global attractor for the solutions to the 3D viscous primitive equations. *Discrete Contin. Dyn. Syst.*, 17(1):159–179, 2007.
45. Ning Ju. On H^2 solutions and z-weak solutions of the 3D primitive equations. *Indiana Univ. Math. J.*, 66(3):973–996, 2017.
46. Tosio Kato. Strong L^p-solutions of the Navier-Stokes equation in $\mathbf{R}^m$, with applications to weak solutions. *Math. Z.*, 187(4):471–480, 1984.
47. Georgy M. Kobelkov. Existence of a solution "in the large" for ocean dynamics equations. *J. Math. Fluid Mech.*, 9(4):588–610, 2007.

48. Igor Kukavica, Nader Masmoudi, Vlad Vicol, and Tak Kwong Wong. On the local well-posedness of the Prandtl and hydrostatic Euler equations with multiple monotonicity regions. *SIAM J. Math. Anal.*, 46(6):3865–3890, 2014.
49. Igor Kukavica, Yuan Pei, Walter Rusin, and Mohammed Ziane. Primitive equations with continuous initial data. *Nonlinearity*, 27(6):1135–1155, 2014.
50. Igor Kukavica and Mohammed Ziane. On the regularity of the primitive equations of the ocean. *Nonlinearity*, 20(12):2739–2753, 2007.
51. Igor Kukavica and Mohammed Ziane. Uniform gradient bounds for the primitive equations of the ocean. *Differential Integral Equations*, 21(9-10):837–849, 2008.
52. Peer C. Kunstmann and Lutz Weis. Maximal L_p-regularity for parabolic equations, Fourier multiplier theorems and H^∞-functional calculus. In *Functional analytic methods for evolution equations*, volume 1855 of *Lecture Notes in Math.*, pages 65–311. Springer, Berlin, 2004.
53. Jinkai Li and Edriss S. Titi. Existence and uniqueness of weak solutions to viscous primitive equations for a certain class of discontinuous initial data. *SIAM J. Math. Anal.*, 49(1):1–28, 2017.
54. Jinkai Li and Edriss S. Titi. *Recent Advances Concerning Certain Class of Geophysical Flows*, pages 933–971. Springer International Publishing, Cham, 2018.
55. Jinkai Li and Edriss S. Titi. The primitive equations as the small aspect ratio limit of the Navier–Stokes equations: Rigorous justification of the hydrostatic approximation. *J. Math. Pures Appl. (9)*, 124:30–58, 2019.
56. J.-L. Lions, R. Temam, and S. Wang. Models for the coupled atmosphere and ocean. (CAO I,II). *Comput. Mech. Adv.*, 1(1):120, 1993.
57. Jacques-Louis Lions, Roger Temam, and Shou Hong Wang. New formulations of the primitive equations of atmosphere and applications. *Nonlinearity*, 5(2):237–288, 1992.
58. Jacques-Louis Lions, Roger Temam, and Shou Hong Wang. On the equations of the large-scale ocean. *Nonlinearity*, 5(5):1007–1053, 1992.
59. Andrew Majda. *Introduction to PDEs and waves for the atmosphere and ocean*, volume 9 of *Courant Lecture Notes in Mathematics*. New York University, Courant Institute of Mathematical Sciences, New York; American Mathematical Society, Providence, RI, 2003.
60. Nader Masmoudi and Tak Kwong Wong. On the H^s theory of hydrostatic Euler equations. *Arch. Ration. Mech. Anal.*, 204(1):231–271, 2012.
61. Kyûya Masuda. On the regularity of solutions of the nonstationary Navier-Stokes equations. In *Approximation methods for Navier-Stokes problems (Proc. Sympos., Univ. Paderborn, Paderborn, 1979)*, volume 771 of *Lecture Notes in Math.*, pages 360–370. Springer, Berlin, 1980.
62. Tobias Nau. L^{p}-theory of cylindrical boundary value problems. Springer Spektrum, Wiesbaden, 2012. An operator-valued Fourier multiplier and functional calculus approach, Dissertation, University of Konstanz, Konstanz, 2012.
63. A. Pazy. *Semigroups of linear operators and applications to partial differential equations*, volume 44 of *Applied Mathematical Sciences*. Springer-Verlag, New York, 1983.
64. J. Pedlosky. *Geophysical Fluid Dynamics, 2nd edition*. Springer, New York, 1987.
65. Madalina Petcu, Roger M. Temam, and Mohammed Ziane. Some mathematical problems in geophysical fluid dynamics. In *Handbook of numerical analysis. Vol. XIV. Special volume: computational methods for the atmosphere and the oceans*, volume 14 of *Handb. Numer. Anal.*, pages 577–750. Elsevier/North-Holland, Amsterdam, 2009.
66. Jan Prüss. Maximal regularity for evolution equations in L_p-spaces. *Conf. Semin. Mat. Univ. Bari*, (285):1–39 (2003), 2002.
67. Jan Prüss and Gieri Simonett. *Moving Interfaces and Quasilinear Parabolic Evolution Equations*, volume 105 of *Monographs in Mathematics*. Birkhäuser/Springer, [Cham], 2016.
68. Jan Prüss and Mathias Wilke. Addendum to the paper "On quasilinear parabolic evolution equations in weighted L_p-spaces II". *J. Evol. Equ.*, 17(4):1381–1388, 2017.
69. Rolf Rannacher and Ridgway Scott. Some optimal error estimates for piecewise linear finite element approximations. *Math. Comp.*, 38(158):437–445, 1982.
70. Martin Saal. Primitive equations with half horizontal viscosity. arXiv:1807.05045, 2018.

71. Hans Triebel. *Interpolation Theory, Function Spaces, Differential Operators.* Johann Ambrosius Barth, Heidelberg, second edition, 1995.
72. Hans Triebel. *Theory of function spaces.* Modern Birkhäuser Classics. Birkhäuser/Springer Basel AG, Basel, 2010. Reprint of 1983 edition, Also published in 1983 by Birkhäuser Verlag.
73. Geoffrey K. Vallis. *Atmospheric and Oceanic Fluid Dynamics. Second Edition.* Cambridge Univ. Press, 2006.
74. W. M. Washington and C. L. Parkinson. *An Introduction to Three Dimensional Climate Modeling. Second Edition.* Oxford University Press, Oxford, 1986.
75. Tak Kwong Wong. Blowup of solutions of the hydrostatic Euler equations. *Proc. Amer. Math. Soc.*, 143(3):1119–1125, 2015.
76. M. Ziane. Regularity results for Stokes type systems related to climatology. *Appl. Math. Lett.*, 8(1):53–58, 1995.
77. Mohammed Ziane. Regularity results for Stokes type systems. *Appl. Anal.*, 58(3-4):263–292, 1995.

Chapter 2
Viscous Compressible Flows Under Pressure

Didier Bresch and Pierre-Emmanuel Jabin

Abstract This chapter deals with the role of pressure in the theory of viscous compressible flows. The pressure state laws and viscosities are described. Special attention is devoted to non-monotone pressure laws and pressure dependent viscosities. The global existence proofs are discussed for approximate systems. Some relevant physical applications are described, including among others the anelastic Euler equations, shallow water model, granular media, or mixture problems.

2.1 Introduction

The existence of global weak solutions to the compressible Navier–Stokes system, in the sense of J. Leray (see [36]), remained a long standing open problem in space dimension strictly greater than one until the first results in [38]. One of the main issue is that the weak bound of the divergence of the velocity field does not *a priori* rule out singular behaviors by the density which may oscillate, concentrate, or even vanish (vacuum state) even if this is not the case initially. This chapter discusses new results related to these questions and should be seen as a "crash-course" on the subject, set for simplicity on periodic domains in dimension 3.

More precisely, we discuss global existence of weak solutions for barotropic compressible flows (without temperature) focusing on pressure dependent viscosities or on non-monotone pressure laws with constant viscosities. Readers interested in the complete mathematical description are referred to the recent preprint [23] (see also the complementary handbook [11] and the Bourbaki paper [47]) and to the explanatory lecture notes [18] and [19] (explaining the published paper [17]) and the references therein.

D. Bresch (✉)
University of Savoie Mont Blanc, Le Bourget du Lac, France
e-mail: Didier.bresch@univ-smb.fr

P.-E. Jabin
University of Maryland, Maryland, MD, USA

T. Bodnár et al. (eds.), *Fluids Under Pressure*, Advances in Mathematical Fluid Mechanics, https://doi.org/10.1007/978-3-030-39639-8_2

Our hope in this volume is to attract young researchers to the topic by emphasizing some of the many remaining problems (or partially solved problems) that remain open after 20 years of contributions on the compressible Navier–Stokes system involving many mathematicians from various countries (China, France, Great Britain, Italy, Poland, Russia, Czech Republic, USA, etc.).

We finally observe that even if the global existence of weak solutions provides few informations on the well posedness of the system, such study has real practical interest. In addition to its physical meaning (turbulent solutions as named by J. Leray), the stability properties of weak solutions on the continuous model help to understand how to construct appropriate stable numerical schemes which do not generally preserve the strong regularity estimates. It also helps to better understand the structure of the equations and mathematical properties of important physical quantities.

Chapter Content Outline

1. Viscous compressible flows: pressure state laws—viscosities expressions
   (a) Compressible Navier–Stokes equations
   (b) Pressure state laws
   (c) Shear and bulk viscosities expressions
   (d) Capillarity versus viscous terms
2. Non-monotone pressure laws
   (a) Appropriate approximate system
   (b) Sketch of the proof of the global existence result
   (c) Some physical situations: anelastic Euler equations, mixture system, etc.
3. Pressure dependent viscosities
   (a) A two-velocity hydrodynamics: approximate system
   (b) Sketch of the proof of the global existence result
   (c) Some physical situations: shallow water, granular media, mixture, etc.
4. Conclusion and comments
5. Some technical important lemmas

2.2 Viscous Compressible Flows: Pressure State Laws—Viscosities Expressions

2.2.1 *Compressible Navier–Stokes Equations*

The barotropic compressible Navier–Stokes equations read

$$\partial_t \rho + \operatorname{div}(\rho u) = 0, \tag{2.1}$$

$$\partial_t(\rho u) + \operatorname{div}(\rho u \otimes u) - 2\operatorname{div}(\mu(\rho)\mathbb{D}(u)) - \nabla(\lambda(\rho)\operatorname{div} u) + \nabla p(\rho) = 0 \tag{2.2}$$

with $\mathbb{D}(u) = (\nabla u + {}^t\nabla u)/2$ the strain tensor and $s \mapsto p(s)$ the given pressure state law. We denote $s \mapsto \lambda(s)$ and $s \mapsto \mu(s)$ the bulk and shear viscosities which, together with the pressure law, completely characterize the physical properties of the fluid within this barotropic framework. The initial conditions naturally read

$$\rho|_{t=0} = \rho_0 \geq 0, \qquad (\rho u)|_{t=0} = m_0.$$

To simplify the presentation, we work in the periodic domain Π^d with typically $d = 3$ though we will also indicate some results for $d = 2$.

The system is dissipative and enjoys the following energy inequality:

$$\begin{aligned}\sup_{t\in[0,T]} \int_\Omega [\frac{1}{2}\rho|u|^2 + \rho\, e(\rho)](t) \\ +2\int_\Omega \mu(\rho)|\mathbb{D}(u) - \frac{1}{3}\operatorname{div} u \operatorname{Id}|^2 + (\lambda(\rho) + \frac{2\mu(\rho)}{3})|\operatorname{div} u|^2 \\ \leq \int_\Omega [\frac{1}{2}\frac{|m_0|^2}{\rho_0} + \rho_0 e(\rho_0)],\end{aligned} \tag{2.3}$$

where $e(\rho) = \int_{\rho^*}^{\rho} p(s)/s^2\, ds$ the potential energy density and ρ^* an arbitrary constant reference density. Note that we assume as usually

$$u_0 = \frac{m_0}{\rho_0} \text{ when } \rho_0 \neq 0 \text{ and } u_0 = 0 \text{ elsewhere }, \tag{2.4}$$

$$\frac{|m_0|^2}{\rho_0} = 0 a.e. \text{ on } \{x \in \Omega : \rho_0(x) = 0\}. \tag{2.5}$$

After the first breakthrough by P. -L. Lions [38] 20 years ago, the existence of global weak solutions for this system has been the object of intensive study with many important contributions including: V.A Vaigant–A.V. Kazhikhov [50], E. Feireisl–A. Novotny–H. Petzeltova [26], P.I. Plotnikov–V.A. Weigant [46], E. Feireisl [27], S. Jiang–Z.P. Xin [32], S. Jiang–Z.P. Xin–P. Zhang [31], D. Bresch–B. Desjardins [8]-[11]–[12], Z. Guo–Q. Jiu–Z. Xin [30], A. Mellet–A. Vasseur [40], A. Vasseur–C. Yu [52], J. Li–Z.P. Xin [37], I. Lacroix/Violet–A. Vasseur [35], D Bresch–B. Desjardins–E. Zatorska [16], P. Mucha–M. Pokorny–E. Zatorska [41], D. Bresch–P.-E. Jabin [17], and D. Bresch–A. Vasseur–C. Yu [23].

The authors have focused their attention in [17] in extending the class of possible pressure law and the possible shear and bulk viscosities. It is important to note that the now classical theory for weak solutions actually has fairly rigid assumptions on the possible state laws, making such an extension non-trivial. A typical example of such assumptions consists in taking

$$\mu \text{ and } \lambda \text{ two constants with } \lambda + 2\mu/3 > 0$$

with, as pressure law, $p(\rho) = a\rho^\gamma$ where $a, \gamma > 0$ are constants. Higher exponents γ are actually easier to handle as they provide more regularity and considerable efforts have been devoted to lowering the minimal value of γ that is needed for global existence. On this point, three breakthroughs have been successively obtained in the multi-dimensional setting: the first by P. -L. Lions in 1998 who succeeded to get global existence of weak solutions for $\gamma > 3d/(d+2)$ and the second by E. Feireisl, A. Novotny, and H. Petzeltova in 2001 who succeeded to reach $\gamma > d/2$ (this last paper is important to cover some physical pressure laws). The case $\gamma = d/2$ has finally been derived by P.I. Plotnikov–V.A. Weigant in 2015.

Within the classical theory, it is possible to relax somewhat the rigid assumption on $p(\rho)$ as long as p remains monotone in ρ. There are of course many classical state laws that are not monotone: The van der Waals pressure law for example. This motivated the article by E. Feireisl in 2002 which can handle non-monotone pressure laws provided that they still become strictly monotone and behave like $a\rho^\gamma$ when $\rho > \bar{\rho}$ for some fixed density $\bar{\rho}$. Only recently have the authors been able to treat general non-monotone pressure laws such as can be obtained by truncated virial pressure laws occurring in cosmology or given by attractive–repulsive pressure laws occurring, for instance, in biology.

The existence of weak solutions for the case where the shear and the bulk viscosities depend on the density had remained completely open until 2004 with the exception of critical work by A.V. Kazkikov–V.A. Vaigant where the shear viscosity is constant and the bulk viscosity depends on the density. In 2004, D. Bresch and B. Desjardins discovered what is now called the BD entropy estimates. It shows that the gradient of a function of the density can be controlled in the standard compressible model (2.1)–(2.2) (i.e., with one velocity field) if the shear and the bulk viscosities satisfy the BD algebraic relation

$$\lambda(\rho) = 2(\mu'(\rho)\rho - \mu(\rho)) \tag{2.6}$$

and if this control is assumed initially. The price to pay is that the viscosities have to vanish when the density vanish. Using this extra control on the density, D. Bresch and B. Desjardins got stability results for compressible Navier–Stokes equations with extra terms: drag terms or singular pressure close to vacuum. Following this breakthrough, the stability and the construction of approximate solutions in space dimension two or three have been investigated but faced considerable difficulties because the corresponding system is a degenerate parabolic system with a complicated mathematical structure which does not allow to define easily an appropriate approximated system. The stability property without extra quantities (drag terms, capillarity, etc.) has been firstly treated in a very important paper by A. Vasseur and A. Mellet. This result is based in 3 dimension on the propagation of an extra integrability on the velocity field when $1 < \gamma < 3$ assuming it is satisfied initially. This extra integrability is usually called the Mellet–Vasseur estimate and it strongly uses the energy and BD entropy estimates. Proposing an approximate system satisfying uniformly the energy estimate, the BD entropy estimate, and the Mellet–Vasseur estimate was for a long-time an open problem.

Concerning the construction of approximate systems, it has been firstly discussed by D. Bresch and B. Desjardins in 2006: appropriate drag terms, cold pressure, or capillarity quantities were the main ingredients. The global existence of weak-entropy solutions (construction of approximate solutions and stability property) has been first solved for

$$\mu(\rho) = \mu\,\rho, \qquad \lambda(\rho) = 0 \text{ with } \mu = \text{cst} > 0$$

and

$$p(\rho) = a\,\rho^\gamma \text{ with } 1 < \gamma < 3$$

in two outstanding papers independently written by A. Vasseur–C. Yu and J. Li–Z.P. Xin in 2015: The solutions satisfied energy estimates, BD entropy, Mellet–Vasseur estimates. We will explain quickly the two approximate systems chosen by these authors in Sect. 2.4.3.

Recently an important paper has been written by I. Lacroix-Violet and A. Vasseur in 2018 where they introduce renormalized formulation in velocity allowing to not use Mellet–Vasseur estimate to conclude and then cover the full range of exponent γ in the pressure law namely

$$p(\rho) = a\,\rho^\gamma \text{ with } \gamma > 1. \tag{2.7}$$

Only recently D. Bresch–A. Vasseur–C. Yu in 2019 solved the general case for a large class of viscosities $\mu(\rho)$ and $\lambda(\rho)$ satisfying the BD relation and (2.7).

Remark 2.2.1 It is interesting to note that if $\mu(\rho)$ is assumed to be constant then $\lambda(\rho)$ satisfying (2.6) has to be constant as well. We will show that it cannot be the case if we also impose that the viscosities satisfy that there exists ε such that $\lambda(\rho) + 2\mu(\rho)/3 \geq \varepsilon\mu(\rho)$ and $\mu(\rho) \geq 0$. In particular this implies λ and μ have to vanish at vacuum if they are to respect those two conditions.

In this note, we discuss the two recent important mathematical results [17] and [23] to make them more accessible to young researchers wishing to start working on viscous compressible flows under pressure. We want to insist that mathematical results often draw on previous breakthroughs as is the case here. In particular to explain the motivation of the two recent works, we start with a more thorough discussion of pressure state laws and shear versus bulk viscosities.

2.2.2 *Pressure State Laws*

As written in [17] and [25], it is in general a non-straightforward question to decide what kind of pressure law should be used given the wide range of possible applications and possible fluids. Well-known equations of state include Dalton's law of partial pressures, the ideal gas law, the van der Waals equation of state, and the virial equation of state. In the barotropic framework, namely with a pressure state which depends only on density, historically the power laws

$$p(\rho) = a\rho^\gamma,$$

with $a, \gamma > 1$ given constants, have first been considered, for instance, by P. -L. Lions with $\gamma > 3d/(d+2)$ (see [38]), E. Feireisl, A. Novotny, and H. Petzeltova. with $\gamma > d/2$ (see [26] and finally P.I. Plotnikov–V.A. Weigant with $\gamma = d/2$ (see [46]). Then in a paper by E. Feireisl (see [27]) a slightly more general assumption (which covers the van der Waals law) is considered namely

$$p \in \mathcal{C}^2([0, +\infty)), \qquad P(0) = 0 \tag{2.8}$$

with

$$a\rho^{\gamma-1} - b \le p'(\rho) \le \frac{1}{a}\rho^{\gamma-1} + b \text{ with } \gamma > d/2, \tag{2.9}$$

where $a, b > 0$ are again given constants. Note that (2.9) necessarily requires that p be increasing after a certain critical value of the density ρ. This monotonicity of the pressure is connected to several well-known difficulties: the monotonicity is required for the stability of the thermodynamical equilibrium and at the level of compressible Euler system non-monotone pressure laws lead to a loss of hyperbolicity in the system. As mentioned before, very recently, D. Bresch–P.E. Jabin have proved in [17] that it is possible to consider a non-monotone pressure law p which is continuous on $[0, +\infty)$, p locally Lipschitz on $(0, +\infty)$ with $p(0) = 0$ such that there exists $C > 0$ with

$$C^{-1}\rho^\gamma - C \le p(\rho) \le C\rho^\gamma + C \tag{2.10}$$

and there exists $\bar{p} > 0$ constant such that for all $s \ge 0$

$$|p'(s)| \le \bar{p}s^{\tilde{\gamma}-1} \tag{2.11}$$

with

$$\gamma > (\max(2, \tilde{\gamma}) + 1)\frac{d}{d+2}. \tag{2.12}$$

It is important to note that such kind of pressure is not thermodynamically stable and does not ensured hyperbolicity for the corresponding compressible Euler equations. We also remark that the authors have in fact considered more general pressure law $p(t, x, \rho)$ which will allow them to consider, for instance, truncated virial pressure laws for the heat conducting Navier–Stokes equations. More precisely the authors hope to consider the heat-conducting compressible Navier–Stokes equations with pressure law under the form

$$p(\rho, \theta) = \rho^\gamma + \theta \sum_{n=0}^{N} B_b(\theta)\rho^n,$$

where $\gamma > 2N \geq 4$. The authors have to assume that the pressure contains a radiative part, namely that B_0 is convex in θ with $C^{-1}\theta^{b-1} \leq B_0(\theta) \leq \theta^{b-1}$ and $C^{-1}\theta^{b-2} \leq B_0'(\theta) \leq \theta^{b-2}$ where $2 \leq b \leq \alpha/2$. For $n \geq 2$, the coefficients B_n can have any sign but a concavity assumption is needed namely

$$\frac{d}{d\theta}(\theta^2 B_n'(\theta)) \leq 0.$$

This ensures, with the assumption on B_0, that $C_v = \partial_\theta e(\rho, \theta) > 0$. The authors require also some specific bounds on the B_n namely that there exists $\bar{B}_N$ and $\varepsilon > 0$ such that

$$|\theta B_n'(\theta)] + |B_n(\theta)| \leq C\theta^{\frac{\gamma-n}{\gamma}b-1-\varepsilon}$$

$$|B_n(\theta)| + [\theta B_n(\theta) - \bar{B}_n| \leq C\theta^{\frac{\alpha}{2}(1-\frac{2n}{\gamma})-\varepsilon}.$$

This important work is under progress in [20] with the stability part on ArXiv. More general pressure laws will be considered in the future work.

2.2.3 Shear and Bulk Viscosities Expressions

Together with the pressure law, the shear and bulk viscosities characterize the fluid under consideration. The theory initiated in the multi-dimensional setting by P.-L. Lions (see [38]) and E. Feireisl–A. Novotny–H. Petzeltova (see [26]) concerns the case

$$\mu \text{ and } \lambda \text{ are constants and } \lambda + 2\mu/3 > 0.$$

The theory is strongly related to the fact that if you take the divergence of the momentum equations, you obtain the following elliptic equation:

$$-\Delta((\lambda + 2\mu)\mathrm{div}u - p(\rho)) = \mathrm{div}[\rho(\dot{u} + f)],$$

where $\dot{u}$ is the total time derivative namely $\dot{u} = \partial_t u + u \cdot \nabla u$. Using the energy estimate and some extra integrability information of the pressure, this allows to get some sort of compactness in space of the effective flux

$$F = (\lambda + 2\mu)\mathrm{div}u - p(\rho)$$

and therefore to replace $\operatorname{div} u$ by $p(\rho)/(2\mu+\lambda)$ if needed. Unfortunately several physical situations involve some anisotropy in the stress tensor: geophysical flows, for instance, use and require to be able to consider, for instance,

$$-2\mathrm{A}\, D(u) - \operatorname{div} u\, \mathrm{B} + p(\rho)\mathrm{C}$$

with A,B,C constant $d \times d$ matrices. Only recently and with very restrictive assumptions on the coefficients, this has been considered by D. Bresch–P.E. Jabin in [17]. More precisely they assume the divergence of the stress tensor as follows:

$$-\operatorname{div}(A(t)\nabla u) - (\mu+\lambda)\nabla \operatorname{div} u + \nabla p(\rho),$$

where $A(t)$ a given smooth and symmetric matrix satisfying

$$A(t) = \mu \mathrm{Id} + \delta A(t), \qquad \mu > 0, \qquad \frac{2}{3}\mu + \lambda - \|\delta A(t)\|_{L^\infty} > 0,$$

where δA is a perturbation around $\mu\mathrm{Id}$. They prove assuming

$$\gamma > \frac{d}{2}\Big[(1+\frac{1}{d}) + \sqrt{1+\frac{1}{d^2}}\Big]$$

that there exists a universal constant $C_* > 0$ such that if

$$\|\delta A\|_{L^\infty} \leq C_*(2\mu+\lambda),$$

then there exist a global weak solution to the compressible Navier–Stokes equations.

The proof involves complex non-local behavior and of course the main problem with general anisotropic diffusion remains fully open. It would be a major breakthrough to be able to treat rather general cases under minimal physical properties such as frame invariance, and definite positivity and symmetry of the stress tensor.

Pressure dependent viscosities for compressible barotropic Navier–Stokes equations are equivalent to density dependent viscosities as long as the pressure is monotone in the density. In space dimension greater than one, no result was obtained before the series of papers by Bresch–Desjardins starting from 2004. Indeed the effective flux property developed by P.-L. Lions and E. Fereisl–A. Novotny–H. Petzeltova is no longer valid when the shear viscosity depends on the density with one important exception if μ is assumed to be constant and only λ depends on the density: This is the case considered in the very nice paper by Vaigant and Kazhikhov (see [50]) with

$$\mu = \mathrm{cst}, \qquad \lambda(\rho) = \rho^\beta \text{ with } \beta > 3.$$

The article [10] identified a quantity related to the gradient of a function of the density if the viscosities satisfy what is called the Bresch–Desjardins constraint namely if the viscosities satisfy (2.6). More precisely it was observed that formally the new velocity

$$v = u + 2\nabla\mu(\rho)/\rho$$

satisfies the following system:

$$\partial_t \rho + \mathrm{div}(\rho u) = 0, \tag{2.13}$$

$$\partial_t(\rho v) + \mathrm{div}(\rho v \otimes u) - 2\mathrm{div}(\mu(\rho)\mathbb{A}(v)) + \nabla p(\rho) = 0, \tag{2.14}$$

where $\mathbb{A}(v) = (\nabla v - {}^t\nabla(v))/2$. It is interesting to note that the obtained system is rather close to the one proposed earlier by H. Brenner. The corresponding energy (called BD entropy) reads

$$\frac{d}{dt}\int_\Omega [\frac{1}{2}\rho|v|^2 + \rho e(\rho)] + 2\int_\Omega \mu(\rho)|\mathbb{A}(u)|^2 + 2\int_\Omega \frac{p'(\rho)\mu'(\rho)}{\rho}|\nabla\rho|^2 = 0.$$

We remark that coupled with the standard energy estimates, this identity provides additional control if the initial data has finite energy and if initially the density ρ^0 satisfies

$$\int_\Omega \frac{|\nabla\mu(\rho^0)|^2}{\rho^0} < +\infty.$$

More precisely, we get control of the following quantities:

$$\nabla\mu(\rho)/\sqrt{\rho} \in L^\infty(0,T;L^2(\Omega)), \qquad \sqrt{\frac{p'(\rho)\mu'(\rho)}{\rho}}\nabla\rho \in L^2(0,T;L^2(\Omega)).$$

Using such additional information, D. Bresch–B. Desjardins firstly and others have been able to get global existence of weak solutions for the compressible Navier–Stokes equations with density dependent viscosities but with extra terms such as drag terms or capillarity quantities: see [2, 4, 9, 12, 14, 16, 41, 51], for instance. Note that the main difficulty for the compressible Navier–Stokes equations with degenerate density dependent viscosities is the limit passage in the nonlinear term $\rho u \otimes u$ contrary to the constant viscosities case where the main difficulty occurs for the limit passage in the pressure law. An important paper by A. Mellet–A. Vasseur (see [40]) has shown that a sequence of approximate solutions of the compressible Navier–Stokes equations satisfies

$$\int_\Omega \rho_n(1 + |u_n|^2)\log(1 + |u_n|^2) < c < +\infty$$

with $c > 0$ independent on n if this estimate is satisfied by the approximate initial data and if the sequence of approximate solutions satisfies uniformly the energy estimate and the BD entropy estimate. Such property is strongly used to get the first stability property without extra terms passing to the limit in $\rho_n u_n \otimes u_n$.

In [52] and [37], respectively, by A. Vasseur–C. Yu and J. Li–Z.P. Xin, real breakthroughs have been obtained proving global existence of entropy-weak solutions without any extra terms (construction of approximate solutions and stability process). The case $\mu(\rho) = \mu\rho$ and $\lambda(\rho) = 0$ with $p(\rho) = a\rho^\gamma$ where $1 < \gamma < 3$ is considered in these two works in completely different manners: The entropy-weak solutions, in these two papers, satisfy the energy estimates, the BD entropy, and the Mellet–Vasseur control. Recently I. Lacroix-Violet and A. Vasseur (see [35]) noticed that, when $\mu(\rho) = \mu\rho$ and $\lambda(\rho) = 0$, it is possible to generalized the renormalization procedure introduced in [52] to not need Mellet–Vasseur estimates to conclude and then to cover a more general pressure law $p(\rho) = a\rho^\gamma$ with $\gamma > 1$. More precisely they introduced the renormalized solution where, for instance, the term $\rho u \otimes u$ is replaced by the quantity $\rho_n \varphi(u_n) \otimes u_n$, where $\varphi \in W^{2,\infty}(\overline{\mathbb{R}}^3)$. Pass to the limit with respect to n in this last quantity is easier and then it remains to choose appropriate nonlinear functions φ_n to recover the global weak formulation after the limit passage with respect to n.

It is important to note that in [37] more general cases have been covered but with a non-symmetric viscous term in the three dimension in space case, namely $-\mathrm{div}(\mu(\rho)\nabla u) - \nabla(\lambda(\rho)\mathrm{div} u)$ with $\lambda(\rho) = (\mu'(\rho)\rho - \mu(\rho))$ and $\mu(\rho) = \mu\rho^\alpha$ where $\alpha \in [3/4, 2)$ with the following assumption on the value γ for the pressure $p(\rho) = a\rho^\gamma$:

$$\text{If } \alpha \in [3/4, 1], \qquad \gamma \in (1, 6\alpha - 3)$$

and

$$\text{if } \alpha \in (1, 2), \qquad \gamma \in [2\alpha - 1, 3\alpha - 1].$$

The main objective was then to cover symmetric viscous terms and to be able to consider shear and bulk viscosities as general as possible, but still satisfying the BD relation. This has been done very recently in [23] by D. Bresch–A.Vasseur–C. Yu . More precisely, they have been able to get global weak solution of the compressible Navier–Stokes equations where

$$p(\rho) = a\rho^\gamma \text{ with } a > 0 \text{ and } \gamma > 1, \tag{2.15}$$

with shear and bulk viscosities satisfying the BD relation with

$$\mu \in \mathcal{C}^0(\mathbb{R}_+; \mathbb{R}_+) \cap \mathcal{C}^2(\mathbb{R}_+^*; \mathbb{R}), \tag{2.16}$$

where $\mathbb{R}_+ = [0, +\infty)$ and $\mathbb{R}_+^* = (0, +\infty)$. They also assume that there exists two positive numbers α_1, α_2 such that

$$\frac{2}{3} < \alpha_1 < \alpha_2 < 4,$$

$$\text{for any } \rho > 0, \qquad 0 < \frac{1}{\alpha_2}\rho\mu'(\rho) \le \mu(\rho) \le \frac{1}{\alpha_1}\rho\mu'(\rho),$$

and that there exists a constant $C > 0$ such that

$$\left|\frac{\rho\mu''(\rho)}{\mu'(\rho)}\right| \le C < +\infty. \tag{2.17}$$

The bulk viscosity is assumed to be given through the BD relation (2.6). Note in particular that (2.6) then implies that there exists ε such that

$$\lambda(\rho) + 2\mu(\rho)/3 \ge \varepsilon\mu(\rho)$$

and

$$\mu(0) = \lambda(0) = 0.$$

Remark 2.2.2 (Open Problem) Note that the authors think that the condition $\alpha_2 < 4$ is not optimal. Such constraint comes from the capillarity quantity which is used to develop the renormalization technique introduced in [35]. It is an interesting open problem to improve such value.

2.2.4 Capillarity Versus Viscous Terms

Relation (2.6) seems to be strange and the authors have found no physical literature and no meaning for such relation but it is interesting to note that such relation occurs, for instance, for capillarity purposes. More precisely the Euler–Korteweg system reads

$$\partial_t\rho + \mathrm{div}(\rho u) = 0, \tag{2.18}$$

$$\partial_t(\rho u) + \mathrm{div}(\rho u \otimes u) - \mathrm{div}\,\mathbb{K} + \nabla p(\rho) = 0, \tag{2.19}$$

where

$$\mathbb{K} = \Big(\rho\mathrm{div}(K(\rho)\nabla\rho) + \frac{1}{2}(K(\rho) - \rho K'(\rho))|\nabla\rho|^2\Big)\mathbb{I}_{\mathbb{R}^d} - K(\rho)\nabla\rho \otimes \nabla\rho$$

with $s \mapsto K(s)$ the given capillarity coefficient behavior. In [7], D. Bresch–F. Couderc–P. Noble–J.P. Vila showed the following interesting relation:

$$
\begin{aligned}
\operatorname{div} \mathbb{K} &= \rho \nabla (\sqrt{K(\rho)} \Delta \int_0^\rho \sqrt{K(s)}\, ds) \\
&= 2\operatorname{div}\big(\widetilde{\mu}(\rho)\mathbb{D}(\widetilde{u})\big) + 2\nabla\big((\widetilde{\mu}'(\rho)\rho - \widetilde{\mu}(\rho))\operatorname{div}\widetilde{u}\big) \qquad (2.20)\\
&= 2\operatorname{div}\big(\widetilde{\mu}(\rho)\mathbb{D}(\widetilde{u})\big) + \nabla\big(\widetilde{\lambda}(\rho)\operatorname{div}\widetilde{u}\big) \qquad (2.21)
\end{aligned}
$$

with $\widetilde{u} = \nabla \widetilde{s}(\rho)$ and where $2\widetilde{\mu}'(\rho) = \sqrt{K(\rho)\rho}$ and $\sqrt{\rho}\, \widetilde{s}'(\rho) = \sqrt{K(\rho)}$ and

$$\widetilde{\lambda}(\rho) = 2(\widetilde{\mu}(\rho)\rho - \widetilde{\mu}(\rho)).$$

In conclusion, the divergence of the stress tensor may be written as a compressible diffusion operator with BD relation between the two viscosities $\widetilde{\mu}$ and $\widetilde{\lambda}$. As an example, if $K(s) = 1/s$ corresponding to the Euler-quantum, we get $\widetilde{\mu}(\rho) = \frac{1}{2}\rho$, $\widetilde{\lambda}(\rho) = 0$, and $\widetilde{s}(\rho) = \log \rho$.

This identity (2.20) plays a crucial role in the proof to construct smooth enough solutions. It defines the appropriate capillarity term to consider in the approximate system namely choosing the capillarity coefficient $K(\rho)$ such that $\widetilde{\mu}(\rho) = \mu(\rho)$. Other identities of $\operatorname{div}\mathbb{K}$ will be used to define the weak solution for the corresponding Navier–Stokes–Korteweg system and to pass to the limit in it namely

$$
\begin{aligned}
2\mu(\rho)\nabla^2(2s(\rho)) + \lambda(\rho)\Delta(2s(\rho)) &= 4\Big[2\sqrt{\mu(\rho)}\nabla\nabla Z(\rho) - \nabla(\sqrt{\mu(\rho)}\nabla Z(\rho)\Big] \\
&\quad + (\frac{2\lambda(\rho)}{\sqrt{\mu(\rho)}} + k(\rho))\Delta Z(\rho)\, \mathrm{Id} - \operatorname{div}[k(\rho)\nabla Z(\rho)]\, \mathrm{Id},
\end{aligned}
\qquad (2.22)
$$

where $Z(\rho) = \int_0^\rho [(\mu(s))^{1/2}\mu'(s)]/s\, ds$ and $k(\rho) = \int_0^\rho \frac{\lambda(s)\mu'(s)}{\mu(s)^{3/2}} ds$. Strong convergence on the viscosity $\mu(\rho)$ and weak convergence on $Z(\rho)$ will be used.

The first mathematical study of Navier–Stokes–Korteweg system with the use of the BD entropy information has been developed in [13]. Note that the case considered in [51, 52] and [35] is related to $\mu(\rho) = \rho$ and $K(\rho) = 4/\rho$ which corresponds to the quantum Navier–Stokes system. We also point out four very interesting papers written by Antonelli–Spirito in [1–4] with one of them without considering Navier–Stokes–Korteweg systems without such relation between the shear viscosity and the capillary coefficient. Note that the quantum Korteweg system has also been considered in the nice paper [24].

Remark 2.2.3 It is interesting to note that a viscous quantum hydrodynamic model can be derived from the Wigner–Fokker–Planck equation by a moment method, see, for instance, the work by S. Brull–F. Mehats, A. Jüngel et al., I. Gamba–A.

Jüngel–A. Vasseur. The viscous regularizations arise from the quantum Fokker–Planck interaction operator. The system reads

$$\partial_t \rho + \operatorname{div}(\rho w) = \varepsilon \Delta \rho,$$

$$\partial_t(\rho w) + \operatorname{div}(\rho w \otimes w) + \nabla p(\rho) - 2\delta^2 \rho \nabla\Big(\frac{1}{\sqrt{\rho}}\Delta\sqrt{\rho}\Big) = \varepsilon \Delta(\rho w).$$

Using the change of variable $u = w - \varepsilon \nabla \log \rho$, the system may be written

$$\partial_t \rho + \operatorname{div}(\rho u) = 0$$

$$\partial_t(\rho u) + \operatorname{div}(\rho u \otimes u) + \nabla p(\rho) - 2(\delta^2 + \varepsilon^2)\rho \nabla\Big(\frac{1}{\sqrt{\rho}}\Delta\sqrt{\rho}\Big) = \varepsilon \operatorname{div}(\rho\, \mathbb{D}u).$$

This shows that physically diffusion quantities and quantum tensor forms are related: Such observation has been done by A. Jüngel.

2.3 Non-monotone Pressure Laws

As mentioned in Sect. 2.2, assuming constant shear and bulk viscosities, barotropic compressible Navier–Stokes equations with non-monotone pressure laws have been considered first by E. Feireisl in [27] and more recently by D. Bresch–P.E. Jabin in [17]. In this last work, the authors get the following existence result for the compressible Navier-Stokes equation with constant viscosities in the following sense:

Definition 2.3.1 For any $T \in (0, +\infty)$, ρ_0, m_0 satisfying some technical assumption, we say that a couple (ρ, u) is a weak renormalized solution with bounded energy if it satisfies the following properties:

- $\rho \in L^\infty(0, T; L^\gamma(\Omega)) \cap \mathcal{C}^0([0, T]; L^\gamma_{\text{weak}}(\Omega))$ with
  $\rho \geq 0$ in $(0, T) \times \Omega$, $\rho|_{t=0} = \rho_0$ a.e. in Ω.
- $u \in L^2(0, T; H^1(\Omega))$ and $\rho|u|^2 \in L^\infty(0, T; L^1(\Omega))$ with
  $\rho u \in \mathcal{C}([0, T]; L^{2\gamma/(\gamma+1)}_{\text{weak}}(\Omega))$ with $\rho u|_{t=0} = m_0$ a.e. in Ω.
- (ρ, u) solves the mass and momentum equations in $\mathcal{D}'((0, T) \times \Omega)$.
- For any smooth b with appropriate monotonicity properties, $b(\rho)$ solves the renormalized equation

$$\partial_t b(\rho) + \operatorname{div}(b(\rho)u) + (b'(\rho)\rho - b(\rho))\operatorname{div} u = 0.$$

- For almost all $\tau \in (0, T)$, (ρ, u) satisfies the energy inequality

$$\int_\Omega (\rho|u|^2 + \rho e(\rho))(\tau) + \int_0^\tau \int_\Omega (\mu|\nabla u|^2 + (\lambda+\mu)|\operatorname{div} u|^2) \leq \int_\Omega (\rho_0|u_0|^2 + \rho_0 e(\rho_0)).$$

More precisely they prove the following theorem:

Theorem 2.3.2 *Assume that the initial data u_0 and ρ_0 satisfies the bound*

$$E_0 = \int_{\Pi^d} (\rho^0 \frac{|u_0|^2}{2} + \rho_0 e(\rho_0))\, dx < +\infty,$$

where $e(\rho) = \int_{\rho^*}^{\rho} p(s)/s^2\, ds$ with the definitions (2.4)–(2.5). *Let the pressure law satisfy* (2.10)–(2.12). *Then there exists a global weak solution of the compressible Navier–Stokes system in the sense of Definition 2.3.1 written above.*

2.3.1 Appropriate Approximate System

For the compressible Navier–Stokes equations with constant viscosities and non-monotone pressure laws, the starting point may be the following system:

$$\begin{aligned} &\partial_t \rho_{k,n} + \operatorname{div}(\rho_{k,n} u_{k,n}) = \varepsilon_n \Delta \rho_{k,n}, \\ &\partial_t(\rho_{k,n} u_{k,n}) + \operatorname{div}(\rho_{k,n} u_{k,n} \otimes u_{k,n}) + \nabla p_k(\rho_{k,n}) \\ &\qquad -\mu \Delta u_{k,n} - (\lambda + \mu) \nabla \operatorname{div} u_{k,n} + \varepsilon_n \nabla \rho_{k,n} \cdot \nabla u_{k,n} = 0, \end{aligned} \tag{2.23}$$

with the initial data

$$\rho_{k,n} u_{k,n}|_{t=0} = m^0, \qquad \rho_{k,n}|_{t=0} = \rho^0.$$

The pressure p_k is defined as follows:

$$p_k(\rho) = p(\rho) \text{ if } \rho \le c_k, \qquad p_k(\rho) = p(c_k) + c(\rho - c_k)^\beta \text{ if } \rho \ge c_k,$$

with large enough β and for $c_k \to \infty$ as $k \to +\infty$ and $\varepsilon_n \to 0$ as $n \to +\infty$. Global existence of solutions for this approximate solution is obtained easily using the parabolic regularization and classical Faedo-Galerkin procedures. Then as usual for compressible Navier–Stokes, one needs to pass to the limit in two steps: First with respect to $n \to +\infty$ to send ε_n to 0, which can be done through variants of stability properties developed by E. Feireisl in [27] for van der Waals type pressure law p_k. The second step consists in taking the limit $k \to \infty$ and $c_k \to \infty$ and this is where new stability results are needed.

2.3.2 Sketch of the Proof of the Global Existence Result

When the viscosities are assumed to be constant, stability estimates are again decomposed in two steps: The first step is to get uniform bounds on ρ and u providing compactness in time on the density and compactness in space on the velocity. The second step is the most difficult and consists in getting compactness in space on the density.

In the other setting considered in this note, when the viscosities depend on the pressure but the pressure law is monotone, the stability will also be decomposed in two different steps with some duality between the two frameworks that will be discussed later on.

We denote ρ_k and u_k the solution obtained from $\rho_{n,k}$ and $u_{n,k}$ by first taking the limit $n \to \infty$. In particular $\rho_k,\ u_k$ solve the system (2.23) but with $\varepsilon_n = 0$.

Uniform Estimates, Compactness in Time of the Density, Compactness on the Velocity From the hypothesis on the pressure law and the energy inequality, one deduces the following uniform bounds:

$$\sup_{t\in[0,T]} \int_\Omega (\rho_k^\gamma + \rho_k\, |u_k|^2)\, dx < C < +\infty,$$

and

$$\int_0^T \int_\Omega |\nabla u_k|^2\, dt\, dx \le C < +\infty,$$

where C does not depend on k.

Then one may improve the integrability of the density, as it was first observed by P. -L. Lions and proved in a more simple way by E. Feireisl–A. Novotny–H. Petzeltova (see [26]). More precisely choosing any smooth, positive function $\chi(t)$ with compact support, and test the momentum equation by $\chi\, g_k = \chi\, \mathcal{B}\rho_k^\theta$ where $\mathcal{B}$ is a Bogovskii linear operator (in x) such that

$$\mathrm{div} g_k = (\rho_k^\theta - \frac{1}{|\Omega|}\int_\Omega \rho_k^\theta), \qquad \|\nabla g_k\|_{L^p} \le C_p \|\rho_k^\theta - \frac{1}{|\Omega|}\int_\Omega \rho_k^\theta\|_{L^p},$$
$$\|\mathcal{B}\rho_k\|_{L^p} \le C_p\|\rho_k\|_{L^p}, \qquad \forall 1 < p < +\infty.$$

By using such test function, one may derive

$$\int_0^T \int_\Omega |\rho_k|^{\gamma+\theta} < C < +\infty \text{ where } 0 < \theta < 2\gamma/d - 1$$

with C again independent on n. Note that there is a gain of integrability for the pressure when $\gamma > d/2$. Moreover if $\gamma > 3d/(d+2)$, one then deduces that the density is square integrable namely $\rho_k \in L^2(0, T; L^2(\Omega))$, which is the case considered by P. -L. Lions.

The compactness in time of the density is derived from the uniform estimate on $\partial_t \rho_k$ given by the continuity equation using the energy estimates. The weak limit of $\rho_k u_k \otimes u_k$ is obtained in two steps: First get uniform integrability on $\rho_k u_k$ using the energy estimates and then use the uniform estimate on $\partial_t(\rho_k u_k)$ obtained from the momentum equation to deduce compactness in $L^2(0, T; H^{-1}(\Omega))$. The key is then to write

$$\int_0^T \int_\Omega \rho_k |u_k|^2 = \int_0^T \langle \rho_k u_k; u_k \rangle_{H^{-1}(\Omega), H^1(\Omega)}$$

to prove convergence in norm using that u_k is uniformly bounded in $L^2(0, T; H^1(\Omega))$ and therefore, up to a subsequence, that u_k weakly converges in this space. Using all this information, it is possible to pass to the limit, by extracting subsequences, and obtain that

$$\begin{aligned}
&\rho_k \rightharpoonup \rho \text{ in } \mathcal{C}^0([0, T]; L^\gamma_{\text{weak}}(\Omega)),\\
&p(\rho_k) \rightharpoonup \overline{p(\rho)} \text{ in } L^{(\gamma+\theta)/\gamma}((0, T) \times \Omega),\\
&\rho_k u_k \rightharpoonup \rho u \text{ in } \mathcal{C}^0([0, T|; L^{2\gamma/(\gamma+1)}(\Omega)),\\
&\rho_k u_k^i u_k^j \rightharpoonup \rho u^i u^j \text{ in } \mathcal{D}'((0, T) \times \Omega) \text{ for } i, j = 1, 2, 3,
\end{aligned}$$

where $\overline{g}$ denotes the weak limit of $\{g_k\}_{k\in\mathbb{N}}$ up to an extraction of a subsequence. In particular we have not identified yet the limit $\overline{p(\rho)}$ of the pressure but we can nevertheless write the system

$$\begin{aligned}
&\partial_t \rho + \operatorname{div}(\rho u) = 0\\
&\partial_t(\rho u) + \operatorname{div}(\rho u \otimes u) - \mu \Delta u - (\lambda + \mu)\nabla \operatorname{div} u + \nabla \overline{p(\rho)} = 0.
\end{aligned} \tag{2.24}$$

The mathematical difficulty consists in proving that (ρ, u) is a renormalized weak solution with bounded energy and the main point is showing that $\overline{p(\rho)} = p(\rho)$ *a.e.* in $(0, T) \times \Omega$.

The main difficulty is this to prove that ρ_k strongly converges to ρ in $L^1((0, T) \times \Omega)$. If the pressure is monotone after a fixed value, the tool which has been used to get the result is to prove commutation of weak convergence with an appropriate strictly convex function. Such a tool unfortunately seems to be ineffective for non-monotone pressure laws. A new method for compactness is therefore needed and has been developed recently. For the reader's convenience, let us explain the main ideas in those two settings.

Pressure Laws That Are Monotone After a Fixed Value In the work by E. Feireisl (see [27]), the pressure law is assumed to be monotone after a fixed value, namely it satisfies (2.8)–(2.9). The author is then able to use tools similar to [26] to prove compactness in space for the density.

For non-monotone pressure laws, we may typically use this approach to cover the first limit $n \to \infty$ in System (2.23), i.e., the limit $\varepsilon_n \to 0$. The presence of the diffusive terms with ε_n complicates the argument and for this reason we present the main idea here for the second limit $k \to \infty$ when one already has $\varepsilon_n = 0$.

More precisely, the method is based on the careful analysis of two defect measures showing

$$osc_{\gamma+1}((0,T)\times\Omega)[\rho_k-\rho] = \sup_{k\geq 1}\Big(\limsup_{n\to+\infty} \|T_k(\rho_k)-T_k(\rho)\|_{L^{\gamma+1}((0,T)\times\Omega)}\Big) < +\infty \tag{2.25}$$

and

$$dft[\rho_k - \rho](t) = \int_\Omega (\overline{\rho\log\rho}(t) - \rho\log\rho(t))\,dx = 0$$

while assuming that this last measure vanishes initially, and where the T_k are cut-off functions defined as

$$T_k(z) = kT\Big(\frac{z}{k}\Big), \qquad k \geq 1,$$

with $T \in \mathcal{C}^1(\mathbb{R})$ and

$$T(z) = z \text{ for } z \leq 1, \qquad T(z) = 2 \text{ for } z \geq 3, \qquad T \text{ concave on } \mathbb{R}.$$

Here again $\overline{b(v)}$ denotes a weak L^1 limit of a sequence $b(v_k)$. More precisely E. Feireisl proves that there exists $C > 0$ such that

$$dft[\rho_k - \rho](t) \leq C \int_0^t dft[\rho_k - \rho](s)ds \text{ for any } t \in [0,T)$$

using the fact that the non-monotone part p_c of the pressure p is such that $p_c \in \mathcal{C}^2([0,+\infty)$ with

$$p_c \geq 0, \qquad p_c(s) \equiv 0 \text{ for all } s \geq \rho^*$$

for a certain fixed value ρ^*. Note that the first defect measure is needed to be able to cover the case $\gamma > d/2$ (following the breakthrough by E. Feireisl-A. Novotny-H. Petzeltova) by using truncation operator. The second defect measure corresponds to the one introduced in the work by P.-L. Lions (see [38]) when $p(\rho) = a\rho^\gamma$ with $\gamma > 3d/(d+2)$.

Let us explain the main idea in this case, where the pressure law is always monotone, the density is square integrable, and truncations are not needed. We can write the two equations

$$\partial_t(\rho \log \rho) + \mathrm{div}(\rho \log \rho u) + \rho \mathrm{div} u = 0,$$

and

$$\partial_t(\overline{\rho \log \rho}) + \mathrm{div}(\overline{\rho \log \rho u}) + \overline{\rho \mathrm{div} u} = 0.$$

Note that if $\rho \in L^2((0,T) \times \Omega$, then using the uniform bound on u_k in $L^2(0,T; H^1(\Omega))$, we have that $\rho\, \mathrm{div} u \in L^1((0,T) \times \Omega)$ and therefore the third quantity is well defined. If $\rho \notin L^2((0,T) \times \Omega$, which is a priori the case when $d/2 < \gamma < 3d/(d+2)$, a truncation procedure is needed and the defect measure (2.25) has been introduced in [26] for this reason.

Subtracting these two equations, and using the compactness in space of the effective flux namely the following property: For all $b \in \mathcal{C}^1([0,+\infty))$ satisfying some increasing properties at infinity

$$\lim_{k\to+\infty} \int_0^T \int_\Omega (p(\rho_k) - (2\mu+\lambda)\mathrm{div} u_k) b(\rho_k)\varphi \tag{2.26}$$
$$= \int_0^T \int_\Omega (\overline{p(\rho)} - (2\mu+\lambda)\mathrm{div} u)\overline{b(\rho)}\varphi.$$

This notably implies that

$$\partial_t(\overline{\rho \log \rho} - \rho \log \rho) + \mathrm{div}((\overline{\rho \log \rho} - \rho \log \rho)u) = \frac{1}{2\mu+\lambda}(\overline{p(\rho)}\rho - \overline{p(\rho)\rho})$$

and using the monotonicity of the pressure, one may prove that

$$dft[\rho_k - \rho](t) \le dft[\rho_k - \rho](0).$$

On the other hand, the strict-convexity of the function $s \mapsto s \log s$ with $s \ge 0$ implies that $dft[\rho_k - \rho](t) \ge 0$. If initially this quantity vanishes, it then vanishes at every later time. The commutation of the weak convergence with a strictly convex function yields compactness of $\{\rho_k\}_k$ in $L^1((0,T) \times \Omega)$.

If the pressure law is monotone after a fixed value, it is possible to modify the argument above to show that $\overline{p(\rho)}\rho - \overline{p(\rho)\rho} \le C\, dft[\rho_k - \rho](t)$ and still conclude by Gronwall lemma.

If we do not impose the pressure law to be monotone after a fixed value, it did not seem feasible to adapt the method and global existence of weak solutions of the compressible Navier–Stokes equations with constant viscosities remained a challenging open problem.

Non-monotone Pressure Laws In the recent paper [17], D. Bresch–P. -E. Jabin introduced a new mathematical tool based on a non-local quantity which encodes the compactness in a quantitative way, namely

$$\int_{\Pi^{2d}} \frac{|\rho_k(t,x)-\rho_k(t,y)|^p}{(h+|x-y|)^k}\,dx\,dy.$$

If $k > d$, this actually controls Besov regularity of ρ_k at order $k-d$ (and hence Sobolev regularity at any lower order). However in the present case, for $0<\theta<1$ and $p \in [1,+\infty)$, one needs to look at exactly the critical case $k=d$

$$\|\rho\|_{p,\theta} = \sup_{h\leq 1//2} |\log h|^{-\theta} \int_{\Pi^{2d}} \frac{|\rho(x)-\rho(y)|^p}{(h+|x-y|)^d}\,dx\,dy$$

and the corresponding spaces

$$W^p_{\log,\theta} = \{\rho \in L^p : \|\rho\|_{p,\theta} < +\infty\}.$$

This still controls the compactness as per the following proposition:

Proposition 2.3.3 *For any $s>0$, $0<\theta<1$ and any $p\in[1,+\infty)$, one has the embeddings $W^{s,p} \subset W^p_{\log,\theta} \subset L^p$ which are compact.*

Such compactness property has been developed, for instance, in the recent paper [5, 6] for continuous transport equation with velocity fields regular enough to study compactness for nonlinear continuity or to design appropriate numerical schemes. Unfortunately, on its own, such regularity cannot hold for solutions ρ of the mass equation with vector fields u such that $\operatorname{div} u \in L^2(0,T;L^2(\Omega))$.

Concerning the compressible Navier–Stokes equations, the authors introduced first a slight variant by considering the singular kernel $\mathcal{K}_{h_0}$ given by

$$\mathcal{K}_{h_0} = \frac{1}{\|K_h\|_{L^1}} \int_{h_0}^1 \frac{K_h}{h}\,dh$$

with appropriate singular kernel K_h as defined in Sect. 2.6 and introduced appropriate $\mathcal{C}^2$ convex function χ to focus on the quantity

$$\int_{\mathbb{T}^d} \mathcal{K}_{h_0}(x-y)\chi(\rho_k(t,x)-\rho_k(t,y)).$$

The main idea by D. Bresch–P. -E. Jabin to propagate quantity is to introduce an auxiliary equation on an appropriate weight $w_k(t,x)$, such that $0\leq w_k(t,x)\leq 1$ and $w_k|_{t=0}\leq 1$, to track good trajectories and to prove the following inequality:

$$\frac{1}{2}\int_{\Pi^{2d}} (w_k(t,x)+w_k(t,y))\mathcal{K}_{h_0}(x-y)\chi(\rho_k(t,x)-\rho_k(t,y))\,dx\,dy$$

$$\leq C\Big(\int_{h_0}^1 \varepsilon(h)\frac{dh}{h} + |\log h_0|^{1/2}\Big) \tag{2.27}$$

with a constant independent on k and where

$$\varepsilon(h) = \frac{1}{\|K_h\|_{L^1(\mathbb{T}^d)}} \int_{\Pi^{2d}} K_h(x-y)\chi(\rho_k^0(x) - \rho_k^0(y))\, dx\, dy \longrightarrow 0 \text{ when } h \to 0$$

as ρ_k^0 is assumed to be compact in L^p with $1 < p < +\infty$. In order to prove (2.27), the small translation property for the square function operator D_h defined in Lemma 2.6.7 is strongly used. In some sense, it helps to exchange value at point y to value at point x for two close points and to encode the cost of such change. To prove the following property on the semi-norm

$$\limsup_{h\to 0} \frac{1}{\|\mathcal{K}_{h_0}\|_{L^1}} \sup_n \int_{\Pi^{2d}} \mathcal{K}_{h_0}(x-y)\chi(\rho_k(t,x) - \rho_k(t,y))\, dx\, dy = 0$$

to get compactness in space of the density using Lemma 2.6.3 given in Sect. 2.6, it remains to derive appropriate properties for the weights to get rid of them without losing too much with respect to h and to be sure that the weights do not vanish too much. The following property will play a crucial role:

$$\int_{\Pi^d} \rho_k(t,x)|\log w_k(t,x)|^\theta \le C < +\infty$$

for some $\theta \in (0,1)$ with C independent of k helping to conclude the following control:

$$\frac{1}{|\log h_0|} \int_{\mathbb{T}^{2d}} \mathcal{K}_{h_0}(x-y)\chi(\rho_k(t,x) - \rho_k(t,y)) \le \frac{c}{|\log(|\log h_0|^{-1/2} + \overline{\varepsilon}(h_0))|^\theta}$$

with $0 < \theta < 1$. Note the important property that if the weight vanishes, the density vanishes almost surely.

To be more precise the weight function is defined as the solution of the following PDE:

$$\partial_t w_k + u_n \cdot \nabla w_k + \lambda\, D_k w_k = 0, \qquad w_k|_{t=0} = \exp(-\lambda \sup \rho_k^0),$$

where $\lambda > 0$ is a constant chosen large enough and $D_k \ge 0$ is the damping quantity linked to the unknowns defined as follows:

$$D_k = \rho_k|\mathrm{div} u_k| + \rho_k^{\widetilde{\gamma}} + M|\nabla u_k| + |\mathrm{div} u_k|,$$

where

$$Mf(x) = \sup_r \frac{1}{|B(x,r)|} \int_{B(x,r)} f(y)\, dy$$

is the maximal function as defined in Sect. 2.6. Note that the weight follows the trajectories and vanishes if the density (equivalently the divergence of the velocity) blows up. The reason for the first term in D_k is to ensure that

$$w_n \le \exp(-\lambda\rho_k)$$

which helps to compensate the penalization in $\rho_k^{\widetilde{\gamma}}$ and to get the property on $\rho_k\,|\log w_n|^\theta$ for some $\theta > 0$. The three last terms are needed to, respectively, counterbalance additional divergence terms in the propagation quantity.

Remark 2.3.4 It is important to note that the flexibility of the method introduced by D. Bresch and P.-E. Jabin lies in the fact that everything is encoded in the choice of appropriate weights to ensure the propagation of the non-local weighted quantity. It always remains of course to derive appropriate properties on such weight to finally remove them and conserve the convergence with respect to the parameter h to get the compactness in space through Lemma 2.6.3 in Sect. 2.6.

The readers who want to see a simple example with the full details is referred to the compressible Stokes system considered, for instance, in [18].

2.3.3 Some Physical Situations: Anelastic System, Mixture System, etc.

In this subsection, we present different mathematical results that have been obtained adapting the method introduced by D. Bresch–P.E. Jabin in [17]. We hope by these examples to show the flexibility of the method. The main difficulty remains to define appropriate unknowns and to introduce appropriate weights.

I. Anelastic Euler Equations In [18] D. Bresch and P.E. Jabin have proposed quantitative estimates and obtained existence of renormalized solution for advective equations with an anelastic constraint in presence of vacuum. More precisely, they consider the following advective equation:

$$a(\partial_t\phi + u\cdot\nabla\phi) = 0 \text{ in } (0,T)\times\Omega \tag{2.28}$$

with a velocity field u such that

$$\operatorname{div}(au) = 0 \text{ in } (0,T)\times\Omega, \qquad au\cdot n|_{(0,T)\times\partial\Omega} = 0, \tag{2.29}$$

and where a is a given non-negative scalar function which depends only on the space variable and is continuous on $\overline{\Omega}$. The initial condition is given by

$$a\phi|_{t=0} = m_0 \text{ in } \Omega. \tag{2.30}$$

Such systems may be obtained through low Mach or low Froude numbers limits in the presence of heterogeneity in the pressure laws. To avoid assuming any regularity on a, the authors impose additional conditions on a namely there exists a measurable non-negative function $\alpha(x)$, $r > 1$, and $q > p^*$ (with as usual $1/p^* + 1/p = 1$) such that

$$\alpha(x) \leq a(x), \tag{2.31}$$
$$A(\alpha, a) = \int_\Omega \big(|\nabla \alpha^{1/p^*}(x)|^q + a(x)(|\log \alpha(x)| + |\nabla \log \alpha(x)|^r)\big) < +\infty.$$

Of course, if $a \in W^{1,p}$ with $p > 1$ and $a|\log a| \in L^1$ then we could just choose $\alpha = a^k$ with $k \geq 1$. Let us now consider a velocity field u such that (with a slight abuse of notation as $\|u\|_a$ is not a norm)

$$\|u\|_a := \|u\|_{L^\infty_t L^p_a} + \int_0^T \int_\Omega a(x)|\nabla u(t,x)| \log(e + |\nabla u(t,x)|)\, dxdt < +\infty \tag{2.32}$$

with p fixed and where the Lebesgue space $L^q_t L^p_a$ and more generally the Sobolev space $L^q_t W^{1,p}_a$ are defined by the norms

$$\|f\|_{L^q_t L^p_a} := \Big\| \Big(\int_\Omega |f(t,x)|^p a(x) dx\Big)^{1/p} \Big\|_{L^q([0,T])} < +\infty$$

$$\|f\|_{L^q_t W^{1,p}_a} := \Big\| \Big(\int_\Omega (|f(t,x)|^p + |\nabla f|^p) a(x) dx\Big)^{1/p} \Big\|_{L^q([0,T])} < +\infty.$$

The main result reads

Theorem 2.3.5

1. For any $\mathcal{C}^1$ sequences a_ε, α_ε, u_ε and a sequence of Lipschitz open domain Ω_ε with

- *a_ε is bounded from below ($\inf_{\Omega_\varepsilon} a_\varepsilon > 0$) and we have the divergence condition*

$$\mathrm{div}(a_\varepsilon u_\varepsilon) = 0$$

- *$a_\varepsilon, \alpha_\varepsilon, u_\varepsilon$ satisfy* (2.29) *and* (2.31)–(2.32) *uniformly with respect to ε:*

$$\sup_\varepsilon A(\alpha_\varepsilon, a_\varepsilon) + \sup \|u_\varepsilon\|_{a_\varepsilon} < +\infty$$

- *Ω_ε converges to Ω for the Hausdorff distance of sets and $\|a_\varepsilon - a\|_{L^1(\Omega_\varepsilon \cap \Omega)} \to 0$ as $\varepsilon \to 0$*

and for any sequence of initial data ϕ_ε^0 *uniformly bounded in* $L^\infty(\mathbb{R}^d)$ *and compact in* $L^1(\mathbb{R}^d)$, *consider the unique Lipschitz solution* ϕ_ε *to*

$$a_\varepsilon(\partial_t \phi_\varepsilon + u_\varepsilon \cdot \phi_\varepsilon) = 0 \text{ in } \Omega_\varepsilon$$

with boundary condition

$$a_\varepsilon u_\varepsilon \cdot n = 0 \text{ on } \partial\Omega_\varepsilon.$$

Then ϕ_ε *is compact in* $L_t^\infty L_{a_\varepsilon}^2$ *and converges to a renormalized solution to* (2.28) *with* (2.30).

2. *Let finally* ϕ_0 *in* L^∞ *and* (a, α, u) *satisfy* (2.29) *and the bounds* (2.31)–(2.32). *Then there exists a renormalized solution* ϕ *to* (2.28) *with initial data* (2.30).

II. Mixture System In [22], D. Bresch–P. Mucha–E. Zatorska proved existence of global in time weak solutions to a compressible two-fluid Stokes system with a single velocity field and algebraic closure for the pressure law. The constitutive relation involves densities of both fluids through an implicit function. Adapting the compactness method explained above in the mono-fluid compressible Navier–Stokes setting, they first prove the weak sequential stability of solutions. Next, they construct weak solutions via unconventional approximation using the Lagrangian formulation, truncations, and stability result of trajectories for rough velocity fields in the spirit of Crippa–De Lellis. This result is the first showing that the method introduced by D. Bresch–P.E. Jabin in [17] may be developed for compressible Navier–Stokes equations with pressure law depending on two phases. The system reads

$$\partial_t(\alpha^\pm \rho^\pm) + \text{div}(\alpha^\pm \rho^\pm u) = 0, \tag{2.33}$$

$$-\mu\Delta u - (\lambda + \mu)\nabla \text{div} u + \nabla p = 0, \tag{2.34}$$

$$\alpha^+ + \alpha^- = 1, \tag{2.35}$$

$$p = p^+ = p^-, \tag{2.36}$$

with constant shear and bulk viscosities μ and λ such that $\lambda + 2\mu > 0$ and $\mu > 0$. The unknowns of the system are the volumetric rates α^+, α^- of presence of fluid $+$ and $-$, respectively, with

$$0 \leq \alpha^\pm \leq 1,$$

the two mass densities ρ^+, ρ^-, and the common velocity field u. By p^+, p^-, we denote the internal barotropic pressures for each fluid with the explicit form:

$$p^+ = a^+(\rho^+)^{\gamma_+}, \qquad p^- = a^-(\rho^-)^{\gamma_-},$$

where $a^\pm > 0$, $\gamma^\pm > 1$. The global existence is obtained on the d-dimensional torus, under the constraint

$$\int_{\mathbb{T}^d} u\,dx = 0$$

and the initial conditions

$$\alpha^+\rho^+|_{t=0} = R_0, \qquad \alpha^-\rho^-|_{t=0} = Q_0, \qquad R_0 \geq 0, \qquad Q_0 \geq 0.$$

Moreover, the following compatibility conditions are required on the initial data:

$$\alpha^+|_{t=0} = \alpha_0^+, \qquad \alpha_-|_{t=0} = \alpha_0^-$$

$$\alpha_0^+ + \alpha_0^- = 1, \qquad \alpha_0^\pm \geq 0, \qquad p^+(\rho_0^+) = p^-(\rho_0^-).$$

The main result reads

Theorem 2.3.6 *Let $\gamma^\pm > 1$ and $a^\pm > 0$ with $\lambda + 2\mu > 0$ and $\mu > 0$ and let the initial data satisfy*

$$\int_{\mathbb{T}^d} (R_0^{\gamma^+} + Q_0^{\gamma^-}) < +\infty, \qquad 0 < \int_{\mathbb{T}^d} R_0\,dx < +\infty, \qquad 0 < \int_{\mathbb{T}^d} Q_0\,dx < +\infty$$

with the compatibility condition

$$\Big(\frac{a^-}{a^+}\Big)^{1/\gamma^-} Q_0 = \Big(1 - \frac{R_0}{Z_0}\Big) Z_0^\gamma \text{ with } R_0 \leq Z_0.$$

Then there exists a global weak solution $(\alpha^\pm, \rho^\pm, u)$ of the system satisfying

$$\alpha^\pm\rho^\pm \in L^\infty(0,T; L^{\gamma^\pm}(\mathbb{T}^d)) \cap L^{2\gamma^\pm}((0,T)\times\mathbb{T}^d) \cap \mathcal{C}([0,T]; L^{\gamma^\pm}(\mathbb{T}^d))$$
$$\rho^\pm \in L^\infty(0,T; L^{\gamma^\pm}(\mathbb{T}^d)) \cap L^{2\gamma^\pm}((0,T)\times\mathbb{T}^d) \cap \mathcal{C}([0,T]; L^{\gamma^\pm}(\mathbb{T}^d)),$$
$$u \in L^2(0,T; H^1(\mathbb{T}^d)),$$

where the mass equations and the momentum equation are satisfied in $\mathcal{D}'((0,T)\times\mathbb{T}^d)$ and the initial condition in $\mathcal{D}'(\mathbb{T}^d)$.

Note that if $a^+ = a^- = 1$ and $\gamma^+ = \gamma^-$, the system is reduced to the stationary Stokes version.

Remark 2.3.7 It is interesting to note that P.-L. Lions and E. Feireisl et al. theory has been extended recently by A. Vasseur–H. Wen–C. Yu (see [53]) to pressure laws depending on two phase (see also D. Maltese & al [39], A. Novotny [42] and A. Novotny–M. Pokorny [43].

2.4 Pressure Dependent Viscosities

Assuming the shear and bulk viscosities satisfying the BD relation, barotropic compressible Navier–Stokes equations with a large class of density dependent shear viscosity (and therefore pressure dependent) have been considered by D. Bresch–A. Vasseur–C. Yu in [23]. In this last work, the authors obtain the global existence of weak solutions in the following sense:

Definition 2.4.1 We say that (ρ, u) is a global weak solution to the compressible Navier–Stokes equations if

- The density satisfies $\rho \in \mathcal{C}^0([0,T]; L^\gamma_{\rm weak}(\Omega))$ with $\rho \geq 0$ in $(0,T)\times\Omega$, $\rho|_{t=0} = \rho_0$ a.e. in Ω with the viscosity $\mu(\rho) \in \mathcal{C}^0([0,T]; L^{3/2}_{\rm weak}(\Omega))$.
- The momentum satisfies $\rho\, u \in \mathcal{C}([0,T]; L^{2\gamma/(\gamma+1)}_{\rm weak}(\Omega))$ with $\rho u|_{t=0} = m_0$.
- The following estimates hold:

$$\|\sqrt{\rho} u\|_{L^\infty(0,T;L^2(\Omega))} < +\infty, \qquad \|\rho\|^\gamma_{L^\infty(0,T;L^\gamma(\Omega))} < +\infty,$$

$$\|\mathbb{T}_\mu\|_{L^2(0,T;L^2(\Omega))} < +\infty, \qquad \kappa(1-\kappa)\|\sqrt{\rho}\nabla s(\rho)\|_{L^\infty(0,T;L^2(\Omega))} < +\infty,$$

$$\kappa\|\sqrt{\mu'(\rho)\rho^{\gamma-2}}\nabla\rho\|_{L^2(0,T;L^2(\Omega))}$$

$$\|\mu(\rho)\|_{L^\infty(0,T;W^{1,1}(\Omega))} + \|\mu(\rho)u\|_{L^\infty(0,T;L^{3/2}(\Omega))\cap L^2(0,T;W^{1,1}(\Omega))} < +\infty$$

$$\|\partial_t\mu(\rho)\|_{L^\infty(0,T;W^{-1,1}(\Omega))} < +\infty,$$

  where $\kappa \in (0,1)$ is a fixed parameter and $s'(\rho) = 2\mu'(\rho)/\rho$.
- The density ρ satisfies the mass equation in $\mathcal{D}'((0,T)\times\Omega)$:

$$\partial_t\rho + {\rm div}(\rho u) = 0.$$

- The velocity satisfies the momentum in $\mathcal{D}'((0,T)\times\Omega)$:

$$\partial_t(\rho u) + {\rm div}(\rho u\otimes u) + \nabla p(\rho) - 2{\rm div}(\sqrt{\mu(\rho)}\mathbb{S}_\mu + \frac{\lambda(\rho)}{2\mu(\rho)}{\rm Tr}(\sqrt{\mu(\rho)}\mathbb{S}_\mu){\rm Id}) = 0,$$

  where Tr denotes the trace operator of the matrix.
- The viscosity satisfies in $\mathcal{D}'((0,T)\times\Omega)$:

$$\partial_t\mu(\rho) + {\rm div}(\mu(\rho)u) + \frac{\lambda(\rho)}{2\mu(\rho)}{\rm Tr}(\sqrt{\mu(\rho)}\mathbb{T}_\mu) = 0, \tag{2.37}$$

  with

$$\mathbb{S}_\mu = \frac{(\mathbb{T}_\mu + \mathbb{T}^t_\mu)}{2},$$

where $\mathbb{T}_\mu$ is defined through

$$\sqrt{\mu(\rho)}\mathbb{T}_\mu = \nabla(\sqrt{\rho}u\frac{\mu(\rho)}{\sqrt{\rho}}) - \sqrt{\rho}u \otimes \sqrt{\rho}\nabla s(\rho) \tag{2.38}$$

$$\frac{\lambda(\rho)}{2\mu(\rho)}\mathrm{Tr}(\sqrt{\mu(\rho)}\mathbb{T}_\mu) \tag{2.39}$$
$$= \Big[\mathrm{div}(\frac{\lambda(\rho)}{\mu(\rho)}\sqrt{\rho}u\frac{\mu(\rho)}{\sqrt{\rho}} - \sqrt{\rho}u \cdot \sqrt{\rho}\nabla s(\rho)\frac{\rho\mu''(\rho)}{\mu'(\rho)}\Big]\mathrm{Id}.$$

More precisely, they prove the following result:

Theorem 2.4.2 *Let $\mu(\rho)$ verify* (2.16)–(2.17) *and $\lambda(\rho)$ given by* (2.6). *Let us assume that the initial data satisfy*

$$\int_\Omega \Big(\frac{1}{2}\rho_0|u_0 + 2\kappa\nabla s(\rho_0)|^2 + \kappa(1-\kappa)\rho_0\frac{|2\nabla s(\rho_0)|^2}{2}\Big)\,dx$$
$$+ \int_\Omega \Big(a\frac{\rho_0^\gamma}{\gamma-1} + \mu(\rho_0)\Big)\,dx \le C < +\infty$$

with $\kappa \in (0,1)$ given and $s'(\rho_0) = \mu'(\rho_0)/\rho_0$. Let T be given such that $0 < T < +\infty$, then, for $p(\rho) = a\rho^\gamma$ with $\gamma > 1$, there exist a global weak solution of the compressible Navier–Stokes equations in the sense of Definition 2.4.1 written above.

Remark 2.4.3 It is important to remark that if we assume μ and λ to satisfy the BD relation with the following constraint that there exists $\varepsilon > 0$ such that

$$\mu(s) \ge 0, \qquad \lambda(s) + 2\mu(s)/3 > \varepsilon\mu(s) \text{ for all } s > 0,$$

then necessarily

$$\mu(0) = \lambda(0) = 0.$$

This implies that the compressible Navier–Stokes equations is a degenerate Hyperbolic-parabolic system close to vacuum. Thus even if we can define u as usually namely

$$u = \frac{\rho u}{\rho} \text{ when } \rho \ne 0 \text{ and } u = 0 \text{ elsewhere},$$

we still cannot talk about ∇u. Thus $\mu(\rho)\mathbb{D}(u)$ and $\lambda(\rho)$divu have to be understood in the weak form written above introducing the tensor $\mathbb{T}_\mu$ which will be in $L^2((0,T)\times\Omega)$.

Remark 2.4.4 Remark that if we take the gradient of equation (2.37) and use the compatibility conditions (2.38)–(2.39), we get the following equation :

$$\partial_t(\rho(2\nabla s(\rho)))+\mathrm{div}(\rho(2\nabla s(\rho)\otimes u)+2\mathrm{div}(\sqrt{\mu(\rho)}{}^t\mathbb{T}_\mu)+\nabla(\frac{\lambda(\rho)}{\mu(\rho)}\mathrm{Tr}(\sqrt{\mu(\rho)}\mathbb{T}_\mu)=0,$$

where $s'(\rho) = \mu'(\rho)/\rho$. Adding this equation to the momentum equation, we obtain the following: on $v = u + 2\nabla s(\rho)$

$$\partial_t(\rho v) + \mathrm{div}(\rho v \otimes u) + \nabla p(\rho) - 2\mathrm{div}(\sqrt{\mu(\rho)}\frac{(\mathbb{T}_\mu - {}^t\mathbb{T}_\mu)}{2}) = 0.$$

This corresponds to equation (2.14) for global weak solutions.

2.4.1 A Two-Velocity Hydrodynamic: Approximate System

In a series of papers over 9 years (2004–2012), H. Brenner (1929–2013) [who was emeritus professor at MIT in chemical engineering] proposed a new theory in compressible fluid mechanics with high gradient of density based on the concept of two different velocities: the mass and the volume velocities. More precisely, the model proposed by H. Brenner written in the barotropic form (i.e., with no temperature) reads

$$\begin{aligned}&\partial_t\rho + \mathrm{div}(\rho u_m) = 0,\\ &\partial_t(\rho u) + \mathrm{div}(\rho u \otimes u_m)\\ &\qquad\qquad -2\,\mathrm{div}(\mu(\rho)\mathbb{D}(u)) - \nabla(\lambda(\rho)\mathrm{div}u) + \nabla p(\rho) = 0,\end{aligned}$$

with

$$u - u_m = K\nabla\log\rho.$$

In other words, a two-velocity hydrodynamic has to be taken into account if we want to consider large gradients of density and this is the case concerning weak regularity solutions. Such system has been studied mathematically in the temperature conducting framework by E. Feireisl and A. Vasseur in [28].

On the other hand, the difficulty to construct approximate solution for the compressible Navier–Stokes system with density dependent viscosities is to ensure that the energy and BD entropy are satisfied uniformly on such approximate system. In [16], D. Bresch–B. Desjardins–E. Zatorska have remarked that it is possible to introduce a parameter $\kappa \in (0, 1)$ in order to define a κ-entropy which is a kind of interpolation between energy and BD entropy: This observation helps to relax the constraints to construct approximate solutions. More precisely, it suffices to remark that

$$(1-\kappa)\rho|u|^2 + \kappa\rho|u + 2\nabla s(\rho)|^2 = \rho|u + 2\kappa\nabla s(\rho)|^2 + (1-\kappa)\kappa\rho|2\nabla s(\rho)|^2.$$

Similar property may be found in real two-fluid hydrodynamic and the interested readers are referred to S. Gavrilyuk–S.M. Shugrin [29] and S.M. Shugrin [48]. This important observation allows to consider an augmented approximate system related to the unknown $(\rho, u + 2\kappa\nabla s(\rho), 2\nabla s(\rho))$ which has a structure similar to a two-velocity hydrodynamic formulation with cross diffusion, namely

$$\partial_t \rho + \mathrm{div}(\rho w) = 2\kappa \Delta\mu(\rho),$$

with

$$\begin{aligned} &\partial_t(\rho w) + \mathrm{div}(\rho w \otimes u) \\ &-2(1-\kappa)\mathrm{div}(\mu(\rho)\mathbb{D}(w)) - 2\kappa \mathrm{div}(\mu(\rho)\mathbb{A}(w)) \\ &-(1-\kappa)\nabla(\lambda(\rho)\mathrm{div}(w-\kappa v)) + \nabla\rho^\gamma + 2(1-\kappa)\kappa\, \mathrm{div}(\mu(\rho)\nabla v) = 0, \end{aligned} \tag{2.40}$$

and

$$\begin{aligned} &\partial_t(\rho v) + \mathrm{div}(\rho v \otimes u) - 2\kappa\, \mathrm{div}(\mu(\rho)\nabla v) \\ &\qquad +2\mathrm{div}(\mu(\rho\nabla^t w) + \nabla(\lambda(\rho)\mathrm{div}(w-\kappa v)) = 0, \end{aligned} \tag{2.41}$$

where

$$v = 2\nabla s(\rho), \qquad w = u + \kappa v.$$

The κ-entropy (energy associated to this system) reads

$$\begin{aligned} &\int_\Omega \Big(\rho\Big(\frac{|w|^2}{2}+(1-\kappa)\kappa\frac{|v|^2}{2}\Big)+\frac{\rho^\gamma}{\gamma-1}\Big)\,dx+2(1-\kappa)\int_0^t\int_\Omega \mu(\rho)|\mathbb{D}w-\kappa\nabla v|^2\,dx\,dt \\ &+(1-\kappa)\int_0^t\int_\Omega \lambda(\rho)(\mathrm{div}w-\kappa\mathrm{div}v)^2\,dx\,dt++2\kappa\int_0^t\int_\Omega \frac{\mu'(\rho)p'(\rho)}{\rho}|\nabla\rho|^2\,dx\,dt \\ &+2\kappa\int_0^t\int_\Omega \mu(\rho)|\mathbb{A}w|^2\,dx\,dt \\ &\le \int_\Omega\Big(\rho_0\Big(\frac{|w_0|^2}{2}+(1-\kappa)\kappa\frac{|v_0|^2}{2}\Big)+\frac{\rho_0^\gamma}{\gamma-1}\Big)\,dx, \end{aligned}$$

where $s' = \mu'(\rho)/\rho$ and $p(\rho) = \rho^\gamma$.

Remark 2.4.5 (Open Problems) The fact that the parameter $\kappa \in (0, 1)$ is actually assumed to be constant leads to some fascinating open questions.

- It could be interesting to couple κ to other physical quantities when performing singular perturbations for compressible Navier–Stokes equations with density dependent viscosities.

- Can one look for κ as an unknown solution of a PDE depending on (ρ, u) and their gradient? This could perhaps help by activating the new unknown $\nabla\mu(\rho)/\sqrt{\rho}$ only where it is necessary.

Remark 2.4.6 It is important to remark that a global weak solution of the compressible Navier–Stokes system in the sense of Definition 2.4.1 satisfies the two-velocity hydrodynamic formulation in a weak sense namely using the tensor $\mathbb{T}_\mu$ and the compatibility conditions.

Remark 2.4.7 Augmented system (two-velocity hydrodynamic formulation) for Euler–Korteweg and Navier–Stokes–Korteweg has been developed in the recent paper [21] to prove weak–strong uniqueness, inviscid limit, or other singular perturbations.

Approximate System The augmented system written above is an over-determined system to be solved in terms of $(\rho, \rho w, \rho v)$. It is necessary to have enough regularity on w to be able to check that $v = 2\nabla s(\rho)$ assuming only $v_0 = 2\nabla s(\rho_0)$. To have smooth enough solutions in the construction, the authors consider the following augmented approximate system:

$$\partial_t \rho + \mathrm{div}(\rho w) = 2\kappa \mathrm{div}(\mu'(\rho)\nabla\rho),$$

with

$$\begin{aligned}
&\partial_t(\rho w) + \mathrm{div}(\rho w \otimes u) + \varepsilon_2[\Delta^{2s} w - \mathrm{div}((1+|\nabla w|^2)\nabla w)] \\
&-2(1-\kappa)\mathrm{div}(\mu(\rho)\mathbb{D}(w)) - 2\kappa \mathrm{div}(\mu(\rho)\mathbb{A}(w)) \qquad (2.42)\\
&-(1-\kappa)\nabla(\lambda(\rho)\mathrm{div}(w-\kappa v)) + \nabla\rho^\gamma + \delta\nabla\rho^{10} + 4(1-\kappa)\kappa \mathrm{div}(\mu(\rho)\nabla^2 s(\rho)) \\
&= -r_0(w - 2\kappa\nabla s(\rho)) - r_1\rho|w - 2\kappa\nabla s(\rho)|(w - 2\kappa\nabla s(\rho)) \\
&-r_2\frac{\rho}{\mu'(\rho)}|w - 2\kappa\nabla s(\rho)|^2(w - 2\kappa\nabla s(\rho)) + r\rho\nabla(\sqrt{k(\rho)}\Delta(\int_0^\rho \sqrt{k(s)}\, ds)),
\end{aligned}$$

and

$$\begin{aligned}
&\partial_t(\rho v) + \mathrm{div}(\rho v \otimes u) - 2\kappa \mathrm{div}(\mu(\rho)\nabla v) \\
&\qquad + 2\mathrm{div}(\mu(\rho)\nabla^t w) + \nabla(\lambda(\rho)\mathrm{div}(w - \kappa v)) = 0, \qquad (2.43)
\end{aligned}$$

where

$$v = 2\nabla s(\rho), \qquad w = u + \kappa v,$$

and

$$k(\rho) = 4(\mu'(\rho))^2/\rho,$$

and finally for $s \geq 2$ and $\inf \mu'(s) = \varepsilon_1 > 0$. Assuming r_1 small enough compared to δ and r_2 small enough compared to r, and denoting

$$E[\rho, u + 2\kappa \nabla s(\rho), \nabla s(\rho)]$$
$$= \int_\Omega \rho \left(\frac{|u + 2\kappa \nabla s(\rho)|^2}{2} + (1-\kappa)\kappa \frac{|\nabla s(\rho)|^2}{2} \right) + \frac{\rho^\gamma}{\gamma - 1} + \frac{\delta \rho^{10}}{9} + G(\rho),$$

the following lengthy κ-entropy inequality holds for the approximate system:

$$E[\rho, u + 2\kappa \nabla \mathbf{s}(\rho), \nabla \mathbf{s}(\rho)](t) + \frac{r}{2} \int_\Omega |\nabla \int_0^\rho \sqrt{K(s)}\, ds|^2 \, dx$$
$$+ \varepsilon_2 \int_0^t \int_\Omega \left(|\delta^s w|^2 + (1 + |\nabla w|^2)|\nabla w|^2 \right) dx\, dt$$
$$+ 2(1-\kappa) \int_0^t \int_\Omega \mu(\rho) |\mathbb{D} u|^2 \, dx\, dt + 2\kappa \int_0^t \int_\Omega \mu(\rho) |\mathbb{A}(u + 2\kappa \nabla s(\rho))|^2 \, dx\, dt$$
$$+ 2(1-\kappa) \int_0^t \int_\Omega (\mu'(\rho)\rho - \mu(\rho))(\mathrm{div} u)^2 \, dx\, dt$$
$$+ 2\kappa \int_0^t \int_\Omega \frac{\mu'(\rho) p'(\rho)}{\rho} |\nabla \rho|^2 \, dx\, dt + 20\kappa \int_0^t \int_\Omega \mu'(\rho) \rho^8 |\nabla \rho|^2 \, dx\, dt$$
$$+ r_0 \int_0^t \int_\Omega |u|^2 \, dx\, dt + r_1 \int_0^t \int_\Omega \rho |u|^3 \, dx\, dt + \frac{r_2}{4} \int_0^t \int_\Omega \frac{\rho}{\mu'(\rho)} |u|^4 \, dx\, dt$$
$$+ \kappa r \int_0^t \int_\Omega \mu(\rho) |2\nabla^2 s(\rho)|^2 \, dx\, dt + \frac{1}{2} \kappa r \int_0^t \int_\Omega \lambda(\rho) |2\Delta s(\rho)|^2 \, dx\, dt$$
$$\leq \int_\Omega \left(\rho_0 \left(\frac{|w_0|^2}{2} + (1-\kappa)\kappa \frac{|v_0|^2}{2} \right) + \frac{\rho_0^\gamma}{\gamma - 1} + \frac{\delta \rho_0^{10}}{9} + \frac{r}{2} |\nabla \int_0^{\rho_0} \sqrt{K(s)} ds|^2 + G(\rho_0) \right) dx$$
$$+ C \frac{r_1}{\delta} \int_\Omega E[\rho, u + 2\kappa \nabla \mathbf{s}(\rho), \nabla \mathbf{s}(\rho)] dx\, dt,$$

where

$$G(\rho) = r_0 \int_1^\rho \int_1^r \frac{\mu'(\xi)}{\xi^2} d\xi\, dr,$$

with for any $0 \leq \rho \leq \varepsilon$

$$4\varepsilon (\ln \rho)_+ \geq G(\rho) \geq -\frac{\varepsilon_1}{4} (\ln \rho)_-.$$

Using Grönwall Lemma, uniform estimates can then be obtained. The regularized quantity for w is added to ensure to solve the augmented system with $v = 2\nabla s(\rho)$.

Proving global existence of weak solutions for such system follows the same lines as in [16] by using a standard Faedo-Galerkin method.

2.4.2 Sketch of the Proof of the Global Existence Result

The proof in [23] yields global existence of weak solutions for the compressible Navier–Stokes equations with density dependent viscosities satisfying the BD relation and it is performed around four main ideas:

- Introduce the equation satisfied by $\mu(\rho)$ in the definition of global weak solutions assuming $\mu(\rho_0)$ in $L^1(\Omega)$ to relax constraints on γ.
- Extend the two-velocities framework by Bresch–Desjardins–Zatorska (see [16]) with capillarity and more general drag terms using a generalization of the quantum Böhm identity developed by Bresch–Couderc–Noble–Vila (see [7]).
- Prove a generalized high-order log-Sobolev dissipation inequality similar to the one used by Jüngel (see [33]) for Navier–Stokes–Quantum system and established by Jüngel–Matthes (see [34]).
- Develop a renormalized framework for viscosities satisfying the BD relation with $\inf \mu'(s) \geq \varepsilon_1$ extending what has been introduced in Lacroix-Violet and Vasseur [35] in the case $\mu(\rho) = \mu\rho$ and $\lambda(\rho) = 0$. This uses the generalized high-order log-Sobolev dissipation inequality. Then one approximates the shear viscosities by a sequence of shear viscosity μ_n such that $\varepsilon_{1n} = \inf \mu_n'(s)$ in the renormalized system. The conclusion holds choosing appropriate functions φ_n in the renormalized formulation and letting n tend to zero through Lebesgue's dominated convergence.

Remark 2.4.8 It is interesting to compare the framework introduced by P.-L. Lions (see [38]) and E. Feireisl–A. Novotny–H. Petzeltova (see [26]) to the one introduced by D. Bresch and B. Desjardins (see [9]).

Constant Viscosities If the shear and bulk viscosities μ and λ are assumed to be constants, the compressible Navier–Stokes system is an hyperbolic-parabolic system. As mentioned compactness in space for the velocity is obtained from the diffusion quantity in energy estimate and compactness in time for the density from the mass equation. To ensure stability in the nonlinear terms

$$\rho u, \qquad \rho u \otimes u, \qquad p(\rho),$$

the authors prove compactness on $\sqrt{\rho}u$ in $L^2(0,T;L^2(\Omega))$ using again the duality product

$$\int_0^T \int_\Omega \rho|u|^2 = \langle \rho u, u\rangle_{L^2(0,T;H^{-1}(\Omega))\times L^2(0,T;H^1(\Omega))}$$

and the strong convergence of $\rho_n u_n$ in $L^2(0, T; H^{-1}(\Omega))$. As explained before, this last convergence uses that $\rho_n u_n$ is bounded in some Lebesgue space and $\partial_t(\rho u)$ is controlled through the momentum equation using the energy estimate informations.

The authors prove compactness in space for the density using the renormalized equation on the density namely the fact that

$$\partial_t \beta(\rho) + \operatorname{div}(\beta(\rho) u \otimes u) + (\beta'(\rho)\rho - \beta(\rho))\operatorname{div} u = 0,$$

choosing $s \mapsto s \log s$ as nonlinear function β. The compactness in space of the effective flux $(\lambda + 2\mu)\operatorname{div} u + p(\rho)$ which is a local quantity is the main property to conclude.

Density Dependent Viscosities If the viscosities are assumed to depend on the density and satisfy the BD relation, then we lose control on the gradient of the velocity field close to the vacuum even if we control the gradient of a function of the density. The pressure term is however easily handled since we have regularity in space of a function of the density and regularity in time of the density using the mass equation. But it is difficult to have control on the velocity. In some sense, this is a kind of duality compared with the constant viscosities case.

The first important lemma to obtain is the following:

Lemma 2.4.9 *Assume that the sequence $\{\rho_n\}_{n\in\mathbb{N}}$ satisfies all estimates uniformly with respect to n. Then, there exists a function $\rho \in L^\infty(0, T; L^\gamma(\Omega))$ such that, up to a subsequence*

$$\mu(\rho_n) \to \mu(\rho) \text{ in } \mathcal{C}([0, T]; L^{3/2}_{\text{weak}}(\Omega)),$$

and

$$\rho_n \to \rho \text{ a.e. in } (0, T) \times \Omega.$$

Moreover

$$\rho_n \to \rho \text{ in } L^{(4\gamma/3)^+}((0, T) \times \Omega),$$

and

$$\nabla p(\rho_n) \rightharpoonup \nabla p(\rho) \text{ in } L^{1+}((0, T) \times \Omega).$$

Another important Lemma, which is a generalization of the one considered in [34, 52] and [35], is the following:

Lemma 2.4.10 *Let $\mu'(\rho) < k\mu(\rho)$ for $2/3 < k < 4$ and*

$$s(\rho) = \int_0^\rho \frac{\mu'(s)}{s}\, ds, \quad Z(\rho) = \int_0^\rho \frac{\sqrt{\mu(s)}}{s}\mu'(s)\, ds, \quad Z_1(\rho) = \int_0^\rho \frac{\mu'(s)}{(\mu(s))^{1/4} s^{1/2}}\, ds.$$

Assume that $\rho > 0$ and $\rho \in L^2(0,T;H^2(\Omega))$ then there exists $\varepsilon(k) > 0$ such that we have the following estimate:

$$\int_0^T \int_\Omega |\nabla^2 Z(\rho)|^2 \,dx\,dt + \varepsilon(k) \int_0^T \int_\Omega |\nabla Z_1(\rho)|^4 \,dx\,dt \leq \frac{C}{\varepsilon(k)} \int_0^T \int_\Omega \mu(\rho) |\nabla^2 s(\rho)|^2 \,dx\,dt,$$

where C is a universal positive constant.

It is a kind of log-Sobolev high-order estimate and the upper constraint on the coefficients α_1, α_2 is needed to conclude. For $\mu(\rho) = \mu\rho$, we take advantage of the inequality developed in [34] and strongly used in [35, 52]. As in [35] (inspired from [52]), we introduce the quantity

$$\int_\Omega \rho\, \varphi_n(\phi_m(\rho)u)\,dx,$$

where ϕ_m is a non-negative cut-off functions defined for any fixed positive m as follows:

$$\phi_m(y) \begin{cases} = 0, & \text{if } 0 \leq y \leq \frac{1}{2m}, \\ = 2my - 1, & \text{if } \frac{1}{2m} \leq y \leq \frac{1}{m}, \\ = 1, & \text{if } \frac{1}{m} \leq y \leq m, \\ = 2 - \frac{y}{m}, & \text{if } m \leq y \leq 2m, \\ = 0, & \text{if } y \geq 2m. \end{cases} \tag{2.44}$$

This allows to define an approximated velocity u_m for the density bounded away from zero and bounded away from infinity by introducing $u_m = u\,\phi_m(\rho)$ for any fixed $m > 0$ and observing that, thanks to Lemma 2.4.10 and the κ entropy estimates, the following control holds:

$$\nabla u_m \in L^2((0,T) \times \Omega).$$

Then using renormalized Lemmas in the spirit of DiPerna–Lions, it is possible to prove that any κ-entropic weak solution of the regularized system is a renormalized solution for any $\varphi \in W^{2,\infty}(\mathbb{R}^d)$. To show such relation, we observe that the regularity of the weak solution is sufficient to justify that

$$\partial_t(\rho u_m) + \operatorname{div}(\rho u_m \otimes u) - \operatorname{div} S_m + F_m = 0,$$

in the distributional sense with

$$S \text{ and } F \in L^1((0,T) \times \Omega).$$

Then we use a test function under the form

$$\Phi(t,x) = [\psi(t)\varphi_n'([u_m]_\varepsilon)]_\varepsilon$$

with $\varphi_n \in W^{2,+\infty}(\mathbb{R}^d)$ (chosen later on) and where $[\cdot]_\varepsilon$ is a regularization by convolution.

We let ε tend to zero and then let m tend to $+\infty$ to get the renormalized formulation. It remains to let n tend to $+\infty$ to get the fact that any global weak solution is a renormalized solution of the Navier–Stokes–Korteweg system using all the informations that we have uniformly with respect to n. Note that the formulation (2.22) for the capillarity term is strongly used to pass to the limit, using Lemma 2.4.10 and Lemma 2.4.9.

Then we let, respectively, r_2, r_1, r_0, δ, and r tend to zero and obtain that there exists a global renormalized solution of the compressible Navier–Stokes system with viscosities such that $\inf \mu'(s) \geq \varepsilon_1$ in the following sense:

Definition 2.4.11 We say that $(\rho, \rho u)$ is a renormalized weak solution in u, if it verifies the κ-entropy estimate uniformly and the estimate on the viscosity $\mu(\rho)$ and for any function $\varphi \in W^{2,\infty}(\mathbb{R}^3)$, there exists two measures R_φ, $\overline{R}_\varphi^1$ with

$$\|R_\varphi\|_{\mathcal{M}(\mathbb{R}^+\times\Omega)} + \|\overline{R}_\varphi^1\|_{\mathcal{M}(\mathbb{R}^+\times\Omega)} \leq C\|\varphi''\|_{L^\infty(\mathbb{R}^3)},$$

where the constant C depends only on the solution $(\rho, \rho u)$, and for any function $\psi \in \mathcal{C}_c^\infty((0,T)\times\Omega)$,

$$\int_0^T \int_\Omega \left(\rho\psi_t + \sqrt{\rho}\sqrt{\rho}u \cdot \nabla\psi\right) dx\, dt = 0,$$

$$\int_0^T \int_\Omega \left(\rho\varphi(u)\psi_t + \rho\varphi(u)\otimes u : \nabla\psi\right) dx\, dt$$
$$- \int_0^T \int_\Omega \left(2(\sqrt{\mu(\rho)}\mathbb{S}_\mu + \frac{\lambda(\rho)}{2\mu(\rho)}\mathrm{Tr}(\sqrt{\mu(\rho)}\mathbb{S}_\mu)\mathrm{Id})\,\varphi'(u)\right) \cdot \nabla\psi\, dxdt$$
$$+\nabla p(\rho)\,\varphi'(u)\psi\, dx\, dt = \langle R_\varphi, \psi\rangle,$$

$$\int_0^T \int_\Omega (\mu(\rho)\psi_t + \frac{\mu(\rho)}{\sqrt{\rho}}\sqrt{\rho}u \cdot \nabla\psi)\, dxdt - \int_0^T \int_\Omega \frac{\lambda(\rho)}{2\mu(\rho)}\mathrm{Tr}(\sqrt{\mu(\rho)}\mathbb{T}_\mu)\psi\, dxdt = 0.$$

For every i, j, k between 1 and d:

$$\sqrt{\mu(\rho)}\varphi_i'(u)[\mathbb{T}_\mu]_{jk} = \partial_j(\mu(\rho)\rho\varphi_i'(u)u_k) - \rho u_k\varphi_i'(u)\sqrt{\rho}\partial_j s(\rho) + \overline{R}_\varphi^1, \tag{2.45}$$

and

$$\|\overline{R}_\varphi^1\|_{\mathcal{M}(\mathbb{R}^+\times\Omega)} + \|R_\varphi\|_{\mathcal{M}(\mathbb{R}^+\times\Omega)} \leq C\|\varphi''\|_{L^\infty}$$

and for any $\overline{\psi} \in C_c^\infty(\Omega)$:

$$\lim_{t\to 0}\int_\Omega \rho(t,x)\overline{\psi}(x)\,dx = \int_\Omega \rho_0(x)\overline{\psi}(x)\,dx,$$

$$\lim_{t\to 0}\int_\Omega \rho(t,x)u(t,x)\overline{\psi}(x)\,dx = \int_\Omega m_0(x)\overline{\psi}(x)\,dx,$$

$$\lim_{t\to 0}\int_\Omega \mu(\rho)(t,x)\overline{\psi}(x)\,dx = \int_\Omega \mu(\rho_0)(x)\overline{\psi}(x)\,dx.$$

After that, we approximate any viscosity μ by a sequence μ_n converging to μ in $\mathcal{C}^0(\mathbb{R}^+)$ and such that $\varepsilon_{1n} = \inf \mu_n' > 0$ to get global existence of renormalized solutions for the compressible Navier–Stokes equations with the shear and the bulk viscosities satisfying the constraint in the theorem. The following lemma plays a crucial role for the various passages to the limit:

Lemma 2.4.12 *For any fixed $2/3 < \alpha_1 < \alpha_2 < 4$, consider sequences δ_n, r_{0n}, r_{1n}, and r_{2n}, such that $r_{i,n} \to r_i \geq 0$ with $i = 0, 1, 2$ and then $\delta_n \to \delta \geq 0$. Consider a family of $\mu_n : \mathbb{R}^+ \to \mathbb{R}^+$ verifying $\inf_s \mu_n'(s) > 0$ and (2.16)–(2.17) for the fixed α_1 and α_2 such that*

$$\mu_n \to \mu \quad \text{in } C^0(\mathbb{R}^+).$$

Then, if (ρ_n, u_n) verifies the uniform estimates written in Definition 2.4.1, up to a subsequence, still denoted n, the following convergences hold:

1. *The sequence $\{\rho_n\}_{n\in\mathbb{N}}$ converges strongly to ρ in $C^0(0,T;L^p(\Omega))$ for any $1 \leq p < \gamma$.*
2. *The sequence $\{\mu_n(\rho_n)\,u_n\}_{n\in\mathbb{N}}$ converges to $\mu(\rho)u$ in $L^\infty(0,T;L^p(\Omega)$ for any $p \in [1, 3/2)$.*
3. *The sequence $\{\sqrt{\mu(\rho_n)}\nabla u_n\}_{n\in\mathbb{N}}$ converges to $\mathbb{T}_\mu$ weakly in $L^2(0,T;L^2(\Omega))$.*
4. *For every function $H \in W^{2,\infty}(\overline{\mathbb{R}^3})$ and $0 < \alpha < 2\gamma/\gamma+1$, we have that $\rho_n^\alpha H(u_n)$ converges to $\rho^\alpha H(u)$ strongly in $L^p(0,T;\Omega)$ for $1 \leq p < \frac{2\gamma}{(\gamma+1)\alpha}$. In particular, the quantity $\sqrt{\mu(\rho_n)}H(u_n)$ converges to $\sqrt{\mu(\rho)}H(u)$ strongly in $L^\infty(0,T;L^2(\Omega))$.*

Then we introduce a non-negative smooth function $\Phi : \mathbb{R} \to \mathbb{R}$ such that Φ has a compact support and $\Phi(s) = 1$ for any $-1 \leq s \leq 1$. Letting $\widetilde{\Phi}(s) = \int_0^s \Phi(r)dr$, we define

$$\varphi_n(y) = n\,\widetilde{\Phi}(\frac{y_1}{n})\,\Phi(\frac{y_2}{n})\,\Phi(\frac{y_3}{n}),$$

for all $y = (y_1, y_2, y_3) \in \mathbb{R}^3$. Note that φ_n is bounded in $W^{2,\infty}(\mathbb{R}^3)$ for any fixed $n > 0$ while $\varphi_n(y)$ converges everywhere to y_1 as n goes to infinity, φ_n' is uniformly bounded in n and converges everywhere to unit vector $(1, 0, 0)$ and

$$\|\varphi_n''\|_{L^\infty} \leq \frac{C}{n} \to 0 \text{ when } n \to +\infty.$$

This allows to control the measures appearing in the renormalized formulation and prove that they converge to zero as $n \to +\infty$. By permuting the directions in φ_n, one may also recover the full vector equation on $\rho\, u$. Lebesgue's dominated convergence is the final ingredient because the couple (ρ, u) is now fixed with all necessary bounds, thanks to the κ-entropy.

2.4.3 Some Physical Situations: Shallow Water, Granular Media, Mixture System

Let us provide here some examples of physical situations where the viscosity may depend on the transported quantity.

I. Viscous Shallow Water Equations in Dimension 2 It is important to note that a typical viscous shallow water system can be written as

$$\begin{aligned}&\partial_t h + \mathrm{div}(hv) = 0,\\&\partial_t (hv) + \mathrm{div}(hv \otimes v) + \nabla \frac{h^2}{2}\\&\qquad\qquad -2\mu \mathrm{div}(h\mathbb{D}(v)) - 2\nu\nabla(h\mathrm{div} v) = 0,\end{aligned}$$

where v is the vertical average of the horizontal velocity and h is the total height of the fluid.

Remark 2.4.13 (Open Problem) To prove global existence of weak solution for such system is a challenging problem because the shear and the bulk viscosities do not satisfy the BD relation.

The shallow water system, without drag term or capillarity terms, has been solved replacing the momentum equation by

$$\partial_t (hu) + \mathrm{div}(hu \otimes u) + \nabla \frac{h^2}{2} - 2\mu \mathrm{div}(h\mathbb{D}(v)) = 0$$

by A. Vasseur–C. Yu [52] and J. Li–Z.P. Xin [37]. These works should be seen as real breakthroughs in the proof of global existence of weak solutions for compressible Navier–Stokes equations with density dependent viscosities. In particular, with [7, 16] and [35], they have been a major source of inspiration to obtain the result in [23] that we have explained previously. For reader's convenience let us give the two approximate systems that has been considered, respectively, in [52] (see also [35]) and [37].

Approximate System by A. Vasseur and C. Yu The construction of approximate solutions in [51] starts with an idea close to the one discussed in [8] but with a well chosen capillarity. More precisely they consider the following system:

$$\begin{aligned}&\partial_t\rho + \mathrm{div}(\rho u) = 0,\\&\partial_t(\rho u) + \mathrm{div}(\rho u\otimes u) + a\nabla\rho^\gamma - 2\mathrm{div}(\rho\mathbb{D}(u))\\&\qquad\qquad + r_0 u + r_1\rho|u|^2u - \kappa\rho\nabla\Big(\frac{1}{\rho^{1/2}}\Delta\rho^{1/2}\Big) = 0.\end{aligned}$$

The capillarity term is often called the quantum pressure. Remark that the Mellet–Vasseur estimate is non-uniformly satisfied with respect to $\kappa > 0$. The strategy by the authors is then to introduce a truncated and regularized quantity to the function $s \mapsto s\log s$ and then pass to the limit $\kappa \to 0$ by relaxing the truncation parameter in an appropriate way. This paper may be seen as the first stone before [35] where they prove that it is possible to not use the Mellet–Vasseur estimate introducing the concept of renormalized solution in velocity. This has been extended by [23] for general viscosities satisfying the BD relation.

Approximate System by J. Li and Z.P. Xin The construction of approximate solutions in [37] starts with the very nice idea to allow regularization on the mass equation and find the appropriate quantities to add in the momentum equation to get BD entropy and Mellet–Vasseur estimates uniformly. The system reads

$$\begin{aligned}&\partial_r\rho + \mathrm{div}(\rho u) = \varepsilon v\Delta v + \varepsilon v\mathrm{div}(|\nabla v|^2\nabla v) + \varepsilon\rho^{-50}\\&\partial_t(\rho u) + \mathrm{div}(\rho u\otimes u) + a\nabla\rho^\gamma - \mathrm{div}(\rho\mathbb{D}(u))\\&\qquad = \sqrt{\varepsilon}\mathrm{div}(\rho\nabla u) + \varepsilon v|\nabla v|^2\nabla v\cdot\nabla u - \varepsilon\rho^{-50}u - \varepsilon\rho|u|^3u,\end{aligned} \tag{2.46}$$

where $v = \rho^{1/2}$ with ε small enough. Note that their approximate system is a parabolic equation for any fixed $\varepsilon > 0$ and hence has smooth effects on the density provided the smooth initial density is strictly away from vacuum. The specific choices of the higher order regularization have several key advantages. First, it can be shown that the smooth solutions to the new system satisfy the energy and the Mellet–Vasseur type estimates. Moreover, after some careful calculations, they find that the most difficult term induced by $\varepsilon v\Delta v + \varepsilon v\mathrm{div}(|\nabla v|^2\nabla v)$ has the right sign which implies that the solutions to our approximate system also satisfy the BD entropy inequality. With all these estimates at hand, they are able to develop De Giorgi-type procedure to bound the density from above and below, in particular, the density is strictly away from vacuum provided the initial one is. It is important to note the regularization that the authors have chosen is related to $\rho^{1/2}$ which has a real meaning from a physical point view. This unknown is linked to the solution of the nonlinear Schrödinger equation using the Madelung transform. Recall that as we mentioned in previous sections, the authors have also considered in 3d a non-symmetric viscous term namely $\mathrm{div}(\mu(\rho)\nabla u)+\nabla(\lambda(\rho)\mathrm{div}u)$ with the bulk viscosity

$\lambda(\rho) = \mu'(\rho)\rho - \mu(\rho)$ and the shear viscosity $\mu(\rho) = \mu\rho^\alpha$ where $\alpha \in [3/4, 2)$ and constraint on γ with respect to α. A regularized system comparable to (2.46) is used.

II. Granular Media It is interesting to note the following system that has been recently derived by C. Perrin (see [44]):

$$\begin{aligned}
&\mathrm{div} u = 0,\\
&\partial_t u + \mathrm{div}(u \otimes u) + \nabla p - 2\mathrm{div}((\mu_0 + \mu_1 \pi)\mathbb{D}(u)) + r_1|u|u = 0,\\
&\partial_t \pi + u \cdot \nabla \pi + \frac{1}{2\mu_1}\pi = \frac{1}{\mu_1} p.
\end{aligned}$$

This system is obtained from a compressible Navier–Stokes equations with singular appropriate pressure law and viscosities depending appropriately from this quantity: The interested reader is referred to [45].

Remark 2.4.14 (Small Open Problem) One should be able to obtain global well posedness of entropy-weak solutions to such system using the BD entropy without the drag term (namely assuming $r_1 = 0$) through the recent framework developed in [23]. Interested readers are encouraged to try to get such result to understand the main steps discussed for compressible Navier–Stokes equations with pressure dependent viscosities. Note that the viscosity depends on the pressure but in a non-local way in space and time, which encodes some memory effects by the material under consideration.

III. Mixture System The formal averaging procedure, for instance, by Ishii or Drew–Passman may lead to a system for mixture written as follows:

$$\begin{aligned}
&\alpha_+ + \alpha_- = 0\\
&\partial_t(\alpha^\pm \rho^\pm) + \mathrm{div}(\alpha^\pm \rho^\pm u^\pm) = 0,\\
&\partial_t(\alpha^\pm \rho^\pm u^\pm) + \mathrm{div}(\alpha^\pm \rho^\pm u^\pm \otimes u^\pm)\\
&\qquad\qquad + \alpha^\pm \nabla p = 2\mathrm{div}(\alpha^\pm \rho^\pm D(u^\pm)),
\end{aligned}$$

with the pressure state closure law

$$p = p^\pm(\rho^\pm).$$

Note that the quantities $R^\pm = \alpha^\pm \rho^\pm$ satisfy mass equations related to the velocities $u^\pm$ and therefore extra regularity may be obtained on $\nabla\sqrt{R^\pm}$ combining in an appropriate manner the corresponding energy and BD entropy estimates.

Remark 2.4.15 (Open Problem) Such non-conservative system may be studied using the framework introduced by D. Bresch and B. Desjardins because we can get physical energy and BD entropy estimates. The interested reader is referred to [15] where extra capillarity forces are added to conclude but it would be interesting to study again this system using the recent results obtained for the mono-fluid compressible Navier–Stokes equations by A. Vasseur–C. Yu and J. Li–Z.P. Xin.

2.5 Conclusion and Comments

In this chapter, we have tried to explain new mathematical results related to compressible Navier–Stokes equations under pressure while emphasizing the historical connections between the various results. Namely we have first sketched the main ingredients to get global existence of weak solutions when the viscosities are assumed to be constants with a non-monotone pressure law (see [17]). Then we presented the main ideas for the global existence of weak solutions when the viscosities depend on the density and satisfy the BD relation (see [23]).

We hope that this will encourage young researchers to try their hand at some of the many remaining problems in this fascinating scientific story that has always involved many international contributors, each of them bringing new stones.

2.6 Some Mathematical Tools

In this section we state technical Lemmas which play crucial roles in the two previous sections.

Some Compactness Lemmas With in particular young researchers in mind, we provide various possible tools allowing to the compactness of sequence depending on appropriate control on time and space oscillations, while emphasizing that the two first lemmas are very classical.

The most famous criterion to get compactness is the Aubin–Lions–Simon–Lemma if we have informations on the time and space derivatives of the sequence. Namely

Lemma 2.6.1 *Let X_0, X, and X_1 be three Banach spaces with $X_0 \subseteq X \subseteq X_1$. Suppose that X_0 is compactly embedded in X and that X is continuously embedded in X_1. For $1 \le p, q \le +\infty$, let*

$$W = \{u \in L^p(0, T; X_0) : \partial_t u \in L^q(0, T; X_1)\}.$$

- *If $p < +\infty$, then the embedding of W into $L^p(0, T; X)$ is compact.*
- *If $p = +\infty$, then the embedding of W into $\mathcal{C}([0, T]; X)$ is compact.*

Another criterion to get compactness, often used for the compressible Navier–Stokes equation with constant viscosities, allows to commute weak convergence and strictly convex function namely

Lemma 2.6.2 *Let $\{\theta_n\}_{n\in\mathbb{N}}$ be a sequence of functions uniformly bounded in $L^p(Q)$ with $1 < p < +\infty$. Let the function $s \mapsto \Phi(s)$ such that*

$$|\Phi(s)| \le C(1 + |s|^p)$$

and Φ is uniformly strictly convex. If

$$\int_Q \overline{\Phi(\theta)} = \int_\Omega \Phi(\theta),$$

where $\bar{\cdot}$ denotes the weak limit, then there exists a subsequence θ_n which converges strongly to θ in $L^1(Q)$.

Let us now present a non-local criterion to get compactness

Proposition 2.6.3 *Let ρ_n be a sequence uniformly bounded in some $L^p((0,T)\times\mathbb{T}^d)$ with $1 \le p < +\infty$. Assume that K_h is a sequence of positive, bounded function such that*

- $\forall \eta > 0 \qquad \sup_h \int_{\mathbb{T}^d} K_h(x) 1_{\{x:\, |x|\le\eta\}} < +\infty$
- $\|K_h\|_{L^1(\mathbb{T}^d)} \to +\infty$ *as* $h \to +\infty$.

If $\partial_t \rho_n \in L^q(0,T; W^{-1,q}(\mathbb{T}^d))$ with $q \ge 1$ uniformly in n and

$$\limsup_n \Big[\frac{1}{\|K_h\|_{L^1}} \int_0^T \int_{\mathbb{T}^d} K_h(x-y)|\rho_n(t,x) - \rho_n(t,y)|^p\, dx\, dt\Big] \to 0 \text{ as } h \to 0,$$

then ρ_n is compact in $L^p((0,T)\times\mathbb{T}^d)$. Conversely if ρ_n is compact in $L^p((0,T)\times\mathbb{T}^d)$, then the above quantity converges to 0 with h.

To sketch the proof, let us just quickly recall that the compactness in space is connected to the classical approximation by convolution. Denote $\overline{\mathcal{K}_h}$ the normalized kernel

$$\overline{\mathcal{K}_h} = \frac{\mathcal{K}_h}{\|\mathcal{K}_h\|_{L^1}}.$$

Write

$$\|\rho_n - \overline{\mathcal{K}_h} \star_x \rho_n\|_{L^p}^p \le \frac{1}{\|\mathcal{K}_h\|_{L^1}^p} \int_{\mathbb{T}^d} \Big(\int_{\mathbb{T}^d} \mathcal{K}_h(x-y)|\rho_n(t,x) - \rho_n(t,y)|\Big)^p$$
$$\le \frac{1}{\|\mathcal{K}_h\|_{L^1}} \int_{\mathbb{T}^d} \mathcal{K}_h(x-y)|\rho_n(t,x) - \rho_n(t,y)|^p$$

which converges to zero as $h \to 0$ uniformly with respect to n by assumption. On the other hand for a fixed h, the sequence $\overline{\mathcal{K}_h} \star_x \rho_n$ in n is compact in x. This would prove the compactness in space. Concerning the compactness in time, we just have to couple everything and use the uniform bound on $\partial_t \rho_n$ as per the usual Aubin–Lions–Simon Lemma.

In the compressible Navier–Stokes equations with non-monotone pressure laws, the following important choice of kernel K_h and its associated $\mathcal{K}_{h_0}$ functions are used:

Definition 2.6.4 Let us define the positive, bounded, and symmetric function K_h such that

$$K_h(x) = \frac{1}{(h + |x|)^a} \text{ for } |x| \le 1/2$$

with some $a > d$ and K_h positive, independent of h for $|x| \ge 2/3$, K_h positive outside $B(0, 3/4)$ and periodized such as to belong in $\mathcal{C}^\infty(\mathbb{T}^d \backslash B(0, 3/4))$. We denote

$$\overline{K_h}(x) = \frac{K_h(x)}{\|K_h\|_{L^1(\mathbb{T}^d)}}.$$

The following important quantity plays a crucial role:

$$\mathcal{K}_{h_0}(x) = \int_{h_0}^1 \overline{K_h}(x) \frac{dh}{h}.$$

Using weights satisfying

$$K_h(x) = K_h(-x), \qquad |x||\nabla K_h|(x) \le C K_h(x)$$

for some constant $C > 0$ then we get that

$$\|\mathcal{K}_{h_0}\|_{L^1} \approx |\log h_0|.$$

Maximal and Square Functions, Translation of Operators The interested reader is referred to [17] and [49] for details and proofs. First we remind the well-known inequality

$$|\Phi(x) - \Phi(y)| \le C\,|x - y|\,(M|\nabla\Phi|(x) + M|\nabla\Phi|(y)), \tag{2.47}$$

where M is the localized maximal operator

$$M f(x) = \sup_{r \le 1} \frac{1}{|B(0,r)|} \int_{B(0,r)} f(x + z)\, dz. \tag{2.48}$$

Let us next mention several other useful mathematical properties that may be proved, see [17]. First one has the following Lemma:

Lemma 2.6.5 *There exists $C > 0$ s.t. for any $u \in W^{1,1}(\mathbb{T}^d)$, one has*

$$|u(x) - u(y)| \le C\,|x-y|\,(D_{|x-y|}u(x) + D_{|x-y|}u(y)),$$

where we denote

$$D_h u(x) = \frac{1}{h}\int_{|z|\le h} \frac{|\nabla u(x+z)|}{|z|^{d-1}}\,dz.$$

Note that this result implies the estimate (2.47) as

Lemma 2.6.6 *There exists $C > 0$, for any $u \in W^{1,p}(\mathbb{T}^d)$ with $p \ge 1$*

$$D_h\, u(x) \le C\, M|\nabla u|(x).$$

The key improvement in using D_h is that small translations of the operator D_h are actually easy to control

Lemma 2.6.7 *Let $u \in H^1(\mathbb{T}^d)$ then have the following estimates:*

$$\int_{h_0}^1 \int_{\mathbb{T}^d} \overline{K}_h(z)\, \|D_{|z|}\, u(.) - D_{|z|}\, u(.+z)\|_{L^2(\mathbb{T}^d)}\, dz\, \frac{dh}{h}$$
$$\le C\,|\log h_0|^{1/2}\, \|u\|_{H^1(\mathbb{T}^d)}. \qquad (2.49)$$

This lemma is critical and explains why we propagate a quantity integrated with respect to h with a weight dh/h namely with the kernel $\mathcal{K}_{h_0}$. It helps to exchange the quantity evaluated at point x to the same quantity but evaluated at point y (when x and y are close enough): This kind of delocalization is actually critical in the proof.

The full proof Lemma 2.6.7 is rather classical and was recalled, for instance, in [17] for any L^p space but we give a brief sketch here (which is simpler as Lemma is L^2 based and we can use Fourier transform).

Sketch of the Proof of Lemma 2.6.7 Note that we can write

$$D_h\, u(x) = L_h \star \nabla u, \quad L(x) = \frac{1_{|x|\le 1}}{|x|^{d-1}}, \quad L_h(z) = h^{-d}\, L(z/h),$$

where L_h is hence a usual convolution operator and $L \in W^{s,1}$ for any $s < 1$. Now

$$\int_{h_0}^1 \int_{\mathbb{T}^d} \overline{K}_h(z)\, \|D_{|z|}\, u(.) - D_{|z|}\, u(. + z)\|_{L^2}\, dz\, \frac{dh}{h}$$

$$\leq C \int_{S^{d-1}} \int_0^1 \|L_r \star \nabla u(.) - L_r \star \nabla u(. + r\,\omega)\|_{L^2}\, \frac{dr}{r + h_0}\, d\omega$$

$$\leq C\, |\log h_0|^{1/2} \left(\int_{S^{d-1}} \int_0^1 \|L_r \star \nabla u(.) - L_r \star \nabla u(. + r\,\omega)\|_{L^2}^2\, \frac{dr}{r + h_0}\, d\omega \right)^{1/2}.$$

For any $\omega \in S^{d-1}$, define $L_r^\omega = L_r(.) - L_r(. + r\,\omega)$. Calculate by Fourier transform

$$\int_0^1 \|L_r \star \nabla u(.) - L_r \star \nabla u(. + r\,\omega)\|_{L^2}^2\, \frac{dr}{r + h_0} = \int_0^1 \sum_{\xi \in \mathbb{T}^d} |\hat{L}_r^\omega|^2(\xi)\, |\xi|^2 |\hat{u}|^2(\xi)\, \frac{dr}{r + h_0}.$$

But $\hat{L}_r^\omega = (1 - e^{i r\, \xi \cdot \omega})\, \hat{L}(r\,\xi)$ and furthermore $|\hat{L}(r\,\xi)| \leq C\,(1 + |r\,\xi|)^{-s}$ for some $s > 0$ since $L \in W^{s,1}$. Therefore

$$\int_0^1 |\hat{L}_r^\omega|^2(\xi)\, \frac{dr}{r + h_0} \leq C,$$

for some constant C independent of ξ, ω, and h_0. This is of course the famous square function calculation and lets us conclude.

Acknowledgements D. Bresch is partially supported by SingFlows project, grand ANR-18-CE40-0027. P.–E. Jabin is partially supported by NSF DMS Grants 161453, 1908739 and NSF Grant RNMS (Ki-Net) 1107444.

Bibliography

1. P. Antonelli, S. Spirito. Global existence of weak solutions to the Navier-Stokes-Korteweg equations. ArXiv:1903.02441 (2019).
2. P. Antonelli, S. Spirito. On the compactness of weak solutions to the Navier-Stokes-Korteweg equations for capillary fluids. ArXiv:1808.03495 (2018).
3. P. Antonelli, S. Spirito. Global Existence of Finite Energy Weak Solutions of Quantum Navier-Stokes Equations. *Archive of Rational Mechanics and Analysis*, 225 (2017), no. 3, 1161–1199.
4. P. Antonelli, S. Spirito. On the compactness of finite energy weak solutions to the Quantum Navier-Stokes equations. *J. of Hyperbolic Differential Equations*, 15 (2018), no. 1, 133–147.
5. F. Ben Belgacem, P.–E. Jabin. Compactness for nonlinear continuity equations. *J. Funct. Anal*, 264, no. 1, 139–168, (2013).

6. F. Ben Belgacem, P.–E. Jabin. Convergence of numerical approximations to non-linear continuity equations with rough force fields. Submitted (2018).
7. D. Bresch, F. Couderc, P. Noble, J.–P. Vila. A generalization of the quantum Bohm identity: hyperbolic CFL condition for Euler-Korteweg equations. *C.R. Acad. Sciences Paris.* volume 354, Issue 1, 39–43, (2016).
8. D. Bresch, B. Desjardins. Existence of global weak solutions for 2D viscous shallow water equations and convergence to the quasi-geostrophic model. *Comm. Math. Phys.,* 238 (2003), no.1-3, 211–223.
9. D. Bresch and B. Desjardins. On the construction of approximate solutions for the 2D viscous shallow water model and for compressible Navier-Stokes models. *J. Math. Pures Appl.* (9) 86 (2006), no. 4, 362–368.
10. D. Bresch, B. Desjardins. Quelques modèles diffusifs capillaires de type Korteweg. *C. R. Acad. Sci. Paris*, section mécanique, **332**, no. 11, 881–886, (2004).
11. D. Bresch, B. Desjardins. Weak solutions via the total energy formulation and their quantitative properties - density dependent viscosities. In: Y. Giga, A. Novotný (éds.) Handbook of Mathematical Analysis in Mechanics of Viscous Fluids. Springer, Berlin (2017).
12. D. Bresch, B. Desjardins. On the existence of global weak solutions to the Navier–Stokes equations for viscous compressible and heat-conducting fluids. *J. Math. Pures et Appl.,* 57–90 (2007).
13. D. Bresch, B. Desjardins, Chi-Kun Lin. On some compressible fluid models: Korteweg, lubrication, and shallow water systems. *Comm. Partial Differential Equations* 28, no. 3-4, 843–868, (2003).
14. D. Bresch, B. Desjardins, D. Gérard-Varet. On compressible Navier-Stokes equations with density dependent viscosities in bounded domains. *J. Math. Pures Appl.* 87 (9), 227–235 (2007).
15. D. Bresch, B. Desjardins, J.–M. Ghidaglia, E. Grenier. Global weak solutions to a generic two-fluid model. *Arch. Rational Mech. Analysis.* 196, Issue 2, 599–629, (2010).
16. D. Bresch, B. Desjardins, E. Zatorska. Two-velocity hydrodynamics in Fluid Mechanics, Part II. Existence of global κ-entropy solutions to compressible Navier-Stokes system with degenerate viscosities. *J. Math. Pures Appl.* Volume 104, Issue 4, 801–836 (2015).
17. D. Bresch, P.-E. Jabin. Global existence of weak solutions for compressible Navier-Stokes equations: thermodynamically unstable pressure and anisotropic viscous stress tensor. *Ann. of Math.* (2) 188, no. 2, 577–684 (2018).
18. D. Bresch, P.–E. Jabin. Quantitative regularity estimates for compressible transport equations. New trends and results in mathematical description of fluid flows. *Birkhauser* 77–113, (2018).
19. D. Bresch, P.–E. Jabin. Quantitative regularity estimates for advective equation with anelastic degenerate constraint. Proc. Int. Cong. of Math, 2161–2186 (2018).
20. D. Bresch, P.–E. Jabin, F. Wang. On heat-conducting Navier-Stokes equations with a truncated virial pressure state law. In preparation (2019).
21. D. Bresch, I. Lacroix-Violet, M. Gisclon. On Navier-Stokes-Korteweg and Euler-Korteweg systems: Application to quantum fluids models. To appear in *Arch. Rational Mech. Anal.* (2019).
22. D. Bresch, P. Mucha, E. Zatorska. Finite-energy solutions for compressible two-fluid Stokes system. *Arch. Rational Mech. Anal.*, 232, Issue 2, 987–1029, (2019).
23. D. Bresch, A. Vasseur, C. Yu. Global existence of entropy-weak solutions to the compressible Navier-Stokes equations with non-linear density dependent viscosities. Submitted (2019).
24. R. Carles, K. Carrapatoso, M. Hillairet. Rigidity results in generalized isothermal fluids. *Annales Henri Lebesgue*, 1, (2018), 47–85.
25. E. Feireisl. Dynamics of viscous compressible fluids. *Oxford Lecture Series in Mathematics and its Applications,* 26. Oxford University Press, Oxford, 2004.
26. E. Feireisl, A. Novotný, H. Petzeltová. On the existence of globally defined weak solutions to the Navier-Stokes equations. *J. Math. Fluid Mech.* **3** (2001), 358–392.

27. E. FEIREISL. Compressible Navier–Stokes Equations with a Non-Monotone Pressure Law. *J. Diff. Eqs* 183, no 1, 97–108, (2002).
28. E. FEIREISL, A. VASSEUR. New perspectives in fluid dynamics: Mathematical analysis of a model proposed by Howard Brenner. New directions in mathematical fluid mechanics, 153–179, *Adv. Math. Fluid Mech.*, Birkhauser Verlag, Basel, 2010.
29. S. GAVRILYUK, S.M. SHUGRIN Media with equations of state that depend on derivatives. *J. Applied Mech. Physics,* vol. 37, No2, (1996).
30. Z. GUO, Q. JIU, Z. XIN. Spherically symmetric isentropic compressible flows with density-dependent viscosity coefficients. *SIAM J. Math. Anal.* 39, no. 5, 1402–1427, (2008).
31. S. JIANG, Z. XIN, P. ZHANG. Global weak solutions to 1D compressible isentropic Navier-Stokes equations with density-dependent viscosity. *Methods Appl. Anal.* 12, no. 3, 239–251, (2005).
32. S. JIANG, P. ZHANG. On spherically symmetric solutions of the compressible isentropic Navier-Stokes equations. *Comm. Math. Phys.* 215, no. 3, 559–581, (2001).
33. A. JÜNGEL. Global weak solutions to compressible Navier-Stokes equations for quantum fluids. *SIAM J. Math. Anal.* 42, no. 3, 1025–1045, (2010).
34. A. JÜNGEL, D. MATTHES. The Derrida–Lebowitz-Speer-Spohn equations: Existence, uniqueness, and Decay rates of the solutions. *SIAM J. Math. Anal.*, 39(6), 1996–2015, (2008).
35. I. LACROIX-VIOLET, A. VASSEUR. Global weak solutions to the compressible quantum Navier-Stokes equation and its semi-classical limit. *J. Math. Pures Appl.* (9) 114, 191–210, (2018).
36. J. LERAY. Sur le mouvement d'un fluide visqueux remplissant l'espace, *Acta Math.* 63, 193–248, (1934).
37. J. LI, Z.P. XIN. Global Existence of Weak Solutions to the Barotropic Compressible Navier-Stokes Flows with Degenerate Viscosities. arXiv:1504.06826 (2015).
38. P.-L. LIONS. *Mathematical topics in fluid mechanics. Vol. 2. Compressible models.* Oxford Lecture Series in Mathematics and its Applications, 10. Oxford Science Publications. The Clarendon Press, Oxford University Press, New York, 1998.
39. D. MALTESE, M. MICHALEK, P. MUCHA, A. NOVOTNY, M. POKORNY, E. ZATORSKA. Existence of weak solutions for compressible Navier-Stokes with entropy transport. *J. Differential Equations*, 261, No. 8, 4448–4485 (2016)
40. A. MELLET, A. VASSEUR. On the barotropic compressible Navier-Stokes equations. *Comm. Partial Differential Equations* 32 (2007), no. 1-3, 431–452.
41. P.B. MUCHA, M. POKORNY, E. ZATORSKA. Approximate solutions to a model of two-component reactive flow. *Discrete Contin. Dyn. Syst. Ser.* S, 7, No. 5 , 1079–1099 (2014).
42. A. NOVOTNY. Weak solutions for a bi-fluid model of a mixture of two compressible non interacting fluids. Submitted (2018).
43. A. NOVOTNY, M. POKORNY. Weak solutions for some compressible multi-component fluid models. Submitted (2018).
44. C. PERRIN. Pressure Dependent Viscosity Model for Granular Media Obtained from Compressible Navier-Stokes Equations. *Appl Math Res Express,* vol. 2016, Iss. 2, p. 289–333 (2016).
45. C. PERRIN. An overview on congestion phenomena in fluid equations. Proceeding Journées EDP 2018. See hal-01994880
46. P.I. PLOTNIKOV, W. WEIGANT. Isothermal Navier-Stokes equations and Radon transform. *SIAM J. Math. Anal.* 47 (2015), no. 1, 626–653.
47. F. ROUSSET. Solutions faibles de l'équation de Navier-Stokes des fluides compressible [d'après A. Vasseur et C. Yu]. Séminaire Bourbaki, 69ème année, 2016–2017, no 1135.
48. S.M. SHUGRIN. Two-velocity hydrodynamics and thermodynamics. *J. Applied Mech and Tech. Physics*, 39, 522–537, (1994).
49. E.M. STEIN *Harmonic Analysis.* Princeton Univ. Press 1995 (second edition).
50. V. A. VAIGANT, A. V. KAZHIKHOV. On the existence of global solutions of two-dimensional Navier-Stokes equations of a compressible viscous fluid. (Russian). *Sibirsk. Mat. Zh.* 36 (1995), no. 6, 1283–1316, ii; translation in *Siberian Math. J.* 36 (1995), no.6, 1108–1141.

51. A. VASSEUR, C. YU. Global weak solutions to compressible quantum Navier-Stokes equations with damping. *SIAM J. Math. Anal.* 48 (2016), no. 2, 1489–1511.
52. A. VASSEUR, C. YU. Existence of Global Weak Solutions for 3D Degenerate Compressible Navier-Stokes Equations. *Inventiones mathematicae* (2016), 1–40.
53. A. VASSEUR, H. WEN, C. YU. Global weak solution to the viscous two-phase model with finite energy. *To appear in J. Math Pures Appl.* (2018).

Chapter 3
Global Well-Posedness for Incompressible–Incompressible Two-Phase Problem

Yoshihiro Shibata and Hirokazu Saito

Abstract This chapter is devoted to some mathematical analysis of the two-phase problem for the viscous incompressible–incompressible capillary flows separated by sharp interface, this problem being called two-phase problem for short, and the Stokes equations with transmission conditions on the sharp interface which is arised from the two-phase problem.

The maximal regularity is a character of the system of equations of parabolic type, and it is a very powerful tool in solving quasilinear equations of parabolic type. The authors of this lecture note have developed a systematic method to derive the maximal regularity theorem for the initial-boundary value problem for the Stokes equations with non-homogeneous boundary conditions, which is based on the $\mathcal{R}$ bounded solution operators theory and L. Weis' operator valued Fourier multiplier theorem. The notion of $\mathcal{R}$ boundedness plays an essential role in the Weis' theory, which takes the place of boundedness in the standard Fourier multiplier theorem of Marcinkiewicz-Mikhilin-Hölmander type.

In this lecture note, we explain how to use the $\mathcal{R}$-bounded solution operators to derive the maximal regularity theorem for the Stokes equations with transmission conditions, and as an application of our maximal regularity theorem, we prove the local well-posedness of the two-phase problem, where the solutions are obtained in the L_p in time and L_q in space maximal regularity class. So far, this framework gives us the best possible regularity class of parabolic quasilinear equations.

Moreover, we prove the global well-posedness for the two-phase problem both in the bounded domain case and the unbounded domain case. A key tool is the decay property of the C^0 analytic semigroup associated with the Stokes equations

Y. Shibata (✉)
Department of Mathematics and Research Institute of Science and Engineering, Waseda University, Tokyo, Japan

Department of Mechanical Engineering and Materials Science, University of Pittsburgh, Pittsburgh, PA, USA

H. Saito
Faculty of Industrial Science and Technology, Tokyo University of Science, Hokkaido, Japan
e-mail: hsaito@rs.tus.ac.jp

T. Bodnár et al. (eds.), *Fluids Under Pressure*, Advances in Mathematical Fluid Mechanics, https://doi.org/10.1007/978-3-030-39639-8_3

with transmission conditions. In the bounded domain case, the decay properties are obtained essentially from the analysis of zero eigenvalue. As a result we prove the exponential stability of our C^0 analytic semigroup in some quotient space, which, together with the conservation of momentum and angular momentum and the maximal regularity theorem, yields the global well-posedness in the case of small initial data and the ball-like reference domain.

On the other hand, in the unbounded domain case, the zero is a continuous spectrum for the Stokes equations with transmission conditions, and so we can prove the polynomial decay properties for the C^0 analytic semigroup only, which, combined with L_p-L_q maximal regularity theorem with suitable choices of p and q, yields the L_p time summability of the L_q space norm of solutions to the nonlinear problem. From this we prove the global well-posedness for the small initial data in the unbounded domain case. Notice that the L_p summability yields L_1 summability in view of the Hölder inequality, which is enough to handle the kinetic equations.

What we want to emphasize here is that our method is based on the construction of $\mathcal{R}$ bounded solution operators and spectral analysis of the zero eigenvalue or generalized eigenvalue of the generalized resolvent problem for the Stokes equations with transmission conditions. The spectral analysis here can be used widely to study the other parabolic linear and quasilinear equations arising from the mathematical study of viscous fluid flows as well as other models in mathematical physics like MHD, multicomponent flows, namatic crystal flows, and so on.

3.1 Introduction

3.1.1 Problem

The maximal regularity is a character of the system of equations of parabolic type, and it is a powerful tool in solving quasilinear equations of parabolic type. The operator valued Fourier multiplier theorem due to L. Weis [40], which is a fruitful result in theory of vector valued harmonic analysis, allows us to prove the maximal regularity theorem for the initial-boundary value problem with non-homogeneous boundary conditions. The authors of this lecture note have developed a deriving method of the maximal regularity theorem in the L_p in time and L_q in space frame work in terms of the $\mathcal{R}$ bounded solution operators for the corresponding resolvent problem. Our original idea goes back to Shibata [27]. The notion of $\mathcal{R}$- boundedness plays an essential role in the Weis operator valued Fourier multiplier theorem, which takes the place of boundedness in the usual Fourier multiplier theorem of Marcinkiewicz-Mikhilin-Hölmander type. And, applying the Weis theorem to the solutions of the resolvent problem represented by using the $\mathcal{R}$- bounded solution operators gives the L_p-L_q maximal regularity of solutions to the time dependent problem. Moreover, the $\mathcal{R}$- boundedness implies the boundedness, so that the existence of $\mathcal{R}$- bounded solution operators yields the generation of C^0 analytic semigroup. Thus, the initial-boundary problem with non-homogeneous boundary conditions can be treated systematically by the use of the $\mathcal{R}$- bounded solution operators, which is not in the case of C^0 analytic semigroup.

In this lecture note, we would like to explain how to derive the L_p-L_q maximal regularity theorem and decay properties of C^0 semigroup associated with some linear problems arising in the study of two-phase problems of the Navier-Stokes equations in terms of $\mathcal{R}$ bounded solution operators and spectral analysis, and how to prove the local and global well-posedness for two-phase problems of the Navier-Stokes equations by using the maximal L_p-L_q regularity theorem and the decay properties of C^0 analytic semigroup associated with the linearized problems.

First of all, we formulate the two-phase problem for the Navier-Stokes equations studied in this lecture note. Let Ω be the N dimensional Euclidean space $\mathbb{R}^N$ ($N \geq 2$) or a bounded domain in $\mathbb{R}^N$ with boundary Γ_- of C^2 compact hypersurface. Let Ω_{t+} be a bounded subdomain of Ω that depends on time $t > 0$ and Γ_t the boundary of Ω_{t+}. We assume that $\Gamma_t \cap \Gamma_- = \emptyset$ when Ω is bounded. Let $\Omega_{t-} = \Omega \setminus (\Omega_{t+} \cup \Gamma_t)$. We consider the situation that $\Omega_{t\pm}$ are occupied by incompressible viscous fluids with mass density $m_\pm$ and viscosity coefficient $\mu_\pm$. We assume that $m_\pm$ are positive constants and $\mu_\pm = \mu_\pm(x)$ are functions defined on $\mathbb{R}^N$ satisfying the conditions (3.3) and (3.4) below. Let $\mathbf{n}_t$ be the unit outer normal to Γ_t pointing from Ω_{t+} to Ω_{t-} and $\dot{\Omega}_t = \Omega_{t+} \cup \Omega_{t-}$. In this section, we consider the two-phase problem of the Navier-Stokes equations separated by sharp interface Γ_t formulated as follows:

$$\begin{cases} m(\partial_t \mathbf{v} + \mathbf{v}\cdot\nabla\mathbf{v}) - \mathrm{Div}\,(\mu\mathbf{D}(\mathbf{v}) - \mathfrak{p}) = 0 & \text{in } \dot{\Omega}_t \text{ for } t \in (0, T), \\ \mathrm{div}\,\mathbf{v} = 0 & \text{in } \dot{\Omega}_t \text{ for } t \in (0, T), \\ [[(\mu\mathbf{D}(\mathbf{v}) - \mathfrak{p}\mathbf{I})\mathbf{n}_t]] = \sigma H(\Gamma_t)\mathbf{n}_t - p_0\mathbf{n}_t & \text{on } \Gamma_t \text{ for } t \in (0, T), \\ [[\mathbf{v}]] = 0 & \text{on } \Gamma_t \text{ for } t \in (0, T), \\ V_n = \mathbf{v}\cdot\mathbf{n}_t & \text{on } \Gamma_t \text{ for } t \in (0, T), \\ \mathbf{v} = 0 & \text{on } \Gamma_- \times (0, T), \\ \dot{\Omega}_t|_{t=0} = \dot{\Omega}_0, \quad \mathbf{v} = \mathbf{v}_0 & \text{in } \dot{\Omega}_0. \end{cases} \tag{3.1}$$

Here, $\partial_t = \partial/\partial t$, $\mathbf{v} = (v_1(x,t), \ldots, v_N(x,t))^\top$ is the velocity field in $\dot{\Omega}_t$, where $M^\top$ denotes the transposed M, $\mathfrak{p} = \mathfrak{p}(x,t)$ the pressure field in $\dot{\Omega}_t$, $\mathbf{D}(\mathbf{v})$ the doubled deformation tensor with $(i,j)^{\text{th}}$ component $\partial_i v_j + \partial_j v_i$, $\partial_i = \partial/\partial x_i$, $\mathbf{I}$ the $N \times N$ identity matrix, σ a positive constant describing the coefficient of the surface tension, $H(\Gamma_t)$ the $N-1$ times fold mean curvature of Γ_t, V_n the evolution velocity of Γ_t along $\mathbf{n}_t$, and p_0 the outside pressure. As for the remaining notation in (3.1), $m = m_+\chi_{\Omega_{t+}} + m_-\chi\Omega_{t-}$ and $\mu = \mu_+\chi_{\Omega_{t+}} + \mu_-\chi_{\Omega_{t-}}$, where $\chi_{\Omega_{t\pm}}$ denotes the characteristic functions of $\Omega_{t\pm}$, that is, $\chi_{\Omega_{t\pm}}(x,t) = 1$ for $(x,t) \in \Omega_{t\pm} \times \{t\}$ and $\chi_{\Omega_{t\pm}}(x,t) = 0$ for $(x,t) \notin \Omega_{t\pm} \times \{t\}$. Moreover, for any function f defined on $\dot{\Omega}_t$, we define the quantity $[[f]]$ by setting

$$[[f]] = [[f]](x_0) = \lim_{\substack{x\to x_0\\ x\in\Omega_{t+}}} f(x) - \lim_{\substack{x\to x_0\\ x\in\Omega_{t-}}} f(x) \quad \text{for } x_0 \in \Gamma_t$$

that is the jump of f across Γ_t; for any matrix field $\mathbf{K}$ with $(i, j)^{\text{th}}$ component K_{ij}, the quantity $\mathrm{Div}\,\mathbf{K}$ is an N-vector with i^{th} component $\sum_{j=1}^N \partial_j K_{ij}$ and for any N-vector field $\mathbf{w} = (w_1, \ldots, w_N)^\top$, we set $\mathrm{div}\,\mathbf{w} = \sum_{j=1}^N \partial_j w_j$. The domains $\Omega_{t\pm}$ and the interface Γ_t are unknown, while $\Omega_{0\pm}$ and Γ_0 are given.

In the equilibrium state, $\mathbf{v} = 0$, and so $\nabla p = 0$, from which $[[p]]$ is a constant on Γ_t. Thus, $\sigma H(\Gamma) = p_0 + \text{constant}$. This constant can be included in the pressure term. Thus, we may assume that

$$p_0 = \sigma H(\Gamma). \tag{3.2}$$

In particular, $\sigma H(\Gamma) = p_0$. Throughout this section we make the following assumptions:

- The interface Γ is a compact hypersurface of C^3 class.
- If Ω is a bounded domain, then Γ_- is a compact hypersurface of C^2 class.
- There exist positive constants $\mu_{\pm 1}$ and $\mu_{\pm 2}$ for which

$$\mu_{1\pm} \leq \mu_\pm(x) \leq \mu_{2\pm} \quad \text{for any } x \in \mathbb{R}^N. \tag{3.3}$$

-

$$\mu_\pm \in H^1_{r,\mathrm{loc}}(\mathbb{R}^N) \quad \text{and} \quad \|\nabla \mu_\pm\|_{L_r(B)} \leq K_{r,\tau} \tag{3.4}$$

for any ball $B \subset \mathbb{R}^N$ with radius $\tau > 0$, where r is an exponent with $N < r < \infty$.

Notice that when $\Omega = \mathbb{R}^N$, $\Gamma_- = \emptyset$, and so we do not impose the boundary condition: $\mathbf{v} = 0$ on Γ_-.

Since $\Omega_{t\pm}$ are unknown, we introduce an unknown function η to describe the surface Γ_t, and by using this function, η, we transform $\Omega_{t\pm}$ to fixed domains $\Omega_\pm$. Let $\dot\Omega = \Omega_+ \cup \Omega_-$, and then the system of linearized equations of the resultant nonlinear problem has the following forms:

$$\begin{cases} \partial_t \mathbf{v} - m^{-1}\mathrm{Div}\,(\mu\mathbf{D}(\mathbf{v}) - \mathfrak{p}\mathbf{I}) = \mathbf{F} & \text{in } \dot\Omega^T, \\ \mathrm{div}\,\mathbf{v} = G = \mathrm{div}\,\mathbf{G} & \text{in } \dot\Omega^T, \\ [[(\mu\mathbf{D}(\mathbf{v}) - \mathfrak{p}\mathbf{I})\mathbf{n}]] - ((\mathcal{B} + \sigma\Delta_\Gamma)\eta)\mathbf{n} = [[\mathbf{H}]], \quad [[\mathbf{v}]] = 0 & \text{on } \Gamma^T, \\ \partial_t \eta + < A_\kappa \mid \nabla_\Gamma \eta > - \mathbf{n}\cdot\mathbf{v} + \mathcal{L}_1\mathbf{v} + \mathcal{L}_2\eta = D & \text{on } \Gamma^T, \\ \mathbf{v} = 0 & \text{on } \Gamma_-^T, \\ (\mathbf{v}, \eta)|_{t=0} = (\mathbf{v}_0, \eta_0) & \text{in } \dot\Omega \times \Gamma. \end{cases} \tag{3.5}$$

Here and in the sequel, m and μ are defined by

$$m = m_+ \chi_{\Omega_+} + m_- \chi_{\Omega_-}, \qquad \mu = \mu_+ \chi_{\Omega_+} + \mu_- \chi_{\Omega_-},$$

where $\chi_{\Omega_\pm}$ are the characteristic function of domains $\Omega_\pm$, $\mathbf{n}$ is the unit outer normal to Γ, $\dot\Omega^T = \dot\Omega \times (0, T)$, $\Gamma^T = \Gamma \times (0, T)$, $\Gamma^T_- = \Gamma_- \times (0, T)$, $\mathcal{B}$, $\mathcal{L}_1$, and $\mathcal{L}_2$ are some linear operators such that

$$\begin{aligned} \|\mathcal{L}_1 \mathbf{v}\|_{W^{2-1/q}_q(\Gamma)} &\leq C\|\mathbf{v}\|_{W^{1-1/q}_q(\Gamma)}, \\ \|\mathcal{L}_2 \eta\|_{W^{2-1/q}_q(\Gamma)} &\leq C\|\eta\|_{W^{2-1/q}_q(\Gamma)}, \\ \|\mathcal{B}\eta\|_{W^{1-1/q}_q(\Gamma)} &\leq C\|\eta\|_{W^{2-1/q}_q(\Gamma)}, \end{aligned} \tag{3.6}$$

Δ_Γ is the Laplace-Beltrami operator on Γ, ∇_Γ the tangential derivatives on Γ, and $< \cdot \mid \cdot >$ denotes the inner-product on Γ. Moreover, A_κ, $\kappa \in [0, 1)$, is an $N - 1$ vector of functions defined on Γ possessing the following properties: $A_0 = 0$ and for any $\kappa \in (0, 1)$

$$\begin{aligned} &|A_\kappa(x)| \leq m_1, \quad |A_\kappa(x) - A_\kappa(y)| \leq m_1|x - y|^a \text{ for any } x, y \in \Gamma, \\ &\|A_\kappa\|_{W^{2-1/r}_r(\Gamma)} \leq m_2\kappa^{-b} \end{aligned} \tag{3.7}$$

for some positive constants m_1, m_2, a, and b independent of $\kappa \in (0, 1)$. We assume that $\Omega_+ \cup \Gamma \cup \Omega_- = \Omega$, that $\Gamma \cap \Gamma_- = \emptyset$, and that $\Omega_+ \cap \Omega_- = \emptyset$.

The main purpose of this lecture note is to prove the local and global well-posedness of problem (3.1), the main theorems for which will be stated in Sects. 3.1.3.1 and 3.1.3.2 below, and the local well-posedness is proved in Sect. 3.5, the global well-posedness in the case where Ω is bounded is proved in Sect. 3.6, and the global well-posedness in the case where $\Omega = \mathbb{R}^N$ is proved in Sect. 3.7. To prove the local well-posedness, the maximal L_p-L_q regularity theorem for equation (3.5) plays an essential role and to prove the global well-posedness, the decay properties of C^0 analytic semigroup associated with equation (3.5), where $\mathbf{F} = G = \mathbf{G} = \mathbf{H} = D = 0$ in the equation (3.5), play an essential role. To prove these properties, we consider the corresponding resolvent problem:

$$\begin{cases} \lambda\mathbf{u} - m^{-1}\mathrm{Div}\,(\mu\mathbf{D}(\mathbf{u}) - \mathfrak{q}\mathbf{I}) = \mathbf{f} & \text{in } \dot\Omega, \\ \mathrm{div}\,\mathbf{u} = g = \mathrm{div}\,\mathbf{g} & \text{in } \dot\Omega, \\ [[(\mu\mathbf{D}(\mathbf{u}) - \mathfrak{q}\mathbf{I})\mathbf{n}]] - ((\mathcal{B} + \sigma\Delta_\Gamma)h)\mathbf{n} = [[\mathbf{h}]], \quad [[\mathbf{u}]] = 0 & \text{on } \Gamma, \\ \lambda h + < A_\kappa \mid \nabla_\Gamma h > - \mathbf{n}\cdot\mathbf{u} + \mathcal{L}_1\mathbf{u} + \mathcal{L}_2 h = d & \text{on } \Gamma, \\ \mathbf{u} = 0 & \text{on } \Gamma_-. \end{cases} \tag{3.8}$$

The maximal L_p-L_q theorems for equation (3.5) are proved by combination of the $\mathcal{R}$ bounded solution operators for equation (3.8) with the Weis operator valued Fourier multiplier theorem [40]. In Sect. 3.2, we will give a detailed proof of the maximal L_p-L_q theorem. Moreover, we will prove the decay properties of solutions to equation (3.8) with $G = \mathbf{G} = \mathbf{H} = 0$ by using spectral analysis in Sect. 3.5.5 when Ω is bounded and in Sect. 3.7 when $\Omega = \mathbb{R}^N$. The proof of the existence of $\mathcal{R}$ bounded solution operators for equation (3.8) is divided into three steps. The main step is the consideration for the model problems in the case where $\Gamma = \mathbb{R}^N_0 := \{x = (x_1, \ldots, x_N) \in \mathbb{R}^N \mid x_N = 0\}$ and $\dot\Omega = \mathbb{R}^N \setminus \mathbb{R}^N_0$, and the detailed proof of the existence of $\mathcal{R}$ bounded solution operators is given in Sect. 3.3 below. The second step is the bent half space case, and in this case employing perturbation argument from the model problem, we can prove the existence of the $\mathcal{R}$ bounded solution operators. The third step is the general domain case and in this case, constructing the parametrix, choosing the spectral parameter λ large enough, and using rather standard argument in theory of parameter elliptic equations where norm is replaced by $\mathcal{R}$ norm, we can prove the existence of $\mathcal{R}$ bounded solution operators. The detailed proofs in the second and third cases are omitted, which are given in Maryani and Saito [20].

In the sequel, concerning the divergence equation: $\operatorname{div} \mathbf{u} = g = \operatorname{div} \mathbf{g}$ in $\dot\Omega$, what $g = \operatorname{div} \mathbf{g}$ means that the data g is given by the relation $g = \operatorname{div} \mathbf{g}$ in $\dot\Omega$ for some N-vector $\mathbf{g}$ in the distribution sense. Moreover, for any $\mathbf{u}$ and $\mathbf{g} \in L_q(\dot\Omega)^N$, $\operatorname{div} \mathbf{u} = \operatorname{div} \mathbf{g}$ means that

$$\mathbf{u} - \mathbf{g} \in J_q(\Omega). \tag{3.9}$$

Here and in the following, $J_q(\Omega)$ denotes the solenoidal space defined by setting

$$J_q(\Omega) = \{\mathbf{v} \in L_q(\dot\Omega) \mid (\mathbf{v}, \nabla\varphi)_{\dot\Omega} = 0 \quad \text{for any } \varphi \in H^1_{q'}(\Omega)\}, \tag{3.10}$$

where

$$\hat H^1_q(\Omega) = \{\varphi \in L_{q,\mathrm{loc}}(\Omega) \mid \nabla\varphi \in L_q(\Omega)\}, \quad q' = q/(q-1).$$

3.1.1.1 Short History

Two-phase problems of two immiscible incompressible viscous fluids have been studied by many mathematicians. This subsection introduces short history of such problems in the following four cases:

1. A bounded domain, in a container, separated by a closed free surface Γ_t;
2. The whole space separated by a closed free surface Γ_t;
3. The whole space separated by a non-compact free surface Γ_t;
4. An infinite layer separated by a non-compact free surface Γ_t.

Case 1 This case deals with the motion of a drop Ω_{+t} inside another liquid, where both fluids are located in a container with a solid surface S and separated by a closed free surface Γ_t. Denisova [8, 9] treated non-capillary fluids and proved, respectively, the existence of classical solutions global in time and the existence of global solutions in a Sobolev space setting based on L_2. Note that the global existence theorems, mentioned above and below, are only proved for small initial data. On the other hand, Denisova and Solonnikov [12] proved the global existence of classical solutions for capillary fluids. In an L_p-setting, Köhne, Prüss, and Wilke proved the global existence of solutions for capillary fluids. Saito, Shibata, and Zhang [25] considered the motion of two immiscible inhomogeneous incompressible viscous fluids without capillarity. They showed the existence of local solutions in general domains containing Cases 1–4, and also the existence of global solutions for Case 1 or the case where the solid surface S is replaced by another closed free surface. Here both the local and global existence theorems in [25] were proved in an L_p-in-time and L_q-in-space setting. Tanaka [36] also treated similar fluids to [25] in Case 1 under the capillarity is taken into account, and proved the global existence of solutions in an L_2-setting. As for problems including the temperature, we refer, e.g., to Tanaka [37], Denisova [7], and Denisova and Nečasová [10].

Case 2 This case describes the motion of a drop Ω_{t+}, bounded by a closed free surface Γ_t, in another liquid $\Omega_{t-} = \mathbb{R}^N \setminus (\Omega_{t+} \cup \Gamma_t)$. Denisova [2] treated capillary fluids and proved a local existence theorem in an L_2-setting. Denisova and Solonnikov [11] also treated capillary fluids, and proved the existence of local classical solutions.

Case 3 Prüss and Simonett [22, 23] proved the existence of strong solutions of equation (3.1) for a given $T > 0$ and small initial data in an L_p-setting in the case where $\mathbb{R}^N$ is separated by a non-compact free surface $\Gamma_t = \{(x', x_N) \mid x_N = h(x', t), x' \in \mathbb{R}^{N-1}\}$ and both the regions $\Omega_{\pm t} = \{(x', x_N) \mid \pm(x_N - h(x', t)) > 0, x' \in \mathbb{R}^{N-1}\}$ are filled with capillary fluids. In addition, Hieber and Saito [15] generalized problems of [22, 23] to ones of non-Newtonian fluids, and proved the existence theorem similar to [22, 23].

Case 4 In this case, both $\Omega_{\pm t}$ are given by infinite layers as follows:

$$\Omega_{+t} = \{(x', x_N) \mid F_-(x', t) < x_N < F_+(x', t), x' \in \mathbb{R}^{N-1}\},$$

$$\Omega_{-t} = \{(x', x_N) \mid b(x') < x_N < F_-(x', t), x' \in \mathbb{R}^{N-1}\},$$

together with the solid boundary $S = \{(x', x_N) \mid x_N = b(x'), x' \in \mathbb{R}^{N-1}\}$ and the two free surfaces $\Gamma_{\pm t}$:

$$\Gamma_{\pm t} = \{(x', x_N) \mid x_N = F_\pm(x', t), x' \in \mathbb{R}^{N-1}\}.$$

Xu and Zhang [41] treated capillary fluids, and proved the local and global existence of solutions in a Sobolev space setting based on L_2. On the other hand, Guo and

Tice [14] and Wang, Tice, and Kim [39] considered the problem in a horizontally periodic setting. They established the global existence of solutions for both non-capillary fluids and capillary fluids in an L_2-setting.

There are also other studies of two-phase problems. For example, two-phase problems of compressible viscous fluids are treated in Denisova [3, 6], Kubo, Shibata, and Soga [19] and Jang, Tice, and Wang [17, 18]; Denisova [4, 5] treated two-phase problems of two fluids of different types, i.e. one fluid is incompressible and another is compressible; Abels [1] considered generalized solutions for two-phase problems of non-Newtonian fluids, and we can find the relevant references therein for the existence of generalized solutions of two-phase problems; Prüss and Simonett [21] proved the Rayleigh-Taylor instability in an L_p-setting.

3.1.1.2 Further Notation

For any scalar function f, we write

$$\nabla f = (\partial_1 f, \ldots, \partial_N f), \quad \bar\nabla f = (f, \partial_1 f, \ldots, \partial_N f),$$
$$\nabla^2 f = (\partial_i\partial_j f \mid i, j = 1, \ldots, N), \quad \bar\nabla^2 f = (\partial_i\partial_j f \mid i, j = 0, 1, \ldots, N),$$

where $\partial_0 f = f$. For any N-vector of function $\mathbf{f} = (f_1, \ldots, f_N)^\top$, we write

$$\nabla\mathbf{f} = (\nabla f_1, \ldots, \nabla f_N), \quad \bar\nabla\mathbf{f} = (\bar\nabla f_1, \ldots, \bar\nabla f_N),$$
$$\nabla^2\mathbf{f} = (\nabla^2 f_1, \ldots, \nabla^2 f_N), \quad \bar\nabla^2\mathbf{f} = (\bar\nabla^2 f_1, \ldots, \bar\nabla^2 f_N).$$

For any m-vector $V = (v_1, \ldots, v_m)$ and n-vector $W = (w_1, \ldots, w_n)$, $V \otimes W$ denotes an (m, n) matrix whose $(i, j)^{\rm th}$ component is $V_i W_j$. For any (mn, N) matrix $A = (A_{ij,k} \mid i = 1, \ldots, m, j = 1, \ldots, n, k = 1, \ldots, N)$, $AV \otimes W$ denotes an N column vector whose $i^{\rm th}$ component is the quantity: $\sum_{j=1}^m \sum_{k=1}^n A_{jk,i} v_j w_k$.

For any domain G, $L_q(G)$, $H^m_q(G)$, and $B^s_{q,p}(G)$ denote respective Lebesgue space, Sobolev space, and Besov space on G, while $\|\cdot\|_{L_q(G)}$, $\|\cdot\|_{H^m_q(G)}$, and $\|\cdot\|_{B^s_{q,p}(G)}$ denote their norms. Let

$$X(\dot\Omega) = \{u \mid u|_{\Omega_\pm} \in X(\Omega_\pm)\}, \quad \|u\|_{X(\dot\Omega)} = \|u|_{\Omega_+}\|_{X(\Omega_+)} + \|u|_{\Omega_-}\|_{X(\Omega_-)}$$

for $X \in \{L_q, H^m_q, B^s_{q,p}\}$. For simplicity, we write $W^s_q(G) = B^s_{q,q}(G)$, while its norm is written by $\|\cdot\|_{W^s_q(G)}$. Let

$$\hat H^1_q(\Omega) = \{u \in L_{q,{\rm loc}}(\Omega) \mid \nabla u \in L_q(\Omega)\}.$$

For any Banach space X with norm $\|\cdot\|_X$, X^N denotes the N product space of X, while the norm of X^N is denoted simply by $\|\cdot\|_X$. Namely,

$$X^N = \{\mathbf{f} = (f_1, \ldots, f_N) \mid f_i \in X\}, \quad \|\mathbf{f}\|_X = \sum_{j=1}^N \|f_j\|_X.$$

For any two Banach spaces X and Y, $\mathcal{L}(X, Y)$ denotes the set of all bounded linear operators from X into Y and $\mathrm{Hol}\,(U, X)$ the set of all X valued holomorphic functions defined on $U \subset \mathbb{C}$, where $\mathbb{C}$ denotes the set of all complex numbers. Set

$$\Sigma_\epsilon = \{\lambda \in \mathbb{C} \setminus \{0\} \mid |\arg\lambda| \leq \pi - \epsilon\}, \quad \Sigma_{\epsilon,\lambda_0} = \{\lambda \in \Sigma_\epsilon \mid |\lambda| \geq \lambda_0\}.$$

For any N-vectors $\mathbf{a}$ and $\mathbf{b}$, we set $\mathbf{a} \cdot \mathbf{b} =< \mathbf{a}, \mathbf{b} >= \sum_{j=1}^N a_i b_i$, where c_i denotes the ith component of $\mathbf{c}$ for $\mathbf{c} \in \{\mathbf{a}, \mathbf{b}\}$. The tangential component $\mathbf{a}_\tau$ with respect to $\mathbf{n}$ is defined by $\mathbf{a}_\tau = \mathbf{a}- < \mathbf{a}, \mathbf{n} > \mathbf{n}$. For any domain D in $\mathbb{R}^N$, and complex valued functions f and g, we set $(f, g)_D = \int_D f(x)\overline{g(x)}\,dx$, where $\overline{g(x)}$ denotes the complex conjugate of $g(x)$. And also, for any hypersurface K, we put $(f, g)_K = \int_K f(x)\overline{g(x)}\,d\omega$, where $d\omega$ denotes the surface element of K. For any two N-vector of functions $\mathbf{f}$ and $\mathbf{g}$, we set $(\mathbf{f}, \mathbf{g})_D = \sum_{j=1}^N (f_j, g_j)$ and $(\mathbf{f}, \mathbf{g})_K = \sum_{j=1}^N (f_j, g_j)_K$, where h_i denotes the ith component of $\mathbf{h}$ for $\mathbf{h} \in \{\mathbf{f}, \mathbf{g}\}$. Throughout this section, the letter C denotes generic constants and $C_{a,b,c,\ldots,}$ denotes that the constant $C_{a,b,c,\ldots}$ depends on the quantities a, b, c, …. The values of C and $C_{a,b,c,\ldots}$ may change from line to line.

3.1.2 Modelling

In this section, we derive the interface conditions so that we assume that $\Omega = \mathbb{R}^N$, that is, $\Gamma_- = \emptyset$ and integration appearing below is finite. We derive the interface conditions which guarantee the conservation of mass and the conservation of momentum. For a while, the mass densities $\rho_\pm = m_\pm$ are assumed to be functions of (x, t). Let $\phi_t : \mathbb{R}^N \to \mathbb{R}^N$ be a bijective map with suitable regularity for each time $t \geq 0$ such that $\phi_t(y)|_{t=0} = y$, and let

$$\Omega_{t\pm} = \{x = \phi_t(y) \mid y \in \Omega_{0\pm}\}.$$

Let $\rho_\pm$, $\mathbf{v}_\pm$, and $\mathfrak{p}_\pm$ satisfy the Navier-Stokes equations:

$$\partial_t \rho_\pm + \mathrm{div}\,(\rho_\pm \mathbf{v}_\pm) = 0 \quad \text{in} \quad \bigcup_{0<t<T} \Omega_{t\pm} \times \{t\}, \tag{3.11}$$

$$\rho_\pm(\partial_t \mathbf{v}_\pm + \mathbf{v}_\pm \cdot \nabla \mathbf{v}_\pm) - \mathrm{Div}\,\mathbf{S}(\mathbf{v}_\pm, \mathfrak{p}_\pm) = 0 \quad \text{in} \quad \bigcup_{0<t<T} \Omega_{t\pm} \times \{t\}. \tag{3.12}$$

We say that Eq. (3.11) is the equation of mass and Eq. (3.12) the equation of momentum. Let

$$\mathbf{S}(\mathbf{v}_\pm, \mathfrak{p}_\pm) = \mu_\pm \mathbf{D}(\mathbf{v}_\pm) + (\nu_\pm - \mu_\pm)\mathrm{div}\, \mathbf{v}_\pm \mathbf{I} - \mathfrak{p}_\pm \mathbf{I}, \tag{3.13}$$

which denotes the stress tensor, where $\mu_\pm$ and $\nu_\pm$ are the first and second viscosity coefficients, respectively. If we assume that $\rho_\pm = m_\pm$ are positive constants, then by (3.11) $\mathrm{div}\, \mathbf{v}_\pm = 0$, which, combined with (3.13), leads to

$$\mathbf{S}(\mathbf{v}_\pm, \mathfrak{p}_\pm) = \mu_\pm \mathbf{D}(\mathbf{v}_\pm) - \mathfrak{p}_\pm \mathbf{I} = \begin{pmatrix} \mathbf{S}_1(\mathbf{v}_\pm, \mathfrak{p}_\pm) \\ \vdots \\ \mathbf{S}_N(\mathbf{v}_\pm, \mathfrak{p}_\pm) \end{pmatrix}.$$

where $\mathbf{S}_i(\mathbf{v}_\pm, \mathfrak{p}_\pm)$ are N row vectors of functions with j-th component

$$\mu_\pm D_{ij}(\mathbf{v}_\pm) + \mathrm{div}\, \mathbf{v}_\pm \delta_{ij} - \mathfrak{p}_\pm \delta_{ij},$$

δ_{ij} being the Kronecker delta symbols, that is, $\delta_{ii} = 1$ and $\delta_{ij} = 0$ $(i \neq j)$. This is the situation in equation (3.1), and then Eq. (3.12) is written componentwise as

$$\begin{aligned} &\rho(\partial_t v_{\pm i} + \sum_{j=1}^N v_{\pm j}\partial_j v_{\pm i}) - \mathrm{div}\, \mathbf{S}_i(\mathbf{v}_\pm, \mathfrak{p}_\pm) \\ &= \rho(\partial_t v_{\pm i} + \sum_{j=1}^N v_{\pm j}\partial_j v_{\pm i}) - \mu_\pm \Delta v_{\pm i} - \nu_\pm \partial_i \mathrm{div}\, \mathbf{v}_\pm + \partial_i \mathfrak{p}_\pm = 0 \end{aligned}$$

for $i = 1, \ldots, N$, where $\Delta v_{\pm i} = \sum_{j=1}^N \dfrac{\partial^2 v_{\pm i}}{\partial x_j^2}$.

To drive interface conditions that preserve the conservation of mass and the conservation of momentum, we use the following theorem.

Theorem 3.1.1 (Reynolds Transport Theorem) *Let $y = \phi_t^{-1}(x)$ be the inverse map of $x = \phi_t(y)$. Let $J = J(y, t)$ be the Jacobian of the transformation: $x = \phi_t(y)$. Then,*

$$\frac{\partial}{\partial t} J(y, t) = (\mathrm{div}_x \mathbf{w}(x, t)) J(y, t),$$

where $\mathbf{w}(x, t) = (\partial_t \phi_t)(\phi_t^{-1}(x))$.

Let $f = f(x, t)$ be a function defined on $\Omega_t = \mathbb{R}^N \setminus \Gamma_t = \Omega_{t+} \cup \Omega_{t-}$, and then

$$\frac{d}{dt} \int_{\Omega_t} f(x, t)\, dx = \int_{\Omega_t} \partial_t f(x, t) + \mathrm{div}\, (f(x, t)\mathbf{w}(x, t))\, dx. \tag{3.14}$$

In fact, by Reynolds transport theorem,

$$\frac{d}{dt}\int_{\Omega_{t\pm}} f(x,t)\,dx = \frac{d}{dt}\int_{\Omega_{0\pm}} f(\phi_t(y),t)J(y,t)\,dy$$
$$= \int_{\Omega_{0\pm}} \{(\partial_t f + (\nabla f)\cdot\partial_t\phi_t)J + f\partial_t J\}\,dy = \int_{\Omega_{t\pm}} (\partial_t f + \mathrm{div}\,(f\mathbf{w}))\,dx.$$

In the sequel, we write

$$\rho = \begin{cases} \rho_+ & \text{for } x \in \Omega_{t+}, \\ \rho_- & \text{for } x \in \Omega_{t-}, \end{cases} \quad \mathbf{v} = \begin{cases} \mathbf{v}_+ & \text{for } x \in \Omega_{t+}, \\ \mathbf{v}_- & \text{for } x \in \Omega_{t-}, \end{cases}$$
$$\mathfrak{p} = \begin{cases} \mathfrak{p}_+ & \text{for } x \in \Omega_{t+}, \\ \mathfrak{p}_- & \text{for } x \in \Omega_{t-}. \end{cases}$$

We first consider the conservation of mass:

$$\frac{d}{dt}\int_{\dot\Omega_t} \rho\,dx = 0. \tag{3.15}$$

By (3.14) and the equation of mass (3.11),

$$\frac{d}{dt}\int_{\dot\Omega_t} \rho\,dx = \int_{\dot\Omega_t} \partial_t\rho + \mathrm{div}\,(\rho\mathbf{w})\,dx = \int_{\dot\Omega_t} \mathrm{div}\,(\rho(\mathbf{w}-\mathbf{v}))\,dx$$
$$= \int_{\Gamma_t} [[\rho(\mathbf{w}-\mathbf{v})\cdot\mathbf{n}_t]]\,d\omega,$$

where $d\omega$ is the area element on Γ_t and we used the fact that $\Gamma_- = \emptyset$. Thus, in order that the mass conservation (3.15) holds, it is sufficient to assume that

$$[[\rho(\mathbf{w}-\mathbf{v})\cdot\mathbf{n}_t]] = 0. \tag{3.16}$$

Assuming that $\rho_+ \neq \rho_-$, by (3.16) we have

$$\mathbf{w}\cdot\mathbf{n}_t = \frac{(\rho_+\mathbf{v}_+ - \rho_-\mathbf{v}_-)\cdot\mathbf{n}_t}{\rho_+ - \rho_-} \quad \text{on } \Gamma_t. \tag{3.17}$$

Let $V_n = \mathbf{w}\cdot\mathbf{n}_t$, which represents the velocity of the evolution of the interface Γ_t, and then (3.17) is written as

$$V_n = \frac{(\rho_+\mathbf{v}_+ - \rho_-\mathbf{v}_-)\cdot\mathbf{n}_t}{\rho_+ - \rho_-} \quad \text{on } \Gamma_t \tag{3.18}$$

for $t \in (0, T)$. This is called **kinematic condition**. In particular, we assume that $[[\mathbf{v}]] = 0$ on Γ_t, that is, $\mathbf{v}_+ = \mathbf{v}_- = \mathbf{v}$ on Γ_t. In this case, kinematic condition reads

$$V_n = \mathbf{v} \cdot \mathbf{n}_t \quad \text{on } \Gamma_t, \tag{3.19}$$

which requires that the interface, although in motion, at all times consists of the same fluid particles. This is the situation of equation (3.1).

Second, we consider the conservation of momentum:

$$\frac{d}{dt} \int_{\dot\Omega_t} \rho v_i \, dx = 0 \quad (i = 1, \dots, N). \tag{3.20}$$

As $\partial_t \rho + \mathrm{div}\,(\rho \mathbf{v}) = 0$ (cf. (3.11)), Eq. (3.12) is rewritten as

$$\partial_t(\rho_\pm v_{\pm i}) + \mathrm{div}\,(\rho_\pm v_{\pm i} \mathbf{v}_\pm) - \mathrm{div}\,\mathbf{S}_i(\mathbf{v}_\pm, \mathfrak{p}_\pm) = 0.$$

And then, by (3.14)

$$\begin{aligned} \frac{d}{dt} \int_{\dot\Omega_t} \rho v_i \, dx &= \int_{\dot\Omega_t} \partial_t(\rho v_i) + \mathrm{div}\,(\rho v_i \mathbf{w}) \, dx \\ &= \int_{\dot\Omega_t} \mathrm{div}\,(\rho v_i (\mathbf{w} - \mathbf{v})) + \mathrm{div}\,\mathbf{S}_i(\mathbf{v}, \mathfrak{p}) \, dx \\ &= \int_{\Gamma_t} [[\rho v_i (\mathbf{w} - \mathbf{v}) \cdot \mathbf{n}_t]] \, d\omega + \int_{\Gamma_t} [[\mathbf{S}_i(\mathbf{v}, \mathfrak{p}) \cdot \mathbf{n}_t]] \, d\omega. \end{aligned}$$

Thus, by (3.16),

$$\frac{d}{dt} \int_{\dot\Omega_t} \rho v_i \, dx = \int_{\Gamma_t} [[\mathbf{S}_i(\mathbf{v}, \mathfrak{p}) \cdot \mathbf{n}_t]] \, d\omega.$$

To obtain (3.20), it suffices to assume that

$$\int_{\Gamma_t} [[\mathbf{S}_i(\mathbf{v}, \mathfrak{p}) \cdot \mathbf{n}_t]] \, d\omega = 0 \quad (i = 1, \dots, N).$$

Let $H(\Gamma_t)$ and Δ_{Γ_t} be the $N-1$ fold mean curvature of Γ_t and the Laplace-Beltrami operator on Γ_t. We know that

$$H(\Gamma_t)\mathbf{n}_t = \Delta_{\Gamma_t} x \quad (x \in \Gamma_t). \tag{3.21}$$

In this lecture note, interface condition is

$$[[\mathbf{S}(\mathbf{v}, \mathfrak{p})\mathbf{n}_t]] = \sigma H(\Gamma_t)\mathbf{n}_t, \quad V_{\Gamma_t} = \mathbf{v}\cdot\mathbf{n}_t \quad \text{on } \Gamma_t \tag{3.22}$$

for $t \in (0, T)$, where $\mathbf{S}(\mathbf{v}_\pm, \mathfrak{p}_\pm)\mathbf{n}_t = (\mathbf{S}_1(\mathbf{v}_\pm, \mathfrak{p}_\pm)\cdot\mathbf{n}_t, \ldots, \mathbf{S}_N(\mathbf{v}_\pm, \mathfrak{p}_\pm)\cdot\mathbf{n}_t)^\top$, and by (3.22) $[[\mathbf{S}_i(\mathbf{v}_\pm, \mathfrak{p}_\pm)\cdot\mathbf{n}_t]] = \sigma\Delta_{\Gamma_t}x_i$ for $i = 1, \ldots, N$. In particular,

$$\int_{\Gamma_t} [[\mathbf{S}_i(\mathbf{v}, \mathfrak{p})\cdot\mathbf{n}_t]]\,d\omega = \int_{\Gamma_t} \Delta_{\Gamma_t}x_i\,d\omega = 0.$$

Thus, the conservation of momentum (3.20) holds.

We will show two more identities. By (3.11), (3.12), and (3.22), we have

$$\frac{d}{dt}\int_{\dot\Omega_t} \rho(x_iv_j - x_jv_i)\,dx = 0, \tag{3.23}$$

$$\frac{d}{dt}\int_{\dot\Omega_t} \rho x_i\,dx = \int_{\dot\Omega_t} \rho v_i\,dx. \tag{3.24}$$

The formula (3.23) is called the conservation of angular momentum. In fact, by (3.14)

$$\begin{aligned}
&\frac{d}{dt}\int_{\dot\Omega_t} \rho(x_iv_j - x_jv_i)\,dx \\
&= \int_{\dot\Omega_t} \{\partial_t(\rho(x_iv_j - x_jv_i)) + \operatorname{div}(\rho(x_iv_j - x_jv_i)\mathbf{w})\}\,dx \\
&= \int_{\dot\Omega_t} \{(\partial_t\rho)(x_iv_j - x_jv_i) + x_i\rho\partial_tv_j - x_j\rho\partial_tv_i + \operatorname{div}(\rho(x_iv_j - x_jv_i)\mathbf{w})\}\,dx \\
&= \int_{\dot\Omega_t} \{-(\operatorname{div}(\rho\mathbf{v}))(x_iv_j - x_jv_i) - \rho(x_i\mathbf{v}\cdot\nabla v_j - x_j\mathbf{v}\cdot\nabla v_i) \\
&\qquad + x_i\operatorname{div}\mathbf{S}_j(\mathbf{v}, \mathfrak{p}) - x_j\operatorname{div}\mathbf{S}_i(\mathbf{v}, \mathfrak{p}) + \operatorname{div}(\rho(x_iv_j - x_jv_i)\mathbf{w})\}\,dx \\
&= \int_{\dot\Omega_t} [\operatorname{div}\{\rho(x_iv_j - x_jv_i)(\mathbf{w} - \mathbf{v})\} + \operatorname{div}(x_i\mathbf{S}_j(\mathbf{v}, \mathfrak{p})) - \operatorname{div}(x_j\mathbf{S}_i(\mathbf{v}, \mathfrak{p})) \\
&\qquad + \sum_{k=1}^N(\delta_{ik}S_{jk}(\mathbf{v}, \mathfrak{p}) - \delta_{jk}S_{ik}(\mathbf{v}, \mathfrak{p})]\,dx,
\end{aligned}$$

where $S_{jk}(\mathbf{v}, \mathfrak{p}) = D_{jk}(\mathbf{v}) + ((\nu - \mu)\operatorname{div}\mathbf{v} - \mathfrak{p})\delta_{jk}$. By (3.16) and $[[\mathbf{v}]] = 0$ on Γ_t, we have

$$\int_{\dot\Omega_t} \mathrm{div}\,\{\rho(x_iv_j - x_jv_i)(\mathbf{w}-\mathbf{v})\}\,dx = \int_{\Gamma_t} \rho(x_iv_j - x_jv_i)[[\mathbf{w}-\mathbf{v})\cdot\mathbf{n}_t]]\,d\omega = 0.$$

By (3.21) and (3.22), we have

$$\begin{aligned}
&\int_{\dot\Omega_t} \mathrm{div}\,(x_i\mathbf{S}_j(\mathbf{v},\mathfrak{p})) - \mathrm{div}\,(x_j\mathbf{S}_i(\mathbf{v},\mathfrak{p}))\,dx \\
&= \int_{\Gamma_t} \{x_i[[\mathbf{S}_j(\mathbf{v},\mathfrak{p})\cdot\mathbf{n}_t]] - x_j[[\mathbf{S}_i(\mathbf{v},\mathfrak{p})\cdot\mathbf{n}_t]]\}\,d\omega \\
&= \sigma\int_{\Gamma_t} \{x_i\Delta_{\Gamma_t}x_j - x_j\Delta_{\Gamma_t}x_i\}\,d\omega \\
&= -\sigma\int_{\Gamma_t} \nabla_{\Gamma_t}x_i\cdot\nabla_{\Gamma_t}x_j - \nabla_{\Gamma_t}x_j\cdot\nabla_{\Gamma_t}x_i\}\,d\omega = 0.
\end{aligned}$$

Since $S_{ij}(\mathbf{v}_\pm,\mathfrak{p}_\pm) = S_{ji}(\mathbf{v}_\pm,\mathfrak{p}_\pm)$, we have

$$\int_{\dot\Omega_t} \{\sum_{k=1}^N(\delta_{ik}S_{jk}(\mathbf{v},\mathfrak{p}) - \delta_{jk}S_{ik}(\mathbf{v},\mathfrak{p}))\,dx = 0.$$

Thus, we have (3.23).

Analogously, by (3.14), (3.11), and (3.16), we have

$$\begin{aligned}
&\frac{d}{dt}\int_{\dot\Omega_t} \rho x_i\,dx = \int_{\dot\Omega_t} \{\partial_t(\rho x_i) + \mathrm{div}\,(\rho x_i\mathbf{w})\}\,dx \\
&= \int_{\dot\Omega_t} \{(\partial_t\rho)x_i + \mathrm{div}\,(\rho x_i\mathbf{w})\}\,dx = \int_{\dot\Omega_t} \{-(\mathrm{div}\,(\rho\mathbf{v}))x_i + \mathrm{div}\,(\rho x_i\mathbf{w})\}\,dx \\
&= \int_{\dot\Omega_t} \{\mathrm{div}\,(x_i\rho(\mathbf{w}-\mathbf{v})) + \rho v_i\}\,dx = \int_{\Gamma_t} x_i[[\rho(\mathbf{w}-\mathbf{v})\cdot\mathbf{n}_t]]\,d\omega + \int_{\dot\Omega_t} \rho v_i\,dx \\
&= \int_{\dot\Omega_t} \rho v_i\,dx.
\end{aligned}$$

Thus, we have (3.24).

Finally, assuming that $\rho_\pm = m_\pm$ are positive constants and adding the initial conditions:

$$\mathbf{v}|_{t=0} = \mathbf{v}_0 \quad \text{in } \dot\Omega_0, \quad \dot\Omega_t|_{t=0} = \dot\Omega_0,$$

we have the set of equations (3.1).

3.1.3 Main Results

3.1.3.1 Local Well-Posedness

In this lecture note, we prove the local and global well-posedness of equation (3.1). In this subsection, we state the local well-posedness. Since Ω_t is unknown, we transform equation (3.1) to nonlinear equations in a fixed domain $\dot\Omega = \Omega \setminus \Gamma$. For this purpose, we introduce the Hanzawa transform. Let $\mathbf{n}$ be the unit outer normal to Γ. Since Γ is a hypersurface of C^3 class, we may assume that $\mathbf{n}$ is defined on $\mathbb{R}^N$ with $\|\mathbf{n}\|_{H^2_\infty(\mathbb{R}^N)} < \infty$ and that there are two tubular neighborhoods of Γ, T^1_u and T^2_u like $T^i_u = \{x \in \mathbb{R}^N \mid \mathrm{dist}\,(x, \Gamma) \leq h_i\}$ such that $\Gamma \subset T^1_u \subset T^2_u \subset \Omega$. Let $\omega(y)$ be a function in $C^\infty(\mathbb{R}^N)$ which equals one in T^1_u and vanishes outside of T^2_u. Let

$$\Gamma_t = \{x = y + \rho(y, t)\mathbf{n} \mid y \in \Gamma\} \tag{3.25}$$

with an unknown function $\rho(y, t)$. Let H_ρ be a suitable extension of $\rho(y, t)$ such that $\rho(y, t) = H_\rho(y, t)$ for $(y, t) \in \Gamma \times (0, T)$ and

$$\begin{aligned} \|H_\rho(\cdot, t)\|_{H^k_q(\mathbb{R}^N)} &\leq C_1\|\rho(\cdot, t)\|_{W^{k-1/q}_q(\Gamma)}, \\ \|\partial_t H_\rho(\cdot, t)\|_{H^\ell_q(\mathbb{R}^N)} &\leq C_1\|\partial_t\rho(\cdot, t)\|_{W^{\ell-1/q}_q(\Gamma)} \end{aligned} \tag{3.26}$$

for $k = 1, 2, 3$ and $\ell = 1, 2$. Let $\Psi_\rho(y, t) = \omega(y)H_\rho(y, t)\mathbf{n}(y)$. The transform:

$$x = y + \Psi_\rho(y, t) \tag{3.27}$$

is called a Hanzawa transform. We assume that

$$\sup_{t\in(0,T)} \|\Psi_\rho(\cdot, t)\|_{H^1_\infty(\mathbb{R}^N)} \leq \delta, \tag{3.28}$$

where δ is a small positive number determined in such a way that several conditions stated below will be satisfied. We choose $\delta \in (0, 1)$, and then the Hanzawa transform defined above is an injective map. In fact, for $x_i = y_i + \Psi_\rho(y_i, t)$ $(i = 1, 2)$,

$$\begin{aligned} |x_1 - x_2| &= |y_1 + \Psi_\rho(y_1, t) - (y_2 + \Psi_\rho(y_2, t))| \\ &\geq |y_1 - y_2| - \|\nabla(\Psi_\rho(\cdot, t))\|_{L_\infty(\mathbb{R}^N)}|y_1 - y_2| \geq (1 - \delta)|y_1 - y_2|, \end{aligned}$$

from which $y_1 \neq y_2$ implies $x_1 \neq x_2$. Let

$$\dot\Omega_t = \{x = y + \Psi_\rho(y, t) \mid y \in \dot\Omega\}, \tag{3.29}$$

and then, the Hanzawa transform maps $\dot\Omega$ onto $\dot\Omega_t$ injectively.

To prove the local well-posedness, using the Hanzawa transform we transform equation (3.1) to the following nonlinear equations:

$$\begin{cases} m\partial_t\mathbf{u} - \mathrm{Div}\,(\mu\mathbf{D}(\mathbf{u}) - \mathfrak{q}\mathbf{I}) = \mathbf{f}(\mathbf{u}, \Psi_\rho) & \text{in } \dot\Omega^T, \\ \mathrm{div}\,\mathbf{u} = g(\mathbf{u}, \Psi_\rho) = \mathrm{div}\,\mathbf{g}(\mathbf{u}, \Psi_\rho) & \text{in } \dot\Omega^T, \\ \partial_t\rho + < \mathbf{u} \mid \nabla'_\Gamma\rho > - \mathbf{u}\cdot\mathbf{n} = d(\mathbf{u}, \Psi_\rho) & \text{on } \Gamma^T, \\ [[(\mu\mathbf{D}(\mathbf{u})\mathbf{n})_\tau]] = [[\mathbf{h}'(\mathbf{u}, \Psi_\rho)]], \quad [[\mathbf{u}]] = 0 & \text{on } \Gamma^T, \\ [[< \mu\mathbf{D}(\mathbf{u})\mathbf{n}, \mathbf{n} > - \mathfrak{q}]] - \sigma(\Delta_\Gamma + \mathbb{B})\rho = [[h_N(\mathbf{u}, \Psi_\rho)]] & \text{on } \Gamma^T, \\ \mathbf{u} = 0 & \text{on } \Gamma_-, \\ \mathbf{u}|_{t=0} = \mathbf{u}_0 \ \text{in } \dot\Omega, \quad \rho|_{t=0} = \rho_0 \ \text{on } \Gamma, \end{cases} \tag{3.30}$$

where setting $\Psi_\rho = \tilde\Psi_\rho = \omega(y)H_\rho(y, t)\mathbf{n}(y)$ and $\alpha = 0$ we define the nonlinear functions $\mathbf{f}(\mathbf{u}, \Psi_\rho)$, $d(\mathbf{u}, \Psi_\rho)$, $g(\mathbf{u}, \Psi_\rho)$, $\mathbf{g}(\mathbf{u}, \Psi_\rho)$, $\mathbf{h}'(\mathbf{u}, \Psi_\rho)$, $h_N(\mathbf{u}, \Psi_\rho)$, and the operator $\mathbb{B}$ by the formulas in respective (3.142), (3.177), (3.136), (3.137), (3.183), (3.196), and (3.193).

Our local well-posedness theorem is stated as follows.

Theorem 3.1.2 *Let $2 < p < \infty$, $N < q < \infty$, $B > 0$ and let Γ be a compact hypersurface of C^3 class. Assume that $2/p + N/q < 1$. Let $\mathbf{u}_0 \in B^{2(1-1/p)}_{q,p}(\dot\Omega)^N$ and $\rho_0 \in B^{3-1/p-1/q}_{q,p}(\Gamma)$ be initial data for which $\|\mathbf{u}_0\|_{B^{2(1-1/p)}_{q,p}(\dot\Omega)} \le B$ holds and the compatibility conditions:*

$$\begin{aligned} \mathbf{u}_0 - \mathbf{g}(\mathbf{u}_0, \Psi_\rho|_{t=0}) \in J_q(\Omega), \quad \mathrm{div}\,\mathbf{u}_0 = g(\mathbf{u}_0, \Psi_\rho|_{t=0}) & \quad \textit{in } \dot\Omega, \\ (\mu\mathbf{D}(\mathbf{u}_0)\mathbf{n})_\tau = \mathbf{h}'(\mathbf{u}_0, \Psi_\rho|_{t=0}), \quad [[\mathbf{u}_0]] = 0 & \quad \textit{on } \Gamma, \\ \mathbf{u}_0 = 0 & \quad \textit{on } \Gamma_-, \end{aligned}$$

are satisfied. Then, there exist a time $T > 0$ and a small number $\epsilon > 0$ depending on R such that if $\|\rho_0\|_{B^{3-1/p-1/q}_{q,p}(\Gamma)} \le \epsilon$, then problem (3.30) admits a unique solution $(\mathbf{u}, \mathfrak{q}, \rho)$ with

$$\begin{aligned} &\mathbf{u} \in L_p((0, T), H^2_q(\dot\Omega)^N) \cap H^1_p((0, T), L_q(\dot\Omega)^N), \\ &\mathfrak{q} \in L_p((0, T), H^1_q(\dot\Omega) + \hat H^1_q(\Omega)), \\ &\rho \in L_p((0, T), W^{3-1/q}_q(\Gamma)) \cap H^1_p((0, T), W^{2-1/q}_q(\Gamma)) \end{aligned}$$

possessing the estimate (3.28) and

$$E_{p,q,T}(\mathbf{u}, \rho) \le CB$$

for some constant C independent of B, where we have set

$$
\begin{aligned}
E_{p,q,T}(\mathbf{u},\rho) &= \|\mathbf{u}\|_{L_p((0,T),H^2_q(\dot\Omega))} + \|\partial_t\mathbf{u}\|_{L_p((0,T),L_q(\dot\Omega))} \\
&\quad + \|\partial_t\rho\|_{L_\infty((0,T),W^{1-1/q}_q(\Gamma))} + \|\partial_t\rho\|_{L_p((0,T),W^{2-1/q}_q(\Omega))} \\
&\quad + \|\rho\|_{L_p((0,T),W^{3-1/p}_q(\Omega))}.
\end{aligned}
$$

Remark 1 From Theorem 3.1.2, the local well-posedness for equation (3.1) follows.

3.1.3.2 Global Well-Posedness

We next consider the global well-posedness. We consider the following two cases: Ω is bounded or $\Omega = \mathbb{R}^N$. The global well-posedness will be studied under the following three conditions:

(A.1) $|\Omega_+| = |B_R| = \dfrac{R^N\omega_N}{N}$, where B_R is the ball in $\mathbb{R}^N$ of radius R centered at the origin, $|D|$ denotes the Lebesgue measure of a Lebesgue measurable set D in $\mathbb{R}^N$, and ω_N is the area of S_1.

(A.2) $\displaystyle\int_{\Omega_{+0}} x\,dx = 0.$

(A.3) Γ_0 is a normal perturbation of S_R given by

$$\Gamma_0 = \{x = y + R^{-1}\rho_0(y)y \mid y \in S_R\}$$

with a given small function $\rho_0(y)$ defined on S_R, where S_R is the sphere in $\mathbb{R}^N$ of radius R.

Notice that $R^{-1}y$ is the unit outer normal to S_R. Let $\rho(y,t)$ be any function defined on S_R such that $\rho(y,t)|_{t=0} = \rho_0(y)$ for $y \in S_R$. Let $\xi(t)$ be the barycenter point of the domain Ω_{t+} defined by

$$\xi(t) = \frac{1}{|\Omega_{t+}|}\int_{\Omega_{t+}} x\,dx.$$

Let Γ_t be given by

$$
\begin{aligned}
\Gamma_t &= \{x = y + \rho(y,t)\mathbf{n} + \xi(t) \mid y \in S_R\} \\
&= \{x = y + R^{-1}\rho(y,t)y + \xi(t) \mid y \in S_R\}.
\end{aligned}
$$

Notice that by (3.15)

$$|\Omega_{t+}| = |\Omega_+| = |B_R|, \tag{3.31}$$

and so we may define $\xi(t)$ by $\xi(t)$ is the barycenter point of the domain Ω_{t+} defined by

$$\xi(t) = \frac{1}{|B_R|}\int_{\Omega_{t+}} x\,dx.$$

It follows from the assumption (A.2) that $\xi(0) = 0$, while the $\xi(t)$ is unknown. By (3.24) and (3.31) we have

$$\xi'(t) = \frac{1}{|B_R|}\int_{\Omega_{t+}} \mathbf{v}(x,t)\,dx.$$

For any function ρ defined on S_R, let H_ρ be a suitable extension of ρ satisfying (3.26), where Γ is replaced by S_R. Below, the method of constructing H_ρ is fixed.

The Case Where Ω Is a Bounded Domain

We first consider the case where Ω is bounded. In this case, let d_0 be a positive number such that $\mathrm{dist}\,(S_R, \Gamma_-) \geq 3d_0$. Let H_ρ be a suitable extension of ρ to $\dot\Omega$ satisfying the estimates:

$$\begin{aligned} \|H_\rho(\cdot,t)\|_{H^k_{r_i}(\dot\Omega)} &\leq C\|\rho(\cdot,t)\|_{W^{k-1/r_i}_{r_i}(S_R)} && \text{for } k = 1,2,3,\\ \|\partial_t H_\rho(\cdot,t)\|_{H^k_{r_i}(\dot\Omega)} &\leq C\|\partial_t\rho(\cdot,t)\|_{W^{k-1/r_i}_{r_i}(S_R)} && \text{for } k = 1,2, \end{aligned} \tag{3.32}$$

for $i = 0, 1, 2$. Let $\zeta(y)$ be a $C^\infty(\mathbb{R}^N)$ function which equals 1 for $|y| < R+d_0$ and 0 for $|y| > R + 2d_0$. We define the Hanzawa transform centered at $\xi(t)$ by (3.27), that is,

$$x = \mathbf{h}_\rho(y,t) = y + \zeta(y)(R^{-1}H_\rho(y,t)y + \xi(t)) \quad \text{for } y \in \Omega. \tag{3.33}$$

Notice that for $y \in B_R$, $\mathbf{h}_\rho(y,t) = y + R^{-1}H_\rho(y,t)y + \xi(t)$. We assume that

$$\sup_{t\in(0,T)} \|\Psi_\rho(\cdot,t)\|_{H^1_\infty(\Omega)} \leq \delta,$$

where $\Psi_\rho = \zeta(y)(R^{-1}\mathbf{n}(y)H_\rho(y,t)y + \xi(t))$. We will choose δ so small that several conditions hold. For example, we choose a $\delta > 0$ so small that the Hanzawa transform is a bijective map from Ω onto Ω with

$$\Gamma_t = \{x = \mathbf{h}_\rho(y,t) \mid y \in \Gamma\}, \quad \Omega_{t\pm} = \{x = \mathbf{h}_\rho(y,t) \mid y \in \Omega_\pm\}$$

for $t \in (0,T)$, where $\Omega_+ = B_R$ and $\Omega_- = \Omega \setminus \{B_R \cup S_R\}$. Let $\mathbf{v}_0(x)$ be an initial data for equation (3.1), and then we set $\mathbf{u}_0(y) = \mathbf{v}_0(\mathbf{h}_{\rho_0}(y))$, where $\mathbf{h}_{\rho_0}(y) =$

$y + \Psi_{\rho_0}(y) = y + \zeta(y)R^{-1}H_{\rho_0}(y)y$ and $H_{\rho_0}(y)$ is a suitable extension of $\rho_0(y)$ such that $H_\rho(y,t)|_{t=0} = H_{\rho_0}(y)$. Notice that $\mathbf{h}_{\rho_0}(y) = \mathbf{h}_\rho(y,0)$. Let $\mathbf{v}$ and $\mathfrak{p}$ satisfy equation (3.1) and we set

$$\mathbf{u}(y,t) = \mathbf{v}(\mathbf{h}_\rho(y,t),t), \quad \mathfrak{q}(y,t) = \mathfrak{p}(\mathbf{h}_\rho(y,t),t) - \frac{\sigma(N-1)}{R}.$$

It then follows from Sects. 3.4.1 and 3.4.2 below that $\mathbf{u}$, $\mathfrak{q}$, and ρ satisfy the following equations:

$$\begin{cases} \partial_t\mathbf{u} - m^{-1}\mathrm{Div}\,(\mu\mathbf{D}(\mathbf{u}) - \mathfrak{q}\mathbf{I}) = \mathbf{f}(\mathbf{u},\Psi_\rho) & \text{in } \dot\Omega^T, \\ \mathrm{div}\,\mathbf{u} = g(\mathbf{u},\Psi_\rho) = \mathrm{div}\,\mathbf{g}(\mathbf{u},\Psi_\rho) & \text{in } \dot\Omega^T, \\ \partial_t\rho - \mathbf{n}\cdot P\mathbf{u} = \tilde d(\mathbf{u},\Psi_\rho) & \text{on } S_R^T, \\ [[(\mu\mathbf{D}(\mathbf{u})\mathbf{n})_\tau]] = [[\mathbf{h}'(\mathbf{u},\Psi_\rho)]] & \text{on } S_R^T, \\ [[< \mu\mathbf{D}(\mathbf{u})\mathbf{n},\mathbf{n} > -\mathfrak{q}]] - \sigma\mathcal{B}\rho = [[h_N(\mathbf{u},\Psi_\rho)]] & \text{on } S_R^T, \\ [[\mathbf{u}]] = 0 & \text{on } S_R^T, \\ \mathbf{u} = 0 & \text{on } \Gamma_-^T, \\ (\mathbf{u},\rho)|_{t=0} = (\mathbf{u}_0,\rho_0) & \text{in } \dot\Omega\times S_R, \end{cases} \tag{3.34}$$

where $\dot\Omega^T = (B_R\cup(\Omega\setminus(B_R\cup S_R)))\times(0,T)$, $S_R^T = S_R\times(0,T)$, $\Gamma_-^T = \Gamma_-\times(0,T)$, $\mathbf{n} = y/|y|$ for $y \in S_R$, $P\mathbf{u} = \mathbf{u} - \dfrac{1}{|B_R|}\displaystyle\int_{B_R}\mathbf{u}\,dy$, $\mathcal{B} = \Delta_{S_R} + \dfrac{N-1}{R^2}$, and Δ_{S_R} is the Laplace-Beltrami operator on S_R. Moreover, setting $\Psi_\rho = \zeta(y)(R^{-1}H_\rho(y,t)y + \xi(t))$ and $\tilde\Psi_\rho = \zeta(y)R^{-1}H_\rho(y,t)y$, we have defined the functions in the right side of (3.34), $\mathbf{f}(\mathbf{u},\Psi_\rho)$, $\tilde d(\mathbf{u},\Psi_\rho)$, $g(\mathbf{u},\Psi_\rho)$, $\mathbf{g}(\mathbf{u},\Psi_\rho)$, $\mathbf{h}'(\mathbf{u},\Psi_\rho)$, and $h_N(\mathbf{u},\Psi_\rho)$ by the formulas in respective (3.142), (3.265), (3.136), (3.136), (3.183), and (3.196).

The global well-posedness theorem for equation (3.34) is stated as follows:

Theorem 3.1.3 *Let p and q be real numbers such that $2 < p < \infty$, $N < q < \infty$ and $2/p + N/q < 1$. Assume that Ω is a bounded domain and Γ_- is a non-empty compact hypersurface of C^2 class. Set*

$$\begin{aligned} \mathcal{S}_{p,q}((a,b)) = \{(\mathbf{u},\mathfrak{q},\rho) \mid\ & \mathbf{u} \in H_p^1((a,b),L_q(\dot\Omega)^N) \cap L_p((a,b),H_q^2(\dot\Omega)^N), \\ & \mathfrak{q} \in L_p((a,b),H_q^1(S_R)), \\ & \rho \in H_p^1((a,b),W_q^{2-1/q}(S_R)) \cap L_p((a,b),W_q^{3-1/q}(S_R))\}. \end{aligned} \tag{3.35}$$

Assume that (A.1), (A.2), and (A.3) hold. Then, there exists a small number $\epsilon > 0$ such that if initial data $\mathbf{u}_0 \in B_{q,p}^{2(1-1/p)}(\dot\Omega)^N$ and $\rho_0 \in B_{q,p}^{3-1/p-/q}(S_R)$ satisfying the smallness condition:

$$\|\mathbf{u}_0\|_{B^{2(1-1/p)}_{q,p}(\dot\Omega)} + \|\rho_0\|_{B^{3-1/p-1/q}_{q,p}(S_R)} \leq \epsilon, \tag{3.36}$$

the compatibility conditions (3.270) *stated in Sect. 3.5.1 below and*

$$\begin{aligned} \mathrm{div}\,\mathbf{u}_0 = g(\mathbf{u}_0, \rho_0) = \mathrm{div}\,\mathbf{g}(\mathbf{u}_0, \rho_0) \quad & \textit{in } \dot\Omega, \\ [[(\mu\mathbf{D}(\mathbf{u}_0)\mathbf{n})_\tau]] = [[\mathbf{h}'(\mathbf{u}_0, \Psi_{\rho_0})]], \quad [[\mathbf{u}_0]] = 0 \quad & \textit{on } S_R, \\ \mathbf{u}_0 = 0 \quad & \textit{on } \Gamma_-, \end{aligned} \tag{3.37}$$

then problem (3.34) *with* $T = \infty$ *admits a unique solution* $(\mathbf{u}, \mathfrak{q}, \rho) \in \mathcal{S}_{p,q}((0,\infty))$ *possessing the estimate:*

$$\begin{aligned} &\|e^{\eta t}\partial_t\mathbf{u}\|_{L_p((0,\infty),L_q(\dot\Omega))} + \|e^{\eta t}\mathbf{u}\|_{L_p((0,\infty),H^2_q(\dot\Omega))} + \|e^{\eta t}\nabla\mathfrak{q}\|_{L_p((0,\infty),L_q(\dot\Omega))} \\ &\quad + \|e^{\eta t}\partial_t\rho\|_{L_p((0,\infty),W^{2-1/q}_q(S_R))} + \|e^{\eta t}\rho\|_{L_p((0,\infty),W^{3-1/q}_q(S_R))} \leq C\epsilon \end{aligned}$$

for some positive constants C and η *independent of* ϵ.

The Case Where $\Omega = \mathbb{R}^N$

In this case, for any function $\rho(y,t)$ defined on $S_R \times (0,T)$, let H_ρ be a suitable extension of ρ such that $\rho(y,t) = H_\rho(y,t)$ for $(y,t) \in S_R \times (0,T)$ and

$$\begin{aligned} \|H_\rho(\cdot,t)\|_{H^k_{r_i}(B_L)} &\leq C\|\rho(\cdot,t)\|_{W^{k-1/r_i}_{r_i}(S_R)} \quad && \text{for } k = 1,2,3, \\ \|\partial_t H_\rho(\cdot,t)\|_{H^k_{r_i}(B_L)} &\leq C\|\partial_t\rho(\cdot,t)\|_{W^{k-1/r_i}_{r_i}(S_R)} \quad && \text{for } k = 1,2, \end{aligned} \tag{3.38}$$

for $i = 0,1,2$, where C is a constant independent of T, $r_0 = q_1/2$, $r_1 = q_1$, and $r_2 = q_2$, and L is a large positive number for which $L > 5R + 2$. Here, q_1 and q_2 are some exponents appearing in Theorem 3.1.4 below. Let $\omega(y)$ be a function in $C^\infty_0(\mathbb{R}^N)$ which equals one for $y \in B_{L-1}$ and zero for $y \notin B_L$. We assume that

$$\sup_{t\in(0,T)} \|\omega(y)R^{-1}H_\rho(y,t)y\|_{H^1_\infty(B_L)} \leq \delta$$

with suitably small positive number δ. Let $\Psi_\rho(y,t) = \omega(y)R^{-1}H_\rho(y,t)y + \xi(t)$ and we define the Hanzawa transform by

$$x = \mathbf{h}_\rho(y,t) = y + \Psi_\rho(y,t). \quad \text{for } y \in \Omega. \tag{3.39}$$

Let $B^R = \{x \in \mathbb{R}^N \mid |x| > R\}$ and set $\Omega_+ = B_R$, $\Omega_- = B^R$, $\dot\Omega = \mathbb{R}^N \setminus S_R = \Omega_+ \cup \Omega_-$,

$$\Gamma_t = \{x = \mathbf{h}_\rho(y,t) \mid y \in S_R\}, \quad \Omega_{t\pm} = \{x = \mathbf{h}_\rho(y,t) \mid y \in \Omega_\pm\}.$$

Let $\mathbf{v}$ and $\mathfrak{p}$ satisfy equation (3.1), and we set

$$\mathbf{u}(y,t) = \mathbf{v}(\mathbf{h}_\rho(y,t),t), \quad \mathfrak{q}(y,t) = \mathfrak{p}(\mathbf{h}_\rho(y,t),t) - \frac{\sigma(N-1)}{R}.$$

Moreover, we set $\mathbf{u}_0(y) = \mathbf{v}_0(\mathbf{h}_{\rho_0}(y))$, where $\mathbf{v}_0$ is the initial data and $\mathbf{h}_{\rho_0}(y) = y + \Psi_\rho|_{t=0}$ with $\rho|_{t=0} = \rho_0$. Then, we see that $\mathbf{u}$, $\mathfrak{q}$, and ρ satisfy the following equations:

$$\begin{cases} \partial_t \mathbf{u} - m^{-1}\mathrm{Div}\,(\mu\mathbf{D}(\mathbf{u}) - \mathfrak{q}\mathbf{I}) = \mathbf{f}(\mathbf{u}, \Psi_\rho) & \text{in } \dot\Omega^T, \\ \mathrm{div}\,\mathbf{u} = g(\mathbf{u}, \Psi_\rho) = \mathrm{div}\,\mathbf{g}(\mathbf{u}, \Psi_\rho) & \text{in } \dot\Omega^T, \\ \partial_t\rho - \mathbf{n}\cdot P\mathbf{u} = \tilde d(\mathbf{u}, \Psi_\rho) & \text{on } S_R^T, \\ [[(\mu\mathbf{D}(\mathbf{u})\mathbf{n})_\tau]] = [[\mathbf{h}'(\mathbf{u}, \Psi_\rho)]] & \text{on } S_R^T, \\ [[< \mu\mathbf{D}(\mathbf{u})\mathbf{n}, \mathbf{n} > -\mathfrak{q}]] - \sigma\mathcal{B}\rho = [[h_N(\mathbf{u}, \Psi_\rho)]] & \text{on } S_R^T, \\ [[\mathbf{u}]] = 0 & \text{on } S_R^T, \\ (\mathbf{u}, \rho)|_{t=0} = (\mathbf{u}_0, \rho_0) & \text{in } \dot\Omega \times S_R, \end{cases} \tag{3.40}$$

where $\dot\Omega^T = (\mathbb{R}^N \setminus S_R) \times (0,T)$, $S_R^T = S_R \times (0,T)$, $\mathbf{n} = y/|y|$ for $y \in S_R$, $P\mathbf{u} = \mathbf{u} - \frac{1}{|B_R|}\int_{B_R} \mathbf{u}\,dy$, $\mathcal{B} = \Delta_{S_R} + \frac{N-1}{R^2}$, and Δ_{S_R} is the Laplace-Beltrami operator on S_R. Setting $\Psi_\rho = \omega(y)R^{-1}H_\rho(y,t)y + \xi(t)$ and $\tilde\Psi_\rho = \omega(y)R^{-1}H_\rho(y,t)y$, we have defined the functions in the right side of (3.40), $\mathbf{f}(\mathbf{u}, \Psi_\rho)$, $\tilde d(\mathbf{u}, \Psi_\rho)$, $g(\mathbf{u}, \Psi_\rho)$, $\mathbf{g}(\mathbf{u}, \Psi_\rho)$, $\mathbf{h}'(\mathbf{u}, \Psi_\rho)$, and $h_N(\mathbf{u}, \Psi_\rho)$ by the formulas in respective (3.142), (3.265), (3.136), (3.137), (3.183), and (3.196).

The global well-posedness for equation (3.40) is stated as follows.

Theorem 3.1.4 *Assume that $\Omega = \mathbb{R}^N$ and that $N \geq 3$. Let L be a large fixed number for which $L > 5R + 2$ and suppose that $\omega(x)$ in Sect. 3.1.3.2 equals one for $x \in B_{L-1}$ and zero for $x \notin B_L$. Let p and q_2 be two exponents satisfying the relations:*

$$2 < p < \infty, \quad N < q_2 < 3N, \quad \frac{2}{p} + \frac{N}{q_2} < 1, \quad \frac{1}{p} + \frac{N}{q_2} > \frac{1}{2}. \tag{3.41}$$

Let q_1 and ζ be two numbers satisfying the relations:

$$\frac{1}{q_1} = \frac{1}{q_2} + \frac{1}{N}, \quad \frac{1}{p} < \zeta < \frac{1}{p} + \frac{N}{q_2}. \tag{3.42}$$

Set $r_0 = q_1/2$, $r_1 = q_1$ *and* $r_2 = q_2$ *and* $X_{p,r_i} = B^{2(1-1/r_i)}_{r_i,p}(\dot\Omega) \times B^{3-1/p-1/r_i}_{r_i,p}(S_R)$. *Let*

$$(\mathbf{u}_0, \rho_0) \in \bigcap_{i=0,1,2} X_{p,r_i}.$$

Suppose that the compatibility conditions:

$$\mathrm{div}\, \mathbf{u}_0 = 0 \quad in\ \dot\Omega, \quad [[(\mathbf{D}(\mathbf{u}_0)\mathbf{n})_\tau]] = [[\mathbf{h}'(\mathbf{u}_0, \Psi_{\rho_0})]], \quad [[\mathbf{u}_0]] = 0 \quad on\ S_R$$

are satisfied. Then, there exists an $\epsilon > 0$ *such that if*

$$\sum_{i=0,1,2} (\|\mathbf{u}_0\|_{B^{2(1-1/r_i)}_{r_i,p}(\dot\Omega)} + \|\rho_0\|_{B^{3-1/p-1/r_i}_{r_i,p}(S_R)}) \leq \epsilon,$$

then equation (3.40) *with* $T = \infty$ *admits a unique solution* $(\mathbf{u}, \rho)$ *with*

$$\mathbf{u} \in \bigcap_{i=0,1,2} (H^1_p((0,\infty), L_{r_i}(\dot\Omega)^N) \cap L_p((0,\infty), H^2_{r_i}(\dot\Omega)^N),$$

$$\rho \in \bigcap_{i=0,1,2} (H^1_p((0,\infty), W^{2-1/r_i}_{r_i}(S_R)) \cap L_p((0,\infty), W^{3-1/r_i}_{r_i}(S_R))$$

possessing the estimate: $E_T(\mathbf{u}, \rho) \leq C\epsilon$ *for any* $T > 0$ *with some constant* C *independent of* ϵ *and* T, *where we have set*

$$\begin{aligned}
&E_T(\mathbf{u}, \rho) \\
&= \| <t>^{\frac{N}{2q_2}+1-\zeta} \rho\|_{L_p((0,T), W^{3-1/q_2}_{q_2}(S_R))} \\
&+ \| <t>^{\frac{N}{2q_2}+1-\zeta} \partial_t\rho\|_{L_p((0,T), W^{2-1/q_2}_{q_2}(S_R))} \\
&+ \| <t>^{\frac{N}{2q_2}+1-\zeta} \mathbf{u}\|_{L_p((0,T), H^2_{q_2}(\dot\Omega))} + \| <t>^{\frac{N}{2q_2}+1-\zeta} \partial_t\mathbf{u}\|_{L_p((0,T), L_{q_2}(\dot\Omega))} \\
&+ \| <t>^{\frac{N}{2q_1}-\zeta} \mathbf{u}\|_{L_p((0,T), H^2_{q_1}(\dot\Omega))} + \| <t>^{\frac{N}{2q_1}-\zeta} \partial_t\mathbf{u}\|_{L_p((0,T), L_{q_1}(\dot\Omega))} \\
&+ \| <t>^{-\zeta} \mathbf{u}\|_{L_p((0,T), H^2_{q_1/2}(\dot\Omega))} + \| <t>^{-\zeta} \partial_t\mathbf{u}\|_{L_p((0,T), L_{q_1/2}(\dot\Omega))} \\
&+ [\mathbf{u}]_{H^1_\infty(\dot\Omega), N/q_1, T} + [\rho]_{W^{2-1/q_2}_{q_2}(S_R), N/q_1, T} + [\partial_t\rho]_{W^{1-1/q_2}_{q_2}(S_R), N/q_1, T} \\
&+ [\mathbf{u}]_{H^1_{q_1/2}(\dot\Omega), 0, T} + [\mathbf{u}]_{H^1_{q_1}(\dot\Omega), N/(2q_1), T} + [\mathbf{u}]_{H^1_{q_2}(B_L), N/(2q_2)+1, T},
\end{aligned}$$

and

$$[\mathbf{v}]_{X,s,T} = \sup_{t \in (0,T)} < t >^s \|\mathbf{v}(\cdot, t)\|_X, \quad < t >= (1+t^2)^{1/2}.$$

Remark 2 If we choose

$$p = \frac{2q_2(1+\sigma)}{q_2 - N}$$

with small positive number σ, then p satisfies the conditions (3.41) and (3.42) with

$$\zeta = \frac{N + q_2 - \sigma}{2q_2(1+\sigma)}.$$

3.2 L_p-L_q Maximal Regularity

3.2.1 $\mathcal{R}$-Bounded Solution Operators

In this section, we study the existence of $\mathcal{R}$-bounded solution operators for equations (3.8). We first introduce the definition of $\mathcal{R}$ boundedness of operator family.

Definition 3.2.1 Let X and Y be two Banach spaces. A family of operators $\mathcal{T} \subset \mathcal{L}(X, Y)$ is called $\mathcal{R}$ bounded on $\mathcal{L}(X, Y)$, if there exist constants $C > 0$ and $p \in [1, \infty)$ such that for each $n \in \mathbb{N}$, $\{T_j\}_{j=1}^n \subset \mathcal{T}$, and $\{f_j\}_{j=1}^n \subset X$, we have

$$\|\sum_{k=1}^n r_k T_k f_k\|_{L_p((0,1),Y)} \leq C\|\sum_{k=1}^n r_k f_k\|_{L_p((0,1),X)}.$$

Here, the Rademacher functions r_k, $k \in \mathbb{N}$ are given by $r_k : [0, 1] \to \{-1, 1\}$ $t \mapsto \text{sign}\,(\sin 2^k \pi t)$. The smallest such C is called $\mathcal{R}$-bound of $\mathcal{T}$ on $\mathcal{L}(X, Y)$, which is denoted by $\mathcal{R}_{\mathcal{L}(X,Y)}\mathcal{T}$.

3.2.1.1 Main Results About $\mathcal{R}$ Bounded Solution Operators

In this subsection, we summarize the results about $\mathcal{R}$ bounded solution operators in the two-phase problem case. We start with the following auxiliary problem:

$$\begin{cases} \lambda \mathbf{u} - m^{-1}\text{Div}\,(\mu \mathbf{D}(\mathbf{u}) - \mathfrak{q}\mathbf{I}) = \mathbf{f} & \text{in } \dot{\Omega}, \\ \text{div}\,\mathbf{u} = g = \text{div}\,\mathbf{g} & \text{in } \dot{\Omega}, \\ [[(\mu \mathbf{D}(\mathbf{u}) - \mathfrak{q}\mathbf{I})\mathbf{n}]] = [[\mathbf{h}]], \quad [[\mathbf{u}]] = 0 & \text{on } \Gamma, \\ \mathbf{u} = 0 & \text{on } \Gamma_-. \end{cases} \tag{3.43}$$

The following theorem follows from the results due to Mariani and Saito [20].

Theorem 3.2.1 *Let* $1 < q < \infty$, $0 < \epsilon < \pi/2$, $N < r < \infty$, *and* $\max(q, q') \leq r$ *with* $q' = q/(q-1)$. *Let*

$$\Sigma_{\epsilon,\lambda_0} = \{\lambda \in \mathbb{C} \mid |\arg\lambda| \leq \pi - \epsilon, \quad |\lambda| \geq \lambda_0\}.$$

Assume that the conditions (3.3) *and* (3.4) *hold and that* Γ *is a compact hypersurface of* C^2 *class. Then, we have the following two assertions:*

(1) **Existence**. *Set*

$$\begin{aligned} X_q = \{(\mathbf{f}, g, \mathbf{g}, \mathbf{h}) \mid \mathbf{f} \in L_q(\dot\Omega)^N, \quad g \in H^1_q(\dot\Omega), \quad \mathbf{g} \in L_q(\dot\Omega)^N,\\ \mathbf{h} \in H^1_q(\dot\Omega)\},\\ \mathcal{X}_q = \{(F_1, \ldots, F_6) \mid F_1, F_4, F_5 \in L_q(\dot\Omega)^N, \quad F_2 \in L_q(\dot\Omega),\\ F_3 \in H^1_q(\dot\Omega), \quad F_6 \in H^1_q(\dot\Omega)^N\}. \end{aligned}$$

Then, there exist a constant $\lambda_0 \geq 1$ *and operator families:*

$$\mathbf{A}(\lambda) \in \mathrm{Hol}\,(\Sigma_{\epsilon,\lambda_0}, \mathcal{L}(\mathcal{X}_q, H^2_q(\dot\Omega)^N)), \quad \mathbf{P}(\lambda) \in \mathrm{Hol}\,(\Sigma_{\epsilon,\lambda_0}, \mathcal{L}(\mathcal{X}_q, H^1_q(\dot\Omega) + \hat H^1_q(\Omega)))$$

such that for any $\lambda \in \Sigma_{\epsilon,\lambda_0}$ *and* $(\mathbf{f}, g, \mathbf{g}, \mathbf{h}) \in X_q(\Omega)$, $\mathbf{u} = \mathbf{A}(\lambda)\mathbf{F}_\lambda$ *and* $\mathfrak{q} = \mathbf{P}(\lambda)\mathbf{F}_\lambda$, *where*

$$\mathbf{F}_\lambda = (\mathbf{f}, \lambda^{1/2}g, g, \lambda\mathbf{g}, \lambda^{1/2}\mathbf{h}, \mathbf{h}),$$

are solutions of equations (3.43), *and moreover*

$$\begin{aligned} \mathcal{R}_{\mathcal{L}(\mathcal{X}_q, H^{2-k}_q(\cdot\Omega)^N}(\{(\tau\partial_\tau)^\ell(\lambda^{k/2}\mathbf{A}(\lambda)) \mid \lambda \in \Sigma_{\epsilon,\lambda_0}\}) \leq r_b \quad (\ell = 0, 1),\\ \mathcal{R}_{\mathcal{L}(\mathcal{X}_q, L_q(\dot\Omega)^N)}(\{(\tau\partial_\tau)^\ell(\nabla\mathbf{P}(\lambda)) \mid \lambda \in \Sigma_{\epsilon,\lambda_0}\}) \leq r_b \quad (\ell = 0, 1) \end{aligned} \tag{3.44}$$

for some positive constant r_b, *where* $\lambda = \gamma + i\tau \in \mathbb{C}$ *and* $k = 0, 1, 2$.

(2) **Uniqueness**. *Let* $\lambda \in \Sigma_{\epsilon,\lambda_0}$. *If* $\mathbf{u}$ *and* $\mathfrak{q}$ *with*

$$\mathbf{u} \in H^2_q(\dot\Omega)^N, \quad \mathfrak{q} \in H^1_q(\dot\Omega) + \hat H^1_q(\Omega)$$

satisfy the homogeneous equations:

$$\begin{cases} \lambda\mathbf{u} - m^{-1}\mathrm{Div}\,(\mu\mathbf{D}(\mathbf{u}) - \mathfrak{q}\mathbf{I}) = 0, & \mathrm{div}\,\mathbf{u} = 0 & \text{in } \dot\Omega,\\ [[(\mu\mathbf{D}(\mathbf{u}) - \mathfrak{q}\mathbf{I})\mathbf{n}]] = 0, & [[\mathbf{u}]] = 0 & \text{on } \Gamma,\\ & \mathbf{u} = 0 & \text{on } \Gamma_-, \end{cases} \tag{3.45}$$

then $\mathbf{u} = 0$ *and* $\mathfrak{q}$ *is a constant in* Ω.

Remark 3 The symbols $F_1, \ldots, F_6$ are corresponding variables to $\mathbf{f}$, $\lambda^{1/2}g$, g, $\lambda\mathbf{g}$, $\lambda^{1/2}\mathbf{h}$, and $\mathbf{h}$, respectively. The norm of space X_q and $\mathcal{X}_q$ are defined by letting

$$\|(\mathbf{f}, g, \mathbf{g}, \mathbf{h})\|_{X_q} = \|(\mathbf{f}, \mathbf{g})\|_{L_q(\dot\Omega)} + \|(g, \mathbf{h})\|_{H^1_q(\dot\Omega)},$$
$$\|(F_1, \ldots, F_6)\|_{\mathcal{X}_q} = \|(F_1, F_2, F_4, F_5)\|_{L_q(\dot\Omega)} + \|(F_3, F_6)\|_{H^1_q(\dot\Omega)}.$$

Thus, we have

$$\|\mathbf{F}_\lambda\|_{\mathcal{X}_q} = \|(\mathbf{f}, \lambda^{1/2}g, \lambda\mathbf{g}, \lambda^{1/2}\mathbf{h})\|_{L_q(\dot\Omega)} + \|(g, \mathbf{h})\|_{H^1_q(\dot\Omega)}. \tag{3.46}$$

We now state a theorem concerning the existence of $\mathcal{R}$-bounded solution operators for equations (3.5)

Theorem 3.2.2 *Let $1 < q < \infty$, $0 < \epsilon < \pi/2$, $N < r < \infty$, and $\max(q, q') \le r$ with $q' = q/(q-1)$. Assume that the conditions* (3.3) *and* (3.4) *hold and that Γ is a compact hypersurfaces of C^3 class. Let $A_0 = 0$ and let A_κ ($\kappa \in (0, 1)$) be an $N-1$ vector of real valued functions satisfying* (3.7). *Let*

$$Y_q = \{\mathbf{F} = (\mathbf{f}, \mathbf{h}, g, \mathbf{g}, d) \mid \mathbf{f} \in L_q(\dot\Omega)^N, \quad \mathbf{g} \in L_q(\dot\Omega)^N, \quad g \in H^1_q(\dot\Omega),$$
$$\mathbf{h} \in H^1_q(\dot\Omega)^N, \quad d \in W^{2-1/q}_q(\Gamma)\},$$
$$\mathcal{Y}_q = \{F = (F_1, F_2, \ldots, F_7) \mid F_1, F_4, F_5 \in L_q(\dot\Omega)^N, \quad F_2 \in L_q(\dot\Omega)^N,$$
$$F_3 \in H^1_q(\dot\Omega), \quad F_6 \in H^1_q(\dot\Omega), \quad F_7 \in W^{2-1/q}_q(\Gamma)\},$$

$$\Lambda_{\kappa,\lambda_0} = \begin{cases} \Sigma_{\epsilon_0,\lambda_0} & \text{for } \kappa = 0, \\ \mathbb{C}_{+,\lambda_0} & \text{for } \kappa \in (0,1), \end{cases} \qquad \ell_\kappa = \begin{cases} 1 & \text{for } \kappa = 0, \\ \kappa^{-b} & \text{for } \kappa \in (0,1), \end{cases}$$

where $\Sigma_{\epsilon,\lambda_0}$ is the same set as in Theorem 3.2.1 and $\mathbb{C}_{+,\lambda_0}$ is a set defined by

$$\mathbb{C}_{+,\lambda_0} = \{\lambda \in \mathbb{C} \mid \mathrm{Re}\,\lambda \ge \lambda_0\}.$$

And then, the following two assertions hold:

(1) **Existence** *There exist a constant $\lambda_0 > 0$ and operator families $\mathcal{A}(\lambda)$, $\mathcal{P}(\lambda)$ and $\mathcal{H}(\lambda)$ with*

$$\mathcal{A}(\lambda) \in \mathrm{Hol}\,(\Lambda_{\kappa,\lambda_0\ell_\kappa}, \mathcal{L}(\mathcal{Y}_q, H^2_q(\dot\Omega)^N)),$$
$$\mathcal{P}(\lambda) \in \mathrm{Hol}\,(\Lambda_{\kappa,\lambda_0\ell_\kappa}, \mathcal{L}(\mathcal{Y}_q, H^1_q(\dot\Omega) + \hat H^1_q(\Omega))),$$
$$\mathcal{H}(\lambda) \in \mathrm{Hol}\,(\Lambda_{\kappa,\lambda_0\ell_\kappa}, \mathcal{L}(\mathcal{Y}_q, W^{3-1/q}_q(\Gamma)))$$

such that for every $\lambda \in \Lambda_{\kappa,\lambda_0\ell_\kappa}$ *and* $(\mathbf{f}, g, \mathbf{g}, \mathbf{h}, d) \in Y_q$, $\mathbf{v} = \mathcal{A}(\mathbf{F}_\lambda, d)$ *and* $\mathfrak{q} = \mathcal{P}(\lambda)(\mathbf{F}_\lambda, d)$, *and* $h = \mathcal{H}(\lambda)(\mathbf{F}_\lambda, d)$ *are solutions of equations* (3.8), *where*

$$(\mathbf{F}_\lambda, d) = (\mathbf{f}, \lambda^{1/2} g, g, \lambda^{1/2}\mathbf{h}, \mathbf{h}, \lambda\mathbf{g}, d).$$

Moreover, we have

$$\mathcal{R}_{\mathcal{L}(\mathcal{Y}_q, H_q^{2-j}(\dot\Omega)^N)}(\{(\tau\partial_\tau)^\ell(\lambda^{j/2}\mathcal{A}(\lambda)) \mid \lambda \in \Lambda_{\kappa,\lambda_0\ell_\kappa}\}) \leq r_b;$$

$$\mathcal{R}_{\mathcal{L}(\mathcal{Y}_q, L_q(\dot\Omega)^N)}(\{(\tau\partial_\tau)^\ell(\nabla\mathcal{P}(\lambda)) \mid \lambda \in \Lambda_{\kappa,\lambda_0\ell_\kappa}\}) \leq r_b;$$

$$\mathcal{R}_{\mathcal{L}(\mathcal{Y}_q, W_q^{3-1/q-k}(\Gamma))}(\{(\tau\partial_\tau)^\ell(\lambda^k\mathcal{H}(\lambda)) \mid \lambda \in \Lambda_{\kappa,\lambda_0\ell_\kappa}\}) \leq r_b$$

for $\ell = 0, 1$, $j = 0, 1, 2$, *and* $k = 0, 1$ *with some constant* $r_b > 0$.

(2) **Uniqueness** *There exists a* $\lambda_0 > 0$ *such that for any* $\lambda \in \Lambda_{\kappa,\lambda_0\ell_\kappa}$, *if* $\mathbf{u}$, $\mathfrak{q}$, *and* h *with*

$$\mathbf{u} \in H_q^2(\dot\Omega)^N, \quad \mathfrak{q} \in H_q^1(\dot\Omega) + \hat H_q^1(\Omega), \quad h \in W_q^{3-1/q}(\Gamma)$$

satisfy the homogeneous equations:

$$\begin{cases} \lambda\mathbf{u} - m^{-1}\mathrm{Div}\,(\mu\mathbf{D}(\mathbf{u}) - \mathfrak{q}\mathbf{I}) = 0, \quad \mathrm{div}\,\mathbf{u} = 0 & \text{in } \dot\Omega, \\ [[(\mu\mathbf{D}(\mathbf{u}) - \mathfrak{q}\mathbf{I})\mathbf{n}]] - ((\mathcal{B} + \sigma\Delta_\Gamma)h)\mathbf{n} = 0, \quad [[\mathbf{u}]] = 0 & \text{on } \Gamma, \\ \lambda h + < A_\kappa \mid \nabla_\Gamma h > - \mathbf{n}\cdot\mathbf{u} + \mathcal{L}_1\mathbf{u} + \mathcal{L}_2 h = 0 & \text{on } \Gamma, \\ \mathbf{u} = 0 & \text{on } \Gamma_-, \end{cases} \tag{3.47}$$

then $\mathbf{u} = 0$, $h = 0$, *and* $\mathfrak{q}$ *is a constant.*

Remark 4 The symbols F_7 is a corresponding variable to d. The norm of spaces Y_q and $\mathcal{Y}_q$ are defined by letting

$$\|(\mathbf{f}, g, \mathbf{g}, \mathbf{h}, d)\|_{Y_q} = \|(\mathbf{f}, g, \mathbf{g}, \mathbf{h})\|_{X_q} + \|d\|_{W_q^{2-1/q}(\Gamma)},$$

$$\|(F_1, \ldots, F_7)\|_{\mathcal{X}_q} = \|(F_1, \ldots, F_6)\|_{\mathcal{X}_q} + \|F_7\|_{W_q^{1-1/q}(\Gamma)}.$$

Thus, we have

$$\|(\mathbf{F}_\lambda, d)\|_{\mathcal{X}_q} = \|(\mathbf{f}, \lambda^{1/2}g, \lambda\mathbf{g}, \lambda^{1/2}\mathbf{h})\|_{L_q(\dot\Omega)} + \|(g, \mathbf{h})\|_{H_q^1(\dot\Omega)} + \|d\|_{W_q^{2-1/q}(\Gamma)}. \tag{3.48}$$

3.2.1.2 Reduced Stokes Equations

Since the pressure term $\mathfrak{p}$ in equations (3.5) has no time evolution, we eliminate $\mathfrak{p}$ to formulate the problem in the semigroup setting. For this purpose we introduce the weak elliptic transmission problem:

$$(m^{-1}\nabla u, \nabla\varphi)_{\dot\Omega} = (\mathbf{f}, \nabla\varphi)_{\dot\Omega} \quad \text{for any } \varphi \in \hat H^1_{q'}(\Omega), \tag{3.49}$$

subject to $[[u]] = g$ on Γ, where $q' = q/(q-1)$. Let $1 < q < \infty$. When Ω is a bounded domain or $\mathbb{R}^N$, we know that for any $\mathbf{f} \in L_q(\dot\Omega)^N$ and $g \in W^{1-1/q}_q(\Gamma)$, problem (3.49) admits a unique solution $u \in H^1_q(\dot\Omega) + \hat H^1_q(\Omega)$ possessing the estimate:

$$\|\nabla u\|_{L_q(\dot\Omega)} \leq C(\|\mathbf{f}\|_{L_q(\dot\Omega)} + \|g\|_{W^{1-1/q}_q(\Gamma)}). \tag{3.50}$$

Using this fact, we define an operator K_0 acting on $\mathbf{u} \in H^2_q(\dot\Omega)^N$ by $u = K_0(\mathbf{u})$, where u is a unique solution of the variational equation:

$$(m^{-1}\nabla u, \nabla\varphi)_{\dot\Omega} = (m^{-1}\mathrm{Div}\,(\mu\mathbf{D}(\mathbf{u})) - \nabla\mathrm{div}\,\mathbf{u}, \nabla\varphi)_{\dot\Omega} \tag{3.51}$$

for any $\varphi \in \hat H^1_{q'}(\Omega)$, subject to $[[u]] =< [[\mu\mathbf{D}(\mathbf{u})\mathbf{n}]], \mathbf{n} > -[[\mathrm{div}\,\mathbf{u}]]$ on Γ. The operator K_0 maps $H^2_q(\dot\Omega)^N$ into $H^1_q(\dot\Omega) + \hat H^1_q(\Omega)$ and possesses the estimate:

$$\|\nabla K_0(\mathbf{u})\|_{L_q(\dot\Omega)} \leq C\|\nabla\mathbf{u}\|_{H^1_q(\dot\Omega)}.$$

And also, we define an operator K_1 acting on $(\mathbf{u}, h) \in H^2_q(\dot\Omega)^N \times W^{3-1/q}_q(\Gamma)$ by $v = K_1(\mathbf{u}, h)$, where v is a unique solution of the variational equation:

$$(m^{-1}\nabla v, \nabla\varphi)_{\dot\Omega} = (m^{-1}\mathrm{Div}\,(\mu\mathbf{D}(\mathbf{u})) - \nabla\mathrm{div}\,\mathbf{u}, \nabla\varphi)_{\dot\Omega} \tag{3.52}$$

for any $\varphi \in \hat H^1_{q'}(\Omega)$, subject to $[[v]] =< [[\mu\mathbf{D}(\mathbf{u})\mathbf{n}]], \mathbf{n} > -[[\mathrm{div}\,\mathbf{u}]] - (\mathcal{B} + \sigma\Delta)h$ on Γ. The operator K_1 maps from $H^2_q(\dot\Omega)^N \times W^{3-1/q}_q(\Gamma)$ into $H^1_q(\dot\Omega) + \hat H^1_q(\Omega)$ and possesses the estimate:

$$\|\nabla K_1(\mathbf{u}, h)\|_{L_q(\dot\Omega)} \leq C(\|\nabla\mathbf{u}\|_{H^1_q(\dot\Omega)} + \|h\|_{W^{3-1/q}_q(\Gamma)}).$$

We call the following equations the reduced Stokes equations with interface conditions:

$$\begin{cases} \lambda\mathbf{v} - m^{-1}\mathrm{Div}\,(\mu\mathbf{D}(\mathbf{v}) - K_0(\mathbf{v})\mathbf{I}) = \mathbf{f} & \text{in } \dot\Omega, \\ [[(\mu\mathbf{D}(\mathbf{v}) - K_0(\mathbf{v})\mathbf{I})\mathbf{n}]] = [[\mathbf{h}]], \quad [[\mathbf{v}]] = 0 & \text{on } \Gamma, \\ \mathbf{v} = 0 & \text{on } \Gamma_-; \end{cases} \tag{3.53}$$

$$\begin{cases} \lambda\mathbf{v} - m^{-1}\mathrm{Div}\,(\mu\mathbf{D}(\mathbf{v}) - K_1(\mathbf{v}, h)\mathbf{I}) = \mathbf{f} & \text{in } \dot\Omega, \\ [[(\mu\mathbf{D}(\mathbf{v}) - K_1(\mathbf{v}, h)\mathbf{I})\mathbf{n}]] - ((\mathcal{B} + \sigma\Delta_\Gamma)h)\mathbf{n} = [[\mathbf{h}]], \ [[\mathbf{v}]] = 0 & \text{on } \Gamma, \\ \lambda h + < A_\kappa \mid \nabla_\Gamma h > - \mathbf{n}\cdot\mathbf{v} + \mathcal{L}_1\mathbf{v} + \mathcal{L}_2 h = d & \text{on } \Gamma, \\ \mathbf{v} = 0 & \text{on } \Gamma_-. \end{cases} \tag{3.54}$$

In equations (3.53) and (3.54), the transmission condition is equivalent to

$$[[(\mu\mathbf{D}(\mathbf{v})\mathbf{n})_\tau]] = [[\mathbf{h}]]_\tau, \quad [[\mathrm{div}\,\mathbf{v}]] = \mathbf{n}\cdot[[\mathbf{h}]], \quad [[\mathbf{v}]] = 0 \quad \text{on } \Gamma, \tag{3.55}$$

where $\mathbf{d}_\tau = \mathbf{d} - < \mathbf{d}, \mathbf{n} > \mathbf{n}$ for any N-vector $\mathbf{d}$.

In the sequel, we shall study the equivalence of equations (3.43) and (3.53) and also the equivalence of equations (3.8) and (3.54). The argument is the same, and so we only consider the equivalence of equations (3.43) and (3.54) below.

We first consider that the unique existence theorem of equations (3.43) implies that of equations (3.54). Suppose that for any $\mathbf{f} \in N^2_q(\dot\Omega)^N$, $g \in H^1_q(\dot\Omega)$, $\mathbf{g} \in L_q(\dot\Omega)^N$, and $\mathbf{h} \in H^1_q(\dot\Omega)$, problem (3.43) admits a unique solution $(\mathbf{u}, \mathfrak{q})$ with

$$\mathbf{u} \in H^2_q(\dot\Omega)^N, \quad \mathfrak{q} \in H^1_q(\dot\Omega) + \hat H^1_q(\Omega).$$

Given $\mathbf{f} \in L_q(\dot\Omega)^N$, let $g \in H^1_q(\dot\Omega)$ be a solution of the resolvent problem for the weak elliptic transmission problem:

$$\lambda(g, \varphi)_{\dot\Omega} + (\nabla g, \nabla\varphi)_{\dot\Omega} = (-\mathbf{f}, \nabla\varphi)_{\dot\Omega} \quad \text{for any } \varphi \in H^1_{q'}(\Omega) \tag{3.56}$$

subject to $[[g]] = [[\mathbf{n}\cdot\mathbf{h}]]$ on Γ. This problem is uniquely solvable for $\lambda \in \Sigma_{\epsilon,\lambda_0}$ with large $\lambda_0 > 0$. Notice that $g = \mathrm{div}\,\mathbf{g}$ with $\mathbf{g} = \lambda^{-1}(\mathbf{f} + \nabla g)$. By the assumption, we have $\mathbf{u} \in H^2_q(\dot\Omega)^N$ and $\mathfrak{q} \in H^1_q(\dot\Omega) + \hat H^1_q(\Omega)$ that solve equations (3.43). Especially, by (3.9), we have

$$\lambda(\mathbf{u}, \nabla\varphi)_{\dot\Omega} = (\mathbf{f} - \nabla g, \nabla\varphi)_{\dot\Omega} \tag{3.57}$$

for any $\varphi \in \hat H^1_{q'}(\Omega)$. We shall prove that $\mathfrak{q} = K_0(\mathbf{u})$. For any $\varphi \in \hat H^1_{q'}(\Omega)$, by (3.43), (3.57), (3.51), and $\mathrm{div}\,\mathbf{u} = g$, we have

$$\begin{aligned}(\mathbf{f}, \nabla\varphi)_{\dot\Omega} &= (\lambda\mathbf{u} - m^{-1}\mathrm{Div}\,(\mu\mathbf{D}(\mathbf{u})) + m^{-1}\nabla\mathfrak{q}, \nabla\varphi)_{\dot\Omega} \\ &= (\mathbf{f} + \nabla g, \nabla\varphi) - (\nabla g, \nabla\varphi)_{\dot\Omega} + (m^{-1}\nabla(\mathfrak{q} - K_0(\mathbf{u})), \nabla\varphi)_{\dot\Omega},\end{aligned}$$

and so we have

$$(m^{-1}\nabla(\mathfrak{q} - K_0(\mathbf{u})), \nabla\varphi)_{\dot\Omega} = 0 \quad \text{for any } \varphi \in \hat H^1_{q'}(\Omega).$$

Moreover, from the interface condition in (3.43) and (3.51) it follows that

$$\begin{aligned}[[\mathfrak{q} - K_0(\mathbf{u})]] &=< [[\mu\mathbf{D}(\mathbf{u})\mathbf{n}]], \mathbf{n} > -[[\mathbf{h}]]\cdot\mathbf{n} - (< [[\mu\mathbf{D}(\mathbf{u})\mathbf{n}]], \mathbf{n} > -[[\operatorname{div}\mathbf{u}]]) \\ &= -[[g]] + [[g]] = 0.\end{aligned}$$

Thus the uniqueness implies that $\mathfrak{q} = K_0(\mathbf{u})$, which implies that $\mathbf{u}$ is a unique solution of equations (3.53).

We assume conversely that problem (3.43) is uniquely solvable. Given $\mathbf{f} \in L_q(\dot\Omega)^N$ and $\mathbf{h} \in H^1_q(\dot\Omega)^N$, let $\theta \in H^1_q(\dot\Omega) + \hat H^1_q(\Omega)$ be a unique solution of the weak elliptic transmission problem:

$$(m^{-1}\nabla\theta, \nabla\varphi)_{\dot\Omega} = (\mathbf{f}, \nabla\varphi)_{\dot\Omega} \quad \text{for any } \varphi \in \hat H^1_{q'}(\Omega)$$

subject to $[[\theta]] = \mathbf{n}\cdot[[\mathbf{h}]]$ on Γ. Setting $\mathfrak{q} = \theta + \mathfrak{p}$ in equations (3.43), we then have

$$\begin{cases} \lambda\mathbf{u} - m^{-1}\operatorname{Div}(\mu\mathbf{D}(\mathbf{u}) - \mathfrak{p}\mathbf{I}) = \mathbf{f} - \nabla\theta & \text{in } \dot\Omega, \\ \operatorname{div}\mathbf{u} = g = \operatorname{div}\mathbf{g} & \text{in } \dot\Omega, \\ [[(\mu\mathbf{D}(\mathbf{u}) - \mathfrak{q}\mathbf{I})\mathbf{n}]] = [[\mathbf{h}]] - < [[\mathbf{h}]], \mathbf{n} > \mathbf{n}, \quad [[\mathbf{u}]] = 0 & \text{on } \Gamma, \\ \mathbf{u} = 0 & \text{on } \Gamma_-. \end{cases} \tag{3.58}$$

Let $\mathbf{f}' = \mathbf{f} - \nabla\theta$ and $\mathbf{h}' = [[\mathbf{h}]] - < [[\mathbf{h}]], \mathbf{n} > \mathbf{n}$, and then

$$\mathbf{h}'\cdot\mathbf{n} = 0 \quad \text{on } \Gamma \quad \text{and} \quad (\mathbf{f}', \nabla\varphi)_{\dot\Omega} = 0 \quad \text{for any } \varphi \in \hat H^1_{q'}(\Omega). \tag{3.59}$$

Given $\mathbf{g} \in L_q(\dot\Omega)^N$ satisfying $g = \operatorname{div}\mathbf{g} \in H^1_q(\dot\Omega)$, let $K \in H^1_q(\dot\Omega) + \hat H^1_{q'}(\dot\Omega)$ be a unique solution of the weak elliptic transmission problem:

$$(m^{-1}\nabla K, \nabla\varphi)_{\dot\Omega} = (\lambda\mathbf{g} - \nabla g, \nabla\varphi)_{\dot\Omega} \quad \text{for any } \varphi \in \hat H^1_{q'}(\Omega) \tag{3.60}$$

subject to $[[K]] = -[[g]]$ on Γ. Let $\mathbf{u} \in H^2_q(\dot\Omega)^N$ be a unique solution of the equations:

$$\begin{cases} \lambda\mathbf{u} - m^{-1}\operatorname{Div}(\mu\mathbf{D}(\mathbf{u}) - K_0(\mathbf{u})\mathbf{I}) = \mathbf{f}' + m^{-1}\nabla K & \text{in } \dot\Omega, \\ [[\mu\mathbf{D}(\mathbf{u}) - K_0(\mathbf{u})\mathbf{I})\mathbf{n}]] = \mathbf{h}'_\tau + [[g]]\mathbf{n}, \quad [[\mathbf{u}]] = 0 & \text{on } \Gamma, \\ \mathbf{u} = 0 & \text{on } \Gamma_-, \end{cases} \tag{3.61}$$

where $\mathbf{h}'_\tau = \mathbf{h}' - < \mathbf{h}, \mathbf{n} > \mathbf{n}$. Using (3.59), (3.60), and (3.61), for any $\varphi \in \hat{H}^1_{q'}(\Omega)$ we have

$$\begin{aligned}(m^{-1}\nabla K, \nabla\varphi)_{\dot\Omega} &= (\mathbf{f}' + m^{-1}\nabla K, \nabla\varphi)_{\dot\Omega} \\ &= (\lambda\mathbf{u} - m^{-1}\mathrm{Div}\,(\mu\mathbf{D}(\mathbf{u}) - K_0(\mathbf{u})\mathbf{I}), \nabla\varphi)_{\dot\Omega} \\ &= (\lambda\mathbf{u}, \nabla\varphi)_{\dot\Omega} - (\nabla\mathrm{div}\,\mathbf{u}, \nabla\varphi)_{\dot\Omega}.\end{aligned} \tag{3.62}$$

Since $H^1_{q'}(\Omega) \subset \hat{H}^1_{q'}(\Omega)$, by (3.60) and (3.62) we have

$$\lambda(\mathrm{div}\,\mathbf{u} - g, \varphi)_{\dot\Omega} + (\nabla(\mathrm{div}\,\mathbf{u} - g), \nabla\varphi)_{\dot\Omega} = 0 \quad \text{for any } \varphi \in H^1_{q'}(\Omega).$$

Here, we have used the fact that $g = \mathrm{div}\,\mathbf{g}$ in $\dot\Omega$. Moreover, by the transmission condition in equations (3.61) and (3.51), we have

$$[[\mathrm{div}\,\mathbf{u} - g]] =< [[\mu\mathbf{D}(\mathbf{u})\mathbf{n}]], \mathbf{n} > -[[K_0(\mathbf{u})]] - [[g]] = 0 \quad \text{on } \Gamma.$$

Thus, the uniqueness implies that $\mathrm{div}\,\mathbf{u} = g$ in $\dot\Omega$, which, combined with (3.60) and (3.62), leads to $\mathbf{u} - \mathbf{g} \in J_q(\Omega)$, that is,

$$(\mathbf{u}, \nabla\varphi)_{\dot\Omega} = (\mathbf{g}, \nabla\varphi)_{\dot\Omega} \quad \text{for any } \varphi \in \hat{H}^1_{q'}(\Omega),$$

because we may assume that $\lambda \neq 0$. Since $[[g]] = -[[K]]$ on Γ, by (3.61) we have

$$\begin{cases} \lambda\mathbf{u} - m^{-1}\mathrm{Div}\,(\mu\mathbf{D}(\mathbf{u}) - (K_0(\mathbf{u}) - K)\mathbf{I}) = \mathbf{f}', \quad \mathrm{div}\,\mathbf{u} = g = \mathrm{div}\,\mathbf{g} & \text{in } \dot\Omega, \\ [[\mu\mathbf{D}(\mathbf{u}) - (K_0(\mathbf{u}) - K)\mathbf{I})\mathbf{n}]] = \mathbf{h}', \quad [[\mathbf{u}]] = 0 & \text{on } \Gamma, \\ \mathbf{u} = 0 & \text{on } \Gamma_-, \end{cases}$$

because $\mathbf{h}' \cdot \mathbf{n} = 0$ on Γ. Thus, $\mathbf{u}$ and $\mathfrak{q} = K_0(\mathbf{u}) - K - \theta$ are required solutions of equation (3.43).

3.2.1.3 $\mathcal{R}$-Bounded Solution Operators for the Reduced Stokes Equations

In this subsection, we state the existence of $\mathcal{R}$ bounded solution operators for (3.8) and (3.43) and also the uniqueness theorems. The following theorem follows from the result due to Mariani and Saito [20].

Theorem 3.2.3 *Let* $1 < q < \infty$, $0 < \epsilon < \pi/2$, $N < r < \infty$, *and* $\max(q, q') \le r$ *with* $q' = q/(q-1)$. *Assume that the conditions* (3.3) *and* (3.4) *hold and that* Γ *is a compact hypersurface of* C^2 *class. Then, we have the following two assertions:*

(1) **Existence**. *Set*

$$X_{r,q} = \{(\mathbf{f}, \mathbf{h}) \mid \mathbf{f} \in L_q(\dot\Omega)^N, \ \ \mathbf{h} \in H^1_q(\dot\Omega)^N\},$$
$$\mathcal{X}_{r,q} = \{(F_1, F_5, F_6) \mid F_1, F_5 \in L_q(\dot\Omega)^N, \ \ F_6 \in H^1_q(\dot\Omega)^N\}.$$

Then, there exists a constant $\lambda_0 \geq 1$ *and an operator family:* $\mathbf{A}_r(\lambda)$ *with*

$$\mathbf{A}_r(\lambda) \in \mathrm{Hol}\,(\Sigma_{\epsilon,\lambda_0}, \mathcal{L}(\mathcal{X}_{r,q}, H^2_q(\dot\Omega)^N))$$

such that for any $\lambda \in \Sigma_{\epsilon,\lambda_0}$ *and* $(\mathbf{f}, \mathbf{h}) \in X_{r,q}(\Omega)$, $\mathbf{u} = \mathbf{A}(\lambda)(\mathbf{f}, \lambda^{1/2}\mathbf{h}, \mathbf{h})$ *is a solution of equations* (3.53), *and moreover*

$$\mathcal{R}_{\mathcal{L}(\mathcal{X}_{r,q}, H^{2-k}_q(\cdot\Omega)^N}(\{(\tau\partial_\tau)^\ell(\lambda^{k/2}\mathbf{A}_r(\lambda)) \mid \lambda \in \Sigma_{\epsilon,\lambda_0}\}) \leq r_b \quad (\ell = 0, 1) \tag{3.63}$$

for $k = 0, 1, 2$ *with some positive constant* r_b, *where* $\lambda = \gamma + i\tau \in \mathbb{C}$.

(2) **Uniqueness**. *Let* $\lambda \in \Sigma_{\epsilon,\lambda_0}$. *If* $\mathbf{u} \in H^2_q(\dot\Omega)^N$ *satisfies the homogeneous equations:*

$$\begin{cases} \lambda\mathbf{u} - m^{-1}\mathrm{Div}\,(\mu\mathbf{D}(\mathbf{u}) - K_0(\mathbf{u})\mathbf{I}) = 0, & \mathrm{div}\,\mathbf{u} = 0 & \text{in } \dot\Omega, \\ [[(\mu\mathbf{D}(\mathbf{u}) - \mathfrak{q}\mathbf{I})\mathbf{n}]] = 0, & [[\mathbf{u}]] = 0 & \text{on } \Gamma, \\ & \mathbf{u} = 0 & \text{on } \Gamma_-, \end{cases} \tag{3.64}$$

then $\mathbf{u} = 0$.

And also, we have the following theorem.

Theorem 3.2.4 *Let* $1 < q < \infty$, $0 < \epsilon < \pi/2$, $N < r < \infty$, *and* $\max(q, q') \leq r$ *with* $q' = q/(q-1)$. *Assume that the conditions* (3.3) *and* (3.4) *hold and that* Γ *is a compact hypersurface of* C^3 *class. Then, we have the following two assertions:*

(1) **Existence**. *Set*

$$Y_{r,q} = \{(\mathbf{f}, \mathbf{h}, d) \mid \mathbf{f} \in L_q(\dot\Omega)^N, \ \ \mathbf{h} \in H^1_q(\dot\Omega)^N, \ \ d \in W^{2-1/q}_q(\Gamma)\},$$
$$\mathcal{Y}_{r,q} = \{(F_1, F_5, F_6, F_7) \mid F_1, F_5 \in L_q(\dot\Omega)^N, \ \ F_6 \in H^1_q(\dot\Omega)^N, \ \ F_7 \in W^{2-1/q}_q(\Gamma)\}.$$

Let $\Lambda_{\kappa,\lambda_0}$ *and* ℓ_κ *be the same set and number as in Theorem 3.2.2. Then, there exist a constant* $\lambda_0 \geq 1$ *and operator families* $\mathcal{A}_r(\lambda)$ *and* $\mathcal{H}_r(\lambda)$ *with*

$$\mathcal{A}_r(\lambda) \in \mathrm{Hol}\,(\Lambda_{\kappa,\lambda_0\ell_\kappa}, \mathcal{L}(\mathcal{Y}_{r,q}, H^2_q(\dot\Omega)^N)), \ \mathcal{H}_r(\lambda) \in \mathrm{Hol}\,(\Lambda_{\kappa,\lambda_0\ell_\kappa}, \mathcal{L}(\mathcal{Y}_{r,q}, H^3_q(\Omega)^N))$$

such that for any $\lambda \in \Lambda_{\kappa,\lambda_0\ell_\kappa}$ *and* $(\mathbf{f}, \mathbf{h}, d) \in Y_{r,q}(\Omega)$, $\mathbf{u} = \mathcal{A}(\lambda)(\mathbf{f}, \lambda^{1/2}\mathbf{h}, \mathbf{h}, d)$ *and* $h = \mathcal{H}(\mathbf{f}, \lambda^{1/2}\mathbf{h}, \mathbf{h}, d)$ *are solutions of equations* (3.54), *and moreover*

$$\mathcal{R}_{\mathcal{L}(\mathcal{Y}_{r,q}, H^{2-j}_q(\cdot\Omega)^N}(\{(\tau\partial_\tau)^\ell(\lambda^{j/2}\mathcal{A}_r(\lambda)) \mid \lambda \in \Lambda_{\kappa,\lambda_0\ell_\kappa}\}) \leq r_b \quad (\ell = 0, 1),$$
$$\mathcal{R}_{\mathcal{L}(\mathcal{Y}_{r,q}, H^{3-k}_q(\cdot\Omega)^N}(\{(\tau\partial_\tau)^\ell(\lambda^{k}\mathcal{H}_r(\lambda)) \mid \lambda \in \Lambda_{\kappa,\lambda_0\ell_\kappa}\}) \leq r_b \quad (\ell = 0, 1) \tag{3.65}$$

for $j = 0, 1, 2$ *and* $k = 0, 1$ *with some positive constant* r_b*, where* $\lambda = \gamma + i\tau \in \mathbb{C}$.

(2) **Uniqueness**. *Let* $\lambda \in \Lambda_{\kappa,\lambda_0\ell_\kappa}$. *If* $\mathbf{v} \in H^2_q(\dot\Omega)^N$ *and* $h \in H^3_q(\Omega)$ *satisfy the homogeneous equations:*

$$\begin{cases} \lambda\mathbf{v} - m^{-1}\mathrm{Div}\,(\mu\mathbf{D}(\mathbf{v}) - K_1(\mathbf{v}, h)\mathbf{I}) = 0, & \mathrm{div}\,\mathbf{u} = 0 \quad \text{in } \dot\Omega, \\ [[(\mu\mathbf{D}(\mathbf{v}) - K_1(\mathbf{v}, h)\mathbf{I})\mathbf{n}]] - ((\mathcal{B} + \sigma\Delta_\Gamma)h)\mathbf{n} = 0, & [[\mathbf{v}]] = 0 \quad \text{on } \Gamma, \\ & \mathbf{v} = 0 \quad \text{on } \Gamma_-, \end{cases} \tag{3.66}$$

then $\mathbf{v} = 0$ *and* $h = 0$.

Remark 5

(1) The norms of spaces $X_{r,q}$, $\mathcal{X}_{r,q}$, $Y_{r,q}$ and $\mathcal{Y}_{r,q}$ are defined by

$$\begin{aligned} \|(\mathbf{f}, \mathbf{h})\|_{X_{r,q}} &= \|\mathbf{f}\|_{L_q(\dot\Omega)} + \|\mathbf{h}\|_{H^1_q(\Omega)}, \\ \|(F_1, F_5, F_6)\|_{\mathcal{X}_{r,q}} &= \|(F_1, F_5)\|_{L_q(\dot\Omega)} + \|F_6\|_{H^1_q(\Omega)}, \\ \|(\mathbf{f}, \mathbf{h}, d)\|_{Y_{r,q}} &= \|\mathbf{f}\|_{L_q(\dot\Omega)} + \|d\|_{W^{2-1/q}_q(\Gamma)} + \|\mathbf{h}\|_{H^1_q(\Omega)}, \\ \|(F_1, F_5, F_6, F_7)\|_{\mathcal{Y}_{r,q}} &= \|(F_1, F_5)\|_{L_q(\dot\Omega)} + \|F_6\|_{H^1_q(\Omega)} + \|F_7\|_{W^{2-1/q}_q(\Gamma)}. \end{aligned}$$

In particular,

$$\begin{aligned} \|(\mathbf{f}, \lambda^{1/2}\mathbf{h}, \mathbf{h})\|_{\mathcal{X}_{r,q}} &= \|\mathbf{f}\|_{L_q(\dot\Omega)} + \|\lambda^{1/2}\mathbf{h}\|_{L_q(\dot\Omega)} + \|\mathbf{h}\|_{H^1_q(\Omega)}, \\ \|(\mathbf{f}, \lambda^{1/2}\mathbf{h}, \mathbf{h}, d)\|_{\mathcal{Y}_{r,q}} &= \|\mathbf{f}\|_{L_q(\dot\Omega)} + \|\lambda^{1/2}\mathbf{h}\|_{L_q(\dot\Omega)} + \|\mathbf{h}\|_{H^1_q(\Omega)} + \|d\|_{W^{2-1/q}_q(\Gamma)}. \end{aligned}$$

(2) To prove Theorem 3.2.4, we reduce the problem to the model problem in $\mathbb{R}^N$, the model problem in the half-space with non-slip condition, and the model problem in $\mathbb{R}^N$ with transmission problem by using the partition of unity. This method is well-known and in the two-phase problem case, the reader can find such arguments, for example, in Maryani and Saito [20]. The model problem in $\mathbb{R}^N$ and in the half-space with non-slip condition are found in Shibata [27, Subsec. 9.3.1 and Subsec. 9.3.3]. The model problem in $\mathbb{R}^N$ with transmission condition was proved in Shibata and Shimizu [34] in the case where $A_\kappa = 0$. In this lecture note, we give a proof of Theorem 3.2.4 in the model problem case with $A_\kappa \neq 0$ in Sect. 3.3 below.

3.2.2 *Generation of C^0 Analytic Semigroup and Maximal L_p-L_q Regularity*

In this section, we consider the L_p-L_q maximal regularity for equations (3.5). As an auxiliary problem, we also consider the evolution equations:

$$\begin{cases} \partial_t \mathbf{v} - m^{-1}\mathrm{Div}\,(\mu\mathbf{D}(\mathbf{v}) - \mathfrak{p}\mathbf{I}) = F & \text{in } \dot{\Omega}^T, \\ \mathrm{div}\,\mathbf{v} = G = \mathrm{div}\,\mathbf{G} & \text{in } \dot{\Omega}^T, \\ [[(\mu\mathbf{D}(\mathbf{v}) - \mathfrak{p}\mathbf{I})\mathbf{n}]] = [[\mathbf{H}]], \quad [[\mathbf{v}]] = 0 & \text{on } \Gamma^T, \\ \mathbf{v} = 0 & \text{on } \Gamma_-^T, \\ \mathbf{v}|_{t=0} = \mathbf{v}_0 & \text{in } \dot{\Omega}. \end{cases} \tag{3.67}$$

3.2.2.1 C^0 Analytic Semigroup Associated with Equation (3.5) and (3.67)

In this subsection, we study the generation of C^0 analytic semigroups corresponding to the following two problems:

$$\begin{cases} \partial_t \mathbf{v} - m^{-1}\mathrm{Div}\,(\mu\mathbf{D}(\mathbf{v}) - \mathfrak{p}\mathbf{I}) = 0 & \text{in } \dot{\Omega}^T, \\ \mathrm{div}\,\mathbf{v} = 0 & \text{in } \dot{\Omega}^T, \\ [[(\mu\mathbf{D}(\mathbf{v}) - \mathfrak{p}\mathbf{I})\mathbf{n}]] = 0, \quad [[\mathbf{v}]] = 0 & \text{on } \Gamma^T, \\ \mathbf{v} = 0 & \text{on } \Gamma_-^T, \\ \mathbf{v}|_{t=0} = \mathbf{v}_0 & \text{in } \dot{\Omega}. \end{cases} \tag{3.68}$$

$$\begin{cases} \partial_t \mathbf{v} - m^{-1}\mathrm{Div}\,(\mu\mathbf{D}(\mathbf{v}) - \mathfrak{p}\mathbf{I}) = 0 & \text{in } \dot{\Omega}^T, \\ \mathrm{div}\,\mathbf{v} = 0 & \text{in } \dot{\Omega}^T, \\ [[(\mu\mathbf{D}(\mathbf{v}) - \mathfrak{p}\mathbf{I})\mathbf{n}]] - ((\mathcal{B} + \sigma\Delta_\Gamma)\eta)\mathbf{n} = 0, \quad [[\mathbf{v}]] = 0 & \text{on } \Gamma^T, \\ \partial_t \eta - \mathbf{n}\cdot\mathbf{v} = 0 & \text{on } \Gamma^T, \\ \mathbf{v} = 0 & \text{on } \Gamma_-^T, \\ (\mathbf{v}, \eta)|_{t=0} = (\mathbf{v}_0, \eta_0) & \text{in } \dot{\Omega}\times\Gamma. \end{cases} \tag{3.69}$$

In order to formulate problems (3.68) and (3.69) in the semigroup setting, we have to eliminate the pressure term $\mathfrak{p}$, because it has no time evolution. For this purpose, we use the operators K_0 and K_1 introduced in Sect. 3.2.1.2. Then, instead of equations (3.68) and (3.69), we consider the following two equations:

$$\begin{cases} \partial_t \mathbf{v} - m^{-1}\mathrm{Div}\,(\mu\mathbf{D}(\mathbf{v}) - K_0(\mathbf{v})\mathbf{I}) = 0 & \text{in } \dot\Omega^T, \\ [[(\mu\mathbf{D}(\mathbf{v}) - K_0(\mathbf{v})\mathbf{I})\mathbf{n}]] = 0, \quad [[\mathbf{v}]] = 0 & \text{on } \Gamma^T, \\ \mathbf{v} = 0 & \text{on } \Gamma_-^T, \\ \mathbf{v}|_{t=0} = \mathbf{v}_0 & \text{in } \dot\Omega. \end{cases} \tag{3.70}$$

$$\begin{cases} \partial_t \mathbf{v} - m^{-1}\mathrm{Div}\,(\mu\mathbf{D}(\mathbf{v}) - K_1(\mathbf{v}, \eta)\mathbf{I}) = 0 & \text{in } \dot\Omega^T, \\ [[(\mu\mathbf{D}(\mathbf{v}) - K_1(\mathbf{v}, \eta)\mathbf{I})\mathbf{n}]] - ((\mathcal{B} + \sigma\Delta_\Gamma)\eta)\mathbf{n} = 0, \quad [[\mathbf{v}]] = 0 & \text{on } \Gamma^T, \\ \partial_t \eta - \mathbf{n}\cdot\mathbf{v} + \mathcal{L}_1\mathbf{v} + \mathcal{L}_2\eta = 0 & \text{on } \Gamma^T, \\ \mathbf{v} = 0 & \text{on } \Gamma_-^T, \\ (\mathbf{v}, \eta)|_{t=0} = (\mathbf{v}_0, \eta_0) & \text{in } \dot\Omega \times \Gamma. \end{cases} \tag{3.71}$$

In view of (3.72), the transmission condition in equations (3.70) and (3.71) reads

$$[[(\mu\mathbf{D}(\mathbf{v})\mathbf{n})_\tau]] = 0, \quad [[\mathrm{div}\,\mathbf{v}]] = 0, \quad [[\mathbf{v}]] = 0 \quad \text{on } \Gamma. \tag{3.72}$$

To formulate problems (3.70) and (3.71) in the semigroup setting, we introduce the following spaces:

$$\begin{aligned} J_q(\dot\Omega) &= \{\mathbf{u} \in L_q(\dot\Omega)^N \mid (\mathbf{u}, \nabla\varphi)_{\dot\Omega} = 0 \quad \text{for any } \varphi \in \hat H^1_{q'}(\Omega)\}, \\ \mathcal{D}_{1,q}(\dot\Omega) &= \{\mathbf{v} \in J_q(\dot\Omega) \cap H^2_q(\dot\Omega)^N \mid \\ &\qquad [[(\mu\mathbf{D}(\mathbf{v})\mathbf{n})_\tau]] = 0, \ [[\mathbf{v}]] = 0 \text{ on } \Gamma, \ \mathbf{v} = 0 \text{ on } \Gamma_-\}, \\ \mathbf{L}_1\mathbf{v} &= m^{-1}\mathrm{Div}\,(\mu\mathbf{D}(\mathbf{v}) - K_0(\mathbf{v})\mathbf{I}) \quad \text{for } \mathbf{v} \in \mathcal{D}_{1,q}; \\ \mathcal{H}_q &= \{(\mathbf{v}, \eta) \mid \mathbf{v} \in J_q(\dot\Omega) \ \eta \in W^{2-1/q}_q(\Gamma)\}, \\ \mathcal{D}_{2,q} &= \{(\mathbf{v}, \eta) \in \mathcal{H}_q \mid \mathbf{v} \in H^2_q(\dot\Omega)^N, \ \eta \in W^{3-1/q}_q(\Gamma), \\ &\qquad [[(\mu\mathbf{D}(\mathbf{v})\mathbf{n})_\tau]] = 0, \ [[\mathbf{v}]] = 0 \text{ on } \Gamma, \ \mathbf{v} = 0 \text{ on } \Gamma_-\}, \\ \mathbf{L}_2(\mathbf{v}, \eta) &= (m^{-1}\mathrm{Div}\,(\mu\mathbf{D}(\mathbf{v}) - K_1(\mathbf{v}, \eta)\mathbf{I}), (\mathbf{n}\cdot\mathbf{v} - \mathcal{L}_1\mathbf{v} - \mathcal{L}_2\eta)|_\Gamma) \end{aligned} \tag{3.73}$$

for $(\mathbf{v}, \eta) \in \mathcal{D}_{2,q}$. Using the symbols defined in (3.73), problem (3.70) is written as

$$\partial_t\mathbf{v} - \mathbf{L}_1\mathbf{v} = 0 \ \text{ for } t > 0, \ \ \mathbf{v}|_{t=0} = \mathbf{v}_0 \tag{3.74}$$

with $\mathbf{v} \in \mathcal{D}_{1,q}$ for any $t > 0$. And, problem (3.71) is written as

$$\partial_t(\mathbf{v}, \eta) - \mathbf{L}_2(\mathbf{v}, \eta) = 0 \ \text{ for } t > 0, \ \ (\mathbf{v}, \eta)|_{t=0} = (\mathbf{v}_0, \eta_0) \tag{3.75}$$

with $(\mathbf{v}, \eta) \in \mathcal{D}_{2,q}$ for any $t > 0$.

Since $\mathcal{R}$-boundedness implies the usual boundedness of the operators, by Theorem 3.2.3 and Theorem 3.2.4, for any $\epsilon \in (0, \pi/2)$ there exists a λ_0 such that both resolvent sets of $\mathbf{L}_1$ and $\mathbf{L}_2$ contain $\Sigma_{\epsilon,\lambda_0}$ and

$$|\lambda|\|(\lambda - \mathbf{L}_1)^{-1}\mathbf{v}_0\|_{L_q(\dot\Omega)} + \|(\lambda - \mathbf{L}_1)^{-1}\mathbf{v}_0\|_{H^2_q(\dot\Omega)} \leq C\|\mathbf{v}_0\|_{L_q(\dot\Omega)};$$

$$|\lambda|\|(\lambda - \mathbf{L}_2)^{-1}(\mathbf{w}_0, \eta_0)\|_{\mathcal{H}_q} + \|(\lambda - \mathbf{L}_2)^{-1}(\mathbf{w}_0, \eta_0)\|_{\mathcal{D}_{2,q}} \leq C\|(\mathbf{w}_0, \eta_0)\|_{\mathcal{H}_q}$$

for any $\lambda \in \Sigma_{\epsilon,\lambda_0}$, $\mathbf{v}_0 \in J_q(\dot\Omega)$ and $(\mathbf{w}_0, \eta_0) \in \mathcal{H}_q$ with some constant C, where we have set

$$\|(\mathbf{w}_0, \eta_0)\|_{\mathcal{H}_q} = \|\mathbf{w}_0\|_{L_q(\dot\Omega)} + \|\eta_0\|_{W^{2-1/q}_q(\Gamma)} \quad \text{for any } (\mathbf{w}_0, \eta_0) \in \mathcal{H}_q;$$

$$\|(\mathbf{w}_0, \eta_0)\|_{\mathcal{D}_{2,q}} = \|\mathbf{w}_0\|_{H^2_q(\dot\Omega)} + \|\eta_0\|_{W^{3-1/q}_q(\Gamma)} \quad \text{for any } (\mathbf{w}_0, \eta_0) \in \mathcal{D}_{2,q}.$$

By analytic semigroup theory, we see that $\mathbf{L}_1$ and $\mathbf{L}_2$ generate C^0 analytic semigroups $\{T_1(t)\}_{t\geq 0}$ and $\{T_2(t)\}_{t\geq 0}$ on $J_q(\dot\Omega)$ and $\mathcal{H}_q$, respectively. Thus. by a standard real-interpolation method (cf. Shibata and Shimizu [33, Theorem 3.9]), we have the following theorems.

Theorem 3.2.5 *Let $1 < p, q < \infty$. Assume that Γ is a compact hypersurface of C^2 class. Let*

$$\mathcal{E}^1_{p,q} = (J_q(\dot\Omega), \mathcal{D}_{1,q}(\dot\Omega))_{1-1/q,p},$$

where $(\cdot, \cdot)_{1-1/p,p}$ denotes a real interpolation functor. Then, for any $\mathbf{v}_0 \in \mathcal{E}^1_{p,q}$, problem (3.70) *admits a unique solution $\mathbf{v}$ with*

$$e^{-\gamma t}\mathbf{v} \in L_p((0, \infty), H^2_q(\dot\Omega)^N) \cap H^1_p((0, T), L_q(\dot\Omega)^N)$$

possessing the estimate:

$$\|e^{-\gamma t}\mathbf{v}\|_{L_p((0,\infty),H^2_q(\dot\Omega))} + \|e^{-\gamma t}\partial_t\mathbf{v}\|_{L_p((0,\infty),L_q(\dot\Omega))} \leq C\|\mathbf{v}_0\|_{B^{2(1-1/p)}_{q,p}(\dot\Omega)}$$

for any $\gamma \geq \gamma_0$ with some constants γ_0 and C. Here, C depends on γ_0 but is independent of γ.

Remark 6

(1) Notice that $\mathcal{E}^1_{p,q} \subset B^{2(1-1/p)}_{q,p}(\dot\Omega)$. Since $\mathbf{v} \in J_q(\dot\Omega)$ for any $t > 0$, $\mathbf{v}$ and $\mathfrak{p} = K_0(\mathbf{v})$ are unique solution of equation (3.68).
(2) $\mathbf{v}_0 \in \mathcal{E}^1_{p,q}$ holds if $\mathbf{v}_0 \in J_q(\dot\Omega)$ and the following conditions hold:

$$\begin{cases} [[(\mu\mathbf{D}(\mathbf{v}_0)\mathbf{n})_\tau]] = 0 \quad \text{on } \Gamma & \text{for } 2/p + 1/q < 1, \\ [[\mathbf{v}_0]] = 0 \quad \text{on } \Gamma, \quad \mathbf{v}_0 = 0 \quad \text{on } \Gamma_- & \text{for } 2/p + 1/q < 2 \end{cases} \tag{3.76}$$

when $2/p + 1/q \neq 1$ and 2.

Theorem 3.2.6 *Let $1 < p, q < \infty$. Assume that Γ is a compact hypersurface of C^3 class. Let*

$$\mathcal{E}^2_{p,q} = (\mathcal{H}_q, \mathcal{D}_{2,q})_{1-1/q,p}.$$

Then, for any $(\mathbf{v}_0, \eta_0) \in \mathcal{E}^2_{p,q}$, problem (3.70) *admits a unique solution $(\mathbf{v}, \eta)$ with*

$$\mathbf{v} \in L_p((0, \infty), H^2_q(\dot\Omega)^N) \cap H^1_p((0, T), L_q(\dot\Omega)^N),$$
$$\eta \in L_p((0, \infty), W^{3-1/q}_q(\Gamma)) \cap H^1_p((0, \infty), W^{2-1/q}_q(\Gamma))$$

possessing the estimate:

$$\|e^{-\gamma t}\mathbf{v}\|_{L_p((0,\infty),H^2_q(\dot\Omega))} + \|e^{-\gamma t}\partial_t\mathbf{v}\|_{L_p((0,\infty),L_q(\dot\Omega))} + \|e^{-\gamma t}\eta\|_{L_p((0,\infty),W^{3-1/q}_q(\Gamma))}$$
$$+ \|e^{-\gamma t}\partial_t\eta\|_{L_p((0,\infty),W^{2-1/q}_q(\Gamma))} \leq C(\|\mathbf{v}_0\|_{B^{2(1-1/p)}_{q,p}(\dot\Omega)} + \|\eta_0\|_{B^{3-1/p-1/q}_{q,p}(\Gamma)})$$

for any $\gamma \geq \gamma_0$ with some constants γ_0 and C. Here, C depends on γ_0 but is independent of γ.

Remark 7

(1) Notice that $\mathcal{E}^2_{p,q} \subset B^{2(1-1/p)}_{q,p}(\dot\Omega) \times B^{3-1/p-1/q}_{q,p}(\Gamma)$. Since $\mathbf{v} \in J_q(\dot\Omega)$ for any $t > 0$, $\mathbf{v}$, $\mathfrak{p} = K_0(\mathbf{v})$ and η are unique solutions of equation (3.69).
(2) $\mathbf{v}_0 \in \mathcal{E}^1_{p,q}$ holds if $\mathbf{v}_0 \in J_q(\dot\Omega)$ and the compatibility condition (3.76) holds when $2/p + 1/q \neq 1, 2$. Notice that there is no compatibility condition for η_0.

3.2.2.2 Maximal L_p-L_q Regularity Theorem

To state our maximal L_p-L_q regularity theorem, we first introduce Bessel potential spaces. Let $\mathcal{F}$ and $\mathcal{F}^{-1}_\tau$ be the Fourier transform and its inverse transform defined by

$$\mathcal{F}[f](\tau) = \int_{\mathbb{R}} e^{-it\tau} f(t)\, dt, \quad \mathcal{F}^{-1}_\tau[g(\tau)](t) = \frac{1}{2\pi}\int_{\mathbb{R}} e^{it\tau} g(\tau)\, d\tau.$$

Let X be a Banach space. For $1 < p < \infty$ and $s > 0$, $H^s_p(\mathbb{R}, x)$ denotes a Bessel potential space defined by

$$H_p^s(\mathbb{R}, X) = \{f \in L_p(\mathbb{R}, X) \mid \\ \|f\|_{H_p^s(\mathbb{R},X)} = \|\mathcal{F}_\tau^{-1}[(1+\tau^2)^{s/2}\mathcal{F}[f](\tau)]\|_{L_p(\mathbb{R},X)} < \infty\},$$

where $L_p(\mathbb{R}, X)$ denotes the X-valued L_p space on $\mathbb{R}$, while $\|\cdot\|_{L_p(\mathbb{R},X)}$ denotes its norm defined by

$$\|f\|_{L_p(\mathbb{R},X)} = \Big(\int_{\mathbb{R}} \|f(t)\|_X^p\,dt\Big)^{1/p}.$$

We now state our maximal L_p-L_q regularity theorems.

Theorem 3.2.7 *Let* $1 < p, q < \infty$ *with* $2/p + 1/q \neq 1, 2$, *and* $T > 0$. *Assume that* Γ *is a compact hypersurface of* C^2 *class. Then, the following two assertions hold.*

(1) **Existence** *There exists a* γ_0 *for which the following assertion holds: Let* $\mathbf{u}_0 \in B_{q,p}^{2(1-1/p)}(\dot\Omega)^N$ *be initial data for equations* (3.67) *and let* $\mathbf{F}$, G, $\mathbf{G}$, *and* $\mathbf{H}$ *be functions in the right side of equations* (3.67) *such that*

$$\mathbf{F} \in L_p((0,T), L_q(\dot\Omega)^N), \quad e^{-\gamma t}\mathbf{G} \in H_p^1(\mathbb{R}, L_q(\dot\Omega)^N),$$
$$e^{-\gamma t}G \in H_p^{1/2}(\mathbb{R}, L_q(\dot\Omega)) \cap L_p(\mathbb{R}, H_q^1(\dot\Omega)),$$
$$e^{-\gamma t}\mathbf{H} \in H_p^{1/2}(\mathbb{R}, L_q(\dot\Omega)^N) \cap L_p(\mathbb{R}, H_q^1(\dot\Omega)^N).$$

for any $\gamma \geq \gamma_0$. *Assume that the compatibility conditions:*

$$\mathbf{u}_0 - \mathbf{G}|_{t=0} \in J_q(\Omega) \text{ and } \operatorname{div}\mathbf{u}_0 = G|_{t=0} \text{ in } \dot\Omega \tag{3.77}$$

hold. In addition, we assume that

$$\begin{aligned} &[[(\mu\mathbf{D}(\mathbf{u}_0)\mathbf{n})_\tau]] = [[\mathbf{H}_\tau]]|_{t=0} \text{ on } \Gamma && \text{for } 2/p + 1/q < 1, \\ &[[\mathbf{u}_0]] = 0 \text{ on } \Gamma, \quad \mathbf{u}_0 = 0 \text{ on } \Gamma_- && \text{for } 2/p + 1/q < 2. \end{aligned} \tag{3.78}$$

Then, problem (3.67) *admits a unique solution* $(\mathbf{v}, \mathfrak{p})$ *with*

$$\mathbf{v} \in L_p((0,T), H_q^2(\dot\Omega)^N) \cap H_p^1((0,T), L_q(\dot\Omega)^N),$$
$$\mathfrak{p} \in L_p((0,T), H_q^1(\dot\Omega) + \hat H_q^1(\Omega))$$

possessing the estimate:

$$\|\mathbf{v}\|_{L_p((0,T),H_q^2(\dot\Omega))} + \|\partial_t\mathbf{v}\|_{L_p((0,T),L_q(\dot\Omega))} + \|\nabla\mathfrak{p}\|_{L_p((0,T),L_q(\dot\Omega))}$$
$$\leq Ce^{\gamma T}\{\|\mathbf{u}_0\|_{B_{q,p}^{2(1-1/p)}(\dot\Omega)} + \|\mathbf{F}\|_{L_p((0,T),L_q(\dot\Omega))} + \|e^{-\gamma t}\partial_t\mathbf{G}\|_{L_p(\mathbb{R},L_q(\dot\Omega))}$$

$$+ \|e^{-\gamma t}G\|_{L_p(\mathbb{R},H^1_q(\dot\Omega))} + \|e^{-\gamma t}\mathbf{H}\|_{L_p(\mathbb{R},H^1_q(\dot\Omega))}$$

$$+ (1+\gamma^{1/2})(\|e^{-\gamma t}G\|_{H^{1/2}_p(\mathbb{R},L_q(\dot\Omega))} + \|e^{-\gamma t}\mathbf{H}\|_{H^{1/2}_p(\mathbb{R},L_q(\dot\Omega))})\}$$

for any $\gamma \geq \gamma_0$ with some constant C independent of γ.

(2) **Uniqueness** *Let $\mathbf{v}$ and $\mathfrak{p}$ with*

$$\mathbf{v} \in L_p((0,T), H^2_q(\dot\Omega)^N) \cap H^1_p((0,T), L_q(\dot\Omega)^N),$$
$$\mathfrak{p} \in L_p((0,T), H^1_q(\dot\Omega) + \hat H^1_q(\dot\Omega))$$

satisfy the homogeneous equations:

$$\begin{cases} \partial_t\mathbf{v} - m^{-1}\mathrm{Div}\,(\mu\mathbf{D}(\mathbf{v}) - \mathfrak{p}\mathbf{I}) = 0 & \text{in } \dot\Omega^T, \\ \mathrm{div}\,\mathbf{v} = 0 & \text{in } \dot\Omega^T, \\ [[(\mu\mathbf{D}(\mathbf{v}) - \mathfrak{p}\mathbf{I})\mathbf{n}]] = 0, \quad [[\mathbf{v}]] = 0 & \text{on } \Gamma^T, \\ \mathbf{v} = 0 & \text{on } \Gamma^T_-, \\ \mathbf{v}|_{t=0} = 0 & \text{in } \dot\Omega, \end{cases} \tag{3.79}$$

then $\mathbf{v} = 0$ and $\mathfrak{p}$ is a constant in Ω.

Theorem 3.2.8 *Let $1 < p, q < \infty$ with $2/p + 1/q \neq 1, 2$, and $T > 0$. Assume that Γ is a compact hypersurface of C^3 class. Then the following two assertions hold:*

(1) **Existence** *There exists a γ_0 such that the following assertion holds: Let $\mathbf{u}_0 \in B^{2(1-1/p)}_{q,p}(\Omega)^N$ and $\eta_0 \in B^{3-1/p-1/q}_{q,p}(\Gamma)$ be initial data for equations* (3.5) *and let $\mathbf{F}$, G, $\mathbf{G}$, D, and $\mathbf{H}$ be functions appearing in the right side of equations* (3.5) *and satisfying the conditions:*

$$\mathbf{F} \in L_p((0,T), L_q(\dot\Omega)^N), \quad D \in L_p((0,T), W^{2-1/q}_q(\Gamma)),$$
$$e^{-\gamma t}G \in L_p(\mathbb{R}, H^1_q(\dot\Omega)) \cap H^{1/2}_p(\mathbb{R}, L_q(\dot\Omega)), \quad e^{-\gamma t}\mathbf{G} \in H^1_p(\mathbb{R}, L_q(\dot\Omega)^N),$$
$$e^{-\gamma t}\mathbf{H} \in L_p(\mathbb{R}, H^1_q(\dot\Omega)^N) \cap H^{1/2}_p(\mathbb{R}, L_q(\dot\Omega)^N)$$

for any $\gamma \geq \gamma_0$ with some γ_0. Assume that the compatibility conditions (3.77) *and* (3.78) *are satisfied. Then, problem* (3.5) *admits a unique solution* $(\mathbf{v}, \mathfrak{p}, \eta)$ *with*

$$\mathbf{v} \in L_p((0,T), H^2_q(\dot\Omega)^N) \cap H^1_p((0,T), L_q(\dot\Omega)^N),$$
$$\mathfrak{p} \in L_p((0,T), H^1_q(\dot\Omega) + H^1_{q,0}(\Omega)),$$
$$\eta \in L_p((0,T), W^{3-1/q}_q(\Gamma)) \cap H^1_p((0,T), W^{2-1/q}_q(\Gamma))$$

possessing the estimate:

$$
\begin{aligned}
&\|\mathbf{v}\|_{L_p((0,T),H^2_q(\dot\Omega))} + \|\partial_t\mathbf{v}\|_{L_p((0,T),L_q(\dot\Omega))}\\
&\quad + \|\eta\|_{L_p((0,T),W^{3-1/q}_q(\Gamma)} + \|\partial_t\eta\|_{L_p((0,T),W^{2-1/q}_q(\Gamma))}\\
&\le Ce^{\gamma\ell_\kappa T}\{\|\mathbf{u}_0\|_{B^{2(1-1/p)}_{q,p}(\Omega)} + \ell_\kappa\|\eta_0\|_{B^{3-1/p-1/q}_{q,p}(\Gamma)}\\
&\quad + \|\mathbf{F}\|_{L_p((0,T),L_q(\dot\Omega))} + \|d\|_{L_p((0,T),W^{2-1/q}_q(\Gamma))} + \|e^{-\gamma t}\partial_t\mathbf{G}\|_{L_p(\mathbb{R},L_q(\dot\Omega))}\\
&\quad + \|e^{-\gamma t}G\|_{L_p(\mathbb{R},H^1_q(\dot\Omega))} + \|e^{-\gamma t}\mathbf{H}\|_{L_p(\mathbb{R},H^1_q(\dot\Omega))}\\
&\quad + (1+\gamma^{1/2})(\|e^{-\gamma t}G\|_{H^{1/2}_p(\mathbb{R},L_q(\dot\Omega))} + \|e^{-\gamma t}\mathbf{H}\|_{H^{1/2}_p(\mathbb{R},L_q(\dot\Omega))})\}
\end{aligned}
$$

for any $\gamma \ge \gamma_0$ *with some constant* C *and* γ_0 *which are independent of* $\kappa \in (0,1)$. *Here,* ℓ_κ *is the same number as in Theorem 3.2.2.*

(2) **Uniqueness** *Let* $\mathbf{v}$, $\mathfrak{p}$, *and* η *with*

$$
\begin{aligned}
&\mathbf{v} \in L_p((0,T),H^2_q(\dot\Omega)^N) \cap H^1_p((0,T),L_q(\dot\Omega)^N),\\
&\mathfrak{p} \in L_p((0,T),H^1_q(\dot\Omega) + \hat H^1_q(\Omega)),\\
&\eta \in L_p((0,T),W^{3-1/q}_q(\Gamma)) \cap H^1_p((0,T),W^{2-1/q}_q(\Gamma))
\end{aligned}
$$

satisfy the homogeneous equations:

$$
\begin{cases}
\partial_t\mathbf{v} - m^{-1}\mathrm{Div}\,(\mu\mathbf{D}(\mathbf{v}) - \mathfrak{p}\mathbf{I}) = 0 & \text{in } \dot\Omega^T,\\
\mathrm{div}\,\mathbf{v} = 0 & \text{in } \dot\Omega^T,\\
[[(\mu\mathbf{D}(\mathbf{v}) - \mathfrak{p}\mathbf{I})\mathbf{n}]] - ((\mathcal{B} + \sigma\Delta_\Gamma)\eta)\mathbf{n} = 0, \quad [[\mathbf{v}]] = 0 & \text{on } \Gamma^T,\\
\lambda\eta + < A_\kappa \mid \nabla_\Gamma\eta > -\mathbf{n}\cdot\mathbf{v} + \mathcal{L}_1\mathbf{v} + \mathcal{L}_2\eta = 0 & \text{on } \Gamma^T,\\
\mathbf{v} = 0 & \text{on } \Gamma^T_-,\\
(\mathbf{v},\eta)|_{t=0} = (0,0) & \text{in } \dot\Omega\times\Gamma,
\end{cases}
\tag{3.80}
$$

then $\mathbf{v} = 0$, $\eta = 0$ *and* $\mathfrak{p}$ *is a constant in* Ω.

Proof of Theorem 3.2.7 We first prove Theorem 3.2.7 with the help of Theorem 3.2.1. The key tool in the proof of our maximal regularity results is the Weis operator valued Fourier multiplier theorem. To state it we need to make a few definitions. For a Banach space, X, $\mathcal{D}(\mathbb{R}, X)$ denotes the space of X-valued $C^\infty(\mathbb{R})$ functions with compact support and $\mathcal{D}'(\mathbb{R}, X) = \mathcal{L}(\mathcal{D}(\mathbb{R}), X)$ the space of X-valued distributions. And also, $\mathcal{S}(\mathbb{R}, X)$ denotes the space of X-valued rapidly decreasing functions and $\mathcal{S}'(\mathbb{R}, X) = \mathcal{L}(\mathcal{S}(\mathbb{R}), X)$ the space of X-valued tempered distributions. Let Y be another Banach space. Then, given $m \in L_{1,\mathrm{loc}}(\mathbb{R}, \mathcal{L}(X, Y))$, we define an operator $T_m : \mathcal{F}^{-1}\mathcal{D}(\mathbb{R}, X) \to \mathcal{S}'(\mathbb{R}, Y)$ by letting

$$T_m\phi = \mathcal{F}^{-1}[m\mathcal{F}[\phi]] \quad \text{for all } \phi \in \mathcal{F}^{-1}\mathcal{D}(\mathbb{R}, X), \tag{3.81}$$

where $\mathcal{F}$ and $\mathcal{F}^{-1}$ denote the respective Fourier transform and Fourier inversion transform. □

Definition 3.2.2 A Banach space X is said to be a UMD Banach space, if the Hilbert transform is bounded on $L_p(\mathbb{R}, X)$ for some (and then all) $p \in (1, \infty)$. Here, the Hilbert transform H operating on $f \in \mathcal{S}(\mathbb{R}, X)$ is defined by

$$[Hf](t) = \frac{1}{\pi} \lim_{\epsilon \to 0+} \int_{|t-s|>\epsilon} \frac{f(s)}{t-s}\, ds \quad (t \in \mathbb{R}).$$

Theorem 3.2.9 (Weis [40]) *Let X and Y be two UMD Banach spaces and $1 < p < \infty$. Let m be a function in $C^1(\mathbb{R} \setminus \{0\}, \mathcal{L}(X, Y))$ such that*

$$\mathcal{R}_{\mathcal{L}(X,Y)}(\{m(\tau) \mid \tau \in \mathbb{R} \setminus \{0\}\}) = \kappa_0 < \infty,$$
$$\mathcal{R}_{\mathcal{L}(X,Y)}(\{\tau m'(\tau) \mid \tau \in \mathbb{R} \setminus \{0\}\}) = \kappa_1 < \infty.$$

Then, the operator T_m defined in (3.81) *is extended to a bounded linear operator from $L_p(\mathbb{R}, X)$ into $L_p(\mathbb{R}, Y)$. Moreover, denoting this extension by T_m, we have*

$$\|T_m f\|_{L_p(\mathbb{R},Y)} \leq C(\kappa_0 + \kappa_1)\|f\|_{L_p(\mathbb{R},X)} \quad \textit{for all } f \in L_p(\mathbb{R}, X)$$

with some positive constant C depending on p.

Let $\mathbf{F}_0$ be the zero extension of $\mathbf{F}$ outside of $(0, T)$, that is, $\mathbf{F}_0(t) = \mathbf{F}(t)$ for $t \in (0, T)$ and $\mathbf{F}_0(t) = 0$ for $t \notin (0, T)$. Notice that $\mathbf{F}_0$, G, $\mathbf{G}$, and $\mathbf{H}$ are defined on the whole line $\mathbb{R}$. Thus, we first consider the equations:

$$\begin{cases} \partial_t \mathbf{u}_1 - m^{-1}\mathrm{Div}\,(\mu\mathbf{D}(\mathbf{u}_1) - \mathfrak{q}_1\mathbf{I}) = \mathbf{F}_0 & \text{in } \dot\Omega \times \mathbb{R}, \\ \mathrm{div}\,\mathbf{u}_1 = G = \mathrm{div}\,\mathbf{G} & \text{in } \dot\Omega \times \mathbb{R}, \\ [[(\mu\mathbf{D}(\mathbf{u}_1) - \mathfrak{q}_1\mathbf{I})\mathbf{n}]] = \mathbf{H}, \quad [[\mathbf{u}_1]] = 0 & \text{on } \Gamma \times \mathbb{R}, \\ \mathbf{u}_1 = 0 & \text{on } \Gamma_- \times \mathbb{R}. \end{cases} \tag{3.82}$$

Let $\mathcal{F}_L$ be the Laplace transform with respect to a time variable t defined by

$$\hat f(\lambda) = \mathcal{F}_L[f](\lambda) = \int_{\mathbb{R}} e^{-\lambda t} f(t)\, dt$$

for $\lambda = \gamma + i\tau \in \mathbb{C}$. Obviously,

$$\mathcal{F}_L[f](\lambda) = \int_{\mathbb{R}} e^{-i\tau t} e^{-\gamma t} f(t)\, dt = \mathcal{F}[e^{-\gamma t} f](\tau).$$

Applying the Laplace transform to equations (3.82) gives

$$\begin{cases} \lambda \hat{\mathbf{u}}_1 - m^{-1}\mathrm{Div}\,(\mu \mathbf{D}(\hat{\mathbf{u}}_1) - \hat{\mathfrak{q}}_1 \mathbf{I}) = \hat{\mathbf{F}}_0 & \text{in } \dot{\Omega}, \\ \mathrm{div}\, \hat{\mathbf{u}}_1 = \hat{G} = \mathrm{div}\, \hat{\mathbf{G}} & \text{in } \dot{\Omega}, \\ [[(\mu \mathbf{D}(\hat{\mathbf{u}}_1) - \hat{\mathfrak{q}}_1 \mathbf{I})\mathbf{n}]] = \hat{\mathbf{H}}, \quad [[\hat{\mathbf{u}}_1]] = 0 & \text{on } \Gamma, \\ \hat{\mathbf{u}}_1 = 0 & \text{on } \Gamma_-. \end{cases} \tag{3.83}$$

Applying Theorem 3.2.1, we have $\hat{\mathbf{u}}_1 = \mathcal{A}(\lambda)\mathbf{F}_\lambda$ and $\hat{\mathfrak{q}}_1 = \mathcal{P}(\lambda)\mathbf{F}_\lambda$ for $\lambda \in \Sigma_{\epsilon_0, \lambda_0}$, where

$$\mathbf{F}_\lambda = (\hat{\mathbf{F}}_0(\lambda), \lambda^{1/2}\hat{G}(\lambda), \hat{G}(\lambda), \lambda\hat{\mathbf{G}}(\lambda), \lambda^{1/2}\hat{\mathbf{H}}(\lambda), \hat{\mathbf{H}}(\lambda)).$$

Let $\mathcal{F}_L^{-1}$ be the inverse Laplace transform defined by

$$\mathcal{F}_L^{-1}[g](t) = \frac{1}{2\pi}\int_{\mathbb{R}} e^{\lambda t} g(\tau)\, d\tau = e^{\gamma t}\frac{1}{2\pi}\int_{\mathbb{R}} e^{i\tau t} g(\tau)\, d\tau$$

for $\lambda = \gamma + i\tau \in \mathbb{C}$. Obviously,

$$\mathcal{F}_L^{-1}[g](t) = e^{\gamma t}\mathcal{F}^{-1}[g](t), \quad \mathcal{F}_L\mathcal{F}_L^{-1} = \mathcal{F}_L^{-1}\mathcal{F}_L = \mathbf{I}.$$

Setting

$$\Lambda_\gamma^{1/2} f = \mathcal{F}_L^{-1}[\lambda^{1/2}\mathcal{F}_L[f]] = e^{\gamma t}\mathcal{F}^{-1}[\lambda^{1/2}\mathcal{F}[e^{-\gamma t} f]],$$

and using the facts that

$$\lambda\hat{\mathbf{G}}(\lambda) = \mathcal{F}_L[\partial_t \mathbf{G}](\lambda), \quad \lambda^{1/2}\hat{f}(\lambda) = \mathcal{F}_L[\Lambda_\gamma^{1/2} f] = \mathcal{F}[e^{-\gamma t}\Lambda_\gamma^{1/2} f]$$

for $f \in \{G, \mathbf{H}\}$ we define $\mathbf{u}_1$ and $\mathfrak{q}_1$ by setting

$$\mathbf{u}_1(\cdot, t) = \mathcal{F}_L[\mathbf{A}(\lambda)\mathbf{F}_\lambda] = e^{\gamma t}\mathcal{F}^{-1}[\mathbf{A}(\lambda)\mathcal{F}[e^{-\gamma t}F(t)](\tau)],$$
$$\mathfrak{q}_1(\cdot, t) = \mathcal{F}_L[\mathbf{P}(\lambda)\mathbf{F}_\lambda] = e^{\gamma t}\mathcal{F}^{-1}[\mathbf{P}(\lambda)\mathcal{F}[e^{-\gamma t}F(t)](\tau)],$$

with $F(t) = (\mathbf{F}_0, \Lambda_\gamma^{1/2}G, G, \partial_t\mathbf{G}, \Lambda_\gamma^{1/2}\mathbf{H}, \mathbf{H})$, where γ is chosen as $\gamma > \lambda_0$, and so $\gamma + i\tau \in \Sigma_{\epsilon_0, \lambda_0}$ for any $\tau \in \mathbb{R}$. By Cauchy's theorem in the theory of one complex variable, $\mathbf{u}_1$ and $\mathfrak{q}_1$ are independent of choice of γ whenever $\gamma > \lambda_0$. Noting that

$$\partial_t \mathbf{u}_1 = \mathcal{F}_L^{-1}[\lambda\mathbf{A}(\lambda)\mathbf{F}_\lambda] = e^{\gamma t}\mathcal{F}^{-1}[\lambda\mathbf{A}(\lambda)\mathcal{F}[e^{-\gamma t}F(t)](\tau)],$$

and applying Theorem 3.2.9, we have

$$\begin{aligned}
&\|e^{-\gamma t}\partial_t \mathbf{u}_1\|_{L_p(\mathbb{R},L_q(\dot\Omega))} + \|e^{-\gamma t}\mathbf{u}_1\|_{L_p(\mathbb{R},H^2_q(\dot\Omega))} + \|e^{-\gamma t}\nabla \mathfrak{q}_1\|_{L_p((\mathbb{R},L_q(\dot\Omega))}\\
&\le Cr_b\|e^{-\gamma t}F\|_{L_p(\mathbb{R},\mathcal{H}_q(\dot\Omega))}\\
&\le Cr_b\{\|e^{-\gamma t}\mathbf{F}\|_{p((0,T),L_q(\dot\Omega))} + \|e^{-\gamma t}G\|_{p(\mathbb{R},H^1_q(\dot\Omega))} + \|e^{-\gamma t}\Lambda^{1/2}_\gamma G\|_{p(\mathbb{R},L_q(\dot\Omega))}\\
&+ \|e^{-\gamma t}\partial_t\mathbf{G}\|_{p(\mathbb{R},L_q(\dot\Omega))} + \|e^{-\gamma t}\mathbf{H}\|_{L_p(\mathbb{R},H^1_q(\Omega))} + \|e^{-\gamma t}\Lambda^{1/2}_\gamma\mathbf{H}\|_{L_p(\mathbb{R},L_q(\Omega))}\}.
\end{aligned} \tag{3.84}$$

Since $|\lambda^{1/2}/(1+\tau^2)^{1/4}| \le C(1+\gamma^{1/2})$, we have

$$\begin{aligned}
&\|e^{-\gamma t}\Lambda^{1/2}_\gamma G\|_{p(\mathbb{R},L_q(\dot\Omega))} + \|e^{-\gamma t}\Lambda^{1/2}_\gamma\mathbf{H}\|_{L_p(\mathbb{R},L_q(\Omega))}\\
&\quad\le C(1+\gamma^{1/2})(\|e^{-\gamma t}G\|_{H^{1/2}_p(\mathbb{R},L_q(\dot\Omega))} + \|e^{-\gamma t}\mathbf{H}\|_{H^{1/2}_p(\mathbb{R},L_q(\Omega))}).
\end{aligned} \tag{3.85}$$

We now write solutions $\mathbf{u}$ and $\mathfrak{q}$ of equations (3.67) by $\mathbf{u} = \mathbf{u}_1 + \mathbf{u}_2$ and $\mathfrak{q} = \mathfrak{q}_1 + \mathfrak{q}_2$, where $\mathbf{u}_2$ and q_2 are solutions of the following equations:

$$\begin{cases}
\partial_t\mathbf{u}_2 - m^{-1}\mathrm{Div}\,(\mu\mathbf{D}(\mathbf{u}_2) - \mathfrak{q}_2\mathbf{I}) = 0, \quad \mathrm{div}\,\mathbf{u}_2 = 0 & \text{in } \dot\Omega\times(0,\infty),\\
[[(\mu\mathbf{D}(\mathbf{u}_2) - \mathfrak{q}_2\mathbf{I})\mathbf{n}]] = 0, \quad [[\mathbf{u}_2]] = 0 & \text{on } \Gamma\times(0,\infty),\\
\mathbf{u}_2 = 0 & \text{on } \Gamma_-\times(0,\infty),\\
\mathbf{u}_2 = \mathbf{u}_0 - \mathbf{u}_1|_{t=0} & \text{in } \dot\Omega.
\end{cases} \tag{3.86}$$

Recall that $\mathrm{div}\,\mathbf{u}_2 = 0$ in $\Omega\times(0,\infty)$ means that $\mathbf{u}_2 \in J_q(\Omega)$ for any $t > 0$. By real interpolation theory, we know that

$$\begin{aligned}
&\sup_{t\in(0,\infty)} e^{-\gamma t}\|\mathbf{u}_1(t)\|_{B^{2(1-1/p)}_{q,p}(\Omega)}\\
&\quad\le C(\|e^{-\gamma t}\mathbf{u}_1\|_{L_p((0,\infty),H^2_q(\Omega))} + \|e^{-\gamma t}\partial_t\mathbf{u}_1\|_{L_p((0,\infty),L_q(\Omega))}).
\end{aligned} \tag{3.87}$$

In fact, this inequality follows from the following theory (cf. Tanabe[p.10][35]): Let X_1 and X_2 be two Banach spaces such that X_2 is a dense subset of X_1, and then

$$L_p((0,\infty),X_2)\cap H^1_p((0,\infty),X_1) \subset C([0,\infty),(X_1,X_2)_{1-1/p,p}), \tag{3.88}$$

and

$$\sup_{t\in(0,\infty)} \|u(t)\|_{(X_1,X_2)_{1-1/p,p}} \le C(\|u\|_{L_p((0,\infty),X_2)} + \|\partial_t u\|_{L_p((0,\infty),X_1)}). \tag{3.89}$$

Since $B^{2(1-1/p)}_{q,p}(\Omega) = (L_p(\Omega), H^2_q(\Omega))_{1-1/p,p}$, we have (3.87). Thus,

$$\mathbf{u}_0 - \mathbf{u}_1|_{t=0} \in B^{2(1-1/p)}_{q,p}(\Omega).$$

By the compatibility condition (3.76) and (3.82), we have

$$(\mathbf{u}_0 - \mathbf{u}_1|_{t=0}, \nabla\varphi) = (\mathbf{u}_0 - \mathbf{G}|_{t=0}, \nabla\varphi) = 0 \quad \text{for any } \varphi \in \hat{H}^1_{q'}(\Omega).$$

Moreover, by the compatibility condition (3.77) and (3.82), we have

$$[[(\mu\mathbf{D}(\mathbf{u}_0 - \mathbf{u}_1|_{t=0})\mathbf{n})_\tau]] = [[(\mu\mathbf{D}(\mathbf{u}_0)\mathbf{n})_\tau]] - [[(\mathbf{H}|_{t=0})_\tau]] = 0 \text{ on } \Gamma \text{ for } 2/p + 1/q < 1,$$
$$[[\mathbf{u}_0 - \mathbf{u}_1|_{t=0}]] = 0 \text{ on } \Gamma, \quad \mathbf{u}_0 - \mathbf{u}_1|_{t=0} = 0 \text{ on } \Gamma_- \text{ for } 2/p + 1/q < 2.$$

From this it follows that if $2/p + /q \neq 1$, then $\mathbf{u}_0 - \mathbf{u}_1|_{t=0} \in \mathcal{D}^1_{q,p}(\Omega)$. Applying Theorem 3.2.5, we see that there exists a $\gamma' > 0$ such that problem (3.86) admits a unique solutions $\mathbf{u}_2$ with some pressure term $\mathfrak{q}_2 = P_1(\mathbf{u}_2)$ and

$$e^{-\gamma' t}\mathbf{u}_2 \in H^1_p((0, \infty), L_q(\Omega)^N) \cap L_p((0, \infty), H^2_q(\Omega)^N) \tag{3.90}$$

possessing the estimate:

$$\begin{aligned} &\|e^{-\gamma' t}\partial_t\mathbf{u}_2\|_{L_p((0,\infty),L_q(\Omega))} + \|e^{-\gamma' t}\mathbf{u}_2\|_{L_p((0,\infty),H^2_q(\Omega))} \\ &\quad\leq C\|\mathbf{u}_0 - \mathbf{u}_1|_{t=0}\|_{B^{2(1-1/p)}_{q,p}(\Omega)}. \end{aligned} \tag{3.91}$$

Thus, setting $\mathbf{u} = \mathbf{u}_1 + \mathbf{u}_2$ and $\mathfrak{q} = \mathfrak{q}_1 + P_1(\mathbf{u}_2)$ and choosing γ_0 in such a way that $\gamma_0 > \max(\lambda_0, \gamma')$, by (3.82), (3.84), (3.85), (3.86), (3.87), (3.90) and (3.91), we see that $\mathbf{u}$ and $\mathfrak{q}$ are required solutions of equations (3.67), which completes the proof of Existence part.

We now prove the uniqueness. Let $\mathbf{v}$ and $\mathfrak{p}$ be solutions to equations (3.79). Let $\mathbf{v}_0$ and $\mathfrak{p}_0$ be the zero extension of $\mathbf{v}$ and $\mathfrak{p}$ to $t < 0$, that is,

$$\mathbf{v}_0(x, t) = \begin{cases} \mathbf{v}(x, t) & \text{for } 0 < t < T, \\ 0 & \text{for } t < 0, \end{cases} \qquad \mathfrak{p}_0(x, t) = \begin{cases} \mathfrak{p}(x, t) & \text{for } 0 < t < T, \\ 0 & \text{for } t < 0. \end{cases}$$

And then, we define $\mathbf{v}_e$ and $\mathfrak{p}_e$ by setting

$$\mathbf{v}_e(x, t) = \begin{cases} \mathbf{v}_0(x, t) & \text{for } t < T, \\ \mathbf{v}_0(x, 2T - t) & \text{for } t > T, \end{cases} \qquad \mathfrak{p}_e(x, t) = \begin{cases} \mathfrak{p}_0(x, t) & \text{for } t < T, \\ \mathfrak{p}_0(x, 2T - t) & \text{for } t > T. \end{cases}$$

Since $\mathbf{v}|_{t=0} = 0$, we see that

$$\mathbf{v}_e \in H^1_p(\mathbb{R}, L_q(\dot\Omega)^N) \cap L_p(\mathbb{R}, H^2_q(\dot\Omega)^N), \quad \mathfrak{p}_e \in L_p(\mathbb{R}, H^1_q(\dot\Omega) + \hat{H}^1_q(\Omega)),$$

and $\mathbf{v}_e$ and $\mathfrak{p}_e$ vanish for $t \notin (0, 2T)$. Let $\hat{\mathbf{v}}_e$ and $\hat{\mathfrak{p}}_e$ be the Laplace transform of $\mathbf{v}_e$ and $\mathfrak{p}_e$ with respect to $t \in \mathbb{R}$, and then $\hat{\mathbf{v}}_e(\cdot, \lambda)$ and $\hat{\mathfrak{p}}_e(\cdot, \lambda)$ are entire functions with respect to $\lambda \in \mathbb{C}$ and satisfy the homogeneous equations:

$$\begin{cases} \lambda\hat{\mathbf{v}}_e - m^{-1}\mathrm{Div}\,(\mu\mathbf{D}(\hat{\mathbf{v}}_e) - \hat{\mathfrak{p}}_e\mathbf{I}) = 0 & \text{in } \dot\Omega, \\ \mathrm{div}\,\hat{\mathbf{v}}_e = 0 & \text{in } \dot\Omega, \\ [[(\mu\mathbf{D}(\hat{\mathbf{v}}_e) - \hat{\mathfrak{p}}_e\mathbf{I})\mathbf{n}]] = 0, \quad [[\hat{\mathbf{v}}_e]] = 0 & \text{on } \Gamma, \\ \hat{\mathbf{v}}_e = 0 & \text{on } \Gamma_-. \end{cases} \tag{3.92}$$

Moreover, by Hölder's inequality we have

$$\begin{aligned} \|\hat{\mathbf{v}}_e(\cdot,\lambda)\|_{H^2_q(\dot\Omega)} &\le \int_{-\infty}^{\infty} e^{|\lambda|t}\|\mathbf{v}_e(\cdot,t)\|_{H^2_q(\dot\Omega)}\,dt \\ &\le 2(p'|\lambda|)^{-1/p'}e^{2|\lambda|T}\|\mathbf{v}\|_{L_p((0,T),H^2_q(\dot\Omega))}. \end{aligned}$$

And also, we have

$$\|\nabla\hat{\mathfrak{p}}_e\|_{L_q(\dot\Omega)} \le 2(p'|\lambda|)^{-1/p'}e^{2|\lambda|T}\|\nabla\mathfrak{p}\|_{L_p((0,T),L_q(\dot\Omega))}.$$

Namely, we have $\hat{\mathbf{v}}_e \in H^2_q(\dot\Omega)^N$ and $\hat{\mathfrak{p}}_e \in H^1_q(\dot\Omega) + \hat{H}^1_q(\Omega)$, and so the uniqueness stated in Theorem 3.2.3 implies that $\hat{\mathbf{v}}_e = 0$ and $\nabla\hat{\mathfrak{p}}_e = 0$ for any $\lambda \in \Sigma_{\epsilon,\lambda_0}$. Thus, we have $\mathbf{v} = 0$ and $\nabla\mathfrak{p} = 0$ in $\dot\Omega$. Since $[[\mathfrak{p}]] = 0$ on Γ, we finally see that $\mathfrak{p}$ is a constant in Ω. This completes the uniqueness, which completes the proof of Theorem 3.2.7.

Proof of Theorem 3.2.8 In view of Theorem 3.2.7, to prove Theorem 3.2.8, first of all, we consider the following auxiliary problem:

$$\begin{cases} \partial_t\mathbf{u} - m^{-1}\mathrm{Div}\,(\mu\mathbf{D}(\mathbf{u}) - \mathfrak{q}\mathbf{I}) = 0, \quad \mathrm{div}\,\mathbf{u} = 0 & \text{in } \dot\Omega^T, \\ [[(\mu\mathbf{D}(\mathbf{u}) - \mathfrak{q}\mathbf{I})\mathbf{n}]] - ((\mathcal{B} + \sigma\Delta_\Gamma)h)\mathbf{n} = 0, \quad [[\mathbf{u}]] = 0 & \text{on } \Gamma^T, \\ \partial_t h + < A_\kappa \mid \nabla_\Gamma h > -\mathbf{n}\cdot\mathbf{u} + \mathcal{L}_1\mathbf{u} + \mathcal{L}_2 h = F & \text{on } \Gamma^T, \\ \mathbf{u} = 0 & \text{on } \Gamma^T_-, \\ (\mathbf{u}, h)|_{t=0} = (0, \eta_0) & \text{in } \dot\Omega \times \Gamma, \end{cases} \tag{3.93}$$

and we prove the following theorem. □

Theorem 3.2.10 *Let* $1 < p, q < \infty$ *and* $T > 0$. *Assume that* Γ *is a compact* C^3 *hypersurface. Then, for any* $\eta_0 \in B^{3-1/p-1/q}_{q,p}(\Gamma)$ *and* $F \in L_p((0,T), W^{2-1/q}_q(\Gamma))$, *problem* (3.93) *admits a solution* $(\mathbf{u}, \mathfrak{q}, h)$ *with*

$$\begin{aligned} &\mathbf{u} \in L_p((0,T), H^2_q(\dot\Omega)^N) \cap H^1_p((0,T), L_q(\dot\Omega)^N), \\ &\mathfrak{q} \in L_p((0,T), H^1_q(\dot\Omega) + \hat{H}^1_q(\Omega)), \end{aligned}$$

$$h \in L_p((0,T), W_q^{3-1/q}(\Gamma)) \cap H_p^1((0,T), W_q^{2-1/q}(\Gamma))$$

possessing the estimate:

$$\begin{aligned}&\|\mathbf{u}\|_{L_p((0,T),H_q^2(\dot\Omega))} + \|\partial_t\mathbf{u}\|_{L_p((0,T),L_q(\dot\Omega))} + \|\nabla\mathfrak{q}\|_{L_p((0,T),L_q(\dot\Omega))}\\ &\quad+ \|h\|_{L_p((0,T),W_q^{3-1/q}(\Gamma))} + \|\partial_t h\|_{L_p((0,T),W_q^{2-1/q}(\Gamma))}\\ &\quad\le Ce^{\gamma\ell_\kappa}\{\ell_\kappa\|\eta_0\|_{B^{3-1/p-1/q}_{q,p}(\Gamma)} + \|F\|_{L_p((0,T),W_q^{2-1/q}(\Gamma))}\}\end{aligned}$$

for any $\gamma \ge \gamma_0$ with some positive constants γ_0 and C. The constant C is independent of γ.

Proof We first consider the case where $\eta_0 = 0$. Let F_0 be the zero extension of F outside of $(0,T)$, that is, $F_0(x,t) = F(x,t)$ for $t \in (0,T)$ and $F_0(x,t) = 0$ for $t \notin (0,T)$. We consider the equations:

$$\begin{cases}\partial_t\mathbf{u}_1 - m^{-1}\mathrm{Div}\,(\mu\mathbf{D}(\mathbf{u}_1) - \mathfrak{q}_1\mathbf{I}) = 0, \quad \mathrm{div}\,\mathbf{u}_1 = 0 & \text{in } \dot\Omega\times\mathbb{R},\\ [[(\mu\mathbf{D}(\mathbf{u}_1) - \mathfrak{q}_1\mathbf{I})\mathbf{n}]] - ((\mathcal{B}+\sigma\Delta_\Gamma)h_1)\mathbf{n} = 0, \quad [[\mathbf{u}_1]] = 0 & \text{on } \Gamma\times\mathbb{R},\\ \partial_t h_1 + < A_\kappa \mid \nabla_\Gamma h_1 > -\mathbf{n}\cdot\mathbf{u}_1 + \mathcal{L}_1\mathbf{u}_1 + \mathcal{L}_2 h_1 = F_0 & \text{on } \Gamma\times\mathbb{R},\\ \mathbf{u}_1 = 0 & \text{on } \Gamma_-\times\mathbb{R}.\end{cases}$$

Applying the Laplace transform and writing $\hat{\mathbf{u}}_1 = \mathcal{L}[\mathbf{u}_1]$, $\hat{\mathfrak{q}}_1 = \mathcal{L}[\mathfrak{q}_1]$, $\hat h_1 = \mathcal{L}[h_1]$, and $\hat F_0 = \mathcal{L}[F_0]$, we have

$$\begin{cases}\lambda\hat{\mathbf{u}}_1 - m^{-1}\mathrm{Div}\,(\mu\mathbf{D}(\hat{\mathbf{u}}_1) - \hat{\mathfrak{q}}_1\mathbf{I}) = 0, \quad \mathrm{div}\,\hat{\mathbf{u}}_1 = 0 & \text{in } \dot\Omega,\\ [[(\mu\mathbf{D}(\hat{\mathbf{u}}_1) - \hat{\mathfrak{q}}_1\mathbf{I})\mathbf{n}]] - ((\mathcal{B}+\sigma\Delta_\Gamma)\hat h_1)\mathbf{n} = 0, \quad [[\hat{\mathbf{u}}_1]] = 0 & \text{on } \Gamma,\\ \lambda\hat h_1 + < A_\kappa \mid \nabla_\Gamma \hat h_1 > -\mathbf{n}\cdot\hat{\mathbf{u}}_1 + \mathcal{L}_1\hat{\mathbf{u}}_1 + \mathcal{L}_2 \hat h_1 = \hat F_0 & \text{on } \Gamma,\\ \hat{\mathbf{u}}_1 = 0 & \text{on } \Gamma_-.\end{cases}$$

Thus, in view of Theorem 3.2.2, we have

$$\begin{aligned}&\mathbf{u}_1 = \mathcal{L}^{-1}[\mathcal{A}(\lambda)(0, \hat F_0(\lambda))], \quad \mathfrak{q}_1 = \mathcal{L}^{-1}[\mathcal{Q}(\lambda)(0, \hat F_0(\lambda))],\\ &h_1 = \mathcal{L}^{-1}[\mathcal{H}(\lambda)(0, \hat F_0(\lambda))].\end{aligned}$$

Applying Theorem 3.2.9 yields that

$$\begin{aligned}&e^{-\gamma t}\mathbf{u}_1 \in L_p(\mathbb{R}, H_q^2(\dot\Omega)^N) \cap H_p^1(\mathbb{R}, L_q(\dot\Omega)^N),\\ &e^{-\gamma t}\mathfrak{q}_1 \in L_p(\mathbb{R}, H_q^1(\dot\Omega) + \hat H_q^1(\Omega)),\end{aligned}$$

$$e^{-\gamma t}h_1 \in L_p(\mathbb{R}, W_q^{3-1/q}(\Gamma)) \cap H_p^1(\mathbb{R}, W_q^{2-1/q}(\Gamma))$$

and the following estimate holds:

$$\begin{aligned}
&\|e^{-\gamma t}\mathbf{u}_1\|_{L_p(\mathbb{R},H_q^1(\dot\Omega))} + \|e^{-\gamma t}\partial_t\mathbf{u}_1\|_{L_p(\mathbb{R},L_q(\dot\Omega))} + \|e^{-\gamma t}\nabla\mathfrak{q}_1\|_{L_p(\mathbb{R},L_q(\dot\Omega))}\\
&\quad + \|e^{-\gamma t}h_1\|_{L_p(\mathbb{R},W_q^{3-1/q}(\Gamma))} + \|e^{-\gamma t}\partial_t h_1\|_{L_p(\mathbb{R},W_q^{2-1/q}(\Gamma))}\\
&\quad \le C\|e^{-\gamma t}F\|_{L_p((0,T),W_q^{2-1/q}(\Gamma))}
\end{aligned} \tag{3.94}$$

for any $\gamma \ge \gamma_0\ell_\kappa$ with some positive constants γ_0 and C, where C is independent of γ.

We now prove that $(\mathbf{u}_1, h_1)|_{t=0} = (0,0)$ on $\dot\Omega \times \Gamma$. Since $|\gamma/\lambda| \le 1$ for any $\lambda \in \Lambda_{\kappa,\lambda_0\ell_\kappa}$, we have

$$\begin{aligned}
\gamma\|e^{-\gamma t}\mathbf{u}_1\|_{L_p(\mathbb{R},L_q(\dot\Omega))} &\le \|e^{-\gamma t}\partial_t\mathbf{u}\|_{L_p(\mathbb{R},L_q(\dot\Omega))},\\
\gamma\|e^{-\gamma t}h_1\|_{L_p(\mathbb{R},W_q^{2-1/q}(\Gamma))} &\le \|e^{-\gamma t}\partial_t h_1\|_{L_p(\mathbb{R},W_q^{2-1/q}(\Gamma))},
\end{aligned}$$

and therefore, by (3.94) we have

$$\begin{aligned}
&\|\mathbf{u}_1\|_{L_p((-\infty),L_q(\dot\Omega))} + \|h_1\|_{L_p((-\infty,0),W_q^{2-1/q}(\Gamma))}\\
&\quad \le \gamma^{-1}(\|e^{-\gamma t}\partial_t\mathbf{u}_1\|_{L_p(\mathbb{R},L_q(\dot\Omega))} + \|e^{-\gamma t}\partial_t h_1\|_{L_p(\mathbb{R},W_q^{2-1/q}(\Gamma))})\\
&\quad \le C\gamma^{-1}\|F\|_{L_p((0,T),L_q(\Gamma))}.
\end{aligned}$$

Letting $\gamma \to \infty$ yields that

$$\|\mathbf{u}_1\|_{L_p((-\infty),L_q(\dot\Omega))} + \|h_1\|_{L_p((-\infty,0),W_q^{2-1/q}(\Gamma))} = 0,$$

which leads to $(\mathbf{u}_1, h_1)|_{t=0} = (0,0)$ on $\dot\Omega \times \Gamma$. In particular, from (3.94) it follows that

$$\begin{aligned}
&\|\mathbf{u}_1\|_{L_p((0,T),H_q^1(\dot\Omega))} + \|\partial_t\mathbf{u}_1\|_{L_p((0,T),L_q(\dot\Omega))} + \|\nabla\mathfrak{q}_1\|_{L_p((0,T),L_q(\dot\Omega))}\\
&\quad + \|h_1\|_{L_p((0,T),W_q^{3-1/q}(\Gamma))} + \|\partial_t h_1\|_{L_p((0,T),W_q^{2-1/q}(\Gamma))}\\
&\quad \le Ce^{\gamma\ell_\kappa T}\|e^{-\gamma\ell_\kappa t}F\|_{L_p((0,T),W_q^{2-1/q}(\Gamma))}
\end{aligned} \tag{3.95}$$

for any $\gamma \ge \gamma_0$ with some positive constants γ_0 and C independent of κ.

We next consider the initial value problem:

$$\begin{cases} \partial_t \mathbf{u}_2 - m^{-1}\mathrm{Div}\,(\mu\mathbf{D}(\mathbf{u}_2) - \mathfrak{q}_2\mathbf{I}) = 0, \quad \mathrm{div}\,\mathbf{u}_2 = 0 & \text{in } \dot{\Omega}^\infty, \\ [[(\mu\mathbf{D}(\mathbf{u}_2) - \mathfrak{q}_1\mathbf{I})\mathbf{n}]] - ((\mathcal{B} + \sigma\Delta_\Gamma)h_2)\mathbf{n} = 0, \quad [[\mathbf{u}_2]] = 0 & \text{on } \Gamma^\infty, \\ \partial_t h_2 + < A_\kappa \mid \nabla_\Gamma h_2 > - \mathbf{n}\cdot\mathbf{u}_2 + \mathcal{L}_1\mathbf{u}_2 + \mathcal{L}_2 h_2 = 0 & \text{on } \Gamma^\infty, \\ \mathbf{u}_2 = 0 & \text{on } \Gamma_-^\infty, \\ (\mathbf{u}_2, h_2)|_{t=0} = (0, \eta_0) & \text{in } \dot{\Omega}\times\Gamma, \end{cases}$$

where we have set $G^\infty = G \times (0, \infty)$ for $G \in \{\dot{\Omega}, \Gamma, \Gamma_-\}$. Let $\mathbf{u}_2$ and h_2 with

$$e^{-\gamma t}\mathbf{u}_2 \in L_p((0,\infty), H^2_q(\dot{\Omega})^N)) \cap H^1_p((0,\infty), L_q(\dot{\Omega})^N),$$
$$e^{-\gamma t}h_2 \in L_p((0,\infty), W^{3-1/q}_q(\Gamma)) \cap H^1_p((0,\infty), W^{2-1/q}_q(\Gamma))$$

be solutions of the equations (3.71) with $\mathbf{v}_0 = 0$. Since $(0, \eta_0) \in \mathcal{E}^2_{p,q}$ when $\eta_0 \in B^{3-1/p-1/q}_{q,p}(\Gamma)$, by Theorem 3.2.6 we know the existence of $\mathbf{u}_2$ and h_2 possessing the estimate:

$$\begin{aligned} &\|e^{-\gamma t}\mathbf{u}_2\|_{L_p((0,\infty),H^1_q(\dot{\Omega}))} + \|e^{-\gamma t}\partial_t\mathbf{u}_2\|_{L_p((0,\infty),L_q(\dot{\Omega}))} \\ &\quad + \|e^{-\gamma t}h_2\|_{L_p((0,\infty),W^{3-1/q}_q(\Gamma))} + \|e^{-\gamma t}\partial_t h_2\|_{L_p((0,\infty),W^{2-1/q}_q(\Gamma))} \\ &\quad \le C\|\eta_0\|_{B^{3-1/p-1/q}_{q,p}(\Gamma)} \end{aligned} \tag{3.96}$$

for any $\gamma \ge \gamma_0$ with some positive constants γ_0 and C, where C is independent of γ. In particular, when $\kappa = 0$, $\mathbf{u} = \mathbf{u}_1 + \mathbf{u}_2$, $\mathfrak{q} = \mathfrak{q}_1 + K(\mathbf{u}_2, h_2)$ and $h = h_1 + h_2$ are required solutions of equations (3.93).

We now consider the case where $\kappa \in (0, 1)$. Let $\mathbf{u}_3$, $\mathfrak{q}_3$, and h_3 be solutions of equations (3.93) with $F = - < A_\kappa \mid \nabla h_2 >$ and $\eta_0 = 0$. We then see the existence of $\mathbf{u}_3$, $\mathfrak{q}_3$, and h_3 with

$$\begin{aligned} &e^{-\gamma t}\mathbf{u}_3 \in L_p(\mathbb{R}, H^2_q(\dot{\Omega})^N) \cap H^1_p(\mathbb{R}, L_q(\dot{\Omega})^N), \\ &e^{-\gamma t}\mathfrak{q}_3 \in L_p(\mathbb{R}, H^1_q(\dot{\Omega}) + \hat{H}^1_q(\Omega)), \\ &e^{-\gamma t}h_3 \in L_p(\mathbb{R}, W^{3-1/q}_q(\Gamma)) \cap H^1_p(\mathbb{R}, W^{2-1/q}_q(\Gamma)) \end{aligned}$$

for any $\gamma \ge \gamma_0\ell_\kappa$ with some constant γ_0 independent of $\kappa \in (0, 1)$. By (3.7) we have

$$\|e^{-\gamma t} < A_\kappa \mid \nabla_\Gamma h_2 > \|_{L_p((0,T),W^{3-1/q}_q(\Gamma))} \le C\kappa^{-b}\|h_2\|_{L_p((0,T),W^{3-1/q}_q(\Gamma))}. \tag{3.97}$$

From (3.96), we have

$$\begin{aligned}
&\|\mathbf{u}_2\|_{L_p((0,T),H^1_q(\dot\Omega))} + \|\partial_t\mathbf{u}_2\|_{L_p((0,T),L_q(\dot\Omega))} \\
&\quad + \|h_2\|_{L_p((0,T),W^{3-1/q}_q(\Gamma))} + \|\partial_t h_2\|_{L_p((0,T),W^{2-1/q}_q(\Gamma))} \\
&\quad\quad \leq Ce^{\gamma\ell_\kappa T}\|\eta_0\|_{B^{3-1/p-1/q}_{q,p}(\Gamma)}
\end{aligned} \tag{3.98}$$

for any $\gamma \geq \gamma_0$, where C and γ_0 are constants independent of $\kappa \in (0, 1)$. Combining (3.95), (3.97), and (3.98) gives

$$\begin{aligned}
&\|\mathbf{u}_3\|_{L_p((0,T),H^1_q(\dot\Omega))} + \|\partial_t\mathbf{u}_3\|_{L_p((0,T),L_q(\dot\Omega))} + \|\nabla\mathfrak{q}_3\|_{L_p((0,T),L_q(\dot\Omega))} \\
&\quad + \|h_3\|_{L_p((0,T),W^{3-1/q}_q(\Gamma))} + \|\partial_t h_3\|_{L_p((0,T),W^{2-1/q}_q(\Gamma))} \\
&\quad \leq Ce^{2\gamma\ell_\kappa T}\kappa^{-b}\|\eta_0\|_{B^{3-1/p-1/q}_{q,p}(\Gamma)}.
\end{aligned} \tag{3.99}$$

Setting $\mathbf{u} = \mathbf{u}_1 + \mathbf{u}_2 + \mathbf{u}_3$, $\mathfrak{q} = \mathfrak{q}_1 + K(\mathbf{u}_2, h_2) + \mathfrak{q}_3$ and $h = h_1 + h_2 + h_3$ and combining (3.95), (3.98), and (3.99), we have Theorem 3.2.10. □

We now complete the proof of Theorem 3.2.8. Let $\mathbf{v}_1$ and $\mathfrak{p}_1$ be solutions of equations (3.67), whose existence is guaranteed by Theorem 3.2.7. In particular, we have

$$\begin{aligned}
&\|\mathbf{v}_1\|_{L_p((0,T),H^2_q(\dot\Omega))} + \|\partial_t\mathbf{v}_1\|_{L_p((0,T),L_q(\dot\Omega))} + \|\nabla\mathfrak{p}_1\|_{L_p((0,T),L_q(\dot\Omega))} \\
&\quad \leq Ce^{\gamma T}\{\|\mathbf{u}_0\|_{B^{2(1-1/p)}_{q,p}(\dot\Omega)} + \|\mathbf{F}\|_{L_p((0,T),L_q(\dot\Omega))} + \|e^{-\gamma t}\partial_t\mathbf{G}\|_{L_p(\mathbb{R},L_q(\dot\Omega))} \\
&\quad + \|e^{-\gamma t}G\|_{L_p(\mathbb{R},H^1_q(\dot\Omega))} + \|e^{-\gamma t}\mathbf{H}\|_{L_p(\mathbb{R},H^1_q(\dot\Omega))} \\
&\quad + (1+\gamma^{1/2})(\|e^{-\gamma t}G\|_{H^{1/2}_p(\mathbb{R},L_q(\dot\Omega))} + \|e^{-\gamma t}\mathbf{H}\|_{H^{1/2}_p(\mathbb{R},L_q(\dot\Omega))})\}
\end{aligned} \tag{3.100}$$

for any $\gamma \geq \gamma_0$ with some constant C independent of γ. Let $\mathbf{v}_2$, $\mathfrak{p}_2$, and h_2 be solutions of equations (3.5) with $\mathbf{F} = G = \mathbf{G} = \mathbf{H} = \mathbf{v}_0 = 0$ and D replaced by $D + \mathbf{n}\cdot\mathbf{v}_1 - \mathcal{L}\mathbf{v}_1$, whose existence is guaranteed by Theorem 3.2.10. Moreover, we have

$$\begin{aligned}
&\|\mathbf{v}_2\|_{L_p((0,T),H^2_q(\dot\Omega))} + \|\partial_t\mathbf{v}_2\|_{L_p((0,T),L_q(\dot\Omega))} + \|\nabla\mathfrak{p}_1\|_{L_p((0,T),L_q(\dot\Omega))} \\
&+ \|h_2\|_{L_p((0,T),W^{3-1/q}_q(\Gamma))} + \|\partial_t h_2\|_{L_p((0,T),W^{2-1/q}_q(\Gamma))} \\
&\leq Ce^{\gamma\ell_\kappa T}\{\ell_\kappa\|\eta_0\|_{B^{3-1/p-1/q}_{q,p}(\Gamma)} + \|D\|_{L_p((0,T),W^{2-1/q}_q(\Gamma))} + \|\mathbf{v}_1\|_{L_p((0,T),H^2_q(\dot\Omega))}\}
\end{aligned} \tag{3.101}$$

for any $\gamma \geq \gamma_0$ with some constants γ_0 and C which are independent of $\kappa \in (0, 1)$. Setting $\mathbf{v} = \mathbf{v}_1 + \mathbf{v}_2$, $\mathfrak{p} = \mathfrak{p}_1 + \mathfrak{p}_2$ and $\eta = h_2$ and combining (3.100) and (3.101), we have the existence part of Theorem 3.2.8.

The uniqueness part of Theorem 3.2.8 can be proved by employing the same argument as in the proof of the uniqueness part of Theorem 3.2.7. This completes the proof of Theorem 3.2.8.

3.3 $\mathcal{R}$-Bounded Solution Operators

3.3.1 Model Problem: $\sigma > 0$

In this section, we consider the following model problem with surface tension $\sigma > 0$:

$$\begin{cases} \lambda\mathbf{u} - m^{-1}\mathrm{Div}\,(\mu\mathbf{D}(\mathbf{u}) - \mathfrak{p}) = \mathbf{f} & \text{in } \dot{\mathbb{R}}^N, \\ \mathrm{div}\,\mathbf{u} = g = \mathrm{div}\,\mathbf{g} & \text{in } \dot{\mathbb{R}}^N, \\ [[(\mu\mathbf{D}(\mathbf{u}) - \mathfrak{p}\mathbf{I})\mathbf{n}_0]] - \sigma(\Delta' h)\mathbf{n}_0 = [[\mathbf{h}]], \quad [[\mathbf{u}]] = 0 & \text{on } \mathbb{R}^N_0, \\ \lambda h - \mathbf{u}\cdot\mathbf{n}_0 = d & \text{on } \mathbb{R}^N_0, \end{cases} \tag{3.102}$$

where $\mathbf{n}_0 = (0, \dots, 0, -1)^\top$, $\dot{\mathbb{R}} = \mathbb{R}^N_+ \cup \mathbb{R}^N_-$ with

$$\mathbb{R}^N_\pm = \{x = (x', x_N) \in \mathbb{R}^N \mid x' = (x_1, \dots, x_{N-1}) \in \mathbb{R}^{N-1}, \pm x_N > 0\},$$

and $\mathbb{R}^N_0$ is a flat interface given by

$$\mathbb{R}^N_0 = \{x = (x', x_N) \in \mathbb{R}^N \mid x' = (x_1, \dots, x_{N-1}) \in \mathbb{R}^{N-1}, x_N = 0\}.$$

Let $(\mathbf{w}, \mathfrak{r})$ be a solution to

$$\begin{cases} \lambda\mathbf{w} - m^{-1}\mathrm{Div}\,(\mu\mathbf{D}(\mathbf{w}) - \mathfrak{r}) = \mathbf{f} & \text{in } \dot{\mathbb{R}}^N, \\ \mathrm{div}\,\mathbf{w} = g = \mathrm{div}\,\mathbf{g} & \text{in } \dot{\mathbb{R}}^N, \\ [[(\mu\mathbf{D}(\mathbf{w}) - \mathfrak{r}\mathbf{I})\mathbf{n}_0]] = [[\mathbf{h}]], \quad [[\mathbf{w}]] = 0 & \text{on } \mathbb{R}^N_0, \end{cases} \tag{3.103}$$

and set $\mathbf{u} = \mathbf{v} + \mathbf{w}$ and $\mathfrak{p} = \mathfrak{q} + \mathfrak{r}$ in (3.102).[1] One then has

[1] We refer to [20] for more details on the system (3.103).

$$\begin{cases} \lambda \mathbf{v} - m^{-1}\mathrm{Div}\,(\mu\mathbf{D}(\mathbf{v}) - \mathfrak{q}) = 0 & \text{in } \dot{\mathbb{R}}^N, \\ \mathrm{div}\,\mathbf{v} = 0 & \text{in } \dot{\mathbb{R}}^N, \\ [[(\mu\mathbf{D}(\mathbf{v}) - \mathfrak{q}\mathbf{I})\mathbf{n}_0]] - \sigma(\Delta' h)\mathbf{n}_0 = 0, \quad [[\mathbf{v}]] = 0 & \text{on } \mathbb{R}^N_0, \\ \lambda h - \mathbf{v}\cdot\mathbf{n}_0 = d & \text{on } \mathbb{R}^N_0, \end{cases} \tag{3.104}$$

where $d + \mathbf{w}\cdot\mathbf{n}_0$ has been replaced by d for simplicity.

The aim of this section is to prove the existence of $\mathcal{R}$-bounded solution operators of (3.104). One starts with deriving representation formulas for solutions of (3.104). Let $j = 1, \dots, N-1$ and $J = 1, \dots, N$ in what follows. The system (3.104) can be written as

$$\begin{cases} m_\pm \lambda v_{\pm J} - \mu_\pm \Delta v_{\pm J} + \nabla \mathfrak{q}_\pm = 0 & \text{in } \mathbb{R}^N_\pm, \\ \mathrm{div}\,\mathbf{v}_\pm = 0 & \text{in } \mathbb{R}^N_\pm, \\ \mu_+(\partial_N v_{+j} + \partial_j v_{+N}) - \mu_-(\partial_N v_{-j} + \partial_j v_{-N}) = 0 & \text{on } \mathbb{R}^N_0, \\ 2\mu_+\partial_N v_{+N} - \mathfrak{q}_+ - (2\mu_-\partial_N v_{-N} - \mathfrak{q}_-) = H & \text{on } \mathbb{R}^N_0, \\ v_{+J} - v_{-J} = 0 & \text{on } \mathbb{R}^N_0, \end{cases} \tag{3.105}$$

where $H = \sigma\Delta' h$, together with

$$\lambda h + v_{-N} = d \quad \text{on } \mathbb{R}^N_0. \tag{3.106}$$

Let $\hat{f}$ be the partial Fourier transform of f defined as

$$\hat{f}(\xi', x_N) = \hat{f}(x_N) = \frac{1}{(2\pi)^{N-1}} \int_{\mathbb{R}^{N-1}} e^{-ix'\cdot\xi'} f(x', x_N)\,dx',$$

and also

$$A = |\xi'|^2, \quad B_\pm = \sqrt{\frac{m_\pm}{\mu_\pm}\lambda + |\xi'|^2} \quad (\mathrm{Re}\, B_\pm > 0).$$

Applying the partial Fourier transform to (3.105) yields for $\pm x_N > 0$

$$m_\pm\lambda\hat{v}_{\pm j}(x_N) - \mu_\pm(\partial_N^2 - A^2)\hat{v}_{\pm j}(x_N) + i\xi_j\hat{\mathfrak{q}}_\pm(x_N) = 0, \tag{3.107}$$

$$m_\pm\lambda\hat{v}_{\pm N}(x_N) - \mu_\pm(\partial_N^2 - A^2)\hat{v}_{\pm N}(x_N) + \partial_N\hat{\mathfrak{q}}_\pm(x_N) = 0, \tag{3.108}$$

$$\sum_{j=1}^{N-1} i\xi_j\hat{v}_{\pm j}(x_N) + \partial_N\hat{v}_{\pm N}(x_N) = 0, \tag{3.109}$$

$$\mu_+(\partial_N\hat{v}_{+j}(0) + i\xi_j\hat{v}_{+N}(0)) - \mu_-(\partial_N\hat{v}_{-j}(0) + i\xi_j\hat{v}_{-N}(0)) = 0, \tag{3.110}$$

$$2\mu_+\partial_N\hat{v}_{+N}(0) - \mathfrak{q}_+(0) - (2\mu_-\partial_N\hat{v}_{-N}(0) - \hat{\mathfrak{q}}_-(0)) = \hat{H}, \tag{3.111}$$

$$\hat{v}_{+J}(0) - \hat{v}_{-J}(0) = 0. \tag{3.112}$$

Analogously, it follows from (3.106) that

$$\lambda\hat{h} + \hat{v}_{-N}(0) = \hat{d}(0). \tag{3.113}$$

One first derives representation formulas, for solutions of (3.107)–(3.112), of the forms:

$$\hat{v}_{\pm J}(x_N) = \alpha_{\pm J}(e^{\mp Ax_N} - e^{\mp B_\pm x_N}) + \beta_{\pm J}e^{\mp B_\pm x_N},$$

$$\hat{\mathfrak{q}}_\pm(x_N) = \gamma_\pm e^{\mp Ax_N}, \quad \pm x_N > 0.$$

Inserting these formulas into (3.107)–(3.109) yields

$$-\mu_\pm(A^2 - B_\pm^2)\alpha_{\pm j} + i\xi_j\gamma_\pm = 0, \tag{3.114}$$

$$-\mu_\pm(A^2 - B_\pm^2)\alpha_{\pm N} \mp A\gamma_\pm = 0, \tag{3.115}$$

$$i\xi' \cdot \alpha'_\pm \mp A\alpha_{\pm N} = 0, \tag{3.116}$$

$$-i\xi' \cdot \alpha'_\pm + i\xi' \cdot \beta'_\pm \pm B_\pm\alpha_{\pm N} \mp B_\pm\beta_{\pm N} = 0, \tag{3.117}$$

where

$$i\xi' \cdot \alpha'_\pm = \sum_{j=1}^{N-1} i\xi_j\alpha_{\pm j}, \quad i\xi' \cdot \beta'_\pm = \sum_{j=1}^{N-1} i\xi_j\beta_{\pm j}.$$

By (3.116) and (3.117),

$$\alpha_{\pm N} = \pm\frac{i\xi' \cdot \beta'_\pm \mp B_\pm\beta_{\pm N}}{A - B_\pm}, \quad i\xi' \cdot \alpha'_\pm = \frac{A(i\xi' \cdot \beta'_\pm \mp B_\pm\beta_{\pm N})}{A - B_\pm}, \tag{3.118}$$

which, combined with (3.115), furnishes

$$\gamma_\pm = -\frac{\mu_\pm(A + B_\pm)(i\xi' \cdot \beta'_\pm \mp B_\pm\beta_{\pm N})}{A}. \tag{3.119}$$

On the other hand, we have from the boundary conditions (3.110)–(3.112)

$$\begin{aligned} &\mu_+((-A + B_+)\alpha_{+j} - B_+\beta_{+j} + i\xi_j\beta_{+N}) \\ &-\mu_-((A - B_-)\alpha_{-j} + B_-\beta_{-j} + i\xi_j\beta_{-N}) = 0, \end{aligned} \tag{3.120}$$

$$2\mu_+((-A + B_+)\alpha_{+N} - B_+\beta_{+N}) - \gamma_+$$

$$-\{2\mu_-((A-B_-)\alpha_{-N}+B_-\beta_{-N})-\gamma_-\}=\hat{H}, \tag{3.121}$$

$$\beta_{+J}-\beta_{-J}=0. \tag{3.122}$$

One then multiplies (3.120) by $i\xi_j$ and sums up the resultant formulas in order to obtain

$$\mu_+((-A+B_+)i\xi'\cdot\alpha'_+-B_+i\xi'\cdot\beta'_+-A^2\beta_{+N})$$
$$-\mu_-((A-B_-)i\xi'\cdot\alpha'_-+B_-i\xi'\cdot\beta'_--A^2\beta_{-N})=0,$$

which, combined with the second formula of (3.118), furnishes

$$\mu_+(-(A+B_+)i\xi'\cdot\beta'_++A(B_+-A)\beta_{+N})$$
$$-\mu_-((A+B_-)i\xi'\cdot\beta'_-+A(B_--A)\beta_{-N})=0. \tag{3.123}$$

In addition, we insert (3.119) and the first formula of (3.118) into (3.121), and multiply the resultant formula by A in order to obtain

$$\mu_+((-A+B_+)i\xi'\cdot\beta'_+-(A+B_+)B_+\beta_{+N})$$
$$-\mu_-((-A+B_-)i\xi'\cdot\beta'_-+(A+B_-)B_-\beta_{-N})=A\hat{H}. \tag{3.124}$$

Note that by (3.122)

$$i\xi'\cdot\beta'_+=i\xi'\cdot\beta'_-,\quad \beta_{+N}=\beta_{-N}. \tag{3.125}$$

These relations, inserted into (3.123) and (3.124), give us the following equations:

$$(-\mu_+(A+B_+)-\mu_-(A+B_-))i\xi'\cdot\beta'_-$$
$$+(\mu_+(B_+-A)-\mu_-(B_--A))A\beta_{-N}=0,$$
$$(\mu_+(-A+B_+)-\mu_-(-A+B_-))i\xi'\cdot\beta'_-$$
$$+(-\mu_+(A+B_+)B_+-\mu_-(A+B_-)B_-)\beta_{-N}=A\hat{H}.$$

Now we have achieved a system of linear equations as follows:

$$\mathbf{L}\begin{pmatrix} i\xi'\cdot\beta'_- \\ \beta_{-N}\end{pmatrix}=\begin{pmatrix}0\\ A\hat{H}\end{pmatrix},$$

where we have set

$$\mathbf{L}=\begin{pmatrix} -\mu_+(A+B_+)-\mu_-(A+B_-) & (\mu_+(B_+-A)-\mu_-(B_--A))A \\ \mu_+(-A+B_+)-\mu_-(-A+B_-) & -\mu_+(A+B_+)B_+-\mu_-(A+B_-)B_- \end{pmatrix}.$$

Let us write the inverse matrix of $\mathbf{L}$ as

$$\mathbf{L}^{-1} = \frac{1}{D(A, B_+, B_-)} \begin{pmatrix} L_{11} & L_{12} \\ L_{21} & L_{22} \end{pmatrix},$$

where

$$\begin{aligned} D(A, B_+, B_-) &= -(\mu_+ - \mu_-)^2 A^3 \\ &\quad + ((3\mu_+ - \mu_-)\mu_+ B_+ + (3\mu_- - \mu_+)\mu_- B_-)A^2 \\ &\quad + ((\mu_+ B_+ + \mu_- B_-)^2 + \mu_+\mu_-(B_+ + B_-)^2)A \\ &\quad + (\mu_+ B_+ + \mu_- B_-)(\mu_+ B_+^2 + \mu_- B_-^2), \\ L_{11} &= -\mu_+(A + B_+)B_+ - \mu_-(A + B_-)B_-, \\ L_{12} &= -(\mu_+(B_+ - A) - \mu_-(B_- - A))A, \\ L_{21} &= -(\mu_+(-A + B_+) - \mu_-(-A + B_-)), \\ L_{22} &= -\mu_+(A + B_+) - \mu_-(A + B_-). \end{aligned}$$

Solving the above system of linear equations, we have

$$\beta_{-N} = \frac{L_{22}}{D(A, B_+, B_-)} A\hat{H} = \frac{-\mu_+(A + B_+) - \mu_-(A + B_-)}{D(A, B_+, B_-)} A\hat{H}. \tag{3.126}$$

The following lemma is proved in Shibata-Shimizu [30, Lemma 5.5].

Lemma 3.3.1 *Let $0 < \epsilon_0 < \pi/2$. Then there exist positive constants C_1 and C_2 such that, for any $(\lambda, \xi') \in \Sigma_{\epsilon_0} \times (\mathbb{R}^{N-1} \times \{0\})$,*

$$\begin{aligned} C_1(|\lambda|^{1/2} + A) \le \operatorname{Re} B \le |B| \le C_2(|\lambda|^{1/2} + A), \\ C_1(|\lambda|^{1/2} + A)^3 \le |D(A, B_+, B_-)| \le C_2(|\lambda|^{1/2} + A)^3. \end{aligned}$$

Next, we consider (3.113). Since $\hat{v}_{-N}(0) = \beta_{-N}$ and since $\hat{H} = -\sigma A^2 \hat{h}$, it holds by (3.113) and (3.126) that

$$\lambda\hat{h} + \frac{\mu_+(A + B_+) + \mu_-(A + B_-)}{D(A, B_+, B_-)} \sigma A^3 \hat{h} = \hat{d}(0).$$

One then solves this equation, and sets

$$E = \lambda D(A, B_+, B_-) + \sigma A^3(\mu_+(A + B_+) + \mu_-(A + B_-))$$

in order to obtain $\hat{h} = E^{-1} D(A, B_+, B_-)\hat{d}(0)$.

At this point, let us prove the following lemma.

Lemma 3.3.2 *Let $0 < \epsilon_0 < \pi/2$ and let E be as above. Then there exist positive constants λ_1 and c such that*

$$|E| \geq c(|\lambda| + A)(|\lambda|^{1/2} + A)^3$$

for any $(\lambda, \xi') \in \Sigma_{\epsilon_0,\lambda_1} \times (\mathbb{R}^{N-1} \setminus \{0\})$.

Proof We first study the case where $|\lambda| \geq R_1 A$ for large $R_1 > 0$. Then, by Lemma 3.3.1,

$$\begin{aligned}|E| &\geq |\lambda||D(A, B_+, B_-)| - C_{\sigma,\mu_+,\mu_-} A^3(|\lambda|^{1/2} + A)\\ &\geq c|\lambda|(|\lambda|^{1/2} + A)^3 - C_{\sigma,\mu_+,\mu_-}(|\lambda|^{1/2} + A)^3(|\lambda|^{1/2} + A)\\ &\geq c|\lambda|(|\lambda|^{1/2} + A)^3 - C_{\sigma,\mu_+,\mu_-}(|\lambda|^{1/2} + A)^3(\lambda_1^{-1/2}|\lambda| + R_1^{-1}|\lambda|)\\ &\geq (c - C_{\sigma,\mu_+,\mu_-}(\lambda_1^{-1/2} + R_1^{-1}))|\lambda|(|\lambda|^{1/2} + A)^3.\end{aligned}$$

Here we choose λ_1 and R_1 large enough so that

$$C_{\sigma,\mu_+,\mu_-}(\lambda_1^{-1/2} + R_1^{-1}) \leq \frac{c}{2},$$

and then we have

$$|E| \geq \frac{c}{2}|\lambda|(|\lambda|^{1/2} + A)^3 \geq \frac{c}{4}(|\lambda| + R_1 A)(|\lambda|^{1/2} + A)^3$$

for $|\lambda| \geq R_1 A$ and $\lambda \in \Sigma_{\epsilon_0,\lambda_1}$.

Next, we consider the case where $|\lambda| \leq R_1 A$ and $\lambda \in \Sigma_{\epsilon_0,\lambda_1}$. In this case, we have $A \geq R_1^{-1}|\lambda| \geq R_1^{-1}\lambda_1^{1/2}|\lambda|^{1/2}$, and so we set $R_2 = R_1^{-1}\lambda_1^{1/2}$. Choosing R_2 large enough, one has

$$B_\pm = A(1 + O(R_2^{-1})),$$

which furnishes

$$D(A, B_+, B_-) = 4(\mu_+ + \mu_-)^2 A^3(1 + O(R_2^{-1})).$$

It thus holds that

$$E = 4\lambda(\mu_+ + \mu_-)^2 A^3(1 + O(R_2^{-1})) + 2\sigma(\mu_+ + \mu_-)A^4(1 + O(R_2^{-1})).$$

Then,

$$\begin{aligned}|E| &\geq |4\lambda(\mu_+ + \mu_-)^2 A^3 + 2\sigma(\mu_+ + \mu_-)A^4|\\ &\quad - (4|\lambda|(\mu_+ + \mu_-)^2 A^3 + 2\sigma(\mu_+ + \mu_-)A^4)O(R_2^{-1}),\end{aligned}$$

which, combined with

$$|\lambda + x| \geq C_\epsilon(|\lambda| + x) \quad (\lambda \in \Sigma_\epsilon, x \geq 0),$$

furnishes

$$\begin{aligned}|E| &\geq C_{\epsilon_0}(4|\lambda|(\mu_+ + \mu_-)^2 A^3 + 2\sigma(\mu_+ + \mu_-)A^4) \\ &\quad - (4|\lambda|(\mu_+ + \mu_-)^2 A^3 + 2\sigma(\mu_+ + \mu_-)A^4) O(R_2^{-1}).\end{aligned}$$

Here we choose R_2 large enough so that

$$C_{\epsilon_0} - O(R_2^{-1}) \geq \frac{C_{\epsilon_0}}{2},$$

and thus we have

$$|E| \geq c(|\lambda| A^3 + A^4) = c(|\lambda| + A)A^3 \geq \frac{c}{8}(|\lambda| + A)(A + R_2|\lambda|^{1/2})^3.$$

This completes the proof of the lemma. □

Let $\varphi(x_N)$ be a function in $C_0^\infty(\mathbb{R})$ such that $\varphi(x_N) = 1$ for $|x_N| \leq 1$ and $\varphi(x_N) = 0$ for $|x_N| \geq 2$. Then $h = h(x)$ is defined by

$$h(x) = \varphi(x_N)\mathcal{F}_{\xi'}^{-1}\Big[e^{-Ax_N}\frac{D(A, B_+, B_-)}{E}\hat{d}(\xi', 0)\Big](x'),$$

where $\mathcal{F}_{\xi'}^{-1}$ is the inverse partial Fourier transform given by

$$\mathcal{F}_{\xi'}^{-1}[f(\xi', x_N)](x') = \int_{\mathbb{R}^{N-1}} e^{ix'\cdot\xi'} f(\xi', x_N)\, d\xi'.$$

By the Volevich trick, we have

$$\begin{aligned}h(x) &= -\varphi(x_N)\int_0^\infty \frac{\partial}{\partial y_N}\mathcal{F}_{\xi'}^{-1}\Big[e^{-A(x_N+y_N)}\frac{D(A, B_+, B_-)}{E}\hat{d}(\xi', y_N)\varphi(y_N)\Big](x')\, dy_N \\ &= \varphi(x_N)\int_0^\infty \mathcal{F}_{\xi'}^{-1}\Big[e^{-A(x_N+y_N)}\frac{AD(A, B_+, B_-)}{E}\hat{d}(\xi', y_N)\varphi(y_N)\Big](x')\, dy_N \\ &\quad - \varphi(x_N)\int_0^\infty \mathcal{F}_{\xi'}^{-1}\Big[e^{-A(x_N+y_N)}\frac{D(A, B_+, B_-)}{E}\partial_N(\hat{d}(\xi', y_N)\varphi(y_N))\Big](x')\, dy_N\end{aligned}$$

$$
\begin{aligned}
&= \varphi(x_N) \int_0^\infty \mathcal{F}_{\xi'}^{-1}\Big[e^{-A(x_N+y_N)} \frac{AD(A, B_+, B_-)}{E(1+A^2)} \\
&\times \mathcal{F}'[(1-\Delta')d](\xi', y_N)\varphi(y_N)\Big](x')\, dy_N \\
&- \varphi(x_N) \int_0^\infty \mathcal{F}_{\xi'}^{-1}\Big[e^{-A(x_N+y_N)} \frac{D(A, B_+, B_-)}{E(1+A^2)} \Big\{\partial_N(\hat{d}(\xi', y_N)\varphi(y_N)) \\
&- \sum_{k=1}^{N-1} i\xi_k \partial_N\Big(\mathcal{F}'[\partial_k d](\xi', y_N)\varphi(y_N)\Big)\Big\}\Big](x')\, dy_N \\
&=: \mathcal{H}'(\lambda)d,
\end{aligned}
$$

where we have set $\mathcal{F}'[f](\xi', y_N) = \hat{f}(\xi', y_N)$. Here we introduce the following lemma proved in [29].

Lemma 3.3.3 *Let Λ be a domain in $\mathbb{C}$ and let $1 < q < \infty$. Let φ and ψ be functions in $C_0^\infty(-2, 2)$. Assume that $m_0 = m_0(\lambda, \xi')$, defined on $\Lambda \times (\mathbb{R}^{N-1} \setminus \{0\})$, satisfies for any multi-index $\alpha' \in \mathbb{N}_0^{N-1}$*

$$
|\partial_{\xi'}^{\alpha'} m_0(\lambda, \xi')| \leq C_{\alpha', \Lambda} |\xi'|^{-|\alpha'|} \quad ((\lambda, \xi') \in \Lambda \times (\mathbb{R}^{N-1} \setminus \{0\}))
$$

with some positive constant $C_{\alpha', \Lambda}$, and define an operator $L(\lambda)$ by

$$
[L(\lambda)g](x) = \varphi(x_N) \int_0^\infty \mathcal{F}_{\xi'}^{-1}\Big[e^{-A(x_N+y_N)} m_0(\lambda, \xi')\hat{g}(\xi', y_N)\psi(y_N)\Big](x')\, dy_N.
$$

Then $L(\lambda) \in \mathrm{Hol}\,(\Lambda, \mathcal{L}(L_q(\mathbb{R}_+^N)))$ and

$$
\mathcal{R}_{\mathcal{L}(L_q(\mathbb{R}_+^N))}(\{L(\lambda) \mid \lambda \in \Lambda\}) \leq r_b
$$

for some positive constant r_b depending on Λ.

Let us define

$$
\mathbf{A} = (i\xi_1, \ldots, i\xi_{N-1}, -A),
$$

and let $\beta \in \mathbb{N}_0^N$ be a multi-index with $|\beta| \leq 2$ and $\gamma \in \mathbb{N}_0^N$ be a multi-index with $|\gamma| \leq 3$. Then, by Bell's formula and Leibniz's rule, we have for any multi-index $\alpha \in \mathbb{N}_0^N$ and for $\ell = 0, 1$

$$
\left|\partial_{\xi'}^{\alpha}(\tau\partial_\tau)^\ell \left(\frac{\lambda \mathbf{A}^\beta D(A, B_+, B_-)}{E(1+A^2)}\right)\right| \leq C_{\lambda_1} A^{-|\alpha|},
$$

$$\left|\partial_{\xi'}^{\alpha}(\tau\partial_\tau)^\ell\left(\frac{\mathbf{A}^\gamma D(A,B_+,B_-)}{E(1+A^2)}\right)\right| \le C_{\lambda_1}A^{-|\alpha|},$$

where $(\lambda,\xi')\in\Sigma_{\epsilon_0,\lambda_1}\times(\mathbb{R}^{N-1}\setminus\{0\})$. Thus applying Lemma 3.3.3 to $\mathcal{H}'(\lambda)$ furnishes that

$$\mathcal{R}_{\mathcal{L}(H_q^2(\mathbb{R}_+^N),H_q^{3-k}(\mathbb{R}_+^N))}\left(\{(\tau\partial_\tau)^\ell(\lambda^k\mathcal{H}'(\lambda))\mid\lambda\in\Sigma_{\epsilon,\lambda_0}\}\right)\le r_b$$

for $k,\ell=0,1$ with some positive constant r_b.

Combining the $\mathcal{R}$-bounded solution operators $\mathcal{H}'(\lambda)$ with [20], we have obtained the following theorem.

Theorem 3.3.4 *Let $1<q<\infty$ and $0<\epsilon_0<\pi/2$. Then, there exist a constant λ_1 and operator families $\mathcal{B}'(\lambda)$, $\mathcal{P}'(\lambda)$, and $\mathcal{H}'(\lambda)$, with*

$$\mathcal{B}'(\lambda)\in\mathrm{Hol}\,(\Sigma_{\epsilon_0,\lambda_1},\mathcal{L}(H_q^2(\mathbb{R}_+^N),H_q^2(\mathbb{R}_+^N)^N)),$$

$$\mathcal{P}'(\lambda)\in\mathrm{Hol}\,(\Sigma_{\epsilon_0,\lambda_1},\mathcal{L}(H_q^2(\mathbb{R}_+^N),\hat{H}_q^1(\mathbb{R}_+^N))),$$

$$\mathcal{H}'(\lambda)\in\mathrm{Hol}\,(\Sigma_{\epsilon_0,\lambda_1},\mathcal{L}(H_q^2(\mathbb{R}_+^N),H_q^3(\mathbb{R}_+^N))),$$

such that, for any $d\in H_q^2(\mathbb{R}_+^N)$, $\mathbf{v}=\mathcal{B}'(\lambda)d$, $\mathfrak{q}=\mathcal{P}'(\lambda)d$, and $h=\mathcal{H}'(\lambda)d$ are unique solutions of (3.104). *Moreover,*

$$\mathcal{R}_{\mathcal{L}(H_q^2(\mathbb{R}_+^N),H_q^{2-j}(\mathbb{R}_+^N)^N)}(\{(\tau\partial_\tau)^\ell(\lambda^{j/2}\mathcal{B}'(\lambda))\mid\lambda\in\Sigma_{\epsilon_0,\lambda_1}\})\le r_b,$$

$$\mathcal{R}_{\mathcal{L}(H_q^2(\mathbb{R}_+^N),L_q(\mathbb{R}_+^N)^N)}(\{(\tau\partial_\tau)^\ell(\nabla\mathcal{P}'(\lambda))\mid\lambda\in\Sigma_{\epsilon_0,\lambda_1}\})\le r_b,$$

$$\mathcal{R}_{\mathcal{L}(H_q^2(\mathbb{R}_+^N),H_q^{3-k}(\mathbb{R}_+^N))}(\{(\tau\partial_\tau)^\ell(\lambda^k\mathcal{H}'(\lambda))\mid\lambda\in\Sigma_{\epsilon_0,\lambda_1}\})\le r_b,$$

for $j=0,1,2$ and $k,\ell=0,1$. Here r_b is a constant depending on λ_1.

Remark 8 Theorem 3.3.4 was essentially proved in Shibata and Shimizu [34].

Using the extension map from $W_q^{m-1/q}(\mathbb{R}_0^N)$, $m\in\mathbb{N}$, to $H_q^m(\mathbb{R}_+^N)$ and the restriction map from $H_q^m(\mathbb{R}_+^N)$ to $W_q^{m-1/q}(\mathbb{R}_0^N)$, we have the following corollary of Theorem 3.3.4.

Corollary 3.3.5 *Let $1<q<\infty$ and $0<\epsilon_0<\pi/2$. Then, there exist a constant λ_1 and operator families $\mathcal{B}'(\lambda)$, $\mathcal{P}'(\lambda)$, and $\mathcal{H}'(\lambda)$, with*

$$\mathcal{B}'(\lambda)\in\mathrm{Hol}\,(\Sigma_{\epsilon_0,\lambda_1},\mathcal{L}(W_q^{2-1/q}(\mathbb{R}_0^N),H_q^2(\mathbb{R}_+^N)^N)),$$

$$\mathcal{P}'(\lambda)\in\mathrm{Hol}\,(\Sigma_{\epsilon_0,\lambda_1},\mathcal{L}(W_q^{2-1/q}(\mathbb{R}_0^N),\hat{H}_q^1(\mathbb{R}_+^N))),$$

$$\mathcal{H}'(\lambda)\in\mathrm{Hol}\,(\Sigma_{\epsilon_0,\lambda_1},\mathcal{L}(W_q^{2-1/q}(\mathbb{R}_0^N),W_q^{3-1/q}(\mathbb{R}_0^N))),$$

such that, for any $d \in W_q^{2-1/q}(\mathbb{R}_0^N)$, $\mathbf{u} = \mathcal{B}'(\lambda)d$, $\mathfrak{q} = \mathcal{P}'(\lambda)d$, *and* $h = \mathcal{H}'(\lambda)d$ *are unique solutions of* (3.104). *Moreover,*

$$\mathcal{R}_{\mathcal{L}(W_q^{2-1/q}(\mathbb{R}_0^N), H_q^{2-j}(\mathbb{R}_+^N)^N)}(\{(\tau\partial_\tau)^\ell(\lambda^{j/2}\mathcal{B}'(\lambda)) \mid \lambda \in \Sigma_{\epsilon_0,\lambda_1}\}) \leq r_b,$$
$$\mathcal{R}_{\mathcal{L}(W_q^{2-1/q}(\mathbb{R}_0^N), L_q(\mathbb{R}_+^N)^N)}(\{(\tau\partial_\tau)^\ell(\nabla\mathcal{P}'(\lambda)) \mid \lambda \in \Sigma_{\epsilon_0,\lambda_1}\}) \leq r_b,$$
$$\mathcal{R}_{\mathcal{L}(W_q^{2-1/q}(\mathbb{R}_0^N), W_q^{3-1/q-k}(\mathbb{R}_0^N))}(\{(\tau\partial_\tau)^\ell(\lambda^k\mathcal{H}'(\lambda)) \mid \lambda \in \Sigma_{\epsilon_0,\lambda_1}\}) \leq r_b,$$

for $j = 0, 1, 2$ *and* $k, \ell = 0, 1$. *Here* r_b *is a constant depending on* λ_1.

3.3.2 Model Problem: $\sigma > 0$ and $\mathbf{a}' \neq 0$

In this section, we consider the following model problem with surface tension $\sigma > 0$ and $\mathbf{a}' = (a_1, \ldots, a_{N-1}) \neq 0$:

$$\begin{cases} \lambda\mathbf{v} - m^{-1}\mathrm{Div}\,(\mu\mathbf{D}(\mathbf{v}) - \mathfrak{q}) = 0 & \text{in } \dot{\mathbb{R}}^N, \\ \mathrm{div}\,\mathbf{v} = 0 & \text{in } \dot{\mathbb{R}}^N, \\ [[(\mu\mathbf{D}(\mathbf{v}) - \mathfrak{q}\mathbf{I})\mathbf{n}_0]] - \sigma(\Delta' h)\mathbf{n}_0 = 0, \quad [[\mathbf{v}]] = 0 & \text{on } \mathbb{R}_0^N, \\ \lambda h + \mathbf{a}' \cdot \nabla' h - \mathbf{v}\cdot\mathbf{n}_0 = d & \text{on } \mathbb{R}_0^N, \end{cases} \tag{3.127}$$

which is a reduced system similar to (3.104). Here we assume for some positive number a_0

$$|\mathbf{a}'| \leq a_0. \tag{3.128}$$

The aim of this section is to prove the existence of $\mathcal{R}$-bounded solution operators of (3.127). One starts with deriving representation formulas for solutions of (3.127). Let $j = 1, \ldots, N-1$ and $J = 1, \ldots, N$ in what follows. The system (3.127) can be written as

$$\begin{cases} m_\pm\lambda v_{\pm J} - \mu_\pm\Delta v_{\pm J} + \nabla\mathfrak{q}_\pm = 0 & \text{in } \mathbb{R}_\pm^N, \\ \mathrm{div}\,\mathbf{v}_\pm = 0 & \text{in } \mathbb{R}_\pm^N, \\ \mu_+(\partial_N v_{+j} + \partial_j v_{+N}) - \mu_-(\partial_N v_{-j} + \partial_j v_{-N}) = 0 & \text{on } \mathbb{R}_0^N, \\ 2\mu_+\partial_N v_{+N} - \mathfrak{q}_+ - (2\mu_-\partial_N v_{-N} - \mathfrak{q}_-) = H & \text{on } \mathbb{R}_0^N, \\ v_{+J} - v_{-J} = 0 & \text{on } \mathbb{R}_0^N, \end{cases}$$

where $H = \sigma\Delta' h$, together with

$$\lambda h + \mathbf{a}' \cdot \nabla' h + v_{-N} = d \quad \text{on } \mathbb{R}^N_0.$$

Applying the partial Fourier transform to this equation yields

$$\lambda \hat{h} + \mathbf{a}' \cdot i\xi' \hat{h} + \hat{v}_{-N}(0) = \hat{d}(0).$$

Since $\hat{v}_{-N}(0) = \beta_{-N}$ with (3.126) as was discussed in the previous section, one has by the last equation

$$(\lambda + \mathbf{a}' \cdot i\xi')\hat{h} + \frac{\mu_+(A+B_+) + \mu_-(A+B_-)}{D(A,B_+,B_-)}\sigma A^3 \hat{h} = \hat{d}(0).$$

One then solves this equation, and sets

$$\tilde{E} = (\lambda + \mathbf{a}' \cdot i\xi')D(A,B_+,B_-) + \sigma A^3(\mu_+(A+B_+) + \mu_-(A+B_-))$$

in order to obtain $\hat{h} = \tilde{E}^{-1}D(A,B_+,B_-)\hat{d}(0)$.

At this point, let us prove the following lemma.

Lemma 3.3.6 *Let $\tilde{E}$ be as above. Then there exist positive numbers λ_1 and c, depending on a_0 given in* (3.128), *such that there holds*

$$|\tilde{E}| \geq c(|\lambda| + A)(|\lambda|^{1/2} + A)^3$$

for any $(\lambda, \xi') \in \mathbb{C}_{+,\lambda_1} \times (\mathbb{R}^{N-1} \setminus \{0\})$. Here $\mathbb{C}_{+,\lambda_1} = \{\lambda \in \mathbb{C} \mid \mathrm{Re}\lambda \geq \lambda_1\}$.

Proof We first study the case where $|\lambda| \geq R_1 A$ for large $R_1 > 0$. Then, by Lemma 3.3.1,

$$\begin{aligned}
|\tilde{E}| &\geq |\lambda||D(A,B_+,B_-)| - |\mathbf{a}'|A|D(A,B_+,B_-)| - C_{\sigma,\mu_+,\mu_-}A^3(|\lambda|^{1/2} + A)\\
&\geq c|\lambda|(|\lambda|^{1/2} + A)^3 - a_0 C_{\mu_+,\mu_-} R_1^{-1}|\lambda|(|\lambda|^{1/2} + A)^3\\
&\quad - C_{\sigma,\mu_+,\mu_-}(|\lambda|^{1/2} + A)^3(|\lambda|^{1/2} + A)\\
&\geq c|\lambda|(|\lambda|^{1/2} + A)^3 - a_0 C_{\mu_+,\mu_-} R_1^{-1}|\lambda|(|\lambda|^{1/2} + A)^3\\
&\quad - C_{\sigma,\mu_+,\mu_-}(|\lambda|^{1/2} + A)^3(\lambda_1^{-1/2}|\lambda| + R_1^{-1}|\lambda|)\\
&\geq c|\lambda|(|\lambda|^{1/2} + A)^3\\
&\quad - \left(a_0 C_{\mu_+,\mu_-} R_1^{-1} + C_{\sigma,\mu_+,\mu_-}\left(\lambda_1^{-1/2} + R_1^{-1}\right)\right)|\lambda|(|\lambda|^{1/2} + A)^3.
\end{aligned}$$

Here we choose R_1 and λ_1 large enough so that

$$a_0 C_{\mu_+,\mu_-} R_1^{-1} + C_{\sigma,\mu_+,\mu_-}\left(\lambda_1^{-1/2} + R_1^{-1}\right) \leq \frac{c}{2},$$

and then we have

$$|E| \geq \frac{c}{2}|\lambda|(|\lambda|^{1/2} + A)^3 \geq \frac{c}{4}(|\lambda| + R_1 A)(|\lambda|^{1/2} + A)^3$$

for $|\lambda| \geq R_1 A$ and $\lambda \in \mathbb{C}_{+,\lambda_1}$.

Next, we study the case where $|\lambda| \leq R_1 A$ and $\lambda \in \mathbb{C}_{+,\lambda_1}$. In this case, we have $A \geq R_1^{-1}|\lambda| \geq R_1^{-1}\lambda_1^{1/2}|\lambda|^{1/2}$, and so we set $R_2 = R_1^{-1}\lambda_1^{1/2}$. Choosing R_2 large enough, one has

$$B_\pm = A(1 + O(R_2^{-1})),$$

which furnishes

$$D(A, B_+, B_-) = 4(\mu_+ + \mu_-)^2 A^3(1 + O(R_2^{-1})).$$

It thus holds that

$$\begin{aligned}\tilde{E} &= (\mathrm{Re}\,\lambda + i(\Im\lambda + \xi' \cdot \mathbf{a}'))4(\mu_+ + \mu_-)^2 A^3(1 + O(R_2^{-1})) \\ &+ \sigma(\mu_+ + \mu_-)2A^4(1 + O(R_2^{-1})).\end{aligned}$$

Taking the real part of this equation and noting $\mathrm{Re}\,\lambda \geq \lambda_1 \geq 1$, we have

$$\begin{aligned}\mathrm{Re}\,\tilde{E} &= (\mathrm{Re}\,\lambda)4(\mu_+ + \mu_-)^2 A^3(1 + O(R_2^{-1})) \\ &+ (\Im\lambda + \xi' \cdot \mathbf{a}')4(\mu_+ + \mu_-)^2 A^3 O(R_2^{-1}) \\ &+ \sigma(\mu_+ + \mu_-)2A^4(1 + O(R_2^{-1})) \\ &\geq 4(\mu_+ + \mu_-)^2 A^3(1 + O(R_2^{-1})) \\ &+ (\Im\lambda + \xi' \cdot \mathbf{a}')4(\mu_+ + \mu_-)^2 A^3 O(R_2^{-1}) \\ &+ \sigma(\mu_+ + \mu_-)2A^4(1 + O(R_2^{-1})).\end{aligned}$$

On the other hand,

$$\begin{aligned}A^3(1 + O(R_2^{-1})) &\leq \lambda_1^{-1/2}|\lambda|^{1/2}A^3(1 + O(R_2^{-1})) \\ &\leq R_2^{-1}A^4(1 + O(R_2^{-1})) \\ &\leq A^4 O(R_2^{-1}), \\ (\Im\lambda + \xi' \cdot \mathbf{a}')A^3 O(R_2^{-1}) &\leq (|\lambda| + a_0 A)A^3 O(R_2^{-1}) \\ &\leq (R_1 A + a_0 A)A^3 O(R_2^{-1}),\end{aligned}$$

which, combined with the above estimate of Re $\tilde{E}$, furnishes

$$
\begin{aligned}
\mathrm{Re}\,\tilde{E} &\geq 2\sigma(\mu_+ + \mu_-)A^4 \\
&\quad - \Big(2\sigma(\mu_+ + \mu_-) + 4(\mu_+ + \mu_-)^2 + 4(\mu_+ + \mu_-)^2(R_1 + a_0)\Big) A^4 O(R_2^{-1}).
\end{aligned}
$$

Here we choose R_2 large enough so that

$$
\begin{aligned}
&\sigma(\mu_+ + \mu_-) \\
&\quad - \Big(2\sigma(\mu_+ + \mu_-) + 4(\mu_+ + \mu_-)^2 + 4(\mu_+ + \mu_-)^2(R_1 + a_0)\Big) O(R_2^{-1}) \geq 0,
\end{aligned}
$$

and thus we have

$$
\begin{aligned}
|\tilde{E}| \geq \mathrm{Re}\tilde{E} &\geq \sigma(\mu_+ + \mu_-)A^4 \\
&\geq c(A + R_1^{-1}|\lambda|)(A + R_2|\lambda|^{1/2})^3.
\end{aligned}
$$

This completes the proof of the lemma. □

Employing the same argument as in the previous section, we obtain the following theorem and its corollary.

Theorem 3.3.7 *Let $1 < q < \infty$. Then there exist a constant λ_1, depending on a_0 given in* (3.128)*, and operator families $\mathcal{B}''(\lambda)$, $\mathcal{P}''(\lambda)$, and $\mathcal{H}''(\lambda)$, with*

$$
\begin{aligned}
\mathcal{B}''(\lambda) &\in \mathrm{Hol}\,(\mathbb{C}_{+,\lambda_1}, \mathcal{L}(H_q^2(\mathbb{R}_+^N), H_q^2(\mathbb{R}_+^N)^N)), \\
\mathcal{P}''(\lambda) &\in \mathrm{Hol}\,(\mathbb{C}_{+,\lambda_1}, \mathcal{L}(H_q^2(\mathbb{R}_+^N), \hat{H}_q^1(\mathbb{R}_+^N))), \\
\mathcal{H}''(\lambda) &\in \mathrm{Hol}\,(\mathbb{C}_{+,\lambda_1}, \mathcal{L}(H_q^2(\mathbb{R}_+^N), H_q^3(\mathbb{R}_+^N))),
\end{aligned}
$$

such that, for any $d \in H_q^2(\mathbb{R}_+^N)$, $\mathbf{v} = \mathcal{B}''(\lambda)d$, $\mathfrak{q} = \mathcal{P}''(\lambda)d$, and $h = \mathcal{H}''(\lambda)d$ are unique solutions of (3.127)*. Moreover,*

$$
\begin{aligned}
\mathcal{R}_{\mathcal{L}(H_q^2(\mathbb{R}_+^N), H_q^{2-j}(\mathbb{R}_+^N)^N)}(\{(\tau\partial_\tau)^\ell(\lambda^{j/2}\mathcal{B}''(\lambda)) \mid \lambda \in \mathbb{C}_{+,\lambda_1}\}) &\leq r_b, \\
\mathcal{R}_{\mathcal{L}(H_q^2(\mathbb{R}_+^N), L_q(\mathbb{R}_+^N)^N)}(\{(\tau\partial_\tau)^\ell(\nabla\mathcal{P}''(\lambda)) \mid \lambda \in \mathbb{C}_{+,\lambda_1}) &\leq r_b, \\
\mathcal{R}_{\mathcal{L}(H_q^2(\mathbb{R}_+^N), H_q^{3-k}(\mathbb{R}_+^N))}(\{(\tau\partial_\tau)^\ell(\lambda^k\mathcal{H}''(\lambda)) \mid \lambda \in \mathbb{C}_{+,\lambda_1}\}) &\leq r_b,
\end{aligned}
$$

for $j = 0, 1, 2$ and $k, \ell = 0, 1$. Here r_b is a constant depending on λ_1 and a_0.

Corollary 3.3.8 *Let $1 < q < \infty$. Then there exist a constant λ_1, depending on a_0 given in* (3.128)*, and operator families $\mathcal{B}''(\lambda)$, $\mathcal{P}''(\lambda)$, and $\mathcal{H}''(\lambda)$, with*

$$\mathcal{B}''(\lambda) \in \mathrm{Hol}\,(\mathbb{C}_{+,\lambda_1}, \mathcal{L}(W_q^{2-1/q}(\mathbb{R}_0^N), H_q^2(\mathbb{R}_+^N)^N)),$$
$$\mathcal{P}''(\lambda) \in \mathrm{Hol}\,(\mathbb{C}_{+,\lambda_1}, \mathcal{L}(W_q^{2-1/q}(\mathbb{R}_0^N), \hat{H}_q^1(\mathbb{R}_+^N))),$$
$$\mathcal{H}''(\lambda) \in \mathrm{Hol}\,(\mathbb{C}_{+,\lambda_1}, \mathcal{L}(W_q^{2-1/q}(\mathbb{R}_0^N), W_q^{3-1/q}(\mathbb{R}_0^N))),$$

such that, for any $d \in W_q^{2-1/q}(\mathbb{R}_0^N)$, $\mathbf{v} = \mathcal{B}''(\lambda)d$, $\mathfrak{q} = \mathcal{P}''(\lambda)d$, *and* $h = \mathcal{H}''(\lambda)d$ *are unique solutions of* (3.127). *Moreover,*

$$\mathcal{R}_{\mathcal{L}(W_q^{2-1/q}(\mathbb{R}_0^N), H_q^{2-j}(\mathbb{R}_+^N)^N)}(\{(\tau\partial_\tau)^\ell(\lambda^{j/2}\mathcal{B}''(\lambda)) \mid \lambda \in \mathbb{C}_{+,\lambda_1}\}) \leq r_b,$$
$$\mathcal{R}_{\mathcal{L}(W_q^{2-1/q}(\mathbb{R}_0^N), L_q(\mathbb{R}_+^N)^N)}(\{(\tau\partial_\tau)^\ell(\nabla\mathcal{P}''(\lambda)) \mid \lambda \in \mathbb{C}_{+,\lambda_1}) \leq r_b,$$
$$\mathcal{R}_{\mathcal{L}(W_q^{2-1/q}(\mathbb{R}_0^N), W_q^{3-1/q-k}(\mathbb{R}_0^N))}(\{(\tau\partial_\tau)^\ell(\lambda^k\mathcal{H}''(\lambda)) \mid \lambda \in \mathbb{C}_{+,\lambda_1}\}) \leq r_b,$$

for $j = 0, 1, 2$ *and* $k, \ell = 0, 1$. *Here* r_b *is a constant depending on* λ_1 *and* a_0.

3.4 Local Well-Posedness

3.4.1 Transformation of Equations and the Divergence Free Condition

In this section, we study how the equations and the divergence free condition are transformed under Hanzawa transformation. Since we use the three different type Hanzawa transforms given in (3.27), (3.33), and (3.39), in this section and next section, let Ψ_ρ denote one of them, that is,

$$\Psi_\rho(y,t) = \omega(y)H_\rho(y,t)\mathbf{n}(y), \quad \zeta(y)(R^{-1}H_\rho(y,t)y + \xi(t)),$$
$$\text{or } \omega(y)R^{-1}H_\rho(y,t)y + \xi(t).$$

Let $\mathbf{v}$ and $\mathfrak{p}$ be solutions of equation (3.1), and let

$$\mathbf{u}(y,t) = \mathbf{v}(y + \Psi_\rho(y,t), t), \quad \mathfrak{q}(y,t) = \mathfrak{p}(y + \Psi_\rho(y,t), t).$$

We now show that the first equation in (3.1) is transformed to

$$m\partial_t\mathbf{u} - \mathrm{Div}\,(\mu\mathbf{D}(\mathbf{u}) - \mathfrak{q}\mathbf{I}) = \mathbf{f}(\mathbf{u}, \Psi_\rho) \quad \text{in } \dot{\Omega}^T, \tag{3.129}$$

and the divergence free condition: $\mathrm{div}\,\mathbf{v} = 0$ in (3.1) is transformed to

$$\mathrm{div}\,\mathbf{u} = g(\mathbf{u}, \Psi_\rho) = \mathrm{div}\,\mathbf{g}(\mathbf{u}, \Psi_\rho) \quad \text{in } \dot{\Omega}^T, \tag{3.130}$$

where $\mathbf{f}(\mathbf{u}, \Psi_\rho)$, $g(\mathbf{u}, \Psi_\rho)$, and $\mathbf{g}(\mathbf{u}, \Psi_\rho)$ are suitable nonlinear functions with respect to $\mathbf{u}$ and $\nabla\Psi_\rho$ given in (3.142), (3.136), and (3.137), below.

Let $\partial x/\partial y$ be the Jacobi matrix of the transformation: $x = y + \Psi_\rho(y, t)$, that is,

$$\frac{\partial x}{\partial y} = \mathbf{I} + \nabla\Psi_\rho(y, t) = (\delta_{ij} + \frac{\partial \Psi_j}{\partial y_i}), \quad \nabla\Psi_\rho = \begin{pmatrix} \partial_1\Psi_1 & \partial_2\Psi_1 & \dots & \partial_N\Psi_1 \\ \partial_1\Psi_2 & \partial_2\Psi_2 & \dots & \partial_N\Psi_2 \\ \vdots & \vdots & \ddots & \vdots \\ \partial_1\Psi_N & \partial_2\Psi_N & \dots & \partial_N\Psi_N \end{pmatrix}$$

where $\Psi_\rho(y, t) = (\Psi_1(y, t), \dots, \Psi_N(y, t))^\top$, and $\partial_i\Psi_j = \dfrac{\partial\Psi_j}{\partial y_i}$. If $0 < \delta < 1$, then

$$\left(\frac{\partial x}{\partial y}\right)^{-1} = \mathbf{I} + \sum_{k=1}^\infty (-\nabla\Psi_\rho(y, t))^k$$

exists, and therefore there exists an $N \times N$ matrix $\mathbf{V}_0(\mathbf{k})$ of C^∞ functions defined on $|\mathbf{k}| < \delta$ such that $\mathbf{V}_0(0) = 0$ and

$$\left(\frac{\partial x}{\partial y}\right)^{-1} = \mathbf{I} + \mathbf{V}_0(\nabla\Psi_\rho(y, t)). \tag{3.131}$$

Here and in the following, $\mathbf{k} = (k_{ij})$ and k_{ij} are the variables corresponding to $\partial_i\Psi_j$.

Let $V_{0ij}(\mathbf{k})$ be the $(i, j)^{\rm th}$ component of $\mathbf{V}_0(\mathbf{k})$. We then have

$$\nabla_x = (\mathbf{I} + \mathbf{V}_0(\mathbf{k}))\nabla_y, \quad \frac{\partial}{\partial x_i} = \sum_{j=1}^N (\delta_{ij} + V_{0ij}(\mathbf{k}))\frac{\partial}{\partial y_j}, \tag{3.132}$$

where $\nabla_z = (\partial/\partial z, \dots, \partial/\partial z_N)^\top$ for $z = x$ and y. By (3.132), we can write $\mathbf{D}(\mathbf{v})$ as $\mathbf{D}(\mathbf{v}) = \mathbf{D}(\mathbf{u}) + \mathcal{D}_\mathbf{D}(\mathbf{k})\nabla\mathbf{u}$ with

$$\begin{aligned} \mathbf{D}(\mathbf{u})_{ij} &= \frac{\partial u_i}{\partial y_j} + \frac{\partial u_j}{\partial y_i}, \\ (\mathcal{D}_\mathbf{D}(\mathbf{k})\nabla\mathbf{u})_{ij} &= \sum_{k=1}^N \Big(V_{0jk}(\mathbf{k})\frac{\partial u_i}{\partial y_k} + V_{0ik}(\mathbf{k})\frac{\partial u_j}{\partial y_k}\Big). \end{aligned} \tag{3.133}$$

We next consider $\operatorname{div}\mathbf{v}$. By (3.132), we have

$$\operatorname{div}_x\mathbf{v} = \sum_{j=1}^N \frac{\partial v_j}{\partial x_j} = \sum_{j,k=1}^N (\delta_{jk} + V_{0jk}(\mathbf{k}))\frac{\partial u_j}{\partial y_k} = \operatorname{div}_y\mathbf{u} + \mathbf{V}_0(\mathbf{k}) : \nabla\mathbf{u}. \tag{3.134}$$

Let J be the Jacobian of the transformation: $x = y + \Psi_\rho(y, t)$. Choosing $\delta > 0$ small enough, we may assume that $J = J(\mathbf{k}) = 1 + J_0(\mathbf{k})$, where $J_0(\mathbf{k})$ is a C^∞ function defined for $|\mathbf{k}| < \sigma$ such that $J_0(0) = 0$.

To obtain another representation formula of $\mathrm{div}_x\mathbf{v}$, we use the inner product $(\cdot, \cdot)_{\dot\Omega_t}$. For any test function $\varphi \in C_0^\infty(\Omega_t)$, we set $\psi(y) = \varphi(x)$. We then have

$$\begin{aligned}(\mathrm{div}_x\mathbf{v}, \varphi)_{\dot\Omega_t} &= -(\mathbf{v}, \nabla\varphi)_{\dot\Omega_t} = -(J\mathbf{u}, (\mathbf{I}+\mathbf{V}_0)\nabla_y\psi)_{\dot\Omega} \\ &= (\mathrm{div}\,((\mathbf{I}+\mathbf{V}_0)^\top J\mathbf{u}), \psi)_{\dot\Omega} = (J^{-1}\mathrm{div}\,((\mathbf{I}+\mathbf{V}_0)^\top J\mathbf{u}), \varphi)_{\dot\Omega_t},\end{aligned}$$

which, combined with (3.134), leads to

$$\mathrm{div}_x\mathbf{v} = \mathrm{div}_y\mathbf{u} + \mathbf{V}_0(\mathbf{k}) : \nabla\mathbf{u} = J^{-1}(\mathrm{div}_y\mathbf{u} + \mathrm{div}_y(J\mathbf{V}_0(\mathbf{k})^\top\mathbf{u})). \tag{3.135}$$

Recalling that $J = 1 + J_0(\mathbf{k})$, we define $g(\mathbf{u}, \Psi_\rho)$ and $\mathbf{g}(\mathbf{u}, \Psi_\rho)$ by letting

$$g(\mathbf{u}, \Psi_\rho) = -(J_0(\mathbf{k})\mathrm{div}\,\mathbf{u} + (1 + J_0(\mathbf{k}))\mathbf{V}_0(\mathbf{k}) : \nabla\mathbf{u}), \tag{3.136}$$

$$\mathbf{g}(\mathbf{u}, \Psi_\rho) = -(1 + J_0(\mathbf{k}))\mathbf{V}_0(\mathbf{k})^\top\mathbf{u}, \tag{3.137}$$

and then by (3.135) we see that the divergence free condition: $\mathrm{div}\,\mathbf{v} = 0$ is transformed to equation (3.130). In particular, it follows from (3.135) that

$$J_0\mathrm{div}\,\mathbf{u} + J\mathbf{V}_0(\mathbf{k}) : \nabla\mathbf{u} = \mathrm{div}\,(J\mathbf{V}_0(\mathbf{k})^\top\mathbf{u}). \tag{3.138}$$

To derive equation (3.129), we first observe that

$$\begin{aligned}&\sum_{j=1}^N \frac{\partial}{\partial x_j}(\mu\mathbf{D}(\mathbf{v})_{ij} - \mathfrak{p}\delta_{ij}) \\ &= \sum_{j,k=1}^N \mu(\delta_{jk} + V_{0jk})\frac{\partial}{\partial y_k}(\mathbf{D}(\mathbf{u})_{ij} + (\mathcal{D}_\mathbf{D}(\mathbf{k})\nabla\mathbf{u})_{ij}) - \sum_{j=1}^N(\delta_{ij} + V_{0ij})\frac{\partial\mathfrak{q}}{\partial y_j},\end{aligned} \tag{3.139}$$

where we have used (3.133). Since

$$\frac{\partial}{\partial t}[v_i(y + \Psi_\rho(y, t), t)] = \frac{\partial v_i}{\partial t}(x, t) + \sum_{j=1}^N \frac{\partial\Psi_j}{\partial t}(y, t)\frac{\partial v_i}{\partial x_j}(x, t),$$

we have

$$\frac{\partial v_i}{\partial t} = \frac{\partial u_i}{\partial t} - \sum_{j,k=1}^N \frac{\partial\Psi_j}{\partial t}(\delta_{jk} + V_{0jk})\frac{\partial u_i}{\partial y_k},$$

and therefore,

$$\frac{\partial v_i}{\partial t} + \sum_{j=1}^{N} v_j \frac{\partial v_i}{\partial x_j} = \frac{\partial u_i}{\partial t} + \sum_{j,k=1}^{N} (u_j - \frac{\partial \Psi_j}{\partial t})(\delta_{jk} + V_{0jk}(\mathbf{k}))\frac{\partial u_i}{\partial y_k}. \tag{3.140}$$

Putting (3.139) and (3.140) together gives

$$\begin{aligned}0 &= \Big(\frac{\partial u_i}{\partial t} + \sum_{j,k=1}^{N} (u_j - \frac{\partial \Psi_j}{\partial t})(\delta_{jk} + V_{0jk}(\mathbf{k}))\frac{\partial u_i}{\partial y_k}\Big) \\ &\quad - \mu \sum_{j,k=1}^{N} (\delta_{jk} + V_{0jk}(\mathbf{k}))\frac{\partial}{\partial y_k}(\mathbf{D}(\mathbf{u})_{ij} + (\mathcal{D}_{\mathbf{D}}(\mathbf{k})\nabla \mathbf{u})_{ij}) \\ &\quad - \sum_{j=1}^{N} (\delta_{ij} + V_{0ij}(\mathbf{k}))\frac{\partial \mathfrak{q}}{\partial y_j}.\end{aligned}$$

Since $(\mathbf{I} + \nabla \Psi_\rho)(\mathbf{I} + \mathbf{V}_0) = (\partial x/\partial y)(\partial y/\partial x) = \mathbf{I}$,

$$\sum_{i=1}^{N} (\delta_{mi} + \partial_m \Psi_i)(\delta_{ij} + V_{0ij}(\mathbf{k})) = \delta_{mj}, \tag{3.141}$$

and so we have

$$\begin{aligned}0 &= \sum_{i=1}^{N} (\delta_{mi} + \partial_m \Psi_i)\Big(\frac{\partial u_i}{\partial t} + \sum_{j,k=1}^{N} (u_j - \frac{\partial \Psi_i}{\partial t})(\delta_{jk} + V_{0jk}(\mathbf{k}))\frac{\partial u_i}{\partial y_k}\Big) \\ &\quad - \mu \sum_{i,j,k=1}^{N} (\delta_{mi} + \partial_m \Psi_i)(\delta_{jk} + V_{0jk}(\mathbf{k}))\frac{\partial}{\partial y_k}(\mathbf{D}(\mathbf{u})_{ij} + (\mathcal{D}_{\mathbf{D}}(\mathbf{k})\nabla \mathbf{u})_{ij}) \\ &\quad - \frac{\partial \mathfrak{q}}{\partial y_m}.\end{aligned}$$

Thus, changing i to ℓ and m to i in the formula above, we define an N-vector of functions $\mathbf{f}(\mathbf{u}, \Psi_\rho)$ by letting

$$\begin{aligned}\mathbf{f}(\mathbf{u}, \Psi_\rho)|_i &= - \sum_{j,k=1}^{N} (u_j - \frac{\partial \Psi_j}{\partial t})(\delta_{jk} + V_{0jk}(\mathbf{k}))\frac{\partial u_i}{\partial y_k} \\ &\quad - \sum_{\ell=1}^{N} \partial_i \Psi_\ell \Big(\frac{\partial u_\ell}{\partial t} + \sum_{j,k=1}^{N} (u_j - \frac{\partial \Psi_j}{\partial t})(\delta_{jk} + V_{0jk}(\mathbf{k}))\frac{\partial u_\ell}{\partial y_k}\Big)\end{aligned}$$

$$+\mu\Big(\sum_{j=1}^N \frac{\partial}{\partial y_j}(\mathcal{D}_{\mathbf{D}}(\mathbf{k})\nabla\mathbf{u})_{ij} + \sum_{j,k=1}^N V_{0jk}(\mathbf{k})\frac{\partial}{\partial y_k}(\mathbf{D}(\mathbf{u})_{ij} + (\mathcal{D}_{\mathbf{D}}(\mathbf{k})\nabla\mathbf{u})_{ij})$$

$$+\sum_{j,k,\ell=1}^N \partial_i\Psi_\ell(\delta_{jk} + V_{0jk}(\mathbf{k}))\frac{\partial}{\partial y_k}(\mathbf{D}(\mathbf{u})_{\ell j} + (\mathcal{D}_{\mathbf{D}}(\mathbf{k})\nabla\mathbf{u})_{\ell j})\Big), \tag{3.142}$$

where $\mathbf{f}(\mathbf{u}, \Psi_\rho)|_i$ denotes the ith component of $\mathbf{f}(\mathbf{u}, \Psi_\rho)$, $\Psi_\rho = (\Psi_1, \ldots, \Psi_N)$, and $\mathbf{k} = (\partial_i\Psi_j)$. We then see that equation (3.12) is transformed to equation (3.129).

3.4.2 Laplace-Beltrami Operator

In this section, we introduce the Laplace-Beltrami operator and some important formulas from differential geometry, which is necessary to know to introduce the surface tension.

Let Γ be a hypersurface of class C^3 in $\mathbb{R}^N$. Locally at $p \in \Gamma$ Γ is parametrized as $p = \phi(\theta) = (\phi_1(\theta), \ldots, \phi_N(\theta))^\top$, where $\theta = (\theta_1, \ldots, \theta_{N-1})$ runs through a domain $\Theta \subset \mathbb{R}^{N-1}$. Let

$$\tau_i = \tau_i(p) = \frac{\partial}{\partial\theta_i}\phi(\theta) = \partial_i\phi \quad (i = 1, \ldots, N-1), \tag{3.143}$$

which forms a basis of the tangent space $T_p\Gamma$ of Γ at p. Let $\mathbf{n} = \mathbf{n}(p)$ denote the outer unit normal of Γ at p. Notice that

$$< \tau_i, \mathbf{n} >= 0. \tag{3.144}$$

Here and in the following, $< \cdot, \cdot >$ denotes a standard inner product in $\mathbb{R}^N$. To introduce the formula for $\mathbf{n}$, we notice that

$$\det\begin{pmatrix} \frac{\partial\phi_1}{\partial\theta_1} & \cdots & \frac{\partial\phi_1}{\partial\theta_{N-1}} & \frac{\partial\phi_1}{\partial\theta_k} \\ \vdots & \ddots & \vdots & \vdots \\ \frac{\partial\phi_N}{\partial\theta_1} & \cdots & \frac{\partial\phi_N}{\partial\theta_{N-1}} & \frac{\partial\phi_N}{\partial\theta_k} \end{pmatrix} = 0$$

for any $k = 1, \ldots, N-1$. Thus, to satisfy (3.144), $\mathbf{n}$ is defined by

$$\mathbf{n} = \frac{(h_1, \ldots, h_N)^\top}{H} \tag{3.145}$$

with $H = \sqrt{\sum_{j=1}^N h_i^2}$ and

$$H_i = \frac{\partial(\phi_1, \ldots, \hat{\phi}_i, \ldots, \phi_N)}{\partial(\theta_1, \ldots, \theta_{N-1})} = (-1)^{N+i} \det \begin{pmatrix} \frac{\partial \phi_1}{\partial \theta_1} & \cdots & \frac{\partial \phi_1}{\partial \theta_{N-1}} \\ \vdots & \ddots & \vdots \\ \frac{\partial \phi_{i-1}}{\partial \theta_1} & \cdots & \frac{\partial \phi_{i-1}}{\partial \theta_{N-1}} \\ \frac{\partial \phi_{i+1}}{\partial \theta_1} & \cdots & \frac{\partial \phi_{i+1}}{\partial \theta_{N-1}} \\ \vdots & \ddots & \vdots \\ \frac{\partial \phi_N}{\partial \theta_1} & \cdots & \frac{\partial \phi_N}{\partial \theta_{N-1}} \end{pmatrix}.$$

For example, when $N = 3$,

$$\mathbf{n} = \frac{\frac{\partial \phi}{\partial \theta_1} \times \frac{\partial \phi}{\partial \theta_2}}{\left|\frac{\partial \phi}{\partial \theta_1} \times \frac{\partial \phi}{\partial \theta_2}\right|} = H^{-1}(h_1, h_2, h_3)^\top \quad h_1 = \frac{\partial \phi_2}{\partial \theta_1}\frac{\partial \phi_3}{\partial \theta_2} - \frac{\partial \phi_3}{\partial \theta_1}\frac{\partial \phi_2}{\partial \theta_2},$$

$$h_2 = \frac{\partial \phi_3}{\partial \theta_1}\frac{\partial \phi_1}{\partial \theta_2} - \frac{\partial \phi_1}{\partial \theta_1}\frac{\partial \phi_3}{\partial \theta_2}, \quad h_3 = \frac{\partial \phi_1}{\partial \theta_1}\frac{\partial \phi_2}{\partial \theta_2} - \frac{\partial \phi_2}{\partial \theta_1}\frac{\partial \phi_1}{\partial \theta_2}, \quad H = \sqrt{h_1^2 + h_2^2 + h_3^2}.$$

Let

$$g_{ij} = g_{ij}(p) =< \tau_i, \tau_j > \quad (i, j = 1, \ldots, N-1),$$

and let G be an $(N-1) \times (N-1)$ matrix whose $(i, j)^{\text{th}}$ components are g_{ij}. The matrix G is called the **first fundamental form** of Γ. In the following, we employ Einstein's summation convention, which means that equal lower and upper indices are to be summed. Since for any $\xi \in \mathbb{R}^{N-1}$ with $\xi \neq 0$, $< G\xi, \xi >= g_{ij}\xi^i\xi^j =< \xi^i\tau_i, \xi^j j\tau_j >= |\xi^i\tau_i|^2 > 0$, G is a positive symmetric matrix, and therefore G^{-1} exists. Let g^{ij} be the $(i, j)^{\text{th}}$ component of G^{-1} and let

$$\tau^i = g^{ij}\tau_j.$$

Using $g^{ik}g_{kj} = g_{ik}g^{kj} = \delta^i_j$, where δ^i_j are the Kronecker delta symbols defined by $\delta^i_i = 1$ and $\delta^i_j = 0$ for $i \neq j$, we have

$$< \tau_i, \tau^j >=< \tau^i, \tau_j >= \delta^i_j. \tag{3.146}$$

In fact, $< \tau_i, \tau^j >=< \tau_i, g^{jk}\tau_k >= g^{jk} < \tau_i, \tau_k >= g^{jk}g_{ki} = \delta^j_i$. Thus, $\{\tau^j\}_{j=1}^{N-1}$ be a dual basis of $\{\tau_i\}_{i=1}^{N-1}$. In particular, we have

$$\tau^i = g^{ij}\tau_j, \quad \tau_i = g_{ij}\tau^j. \tag{3.147}$$

For any $\mathbf{a} \in T_p\Gamma$, we write

$$\mathbf{a} = a^i\tau_i = a_i\tau^j$$

By (3.146), we have $< \mathbf{a}, \tau^i >=< a^j\tau_j, \tau^i >= a^j < \tau_j, \tau^i >= a^i$ and $< \mathbf{a}, \tau_i >= a_j < \tau^j, \tau_i >= a_i$, and so

$$\mathbf{a} =< \mathbf{a}, \tau^i > \tau_i =< \mathbf{a}, \tau_i > \tau^i. \tag{3.148}$$

In particular, by (3.146) and (3.147), we have

$$a_i = g_{ij}a^j, \quad a^i = g^{ij}a_j$$

with $a_i =< \mathbf{a}, \tau_i >$ and $a^i =< \mathbf{a}, \tau^i >$. Notice that $\{\tau_1, \ldots, \tau_{N-1}, \mathbf{n}\}$ forms a basis of $\mathbb{R}^N$. Namely, for any N vector $\mathbf{b} \in \mathbb{R}^N$ we have

$$\mathbf{b} = b^i\tau_i+ < \mathbf{b}, \mathbf{n} > \mathbf{n} = b_i\tau^i+ < \mathbf{b}, \mathbf{n} > \mathbf{n}$$

with $b^i =< \mathbf{b}, \tau^i >$ and $b_i =< \mathbf{b}, \tau_i >$.

We next consider $\tau_{ij} = \partial_i\partial_j\phi = \partial_j\tau_i$. Notice that $\tau_{ij} = \tau_{ji}$. Let

$$\Lambda^k_{ij} =< \tau_{ij}, \tau^k >, \quad \ell_{ij} =< \tau_{ij}, \mathbf{n} > . \tag{3.149}$$

and then, we have

$$\tau_{ij} = \Lambda^k_{ij}\tau_k + \ell_{ij}\mathbf{n}. \tag{3.150}$$

Let L be an $N - 1 \times N - 1$ matrix whose $(i, j)^{\text{th}}$ component is ℓ_{ij} which is called the **second fundamental form** of Γ. Let

$$\mathcal{H}(\Gamma) = \frac{1}{N-1}\text{tr}\,(G^{-1}L) = \frac{1}{N-1}g^{ij}\ell_{ij}. \tag{3.151}$$

The $\mathcal{H}(\Gamma)$ is called the **mean curvature** of Γ.

Let $g = \det G$. One of the most important formulas in this section is

$$\partial_i(\sqrt{g}g^{ij}\tau_j) = \sqrt{g}g^{ij}\ell_{ij}\mathbf{n}. \tag{3.152}$$

This formula will be proved below.

The Λ^k_{ij} is called **Christoffel symbols**. We know the formula:

$$\Lambda^r_{ij} = \frac{1}{2}g^{rk}(\partial_i g_{jk} + \partial_j g_{ki} - \partial_k g_{ij}). \tag{3.153}$$

In fact,

$$\partial_i g_{jk} = \partial_i < \tau_j, \tau_k >=< \tau_{ij}, \tau_k > + < \tau_j, \tau_{ki} >,$$
$$\partial_j g_{ki} = \partial_j < \tau_k, \tau_i >=< \tau_{jk}, \tau_i > + < \tau_k, \tau_{ij} >,$$

$$\partial_k g_{ij} = \partial_k < \tau_i, \tau_j > = < \tau_{ki}, \tau_j > + < \tau_i, \tau_{jk} >,$$

and so we have

$$\partial_i g_{jk} + \partial_j g_{ki} - \partial_k g_{ij} = 2 < \tau_{ij}, \tau_k >,$$

which implies (3.153).

We now prove (3.152) in several steps. We first prove

$$\partial_k g_{ij} = g_{jr}\Lambda^r_{ki} + g_{ir}\Lambda^r_{kj}. \tag{3.154}$$

In fact, by (3.150) and $< \tau_i, \mathbf{n} > = 0$,

$$\begin{aligned}\partial_k g_{ij} &= < \tau_{ki}, \tau_j > + < \tau_i, \tau_{kj} > = < \Lambda^r_{ki}\tau_r, \tau_j > + < \tau_i, \Lambda^r_{kj}\tau_r > \\ &= g_{jr}\Lambda^r_{ki} + g_{ir}\Lambda^r_{kj}.\end{aligned}$$

We next prove

$$\partial_k g^{ij} = -g^{ir}\Lambda^j_{rk} - g^{jr}\Lambda^i_{rk}, \tag{3.155}$$

In fact, by $g^{ij}g_{j\ell} = \delta^i_\ell$, we have

$$0 = \partial_k(g^{ij}g_{jk}) = (\partial_k g^{ij})g_{j\ell} + g^{ij}\partial_k g_{j\ell}.$$

Thus, using (3.154), we have

$$\begin{aligned}(\partial_k g^{ij})g_{j\ell} &= -g^{ij}\partial_k g_{j\ell} = -g^{ij}(g_{\ell r}\Lambda^r_{kj} + g_{jr}\Lambda^r_{k\ell}) \\ &= -g^{ij}g_{\ell r}\Lambda^r_{kj} - \delta^i_r\Lambda^r_{k\ell} = -g^{ij}g_{\ell r}\Lambda^r_{kj} - \Lambda^i_{k\ell}.\end{aligned}$$

From this formula it follows that

$$\begin{aligned}\partial_k g^{im} &= (\partial_k g^{ij})\delta^m_j = (\partial_k g^{ij})g_{j\ell}g^{\ell m} = -g^{\ell m}(g^{ij}g_{\ell r}\Lambda^r_{kj} + \Lambda^i_{k\ell}) \\ &= -\delta^m_r g^{ij}\Lambda^r_{kj} - g^{\ell m}\Lambda^i_{k\ell} = -g^{ij}\Lambda^m_{kj} - g^{m\ell}\Lambda^i_{k\ell} = -g^{i\ell}\Lambda^m_{k\ell} - g^{m\ell}\Lambda^i_{k\ell},\end{aligned}$$

and therefore, setting $m = j$ and $\ell = r$ and using the identities: $\Lambda^m_{k\ell} = \Lambda^m_{\ell k}$ and $\Lambda^i_{k\ell} = \Lambda^i_{\ell k}$, we have (3.155).

We next prove

$$\partial_i g = 2\Lambda^j_{ij}g. \tag{3.156}$$

Recall that $g = \det G$ and the $(i, j)^{\text{th}}$ component of G is $< \tau_i, \tau_j >$. From the definition of differentiation, we have

$$\begin{aligned}
&\frac{\det G(x+\mathbf{e}_i\Delta x_i)-\det G(x)}{\Delta x_i}\\
&=\frac{\det(G(x)+\partial_i G(x)\Delta x_i)-\det G(x)}{\Delta x_i}+O(\Delta x_i)\\
&=\frac{\det G(x)(\det(I+\partial_i G(x)G^{-1}(x)\Delta x_i)-1)}{\Delta x_i}+O(\Delta x_i)\\
&=\det G(x)\,\mathrm{tr}\,(\partial_i G(x)G^{-1}(x))+O(\Delta x_i).
\end{aligned}$$

Thus, we have

$$\partial_i g=\det G\,\mathrm{tr}\,(\partial_i GG^{-1}).$$

Using (3.150) and $<\tau_i,\mathbf{n}>=0$, we have

$$\begin{aligned}
\mathrm{tr}\,(\partial_i GG^{-1})&=\partial_i(<\tau_j,\tau_k>g^{kj})=(<\tau_{ij},\tau_k>+<\tau_j,\tau_{ik}>)g^{kj}\\
&=((\Lambda^r_{ij}\tau_r,\tau_k>+<\Lambda^r_{ik}\tau_r,\tau_j>)g^{kj}=(g_{rk}\Lambda^r_{ij}+g_{rj}\Lambda^r_{ik})g^{kj}\\
&=2\Lambda^j_{ij}
\end{aligned}$$

Putting these two formulas gives (3.156).

We next prove

$$\partial_i(\sqrt{g}g^{ij})=-\sqrt{g}g^{ik}\Lambda^j_{ik}. \tag{3.157}$$

In fact, by (3.155) and (3.156),

$$\begin{aligned}
\frac{1}{\sqrt{g}}\partial_i(\sqrt{g}g^{ij})&=\frac{1}{2g}(\partial_i g)g^{ij}+\partial_i g^{ij}=\frac{1}{2g}2\Lambda^\ell_{i\ell}gg^{ij}-g^{ir}\Lambda^j_{ri}-g^{jr}\Lambda^i_{ri}\\
&=g^{jr}\Lambda^\ell_{r\ell}-g^{ir}\Lambda^j_{ri}-g^{jr}\Lambda^\ell_{\ell r}=-g^{ik}\Lambda^j_{ik}.
\end{aligned}$$

Thus, we have (3.157).

We now prove (3.152). By (3.150) and (3.157),

$$\begin{aligned}
\partial_i(\sqrt{g}g^{ij}\tau_j)&=\partial_i(\sqrt{g}g^{ij})\tau_j+\sqrt{g}g^{ij}\partial_i\tau_{ij}\\
&=-\sqrt{g}g^{ik}\Lambda^j_{ik}\tau_j+\sqrt{g}g^{ij}(\Lambda^k_{ij}\tau_k+\ell_{ij}\mathbf{n})\\
&=-\sqrt{g}g^{ij}\Lambda^k_{ij}\tau_k+\sqrt{g}g^{ij}\Lambda^k_{ij}\tau_k+\sqrt{g}g^{ij}\ell_{ij}\mathbf{n}\\
&=\sqrt{g}g^{ij}\ell_{ij}\mathbf{n}.
\end{aligned}$$

Thus, we have(3.152).

We now introduce the **Laplace-Beltrami operator** Δ_Γ on Γ, which is defined by

$$\Delta_\Gamma f = \frac{1}{\sqrt{g}}\partial_i(\sqrt{g}g^{ij}\partial_j f).$$

By (3.157), we have

$$\Delta_\Gamma f = g^{ij}\partial_i\partial_j f - g^{ik}\Lambda^j_{ik}\partial_j f. \tag{3.158}$$

By (3.151) and (3.152), we have

$$\Delta_\Gamma \phi = (N-1)\mathcal{H}(\Gamma)\mathbf{n}.$$

Usually, we put $H(\Gamma) = (N-1)\mathcal{H}(\Gamma)$, and so we have

$$\Delta_\Gamma x = H(\Gamma)\mathbf{n} \quad \text{for } x \in \Gamma.$$

3.4.3 *Transformation of the Boundary Conditions*

Since Γ is a compact hypersurface of C^3 class, there exist a positive number d, a finite number of points $x_\ell \in \Gamma$, and N-vector of functions Φ_ℓ of C^3 class defined on $\mathbb{R}^N$ ($\ell = 1, \ldots, \ell_0$) such that $\Gamma \subset \cup_{\ell=1}^{\ell_0} B_\ell$ and

$$\Gamma \cap B_\ell = \{y = \Phi_\ell(p', 0) \mid p' = (p_1, \ldots, p_{N-1}) \in \mathbb{R}^{N-1}\} \cap B_\ell, \tag{3.159}$$

where $B_\ell = B_d(x_\ell)$ is the ball with center x_ℓ and radius d. Let $\{\zeta_\ell\}_{\ell=1}^{\ell_0}$ be a partition of unity such that supp $\zeta_\ell \subset B_\ell$ and $\sum_{\ell=1}^{\ell_0} \zeta_\ell = 1$ on Γ. In the following we use the formula:

$$f = \sum_{\ell=1}^{\ell_0} \zeta_\ell f \quad \text{in } \Gamma$$

for any function, f, defined on Γ.

We write $\rho = \rho(y(p_1, \ldots, p_{N-1}, 0), t)$ in the following. Recall that $\Psi_\rho = \rho$ or $= \rho + \xi(t)$ on Γ. Thus, the derivatives of Ψ_ρ coincide the derivatives of ρ on Γ. From this fact, we set $\tilde\Psi_\rho = \omega(y)\Psi_\rho(y, t)$, $\omega(y)R^{-1}H_\rho(y, t)y$ or $\omega(y)R^{-1}H_\rho(y, t)y$. By the chain rule, we have

$$\frac{\partial\rho}{\partial p_i} = \frac{\partial}{\partial p_i}\tilde\Psi_\rho(\Phi_\ell(p_1, \ldots, p_{N-1}, 0), t) = \sum_{m=1}^N \frac{\partial\tilde\Psi_\rho}{\partial y_m}\frac{\partial\Phi_{\ell,m}}{\partial p_i}|_{p_N=0}, \tag{3.160}$$

where we have set $\Phi_\ell = (\Phi_{\ell,1}, \ldots, \Phi_{\ell,N})^\top$, and so, $\partial\rho/\partial p_i$ is defined in B_ℓ by letting

$$\frac{\partial\rho}{\partial p_i} = \sum_{m=1}^{N} \frac{\partial\tilde{\Psi}_\rho}{\partial y_m} \circ \Phi_\ell \frac{\partial\Phi_{\ell,m}}{\partial p_i}. \tag{3.161}$$

We first represent $\mathbf{n}_t$. Recall that Γ_t is given by $x = y + \rho(y,t)\mathbf{n} + \xi(t)$ for $y \in \Gamma$ (cf. (3.25)). Let

$$\mathbf{n}_t = a(\mathbf{n} + \sum_{i=1}^{N-1} b_i\tau_i) \quad \text{with } \tau_i = \frac{\partial}{\partial p_i} y = \frac{\partial}{\partial p_i}\Phi_\ell(p', 0).$$

These vectors τ_i ($i = 1, \ldots, N-1$) form a basis of the tangent space of Γ at $y = y(p_1, \ldots, y_{N-1}, 0)$. Since $|\mathbf{n}_t|^2 = 1$, we have

$$1 = a^2(1 + \sum_{i,j=1}^{N-1} g_{ij}b_ib_j) \quad \text{with } g_{ij} = \tau_i \cdot \tau_j, \tag{3.162}$$

because $\tau_i \cdot \mathbf{n} = 0$. The vectors $\dfrac{\partial x}{\partial p_i}$ ($i = 1, \ldots, N-1$) form a basis of the tangent space of Γ_t, and so $\dfrac{\partial x}{\partial p_i} \cdot \mathbf{n}_t = 0$. Thus, we have

$$0 = a(\mathbf{n} + \sum_{j=1}^{N-1} b_j\tau_j) \cdot (\frac{\partial y}{\partial p_i} + \frac{\partial\rho}{\partial p_i}\mathbf{n} + \rho\frac{\partial\mathbf{n}}{\partial p_i}).$$

Since $\mathbf{n} \cdot \dfrac{\partial y}{\partial p_i} = \mathbf{n} \cdot \tau_i = 0$, $\dfrac{\partial\mathbf{n}}{\partial p_i} \cdot \mathbf{n} = 0$ (because of $|\mathbf{n}|^2 = 1$), and $\dfrac{\partial y}{\partial p_i} \cdot \dfrac{\partial y}{\partial p_j} = \tau_i \cdot \tau_j = g_{ij}$, we have

$$\frac{\partial\rho}{\partial p_i} + \sum_{i=1}^{N-1}(g_{ij} + \rho\frac{\partial\mathbf{n}}{\partial p_i} \cdot \tau_j)b_j = 0. \tag{3.163}$$

Let H be an $(N-1) \times (N-1)$ matrix whose $(i, j)^{\text{th}}$ component is $\dfrac{\partial\mathbf{n}}{\partial p_i} \cdot \tau_j$. Since $\mathbf{n}$ is defined in $\mathbb{R}^N$ as an N vector of C^2 functions with $\|\mathbf{n}\|_{H^2_\infty(\mathbb{R}^N)} < \infty$, we can write

$$\frac{\partial\mathbf{n}}{\partial p_i} \cdot \tau_j = \sum_{m=1}^{N} \frac{\partial\mathbf{n}}{\partial y_m} \circ \Phi_\ell \frac{\partial\Phi_{\ell,m}}{\partial p_i} \cdot \frac{\partial\Phi_\ell}{\partial p_j}, \tag{3.164}$$

Thus, letting

$$M_{ij} = \sum_{m=1}^{N} \frac{\partial \mathbf{n}}{\partial y_m} \circ \Phi_\ell \frac{\partial \Phi_{\ell,m}}{\partial p_i} \cdot \frac{\partial \Phi_\ell}{\partial p_j}.$$

The $(N-1) \times (N-1)$ matrix M with $(i,j)^{\rm th}$ component M_{ij} is defined in $\mathbb{R}^N$ and $\|M\|_{H^2_\infty(\mathbb{R}^N)} \leq C$ with some constant C independent of $\ell \in \mathbb{N}$. Under the assumption (3.28), we may assume that the inverse of $\mathbf{I}+\rho M G^{-1}$ exists. Let $\nabla'_\Gamma \rho = (\frac{\partial \rho}{\partial p_1}, \ldots, \frac{\partial \rho}{\partial p_{N-1}})$ with $\frac{\partial \rho}{\partial p_i} = \frac{\partial}{\partial p_i}\rho(\Phi_\ell(p', 0))$, and then by (3.163) we have

$$b = -(G+\rho M)^{-1}\nabla'_\Gamma \rho = -G^{-1}(\mathbf{I}+\rho M G^{-1})^{-1}\nabla'_\Gamma\rho. \tag{3.165}$$

Putting (3.165) and (3.162) together gives

$$\begin{aligned} a &= (1+ < Gb, b >)^{-1/2} \\ &= (1+ < \mathbf{I}+\rho M G^{-1})^{-1}\nabla'_\Gamma\rho, G^{-1}(\mathbf{I}+\rho M G^{-1})^{-1}\nabla'_\Gamma\rho >)^{-1/2}. \end{aligned}$$

Thus, we have

$$\mathbf{n}_t = \mathbf{n} - \sum_{i,j=1}^{N-1} g^{ij}\tau_i \frac{\partial \rho}{\partial p_j} + \mathbf{V}_\Gamma(\rho, \nabla'_\Gamma \rho) \tag{3.166}$$

where we have set

$$\begin{aligned} \mathbf{V}_\Gamma(\rho, \nabla'_\Gamma\rho) &= - < G^{-1}((\mathbf{I}+\rho M G^{-1})^{-1} - \mathbf{I})\nabla'_\Gamma\rho, \tau > \\ &+ \{(1+ < (\mathbf{I}+\rho M G^{-1})^{-1}\nabla'_\Gamma\rho, G^{-1}(\mathbf{I}+\rho M G^{-1})^{-1}\nabla'_\Gamma\rho >)^{-1/2} - 1\} \\ &\qquad \times (\mathbf{n} - < G^{-1}(\mathbf{I}+\rho M G^{-1})^{-1}\nabla'_\Gamma\rho, \tau >). \end{aligned}$$

From (3.161), $\nabla'_\Gamma\rho$ is extended to $\mathbb{R}^N$ by letting $\nabla'_\Gamma\rho = (\nabla\Phi_\ell)\nabla\tilde\Psi_\rho \circ \Phi_\ell := \Psi_{\rho,\ell}$, and so $\mathbf{V}_\Gamma(\rho, \nabla'_\Gamma\rho)$ can be extended to B_ℓ by

$$\begin{aligned} \mathbf{V}_\Gamma(\rho, \nabla'_\Gamma\rho) &= - < G^{-1}((\mathbf{I}+\tilde\Psi_\rho M G^{-1})^{-1} - \mathbf{I})\Psi_{\rho,\ell}, \tau > \\ &+ \{(1+ < (\mathbf{I}+\tilde\Psi_\rho M G^{-1})^{-1}\Psi_{\rho,\ell}, G^{-1}(\mathbf{I}+\rho M G^{-1})^{-1}\Psi_{\rho,\ell} >)^{-1/2} - 1\} \\ &\quad \times (\mathbf{n} - < G^{-1}(\mathbf{I}+\rho M G^{-1})^{-1}\Psi_{\rho,\ell}, \tau >), \end{aligned} \tag{3.167}$$

Thus, we may write

$$\mathbf{V}_\Gamma(\rho, \nabla'_\Gamma\rho) = \mathbf{V}_{\Gamma,\ell}(\bar{\mathbf{k}})\bar\nabla\tilde\Psi_\rho \otimes \bar\nabla\tilde\Psi_\rho$$

on B_ℓ with some function $\mathbf{V}_{\Gamma,\ell}(\bar{\mathbf{k}}) = \mathbf{V}_{\Gamma,\ell}(y, \bar{\mathbf{k}})$ defined on $B_\ell \times \{\bar{\mathbf{k}} \mid |\bar{\mathbf{k}}| \leq \delta\}$ with $\mathbf{V}_{\Gamma,\ell}(0) = 0$ possessing the estimate

$$\|(\mathbf{V}_{\Gamma,\ell}(\bar{\mathbf{k}}), \partial_{\bar{\mathbf{k}}}\mathbf{V}_{\Gamma,\ell}(\cdot, \mathbf{k}))\|_{H^1_\infty(B_\ell)} \leq C$$

with some constant C independent of ℓ. Here and in the following $\bar{\mathbf{k}}$ are the corresponding variables to $\bar{\nabla}\tilde{\Psi}_\rho = (\tilde{\Psi}_\rho, \nabla\tilde{\Psi}_\rho)$. In view of (3.166), we have

$$\mathbf{n}_t = \mathbf{n} - \sum_{i,j=1}^{N-1} g^{ij}\tau_i \frac{\partial \rho}{\partial y_j} + \mathbf{V}_{\Gamma,\ell}(\bar{\mathbf{k}})\bar{\nabla}\tilde{\Psi}_\rho \otimes \bar{\nabla}\tilde{\Psi}_\rho \quad \text{on } B_\ell \cap \Gamma. \tag{3.168}$$

Let

$$\mathbf{V}_\Gamma(\bar{\mathbf{k}}) = \sum_{\ell=1}^{\ell_0} \zeta_\ell \mathbf{V}_{\Gamma,\ell}(\bar{\mathbf{k}}), \tag{3.169}$$

and then we have

$$\|(\mathbf{V}_\Gamma(\bar{\mathbf{k}}), \partial_{\bar{\mathbf{k}}}\mathbf{V}_\Gamma(\cdot, \mathbf{k}))\|_{H^1_\infty(\Omega)} \leq C \tag{3.170}$$

for $|\bar{\mathbf{k}}| \leq \delta$ with some constant $C > 0$.

In view of (3.161) and (3.166), the unit outer normal $\mathbf{n}_t$ is also represented by

$$\mathbf{n}_t = \mathbf{n} - \sum_{i,j=1}^{N-1}\sum_{m=1}^{N} g^{ij}\tau_i \frac{\partial \tilde{\Psi}_\rho}{\partial y_m} \circ \Phi_\ell \frac{\partial \Phi_{\ell,m}}{\partial p_i} + \mathbf{V}_{\Gamma,\ell}(\bar{\mathbf{k}})\bar{\nabla}\tilde{\Psi}_\rho$$

on B_ℓ, and so we may write

$$\mathbf{n}_t = \mathbf{n} + \tilde{\mathbf{V}}_{\Gamma,\ell}(\bar{\nabla}\tilde{\Psi}_\rho)\bar{\nabla}\tilde{\Psi}_\rho$$

for some functions $\tilde{\mathbf{V}}_{\Gamma,\ell}(\bar{\mathbf{k}}) = \tilde{\mathbf{V}}_{\Gamma,\ell}(y, \bar{\mathbf{k}})$ defined on $B_\ell \times \{\bar{\mathbf{k}} \mid |\bar{\mathbf{k}}| \leq \delta\}$ possessing the estimate:

$$\|(\tilde{\mathbf{V}}_{\Gamma,\ell}, \partial_{\bar{\mathbf{k}}}\tilde{\mathbf{V}}_{\Gamma,\ell})(\cdot, \bar{\mathbf{k}})\|_{H^1_\infty(B_\ell)} \leq C \quad \text{for } |\bar{\mathbf{k}}| \leq \delta \tag{3.171}$$

with some constant C independent of $\ell \in \mathbb{N}$. Thus, setting

$$\tilde{\mathbf{V}}_\Gamma(\bar{\mathbf{k}}) = \sum_{\ell=1}^{\ell_0} \zeta_\ell \tilde{\mathbf{V}}_{\Gamma,\ell}(\bar{\mathbf{k}}), \tag{3.172}$$

we have

$$\mathbf{n}_t = \mathbf{n} + \tilde{\mathbf{V}}_\Gamma(\bar\nabla\tilde\Psi_\rho)\bar\nabla\tilde\Psi_\rho, \tag{3.173}$$

and

$$\|(\tilde{\mathbf{V}}_\Gamma(\cdot, \bar{\mathbf{k}}), \partial_{\bar{\mathbf{k}}}\mathbf{V}_\Gamma(\cdot, \bar{\mathbf{k}}))\|_{H^1_\infty(\Omega)} \leq C \quad \text{for } |\bar{\mathbf{k}}| \leq \delta. \tag{3.174}$$

We now consider the kinematic equation: $V_{\Gamma_t} = \mathbf{v} \cdot \mathbf{n}_t$ in (3.1). We have to consider the cases where $x = y + \rho(y, t)\mathbf{n}$ in (3.27) and $x = y + \rho(y, t)\mathbf{n} + \xi(t)$ in (3.33) and (3.39), and so for simplicity $x = y + \rho(y, t)\mathbf{n} + \alpha\xi(t)$, where $\alpha = 0$ in case of (3.27) and $\alpha = 1$ in case of (3.33) and (3.39). By (3.168) we have

$$\begin{aligned}
V_{\Gamma_t} &= \frac{\partial x}{\partial t} \cdot \mathbf{n}_t \\
&= \sum_{\ell=1}^{\ell_0} \zeta_\ell < (\partial_t\rho)\mathbf{n} + \alpha\xi'(t), \mathbf{n} - \sum_{i,j=1}^{N-1} g^{ij}\tau_i \frac{\partial\rho}{\partial p_j} + \mathbf{V}_{\Gamma,\ell}(\bar{\mathbf{k}})(\bar\nabla\tilde\Psi_\rho, \bar\nabla\tilde\Psi_\rho) > \\
&= \partial_t\rho + \alpha\xi'(t) \cdot \mathbf{n} + \partial_t\rho < \mathbf{n}, \mathbf{V}_\Gamma(\bar{\mathbf{k}})(\bar\nabla\tilde\Psi_\rho, \bar\nabla\tilde\Psi_\rho) > \\
&- < \alpha\xi'(t) \mid \bar\nabla'_\Gamma\rho > + < \alpha\xi'(t), \mathbf{V}_\Gamma(\bar{\mathbf{k}})(\bar\nabla\tilde\Psi_\rho, \bar\nabla\tilde\Psi_\rho) >,
\end{aligned}$$

where, for any N-vector function $\mathbf{d}$, we have set

$$< \mathbf{d} \mid \nabla'_\Gamma\rho >= \sum_{i,j=1}^{N-1} g^{ij} < \tau_i, \mathbf{d} > \frac{\partial\rho}{\partial p_j}. \tag{3.175}$$

On the other hand,

$$\begin{aligned}
\mathbf{v} \cdot \mathbf{n}_t &= \sum_{\ell=1}^{\ell_0} \zeta_\ell \mathbf{u} \cdot (\mathbf{n} - \sum_{i,j=1}^{N-1} g^{ij}\tau_i \frac{\partial\rho}{\partial p_j} + \mathbf{V}_{\Gamma,\ell}(\bar{\mathbf{k}})\bar\nabla\tilde\Psi_\rho) \\
&= \mathbf{n} \cdot \mathbf{u} - < \mathbf{u} \mid \nabla'_\Gamma\rho > + \mathbf{u} \cdot \mathbf{V}_\Gamma(\bar{\mathbf{k}})\bar\nabla\tilde\Psi_\rho \otimes \bar\nabla\tilde\Psi_\rho,
\end{aligned}$$

where we have used $[[\mathbf{v}]] = 0$ on Γ. From the consideration above, the kinematic equation is transformed to the following equation:

$$\partial_t\rho + < \mathbf{u} \mid \nabla'_\Gamma\rho > - \mathbf{n} \cdot \mathbf{u} + \alpha\xi'(t) \cdot \mathbf{n} = d(\mathbf{u}, \tilde\Psi_\rho) \quad \text{on } \Gamma \times (0, T) \tag{3.176}$$

with

$$
\begin{aligned}
d(\mathbf{u}, \tilde{\Psi}_\rho) &= \mathbf{u} \cdot \mathbf{V}_\Gamma(\bar{\mathbf{k}})\bar{\nabla}\tilde{\Psi}_\rho \otimes \bar{\nabla}\tilde{\Psi}_\rho - \partial_t \rho < \mathbf{n}, \mathbf{V}_\Gamma(\bar{\mathbf{k}})\bar{\nabla}\tilde{\Psi}_\rho \otimes \bar{\nabla}\tilde{\Psi}_\rho > \\
&\quad + < \alpha\xi'(t) \mid \nabla'_\Gamma \rho > - < \alpha\xi'(t), \mathbf{V}_\Gamma(\bar{\mathbf{k}})\bar{\nabla}\tilde{\Psi}_\rho \otimes \bar{\nabla}\tilde{\Psi}_\rho > .
\end{aligned} \tag{3.177}
$$

We next consider the boundary condition:

$$
[[(\mu\mathbf{D}(\mathbf{v}) - \mathfrak{p}\mathbf{I})\mathbf{n}_t]] = \sigma H(\Gamma_t)\mathbf{n}_t \quad \text{on } \Gamma_t \tag{3.178}
$$

for $0 < t < T$ in equations (3.1). It is convenient to divide the formula in (3.178) into the tangential part and normal part on Γ_t as follows:

$$
[[\Pi_t \mu\mathbf{D}(\mathbf{v})\mathbf{n}_t]] = 0, \tag{3.179}
$$

$$
[[< \mu\mathbf{D}(\mathbf{v})\mathbf{n}_t, \mathbf{n}_t > - \mathfrak{p}]] = \sigma < H(\Gamma_t)\mathbf{n}_t, \mathbf{n}_t >, \tag{3.180}
$$

where Π_t is defined by $\Pi_t \mathbf{d} = \mathbf{d} - < \mathbf{d}, \mathbf{n}_t > \mathbf{n}_t$ for any N vector $\mathbf{d}$.

By (3.173) we can write

$$
\begin{aligned}
\Pi_t \mathbf{d} &= \mathbf{d}_\tau - < \mathbf{d}, \mathbf{n} > \tilde{\mathbf{V}}_\Gamma(\bar{\nabla}\tilde{\Psi}_\rho) - < \mathbf{d}, \tilde{\mathbf{V}}_\Gamma(\bar{\nabla}\tilde{\Psi}_\rho)\bar{\nabla}\tilde{\Psi}_\rho > \mathbf{n} \\
&\quad - < \mathbf{d}, \tilde{\mathbf{V}}_\Gamma(\bar{\nabla}\tilde{\Psi}_\rho)\bar{\nabla}\tilde{\Psi}_\rho > \tilde{\mathbf{V}}_\Gamma(\bar{\nabla}\tilde{\Psi}_\rho)\bar{\nabla}\tilde{\Psi}_\rho,
\end{aligned} \tag{3.181}
$$

for any $\mathbf{d} \in \mathbb{R}^N$. Thus, recalling (3.133), we see that the boundary condition (3.180) is transformed to the following formula:

$$
[[(\mu\mathbf{D}(\mathbf{u})\mathbf{n})_\tau]] = [[\mathbf{h}'(\mathbf{u}, \Psi_\rho)]] \quad \text{on } \Gamma \times (0, T), \tag{3.182}
$$

where we have set

$$
\begin{aligned}
\mathbf{h}'(\mathbf{u}, \Psi_\rho) &= -\mu\{\Pi_0 \mathcal{D}_\mathbf{D}(\mathbf{k})\nabla\mathbf{u} + < \tilde{\mathbf{V}}_\Gamma(\bar{\mathbf{k}})\bar{\mathbf{k}}, \mathbf{D}(\mathbf{u}) + \mathcal{D}_\mathbf{D}(\mathbf{k})\nabla\mathbf{u} > \mathbf{n} \\
&\quad + < \mathbf{n}, \mathbf{D}(\mathbf{u}) + \mathcal{D}_\mathbf{D}(\mathbf{k})\nabla\mathbf{u} > \tilde{\mathbf{V}}_\Gamma(\bar{\mathbf{k}})\bar{\mathbf{k}} \\
&\quad + < \tilde{\mathbf{V}}_\Gamma(\bar{\mathbf{k}})\bar{\mathbf{k}}, \mathbf{D}(\mathbf{u}) + \mathcal{D}_\mathbf{D}(\mathbf{k})\nabla\mathbf{u} > \tilde{\mathbf{V}}_{\Gamma,\ell}(\bar{\mathbf{k}})\bar{\mathbf{k}}\}
\end{aligned} \tag{3.183}
$$

with $\mathbf{k} = \nabla\Psi_\rho$ and $\bar{\mathbf{k}} = (\tilde{\Psi}_\rho, \nabla\tilde{\Psi}_\rho)$.

To consider the transformation of the boundary condition (3.180), we first consider the Laplace-Beltrami operator on Γ_t. From (3.158), we have

$$
\Delta_{\Gamma_t} f = g_t^{ij}\partial_i\partial_j f - g_t^{ij}\Lambda_{tik}^j \partial_j f, \tag{3.184}
$$

where Λ_{tik}^j are Christoffel symbols of Γ_t defined by (3.149). Recall $x = y + \rho\mathbf{n} + \xi(t)$. The first fundamental form of $\Gamma_t = (g_{tij})$ is given by

$$
g_{tij} = < \frac{\partial x}{\partial x_i}, \frac{\partial x}{\partial x_j} > = g_{ij} + \tilde{g}_{ij}\rho + \frac{\partial\rho}{\partial p_i}\frac{\partial\rho}{\partial p_j} \tag{3.185}
$$

with

$$\tilde{g}_{ij} =< \tau_i, \frac{\partial \mathbf{n}}{\partial p_j} > + < \tau_j, \frac{\partial \mathbf{n}}{\partial p_i} > + < \frac{\partial \mathbf{n}}{\partial p_i}, \frac{\partial \mathbf{n}}{\partial p_j} > .$$

In view of (3.159), we can write τ_i and $\partial_j \mathbf{n}$ as

$$\tau_i = \frac{\partial y}{\partial p_i} = \sum_{j=1}^{N-1} \frac{\partial \Phi_\ell(p)}{\partial p_j}, \quad \frac{\partial \mathbf{n}}{\partial p_j} = \sum_{k=1}^{N} \frac{\partial \mathbf{n}}{\partial y_k} \frac{\partial \Phi_{\ell k}(p)}{\partial p_j}. \tag{3.186}$$

Let Q be an $(N-1) \times (N-1)$ matrix whose $(i, j)^{\text{th}}$ components are $\tilde{g}_{ij}$, and then

$$G_t = G + \rho Q + \nabla_p \rho \otimes \nabla_p \rho = G(\mathbf{I} + \rho G^{-1} Q + G^{-1} \nabla_p \rho \otimes \nabla_p \rho),$$

where $\nabla_p \rho = (\frac{\partial \rho}{\partial p_1}, \ldots, \frac{\partial \rho}{\partial p_{N-1}})$. Choosing $\delta > 0$ small enough in (3.28), we know that the inverse of $\mathbf{I} + \rho G^{-1} Q + \nabla_p \rho \otimes \nabla_p \rho$ exists, and so

$$G_t^{-1} = (\mathbf{I} + \rho G^{-1} Q + G^{-1} \nabla_p \rho \otimes \nabla_p \rho))^{-1} G^{-1} = G^{-1} - \rho G^{-1} Q G^{-1} + O_2,$$

that is,

$$g_t^{ij} = g^{ij} + \rho h^{ij} + O_2,$$

where h^{ij} denotes the $(i, j)^{\text{th}}$ component of $-G^{-1} Q G^{-1}$. Here and in the following, O_2 denotes some nonlinear function with respect to ρ and $\nabla_p \rho$ of the form:

$$O_2 = a_0 \tilde{\Psi}_\rho^2 + \sum_{j=1}^{N} a_j \tilde{\Psi}_\rho \frac{\partial \tilde{\Psi}_\rho}{\partial y_j} + \sum_{j,k=1}^{N} b_{jk} \frac{\partial \tilde{\Psi}_\rho}{\partial y_j} \frac{\partial \tilde{\Psi}_\rho}{\partial y_k} \tag{3.187}$$

with suitable functions a_j, and b_{jk} possessing the estimates

$$|(a_j, b_{jk})| \leq C, \quad |\nabla(a_j, b_{jk})| \leq C(|\nabla \tilde{\Psi}_\rho| + |\nabla^2 \tilde{\Psi}_\rho|), \tag{3.188}$$

provided that (3.28) holds. Moreover, the Christoffel symbols are given by $\Lambda_{tij}^k = g_t^{k\ell} < \tau_{tij}, \tau_{t\ell} >$ with

$$\tau_{ti} = \frac{\partial x}{\partial p_i}, \quad \tau_{tij} = \frac{\partial^2 x}{\partial p_i \partial p_j}.$$

Since

$$\tau_{tij} = \tau_{ij} + \partial_i \partial_j (\rho \mathbf{n}) = \tau_{ij} + \rho \partial_i \partial_j \mathbf{n} + \partial_i \partial_j \mathbf{n} + \partial_j \rho \partial_i \mathbf{n} + \partial_i \partial_j \rho \mathbf{n},$$

we have

$$< \tau_{tij}, \tau_{t\ell} > = < \tau_{ij}, \tau_\ell > + \rho(< \partial_i\partial_j\mathbf{n}, \tau_\ell > + < \tau_{ij}, \partial_\ell\mathbf{n} >) \\ + \partial_j\rho < \partial_i\mathbf{n}, \tau_\ell > + \partial_i\rho < \partial_j\mathbf{n}, \tau_\ell > .$$

Thus,

$$\begin{aligned} \Lambda^k_{tij} &= < g^{k\ell} + h^{k\ell}\rho + O_2, < \tau_{ij}, \tau_\ell > + \rho(< \partial_i\partial_j\mathbf{n}, \tau_\ell > + < \tau_{ij}, \partial_\ell\mathbf{n} >) \\ &\quad + \partial_j\rho < \partial_i\mathbf{n}, \tau_\ell > + \partial_i\rho < \partial_j\mathbf{n}, \tau_\ell >> \\ &= \Lambda^k_{ij} + \rho g^{k\ell}(< \partial_i\partial_j\mathbf{n}, \tau_\ell > + < \tau_{ij}, \partial_\ell\mathbf{n} >) \\ &\quad + \partial_j\rho\, g^{k\ell} < \partial_i\mathbf{n}, \tau_\ell > + \partial_i\rho\, g^{k\ell} < \partial_j\mathbf{n}, \tau_\ell > + O_2, \end{aligned}$$

and so we may write

$$\Lambda^k_{tij} = \Lambda^k_{ij} + \rho A^k_{ij} + \partial_i\rho B^k_j + \partial_j\rho B^k_i + O_2 \tag{3.189}$$

with

$$A^k_{ij} = g^{k\ell}(< \partial_i\partial_j\mathbf{n}, \tau_\ell > + < \tau_{ij}, \partial_\ell\mathbf{n} >), \\ B^k_j = g^{k\ell} < \partial_i\mathbf{n}, \tau_\ell >, \quad B^k_i = g^{k\ell} < \partial_j\mathbf{n}, \tau_\ell > .$$

Combining these formulas with (3.184) gives

$$\Delta_{\Gamma_t} f = \Delta_\Gamma f + \sum_{i,j=1}^{N-1} (h^{ij}\rho + O_2)\partial_i\partial_j f + \sum_{k=1}^{N-1} (h^k + O_2)\partial_k f \tag{3.190}$$

with

$$h^k = -\sum_{i,j=1}^{N-1} (g^{ij} A^k_{ij}\rho + g^{ij} B^k_j\partial_i\rho + g^{ij} B^k_i\partial_j\rho + \Lambda^k_{ij}h^{ij}\rho).$$

Thus, we have

$$\begin{aligned} H(\Gamma_t)\mathbf{n}_t &= \Delta_{\Gamma_t}(y + \rho\mathbf{n}) \\ &= \Delta_\Gamma y + \rho\Delta_\Gamma\mathbf{n} + \mathbf{n}\Delta_\Gamma\rho + g^{ij}(\partial_i\rho\partial_j\mathbf{n} + \partial_j\rho\partial_i\mathbf{n}) \\ &\quad + \rho h^{ij}\partial_i\partial_j y + (\rho h^{ij}\partial_i\partial_j\rho)\mathbf{n} + h^k\partial_k y + O_2\partial_i\partial_j\rho + O_2, \end{aligned}$$

which, combined with (3.166) and $< \mathbf{n}, \partial_j\mathbf{n} >= 0$, leads to

$$< H(\Gamma_t)\mathbf{n}_t, \mathbf{n}_t > =< \Delta_\Gamma y, \mathbf{n} > +\rho < \Delta_\Gamma \mathbf{n}, \mathbf{n} > +\Delta_\Gamma \rho + \rho h^{ij} < \partial_i\partial_j y, \mathbf{n} >$$
$$+ (h^{ij}\rho + O_2)\partial_i\partial_j\rho - g^{ij} < \Delta_\Gamma y, \tau_i > \partial_j\rho + O_2.$$

Noting that $\Delta_\Gamma y = H(\Gamma)\mathbf{n}$, we have

$$< H(\Gamma_t)\mathbf{n}_t, \mathbf{n}_t > = \Delta_\Gamma\rho + H(\Gamma) + \rho < \Delta_\Gamma\mathbf{n}, \mathbf{n} >$$
$$+ \rho h^{ij} < \partial_i\partial_j y, \mathbf{n} > +(h^{ij}\rho + O_2)\partial_i\partial_j\rho + O_2.$$

Recalling that (3.187) and (3.188), we see that there exists a function $\mathbf{V}'_\Gamma(\bar{\mathbf{k}})$ defined on $\overline{\Omega} \times \{\bar{\mathbf{k}} \mid |\bar{\mathbf{k}}| \leq C\}$ satisfying the estimate:

$$\sup_{|\bar{\mathbf{k}}|\leq\delta} \|(\mathbf{V}'_\Gamma(\cdot, \bar{\mathbf{k}}), \partial_{\bar{\mathbf{k}}}\mathbf{V}'_\Gamma(\cdot, \bar{\mathbf{k}}))\|_{H^1_\infty(\Omega)} \leq C$$

with some constant C, where $\partial_{\bar{\mathbf{k}}}$ denotes the partial derivatives with respect to variables $\bar{\mathbf{k}}$, for which

$$\sum_{\ell=1}^{\ell_0} \zeta_\ell((h^{ij}\rho + O_2)\partial_i\partial_j\rho + O_2) = \mathbf{V}'_\Gamma(\bar{\mathbf{k}})\bar{\mathbf{k}} \otimes \bar{\bar{\mathbf{k}}}, \tag{3.191}$$

where $\bar{\mathbf{k}} = (\tilde{\Psi}_\rho, \nabla\tilde{\Psi}_\rho)$ and $\bar{\bar{\mathbf{k}}} = (\tilde{\Psi}_\rho, \nabla\tilde{\Psi}_\rho, \nabla^2\tilde{\Psi}_\rho)$. Therefore, we have

$$< H(\Gamma_t)\mathbf{n}_t, \mathbf{n}_t > = \Delta_\Gamma\rho + H(\Gamma) + \mathbb{B}\rho + \mathbf{V}'_\Gamma(\bar{\nabla}\tilde{\Psi}_\rho)\bar{\nabla}\tilde{\Psi}_\rho \otimes \bar{\nabla}^2\tilde{\Psi}_\rho, \tag{3.192}$$

where $\bar{\nabla}^2\tilde{\Psi}_\rho = (\tilde{\Psi}_\rho, \nabla\tilde{\Psi}_\rho, \nabla^2\tilde{\Psi}_\rho)$ and we have set

$$\mathbb{B}\rho = \{< \Delta_\Gamma\mathbf{n}, \mathbf{n} > + \sum_{\ell=1}^{\ell_0} \zeta_\ell(\sum_{i,j=1}^{N-1} h^{ij} < \partial_i\partial_j y, \mathbf{n} >)\}\rho. \tag{3.193}$$

To turn to equation (3.180), in view of (3.133), (3.173), and (3.172), we write

$$\begin{aligned} < \mathbf{n}, \mu\mathbf{D}(\mathbf{v})\mathbf{n}_t > &=< \mathbf{n}, \mu\mathbf{D}(\mathbf{u})\mathbf{n} > + < \mathbf{n}, \mu\mathbf{D}(\mathbf{u})\mathbf{V}_\Gamma(\bar{\mathbf{k}})\bar{\mathbf{k}} > \\ &+ < \mathbf{n}, \mu(\mathcal{D}_\mathbf{D}(\mathbf{k})\nabla\mathbf{u})(\mathbf{n} + \mathbf{V}_\Gamma(\bar{\mathbf{k}})\bar{\mathbf{k}}) >, \end{aligned} \tag{3.194}$$

where $\mathbf{k} = \nabla\Psi_\rho$ and $\bar{\mathbf{k}} = (\tilde{\Psi}_\rho, \nabla\tilde{\Psi}_\rho)$. Thus, from (3.192) and (3.194) it follows that the boundary condition (3.180) is transformed to

$$[[< \mathbf{n}, \mu\mathbf{D}(\mathbf{u})\mathbf{n} > -\mathfrak{q}]] - \sigma(\Delta_\Gamma\rho + \mathbb{B}\rho) = [[h_N(\mathbf{u}, \Psi_\rho)]], \tag{3.195}$$

where we have set

$$
\begin{aligned}
h_N(\mathbf{u}, \Psi_\rho) &= - < \mathbf{n}, \mu\mathbf{D}(\mathbf{u})\mathbf{V}_\Gamma(\bar{\mathbf{k}})\bar{\mathbf{k}} > \\
&- < \mathbf{n}, \mu(\mathcal{D}_\mathbf{D}(\mathbf{k})\nabla\mathbf{u})(\mathbf{n} + \mathbf{V}_\Gamma(\bar{\mathbf{k}})\bar{\mathbf{k}}) > + \sigma\mathbf{V}'_\Gamma(\bar{\mathbf{k}})(\bar{\mathbf{k}}, \bar{\bar{\mathbf{k}}})_+,
\end{aligned}
\tag{3.196}
$$

where $\mathbf{k} = \nabla\Psi_\rho$, $\bar{\mathbf{k}} = \bar{\nabla}\tilde{\Psi}_\rho$, $\bar{\bar{\mathbf{k}}} = \bar{\nabla}^2\tilde{\Psi}_\rho$, and

$$
\mathbf{V}'_\Gamma(\bar{\mathbf{k}})(\bar{\mathbf{k}}, \bar{\bar{\mathbf{k}}})_+ = \begin{cases} \mathbf{V}'_\Gamma(\bar{\mathbf{k}})\bar{\mathbf{k}} \otimes \bar{\bar{\mathbf{k}}} & \text{for } \Omega_+, \\ 0 & \text{for } \Omega_-. \end{cases}
$$

Here, we have used (3.2).

3.4.4 Local Well-Posedness

Let ρ, H_ρ, and ω be the same functions as in Sect. 3.1.3.1. Let $\mathbf{n}$ be the unit outer normal to Γ and the extension of $\mathbf{n}$ to $\mathbb{R}^N$ is also written as $\mathbf{n}$. The Γ_t is a set defined by (3.25) and the H_ρ is assumed to satisfy (3.26). Let $\Psi_\rho(y, t) = \omega(y)H_\rho(y, t)\mathbf{n}(y)$ and we consider the transform (3.27). Then, equation (3.1) is transformed to the following equations:

$$
\begin{cases}
m\partial_t\mathbf{u} - \mathrm{Div}\,(\mu\mathbf{D}(\mathbf{u}) - \mathfrak{q}\mathbf{I}) = \mathbf{f}(\mathbf{u}, \Psi_\rho) & \text{in } \dot{\Omega}^T, \\
\mathrm{div}\,\mathbf{u} = g(\mathbf{u}, \Psi_\rho) = \mathrm{div}\,\mathbf{g}(\mathbf{u}, \Psi_\rho) & \text{in } \dot{\Omega}^T, \\
\partial_t\rho + < \mathbf{u} \mid \nabla'_\Gamma\rho > - \mathbf{u}\cdot\mathbf{n} = d(\mathbf{u}, \Psi_\rho) & \text{on } \Gamma^T, \\
[[(\mu\mathbf{D}(\mathbf{u})\mathbf{n})_\tau]] = [[\mathbf{h}'(\mathbf{u}, \Psi_\rho)]], \quad [[\mathbf{u}]] = 0 & \text{on } \Gamma^T, \\
[[< \mu\mathbf{D}(\mathbf{u})\mathbf{n}, \mathbf{n} > - \mathfrak{q}]] - \sigma(\Delta_\Gamma + \mathbb{B})\rho = [[h_N(\mathbf{u}, \Psi_\rho)]] & \text{on } \Gamma^T, \\
\mathbf{u} = 0 & \text{on } \Gamma_-^T, \\
\mathbf{u}|_{t=0} = \mathbf{u}_0 \ \text{in } \dot{\Omega}, \quad \rho|_{t=0} = \rho_0 \ \text{on } \Gamma,
\end{cases}
\tag{3.197}
$$

where setting $\Psi_\rho(y, t) = \tilde{\Psi}_\rho(y, t) = \omega(y)H_\rho(y, t)\mathbf{n}(y)$, we have defined $\mathbf{f}(\mathbf{u}, \Psi_\rho)$, $d(\mathbf{u}, \Psi_\rho)$, $g(\mathbf{u}, \Psi_\rho)$, $\mathbf{g}(\mathbf{u}, \Psi_\rho)$, $\mathbf{h}'(\mathbf{u}, \Psi_\rho)$, and $h_N(\mathbf{u}, \Psi_\rho)$ by the formulas in respective (3.142), (3.177), (3.136), and (3.137), (3.183), (3.196). And, the operator $\mathbb{B}$ is given in (3.193). If we move the term $< \mathbf{u} \mid \nabla'_\Gamma\rho >$ to the right-hand side and use the usual fixed point argument based on Theorem 3.2.8 with $\kappa = 0$, then we have to assume that both initial data $\mathbf{u}_0$ and ρ_0 are small. But, this is not satisfactory. We have to treat at least the case where the initial velocity $\mathbf{u}_0$ is arbitrarily large. For this purpose, we approximate $\mathbf{u}$ as follows: Let $\mathbf{u}_0$ be an initial data belonging to $B^{2(1-1/p)}_{q,p}(\dot{\Omega})^N$ and set $\mathbf{u}_{0+} = \mathbf{u}|_{\Omega_+}$. Notice that $[[\mathbf{u}_0]] = 0$ on Γ, which is one of the compatibility conditions for the initial data. Let $\tilde{\mathbf{u}}_{0+}$ be an extension of $\mathbf{u}_{0+}$ to $\mathbb{R}^N$ such that $\tilde{\mathbf{u}}_{0+} = \mathbf{u}_{0+}$ in Ω_+ and

$$\|\tilde{\mathbf{u}}_{0+}\|_{B^{2(1-1/p)}_{q,p}(\mathbb{R}^N)} \leq C\|\mathbf{u}_{0+}\|_{B^{2(1-1/p)}_{q,p}(\Omega_+)}. \tag{3.198}$$

Let

$$\mathbf{u}_\kappa = \frac{1}{\kappa}\int_0^\kappa T_0(s)\tilde{\mathbf{u}}_{0+}\,ds,$$

where $T_0(s)$ is the heat semigroup on $\mathbb{R}^N$ which is defined by letting

$$T_0(s)f = \frac{1}{(2\pi)^N}\int_{\mathbb{R}^N} e^{ix\cdot\xi}e^{-|\xi|^2 s}\mathcal{F}[f](\xi)\,d\xi.$$

Since $T_0(s)$ is a bounded C^0 analytic semigroup, we have

$$\begin{aligned}
&\|T_0(\cdot)\tilde{\mathbf{u}}_{0+}\|_{L_p((0,\infty),H^2_q(\mathbb{R}^N))} + \|\partial_t T_0(\cdot)\tilde{\mathbf{u}}_{0+}\|_{L_p((0,\infty),L_q(\mathbb{R}^N))}\\
&\quad + \|T_0(\cdot)\tilde{\mathbf{u}}_{0+}\|_{L_\infty((0,\infty),B^{2(1-1/p)}_{q,p}(\mathbb{R}^N))} \leq C\|\mathbf{u}_{0+}\|_{B^{2(1-1/p)}_{q,p}(\Omega_+)},
\end{aligned}$$

which yields

$$\begin{aligned}
\|\mathbf{u}_\kappa\|_{B^{2(1-1/p)}_{q,p}(\mathbb{R}^N)} &\leq C\|\mathbf{u}_{0+}\|_{B^{2(1-1/p)}_{q,p}(\Omega_+)},\\
\|\mathbf{u}_\kappa\|_{H^2_q(\mathbb{R}^N)} &\leq C\kappa^{-1/p}\|\mathbf{u}_{0+}\|_{B^{2(1-1/p)}_{q,p}(\Omega_+)},
\end{aligned} \tag{3.199}$$

where $1 < p, q < \infty$.

Instead of equation (3.197), we shall solve the following equations:

$$\begin{cases}
m\partial_t\mathbf{u} - \mathrm{Div}\,(\mu\mathbf{D}(\mathbf{u}) - \mathfrak{q}\mathbf{I}) = \mathbf{f}(\mathbf{u},\Psi_\rho) & \text{in } \dot\Omega^T,\\
\mathrm{div}\,\mathbf{u} = g(\mathbf{u},\Psi_\rho) = \mathrm{div}\,\mathbf{g}(\mathbf{u},\Psi_\rho) & \text{in } \dot\Omega^T,\\
\partial_t\rho + <\mathbf{u}_\kappa \mid \nabla'_\Gamma\rho> - \mathbf{u}\cdot\mathbf{n} = d(\mathbf{u},\Psi_\rho) + <\mathbf{u}_\kappa - \mathbf{u} \mid \nabla'_\Gamma\rho> & \text{on } \Gamma^T,\\
[[\mu\mathbf{D}(\mathbf{u})\mathbf{n}]]_\tau = [[\mathbf{h}'(\mathbf{u},\Psi_\rho)]], \quad [[\mathbf{u}]] = 0 & \text{on } \Gamma^T,\\
[[<\mu\mathbf{D}(\mathbf{u})\mathbf{n},\mathbf{n}> - \mathfrak{q}]] - \sigma(\Delta_\Gamma + \mathbb{B})\rho = [[h_N(\mathbf{u},\Psi_\rho)]] & \text{on } \Gamma^T,\\
\mathbf{u} = 0 & \text{on } \Gamma_-^T,\\
\mathbf{u}|_{t=0} = \mathbf{u}_0 \text{ in } \dot\Omega, \quad \rho|_{t=0} = \rho_0 \text{ on } \Gamma.
\end{cases} \tag{3.200}$$

This is our linearization principle for equation (3.1). And then, we have Theorem 3.1.2, which was stated in Sect. 3.1.3.1 and will be proved in Sect. 3.4.5 below.

3.4.5 Proof of Theorem 3.1.2

Since $N < q < \infty$, by Sobolev's inequality we have the following estimates:

$$\|f\|_{L_\infty(\Omega_\pm)} \leq C\|f\|_{H^1_q(\Omega_\pm)}, \quad \|fg\|_{H^1_q(\Omega_\pm)} \leq C\|f\|_{H^1_q(\Omega_\pm)}\|g\|_{H^1_q(\Omega_\pm)},$$
$$\|fg\|_{H^2_q(\Omega_\pm)} \leq C(\|f\|_{H^2_q(\Omega_\pm)}\|g\|_{H^1_q(\Omega_\pm)} + \|f\|_{H^1_q(\Omega_\pm)}\|g\|_{H^2_q(\Omega_\pm)}),$$
$$\|fg\|_{W^{1-1/q}_q(\Gamma)} \leq C\|f\|_{W^{1-1/q}_q(\Gamma)}\|g\|_{W^{1-1/q}_q(\Gamma)}, \tag{3.201}$$
$$\|fg\|_{W^{2-1/q}_q(\Gamma)} \leq C(\|f\|_{W^{2-1/q}_q(\Gamma)}\|g\|_{W^{1-1/q}_q(\Gamma)} + \|f\|_{W^{1-1/q}_q(\Gamma)}\|g\|_{W^{2-1/q}_q(\Gamma)}).$$

Moreover, since $2/p + N/q < 1$, we have

$$\|f\|_{H^1_\infty(\Omega_\pm)} \leq C\|f\|_{B^{2(1-1/p)}_{q,p}(\Omega_\pm)}, \quad \|f\|_{H^2_\infty(\Gamma)} \leq C\|f\|_{B^{3-1/p-1/q}_{q,p}(\Gamma)}. \tag{3.202}$$

We define an underlying space $\mathbf{U}_T$ by letting

$$\begin{aligned}\mathbf{U}_T = \{(\mathbf{u}, \rho) \mid\ & \mathbf{u} \in H^1_p((0,T), L_q(\dot\Omega)^N) \cap L_p((0,T), H^2_q(\dot\Omega)^N,\\ & \rho \in H^1_p((0,T), W^{2-1/q}_q(\Gamma)) \cap L_p((0,T), W^{3-1/q}_q(\Gamma)),\\ & \mathbf{u}|_{t=0} = \mathbf{u}_0 \quad \text{in } \dot\Omega, \quad \rho|_{t=0} = \rho_0 \quad \text{on } \Gamma,\\ & E_{p,q,T}(\mathbf{u}, \rho) \leq L, \quad \sup_{t\in(0,T)} \|\rho(\cdot,t)\|_{H^1_\infty(\Gamma)} \leq \delta\},\end{aligned}$$

where L is a number determined later. Since L is chosen large and ϵ small eventually, we may assume that $0 < \epsilon < 1 < L$ in the following. Thus, for example, we will use the inequality: $1 + \epsilon + L < 3L$ below.

Let $(\mathbf{v}, h) \in \mathbf{U}_T$ and let $\mathbf{u}$, $\mathfrak{q}$, and ρ be solutions of the linear equations:

$$\begin{cases} \partial_t\mathbf{u} - m^{-1}\mathrm{Div}\,(\mu\mathbf{D}(\mathbf{u}) - \mathfrak{q}\mathbf{I}) = \mathbf{f}(\mathbf{v}, \Psi_h) & \text{in } \dot\Omega^T,\\ \mathrm{div}\,\mathbf{u} = g(\mathbf{u}, \Psi_h) = \mathrm{div}\,\mathbf{g}(\mathbf{v}, \Psi_h) & \text{in } \dot\Omega^T,\\ \partial_t\rho + < \mathbf{u}_\kappa \mid \nabla'_\Gamma\rho > -\mathbf{u}\cdot\mathbf{n} = d_\kappa(\mathbf{v}, \Psi_h) & \text{on } \Gamma^T,\\ [[(\mu\mathbf{D}(\mathbf{u})\mathbf{n})]]_\tau = [[\mathbf{h}'(\mathbf{v}, \Psi_h)]], \quad [[\mathbf{u}]] = 0 & \text{on } \Gamma^T,\\ [[< \mu\mathbf{D}(\mathbf{u})\mathbf{n}, \mathbf{n} > -\mathfrak{q}]] - \sigma(\Delta_\Gamma\rho + \mathbb{B}\rho) = [[h_N(\mathbf{v}, \Psi_h)]] & \text{on } \Gamma^T,\\ \mathbf{u} = 0 & \text{on } \Gamma_-,\\ (\mathbf{u}, \rho)|_{t=0} = (\mathbf{u}_0, \rho_0) & \text{in } \dot\Omega \times \Gamma, \end{cases} \tag{3.203}$$

where we have set

$$d_\kappa(\mathbf{v}, \Psi_h) = d(\mathbf{v}, \Psi_h) + < \mathbf{u}_\kappa - \mathbf{v} \mid \nabla'_\Gamma h > .$$

Here and in the following, we set $\Psi_h = \omega(y)H_h(y,t)\mathbf{n}(y)$. In view of (3.26), we may assume that

$$\sup_{t\in(0,T)} \|\Psi_h(\cdot,t)\|_{H^1_\infty(\Omega)} \leq \delta, \tag{3.204}$$

because we have assumed that $\sup_{t\in(0,T)} \|h(\cdot,t)\|_{H^1_\infty(\Gamma)} \leq \delta$. Since

$$E_{p,q,T}(\mathbf{v}, h) \leq L, \tag{3.205}$$

we have

$$\begin{aligned}
&\|\partial_t h\|_{L_\infty((0,T),W^{1-1/q}_q(\Gamma))} + \|\partial_t h\|_{L_p((0,T),W^{2-1/q}_q(\Gamma))} + \|h\|_{L_p((0,T),W^{3-1/q}_q(\Gamma))} \\
&\quad + \|\partial_t \mathbf{v}\|_{L_p((,T),L_q(\dot\Omega))} + \|\mathbf{v}\|_{L_p((0,T),H^2_q(\dot\Omega))} \leq L.
\end{aligned} \tag{3.206}$$

Moreover, we use the assumption:

$$\|\mathbf{u}_0\|_{B^{2(1-1/p)}_{q,p}(\dot\Omega)} \leq B, \quad \|\rho_0\|_{B^{3-1/p-1/q}_{q,p}(\Gamma)} \leq \epsilon, \tag{3.207}$$

where ϵ is a small constant determined later.

To obtain the estimates of $\mathbf{u}$ and ρ, we shall use Theorem 3.2.8. For this purpose, we shall estimate the nonlinear functions appearing on the right-hand side of equations (3.203). We first prove that

$$\|\mathbf{f}(\mathbf{v}, \Psi_h)\|_{L_p((0,T),L_q(\dot\Omega))} \leq C\{T^{1/p}(L+B)^2 + (\epsilon + T^{1/p'}L)L\}. \tag{3.208}$$

The definition of $\mathbf{f}(\mathbf{v}, \Psi_h)$ is given by replacing $\xi(t)$, ρ and $\mathbf{u}$ by 0, h, and $\mathbf{v}$ in (3.142). Since $|\mathbf{V}_0(\mathbf{k})| \leq C|\mathbf{k}|$ when $|\mathbf{k}| \leq \delta$, by (3.204) we have

$$\begin{aligned}
\|\mathbf{f}(\mathbf{v}, \Psi_h)\|_{L_q(\dot\Omega)} &\leq C\{\|\mathbf{v}\|_{L_\infty(\dot\Omega)}\|\nabla\mathbf{v}\|_{L_q(\dot\Omega)} + \|\partial_t\Psi_h\|_{L_\infty(\dot\Omega)}\|\nabla\mathbf{v}\|_{L_q(\dot\Omega)} \\
&\quad + \|\nabla\Psi_h\|_{L_\infty(\dot\Omega)}\|\partial_t\mathbf{v}\|_{L_q(\dot\Omega)} + \|\nabla\Psi_h\|_{L_\infty(\dot\Omega)}\|\nabla^2\mathbf{v}\|_{L_q(\dot\Omega)} \\
&\quad + \|\nabla^2\Psi_h\|_{L_q(\dot\Omega)}\|\mathbf{v}\|_{L_\infty(\dot\Omega)}\}.
\end{aligned} \tag{3.209}$$

By (3.26) and (3.201), we have

$$
\begin{aligned}
\|\mathbf{v}(\cdot,t)\|_{L_\infty(\dot\Omega)} &\leq C\|\mathbf{v}(\cdot,t)\|_{H^1_q(\dot\Omega)},\\
\|\partial_t\Psi_h(\cdot,t)\|_{L_\infty(\dot\Omega)} &\leq C\|\partial_t h(\cdot,t)\|_{W^{1-1/q}_q(\Gamma)},\\
\|\nabla\Psi_h(\cdot,t)\|_{L_\infty(\dot\Omega)} &\leq C\|h(\cdot,t)\|_{W^{2-1/q}_q(\Gamma)},\\
\|\nabla\mathbf{v}(\cdot,t)\|_{L_\infty(\dot\Omega)} &\leq C\|\mathbf{v}(\cdot,t)\|_{H^2_q(\dot\Omega)},
\end{aligned} \tag{3.210}
$$

and so,

$$
\begin{aligned}
&\|\mathbf{f}(\mathbf{v},\Psi_h)\|_{L_p((0,T),L_q(\dot\Omega))} \leq C\{\|\mathbf{v}\|^2_{L_\infty((0,T),H^1_q(\dot\Omega))}T^{1/p}\\
&+\|\mathbf{v}\|_{L_\infty((0,T),H^1_q(\dot\Omega))}\|\partial_t h\|_{L_\infty((0,T),W^{1-1/q}_q(\Gamma))}T^{1/p}\\
&+\|h\|_{L_\infty((0,T),W^{2-1/q}_q(\Gamma))}(\|\partial_t\mathbf{v}\|_{L_p((0,T),L_q(\dot\Omega))}+\|\mathbf{v}\|_{L_p((0,T),H^2_q(\dot\Omega))})\}.
\end{aligned} \tag{3.211}
$$

In what follows, we shall use the following inequalities:

$$
\begin{aligned}
&\|\mathbf{v}\|_{L_\infty((0,T),B^{2(1-1/p)}_{q,p}(\dot\Omega))} \leq C\{\|\mathbf{u}_0\|_{B^{2(1-1/p)}_{q,p}(\dot\Omega)}\\
&\quad+\|\mathbf{v}\|_{L_p((0,T),H^2_q(\dot\Omega))}+\|\partial_t\mathbf{v}\|_{L_p((0,T),L_q(\dot\Omega))}\},
\end{aligned} \tag{3.212}
$$

$$
\begin{aligned}
&\|h\|_{L_\infty((0,T),B^{3-1/p-1/q}_{q,p}(\Gamma))} \leq C\{\|\rho_0\|_{B^{3-1/p-1/q}_{q,p}(\Gamma)}\\
&\quad+\|h\|_{L_p((0,T),W^{3-1/q}_q(\Gamma))}+\|\partial_t h\|_{L_p((0,T),W^{2-1/q}_q(\Gamma))}\},
\end{aligned} \tag{3.213}
$$

$$
\begin{aligned}
&\|h(\cdot,t)\|_{L_\infty((0,T),W^{2-1/q}_q(\Gamma))}\\
&\leq \|\rho_0\|_{W^{2-1/q}_q(\Gamma)}+\int_0^T\|\partial_s h(\cdot,s)\|_{W^{2-1/q}_q(\Gamma)}\,ds\\
&\leq \|\rho_0\|_{W^{2-1/q}_q(\Gamma)}+T^{1/p'}L,
\end{aligned} \tag{3.214}
$$

for some constant C independent of T. The inequalities: (3.212) and (3.213) will be proved later. Combining (3.211) with (3.212), (3.214), (3.206) and (3.207) gives (3.208).

We next consider $d_\kappa(\mathbf{u},\Psi_h)$. Since $\xi(t)=0$ in (3.27), by (3.177) we have

$$
d(\mathbf{v},\Psi_h)=\mathbf{v}\cdot\mathbf{V}_\Gamma(\bar\nabla\Psi_h)(\bar\nabla\Psi_h,\bar\nabla\Psi_h)-\partial_t h<\mathbf{n},\mathbf{V}_\Gamma(\bar\nabla\Psi_h)\bar\nabla\Psi_h\otimes\bar\nabla\Psi_h>. \tag{3.215}
$$

Applying (3.170), (3.204) and (3.201), we have

$$
\begin{aligned}
&\|\mathbf{V}_\Gamma(\bar\nabla\Psi_h)\bar\nabla\Psi_h\otimes\bar\nabla\Psi_h\|_{W^{1-1/q}_q(\Gamma)} \leq C\|h(\cdot,t)\|^2_{W^{2-1/q}_q(\Gamma)},\\
&\|\mathbf{V}_\Gamma(\bar\nabla\Psi_h)\bar\nabla\Psi_h\otimes\bar\nabla\Psi_h\|_{W^{2-1/q}_q(\Gamma)}
\end{aligned}
$$

$$\leq C\|h(\cdot,t)\|_{W^{2-1/q}_q(\Gamma)}\|h(\cdot,t)\|_{W^{3-1/q}_q(\Gamma)},$$

and so, by (3.210) and (3.201)

$$\begin{aligned}
&\|d(\mathbf{v},\Psi_h)\|_{W^{1-1/q}_q(\Gamma)}\\
&\leq C(\|\partial_t h(\cdot,t)\|_{W^{1-1/q}_q(\Gamma)}+\|\mathbf{v}(\cdot,t)\|_{H^1_q(\dot\Omega)})\|h(\cdot,t)\|_{W^{2-1/q}_q(\Gamma)},\\
&\|d(\mathbf{v},\Psi_h)\|_{W^{2-1/q}_q(\Gamma)}\\
&\leq C\{(\|\partial_t h(\cdot,t)\|_{W^{2-1/q}_q(\Gamma)}+\|\mathbf{v}(\cdot,t)\|_{H^2_q(\dot\Omega)})\|h(\cdot,t)\|_{W^{2-1/q}_q(\Gamma)}\\
&+(\|\partial_t h(\cdot,t)\|_{W^{1-1/q}_q(\Gamma)}+\|\mathbf{v}(\cdot,t)\|_{H^1_q(\dot\Omega)})\|h(\cdot,t)\|_{W^{2-1/q}_q(\Gamma)}\|h(\cdot,t)\|_{W^{3-1/q}_q(\Gamma)}\}.
\end{aligned} \tag{3.216}$$

Thus, by (3.206) and (3.207), we have

$$\begin{aligned}
\sup_{t\in(0,T)}\|d(\mathbf{v},\Psi_h)\|_{W^{1-1/q}_q(\Gamma)}&\leq C(L+B)(\epsilon+T^{1/p'}L),\\
\|d(\mathbf{v},\Psi_h)\|_{L_p((0,T),W^{2-1/q}_q(\Gamma))}&\leq C(L+B)L(\epsilon+T^{1/p'}L).
\end{aligned} \tag{3.217}$$

By (3.199) and (3.201), we have

$$\begin{aligned}
&\|<\mathbf{u}_\kappa-\mathbf{v}\mid\nabla'_\Gamma h>\|_{W^{1-1/q}_q(\Gamma)}\\
&\quad\leq C(\|\mathbf{u}_{0+}\|_{B^{2(1-1/p)}_{q,p}(\Omega_+)}+\|\mathbf{v}_+(\cdot,t)\|_{H^1_q(\Omega_+)})\|h(\cdot,t)\|_{W^{2-1/q}_q(\Gamma)},\\
&\|<\mathbf{u}_\kappa-\mathbf{v}\mid\nabla'_\Gamma h>\|_{W^{2-1/q}_q(\Gamma)}\\
&\quad\leq C(\|\mathbf{u}_\kappa\|_{H^2_q(\Omega_+)}+\|\mathbf{v}(\cdot,t)\|_{H^2_q(\Omega_+)})\|h(\cdot,t)\|_{W^{2-1/q}_q(\Gamma)}\\
&\qquad+\|\mathbf{u}_\kappa-\mathbf{v}_+(\cdot,t)\|_{H^1_q(\Omega_+)}\|h(\cdot,t)\|_{W^{3-1/q}_q(\Gamma)}).
\end{aligned} \tag{3.218}$$

By (3.212), (3.214), (3.206), and (3.207), we have

$$\|<\mathbf{u}_\kappa-\mathbf{v}\mid\nabla'_\Gamma h>\|_{L_\infty((0,T),W^{1-1/q}_q(\Gamma))}\leq C(B+L)(\epsilon+T^{1/p'}L). \tag{3.219}$$

By the definition of $\mathbf{u}_\kappa$, we have

$$\mathbf{u}_\kappa-\mathbf{u}_0=\frac{1}{\kappa}\int_0^\kappa(T(s)\tilde{\mathbf{u}}_0-\mathbf{u}_0)\,ds,$$

and so, we have

$$\|\mathbf{u}_\kappa - \mathbf{u}_0\|_{L_q(\Omega_+)} \leq \frac{1}{\kappa}\int_0^\kappa \Big(\int_0^s \|\partial_s T(r)\tilde{\mathbf{u}}_0\|_{L_q(\Omega_+)}\,dr\Big)\,ds$$
$$\leq C\kappa^{1/p'}\|\mathbf{u}_0\|_{B^{2(1-1/p)}_{q,p}(\Omega_+)}.$$

Let s be a positive number such that $1 < s < 2(1-1/p)$, and then by interpolation theory and (3.199) we have

$$\begin{aligned}\|\mathbf{u}_\kappa - \mathbf{u}_{0+}\|_{H^1_q(\Omega_+)} &\leq C\|\mathbf{u}_\kappa - \mathbf{u}_{0+}\|^{1-1/s}_{L_q(\Omega_+)}\|\mathbf{u}_\kappa - \mathbf{u}_{0+}\|^{1/s}_{W^s_q(\Omega_+)} \\ &\leq C\kappa^{(1-1/s)/p'}\|\mathbf{u}_{0+}\|_{B^{2(1-1/p)}_{q,p}(\Omega_+)}.\end{aligned} \tag{3.220}$$

Writing $\mathbf{v}_+(\cdot,t) - \mathbf{u}_{0+} = \int_0^t \partial_r\mathbf{v}_+(\cdot,r)\,dr$, we have

$$\|\mathbf{v}_+(\cdot,t) - \mathbf{u}_{0+}\|_{L_q(\Omega_+)} \leq LT^{1/p'}.$$

On the other hand,

$$\|\mathbf{v}_+(\cdot,t) - \mathbf{u}_{0+}\|_{B^{2(1-1/p)}_{q,p}(\Omega_+)} \leq L + \|\mathbf{u}_0\|_{B^{2(1-1/p)}_{q,p}(\dot\Omega)},$$

and so,

$$\|\mathbf{v}_+(\cdot,t) - \mathbf{u}_0\|_{H^1_q(\Omega_+)} \leq (L+B)T^{(1-1/s)/p'}. \tag{3.221}$$

Putting (3.220) and (3.221) together gives

$$\sup_{t\in(0,T)}\|\mathbf{v}_+(\cdot,t) - \mathbf{u}_\kappa\|_{H^1_q(\Omega_+)} \leq C(\kappa^{(1-1/s)/p'} + T^{(1-1/s)/p'})(L+B). \tag{3.222}$$

By (3.218)

$$\begin{aligned}&\| < \mathbf{u}_\kappa - \mathbf{v} \mid \nabla'_\Gamma h > \|_{L_p((0,T),W^{2-1/q}_q(\Gamma)} \\ &\leq C\{(\|\mathbf{u}_\kappa\|_{H^2_q(\Omega_+)}T^{1/p} + \|\mathbf{v}_+\|_{L_p((0,T),H^2_q(\Omega_+))})\|h\|_{L_\infty((0,T),W^{2-1/q}_q(\Gamma))} \\ &\qquad + \|\mathbf{u}_\kappa - \mathbf{v}_+\|_{L_\infty((0,T),H^1_q(\Omega_+))}\|h\|_{L_p((0,T),W^{3-1/q}_q(\Gamma))}\}.\end{aligned}$$

Thus, by (3.199), (3.206), (3.207), and (3.214), we have

$$\begin{aligned}&\| < \mathbf{u}_\kappa - \mathbf{v} \mid \nabla'_\Gamma h > \|_{L_p((0,T),W^{2-1/q}_q(\Gamma))} \\ &\leq C\{(B\kappa^{-1/p}T^{1/p} + L)(\epsilon + T^{1/p'}L) \\ &\quad + L(L+B)(\kappa^{(1-1/s)/p'} + T^{(1-1/s)/p'})\}\end{aligned} \tag{3.223}$$

for some constant $s \in (1, 2(1-1/p))$. Putting (3.217), (3.219), (3.222), and (3.223) together gives

$$
\begin{aligned}
&\|\tilde{d}_\kappa(\mathbf{v}, \Psi_h)\|_{L_\infty((0,T), W_q^{1-1/q}(\Gamma))} \\
&\quad \le C(L+B)(\epsilon + T^{1/p'}L + \kappa^{(1-1/s)p'} + T^{1/p'}), \\
&\|\tilde{d}_\kappa(\mathbf{v}, \Psi_h)\|_{L_p((0,T), W_q^{2-1/q}(\Gamma))} \le C\{(L + B\kappa^{-1/p}T^{1/p})(\epsilon + T^{1/p'}L) \\
&\quad + L(L+B)(\epsilon + T^{1/p'}L + \kappa^{(1-1/s)/p'} + T^{(1-1/s)/p'})\}, \qquad (3.224)
\end{aligned}
$$

where s is a constant $\in (1, 2(1-1/p))$.

We now estimate $g(\mathbf{v}, \Psi_h)$ and $\mathbf{g}(\mathbf{v}, \Psi_h)$, which are defined in (3.136) and (3.137). For this purpose, we have to extend $g(\mathbf{v}, \Psi_h)$ and $\mathbf{g}(\mathbf{v}, \Psi_h)$ to the whole time interval $\mathbb{R}$. Before turning to the extension of these functions, we make a few definitions. Let $\tilde{\mathbf{u}}_{0\pm} \in B^{2(1-1/p)}_{q,p}(\mathbb{R}^N)$ be an extension of $\mathbf{u}_0|_{\Omega_\pm} \in B^{2(1-1/p)}_{q,p}(\Omega_\pm)$ to $\mathbb{R}^N$ such that

$$
\mathbf{u}_0|_{\Omega_\pm} = \tilde{\mathbf{u}}_{0\pm} \quad \text{in } \Omega_\pm, \quad \|\tilde{\mathbf{u}}_{0\pm}\|_{X(\mathbb{R}^N)} \le C\|\mathbf{u}_0\|_{X(\Omega_\pm)}
$$

for $X \in \{H^k_q, B^{2(1-1/p)}_{q,p}\}$. Let $T_{v\pm}(t)\mathbf{u}_0$ be defined by setting

$$
T_{v\pm}(t)\mathbf{u}_0 = e^{-(2-\Delta)t}\tilde{\mathbf{u}}_{0\pm} = \mathcal{F}^{-1}[e^{-(|\xi|^2+2)t}\mathcal{F}[\tilde{\mathbf{u}}_{0\pm}](\xi)] \qquad (3.225)
$$

and set $T_v(t)\mathbf{u}_0 = T_{v\pm}(t)\mathbf{u}_0|_{\Omega_\pm}$ for $x \in \Omega_\pm$. And then, we have $T_v(0)\mathbf{u}_0 = \mathbf{u}_0$ in $\dot\Omega$ and

$$
\begin{aligned}
\|e^t T_v(t)\mathbf{u}_0\|_{X(\dot\Omega)} &\le C\|\mathbf{u}_0\|_{X(\dot\Omega)}, \\
\|e^t T_v(\cdot)\mathbf{u}_0\|_{H^1_p((0,\infty), L_q(\dot\Omega))} &+ \|e^t T_v(\cdot)\mathbf{u}_0\|_{L_p((0,\infty), H^2_q(\dot\Omega))} \\
&\le C\|\mathbf{u}_0\|_{B^{2(1-1/p)}_{q,p}(\dot\Omega)}. \qquad (3.226)
\end{aligned}
$$

Let $\mathbf{W}$, P, and Ξ be solutions of the equations:

$$
\begin{cases}
\partial_t \mathbf{W} + \lambda_0 \mathbf{W} - m^{-1}\mathrm{Div}\,(\mu\mathbf{D}(\mathbf{W}) - P\mathbf{I}) = 0, \quad \mathrm{div}\,\mathbf{W} = 0 & \text{in } \dot\Omega \times (0,\infty), \\
\partial_t \Xi + \lambda_0 \Xi - \mathbf{W}\cdot\mathbf{n} = 0 & \text{on } \Gamma \times (0,\infty), \\
[[(\mu\mathbf{D}(\mathbf{W}) - P\mathbf{I})\mathbf{n}]] - \sigma(\Delta_\Gamma \Xi + < \mathbf{n}, \Delta_\Gamma \mathbf{n} > \Xi) = 0 & \text{on } \Gamma \times (0,\infty), \\
[[\mathbf{W}]] = 0 & \text{on } \Gamma \times (0,\infty), \\
\mathbf{W} = 0 & \text{on } \Gamma_- \times (0,\infty), \\
(\mathbf{W}, \Xi)|_{t=0} = (0, \rho_0) & \text{in } \dot\Omega \times \Gamma.
\end{cases}
$$

If we choose a positive number λ_0 large enough, then by Theorem 3.2.8 with $\kappa = 0$, we know the existence of $\mathbf{W}$ and Ξ with

$$\mathbf{W} \in L_p((0,\infty), H^2_q(\dot\Omega)^N) \cap H^1_p((0,\infty), L_q(\dot\Omega)^N),$$
$$\Xi \in L_p((0,\infty), W^{3-1/q}_q(\Gamma)) \cap H^1_p((0,\infty), W^{2-1/q}_q(\Gamma))$$

possessing the estimates:

$$\begin{aligned}
&\|e^t\mathbf{W}\|_{L_\infty((0,\infty),B^{2(1-1/p)}_{q,p}(\dot\Omega))} + \|e^t\mathbf{W}\|_{L_p((0,\infty),H^2_q(\dot\Omega))} + \|e^t\partial_t\mathbf{W}\|_{L_p((0,\infty),L_q(\dot\Omega))}\\
&\quad + \|e^t\,\Xi\|_{L_\infty((0,\infty),B^{3-1/p-1/q}_{q,p}(\Gamma))} + \|e^t\,\Xi\|_{L_p((0,\infty),W^{3-1/q}_q(\Gamma))}\\
&\quad + \|e^t\partial_t\,\Xi\|_{L_p((0,\infty),W^{1-1/q}_q(\Gamma))} \leq C\|\rho_0\|_{B^{3-1/p-1/q}_{q,p}(\Gamma)},
\end{aligned}$$

where we have used the inequalities:

$$\begin{aligned}
&\|e^t\mathbf{W}\|_{L_\infty((0,\infty),B^{2(1-1/p)}_{q,p}(\dot\Omega))}\\
&\quad\leq C(\|e^t\mathbf{W}\|_{L_p((0,\infty),H^2_q(\dot\Omega))} + \|e^t\partial_t\mathbf{W}\|_{L_p((0,\infty),L_q(\dot\Omega))}),\\
&\|e^t\,\Xi\|_{L_\infty((0,\infty),B^{3-1/p-1/q}_{q,p}(\Gamma))}\\
&\quad\leq C(\|e^t\,\Xi\|_{L_p((0,\infty),W^{3-1/q}_q(\Gamma))} + \|e^t\partial_t\,\Xi\|_{L_p((0,\infty),W^{2-1/q}_q(\Gamma))})
\end{aligned}$$

which follow from real interpolation (3.88) and (3.89). Moreover, by the trace theorem and the kinematic equation, we have

$$\begin{aligned}
&\|e^t\partial_t\,\Xi\|_{L_\infty((0,\infty),W^{1-1/q}_q(\Gamma))}\\
&\quad\leq \lambda_0\|e^t\,\Xi\|_{L_\infty((0,\infty),W^{1-1/q}_q(\Gamma))} + \|e^t\mathbf{n}\cdot\mathbf{W}\|_{L_\infty((0,\infty),W^{1-1/q}_q(\Gamma))}\\
&\quad\leq C\|\rho_0\|_{W^{3-1/p-1/q}_{q,p}(\Gamma)}.
\end{aligned}$$

Setting $T_h(t)\rho_0 = H_\Xi(x,t)$, we have $T_h(0)\rho_0 = \rho_0$ in Γ, and therefore, we may assume that $T_h(0)\rho_0 = H_h|_{t=0}$. Moreover, in view of (3.26), we have

$$\begin{aligned}
&\|e^tT_h(\cdot)\rho_0\|_{L_\infty(0,\infty),B^{3-1/p}_{q,p}(\dot\Omega))} + \|e^t\partial_tT_h(\cdot)\rho_0\|_{L_\infty((0,\infty),H^1_q(\dot\Omega))}\\
&\quad + \|e^tT_h(\cdot)\rho_0\|_{L_p((0,\infty),H^3_q(\dot\Omega))} + \|e^t\partial_tT_h(\cdot)\rho_0\|_{L_p((0,\infty),H^2_q(\dot\Omega))}\\
&\quad\leq C\|\rho_0\|_{B^{3-1/p-1/q}_{q,p}(\Gamma)}.
\end{aligned}\tag{3.227}$$

Given a function, $f(t)$, defined on $(0, T)$, the extension, $e_T[f]$, of f is defined by letting

$$e_T[f](t) = \begin{cases} 0 & \text{for } t < 0, \\ f(t) & \text{for } 0 < t < T, \\ f(2T - t) & \text{for } T < t < 2T, \\ 0 & \text{for } t > 2T. \end{cases} \tag{3.228}$$

Obviously, $e_T[f](t) = f(t)$ for $t \in (0, T)$ and $e_T[f](t) = 0$ for $t \notin (0, 2T)$. Moreover, if $f|_{t=0} = 0$, then

$$\partial_t e_T[f](t) = \begin{cases} 0 & \text{for } t < 0, \\ (\partial_t f)(t) & \text{for } 0 < t < T, \\ -(\partial_t f)(2T - t) & \text{for} T < t < 2T, \\ 0 & \text{for } t > 2T. \end{cases} \tag{3.229}$$

Let $\psi(t)$ be a function in $C^\infty(\mathbb{R})$ which equals one for $t > -1$ and zero for $t < -2$.

Under these preparations, we now define the extensions, $\mathcal{E}_1[\mathbf{v}]$, and $\mathcal{E}_2[\Psi_h]$, of $\mathbf{v}$ and Ψ_h to $\mathbb{R}$ by letting

$$\begin{aligned} \mathcal{E}_1[\mathbf{v}] &= e_T[\mathbf{v} - T_v(t)\mathbf{u}_0] + \psi(t)T_v(|t|)\mathbf{u}_0, \\ \mathcal{E}_2[\Psi_h] &= e_T[\omega(y)(H_h(y, t) - T_h(t)\rho_0)\mathbf{n}] + \psi(t)T_h(|t|)\rho_0, \end{aligned} \tag{3.230}$$

where we notice that $\mathbf{v}|_{t=0} = T_v(0)\mathbf{u}_0 = \mathbf{u}_0$ and $\Psi_h|_{t=0} = T_h(0)\rho_0$. Obviously, we have

$$\mathcal{E}_1[\mathbf{v}] = \mathbf{v}, \quad \mathcal{E}_2[\Psi_h] = \Psi_h \quad \text{in } \dot{\Omega}^T. \tag{3.231}$$

Let

$$\begin{aligned} \tilde{g}(\mathbf{v}, \Psi_h) &= -(J_0(\nabla\mathcal{E}_2[\Psi_h])\operatorname{div}\mathcal{E}_1[\mathbf{v}] \\ &\qquad + (1 + J_0(\nabla\mathcal{E}_2[\Psi_h]))\mathbf{V}_0(\nabla\mathcal{E}_2[\Psi_h]) : \nabla\mathcal{E}_1[\mathbf{v}]), \\ \tilde{\mathbf{g}}(\mathbf{v}, \Psi_h) &= -(1 + J_0(\nabla\mathcal{E}_2[\Psi_h]))^\top\mathbf{V}_0(\nabla\mathcal{E}_2[\Psi_h])\mathcal{E}_1[\mathbf{v}], \end{aligned} \tag{3.232}$$

and then, considering the transformation: $x = y + \mathcal{E}_2[\Psi_h]$ instead of $x = y + \Psi_h$, by (3.136) and (3.137), (3.138), and (3.231), we have

$$\begin{aligned} \tilde{g}(\mathbf{v}, \Psi_h) = g(v, \Psi_h), \quad \tilde{\mathbf{g}}(\mathbf{v}, \Psi_h) &= \mathbf{g}(\mathbf{v}, \Psi_h) \quad \text{in } \dot{\Omega}^T, \\ \operatorname{div}\tilde{\mathbf{g}}(\mathbf{v}, \Psi_h) &= \tilde{g}(\mathbf{v}, \Psi_h) \quad \text{in } \dot{\Omega} \times \mathbb{R}. \end{aligned} \tag{3.233}$$

By (3.227) and (3.214), we have

$$\sup_{t\in\mathbb{R}} \|\mathcal{E}_2[\Psi_h]\|_{H^1_\infty(\dot\Omega)} \leq C(\sup_{t\in(0,T)} \|\Psi_h\|_{H^2_q(\dot\Omega)} + \|T(\cdot)\rho_0\|_{L_\infty, H^2_q(\dot\Omega)})$$
$$\leq C(\|\rho_0\|_{W^{3-1/p1/q}_q(\Gamma)} + T^{1/p'}L).$$

Thus, we choose T and $\|\rho\|_{W^{3-1/p-1/q}_{q,p}(\dot\Omega)}$ so small that

$$\sup_{t\in\mathbb{R}} \|\mathcal{E}_2[\Psi_h]\|_{H^1_\infty(\dot\Omega)} \leq \delta. \tag{3.234}$$

Since $\mathbf{V}_0(0) = 0$, in view of (3.232), we may simply write

$$\tilde{\mathbf{g}}(\mathbf{v}, \Psi_h) = \mathbf{V}_{\mathbf{g}}(\nabla\mathcal{E}_2[\Psi_h])\nabla\mathcal{E}_2[\Psi_h] \otimes \mathcal{E}_1(\mathbf{v}).$$

Since

$$\partial_t\tilde{\mathbf{g}}(\mathbf{v}, \Psi_h) = \mathbf{V}_{\mathbf{g}}(\nabla\mathcal{E}_2[\Psi_h])\partial_t\nabla\mathcal{E}_2[\Psi_h] \otimes \mathcal{E}_1(\mathbf{v})$$
$$+ \mathbf{V}_{\mathbf{g}}(\nabla\mathcal{E}_2[\Psi_h])\nabla\mathcal{E}_2[\Psi_h] \otimes \partial_t\mathcal{E}_1(\mathbf{v})$$
$$+ \mathbf{V}'_{\mathbf{g}}(\nabla\mathcal{E}_2[\Psi_h])\partial_t\nabla\mathcal{E}_2[\Psi_h]\nabla\mathcal{E}_2[\Psi_h] \otimes \mathcal{E}_1(\mathbf{v})$$

where $\mathbf{V}'_{\mathbf{g}}$ denotes the derivative of $\mathbf{V}_{\mathbf{g}}$ with respect to $\mathbf{k}$, and so by (3.234) we have

$$\begin{aligned}\|\partial_t\tilde{\mathbf{g}}(\mathbf{v}, \Psi_h)\|_{L_q(\dot\Omega)} &\leq C\{\|\nabla\mathcal{E}_2[\Psi_h]\|_{H^1_q(\dot\Omega)}\|\partial_t\mathcal{E}_1[\mathbf{v}]\|_{L_q(\dot\Omega)} \\ &\quad + \|\partial_t\nabla\mathcal{E}_2[\Psi_h]\|_{L_q(\dot\Omega)}\|\mathcal{E}_1[\mathbf{v}]\|_{H^1_q(\dot\Omega)}\}.\end{aligned} \tag{3.235}$$

Using (3.229), (3.201), (3.214), (3.211), (3.206), (3.207), (3.226), and (3.227), for any $\gamma > 0$, we have

$$\begin{aligned}&\|e^{-\gamma t}\partial_t\tilde{\mathbf{g}}(\mathbf{v}, \Psi_h)\|_{L_p(\mathbb{R}, L_q(\dot\Omega))} \\ &\leq C(\|\Psi_h\|_{L_\infty((0,T), H^2_q(\dot\Omega))} + \|T_h(\cdot)\rho_0\|_{L_\infty((0,\infty), H^2_q(\dot\Omega))}) \\ &\qquad \times (\|\partial_t\mathbf{v}\|_{L_p((0,T), L_q(\dot\Omega))} + \|e^{-\gamma t}T_v(|\cdot|)\mathbf{u}_0\|_{H^1_p((-2,\infty), L_q(\dot\Omega))}) \\ &\quad + (T^{1/p}\|\partial_t\Psi_h\|_{L_\infty((0,\infty), H^1_q(\dot\Omega))} + \|\partial_t T_h(|\cdot|)\rho_0\|_{L_p((-2,\infty), H^1_q(\dot\Omega))}) \\ &\qquad \times (\|\mathbf{v}\|_{L_\infty((0,T), H^1_q(\dot\Omega))} + \|T_v(\cdot)\mathbf{u}_0\|_{L_\infty((0,\infty), H^1_q(\dot\Omega))}) \\ &\leq C(\epsilon + L(T^{1/p'} + T^{1/p}))(L + e^{2\gamma}B).\end{aligned} \tag{3.236}$$

We next prove that

$$
\begin{aligned}
&\|e^{-\gamma t}\tilde g(\mathbf{v}, \Psi_h)\|_{H^{1/2}_p(\mathbb{R}, L_q(\dot\Omega))} + \|e^{-\gamma t}\tilde g(\mathbf{v}, \Psi_h)\|_{L_p(\mathbb{R}, H^1_q(\dot\Omega))} \\
&\quad\leq C(\epsilon + L(T^{1/p'} + T^{(q-N)/(pq)}))(L + e^{2\gamma}B)
\end{aligned}
\tag{3.237}
$$

for any $\gamma \geq 0$. To prove (3.237), we use the following three lemmata.

Lemma 3.4.1 *Let $1 < p < \infty$, $N < q < \infty$ and $0 < T \leq 1$. Let $f \in H^1_p(\mathbb{R}, H^1_q(\dot\Omega))$ and $g \in H^{1/q}_p(\mathbb{R}, L_q(\dot\Omega)) \cap L_p(\mathbb{R}, H^1_q(\dot\Omega))$. If f vanishes for $t \notin (0, 2T)$, then we have*

$$
\begin{aligned}
&\|fg\|_{H^{1/2}_p(\mathbb{R}, L_q(\dot\Omega))} + \|fg\|_{L_p(\mathbb{R}, H^1_q(\dot\Omega))} \\
&\leq C\{\|f\|_{L_\infty(\mathbb{R}, H^1_q(\dot\Omega))} + T^{(q-N)/(pq)}\|\partial_t f\|^{1-N/(2q)}_{L_\infty(\mathbb{R}, L_q(\dot\Omega))}\|\partial_t f\|^{N/(2q)}_{L_p(\mathbb{R}, H^1_q(\dot\Omega))}\} \\
&\quad\times(\|g\|_{H^{1/2}_p(\mathbb{R}, L_q(\dot\Omega))} + \|g\|_{L_p(\mathbb{R}, H^1_q(\dot\Omega))}).
\end{aligned}
$$

Proof For a proof, see Shibata-Shimizu [32, Lemma 2.6]. □

Remark 9 We replace $\|f\|_{L_\infty(\mathbb{R}, H^1_q(\dot\Omega))}$ by $T^{1/p'}\|\partial_t f\|_{L_p(\mathbb{R}, H^1_q(\dot\Omega))}$ in Lemma 3.4.1. In fact, since $f|_{t=0} = 0$, representing $f(t) = \int_0^t \partial_s f(\cdot, s)\,ds$ and applying Hölder's inequality, we have

$$
\|f\|_{L_\infty(\mathbb{R}, H^1_q(\dot\Omega))} \leq T^{1/p'}\|\partial_t f\|_{L_p(\mathbb{R}, H^1_q(\dot\Omega))}.
$$

Lemma 3.4.2 *Let $1 < p < \infty$ and $N < q < \infty$. Let*

$$
\begin{aligned}
&f \in L_\infty(\mathbb{R}, H^1_q(\dot\Omega)) \cap H^1_\infty(\mathbb{R}, L_q(\dot\Omega)), \\
&g \in H^{1/2}_p(\mathbb{R}, L_q(\dot\Omega)) \cap L_p(\mathbb{R}, H^1_q(\dot\Omega)).
\end{aligned}
$$

Then, we have

$$
\begin{aligned}
&\|fg\|_{H^{1/2}_p(\mathbb{R}, L_q(\dot\Omega))} + \|fg\|_{L_p(\mathbb{R}, H^1_q(\dot\Omega))} \\
&\leq C(\|f\|_{L_\infty(\mathbb{R}, H^1_q(\dot\Omega))} + \|f\|_{H^1_\infty(\mathbb{R}, L_q(\dot\Omega))}) \\
&\quad\times(\|g\|_{H^{1/2}_p(\mathbb{R}, L_q(\dot\Omega))} + \|g\|_{L_p(\mathbb{R}, H^1_q(\dot\Omega))}).
\end{aligned}
\tag{3.238}
$$

Proof To prove the lemma, we use a complex interpolation:

$$
\begin{aligned}
&H^{1/2}_p(\mathbb{R}, L_q(\dot\Omega)) \cap L_p(\mathbb{R}, H^{1/2}_q(\dot\Omega)) \\
&\quad= (L_p(\mathbb{R}, L_q(\dot\Omega)), H^1_p(\mathbb{R}, L_q(\dot\Omega)) \cap L_p(\mathbb{R}, H^1_q(\dot\Omega)))_{[1/2]},
\end{aligned}
$$

where $(\cdot, \cdot)_{[1/2]}$ denotes a complex interpolation functor. Thus, we observe that

$$\begin{aligned}
&\|fg\|_{H^1_p(\mathbb{R},L_q(\dot\Omega))} + \|fg\|_{L_p(\mathbb{R},H^1_q(\dot\Omega))}\\
&\quad\le C(\|f\|_{H^1_\infty(\mathbb{R},L_q(\dot\Omega))}\|g\|_{L_p(\mathbb{R},H^1_q(\dot\Omega))} + \|f\|_{L_\infty(\mathbb{R},H^1_q(\dot\Omega))}\|g\|_{H^1_p(\mathbb{R},L_q(\dot\Omega))})\\
&\quad+ C\|f\|_{L_\infty(\mathbb{R},H^1_q(\dot\Omega))}\|g\|_{L_p(\mathbb{R},H^1_q(\dot\Omega))}\\
&\quad\le (\|f\|_{H^1_\infty(\mathbb{R},L_q(\dot\Omega))} + \|f\|_{L_\infty(\mathbb{R},H^1_q(\dot\Omega))})(\|g\|_{H^1_p(\mathbb{R},L_q(\dot\Omega))} + \|g\|_{L_p(\mathbb{R},H^1_q(\dot\Omega))}),
\end{aligned}$$

where we have used Sobolev's inequality: $\|v\|_{L_\infty(\dot\Omega)} \le C\|v\|_{H^1_q(\dot\Omega)}$. We also use Sobolev's inequality to obtain

$$\begin{aligned}
\|fg\|_{L_p(\mathbb{R},L_q(\dot\Omega))} &\le \|f\|_{L_\infty(\mathbb{R},H^1_q(\dot\Omega))}\|g\|_{L_p(\mathbb{R},L_q(\dot\Omega))}\\
&\le (\|f\|_{H^1_\infty(\mathbb{R},L_q(\dot\Omega))} + \|f\|_{L_\infty(\mathbb{R},H^1_q(\dot\Omega))})\|g\|_{L_p(\mathbb{R},L_q(\dot\Omega))}.
\end{aligned}$$

Interpolating these two inequalities yields

$$\begin{aligned}
\|fg\|_{H^{1/2}_p(\mathbb{R},L_q(\dot\Omega))} \le&(\|f\|_{H^1_\infty(\mathbb{R},L_q(\dot\Omega))} + \|f\|_{L_\infty(\mathbb{R},H^1_q(\dot\Omega))})\\
&\times (\|g\|_{H^{1/2}_p(\mathbb{R},L_q(\dot\Omega))} + \|g\|_{L_p(\mathbb{R},H^{1/2}_q(\dot\Omega))}).
\end{aligned}$$

Since $\|g\|_{L_p(\mathbb{R},H^{1/2}_q(\dot\Omega))} \le \|g\|_{L_p(\mathbb{R},H^1_q(\dot\Omega))}$, we have (3.238), which completes the proof of Lemma 3.4.2. □

Lemma 3.4.3 *Let* $1 < p, q < \infty$. *Then,*

$$H^1_p(\mathbb{R}, L_q(\dot\Omega)) \cap L_p(\mathbb{R}, H^2_q(\dot\Omega)) \subset H^{1/2}_p(\mathbb{R}, H^1_q(\dot\Omega))$$

and

$$\|u\|_{H^{1/2}_p(\mathbb{R},H^1_q(\dot\Omega))} \le C\{\|u\|_{L_p(\mathbb{R},H^2_q(\dot\Omega))} + \|\partial_t u\|_{L_p(\mathbb{R},L_q(\dot\Omega))}\}.$$

Since $J_0(0) = 0$ and $\mathbf{V}_0(0) = 0$, by (3.234) we may write

$$\tilde g(\mathbf{v}, \Psi_h) = V_g(\nabla\mathcal{E}_2[\Psi_h])\nabla\mathcal{E}_2[\Psi_h]) \otimes \nabla\mathcal{E}_1[\mathbf{v}] \tag{3.239}$$

symbolically with some matrix of C^1 functions $V_g(\mathbf{k})$ defined on $|\mathbf{k}| \le \delta$. By (3.201), (3.26), (3.234), (3.230), (3.227), (3.214), and (3.207), we have

$$\begin{aligned}
&\|V_g(\nabla\mathcal{E}_2[\Psi_h])\nabla\mathcal{E}_2[\Psi_h]\|_{L_\infty(\mathbb{R},H^1_q(\dot\Omega))} \le C\|\mathcal{E}_2[\Psi_h]\|_{L_\infty(\mathbb{R},H^2_q(\dot\Omega))}\\
&\quad\le C(\|T_h(\cdot)\rho_0\|_{L_\infty(\mathbb{R},H^2_q(\dot\Omega))} + \|\Psi_h\|_{L_\infty(\mathbb{R},H^2_q(\dot\Omega))}) \le C(\epsilon + T^{1/p'}L);\\
&\|\partial_t(V_g(\nabla\mathcal{E}_2[\Psi_h])\nabla\mathcal{E}_2[\Psi_h])\|_{L_\infty(\mathbb{R},H^1_q(\dot\Omega))} \le C\|\partial_t\nabla\mathcal{E}_2[\Psi_h]\|_{L_\infty(\mathbb{R},L_\infty))}
\end{aligned}$$

$$
\begin{aligned}
&\leq C(\epsilon + L) \leq 2CL;\\
&\|\partial_t(V_g(\nabla\mathcal{E}_2[\Psi_h])\nabla\mathcal{E}_2[\Psi_h])\|_{L_p(\mathbb{R},H^1_q(\dot\Omega))} \leq C\|\partial_t\nabla\mathcal{E}_2[\Psi_h]\|_{L_p((\mathbb{R},H^1_q(\dot\Omega))}\\
&\leq C(\epsilon + L) \leq 2CL.
\end{aligned}
$$

Thus, by Lemma 3.4.1 and Lemma 3.4.3 we have

$$
\begin{aligned}
&\|e^{-\gamma t}\tilde g(\mathbf{v}, \Psi_h)\|_{H^{1/2}_p(\mathbb{R},L_q(\dot\Omega))} + \|e^{-\gamma t}\tilde g(\mathbf{v}, \Psi_h)\|_{L_p(\mathbb{R},H^1_q(\dot\Omega))}\\
&\leq C(\epsilon + T^{1/p'}L + T^{(q-N)/(pq)}L)\\
&\qquad\qquad \times (\|e^{-\gamma t}\mathcal{E}_1[\mathbf{v}]\|_{H^1_p(\mathbb{R},L_q(\dot\Omega))} + \|e^{-\gamma t}\mathcal{E}_1[\mathbf{v}]\|_{L_p(\mathbb{R},H^2_q(\dot\Omega))})
\end{aligned}
$$

for any $\gamma \geq 0$, which, combined with (3.206), (3.207), and (3.226), leads to (3.237).

We finally estimate $\mathbf{h}'(\mathbf{v}, \Psi_h)$ and $h_N(\mathbf{v}, h)$ given in (3.183) and (3.196). In view of (3.183), we may symbolically write $\mathbf{h}'(\mathbf{v}, \Psi_h)$ as

$$
\mathbf{h}'(\mathbf{v}, \Psi_h) = \mathbf{V}'_{\mathbf{h}}(\bar\nabla\Psi_h))\bar\nabla\Psi_h \otimes \nabla\mathbf{v} \tag{3.240}
$$

with some matrix of C^1 functions $\mathbf{V}'_{\mathbf{h}}(\bar{\mathbf{k}}) = \mathbf{V}'_{\mathbf{h}}(y, \bar{\mathbf{k}})$ defined on $\dot\Omega \times \{\bar{\mathbf{k}} \mid |\bar{\mathbf{k}}| \leq \delta\}$ possessing the estimate:

$$
\sup_{|\bar{\mathbf{k}}|\leq\delta} \|(\mathbf{V}'_{\mathbf{h}}(\cdot, \bar{\mathbf{k}}), \partial_{\bar{\mathbf{k}}}\mathbf{V}'_{\mathbf{h}}(\cdot, \bar{\mathbf{k}}))\|_{H^1_\infty(\dot\Omega)} \leq C
$$

with some constant C. We extend $\mathbf{h}'(\mathbf{v}, \Psi_h)$ to the whole time interval $\mathbb{R}$ by letting

$$
\tilde{\mathbf{h}}'(\mathbf{v}, \Psi_h) = \mathbf{V}'_{\mathbf{h}}(\bar\nabla\mathcal{E}_2[\Psi_h])\nabla\mathcal{E}_2[\Psi_h] \otimes \nabla\mathcal{E}_1[\mathbf{v}]. \tag{3.241}
$$

Employing the same argument as in the proof of (3.237), for any $\gamma > 0$ we have

$$
\begin{aligned}
&\|e^{-\gamma t}\tilde{\mathbf{h}}'(\mathbf{v}, \Psi_h)\|_{H^{1/2}_p(\mathbb{R},L_q(\dot\Omega))} + \|e^{-\gamma t}\tilde{\mathbf{h}}'(\mathbf{v}, \Psi_h)\|_{L_p(\mathbb{R},H^1_q(\dot\Omega))}\\
&\leq C(L(T^{1/p'} + T^{(q-N)/(pq)}) + \epsilon)(L + e^{2\gamma}B).
\end{aligned} \tag{3.242}
$$

In (3.196) we may write

$$
\begin{aligned}
&- < \mathbf{n}, \mu\mathbf{D}(\mathbf{v})\mathbf{V}_\Gamma(\bar{\mathbf{k}})\bar{\mathbf{k}} > - < \mathbf{n}, \mu(\mathcal{D}_{\mathbf{D}}(\mathbf{k})\nabla\mathbf{v})(\mathbf{n} + \mathbf{V}_\Gamma(\bar{\mathbf{k}})\bar{\mathbf{k}}) >\\
&= \mathbf{V}_{\mathbf{h},N}(\bar{\mathbf{k}})(\nabla\Psi_h, \nabla\mathbf{v})
\end{aligned}
$$

with some function $\mathbf{V}_{\mathbf{h},N}(\bar{\mathbf{k}}) = \mathbf{V}_{\mathbf{h},N}(y, \bar{\mathbf{k}})$ defined on $\dot\Omega \times \{\mathbf{k} \mid |\mathbf{k}| \leq \delta\}$ possessing the estimate:

$$\sup_{|\bar{\mathbf{k}}|\leq\delta} \|(\mathbf{V}_{\mathbf{h},N}(\cdot,\bar{\mathbf{k}}), \partial_{\bar{\mathbf{k}}}\mathbf{V}_{\mathbf{h},N}(\cdot,\bar{\mathbf{k}}))\|_{H^1_\infty(\dot\Omega)} \leq C.$$

Thus, we may write $h_N(\mathbf{v}, \Psi_h)$ as

$$h_N(\mathbf{v}, \Psi_h) = \mathbf{V}_{\mathbf{h},N}(\bar\nabla\Psi_h)\nabla\Psi_h \otimes \nabla\mathbf{v} + \sigma\mathbf{V}'_\Gamma(\bar\nabla\Psi_h)\bar\nabla\Psi_h \otimes \bar\nabla^2\Psi_h, \tag{3.243}$$

and so, we can define the extension of $h_N(\mathbf{v}, \Psi_h)$ by letting

$$\begin{aligned} h_N(\mathbf{v}, \Psi_h) = \mathbf{V}_{\mathbf{h},N}(\bar\nabla\mathcal{E}_2[\Psi_h])\bar\nabla\mathcal{E}_2[\Psi_h] \otimes \nabla\mathcal{E}_1[\mathbf{v}] \\ + \sigma V^1_\Gamma(\bar\nabla\mathcal{E}_2[\Psi_h])\bar\nabla\mathcal{E}_2[\Psi_h] \otimes \bar\nabla^2\mathcal{E}_2[\Psi_h]. \end{aligned}$$

Using Lemma 3.4.1, Lemma 3.4.3, (3.226), and (3.227), we have

$$\begin{aligned} &\|e^{-\gamma t}\tilde h_N(\mathbf{v}, \Psi_h)\|_{L_p(\mathbb{R}, H^1_q(\dot\Omega))} + \|e^{-\gamma t}\tilde h_N(\mathbf{v}, \Psi_h)\|_{H^{1/2}_p(\mathbb{R}, L_q(\dot\Omega))} \\ &\leq C(\epsilon + (T^{1/p'} + T^{(q-N)/(pq)})L)(\|e^{-\gamma t}\mathcal{E}_1[\mathbf{v}]\|_{L_p(\mathbb{R}, H^2_q(\dot\Omega))} \\ &\quad + \|e^{-\gamma t}\mathcal{E}_1[\mathbf{v}]\|_{H^1_p(\mathbb{R}, L_q(\dot\Omega))} + \|e^{-\gamma t}\bar\nabla^2\mathcal{E}_2[\Psi_h]\|_{L_p(\mathbb{R}, H^1_q(\dot\Omega))} \\ &\quad + \|e^{-\gamma t}\bar\nabla^2\mathcal{E}_2[\Psi_h]\|_{H^{1/2}_p(\mathbb{R}, L_q(\dot\Omega))}). \end{aligned}$$

By the fact that $H^1_p(\mathbb{R}, L_q(\dot\Omega)) \subset H^{1/2}_p(\mathbb{R}, L_q(\dot\Omega))$, we have

$$\|e^{-\gamma t}\bar\nabla^2\mathcal{E}_2[\Psi_h]\|_{H^{1/2}_p(\mathbb{R}, L_q(\dot\Omega))} \leq C\|e^{-\gamma t}\bar\nabla^2\mathcal{E}_2[\Psi_h])\|_{H^1_p(\mathbb{R}, L_q(\dot\Omega))}.$$

Therefore, by (3.226), (3.227), (3.206), and (3.207), we have

$$\begin{aligned} &\|\tilde h_N(\mathbf{v}, \Psi_h)\|_{L_p(\mathbb{R}, H^1_q(\dot\Omega))} + \|\tilde h_N(\mathbf{v}, \Psi_h)\|_{H^{1/2}_p(\mathbb{R}, L_q(\dot\Omega))} \\ &\leq C(\epsilon + (T^{1/p'} + T^{(q-N)/(pq)})L)(e^{2\gamma}B + L). \end{aligned} \tag{3.244}$$

Notice that

$$\begin{aligned} \|\mathbf{u}_\kappa\|_{H^1_\infty(\Omega_+)} &\leq C\|\mathbf{u}_\kappa\|_{B^{2(1-1/p)}_{q,p}(\Omega_+)} \leq C\|\mathbf{u}_0\|_{B^{2(1-1/p)}_{q,p}(\dot\Omega)} \leq CB, \\ \|\mathbf{u}_\kappa\|_{H^2_q(\Omega_+)} &\leq \frac{1}{\kappa}\int_0^\kappa \|T(s)\tilde{\mathbf{u}}_0\|_{H^2_q(\dot\Omega)}\,ds \\ &\leq \frac{1}{\kappa}\kappa^{1/p'}\Big(\int_0^\kappa \|T(s)\tilde{\mathbf{u}}_0\|^p_{H^2_q(\dot\Omega)}\,ds\Big)^{1/p} \leq CB\kappa^{-1/p}. \end{aligned}$$

Applying Theorem 3.2.7 and the estimates (3.212), (3.213), (3.208), (3.224), (3.236), (3.242), (3.244), and (3.207), we have

$$
\begin{aligned}
&\sup_{t\in(0,T)} (\|\mathbf{u}(\cdot,t)\|_{B^{2(1-1/p)}_{q,p}(\dot\Omega)} + \|\rho(\cdot,t)\|_{B^{3-1/p-1/q}_{q,p}(\Gamma)}) \\
&\quad + \|\mathbf{u}\|_{L_p((0,T),H^2_q(\dot\Omega))} + \|\partial_t\mathbf{u}\|_{L_p((0,T),L_q(\dot\Omega))} \\
&\quad + \|\rho\|_{L_p((0,T),W^{3-1/q}_q(\Gamma))} + \|\partial_t\rho\|_{L_p((0,T),W^{2-1/q}_q(\Gamma))} \\
&\leq Ce^{\gamma\kappa^{-b}T}\{B + \kappa^{-b}\epsilon + T^{1/p}(L+B)^2 \\
&\quad + (\epsilon + T^{1/p'}L)L + (L + B\kappa^{-1/p}T^{1/p})(\epsilon + T^{1/p'}L) \\
&\quad + L(L+B)(\epsilon + T^{1/p'}L + \kappa^{(1-1/s)/p'} + T^{(1-1/s)/p'}) \\
&\quad + (\epsilon + L(T^{1/p'} + T^{(1-1/s)/p'} + T^{(q-N)/(pq)}))(L + e^{2\gamma}B)\}
\end{aligned}
\tag{3.245}
$$

for some positive constants γ and C independent of B, ϵ and T. By the kinetic equation in equations (3.203), (3.199), (3.217), and (3.245), we have

$$
\begin{aligned}
&\sup_{t\in(0,T)} \|\partial_t\rho(\cdot,t)\|_{W^{1-1/q}_q(\Gamma)} \\
&\quad \leq \|\mathbf{u}_\kappa\|_{H^1_q(\Omega_+)} \sup_{t\in(0,T)} \|\rho(\cdot,t)\|_{W^{2-1/q}_q(\Gamma)} \\
&\qquad + C\sup_{t\in(0,T)} \|\mathbf{u}(\cdot,t)\|_{H^1_q(\dot\Omega)} + \sup_{t\in(0,T)} \|d(\cdot,t)\|_{W^{1-1/q}_q(\Gamma)} \\
&\quad \leq CB\sup_{t\in(0,T)} \|\rho(\cdot,t)\|_{W^{3-1/p-1/q}_q(\Gamma)} + C\sup_{t\in(0,T)} \|\mathbf{u}(\cdot,t)\|_{B^{2(1-1/p)}_{q,p}(\dot\Omega)} \\
&\quad + C\tilde{E}_{p,q,T}(\mathbf{u},\rho),
\end{aligned}
\tag{3.246}
$$

where we have set

$$
\begin{aligned}
\tilde{E}_{p,q,T} &= \|\mathbf{u}\|_{L_p((0,T),H^2_q(\dot\Omega))} + \|\partial_t\mathbf{u}\|_{L_p((0,T),L(\dot\Omega))} \\
&\quad + \|\rho\|_{L_p((0,T),W^{3-1/q}_q(\Gamma))} + \|\partial_t\rho\|_{L_p((0,T),W^{2-1/q}_q(\Gamma))}.
\end{aligned}
$$

In (3.245) and (3.246), we set $\kappa = \epsilon = T$ and choose $T \in (0,1)$ so small that $\gamma BT^{1/p'} \leq 1$, and then by (3.245), (3.246), and (3.212), we have

$$
\begin{aligned}
E_{p,q,T}(\mathbf{u},\rho) &\leq Ce^{\gamma\kappa^{-b}T}(1+B)\{B + \kappa^{-b}\epsilon + T^{1/p}(L+B)^2 \\
&\quad + (\epsilon + T^{1/p'}L)L + (L + B\kappa^{-1/p}T^{1/p})(\epsilon + T^{1/p'}L) \\
&\quad + L(L+B)(\epsilon + T^{1/p'}L + \kappa^{(1-1/s)/p'} + T^{(1-1/s)/p'}) \\
&\quad + (\epsilon + L(T^{1/p'} + T^{1/p} + T^{(q-N)/(pq)}))(L + e^{2\gamma}B)\}.
\end{aligned}
\tag{3.247}
$$

Note that $1 - 2/p > 0$. Choosing κ, ϵ and T in such a way that

$$\begin{aligned}
&\gamma\kappa^{-b}T \leq 1, \quad \kappa^{-b}\epsilon \leq B,\\
&T^{1/p}(L+B)^2 + (\epsilon + T^{1/p'}L)L + (L + B\kappa^{-1/p}T^{1/p})(\epsilon + T^{1/p'}L)\\
&\quad + L(L+B)(\epsilon + T^{1/p'}L + \kappa^{(1-1/s)/p'} + T^{(1-1/s)/p'})\\
&\quad + (\epsilon + L(T^{1/p'} + T^{1/p} + T^{(q-N)/(pq)}))(L + e^{2\gamma}B) \leq B,
\end{aligned}$$

we have

$$E_{p,q,T}(\mathbf{u}, \rho) \leq 3Ce(1+B)B,$$

and so setting $L = 3Ce(1+B)B$, we have

$$E_{p,q,T}(\mathbf{u}, \rho) \leq L. \tag{3.248}$$

Let $\mathcal{M}$ be a map defined by letting $\mathcal{M}(\mathbf{v}, h) = (\mathbf{u}, \rho)$, and then by (3.248) $\mathcal{M}$ maps $\mathbf{U}_T$ into itself. We can also prove that $\mathcal{M}$ is a contraction map. Namely, choosing $\kappa = \epsilon = T$ smaller if necessary, we can show that for any $(\mathbf{v}_i, h_i) \in \mathbf{U}_T$ $(i = 1, 2)$,

$$E_{p,q,T}(\mathcal{M}(\mathbf{v}_1, h_1) - \mathcal{M}(\mathbf{v}_2, h_2)) \leq (1/2)E_{p,q,T}((\mathbf{v}_1, \rho_1) - (\mathbf{v}_2, \rho_2)).$$

Thus, by the contraction mapping principle, we have Theorem 3.1.2, which completes the proof of Theorem 3.1.2.

We finally prove the inequalities (3.212) and (3.213). Let $\mathcal{E}_1[\mathbf{v}]$ be the function given in (3.230). By (3.231) and (3.89), we have

$$\begin{aligned}
&\|\mathbf{v}\|_{L_\infty((0,T), B^{2(1-1/p)}_{q,p}(\dot\Omega))} \leq \|\mathcal{E}_1[\mathbf{v}]\|_{L_\infty((0,T), B^{2(1-1/p)}_{q,p}(\dot\Omega))}\\
&\leq C\{\|\mathcal{E}_1[\mathbf{v}]\|_{L_p((0,\infty), H^2_q(\dot\Omega))} + \|\partial_t\mathcal{E}_1[\mathbf{v}]\|_{L_p((0,\infty), L_q(\dot\Omega))}\},
\end{aligned}$$

which, combined with (3.226), leads to the inequality (3.212). Analogously, using $\mathcal{E}_2[\Psi_h]$ given in (3.230) and (3.89), we have

$$\begin{aligned}
&\|\Psi_h\|_{L_\infty((0,T), B^{3-1/p-1/q}_{q,p}(\Gamma))}\\
&\quad\leq C\{\|\Psi_h\|_{L_p((0,T), H^3_q(\dot\Omega))} + \|\partial_t\Psi_h\|_{L_p((0,T), H^2_q(\dot\Omega))}\},
\end{aligned}$$

which, combined with (3.227) and (3.26), leads to the inequality in (3.213).

3.5 Global Well-Posedness in the Case That Ω Is a Bounded Domain

3.5.1 Formulation of the Problem

In this section, we study the global well-posedness of equation (3.1) in the case where Ω is a bounded domain. We already formulated the problem in Subsec 3.1.3.2. But, we repeat the formulation below. Let B_R be the ball of radius R centered at the origin and let S_R be the sphere of radius R centered at the origin. We assume that

(A.1) $|\Omega_+| = |B_R| = \dfrac{R^N \omega_N}{N}$, where $|D|$ denotes the Lebesgue measure of a Lebesgue measurable set D in $\mathbb{R}^N$ and ω_N is the area of S_1.

(A.2) $\displaystyle\int_{\Omega_+} x\,dx = 0.$

(A.3) Γ is a normal perturbation of S_R given by

$$\Gamma = \{x = y + R^{-1}\rho_0(y)y \mid y \in S_R\}$$

with a given small function $\rho_0(y)$ defined on S_R.

Notice that $R^{-1}y$ is the unit outer normal to S_R. Let Γ_t be given by

$$\begin{aligned}\Gamma_t &= \{x = y + \rho(y,t)\mathbf{n} + \xi(t) \mid y \in S_R\}\\ &= \{x = y + R^{-1}\rho(y,t)y + \xi(t) \mid y \in S_R\}\end{aligned} \tag{3.249}$$

where $\rho(y,t)$ is an unknown function with $\rho(y,0) = \rho_0(y)$ for $y \in S_R$ and $\xi(t)$ is the barycenter point of the domain Ω_{t+} defined by

$$\xi(t) = \frac{1}{|\Omega_{t+}|}\int_{\Omega_{t+}} x\,dx. \tag{3.250}$$

Notice that by (3.15)

$$|\Omega_{t+}| = |\Omega_+| = |B_R|. \tag{3.251}$$

It follows from the assumption (A.2) that $\xi(0) = 0$, while the $\xi(t)$ is also unknown. Moreover, by (3.24) and (3.251) we have

$$\xi'(t) = \frac{1}{|B_R|}\int_{\Omega_{t+}} \mathbf{v}(x,t)\,dx. \tag{3.252}$$

For any function ρ defined on S_R, let H_ρ be a suitable extension of ρ satisfying the condition:

$$\begin{aligned}\|H_\rho(\cdot,t)\|_{H^k_q(\mathbb{R}^N)} &\le C_1\|\rho(\cdot,t)\|_{W^{k-1/q}_q(S_R)},\\ \|\partial_t H_\rho(\cdot,t)\|_{H^\ell_q(\mathbb{R}^N)} &\le C_1\|\partial_t\rho(\cdot,t)\|_{W^{\ell-1/q}_q(S_R)}\end{aligned} \tag{3.253}$$

for $k=1,2,3$ and $\ell=1,2$. Let d_0 be a positive number such that $\mathrm{dist}\,(S_R,\Gamma_-)\ge 3d_0$. Let $\zeta(y)$ be a $C^\infty(\mathbb{R}^N)$ function which equals 1 for $|y|<R+d_0$ and 0 for $|y|>R+2d_0$. We define the Hanzawa transform centered at $\xi(t)$ by (3.27), that is,

$$x=\mathbf{h}_\rho(y,t)=y+\zeta(y)(R^{-1}H_\rho(y,t)y+\xi(t))\quad\text{for } y\in\Omega. \tag{3.254}$$

Notice that for $y\in B_R$, $\mathbf{h}_\rho(y,t)=y+R^{-1}H_\rho(y,t)y+\xi(t)$. In the following, we set $\Psi_\rho=\zeta(y)(R^{-1}\mathbf{n}(y)H_\rho(y,t)y+\xi(t))$ and we assume that

$$\sup_{t\in(0,T)}\|\Psi_\rho(\cdot,t)\|_{H^1_\infty(\Omega)}\le\delta. \tag{3.255}$$

In this case,

$$\begin{aligned}\nabla\Psi_\rho&=\nabla(\zeta(y)R^{-1}\mathbf{n}(y)H_\rho(y,t)y)+\nabla\zeta(y)\xi(t),\\ \nabla\tilde\Psi_\rho&=\nabla(\zeta(y)R^{-1}\mathbf{n}(y)H_\rho(y,t)y),\\ \partial_t\Psi_\rho&=\nabla(\zeta(y)R^{-1}\mathbf{n}\partial_tH_\rho(y,t)y+\xi'(t)).\end{aligned} \tag{3.256}$$

We will choose δ so small that several conditions hold. For example, we choose a $\delta>0$ so small that the Hanzawa transform is a bijective map from Ω onto Ω with

$$\Gamma_t=\{x=\mathbf{h}_\rho(y,t)\mid y\in\Gamma\},\quad \Omega_{t\pm}=\{x=\mathbf{h}_\rho(y,t)\mid y\in\Omega_\pm\} \tag{3.257}$$

for $t\in(0,T)$, where $\Omega_+=B_R$ and $\Omega_-=\Omega\setminus\{B_R\cup S_R\}$. Let $\mathbf{v}_0(x)$ be an initial data for equation (3.1), and then we set $\mathbf{u}_0(y)=\mathbf{v}_0(\mathbf{h}_{\rho_0}(y))$, where $\mathbf{h}_{\rho_0}(y)=y+\Psi_{\rho_0}=y+\zeta(y)R^{-1}H_{\rho_0}(y)y$. Notice that $\mathbf{h}_{\rho_0}(y)=\mathbf{h}_\rho(y,0)$ if $\rho|_{t=0}=\rho_0$. Let $\mathbf{v}$ and $\mathfrak{p}$ satisfy equation (3.1) and we set

$$\mathbf{u}(y,t)=\mathbf{v}(\mathbf{h}_\rho(y,t),t),\quad \mathfrak{q}(y,t)=\mathfrak{p}(\mathbf{h}_\rho(y,t),t)-\frac{\sigma(N-1)}{R}. \tag{3.258}$$

It then follows from Sect. 3.4.1 and Sect. 3.4.2 that $\mathbf{u}$, $\mathfrak{q}$, and ρ satisfy equations (3.34) in Sect. 3.1.3.2.

Since $dx=dy+J_0(\mathbf{k})dy$ with

$$J_0(\mathbf{k})=\det\begin{pmatrix}\frac{\partial\Psi_{\rho 1}(y,t)}{\partial y_1}&\cdots&\frac{\partial\Psi_{\rho 1}(y,t)}{\partial y_N}\\ \vdots&\ddots&\vdots\\ \frac{\partial\Psi_{\rho N}(y,t)}{\partial y_1}&\cdots&\frac{\partial\Psi_{\rho N}(y,t)}{\partial y_N}\end{pmatrix}$$

for $\Psi_\rho = {}^\top(\Psi_{\rho 1}, \ldots, \Psi_{\rho N})$, by (3.252) we have $\xi'(t) = \xi'_{\mathbf{u},\rho}(t)$, where $\xi'_{\mathbf{u},\rho}(t)$ is defined by

$$\xi'_{\mathbf{u},\rho}(t) := \frac{1}{|B_R|}\int_{B_R} \mathbf{u}(y,t)\,dy + \frac{1}{|B_R|}\int_{B_R} \mathbf{u}(y,t)J_0(\mathbf{k})\,dy. \tag{3.259}$$

In particular,

$$\frac{\partial \Phi_{\rho i}}{\partial y_j} = R^{-1}\frac{\partial}{\partial y_j}(H_\rho(y,t)y_i) = R^{-1}(\delta_{ij}H_\rho(y,t) + \frac{\partial H_\rho(y,t)}{\partial y_j}y_i)$$

for $y \in B_R$, because $\zeta(y) = 1$ on B_R. Thus, we may write

$$J_0(\mathbf{k}) = \mathfrak{J}(H_\rho, \nabla H_\rho)(H_\rho, \nabla H_\rho) \quad \text{on } B_R$$

with some matrix $\mathfrak{J}$ of smooth functions such that

$$\|(\mathfrak{J}(\mathbf{i}), \partial_{\mathbf{i}}\mathfrak{J}(\mathbf{i}))\|_{L_\infty(L_1)} \leq C$$

for some constant $C > 0$, where $\mathbf{i}$ is a variable corresponding to $(H_\rho, \nabla H_\rho)$ and $L_1 = \{\mathbf{i} \in \mathbb{R}^{N+1} \mid |\mathbf{i}| \leq 1\}$. By Sobolev's inequality,

$$|J_0(\mathbf{k})| \leq C\|H_\rho(\cdot,t)\|_{H^2_q(B_R)}, \quad |\xi'_{\mathbf{u},\rho}(t)| \leq C_R\|\mathbf{u}(\cdot,t)\|_{L_q(B_R)} \tag{3.260}$$

provided that

$$\|H_\rho(\cdot,t)\|_{H^1_\infty(\Omega)} \leq 1. \tag{3.261}$$

Since $\xi(0) = 0$, we define $\xi_{\mathbf{u},\rho}(t)$ by setting

$$\begin{aligned}\xi_{\mathbf{u},\rho}(t) &= \int_0^t \xi'_{\mathbf{u},\rho}(\tau)\,d\tau \\ &= \frac{1}{|B_R|}\int_0^t\int_{B_R} \mathbf{u}(y,\tau)\,dyd\tau + \frac{1}{|B_R|}\int_0^t\int_{B_R} \mathbf{u}(y,\tau)J_0(\mathbf{k})\,dyd\tau.\end{aligned}$$

In particular, by Hölder's inequality and (3.260) we have

$$\begin{aligned}\sup_{0\leq t<T} |\xi_{\mathbf{u},\rho}(t)| &\leq C_R\int_0^T \|\mathbf{u}(\cdot,\tau)\|_{L_q(B_R)}\,d\tau \\ &\leq C_R\Big(\int_0^T e^{-\eta\tau p'}\,d\tau\Big)^{1/p'}\|e^{\eta t}\mathbf{u}\|_{L_p((0,T),L_q(B_R))} \\ &\leq C_R(\eta p')^{1/p'}\|e^{\eta t}\mathbf{u}\|_{L_p((0,T),L_q(B_R))}\end{aligned} \tag{3.262}$$

provided that (3.261) holds. By (3.260) and (3.261),

$$\|\Psi_\rho(\cdot, t)\|_{H^1_\infty(\dot\Omega)} \leq M_0(\|H_\rho(\cdot, t)\|_{H^1_\infty(\dot\Omega)} + |\xi_{\mathbf{u},\rho}(t)|) \tag{3.263}$$

provided that $\|H_\rho(\cdot, t)\|_{H^1_\infty(\dot\Omega)} \leq 1$, where M_0 is a constant depending on Ω, R, p and η, but independent of T. Let δ_1 be a positive number for which $2M_0\delta_1 \leq \delta$, and then by (3.263) we have $\|\Psi_\rho(\cdot, t)\|_{H^1_\infty(\dot\Omega)} \leq \delta$. From this point of view, instead of (3.255), we assume that

$$\|H_\rho(\cdot, t)\|_{H^1_\infty(\dot\Omega)} \leq \delta_1, \quad |\xi_{\mathbf{u},\rho}(t)| \leq \delta_1. \tag{3.264}$$

By (3.259), (3.176), and (3.177), we have

$$\partial_t \rho - \mathbf{n} \cdot P\mathbf{u} = \tilde d(\mathbf{u}, \Psi_\rho),$$

where $\tilde d(\mathbf{u}, \Psi_\rho)$ is given by letting

$$\tilde d(\mathbf{u}, \Psi_\rho) = d(\mathbf{u}, \Psi_\rho) - < \mathbf{u} \mid \nabla'_\Gamma \rho > + \frac{1}{|B_R|}\int_{B_R} \mathbf{u}(y, t) J_0(\mathbf{k})\, dy, \tag{3.265}$$

where $d(\mathbf{u}, \Psi_\rho)$ is a nonlinear function given in (3.177) with $\alpha = 1$ and $\tilde\Psi_\rho = \zeta(y)R^{-1}H_\rho(y, t)y$. In this section, $\mathbf{n} = y/|y|$ for $y \in \mathbb{R}^N \setminus \{0\}$.

Since $-(N-1)/R^2$ is the first eigenvalue of the Laplace-Beltrami operator Δ_{S_R} with eigenfunctions $x_1, \ldots, x_N$, we modify the kinematic equation in equations (3.34). From (3.250) and (3.249) it follows that

$$\begin{aligned} 0 &= \int_{\Omega_{t+}} (x - \xi(t))\, dx = \int_{S_R} \Big(\int_0^{1+R^{-1}\rho} r^N \omega\, dr\Big)\, d\omega \\ &= \frac{1}{N+1}\int_{S_R} (1 + R^{-1}\rho)^{N+1} \omega\, d\omega \\ &= R^{-1}\int_{|\omega|=R} \rho\omega\, d\omega + \sum_{\ell=2}^{N+1} \frac{{}_{N+1}C_\ell}{N+1}\int_{S_R} (R^{-1}\rho)^\ell \omega\, d\omega, \end{aligned} \tag{3.266}$$

Moreover, from (3.251) and (3.249) it follows that

$$\begin{aligned} 0 &= \int_{\Omega_{t+}} dx - |B_R| = \int_{S_R} \Big(\int_0^{1+R^{-1}\rho} r^{N-1}\, dr\Big)\, d\omega - |B_R| \\ &= R^{-1}\int_{|\omega|=R} \rho\, d\omega + \sum_{\ell=2}^{N} \frac{{}_NC_\ell}{N}\int_{S_R} (R^{-1}\rho)^\ell\, d\omega. \end{aligned} \tag{3.267}$$

Inserting these formulas into the kinematic equation yields that

$$\partial_t \rho + \int_{S_R} \rho\, d\omega + < \int_{S_R} \rho\omega\, d\omega, \mathbf{n} > - \mathbf{n}\cdot P\mathbf{u} = \bar{d}(\mathbf{u}, \Psi_\rho) \quad \text{on } S_R^T, \tag{3.268}$$

with

$$\begin{aligned}\bar{d}(\mathbf{u}, \Psi_\rho) = \tilde{d}(\mathbf{u}, \Psi_\rho) &- \sum_{\ell=2}^{N} \frac{{}_N C_\ell}{N} R^{-\ell+1} \int_{S_R} \rho^\ell\, d\omega \\ &- \sum_{\ell=2}^{N+1} \frac{{}_{N+1}C_\ell}{N+1} R^{-\ell+1} < \int_{S_R} \rho^\ell \omega\, d\omega, \mathbf{n} > .\end{aligned} \tag{3.269}$$

Notice that $\mathbf{n} = R^{-1}\omega$ with $\omega \in S_R$. Under the conditions (3.250) and (3.251), the following two equations are equivalent on S_R^T:

- $\rho_t - \mathbf{n}\cdot P\mathbf{u} = \tilde{d}(\mathbf{u}, \Psi_\rho)$;
- $\partial_t \rho + \int_{|\omega|=R} \rho\, d\omega + < \int_{|\omega|=R} \rho\omega\, d\omega, x > - \mathbf{n}\cdot P\mathbf{u} = \bar{d}(\mathbf{u}, \Psi_\rho)$

In particular, we require that the initial data ρ_0 satisfies the conditions:

$$\int_{|\omega|=R} (1 + R^{-1}\rho_0)^N\, d\omega = |S_R|, \quad \int_{S_R} (1 + R^{-1}\rho_0)^{N+1}\omega\, d\omega = 0, \tag{3.270}$$

which follows from (A.1) and (A.2).

In what follows, we consider the equations:

$$\begin{cases} \partial_t \mathbf{u} - m^{-1}\mathrm{Div}\,(\mu\mathbf{D}(\mathbf{u}) - \mathfrak{q}\mathbf{I}) = \mathbf{f}(\mathbf{u}, \Psi_\rho) & \text{in } \dot{\Omega}^T, \\ \mathrm{div}\,\mathbf{u} = g(\mathbf{u}, H_\rho) = \mathrm{div}\,\mathbf{g}(\mathbf{u}, \Psi_\rho) & \text{in } \dot{\Omega}^T, \\ \partial_t \rho + \int_{S_R} \rho\, d\omega + < \int_{S_R} \rho\omega\, d\omega, \mathbf{n} > - \mathbf{n}\cdot P\mathbf{u} = \bar{d}(\mathbf{u}, \Psi_\rho) & \text{on } S_R^T, \\ [[(\mu\mathbf{D}(\mathbf{u})\mathbf{n})_\tau]] = [[\mathbf{h}'(\mathbf{u}, \Psi_\rho)]] & \text{on } S_R^T, \\ [[< \mu\mathbf{D}(\mathbf{u})\mathbf{n}, \mathbf{n} > - \mathfrak{q}]] - \sigma\mathcal{B}\rho = [[h_N(\mathbf{u}, \Psi_\rho)]] & \text{on } S_R^T, \\ [[\mathbf{u}]] = 0 & \text{on } S_R^T, \\ \mathbf{u} = 0 & \text{on } \Gamma_-^T, \\ (\mathbf{u}, \rho)|_{t=0} = (\mathbf{u}_0, \rho_0) & \text{in } \dot{\Omega} \times S_R. \end{cases} \tag{3.271}$$

Our main result of this section is Theorem 3.1.3 stated in Sect. 3.1.3.2 and we shall prove it in the following.

3.5.2 Derivation of Nonlinear Term $h_N(\mathbf{u}, \Psi_\rho)$ in Equation (3.271)

In this section, we derive the nonlinear term $h_N(\mathbf{u}, \Psi_\rho)$ in (3.271). Let $\mathbf{n} \in S_1$ be represented by $\mathbf{n} = \mathbf{n}(p_1, \ldots, p_{N-1})$ with a local coordinate $(p_1, \ldots, p_{N-1})$, and then for $x = (R+\rho)\mathbf{n} + \xi(t) \in \Gamma_t$, we have

$$\frac{\partial x}{\partial p_j} = (R+\rho)\tau_j + \frac{\partial \rho}{\partial p_j}\mathbf{n}$$

where $\tau_j = \frac{\partial \mathbf{n}}{\partial p_j}$. Since $\tau_j \cdot \mathbf{n} = 0$, the $(i,j)^{\text{th}}$ component of the first fundamental form G_t of Γ_t is given by

$$g_{tij} = \frac{\partial x}{\partial p_i} \cdot \frac{\partial x}{\partial p_j} = (R+\rho)^2 g_{ij} + \frac{\partial \rho}{\partial p_i}\frac{\partial \rho}{\partial p_j},$$

where $g_{ij} = \tau_i \cdot \tau_j$ are the $(i,j)^{\text{th}}$ elements of the first fundamental forms of S_1, and so

$$\begin{aligned} G_t &= (R+\rho)^2(G + (R+\rho)^{-2}\nabla_p\rho \otimes \nabla_p\rho) \\ &= (R+\rho)^2 G(\mathbf{I} + (R+\rho)^{-2}(G^{-1}\nabla_p\rho) \otimes \nabla_p\rho), \end{aligned}$$

where $\nabla_p\rho = {}^\top(\partial\rho/\partial p_1, \ldots, \partial\rho/\partial p_{N-1})$. Since

$$\det(\mathbf{I} + \mathbf{a}' \otimes \mathbf{b}') = 1 + \mathbf{a}' \cdot \mathbf{b}', \quad (\mathbf{I} + \mathbf{a}' \otimes \mathbf{b}')^{-1} = \mathbf{I} - \frac{\mathbf{a}' \otimes \mathbf{b}'}{1 + \mathbf{a}' \cdot \mathbf{b}'} \tag{3.272}$$

for any $N-1$ vectors $\mathbf{a}'$ and $\mathbf{b}' \in \mathbb{R}^{N-1}$, we have

$$\begin{aligned} G_t^{-1} &= (R+\rho)^{-2}\Big(\mathbf{I} - \frac{(R+\rho)^{-2}(G^{-1}\nabla_p\rho) \otimes \nabla_p\rho}{1 + (R+\rho)^{-2} < G^{-1}\nabla_p\rho, \nabla_p\rho >}\Big)G^{-1} \\ &= (R+\rho)^{-2}G^{-1} + O_2, \end{aligned}$$

where O_2 denotes the same symbol as in (3.187). Namely,

$$g_t^{ij} = (R+\rho)^{-2}g^{ij} + O_2,$$

componentwise.

We next calculate the Christoffel symbols of Γ_t. Since

$$\tau_{ti} = (R+\rho)\tau_i + \frac{\partial \rho}{\partial p_i}\mathbf{n},$$

$$\tau_{tij} = (R+\rho)\tau_{ij} + \frac{\partial\rho}{\partial p_j}\tau_i + \frac{\partial\rho}{\partial p_i}\tau_j + \frac{\partial^2\rho}{\partial p_i\partial p_j}\mathbf{n},$$

we have

$$< \tau_{tij}, \tau_{t\ell} > = (R+\rho)^2 < \tau_{ij}, \tau_\ell > +(R+\rho)(\frac{\partial\rho}{\partial p_\ell}\ell_{ij} + g_{i\ell}\frac{\partial\rho}{\partial p_j} + g_{j\ell}\frac{\partial\rho}{\partial p_i})$$
$$+ \frac{\partial^2\rho}{\partial p_i\partial p_j}\frac{\partial\rho}{\partial p_\ell},$$

and so

$$\begin{aligned}
\Lambda^k_{tij} &= g_t^{k\ell} < \tau_{tij}, \tau_{t\ell} > \\
&=< (R+\rho)^{-2}g^{k\ell} + O_2, (R+\rho)^2 < \tau_{ij}, \tau_\ell > \\
&+ (R+\rho)(\frac{\partial\rho}{\partial p_\ell}\ell_{ij} + g_{i\ell}\frac{\partial\rho}{\partial p_j} + g_{j\ell}\frac{\partial\rho}{\partial p_i}) + \frac{\partial^2\rho}{\partial p_i\partial p_j}\frac{\partial\rho}{\partial p_\ell} > \\
&= \Lambda^k_{ij} + (R+\rho)^{-1}(g^{k\ell}\frac{\partial\rho}{\partial p_\ell}\ell_{ij} + \delta^k_i\frac{\partial\rho}{\partial p_j} + \delta^k_j\frac{\partial\rho}{\partial p_i}) \\
&+ ((R+\rho)^{-2}g^{k\ell}\frac{\partial\rho}{\partial p_\ell} + O_2)\frac{\partial^2\rho}{\partial p_i\partial p_j} + O_2.
\end{aligned}$$

Thus,

$$\begin{aligned}
\Delta_{\Gamma_t} f &= g_t^{ij}(\partial_i\partial_j f - \Lambda^k_{tij}\partial_k f) \\
&= (R+\rho)^{-2}g^{ij}(\partial_i\partial_j f - \Lambda^k_{ij}\partial_k f) + (A^k\nabla_p^2\rho + O_2)\partial_k f
\end{aligned}$$

with

$$A^k\nabla_p^2\rho = ((R+\rho)^{-4}g^{k\ell}g^{ij}\frac{\partial\rho}{\partial p_\ell} + O_2)\frac{\partial^2\rho}{\partial p_i\partial p_j},$$

and so

$$\begin{aligned}
&H(\Gamma_t)\mathbf{n}_t = \Delta_{\Gamma_t}[(R+\rho)\mathbf{n} + \xi(t)] \\
&= (R+\rho)^{-2}g^{ij}(\partial_i\partial_j - \Lambda^k_{ij}\partial_k)((R+\rho)\mathbf{n}) + (A^k\nabla_p^2\rho + O_2)\partial_k((R+\rho)\mathbf{n}) \\
&= (R+\rho)^{-1}g^{ij}(\partial_i\partial_j\mathbf{n} - \Lambda^k_{ij}\partial_k\mathbf{n}) + (R+\rho)^{-2}g^{ij}(\partial_i\rho\partial_j\mathbf{n} + \partial_j\rho\partial_i\mathbf{n}) \\
&\quad + (R+\rho)^{-2}g^{ij}(\partial_i\partial_j\rho - \Lambda^k_{ij}\partial_k\rho)\mathbf{n} + ((A^k\nabla_p^2\rho + O_2)\partial_k\rho)\mathbf{n} \\
&\quad + (A^k\nabla_p^2\rho + O_2)(R+\rho)\partial_k\mathbf{n}.
\end{aligned}$$

Combining this formula with (3.166) and using $< \partial_i \mathbf{n}, \mathbf{n} >= 0$ gives

$$
\begin{aligned}
&< H(\Gamma_t)\mathbf{n}_t\, \mathbf{n}_t >=< H(\Gamma_t)\mathbf{n}_t, \mathbf{n} - g^{k\ell}(\partial_\ell \rho)\tau_k + O_2 > \\
&= (R+\rho)^{-1} < \Delta_{S_1}\mathbf{n}, \mathbf{n} > +(R+\rho)^{-2}\Delta_{S_1}\rho + (A^k \nabla_p^2 \rho + O_2)\partial_k \rho \\
&+ (R+\rho)^{-1} g^{k\ell}\partial_\ell \rho < \Delta_{S_1}\mathbf{n}, \tau_\ell > +(A^m \nabla_p^2 \rho)\partial_m \rho + O_2 \nabla_p^2 \rho + O_2.
\end{aligned}
$$

Since $\Delta_{S_1}\mathbf{n} = -(N-1)\mathbf{n}$, we have

$$
\begin{aligned}
< H(\Gamma_t)\mathbf{n}_t, \mathbf{n}_t > &= -(R+\rho)^{-1}(N-1) + (R+\rho)^{-2}\Delta_{S_1}\rho \\
&+ (A^m \nabla_p^2 \rho)\partial_m \rho + O_2 \nabla_p^2 \rho + O_2.
\end{aligned}
$$

Since

$$
\begin{aligned}
(R+\rho)^{-1} &= R^{-1} - \rho R^{-2} + O(\rho^2), \\
(R+\rho)^{-2}\Delta_{S_1}\rho &= R^{-2}\Delta_{S_1}\rho + 2R^{-3}\rho\Delta_{S_1}\rho + O_2\nabla_p^2\rho,
\end{aligned}
$$

we have

$$
\begin{aligned}
&< H(\Gamma_t)\mathbf{n}_t, \mathbf{n}_t > \\
&= -\frac{N-1}{R} + \mathcal{B}\rho + 2R^{-3}\rho\Delta_{S_1}\rho + (A^m \nabla_p^2 \rho)\partial_m \rho + O_2 \nabla_p^2 \rho + O_2.
\end{aligned}
\tag{3.273}
$$

Replacing the pressure term $\mathfrak{q}$ by $\mathfrak{q} + \sigma\frac{N-1}{R}$, we have

$$
< \mu \mathbf{D}(\mathbf{u})\mathbf{n}, \mathbf{n} > -\mathfrak{q} - \sigma\mathcal{B}\rho = h_N(\mathbf{u}, \Psi_\rho)
$$

on S_R^T. In view of (3.191), (3.194), and (3.243), $h_N(\mathbf{u}, \Psi_\rho)$ may be defined by letting

$$
h_N(\mathbf{u}, \Psi_\rho) = \mathbf{V}_{h,N}(\bar\nabla H_\rho)\bar\nabla H_\rho \otimes \nabla \mathbf{u} + \sigma \tilde{\mathbf{V}}'_\Gamma(\bar\nabla H_\rho)\bar\nabla H_\rho \otimes \bar\nabla^2 H_\rho, \tag{3.274}
$$

where $\mathbf{V}_{h,N}(\bar{\mathbf{k}})$ and $\tilde{\mathbf{V}}'_\Gamma(\bar{\mathbf{k}})$ are some matrices of functions defined on $\Omega \times \{\bar{\mathbf{k}} \mid |\bar{\mathbf{k}}| \le \delta\}$ possessing the estimate:

$$
\begin{aligned}
&\sup_{|\bar{\mathbf{k}}| \le \delta} \|(\mathbf{V}_{h,N}(\cdot, \bar{\mathbf{k}}), \partial_{\bar{\mathbf{k}}}\mathbf{V}_{h,N}(\cdot, \bar{\mathbf{k}}))\|_{H^1_\infty(\Omega)} \le C, \\
&\sup_{|\bar{\mathbf{k}}| \le \delta} \|(\tilde{\mathbf{V}}'_\Gamma(\cdot, \bar{\mathbf{k}}), \partial_{\bar{\mathbf{k}}}\tilde{\mathbf{V}}'_\Gamma(\cdot, \bar{\mathbf{k}}))\|_{H^1_\infty(\Omega)} \le C,
\end{aligned}
$$

for some constant C.

3.5.3 Local Well-Posedness

In this section, we state the local well-posedness of equations (3.271). The following theorem holds both in the cases that Ω is a bounded domain and that $\Omega = \mathbb{R}^N$.

Theorem 3.5.1 *Let $N < q < \infty$, $2 < p < \infty$ and $T > 0$. Assume that Γ_- is a compact hypersurface of C^2. Assume that $2/p + N/q < 1$. Then, there exists a constant $\epsilon > 0$ depending on T such that if initial data $\mathbf{u}_0 \in B^{2(1-1/p)}_{q,p}(\dot\Omega)$ and $\rho_0 \in B^{3-1/p-1/q}_{q,p}(S_R)$ satisfy the smallness condition:*

$$\|\mathbf{u}_0\|_{B^{2(1-1/p)}_{q,p}(\dot\Omega)} + \|\rho_0\|_{B^{3-1/p-1/q}_{q,p}(S_R)} \leq \epsilon^2 \tag{3.275}$$

and the compatibility conditions (3.270) *and*

$$\begin{aligned} \operatorname{div} \mathbf{u}_0 &= g(\mathbf{u}_0, \rho_0) \quad && \text{in } \dot\Omega, \\ [[(\mu \mathbf{D}(\mathbf{u}_0)\mathbf{n})_\tau]] = [[\mathbf{h}'(\mathbf{u}_0, \rho_0)]], \quad [[\mathbf{u}_0]] &= 0 \quad && \text{on } S_R, \\ \mathbf{u}_0 &= 0 \quad && \text{on } \Gamma_1, \end{aligned} \tag{3.276}$$

then problem (3.271) *admits a unique solution $(\mathbf{u}, \mathfrak{q}, \rho) \in \mathcal{S}_{p,q}((0, T))$ possessing the estimates:*

$$\sup_{0<t<T} \|H_\rho(\cdot, t)\|_{H^1_\infty(\dot\Omega)} \leq \delta_1, \qquad \sup_{0<t<T} |\xi_{\mathbf{u},\rho}(t)| \leq \delta_1, \qquad E_{p,q,T}(\mathbf{u}, \rho) \leq \epsilon.$$

Here and in the following, for $0 \leq a < b \leq \infty$, we set

$$\begin{aligned} E_{p,q,T}(\mathbf{u}, \rho) = {} & \|\partial_t \mathbf{u}\|_{L_p((0,T),L_q(\dot\Omega))} + \|\mathbf{u}\|_{L_p((0,T),H^2_q(\dot\Omega))} \\ & + \|\partial_t \rho\|_{L_p((0,T),W^{2-1/q}_q(S_R))} + \|\rho\|_{L_p((0,T),W^{3-1/q}_q(S_R))} \\ & + \|\partial_t \rho\|_{L_\infty((0,T),W^{1-1/q}_q(S_R))}). \end{aligned}$$

Remark 10 Moving the terms:

$$\frac{1}{|B_R|}\int_{B_R} \mathbf{u}\, dx, \qquad \int_{|\omega|=R} \rho\, d\omega, \qquad < \int_{|\omega|=R} \rho\omega\, d\omega, \mathbf{n} >$$

to the right-hand side, employing the similar argument to that in the proof of Theorem 3.1.2 and using Theorem 3.2.8 with $\kappa = 0$, we can prove Theorem 3.5.1, and so we may omit its proof.

3.5.4 Decay Estimates of Solutions for the Linearized Equations

To prove Theorem 3.1.3, the key tool is decay properties of solutions of the Stokes equations:

$$\begin{cases} \partial_t \mathbf{u} - m^{-1}\mathrm{Div}\,(\mu\mathbf{D}(\mathbf{u}) - \mathfrak{q}\mathbf{I}) = \mathbf{f} & \text{in } \dot\Omega^T, \\ \mathrm{div}\,\mathbf{u} = g = \mathrm{div}\,\mathbf{g} & \text{in } \dot\Omega^T, \\ \partial_t\rho + \mathcal{L}_3\rho - \mathbf{n}\cdot P\mathbf{u} = d & \text{on } S_R^T, \\ [[(\mu\mathbf{D}(\mathbf{u}) - \mathfrak{q}\mathbf{I})\mathbf{n}]] - \sigma(\mathcal{B}\rho)\mathbf{n} = [[\mathbf{h}]] & \text{on } S_R^T, \\ [[\mathbf{u}]] = 0 & \text{on } S_R^T, \\ \mathbf{u} = 0 & \text{on } \Gamma_-^T, \\ (\mathbf{u}, \rho)|_{t=0} = (\mathbf{u}_0, \rho_0) & \text{in } \dot\Omega \times S_R. \end{cases} \tag{3.277}$$

Here and in the following, $\mathcal{L}_3$ denotes an operator defined by setting

$$\mathcal{L}_3\rho = \int_{S_R} \rho\, d\omega + < \int_{S_R} \rho\omega\, d\omega, \mathbf{n} > . \tag{3.278}$$

Here and in the following, $d\omega$ denotes the surface element of S_R and $\omega = (\omega_1, \ldots, \omega_N)$ denotes the point of S_R. Note that $\mathbf{n} = \omega/R$. Obviously, we have

$$\|\mathcal{L}_3\rho\|_{W_q^{2-1/q}(S_R)} \leq C\|\rho\|_{W_q^{2-1/q}(S_R)}.$$

Thus, we can apply Theorem 3.2.8 and Theorem 3.2.4 below.

We will prove the following theorem.

Theorem 3.5.2 *Let $1 < p, q < \infty$, $2/p + 1/q \neq 1$, and $2/p + 1/q \neq 2$. Assume that Γ_- is a non-empty compact hypersurfaces of C^2 class. Then, there exists an $\eta > 0$ such that the following assertion holds: Let $\mathbf{u}_0 \in B_{q,p}^{2(1-1/p)}(\dot\Omega)^N$ and $\rho_0 \in B_{q,p}^{3-1/p-1/q}(S_R)$ be initial data for equations* (3.277) *and let $\mathbf{f}$, g, $\mathbf{g}$, d, $\mathbf{h}$ be given functions on the right side of equations* (3.277)*. Assume that*

$$\mathbf{f} \in L_p((0, T), L_q(\dot\Omega)^N), \quad d \in L_p((0, T), W_q^{2-1/q}(S_R)).$$

Moreover, we assume that there exist g_η, $\mathbf{g}_\eta$, and $\mathbf{h}_\eta$ such that $g_\eta = e^{\eta t}g$, $\mathbf{g}_\eta = e^{\eta t}\mathbf{g}$, and $\mathbf{h}_\eta = e^{\eta t}\mathbf{h}$ for $t \in (0, T)$ and

$$g_\eta \in H_p^{1/2}(\mathbb{R}, L_q(\dot\Omega)) \cap L_p(\mathbb{R}, H_q^1(\dot\Omega)), \quad \mathbf{g}_\eta \in H_p^1(\mathbb{R}, L_q(\dot\Omega)^N), \quad \mathrm{div}\,\mathbf{g}_\eta = g_\eta,$$

$$\mathbf{h}_\eta \in H^{1/2}_p(\mathbb{R}, L_q(\dot\Omega)^N) \cap L_p(\mathbb{R}, H^1_q(\dot\Omega)^N).$$

We also assume that the following compatibility conditions hold:

$$\begin{aligned}
\operatorname{div}\mathbf{u}_0 &= g|_{t=0} \quad \text{in } \dot\Omega\\
[[(\mu\mathbf{D}(\mathbf{u}_0)\mathbf{n})_\tau]] &= [[(\mathbf{h}|_{t=0})_\tau]] \quad \text{on } S_R \text{ provided } 2/p + 1/q < 1,\\
[[\mathbf{u}_0]] = 0 \quad \text{on } S_R, \quad \mathbf{u}_0 &= 0 \quad \text{on } \Gamma_- \text{ provided } 2/p + 1/q < 2.
\end{aligned}$$

Then, problem (3.277) *admits a unique solution* $(\mathbf{u}, \mathfrak{q}, \rho) \in \mathcal{S}_{p,q}((0,\infty))$ *possessing the estimate:*

$$\mathcal{I}_{p,q,T}(\mathbf{u}, \rho; \eta) \leq C\mathcal{J}_{p,q,T}(\mathbf{u}_0, \rho_0, \mathbf{f}, g, \mathbf{g}, d, \mathbf{h}; \eta) \tag{3.279}$$

for some constant C independent of η. *Here and in the following, we set*

$$\begin{aligned}
\mathcal{I}_{p,q,T}(\mathbf{u}, \rho; \eta) &= \|e^{\eta t}\mathbf{u}\|_{L_p((0,T),H^2_q(\dot\Omega))} + \|e^{\eta t}\partial_t\mathbf{u}\|_{L_p((0,T),L_q(\dot\Omega))}\\
&+ \|e^{\eta t}\rho\|_{L_p((0,T),W^{3-1/q}_q(S_R))} + \|e^{\eta t}\partial_t\rho\|_{L_p((0,T),W^{2-1/q}_q(S_R))};\\
\mathcal{J}_{p,q,T}(\mathbf{u}_0, \rho_0, \mathbf{f}, g, \mathbf{g}, d, \mathbf{h}; \eta) &= \|\mathbf{u}_0\|_{B^{2(1-1/p)}_{q,p}(\dot\Omega)} + \|\rho_0\|_{B^{3-1/p-1/q}_{q,p}(S_R)}\\
&+ \|e^{\eta t}\mathbf{f}\|_{L_p((0,T),L_q(\dot\Omega))} + \|e^{\eta t}d\|_{L_p((0,T),W^{2-1/q}_q(S_R))} + \|\partial_t\mathbf{g}_\eta\|_{L_p(\mathbb{R},L_q(\dot\Omega))}\\
&+ \|(g_\eta, \mathbf{h}_\eta)\|_{H^{1/2}_p(\mathbb{R},L_q(\dot\Omega))} + \|(g_\eta, \mathbf{h}_\eta)\|_{L_p(\mathbb{R},H^1_q(\dot\Omega))}
\end{aligned}$$

To prove Theorem 3.5.2, we first consider the following shifted equations:

$$\begin{cases}
\partial_t\mathbf{u}_1 + \lambda_1\mathbf{u}_1 - m^{-1}\operatorname{Div}(\mu\mathbf{D}(\mathbf{u}_1) - \mathfrak{p}_1\mathbf{I}) = \mathbf{f} & \text{in } \dot\Omega^T,\\
\operatorname{div}\mathbf{u}_1 = g = \operatorname{div}\mathbf{g} & \text{in } \dot\Omega^T,\\
\partial_t\rho_1 + \lambda_1\rho_1 + \mathcal{L}_3\rho_1 - \mathbf{n}\cdot P\mathbf{u}_1 = d & \text{on } S_R^T,\\
[[(\mu\mathbf{D}(\mathbf{u}_1) - \mathfrak{p}_1\mathbf{I})\mathbf{n}]] - \sigma(\mathcal{B}\rho_1)\mathbf{n} = \mathbf{h} & \text{on } S_R^T,\\
[[\mathbf{u}_1]] = 0 & \text{on } S_T^R,\\
\mathbf{u}_1 = 0 & \text{on } \Gamma_1^T,\\
(\mathbf{u}_1, \rho_1)|_{t=0} = (\mathbf{u}_0, \rho_0) & \text{on } \dot\Omega \times S_R.
\end{cases} \tag{3.280}$$

For the shifted equation (3.280), the following theorem holds both in the cases that Ω is a bounded domain and that $\Omega = \mathbb{R}^N$.

Theorem 3.5.3 *Let* $1 < p, q < \infty$ *and* $T > 0$. *Assume that* $2/p + 1/q \neq 1, 2$. *Then, there exists a positive constant* $\lambda_2 > 0$ *such that if* $\lambda_1 \geq \lambda_2$, *then the following* *assertion holds: Let* $\mathbf{u}_0 \in B^{2(1-1/p)}_{q,p}(\dot\Omega)^N$ *and* $\rho_0 \in B^{3-1/p-1/q}_{q,p}(S_R)$ *be initial data for equations* (3.280) *and let* $\mathbf{f}$, g, $\mathbf{g}$, d, $\mathbf{h}$ *be given functions in the right side of equations* (3.280) *with*

$$\begin{aligned}
&\mathbf{f} \in L_p(\mathbb{R}, L_q(\dot\Omega)^N), \quad g \in H^1_p(\mathbb{R}, H^1_q(\dot\Omega)) \cap H^{1/2}_p(\mathbb{R}, L_q(\dot\Omega)),\\
&\mathbf{g} \in H^1_p(\mathbb{R}, L_q(\dot\Omega)^N), \quad d \in L_p(\mathbb{R}, W^{2-1/q}_q(S_R)),\\
&\mathbf{h} \in H^1_p(\mathbb{R}, H^1_q(\dot\Omega)^N) \cap H^{1/2}_p(\mathbb{R}, L_q(\dot\Omega)^N).
\end{aligned}$$

Assume that the compatibility condition:

$$\begin{aligned}
\operatorname{div}\mathbf{u}_0 = g|_{t=0} \quad &\text{in } \dot\Omega\\
[[(\mu\mathbf{D}(\mathbf{u}_0)\mathbf{n})_\tau]] = [[(\mathbf{h}|_{t=0})_\tau]] \quad &\text{on } S_R \text{ provided } 2/p + 1/q < 1,\\
[[\mathbf{u}_0]] = 0 \quad \text{on } S_R, \quad \mathbf{u}_0 = 0 \quad &\text{on } \Gamma_- \text{ provided } 2/p + 1/q < 2.
\end{aligned}$$

Then, problem (3.280) *admits a unique solution* $(\mathbf{u}, \mathfrak{q}, \rho) \in \mathcal{S}_{p,q}((0,\infty))$ *possessing the estimate:*

$$\mathcal{I}_{p,q,T}(\mathbf{u}_1, \rho_1; 0) \leq C\mathcal{J}_{p,q,T}(\mathbf{u}_0, \rho_0, \mathbf{f}, g, \mathbf{g}, d, \mathbf{h}; 0)$$

for some constant C.

Proof Let $\mathcal{A}(\lambda)$, $\mathcal{P}(\lambda)$, and $\mathcal{H}(\lambda)$ be operators given in Theorem 3.2.2 in the case where $\kappa = 0$. And then, $\mathcal{A}(\lambda + \lambda_1)$, $\mathcal{P}(\lambda + \lambda_1)$, and $\mathcal{H}(\lambda + \lambda_1)$ are $\mathcal{R}$ bounded solution operators for the generalized resolvent problem corresponding to equations (3.280). Let $-(\lambda_1 - \lambda_0) + \Sigma_{\epsilon_0} = \{\lambda \in \mathbb{C} \mid \lambda + (\lambda_1 - \lambda_0) \in \Sigma_{\epsilon_0}\}$. We then see that there exists a $\lambda_2 > 0$ such that $-\lambda_1 + \Sigma_{\epsilon_0,\lambda_0} \supset \Sigma_{\epsilon_0} - (\lambda_1 - \lambda_0)$ provided $\lambda_1 \geq \lambda_2$. So, by Theorem 3.2.2 we have

$$\begin{aligned}
\mathcal{R}_{\mathcal{L}(\mathcal{X}_q(\Omega), H^{2-j}_q(\Omega)^N)}(\{(\tau\partial_\tau)^\ell(\lambda^{j/2}\mathcal{A}(\lambda+\lambda_1)) \mid \lambda \in -(\lambda_1 - \lambda_0) + \Sigma_{\epsilon_0}\}) \leq r_b;\\
\mathcal{R}_{\mathcal{L}(\mathcal{X}_q(\Omega), L_q(\Omega)^N)}(\{(\tau\partial_\tau)^\ell(\nabla\mathcal{P}(\lambda+\lambda_1)) \mid \lambda \in -(\lambda_1 - \lambda_0) + \Sigma_{\epsilon_0}\}) \leq r_b;\\
\mathcal{R}_{\mathcal{L}(\mathcal{X}_q(\Omega), W^{3-k}_q(\Gamma))}(\{(\tau\partial_\tau)^\ell(\lambda^k\mathcal{H}(\lambda+\lambda_1)) \mid \lambda \in -(\lambda_1 - \lambda_0) + \Sigma_{\epsilon_0}\}) \leq r_b
\end{aligned}$$

for $\ell = 0, 1$, $j = 0, 1, 2$, and $k = 0, 1$. Thus, we can choose $\gamma = 0$ in the argument in Sect. 3.2.2.2, and so we have the theorem. □

For any $\eta > 0$, $e^{\eta t}\mathbf{u}$, $e^{\eta t}\mathfrak{q}$ and $e^{\eta t}\rho$ satisfy the equations:

$$\left.\begin{aligned}
\partial_t(e^{\eta t}\mathbf{u}_1) + (\lambda_1 - \eta)e^{\eta t}\mathbf{u}_1 - m^{-1}\operatorname{Div}(\mu\mathbf{D}(e^{\eta t}\mathbf{u}_1) - e^{\eta t}\mathfrak{p}_1\mathbf{I}) = e^{\eta t}\mathbf{f}\\
\operatorname{div} e^{\eta t}\mathbf{u}_1 = e^{\eta t}g = \operatorname{div} e^{\eta t}\mathbf{g}
\end{aligned}\right\} \quad \text{in } \dot\Omega^T,$$

$$\left.\begin{aligned} \partial_t(e^{\eta t}\rho_1) + (\lambda_1 - \eta)e^{\eta t}\rho_1 + \mathcal{L}_3(e^{\eta t}\rho_1) - \mathbf{n}\cdot P(e^{\eta t}\mathbf{u}_1) &= e^{\eta t}d \\ (\mu\mathbf{D}(e^{\eta t}\mathbf{u}_1) - e^{\eta t}\mathfrak{p}_1\mathbf{I})\mathbf{n}) - \sigma(\mathcal{B}(e^{\eta t}\rho_1))\mathbf{n} &= e^{\eta t}\mathbf{h}, \\ [[e^{\eta t}\mathbf{u}_1]] &= 0 \end{aligned}\right\} \text{ on } S_R^T,$$

$$e^{\eta t}\mathbf{u}_1 = 0 \text{ on } \Gamma_-^T,$$

$$(e^{\eta t}\mathbf{u}_1, e^{\eta t}\rho_1)|_{t=0} = (\mathbf{u}_0, \rho_0) \text{ on } \dot\Omega \times S_R.$$

Choosing $\lambda_1 > 0$ and $\eta > 0$ in such a way that $\lambda_1 - \eta \geq \lambda_2$, by Theorem 3.5.3, we have the following corollary.

Corollary 3.5.4 *Let $1 < p, q < \infty$ and $T > 0$. Assume that $2/p + 1/q \neq 1, 2$. Then, there exists a positive constant $\eta_0 > 0$ such that if $0 < \eta \leq \eta_0$, then the following assertion holds: Let $\mathbf{u}_0 \in B^{2(1-1/p)}_{q,p}(\dot\Omega)^N$ and $\rho_0 \in B^{3-1/p-1/q}_{q,p}(S_R)$ be initial data for equations* (3.277) *and let $\mathbf{f}$, g, $\mathbf{g}$, d, $\mathbf{h}$ be given functions in the right side of equations* (3.277). *Assume that*

$$\mathbf{f} \in L_p((0,T), L_q(\dot\Omega)^N), \quad d \in L_p((0,T), W_q^{2-1/q}(S_R)).$$

Moreover, we assume that there exist g_η, $\mathbf{g}_\eta$, and $\mathbf{h}_\eta$ such that $g_\eta = e^{\eta t}g$, $\mathbf{g}_\eta = e^{\eta t}\mathbf{g}$, and $\mathbf{h}_\eta = e^{\eta t}\mathbf{h}$ for $t \in (0, T)$ and

$$g_\eta \in H_p^{1/2}(\mathbb{R}, L_q(\dot\Omega)) \cap L_p(\mathbb{R}, H_q^1(\dot\Omega)), \quad \mathbf{g}_\eta \in H_p^1(\mathbb{R}, L_q(\dot\Omega)^N), \quad \operatorname{div}\mathbf{g}_\eta = g_\eta,$$
$$\mathbf{h}_\eta \in H_p^{1/2}(\mathbb{R}, L_q(\dot\Omega)^N) \cap L_p(\mathbb{R}, H_q^1(\dot\Omega)^N).$$

We also assume that the following compatibility conditions hold:

$$\operatorname{div}\mathbf{u}_0 = g|_{t=0} \quad \text{in } \dot\Omega$$
$$[[(\mu\mathbf{D}(\mathbf{u}_0)\mathbf{n})_\tau]] = [[(\mathbf{h}|_{t=0})_\tau]] \quad \text{on } S_R \text{ provided } 2/p + 1/q < 1,$$
$$[[\mathbf{u}_0]] = 0 \quad \text{on } S_R, \quad \mathbf{u}_0 = 0 \quad \text{on } \Gamma_- \text{ provided } 2/p + 1/q < 2.$$

Then, problem (3.280) *admits a unique solution $(\mathbf{u}, \mathfrak{q}, \rho) \in \mathcal{S}_{p,q}((0,\infty))$ possessing the estimate:*

$$\mathcal{I}_{p,q,T}(\mathbf{u}_1, \rho_1; \eta) \leq C\mathcal{J}_{p,q,T}(\mathbf{u}_0, \rho_0, \mathbf{f}, g, \mathbf{g}, d, \mathbf{h}; \eta) \tag{3.281}$$

for some constant C.

We consider the solutions $\mathbf{u}$, $\mathfrak{p}$, and ρ of problem (3.277) of the form: $\mathbf{u} = \mathbf{u}_1 + \mathbf{v}$, $\mathfrak{p} = \mathfrak{p}_1 + \mathfrak{q}$, and $\rho = \rho_1 + h$, where $\mathbf{u}_1$, $\mathfrak{p}_1$, and ρ_1 are solutions of the shifted equation (3.280), and then $\mathbf{v}$, $\mathfrak{q}$, and h should satisfy the equations:

$$
\begin{cases}
\partial_t \mathbf{v} - m^{-1}\mathrm{Div}\,(\mu\mathbf{D}(\mathbf{v}) - \mathfrak{q}\mathbf{I}) = -\lambda_1\mathbf{u}_1, \quad \mathrm{div}\,\mathbf{v} = 0 & \text{in } \dot\Omega^T, \\
\partial_t h + \mathcal{L}_3 h - \mathbf{n}\cdot P\mathbf{v} = -\lambda_1\rho_1 & \text{on } S_R^T, \\
[[(\mu\mathbf{D}(\mathbf{v}) - \mathfrak{q}\mathbf{I})\mathbf{n}]] - \sigma(\mathcal{B}h)\mathbf{n} = 0, \quad [[\mathbf{v}]] = 0 & \text{on } S_R^T, \\
\mathbf{v} = 0 & \text{on } \Gamma_-^T, \\
(\mathbf{v}, h)|_{t=0} = (0, 0) & \text{on } \dot\Omega \times S_R.
\end{cases}
\tag{3.282}
$$

Recall the definition of $J_q(\dot\Omega)$ given in (3.73) in Sect. 3.2.2.1, that is,

$$
J_q(\dot\Omega) = \{\mathbf{f} \in L_q(\dot\Omega)^N \mid (\mathbf{f}, \nabla\varphi)_{\dot\Omega} = 0 \quad \text{for any } \varphi \in \hat H^1_{q'}(\Omega)\}.
$$

Since $C_0^\infty(\Omega) \subset \hat H^1_{q'}(\Omega)$, $\mathbf{u} \in J_q(\dot\Omega)$ satisfies $\mathrm{div}\,\mathbf{u} = 0$ in $\dot\Omega$. Let $\psi \in \hat H^1_{q,0}(\dot\Omega)$ be a solution of the variational equation:

$$
(\nabla\psi, \nabla\varphi)_{\dot\Omega} = (\mathbf{u}_1, \nabla\varphi)_{\dot\Omega} \quad \text{for any } \varphi \in \hat H^1_{q'}(\Omega), \tag{3.283}
$$

and let $\mathbf{w} = \mathbf{u}_1 - \nabla\psi$. Then, $\mathbf{w} \in J_q(\dot\Omega)$ and

$$
\|\mathbf{w}\|_{L_q(\dot\Omega)} + \|\nabla\psi\|_{L_q(\dot\Omega)} \leq C\|\mathbf{u}_1\|_{L_q(\dot\Omega)}. \tag{3.284}
$$

Using $\mathbf{w}$ and ψ , we can rewrite the first equation in (3.282) as follows:

$$
\partial_t \mathbf{v} - m^{-1}\mathrm{Div}\,(\mu\mathbf{D}(\mathbf{v}) - (\mathfrak{q} + \lambda_1\psi)\mathbf{I}) = -\lambda_1\mathbf{w}, \quad \mathrm{div}\,\mathbf{v} = 0 \quad \text{in } \dot\Omega^T.
$$

Thus, in what follows we may assume that

$$
\mathbf{u}_1 \in H^1_p((0, T), J_q(\dot\Omega)) \cap L_p((0, T), H^2_q(\dot\Omega)^N). \tag{3.285}
$$

Let $K(\mathbf{u}, h) \in H^1_q(\dot\Omega) + \hat H^1_q(\Omega)$ be a unique solution of the weak Dirichlet problem

$$
(m^{-1}\nabla K(\mathbf{u}, h), \nabla\varphi)_{\dot\Omega} = (m^{-1}\mathrm{Div}\,(\mu\mathbf{D}(\mathbf{u})) - \nabla\mathrm{div}\,\mathbf{u}, \nabla\varphi)_{\dot\Omega} \tag{3.286}
$$

for any $\varphi \in \hat H^1_{q'}(\Omega)$, subject to

$$
[[K(\mathbf{u}, h)]] = [[< \mu\mathbf{D}(\mathbf{u})\mathbf{n}, \mathbf{n} >]] - \sigma(\mathcal{B}h) - [[\mathrm{div}\,\mathbf{u}]] \quad \text{on } S_R. \tag{3.287}
$$

And then, to handle problem (3.282) in the semigroup setting, we consider the initial value problem:

$$\begin{cases} \partial_t \mathbf{v} - m^{-1}\mathrm{Div}\,(\mu\mathbf{D}(\mathbf{v}) - K(\mathbf{v}, h)\mathbf{I}) = 0 & \text{in } B_R \times (0, \infty), \\ \partial_t h + \mathcal{L}_3 h - \mathbf{n}\cdot P\mathbf{v} = 0 & \text{on } S_R \times (0, \infty), \\ [[(\mu\mathbf{D}(\mathbf{v}) - K(\mathbf{v}, h)\mathbf{I})\mathbf{n}]] - \sigma(\mathcal{B}h)\mathbf{n} = 0, \quad [[\mathbf{v}]] = 0 & \text{on } S_R \times (0, \infty), \\ \mathbf{v} = 0 & \text{on } \Gamma_- \times (0, \infty), \\ (\mathbf{v}, h)|_{t=0} = (\mathbf{v}_0, \rho_0) & \text{on } \dot\Omega \times S_R. \end{cases} \tag{3.288}$$

Note that $[[(\mu\mathbf{D}(\mathbf{v}) - K(\mathbf{v}, h)\mathbf{I})\mathbf{n}]] - \sigma(\mathcal{B}h)\mathbf{n} = 0$ on $S_R \times (0, \infty)$ is equivalent to

$$[[(\mathbf{D}(\mathbf{v})\mathbf{n})_\tau]] = 0, \quad [[\mathrm{div}\,\mathbf{v}]] = 0 \quad \text{on } S_R \times (0, \infty). \tag{3.289}$$

Defining $\mathcal{H}_q(B_R)$, $\mathcal{D}_q(B_R)$, and $\mathcal{A}_q(\mathbf{u}, h)$ by

$$\begin{aligned} \mathcal{H}_q &= \{(\mathbf{v}, h) \mid \mathbf{v} \in J_q(\dot\Omega), \quad h \in W^{2-1/q}_q(S_R)\}, \\ \mathcal{D}_q &= \{(\mathbf{v}, h) \in \mathcal{H}_q \mid \mathbf{v} \in H^2_q(\dot\Omega)^N, \ h \in W^{3-1/q}_q(S_R), \\ &\qquad [[(\mu\mathbf{D}(\mathbf{v})\mathbf{n})_\tau]] = 0, \quad [[\mathbf{v}]] = 0 \text{ on } S_R, \quad \mathbf{v} = 0 \quad \text{on } \Gamma_-\}, \\ \mathcal{A}_q(\mathbf{v}, h) &= (m^{-1}\mathrm{Div}\,(\mathbf{D}(\mathbf{v}) - K(\mathbf{v}, h)\mathbf{I}), (-\mathcal{L}_3 h + \mathbf{n}\cdot P\mathbf{v})|_{S_R}) \end{aligned} \tag{3.290}$$

for $(\mathbf{v}, h) \in \mathcal{D}_q(B_R)$, we see that equations (3.288) is formulated by

$$\partial_t U = \mathcal{A}_q U \quad (t > 0), \quad U|_{t=0} = U_0 \tag{3.291}$$

with $U = (\mathbf{v}, h) \in \mathcal{D}_q$ for $t > 0$ and $U_0 = (\mathbf{u}_0, \rho_0) \in \mathcal{H}_q$. According to Theorem 3.2.4, we see that $\mathcal{A}_q$ generates a C^0 analytic semigroup $\{T(t)\}_{t\geq 0}$ on $\mathcal{H}_q(B_R)$. Moreover, we have

Theorem 3.5.5 *Let $1 < q < \infty$. Then, $\{T(t)\}_{t\geq 0}$ is exponentially stable, that is,*

$$\|T(t)(\mathbf{f}, g)\|_{\mathcal{H}_q} \leq Ce^{-\eta_1 t}\|(\mathbf{f}, g)\|_{\mathcal{H}_q} \tag{3.292}$$

for any $t > 0$ and $(\mathbf{f}, g) \in \mathcal{H}_q(B_R)$ with some positive constants C and η_1, where we have set

$$\|(\mathbf{f}, g)\|_{\mathcal{H}_1} = \|\mathbf{f}\|_{L_q(\dot\Omega)} + \|g\|_{W^{2-1/q}_q(S_R)}.$$

Postponing the proof of Theorem 3.5.5 to the next section, we continue to prove Theorem 3.5.2. In view of Duhamel's principle, solutions $\mathbf{v}$ and h of equations (3.282) are represented uniquely by

$$(\mathbf{v}, h)(\cdot, s) = \int_0^s T(s - r)(-\lambda_1\mathbf{u}_1(\cdot, r), -\lambda_1\rho_1(\cdot, r))\,dr.$$

By (3.292),

$$
\begin{aligned}
\|(\mathbf{v}, h)(\cdot, s)\|_{\mathcal{H}_q} &\le C\int_0^s e^{-\eta_1(s-r)}\|(\mathbf{u}_1(\cdot, r), \rho_1(\cdot, r))\|_{\mathcal{H}_q}\,dr \\
&\le C\Big(\int_0^s e^{-\eta_1(s-r)}\,dr\Big)^{1/p'}\Big(\int_0^s e^{-\eta_1(s-r)}\|(\mathbf{u}_1(\cdot, r), \rho_1(\cdot, r))\|_{\mathcal{H}_q}^p\,dr\Big)^{1/p}.
\end{aligned}
$$

Choosing $\eta > 0$ smaller if necessary, we may assume that $0 < \eta p < \eta_1$ without loss of generality. Thus, by the inequality above we have

$$
\begin{aligned}
&\int_0^t (e^{\eta s}\|(\mathbf{v}, h)(\cdot, s)\|_{\mathcal{H}_q})^p\,ds \\
&\quad\le C\int_0^t\Big(\int_0^s e^{\eta sp}e^{-\eta_1(s-r)}\|(\mathbf{u}_1(\cdot, r), \rho_1(\cdot, r))\|_{\mathcal{H}_q}^p\,dr\Big)\,ds \\
&\quad= \int_0^t\Big(\int_0^s e^{-(\eta_1-p\eta)(s-r)}(e^{\eta r}\|(\mathbf{u}_1(\cdot, r), \rho_1(\cdot, r))\|_{\mathcal{H}_q})^p\,dr\Big)\,ds \\
&\quad= \int_0^t (e^{\eta r}\|(\mathbf{u}_1(\cdot, r), \rho_1(\cdot, r))\|_{\mathcal{H}_q})^p\Big(\int_r^t e^{-(\eta_1-p\eta)(s-r)}\,ds\Big)\,dr \\
&\quad\le (\eta_1 - p\eta)^{-1}\int_0^T (e^{\eta r}\|(\mathbf{u}_1(\cdot, r), \rho_1(\cdot, r))\|_{\mathcal{H}_q})^p\,dr,
\end{aligned}
$$

which, combined with (3.281), leads to

$$
\|e^{\eta s}(\mathbf{v}, h)\|_{L_p((0,t),\mathcal{H}_q)} \le C\mathcal{J}_{p,q,T}(\mathbf{u}_0, \rho_0, \mathbf{f}, g, \mathbf{g}, d, \mathbf{h}; \eta) \tag{3.293}
$$

for any $t \in (0, T)$. In view of equations (3.282), we see that $\mathbf{v}$ and h satisfy the shifted equations:

$$
\begin{cases}
\partial_t\mathbf{v} + \lambda_1\mathbf{v} - m^{-1}\mathrm{Div}\,(\mu\mathbf{D}(\tilde{\mathbf{v}}) - K(\mathbf{v}, h)\mathbf{I}) = \lambda_1\mathbf{v} - \lambda_1\mathbf{u}_1, \quad \mathrm{div}\,\mathbf{v} = 0 & \text{in } \dot{\Omega}^T, \\
\partial_t h + \lambda_1 h + \mathcal{L}_3 h - \mathbf{n}\cdot P\mathbf{v} = \lambda_1 h - \lambda_1\rho_1 & \text{on } S_R^T, \\
[[(\mu\mathbf{D}(\mathbf{v}) - K(\mathbf{v}, h)\mathbf{I})\mathbf{n}]] - \sigma(\mathcal{B}h)\mathbf{n} = 0, \quad [[\mathbf{v}]] = 0 & \text{on } S_R^T, \\
\qquad\mathbf{v} = 0 & \text{on } \Gamma_-^T, \\
\qquad(\mathbf{v}, h)|_{t=0} = (0, 0) \quad \text{on } \dot{\Omega}\times S_R,
\end{cases}
$$

by (3.281) and (3.293), we have

$$
\mathcal{I}_{p,q,T}(\mathbf{v}, h; \eta) \le C\mathcal{J}_{p,q,T}(\mathbf{u}_0, \rho_0, \mathbf{f}, f_d, \mathbf{f}_d, g, \mathbf{h}; \eta).
$$

Thus, $\mathbf{u} = \mathbf{u}_1 + \mathbf{v}$ and $\rho = \rho_1 + h$ are required solutions of equations (3.277), which completes the proof of Theorem 3.5.2.

3.5.5 Exponential Stability of Continuous Analytic Semigroup Associated with Equations (3.288)

In this section, we shall prove Theorem 3.5.5 stated in the previous section. For this purpose, we consider the equations:

$$(\lambda \mathbf{I} - \mathcal{A}_q)U = F \tag{3.294}$$

for $F = (\mathbf{f}, g) \in J_q(\Omega)$ and $U = (\mathbf{v}, h) \in \mathcal{D}_q \cap \dot{J}_q(B_R)$, which is the resolvent problem corresponding to equations (3.291). Here, $\mathbf{I}$ is the identity operator, $J_q(B_R)$ is the space defined in (3.73), and $\mathcal{D}_q$ and $\mathcal{A}_q$ are the domain and the operator defined in (3.290). Since $\mathcal{R}$ boundedness implies the usual boundedness of operator families, by Theorem 3.2.4 and a result in Shibata and Shimizu [30], we have the following theorem.

Theorem 3.5.6 *Let $1 < q < \infty$ and $0 < \epsilon_0 < \pi/2$. Then, there exists a $\lambda_0 > 0$ such that for any $\lambda \in \Sigma_{\epsilon_0,\lambda_0}$ and $F \in \mathcal{H}_q$, equations (3.294) admits a unique solution $U \in \mathcal{D}_q$ possessing the estimate:*

$$|\lambda|\|U\|_{\mathcal{H}_q} + \|U\|_{\mathcal{D}_q} \leq C\|F\|_{\mathcal{H}_q} \tag{3.295}$$

for some constant $C > 0$. Here, $\mathcal{H}_q$ is the space defined in (3.290).

Our task in this section is to prove the following theorem.

Theorem 3.5.7 *Let $1 < q < \infty$ and let λ_0 be a number appearing in Theorem 3.5.6. Let*

$$\Lambda_{\lambda_0} = \{\lambda \in \mathbb{C} \mid \mathrm{Re}\,\lambda \geq 0, \quad |\lambda| \leq \lambda_0\}.$$

Then, for any $\lambda \in \Lambda_{\lambda_0}$ and $F \in \dot{\mathcal{H}}_q$, problem (3.294) admits a unique solution $U \in \mathcal{D}_q$ possessing the estimate:

$$\|U\|_{\mathcal{D}_q} \leq C\|F\|_{\mathcal{H}_q} \tag{3.296}$$

for some constant C.

Remark 11 Combining Theorem 3.5.6 and Theorem 3.5.7 yields immediately Theorem 3.5.5. Thus, this section mainly devotes to proving Theorem 3.5.7

Proof In view of Theorem 3.5.6, $(\lambda_0 - \mathcal{A}_q)^{-1}$ exists. For any $\lambda \in \Lambda_{\lambda_0}$, we write

$$\lambda - \mathcal{A}_q = \lambda - \lambda_0 + \lambda_0 - \mathcal{A}_q = (\mathbf{I} + (\lambda - \lambda_0)(\lambda_0 - \mathcal{A}_q)^{-1})(\lambda_0 - \mathcal{A}_q),$$

and therefore,

$$(\lambda - \mathcal{A}_q)^{-1} = (\lambda_0 - \mathcal{A}_q)^{-1}(\mathbf{I} + (\lambda - \lambda_0)(\lambda_0 - \mathcal{A}_q)^{-1})^{-1}$$

provided that $(\mathbf{I} + (\lambda - \lambda_0)(\lambda_0 - \mathcal{A}_q)^{-1})^{-1}$ exists. Since Ω is a bounded domain, by Rellich's compactness theorem $\mathcal{D}_q$ is compactly embedded into $\mathcal{H}_q$, and so $(\lambda_0 - \mathcal{A}_q)^{-1}$ is a compact operator on $\mathcal{H}_q$. Thus, by the Riesz-Schauder theorem, especially the Fredholm alternative principle, the triviality of the kernel of $\mathbf{I} + (\lambda - \lambda_0)(\lambda_0 - \mathcal{A}_q)^{-1}$ yields the existence of inverse operator, $(\mathbf{I}+(\lambda-\lambda_0)(\lambda_0-\mathcal{A}_q)^{-1})^{-1}$. Thus, let $F \in \mathcal{H}_q$ satisfy the equation: $(\mathbf{I} + (\lambda - \lambda_0)(\lambda_0 - \mathcal{A}_q)^{-1})F = 0$. Notice that $F = -(\lambda - \lambda_0)(\lambda_0 - \mathcal{A}_q)^{-1}F \in \mathcal{D}_q$. Moreover,

$$(\lambda - \mathcal{A}_q)F = (\lambda_0 - \mathcal{A}_q + (\lambda - \lambda_0))F = -(\lambda - \lambda_0)F + (\lambda - \lambda_0)F = 0.$$

Thus, setting $(\mathbf{u}, \rho) = F$, in view of (3.290) we see that $(\mathbf{u}, \rho) \in \mathcal{D}_q$ and $(\mathbf{u}, \rho)$ satisfy the homogeneous equations:

$$\begin{cases} \lambda\mathbf{u} - m^{-1}\mathrm{Div}\,(\mu\mathbf{D}(\mathbf{u}) - K(\mathbf{u}, \rho)\mathbf{I}) = 0, & \text{in } \dot\Omega, \\ \lambda\rho + \mathcal{L}_3\rho - \mathbf{n}\cdot P\mathbf{u} = 0 & \text{on } S_R, \\ [[(\mu\mathbf{D}(\mathbf{u}) - K(\mathbf{u}, \rho)\mathbf{I})\mathbf{n}]] - \sigma(\mathcal{B}\rho)\mathbf{n} = 0, \quad [[\mathbf{u}]] = 0 & \text{on } S_R, \\ \mathbf{u} = 0 & \text{on } \Gamma_-, \end{cases} \tag{3.297}$$

where we have used (3.287). In fact, it follows from the boundedness of $\dot\Omega$ and $\mathbf{u} \in J_q(\dot\Omega) \cap H^2_q(\dot\Omega)^N$ that $\mathrm{div}\,\mathbf{u} = 0$ in $\dot\Omega$. In particular, $[[\mathrm{div}\,\mathbf{u}]] = 0$ on S_R. Thus, by (3.287) $[[< \mu\mathbf{D}(\mathbf{u}) - K(\mathbf{u}, \rho)\mathbf{I})\mathbf{n}, \mathbf{n} >]] - \sigma(\mathcal{B}\rho) = 0$ on S_R, which, combined with $[[\mu\mathbf{D}(\mathbf{u})\mathbf{n})_\tau]] = 0$ on S_R, leads to

$$[[(\mu\mathbf{D}(\mathbf{u}) - K(\mathbf{u}, \rho)\mathbf{I})\mathbf{n}]] - \sigma(\mathcal{B}\rho)\mathbf{n} = 0 \quad \text{on } S_R.$$

By the second equation in equations (3.297) and the divergence theorem of Gauß,

$$\begin{aligned} 0 &= (\lambda\rho + \mathcal{L}_3\rho - \mathbf{n}\cdot P\mathbf{u}, 1)_{S_R} = (\lambda + |S_R|)\int_{S_R} \rho\,d\omega - (\mathrm{div}\,P\mathbf{u}, 1)_{B_R}, \\ 0 &= (\lambda\rho + \mathcal{L}_3\rho - \mathbf{n}\cdot P\mathbf{u}, x_i)_{S_R} \\ &= (\lambda + R^{-1}\int_{S_R} \omega_i^2\,d\omega)\int_{S_R} \omega_i\rho\,d\omega - (\mathrm{div}\,(x_i P\mathbf{u}), 1)_{B_R}, \end{aligned}$$

where $|S_R|$ denotes the area of S_R. Since $\mathrm{div}\,\mathbf{u} = 0$ on $\dot\Omega$, we have

$$(\mathrm{div}\, P\mathbf{u}, 1)_{B_R} = 0, \quad (\mathrm{div}\,(x_i P\mathbf{u}), 1)_{B_R} = \int_{B_R} \Big(u_i - \frac{1}{|B_R|}\int_{B_R} u_i\, dx\Big)\, dx = 0.$$

Thus, we have

$$\int_{S_R} \rho\, d\omega = 0, \quad \int_{S_R} \omega_i \rho\, d\omega = 0 \ \ (i = 1, \ldots, N), \tag{3.298}$$

because $\lambda + |S_R| \neq 0$ and $\lambda + \int_{S_R} \omega_i^2\, d\omega_i \neq 0$ for $\mathrm{Re}\,\lambda \geq 0$.

We first consider the case that $2 \leq q < \infty$. In this case, $(\mathbf{u}, \rho) \in \mathcal{D}_2$, because $\dot\Omega$ is bounded. Multiplying the first equation of (3.297) with $\overline{\mathbf{u}}$, integrating the resultant formula over $\dot\Omega$ and using the divergence theorem of Gauß and the facts that $[[\mathbf{u}]] = 0$ on S_R and $\mathbf{u} = 0$ on Γ_-, we have

$$\begin{aligned} 0 &= \lambda(m\mathbf{u}, \mathbf{u})_{\dot\Omega} - ([[\mu\mathbf{D}(\mathbf{u}) - K(\mathbf{u}, \rho)\mathbf{I})\mathbf{n}]], \mathbf{u})_{S_R} \\ &\quad + \frac{1}{2}(\mu\mathbf{D}(\mathbf{u}), \mathbf{D}(\mathbf{u}))_{\dot\Omega} - (K(\mathbf{u}, \rho), \mathrm{div}\,\mathbf{u})_{\dot\Omega}. \end{aligned}$$

By the interface condition of equations (3.297),

$$\begin{aligned} &([[\mu\mathbf{D}(\mathbf{u}) - K(\mathbf{u}, \rho)\mathbf{I})\mathbf{n}]], \mathbf{u})_{S_R} = ([[< \mu\mathbf{D}(\mathbf{u})\mathbf{n}, \mathbf{n} > - K(\mathbf{u}, \rho)]], \mathbf{u}\cdot\mathbf{n})_{S_R} \\ &= \sigma(\mathcal{B}\rho, \mathbf{u}\cdot\mathbf{n})_{S_R} = \sigma(\mathcal{B}\rho, \lambda\rho + \mathcal{L}_3\rho + |B_R|^{-1}\mathbf{n}\cdot\int_{B_R} \mathbf{u}\, dx)_{S_R}. \end{aligned}$$

Since ω_i $(i = 1, \ldots, N)$ are eigenfunctions of the Laplace-Beltrami operator Δ_{S_R}, we have $\mathcal{B}\omega_i = 0$ on S_R, and so

$$([[\mu\mathbf{D}(\mathbf{u}) - K(\mathbf{u}, \rho)\mathbf{I})\mathbf{n}]], \mathbf{u})_{S_R} = \bar\lambda\sigma(\mathcal{B}\rho, \rho)_{S_R} + \frac{\sigma(N-1)}{R^2}\Big|\int_{S_R} \rho\, d\omega\Big|^2.$$

Thus, using the fact that $\mathrm{div}\,\mathbf{u} = 0$ in $\dot\Omega$ and (3.298), we have

$$([[\mu\mathbf{D}(\mathbf{u}) - K(\mathbf{u}, \rho)\mathbf{I})\mathbf{n}]], \mathbf{u})_{S_R} = \bar\lambda\sigma(\mathcal{B}\rho, \rho)_{S_R}.$$

Summing up, we have obtained

$$0 = \lambda(m\mathbf{u}, \mathbf{u})_{\dot\Omega} + \frac{1}{2}(\mu\mathbf{D}(\mathbf{u}), \mathbf{D}(\mathbf{u}))_{\dot\Omega} - \bar\lambda\sigma(\mathcal{B}\rho, \rho)_{S_R}. \tag{3.299}$$

To handle $(\mathcal{B}\rho, \rho)_{S_R}$, we use the following lemma. □

Lemma 3.5.8 *Let*

$$\dot H_2^2(S_R) = \{h \in H_2^2(S_R) \mid (h, 1)_{S_R} = 0, \quad (h, x_j)_{S_R} = 0 \ \ (j = 1, \ldots, N)\}.$$

Then,

$$-(\mathcal{B}h, h) \geq c\|h\|^2_{L_2(S_R)} \tag{3.300}$$

for any $h \in \dot{H}^2_2(S_R)$ *with some constant* $c > 0$.

Postponing the proof of Lemma 3.5.8, we continue the proof of Theorem 3.5.6. By (3.298) and Lemma 3.5.8, we have

$$-(\mathcal{B}\rho, \rho)_{S_R} \geq c\|\rho\|^2_{L_2(S_R)}.$$

Thus, taking the real part of the formula (3.299), we have

$$0 \geq \mathrm{Re}\,\lambda(m\mathbf{u}, \mathbf{u})_{\dot{\Omega}} + \frac{1}{2}(\mu\mathbf{D}(\mathbf{u}), \mathbf{D}(\mathbf{u}))_{\dot{\Omega}} + \mathrm{Re}\,\lambda\,\sigma c\|\rho\|_{L_2(S_R)},$$

which leads to $\mathbf{D}(\mathbf{u}) = 0$ in $\dot{\Omega}$. Since $[[\mathbf{u}]] = 0$ on S_R and $\mathbf{u} = 0$ on Γ_-, we have $\mathbf{u} = 0$ in $\dot{\Omega}$. By the first equation of equations (3.297), we have $\nabla K(\mathbf{u}, \rho) = 0$ in $\dot{\Omega}$, which yields that $K(\mathbf{u}, \rho)$ is a constant in B_R and $\dot{\Omega} \setminus B_R$. Thus, $[[K(\mathbf{u}, \rho)]]$ is a constant. By the interface condition of equations (3.297), we have

$$[[K(\mathbf{u}, \rho)]] - \sigma\mathcal{B}\rho = 0 \quad \text{on } S_R.$$

Integrating this formula on S_R and using (3.298), we have $[[K(\mathbf{u}, \rho)]] = 0$ on S_R. Thus, $\mathcal{B}\rho = 0$ on S_R, which, combined with (3.298) and Lemma 3.5.8, leads to $\rho = 0$. This completes the proof of Theorem 3.5.7 in the case that $2 \leq q < \infty$.

We next consider the case that $1 < q < 2$. Given any $\mathbf{f} \in J_{q'}(\dot{\Omega})$, let $(\mathbf{v}, h) \in \mathcal{D}_{q'}$ be solutions of the equations:

$$\begin{cases} \bar{\lambda}\mathbf{v} - m^{-1}\mathrm{Div}\,(\mu\mathbf{D}(\mathbf{v}) - K(\mathbf{v}, h)\mathbf{I}) = \mathbf{f}, & \text{in } \dot{\Omega}, \\ \bar{\lambda}h + \mathcal{L}_3 h - \mathbf{n}\cdot P\mathbf{v} = 0 & \text{on } S_R, \\ [[(\mu\mathbf{D}(\mathbf{v}) - K(\mathbf{v}, h)\mathbf{I})\mathbf{n}]] - \sigma(\mathcal{B}h)\mathbf{n} = 0, \quad [[\mathbf{u}]] = 0 & \text{on } S_R, \\ \mathbf{v} = 0 & \text{on } \Gamma_-. \end{cases} \tag{3.301}$$

Since $\mathbf{v} \in J_q(\dot{\Omega}) \cap H^2_q(\dot{\Omega})^N$, we have $\mathrm{div}\,\mathbf{v} = 0$ in $\dot{\Omega}$. Thus, employing the same argument as that in the proof of (3.298), by the second equation of (3.301) we have

$$\int_{S_R} h\,d\omega = 0, \quad \int_{S_R} \omega_i h\,d\omega = 0 \ \ (i = 1, \ldots, N). \tag{3.302}$$

By the divergence theorem of Gauß, (3.298) and (3.302),

$$
\begin{aligned}
0 &= (m\lambda\mathbf{u} - \mathrm{Div}\,(\mu\mathbf{D}(\mathbf{u}) - K(\mathbf{u},\rho)\mathbf{I}), \mathbf{v})_{\dot\Omega} \\
&= \lambda(m\mathbf{u}, \mathbf{v})_{\dot\Omega} - \lambda\sigma(\mathcal{B}\rho, h)_{S_R} + \frac{1}{2}(\mu\mathbf{D}(\mathbf{u}), \mathbf{D}(\mathbf{v}))_{\dot\Omega}, \\
(\mathbf{u}, \mathbf{f})_{\dot\Omega} &= (\mathbf{u}, m\lambda\mathbf{v})_{\dot\Omega} - (\mathbf{u}, \mathrm{Div}\,(\mu\mathbf{D}(\mathbf{v}) - K(\mathbf{v}, h)\mathbf{I}))_{\dot\Omega} \\
&= m\lambda(\mathbf{u}, \mathbf{v})_{\dot\Omega} - \lambda\sigma(\rho, \mathcal{B}h)_{S_R} + \frac{1}{2}(\mathbf{D}(\mathbf{u}), \mu\mathbf{D}(\mathbf{v}))_{\dot\Omega}.
\end{aligned}
\tag{3.303}
$$

Since

$$
(\mathcal{B}\rho, h)_{S_R} = -(\nabla_\Gamma\rho, \nabla_\Gamma h)_{S_R} + \frac{N-1}{R^2}(\rho, h)_{S_R} = (\rho, \mathcal{B}h)_{S_R},
$$

we have

$$
(\mathbf{u}, \mathbf{f})_{\dot\Omega} = 0 \quad \text{for any } \mathbf{f} \in J_{q'}(\dot\Omega). \tag{3.304}
$$

Given any $\mathbf{g} \in L_{q'}(\dot\Omega)^N$, let $\psi \in \hat H^1_{q'}(\dot\Omega)$ be a solution of the variational equation:

$$
(m^{-1}\nabla\psi, \nabla\varphi)_{\dot\Omega} = (\mathbf{g}, \nabla\varphi)_{\dot\Omega} \quad \text{for any } \varphi \in \hat H^1_q(\Omega).
$$

And then, $\mathbf{g} = \mathbf{f} + \nabla\psi$ with $\mathbf{f} = \mathbf{g} - \nabla\psi \in J_q(\dot\Omega)$. Since $\mathbf{u} \in J_q(\dot\Omega)$, we have $(\mathbf{u}, \nabla\psi)_{\dot\Omega} = 0$, and so by (3.304) we have $(\mathbf{u}, \mathbf{g})_{\dot\Omega} = 0$ for any $\mathbf{g} \in L_{q'}(\dot\Omega)^N$, which yields that $\mathbf{u} = 0$. By the second equation of equation (3.297) and (3.298), we have $\lambda\rho = 0$ on S_R, and therefore $\rho = 0$ when $\lambda \neq 0$. When $\lambda = 0$, by the first equation of (3.297), we have $\nabla K(\mathbf{u}, \rho) = 0$ in $\dot\Omega$, which yields that $[[K(\mathbf{u}, \rho)]]$ is a constant. Thus, by the interface condition of equation (3.297), we have

$$
[[K(\mathbf{u}, \rho]] - \sigma\mathcal{B}\rho = 0 \quad \text{on } S_R.
$$

Integrating this formula on S_R and using (3.298), we have $[[K(\mathbf{u}, \rho)]] = 0$ on S_R. Thus, we have $\mathcal{B}\rho = 0$ on S_R. By the hypoellipticity of the operator Δ_{S_R}, we see that $\rho \in H^2_2(S_R)$, and therefore combining (3.298) and Lemma 3.5.8 yields that $\rho = 0$. This completes the proof of Theorem 3.5.7.

3.5.6 A Proof of Theorem 3.1.3

In this section, we prove Theorem 3.1.3. Assume that the initial data $\mathbf{u}_0$ and ρ_0 with

$$
\mathbf{u}_0 \in B^{2-2/p}_{q,p}(\dot\Omega)^N, \quad \rho_0 \in B^{3-1/p-1/q}_{q,p}(S_R)
$$

satisfy the smallness condition:

$$\|\mathbf{u}_0\|_{H^2_q(\dot\Omega)} + \|\rho_0\|_{B^{3-1/p-1/q}_{q,p}(S_R)} \leq \epsilon \tag{3.305}$$

with small constant $\epsilon > 0$ as well as the compatibility conditions (3.37) and (3.269). For the notational simplicity, we write

$$\begin{aligned}
\mathcal{I} &= \|\mathbf{u}_0\|_{B^{2(1-1/p)}_{q,p}(\dot\Omega)} + \|\rho_0\|_{B^{3-1/p-1/q}_{q,p}(S_R)},\\
\mathcal{I}_{p,q,T}(\mathbf{u}, \rho; \eta) &= \|e^{\eta t}\mathbf{u}\|_{L_p((0,T),H^2_q(\dot\Omega))} + \|e^{\eta t}\partial_t\mathbf{u}\|_{L_p((0,T),L_q(\dot\Omega))}\\
&\quad + \|e^{\eta t}\rho\|_{L_p((0,T),W^{3-1/q}_q(S_R))} + \|e^{\eta t}\partial_t\rho\|_{L_p((0,T),W^{2-1/q}_q(S_R))},\\
E_{p,q,T}(\mathbf{u}, \rho; \eta) &= \mathcal{I}_{p,q,T}(\mathbf{u}, \rho; \eta) + \|e^{\eta t}\partial_t\rho\|_{L_\infty((0,T),W^{1-1/q}_q(S_R))}.
\end{aligned}$$

Notice that

$$\begin{aligned}
&\|e^{\eta t}\mathbf{u}\|_{L_\infty((0,T),B^{2(1-1/p)}_{q,p}(\dot\Omega))} + \|e^{\eta t}\rho\|_{L_\infty((0,T),B^{3-1/p-1/q}_{q,p}(S_R))}\\
&\qquad\leq C(\mathcal{I} + E_{p,q,T}(\mathbf{u}, \rho; \eta)),
\end{aligned}$$

which follows from (3.212) and (3.213). Since we choose ϵ small enough eventually, we may assume that $0 < \epsilon < 1$.

Let $T_0 > 2$ be a given number. In view of Theorem 3.6.1, there exists a small number $\epsilon_1 > 0$ such that equations (3.271) admits a unique solution $(\mathbf{u}, \mathfrak{q}, \rho) \in \mathcal{S}_{p,q}(0, 2)$ satisfying the conditions:

$$\sup_{0<t<2} \|H_\rho(\cdot, t)\|_{H^1_\infty(\dot\Omega)} \leq \delta_1, \quad \sup_{0<t<2} |\xi_{\mathbf{u},\rho}(t)| \leq \delta_1, \quad E_{p,q,2}(\mathbf{u}, \rho; 0) \leq \epsilon_1 \tag{3.306}$$

provided that $\mathcal{I} \leq \epsilon_1^2$, where ϵ_1 is some small positive number, δ_1 is the same constant as in (3.264), $\xi'_{\mathbf{u},\rho}(t)$ is the function defined in (3.259), and we have set

$$\xi_{\mathbf{u},\rho}(t) = \int_0^t \xi'_{\mathbf{u},\rho}(s)\,ds.$$

In particular, by (3.263) we have

$$\sup_{0<t<T_0} \|\Psi_\rho(\cdot, t)\|_{H^1_\infty(\dot\Omega)} \leq \delta.$$

We now assume that the existence of unique solution $(\mathbf{u}, \mathfrak{q}, \rho) \in \mathcal{S}_{p,q}(0, T_0)$ of equations (3.271). Moreover, in view of (3.263), we assume that $(\mathbf{u}, \rho)$ satisfies the condition:

$$\sup_{0<t<T_0} \|H_\rho(\cdot, t)\|_{H^1_\infty(\dot\Omega)} \leq \delta_1, \quad \sup_{0<t<T_0} |\xi_{\mathbf{u},\rho}(t)| \leq \delta_1, \quad E_{p,q,T_0}(\mathbf{u}, \rho, 0) \leq \epsilon_1. \tag{3.307}$$

Since $2M_0\delta_1 \leq \delta$, by (3.263) we have

$$\sup_{0<t<T_0} \|\Psi_\rho(\cdot, t)\|_{H^1_\infty(\dot\Omega)} \leq \delta.$$

We shall prove that $\mathbf{u}$, $\mathfrak{q}$, and ρ can be prolonged beyond T keeping the condition (3.307) provided that $\epsilon > 0$ is small enough. For this purpose, it suffices to prove that the inequality:

$$E_{p,q,T}(\mathbf{u}, \rho; \eta) \leq M(\mathcal{I} + E_{p,q,T}(\mathbf{u}, \rho; \eta)^2 + E_{p,q,T}(\mathbf{u}, \rho; \eta)^3) \tag{3.308}$$

holds for any $T \in (0, T_0)$ with some constant $M > 0$ independent of ϵ, ϵ_1, T, and T_0, where η is the same positive constant as in Theorem 3.5.2. In fact, let $r_0(\epsilon)$ and $r_\pm(\epsilon)$ be three different solutions of the algebraic equation: $x^3+x^2+\epsilon-M^{-1}x = 0$ with $r_0(\epsilon) = M\epsilon + O(\epsilon^2)$, $r_+(\epsilon) = M^{-1} + O(M^{-2}) + O(\epsilon)$, and $r_-(\epsilon) = -1 - M^{-1} + O(M^{-2}) + O(\epsilon)$ as $M \to \infty$ and $\epsilon \to 0$. Since $E_{p,q,T}(\mathbf{u}, \rho; \eta) \geq 0 > r_-(\epsilon)$, by (3.308) one of the following cases holds:

$$E_{p,q,T}(\mathbf{u}, \rho; \eta) \leq r_0(\epsilon), \quad E_{p,q,T}(\mathbf{u}, \rho; \eta) \geq r_+(\epsilon).$$

Since we may change M larger for (3.308) to hold if necessary, and since we choose $\epsilon_1 > 0$ relatively small, we may assume that

$$e\epsilon_1 < M/2. \tag{3.309}$$

By (3.309), $E_{p,q,1/\eta}(\mathbf{u}, \rho; \eta) \leq e\epsilon_1 < M/2 < r_+(\epsilon)$, and therefore,

$$E_{p,q,T}(\mathbf{u}, \rho; \eta) \leq r_0(\epsilon) \quad \text{for } T \in (0, 1/\eta).$$

But, $E_{p,q,T}(\mathbf{u}, \rho; \eta)$ is a continuous function with respect to $T \in (0, T_0)$, which yields that

$$E_{p,q,T}(\mathbf{u}, \rho; \eta) \leq r_0(\epsilon) \quad \text{for any } T \in (0, T_0). \tag{3.310}$$

Especially, setting $T_1 = T_0 - 1/2$, we have

$$\|\mathbf{u}(\cdot, T_1)\|_{B^{2-2/p}_q(\dot\Omega)} + \|\rho(\cdot, T_1)\|_{W^{3-1/p-1/q}_{q,p}(S_R)} \leq r_0(\epsilon).$$

Thus, choosing $\epsilon > 0$ small enough and employing the same argument as that in proving Theorem 3.5.1, we see that there exists a unique solution $(\mathbf{v}, \mathfrak{q}, h) \in \mathcal{S}_{p,q}((T_1, T_1 + 1))$ of the equations:

$$
\begin{cases}
\partial_t \mathbf{v} - \mathrm{Div}\,(\mu \mathbf{D}(\mathbf{v}) - \mathfrak{p}\mathbf{I}) = \mathbf{f}(\mathbf{v}, \Psi_h) & \text{in } \dot\Omega \times (T_1, T_1+1), \\
\mathrm{div}\,\mathbf{v} = g(\mathbf{v}, \Psi_h) = \mathrm{div}\,\mathbf{g}(\mathbf{v}, \Psi_h) & \text{in } \dot\Omega \times (T_1, T_1+1), \\
\partial_t h + \mathcal{L}_3 h - \mathbf{n}\cdot P\mathbf{v} = \bar d(\mathbf{v}, \Psi_h) & \text{on } S_R \times (T_1, T_1+1), \\
[[(\mu \mathbf{D}(\mathbf{v})\mathbf{n})_\tau]] = [[\mathbf{h}'(\mathbf{v}, \Psi_h) & \text{on } S_R \times (T_1, T_1+1), \\
[[< \mu \mathbf{D}(\mathbf{v})\mathbf{n}, \mathbf{n} > -\mathfrak{p}]] - \sigma \mathcal{B} h = [[h_N(\mathbf{v}, \Psi_h)]] & \text{on } S_R \times (T_1, T_1+1), \\
[[\mathbf{v}]] = 0 & \text{on } S_R \times (T_1, T_1+1), \\
\mathbf{v} = 0 & \text{on } \Gamma_- \times (T_1, T_1+1), \\
(\mathbf{v}, h)|_{t=T_1} = (\mathbf{u}(\cdot, T_1), \rho(\cdot, T_1)) & \text{in } \dot\Omega \times S_R,
\end{cases}
\tag{3.311}
$$

which satisfies the condition:

$$
\sup_{T_1<t<T_1+1} \|H_h(\cdot, t)\|_{H^1_\infty(\dot\Omega)} \leq \delta_1, \qquad \sup_{T_1<t<T_1+1} |\xi'_{\mathbf{v},h}(t)| \leq \delta_1/2,
$$
$$
E_{p,q,(T_1,T_1+1)}(\mathbf{u}, \rho) \leq \epsilon_1
$$

with

$$
\begin{aligned}
&E_{p,q,(T_1,T_1+1)}(\mathbf{u}, \rho) \\
&\quad = \|\partial_t \mathbf{u}\|_{L_p(T_1,T_1+1),L_q(\dot\Omega))} + \|\mathbf{u}\|_{L_p((T_1,T_1+1),H^2_q(\dot\Omega))} \\
&\quad + \|\partial_t \rho\|_{L_p((T_1,T_1+1),W^{2-1/q}_q(S_R))} + \|\rho\|_{L_p((T_1,T_1+1),W^{3-1/q}_q(S_R))} \\
&\quad + \|\partial_t \rho\|_{L_\infty((T_1,T_1+1),W^{1-1/q}_q(S_R)}).
\end{aligned}
$$

Let

$$
\mathbf{u}_1 = \begin{cases} \mathbf{u} & (0 < t \leq T_1), \\ \mathbf{v} & (T_1 < t < T_1+1), \end{cases} \qquad \mathfrak{q}_1 = \begin{cases} \mathfrak{q} & (0 < t \leq T_1), \\ \mathfrak{p} & (T_1 < t < T_1+1), \end{cases}
$$
$$
\rho_1 = \begin{cases} \rho & (0 < t \leq T_1), \\ h & (T_1 < t < T_1+1), \end{cases}
$$

and then $(\mathbf{u}_1, \mathfrak{q}_1, \rho_1)$ belongs to $\mathcal{S}_{p,q}((0, T_1+1))$ and satisfies the condition:

$$
\sup_{0<t<T_1+1} \|H_{\rho_1}(\cdot, t)\|_{H^1_\infty(\dot\Omega)} \leq \delta_1, \quad E_{p,q,T_1+1}(\mathbf{u}_1, \rho_1, 0) \leq \epsilon_1.
$$

Moreover, letting

$$\xi_{\mathbf{u}_1,\rho_1}(t) = \int_0^t \xi'_{\mathbf{u}_1,\rho_1}(s)\,ds,$$

and using (3.262) we have

$$\begin{aligned}\sup_{0<t<T_1+1} |\xi_{\mathbf{u}_1,\rho_1}(t)| &\leq \int_0^{T_1} |\xi'_{\mathbf{u},\rho}(s)|\,ds + \int_{T_1}^{T_1+1} |\xi'_{\mathbf{v},h}(s)|\,ds \\ &\leq C_R(\eta p')^{1/p'} \|e^{\eta t}\mathbf{u}\|_{L_p((0,T_0),L_q(B_R))} + \delta_1/2 \\ &\leq C_R(\eta p')^{1/p'} r_0(\epsilon) + \delta_1/2.\end{aligned}$$

Choosing $\epsilon > 0$ so small that $C_R(\eta p')^{1/p'} r_0(\epsilon) \leq \delta_1/2$ yields that

$$\sup_{0<t<T_1+1} |\xi_{\mathbf{u}_1,\rho_1}(t)| \leq \delta_1.$$

Since $T_1 + 1 = T_0 + 1/2$, repeating this argument, we can prolong $\mathbf{u}$, $\mathfrak{q}$, and ρ to time interval $(0, \infty)$. This completes the proof of Theorem 3.1.3.

Below, we prove (3.308). Since $\mathcal{I}$ will be chosen small enough eventually, we may assume that $0 < \mathcal{I} < 1$ below. In particular, we have

$$\begin{aligned}(\mathcal{I} + E_{p,q,T}(\mathbf{u}, \rho, \eta))^2 &\leq C(\mathcal{I} + E_{p,q,T}(\mathbf{u}, \rho; \eta)^2), \\ (\mathcal{I} + E_{p,q,T}(\mathbf{u}, \rho, \eta))^3 &\leq C(\mathcal{I} + E_{p,q,T}(\mathbf{u}, \rho; \eta)^2 + E_{p,q,T}(\mathbf{u}, \rho; \eta)^3).\end{aligned}$$

Let $(\mathbf{u}, \mathfrak{q}, \rho) \in \mathcal{S}_{p,q}((0, T))$ satisfy equations (3.271) and condition (3.307). We first estimate the nonlinear terms $\mathbf{f}(\mathbf{u}, \Psi_\rho)$ and $\bar{d}(\mathbf{u}, \Psi_\rho)$. Let H_ρ be a suitable extension of ρ to $\dot\Omega$ such that $H_\rho = \rho$ on S_R, and (3.253) and (3.306) hold. Since $p > 2$ and $1/p + 1/q < 1$, noting (3.253), we have

$$\begin{aligned}\|\mathbf{u}(\cdot, t)\|_{H^1_q(\dot\Omega)} &\leq C\|\mathbf{u}(\cdot, t)\|_{B^{2(1-1/p)}_{q,p}(\dot\Omega)}, \\ \|H_\rho(\cdot, t)\|_{H^2_q(\dot\Omega)} &\leq C\|\rho(\cdot, t)\|_{B^{3-1/p-1/q}_{q,p}(S_R)}.\end{aligned} \tag{3.312}$$

Moreover, by (3.212) and (3.213) we have

$$\begin{aligned}\sup_{t\in(0,T)} e^{\eta t}\|\mathbf{u}(\cdot, t)\|_{B^{2(1-1/p)}_{q,p}(\dot\Omega)} &\leq C(\mathcal{I} + E_{p,q,T}(\mathbf{u}, \rho; \eta)) \\ \sup_{t\in(0,T)} e^{\eta t}\|\rho(\cdot, t)\|_{B^{3-1/p-1/q}_{q,p}(S_R)} &\leq C(\mathcal{I} + E_{p,q,T}(\mathbf{u}, \rho; \eta)),\end{aligned} \tag{3.313}$$

which, combined with (3.312), leads to

$$\sup_{t\in(0,T)} e^{\eta t}\|\mathbf{u}(\cdot,t)\|_{H^1_q(\dot\Omega)} \leq C(\mathcal{I}+E_{p,q,T}(\mathbf{u},\rho;\eta)),$$
$$\sup_{t\in(0,T)} e^{\eta t}\|\rho(\cdot,t)\|_{W^{2-1/q}_q(S_R)} \leq C(\mathcal{I}+E_{p,q,T}(\mathbf{u},\rho;\eta)). \tag{3.314}$$

By (3.260) and (3.261), we have

$$\sup_{t\in(0,T)} |\xi_{\mathbf{u},\rho}(t)| \leq C(\mathcal{I}+E_{p,q,T}(\mathbf{u},\rho,\eta)),$$
$$\sup_{t\in(0,T)} e^{\eta t}|\xi'_{\mathbf{u},\rho}(t)| \leq C(\mathcal{I}+E_{p,q,T}(\mathbf{u},\rho,\eta)). \tag{3.315}$$

In the following, we use Sobolev's inequality especially to obtain the estimates:

$$\|v\|_{L_\infty(\dot\Omega)} \leq C\|v\|_{H^1_q(\dot\Omega)}, \quad \|v\|_{L_\infty(S_R)} \leq C\|v\|_{W^{1-1/q}_q(S_R)}.$$

From (3.209), (3.253), and Sobolev's inequality, we have

$$\begin{aligned}
&\|\mathbf{f}(\mathbf{u},\Psi_\rho)\|_{L_q(\dot\Omega)}\\
&\leq C\{(\|\mathbf{u}(\cdot,t)\|_{H^1_q(\dot\Omega)} + \|\partial_t H_\rho(\cdot,t)\|_{H^1_q(\dot\Omega)} + |\xi'_{\mathbf{u},\rho)}(t)|)\|\nabla\mathbf{u}(\cdot,t)\|_{L_q(\dot\Omega)}\\
&\quad + (\|H_\rho(\cdot,t)\|_{H^1_\infty(\dot\Omega)} + |\xi_{\mathbf{u},\rho}(t)|)(\|\mathbf{u}(\cdot,t)\|_{H^2_q(\dot\Omega)} + \|\partial_t\mathbf{u}(\cdot,t)\|_{L_q(\dot\Omega)})\\
&\quad + (\|H_\rho(\cdot,t)\|_{H^2_q(\dot\Omega)} + |\xi_{\mathbf{u},\rho}(t)|)\|\mathbf{u}(\cdot,t)\|_{H^1_q(\dot\Omega)}\}.
\end{aligned} \tag{3.316}$$

By (3.314), (3.315), and (3.316), we have

$$\|e^{\eta t}\mathbf{f}(\mathbf{u},\Psi_\rho)\|_{L_p((0,T),L_q(\dot\Omega))} \leq C(\mathcal{I}+E_{p,q,T}(\mathbf{u},\rho,\eta))E_{p,q,T}(\mathbf{u},\rho,\eta). \tag{3.317}$$

Here and in the following, C denotes a generic constant independent of ϵ.

We next consider $\bar d(\mathbf{u},\Psi_\rho)$. Let $d(\mathbf{u},\Psi_\rho)$ be the function defined in (3.177). Since $\bar\Psi_\rho = \zeta(y)R^{-1}H_\rho(y,t)y$, recalling that $\bar{\mathbf{k}} = \bar\nabla\Psi_\rho$ and using (3.307), (3.32), and (3.201), we have

$$\begin{aligned}
&\| < \mathbf{V}_\Gamma(\bar{\mathbf{k}})\bar\nabla\Psi_\rho, \bar\nabla\Psi_\rho > \|_{W^{1-1/q}_q(S_R)} \leq C\|\rho(\cdot,t)\|_{W^{2-1/q}_q(S_R)},\\
&\| < \mathbf{V}_\Gamma(\bar{\mathbf{k}})\bar\nabla\Psi_\rho, \bar\nabla\Psi_\rho > \|_{W^{2-1/q}_q(S_R)} \leq C\|\rho(\cdot,t)\|_{W^{2-1/q}_q(S_R)}\|\rho(\cdot,t)\|_{W^{3-1/q}_q(S_R)},\\
&\|\partial_t\rho(\cdot,t)\|_{L_\infty(S_R)} \leq C\|\partial_t\rho(\cdot,t)\|_{W^{1-1/q}_q(S_R)}.
\end{aligned} \tag{3.318}$$

Applying (3.260) and (3.318) to the formula in (3.177), we have

$$\|d(\mathbf{u},\Psi_\rho)(\cdot,t)\|_{W^{1-1/q}_q(S_R)}$$

$$\leq C\|\rho(\cdot,t)\|_{W_q^{2-1/q}(S_R)}(\|\partial_t\rho(\cdot,t)\|_{W_q^{1-1/q}(S_R)} + \|\mathbf{u}(\cdot,t)\|_{H_q^1(\dot\Omega)}),$$

$$\|d(\mathbf{u},\Psi_\rho)(\cdot,t)\|_{W_q^{2-1/q}(S_R)}$$
$$\leq C\{\|\rho(\cdot,t)\|_{W_q^{2-1/q}(S_R)}(\|\partial_t\rho(\cdot,t)\|_{W_q^{2-1/q}(S_R)} + \|\mathbf{u}(\cdot,t)\|_{H_q^2(\dot\Omega)})$$
$$+ (\|\partial_t\rho(\cdot,t)\|_{W_q^{1-1/q}(S_R)} + \|\mathbf{u}(\cdot,t)\|_{H_q^1(\dot\Omega)})\|\rho(\cdot,t)\|_{W_q^{2-1/q}(S_R)}\|\rho(\cdot,t)\|_{W_q^{3-1/q}(S_R)}$$
$$+ \|\mathbf{u}(\cdot,t)\|_{H_q^1(\dot\Omega)}\|\rho(\cdot,t)\|_{W_q^{3-1/q}(S_R)}\}. \tag{3.319}$$

By (3.201) and Sobolev's inequality, we have

$$\| <\mathbf{u} \mid \nabla'_\Gamma\rho> \|_{W_q^{1-1/q}(S_R))} \leq C\|\mathbf{u}(\cdot,t)\|_{H_q^1(B_R))}\|\rho(\cdot,t)\|_{W_q^{2-1/q}(S_R))}$$
$$\| <\mathbf{u} \mid \nabla'_\Gamma\rho> \|_{W_q^{2-1/q}(S_R))} \leq C\{\|\mathbf{u}(\cdot,t)\|_{H_q^2(B_R))}\|\rho(\cdot,t)\|_{W_q^{2-1/q}(S_R))} \tag{3.320}$$
$$+ \|\mathbf{u}(\cdot,t)\|_{H_q^1(B_R))}\|\rho(\cdot,t)\|_{W_q^{3-1/q}(S_R))}\}.$$

Moreover, noting (3.307), we have

$$\begin{aligned}\|\int_{S_R}\rho^\ell\,d\omega\|_{W_q^{2-1/q}(S_R)} &\leq C\|\rho(\cdot,t)\|^2_{W_q^{1-1/q}(S_R)},\\ \|\int_{S_R}\rho^\ell\omega\,d\omega\|_{W_q^{2-1/q}(S_R)} &\leq C\|\rho(\cdot,t)\|^2_{W_q^{1-1/q}(S_R)},\end{aligned} \tag{3.321}$$

for $\ell \geq 2$. Thus, by (3.319), (3.320), (3.321), and (3.314), we have

$$\|e^{\eta t}\bar d(\mathbf{u},\Psi_\rho)\|_{L_p((0,T),W_q^{1-1/q}(S_R))} \leq C(\mathcal{I} + E_{p,q,T}(\mathbf{u},\rho;\eta))E_{p,q,T}(\mathbf{u},\rho;\eta),$$
$$\|e^{\eta t}\bar d(\mathbf{u},\Psi_\rho)\|_{L_p((0,T),W_q^{2-1/q}(S_R))} \leq C\{(\mathcal{I} + E_{p,q,T}(\mathbf{u},\rho;\eta))E_{p,q,T}(\mathbf{u},\rho;\eta)$$
$$+ (\mathcal{I} + E_{p,q,T}(\mathbf{u},\rho;\eta))^2E_{p,q,T}(\mathbf{u},\rho;\eta)\}. \tag{3.322}$$

We now consider $g(\mathbf{u},\Psi_\rho)$ and $\mathbf{g}(\mathbf{u},\Psi_\rho)$. We want to extend $g(\mathbf{u},\Psi_\rho)$ and $\mathbf{g}(\mathbf{u},\Psi_\rho)$ to $t \in \mathbb{R}$. For this purpose, we extend Ψ_ρ for all $t \in \mathbb{R}$. Recall that $\Psi_\rho = \zeta(y)R^{-1}H_\rho(y,t)y + \xi_{\mathbf{u},\rho}(t)$. Since $\xi_{\mathbf{u},\rho}(0) = 0$, we define $\bar\xi_{\mathbf{u},\rho}(t) = e_T[\xi_{\mathbf{u},\rho}](t)$, where e_T is the same extension operator as in (3.228). Let $T_v(t)\rho_0$ and $T_h(t)\rho_0$ be the same functions as in Sect. 3.4.5. Let $\psi(t)$ be a function in $C^\infty(\mathbb{R})$ which equals 1 for $t > -1$ and 0 for $t < -2$. Let

$$\mathcal{E}_1[e^{\eta t}\mathbf{u}] = e_T[e^{\eta t}\mathbf{u}(y,t) - T_v(t)\mathbf{u}_0] + \psi(t)T_v(t)],$$
$$\bar{\mathcal{E}}_2[\Psi_\rho] = \zeta(y)(R^{-1}(e_T[H_\rho(y,t) - T_h(t)\rho_0] + \psi(t)T_h(t)\rho_0)y + \bar\xi_{\mathbf{u},\rho}(t)).$$

Since $(e^{\eta t}\mathbf{u}(y,t) - T_v(t)\mathbf{u}_0)|_{t=0} = \mathbf{u}_0 - \mathbf{u}_0 = 0$ and $(H_\rho(y,t) - T_h(t)\rho_0)|_{t=0} = H_{\rho_0} - H_{\rho_0} = 0$, we can use (3.229) for the time derivative of $\mathcal{E}_1[e^{\eta t}\mathbf{u}]$ and $\bar{\mathcal{E}}_2[\Psi_\rho]$. Notice that $\mathcal{E}_1[e^{\eta t}\mathbf{u}] = e^{\eta t}\mathbf{u}$ and $\bar{\mathcal{E}}_2[\Psi_\rho] = \Psi_\rho$ for $t \in (0, T)$. Choosing δ_1 smaller if necessary, we may assume that

$$\sup_{t\in\mathbb{R}} \|\bar{\mathcal{E}}_2[\Psi_\rho]\|_{H^1_\infty(\dot\Omega)} \leq \delta. \tag{3.323}$$

We first extend $g(\mathbf{u}, \Psi_\rho)$ and $\mathbf{g}(\mathbf{u}, \Psi_\rho)$ by using $\mathcal{E}_1[e^{\eta t}\mathbf{u}]$ and $\bar{\mathcal{E}}_2[\Psi_\rho]$. In view of (3.232), we may write

$$g(\mathbf{u}, \Psi_\rho) = V_g(\nabla\Psi_\rho)\nabla\Psi_\rho \otimes \nabla\mathbf{u}, \quad \mathbf{g}(\mathbf{u}, \Psi_\rho) = V_{\mathbf{g}}(\nabla\Psi_\rho)\nabla\Psi \otimes \mathbf{u}.$$

Let

$$\begin{aligned} g_\eta(\mathbf{u}, \Psi_\rho) &= V_g(\nabla\bar{\mathcal{E}}_2[\Psi_\rho])\nabla\bar{\mathcal{E}}_2[\Psi_\rho] \otimes \nabla\mathcal{E}_1[e^{\eta t}\mathbf{u}], \\ \mathbf{g}_\eta(\mathbf{u}, \Psi_\rho) &= V_{\mathbf{g}}(\nabla\bar{\mathcal{E}}_2[\Psi_\rho])\nabla\bar{\mathcal{E}}_2[\Psi_\rho] \otimes \mathcal{E}_1[e^{\eta t}\mathbf{u}]. \end{aligned}$$

We have

$$g_\eta(\mathbf{u}, \Psi_\rho) = e^{\eta t} g(\mathbf{u}, \Psi_\rho), \quad \mathbf{g}_\eta(\mathbf{u}, \Psi_\rho) = e^{\eta t}\mathbf{g}(\mathbf{u}, \Psi_\rho)$$

for $t \in (0, T)$, and therefore g_η and $\mathbf{g}_\eta$ are the extension of $e^{\eta t} g$ and $e^{\eta t}\mathbf{g}$ to the whole time interval $\mathbb{R}$. Moreover, we see that $\operatorname{div}\mathbf{g}_\eta(\mathbf{u}, \Psi_\rho) = g_\eta(\mathbf{u}, \Psi_\rho)$. Since

$$\begin{aligned} \partial_t\mathbf{g}_\eta(\mathbf{u}, \Psi_\rho) &= V_{\mathbf{g}}(\nabla\bar{\mathcal{E}}_2[\Psi_\rho])\partial_t\nabla\bar{\mathcal{E}}_2[\Psi_\rho] \otimes \mathcal{E}_1[e^{\eta t}\mathbf{u}] \\ &\quad + V_{\mathbf{g}}(\nabla\bar{\mathcal{E}}_2[\Psi_\rho])\nabla\bar{\mathcal{E}}_2[\Psi_\rho] \otimes \partial_t\mathcal{E}_1[e^{\eta t}\mathbf{u}] \\ &\quad + V'_{\mathbf{g}}(\nabla\bar{\mathcal{E}}_2[\Psi_\rho])\partial_t\nabla\bar{\mathcal{E}}_2[\Psi_\rho]\nabla\bar{\mathcal{E}}_2[\Psi_\rho] \otimes \mathcal{E}_1[e^{\eta t}\mathbf{u}]), \end{aligned}$$

by (3.323), Sobolev's inequality, and (3.253), we have

$$\begin{aligned} &\|\partial_t\mathbf{g}_\eta(\mathbf{u}, \Psi_\rho)\|_{L_p(\mathbb{R}, L_q(\dot\Omega))} \\ &\leq C\{(\mathcal{J} + \|\partial_t\rho\|_{L_p((0,T), W^{2-1/q}_q(S_R))})(\mathcal{J} + \|e^{\eta t}\mathbf{u}\|_{L_\infty((0,T), H^1_q(\dot\Omega))}) \\ &+ \sup_{0<t<T} |\xi'_{\mathbf{u},\rho}(t)|(\mathcal{J} + \|e^{\eta t}\mathbf{u}\|_{L_p((0,T), H^1_q(\dot\Omega))}) \\ &+ (\mathcal{J} + \|\rho\|_{L_\infty((0,T), W^{2-1/q}_q(S_R))} + \sup_{0<t<T} |\xi_{\mathbf{u},\rho}(t)|)(\mathcal{J} + \|e^{\eta t}\partial_t\mathbf{u}\|_{L_p((0,T), L_q(\dot\Omega))})\}. \end{aligned}$$

Thus, by (3.314), (3.315) and the fact that $0 < \mathcal{J} < 1$, we have

$$\|\partial_t\mathbf{g}_\eta(\mathbf{u}, \Psi_\rho)\|_{L_p(\mathbb{R}, L_q(\dot\Omega))} \leq C(\mathcal{J} + E_{p,q,T}(\mathbf{u}, \Psi_\rho)^2 + E_{p,q,T}(\mathbf{u}, \Psi_\rho)^3). \tag{3.324}$$

To estimate $g_\eta(\mathbf{u}, \Psi_\rho)$, we use Lemma 3.4.2. By (3.323), (3.227), and (3.253), we have

$$\begin{aligned}
&\|V_g(\nabla\bar{\mathcal{E}}_2[\Psi_\rho])\nabla\bar{\mathcal{E}}_2[\Psi_\rho]\|_{H^1_\infty(\mathbb{R},L_q(\dot\Omega))} \\
&\quad\le C(\mathcal{I}+\|\partial_t\rho\|_{L_\infty((0,T),W^{1-1/q}_q(S_R))}+\sup_{0<t<T}|\xi'_{\mathbf{u},\rho}(t)|), \\
&\|V_g(\nabla\bar{\mathcal{E}}_2[\Psi_\rho])\nabla\bar{\mathcal{E}}_2[\Psi_\rho]\|_{L_\infty(\mathbb{R},H^1_q(\dot\Omega))} \\
&\quad\le C(\mathcal{I}+\|\rho\|_{L_\infty((0,T),W^{2-1/q}_q(S_R))}+\sup_{0<t<T}|\xi_{\mathbf{u},\rho}(t)|).
\end{aligned} \tag{3.325}$$

Thus, by Lemma 3.4.2, we have

$$\begin{aligned}
&\|g_\eta(\mathbf{u}, \Psi_\rho)\|_{H^{1/2}_p(\mathbb{R},L_q(\dot\Omega))}+\|g_\eta(\mathbf{u}, \Psi_\rho)\|_{L_p(\mathbb{R},H^1_q(\dot\Omega))} \\
&\le C(\mathcal{I}+\|\partial_t\rho\|_{L_\infty((0,T),W^{1-1/q}_q(S_R))}+\sup_{0<t<T}|\xi'_{\mathbf{u},\rho}(t)| \\
&+\|\rho\|_{L_\infty((0,T),W^{2-1/q}_q(S_R))}+\sup_{0<t<T}|\xi_{\mathbf{u},\rho}(t)|) \\
&\times(\|\nabla\mathcal{E}_1[e^{\eta t}\mathbf{u}]\|_{H^{1/2}_p(\mathbb{R},L_q(\dot\Omega))}+\|\nabla\mathcal{E}_1[e^{\eta t}\mathbf{u}]\|_{L_p(\mathbb{R},H^1_q(\dot\Omega))}).
\end{aligned} \tag{3.326}$$

Applying Lemma 3.4.3, we have

$$\begin{aligned}
&\|\nabla\mathcal{E}_1[e^{\eta t}\mathbf{u}]\|_{H^{1/2}_p(\mathbb{R},L_q(\dot\Omega))}+\|\nabla\mathcal{E}_1[e^{\eta t}\mathbf{u}]\|_{L_p(\mathbb{R},H^1_q(\dot\Omega))} \\
&\quad\le C(\mathcal{I}+\|e^{\eta t}\partial_t\mathbf{u}\|_{L_p((0,T),L_q(\dot\Omega))}+\|e^{\eta t}\mathbf{u}\|_{L_p((0,T),H^2_q(\dot\Omega))}),
\end{aligned}$$

which, combined with (3.326) and (3.315), leads to

$$\begin{aligned}
&\|g_\eta(\mathbf{u}, \Psi_\rho)\|_{H^{1/2}_p(\mathbb{R},L_q(\dot\Omega))}+\|g_\eta(\mathbf{u}, \Psi_\rho)\|_{L_p(\mathbb{R},H^1_q(\dot\Omega))} \\
&\quad\le C(\mathcal{I}+E_{p,q,T}(\mathbf{u}, \rho; \eta)^2).
\end{aligned} \tag{3.327}$$

We next consider $\mathbf{h}'(\mathbf{u}, \Psi_\rho)$. Instead of $\bar{\mathcal{E}}_2[\Psi_\rho]$, we use $\tilde{\mathcal{E}}_2[\Psi_\rho]$, where $\tilde{\mathcal{E}}_2[\Psi_\rho]$ is defined by setting

$$\tilde{\mathcal{E}}_2[\Psi_\rho]=\zeta(y)R^{-1}(e_T[H_\rho(y,t)-T_h(t)\rho_0]+\psi(t)T_h(|t|)\rho_0)y.$$

We may assume that

$$\sup_{t\in\mathbb{R}}\|\tilde{\mathcal{E}}_2[\Psi_\rho](\cdot, t)\|_{H^1_\infty(\dot\Omega)}\le\delta. \tag{3.328}$$

Note that $\tilde\Psi_\rho=\tilde{\mathcal{E}}_2[\Psi_\rho]$ for $t\in(0, T)$. In view of (3.183), we may write

$$\mathbf{h}'(\mathbf{u}, \Psi_\rho) = V_{\mathbf{h}'}(\bar\nabla\tilde\Psi_\rho)\bar\nabla\tilde\Psi_\rho \otimes \nabla\mathbf{u}.$$

Thus, we define the extension of $e^{\eta t}\mathbf{h}'(\mathbf{u}, \Psi_\rho)$ by setting

$$\mathbf{h}'_\eta = V_{\mathbf{h}'}(\bar\nabla\tilde{\mathcal{E}}_2[\Psi_\rho])\bar\nabla\tilde{\mathcal{E}}_2[\Psi_\rho] \otimes \nabla\mathcal{E}_1[e^{\eta t}\mathbf{u}].$$

And then, employing the same argument as obtaining (3.327), we have

$$\begin{aligned} &\|\mathbf{h}'_\eta(\mathbf{u}, \Psi_\rho)\|_{H^{1/2}_p(\mathbb{R}, L_q(\dot\Omega))} + \|\mathbf{h}'_\eta(\mathbf{u}, \Psi_\rho)\|_{L_p(\mathbb{R}, H^1_q(\dot\Omega))} \\ &\quad \leq C(\mathcal{I} + E_{p,q,T}(\mathbf{u}, \rho; \eta)^2). \end{aligned} \tag{3.329}$$

We finally consider $h_N(\mathbf{u}, \Psi_\rho)$ given in (3.274). In view of (3.274), we define the extension of $h_N(\mathbf{u}, \Psi_\rho)$ to the whole time interval $\mathbb{R}$ by letting

$$\begin{aligned} h_{\eta,N}(\mathbf{u}, \Psi_\rho) &= V_{h,N}(\bar\nabla\tilde{\mathcal{E}}_2[\Psi_\rho])\bar\nabla\tilde{\mathcal{E}}_2[\Psi_\rho] \otimes \nabla\mathcal{E}_1[e^{\eta t}\mathbf{u}]) \\ &\quad + \sigma\tilde V'_\Gamma(\bar\nabla\tilde{\mathcal{E}}_2[\Psi_\rho])\bar\nabla\tilde{\mathcal{E}}_2[\Psi_\rho] \otimes \bar\nabla^2\tilde{\mathcal{E}}_2[e^{\eta t}\Psi_\rho]). \end{aligned}$$

Notice that $e^{\eta t}\Psi_\rho = \Psi_{e^{\eta t}\rho}$. Employing the same argument as in obtaining (3.326), we have

$$\begin{aligned} &\|h_{\eta,N}(\mathbf{u}, \Psi_\rho)\|_{H^{1/2}_p(\mathbb{R}, L_q(\dot\Omega))} + \|h_{\eta,N}(\mathbf{u}, \Psi_\rho)\|_{L_p(\mathbb{R}, H^1_q(\dot\Omega))} \\ &\quad \leq C(\mathcal{I} + \|\partial_t\rho\|_{L_\infty((0,T), W^{1-1/q}_q(S_R))} + \|\Psi_\rho\|_{L_\infty((0,T), H^2_q(\dot\Omega))}) \\ &\quad \times (\|\nabla\mathcal{E}_1[e^{\eta t}\mathbf{u}]\|_{H^{1/2}_p(\mathbb{R}, L_q(\dot\Omega))} + \|\nabla\mathcal{E}_1[e^{\eta t}\mathbf{u}]\|_{L_p(\mathbb{R}, H^1_q(\dot\Omega))} \\ &\quad + \|\bar\nabla^2\tilde{\mathcal{E}}_2[e^{\eta t}\Psi_\rho]\|_{H^{1/2}_p(\mathbb{R}, L_q(\dot\Omega))} + \|\bar\nabla^2\tilde{\mathcal{E}}_2[e^{\eta t}\Psi_\rho]\|_{L_p(\mathbb{R}, H^1_q(\dot\Omega))}) \end{aligned}$$

Since

$$\begin{aligned} \|e^{\eta t}\bar\nabla^2\tilde{\mathcal{E}}_2[\Psi_\rho]\|_{H^{1/2}_p(\mathbb{R}, L_q(\dot\Omega))} &\leq \|e^{\eta t}\bar\nabla^2\tilde{\mathcal{E}}_2[\Psi_\rho]\|_{H^1_p(\mathbb{R}, L_q(\dot\Omega))} \\ &\leq C(\mathcal{I} + \|e^{\eta t}\Psi_\rho\|_{H^1_p((0,T), H^2_q(\dot\Omega))}); \\ \|\bar\nabla^2\tilde{\mathcal{E}}_2[e^{\eta t}\Psi_\rho]\|_{L_p(\mathbb{R}, H^1_q(\dot\Omega))}) &\leq C(\mathcal{I} + \|e^{\eta t}\Psi_\rho\|_{L_p((0,T), H^3_q(\dot\Omega))}), \end{aligned}$$

using (3.253), we have

$$\begin{aligned} &\|h_{\eta,N}(\mathbf{u}, \Psi_\rho)\|_{H^{1/2}_p(\mathbb{R}, L_q(\dot\Omega))} + \|h_{\eta,N}(\mathbf{u}, \Psi_\rho)\|_{L_p(\mathbb{R}, H^1_q(\dot\Omega))} \\ &\quad \leq C(\mathcal{I} + E_{p,q,T}(\mathbf{u}, \rho; \eta))^2. \end{aligned} \tag{3.330}$$

Applying Theorem 3.5.2 to equations (3.271) and using (3.317), (3.322), (3.324), (3.327), (3.329), and (3.330), we have

$$\begin{aligned}
&\|e^{\eta t}\mathbf{u}\|_{L_p((0,T),H_q^2(\dot\Omega))} + \|e^{\eta t}\partial_t\mathbf{u}\|_{L_p((0,T),L_q(\dot\Omega))} \\
&\quad + \|e^{\eta t}\rho\|_{L_p((0,T),W_q^{3-1/q}(S_R))} + \|e^{\eta t}\partial_t\rho\|_{L_p((0,T),W_q^{2-1/q}(S_R))} \\
&\quad \leq C(\mathfrak{I} + E_{p,q,T}(\mathbf{u},\eta;\eta)^2 + E_{p,q,T}(\mathbf{u},\rho;\eta)^3)
\end{aligned} \tag{3.331}$$

for some positive constants C. Moreover, by the third equation in equation (3.271), we have

$$\begin{aligned}
&\|e^{\eta t}\partial_t\rho\|_{L_\infty((0,T),W_q^{1-1/q}(S_R))} \\
&\quad \leq C(\|e^{\eta t}\mathbf{u}\|_{L_\infty((0,T),H_q^1(\dot\Omega))} + \|e^{\eta t}\tilde d(\mathbf{v},\Psi_h)\|_{L_\infty((0,T),W_q^{1-1/q}(S_R))}),
\end{aligned}$$

which, combined with (3.212), (3.213), and (3.331), yields

$$\|e^{\eta t}\partial_t\rho\|_{L_\infty((0,T),W_q^{1-1/q}(S_R))} \leq C(\mathfrak{I} + E_{p,q,T}(\mathbf{u},\rho;\eta)^2 + E_{p,q,T}(\mathbf{u},\rho;\eta)^3).$$

Combining this inequality with (3.331) gives

$$E_{p,q,T}(\mathbf{u},\rho) \leq C(\mathfrak{I} + E_{p,q,T}(\mathbf{u},\rho;\eta)^2 + E_{p,q,T}(\mathbf{u},\rho;\eta)^3) \tag{3.332}$$

for some positive constants C, and therefore we have (3.308).

Finally, we define Γ_t and $\Omega_{t\pm}$ by

$$\begin{aligned}
\Gamma_t &= \{x = (1 + R^{-1}\rho(y,t))y + \xi_{\mathbf{u},\rho}(t) \mid y \in S_R\}, \\
\Omega_{t+} &= \{x = y + \Psi_\rho(y,t) \mid y \in B_R\}, \\
\Omega_{t-} &= \{x = y + \Psi_\rho(y,t) \mid y \in \Omega \setminus (B_R \cup S_R)\},
\end{aligned}$$

where $\Psi_\rho(y,t) = \eta(y)(R^{-1}H_\rho(y,t)y + \xi_{\mathbf{u},\rho}(t))$. Note that $\Omega = \Omega_{t+} \cup \Gamma_t \cup \Omega_{t-}$ (cf. (3.249), (3.254), and (3.257)). Let

$$\xi(t) = \frac{1}{|\Omega_+|}\int_{\Omega_{t+}} x\,dx.$$

Let $\mathbf{v}(x,t) = \mathbf{u}(h_\rho^{-1}(x,t),t)$ and $\mathfrak{p}(x,t) = \frac{\sigma(N-1)}{R} + \mathfrak{q}(h_\rho^{-1}(x,t),t)$, and then by (3.271), $\operatorname{div}\mathbf{v} = 0$, and so $|\Omega_{t+}| = |\Omega_+| = |B_R|$ by (A.1). Thus, by (3.252) we have $\xi'(t) = \xi'_{\mathbf{u},\rho}(t)$. By (A.2), $\xi(0) = 0$, and so $\xi(t) = \xi_{\mathbf{u},\rho}(t)$, that is,

$$\xi_{\mathbf{u},\rho}(t) = \frac{1}{|\Omega_t|}\int_{\Omega_t} x\,dx.$$

Thus, by (3.266), (3.267), (3.268), and (3.269), we have

$$\partial_t \rho - \mathbf{n} \cdot P\mathbf{u} = \tilde{d}(\mathbf{u}, \Psi_\rho).$$

Thus, $\mathbf{u}$, $\mathfrak{q}$, and ρ satisfy Eq. (3.271). This completes the proof of Theorem 3.1.3.

3.6 Global Well-Posedness in the Case That $\Omega = \mathbb{R}^N$ with $N \geq 3$

3.6.1 *Reformulation of Problem (3.40)*

In this section, we shall prove the global well-posedness in the case where $\Omega = \mathbb{R}^N$. We first repeat the formulation of the problem, although it was given in the corresponding part of Subsec: 3.1.3.2. As was stated in Sect. 3.5.1, under the assumptions (A.1), (A.2), and (A.3), we introduce the barycenter $\xi(t)$ of Ω_{t+}, which is defined by (3.250). Since $\operatorname{div} \mathbf{v} = 0$ in Ω_{t+}, we have (3.251). By the assumption (A.2), we have $\xi(0) = 0$. Moreover, we have (3.252). For any function $\rho(y, t)$ defined on $S_R \times (0, T)$, let H_ρ be a suitable extension of ρ such that $\rho(y, t) = H_\rho(y, t)$ for $(y, t) \in S_R \times (0, T)$ and

$$\begin{aligned}
&\|H_\rho(\cdot, t)\|_{H^k_{r_i}(B_L)} \leq C\|\rho(\cdot, t)\|_{W^{k-1/r_i}_{r_i}(S_R)} && \text{for } k = 1, 2, 3, \\
&\|\partial_t H_\rho(\cdot, t)\|_{H^k_{r_i}(B_L)} \leq C\|\partial_t \rho(\cdot, t)\|_{W^{k-1/r_i}_{r_i}(S_R)} && \text{for } k = 1, 2,
\end{aligned} \tag{3.333}$$

for $i = 0, 1, 2$, where C is a constant independent of T, $r_0 = q_1/2$, $r_1 = q_1$, and $r_2 = q_2$, and L is a large positive number for which $L > 5R + 2$. Let $\omega(y)$ be a function in $C^\infty_0(\mathbb{R}^N)$ which equals one for $y \in B_{L-1}$ and zero for $y \notin B_L$. In the following, we set

$$\Psi_\rho(y, t) = R^{-1} H_\rho(y, t)\omega(y)y$$

and we assume that

$$\sup_{t \in (0,T)} \|\Psi_\rho(\cdot, t)\|_{H^1_\infty(B_L)} \leq \delta \tag{3.334}$$

with suitably small positive number δ. We define the Hanzawa transform by setting

$$\mathbf{h}_\rho(y, t) = y + \Psi_\rho(y, t) + \xi(t).$$

Choosing $\delta > 0$ small enough, we see that the correspondence: $x = \mathbf{h}_\rho(y, t)$ is injective for any $t \in (0, T)$. Let $B^R = \{x \in \mathbb{R}^N \mid |x| > R\}$ and set $\dot{\Omega} = \Omega_+ \cup \Omega_- =$

$\mathbb{R}^N \setminus S_R$ with $\Omega_+ = B_R$ and $\Omega_- = B^R$. Moreover, we set

$$\Gamma_t = \{x = \mathbf{h}_\rho(y,t) \mid y \in S_R\}, \quad \Omega_{t\pm} = \{x = \mathbf{h}_\rho(y,t) \mid y \in \Omega_\pm\}.$$

Let $\mathbf{v}$ and $\mathfrak{p}$ satisfy equations (3.1), and we set

$$\mathbf{u}(y,t) = \mathbf{v}(\mathbf{h}_\rho(y,t),t), \quad \mathfrak{q}(y,t) = \mathfrak{p}(\mathbf{h}_\rho(y,t),t) - \frac{\sigma(N-1)}{R}.$$

Moreover, we set $\mathbf{u}_0(y) = \mathbf{v}_0(\mathbf{h}_{\rho_0}(y))$, where $\mathbf{v}_0$ is the initial data and $\mathbf{h}_{\rho_0}(y) = y + \Psi_\rho|_{t=0}$ with $\rho|_{t=0} = \rho_0$. Then, we see that $\mathbf{u}$, $\mathfrak{q}$, and ρ satisfy equations (3.40)

To handle the zero and first eigenvalues of the Laplace-Beltrami operator Δ_{S_R}, we introduce the relations (3.266) and (3.267), and so our problem is finally the following equations:

$$\begin{cases} \partial_t\mathbf{u} - m^{-1}\mathrm{Div}\,(\mu\mathbf{D}(\mathbf{u}) - \mathfrak{q}\mathbf{I}) = \mathbf{f}(\mathbf{u}, \Psi_\rho) & \text{in } \dot\Omega^T, \\ \mathrm{div}\,\mathbf{u} = g(\mathbf{u}, H_\rho) = \mathrm{div}\,\mathbf{g}(\mathbf{u}, \Psi_\rho) & \text{in } \dot\Omega^T, \\ \partial_t\rho + \mathcal{L}_3\rho - \mathbf{n}\cdot P\mathbf{u} = \bar d(\mathbf{u}, \Psi_\rho) & \text{on } S_R^T, \\ [[(\mu\mathbf{D}(\mathbf{u})\mathbf{n})_\tau]] = [[\mathbf{h}'(\mathbf{u}, \Psi_\rho)]] & \text{on } S_R^T, \\ [[< \mu\mathbf{D}(\mathbf{u})\mathbf{n}, \mathbf{n}> - \mathfrak{q}]] - \sigma\mathcal{B}\rho = [[h_N(\mathbf{u}, \Psi_\rho)]] & \text{on } S_R^T, \\ [[\mathbf{u}]] = 0 & \text{on } S_R^T, \\ (\mathbf{u}, \rho)|_{t=0} = (\mathbf{u}_0, \rho_0) & \text{in } \dot\Omega \times S_R, \end{cases} \tag{3.335}$$

where $\mathcal{L}_3$ is the operator given in (3.278) and $\mathbf{n} = y/|y|$ for $y \in \mathbb{R}^N \setminus \{0\}$.

The purpose of this section is to prove the unique existence of solutions of equation (3.335) globally in time. Namely, we shall prove Theorem 3.1.4. In view of Theorem 3.1.2, we may assume that problem (3.335) admits a unique solution up to certain time $T > 0$. Choosing the initial data small enough, we may assume that $T > 6$. And then, we shall prove that local in time solution can be prolonged beyond T.

Let r_i and p be the numbers given in Theorem 3.1.4. Let $\mathbf{u}$, $\mathfrak{q}$, and ρ be solutions of equations (3.335) satisfying (3.334) with

$$\mathbf{u} \in \bigcap_{i=0,1,2} (H^1_p((0,T), L_{r_i}(\dot\Omega)) \cap L_p((0,\infty), H^2_{r_i}(\dot\Omega)),$$

$$\rho \in \bigcap_{i=0,1,2} (H^1_p((0,T), W^{2-1/r_i}_{r_i}(S_R)) \cap L_p((0,\infty), W^{3-1/r_i}_{r_i}(S_R)).$$

As was stated in the first part of Sect. 3.5.6, to prolong such local solutions beyond T, it suffices to prove that

$$E_T(\mathbf{u}, \rho) \leq M(\mathcal{I} + E_T(\mathbf{u}, \rho)^2 + E_T(\mathbf{u}, \rho)^3) \tag{3.336}$$

for some constant $M > 0$ independent of ϵ and T, where $E_T(\mathbf{u}, \rho)$ is the quantity given in Theorem 3.1.4, and we have set

$$\mathcal{I} = \sum_{i=0,1,2} (\|\mathbf{u}_0\|_{B^{2(1-1/r_i)}_{r_i,p}(\dot{\Omega})} + \|\rho_0\|_{B^{3-1/p-1/r_i}_{r_i,p}(S_R)}).$$

Since we choose $\mathcal{I}$ is small enough eventually, we may assume that $\mathcal{I} \leq 1$. Below, we shall show (3.336).

3.6.2 Decay Properties of Solutions for Some Linear Equations

In this section, we shall study decay properties of solutions for some linear equations. We first consider the time-shifted equations:

$$\begin{cases} \partial_t \mathbf{u} + \lambda_1 \mathbf{u} - m^{-1}\mathrm{Div}\,(\mu\mathbf{D}(\mathbf{u}) - \mathfrak{p}\mathbf{I}) = \mathbf{f} & \text{in } \dot{\Omega}^T, \\ \mathrm{div}\,\mathbf{u} = g = \mathrm{div}\,\mathbf{g} & \text{in } \dot{\Omega}^T, \\ \partial_t \rho + \lambda_1 \rho + \mathcal{L}_3 \rho - \mathbf{n} \cdot P\mathbf{u} = d & \text{on } S_R^T, \\ [[(\mu\mathbf{D}(\mathbf{u}) - \mathfrak{p}\mathbf{I})\mathbf{n}]] - \sigma(\mathcal{B}\rho)\mathbf{n} = \mathbf{h} & \text{on } S_R^T, \\ [[\mathbf{u}]] = 0 & \text{on } S_R^T, \\ (\mathbf{u}, \rho)|_{t=0} = (\mathbf{u}_0, \rho_0) & \text{on } \dot{\Omega} \times S_R. \end{cases} \tag{3.337}$$

For the time-shifted equations (3.337), the following theorem holds.

Theorem 3.6.1 *Let $1 < p, q < \infty$ and $T > 0$. Assume that $2/p + 1/q \neq 1, 2$. Let $a \geq 0$. Then, there exist positive constants $\lambda_2 > 0$ such that if $\lambda_1 \geq \lambda_2$, then the following assertion holds: Let $\mathbf{u}_0 \in B^{2(1-1/p)}_{q,p}(\dot{\Omega})^N$ and $\rho_0 \in B^{3-1/p-1/q}_{q,p}(S_R)$ be initial data for equations* (3.280) *and let $\mathbf{f}$, g, $\mathbf{g}$, d, $\mathbf{h}$ be given functions in the right side of equations* (3.280). *Assume that*

$$\mathbf{f} \in L_p(\mathbb{R}, L_q(\dot{\Omega})^N), \quad d \in L_p(\mathbb{R}, W^{2-1/q}_q(S_R)).$$

Moreover, there exist $\tilde{g}$, $\tilde{\mathbf{g}}$, and $\tilde{\mathbf{h}}$ such that $< t >^a g = \tilde{g}$, $< t >^a \mathbf{g} = \tilde{\mathbf{g}}$, $< t >^a \mathbf{h} = \tilde{\mathbf{h}}$, for $t \in (0, T)$, and

$$\tilde{g} \in H^1_p(\mathbb{R}, H^1_q(\dot{\Omega})) \cap H^{1/2}_p(\mathbb{R}, L_q(\dot{\Omega})),$$
$$\tilde{\mathbf{g}} \in H^1_p(\mathbb{R}, L_q(\dot{\Omega})^N), \quad \mathrm{div}\,\tilde{\mathbf{g}} = \tilde{g},$$

$$\mathbf{h} \in H^1_p(\mathbb{R}, H^1_q(\dot\Omega)^N) \cap H^{1/2}_p(\mathbb{R}, L_q(\dot\Omega)^N).$$

Assume that the compatibility condition:

$$\begin{aligned} \operatorname{div} \mathbf{u}_0 = g|_{t=0} \quad &\text{in } \dot\Omega \\ [[(\mu\mathbf{D}(\mathbf{u}_0)\mathbf{n})_\tau]] = [[(\mathbf{h}|_{t=0})_\tau]] \quad &\text{on } S_R \text{ provided } 2/p + 1/q < 1, \\ [[\mathbf{u}_0]] = 0 \quad \text{on } S_R, \quad \mathbf{u}_0 = 0 \quad &\text{on } \Gamma_- \text{ provided } 2/p + 1/q < 2. \end{aligned}$$

Then, problem (3.280) *admits a unique solution* $(\mathbf{u}, \mathfrak{q}, \rho) \in \mathcal{S}_{p,q}((0, T))$ *possessing the estimate:*

$$\mathcal{E}_T(\mathbf{u}, \rho, a) \le C(\mathcal{I}(\mathbf{u}_0, \rho_0) + \mathcal{F}_T(\mathbf{f}, g, \mathbf{g}, \mathbf{h}, a)) \tag{3.338}$$

with some constant $C > 0$ *independent of* T, *where we have set*

$$\begin{aligned} \mathcal{I}(\mathbf{u}_0, \rho_0) &= \|\mathbf{u}_0\|_{B^{2(1-1/p)}_{q,p}(\dot\Omega)} + \|\rho\|_{B^{3-1/p-1/q}_{q,p}(S_R)}, \\ \mathcal{E}_T(\mathbf{u}, \rho, a) &= \| < t >^a \partial_t \mathbf{u}\|_{L_p((0,T),L_q(\dot\Omega))} + \| < t >^a \mathbf{u}\|_{L_p((0,T),H^2_q(\dot\Omega))} \\ &\quad + \| < t >^a \partial_t \rho\|_{L_p((0,T),W^{2-1/q}_q(S_R))} + \| < t >^a \rho\|_{L_p((0,T),W^{3-1/q}_q(S_R))}, \\ \mathcal{F}_T(\mathbf{f}, g, \mathbf{g}, \mathbf{h}, a) &= \| < t >^a \mathbf{f}\|_{L_p((0,T),L_q(\dot\Omega))} + \| < t >^a d\|_{L_p((0,T),W^{2-1/q}_q(S_R))} \\ &\quad + \|\partial_t \tilde{\mathbf{g}}\|_{L_p(\mathbb{R},L_q(\dot\Omega))} + \|(\tilde g, \tilde{\mathbf{h}})\|_{L_p(\mathbb{R},H^1_q(\dot\Omega))} + \|(\tilde g, \tilde{\mathbf{h}})\|_{H^{1/2}_p(\mathbb{R},L_q(\dot\Omega))} \end{aligned}$$

for some constant C *independent of* T.

Proof When $a = 0$, by Theorem 3.5.3 problem (3.337) admits a unique solution $(\mathbf{u}, \mathfrak{q}, \rho) \in \mathcal{S}_{p,q}((0, T))$ possessing the estimate:

$$\mathcal{E}_T(\mathbf{u}, \rho, 0) \le C(\mathcal{I}(\mathbf{u}_0, \rho_0) + \mathcal{F}_T(\mathbf{f}, g, \mathbf{g}, \mathbf{h}, 0)). \tag{3.339}$$

When $0 < a \le 1$, let $< t >^a \mathbf{u} = \mathbf{v}$, $< t >^a \mathfrak{p} = \mathfrak{q}$ and $< t >^a \rho = h$, and then $\mathbf{v}$, $\mathfrak{q}$, and h satisfy the equations:

$$\begin{aligned} 2\partial_t \mathbf{v} + \lambda_1 \mathbf{v} - m^{-1}\mathrm{Div}\,(\mu\mathbf{D}(\mathbf{v}) - \mathfrak{q}\mathbf{I}) = 2a < t >^{a-1} t\mathbf{u} + < t >^a \mathbf{f} \quad &\text{in } \dot\Omega^T, \\ \operatorname{div} \mathbf{v} = < t >^a g = \operatorname{div}(< t >^a \mathbf{g}) \quad &\text{in } \dot\Omega^T, \\ \partial_t h + \lambda_1 h + \mathcal{L}_3 h - \mathbf{n} \cdot P\mathbf{v} = 2a < t >^{a-1} t\rho + < t >^a d \quad &\text{on } S^T_R, \\ [[(\mu\mathbf{D}(\mathbf{v}) - \mathfrak{q}\mathbf{I})\mathbf{n}]] - \sigma(\mathcal{B}h)\mathbf{n} = < t >^a \mathbf{h} \quad &\text{on } S^T_R, \\ [[\mathbf{v}]] = 0 \quad &\text{on } S^R_T, \\ (\mathbf{v}, h)|_{t=0} = (\mathbf{u}_0, \rho_0) \quad &\text{on } \dot\Omega \times S_R. \end{aligned}$$

Then, by Theorem 3.5.3, $< t >^a$ **u**, $< t >^a$ $\mathfrak{p} = \mathfrak{q}$ and $< t >^a$ $\mathbf{u} = \mathbf{v}$ are unique solutions of equations (3.337) possessing the estimate:

$$\begin{aligned}\mathcal{E}_T(\mathbf{u}, \rho, a) \leq C(\mathcal{I}(\mathbf{u}_0, \rho_0) + \mathcal{F}_T(\mathbf{f}, g, \mathbf{g}, \mathbf{h}, a)\\ + \|\mathbf{u}\|_{L_p((0,T),L_q(\dot\Omega))} + \|\rho\|_{L_p((0,T),W^{2-1/q}_q(S_R))}).\end{aligned} \tag{3.340}$$

Here, we have used that $| < t >^{a-1} | \leq 1$ for any $t \in \mathbb{R}$. Thus, combining (3.339) and (3.340) yields (3.338).

When $a > 1$, repeated use of the argument above yields (3.338). This completes the proof of Theorem 3.6.1. □

We next consider the equations:

$$\begin{cases} \partial_t\mathbf{u} - m^{-1}\mathrm{Div}\,(\mu\mathbf{D}(\mathbf{u}) - \mathfrak{p}\mathbf{I}) = \mathbf{f} & \text{in } \dot\Omega^T,\\ \mathrm{div}\,\mathbf{u} = 0 & \text{in } \dot\Omega^T,\\ \partial_t\rho + \mathcal{L}_3\rho - \mathbf{n}\cdot P\mathbf{u} = d & \text{on } S_R^T,\\ [[(\mu\mathbf{D}(\mathbf{u}) - \mathfrak{p}\mathbf{I})\mathbf{n}]] - \sigma(\mathcal{B}\rho)\mathbf{n} = 0 & \text{on } S_R^T,\\ [[\mathbf{u}]] = 0 & \text{on } S_T^R,\\ (\mathbf{u}, \rho)|_{t=0} = (0, 0) & \text{on } \dot\Omega \times S_R. \end{cases} \tag{3.341}$$

To solve equations (3.341), we formulate the problem in the analytic semigroup setting. Namely, we consider the equations:

$$\begin{cases} \partial_t\mathbf{v} - m^{-1}\mathrm{Div}\,(\mu\mathbf{D}(\mathbf{v}) - K(\mathbf{v}, \eta)\mathbf{I}) = 0 & \text{in } \dot\Omega^T,\\ [[(\mu\mathbf{D}(\mathbf{v}) - K(\mathbf{v}, \eta)\mathbf{I})\mathbf{n}]] - (\mathcal{B}\eta)\mathbf{n} = 0, \quad [[\mathbf{v}]] = 0 & \text{on } S_R^T,\\ \partial_t\eta + \mathcal{L}_3\eta - \mathbf{n}\cdot P\mathbf{v} = 0 & \text{on } S_R^T,\\ (\mathbf{v}, \eta)|_{t=0} = (\mathbf{v}_0, \eta_0) & \text{in } \dot\Omega \times \Gamma, \end{cases} \tag{3.342}$$

where $v = K(\mathbf{v}, \eta)$ be a unique solution of the variational problem:

$$(m^{-1}\nabla v, \nabla\varphi)_{\dot\Omega} = (m^{-1}\mathrm{Div}\,(\mu\mathbf{D}(\mathbf{v})) - \nabla\mathrm{div}\,\mathbf{v}, \nabla\varphi)_{\dot\Omega}$$

for any $\varphi \in \hat H^1_{q'}(\mathbb{R}^N)$ subject to

$$[[v]] =< [[\mu\mathbf{D}(\mathbf{v})\mathbf{n}]], \mathbf{n} > -[[\mathrm{div}\,\mathbf{v}]] - \mathcal{B}\eta \quad \text{on } S_R.$$

Let

$$
\begin{aligned}
J_q(\dot\Omega) &= \{\mathbf{f} \in L_q(\dot\Omega)^N \mid (\mathbf{f}, \nabla\varphi)_{\dot\Omega} = 0 \quad \text{for any } \varphi \in \hat H^1_{q'}(\mathbb{R}^N)\},\\
\mathcal{H}_q &= \{(\mathbf{v}, \eta) \mid \mathbf{v} \in J_q(\dot\Omega), \quad \eta \in W^{2-1/q}_q(S_R)\},\\
\mathcal{D}_q &= \{(\mathbf{v}, \eta) \in \mathcal{H}_q \mid \mathbf{v} \in H^2_q(\dot\Omega)^N, \quad \eta \in W^{3-1/q}_q(S_R)\\
&\qquad [[(\mu\mathbf{D}(\mathbf{v})\mathbf{n})_\tau]] = 0, \quad [[\mathbf{v}]] = 0 \quad \text{on } S_R\},\\
\mathcal{A}_q(\mathbf{v}, \eta) &= (m^{-1}\mathrm{Div}\,(\mu\mathbf{D}(\mathbf{v}) - K(\mathbf{v}, \eta)\mathbf{I}), (-\mathcal{L}_3\eta + \mathbf{n}\cdot P\mathbf{v})|_{S_R})
\end{aligned}
\tag{3.343}
$$

for $(\mathbf{v}, \eta) \in \mathcal{D}_q$. In the following, we set

$$
\begin{aligned}
\|(\mathbf{v}, \eta)\|_{\mathcal{H}_q} &= \|\mathbf{v}\|_{L_q(\dot\Omega)} + \|\eta\|_{W^{2-1/q}_q(S_R)},\\
\|(\mathbf{v}, \eta)\|_{\mathcal{D}_q} &= \|\mathbf{v}\|_{H_q(\dot\Omega)} + \|\eta\|_{W^{3-1/q}_q(S_R)}.
\end{aligned}
\tag{3.344}
$$

Using the symbols defined in (3.343), problem (3.342) is written as

$$\partial_t(\mathbf{v}, \eta) - \mathcal{A}_q(\mathbf{v}, \eta) = (0, 0) \quad \text{for } t > 0, \quad (\mathbf{v}, \eta)|_{t=0} = (\mathbf{v}_0, \eta_0).$$

As was mentioned in Sect. 3.2.2.1, the operator $\mathcal{A}_q$ generates a C^0 analytic semigroup $\{T(t)\}_{t\geq 0}$. To represent solutions $\mathbf{u}$ and ρ by using $\{T(t)\}_{t\geq 0}$, we use the solenoidal decomposition for $\mathbf{f}$. For any $\mathbf{f} \in L_p((0, T), L_q(\dot\Omega)^N)$ let $\psi \in L_p((0, T), \hat H^1_q(\mathbb{R}^N))$ be a solution of the variational problem:

$$(m^{-1}\nabla\psi, \nabla\varphi)_{\dot\Omega} = (\mathbf{f}, \nabla\varphi)_{\dot\Omega} \quad \text{for any } \varphi \in \hat H^1_{q'}(\mathbb{R}^N)$$

and then the projection $\mathcal{P}$ and $\mathcal{Q}$ are defined by setting

$$\mathcal{P}\mathbf{f} = \mathbf{f} - m^{-1}\nabla\psi, \quad \mathcal{Q}\mathbf{f} = m^{-1}\psi.$$

We know that

$$\|\mathcal{P}\mathbf{f}\|_{L_q(\dot\Omega)} \leq C\|\mathbf{f}\|_{L_q(\dot\Omega)}. \tag{3.345}$$

And then, the solutions $\mathbf{u}$, $\mathfrak{p}$, and ρ of equations (3.341) are written as

$$(\mathbf{u}, \rho) = \int_0^t T(t-s)(\mathcal{P}\mathbf{f}(\cdot, t), d(\cdot, s))\,ds, \quad \mathfrak{p} = \mathcal{Q}\mathbf{f} + K(\mathbf{u}, \rho). \tag{3.346}$$

Let $(\mathbf{v}, \eta) = T(t)(\mathbf{v}_0, \eta_0)$. Then, from Theorem 3.7.1 proved in Sect. 3.7 below, we know the following decay properties of $\mathbf{v}$ and η: For any $t \geq 3$, we have

$$\begin{aligned} \|\mathbf{v}(\cdot, t)\|_{L_r(\dot\Omega)} &\leq C_{r,q} t^{-\frac{N}{2}\left(\frac{1}{q}-\frac{1}{r}\right)} \|(\mathbf{v}_0, \eta_0)\|_{\mathcal{H}_q}, \\ \|\nabla \mathbf{v}(\cdot, t)\|_{L_r(\dot\Omega)} &\leq C_{r,q} t^{-\min(\frac{1}{2}+\frac{N}{2}\left(\frac{1}{q}-\frac{1}{2}\right), \frac{N}{2q})} \|(\mathbf{v}_0, \eta_0)\|_{\mathcal{H}_q} \end{aligned} \tag{3.347}$$

provided that $1 < q \leq r \leq \infty$ and $q \neq \infty$. And, for any $t \geq 3$, we have

$$\|\partial_t^m (\mathbf{v}(\cdot, t), \eta(\cdot, t))\|_{H^2_r(\dot B_L) \times W^{3-1/r}_r(S_R)} \leq C_{r,q} t^{-\frac{N}{2q}} \|(\mathbf{v}_0, \eta_0)\|_{\mathcal{H}_q} \tag{3.348}$$

provided that $1 < q \leq r < \infty$. Here and in the following, $\dot B_L = B_L \setminus S_R$.

Using (3.345), (3.346), Theorem 3.6.1 and (3.348), we prove the following decay theorem.

Theorem 3.6.2 *Let $T > 0$ and let q_1, q_2, p, and ζ be the constants given in Theorem 3.1.4 satisfying the conditions* (3.41) *and* (3.42). *Let*

$$a_0 = 0, \quad a_1 = \frac{N}{2q_1}, \quad a_2 = \frac{N}{2q_2} + 1 \quad r_0 = q_1/2, \quad r_1 = q_1, \quad r_2 = q_2. \tag{3.349}$$

Assume that $(\mathbf{f}, d) \in \cap_{i=0}^2 L_p((0, T), \mathcal{H}_{r_i})$. Then, problem (3.341) *admits a unique solution $(\mathbf{u}, \mathfrak{p}, \rho)$ with*

$$\begin{aligned} &\mathbf{u} \in \cap_{i=0}^2 (H^1_p((0, T), L_{r_i}(\dot\Omega)^N) \cap L_p((0, T), H^2_{r_i}(\dot\Omega)^N)), \\ &\mathfrak{p} \in L_p((0, T), H^1_{r_i}(\dot\Omega) + \hat H^1_{r_i}(\mathbb{R}^N)), \\ &\rho \in \cap_{i=0}^2 (H^1_p((0, T), W^{2-1/r_i}_{r_i,p}(S_R)) \cap L_p((0, T), W^{3-1/r_i}_{r_i,p}(S_R))) \end{aligned}$$

possessing the estimate:

$$\begin{aligned} E_T(\mathbf{u}, \rho) \leq C(&\sum_{i=0}^2 \| < t >^{\frac{N}{q_1}-\zeta+a_i} (\mathbf{f}, d)\|_{L_p((0,T),\mathcal{H}_{r_i})} \\ &+ [d]_{W^{2-1/q_2}_{q_2}(S_R), N/q_1, T}) \end{aligned} \tag{3.350}$$

for some constant $C > 0$, where ζ is the number given in Theorem 3.1.4.

Proof We first prove that

$$
\begin{aligned}
\|\mathbf{u}(\cdot,t)\|_{H^1_\infty(\dot\Omega)} &\le Ct^{-\frac{N}{q_1}} J_T(\mathbf{f},d);\\
\|\rho(\cdot,t)\|_{W^{2-1/q_2}_{q_2}(S_R)} &\le Ct^{-\frac{N}{q_1}} J_T(\mathbf{f},d) \quad (m=0,1);\\
\|\mathbf{u}(\cdot,t)\|_{H^1_{q_1/2}(\dot\Omega)} &\le CJ_T(\mathbf{f},d);\\
\|\mathbf{u}(\cdot,t)\|_{H^1_{q_1}(\dot\Omega)} &\le Ct^{-\frac{N}{2q_1}} J_T(\mathbf{f},d);\\
\|\mathbf{u}(\cdot,t)\|_{H^1_{q_2}(\dot\Omega)} &\le Ct^{-\frac{N}{2q_2}-1} J_T(\mathbf{f},d)
\end{aligned}
\tag{3.351}
$$

when $t > 6$, where we have set

$$
J_T(\mathbf{f},d) = \sum_{i=0}^{2} \| < t >^{\frac{N}{q_1}-\zeta+a_i} (\mathbf{f},d)\|_{L_p((0,T),\mathcal{H}_{r_i})}.
$$

Applying (3.348) with $r=\infty$ and $q=q_1/r$ to (3.346) and $r=\infty$ and $q=q_2$, we have

$$
\|\mathbf{u}(\cdot,t)\|_{H^1_\infty(\dot\Omega)} \le C(I_{\infty\mathbf{u}} + II_{\infty,\mathbf{u}} + III_{\infty,\mathbf{u}})
$$

with

$$
\begin{aligned}
I_{\infty,\mathbf{u}} &= \int_0^{t/2} (t-s)^{-\frac{N}{q_1}} \|F(\cdot,s)\|_{\mathcal{H}_{q_1/2}}\,ds;\\
II_{\infty,\mathbf{u}} &= \int_{t/2}^{t-1} (t-s)^{-\frac{N}{2q_2}} \|F(\cdot,s)\|_{\mathcal{H}_{q_2}}\,ds;\\
III_{\infty,\mathbf{u}} &= \int_{t-1}^{t} (t-s)^{-\frac{1}{2}\left(1+\frac{N}{q_2}+\kappa\right)} \|F(\cdot,s)\|_{\mathcal{H}_{q_2}}\,ds,
\end{aligned}
$$

where $F = (\mathcal{P}\mathbf{f}, d)$ and κ is a very small positive number. We shall explain how to obtain $III_{\infty,\mathbf{u}}$. We first use the fact that $W^{1+\frac{N}{q_2}+\kappa}_{q_2}(\dot\Omega)$ is continuously embedded into $H^1_\infty(\dot\Omega)$ to obtain

$$
\left\|\int_{t-1}^t T(t-s)F(\cdot,s)\,ds\right\|_{H^1_\infty(\dot\Omega)} \le C\int_{t-1}^t \|T(t-s)F(\cdot,s)\|_{W^{1+\frac{N}{q_2}+\kappa}_{q_2}(\dot\Omega)}\,ds.
$$

By theory of C^0 analytic semigroup, we have

$$
\|T(t)F_0\|_{\mathcal{D}_{q_2}} \le Ct^{-1}\|F_0\|_{\mathcal{H}_{q_2}}, \quad \|T(t)F_0\|_{\mathcal{H}_{q_2}} \le C\|F_0\|_{\mathcal{H}_{q_2}}
\tag{3.352}
$$

for any $t \in (0, 1)$ with some constant $C > 0$. Notice that

$$W_{q_2}^{1+\frac{N}{q_2}+\kappa}(\dot\Omega) \times W_{q_2}^{\frac{5}{2}+\frac{N}{2q_2}+\kappa-\frac{1}{q_2}}(S_R) = (\mathcal{H}_{q_2}, \mathcal{D}_{q_2})_{\theta,q_2} \tag{3.353}$$

with $\theta = \frac{1}{2}(1 + \frac{N}{q_2} + \kappa)$, where $(\cdot, \cdot)_{\theta,q_2}$ is a real interpolation functor. Combining (3.352) and (3.353) yields

$$\|T(t)F_0\|_{(\mathcal{H}_q,\mathcal{D}_q)_{\theta,q_2}} \leq Ct^{-\frac{1}{2}\left(1+\frac{N}{q_2}+\kappa\right)}\|F_0\|_{\mathcal{H}_{q_2}},$$

which leads to $III_{\infty,\mathbf{u}}$.

Below, for notational simplicity, we set

$$J_{T,r_i}(\mathbf{f}, d) = \| < t >^{\frac{N}{q_1}-\zeta+a_i} (\mathbf{f}, d)\|_{L_p((0,T),\mathcal{H}_{r_i})}.$$

By Hölder's inequality and (3.345), we have

$$\begin{aligned} I_{\infty,\mathbf{u}} &\leq (t/2)^{-\frac{N}{q_1}} \int_0^{t/2} \|F(\cdot, s)\|_{\mathcal{H}_{q_1/2}}\, ds \\ &\leq (t/2)^{-\frac{N}{q_1}} \left(\int_0^{t/2} < s >^{-\left(\frac{N}{q_1}-\zeta+a_0\right)p'} ds\right)^{1/p'} J_{T,r_0}(\mathbf{f}, d). \end{aligned}$$

Since

$$\frac{N}{q_1} - \zeta + a_0 - \frac{1}{p'} = \frac{N}{q_2} - \zeta + \frac{1}{p} > 0 \tag{3.354}$$

as follows from (3.42), we have $(\frac{N}{q_1} - \zeta + a_0)p' > 1$, and so we have

$$I_{\infty,\mathbf{u}} \leq C < t >^{-\frac{N}{q_1}} \| < t >^{\frac{N}{q_1}-\zeta+a_0} (\mathbf{f}, d)\|_{L_p((0,T),\mathcal{H}_{q_1/2})}.$$

Analogously, we have

$$\begin{aligned} &II_{\infty,\mathbf{u}} \\ &\leq \int_{t/2}^{t-1} (t-s)^{-\frac{N}{2q_2}} < s >^{-\frac{N}{q_1}+\zeta-a_2} < s >^{\frac{N}{q_1}-\zeta+a_2} \|(\mathbf{f}(\cdot, s), d(\cdot, s))\|_{\mathcal{H}_{q_2}}\, ds \\ &\leq < t/2 >^{-\frac{N}{q_1}+\zeta-a_2} \left(\int_{t/2}^{t-1} (t-s)^{-p'\frac{N}{2q_2}} ds\right)^{1/p'} J_{T,r_2}(\mathbf{f}, d). \end{aligned}$$

We proceed as

$$< t/2 >^{-\frac{N}{q_1}+\zeta-a_2}\Big(\int_{t/2}^{t-1}(t-s)^{-p'\frac{N}{2q_2}}\,ds\Big)^{1/p'} \leq Ct^{-\frac{N}{q_1}-\zeta+a_2}t^{-\frac{N}{2q_2}+\frac{1}{p'}}.$$

Since

$$\frac{N}{q_1}-\zeta+a_2+\frac{N}{2q_2}-\frac{1}{p'}=\frac{N}{q_1}+\frac{N}{q_2}+\frac{1}{p}-\zeta>\frac{N}{q_1} \tag{3.355}$$

as follows from (3.42), we have

$$II_{\infty,\mathbf{u}} \leq C < t >^{-\frac{N}{q_1}} J_{T,r_2}(\mathbf{f}, d).$$

By Hölder's inequality, we have

$$III_{\infty,\mathbf{u}} \leq < t-1 >^{-\frac{N}{q_1}+\zeta-a_2}\Big(\int_{t-1}^{t}(t-s)^{-\frac{p'}{2}\left(1+\frac{N}{q_2}+\kappa\right)}\,ds\Big)\,J_{T,r_2}(\mathbf{f}, t).$$

Since

$$1+\frac{N}{q_2}+\kappa-\frac{2}{p'}=\frac{N}{q_2}+\frac{2}{p}-1+\kappa<0 \tag{3.356}$$

with some very small number κ as follows from (3.41) and since

$$a_2-\zeta=\frac{N}{2q_2}+1-\zeta>\frac{N}{2q_2}+\frac{1}{p}-\zeta>0 \tag{3.357}$$

as follows from (3.42), we have

$$III_{\infty,\mathbf{u}} \leq C < t >^{-\frac{N}{q_1}} J_{T,r_2}(\mathbf{f}, d).$$

Putting the estimates for $I_{\mathbf{u},\infty}$, $II_{\infty,\mathbf{u}}$, and $III_{\infty,\mathbf{u}}$ together yields that

$$\|\mathbf{u}(\cdot, t)\|_{H^1_\infty(\dot\Omega)} \leq Ct^{-\frac{N}{q_1}}(J_{T,r_0}(\mathbf{f}, d)+J_{T,r_2}(\mathbf{f}, d)).$$

We next estimate $\|\rho(\cdot, t)\|_{W^{2-\frac{1}{q_2}}_{q_2}(S_R)}$. Applying (3.348) to (3.346) and using (3.353), we have

$$\|\rho(\cdot, t)\|_{W^{2-\frac{1}{q_2}}_{q_2}(S_R)} \leq C(I_\rho, +II_\rho+III_\rho)$$

with

$$I_\rho = \int_0^{t/2} (t-s)^{-\frac{N}{q_1}} \|F(\cdot,s)\|_{\mathcal{H}_{q_1/2}}\,ds;$$
$$II_\rho = \int_{t/2}^{t-1} (t-s)^{-\frac{N}{2q_2}} \|F(\cdot,s)\|_{\mathcal{H}_{q_2}}\,ds;$$
$$III_\rho = \int_{t-1}^{t} \|F(\cdot,s)\|_{\mathcal{H}_{q_2}}\,ds.$$

Employing the same argument as in obtaining the first inequality in (3.351),

$$\|\rho(\cdot,t)\|_{W_{q_2}^{2-\frac{1}{q_2}}(S_R)} \le Ct^{-\frac{N}{q_1}}(J_{T,r_0}(\mathbf{f},d) + J_{T,r_2}(\mathbf{f},d)).$$

We next estimate $\|\mathbf{u}(\cdot,t)\|_{H^1_{q_1/2}}$. Applying (3.347) with $r = q = q_1/2$ to (3.346) and using (3.345)

$$\begin{aligned}\|\mathbf{u}(\cdot,t)\|_{H^1_{q_1/2}(\dot\Omega)} &\le C\int_0^t \|F(\cdot,s)\|_{\mathcal{H}_{q_1/2}}\,ds \\ &\le C\Big(\int_0^t <s>^{-(\frac{N}{q_1}-\zeta+a_0)p'}\,ds\Big)^{1/p'} J_{T,q_1/2}(\mathbf{f},\rho).\end{aligned}$$

From (3.354) it follows that

$$\Big(\int_0^t <s>^{-(\frac{N}{q_1}-\zeta+a_0)p'}\,ds\Big)^{1/p'} < \infty.$$

Thus, we have

$$\|\mathbf{u}(\cdot,t)\|_{H^1_{q_1/2}(\dot\Omega)} \le CJ_{T,q_1/2}(\mathbf{f},d).$$

We next estimate $\|\mathbf{u}(\cdot,t)\|_{H^1_{q_1}(\dot\Omega)}$. Applying (3.347) with $r = q_1$ an $q = q_1/2$ to (3.346) and using (3.345) and the estimate for C^0 analytic semigroup for $t \in (0,1)$, we have

$$\|\mathbf{u}(\cdot,t)\|_{H^1_{q_1}(\dot\Omega)} \le C(I_{q_1} + II_{q_1} + III_{q_1})$$

with

$$I_{q_1} = \int_0^{t/2} (t-s)^{-\frac{N}{2}\left(\frac{1}{q_1/2}-\frac{1}{q_1}\right)} \|F(\cdot,s)\|_{\mathcal{H}_{q_1/2}}\,ds,$$
$$II_{q_1} = \int_{t/2}^{t-1} (t-s)^{-\frac{N}{2}\left(\frac{1}{q_1/2}-\frac{1}{q_1}\right)} \|F(\cdot,s)\|_{\mathcal{H}_{q_1/2}}\,ds,$$

$$III_{q_1} = \int_{t-1}^{t} (t-s)^{-\frac{1}{2}} \|F(\cdot, s)\|_{\mathcal{H}_{q_1}}\, ds.$$

Noting that $\frac{N}{2}(\frac{1}{q_1/2} - \frac{1}{q_1}) = \frac{N}{2q_1}$, by Hölder's inequality we have

$$I_{q_1} \leq <t/2>^{-\frac{N}{2q_1}} \left(\int_0^{t/2} <s>^{-\left(\frac{N}{q_1}+a_1-\zeta\right)p'}\, ds\right)^{1/p'} J_{T,r_1}(\mathbf{f}, d).$$

Since

$$\frac{N}{q_1} + a_1 - \zeta - \frac{1}{p'} = \frac{3N}{2q_1} - \zeta - \frac{1}{p'} = \frac{N}{2q_1} + \frac{N}{q_2} + \frac{1}{p} - \zeta > \frac{N}{2q_1} > 0 \tag{3.358}$$

as follows from (3.42), we have

$$\int_0^{t/2} <s>^{-\left(\frac{N}{q_1}+a_1-\zeta\right)p'}\, ds < \infty,$$

which yields that

$$I_{q_1} \leq Ct^{-\frac{N}{2q_1}} J_{T,r_1}(\mathbf{f}, d).$$

Analogously, by Hölder's inequality and $\frac{N}{2}(\frac{1}{q_1/2} - \frac{1}{q_1}) = \frac{N}{2q_1}$, we have

$$\begin{aligned} II_{q_1} &\leq <t/2>^{-\left(\frac{N}{q_1}+a_1-\zeta\right)} \left(\int_{t/2}^{t-1} (t-s)^{-\frac{p'N}{2q_1}}\, ds\right)^{1/p'} J_{T,r_1}(\mathbf{f}, d) \\ &\leq Ct^{-\left(\frac{N}{q_1}+a_1-\zeta\right)} t^{-\frac{N}{2q_1}+\frac{1}{p'}} J_{T,r_1}(\mathbf{f}, d). \end{aligned}$$

Since

$$\frac{N}{q_1} + a_1 - \zeta + \frac{N}{2q_1} - \frac{1}{p'} > \frac{N}{q_1}, \tag{3.359}$$

as follows from (3.358), we have

$$II_{q_1} \leq Ct^{-\frac{N}{q_1}} J_{T,r_1}(\mathbf{f}, d).$$

Since $p > 2$, we have $p' < 2$, and so by Hölder's inequality we have

$$III_{q_1} \leq <t>^{-\left(\frac{N}{q_1}-\zeta+a_1\right)} \left(\int_{t-1}^{t} (t-s)^{\frac{p'}{2}}\, ds\right)^{1/p'} J_{T,r_1}(\mathbf{f}, d) \leq Ct^{-\frac{N}{2q_1}} J_{T,r_1}(\mathbf{f}, d),$$

where we have used the fact that $\frac{N}{q_1} + a_1 - \zeta > \frac{N}{2q_1}$ which follows from (3.359). Therefore, we have

$$\|\mathbf{u}(\cdot, t)\|_{H^1_{q_1}(\dot\Omega)} \leq Ct^{-\frac{N}{2q_1}} J_{T,r_1}(\mathbf{f}, d).$$

We finally estimate $\|\mathbf{u}(\cdot, t)\|_{H^1_{q_2}(\dot\Omega)}$. Applying (3.347) with $r = q_2$ an $q = q_1/2$ to (3.346) and using (3.345) and the estimate for C^0 analytic semigroup for $t \in (0, 1)$, we have

$$\|\mathbf{u}(\cdot, t)\|_{H^1_{q_2}(\dot\Omega)} \leq C(I_{q_2} + II_{q_2} + III_{q_2})$$

with

$$\begin{aligned} I_{q_2} &= \int_0^{t/2} (t-s)^{-\frac{N}{2}\left(\frac{1}{q_1/2} - \frac{1}{q_2}\right)} \|F(\cdot, s)\|_{\mathcal{H}_{q_1/2}}\, ds, \\ II_{q_2} &= \int_{t/2}^{t-1} \|F(\cdot, s)\|_{\mathcal{H}_{q_2}}\, ds, \\ III_{q_1} &= \int_{t-1}^{t} (t-s)^{-\frac{1}{2}} \|F(\cdot, s)\|_{\mathcal{H}_{q_2}}\, ds. \end{aligned}$$

Since

$$\frac{N}{2}\left(\frac{1}{2q_1} - \frac{1}{q_2}\right) = \frac{N}{q_1} - \frac{N}{2q_2} = \frac{N}{2q_2} + 1,$$

by Hölder's inequality we have

$$I_{q_2} \leq (t/2)^{-\frac{N}{2q_2}+1} \left(\int_0^{t/2} < s >^{-\left(\frac{N}{q_1} - \zeta + a_0\right)p'} ds\right)^{1/p'} J_{T,r_0}(\mathbf{f}, d).$$

By (3.354) we have

$$\left(\int_0^{t/2} < s >^{-\left(\frac{N}{q_1} - \zeta + a_0\right)p'} ds\right)^{1/p'} < \infty,$$

and so we have

$$I_{q_2} \leq Ct^{-\frac{N}{q_1}} J_{T,r_0}(\mathbf{f}, d).$$

By Hölder's inequality, we have

$$
\begin{aligned}
II_{q_2} &\leq < t/2 >^{-\left(\frac{N}{q_1}+a_2-\zeta\right)} \left(\int_{t/2}^{t-1} ds\right)^{1/p'} J_{T,r_2}(\mathbf{f}, d) \\
&\leq Ct^{-\left(\frac{N}{q_1}+a_2-\zeta\right)} t^{\frac{1}{p'}} J_{T,r_2}(\mathbf{f}, d).
\end{aligned}
$$

Since

$$
\frac{N}{q_1} + a_2 - \zeta - \frac{1}{p'} = \frac{N}{q_2} + 1 - 1 + \frac{1}{p} - \zeta + \frac{N}{2q_2} + 1 > \frac{N}{2q_2} + 1
$$

as follows from (3.42), we have

$$
II_{q_2} \leq Ct^{-\frac{N}{2q_2}+1} J_{T,r_2}(\mathbf{f}, d).
$$

By Hölder's inequality, $p'/2 < 1$ and the fact that $\frac{N}{q_1} - \zeta > 0$, we have

$$
\begin{aligned}
III_{q_2} &\leq < t - 1 >^{-\left(\frac{N}{q_1}-\zeta+a_2\right)} \left(\int_{t-1}^{t} (t-s)^{-\frac{p'}{2}} ds\right)^{1/p'} J_{T,r_2}(\mathbf{f}, d) \\
&\leq Ct^{-\left(\frac{N}{2q_2}+1\right)} J_{T,r_2}(\mathbf{f}, d).
\end{aligned}
$$

Summing up, we have obtained

$$
\|\mathbf{u}(\cdot, t)\|_{H^1_{q_2}(\dot{\Omega})} \leq Ct^{-\left(\frac{N}{2q_2}+1\right)} (J_{T,r_0}(\mathbf{f}, d) + J_{T,r_2}(\mathbf{f}, d)).
$$

Therefore, we have (3.351).

We next consider the case where $0 < S \leq 6$. Applying Theorem 3.2.8 to equation (3.341), we have

$$
\begin{aligned}
&\|\mathbf{u}\|_{L_p((0,S),H^2_{r_i}(\dot{\Omega}))} + \|\partial_t \mathbf{u}\|_{L_p((0,S),L_{r_i}(\dot{\Omega}))} \\
&\quad + \|\rho\|_{L_p((0,S),W^{3-1/r_i}_{r_i}(S_R))} + \|\partial_t \rho\|_{L_p((0,S),W^{2-r_i}_{r_i}(S_R))} \\
&\quad \leq C(\|\mathbf{f}\|_{L_p((0,T),L_{r_i}(\dot{\Omega}))} + \|d\|_{L_p((0,S),W^{2-r_i}_{r_i}(S_R))}).
\end{aligned}
\tag{3.360}
$$

Thus, by (3.212), (3.213), and (3.351), we have

$$
\begin{aligned}
&[\mathbf{u}]_{H^1_\infty(\dot{\Omega}),N/q_1,T} + [\rho]_{W^{2-\frac{1}{q_2}}_{q_2}(S_R),N/q_1,T} + [\mathbf{u}]_{H^1_{q_2}(\dot{\Omega}),N/(2q_2)+1,T} \\
&\quad + [\mathbf{u}]_{H^1_{q_1}(\dot{\Omega}),N/(2q_1),T} + [\mathbf{u}]_{H^1_{q_1/2}(\dot{\Omega}),0,T} \leq J_T(\mathbf{f}, d).
\end{aligned}
\tag{3.361}
$$

By the third equation of equations (3.341) and Sobolev's imbedding theorem, we have

$$[\partial_t \rho]_{W^{1-1/q_2}_{q_2}, N/q_1, T} \leq C([\mathbf{u}]_{H^1_{q_1}(B_R), N/q_1, T} + [d]_{W^{1-1/q_2}_{q_2}(S_R), N/q_1, T}).$$

Since $[\mathbf{u}]_{H^1_{q_1}(B_R), N/q_1, T} \leq C_R[\mathbf{u}]_{H^1_\infty(B_R), N/q_1, T}$, by (3.361) we have

$$[\partial_t \rho]_{W^{1-1/q_2}_{q_2}, N/q_1, T} \leq J_T(\mathbf{f}, d) + [d]_{W^{1-1/q_2}_{q_2}(S_R), N/q_1, T}.$$

By (3.333), we have

$$[\Psi_\rho]_{H^2_{q_2}(B_L), N/q_1, T} \leq C[\rho]_{W^{2-1/q_2}_{q_2}(S_R), N/q_1, T},$$

$$[\partial_t \Psi_\rho]_{H^1_{q_2}(B_L), N/q_1, T} \leq C[\partial_t \rho]_{W^{1-1/q_2}_{q_2}(S_R), N/q_1, T}.$$

This completes the proof of Theorem 3.6.2. □

3.6.3 Estimate of Nonlinear Terms and a Proof of Theorem 3.1.4

In this section, we first estimate nonlinear terms appearing on the right side of equations (3.34). Notice that

$$r_0 < r_1 < r_2, \quad a_0 < a_1 < a_2. \tag{3.362}$$

We first estimate $\mathbf{f} = \mathbf{f}(\mathbf{u}, \Psi_\rho)$ given in (3.142). We shall prove that

$$\| < t >^{\frac{N}{q_1} - \zeta + a_i} \mathbf{f}(\mathbf{u}, \Psi_\rho)\|_{L_p((0,T), L_{r_i}(\dot\Omega))} \leq C E_T(\mathbf{u}, \rho)^2 \quad (i = 0, 1, 2). \tag{3.363}$$

Recall that $\Psi_\rho = \omega(y)R^{-1}H_\rho(y, t)y + \xi(t)$, and so $\nabla\Psi_\rho = \nabla(\omega(y)R^{-1}H_\rho(y, t)y)$. Using (3.334), supp $\Psi_\rho(x, t) \subset B_L \times (0, T)$, we have

$$\begin{aligned}
&\|\mathbf{f}(\mathbf{u}, \Psi_\rho)\|_{L_q(\dot\Omega)} \\
&\leq C\{\|\mathbf{u}(\cdot, t) \cdot \nabla\mathbf{u}(\cdot, t)\|_{L_q(\dot\Omega)} + \|\partial_t \Psi_\rho(\cdot, t)\nabla\mathbf{u}(\cdot, t)\|_{L_q(\dot\Omega)} + |\xi'(t)|\|\nabla\mathbf{u}(\cdot, t)\|_{L_q(\dot\Omega)} \\
&\quad + \|\nabla\Psi_\rho(\cdot t)(\partial_t\mathbf{u}(\cdot, t), \nabla^2\mathbf{u}(\cdot, t))\|_{L_q(\dot\Omega)} + \|\nabla^2\Psi_\rho(\cdot, t)\nabla\mathbf{u}(\cdot, t)\|_{L_q(\dot\Omega)}\}.
\end{aligned}$$

By Hölder's inequality, we have

$$
\begin{aligned}
&\| < t >^{\frac{N}{q_1}-\zeta+a_i} \mathbf{u}\cdot\nabla\mathbf{u}\|_{L_p((0,T),L_{q_1/2}(\dot\Omega))}\\
&\quad \leq [\mathbf{u}]_{H^1_{q_1}(\dot\Omega),N/(2q_1),T}\| < t >^{\frac{N}{2q_1}-\zeta} \mathbf{u}\|_{L_p((0,T),H^2_{q_1}(\dot\Omega))};\\
&\| < t >^{\frac{3N}{2q_1}-\zeta} \mathbf{u}\cdot\nabla\mathbf{u}\|_{L_p((0,T),L_{q_1}(\dot\Omega))}\\
&\quad \leq [\mathbf{u}]_{H^1_\infty(\dot\Omega),N/q_1,T}\| < t >^{\frac{N}{2q_1}-\zeta} \mathbf{u}\|_{L_p((0,T),H^2_{q_1}(\dot\Omega))};\\
&\| < t >^{\frac{N}{q_1}+\frac{N}{2q_2}+1-\zeta} \mathbf{u}\cdot\nabla\mathbf{u}\|_{L_p((0,T),L_{q_2}(\dot\Omega))}\\
&\quad \leq [\mathbf{u}]_{H^1_{q_1}(\dot\Omega),N/(2q_1),T}\| < t >^{\frac{N}{2q_2}+1-\zeta} \mathbf{u}\|_{L_p((0,T),H^2_{q_2}(\dot\Omega))}.
\end{aligned}
\tag{3.364}
$$

Noting that $\mathrm{supp}\,\Psi_\rho \subset B_L$ and using (3.362) and Sobolev's inequality:

$$
\|v\|_{L_\infty(B_R)} \leq C\|v\|_{H^1_{q_2}(B_R)}, \tag{3.365}
$$

we have

$$
\begin{aligned}
&\| < t >^{\frac{N}{q_1}-\zeta+a_i} \partial_t\Psi_\rho\nabla\mathbf{u}\|_{L_p((0,T),L_{r_i}(\dot\Omega))}\\
&\quad \leq [\partial_t\rho]_{W^{1-1/q_2}_{q_2}(S_R),N/q_1,T}\| < t >^{a_2-\zeta} \mathbf{u}\|_{L_p((0,T),H^2_{q_2}(\dot\Omega))}\\
&\| < t >^{\frac{N}{q_1}-\zeta+a_i} \nabla\Psi_\rho(\partial_t\mathbf{u}, \nabla^2\mathbf{u})\|_{L_p((0,T),L_{r_i}(\dot\Omega))}\\
&\quad \leq [\rho]_{W^{2-1/q_2}_{q_2}(S_R),N/q_1,T}\| < t >^{a_2-\zeta} (\partial_t\mathbf{u}, \nabla^2\mathbf{u})\|_{L_p((0,T),L_{q_2}(\dot\Omega))};\\
&\| < t >^{\frac{N}{q_1}-\zeta+a_i} \nabla^2\Psi_\rho\nabla\mathbf{u}\|_{L_p((0,T),L_{r_i}(\dot\Omega))}\\
&\quad \leq [\rho]_{W^{2-q_2}_{q_2}(S_R),N/q_1,T}\| < t >^{a_i-\zeta} \mathbf{u}\|_{L_p((0,T),H^2_{q_2}(\dot\Omega))}
\end{aligned}
\tag{3.366}
$$

for $i = 0, 1, 2$. Using the estimate: $|\xi'(t)| \leq C_R\|\mathbf{u}(\cdot, t)\|_{L_\infty(B_R)}$, we have

$$
\begin{aligned}
&\| < t >^{\frac{N}{q_1}-\zeta+a_i} |\xi'(\cdot)|\nabla\mathbf{u}\|_{L_p((0,T),L_{r_i}(\dot\Omega))}\\
&\quad \leq C_L\|\mathbf{u}\|_{H^1_\infty(\dot\Omega),N/q_1,T}\| < t >^{a_i-\zeta} \mathbf{u}\|_{L_p((0,T),H^2_{r_i}(\dot\Omega))}.
\end{aligned}
\tag{3.367}
$$

Putting (3.364), (3.366), and (3.367) together, we have obtained (3.363).

We next consider $\bar d(\mathbf{u}, \psi_\rho)$. We shall prove that

$$
\| < t >^{\frac{N}{q_1}-\zeta+a_i} \bar d(\mathbf{u}, \Psi_\rho)\|_{L_p((0,T),W^{2-1/r_i}_{r_i}(S_R))} \leq C(E_T(\mathbf{u}, \rho)^2 + E_T(\mathbf{u}, \rho)^3)
\tag{3.368}
$$

for $i = 0, 1, 2$. To prove (3.368), in view of the fact: $\|\mathbf{v}\|_{W^{\ell-1/r_i}_{r_i}(S_R)} \leq C\|\mathbf{v}\|_{H^2_{r_2}(B_R)}$ for $\ell = 1, 2$ and $i = 0, 1, 2$ and (3.362), it is sufficient to prove that

$$\| < t >^{\frac{N}{q_1}-\zeta+a_2} \bar{d}(\mathbf{u}, \rho)\|_{L_p((0,T),H^2_{q_2}(B_R))} \leq C(E_T(\mathbf{u}, \rho)^2 + E_T(\mathbf{u}, \rho)^3). \tag{3.369}$$

Recall that $d(\mathbf{u}, \Psi_\rho)$ is defined by replacing $\mathbf{v}$ and h by $\mathbf{u}$ and ρ in (3.215). And then, using (3.334) and Sobolev's inequality: (3.365), we have

$$\begin{aligned}
&\| < t >^{\frac{N}{q_1}-\zeta+a_2} d(\mathbf{u}, \Psi_\rho)\|_{L_p((0,T),H^2_{q_2}(B_R))} \\
&\quad \leq C([\rho]_{W^{2-1/q_2}_{q_2}(S_R),N/q_1,T} + [\partial_t\rho]_{W^{1-1/q_2}_{q_2}(S_R),N/q_1,T} \\
&\qquad\qquad + [\rho]_{W^{2-1/q_2}_{q_2}(S_R),N/q_1,T}[\partial_t\rho]_{W^{1-1/q_2}_{q_2}(S_R),N/q_1,T}) \\
&\quad \times (\| < t >^{a_2-\zeta} \mathbf{u}\|_{L_p((0,T),H^2_{q_2}(\dot\Omega))} + \| < t >^{a_2-\zeta} \partial_t\rho\|_{L_p((0,T),W^{2-1/q_2}_{q_2}(\dot\Omega))} \\
&\qquad\qquad + \| < t >^{a_2-\zeta} \rho\|_{L_p((0,T),W^{3-1/q_2}_{q_2}(\dot\Omega))}) \\
&\quad \leq C(E_T(\mathbf{u}, \rho)^2 + E_T(\mathbf{u}, \rho)^3)).
\end{aligned}$$

Using (3.334), Sobolev's inequality (3.365), and the fact that

$$\frac{N}{q_1} - \frac{N}{2q_2} - 1 + \zeta = \frac{N}{2q_2} + \zeta > \zeta > 1/p, \tag{3.370}$$

we have

$$\begin{aligned}
&\| < t >^{\frac{N}{q_1}-\zeta+a_2} < \mathbf{u}, \mid \nabla'_\Gamma\rho > \|_{L_p((0,T),H^2_{q_2}(B_R))} \\
&\quad \leq C([\rho]_{W^{2-1/q_2}_{q_2}(S_R),N/q_1,T}\| < t >^{a_2-\zeta} \mathbf{u}\|_{L_p((0,T),H^2_{q_2}(\dot\Omega))} \\
&\qquad + [\mathbf{u}]_{H^1_\infty(\dot\Omega),N/q_1,T}\| < t >^{a_2-\zeta} \rho\|_{L_p((0,T),W^{3-1/q_2}_{q_2}(S_R))}) \\
&\quad \leq CE_T(\mathbf{u}, \rho)^2; \\
&\| < t >^{\frac{N}{q_1}-\zeta+a_2} \int_{B_R} \mathbf{u}J_0(\mathbf{k})\,dy\|_{L_p((0,T),H^2_{q_2}(B_R))} \\
&\quad \leq C_R[\mathbf{u}]_{H^1_\infty(\dot\Omega),N/q_1,T}[\rho]_{W^{2-1/q_2}_{q_2}(S_R),N/q_1,T}\| < t >^{-(\frac{N}{q_1}-\frac{N}{2q_2}-1+\zeta)} \|_{L_p((0,T))} \\
&\quad \leq CE_T(\mathbf{u}, \rho)^2; \\
&\| < t >^{\frac{N}{q_1}-\zeta+a_2} \int_{S_R} \rho^\ell\,d\omega\|_{L_p((0,T),H^2_{q_2}(B_R))}
\end{aligned}$$

$$
\begin{aligned}
&\leq [\rho]^2_{W^{2-1/q_2}_{q_2}(S_R),N/q_1,T} \| < t >^{-(\frac{N}{q_1}-\frac{N}{2q_2}-1+\zeta)} \|_{L_p((0,T))} \\
&\leq CE_T(\mathbf{u},\rho)^2; \\
\| &< t >^{\frac{N}{q_1}-\zeta+a_2} \int_{S_R} \rho^\ell \mathbf{n}_i\,d\omega \|_{L_p((0,T),H^2_{q_2}(B_R))} \\
&\leq [\rho]^2_{W^{2-1/q_2}_{q_2}(S_R),N/q_1,T} \| < t >^{-(\frac{N}{q_1}-\frac{N}{2q_2}-1+\zeta)} \|_{L_p((0,T))} \\
&\leq CE_T(\mathbf{u},\rho)^2,
\end{aligned}
$$

where $\ell \geq 2$. Summing up, we have obtained (3.368).

We next consider $< t >^{\frac{N}{q_1}-\zeta+a_i} g(\mathbf{u},\Psi_\rho)$ and $< t >^{\frac{N}{q_1}-\zeta+a_i} \mathbf{g}(\mathbf{u},\Psi_\rho)$. According to the definition of $g(\mathbf{u},\Psi_\rho)$ and $\mathbf{g}(\mathbf{u},\Psi_\rho)$ given in (3.136) and (3.137), respectively, we may write

$$
g(\mathbf{u},\Psi_\rho) = V_g(\nabla\Psi_\rho)\nabla\Psi_\rho \otimes \nabla\mathbf{u}, \quad \mathbf{g}(\mathbf{u},\Psi_\rho) = V_{\mathbf{g}}(\nabla\Psi_\rho)\nabla\Psi_\rho \otimes \mathbf{u}.
$$

Thus, we write

$$
\begin{aligned}
< t >^{\frac{N}{q_1}-\zeta+a_i} g(\mathbf{u},\psi_\rho) &= V_g(\nabla\Psi_\rho)\nabla(< t >^{\frac{N}{q_1}} \Psi_\rho) \otimes \nabla(< t >^{a_i-\zeta} \mathbf{u}), \\
< t >^{\frac{N}{q_1}-\zeta+a_i} \mathbf{g}(\mathbf{u},\psi_\rho) &= V_{\mathbf{g}}(\nabla\Psi_\rho)\nabla(< t >^{\frac{N}{q_1}} \Psi_\rho) \otimes < t >^{a_i-\zeta} \mathbf{u}.
\end{aligned}
$$

Using the symbols given in (3.230) and noting the relations:

$$
< t >^a \Psi_\rho = \Psi_{<t>^a\rho}, \quad \Psi_{<t>^a\rho}|_{t=0} = \Psi_{\rho_0}, \quad < t >^a \mathbf{u}|_{t=0} = \mathbf{u}_0,
$$

we define $g_{a_i}(\mathbf{u},\Psi_\rho)$ and $\mathbf{g}_{a_i}(\mathbf{u},\Psi_\rho)$ by letting

$$
\begin{aligned}
g_{a_i}(\mathbf{u},\Psi_\rho) &= V_g(\nabla\mathcal{E}_2[\Psi_\rho])\nabla\mathcal{E}_2[\Psi_{<t>^{N/q_1}\rho}] \otimes \nabla\mathcal{E}_1[< t >^{a_i-\zeta} \mathbf{u}], \\
\mathbf{g}_{a_i}(\mathbf{u},\Psi_\rho) &= V_{\mathbf{g}}(\nabla\mathcal{E}_2[\Psi_\rho])\nabla\mathcal{E}_2[\Psi_{<t>^{N/q_1}\rho}] \otimes \mathcal{E}_1[< t >^{a_i-\zeta} \mathbf{u}].
\end{aligned}
$$

Let $\tilde{g}(\mathbf{u},\Psi_\rho)$ and $\tilde{\mathbf{g}}(\mathbf{u},\Psi_\rho)$ be the extensions of $g(\mathbf{u},\Psi_\rho)$ and $\mathbf{g}(\mathbf{u},\Psi_\rho)$ defined in (3.232) by replacing $\mathbf{v}$ and Ψ_h by $\mathbf{u}$ and Ψ_ρ. Notice that

$$
g_{a_i}(\mathbf{u},\Psi_\rho) = < t >^{\frac{N}{q_1}-\zeta+a_i} g(\mathbf{u},\psi_\rho), \quad \mathbf{g}_{a_i}(\mathbf{u},\Psi_\rho) = < t >^{\frac{N}{q_1}-\zeta+a_i} \mathbf{g}(\mathbf{u},\psi_\rho).
$$

We shall prove that

$$
\|\partial_t \mathbf{g}_{a_i}(\mathbf{u},\Psi_\rho)\|_{L_p((0,T),L_{r_i}(\dot\Omega))} \leq C((\mathcal{I} + E_T(\mathbf{u},\rho))^2 + (\mathcal{I} + E_T(\mathbf{u},\rho))^3), \tag{3.371}
$$

$$\|g_{a_i}(\mathbf{u}, \Psi_\rho)\|_{H^{1/2}_p((0,T),L_{r_i}(\dot\Omega))} \leq C((\mathfrak{I} + E_T(\mathbf{u}, \rho))^2 + (\mathfrak{I} + E_T(\mathbf{u}, \rho))^3), \tag{3.372}$$

$$\|g_{a_i}(\mathbf{u}, \Psi_\rho)\|_{L_p((0,T),H^1_{r_i}(\dot\Omega))} \leq C((\mathfrak{I} + E_T(\mathbf{u}, \rho))^2 + (\mathfrak{I} + E_T(\mathbf{u}, \rho))^3) \tag{3.373}$$

for $i = 0, 1, 2$.

We first prove (3.371). We have

$$\begin{aligned}
&\partial_t \mathbf{g}_{a_i}(\mathbf{u}, \Psi_\rho) \\
&= V_{\mathbf{g}}(\nabla\mathcal{E}_2[\Psi_\rho])\nabla\mathcal{E}_2[\Psi_{<t>^{N/q_1}\rho}] \otimes \partial_t\mathcal{E}_1[<t>^{a_i-\zeta}\mathbf{u}]) \\
&+ V_{\mathbf{g}}(\nabla\mathcal{E}_2[\Psi_\rho])\partial_t\nabla\mathcal{E}_2[\Psi_{<t>^{N/q_1}\rho}] \otimes \mathcal{E}_1[<t>^{a_i-\zeta}\mathbf{u}]) \\
&+ V'_{\mathbf{g}}(\nabla\mathcal{E}_2[\Psi_\rho])\partial_t\nabla\mathcal{E}_2[\Psi_\rho]\nabla\mathcal{E}_2[\Psi_{<t>^{N/q_1}\rho}] \otimes \mathcal{E}_1[<t>^{a_i-\zeta}\mathbf{u}]).
\end{aligned}$$

Using (3.227), (3.234) and Sobolev's inequality (3.365), and noting that

$$\operatorname{supp}\nabla\mathcal{E}_2[\Psi_{<t>^{N/q_1}\rho}], \quad \operatorname{supp}\nabla\mathcal{E}_2[\Psi_\rho] \subset B_L \tag{3.374}$$

we have

$$\begin{aligned}
&\|\partial_t \mathbf{g}_{a_i}(\mathbf{u}, \Psi_\rho)\|_{L_{r_i}(\dot\Omega)} \\
&\leq C_L\{(\mathfrak{I} + [\rho]_{W^{2-1/q_2}_{q_2}(B_L),N/q_1,T})\|\partial_t\mathcal{E}_1[<t>^{a_i-\zeta}\mathbf{u}]\|_{L_{r_i}(\dot\Omega)} \\
&+ (\mathfrak{I} + [\partial_t\rho]_{W^{1-1/q_2}_{q_2}(B_L),N/q_1,T})\|\mathcal{E}_1[<t>^{a_i-\zeta}\mathbf{u}]\|_{H^1_{q_2}(\dot\Omega)} \\
&+ (\mathfrak{I} + [\partial_t\rho]_{W^{1-1/q_2}_{q_2}(B_L),N/q_2,T})(\mathfrak{I} + [\rho]_{W^{2-1/q_2}_\infty(B_L),N/q_1,T})^2 \\
&\qquad \times \|\mathcal{E}_1[<t>^{a_i-\zeta}\mathbf{u}]\|_{H^1_{q_2}(\dot\Omega)}\}.
\end{aligned}$$

Thus, noting (3.362), by (3.226) we have (3.371).

By (3.374), Hölder's inequality, and (3.362), we have

$$\begin{aligned}
\|g_{a_i}(\mathbf{u}, \Psi_\rho)\|_{H^{1/2}_p(\mathbb{R},L_{r_i}(\dot\Omega))} &\leq C_L\|g_{a_i}(\mathbf{u}, \Psi_\rho)\|_{H^{1/2}_p(\mathbb{R},L_{q_2}(\dot\Omega))}, \\
\|g_{a_i}(\mathbf{u}, \Psi_\rho)\|_{L_p(\mathbb{R},H^1_{r_i}(\dot\Omega))} &\leq C_L\|g_{a_i}(\mathbf{u}, \Psi_\rho)\|_{H^{1/2}_p(\mathbb{R},L_{q_2}(\dot\Omega))}.
\end{aligned}$$

Thus, it suffices to prove (3.372) and (3.373) in the case where $r_i = q_2$. To prove inequalities in (3.372) and (3.373) in the case where $r_i = q_2$, we use Lemma 3.4.2 with $q = q_2$. By (3.227), (3.234), (3.374) and Sobolev's inequality (3.365), we have

$$\begin{aligned}
&\|V_g(\nabla\mathcal{E}_2[\Psi_\rho])\nabla\mathcal{E}_2[\Psi_{<t>^{N/q_1}\rho}]\|_{H^1_\infty(\mathbb{R},L_{q_2}(\dot\Omega))}\\
&\quad\leq C\{\|\partial_t\nabla\mathcal{E}_2[\Psi_{<t>^{N/q_1}\rho}]\|_{L_\infty(\mathbb{R},L_{q_2}(B_L))}\\
&\qquad+\|\partial_t\nabla\mathcal{E}_2[\Psi_\rho]\|_{L_\infty(\mathbb{R},L_{q_2}(B_L))}\|\nabla\mathcal{E}_2[\Psi_{<t>^{N/q_1}\rho}]\|_{L_\infty(\mathbb{R},H^1_{q_2}(B_L))}\}\\
&\quad\leq C(\mathcal{I}+E_T(\mathbf{u},\rho)+(\mathcal{I}+E_T(\mathbf{u},\rho))^2).
\end{aligned}$$

Thus, by Lemma 3.4.2

$$\begin{aligned}
&\|g_{a_i}(\mathbf{u},\rho)\|_{H^1_p(\mathbb{R},L_{q_2}(\dot\Omega))}+\|g_{a_i}(\mathbf{u},\rho)\|_{L_p(\mathbb{R},H^1_{q_2}(\dot\Omega))}\\
&\quad\leq C(\mathcal{I}+E_T(\mathbf{u},\rho)+(\mathcal{I}+E_T(\mathbf{u},\rho))^2)\\
&\qquad\times(\|\nabla\mathcal{E}_1[<t>^{a_i-\zeta}\mathbf{u}]\|_{H^{1/2}_p(\mathbb{R},L_{q_2}(\dot\Omega))}+\|\nabla\mathcal{E}_1[<t>^{a_i-\zeta}\mathbf{u}]\|_{L_p(\mathbb{R},H^1_{q_2}(\dot\Omega))}).
\end{aligned}$$

By Lemma 3.4.3, we have

$$\begin{aligned}
&\|\nabla\mathcal{E}_1[<t>^{a_i-\zeta}\mathbf{u}]\|_{H^{1/2}_p(\mathbb{R},L_{q_2}(\dot\Omega))}+\|\nabla\mathcal{E}_1[<t>^{a_i-\zeta}\mathbf{u}]\|_{L_p(\mathbb{R},H^1_{q_2}(\dot\Omega))}\\
&\leq C(\|\nabla\mathcal{E}_1[<t>^{a_i-\zeta}\mathbf{u}]\|_{H^1_p(\mathbb{R},L_{q_2}(\dot\Omega))}+\|\nabla\mathcal{E}_1[<t>^{a_i-\zeta}\mathbf{u}]\|_{L_p(\mathbb{R},H^2_{q_2}(\dot\Omega))})\\
&\leq C(\mathcal{I}+\|<t>^{a_i-\zeta}\partial_t\mathbf{u}\|_{L_p((0,T),L_{q_2}(\dot\Omega))}+\|<t>^{a_i-\zeta}\mathbf{u}\|_{L_p((0,T),H^2_{q_2}(\dot\Omega))})\\
&\leq C(\mathcal{I}+\|<t>^{a_2-\zeta}\partial_t\mathbf{u}\|_{L_p((0,T),L_{q_2}(\dot\Omega))}+\|<t>^{a_2-\zeta}\mathbf{u}\|_{L_p((0,T),H^2_{q_2}(\dot\Omega))})
\end{aligned}$$

where we have used the fact that $a_0 < a_1 < a_2$ in the last step. Putting these inequalities together yields inequalities in (3.372) and (3.373).

We now consider $<t>^{\frac{N}{q_1}-\zeta+a_i}\mathbf{h}'(\mathbf{u},\Psi_\rho)$. As was mentioned in (3.237), we may write

$$<t>^{\frac{N}{q_1}-\zeta+a_i}\mathbf{h}'(\mathbf{u},\Psi_\rho)=\mathbf{V}_{\mathbf{h}'}(\bar\nabla\Psi_\rho)\bar\nabla(<t>^{\frac{N}{q_1}}\Psi_\rho)\otimes\nabla(<t>^{a_i-\zeta}\mathbf{u})).$$

Thus, the extension of $<t>^{\frac{N}{q_1}-\zeta+a_i}\mathbf{h}'(\mathbf{u},\Psi_\rho)$ to the whole time interval $\mathbb{R}$ is defined by letting

$$\mathbf{h}'_{a_i}(\mathbf{u},\Psi_\rho)=\mathbf{V}_{\mathbf{h}'}(\bar\nabla\mathcal{E}_2[\Psi_\rho])\bar\nabla\mathcal{E}_2[\Psi_{<t>^{N/q_1}\rho}]\otimes\nabla\mathcal{E}_1[<t>^{a_i-\zeta}\mathbf{u}].$$

Employing the same argument as in the proof of inequalities in (3.372) and (3.373), we have

$$\begin{aligned}
&\|\mathbf{h}'_{a_i}(\mathbf{u},\Psi_\rho)\|_{H^{1/2}_p(\mathbb{R},L_{r_i}(\dot\Omega))}+\|\mathbf{h}'_{a_i}(\mathbf{u},\Psi_\rho)\|_{L_p(\mathbb{R},H^1_{r_i}(\dot\Omega))}\\
&\quad\leq C((\mathcal{I}+E_T(\mathbf{u},\rho))^2+(\mathcal{I}+E_T(\mathbf{u},\rho))^3)
\end{aligned}\tag{3.375}$$

for $i = 0, 1, 2$.

We finally consider $<t>^{\frac{N}{q_1}-\zeta+a_i} h_N(\mathbf{u}, \Psi_\rho)$. In view of (3.274), we may write

$$<t>^{\frac{N}{q_1}-\zeta+a_i} h_N(\mathbf{u}, \Psi_\rho) = \mathbf{V}_{h,N}(\bar\nabla\Psi_\rho)\bar\nabla(<t>^{\frac{N}{q_1}} \Psi_\rho) \otimes \nabla(<t>^{a_i-\zeta} \mathbf{u})$$
$$+ \sigma\tilde{\mathbf{V}}'_\Gamma(\bar\nabla\Psi_\rho)\bar\nabla(<t>^{\frac{N}{q_1}} \Psi_\rho) \otimes \bar\nabla^2(<t>^{a_i-\zeta} \Psi_\rho).$$

Thus, the extension of $<t>^{\frac{N}{q_1}-\zeta+a_i} h_N(\mathbf{u}, \Psi_\rho)$ to the whole time interval $\mathbb{R}$ is defined by letting

$$h_{N,a_i}(\mathbf{u}, \Psi_\rho) = \mathbf{V}_{h,N}(\bar\nabla\mathcal{E}_2[\Psi_\rho])\bar\nabla\mathcal{E}_2[\Psi_{<t>^{N/q_1}\rho}] \otimes \nabla(<t>^{a_i-\zeta} \mathbf{u})$$
$$+ \sigma\tilde{\mathbf{V}}'_\Gamma(\bar\nabla\mathcal{E}_2[\Psi_\rho])\bar\nabla\mathcal{E}_2[\Psi_{<t>^{N/q_1}\rho}] \otimes \bar\nabla^2\mathcal{E}_2[\Psi_{<t>^{N/q_1}\rho}].$$

Employing the same argument as in the proof of inequalities in (3.372) and (3.373), we have

$$\|h_{N,a_i}(\mathbf{u}, \Psi_\rho)\|_{H^{1/2}_p(\mathbb{R}, L_{q_2}(\dot\Omega))} + \|h_{N,a_i}(\mathbf{u}, \Psi_\rho)\|_{L_p(\mathbb{R}, H^1_{q_2}(\dot\Omega))}$$
$$\leq C(\mathcal{I} + E_T(\mathbf{u}, \rho) + (\mathcal{I} + E_T(\mathbf{u}, \rho))^2)$$
$$\times (\|\nabla\mathcal{E}_1[<t>^{a_i-\zeta} \mathbf{u}]\|_{H^{1/2}_p(\mathbb{R}, L_{q_2}(\dot\Omega))} + \|\nabla\mathcal{E}_1[<t>^{a_i-\zeta} \mathbf{u}]\|_{L_p(\mathbb{R}, H^1_{q_2}(\dot\Omega))}$$
$$+ \|\bar\nabla^2\mathcal{E}_2[\Psi_{<t>^{N/q_1}\rho}]\|_{H^{1/2}_p(\mathbb{R}, L_{q_2}(\dot\Omega))} + \|\bar\nabla^2\mathcal{E}_2[\Psi_{<t>^{N/q_1}\rho}]\|_{L_p(\mathbb{R}, H^1_{q_2}(\dot\Omega))}).$$

By Lemma 3.4.3 and (3.226), we have

$$\|\nabla\mathcal{E}_1[<t>^{a_i-\zeta} \mathbf{u}]\|_{H^{1/2}_p(\mathbb{R}, L_{q_2}(\dot\Omega))} + \|\nabla\mathcal{E}_1[<t>^{a_i-\zeta} \mathbf{u}]\|_{L_p(\mathbb{R}, H^1_{q_2}(\dot\Omega))}$$
$$\leq C(\|\mathcal{E}_1[<t>^{a_i-\zeta} \mathbf{u}]\|_{H^1_p(\mathbb{R}, L_{q_2}(\dot\Omega))} + \|\mathcal{E}_1[<t>^{a_i-\zeta} \mathbf{u}]\|_{L_p(\mathbb{R}, H^2_{q_2}(\dot\Omega))})$$
$$\leq C(\mathcal{I} + \| <t>^{a_i-\zeta} \partial_t\mathbf{u}\|_{L_p((0,T), L_{q_2}(\dot\Omega))} + \| <t>^{a_i-\zeta} \mathbf{u}\|_{L_p((0,T), H^2_{q_2}(\dot\Omega))})$$
$$\leq C(\mathcal{I} + \| <t>^{a_2-\zeta} \partial_t\mathbf{u}\|_{L_p((0,T), L_{q_2}(\dot\Omega))} + \| <t>^{a_2-\zeta} \mathbf{u}\|_{L_p((0,T), H^2_{q_2}(\dot\Omega))})$$
$$\leq C(\mathcal{I} + E_T(\mathbf{u}, \rho)),$$

where we have used the facts that $a_0 < a_1 < a_2$ in the last step. Moreover, using (3.227), we have

$$\|\bar\nabla^2\mathcal{E}_2[\Psi_{<t>^{N/q_1}\rho}]\|_{H^{1/2}_p(\mathbb{R}, L_{q_2}(\dot\Omega))} + \|\bar\nabla^2\mathcal{E}_2[\Psi_{<t>^{N/q_1}\rho}]\|_{L_p(\mathbb{R}, H^1_{q_2}(\dot\Omega))}$$
$$\leq \|\mathcal{E}_2[\Psi_{<t>^{N/q_1}\rho}]\|_{H^1_p(\mathbb{R}, H^2_{q_2}(\dot\Omega))} + \|\mathcal{E}_2[\Psi_{<t>^{N/q_1}\rho}]\|_{L_p(\mathbb{R}, H^3_{q_2}(\dot\Omega))}$$

$$\leq C(\mathcal{I} + \| <t>^{a_i-\zeta} \partial_t \Psi_\rho \|_{L_p((0,T),H^2_{q_2}(\dot\Omega))} + \| <t>^{a_i-\zeta} \Psi_\rho \|_{L_p((0,T),H^3_{q_3}(\dot\Omega))})$$

$$\leq C(\mathcal{I} + \| <t>^{a_2-\zeta} \partial_t \Psi_\rho \|_{L_p((0,T),H^2_{q_2}(\dot\Omega))} + \| <t>^{a_2-\zeta} \Psi_\rho \|_{L_p((0,T),H^3_{q_3}(\dot\Omega))})$$

$$\leq C(\mathcal{I} + E_T(\mathbf{u}, \rho)),$$

where we have used the facts that $a_0 < a_1 < a_2$ in the last step. Putting inequalities obtained above together yields

$$\begin{aligned} &\|h_{N,a_i}(\mathbf{u}, \Psi_\rho)\|_{H^{1/2}_p(\mathbb{R},L_{q_2}(\dot\Omega))} + \|h_{N,a_i}(\mathbf{u}, \Psi_\rho)\|_{L_p(\mathbb{R},H^1_{q_2}(\dot\Omega))} \\ &\quad\leq C((\mathcal{I} + E_T(\mathbf{u}, \rho))^2 + (\mathcal{I} + E_T(\mathbf{u}, \rho))^3) \end{aligned} \tag{3.376}$$

for $i = 0, 1, 2$. By (3.374) and Hölder's inequality, we have

$$\begin{aligned} &\|h_{N,a_i}(\mathbf{u}, \Psi_\rho)\|_{H^{1/2}_p(\mathbb{R},L_{r_i}(\dot\Omega))} + \|h_{N,a_i}(\mathbf{u}, \Psi_\rho)\|_{L_p(\mathbb{R},H^1_{r_i}(\dot\Omega))} \\ &\leq C_L(\|h_{N,a_i}(\mathbf{u}, \Psi_\rho)\|_{H^{1/2}_p(\mathbb{R},L_{q_2}(\dot\Omega))} + \|h_{N,a_i}(\mathbf{u}, \Psi_\rho)\|_{L_p(\mathbb{R},H^1_{q_2}(\dot\Omega))}), \end{aligned}$$

which, combined with (3.376), leads to

$$\begin{aligned} &\|h_{N,a_i}(\mathbf{u}, \Psi_\rho)\|_{H^{1/2}_p(\mathbb{R},L_{r_i}(\dot\Omega))} + \|h_{N,a_i}(\mathbf{u}, \Psi_\rho)\|_{L_p(\mathbb{R},H^1_{r_i}(\dot\Omega))} \\ &\quad\leq C((\mathcal{I} + E_T(\mathbf{u}, \rho))^2 + (\mathcal{I} + E_T(\mathbf{u}, \rho))^3) \end{aligned} \tag{3.377}$$

for $i = 0, 1, 2$.

A Proof of Theorem 3.1.4 We now prove Theorem 3.1.4. As was seen in Sect. 3.5.6, to prove Theorem 3.1.4 it suffices to prove (3.336). Let $\mathbf{v}$, $\mathfrak{p}$, and h be unique solutions of the shifted equations:

$$\begin{cases} \partial_t\mathbf{v} + \lambda_0\mathbf{v} - m^{-1}\mathrm{Div}\,(\mu\mathbf{D}(\mathbf{v}) - \mathfrak{p}\mathbf{I}) = \mathbf{f}(\mathbf{u}, \Psi_\rho) & \text{in } \dot\Omega^T, \\ \mathrm{div}\,\mathbf{v} = g(\mathbf{u}, H_\rho) = \mathrm{div}\,\mathbf{g}(\mathbf{u}, \Psi_\rho) & \text{in } \dot\Omega^T, \\ \partial_t h + \lambda_0 h + \mathcal{L}_3 h - \mathbf{n}\cdot P\mathbf{v} = \bar d(\mathbf{u}, \Psi_\rho) & \text{on } S_R^T, \\ [[(\mu\mathbf{D}(\mathbf{v})\mathbf{n})_\tau]] = [[\mathbf{h}'(\mathbf{u}, \Psi_\rho)]] & \text{on } S_R^T, \\ [[<\mu\mathbf{D}(\mathbf{v})\mathbf{n}, \mathbf{n}> -\mathfrak{p}]] - \sigma\mathcal{B}\rho = [[h_N(\mathbf{u}, \Psi_\rho)]] & \text{on } S_R^T, \\ [[\mathbf{v}]] = 0 & \text{on } S_R^T, \\ (\mathbf{v}, h)|_{t=0} = (\mathbf{u}_0, \rho_0) \quad \text{in } \dot\Omega \times S_R. \end{cases} \tag{3.378}$$

Choosing $\lambda_0 > 0$ large enough, by Theorem 3.6.1 and estimates: (3.363), (3.368), (3.371), (3.372), (3.373), (3.375), and (3.377) we see that

$$(\mathbf{v}, h) \in \cap_{i=0}^2 (H_p^1((0,T), \mathcal{H}_{r_i}) \cap L_p((0,T), \mathcal{D}_{r_i})),$$

and $\mathbf{v}$ and h possess the estimate:

$$\begin{aligned}\sum_{i=0}^2 \{ \| < t >^{\frac{N}{q_1}-\zeta+a_i} \partial_t(\mathbf{v}, h)\|_{L_p((0,T),\mathcal{H}_{r_i})} \\ + \| < t >^{\frac{N}{q_1}-\zeta+a_i} (\mathbf{v}, h)\|_{L_p((0,T),\mathcal{D}_{r_i})}\} \\ \leq C(\mathcal{I} + E_T(\mathbf{u}, \rho)^2 + E_T(\mathbf{u}, \rho)^3).\end{aligned} \tag{3.379}$$

Here and in the following, we use the estimate:

$$(\mathcal{I} + E_T(\mathbf{u}, \rho))^2 + (\mathcal{I} + E_T(\mathbf{u}, \rho))^3 \leq C(\mathcal{I} + E_T(\mathbf{u}, \rho)^2 + E_T(\mathbf{u}, \rho)^3)$$

for some constant, because $\mathcal{I}$ is assumed to be less than 1.

Let $\mathbf{u}$, $\mathfrak{q}$, and ρ be unique solutions of equation (3.335) and we set

$$\mathbf{u} = \mathbf{v} + \mathbf{w}, \quad \mathfrak{q} = \mathfrak{q} + \mathfrak{r}, \quad \rho = h + \eta.$$

Then, $\mathbf{w}$, $\mathfrak{r}$, and η satisfy the equations:

$$\begin{cases} \partial_t \mathbf{w} - m^{-1}\mathrm{Div}\,(\mu\mathbf{D}(\mathbf{w}) - \mathfrak{r}\mathbf{I}) = -\lambda_0 \mathbf{v} & \text{in } \dot\Omega^T, \\ \mathrm{div}\,\mathbf{w} = 0 & \text{in } \dot\Omega^T, \\ \partial_t\eta + \lambda_0\eta + \mathcal{L}_3\eta - \mathbf{n}\cdot P\mathbf{w} = -\lambda_0 h & \text{on } S_R^T, \\ [[(\mu\mathbf{D}(\mathbf{w})\mathbf{n})_\tau]] = 0 & \text{on } S_R^T, \\ [[< \mu\mathbf{D}(\mathbf{w})\mathbf{n}, \mathbf{n} > -\mathfrak{r}]] - \sigma\mathcal{B}\eta = 0 & \text{on } S_R^T, \\ [[\mathbf{w}]] = 0 & \text{on } S_R^T, \\ (\mathbf{w}, \eta)|_{t=0} = (0,0) & \text{in } \dot\Omega \times S_R. \end{cases} \tag{3.380}$$

By (3.379) and (3.213), we have

$$\begin{aligned}[h]_{W_{q_2,p}^{2-1/q_2}(S_R), N/q_1, T} \leq C\{\| < t >^{\frac{N}{q_1}} \partial_t\rho\|_{L_p((0,T),W_{q_2}^{2-1/q_2}(S_R))} \\ + \| < t >^{\frac{N}{q_1}} \rho\|_{L_p((0,T),W_{q_2}^{3-1/q_2}(S_R))} + \|\rho_0\|_{W_{q_2,p}^{3-1/p-1/q_2}(S_R)}\} \\ \leq C(\mathcal{I} + E_T(\mathbf{u}, \rho)^2 + E_T(\mathbf{u}, \rho)^3).\end{aligned} \tag{3.381}$$

By Theorem 3.6.2, (3.379), and (3.381), we see that

$$(\mathbf{w}, \eta) \in \cap_{i=0}^{2}(H_p^1((0, T), \mathcal{H}_{r_i}) \cap L_p((0, T), \mathcal{D}_{r_i}))$$

and w and θ possess the estimate:

$$E_T[\mathbf{w}, \theta] \leq C(\mathcal{I} + E_T(\mathbf{u}, \rho)^2 + E_T(\mathbf{u}, \rho)^3) \tag{3.382}$$

for some constant $C > 0$. From (3.379), (3.212), and (3.213) it follows that

$$\begin{aligned}
&[\rho]_{B^{3-1/p-1/q_2}_{q_2,p}(S_R), N/q_1, T} + [\mathbf{v}]_{H^1_\infty(\dot\Omega), N/q_1, T} + [\mathbf{v}]_{H^1_{q_1/2}(\dot\Omega), 0, T} \\
&\quad + [\mathbf{v}]_{H^1_{q_1}(\dot\Omega), N/(2q_1), T} + [\mathbf{v}]_{H^1_{q_2}(\dot\Omega), N/(2q_2)+1, T} \\
&\quad \leq C(\mathcal{I} + E_T(\mathbf{u}, \rho)^2 + E_T(\mathbf{u}, \rho)^3),
\end{aligned} \tag{3.383}$$

because $N/q_1 - \zeta + a_i \geq N/q_1$ $(i = 0, 1, 2)$, and $1 + N/q_2 < 2(1 - 1/p)$. By the third equation of equations (3.335), we have

$$\begin{aligned}
&[\partial_t \rho]_{W^{1-1/q_2}_{q_2}(S_R), N/q_1, T} \\
&\leq C([\mathbf{v}]_{H^1_{q_2}(\dot\Omega), N/q_1, T} + [\rho]_{L_{q_2}(S_R), N/q_1, T} + [\bar d(\mathbf{u}, \Psi_\rho]_{W^{1-1/q_2}_{q_2}(S_R), N/q_1, T}).
\end{aligned} \tag{3.384}$$

By (3.215), (3.201), (3.333), and (3.334), we have

$$\begin{aligned}
[d(\mathbf{u}, \Psi_\rho)]_{W^{1-1/q_2}_{q_2}(S_R), N/q_2, T} &\leq C([\mathbf{u}]_{H^1_{q_2}(\dot\Omega), N/q_1, T}[\rho]_{W^{2-1/q_2}_{q_2}(S_R), N/q_2, T} \\
&\quad + [\partial_t \rho]_{W^{1-1/q_2}_{q_2}(S_R), N/q_1, T}[\rho]_{W^{2-1/q_2}_{q_2}(S_R), N/q_1, T}), \\
[< \mathbf{u} \mid \nabla'_\Gamma \rho >]_{W^{1-1/q_2}_{q_2}(S_R), N/q_1, T} &\leq C[\mathbf{u}]_{H^1_{q_2}(\dot\Omega), N/q_2, T}[\rho]_{W^{1-1/q_2}_{q_2}(S_R), N/q_1, T}, \\
[\int_{B_R} \mathbf{u} J_0(\mathbf{k}), dy]_{W^{1-1/q_2}_{q_2}(S_R), N/q_1, T} &\leq C[\mathbf{u}]_{H^1_{q_2}(\dot\Omega), N/q_1, T}[\rho]_{W^{2-1/q_2}_{q_2}(S_R), N/q_1, T}, \\
[\int_{S_R} \rho^\ell \, d\omega]_{W^{1-1/q_2}_{q_2}(S_R), N/q_1, T} &\leq C[\rho]^2_{W^{2-1/q_2}_{q_2}(S_R), N/q_1, T} \qquad (\ell \geq 2), \\
[\int_{S_R} \rho^\ell \mathbf{n}_i \, d\omega]_{W^{1-1/q_2}_{q_2}(S_R), N/q_1, T} &\leq C[\rho]^2_{W^{2-1/q_2}_{q_2}(S_R), N/q_1, T} \qquad (\ell \geq 2).
\end{aligned}$$

Thus, we have

$$[\bar d(\mathbf{u}, \Psi_\rho)]_{W^{1-1/q_2}_{q_2}(S_R), N/q_1, T} \leq C E_T(\mathbf{u}, \rho)^2. \tag{3.385}$$

Combining (3.379), (3.383), (3.384), (3.385), and (3.333), we have

$$E_T(\mathbf{v}, h) \leq C(\mathcal{I} + E_T(\mathbf{u}, \rho)^2 + E_T(\mathbf{u}, \rho)^3),$$

which, combined with (3.382), yields

$$E_T(\mathbf{u}, \rho) \le C(\mathcal{I} + E_T(\mathbf{u}, \rho)^2 + E_T(\mathbf{u}, \rho)^3).$$

Namely, we have proved (3.336), which completes the proof of Theorem 3.1.4. The rest of the proof is to prove (3.347) and (3.348), which will be given in the next section. □

3.7 On the Decay Properties of Stokes Semigroup Associated with Two-Phase Problem

3.7.1 Introduction

In this section, we shall prove (3.347) and (3.348). To make this section self-contained as much as possible, we formulate the problem again. We consider the following linear equations:

$$\begin{cases} \partial_t \mathbf{v} - m^{-1}\mathrm{Div}\,(\mu\mathbf{D}(\mathbf{v}) - \mathfrak{q}\mathbf{I}) = 0 & \text{in } \dot\Omega^T, \\ [[(\mu\mathbf{D}(\mathbf{v}) - \mathfrak{q}\mathbf{I})\mathbf{n}]] - (\mathcal{B}\eta)\mathbf{n} = 0, \quad [[\mathbf{v}]] = 0 & \text{on } S_R^T, \\ \partial_t\eta + \mathcal{L}_3\eta - \mathbf{n}\cdot P\mathbf{v} = 0 & \text{on } S_R^T, \\ (\mathbf{v}, \eta)|_{t=0} = (\mathbf{v}_0, \eta_0) & \text{in } \dot\Omega \times S_R. \end{cases} \tag{3.386}$$

In this section, $\dot\Omega = \mathbb{R}^N \setminus S_R$, $\mathcal{B}$ and P are the same operators as in equations (3.34) in Subsec 3.1.3.2 and $\mathcal{L}_3$ is the operator defined in (3.278) in Sect. 3.5.4. To formulate problem (3.386) in the semi-group setting, we eliminate the pressure term $\mathfrak{q}$. Let $K(\mathbf{u}, \rho) \in H^1_q(\dot\Omega) + \hat H^1_q(\mathbb{R}^N)$ be a unique solution of the variational problem:

$$(\nabla K(\mathbf{u}, \rho), \nabla\psi)_{\dot\Omega} = (m^{-1}\mathrm{Div}\,(\mu\mathbf{D}(\mathbf{u})) - \nabla\mathrm{div}\,\mathbf{u}, \nabla\psi)_{\dot\Omega} \tag{3.387}$$

for $\psi \in \hat H^1_{q}(\mathbb{R}^N)$, subject to

$$K(\mathbf{u}, \rho) =< [[\mu\mathbf{D}(\mathbf{u})]]\mathbf{n}, \mathbf{n} > -\sigma(\mathcal{B}\rho)\mathbf{n} - [[\mathrm{div}\,\mathbf{u}]] \quad \text{on } S_R, \tag{3.388}$$

where we have set $q' = q/(q-1)$, and

$$\hat H^1_q(\mathbb{R}^N) = \{\psi \in L_{q,\mathrm{loc}}(\mathbb{R}^N) \mid \nabla\psi \in L_q(\mathbb{R}^N)\}.$$

The existence of such $K(\mathbf{u}, \rho)$ possessing the estimate:

$$\|\nabla K(\mathbf{u}, \rho)\|_{L_q(\dot\Omega)} \le C(\|\mathbf{u}\|_{H^2_q(\dot\Omega)} + \|\rho\|_{W^{3-1/q}_q(S_R)}) \tag{3.389}$$

is proved in Saito and Zhang [24]. We now consider the reduced Stokes problem:

$$\begin{cases} \partial_t \mathbf{u} - m^{-1}\mathrm{Div}\,(\mu\mathbf{D}(\mathbf{u}) - K(\mathbf{u},\rho)\mathbf{I}) = 0 & \text{in } \dot\Omega^T, \\ \partial_t\rho + \mathcal{L}_3\rho - \mathbf{n}\cdot P\mathbf{u} = 0 & \text{on } S_R^T, \\ [[\mu\mathbf{D}(\mathbf{u}) - K(\mathbf{u},\rho)\mathbf{I}]]\mathbf{n} - \sigma(\mathcal{B}\rho)\mathbf{n} = 0 & \text{on } S_T^R, \\ [[\mathbf{u}]] = 0 & \text{on } S_T^R, \\ (\mathbf{u},\rho)|_{t=0} = (\mathbf{u}_0,\rho_0) & \text{in } \dot\Omega\times S_R. \end{cases} \tag{3.390}$$

Notice that the transmission conditions: $[[\mu\mathbf{D}(\mathbf{u}) - K(\mathbf{u},\rho)\mathbf{I}]]\mathbf{n} - \sigma(\mathcal{B}\rho)\mathbf{n} = 0$ on S_R are equivalent to

$$([[\mu\mathbf{D}(\mathbf{u})]]\mathbf{n})_\tau = 0 \text{ and } [[\mathrm{div}\,\mathbf{u}]] = 0 \tag{3.391}$$

on S_R, where, for any N-vector $\mathbf{d}$, we have set $\mathbf{d}_\tau = \mathbf{d} - < \mathbf{d}, \mathbf{n} > \mathbf{n}$. Given $\mathbf{f} \in L_q(\dot\Omega)^N$, let $\mathcal{Q}\mathbf{f} \in \hat H^1_q(\mathbb{R}^N)$ be a unique solution of the variational problem:

$$(m^{-1}\nabla\mathcal{Q}\mathbf{f}, \nabla\psi)_{\dot\Omega} = (\mathbf{f}, \nabla\psi)_{\dot\Omega} \tag{3.392}$$

for any $\psi \in \hat H^1_{q'}(\mathbb{R}^N)$ with $[[\mathcal{Q}\mathbf{f}]] = 0$ on S_R. The existence of such $\mathcal{Q}\mathbf{f}$ possessing the estimate:

$$\|\nabla\mathcal{Q}\mathbf{f}\|_{L_q(\mathbb{R}^N)} \leq C\|\mathbf{f}\|_{L_q(\dot\Omega)}$$

is discussed in Saito and Zhang [24]. Let $\mathcal{P}\mathbf{f} = \mathbf{f} - m^{-1}\mathcal{Q}\mathbf{f}$, and then

$$\mathcal{P}\mathbf{f} \in J_q(\dot\Omega), \quad \|\mathcal{P}\mathbf{f}\|_{L_q(\dot\Omega)} \leq C\|\mathbf{f}\|_{L_q(\dot\Omega)}, \tag{3.393}$$

where $J_q(\dot\Omega)$ is a solenoidal space defined by

$$J_q(\dot\Omega) = \{\mathbf{f} \in L_q(\dot\Omega)^N \mid (\mathbf{f}, \nabla\psi)_{\dot\Omega} = 0 \quad \text{for any } \psi \in \hat H^1_{q'}(\mathbb{R}^N)\}. \tag{3.394}$$

If $\mathbf{f} \in H^1_q(\dot\Omega)^N$, $[[\mathbf{f}]] = 0$ on S_R, and $\mathbf{f} \in J_q(\dot\Omega)$, then $\mathrm{div}\,\mathbf{f} = 0$ in $\mathbb{R}^N$. But, we can prove the opposite direction (cf. Shibata [28, Lemma 14]), that is, if $\mathbf{f} \in H^1_q(\dot\Omega)^N$ satisfies $[[\mathbf{f}]] = 0$ on S_R and $\mathrm{div}\,\mathbf{f} = 0$ in $\dot\Omega$, then $\mathbf{f} \in J_q(\dot\Omega)$.

Let

$$\begin{aligned}
\mathcal{H}_q =&\{(\mathbf{f}, g) \mid \mathbf{f} \in L_q(\dot\Omega)^N, \quad g \in W_q^{2-1/q}(S_R)\},\\
&\|(\mathbf{f}, g)\|_{\mathcal{H}_q} = \|\mathbf{f}\|_{L_q(\dot\Omega)} + \|g\|_{W_q^{2-1/q}(S_R)},\\
\mathcal{D}_q =&\{(\mathbf{u}, \rho) \in \mathcal{H}_q \mid \mathbf{u} \in H_q^2(\dot\Omega)^N, \quad \rho \in W_q^{3-1/q}(S_R),\\
&([[\mu\mathbf{D}(\mathbf{u})]]\mathbf{n})_\tau = 0, \;\; [[\operatorname{div}\mathbf{u}]] = 0, \;\; [[\mathbf{u}]] = 0 \quad \text{on } S_R\},\\
&\|(\mathbf{u}, \rho)\|_{\mathcal{D}_q} = \|\mathbf{u}\|_{H_q^2(\dot\Omega)} + \|\rho\|_{W_q^{3-1/q}(S_R)},\\
\mathcal{A}_q(\mathbf{u}, \rho) =&(m^{-1}\mathrm{Div}\,(\mu\mathbf{D}(\mathbf{u}) - K(\mathbf{u}, \rho)\mathbf{I}), -\mathcal{L}_3\rho + \mathbf{n}\cdot P\mathbf{u}|_{S_R})
\end{aligned} \tag{3.395}$$

for $(\mathbf{u}, \rho) \in \mathcal{D}_q$ where we have used (3.391) to define the domain $\mathcal{D}_q$. Obviously, problem (3.390) is formulated by

$$\partial_t(\mathbf{u}, \rho) - \mathcal{A}_q(\mathbf{u}, \rho) = 0 \quad \text{in } Q_\infty, \quad (\mathbf{u}, \rho)|_{t=0} = (\mathbf{u}_0, \rho_0) \quad \text{in } \dot\Omega \times S_R, \tag{3.396}$$

where $Q_\infty = \dot\Omega \times (0, \infty)$. According to Theorem 3.2.4 in Sect. 3.2.1.3, we know that for any $\epsilon \in (0, \pi/2)$ there exists a $\lambda_0 > 0$ such that the corresponding resolvent problem:

$$\lambda(\mathbf{v}, h) - \mathcal{A}_q(\mathbf{v}, h) = (\mathbf{f}, g) \quad \text{in } \dot\Omega \tag{3.397}$$

admits a unique solution $(\mathbf{v}, h) \in \mathcal{D}_q$ for any $(\mathbf{f}, g) \in \mathcal{H}_q$ and $\lambda \in \Sigma_{\epsilon,\lambda_0}$. Moreover, these $\mathbf{v}$ and h possess the estimate:

$$|\lambda|\|(\mathbf{v}, h)\|_{\mathcal{H}_q} + \|(\mathbf{v}, h)\|_{\mathcal{D}_q} \le C\|(\mathbf{f}, g)\|_{\mathcal{H}_q}. \tag{3.398}$$

By semigroup theory, the operator $\mathcal{A}_q$ generates a C_0 analytic semigroup $\{T(t)\}_{t\ge0}$ on $\mathcal{H}_q$. Moreover, the solution $\mathbf{u}$ and ρ of equation (3.396) are represented as

$$(\mathbf{u}(\cdot, t), \rho(\cdot, t)) = T(t)(\mathcal{P}\mathbf{u}_0, \rho_0), \tag{3.399}$$

with $\mathfrak{q} = K(\mathbf{u}, \rho) + \mathcal{Q}\mathbf{f}$.

In this section, we prove the following theorem.

Theorem 3.7.1 *Assume that* $N \ge 3$. *For any* $(\mathbf{u}_0, \rho_0) \in \mathcal{H}_q$, *we set* $(\mathbf{u}, \rho) = T(t)(\mathbf{u}_0, \rho_0)$, *and then, for any* $t \ge 3$, *and* p *and* q *with* $1 < q \le p \le \infty$ *and* $q \ne \infty$, *we have*

$$\|\mathbf{u}(\cdot, t)\|_{L_p(\dot\Omega)} \le C_{p,q} t^{-\frac{N}{2}\left(\frac{1}{q}-\frac{1}{p}\right)}\|(\mathbf{u}_0, \rho_0)\|_{\mathcal{H}_q};$$

$$\|\nabla\mathbf{u}(\cdot, t)\|_{L_p(\dot\Omega)} \le C_{p,q} t^{-\min\left(\frac{1}{2}+\frac{N}{2}\left(\frac{1}{q}-\frac{1}{p}\right), \frac{N}{2q}\right)}\|(\mathbf{u}_0, \rho_0)\|_{\mathcal{H}_q}.$$

Moreover, for $t \geq 3$, $m \in \mathbb{N}_0$, $L \geq 5R$, *and* $1 < q \leq p < \infty$, *we have*

$$\|\partial_t^m(\mathbf{u}(\cdot,t),\rho(\cdot,t))\|_{H^2_p(\dot B_L)\times W^{3-1/p}_p(S_R)} \leq Ct^{-\frac{N}{2q}}\|(\mathbf{u}_0,\rho_0)\|_{\mathcal{H}_q}$$

for some constant C *depending on* p, q, m, *and* L. *Here and in the following, we set* $\dot B_L = B_R \cup \{x \in \mathbb{R}^N \mid R < |x| < L\}$ *for* $L > R$.

To prove Theorem 3.7.1, the main step is to prove the local energy decay estimate stated as in the following theorem.

Theorem 3.7.2 *Assume that* $N \geq 3$. *Let* $1 < q < \infty$. *Let*

$$L_{q,5R}(\dot\Omega) = \{f \in L_q(\dot\Omega) \mid \operatorname{supp} f \subset B_R\}.$$

Then, the semigroup $\{T(t)\}_{t\geq 0}$ *has the following property:*

$$\begin{aligned}&\|\partial_t^m T(t)(\mathcal{P}\mathbf{u}_0,\rho_0)\|_{H^2_q(\dot B_L)\times W^{3-1/q}_q(S_R)}\\ &\quad\leq C_{L,m}t^{-m-\frac{N}{2}}\|(\mathbf{u}_0,\rho_0)\|_{\mathcal{H}_q}\end{aligned} \tag{3.400}$$

for any $t \geq 1$, $m \in \mathbb{N}_0$, $L > 5R$, *and* $(\mathbf{u}_0,\rho_0) \in L_{q,5R}(\dot\Omega)^N \times W^{2-1/q}_q(S_R)$.

Notation In this section, in addition, we use the following symbols. $\mathbb{C}$, $\mathbb{R}$, and $\mathbb{N}$ denote the sets of all complex numbers, real numbers, and natural numbers. Let

$$L_{q,5R}(\dot\Omega) = \{f \in L_q(\dot\Omega) \mid \operatorname{supp} f \subset B_R\}$$

$U_r = \{\lambda \in \mathbb{C} \mid |\lambda| < r\}$ and $\dot U_r = \{\lambda \in U_r \mid \lambda \notin (-\infty, 0]\}$. Let $\mathrm{Hol}(U, X)$ denote the set of all X-valued holomorphic functions defined in a domain U of $\mathbb{C}$, let $\mathcal{B}(U, X)$ denote the set of all X-valued bounded holomorphic functions defined in U, and let $\mathcal{L}(X, Y)$ denote the set of all bounded linear operators from X into Y. $\mathcal{L}(X, X)$ is simply written as $\mathcal{L}(X)$.

3.7.2 Analysis in a Bounded Domain

Let $\dot B_{5R} = B_{5R} \setminus S_R$. In this section, we consider the following resolvent problem:

$$\begin{cases}\lambda\mathbf{u} - m^{-1}\mathrm{Div}\,(\mu\mathbf{D}(\mathbf{u}) - \mathbf{q}\mathbf{I}) = \mathbf{f}, \quad \mathrm{div}\,\mathbf{u} = 0 & \text{in } \dot B_{5R},\\ \lambda\rho + \mathcal{L}_3\rho - \mathbf{n}\cdot P\mathbf{u} = g & \text{on } S_R,\\ [[\mu\mathbf{D}(\mathbf{u}) - \mathbf{q}\mathbf{I}]]\mathbf{n} - \sigma(\mathcal{B}\rho)\mathbf{n} = 0, \quad [[\mathbf{u}]] = 0 & \text{on } S_R,\\ \mathbf{u} = 0 & \text{on } S_{5R}.\end{cases} \tag{3.401}$$

From Theorem 3.5.6 and Theorem 3.5.7, we have the following theorem.

Theorem 3.7.3 *Let* $1 < q < \infty$. *Then, there exist an* $\delta_0 > 0$ *and* $\epsilon_0 \in (0, \pi/2)$ *such that for any* $\lambda \in -\delta_0 + \Sigma_{\epsilon_0}$, $\mathbf{f} \in L_q(\dot{B}_{5R})^N$ *and* $g \in W_q^{2-1/q}(S_R)$, *problem* (3.401) *admits unique solution* $(\mathbf{u}, \mathfrak{q}, \rho)$ *with*

$$\mathbf{u} \in H_q^2(\dot{B}_{5R})^N, \quad \mathfrak{q} \in H_q^1(\dot{B}_{5R}), \quad \rho \in W_q^{3-1/q}(S_R)$$

possessing the estimate:

$$\begin{aligned}&|\lambda|\|\mathbf{u}\|_{L_q(B_{5R})} + \|\mathbf{u}\|_{H_q^2(\dot{B}_{5R})} + \|\mathfrak{q}\|_{H_q^1(\dot{B}_{5R})} + |\lambda|\|\rho\|_{W_q^{2-1/q}(S_R)} + \|\rho\|_{W_q^{3-1/q}(S_R)}\\ &\leq C(\|\mathbf{f}\|_{L_q(B_{5R})} + \|g\|_{W_q^{2-1/q}(S_R)}),\end{aligned}$$

where the constant C *is a constant independent of* $\lambda \in \Sigma_\epsilon - \lambda_0$.

3.7.3 Analysis in the Whole Space Near $\lambda = 0$

In this section, we consider the resolvent problem for the following Stokes system:

$$\lambda\mathbf{u} - m_-^{-1}\mathrm{Div}\,(\mu_-\mathbf{D}(\mathbf{u}) - \mathfrak{p}\mathbf{I}) = \mathbf{f}, \quad \mathrm{div}\,\mathbf{u} = 0 \quad \text{in } \mathbb{R}^N. \tag{3.402}$$

Recall that $m = m_-$ in $B^R = \{x \in \mathbb{R}^N \mid |x| > R\} = \Omega_-$. We define the Fourier transform and the inverse Fourier transform by setting

$$\hat{f}(\xi)=\mathcal{F}[f](\xi) = \int_{\mathbb{R}^N} e^{-ix\cdot\xi} f(x)\,dx, \ \mathcal{F}_\xi^{-1}[g(\xi)](x) = \frac{1}{(2\pi)^N}\int_{\mathbb{R}^N} e^{ix\cdot\xi} g(\xi)\,d\xi.$$

Applying the Fourier transform and the inverse transform gives the representation formulas of $\mathbf{u}$ and $\mathfrak{p}$ as follows:

$$R_0(\lambda)\mathbf{f} = \mathcal{F}^{-1}\Big[\frac{P(\xi)\hat{\mathbf{f}}(\xi)}{\lambda + m_-^{-1}\mu_-|\xi|^2}\Big](x), \quad \Pi\mathbf{f} = -m_-\mathcal{F}^{-1}\Big[\frac{i\xi\cdot\hat{\mathbf{f}}(xi)}{|\xi|^2}\Big](x), \tag{3.403}$$

where $P(\xi)$ is an $N \times N$ matrix with $(i, j)^{\text{th}}$ component $\delta_{ij} - \xi_i\xi_j|\xi|^{-2}$, δ_{ij} being Kronecker's delta symbols. The following theorem was proved by Shibata and Shimizu [31, Theorem 3.1].

Theorem 3.7.4 *Let* $1 < q < \infty$ *and* $0 < \epsilon < \pi/2$. *Assume that* $N \geq 3$. *Let*

$$L_{q,5R}(\mathbb{R}^N) = \{f \in L_q(\mathbb{R}^N) \mid f(x) = 0 \text{ for } x \notin B_{5R}\}.$$

Then, there exist operators $G_j(\lambda)$ $(j = 0, 1)$ *with*

$$G_j(\lambda) \in \mathrm{Hol}\,(U_{1/2}, \mathcal{L}(L_{q,5R}(\mathbb{R}^N)^N, H^2_{q,\mathrm{loc}}(\mathbb{R}^N)^N))$$

such that $R_0(\lambda)$ has the expansion formula:

$$R_0(\lambda)\mathbf{f} = \lambda^{\frac{N}{2}-1}(\log\lambda)^{\sigma(N)}G_1(\lambda)\mathbf{f} + G_2(\lambda)\mathbf{f} \quad \text{in } B_L \tag{3.404}$$

for any $L > 0$, $\lambda \in \dot{U}_{1/2}$ and $\mathbf{f} \in L_{q,5R}(\mathbb{R}^N)^N$, where $\sigma(N)$ is a number defined by setting

$$\sigma(N) = \begin{cases} 1 & \text{provided } N \text{ is even,} \\ 0 & \text{provided } N \text{ is odd}. \end{cases}$$

3.7.4 Analysis in $\dot\Omega$ Near $\lambda = 0$

In this section and next section, we consider the following resolvent problem:

$$\begin{cases} \lambda\mathbf{u} - m^{-1}\mathrm{Div}\,(\mu\mathbf{D}(\mathbf{u}) - \mathfrak{q}\mathbf{I}) = \mathbf{f}, \quad \mathrm{div}\,\mathbf{u} = 0 & \text{in } \dot\Omega, \\ \lambda\rho + \mathcal{L}_3\rho - \mathbf{n}\cdot P\mathbf{u} = g & \text{on } S_R, \\ [[\mu\mathbf{D}(\mathbf{u}) - \mathfrak{q}\mathbf{I}]]\mathbf{n} - \sigma(\mathcal{B}\rho)\mathbf{n} = 0, \quad [[\mathbf{u}]] = 0 & \text{on } S_R. \end{cases} \tag{3.405}$$

From Theorem 3.2.2, we have the following theorem.

Theorem 3.7.5 *Let $1 < q < \infty$ and $0 < \epsilon < \pi/2$. Then, there exists a $\lambda_1 > 0$ for which the following assertion holds: Given $\lambda \in \Sigma_{\epsilon,\lambda_1}$, $(\mathbf{f}, g) \in \mathcal{H}_q$, problem* (3.405) *admits a unique solution $(\mathbf{u}, \rho) \in \mathcal{D}_q$ with some pressure term $\mathfrak{q} \in H^1_q(\dot\Omega) + \hat{H}^1_q(\mathbb{R}^N)$, possessing the estimate:*

$$|\lambda|\|(\mathbf{u}, \rho)\|_{\mathcal{H}_q} + \|(\mathbf{u}, \rho)\|_{\mathcal{D}_q} \leq C\|(\mathbf{f}, g)\|_{\mathcal{H}_q}$$

for some constant $C > 0$.

To analyze the case where λ ranges near 0, we assume that $\mathbf{f} \in L_{q,5R}(\dot\Omega)$. We shall prove the following theorem.

Theorem 3.7.6 *Let $1 < q < \infty$ and $0 < \epsilon < \pi/2$. Then, there exist a positive constant d and a solution operator $\mathcal{S}(\lambda)$ defined on $\dot{U}_d$ such that for any $\lambda \in \dot{U}_d$ and $(\mathbf{f}, g) \in L_{q,5R}(\dot\Omega)^N \times W^{2-1/q}_q(S_R)$, $(\mathbf{u}, \mathfrak{p}, \rho) = \mathcal{S}(\lambda)(\mathbf{f}, g)$ is a unique solution of equations* (3.405).

Moreover, $\mathcal{S}(\lambda)$ possesses the following properties:

(1) $\mathcal{S}(\lambda) \in \mathrm{Hol}\,(\Sigma_\epsilon \cap U_d, \mathcal{L}_q(\dot\Omega))$, where we have set

$$\mathcal{L}_q(\dot\Omega) = \mathcal{L}(L_{q,5R}(\dot\Omega)^N \times W_q^{2-1/q}(S_R), H_q^2(\dot\Omega) \times \hat H_q^1(\dot\Omega) \times W_q^{3-1/q}(S_R)),$$

and for any $\lambda \in \Sigma_\epsilon \cap U_d$ *and* $(\mathbf{f}, g) \in L_{q,5R}(\dot\Omega)^N \times W_q^{2-1/q}(S_R)$, $(\mathbf{u}, \mathfrak{p}, \rho) = \mathcal{S}(\lambda)(\mathbf{f}, g)$ *satisfies the following estimate:*

$$\|(\lambda\mathbf{u}, \lambda^{1/2}\nabla\mathbf{u}, \nabla^2\mathbf{u})\|_{L_q(\dot\Omega)} + \|\nabla\mathfrak{p}\|_{L_q(\dot\Omega)} + \|\rho\|_{W_q^{3-1/q}(S_R)} \leq C(\|\mathbf{f}\|_{L_q(\dot\Omega)} + \|g\|_{W_q^{3-1/q}(S_R)}).$$

Moreover, when $\lambda = 0$, $\mathbf{u}$ *and* $\mathfrak{p}$ *satisfy the following estimate:*

$$\begin{aligned} |\mathbf{u}(x)| &\leq C|x|^{-(N-2)}(\|\mathbf{f}\|_{L_q(\dot\Omega)} + \|g\|_{W_q^{2-1/q}(S_R)}), \\ |\nabla\mathbf{u}(x)| &\leq C|x|^{-(N-1)}(\|\mathbf{f}\|_{L_q(\dot\Omega)} + \|g\|_{W_q^{2-1/q}(S_R)}), \\ |\mathfrak{p}(x)| &\leq C|x|^{-(N-1)}(\|\mathbf{f}\|_{L_q(\dot\Omega)} + \|g\|_{W_q^{2-1/q}(S_R)}). \end{aligned} \tag{3.406}$$

(2) There exist $\mathcal{S}_i(\lambda)$ $(i = 0, 1)$ *and* $\mathcal{S}_2(\lambda)$ *with*

$$\mathcal{S}_i(\lambda) \in \mathrm{Hol}\,(U_d, \mathcal{L}_{q,L}(\dot\Omega)), \quad \mathcal{S}_2(\lambda) \in \mathcal{B}(\dot U_d, \mathcal{L}_{q,L}(\dot\Omega))$$

such that

$$\mathcal{S}(\lambda)(\mathbf{f}, g) = \lambda^{\frac{N}{2}-1}(\log\lambda)^{\sigma(N)}\mathcal{S}_0(\lambda)(\mathbf{f}, g) + \mathcal{S}_1(\lambda)(\mathbf{f}, g) + (\lambda^{\frac{N}{2}-1}(\log\lambda)^{\sigma(N)})^2\mathcal{S}_2(\lambda)(\mathbf{f}, g)$$

in B_L,

where we have set

$$\mathcal{L}_{q,L}(\dot\Omega) = \mathcal{L}(L_{q,5R}(\dot\Omega)^N \times W_q^{2-1/q}(S_R), H_q^2(\dot B_L)^N \times \hat H_q^1(\dot B_L) \times W_q^{3-1/q}(S_R)).$$

In the following, we shall prove Theorem 3.7.6. We first introduce the so-called Bogovskiĭ's operator which plays an important role to construct a parametrix of equations (3.405).

Lemma 3.7.7 *Let* $1 < q < \infty$, $m \in \mathbb{N} \cup \{0\}$, *and let* G *be a bounded domain whose boundary* ∂G *is a hypersurface of* C^{m+1} *class. Let*

$$H^0_{q,0}(G) = L_q(G), \quad H^m_{q,0}(G) = \{f \in H^m_q(G) \mid \partial_x^\alpha f|_{\partial G} = 0 \quad (|\alpha| \leq m-1)\}$$

for $m \geq 1$. *Then, there exists a linear map* $\mathbb{B} : H^m_{q,0}(G) \to H^{m+1}_{q,0}(G)^N$ *having the following properties:*

(1) $\operatorname{div} \mathbb{B}[f] = f - \rho \int_G f\,dx$ *for some* $\rho \in C_0^\infty(G)$ *such that* $\rho \geq 0$ *and* $\int_G \rho\,dx = 1$. *In particular, if* $\int_G f\,dx = 0$*, then* $\operatorname{div} \mathbb{B}[f] = f$.

(2) We have the estimate:

$$\|\mathbb{B}[f]\|_{H_q^{k+1}(G)} \leq C_{q,k,G} \|f\|_{H_q^k(G)} \quad (k = 0, \ldots, m).$$

(3) If $f = \partial g / \partial x_i$ *with some* $g \in H_{q,0}^{m+1}(G)$ *and* $i \in \{1, \ldots, N\}$*, then*

$$\|\mathbb{B}[f]\|_{H_q^k(G)} \leq C_{q,k,G} \|g\|_{H_q^k(G)} \quad (k = 0, \ldots, m).$$

Remark 12

(1) $\mathbb{B}$ is called a Bogovskiĭ operator.
(2) Since $\mathbb{B}[f] \in H_{q,0}^{m+1}(G)^N$, the zero extension of $\mathbb{B}[f]$ to $\mathbb{R}^N$ is also written by $\mathbb{B}[f]$. Thus, $\mathbb{B}[f] \in H_q^{m+1}(\mathbb{R}^N)^N$ and $\operatorname{supp} \mathbb{B}[f] \subset G$.

Proof For a proof, see Galdi [13]. □

To apply Lemma 3.7.7, we use the following lemma.

Lemma 3.7.8 *Let* $1 < q < \infty$ *and* $2R < L_1 < L_2 < L_3 < L_4 < 5R$*. Let* χ *be a function in* $C^\infty(\mathbb{R}^N)$ *which equals one for* $x \in B_{L_2}$ *and zero for* $x \notin B_{L_3}$*. If* $\mathbf{v} \in H_q^2(G)^N$*,* $G \in \{\mathbb{R}^N, B_{5R}\}$*, satisfies* $\operatorname{div} \mathbf{v} = 0$ *in* D_{L_1,L_4}*, then there exists a* $\mathbf{u} \in H_q^2(\mathbb{R}^N)^N$ *such that* $\operatorname{supp} \mathbf{u} \subset D_{L_2,L_5}$*,* $\operatorname{div} \mathbf{u} = 0$ *in* D_{L_1,L_4} *in* $\mathbb{R}^N$ *and* $(\nabla\chi) \cdot \mathbf{u} = (\nabla\chi) \cdot \mathbf{v}$ *in* $\mathbb{R}^N$*. As a consequence,* $(\nabla\chi) \cdot \mathbf{v} \in \dot{H}_{q,a}^3(D_{L_2,L_3})$*, where we have set* $D_{A,B} = \{x \in \mathbb{R}^N \mid A \leq |x| \leq B\}$*, and*

$$\dot{H}_{q,a}^3(D_{L_2,L_3}) = \{f \in H_{q,0}^3(D_{L_2,L_3}) \mid \int_{D_{L_2,L_3}} f\,dx = 0\}.$$

Proof For a proof, see Shibata [28, Lemma 5, Lemma 6]. □

We first consider the case where $g = 0$ in equations (3.405). Given $\mathbf{g} \in L_q(B_{5R})^N$, let $\mathbf{u} \in H_q^2(\dot{B}_{5R})$, $(\mathfrak{p}, \rho)$ with

$$\mathfrak{p} \in H_q^1(\dot{B}_{5R}), \quad \rho \in W_q^{3-1/q}(S_R)$$

be a unique solution of the following equations:

$$\begin{cases} \lambda\mathbf{u} - m^{-1}\operatorname{Div}(\mu\mathbf{D}(\mathbf{u}) - \mathfrak{p}\mathbf{I}) = \mathbf{g}, \quad \operatorname{div}\mathbf{u} = 0 & \text{in } \dot{B}_{5R}, \\ \lambda\rho + \mathcal{L}_3\rho - \mathbf{n} \cdot P\mathbf{u} = 0 & \text{on } S_R, \\ [[\mu\mathbf{D}(\mathbf{u}) - \mathfrak{p}\mathbf{I}]]\mathbf{n} - (\sigma\mathcal{B}\rho)\mathbf{n} = 0, \quad [[\mathbf{u}]] = 0 & \text{on } S_R, \\ \mathbf{u} = 0 & \text{on } S_{5R}. \end{cases} \tag{3.407}$$

The unique existence of $\mathbf{u}$, $\mathfrak{p}$, and ρ are guaranteed by Theorem 3.7.3. Let $\mathbf{A}(\lambda)$, $B(\lambda)$, and $H(\lambda)$ be operators acting on $\mathbf{g} \in L_q(\dot{B}_{5R})^N$ defined by setting $\mathbf{A}(\lambda)\mathbf{g} = \mathbf{u}$, $B(\lambda)\mathbf{g} = \mathfrak{p}$, and $H(\lambda)\mathbf{g} = \rho$. We know that

$$\begin{aligned}
&\mathbf{A}(\lambda) \in \mathrm{Hol}\,(-\delta_0 + \Sigma_{\epsilon_0}, \mathcal{L}(L_q(\dot{B}_{5R}), H^2_q(\dot{B}_{5R})^N)),\\
&B(\lambda) \in \mathrm{Hol}\,(-\delta_0 + \Sigma_{\epsilon_0}, \mathcal{L}(L_q(\dot{B}_{5R}), H^1_q(\dot{B}_{5R}))),\\
&H(\lambda) \in \mathrm{Hol}\,(-\delta_0 + \Sigma_{\epsilon_0}, \mathcal{L}(L_q(\dot{B}_{5R}), W^{3-1/q}_q(S_R))),
\end{aligned} \tag{3.408}$$

where δ_0 and ϵ_0 are constants given in Theorem 3.7.3. Let φ be a function in $C^\infty(\mathbb{R}^N)$ which equals to 1 for $x \in B_{3R}$ and 0 for $x \notin B_{4R}$. Let $R_0(\lambda)$ and Π be operators given in (3.403). Let $\mathbb{B}$ be the Bogovskiĭ operator. For given f defined on $\dot{\Omega}$, $r_R f$ denotes the restriction of f to $\dot{B}_{5R}$. We let define operators $\Phi(\lambda)$, $\Psi(\lambda)$ and $\Theta(\lambda)$ acting on $\mathbf{f} \in L_q(\dot{B}_{5R})^N$ by letting

$$\begin{aligned}
&\Phi(\lambda)\mathbf{f} = (1-\varphi)R_0(\lambda)\mathbf{f} + \varphi\mathbf{A}(\lambda)r_R\mathbf{f} + \mathbb{B}[(\nabla\varphi)\cdot(R_0(\lambda)\mathbf{f} - \mathbf{A}(\lambda)r_R\mathbf{f})],\\
&\Psi(\lambda)\mathbf{f} = (1-\varphi)\Pi\mathbf{f} + \varphi B(\lambda)r_R\mathbf{f},\\
&\Theta(\lambda)\mathbf{f} = H(\lambda)r_R\mathbf{f}.
\end{aligned} \tag{3.409}$$

For any constant c, $\nabla(B(\lambda)\mathbf{g}+c) = \nabla B(\lambda)\mathbf{g}$ in $\dot{B}_{5R}$ and $[[B(\lambda)\mathbf{g}+c]] = [[B(\lambda)\mathbf{g}]]$ on S_R. So, for $B(\lambda)$ given in Theorem 3.7.3, we choose c in such a way that

$$\int_{D_{3R,4R}} (\Pi\mathbf{f} - (B(0)r_R\mathbf{f} + c))\,dx = 0.$$

Thus, without loss of generality, we may assume that

$$\int_{D_{3R,4R}} (\Pi\mathbf{f} - B(0)r_R\mathbf{f})\,dx = 0. \tag{3.410}$$

Substituting $\Phi(\lambda)\mathbf{f}$, $\Psi(\lambda)\mathbf{f}$ and $\Theta(\lambda)\mathbf{f}$ into equation (3.405) with $g = 0$ and noting that $\varphi = 1$ on B_{3R} and $S_R \subset B_{3R}$, we have

$$\begin{cases}
\lambda\Phi(\lambda)\mathbf{f} - m^{-1}\mathrm{Div}\,(\mu\mathbf{D}(\Phi(\lambda)\mathbf{f}) - (\Psi(\lambda)\mathbf{f})\mathbf{I}) = \mathbf{f} + T(\lambda)\mathbf{f} & \text{in } \dot{\Omega},\\
\mathrm{div}\,\Phi(\lambda)\mathbf{f} = 0 & \text{in } \dot{\Omega},\\
\lambda\Theta(\lambda)\mathbf{f} + \mathcal{L}_3\Theta(\lambda)\mathbf{f} - \mathbf{n}\cdot P\Phi(\lambda)\mathbf{f} = 0 & \text{on } S_R,\\
[[\mu\mathbf{D}(\Phi(\lambda)\mathbf{f}) - (\Psi(\lambda)\mathbf{f})\mathbf{I}]]\mathbf{n} - \sigma(\mathcal{B}\Theta(\lambda)\mathbf{f})\mathbf{n} = 0 & \text{on } S_R,\\
[[\Phi(\lambda)\mathbf{f}]] = 0 & \text{on } S_R,
\end{cases} \tag{3.411}$$

where $T(\lambda)$ is an operator acting on $\mathbf{f} \in L_q(B_{5R})$ defined by setting

$$T(\lambda)\mathbf{f} = m_-^{-1}\mu_-\{2(\nabla\varphi)\cdot\nabla(R_0(\lambda)\mathbf{f} - \mathbf{A}(\lambda)r_R\mathbf{f}) + (\Delta\varphi)(R_0(\lambda)\mathbf{f} - \mathbf{A}(\lambda)r_R\mathbf{f})\}$$
$$+ \lambda\mathbb{B}[(\nabla\varphi)\cdot(R_0(\lambda)\mathbf{f} - \mathbf{A}(\lambda)r_R\mathbf{f})] - m_-^{-1}\mu_-\Delta\mathbb{B}[(\nabla\varphi)\cdot(R_0(\lambda)\mathbf{f} - \mathbf{A}(\lambda)r_R\mathbf{f})]$$
$$- m_-^{-1}(\nabla\varphi)(\Pi\mathbf{f} - B(\lambda)r_R\mathbf{f}).$$

In particular,

$$T(\lambda)\mathbf{f} - T(0)\mathbf{f} \tag{3.412}$$
$$= m_-^{-1}\mu_-\{2(\nabla\varphi)\cdot\nabla(R_0(\lambda)\mathbf{f} - R_0(0)\mathbf{f}) + (\Delta\varphi)(R_0(\lambda)\mathbf{f} - R_0(0)\mathbf{f})$$
$$- (2\nabla\varphi)\cdot\nabla(\mathbf{A}(\lambda)r_R\mathbf{f} - \mathbf{A}(0)r_R\mathbf{f}) - (\Delta\varphi)(\mathbf{A}(\lambda)r_R\mathbf{f} - \mathbf{A}(0)r_R\mathbf{f})\}$$
$$+ \lambda\mathbb{B}[(\nabla\varphi)\cdot(R_0(\lambda)\mathbf{f} - R_0(0)\mathbf{f})] - \lambda\mathbb{B}[(\nabla\varphi)\cdot(\mathbf{A}(\lambda)r_R\mathbf{f} + \mathbf{A}(0)r_R\mathbf{f})]$$
$$- m_-^{-1}\mu_-(\Delta\mathbb{B}[(\nabla\varphi)\cdot(R_0(\lambda)\mathbf{f} - R_0(0)\mathbf{f})] - \Delta\mathbb{B}[(\nabla\varphi)\cdot(\mathbf{A}(\lambda)r_R\mathbf{f} - \mathbf{A}(0)r_R\mathbf{f})])$$
$$+ m_-^{-1}(\nabla\varphi)(B(\lambda)r_R\mathbf{f} - B(0)r_R\mathbf{f}), \tag{3.413}$$
$$T(0)\mathbf{f} = m_-^{-1}\mu_-\{2(\nabla\varphi)\cdot\nabla(R_0(0)\mathbf{f} - \mathbf{A}(0)r_R\mathbf{f}) + (\Delta\varphi)(R_0(0)\mathbf{f} - \mathbf{A}(0)r_R\mathbf{f})\}$$
$$+ \lambda\mathbb{B}[(\nabla\varphi)\cdot(R_0(0)\mathbf{f} - \mathbf{A}(0)r_R\mathbf{f})] - m_-^{-1}\mu_-\Delta\mathbb{B}[(\nabla\varphi)\cdot(R_0(0)\mathbf{f} - \mathbf{A}(0)r_R\mathbf{f})]$$
$$- m_-^{-1}(\nabla\varphi)(\Pi\mathbf{f} - B(0)r_R\mathbf{f}). \tag{3.414}$$

Let $\mathbf{u} = \Phi(0)\mathbf{f}$, $\mathfrak{p} = \Psi(0)\mathbf{f}$ and $\rho = H(0)\mathbf{f}$, and then we have

$$\begin{cases} -m^{-1}\mathrm{Div}\,(\mu\mathbf{D}(\mathbf{u}) - \mathfrak{p}\mathbf{I}) = (\mathbf{I} + T(0))\mathbf{f}, \quad \mathrm{div}\,\mathbf{u} = 0 & \text{in } \dot\Omega, \\ \mathcal{L}_3\rho - \mathbf{n}\cdot P\mathbf{u} = 0 & \text{on } S_R, \\ [[\mu\mathbf{D}(\mathbf{u}) - \mathfrak{p}\mathbf{I}]]\mathbf{n} - \sigma(\mathcal{B}\rho)\mathbf{n} = 0, \quad [[\mathbf{u}]] = 0 & \text{on } S_R. \end{cases} \tag{3.415}$$

Lemma 3.7.9 *Let $1 < q < \infty$. Then, the inverse of $\mathbf{I} + T(0)$ exists as an element of $\mathcal{L}(L_{q,5R}(\dot\Omega)^N)$.*

Proof Since $T(0)\mathbf{f} \in H^1_q(\mathbb{R}^N)^N$ and $\mathrm{supp}\,T(0)\mathbf{f} \subset D_{3R,4R}$, by the Rellich compactness theorem, we see that $T(0)$ is a compact operator on $L_{q,5R}(\dot\Omega)^N$. Thus, to prove the existence of the inverse operator $(\mathbf{I} + T(0))^{-1}$, in view of Riesz-Schauder theory, Fredhold's alternative principle, it is sufficient to prove that $\mathrm{Ker}\,(\mathbf{I} + T(0)) = \{0\}$. In what follows, we shall prove this fact. Let $\mathbf{f}$ be an element of $L_{q,5R}(\dot\Omega)^N$ for which $(\mathbf{I} + T(0))\mathbf{f} = 0$. Since $\mathbf{f} = -T(0)\mathbf{f}$, we see that

$$\mathrm{supp}\,\mathbf{f} \subset D_{3R,4R}. \tag{3.416}$$

Let $\mathbf{u} = \Phi(0)\mathbf{f}$, $\mathfrak{p} = \Psi(0)\mathbf{f}$ and $\rho = \Xi(0)\mathbf{f} = H(0)r_R\mathbf{f}$, and then by (3.415) $\mathbf{u}$, $\mathfrak{p}$, and ρ satisfy the homogeneous equations:

$$\begin{cases} -m^{-1}\mathrm{Div}\,(\mu\mathbf{D}(\mathbf{u}) - \mathfrak{p}\mathbf{I}) = 0, \quad \mathrm{div}\,\mathbf{u} = 0 & \text{in } \dot\Omega, \\ \mathcal{L}_3\rho - \mathbf{n}\cdot P\mathbf{u} = 0 & \text{on } S_R, \\ [[\mu\mathbf{D}(\mathbf{u}) - \mathfrak{p}\mathbf{I}]]\mathbf{n} - \sigma(\mathcal{B}\rho)\mathbf{n} = 0, \quad [[\mathbf{u}]] = 0 & \text{on } S_R. \end{cases} \tag{3.417}$$

As was seen in (3.298), from the second equation in (3.417) and $\mathrm{div}\,\mathbf{u} = 0$ in B_R, we have

$$(\rho, 1)_{S_R} = 0, \quad (\rho, x_i)_{S_R} = 0 \quad (i = 1, \ldots, N). \tag{3.418}$$

In particular, $\mathcal{L}_3\rho = 0$, and therefore,

$$\mathbf{n}\cdot\mathbf{u} = \frac{1}{|B_R|}\int_{B_R} \mathbf{u}\,dx = \mathbf{n} \quad \text{on } S_R. \tag{3.419}$$

We first assume that

$$\mathbf{u} = 0, \quad \mathfrak{p} = d \quad \text{in } \dot\Omega \tag{3.420}$$

where d is some constant. In this case, by the boundary condition, $\mathcal{B}\rho = 0$ on S_R. Thus, ρ belongs to the eigenspace of the first eigenvalue of the Laplace-Beltrami operator Δ_{S_R} on S_R. But, ρ satisfies (3.418), and so $\rho = 0$. By (3.420) and (3.409)

$$\begin{aligned} 0 &= (1-\varphi)R_0(0)\mathbf{f} + \varphi\mathbf{A}(0)r_R\mathbf{f} + \mathbb{B}[(\nabla\varphi)(\cdot(R_0(0)\mathbf{f} - \mathbf{A}(0)r_R\mathbf{f})], \\ d &= (1-\varphi)\Pi\mathbf{f} + \varphi B(0)r_R\mathbf{f} \end{aligned} \tag{3.421}$$

in $\dot\Omega$. Since $\mathrm{supp}\,\mathbb{B}[(\nabla\varphi)\cdot(R_0(0)\mathbf{f} - \mathbf{A}(0)r_R\mathbf{f})] \subset D_{3R,4R}$, by (3.421) we have

$$\begin{aligned} R_0(0)\mathbf{f} &= 0, & \Pi\mathbf{f} &= d & &\text{for } x \notin B_{4R}, \\ \mathbf{A}(0)r_R\mathbf{f} &= 0, & B(0)r_R\mathbf{f} &= d & &\text{for } x \in \dot B_{3R}. \end{aligned} \tag{3.422}$$

If we define $\mathbf{w}$ and $\mathfrak{q}$ by setting

$$\mathbf{w} = \begin{cases} \mathbf{A}(0)r_R\mathbf{f} & \text{for } x \in D_{3R,5R}, \\ 0 & \text{for } x \in B_{3R}, \end{cases} \quad \mathfrak{q} = \begin{cases} B(0)r_R\mathbf{f} & \text{for } x \in D_{3R,5R}, \\ d & \text{for } x \in B_{3R}, \end{cases}$$

then noting (3.416), we see that $\mathbf{w} \in H^2_q(B_{5R})$, $\mathfrak{q} \in H^1_q(B_{5R})$ and $\mathbf{w}$ and $\mathfrak{q}$ satisfy the equations:

$$-m_-^{-1}\mu_-\Delta\mathbf{w} + m_-^{-1}\nabla q = \mathbf{f}, \quad \mathrm{div}\,\mathbf{w} = 0 \quad \text{in } B_{5R}, \quad \mathbf{w}|_{S_{5R}} = 0. \tag{3.423}$$

In view of (3.422), $R(0)\mathbf{f}$ and $\Pi\mathbf{f}$ also satisfy (3.423), and so the uniqueness yields that $R(0)\mathbf{f} = \mathbf{w}$ in B_{5R} and $\Pi\mathbf{f} - \mathfrak{q} = c$ with some constant c in B_{5R}. But, by (3.410)

$$\int_{D_{3R,4R}} c\,dx = \int_{3R,4R} (\Pi\mathbf{f} - \mathfrak{q})\,dx = \int_{3R,4R} (\Pi\mathbf{f} - B(0)r_R\mathbf{f})\,dx = 0,$$

which leads to $c = 0$. Thus, we have

$$R_0(0)\mathbf{f} = \mathbf{w} = \mathbf{A}(0)r_R\mathbf{f}, \quad \Pi\mathbf{f} = \mathfrak{q} = B(0)r_R\mathbf{f} \quad \text{in } \dot{B}_{5R}. \tag{3.424}$$

By (3.424), $(\nabla\varphi)\cdot(R_0(0)\mathbf{f}-\mathbf{A}(0)r_R\mathbf{f}) = 0$, and so $\mathbb{B}[(\nabla\varphi)\cdot(R_0(0)\mathbf{f}-\mathbf{A}(0)r_R\mathbf{f})] = 0$. Thus, by (3.421)

$$0 = R_0(0)\mathbf{f} + \varphi(\mathbf{A}(0)r_R\mathbf{f} - R_0(0)\mathbf{f}) = R_0(0)\mathbf{f}, \quad d = \Pi\mathbf{f} + \varphi(B(0)r_R\mathbf{f} - \Pi\mathbf{f}) = \Pi\mathbf{f}$$

in $\dot{\Omega}$, which implies that

$$\mathbf{f} = m_-^{-1}\mu_-\Delta R_0(0)\mathbf{f} + m_-^{-1}\nabla\Pi\mathbf{f} = 0 \quad \text{in } \dot{\Omega}.$$

But, (3.416) holds, and so $\mathbf{f} = 0$.

To complete the proof of Lemma 3.7.9, it suffices to prove (3.420). Since (3.406) hold, (3.420) follows from the following lemma. □

Lemma 3.7.10 *Let* $1 < q < \infty$. *Let* $\mathbf{u} \in H^2_q(\dot{\Omega})^N$, $\mathfrak{p} \in \hat{H}^1_q(\dot{\Omega})$ *and* $\rho \in W^{3-1/q}_q(S_R)$ *satisfy the homogeneous equations:*

$$\begin{cases} -m^{-1}\mathrm{Div}\,(\mu\mathbf{D}(\mathbf{u}) - \mathfrak{p}\pi) = 0, \quad \mathrm{div}\,\mathbf{u} = 0 & \text{in } \dot{\Omega}, \\ \mathcal{L}_3\rho - \mathbf{n}\cdot P\mathbf{u} = 0 & \text{on } S_R, \\ [[\mu\mathbf{D}(\mathbf{u}) - \mathfrak{p}\mathbf{I}]]\mathbf{n} - \sigma(\mathcal{B}\rho)\mathbf{n} = 0, \quad [[\mathbf{u}]] = 0 & \text{on } S_R. \end{cases} \tag{3.425}$$

Moreover, $\mathbf{u}$ *and* $\mathfrak{p}$ *satisfy the decay condition:*

$$|\mathbf{u}(x)| \le C|x|^{-(N-2)}, \quad |\nabla\mathbf{u}(x)| \le C|x|^{-(N-1)}, \quad |\mathfrak{p}(x)| \le C|x|^{-(N-1)} \tag{3.426}$$

for $x \notin B_{6R}$. *Then,* $\mathbf{u} = 0$, $\mathfrak{p} = d$, *and* $\rho = 0$, *where* d *is some constant and* $\hat{H}^1_q(\dot{\Omega})$ *is a homogeneous space defined by setting*

$$\hat{H}^1_q(\dot{\Omega}) = \{\theta \in L_q(\dot{\Omega}) \mid \nabla\theta \in L_q(\dot{\Omega})^N\}.$$

Proof First, we consider the case where $2 \le q < \infty$. Let ψ be a function in $C^\infty(\mathbb{R}^N)$ which equals 1 for $x \in B_1$ and 0 for $x \notin B_2$. Set $\psi_L(x) = \psi(x/L)$. Notice that $\mathbf{u} \in H^2_{q,\mathrm{loc}}(\dot{\Omega}) \subset H^2_{2,\mathrm{loc}}(\dot{\Omega})$ and $\mathfrak{p} \in \hat{H}^1_q(\dot{\Omega}) \subset H^1_{2,\mathrm{loc}}(\dot{\Omega})$. By the divergence theorem of Gauss,

$$0 = (-m^{-1}\mathrm{Div}\,(\mu\mathbf{D}(\mathbf{u}) - \mathfrak{p}\mathbf{I}), \psi_L m u)_{\dot{\Omega}}$$

$$= -([[\mu\mathbf{D}(\mathbf{u}) - \mathfrak{p}\mathbf{I}]]\mathbf{n}, \mathbf{u})_{S_R} + (\frac{\mu}{2}\mathbf{D}(\mathbf{u}), \mathbf{D}(\psi_L\mathbf{u}))_{\dot\Omega} - (\mathfrak{p}, \mathrm{div}\,(\psi_L\mathbf{u}))_{\dot\Omega}.$$

Since

$$\mathbf{n}\cdot P\mathbf{u} = \mathbf{n}\cdot\mathbf{u} - \frac{1}{|B_R|}\int_{B_R}\mathbf{u}\,dx = 0,$$

we have

$$\begin{aligned}([[\mu\mathbf{D}(\mathbf{u}) - \mathfrak{p}\mathbf{I}]]\mathbf{n}, \mathbf{u})_{S_R} &= (\sigma(\mathcal{B}\rho), \mathbf{n}\cdot\mathbf{u})_{S_R}\\ &= \sum_{j=1}^N \sigma(\mathcal{B}\rho, \omega_j)_{S_R}\frac{1}{|B_R|}\int_{B_R} u_j\,dx = 0,\end{aligned}$$

because $\mathcal{B}\omega_j = R^{-1}\mathcal{B}x_j = 0$ for $j = 1, \ldots, N$. And also, we have

$$\begin{aligned}&(\frac{\mu}{2}\mathbf{D}(\mathbf{u}), \mathbf{D}(\psi_L\mathbf{u}))_{\dot\Omega}\\ &= (\frac{\mu}{2}\mathbf{D}(\mathbf{u}), \mathbf{D}(\mathbf{u})\psi_L)_{\dot\Omega} + \sum_{i,j=1}^N(\frac{\mu}{2}D_{ij}(\mathbf{u}), (\frac{\partial\psi_L}{\partial x_i}u_j + \frac{\partial\psi_L}{\partial x_j}u_i))_{\dot\Omega}.\end{aligned}$$

By (3.426) and the assumption that $N \geq 3$, we have

$$|(\frac{\mu}{2}D_{ij}(\mathbf{u}), \frac{\partial\psi_L}{\partial x_i}u_j)_{\dot\Omega}| \leq CL^{-1}\int_{D_{L,2L}}|\nabla\mathbf{u}(x)||\mathbf{u}(x)|\,dx \leq CL^{-2N+2} \to 0$$

as $L \to \infty$. In the same manner, by (3.426), $\mathrm{div}\,\mathbf{u} = 0$, and the assumption that $N \geq 3$, we have

$$\lim_{L\to\infty}(\mathfrak{p}, \mathrm{div}\,(\psi_L\mathbf{u}))_{\dot\Omega} = 0.$$

Thus, we have

$$0 = (\frac{\mu}{2}\mathbf{D}(\mathbf{u}), \mathbf{D}(\mathbf{u}))_{\dot\Omega},$$

which leads to $\mathbf{D}(\mathbf{u}) = 0$ in $\dot\Omega$. Since $[[\mathbf{u}]] = 0$ on S_R, we have $\mathbf{D}(\mathbf{u}) = 0$ in $\mathbb{R}^N$. Thus, $\mathbf{u}$ is written as $\mathbf{u}(x) = Ax + \mathbf{b}$ for some anti-symmetric matrix A and constant vector $\mathbf{b} \in \mathbb{R}^N$. Thus, by (3.426), we have $\mathbf{u} = 0$, which, combined with the equation, yields that $\nabla\mathfrak{p} = 0$. In particular, $[[\mathfrak{p}]]$ is constant. Since ρ satisfies (3.418), by the jump condition and $\mathbf{u} = 0$, we have

$$0 = [[\mathfrak{p}]] \int_{S_R} d\tau - \sigma(\mathcal{B}\rho, 1)_{S_R} = [[\mathfrak{p}]]|S_R|,$$

where τ is the area element of S_R, which leads to $[[\mathfrak{p}]] = 0$. Thus, $\mathfrak{p} = d$ for some constant d in $\mathbb{R}^N$. By $\mathcal{B}\rho = 0$ on S_R, ρ is an eigenfunction corresponding to the first eigenvalue $(N-1)/R^2$ of the Laplace-Beltrami operator S_R. But, $(\rho, 1)_{S_R} = 0$, and $(\rho, x_i)_{S_R} = 0$ for $i = 1, \ldots, N$ as follows from (3.418), and so $\rho = 0$. Thus, we have proved the lemma in the case where $2 \leq q < \infty$. In particular, we have proved Lemma 3.7.9 in the case where $2 \leq q < \infty$. In view of equations (3.411), we have proved the existence of solutions, $\mathbf{u} = \Phi(0)(\mathbf{I} + T(0))^{-1}\mathbf{f}$, $\mathfrak{p} = \Psi(0)(\mathbf{I} + T(0))^{-1}\mathbf{f}$ and $\rho = \Xi(0)(\mathbf{I} + T(0))^{-1}\mathbf{f}$ to equations (3.405) with $\lambda = 0$ and $g = 0$ in the case where $2 \leq q < \infty$.

We next consider the case where $1 < q < 2$. Given any $\mathbf{f} \in L_{q,5R}(\dot\Omega)$, let $\mathbf{v} = \Phi(0)(\mathbf{I} + T(0))^{-1}\mathbf{f}$, $\mathfrak{q} = \Psi(0)(\mathbf{I} + T(0))^{-1}\mathbf{f}$ and $h = \Xi(0)(\mathbf{I} + T(0))^{-1}\mathbf{f}$ be solutions of the equations:

$$\begin{cases} -m^{-1}\mathrm{Div}\,(\mu\mathbf{D}(\mathbf{v}) - \mathfrak{q}\mathbf{I}) = \mathbf{f}, \quad \mathrm{div}\,\mathbf{v} = 0 & \text{in } \dot\Omega, \\ \mathcal{L}_3 h - \mathbf{n}\cdot P\mathbf{v} = 0 & \text{on } S_R, \\ [[\mu\mathbf{D}(\mathbf{v}) - \mathfrak{q}\mathbf{I}]]\mathbf{n} - \sigma(\mathcal{B}h) = 0, \quad [[\mathbf{v}]] = 0 & \text{on } S_R. \end{cases} \tag{3.427}$$

In particular, by (3.406), we know that $\mathbf{v}$ and $\mathfrak{q}$ satisfy the decay condition (3.426). Since $\mathbf{u}$ and $\mathfrak{p}$ also satisfy the decay condition (3.426), using ψ_L and Green's formula, and letting $L \to \infty$, we have

$$(\mathbf{u}, m^{-1}\mathbf{f})_{\dot\Omega} = -\sigma(\mathbf{u}\cdot\mathbf{n}, \mathcal{B}h)_{S_R} + (\frac{\mu}{2}\mathbf{D}(\mathbf{u}), \mathbf{D}(\mathbf{v}))_{\dot\Omega}.$$

Since

$$\mathbf{u}\cdot\mathbf{n} = \frac{1}{|B_R|}\int_{B_R} \mathbf{u}\,dx \cdot \mathbf{n},$$

as follows from (3.419), we have

$$(\mathbf{u}\cdot\mathbf{n}, \mathcal{B}h)_{S_R} = \sum_{j=1}^N \frac{1}{|B_R|}\int_{B_R} u_j\,dx(\omega_j, \mathcal{B}h)_{S_R} = 0,$$

because $(\omega_j, \mathcal{B}h)_{S_R} = R^{-1}(\mathcal{B}x_j, h)_{S_R} = 0$ for $j = 1, \ldots, N$. Thus, we have

$$(\mathbf{u}, m^{-1}\mathbf{f})_{\dot\Omega} = (\frac{\mu}{2}\mathbf{D}(\mathbf{u}), \mathbf{D}(\mathbf{v}))_{\dot\Omega}.$$

Analogously, we have

$$0 = (m^{-1}\mathrm{Div}\,(\mu\mathbf{D}(\mathbf{u}) - \mathfrak{p}\mathbf{I}), m\mathbf{v})_{\dot\Omega} = (\frac{\mu}{2}\mathbf{D}(\mathbf{u}), \mathbf{D}(\mathbf{v}))_{\dot\Omega}.$$

And therefore, $(m\mathbf{u}, \mathbf{f})_{\dot\Omega} = 0$ for any $\mathbf{f} \in L_{q,5R}(\dot\Omega)$, which yields that $m\mathbf{u} = 0$ on $\dot B_{5R}$. By the first equation of (3.425), $\nabla\mathfrak{p} = 0$ in $\dot B_{5R}$. Thus, $[[\mathfrak{p}]]$ is a constant. By the jump condition in (3.425), we have

$$[[\mathfrak{p}]] - \sigma(\mathcal{B}\rho) = 0 \quad \text{on } S_R.$$

Since $\rho \in W_q^{3-1/q}(S_R)$, ρ satisfies (3.418). In particular,

$$([[\mathfrak{p}]], 1)_{S_R} = (\mathcal{B}\rho, 1)_{S_R} = \frac{N-1}{R^2}(\rho, 1)_{S_R} = 0,$$

which leads to $[[\mathfrak{p}]] = 0$. Thus, $\mathfrak{p} = c$ in B_{5R} for some constant c. Moreover, combining the fact that $\mathcal{B}\rho = 0$ on S_R and (3.418), we have $\rho = 0$.

Since $\mathbf{u} = 0$ in B_{5R}, we see that $\mathbf{u}|_{S_{5R}} = 0$. Thus, by (3.417), $\mathbf{u}$ and $\mathfrak{p}$ with

$$\mathbf{u} \in H^2_{q,\mathrm{loc}}(B^{5R})^N, \quad \mathfrak{p} \in H^1_{q,\mathrm{loc}}(B^{5R})$$

satisfy the equations:

$$-m_-^{-1}\mu_-\Delta\mathbf{u} + m_-^{-1}\nabla\mathfrak{p} = 0, \quad \mathrm{div}\,\mathbf{u} = 0 \quad \text{in } B^{5R}, \quad \mathbf{u}|_{S_{5R}} = 0, \tag{3.428}$$

and the radiation condition (3.426). By the hypo-ellipticity, we see that

$$\mathbf{u} \in H^2_{2,\mathrm{loc}}(\overline{B^{5R}}), \quad \mathfrak{p} \in H^1_{2,\mathrm{loc}}(\overline{B^{5R}}).$$

Thus, multiplying the first equation in (3.428) with $\psi_L\mathbf{u}$, using (3.426) and the divergence theorem of Gauss, and letting $L \to \infty$, we have $\|\nabla\mathbf{u}\|_{L_2(B^{5R})} = 0$, which, combined with $\mathbf{u}|_{S_{5R}} = 0$, leads to $\mathbf{u} = 0$ in B^{5R}. Thus, $\nabla\mathfrak{p} = 0$ in B^{5R}, and so $\mathfrak{p}$ is a constant. But, $|\mathfrak{p}(x)| \leq C|x|^{-(N-1)}$ for $|x| \to \infty$, and so $\mathfrak{p} = 0$ in B^{5R}. Thus, we have proved (3.420). This completes the proofs of Lemma 3.7.10 and Lemma 3.7.9. □

To prove the expansion formula of $(\mathbf{I}+T(\lambda))^{-1}$ near $\lambda = 0$, we use the following lemma.

Lemma 3.7.11 *Let X be a Banach space with norm $\|\cdot\|$, and $\mathbf{I}$ be the identity operator on X. Let $f_1(\lambda)$ and $f_2(\lambda)$ be two X-valued bounded holomorphic functions defined on U_{d_0} for some $d_0 > 0$. Then, there exist a constant $d_1 \in (0, d_0)$ and infinitely many X-valued bounded holomorphic functions $g_j(\lambda)$ defined on U_{d_1} such that*

$$(\mathbf{I} - (zf_1(\lambda) + \lambda f_2(\lambda)))^{-1} = \sum_{j=0}^{\infty} g_j(\lambda) z^j$$

in U_{d_1} *for any* $z \in U_{d_1}$, *and*

$$\sup_{\lambda \in U_{d_1}} \|g_j(\lambda)\| \le d_1^{-j} M$$

for some constant M *depending on* $\sup_{\lambda \in U_{d_0}} \|f_i(\lambda)\|_X$ $(i = 1, 2)$.

Proof Let K be a positive number for which

$$\sup_{\lambda \in U_{d_0}} \|f_i(\lambda)\|_X \le K \quad (i = 1, 2).$$

Let $f_{m,k}(\lambda)$ be X-valued holomorphic functions defined by the formula:

$$(zf_1(\lambda) + \lambda f_2(\lambda))^m = \sum_{k=0}^{m} z^k \lambda^{m-k} f_{m,k}(\lambda).$$

We see that

$$\sup_{\lambda \in U_{d_0}} \|f_{m,k}(\lambda)\| \le {}_mC_k K^m$$

where ${}_mC_k = m!/(k!(m-k)!)$. We may assume that $1/4K \in (0, d_0)$. Since

$$\begin{aligned}
&\sum_{k=0}^{\infty} \Big(\sum_{\ell=0}^{\infty} \sup_{\lambda \in U_{1/4K}} \|f_{k+\ell,k}(\lambda)\lambda^{\ell}\| \Big) (1/4K)^k \\
&= \sum_{k=0}^{\infty} \Big(\sum_{m=k}^{\infty} \sup_{\lambda \in U_{1/4K}} \|f_{m,k}(\lambda)\lambda^{m-k}\| \Big) (1/4K)^k \\
&= \sum_{m=0}^{\infty} \Big(\sum_{k=0}^{m} \sup_{\lambda \in U_{1/4K}} \|f_{m,k}(\lambda)\lambda^{m-k}\| (1/4K)^k \Big) \\
&\le \sum_{m=0}^{\infty} (1/4)^m \sum_{k=0}^{m} {}_mC_k = \sum_{m=0}^{\infty} (1/2)^m = 2,
\end{aligned}$$

we have

$$\Big(\sum_{\ell=0}^{\infty} \sup_{\lambda \in U_{1/4K}} \|f_{k+\ell,k}(\lambda)\lambda^{\ell}\| \Big) (1/4K)^k \le M$$

for any $k \geq 0$ with some constant M, which depends on K. Let

$$g_k(\lambda) = \sum_{\ell=0}^{\infty} f_{k+\ell,k}(\lambda)\lambda^{\ell},$$

and then, $g_k(\lambda) \in \mathrm{Hol}\,(U_{1/4K}, X)$ and

$$\sup_{\lambda \in U_{1/4K}} \|g_k(\lambda)\| \leq \sum_{\ell=0}^{\infty} \sup_{\lambda \in U_{1/4K}} \|f_{k+\ell,k}(\lambda)\lambda^{\ell}\| \leq (4K)^k M.$$

Moreover,

$$\sum_{k=0}^{\infty} g_k(\lambda)z^k = \sum_{k=0}^{\infty}(\sum_{\ell=0}^{\infty} f_{k+\ell,k}(\lambda)\lambda^{\ell})z^k = \sum_{m=0}^{\infty}(zf_1(\lambda) + \lambda f_2(\lambda))^m.$$

This completes the proof of Lemma 3.7.11. □

Proof of Theorem 3.7.6 By Theorem 3.7.4 we have

$$(\mathbf{I} + T(0))^{-1}(T(\lambda) - T(0)) = \lambda^{\frac{N}{2}-1}(\log \lambda)^{\sigma(N)} f_1(\lambda) + f_2(\lambda)\lambda$$

for some $f_i \in \mathrm{Hol}\,(U_{1/2}, \mathcal{L}(L_{q,5R}(\dot{\Omega})^N))$ $(i = 1, 2)$. Setting $z = \lambda^{\frac{N}{2}-1}(\log \lambda)^{\sigma(N)}$, by Lemma 3.7.11 there exist a constant $d \in (0, 1/2)$ and infinitely many $g_j(\lambda) \in \mathrm{Hol}\,(U_d, \mathcal{L}(L_{q,5R}(\dot{\Omega})))$ such that

$$(\mathbf{I} + (\mathbf{I} + T(0))^{-1}(T(\lambda) - T(0)))^{-1} = \sum_{j=0}^{\infty} g_j(\lambda)(\lambda^{\frac{N}{2}-1}(\log \lambda)^{\sigma(N)})^j \tag{3.429}$$

for $\lambda \in U_d$, and

$$\sup_{\lambda \in U_d} \|g_j(\lambda)\|_{\mathcal{L}(L_{q,5R}(\dot{\Omega}))} \leq d^j M,$$

where M is a constant depending only on $\sup_{\lambda \in U_{1/2}} \|f_i(\lambda)\|_{\mathcal{L}(L_{q,5R}(\dot{\Omega}))}$. Thus, in view of (3.411), we define an operator $\mathcal{T}(\lambda)$ acting on $\mathbf{f} \in L_{q,5R}(\dot{\Omega})^N$ by letting

$$\mathcal{T}(\lambda)\mathbf{f} = (\Phi(\lambda), \Psi(\lambda), \Theta(\lambda))(\mathbf{I} + T(\lambda))^{-1}\mathbf{f}.$$

Since

$$(\mathbf{I} + T(\lambda))^{-1} = (\mathbf{I} + (\mathbf{I} + T(0))^{-1}(T(\lambda) - T(0)))^{-1}(\mathbf{I} + T(0))^{-1}\mathbf{f},$$

by Theorem 3.7.4 and (3.429), we have

$$\mathcal{T}(\lambda) = \mathcal{T}_1(\lambda) + \lambda^{\frac{N}{2}-1}(\log\lambda)^{\sigma(N)}\mathcal{T}_2(\lambda) + (\lambda^{\frac{N}{2}-1}(\log\lambda)^{\sigma(N)})^2\mathcal{T}_3(\lambda) \tag{3.430}$$

for $\lambda \in U_d$, where

$$\mathcal{T}_i(\lambda) \in \text{Hol}, (U_d, \mathcal{L}(L_{q,5R}(\dot\Omega)^N)) \quad (i = 1, 2),$$
$$\mathcal{T}_3(\lambda) \in \mathcal{B}(\dot U_d, \mathcal{L}(L_{q,5R}(\dot\Omega), \mathcal{L}_{q,L}(\dot\Omega))),$$
$$\sup_{\lambda\in U_d} \|\mathcal{T}_i(\lambda)\|_{\mathcal{L}(L_{q,5R}(\dot\Omega),\mathcal{L}_{q,L}(\dot\Omega))} \le M_L \quad (i = 1, 2, 3)$$

for any $L > 0$ with some constant M_L depending on L, $m_\pm$, $\mu_\pm$, and q.

In view of Theorem 3.7.3, there exists an operator $\mathcal{W}(\lambda)$ with

$$\mathcal{W}(\lambda) \in \text{Hol}\,(-\delta_0 + \Sigma_{\epsilon_0}, \mathcal{L}(W^{2-1/q}_q(S_R), \mathcal{M}_q(\dot B_{5R}))),$$

where $\mathcal{M}_q(\dot B_{5R}) = H^2_q(\dot B_{5R})^N \times H^1_q(\dot B_{5R}) \times W^{3-1/q}_q(S_R)$, such that for any $\lambda \in -\delta_0 + \Sigma_{\epsilon_0}$ and $g \in W^{2-1/q}_q(S_R)$, $(\mathbf{v}, \mathfrak{p}, h) = \mathcal{W}(\lambda)g$ is a unique solution of the equations:

$$\begin{cases} \lambda\mathbf{v} - m^{-1}\text{Div}\,(\mu\mathbf{D}(\mathbf{v}) - \mathfrak{p}\mathbf{I}) = 0, \quad \text{div}\,\mathbf{v} = 0 & \text{in } \dot B_{5R}, \\ \lambda h + \mathcal{L}_3 h - \mathbf{u}\cdot\mathbf{n} = g & \text{on } S_R, \\ [[\mu\mathbf{D}(\mathbf{v}) - \mathfrak{p}\mathbf{I}]]\mathbf{n} - \sigma(\mathcal{B}h)\mathbf{n} = 0, \quad [[\mathbf{v}]] = 0 & \text{on } S_R. \end{cases}$$

For solutions $\mathbf{u}$, $\mathfrak{q}$, and ρ of equation (3.405), we set $\mathbf{u} = \varphi\mathbf{v} - \mathbb{B}[(\nabla\varphi)\cdot\mathbf{v}] + \mathbf{w}$, $\mathfrak{q} = \varphi\mathfrak{p} + \mathfrak{r}$, and $\rho = h + \vartheta$, and then, noting that $\varphi = 1$ for $x \in B_{3R}$, we see that $\mathbf{w}$, $\mathfrak{r}$, and ϑ satisfy the equations:

$$\begin{cases} \lambda\mathbf{w} - m^{-1}\text{Div}\,(\mu\mathbf{D}(\mathbf{w}) - \mathfrak{r}\mathbf{I}) = \mathbf{g}, \quad \text{div}\,\mathbf{v} = 0 & \text{in } \dot\Omega, \\ \lambda\vartheta + \mathcal{L}_3\vartheta - \mathbf{w}\cdot\mathbf{n} = 0 & \text{on } S_R, \\ [[\mu\mathbf{D}(\mathbf{w}) - \mathfrak{r}\mathbf{I}]]\mathbf{n} - \sigma(\mathcal{B}\vartheta)\mathbf{n} = 0, \quad [[\mathbf{w}]] = 0 & \text{on } S_R, \end{cases}$$

with

$$\mathbf{g} = \mathbf{f} + \lambda\mathbb{B}[(\nabla\varphi)\cdot\mathbf{v}] + m^{-1}\{\text{Div}\,(\mu\mathbf{D}(\varphi\mathbf{v})) - \varphi\text{Div}\,(\mu\mathbf{D}(\mathbf{v}))\}$$
$$- m^{-1}(\nabla\varphi)\mathfrak{p} - m^{-1}\text{Div}\,(\mu\mathbf{D}(\mathbb{B}[(\nabla\varphi)\cdot\mathbf{v}])).$$

Since $(\mathbf{w}, \mathfrak{r}, \vartheta) = \mathcal{T}(\lambda)\mathbf{g}$, by (3.430) we have Theorem 3.7.6. This completes the proof of Theorem 3.7.6. □

3.7.5 Analysis in the Middle Range

Let λ_1 be a positive number given in Theorem 3.7.5. Given any $\delta > 0$, we set

$$\begin{aligned}\Xi_\delta =&\{\lambda \in \mathbb{C} \mid \delta \leq \mathrm{Re}\,\lambda \leq \lambda_1, \quad |\mathrm{Im}\,\lambda| \leq \delta\}\\ &\cup\{\lambda \in \mathbb{C} \mid 0 \leq \mathrm{Re}\,\lambda \leq \lambda_1, \quad \delta \leq |\mathrm{Im}\,\lambda| \leq \lambda_1\}.\end{aligned}$$

In this section, for $\lambda \in \Xi_\delta$ we consider equation (3.405). We shall prove the following theorem.

Theorem 3.7.12 *Let $1 < q < \infty$ and let λ_1 be a positive constant given in Theorem 3.7.5. Let $\delta > 0$ be any small positive number. Then, for any $\lambda \in \Xi_\delta$, $(\mathbf{f}, g) \in \mathcal{H}_q$, problem* (3.405) *admits a unique solution $(\mathbf{u}, \rho) \in \mathcal{D}_q$ with some pressure term $\mathfrak{p} \in \hat{H}^1_q(\dot\Omega)$ possessing the estimate:*

$$\|(\mathbf{u}, \rho)\|_{\mathcal{D}_q} + \|\nabla \mathfrak{p}\|_{L_q(\mathbb{R}^N)} \leq C\|(\mathbf{f}, g)\|_{\mathcal{H}_q}. \tag{3.431}$$

From Theorem 3.7.12, immediately we have the following corollary.

Corollary 3.7.13 *Let $1 < q < \infty$ and let λ_1 and δ be the same positive numbers as in Theorem 3.7.12. For any $\tilde\delta > 0$, we set*

$$\begin{aligned}\tilde\Xi_{\tilde\delta} =&\{\lambda \in \mathbb{C} \mid \delta \leq |\mathrm{Im}\,\lambda| \leq \lambda_1, \quad -\tilde\delta \leq \mathrm{Re}\,\lambda \leq \lambda_1\}\\ &\cup\{\lambda \in \mathbb{C} \mid \delta \leq \mathrm{Re}\,\lambda \leq \lambda_1, \quad |\mathrm{Im}\,\lambda| \leq \delta\}.\end{aligned}$$

Then, there exists a $\tilde\delta > 0$ for which the following assertion holds: For any $\lambda \in \tilde\Xi_{\tilde\delta}$, $(\mathbf{f}, g) \in \mathcal{H}_q$, problem (3.405) *admits a unique solution $(\mathbf{u}, \rho) \in \mathcal{D}_q$ with some pressure term $\mathfrak{p} \in \hat{H}^1_q(\dot\Omega)$ possessing the estimate:*

$$\|(\mathbf{u}, \rho)\|_{\mathcal{D}_q} + \|\nabla \mathfrak{p}\|_{L_q(\mathbb{R}^N)} \leq C\|(\mathbf{f}, g)\|_{\mathcal{H}_q}.$$

We now prove Theorem 3.7.12. Since Ξ_δ is a compact set, if we prove the unique existence of solutions, then the uniform estimate (3.431) of solutions follows automatically. Let $\mathbf{u}_1 \in H^2_q(\dot B_{5R})$, $\mathfrak{p}_1 \in H^1_q(\dot B_{5R})$ and $\rho_1 \in W^{3-1/q}_q(S_R)$ be solutions of the equations:

$$\begin{cases} \lambda\mathbf{u}_1 - m^{-1}\mathrm{Div}\,(\mu\mathbf{D}(\mathbf{u}_1) - \mathfrak{p}_1\mathbf{I}) = r_R\mathbf{f}, \quad \mathrm{div}\,\mathbf{u} = 0 & \text{in } \dot B_{5R},\\ \lambda\rho_1 + \mathcal{L}_3\rho_1 - \mathbf{n}\cdot P\mathbf{u}_1 = g & \text{on } S_R,\\ [[\mu\mathbf{D}(\mathbf{u}_1) - \mathfrak{p}_1\mathbf{I}]]\mathbf{n} - \sigma(\mathcal{B}\rho_1)\mathbf{n} = 0, \quad [[\mathbf{u}_1]] = 0 & \text{on } S_R,\\ \mathbf{u}_1 = 0 & \text{on } S_{5R}, \end{cases} \tag{3.432}$$

where $r_R\mathbf{f}$ is the restriction of $\mathbf{f}$ to B_{5R}. Let φ be a function in C^∞ which equals 1 for $x \in B_{3R}$ and 0 for $x \notin B_{4R}$. Let

$$\mathbf{u} = \mathbf{v} + \varphi\mathbf{u}_1 - \mathbb{B}[(\nabla\varphi)\cdot\mathbf{u}_1], \quad \mathfrak{p} = \mathfrak{q} + \varphi\mathfrak{p}, \quad \rho = \rho_1 + h$$

in equations (3.405), and then $\mathbf{v}$, $\mathfrak{q}$, and h satisfy the following equations:

$$\begin{cases} \lambda\mathbf{v} - m^{-1}\mathrm{Div}\,(\mu\mathbf{D}(\mathbf{v}) - \mathfrak{q}\mathbf{I}) = \mathbf{g}, \quad \mathrm{div}\,\mathbf{v} = 0 & \text{in } \dot\Omega, \\ \lambda h + \mathcal{L}_3 h - \mathbf{n}\cdot P\mathbf{v} = 0 & \text{on } S_R, \\ [[\mu\mathbf{D}(\mathbf{v}) - \mathfrak{q}\mathbf{I}]]\mathbf{n} - \sigma(\mathcal{B}h)\mathbf{n} = 0, \quad [[\mathbf{v}]] = 0 & \text{on } S_R \end{cases} \tag{3.433}$$

with

$$\begin{aligned} \mathbf{g} &= (1-\varphi)\mathbf{f} + m^{-1}\{\mathrm{Div}\,(\mu\mathbf{D}(\varphi\mathbf{u}_1)) - \varphi\mathrm{Div}\,(\mu\mathbf{D}(\mathbf{u}_1))\} + \lambda\mathbb{B}[(\nabla\varphi)\cdot\mathbf{u}_1] \\ &\quad - m^{-1}\mathrm{Div}\,(\mu\mathbf{D}(\mathbb{B}[(\nabla\varphi)\cdot\mathbf{u}_1])) - m^{-1}(\nabla\varphi)\mathfrak{p}_1. \end{aligned}$$

Noting $|\lambda| \leq \lambda_0$, by Theorem 3.7.3 we have

$$\|(\mathbf{u}_1, \rho_1)\|_{H^2_q(\dot B_{5R})\times W^{3-1/q}_q(S_R)} + \|\mathfrak{p}_1\|_{H^1_q(\dot B_{5R})} \leq C\|(\mathbf{f}, g)\|_{L_q(\mathcal{D}_q)}. \tag{3.434}$$

Using Lemma 3.7.7 to estimate $\mathbb{B}[(\nabla\varphi)\cdot\mathbf{u}_1]$, by (3.434) we have

$$\|\mathbf{g}\|_{L_q(\dot\Omega)} \leq C\|(\mathbf{f}, g)\|_{\mathcal{H}_q}.$$

In what follows, we study equations (3.433) for $\mathbf{g} \in L_q(\dot\Omega)^N$. As was seen in Sect. 3.7.4, we analyze equations (3.433) by constructing a parametrix. For this purpose, we use the same symbols as in Sect. 3.7.4. Let

$$\begin{aligned} \Phi(\lambda)\mathbf{g} &= (1-\varphi)R_0(\lambda)\mathbf{g} + \varphi\mathbf{A}(\lambda)r_R\mathbf{g} + \mathbb{B}[(\nabla\varphi)\cdot(R_0(\lambda)\mathbf{g} - \mathbf{A}(\lambda)r_R\mathbf{g})], \\ \Psi(\lambda)\mathbf{g} &= (1-\varphi)\Pi\mathbf{g} + \varphi B(\lambda)r_R\mathbf{g}, \\ \Theta(\lambda)\mathbf{g} &= H(\lambda)r_R\mathbf{g}. \end{aligned} \tag{3.435}$$

Adding some constant if necessary, without loss of generality, we may assume that

$$\int_{D_{3R,4R}} (\Pi\mathbf{g} - B(\lambda)r_R\mathbf{g})\,dx = 0. \tag{3.436}$$

Substituting the formulas in equations (3.435) into equations (3.433), we have

$$\begin{cases} \lambda\Phi(\lambda)\mathbf{g} - m^{-1}\mathrm{Div}\,(\mu\mathbf{D}(\Phi(\lambda)\mathbf{g}) - (\Psi(\lambda)\mathbf{g})\mathbf{I}) = \mathbf{g} + T(\lambda)\mathbf{g} & \text{in } \dot\Omega, \\ \mathrm{div}\,\Phi(\lambda)\mathbf{g} = 0 & \text{in } \dot\Omega, \\ \lambda\Theta(\lambda)\mathbf{g} + \mathcal{L}_3\Theta(\lambda)\mathbf{g} - \mathbf{n}\cdot P\Phi(\lambda)\mathbf{g} = 0 & \text{on } S_R, \\ [[\mu\mathbf{D}(\Phi(\lambda)\mathbf{g}) - (\Psi(\lambda)\mathbf{g})\mathbf{I}]]\mathbf{n} - \sigma(\mathcal{B}\Theta(\lambda)\mathbf{g})\mathbf{n} = 0 & \text{on } S_R, \\ [[\Phi(\lambda)\mathbf{g}]] = 0 & \text{on } S_R, \end{cases} \tag{3.437}$$

where $T(\lambda)$ is an operator acting on $\mathbf{g} \in L_q(B_{5R})^N$ defined by setting

$$\begin{aligned} T(\lambda)\mathbf{g} &= m_-^{-1}\mu_-\{2(\nabla\varphi)\cdot\nabla(R_0(\lambda)\mathbf{g} - \mathbf{A}(\lambda)r_R\mathbf{g}) + (\Delta\varphi)(R_0(\lambda)\mathbf{g} - \mathbf{A}(\lambda)r_R\mathbf{g})\} \\ &+ \lambda\mathbb{B}[(\nabla\varphi)\cdot(R_0(\lambda)\mathbf{g} - \mathbf{A}(\lambda)r_R\mathbf{g})] - m_-^{-1}\mu_-\Delta\mathbb{B}[(\nabla\varphi)\cdot(R_0(\lambda)\mathbf{g} - \mathbf{A}(\lambda)r_R\mathbf{g})] \\ &- m_-^{-1}(\nabla\varphi)(\Pi\mathbf{g} - B(\lambda)r_R\mathbf{g}). \end{aligned}$$

Since $T(\lambda)\mathbf{g} \in H_q^1(\dot\Omega)^N$ and $\mathrm{supp}\,T(\lambda)\mathbf{g} \subset D_{3R,4R}$, by Rellich's compactness theorem $T(\lambda)$ is a compact operator on $L_q(\dot\Omega)^N$. By using this fact, we shall prove the following lemma.

Lemma 3.7.14 *Let* $1 < q < \infty$. *For any* $\lambda \in \Xi_\delta$, *the inverse of* $\mathbf{I} + T(\lambda)$ *exists in* $\mathcal{L}(L_q(\dot\Omega)^N)$.

Proof Since $T(\lambda)$ is a compact operator on $L_q(\dot\Omega)^N$, to prove the existence of $(\mathbf{I} + T(\lambda))^{-1} \in \mathcal{L}(L_q(\dot\Omega)^N)$, in view of Riesz-Schauder theory, especially Fredholm alternative principle, it is sufficient to prove that $\mathrm{Ker}\,(\mathbf{I} + T(\lambda)) = \{0\}$. Thus, we take $\mathbf{g} \in L_q(\dot\Omega)^N$ such that $\mathbf{g} + T(\lambda))\mathbf{g} = 0$. From $\mathbf{g} = -T(\lambda)\mathbf{g}$ it follows that

$$\mathrm{supp}\,\mathbf{g} \subset D_{3R,4R}. \tag{3.438}$$

Let $\mathbf{u} = \Phi(\lambda)\mathbf{g}$, $\mathfrak{p} = \Psi(\lambda)\mathbf{g}$, and $h = \Theta(\lambda)\mathbf{g}$, and then by (3.411), $\mathbf{u} \in H_q^2(\dot\Omega)^N$, $\mathfrak{p} \in \hat H_q^1(\dot\Omega)$ and $h \in W_q^{3-1/q}(S_R)$ satisfy the homogeneous equations:

$$\begin{cases} \lambda\mathbf{u} - m^{-1}\mathrm{Div}\,(\mu\mathbf{D}(\mathbf{u}) - \mathfrak{p}\mathbf{I}) = 0, \quad \mathrm{div}\,\mathbf{u} = 0 & \text{in } \dot\Omega, \\ \lambda h + \mathcal{L}_3 h - \mathbf{n}\cdot P\mathbf{u} = 0 & \text{on } S_R, \\ [[\mu\mathbf{D}(\mathbf{u}) - \mathfrak{p}\mathbf{I}]]\mathbf{n} - \sigma(\mathcal{B}h)\mathbf{n} = 0, \quad [[\mathbf{u}]] = 0 & \text{on } S_R. \end{cases} \tag{3.439}$$

By the first equation on S_R and $\mathrm{div}\,\mathbf{u} = 0$ in B_R, we have

$$(h, 1)_{S_R} = 0, \quad (h, x_i)_{S_R} = 0 \quad (i = 1, \ldots, N), \tag{3.440}$$

and so

$$\mathcal{L}_3 h = 0. \tag{3.441}$$

By (3.438), we have

$$|\mathfrak{p}(x)| \leq C|x|^{-(N-1)} \quad \text{for } x \notin B_{6R}. \tag{3.442}$$

We first assume that

$$\mathbf{u} = 0, \quad \mathfrak{p} = 0. \tag{3.443}$$

Since $\lambda \neq 0$, by the first equation on S_R and (3.441), we have $h = 0$. Putting (3.443) and the definition of $\Phi(\lambda)\mathbf{g}$ in (3.409) together gives

$$\begin{aligned} 0 &= (1-\varphi)R_0(\lambda)\mathbf{g} + \varphi\mathbf{A}(\lambda)r_R\mathbf{g} + \mathbb{B}[(\nabla\varphi)\cdot(R_0(\lambda)\mathbf{g} - \mathbf{A}(\lambda)r_R\mathbf{g})], \\ 0 &= (1-\varphi)\Pi\mathbf{g} + \varphi B(\lambda)r_R\mathbf{g}. \end{aligned} \tag{3.444}$$

Since $\mathbb{B}[(\nabla\varphi)\cdot(R_0(\lambda)\mathbf{g} - \mathbf{A}(\lambda)r_R\mathbf{g})] \subset D_{3R,4R}$, by (3.444) we have

$$\begin{aligned} R_0(\lambda)\mathbf{g} &= 0 \qquad & \Pi\mathbf{g} &= 0 \qquad & &\text{for } x \notin B_{4R}, \\ \mathbf{A}(\lambda)\mathbf{g} &= 0 \qquad & B(\lambda)r_R\mathbf{g} &= 0 \qquad & &\text{for } x \in B_{3R}. \end{aligned} \tag{3.445}$$

Set

$$\mathbf{v} = \begin{cases} \mathbf{A}(\lambda)r_R\mathbf{g} & \text{for } x \in D_{2R,5R}, \\ 0 & \text{for } x \in B_{2R}, \end{cases} \qquad \mathfrak{q} = \begin{cases} B(\lambda)r_R\mathbf{g} & \text{for } x \in D_{2R,5R}, \\ 0 & \text{for } x \in B_{2R}. \end{cases}$$

In view of (3.445), $\mathbf{v} \in H^2_q(B_{5R})^N$ and $\mathfrak{q} \in H^1_q(B_{5R})$. Moreover, by (3.438) and equations (3.401), $\mathbf{v}$ and $\mathfrak{q}$ satisfy the equations:

$$\lambda\mathbf{v} - m_-^{-1}\mathrm{Div}\,(\mu_-\mathbf{D}(\mathbf{v}) - \mathfrak{q}\mathbf{I}) = \mathbf{g}, \quad \mathrm{div}\,\mathbf{v} = 0 \quad \text{in } B_{5R}, \quad \mathbf{v}|_{5R} = 0. \tag{3.446}$$

Moreover, since $R_0(\lambda)g = 0$ on S_{5R} as follows from (3.445), we see that $R_0(\lambda)\mathbf{g}$ and $\Pi\mathbf{g}$ also satisfy equations (3.446), and so the uniqueness yields that $\mathbf{v} = R_0(\lambda)\mathbf{g}$ and $\mathfrak{q} - \Pi\mathbf{g} = c$ with some constant c. But, by (3.410),

$$\int_{D_{3R,4R}} c\,dx = \int_{D_{3R,4R}} (\Pi\mathbf{g} - B(\lambda)r_R\mathbf{g})\,dx = 0,$$

which implies that $c = 0$. Noting that $\mathbf{A}(\lambda)r_R\mathbf{g} = 0$ and $B(\lambda)r_R\mathbf{g} = 0$ for $x \in B_{3R}$ (cf. (3.445)), we have

$$R_0(\lambda)\mathbf{g} = \mathbf{A}(\lambda)r_R\mathbf{g}, \quad \Pi\mathbf{g} = B(\lambda)r_R\mathbf{g} \quad \text{for } x \in B_{5R},$$

which leads to $\varphi(R_0(\mathbf{f}) - \mathbf{A}(\lambda)r_R\mathbf{f}) = 0$, $(\nabla\varphi)\cdot(R_0(\mathbf{f}) - \mathbf{A}(\lambda)r_R\mathbf{f}) = 0$, and $\varphi(\Pi\mathbf{g} - B(\lambda)r_R\mathbf{g}) = 0$. Thus, by (3.444) we have $R_0(\lambda)\mathbf{g} = 0$ and $\Pi\mathbf{g} = 0$ in B_{5R}, which yields that $\mathbf{g} = 0$ in B_{5R}. Since (3.438) holds, we have $\mathbf{g} = 0$. Thus, $(\mathbf{I} + T(\lambda))^{-1} \in \mathcal{L}(L_q(\dot\Omega)^N)$ exists. In particular, $\mathbf{u} = \Phi(\lambda)(\mathbf{I} + T(\lambda))^{-1}\mathbf{g}$, $\mathfrak{p} = \Psi(\lambda)(\mathbf{I} + T(\lambda))^{-1}\mathbf{g}$ and $\rho = \Theta(\lambda)(\mathbf{I} + T(\lambda))^{-1}\mathbf{g}$ are solutions of equations (3.405).

We now prove (3.443). We first consider the case $2 \leq q < \infty$. Since $\mathbf{g} \in L_q(\dot\Omega)^N$ with $2 \leq q < \infty$ and $\mathbf{g}$ has a compact support (cf. (3.438)), $\mathbf{g} \in L_2(\mathbb{R}^N)^N$. Thus, $R_0(\lambda)\mathbf{g} \in H^2_q(\mathbb{R}^N)^N$ and $\nabla\Pi\mathbf{g} \in L_q(\mathbb{R}^N)^N$. Since $\mathbf{A}(\lambda)r_R\mathbf{g} \in H^2_q(\dot B_{5R}) \subset H^2_2(\dot B_{5R})$, and $B(\lambda)r_R\mathbf{g} \in H^1_q(\dot B_{5R}) \subset H^1_2(\dot B_{5R})$, we have $\mathbf{u} \in H^2_q(\dot\Omega)^N$ and $\nabla\mathfrak{p} \in L_2(\dot\Omega)^N$. Moreover, since $N \geq 3$, by (3.442) and the fact that $\mathfrak{p} \in L_{2,\mathrm{loc}}(\mathbb{R}^N)$, we have $\mathfrak{p} \in L_2(\mathbb{R}^N)$. Thus, multiplying the first equation of (3.439) with $m\mathbf{u}$, using the transmission conditions on S_R in (3.439), $\mathcal{B}x_i = 0$ on S_R, (3.441), and the divergence theorem of Gauss, we have

$$0 = \lambda(m\mathbf{u}, \mathbf{u})_{\dot\Omega} - \bar\lambda\sigma(\mathcal{B}h, h)_{S_R} + \frac{1}{2}(\mu\mathbf{D}(\mathbf{u}), \mathbf{D}(\mathbf{u}))_{\dot\Omega}.$$

Taking the real part yields that

$$0 = (\mathrm{Re}\,\lambda)\{(m\mathbf{u}, \mathbf{u})\|_{\dot\Omega} - \sigma(\mathcal{B}h, h)_{S_R}\} + \frac{1}{2}(\mu\mathbf{D}(\mathbf{u}), \mathbf{D}(\mathbf{u}))_{\dot\Omega}.$$

Since (3.440) holds, by Lemma 3.5.8 we have

$$-(\mathcal{B}h, h)_{S_R} \geq c\|h\|^2_{L_2(S_R)} \tag{3.447}$$

with some constant $c > 0$. Thus, we have

$$0 \geq (\mathrm{Re}\,\lambda)\{(m\mathbf{u}, \mathbf{u})\|_{\dot\Omega} + c\|h\|^2_{L_2(S_R)}\} + \frac{1}{2}(\mu\mathbf{D}(\mathbf{u}), \mathbf{D}(\mathbf{u}))_{\dot\Omega}.$$

which leads to $\mathbf{D}(\mathbf{u}) = 0$ in $\dot\Omega$, because $\mathrm{Re}\,\lambda \geq 0$. Since $[[\mathbf{u}]] = 0$ on S_R, $\mathbf{u} \in H^1_q(\mathbb{R}^N)^N$ and $\mathbf{D}(\mathbf{u}) = 0$. Thus, $\mathbf{u} = 0$. By the condition on S_R in (3.439) and (3.441), $h = 0$, because $\lambda \neq 0$. By the equations in $\dot\Omega$ and the transmission condition on S_R in (3.439), we have $\nabla\mathfrak{p} = 0$ in $\dot\Omega$ and $[[\mathfrak{p}]] = 0$ on S_R, which leads to $\mathfrak{p} = c$ in $\mathbb{R}^N$ with some constant c. Thus, by (3.442), $\mathfrak{p} = 0$. This completes the proof of (3.443), and therefore, we have proved Theorem 3.7.12 in the case where $2 \leq q < \infty$.

Finally, we prove (3.443) in the case where $1 < q < 2$. For any $\mathbf{f} \in L_{q',5R}(\dot\Omega)^N$, let $\mathbf{v} \in H^2_{q'}(\dot\Omega)^N$, $\mathfrak{q} \in \hat H^1_{q'}(\dot\Omega)$, and $\rho \in W^{3-1/q}_q(S_R)$ be solutions of the equations:

$$\begin{cases} \bar\lambda\mathbf{v} - m^{-1}\mathrm{Div}\,(\mu\mathbf{D}(\mathbf{v}) - \mathfrak{q}\mathbf{I}) = \mathbf{f}, \quad \mathrm{div}\,\mathbf{v} = 0 & \text{in } \dot\Omega, \\ \lambda\rho + \mathcal{L}_3\rho - \mathbf{n}\cdot P\mathbf{v} = 0 & \text{on } S_R, \\ [[\mu\mathbf{D}(\mathbf{v}) - \mathfrak{q}\mathbf{I}]]\mathbf{n} - \sigma(\mathcal{B}\rho)\mathbf{n} = 0, \quad [[\mathbf{v}]] = 0 & \text{on } S_R. \end{cases}$$

By the first equation on S_R and $\operatorname{div}\mathbf{v} = 0$ on S_R, we have

$$(\rho, 1)_{S_R} = 0, \quad (\rho, x_i)_{S_R} = 0 \quad (i = 1, \ldots, N). \tag{3.448}$$

In particular,

$$\mathcal{L}_3\rho = 0. \tag{3.449}$$

Noting that $\mathcal{B}x_i = 0$ on S_R, by Green's formula, we have

$$(\mathbf{u}, m\mathbf{f})_{\dot\Omega} = \lambda(\mathbf{u}, m\mathbf{v}) - (\mathbf{u}, [[\mu\mathbf{D}(\mathbf{v}) - \mathfrak{q}\mathbf{I}]]\mathbf{n})_{S_R} + \frac{\mu}{2}(\mathbf{D}(\mathbf{u}), \mathbf{D}(\mathbf{v}))_{\dot\Omega},$$

where we have used the following lemma.

Lemma 3.7.15 *Let* $1 < q < \infty$, $\mathfrak{p} \in \hat H^1_q(\dot\Omega)$ *and* $\mathbf{v} \in H^1_{q'}(\dot\Omega)^N$, $q' = q/(q-1)$. *If* $\operatorname{div}\mathbf{v} = 0$ *in* $\dot\Omega$ *and* $[[\mathbf{v}]] = 0$ *on* S_R, *then*

$$(\mathbf{v}, \nabla\mathfrak{p})_{\dot\Omega} = (\mathbf{v}, [[\mathfrak{p}]]\mathbf{n})_{S_R}. \tag{3.450}$$

Proof Let $\tilde{\mathfrak{p}}_-$ be an extension of $\mathfrak{p}_-$ such that $\tilde{\mathfrak{p}}_- = \mathfrak{p}_-$ in B^R and

$$\|(\tilde{\mathfrak{p}}_- - c)w^{-1}\|_{L_q(\mathbb{R}^N)} \leq C\|\nabla\tilde{\mathfrak{p}}_-\|_{L_q(\mathbb{R}^N)} \tag{3.451}$$

for some constants c and C, where w is a weight function defined by letting

$$w(x) = \begin{cases} (1+|x|^2)^{1/2} & \text{for } q \neq N, \\ (1+|x|^2)^{1/2}\log(2+|x|^2) & \text{for } q = N. \end{cases}$$

Such $\tilde{\mathfrak{p}}_-$ can be constructed by using the method given in Galdi [13, Chapter II]). Let $\psi(t)$ be a function in $C^\infty(\mathbb{R})$ which equals one for $|t| \leq 1/2$ and zero for $|t| \geq 1$. For large $L > 0$, let

$$\psi_L(x) = \psi\Big(\frac{\ln\ln|x|}{\ln\ln L}\Big).$$

We see that $\psi_L(x) = 1$ for $|x| \leq \exp(\ln L)^{1/2}$ and $\psi_L(x) = 0$ for $|x| \geq L$,

$$|\nabla\psi_L(x)| \leq (\ln\ln L)^{-1}(|x|\ln|x|)^{-1}, \quad \operatorname{supp}\nabla\psi_L \subset \tilde D_L \tag{3.452}$$

where $\tilde D_L = D_{\exp(\ln L)^{1/2}, L}$. By the divergence theorem of Gauss, we have

$$(\psi_L\mathbf{v}, \nabla\mathfrak{p})_{\dot\Omega} = (\mathbf{v}, [[\mathfrak{p}]]\mathbf{n})_{S_R} - ((\nabla\psi_L)\cdot\mathbf{v}, \tilde{\mathfrak{p}}_-)_{\mathbb{R}^N}$$

for any L with $(\ln L)^{1/2} > R$. By (3.452), we have

$$((\nabla\psi_L)\cdot\mathbf{v}, \mathfrak{p}_-)_{\mathbb{R}^N} = (w(\nabla\psi_L)\cdot\mathbf{v}, (\tilde{\mathfrak{p}}_- - c)w^{-1})_{\tilde{D}} + ((\nabla\psi_L)\cdot\mathbf{v}, c)_{\mathbb{R}^N},$$

where c is the constant given in (3.451). By (3.451)

$$|(\mathbf{v}, (\tilde{\mathfrak{p}}_- - c)w^{-1})_{\tilde{D}_L}| \leq C\|\mathbf{v}\|_{L_{q'}(\tilde{D}_L)}\|\nabla\tilde{\mathfrak{p}}_-\|_{L_q(\mathbb{R}^N)} \to 0$$

as $L \to \infty$. Since $((\nabla\psi_L)\cdot\mathbf{v}, c)_{\mathbb{R}^N} = 0$ as follows from Lemma 3.7.8, we have (3.450), which completes the proof of Lemma 3.7.15. □

We continue the proof of Theorem 3.7.12. By the conditions on S_R and (3.449), we have

$$\begin{aligned}(\mathbf{u}, [[\mu\mathbf{D}(\mathbf{v}) - \mathfrak{q}\mathbf{I}]]\mathbf{n})_{S_R} &= (\mathbf{u}\cdot\mathbf{n}, \sigma\mathcal{B}\rho)_{S_R} \\ &= \sigma(\lambda h, \mathcal{B}\rho)_{S_R} - \sum_{j=1}^N \frac{1}{R|B_R|}u_j\,dx(x_j, \mathcal{B}\rho)_{S_R} \\ &= -\sigma\lambda\{(\nabla_{S_R}h, \nabla_{S_R}\rho)_{S_R} - \frac{N-1}{R^2}(h, \rho)_{S_R}\},\end{aligned}$$

where we have used

$$(x_j, \mathcal{B}\rho)_{S_R} = (\mathcal{B}x_j, \rho)_{S_R} = 0.$$

Thus, we have

$$\begin{aligned}&(\mathbf{u}, m\mathbf{f})_{\dot\Omega} \\ &= \lambda(\mathbf{u}, m\mathbf{v}) + \sigma\lambda\{(\nabla_{S_R}h, \nabla_{S_R}\rho)_{S_R} - \frac{N-1}{R^2}(h, \rho)_{S_R}\} + \frac{\mu}{2}(\mathbf{D}(\mathbf{u}), \mathbf{D}(\mathbf{v}))_{\dot\Omega}.\end{aligned}$$

Analogously, by (3.439) we have

$$\begin{aligned}0 &= (\lambda\mathbf{u} - m^{-1}\mathrm{Div}\,(\mu\mathbf{D}(\mathbf{u}) - \mathfrak{p}\mathbf{I}), m\mathbf{v})_{\dot\Omega} \\ &= \lambda(\mathbf{u}, m\mathbf{v}) + \sigma\lambda\{(\nabla_{S_R}h, \nabla_{S_R}\rho)_{S_R} - \frac{N-1}{R^2}(h, \rho)_{S_R}\} + \frac{\mu}{2}(\mathbf{D}(\mathbf{u}), \mathbf{D}(\mathbf{v}))_{\dot\Omega}.\end{aligned}$$

Combining these two equalities gives that $(m\mathbf{u}, \mathbf{f})_{\dot\Omega} = 0$ for any $\mathbf{f} \in L_q(\dot\Omega)^N$, which implies that $\mathbf{u} = 0$ in $\dot\Omega$. By the equations in (3.439) and (3.449), we have $\nabla\mathfrak{p} = 0$ in $\dot\Omega$, $\lambda h = 0$ on S_R, and $[[\mathfrak{p}]] + \sigma(\mathcal{B}h) = 0$ on S_R. Since $\lambda \neq 0$, we have h, and so $[[\mathfrak{p}]] = 0$ on S_R. Since $\nabla\mathfrak{p} = 0$ in $\dot\Omega$ and $[[\mathfrak{p}]] = 0$ on S_R, $\mathfrak{p}$ is a constant in $\mathbb{R}^N$. By (3.442), we have $\mathfrak{p} = 0$. Thus, we have (3.443). This completes the proof of Theorem 3.7.12. □

3.7.6 Local Energy Decay Estimates

In this section, we shall prove Theorem 3.7.2. Let ϵ be chosen in such a way that $0 < \epsilon < \pi/5$. Let d be a positive number given in Theorem 3.7.6. We choose $\delta > 0$ in Theorem 3.7.12 in such a way that $0 < \delta < d/2$. Let $\tilde\delta$ be a positive number given in Corollary 3.7.13. Let λ_1 be the constant given in Theorem 3.7.5. Let

$$\Gamma = \{\lambda \in \mathbb{C} \mid \lambda = e^{\pm i3\pi/4}\ell \mid |\ell| \geq 2\lambda_1\} \cup \{\lambda = 2\lambda_1 e^{i\theta} \mid -\frac{3\pi}{4} \leq \theta \leq \frac{3\pi}{4}\},$$

From theory of C^0 analytic semigroup, we know that $\{T(t)\}_{t\geq 0}$ is represented as

$$T(t)(\mathcal{P}\mathbf{f}, g) = \frac{1}{2\pi}\int_\Gamma e^{\lambda t}(\lambda + \mathcal{A}_q)^{-1}(\mathcal{P}\mathbf{f}, g)\,d\lambda$$

for $t > 0$. Let

$$\Gamma_1^\pm = \{\lambda = e^{\pm i3\pi/4}\ell \mid |\ell| \geq 2\lambda_1\} \cup \{\lambda = \pm\ell_0 + is \mid -\ell_1 \leq s - \tilde\delta/2\};$$
$$\Gamma_2^\pm = \{\lambda = -\tilde\delta/2 \pm is \mid 0 \leq s \leq \ell_0\};$$

and let Γ_3 be a smooth loop joining the points $\lambda = (\tilde\delta/2)e^{i\pi}$ and $\lambda = (\tilde\delta/2)e^{-i\pi}$ and going around the cut in $\dot U_d$ and connecting Γ_2^+ and Γ_2^-. In view of Theorem 3.7.5 and Corollary 3.7.13, $(\lambda+\mathcal{A}_q)^{-1}(\mathcal{P}\mathbf{f}, g)$ are holomorphic in $\Sigma_{\epsilon,\lambda_0}\cup\tilde\Xi_\delta$. Moreover, in view of Theorem 3.7.6, $(\lambda+\mathcal{A}_q)^{-1}(\mathcal{P}\mathbf{f}, g) = \mathcal{S}(\lambda)(\mathbf{f}, g)$ for $\lambda \in \tilde\Xi_\delta \cap \dot U_d$ and $(\mathbf{f}, g) \in L_{q,5R}(\dot\Omega)^N \times W_q^{2-1/q}(S_R)$. Since $\mathcal{S}(\lambda)$ is also holomorphic in $\dot U_d$, by Cauchy's integral formula, we have

$$T(t)(\mathcal{P}\mathbf{f}, g) = I_1^+(t) + I_2^+(t) + I_3(t) + I_2^-(t) + I_1^-(t),$$

where we have set

$$I_1^\pm(t) = \frac{1}{2\pi i}\int_{\Gamma_1^\pm} e^{\lambda t}(\lambda + \mathcal{A}_q)^{-1}(\mathcal{P}\mathbf{f}, g)\,d\lambda;$$
$$I_2^\pm(t) = \frac{1}{2\pi i}\int_{\Gamma_2^\pm} e^{\lambda t}(\lambda + \mathcal{A}_q)^{-1}(\mathcal{P}\mathbf{f}, g)\,d\lambda;$$
$$I_3(t) = \frac{1}{2\pi i}\int_{\Gamma_3} e^{\lambda t}\mathcal{S}(\lambda)(\mathbf{f}, g)\,d\lambda.$$

We assume that $t \geq 1$. By Theorem 3.7.5 and Corollary 3.7.13, we have

$$\|\partial_t^m I_i^\pm(t)\|_{H_q^2(\dot\Omega)\times W_q^{3-1/q}(S_R)} \leq C_m e^{-ct}(\|\mathbf{f}\|_{L_q(\dot\Omega)} + \|g\|_{W_q^{2-1/q}(S_R)})$$

for $i = 1, 2$ and any $m \in \mathbb{N}_0$ with some positive constants C_m and c provided that $t \geq 1$. Here, c is independent of m. Moreover, using Theorem 3.7.3 and applying Lemma 7 in Vainberg [38, p.369] to I_3, we have

$$\|\partial_t^m I_3(t)\|_{H^2_q(\dot{B}_{5R})\times W^{3-1/q}_q(S_R)} \leq C_m t^{-m-\frac{N}{2}}(\|\mathbf{f}\|_{L_q(\dot{\Omega})} + \|g\|_{W^{2-1/q}_q(S_R)})$$

for any $m \in \mathbb{N}_0$ with some positive constants C_m and c provided that $t \geq 1$. In the two inequalities above, the constant C_m depends on m. This completes the proof of Theorem 3.7.2.

3.7.7 Decay Estimate for Semigroup $\{T(t)\}_{t\geq 0}$

In this section, we prove Theorem 3.7.1. Let $\mathbf{u}_0 \in J_q(\dot{\Omega})$ and $\rho_0 \in W^{2-1/q}_q(S_R)$. As was pointed out in Iwashita [16], representing $T(t)(\mathbf{u}_0, \rho_0) = T(t-1)T(1)(\mathbf{u}_0, \rho_0)$ and setting $(\mathbf{f}, g) = T(1)(\mathbf{u}_0, \rho_0)$, we have $(\mathbf{f}, g) \in (J_q(\dot{\Omega}) \cap \mathcal{D}_q) \times W^{3-1/q}_q(S_R)$ and $T(t)(\mathbf{u}_0, g) = T(t-1)(\mathbf{f}, g)$. Since we consider the decay estimate for large $t > 3$, we consider $(\mathbf{u}(\cdot, t), \rho(\cdot, t)) = T(t-1)(\mathbf{f}, g)$ for $t \geq 2$. Notice that

$$\|(\mathbf{f}, g)\|_{\mathcal{D}_q} \leq C\|(\mathbf{u}_0, \rho_0)\|_{\mathcal{H}_q}.$$

Moreover, $(\mathbf{u}, \rho)$ satisfy the equations:

$$\begin{cases} \partial_t \mathbf{u} - m^{-1}\mathrm{Div}\,(\mu\mathbf{D}(\mathbf{u}) - \mathfrak{p}\mathbf{I}) = 0 & \text{in } Q_\infty, \\ \mathrm{div}\,\mathbf{u} = 0 & \text{in } Q_\infty, \\ \partial_t\rho + \mathcal{L}_3\rho - \mathbf{n}\cdot P\mathbf{u} = 0 & \text{on } \partial Q_\infty, \\ [[\mu\mathbf{D}(\mathbf{u}) - \mathfrak{q}\mathbf{I}]]\mathbf{n} - \sigma(\mathcal{B}\rho)\mathbf{n} = 0 & \text{on } \partial Q_\infty, \\ [[\mathbf{u}]] = 0 & \text{on } \partial Q_\infty, \\ (\mathbf{u}, \rho)|_{t=1} = (\mathbf{f}, g) & \text{in } \dot{\Omega}\times S_R, \end{cases} \tag{3.453}$$

where $Q_\infty = \dot{\Omega} \times (0, \infty)$ and $\partial Q_\infty = S_R \times (0, \infty)$. Let L be a large number such that $L > 6R + 5$. Since $\mathfrak{p} + c$ satisfies equation (3.453) for any constant c, choosing c suitably and writing $\mathfrak{p} + c$ by $\mathfrak{p}$, we may assume that

$$\int_{D_{R,L}} \mathfrak{p}(x, t)\,dx = 0. \tag{3.454}$$

By Poincares' inequality, we have

$$\begin{aligned}\int_{D_{R,L}} |\mathfrak{p}(x,t)|^q\,dx &\leq C\int_{D_{R,L}} |\nabla\mathfrak{p}(x,t)|^q\,dx \\ &\leq C(\|\partial_t\mathbf{u}(t,\cdot)\|_{L_q(D_{R,L})} + \|\mathbf{u}(\cdot,t)\|_{H^2_q(D_{R,L})})\end{aligned} \tag{3.455}$$

for $t > 0$. Let ψ be a function in $C^\infty(\mathbb{R}^N)$ which equals 1 for $x \notin B_{6R}$ and equals 0 for $x \in B_{5R}$ and let $\tilde{\psi}$ be a function in $C^\infty(\mathbb{R}^N)$ which equals 1 for $x \notin B_{5R}$ and equals 0 for $x \in B_{4R}$. Let $\mathbf{v}_0 = \psi\mathbf{f} - \mathbb{B}[(\nabla\psi)\cdot\mathbf{f}]$, and then $\mathbf{v}_0 = \mathbf{f}$ for $x \notin B_{6R}$ and $\operatorname{div}\mathbf{v}_0 = 0$ in $\mathbb{R}^N$. In particular, $\mathbf{v}_0 \in J_q(\mathbb{R}^N) \cap H^2_q(\mathbb{R}^N)^N$. Let $\mathbf{v}(\cdot,t) = e^{m_-^{-1}\mu_-\Delta t}\mathbf{v}_0$, and then $\mathbf{v}$ satisfies the Stokes equations in $\mathbb{R}^N$:

$$\begin{cases} \partial_t\mathbf{v} - m_-^{-1}\mu_-\Delta\mathbf{v} = 0, \quad \operatorname{div}\mathbf{v} = 0 & \text{in } \mathbb{R}^N \times (0,\infty), \\ \mathbf{v}|_{t=0} = \mathbf{v}_0 & \text{in } \mathbb{R}^N. \end{cases} \tag{3.456}$$

We know the following L_p-L_q estimates:

$$\|\partial_t^j\partial_x^\alpha\mathbf{v}(\cdot,t)\|_{L_p(\mathbb{R}^N)} \leq C_{p,q}t^{-j-\frac{|\alpha|}{2}-\frac{N}{2}\left(\frac{1}{q}-\frac{1}{p}\right)}\|\mathbf{v}_0\|_{L_q(\mathbb{R}^N)} \tag{3.457}$$

for any $t > 0$ provided $1 \leq q \leq p \leq \infty$. If we set

$$\tilde{\mathbf{v}} = \tilde{\psi}\mathbf{v} - \mathbb{B}[(\nabla\tilde{\psi})\cdot\mathbf{v}],$$

by (3.456), we have

$$\begin{cases} \partial_t\tilde{\mathbf{v}} - m^{-1}\operatorname{Div}(\mu\mathbf{D}(\tilde{\mathbf{v}})) = \mathbf{g}, \quad \operatorname{div}\tilde{\mathbf{v}} = 0 & \text{in } Q_\infty, \\ \tilde{\mathbf{v}}|_{t=0} = \mathbf{v}_0 & \text{in } \dot{\Omega}. \end{cases} \tag{3.458}$$

In fact, using the fact

$$\nabla\tilde{\psi} = 0 \quad \text{for } x \in \operatorname{supp}\mathbf{v}_0, \qquad \tilde{\psi} = 1 \ \text{ for } |x| > 5R, \qquad \tilde{\psi} = 0 \ \text{ for } x \in B_{4R},$$

we have

$$\tilde{\mathbf{v}}|_{t=0} = \tilde{\psi}\mathbf{v}_0 - \mathbb{B}[(\nabla\tilde{\psi})\cdot\mathbf{v}_0] = \tilde{\psi}(\psi\mathbf{f} - \mathbb{B}[(\nabla\psi)\cdot\mathbf{f}]) = \psi\mathbf{f} - \mathbb{B}[(\nabla\psi)\cdot\mathbf{f}] = \mathbf{v}_0.$$

Moreover,

$$\tilde{\mathbf{v}} = 0 \quad \text{for } x \in B_{4R}. \tag{3.459}$$

To find $\mathbf{g}$, we set $\mathbf{g} = \partial_t\tilde{\mathbf{v}} - m^{-1}\operatorname{Div}(\mu\mathbf{D}(\tilde{\mathbf{v}}))$. Since $\operatorname{div}\tilde{\mathbf{v}} = 0$, we have $\operatorname{div}\mathbf{g} = 0$. Moreover,

$$\mathbf{g} = \psi(\partial_t \mathbf{v} - m^{-1}\mu_- \Delta \mathbf{v}) - m_-^{-1}\mu_-(2(\nabla\tilde{\psi}) : \nabla \mathbf{v} + (\Delta\tilde{\psi})\mathbf{v})$$
$$- \partial_t \mathbb{B}[(\nabla\tilde{\psi}) \cdot \mathbf{v}(t)] + m^{-1}\mu_- \Delta \mathbb{B}[(\nabla\tilde{\psi}) \cdot \mathbf{v}].$$

Since $\partial_t \mathbf{v} - m^{-1}\mu_- \Delta \mathbf{v} = 0$, we have

$$\mathbf{g} = -m_-^{-1}\mu_-(2(\nabla\tilde{\psi}) : \nabla \mathbf{v} + (\Delta\tilde{\psi})\mathbf{v}) - \partial_t \mathbb{B}[(\nabla\tilde{\psi}) \cdot \mathbf{v}(t)] + m^{-1}\mu_- \Delta \mathbb{B}[(\nabla\tilde{\psi}) \cdot \mathbf{v}].$$

Notice that

$$\partial_t \mathbb{B}[(\nabla\tilde{\psi}) \cdot \mathbf{v}] = \mathbb{B}[(\nabla\tilde{\psi}) \cdot \mathbf{v}_t]] = -m_-^{-1}\mu_- \mathbb{B}[(\nabla\tilde{\psi}) \cdot \Delta \mathbf{v}].$$

Using Lemma 3.7.7 and (3.457), we have

$$\begin{aligned} \|\tilde{\mathbf{v}}(\cdot, t)\|_{L_p(\mathbb{R}^N)} &\leq C_{p,q} t^{-\frac{N}{2}\left(\frac{1}{q}-\frac{1}{p}\right)} D, \\ \|\nabla\tilde{\mathbf{v}}(\cdot, t)\|_{L_p(\mathbb{R}^N)} &\leq C_{p,q} t^{-\min(\frac{1}{2}+\frac{N}{2}\left(\frac{1}{q}-\frac{1}{p}\right), \frac{N}{2q})} D, \\ \|\partial_t^m \tilde{\mathbf{v}}(\cdot, t)\|_{H_q^2(B_L)} &\leq C_{p,q,L,m} t^{-m-\frac{N}{2q}} D \end{aligned} \tag{3.460}$$

for any $t \geq 1$ and $m \in \mathbb{N}_0$ provided $1 < q \leq p \leq \infty$ and $q \neq \infty$, where we have set $D = \|(\mathbf{f}, g)\|_{\mathcal{D}_q}$. In fact, by (3.457) we see easily that

$$\begin{aligned} &\|\tilde{\psi}\mathbf{v}(\cdot, t)\|_{L_p(\mathbb{R}^N)} \leq \|\tilde{\psi}\|_{L_\infty(\mathbb{R}^N)} C_{p,q} t^{-\frac{N}{2}\left(\frac{1}{q}-\frac{1}{p}\right)} \|\mathbf{v}_0\|_{L_q(\mathbb{R}^N)}, \\ &\|\nabla(\tilde{\psi}\mathbf{v})\|_{L_p(\mathbb{R}^N)} \leq \|\tilde{\psi}\|_{L_\infty(\mathbb{R}^N)} C_{p,q} \|\nabla \mathbf{v}(\cdot, t)\|_{L_p(\mathbb{R}^N)} + \|\nabla\tilde{\psi}\|_{L_p(\mathbb{R}^N)} \|\mathbf{v}\|_{L_\infty(\mathbb{R}^N)} \\ &\quad \leq C_{p,q,\tilde{\psi}} (t^{-\frac{1}{2}-\frac{N}{2}\left(\frac{1}{q}-\frac{1}{p}\right)} + t^{-\frac{N}{2q}}) \|\mathbf{v}_0\|_{L_q(\mathbb{R}^N)}; \\ &\|\nabla^2(\tilde{\psi}\mathbf{v})\|_{L_q(B_L)} \leq C_{p,q} \|\tilde{\psi}\|_{H_q^2(B_L)} \|\mathbf{v}\|_{H_\infty^2(\mathbb{R}^N)} \leq C_{p,q,\tilde{\psi}} t^{-\frac{N}{2q}} \|\mathbf{v}_0\|_{L_q(\mathbb{R}^N)} \end{aligned}$$

for $t > 1$. By Lemma 3.7.7 and (3.457), we have

$$\begin{aligned} &\|\mathbb{B}[(\nabla\tilde{\psi}) \cdot \Delta \mathbf{v}]\|_{H_q^1(\mathbb{R}^N)} + \|\Delta \mathbb{B}[(\nabla\tilde{\psi}) \cdot \mathbf{v}]\|_{H_q^1(\mathbb{R}^N)} \leq C\|\mathbf{v}(\cdot, t)\|_{H_q^2(B_L)} \\ &\quad \leq C|B_L|^{1/q} \|\mathbf{v}(\cdot, t)\|_{H_\infty^2(\mathbb{R}^N)} \leq C_L t^{-\frac{N}{2q}} \|\mathbf{v}_0\|_{L_q(\mathbb{R}^N)} \end{aligned}$$

for $t > 1$, where $|B_L|$ denotes the volume of B_L. Since $\|\mathbf{v}_0\|_{L_q(\mathbb{R}^N)} \leq CD$, combining the estimates above gives (3.460).

Analogously, by Lemma 3.7.7 and (3.457), we have

$$
\begin{aligned}
\|\mathbf{g}(\cdot, t)\|_{H^1_q(\mathbb{R}^N)} &\leq C\|\mathbf{v}(\cdot, t)\|_{H^1_\infty(\mathbb{R}^N)} \leq Ct^{-\frac{N}{2q}}\|\mathbf{v}_0\|_{L_q(\mathbb{R}^N)} \\
&\leq Ct^{-\frac{N}{2q}}D \quad \text{for } t > 2, \\
\|\mathbf{g}(\cdot, t)\|_{H^1_q(\mathbb{R}^N)} &\leq C\|\mathbf{v}(\cdot, t)\|_{H^2_q(\mathbb{R}^N)} \leq C\|\mathbf{v}_0\|_{H^2_q(\mathbb{R}^N)} \\
&\leq CD \qquad \text{for } 0 < t < 2.
\end{aligned}
\tag{3.461}
$$

We now set $\mathbf{u} = \tilde{\mathbf{v}} + \mathbf{w}$ in equation (3.453), and then $\mathbf{w}$ satisfies the equations:

$$
\begin{cases}
\partial_t \mathbf{w} - m^{-1}\mathrm{Div}\,(\mu\mathbf{D}(\mathbf{w}) - \mathfrak{p}\mathbf{I}) = -\mathbf{g} & \text{in } Q_\infty, \\
\mathrm{div}\,\mathbf{w} = 0 & \text{in } Q_\infty, \\
\partial_t\rho + \mathcal{L}_3\rho - \mathbf{n}\cdot P\mathbf{w} = 0 & \text{on } \partial Q_\infty, \\
[[\mu\mathbf{D}(\mathbf{w}) - \mathfrak{p}\mathbf{I}]]\mathbf{n} - \sigma(\mathcal{B}\rho)\mathbf{n} = 0 & \text{on } \partial Q_\infty, \\
[[\mathbf{w}]] = 0 & \text{on } \partial Q_\infty, \\
(\mathbf{w}, \rho)|_{t=0} = (\mathbf{f} - \mathbf{v}_0, g) & \text{in } \dot\Omega \times S_R.
\end{cases}
\tag{3.462}
$$

Notice that $\mathbf{w} \in \mathcal{D}_q \cap J_q(\dot\Omega)$. Let $\mathbf{w}_0 = \mathbf{f} - \mathbf{v}_0$. Recalling that $\mathbf{g}(\cdot, t) \in J_q(\dot\Omega)$ for all $t > 0$, by Duhamel's principle we have

$$
(\mathbf{w}(\cdot, t), \rho(\cdot, t)) = T(t)(\mathbf{w}_0, \rho_0) - \int_0^t T(t-s)(\mathbf{g}(\cdot, s), 0)\,ds.
$$

Since $\mathrm{supp}\,\mathbf{w}_0 \subset B_{5R}$ and $\mathrm{supp}\,\mathbf{g}(\cdot, t) \subset B_{5R}$ for any $t > 0$, by Theorem 3.7.2 we have

$$
\begin{aligned}
&\|\partial_t^m(\mathbf{w}(\cdot, t), \rho(\cdot, t))\|_{H^2_q(\dot B_L)\times W^{3-1/q}_q(S_R)} \\
&\quad \leq C_m t^{-m-\frac{N}{2}}\|(\mathbf{w}_0, \rho_0)\|_{\mathcal{H}_q} + CI(t)
\end{aligned}
$$

with

$$
\begin{aligned}
I(t) = &\int_0^1 (t-s)^{-\frac{N}{2}}\|\mathbf{g}(\cdot, s)\|_{L_q(\mathbb{R}^N)}\,ds + \int_1^{t-1}(t-s)^{-\frac{N}{2}}\|\mathbf{g}(\cdot, s)\|_{L_q(\mathbb{R}^N)}\,ds \\
&+ \int_{t-1}^t (t-s)^{-1/2}\|\mathbf{g}(\cdot, s)\|_{H^1_q(\mathbb{R}^N)}\,ds.
\end{aligned}
$$

Since $N/2 > 1$, using (3.461) we have

$$
\|\partial_t^m(\mathbf{w}(\cdot, t), \rho(\cdot, t))\|_{H^2_q(\dot B_L)\times W^{3-1/q}_q(S_R)} \leq Ct^{-\frac{N}{2q}}D.
\tag{3.463}
$$

Since $\{T(t)\}_{t\geq 0}$ is a C^0 analytic semigroup, we know that

$$\|T(t)(\mathbf{u}_0, \rho_0)\|_{\mathcal{D}_q} \leq C\|(\mathbf{u}_0, \rho_0)\|_{\mathcal{D}_q},$$
$$\|T(t)(\mathbf{u}_0, \rho_0)\|_{\mathcal{D}_q} \leq Ct^{-1}\|(\mathbf{u}_0, \rho_0)\|_{\mathcal{D}_q}$$

for $0 < t \leq 2$. By real interpolation, we also have

$$\|T(t)(\mathbf{u}_0, \rho_0)\|_{\mathcal{D}_q} \leq Ct^{-1/2}\|(\mathbf{u}_0, \rho_0)\|_{H^1_q(\dot\Omega)\times W^{5/2}_q(S_R)},$$

for $0 < t \leq 2$. Thus, we have

$$\|\mathbf{w}\|_{H^2_q(\dot\Omega)} \leq C(\|\mathbf{w}_0, \rho_0)\|_{\mathcal{D}_q} + \int_0^t (t-s)^{-1/2}\|\mathbf{g}(\cdot, s)\|_{H^1_q(\dot\Omega)}\,ds) \leq CD \tag{3.464}$$

for $0 < t < 2$. By (3.457), we have

$$\|\partial_t\tilde{\mathbf{v}}(t,\cdot)\|_{L_q(D_{R,L})} + \|\tilde{\mathbf{v}}(\cdot,t)\|_{H^2_q(D_{R,L})} \leq Ct^{-\frac{N}{2q}}\|\mathbf{v}_0\|_{L_q(\mathbb{R}^N)} \leq Ct^{-\frac{N}{2q}}D \quad \text{for } t > 2,$$
$$\|\partial_t\tilde{\mathbf{v}}(t,\cdot)\|_{L_q(D_{R,L})} + \|\tilde{\mathbf{v}}(\cdot,t)\|_{H^2_q(D_{R,L})} \leq C\|(\mathbf{f}, g)\|_{\mathcal{D}_q} \leq CD \quad \text{for } 0 < t < 2.$$

Combining these inequalities with (3.455), (3.463), and (3.464) gives

$$\|\mathfrak{p}(\cdot,t)\|_{H^1_q(D_{R,L})} \leq C\begin{cases} t^{-\frac{N}{2q}}D & (t > 2), \\ D & (0 < t < 2). \end{cases} \tag{3.465}$$

Let κ be a function in $C^\infty(\mathbb{R}^N)$ which equals 1 for $|x| > L-1$ and equals 0 for $|x| < L-2$ and let

$$\mathbf{z} = \kappa\mathbf{w} - \mathbb{B}[(\nabla\kappa)\cdot\mathbf{w}].$$

Then, $\mathbf{z}$ satisfies the Cauchy problem:

$$\begin{cases} \partial_t\mathbf{z} - m_-^{-1}\mu_-\Delta\mathbf{z} + \nabla(\kappa\mathfrak{p}) = \mathbf{h} & \text{in } \mathbb{R}^N\times(0,\infty), \\ \operatorname{div}\mathbf{z} = 0 & \text{in } \mathbb{R}^N\times(0,\infty), \\ \mathbf{z}|_{t=0} = 0 & \text{in } \mathbb{R}^N. \end{cases} \tag{3.466}$$

Here,

$$\mathbf{z}|_{t=0} = \kappa\mathbf{w}_0 - \mathbb{B}[(\nabla\kappa)\cdot\mathbf{w}_0] = 0$$

because $\operatorname{supp}\mathbf{w}_0 \subset B_{6R}$ and $L-2 > 6R$. And

$$\begin{aligned}\mathbf{h} &= \partial_t\mathbf{z} - m_-^{-1}\mu_-\Delta\mathbf{z} + \nabla(\kappa\mathfrak{p})\\ &= \kappa(\partial_t\mathbf{w} - m_-^{-1}\mu_-\Delta\mathbf{w} + \nabla\mathfrak{p}) - m_-^{-1}\mu_-(2(\nabla\kappa):\nabla\mathbf{w} + (\Delta\kappa)\mathbf{w})\\ &\quad - m_-^{-1}(\nabla\kappa)\mathfrak{p} - \partial_t\mathbb{B}[(\nabla\kappa)\cdot\mathbf{w}] + m_-^{-1}\mu_-\Delta\mathbb{B}[(\nabla\kappa)\cdot\mathbf{w}].\end{aligned}$$

Since $\operatorname{supp}\mathbf{g}(\cdot, t) \subset B_{5R}$, we have $\kappa(\partial_t\mathbf{w} - m_-^{-1}\mu_-\Delta\mathbf{w} + \nabla\mathfrak{p}) = 0$, and so

$$\begin{aligned}\mathbf{h} = &-m_-^{-1}\mu_-(2(\nabla\kappa):\nabla\mathbf{w} + (\Delta\kappa)\mathbf{w}) - m_-^{-1}(\nabla\kappa)\mathfrak{p}\\ &- \mathbb{B}[(\nabla\kappa)\cdot\mathbf{w}_t] + m_-^{-1}\mu_-\Delta\mathbb{B}[(\nabla\kappa)\cdot\mathbf{w}].\end{aligned}$$

Notice that

$$\mathbb{B}[(\nabla\kappa)\cdot\mathbf{w}_t] = m_-^{-1}\mu_-\mathbb{B}[(\nabla\kappa)\cdot\Delta\mathbf{w}] + m_-^{-1}\mathbb{B}[(\nabla\kappa)\cdot\nabla\mathfrak{p}].$$

Since $\operatorname{supp}\mathbf{h}(\cdot, s) \subset D_{L-2,L-1}$, by Lemma 3.7.7, (3.463), (3.464), and (3.465), we have

$$\|\mathbf{h}(\cdot, s)\|_{L_r(\mathbb{R}^N)} \leq C_{q,r}\|\mathbf{h}(\cdot, s)\|_{L_q(D_{L-2,L-1})} \leq CD\begin{cases} s^{-\frac{N}{2q}} & (s > 2),\\ 1 & (0 < s < 2),\end{cases} \tag{3.467}$$

where r is any index with $1 < r \leq q$. Let $\{S(t)\}_{t\geq 0}$ be the Stokes semi-group in $\mathbb{R}^N$, and then

$$\mathbf{z}(\cdot, t) = \int_0^t S(t-s)P_q\mathbf{h}(\cdot, s)\,ds,$$

where P_q denotes the Helmholtz decomposition in $\mathbb{R}^N$. We know the following estimates:

$$\begin{aligned}\|S(t)P_q\mathbf{w}\|_{L_p(\mathbb{R}^N)} &\leq Ct^{-\frac{N}{2}\left(\frac{1}{q}-\frac{1}{p}\right)}\|\mathbf{w}\|_{L_q(\mathbb{R}^N)},\\ \|\nabla S(t)P_q\mathbf{w}\|_{L_p(\mathbb{R}^N)} &\leq Ct^{-\frac{1}{2}-\frac{N}{2}\left(\frac{1}{q}-\frac{1}{p}\right)}\|\mathbf{w}\|_{L_q(\mathbb{R}^N)}\end{aligned} \tag{3.468}$$

for any $t > 0$ provided that $1 < q \leq p \leq \infty$ and $q \neq \infty$. Applying (3.468), we have

$$\begin{aligned}\|\mathbf{z}(\cdot, t)\|_{L_p(\mathbb{R}^N)} \leq C\Big\{&\int_0^1 (t-s)^{-\frac{N}{2}\left(\frac{1}{q}-\frac{1}{p}\right)}\|\mathbf{h}(\cdot, s)\|_{L_q(\mathbb{R}^N)}\,ds\\ &+ \int_1^{t-1}(t-s)^{-\frac{N}{2}\left(\frac{1}{r}-\frac{1}{p}\right)}\|\mathbf{h}(\cdot, s)\|_{L_r(\mathbb{R}^N)}\,ds\end{aligned}$$

$$+\int_{t-1}^{t} (t-s)^{-\frac{N}{2}\left(\frac{1}{q}-\frac{1}{p}\right)} \|\mathbf{h}(\cdot, s)\|_{L_q(\mathbb{R}^N)}\, ds\Big\}.$$

We first consider the case where $\frac{N}{2}(\frac{1}{q}-\frac{1}{p}) < 1$. By (3.467),

$$\begin{aligned}&\int_0^1 (t-s)^{-\frac{N}{2}\left(\frac{1}{q}-\frac{1}{p}\right)} \|\mathbf{h}(\cdot, s)\|_{L_q(\mathbb{R}^N)}\, ds\\ &\quad\le CD(t-1)^{-\frac{N}{2}\left(\frac{1}{q}-\frac{1}{p}\right)} \int_0^1 ds \le CDt^{-\frac{N}{2}\left(\frac{1}{q}-\frac{1}{p}\right)}.\end{aligned}$$

Since we assume that $\frac{N}{2}(\frac{1}{q}-\frac{1}{p}) < 1$, we have

$$\begin{aligned}&\int_{t-1}^{t} (t-s)^{-\frac{N}{2}\left(\frac{1}{q}-\frac{1}{p}\right)} \|\mathbf{h}(\cdot, s)\|_{L_q(\mathbb{R}^N)}\, ds\\ &\quad\le CD(t-1)^{-\frac{N}{2q}} \int_{t-1}^{t} (t-s)^{-\frac{N}{2}\left(\frac{1}{q}-\frac{1}{p}\right)} ds \le Cd_{pq}Dt^{-\frac{N}{2q}},\end{aligned}$$

where we have set

$$d_{pq} = \Big(1-\frac{N}{2}\Big(\frac{1}{q}-\frac{1}{p}\Big)\Big)^{-1}.$$

By (3.467), we have

$$\begin{aligned}&\int_1^{t-1} (t-s)^{-\frac{N}{2}\left(\frac{1}{r}-\frac{1}{p}\right)} \|\mathbf{h}(\cdot, s)\|_{L_r(\mathbb{R}^N)}\, ds\\ &\quad\le CD\int_1^{t-1} (t-s)^{-\frac{N}{2}\left(\frac{1}{r}-\frac{1}{p}\right)} s^{-\frac{N}{2q}}\, ds\\ &\quad= CD\int_1^{t/2} (t-s)^{-\frac{N}{2}\left(\frac{1}{r}-\frac{1}{p}\right)} s^{-\frac{N}{2q}}\, ds + CD\int_1^{t/2} s^{-\frac{N}{2}\left(\frac{1}{r}-\frac{1}{p}\right)} (t-s)^{-\frac{N}{2q}}\, ds.\end{aligned}$$

For $0 < s < t/2$, we have $t-s > s$, and so

$$\begin{aligned}(t-s)^{-\frac{N}{2}\left(\frac{1}{r}-\frac{1}{p}\right)} &= (t-s)^{-\frac{N}{2}\left(\frac{1}{q}-\frac{1}{p}\right)}(t-s)^{-\frac{N}{2}\left(\frac{1}{r}-\frac{1}{q}\right)}\\ &\le (t-s)^{-\frac{N}{2}\left(\frac{1}{q}-\frac{1}{p}\right)} s^{-\frac{N}{2}\left(\frac{1}{r}-\frac{1}{q}\right)}.\end{aligned}$$

Thus, choosing r in such a way that $N/(2r) > 1$ and $1 < r \le q$, we have

$$\int_1^{t/2} (t-s)^{-\frac{N}{2}\left(\frac{1}{r}-\frac{1}{p}\right)} s^{-\frac{N}{2q}}\, ds \le \int_1^{t/2} (t-s)^{-\frac{N}{2}\left(\frac{1}{q}-\frac{1}{p}\right)} s^{-\frac{N}{2r}}\, ds \le d_r(t/2)^{-\frac{N}{2}\left(\frac{1}{q}-\frac{1}{p}\right)},$$

where we have set

$$d_r = (1 - N/(2r))^{-1}.$$

Analogously,

$$s^{-\frac{N}{2}\left(\frac{1}{r}-\frac{1}{p}\right)}(t-s)^{-\frac{N}{2q}} = s^{-\frac{N}{2r}}s^{\frac{N}{2p}}(t-s)^{-\frac{N}{2q}}$$
$$\leq s^{-\frac{N}{2r}}(t-s)^{\frac{N}{2p}}(t-s)^{-\frac{N}{2q}} = s^{-\frac{N}{2r}}(t-s)^{-\frac{N}{2}\left(\frac{1}{q}-\frac{1}{p}\right)}$$

and so,

$$\int_1^{t/2} s^{-\frac{N}{2}\left(\frac{1}{r}-\frac{1}{p}\right)}(t-s)^{-\frac{N}{2q}} \leq d_r(t/2)^{-\frac{N}{2}\left(\frac{1}{q}-\frac{1}{p}\right)}.$$

Summing up, we have obtained

$$\|\mathbf{z}(\cdot,t)\|_{L_p(\mathbb{R}^N)} \leq C_{p,q}t^{-\frac{N}{2}\left(\frac{1}{q}-\frac{1}{p}\right)}D$$

provided $1 < q \leq p \leq \infty, q \neq \infty$ and $\frac{N}{2}\left(\frac{1}{q}-\frac{1}{p}\right) < 1$.

We next consider $\nabla\mathbf{z}$. We also assume that $\frac{N}{2}\left(\frac{1}{q}-\frac{1}{p}\right) < 1$. By (3.468), we have

$$\begin{aligned}\|\nabla\mathbf{z}(\cdot,t)\|_{L_p(\mathbb{R}^N)} \leq C\Big\{&\int_0^1 (t-s)^{-\frac{1}{2}-\frac{N}{2}\left(\frac{1}{q}-\frac{1}{p}\right)}\|\mathbf{h}(\cdot,s)\|_{L_q(\mathbb{R}^N)}\,ds \\ &+\int_1^{t-1} (t-s)^{-\frac{1}{2}-\frac{N}{2}\left(\frac{1}{r}-\frac{1}{p}\right)}\|\mathbf{h}(\cdot,s)\|_{L_r(\mathbb{R}^N)}\,ds \\ &+\int_{t-1}^{t} (t-s)^{-\frac{N}{2}\left(\frac{1}{q}-\frac{1}{p}\right)}\|\nabla\mathbf{h}(\cdot,s)\|_{L_q(\mathbb{R}^N)}\,ds\Big\}.\end{aligned}$$

By (3.467)

$$\begin{aligned}\int_0^1 (t-s)^{-\frac{1}{2}-\frac{N}{2}\left(\frac{1}{q}-\frac{1}{p}\right)}\|\mathbf{h}(\cdot,s)\|_{L_q(\mathbb{R}^N)}\,ds &\leq CD\int_0^1 (t-s)^{-\frac{1}{2}-\frac{N}{2}\left(\frac{1}{q}-\frac{1}{p}\right)}\,ds \\ &\leq CDt^{-\frac{1}{2}-\frac{N}{2}\left(\frac{1}{q}-\frac{1}{p}\right)}.\end{aligned}$$

By (3.467), we also have

$$\begin{aligned}\int_{t-1}^t (t-s)^{-\frac{N}{2}\left(\frac{1}{q}-\frac{1}{p}\right)}\|\nabla\mathbf{h}(\cdot,s)\|_{L_q(\mathbb{R}^N)}\,ds &\leq CD\int_{t-1}^t (t-s)^{-\frac{N}{2}\left(\frac{1}{q}-\frac{1}{p}\right)}s^{-\frac{N}{2q}}\,ds \\ &\leq Cd_{pq}Dt^{-\frac{N}{2q}}.\end{aligned}$$

Analogously to the estimate of **z**, we have

$$\int_1^{t-1}(t-s)^{-\frac12-\frac N2\left(\frac1r-\frac1p\right)}\|\mathbf{h}(\cdot,s)\|_{L_r(\mathbb{R}^N)}\,ds$$
$$\leq CD\int_1^{t/2}(t-s)^{-\frac12-\frac N2\left(\frac1r-\frac1p\right)}s^{-\frac{N}{2q}}\,ds+CD\int_1^{t/2}s^{-\frac12-\frac N2\left(\frac1r-\frac1p\right)}(t-s)^{-\frac{N}{2q}}\,ds.$$

For $0 < s < t/2$, we have

$$(t-s)^{-\frac12-\frac N2\left(\frac1r-\frac1p\right)}s^{-\frac{N}{2q}}\leq(t-s)^{-\frac12-\frac N2\left(\frac1q-\frac1p\right)}s^{-\frac{N}{2r}},$$

and so

$$\int_1^{t/2}(t-s)^{-\frac12-\frac N2\left(\frac1r-\frac1p\right)}s^{-\frac{N}{2q}}\,ds\leq d_r(t/2)^{-\frac12-\frac N2\left(\frac1q-\frac1p\right)}.$$

If we choose r in such a way that $\frac12+\frac N2\left(\frac1r-\frac1p\right)>1$, then we have

$$\int_1^{t/2}s^{-\frac12-\frac N2\left(\frac1r-\frac1p\right)}(t-s)^{-\frac{N}{2q}}\,ds\leq\tilde d_{pq}(t/2)^{-\frac{N}{2q}},$$

where we have set

$$\tilde d_{pq}=\left(\frac N2\left(\frac1r-\frac1p\right)-\frac12\right)^{-1}.$$

Moreover, if we choose r only in such a way that $\frac12+\frac N2\left(\frac1r-\frac1p\right)<1$, then

$$\int_1^{t/2}s^{-\frac12-\frac N2\left(\frac1r-\frac1p\right)}(t-s)^{-\frac{N}{2q}}\,ds\leq(t/2)^{-\frac{N}{2q}}\int_1^{t/2}s^{-\frac12-\frac N2\left(\frac1r-\frac1p\right)}\,ds$$
$$\leq C(t/2)^{-\frac{N}{2q}}(t/2)^{\frac12-\frac N2\left(\frac1r-\frac1p\right)}\leq Ct^{-\frac{N}{2q}+\frac12-\frac N2\left(\frac1r-\frac1p\right)}=Ct^{-\frac12-\frac N2\left(\frac1q-\frac1p\right)}t^{1-\frac{N}{2r}}.$$

Since $N\geq3$, we can choose r in such a way that $N/(2r)>1$ and $1<r<q$, and so we have

$$\int_1^{t/2}s^{-\frac12-\frac N2\left(\frac1r-\frac1p\right)}(t-s)^{-\frac{N}{2q}}\,ds\leq Ct^{-\frac12-\frac N2\left(\frac1q-\frac1p\right)}.$$

Summing up, we have obtained

$$\|\nabla\mathbf{z}(\cdot,t)\|_{L_p(\mathbb{R}^N)}\leq C_{p,q}t^{-\min\left(\frac12+\frac N2\left(\frac1q-\frac1p\right),\frac{N}{2q}\right)}D$$

provided $1 < q \leq p \leq \infty$, $q \neq \infty$ and $\frac{N}{2}\left(\frac{1}{q} - \frac{1}{p}\right) < 1$. Setting $(\mathbf{u}(\cdot, t), \rho(\cdot, t)) = T(t)(\mathbf{u}_0, \rho_0)$ and recalling (3.460) and (3.463), we have

$$\begin{aligned} \|\mathbf{u}(\cdot, t)\|_{L_p(\dot\Omega)} &\leq C_{p,q} t^{-\frac{N}{2}\left(\frac{1}{q}-\frac{1}{p}\right)} \|(\mathbf{u}_0, \rho_0)\|_{\mathcal{H}_q} \\ \|\nabla \mathbf{u}(\cdot, t)\|_{L_p(\dot\Omega)} &\leq C_{p,q} t^{-\min\left(\frac{1}{2}+\frac{N}{2}\left(\frac{1}{q}-\frac{1}{p}\right), \frac{N}{2q}\right)} \|(\mathbf{u}_0, \rho_0)\|_{\mathcal{H}_q} \end{aligned} \tag{3.469}$$

for any $(\mathbf{u}_0, \rho_0) \in J_q(\dot\Omega) \times W_q^{2-1/q}(S_R)$ and $t > 2$ provided $1 < q \leq p \leq \infty$, $q \neq \infty$, and $\frac{N}{2}\left(\frac{1}{q} - \frac{1}{p}\right) < 1$. Moreover, by (3.460) and (3.463),

$$\|\partial_t^m (\mathbf{u}(\cdot, t), \rho(\cdot, t))\|_{H_q^2(\dot B_L) \times W_q^{3-1/q}(S_R)} \leq C_q t^{-\frac{N}{2q}} \|(\mathbf{u}_0, \rho_0)\|_{\mathcal{H}_q} \tag{3.470}$$

for $t > 2$. By Sobolev's imbedding theorem, $H_q^3(\mathbb{R}^N) \subset H_p^2(\mathbb{R}^N)$ for any $1 \leq q \leq p \leq \infty$ with $N\left(\frac{1}{q} - \frac{1}{p}\right) < 1$, where the inclusion is continuous, which combined with the trace theorem, leads to

$$\|\rho(\cdot, t)\|_{W_p^{2-1/p}(S_R)} \leq C_{p,q} \|\rho(\cdot, t)\|_{W_q^{3-1/q}(S_R)}$$

provided that $N\left(\frac{1}{q} - \frac{1}{p}\right) < 1$. Thus, by (3.470),

$$\|\rho(\cdot, t)\|_{W_p^{2-1/p}(S_R)} \leq C_{p,q} t^{-\frac{N}{2q}} \|(\mathbf{u}_0, \rho_0)\|_{\mathcal{H}_q} \leq C_{p,q} t^{-\frac{N}{2}\left(\frac{1}{q}-\frac{1}{p}\right)} \|(\mathbf{u}_0, \rho_0)\|_{\mathcal{H}_q},$$

which, combined with (3.469), leads to

$$\|T(t)(\mathbf{u}_0, \rho_0)\|_{\mathcal{H}_p} \leq C_{p,q} t^{-\frac{N}{2}\left(\frac{1}{q}-\frac{1}{p}\right)} \|(\mathbf{u}_0, \rho_0)\|_{\mathcal{H}_q} \tag{3.471}$$

for $t > 3$ and $(\mathbf{u}_0, \rho_0) \in J_q(\dot\Omega) \times W_q^{2-1/q}(S_R)$, provided $1 < q \leq p \leq \infty$, $q \neq \infty$ and $N\left(\frac{1}{q} - \frac{1}{p}\right) < 1$, because $(\mathbf{u}(\cdot, t), \rho(\cdot, t)) = T(t+1)(\mathbf{u}_0, \rho_0)$.

We now consider the case where $1 \leq N\left(\frac{1}{q} - \frac{1}{p}\right) < 2$. Notice that

$$T(t)(\mathbf{u}_0, \rho_0) \in J_q(\dot\Omega) \times W_q^2(S_R) \quad \text{for } (\mathbf{u}_0, \rho_0) \in J_q(\dot\Omega) \times W_q^2(S_R).$$

Since $\{T(t)\}_{t \geq 0}$ is a semi-group, we have

$$T(t)(\mathbf{u}_0, \rho_0) = T(t/2)T(t/2)(\mathbf{u}_0, \rho_0).$$

Choosing q_1 in such a way that $N\left(\frac{1}{q_1}-\frac{1}{p}\right)<1$, $N\left(\frac{1}{q}-\frac{1}{q_1}\right)<1$, $1<q<q_1\leq p\leq\infty$ and $q_1\neq\infty$, we have

$$\begin{aligned}\|T(t)(\mathbf{u}_0,\rho_0)\|_{\mathcal{H}_p}&=\|T(t/2)T(t/2)(\mathbf{u}_0,\rho_0)\|_{\mathcal{H}_p}\\&\leq C_{p,q_1}(t/2)^{-\frac{N}{2}\left(\frac{1}{q_1}-\frac{1}{p}\right)}\|T(t/2)(\mathbf{u}_0,\rho_0)\|_{\mathcal{H}_{q_1}}\\&\leq C_{p,q_1}C_{q_1,q}(t/2)^{-\frac{N}{2}\left(\frac{1}{q_1}-\frac{1}{p}\right)}(t/2)^{-\frac{N}{2}\left(\frac{1}{q}-\frac{1}{q_1}\right)}\|(\mathbf{u}_0,\rho_0)\|_{\mathcal{H}_q}\\&\leq C_{p,q}t^{-\frac{N}{2}\left(\frac{1}{q}-\frac{1}{p}\right)}\|(\mathbf{u}_0,\rho_0)\|_{\mathcal{H}_q}\end{aligned}$$

for $t>6$. Repeating this argument at most N times, we have

$$\|T(t)(\mathbf{u}_0,\rho_0)\|_{\mathcal{H}_p}\leq C_{p,q}t^{-\frac{N}{2}\left(\frac{1}{q}-\frac{1}{p}\right)}\|(\mathbf{u}_0,\rho_0)\|_{\mathcal{H}_q}\tag{3.472}$$

for $t>3N$ provided $1<q\leq p\leq\infty$ and $q\neq\infty$.

Since $\mathcal{A}_q^m$ maps $\mathcal{D}_q^m$ into $\mathcal{H}_q$, where $\mathcal{D}_q^m\subset H_q^{2m}(\dot\Omega)\times W_q^{2+m-1/q}(S_R)$, by semigroup theory, we have

$$\|T(t)(\mathbf{u}_0,\rho_0)\|_{H_q^{2m}(\dot\Omega)\times W_q^{2+m-1/q}(S_R)}\leq Ce^{ct}t^{-m}\|(\mathbf{u}_0,\rho)\|_{\mathcal{H}_q}$$

for any $t>0$ and $m\in\mathbb{N}$ with some constant C_m and c, where C_m is a constant depending on m and c is a constant independent of m. By Sobolev's imbedding theorem, $W^{N\left(\frac{1}{q}-\frac{1}{p}\right)}(\dot\Omega)\times W_q^{2+N\left(\frac{1}{q}-\frac{1}{p}\right)-1/q}(S_R)$ is continuously imbedded into $L_p(\dot\Omega)\times W_p^{2-1/p}(S_R)$, and so

$$\|T(t)(\mathbf{u}_0,\rho_0)\|_{\mathcal{H}_p}\leq Ce^{ct}t^{-N\left(\frac{1}{q}-\frac{1}{p}\right)}\|(\mathbf{u}_0,\rho_0)\|_{\mathcal{H}_q}.$$

for any $t>0$, which, combined with (3.472), leads to

$$\|T(t)(\mathbf{u}_0,\rho_0)\|_{\mathcal{H}_p}\leq C_{p,q}t^{-\frac{N}{2}\left(\frac{1}{q}-\frac{1}{p}\right)}\|(\mathbf{u}_0,\rho_0)\|_{\mathcal{H}_q}\tag{3.473}$$

for any $t\geq1$ and $(\mathbf{u}_0,\rho_0)\in J_q(\dot\Omega)\times W_q^{2-1/q}(S_R)$, provided $1<q\leq p\leq\infty$ and $q\neq\infty$.

When $1\leq N\left(\frac{1}{q}-\frac{1}{p}\right)<2$, choosing q_1 in such a way that $1<q<q_1<p\leq\infty$, $N\left(\frac{1}{q_1}-\frac{1}{p}\right)<1$, and $N\left(\frac{1}{q}-\frac{1}{q_1}\right)<1$, by (3.469), (3.473) and the semigroup property, we have

$$\|\nabla\mathbf{u}(\cdot,t)\|_{L_p(\dot\Omega)}$$

$$
\begin{aligned}
&\le C_{p,q_1}(t/2)^{-\min(\frac12+\frac N2\left(\frac1{q_1}-\frac1p\right),\frac{N}{2q_1})}\|T(t/2+1)(\mathbf{u}_0,\rho_0)\|_{\mathcal{H}_{q_1}}\\
&\le C_{p,q_1}C_{q_1,q}(t/2)^{-\min(\frac12+\frac N2\left(\frac1{q_1}-\frac1p\right),\frac{N}{2q_1})}(t/2)^{-\frac N2\left(\frac1{q_1}-\frac1p\right)}\|T(1)(\mathbf{u}_0,\rho_0)\|_{\mathcal{H}_q}\\
&\le C_{p,q}t^{-\min(\frac12+\frac N2\left(\frac1q-\frac1p\right),\frac{N}{2q})}\|(\mathbf{u}_0,\rho_0)\|_{\mathcal{H}_q}
\end{aligned}
$$

for any $t \ge 1$ and $(\mathbf{u}_0, \rho_0) \in J_q(\dot\Omega) \times W_q^{2-1/q}(S_R)$ provided $1 < q \le p \le \infty$ and $q \ne \infty$.

Analogously, by (3.470), (3.473) and the semigroup property, we have

$$
\begin{aligned}
&\|\partial_t^m(\mathbf{u}(\cdot,t),\rho(\cdot,t))\|_{H^2_p(\dot B_L)\times W_p^{3-1/p}(S_R)}\\
&\le C_p(t/2)^{-\frac{N}{2p}}\|T(t/2+1)(\mathbf{u}_0,\rho_0)\|_{\mathcal{H}_p}\\
&\le C_pC_{p,q}(t/2)^{-\frac{N}{2p}}(t/2)^{-\frac N2\left(\frac1q-\frac1p\right)}\|(\mathbf{u}_0,\rho_0)\|_{\mathcal{H}_q}\\
&\le C_{p,q}t^{-\frac{N}{2q}}\|(\mathbf{u}_0,\rho_0)\|_{\mathcal{H}_q}
\end{aligned}
$$

for any $t \ge 3$ and $(\mathbf{u}_0, \rho_0) \in J_q(\dot\Omega) \times W_q^{2-1/q}(S_R)$ provided $1 < q \le p < \infty$. This completes the proof of Theorem 3.7.1.

Acknowledgements Yoshihiro Shibata was partially supported by JSPS Grant-in-aid for Scientific Research (A) 17H0109, and Top Global University Project. Hirokazu Saito was partially supported by JSPS Grant-in-aid for Young Scientists (B) 17K14224.

Bibliography

1. H. Abels, *On generalized solutions of two-phase flows for viscous incompressible fluids*, Interfaces Free Bound. **9** (1) (2007), 31–65.
2. I. V. Denisova, *Problem of the motion of two viscous incompressible fluids separated by a closed free interface*, Acta Appl. Math., **37** (1-2) (1994), 31–40.
3. I. V. Denisova, *The problem of the motion of two compressible fluids separated by a closed free surface*, Zap. Nauchn. Sem. S.-Peterburg. Otdel. Mat. Inst. Steklov. (POMI), **243** (1997), 61–86; English transl.: J. Math. Sci. (N. Y.), **99** (1) (2000), 837–853.
4. I. V. Denisova, *Evolution of compressible and incompressible fluids separated by a closed interface*, Interfaces Free Bound. **2** (3) (2000), 283–312.
5. I. V. Denisova, *Evolution of a closed interface between two liquids of different types*, European Congress of Mathematics, Vol. II (Barcelona, 2000), 263–272, Progr. Math., **202**, Birkhäuser, Basel, 2001.
6. I. V. Denisova, *Solvability in weighted Hölder spaces for a problem governing the evolution of two compressible fluids*, Zap. Nauchn. Sem. S.-Peterburg. Otdel. Mat. Inst. Steklov. (POMI), **295** (2003), 57–89; English transl.: J. Math. Sci. (N. Y.) **127** (2) (2005), 1849–1868.
7. I. V. Denisova, *On the problem of thermocapillary convection for two incompressible fluids separated by a closed interface*, In Trends in Partial Differential Equations of Mathematical

Physics, Volume 61 of Progr. Nonlinear Differential Equations Appl., pages 45–64. Birkhäuser, Basel, 2005.
8. I. V. Denisova, *Global solvability of a problem on two fluid motion without surface tension*, Zap. Nauchn. Sem. S.-Peterburg. Otdel. Mat. Inst. Steklov. (POMI), **348** (2007), 19–39; English transl.: J. Math. Sci. (N.Y.) **152** (5) (2008), 625–637.
9. I. V. Denisova, *Global L_2-solvability of a problem governing two-phase fluid motion without surface tension*, Port. Math., **71** (1) (2014), 1–24.
10. I. V. Denisova and Š. Nečasová, *Oberbeck-Boussinesq approximation for the motion of two incompressible fluids*, Zap. Nauchn. Sem. S.-Peterburg. Otdel. Mat. Inst. Steklov. (POMI), **362** (2008), 92–119; English transl.: J. Math. Sci. (N. Y.), **159** (4) (2009), 436–451.
11. I. V. Denisova and V. A. Solonnikov, *Classical solvability of the problem of the motion of two viscous incompressible fluids*, Algebra i Analiz, **7** (5) (1995), 101–142; English transl.: St. Petersburg Math. J., **7** (5) (1996), 755–786.
12. I. V. Denisova and V. A. Solonnikov, *Global solvability of the problem of the motion of two incompressible capillary fluids in a container*, Zap. Nauchn. Sem. S.-Peterburg. Otdel. Mat. Inst. Steklov. (POMI), **397** (2011), 20–52; English transl.: J. Math. Sci. (N. Y.), **185** (5) (2012), 668–686.
13. G. P. Galdi, An Introduction to the Mathematical Theory of the Navier-Stokes Equations, Steady-State Problems, Second Edition, *Springer Monographs in Mathematics*, Springer 2011.
14. Y. Guo and I. Tice, *Almost exponential decay of periodic viscous surface waves without surface tension*, Arch. Ration. Mech. Anal. **207** (2) (2013), 459–531.
15. M. Hieber and H. Saito, *Strong solutions for two-phase free boundary problems for a class of non-Newtonian fluids*, J. Evol. Equ., **17** (1) (2017), 335–358.
16. H. Iwashita, *L_q-L_r estimates for solutions of the nonstationary Stokes equations in an exterior domain and the Navier-Stokes initial value problems in L_q spaces*, Math. Ann. **285** (1989), 265–288.
17. J. Jang, I. Tice, and Y. Wang, *The compressible viscous surface-internal wave problem: stability and vanishing surface tension limit*, Comm. Math. Phys. **343** (3) (2016), 1039–1113.
18. J. Jang, I. Tice, and Y. Wang, *The compressible viscous surface-internal wave problem: local well-posedness*, SIAM J. Math. Anal. **48** (4) (2016), 2602–2673.
19. T. Kubo, Y. Shibata, and K. Soga, *On some two-phase problem for compressible and compressible viscous fluid flow separated by sharp interface*, Discrete Contin. Dyn. Syst. **36** (7) (2016), 3741–3774.
20. S. Maryani and H. Saito, *On the R-boundedness of solution operator families for two-phase Stokes resolvent equations*, Differential Integral Equations **30** (1-2) (2017), 1–52.
21. J. Prüss and G. Simonett, *On the Rayleigh-Taylor instability for the two-phase Navier-Stokes equations*, Indiana Univ. Math. J., **59** (6) (2010), 1853–1871.
22. J. Prüss and G. Simonett, *On the two-phase Navier-Stokes equations with surface tension*, Interfaces Free Bound., **12** (3) (2010), 311–345.
23. J. Prüss and G. Simonett, *Analytic solutions for the two-phase Navier-Stokes equations with surface tension and gravity*, In Parabolic Problems, Volume 80 of Progr. Nonlinear Differential Equations Appl.,
24. H. Saito and X. Zhang, *Unique solvability of weak problems associated with two-phase incompressible flows*, Preprint in 2018.
25. H. Saito, Y. Shibata, and X. Zhang, *Some free boundary problem for two phase inhomogeneous incompressible flow*, submitted. arXiv:1811.02179
26. Y. Shibata, *On the $\mathcal{R}$-boundedness of solution operators for the Stokes equations with free boundary condition*, Differential Integral Equations, **27**(3-4) (2014), 313–368.
27. Y. Shibata, *On the $\mathcal{R}$-bounded solution operator and the maximal L_p-L_q regularity of the Stokes equations with free boundary condition*, Springer Proceedings in Mathematics & Statistics Vol. 183 2016, Mathematical Fluid Dynamics, Present and Future, Tokyo, Japan, November 204, ed. Y. Shibata and Y. Suzuki, pp.203–285.

28. Y. Shibata, *On the L_p-L_q decay estimate for the Stokes equations with free boundary conditions in an exterior domain*, Asymptotic Analysis **107** (1-2) (2018), 33–72, DOI: 10.3233/ASY-171449
29. Y. Shibata, *$\mathcal{R}$ boundedness, Maximal Regularity and Free Boundary Problems for the Navier Stokes Equations*, Lecture Notes in Mathematics 2254 (2020), Springer.
30. Y. Shibata and S. Shimizu, *On a resolvent estimate of the interface problem for the Stokes system in a bounded domain*, J. Differential Equations, 191(2) (2003), 408–444.
31. Y. Shibata and S. Shimizu, *Decay properties of the Stokes semigroup in exterior domains with Neumann boundary condition*, J. Math. Soc. Japan, **59** (1) (2007), 1–34.
32. Y. Shibata and S. Shimizu, *On a free boundary problem for the Navier-Stokes equations*, Differential Integral Equations, **20**(3) (2007), 241–276.
33. Y. Shibata and S. Shimizu, *On the L_p-L_q maximal regularity of the Neumann problem for the Stokes equations in a bounded domain*, J. reine angew. Math. **615** (2008), 157–209. DOI 10.1515/CRELLE.2008.013
34. Y. Shibata and S. Shimizu, *Maximal L_p-L_q regularity for the two-phase Stokes equations*; Model problems, J. Differential Equations, 251(2) (2011), 373–419.
35. H. Tanabe, *Functional analytic methods for partial differential equations*, Pure and Appl. Math.. A series of monographs and text books. Marcel Dekker, New York and Basel, 1997. ISBN 0-8247-9774-4.
36. N. Tanaka, *Global existence of two phase nonhomogeneous viscous incompressible fluid flow*, Comm. Partial Differential Equations, **18** (1-2) (1993), 41–81.
37. N. Tanaka, *Two-phase free boundary problem for viscous incompressible thermocapillary convection*, Japan. J. Math. (N. S.), **21** (1) (1995), 1–42.
38. B. Vainberg, *Asymptotic Methods in Equations of Mathematical Physics*, In Russian Moscow Univ. Press, 182; Gordon and Breach Publishers, New York, London, Paris, Montreux, Tokyo, 1989; English translation.
39. Y. Wang, I. Tice, and C. Kim, *The viscous surface-internal wave problem: global well-posedness and decay*, Arch. Rational Mech. Anal.,**212** 81) (2014), 1–92.
40. L. Weis, *Operator-valued Fourier multiplier theorems and maximal L_p-regularity*. Math. Ann. **319** (2001), 735–758.
41. L. Xu and Z. Zhang, *On the free boundary problem to the two viscous immiscible fluids*, J. Differential Equations **248** (5) (2010), 1044–1111.

Chapter 4
The Role of Pressure in the Theory of Weak Solutions to the Navier-Stokes Equations

Jiří Neustupa

Abstract Sections 4.1 and 4.2 contain an introduction, notation and definitions and basic properties of used function spaces and operators. A pressure, associated with a weak solution to the Navier-Stokes equations for incompressible fluid, is constructed in Sect. 4.3. The interior regularity of the pressure in regions, where the velocity satisfies Serrin's integrability conditions, is studied in Sect. 4.4. Finally, Sect. 4.5 is devoted to criteria of regularity for weak solutions to the Navier-Stokes equations, formulated in terms of the pressure.

Keywords Viscous incompressible fluid · Navier-Stokes equations · Weak solutions · Associated pressure · Regularity

MSC2010: 35Q30, 35Q35, 76D03, 76D05

4.1 Introduction

Although the pressure is one of the unknowns in the Navier-Stokes equations for an incompressible fluid, it does not explicitly appear in the weak formulation of the initial-boundary value problem for these equations. We explain how an appropriate pressure, as a distribution, can be associated with a weak solution of the Navier-Stokes equations. If the equations are considered in a "smooth" spatial domain Ω, then the associated pressure is a function with a certain rate of the space-time integrability. We give a brief description of related results with main ideas of the proofs in Sect. 4.3.3. Moreover, we also show in Sect. 4.3.4 how a certain pressure can be associated with a weak solution to more general classes of equations, including, e.g., equations of motion of some non-Newtonian incompressible fluids.

J. Neustupa (✉)
Institute of Mathematics CAS, Prague, Czech Republic
e-mail: neustupa@math.cas.cz

T. Bodnár et al. (eds.), *Fluids Under Pressure*, Advances in Mathematical Fluid Mechanics, https://doi.org/10.1007/978-3-030-39639-8_4

The classical Ohyama's and Serrin's results from the 60s of the 20th century provide an information on the interior spatial regularity of a weak solution to the Navier-Stokes equations in regions, where the solution satisfies a certain additional condition of integrability (the so-called Serrin's condition). In Sect. 4.4, we explain what is known about the regularity of the associated pressure in these regions, how the regularity of pressure is related to the regularity of the time-derivative of velocity, and we also show that the question of interior regularity of pressure is closely connected with the type of boundary conditions, satisfied by velocity.

One of the main open problems in the theory of the Navier-Stokes equations is the question of regularity of a weak solution. There exist many results (the so-called regularity criteria), showing that the weak solution is regular under some additional (a posteriori) assumptions. We give a survey of criteria that impose a posteriori conditions on the associated pressure and we describe main ideas of the proofs of some of them in Sect. 4.5.

There still remain several other fields in the theory of the Navier-Stokes or closely related equations where the pressure plays an important role, but we do not touch them in this text. One of these fields is the use of pressure in boundary conditions. Interested readers can find many papers on numerical solution of the Navier-Stokes equations with the so called "pressure boundary conditions" in literature. Of theoretically oriented papers on boundary conditions involving the pressure, we cite, e.g., [53] (by S. Marušić—the unsteady case) and [2] (by Ch. Amrouche and N. Seloula—the steady case). The pressure has also been used in outflow conditions on artificial boundaries, see, e.g., [43] (by P. Kučera and Z. Skalák—steady flow in a channel), [44] (by P. Kučera—unsteady flow in a channel), [25, 26] (by M. Feistauer and T. Neustupa—flows through profile cascades) and [41, 42] (by S. Kračmar and J. Neustupa—steady and unsteady flow in a channel, modeled by means of variational inequalities of the Navier-Stokes type). Another interesting field are equations of motion of an incompressible fluid with a pressure dependent viscosity. Of a series of papers on this theme, we cite, e.g., [32] and [33] (by F. Gazzola, respectively, F. Gazzola and P. Secchi—opening qualitative results), [51] (by J. Málek, J. Nečas and K. R. Rajagopal–first qualitative results on fluids with pressure and shear-rate dependent viscosity), [52] (by J. Málek and K. R. Rajagopal—survey chapter in a handbook, containing derivation as well as mathematical analysis of the models) and [48] (by M. Lanzendörfer and J. Stebel—pressure and shear-rate dependent viscosity in combination with pressure boundary conditions). One can observe in scientific literature that a growing attention is being devoted in recent years to mathematical models, based on the Navier-Stokes equations, of flows in domains with time-varying boundaries, e.g., around rotating bodies. Naturally, the pressure and its estimates play a fundamental role in these models. Related questions are discussed, e.g., in [31] (by G. P. Galdi and J. Neustupa—motion of the Navier-Stokes fluid around a moving body) and [56] (by Š. Nečasová and J. Wolf—motion of a rigid body in a viscous fluid). Here, the readers can also find an extensive list of further references.

4.2 Notation and Basic Properties of Relevant Function Spaces and Their Duals

Let Ω be a domain in $\mathbb{R}^3$. Although most of the notions and results, presented in this section, also make sense or hold if $\Omega \subset \mathbb{R}^N$ (for $N \geq 2$), we focus on the case $N = 3$ mainly because we study the pressure associated with a weak solution to the Navier-Stokes equations, and the weak solution and its properties are especially important in three space dimensions. This is caused by the fact the Navier-Stokes equations are physically meaningful if $N = 2$ or $N = 3$, and on the other hand, if $N \geq 3$ then the weak solution is the only type of solution that is known to exist on an arbitrarily long time interval without any restriction on the size of given data. (The question of regularity of a weak solution, i.e. the question whether the weak solution coincides with a strong or classical solution if all the given data are sufficiently "smooth," is still open, see, e.g., [30].)

A) Notation We use this notation of functions, function spaces, dual spaces, etc.:

- The outer normal vector to $\partial\Omega$ is denoted by $\mathbf{n}$.
- Vector functions and spaces of vector functions are denoted by boldface letters.
- $\mathbf{C}^{\infty}_{0,\sigma}(\Omega)$ denotes the linear space of infinitely differentiable divergence-free vector functions in Ω, with a compact support in Ω.
- Let $1 < q < \infty$. We denote by $\mathbf{L}^q_\sigma(\Omega)$ the closure of $\mathbf{C}^{\infty}_{0,\sigma}(\Omega)$ in $\mathbf{L}^q(\Omega)$.
- $\mathbf{W}^{1,q}_{0,\sigma}(\Omega)$ denotes the closure of $\mathbf{C}^{\infty}_{0,\sigma}(\Omega)$ in $\mathbf{W}^{1,q}(\Omega)$. (The space $\mathbf{W}^{1,q}_{0,\sigma}(\Omega)$ generally satisfies the inclusion $\mathbf{W}^{1,q}_{0,\sigma}(\Omega) \subset \{\mathbf{v} \in \mathbf{W}^{1,q}_0(\Omega);\ \operatorname{div}\mathbf{v} = 0 \text{ a.e. in } \Omega\}$. The equality holds, e.g., if Ω has a bounded locally Lipschitz boundary or Ω is a half-space, see [29, Sec. III.4] for more details.)
- The norm in $L^q(\Omega)$ and in $\mathbf{L}^q(\Omega)$ is denoted by $\|\,.\,\|_q$. The norm in $W^{k,q}(\Omega)$ and in $\mathbf{W}^{k,q}(\Omega)$ (for $k \in \mathbb{N}$) is denoted by $\|\,.\,\|_{k,q}$. If the considered domain differs from Ω, then we use, e.g., the notation $\|\,.\,\|_{q;\,\Omega'}$, etc. The scalar product in $L^2(\Omega)$ and in $\mathbf{L}^2(\Omega)$ is denoted by $(\,.\,,\,.\,)_2$ and the scalar product in $W^{1,2}(\Omega)$ and in $\mathbf{W}^{1,2}(\Omega)$ is denoted by $(\,.\,,\,.\,)_{1,2}$.
- The conjugate exponent is denoted by prime, so that, e.g., $q' = q/(q-1)$. $\mathbf{W}^{-1,q'}_0(\Omega)$ denotes the dual space to $\mathbf{W}^{1,q}_0(\Omega)$ and $\mathbf{W}^{-1,q'}_{0,\sigma}(\Omega)$ denotes the dual space to $\mathbf{W}^{1,q}_{0,\sigma}(\Omega)$. The norm in $\mathbf{W}^{-1,q'}_0(\Omega)$, respectively, $\mathbf{W}^{-1,q'}_{0,\sigma}(\Omega)$, is denoted by $\|\,.\,\|_{-1,q'}$, respectively, by $\|\,.\,\|_{-1,q';\,\sigma}$.
- The duality between elements of $\mathbf{W}^{-1,q'}_0(\Omega)$ and $\mathbf{W}^{1,q}_0(\Omega)$ is denoted by $\langle\,.\,,\,.\,\rangle_\Omega$ and the duality between elements of $\mathbf{W}^{-1,q'}_{0,\sigma}(\Omega)$ and $\mathbf{W}^{1,q}_{0,\sigma}(\Omega)$ is denoted by $\langle\,.\,,\,.\,\rangle_{\Omega,\sigma}$.
- $\mathbf{W}^{1,q}_{0,\sigma}(\Omega)^\perp$ denotes the space of annihilators of $\mathbf{W}^{1,q}_{0,\sigma}(\Omega)$ in $\mathbf{W}^{-1,q'}_0(\Omega)$. i.e. the space $\big\{\mathbf{f} \in \mathbf{W}^{-1,q'}_0(\Omega);\ \forall \boldsymbol{\varphi} \in \mathbf{W}^{1,q}_{0,\sigma}(\Omega) : \langle \mathbf{f}, \boldsymbol{\varphi}\rangle_\Omega = 0\big\}$.

B) $\mathbf{L}^{q'}(\Omega)$ and $\mathbf{L}^{q'}_\sigma(\Omega)$ as Subspaces of $\mathbf{W}_0^{-1,q'}(\Omega)$ and $\mathbf{W}_{0,\sigma}^{-1,q'}(\Omega)$, Respectively

The Lebesgue space $\mathbf{L}^{q'}(\Omega)$ can be identified with a subspace of $\mathbf{W}_0^{-1,q'}(\Omega)$ so that if $\mathbf{f} \in \mathbf{L}^{q'}(\Omega)$ then

$$\langle \mathbf{f}, \boldsymbol{\varphi} \rangle_\Omega := \int_\Omega \mathbf{f} \cdot \boldsymbol{\varphi} \, \mathrm{d}\mathbf{x} \tag{4.1}$$

for all $\boldsymbol{\varphi} \in \mathbf{W}_0^{1,q}(\Omega)$. Similarly, $\mathbf{L}^{q'}_\sigma(\Omega)$ can be identified with a subspace of $\mathbf{W}_{0,\sigma}^{-1,q'}(\Omega)$ so that if $\mathbf{f} \in \mathbf{L}^{q'}_\sigma(\Omega)$ then

$$\langle \mathbf{f}, \boldsymbol{\varphi} \rangle_{\Omega,\sigma} := \int_\Omega \mathbf{f} \cdot \boldsymbol{\varphi} \, \mathrm{d}\mathbf{x} \tag{4.2}$$

for all $\boldsymbol{\varphi} \in \mathbf{W}_{0,\sigma}^{1,q}(\Omega)$. We observe that if $\mathbf{f} \in \mathbf{L}^{q'}_\sigma(\Omega)$ and $\boldsymbol{\varphi} \in \mathbf{W}_{0,\sigma}^{1,q}(\Omega)$ then the dualities $\langle \mathbf{f}, \boldsymbol{\varphi} \rangle_\Omega$ and $\langle \mathbf{f}, \boldsymbol{\varphi} \rangle_{\Omega,\sigma}$ coincide, because they are expressed by the same integral.

Note that if $\mathbf{f} \in \mathbf{L}^{q'}(\Omega)$ then the integral on the right-hand side of (4.1) also defines a bounded linear functional on $\mathbf{W}_{0,\sigma}^{1,q}(\Omega)$. This, however, does not mean that $\mathbf{L}^{q'}(\Omega)$ can be identified with a subspace of $\mathbf{W}_{0,\sigma}^{-1,q'}(\Omega)$. The reason is, for instance, that the spaces $\mathbf{L}^{q'}(\Omega)$ and $\mathbf{W}_{0,\sigma}^{-1,q'}(\Omega)$ do not have the same zero element. (If ψ is a non-constant function in $C_0^\infty(\Omega)$, then $\nabla\psi$ is a non-zero element of $\mathbf{L}^{q'}(\Omega)$, but it induces the zero element of $\mathbf{W}_{0,\sigma}^{-1,q'}(\Omega)$.)

C) Definition and Properties of Operator $\mathcal{P}_{q'}$ $\mathbf{W}_{0,\sigma}^{1,q}(\Omega)$ is a closed subspace of $\mathbf{W}_0^{1,q}(\Omega)$. If $\mathbf{f} \in \mathbf{W}_0^{-1,q'}(\Omega)$ (i.e., $\mathbf{f}$ is a bounded linear functional on $\mathbf{W}_0^{1,q}(\Omega)$), then we denote by $\mathcal{P}_{q'}\mathbf{f}$ the restriction of $\mathbf{f}$ to $\mathbf{W}_{0,\sigma}^{1,q}(\Omega)$. Thus, $\mathcal{P}_{q'}\mathbf{f}$ is an element of $\mathbf{W}_{0,\sigma}^{-1,q'}(\Omega)$, defined by the equation

$$\langle \mathcal{P}_{q'}\mathbf{f}, \boldsymbol{\varphi} \rangle_{\Omega,\sigma} := \langle \mathbf{f}, \boldsymbol{\varphi} \rangle_\Omega \qquad \text{for all } \boldsymbol{\varphi} \in \mathbf{W}_{0,\sigma}^{1,q}(\Omega). \tag{4.3}$$

Note that some authors denote the restriction of $\mathbf{f}$ to $\mathbf{W}_{0,\sigma}^{1,q}(\Omega)$ (which is an element of $\mathbf{W}_{0,\sigma}^{-1,q'}(\Omega)$) by the same symbol $\mathbf{f}$. However, since this may lead to inaccuracies, we prefer to use a different notation for $\mathbf{f}$ and the restriction. Due to similar reasons, we also use a different notation for the duality between $\mathbf{W}_0^{-1,q'}(\Omega)$ and $\mathbf{W}_0^{1,q}(\Omega)$ and the duality between $\mathbf{W}_{0,\sigma}^{-1,q'}(\Omega)$ and $\mathbf{W}_{0,\sigma}^{1,q}(\Omega)$.

Obviously, $\mathcal{P}_{q'}$ is a linear operator from $\mathbf{W}_0^{-1,q'}(\Omega)$ to $\mathbf{W}_{0,\sigma}^{-1,q'}(\Omega)$, whose domain is the whole space $\mathbf{W}_0^{-1,q'}(\Omega)$.

Lemma 4.2.1 *The operator $\mathcal{P}_{q'}$ is bounded, its range is $\mathbf{W}_{0,\sigma}^{-1,q'}(\Omega)$ and $\mathcal{P}_{q'}$ is not one-to-one.*

Proof The proof of the boundedness of $\mathcal{P}_{q'}$ is a simple exercise: let $\mathbf{f} \in \mathbf{W}_0^{-1,q'}(\Omega)$. Then

$$\begin{aligned} \|\mathcal{P}_{q'}\mathbf{f}\|_{-1,q';\sigma} &= \sup_{\boldsymbol{\varphi}\in\mathbf{W}_{0,\sigma}^{1,q}(\Omega);\ \boldsymbol{\varphi}\neq\mathbf{0}} \frac{|\langle\mathcal{P}_{q'}\mathbf{f},\boldsymbol{\varphi}\rangle_{\Omega,\sigma}|}{\|\boldsymbol{\varphi}\|_{1,q}} = \sup_{\boldsymbol{\varphi}\in\mathbf{W}_{0,\sigma}^{1,q}(\Omega);\ \boldsymbol{\varphi}\neq\mathbf{0}} \frac{|\langle\mathbf{f},\boldsymbol{\varphi}\rangle_{\Omega}|}{\|\boldsymbol{\varphi}\|_{1,q}} \\ &\leq \sup_{\boldsymbol{\varphi}\in\mathbf{W}_{0}^{1,q}(\Omega);\ \boldsymbol{\varphi}\neq\mathbf{0}} \frac{|\langle\mathbf{f},\boldsymbol{\varphi}\rangle_{\Omega}|}{\|\boldsymbol{\varphi}\|_{1,q}} = \|\mathbf{f}\|_{-1,q'}. \end{aligned}$$

This implies the boundedness of $\mathcal{P}_{q'}$.

Let $\mathbf{g} \in \mathbf{W}_{0,\sigma}^{-1,q'}(\Omega)$. There exists (by the Hahn-Banach theorem) an extension of $\mathbf{g}$ from $\mathbf{W}_{0,\sigma}^{1,q}(\Omega)$ to $\mathbf{W}_0^{1,q}(\Omega)$, which we denote by $\mathbf{g}_{\text{ext}}$. The extension is an element of $\mathbf{W}_0^{-1,q'}(\Omega)$, satisfying $\|\mathbf{g}_{\text{ext}}\|_{-1,q'} = \|\mathbf{g}\|_{-1,q';\sigma}$ and

$$\langle\mathbf{g}_{\text{ext}},\boldsymbol{\varphi}\rangle_{\Omega} = \langle\mathbf{g},\boldsymbol{\varphi}\rangle_{\Omega,\sigma} \tag{4.4}$$

for all $\boldsymbol{\varphi} \in \mathbf{W}_{0,\sigma}^{1,q}(\Omega)$. This shows that $\mathbf{g} = \mathcal{P}_{q'}(\mathbf{g}_{\text{ext}})$. Consequently, $\mathbf{W}_{0,\sigma}^{-1,q'}(\Omega) = R(\mathcal{P}_{q'})$ (the range of $\mathcal{P}_{q'}$).

Finally, taking $\mathbf{f} = \nabla g$ for $g \in C_0^{\infty}(\Omega)$, we get

$$\langle\mathcal{P}_{q'}\mathbf{f},\boldsymbol{\varphi}\rangle_{\Omega,\sigma} = \langle\mathbf{f},\boldsymbol{\varphi}\rangle_{\Omega} = \int_{\Omega}\nabla q\cdot\boldsymbol{\varphi}\,\mathrm{d}\mathbf{x} = 0$$

for all $\boldsymbol{\varphi} \in \mathbf{W}_{0,\sigma}^{1,q}(\Omega)$. This shows that the operator $\mathcal{P}_{q'}$ is not one-to-one. □

In order to derive an explicit formula for $\mathcal{P}_{q'}$, we define the mapping $\mathcal{S}_{q'}$: $\mathbf{W}_{0,\sigma}^{-1,q'}(\Omega) \to \mathbf{W}_0^{-1,q'}(\Omega)\big|_{\mathbf{W}_{0,\sigma}^{1,q}(\Omega)^{\perp}}$ (the quotient space) by the equation

$$\mathcal{S}_{q'}(\mathbf{g}) := \mathbf{g}_{\text{ext}} + \mathbf{W}_{0,\sigma}^{1,q}(\Omega)^{\perp}$$

(for $\mathbf{g} \in \mathbf{W}_{0,\sigma}^{1,q}(\Omega)^{\perp}$). The definition of $\mathcal{S}_{q'}$ is independent of a concrete extension of $\mathbf{g} \in \mathbf{W}_{0,\sigma}^{1,q}(\Omega)$ to $\mathbf{g}_{\text{ext}} \in \mathbf{W}_0^{-1,q'}(\Omega)$ due to these reasons: let $\mathbf{g}'_{\text{ext}}$ and $\mathbf{g}''_{\text{ext}}$ be two such extensions. They coincide on $\mathbf{W}_{0,\sigma}^{1,q}(\Omega)$, hence $\mathbf{g}'_{\text{ext}} - \mathbf{g}''_{\text{ext}} \in \mathbf{W}_{0,\sigma}^{1,q}(\Omega)^{\perp}$. Denote by $\mathcal{S}'_{q'}$ the mapping defined by means of extension $\mathbf{g}'_{\text{ext}}$ and by $\mathcal{S}''_{q'}$ the mapping defined by means of extension $\mathbf{g}''_{\text{ext}}$. Then, for $\mathbf{g} \in \mathbf{W}_{0,\sigma}^{-1,q'}(\Omega)$, we have

$$\mathcal{S}'_{q'}(\mathbf{g}) - \mathcal{S}''_{q'}(\mathbf{g}) = (\mathbf{g}'_{\text{ext}} - \mathbf{g}''_{\text{ext}}) + \mathbf{W}_{0,\sigma}^{1,q}(\Omega)^{\perp} = \mathbf{W}_{0,\sigma}^{1,q}(\Omega)^{\perp},$$

which is the zero element of the quotient space $\mathbf{W}_0^{-1,q'}(\Omega)\big|_{\mathbf{W}_{0,\sigma}^{1,q}(\Omega)^\perp}$.

The next lemma directly follows from Theorem 4.9 in [64]:

Lemma 4.2.2 *The mapping $\mathcal{S}_{q'}$ is an isometric isomorphism of $\mathbf{W}_{0,\sigma}^{-1,q'}(\Omega)$ onto $\mathbf{W}_0^{-1,q'}(\Omega)\big|_{\mathbf{W}_{0,\sigma}^{1,q}(\Omega)^\perp}$.*

Recall that the so-called *quotient mapping* $\mathcal{Q}_{q'}$ of $\mathbf{W}_0^{-1,q'}(\Omega)$ to $\mathbf{W}_0^{-1,q'}(\Omega)\big|_{\mathbf{W}_{0,\sigma}^{1,q}(\Omega)^\perp}$ is defined by the equation

$$\mathcal{Q}_{q'}(\mathbf{f}) := \mathbf{f} + \mathbf{W}_{0,\sigma}^{1,q}(\Omega)^\perp \qquad \text{for } \mathbf{f} \in \mathbf{W}_0^{-1,q'}(\Omega).$$

Naturally, the range of the mapping $\mathcal{Q}_{q'}$ is the whole quotient space $\mathbf{W}_0^{-1,q'}(\Omega)\big|_{\mathbf{W}_{0,\sigma}^{1,q}(\Omega)^\perp}$. Moreover, we also have $\mathbf{W}_0^{-1,q'}(\Omega)\big|_{\mathbf{W}_{0,\sigma}^{1,q}(\Omega)^\perp} = \mathcal{S}_{q'}(\mathbf{W}_{0,\sigma}^{-1,q'}(\Omega))$. Hence $\mathcal{S}_{q'}^{-1}\mathcal{Q}_{q'}$ maps $\mathbf{W}_0^{-1,q'}(\Omega)$ onto $\mathbf{W}_{0,\sigma}^{-1,q'}(\Omega)$. Concretely, if $\mathbf{f} \in \mathbf{W}_0^{-1,q'}(\Omega)$, then $\mathcal{S}_{q'}^{-1}\mathcal{Q}_{q'}(\mathbf{f})$ is an element $\mathbf{g} \in \mathbf{W}_{0,\sigma}^{-1,q'}(\Omega)$ such that $\mathcal{S}_{q'}(\mathbf{g}) = \mathcal{Q}_{q'}(\mathbf{f})$, which means that

$$\mathbf{g}_{\text{ext}} + \mathbf{W}_{0,\sigma}^{1,q}(\Omega)^\perp = \mathbf{f} + \mathbf{W}_{0,\sigma}^{1,q'}(\Omega)^\perp.$$

Hence $\mathbf{g}_{\text{ext}} - \mathbf{f} \in \mathbf{W}_{0,\sigma}^{1,q}(\Omega)^\perp$. Consequently,

$$\langle \mathbf{f}, \boldsymbol{\varphi} \rangle_\Omega = \langle \mathbf{g}_{\text{ext}}, \boldsymbol{\varphi} \rangle_\Omega = \langle \mathbf{g}, \boldsymbol{\varphi} \rangle_{\Omega,\sigma},$$

for all $\boldsymbol{\varphi} \in \mathbf{W}_{0,\sigma}^{1,q}(\Omega)$. This yields $\mathbf{g} = \mathcal{P}_{q'}\mathbf{f}$. Since $\mathbf{g}$ also equals $\mathcal{S}_{q'}^{-1}\mathcal{Q}_{q'}(\mathbf{f})$, we have proven the lemma:

Lemma 4.2.3 *The operator $\mathcal{P}_{q'}$ can be explicitly expressed by the formula*

$$\mathcal{P}_{q'} = \mathcal{S}_{q'}^{-1}\mathcal{Q}_{q'}. \tag{4.5}$$

D) More on the Space $\mathbf{W}_{0,\sigma}^{1,q}(\Omega)^\perp$ Put $\mathbf{G}_{q'}(\Omega) := \{\nabla\psi \in \mathbf{L}^{q'}(\Omega);\ \psi \in W_{loc}^{1,q'}(\Omega)\}$. $\mathbf{G}_{q'}(\Omega)$ is a closed subspace of $\mathbf{L}^{q'}(\Omega)$, see [29, Exercise III.1.2]. Identifying $\mathbf{G}_{q'}(\Omega)$ with a subspace of $\mathbf{W}_0^{-1,q'}(\Omega)$ in the sense explained in Sect. 4.2B, *we denote by ${}^\perp\mathbf{G}_{q'}(\Omega)$ the linear space $\{\boldsymbol{\varphi} \in \mathbf{W}_0^{1,q}(\Omega);\ \forall \mathbf{f} \in \mathbf{G}_{q'}(\Omega) : \langle \mathbf{f}, \boldsymbol{\varphi} \rangle_\Omega = 0\}$.* It follows from [29, Lemma III.2.1] that $\mathbf{W}_{0,\sigma}^{1,q}(\Omega) = {}^\perp\mathbf{G}_{q'}(\Omega)$. Hence $\mathbf{W}_{0,\sigma}^{1,q}(\Omega)^\perp = ({}^\perp\mathbf{G}_{q'}(\Omega))^\perp$ and applying Theorem 4.7 in [64], we can characterize $\mathbf{W}_{0,\sigma}^{1,q}(\Omega)^\perp$ as a closure of $\mathbf{G}_{q'}(\Omega)$ in the weak-$*$ topology of $\mathbf{W}_0^{-1,q'}(\Omega)$.

The next lemma tells us more on elements of $\mathbf{W}^{1,q}_{0,\sigma}(\Omega)^{\perp}$ in the case of a bounded Lipschitz domain Ω. It comes from [75, Lemma II.2.2.2].

Lemma 4.2.4 *Let Ω be a bounded Lipschitz domain in $\mathbb{R}^3$, Ω_0 be a nonempty sub-domain of Ω, $1 < q < \infty$ and $\mathbf{f}$ be a bounded linear functional on $\mathbf{W}^{1,q}_0(\Omega)$ (i.e., an element of $\mathbf{W}^{-1,q'}_0(\Omega)$) that vanishes on $\mathbf{W}^{1,q}_{0,\sigma}(\Omega)$ (which means that $\mathbf{f} \in \mathbf{W}^{1,q}_{0,\sigma}(\Omega)^{\perp}$). Then there exists a unique function $p \in L^{q'}(\Omega)$ such that $\int_{\Omega_0} p \, \mathrm{d}\mathbf{x} = 0$,*

$$\langle \mathbf{f}, \boldsymbol{\psi} \rangle_{\Omega} = -\int_{\Omega} p \, \mathrm{div}\, \boldsymbol{\psi} \, \mathrm{d}\mathbf{x} \tag{4.6}$$

for all $\boldsymbol{\psi} \in \mathbf{W}^{1,q}_0(\Omega)$ and

$$\|p\|_{q'} \leq c \, \|\mathbf{f}\|_{-1,q'} \tag{4.7}$$

where $c = c(q, \Omega_0, \Omega)$.

Formula (4.6) shows that $\mathbf{f} = \nabla p$, where operator ∇ acts on p in the sense of distributions. Thus, we may symbolically write $\mathbf{W}^{1,q}_{0,\sigma}(\Omega)^{\perp} = \nabla(L^{q'}(\Omega))$.

Denote by $L^{q'}_{pot}(\Omega)$ the set of all $p \in L^{q'}_{loc}(\Omega)$ such that $\nabla p \in \mathbf{W}^{-1,q'}_0(\Omega)$. The next lemma follows from Lemma 4.2.4 and it provides more information on $\mathbf{W}^{1,q}_{0,\sigma}(\Omega)^{\perp}$ in the case of an arbitrary domain Ω in $\mathbb{R}^3$.

Lemma 4.2.5 *If Ω is an arbitrary domain in $\mathbb{R}^3$, $\mathbf{f} \in \mathbf{W}^{1,q}_{0,\sigma}(\Omega)^{\perp}$ and $\Omega_0 \subset\subset \Omega$ is a nonempty sub-domain of Ω, then there is a unique $p \in L^{q'}_{pot}(\Omega)$ such that $\mathbf{f} = \nabla p$ (the distributional gradient of p) and $\int_{\Omega_0} p \, \mathrm{d}\mathbf{x} = 0$.*

(Here and further on, $\Omega_0 \subset\subset \Omega$ means that Ω_0 is a bounded sub-domain of Ω such that $\overline{\Omega_0} \subset \Omega$.)

Lemma 4.2.5 shows that $\mathbf{W}^{1,q}_{0,\sigma}(\Omega)^{\perp} \subset \nabla(L^{q'}_{pot}(\Omega))$. Since the opposite inclusion is obvious, we obtain the equality $\mathbf{W}^{1,q}_{0,\sigma}(\Omega)^{\perp} = \nabla(L^{q'}_{pot}(\Omega))$.

Proof of Lemma 4.2.5 The domain Ω can be expressed as a countable union $\Omega = \bigcup_{k=1}^{\infty} \Omega_k$, where Ω_k are bounded Lipschitz domains, such that they create an "increasing" sequence in the sense that $\Omega_k \subset \Omega_{k+1}$ for all $k \in \mathbb{N}$. (See [75, Lemma II.1.4.1].) We can assume without loss of generality that $\Omega_0 \subset \Omega_1$. Since $\mathbf{W}^{1,q}_{0,\sigma}(\Omega_k) \hookrightarrow \mathbf{W}^{1,q}_{0,\sigma}(\Omega)$, we have $\mathbf{f} \in \mathbf{W}^{1,q}_{0,\sigma}(\Omega_k)^{\perp}$. Thus, due to Lemma 4.2.4, there exists a unique $p_k \in L^{q'}(\Omega_k)$ such that $\int_{\Omega_0} p_k \, \mathrm{d}\mathbf{x} = 0$ and

$$\langle \mathbf{f}, \boldsymbol{\psi} \rangle_{\Omega} = -\int_{\Omega_k} p_k \, \mathrm{div}\, \boldsymbol{\psi} \, \mathrm{d}\mathbf{x}$$

for all $\boldsymbol{\psi} \in \mathbf{W}_0^{1,q}(\Omega_k)$. (Function $\boldsymbol{\psi}$, extended by zero to $\Omega \setminus \Omega_k$, is an element of $\mathbf{W}_0^{1,q}(\Omega)$, hence the duality $\langle \mathbf{f}, \boldsymbol{\psi} \rangle_\Omega$ has a sense.) As the same formula also holds with $k + 1$ instead of k we deduce that $p_{k+1} = p_k$ in Ω_k. Setting $p := p_k$ for $k \in \Omega_k$, we observe that $p \in L^{q'}_{loc}(\Omega)$, $\int_{\Omega_0} p \, \mathrm{d}\mathbf{x} = 0$ and $\nabla p = \mathbf{f}$ in Ω in the sense of distributions. □

E) The Helmholtz Decomposition and the Relation Between the Helmholtz Projection and Operator $\mathcal{P}_{q'}$ If each function $\mathbf{g} \in \mathbf{L}^{q'}(\Omega)$ can be uniquely expressed in the form $\mathbf{g} = \mathbf{v} + \nabla\psi$ for some $\mathbf{v} \in \mathbf{L}^{q'}_\sigma(\Omega)$ and $\nabla\psi \in \mathbf{G}_{q'}(\Omega)$, which is equivalent to the validity of the decomposition

$$\mathbf{L}^{q'}(\Omega) = \mathbf{L}^{q'}_\sigma(\Omega) \oplus \mathbf{G}_{q'}(\Omega), \tag{4.8}$$

then we write $\mathbf{v} = P_{q'}\mathbf{g}$. Decomposition (4.8) is called the *Helmholtz decomposition* (some authors also use the name *Helmholtz-Weyl decomposition*) and the operator $P_{q'}$ is called the *Helmholtz projection.* The existence of the Helmholtz decomposition is a subtle problem and it depends on exponent q' and the shape of domain Ω. It is known that if $q' = 2$ then the Helmholtz decomposition exists on an arbitrary domain Ω and P_2, respectively, $I - P_2$, is an orthogonal projection of $\mathbf{L}^2(\Omega)$ onto $\mathbf{L}^2_\sigma(\Omega)$, respectively, onto $\mathbf{G}_2(\Omega)$. If $q' \neq 2$, then the Helmholtz decomposition exists, e.g., if Ω is any domain in $\mathbb{R}^3$ of the class C^2 (see [29, p. 152]) or a bounded or exterior domain of the class C^1 (see [70]) or a bounded convex domain (see [34]). Interested readers can also find further information on the Helmholtz decomposition and projection, e.g., in [22, 27] and [40].

In the rest of this paragraph, we assume that the Helmholtz decomposition of $\mathbf{L}^{q'}(\Omega)$ exists and we explain the relation between the operators $\mathcal{P}_{q'}$ and $P_{q'}$. Let $\mathbf{g} \in \mathbf{L}^{q'}(\Omega)$. Treating $\mathbf{g}$ as an element of $\mathbf{W}_0^{-1,q'}(\Omega)$, we have

$$\langle \mathcal{P}_{q'}\mathbf{g}, \boldsymbol{\varphi} \rangle_{\Omega,\sigma} = \langle \mathbf{g}, \boldsymbol{\varphi} \rangle_\Omega$$

for all $\boldsymbol{\varphi} \in \mathbf{W}^{1,q}_{0,\sigma}(\Omega)$. Writing $\mathbf{g} = P_{q'}\mathbf{g} + (I - P_{q'})\mathbf{g}$, we also get

$$\langle \mathbf{g}, \boldsymbol{\varphi} \rangle_\Omega = \big\langle P_{q'}\mathbf{g} + (I - P_{q'})\mathbf{g}, \boldsymbol{\varphi} \big\rangle_\Omega = \big\langle P_{q'}\mathbf{g}, \boldsymbol{\varphi} \big\rangle_\Omega = \big\langle P_{q'}\mathbf{g}, \boldsymbol{\varphi} \big\rangle_{\Omega,\sigma}$$

for all $\boldsymbol{\varphi} \in \mathbf{W}^{1,q}_{0,\sigma}(\Omega)$. (The second equality holds because $(I - P_{q'})\mathbf{g} \in \mathbf{G}_{q'}(\Omega)$ and it therefore annihilates $\boldsymbol{\varphi}$. The last equality follows from the inclusions $P_{q'}\mathbf{g} \in \mathbf{L}^{q'}_\sigma(\Omega)$, $\boldsymbol{\varphi} \in \mathbf{W}^{1,q}_{0,\sigma}(\Omega)$ and formulas (4.1) and (4.2).) Consequently, $\langle \mathcal{P}_{q'}\mathbf{g}, \boldsymbol{\varphi} \rangle_{\Omega,\sigma} = \langle P_{q'}\mathbf{g}, \boldsymbol{\varphi} \rangle_{\Omega,\sigma}$ for all $\boldsymbol{\varphi} \in \mathbf{W}^{1,q}_{0,\sigma}(\Omega)$. Hence $\mathcal{P}_{q'}\mathbf{g}$ and $P_{q'}\mathbf{g}$ represent the same element of $\mathbf{W}^{-1,q'}_{0,\sigma}(\Omega)$. As $P_{q'}\mathbf{g} \in \mathbf{L}^{q'}_\sigma(\Omega)$, $\mathcal{P}_{q'}\mathbf{g}$ can also be considered to be an element of $\mathbf{L}^{q'}_\sigma(\Omega)$ (which induces a functional in $\mathbf{W}^{-1,q}_{0,\sigma}(\Omega)$ in the way explained in Sect. 4.2B). In this sense, we observe that *the Helmholtz projection $P_{q'}$ coincides with the restriction of $\mathcal{P}_{q'}$ to $\mathbf{L}^{q'}(\Omega)$.*

4.3 A Pressure Associated with a Weak Solution to the Navier-Stokes Equations

4.3.1 The Navier-Stokes Initial-Boundary Value Problem and Its Weak Formulation

Let Ω be an arbitrary domain in $\mathbb{R}^3$ and $T > 0$. We denote $Q_T := \Omega \times (0, T)$ and $\Gamma_T := \partial\Omega \times (0, T)$.

A) A Classical Form of the Navier-Stokes IBVP We consider the Navier-Stokes IBVP (i.e., the initial-boundary value problem), given by the equations

$$\partial_t \mathbf{u} + \mathbf{u} \cdot \nabla \mathbf{u} + \nabla p = \nu \Delta \mathbf{u} + \mathbf{f} \qquad \text{in } Q_T, \tag{4.9}$$

$$\operatorname{div} \mathbf{u} = 0 \qquad \text{in } Q_T, \tag{4.10}$$

the boundary condition

$$\mathbf{u} = \mathbf{0} \qquad \text{on } \Gamma_T \tag{4.11}$$

and the initial condition

$$\mathbf{u} = \mathbf{u}_0 \qquad \text{in } \Omega \times \{0\}. \tag{4.12}$$

The equations (4.9) and (4.10) describe the motion of an incompressible Newtonian fluid. The unknowns are $\mathbf{u}$ (velocity) and p (pressure). Function $\mathbf{f}$ represents an external body force and ν is the kinematic coefficient of viscosity. It is supposed to be a positive constant.

Note that ∇, Δ, and div always denote operators, acting only in the spatial variables.

B) The 1st Weak Formulation of the Navier-Stokes IBVP (4.9)–(4.12) *Given* $\mathbf{u}_0 \in \mathbf{L}^2_\sigma(\Omega)$ *and* $\mathbf{f} \in L^2(0, T;\ \mathbf{W}_0^{-1,2}(\Omega))$. *A function* $\mathbf{u} \in L^\infty(0, T;\ \mathbf{L}^2_\sigma(\Omega)) \cap L^2(0, T;\ \mathbf{W}^{1,2}_{0,\sigma}(\Omega))$ *is said to be a weak solution to the problem (4.9)–(4.12) if*

$$\int_0^T \int_\Omega \big[-\mathbf{u} \cdot \partial_t \boldsymbol{\phi} + \nu \nabla \mathbf{u} : \nabla \boldsymbol{\phi} + \mathbf{u} \cdot \nabla \mathbf{u} \cdot \boldsymbol{\phi} \big] \, \mathrm{d}\mathbf{x} \, \mathrm{d}t$$
$$= \int_\Omega \mathbf{u}_0 \cdot \boldsymbol{\phi}(\mathbf{x}, 0) \, \mathrm{d}\mathbf{x} + \int_0^T \langle \mathbf{f}, \boldsymbol{\phi} \rangle_\Omega \, \mathrm{d}t \tag{4.13}$$

for all $\boldsymbol{\phi} \in C^\infty\big([0, T];\ \mathbf{W}^{1,2}_{0,\sigma}(\Omega)\big)$ *such that* $\boldsymbol{\phi}(T) = \mathbf{0}$.

Equation (4.13) follows from (4.9), (4.10) if one formally multiplies equation (4.9) by the test function $\boldsymbol{\phi}$ with the aforementioned properties, applies the integration by parts and uses the boundary condition (4.11) and the initial condition (4.12). As the integral of $\nabla p \cdot \boldsymbol{\phi}$ vanishes, the pressure p does not explicitly appear in (4.13).

On the other hand, if $\mathbf{f} \in \mathbf{L}^2(Q_T)$ and $\mathbf{u}$ is a weak solution with the additional properties $\partial_t \mathbf{u} \in \mathbf{L}^2(Q_T)$ and $\mathbf{u} \in L^2(0, T;\, \mathbf{W}^{2,2}(\Omega))$ then, considering the test functions $\boldsymbol{\phi}$ in (4.13) of the form $\boldsymbol{\phi}(\mathbf{x}, t) = \boldsymbol{\varphi}(\mathbf{x})\, \vartheta(t)$ where $\boldsymbol{\varphi} \in \mathbf{C}^\infty_{0,\sigma}(\Omega)$ and $\vartheta \in C^\infty_0((0, T))$, and applying the backward integration by parts, one obtains the equation

$$\int_\Omega \big[\partial_t \mathbf{u} + \mathbf{u} \cdot \nabla \mathbf{u} - \nu \Delta \mathbf{u} - \mathbf{f}\big] \cdot \boldsymbol{\varphi} \, \mathrm{d}\mathbf{x} \, = \, 0$$

for a.a. $t \in (0, T)$. As $\mathbf{C}^\infty_{0,\sigma}(\Omega)$ is dense in $\mathbf{L}^2_\sigma(\Omega)$, this equation shows that $P_2[\partial_t \mathbf{u} + \mathbf{u} \cdot \nabla \mathbf{u} - \nu \Delta \mathbf{u} - \mathbf{f}] = \mathbf{0}$ at a.a. time instants $t \in (0, T)$. Consequently, to a.a. $t \in (0, T)$, there exists $p \in W^{1,2}_{loc}(\Omega)$ such that $\nabla p = (I - P_2)[\partial_t \mathbf{u} + \mathbf{u} \cdot \nabla \mathbf{u} - \nu \Delta \mathbf{u} - \mathbf{f}]$ and the functions $\mathbf{u}$ and p satisfy equation (4.9) (as an equation in $\mathbf{L}^2(\Omega)$) at a.a. time instants $t \in (0, T)$. It follows from the boundedness of projection P_2 in $\mathbf{L}^2(\Omega)$ and the assumed properties of functions $\mathbf{u}$ and $\mathbf{f}$ that $\nabla p \in \mathbf{L}^2(Q_T)$. This is a "naive" way one can reconstruct the pressure from the weak solution to the IBVP (4.9)–(4.12). We call it "naïve," because it requires the additional assumptions on the smoothness of the weak solution $\mathbf{u}$ and function $\mathbf{f}$. On a general level, however, it is not known whether these assumptions (especially the assumptions that concern $\mathbf{u}$) are satisfied, even if all the input data in the weak formulation of the IBVP (4.9)–(4.12) are arbitrarily smooth. (This is connected with the famous open problem of regularity of weak solutions to the Navier-Stokes equations.) Nevertheless, we show in Sect. 4.3.2C that one can always assign a certain pressure to the weak solution to the IBVP (4.9)–(4.12). The pressure generally exists only as a distribution and it can be represented by a function with an appropriate rate of integrability if domain Ω is "smooth," see Sect. 4.3.3.

C) The 2nd Weak Formulation of the Navier-Stokes IBVP (4.9)–(4.12) We define the operators $\mathcal{A} : \mathbf{W}^{1,2}_0(\Omega) \to \mathbf{W}^{-1,2}_0(\Omega)$ and $\mathcal{B} : \big[\mathbf{W}^{1,2}_0(\Omega)\big]^2 \to \mathbf{W}^{-1,2}_0(\Omega)$ by the equations

$$\big\langle \mathcal{A}\mathbf{v}, \boldsymbol{\varphi} \big\rangle_\Omega \, := \int_\Omega \nabla \mathbf{v} : \nabla \boldsymbol{\varphi} \, \mathrm{d}\mathbf{x} \qquad \text{for } \mathbf{v}, \boldsymbol{\varphi} \in \mathbf{W}^{1,2}_0(\Omega),$$

$$\big\langle \mathcal{B}(\mathbf{v}, \mathbf{w}), \boldsymbol{\varphi} \big\rangle_\Omega \, := \int_\Omega \mathbf{v} \cdot \nabla \mathbf{w} \cdot \boldsymbol{\varphi} \, \mathrm{d}\mathbf{x} \qquad \text{for } \mathbf{v}, \mathbf{w}, \boldsymbol{\varphi} \in \mathbf{W}^{1,2}_0(\Omega).$$

Obviously, $\mathcal{A}$ is one-to-one and

$$\|\mathcal{A}\mathbf{v}\|_{-1,2} \, \leq \, \|\nabla \mathbf{v}\|_2. \tag{4.14}$$

If $k > 0$, then the range of $\mathcal{A}+kI$ is the whole space $\mathbf{W}^{-1,2}_0(\Omega)$ (by the Lax-Milgram theorem) and $(\mathcal{A} + kI)^{-1}$ is a bounded operator from $\mathbf{W}^{-1,2}_0(\Omega)$ onto $\mathbf{W}^{1,2}_0(\Omega)$. If Ω is bounded, then the same statements also hold for $k = 0$. The bilinear operator $\mathcal{B}$ satisfies

$$
\begin{aligned}
\|\mathcal{B}(\mathbf{v},\mathbf{w})\|_{-1,2} &= \sup_{\boldsymbol{\varphi}\in \mathbf{W}_0^{1,2}(\Omega),\ \boldsymbol{\varphi}\neq \mathbf{0}} \frac{|\langle \mathcal{B}(\mathbf{v},\mathbf{w}),\boldsymbol{\varphi}\rangle_\Omega|}{\|\boldsymbol{\varphi}\|_{1,2}} \\
&= \sup_{\boldsymbol{\varphi}\in \mathbf{W}_0^{1,2}(\Omega),\ \boldsymbol{\varphi}\neq \mathbf{0}} \frac{|(\mathbf{v}\cdot\nabla\mathbf{w},\,\boldsymbol{\varphi})_2|}{\|\boldsymbol{\varphi}\|_{1,2}} \leq \sup_{\boldsymbol{\varphi}\in \mathbf{W}_0^{1,2}(\Omega),\ \boldsymbol{\varphi}\neq \mathbf{0}} \frac{\|\mathbf{v}\|_2^{1/2}\,\|\mathbf{v}\|_6^{1/2}\,\|\nabla\mathbf{w}\|_2\,\|\boldsymbol{\varphi}\|_6}{\|\boldsymbol{\varphi}\|_{1,2}} \\
&\leq c\,\|\mathbf{v}\|_2^{1/2}\,\|\nabla\mathbf{v}\|_2^{1/2}\,\|\nabla\mathbf{w}\|_2.
\end{aligned} \tag{4.15}
$$

(We have used the Sobolev inequality $\|\mathbf{v}\|_6 \leq c\,\|\nabla\mathbf{v}\|_2$, valid for all $\mathbf{v}\in\mathbf{W}_0^{1,2}(\Omega)$, see, e.g., [29, p. 54]. Here and further on, c denotes the generic constant.)

Let $\mathbf{u}$ be a weak solution of the IBVP (4.9)–(4.12) in the sense of definition 4.3.1.B. It follows from the estimates (4.14) and (4.15) that

$$
\mathcal{A}\mathbf{u}\in L^2(0,T;\ \mathbf{W}_0^{-1,2}(\Omega)) \quad\text{and}\quad \mathcal{B}(\mathbf{u},\mathbf{u})\in L^{4/3}(0,T;\ \mathbf{W}_0^{-1,2}(\Omega)). \tag{4.16}
$$

Considering $\boldsymbol{\phi}$ in (4.13) in the form $\boldsymbol{\phi}(\mathbf{x},t)=\boldsymbol{\varphi}(\mathbf{x})\,\vartheta(t)$ where $\boldsymbol{\varphi}\in\mathbf{W}_{0,\sigma}^{1,2}(\Omega)$ and $\vartheta\in C_0^\infty((0,T))$, we deduce that $\mathbf{u}$ satisfies the equation

$$
\frac{\mathrm{d}}{\mathrm{d}t}(\mathbf{u},\boldsymbol{\varphi})_2+\nu\,\big\langle\mathcal{A}\mathbf{u},\boldsymbol{\varphi}\big\rangle_\Omega+\big\langle\mathcal{B}(\mathbf{u},\mathbf{u}),\boldsymbol{\varphi}\big\rangle_\Omega = \langle\mathbf{f},\boldsymbol{\varphi}\rangle_\Omega \tag{4.17}
$$

a.e. in $(0,T)$, where the derivative of $(\mathbf{u},\boldsymbol{\varphi})_2$ means the derivative in the sense of distributions. It follows from (4.16) that $\langle\mathcal{A}\mathbf{u},\boldsymbol{\varphi}\rangle_\Omega\in L^2(0,T)$ and $\langle\mathcal{B}(\mathbf{u},\mathbf{u}),\boldsymbol{\varphi}\rangle_\Omega\in L^{4/3}(0,T)$. Since $\langle\mathbf{f},\boldsymbol{\varphi}\rangle_\Omega\in L^2(0,T)$, we observe from (4.17) that the distributional derivative of $(\mathbf{u},\boldsymbol{\varphi})_2$ with respect to t is in $L^{4/3}(0,T)$. Hence $(\mathbf{u},\boldsymbol{\varphi})_2$ is a.e. in $[0,T)$ equal to a continuous function and the weak solution $\mathbf{u}$ is (after a possible redefinition on a set of measure zero) a weakly continuous function from $[0,T)$ to $\mathbf{L}^2_\sigma(\Omega)$. Now, one can easily deduce from (4.13) that $\mathbf{u}$ satisfies the initial condition (4.12) in the sense that

$$
(\mathbf{u},\boldsymbol{\varphi})_2\big|_{t=0} = (\mathbf{u}_0,\boldsymbol{\varphi})_2 \tag{4.18}
$$

for all $\boldsymbol{\varphi}\in\mathbf{W}_{0,\sigma}^{1,2}(\Omega)$. Thus, we come to the 2nd weak formulation of the IBVP (4.9)–(4.12):

Given $\mathbf{u}_0\in\mathbf{L}^2_\sigma(\Omega)$ *and* $\mathbf{f}\in L^2(0,T;\ \mathbf{W}_0^{-1,2}(\Omega))$. *Find* $\mathbf{u}\in L^\infty(0,T;\ \mathbf{L}^2_\sigma(\Omega))\cap L^2(0,T;\ \mathbf{W}_{0,\sigma}^{1,2}(\Omega))$ *(called the weak solution) such that* $\mathbf{u}$ *satisfies equation (4.17) a.e. in* $(0,T)$ *and the initial condition (4.18) for all* $\boldsymbol{\varphi}\in\mathbf{W}_{0,\sigma}^{1,2}(\Omega)$.

We have shown that if $\mathbf{u}$ is a weak solution of the IBVP (4.9)–(4.12) in the sense of the 1st definition then it also satisfies the 2nd definition. One can also show the opposite, i.e. if $\mathbf{u}$ satisfies the 2nd definition, then it also satisfies the 1st definition. For that purpose, it is sufficient to take into account that any test function $\boldsymbol{\phi}$ from (4.13) can be approximated with an arbitrarily small error (measured in the norm of $C^1\big([0,T];\ \mathbf{W}_{0,\sigma}^{1,2}(\Omega)\big)$) by a finite linear combination of functions of the type $\boldsymbol{\varphi}(\mathbf{x})\,\vartheta(t)$, where $\boldsymbol{\varphi} \in \mathbf{W}^{1,2}_{0,\sigma}(\Omega)$ and $\vartheta \in C^\infty([0,T])$, $\vartheta(T) = 0$, and that each such pair $\boldsymbol{\varphi}$, ϑ satisfies the equation

$$\int_0^T \big[-(\mathbf{u},\boldsymbol{\varphi})_2\,\vartheta'(t) + \nu\,\langle \mathcal{A}\mathbf{u},\boldsymbol{\varphi}\rangle_\Omega\,\vartheta(t) + \langle \mathcal{B}(\mathbf{u},\mathbf{u}),\boldsymbol{\varphi}\rangle_\Omega\,\vartheta(t)\big]\,\mathrm{d}t$$
$$= (\mathbf{u}_0,\boldsymbol{\varphi})_2\,\vartheta(0) + \int_0^T \langle \mathbf{f},\boldsymbol{\varphi}\rangle_\Omega\,\vartheta(t)\,\mathrm{d}t,$$

which follows from (4.17).

Before we proceed with another weak formulation of the IBVP (4.9)–(4.12), we present a lemma which plays an important role in studies of non-steady problems. It coincides with Lemma III.1.1 in [79].

Lemma 4.3.1 *Let* $\mathbf{X}$ *be a Banach space with the dual* $\mathbf{X}^*$, $\langle\,.\,,\,.\,\rangle$ *be the duality between* $\mathbf{X}^*$ *and* $\mathbf{X}$, $-\infty < a < b < \infty$ *and* $\mathbf{u},\ \mathbf{g} \in L^1(a,b;\ \mathbf{X})$. *Then the following three conditions are equivalent:*

1) $\mathbf{u}$ *is a.e. in* (a,b) *equal to a primitive function of* $\mathbf{g}$, *which means that*

$$\mathbf{u}(t) = \boldsymbol{\xi} + \int_a^t \mathbf{g}(s)\,\mathrm{d}s \quad \textit{for some } \boldsymbol{\xi} \in \mathbf{X} \textit{ and a.a. } t \in (a,b),$$

2) $\displaystyle\int_a^b \vartheta'(t)\,\mathbf{u}(t)\,\mathrm{d}t = -\int_a^b \vartheta(t)\,\mathbf{g}(t)\,\mathrm{d}t \quad$ *for all* $\vartheta \in C_0^\infty((a,b))$,

3) $\displaystyle\frac{\mathrm{d}}{\mathrm{d}t}\,\langle \boldsymbol{\eta},\mathbf{u}\rangle = \langle \boldsymbol{\eta},\mathbf{g}\rangle \quad$ *in the sense of distributions in* (a,b) *for each* $\boldsymbol{\eta} \in \mathbf{X}^*$.

If the conditions 1)–3) are fulfilled, then $\mathbf{u}$ *is a.e. in* (a,b) *equal to a continuous function from* $[a,b]$ *to* $\mathbf{X}$.

Note that if functions $\mathbf{u}$ and $\mathbf{g}$ are related as in item 2) then $\mathbf{g}$ is called the *distributional derivative* of $\mathbf{u}$ with respect to t and it is usually denoted by $\mathbf{u}'$.

D) The 3rd Weak Formulation of the Navier-Stokes IBVP (4.9)–(4.12) Equation (4.17) can also be written in the equivalent form

$$\frac{\mathrm{d}}{\mathrm{d}t}\,(\mathbf{u},\boldsymbol{\varphi})_2 + \nu\,\big\langle \mathcal{P}_2\mathcal{A}\mathbf{u},\boldsymbol{\varphi}\big\rangle_{\Omega,\sigma} + \big\langle \mathcal{P}_2\mathcal{B}(\mathbf{u},\mathbf{u}),\boldsymbol{\varphi}\big\rangle_{\Omega,\sigma} = \big\langle \mathcal{P}_2\mathbf{f},\boldsymbol{\varphi}\big\rangle_{\Omega,\sigma}. \tag{4.19}$$

Let us denote by $(\mathbf{u}')_\sigma$ the distributional derivative with respect to t of $\mathbf{u}$, as a function from $(0,T)$ to $\mathbf{W}^{-1,2}_{0,\sigma}(\Omega)$. (We explain in Sect. 4.3.1E why we use the notation $(\mathbf{u}')_\sigma$ and not just $\mathbf{u}'$.) Applying Lemma 4.3.1 (with $\mathbf{X} = \mathbf{W}^{-1,2}_{0,\sigma}(\Omega)$ and $\mathbf{X}^* = \mathbf{W}^{1,2}_{0,\sigma}(\Omega)$), we deduce that equation (4.19) is equivalent to

$$(\mathbf{u}')_\sigma + \nu\mathcal{P}_2\mathcal{A}\mathbf{u} + \mathcal{P}_2\mathcal{B}(\mathbf{u},\mathbf{u}) = \mathcal{P}_2\mathbf{f}, \tag{4.20}$$

which is an equation in $\mathbf{W}_{0,\sigma}^{-1,2}(\Omega)$, satisfied a.e. in the time interval $(0, T)$. Due to Lemma 4.2.1 and (4.16), $(\mathbf{u}')_\sigma \in L^{4/3}(0, T;\ \mathbf{W}_{0,\sigma}^{-1,2}(\Omega))$. Hence $\mathbf{u}$ coincides a.e. in $(0, T)$ with a continuous function from $[0, T)$ to $\mathbf{W}_{0,\sigma}^{-1,2}(\Omega)$. Thus, we obtain the 3rd equivalent definition of a weak solution to the IBVP (4.9)–(4.12):

Given $\mathbf{u}_0 \in \mathbf{L}^2_\sigma(\Omega)$ *and* $\mathbf{f} \in L^2(0, T;\ \mathbf{W}_0^{-1,2}(\Omega))$. *Function* $\mathbf{u} \in L^\infty(0, T;\ \mathbf{L}^2_\sigma(\Omega)) \cap L^2(0, T;\ \mathbf{W}_{0,\sigma}^{1,2}(\Omega))$ *is called a weak solution to the IBVP (4.9)–(4.12) if* $\mathbf{u}$ *satisfies equation (4.20) a.e. in the interval* $(0, T)$ *and the initial condition (4.12), where* $\mathbf{u}|_{t=0}$ *is the value of the aforementioned continuous function at time* $t = 0$.

We have explained that if $\mathbf{u}$ is a weak solution in the sense of the 2nd definition then it satisfies the 3rd definition. The validity of the opposite implication can be simply verified by means of Lemma 4.3.1.

E) Remark We have shown that $\mathbf{u}$ coincides a.e. in $(0, T)$ with a continuous function from $[0, T)$ to $\mathbf{W}_{0,\sigma}^{-1,2}(\Omega)$. This, however, does not imply that $\mathbf{u}$ coincides a.e. in $(0, T)$ with a continuous function from $[0, T)$ to $\mathbf{W}_0^{-1,2}(\Omega)$. It is connected with the fact that the derivative $(\mathbf{u}')_\sigma$ in equation (4.20) is the distributional derivative with respect to t of $\mathbf{u}$, as a function from $(0, T)$ to $\mathbf{W}_{0,\sigma}^{-1,2}(\Omega)$ and not the distributional derivative with respect to t of $\mathbf{u}$, as a function from $(0, T)$ to $\mathbf{W}_0^{-1,2}(\Omega)$. As it is important to distinguish between these two derivatives, we use the different notation: while the first derivative is denoted by $(\mathbf{u}')_\sigma$, the second is denoted just by $\mathbf{u}'$. We can formally write $(\mathbf{u}')_\sigma = \mathcal{P}_2\mathbf{u}'$.

4.3.2 An Associated Pressure and Its Existence

A) Projections $E^{1,2}$ and $E^{-1,2}$ By analogy with $\mathbf{W}_0^{1,2}(\Omega)$, which is a Hilbert space with the scalar product $(\,.\,,\,.\,)_{1,2} = \langle(\mathcal{A} + I)\,.\,,\,.\,\rangle_\Omega$, $\mathbf{W}_0^{-1,2}(\Omega)$ is a Hilbert space with the scalar product

$$(\mathbf{f}, \mathbf{g})_{-1,2} := \langle\mathbf{f}, (\mathcal{A} + I)^{-1}\mathbf{g}\rangle_\Omega = \big((\mathcal{A} + I)^{-1}\mathbf{f}, (\mathcal{A} + I)^{-1}\mathbf{g}\big)_{1,2}. \tag{4.21}$$

Denote by $E^{1,2}$ the orthogonal projection in $\mathbf{W}_0^{1,2}(\Omega)$, such that

$$\ker E^{1,2} = \mathbf{W}_{0,\sigma}^{1,2}(\Omega), \tag{4.22}$$

and by $E^{-1,2}$ the adjoint projection in $\mathbf{W}_0^{-1,2}(\Omega)$. It follows from (4.22) that the range of $E^{-1,2}$ is $\mathbf{W}_{0,\sigma}^{1,2}(\Omega)^\perp$. Due to (4.21) and the orthogonality of $E^{1,2}$, we have

$$\langle\mathbf{f}, E^{1,2}\boldsymbol{\psi}\rangle_\Omega = \big((\mathcal{A} + I)^{-1}\mathbf{f}, E^{1,2}\boldsymbol{\psi}\big)_{1,2} = \big(E^{1,2}(\mathcal{A} + I)^{-1}\mathbf{f}, \boldsymbol{\psi}\big)_{1,2}$$

for all $\mathbf{f} \in \mathbf{W}_0^{-1,2}(\Omega)$ and $\boldsymbol{\psi} \in \mathbf{W}_0^{1,2}(\Omega)$. However, using (4.21) and the fact that $E^{-1,2}$ is adjoint to $E^{1,2}$, we also have

$$\langle \mathbf{f}, E^{1,2}\boldsymbol{\psi}\rangle_\Omega = \langle E^{-1,2}\mathbf{f}, \boldsymbol{\psi}\rangle_\Omega = \big((\mathcal{A}+I)^{-1}E^{-1,2}\mathbf{f}, \boldsymbol{\psi}\big)_{1,2}$$

for all $\mathbf{f} \in \mathbf{W}_0^{-1,2}(\Omega)$ and $\boldsymbol{\psi} \in \mathbf{W}_0^{1,2}(\Omega)$. This shows that

$$E^{1,2}(\mathcal{A}+I)^{-1} = (\mathcal{A}+I)^{-1}E^{-1,2}. \tag{4.23}$$

Applying this identity and the orthogonality of projection $E^{1,2}$, we get

$$\begin{aligned}\big(E^{-1,2}\mathbf{f}, \mathbf{g}\big)_{-1,2} &= \big((\mathcal{A}+I)^{-1}E^{-1,2}\mathbf{f}, (\mathcal{A}+I)^{-1}\mathbf{g}\big)_{1,2} \\ &= \big(E^{1,2}(\mathcal{A}+I)^{-1}\mathbf{f}, (\mathcal{A}+I)^{-1}\mathbf{g}\big)_{1,2} = \big((\mathcal{A}+I)^{-1}\mathbf{f}, E^{1,2}(\mathcal{A}+I)^{-1}\mathbf{g}\big)_{1,2} \\ &= \big((\mathcal{A}+I)^{-1}\mathbf{f}, (\mathcal{A}+I)^{-1}E^{-1,2}\mathbf{g}\big)_{1,2} = \big(\mathbf{f}, E^{-1,2}\mathbf{g}\big)_{-1,2}\end{aligned}$$

for all $\mathbf{f}, \mathbf{g} \in \mathbf{W}_0^{-1,2}(\Omega)$, which shows that projection $E^{-1,2}$ is orthogonal, too.

Finally, let $\phi \in C_0^\infty(\Omega)$. Then $(\mathcal{A}+I)\nabla\phi \equiv \nabla(-\Delta+I)\phi \in \mathbf{W}_{0,\sigma}^{1,2}(\Omega)^\perp$. Hence $E^{-1,2}(\mathcal{A}+I)\nabla\phi = (\mathcal{A}+I)\nabla\phi$. Moreover, applying (4.23), we also have $E^{-1,2}(\mathcal{A}+I)\nabla\phi = (\mathcal{A}+I)E^{1,2}\nabla\phi$. As $\mathcal{A}+I$ is a one-to-one operator from $\mathbf{W}_0^{1,2}(\Omega)$ to $\mathbf{W}_0^{-1,2}(\Omega)$, the last two equalities show that

$$E^{1,2}\nabla\phi = \nabla\phi \qquad \text{for all } \phi \in C_0^\infty(\Omega). \tag{4.24}$$

B) An Associated Pressure The external body force $\mathbf{f}$ is supposed to be an element of $L^2(0,T;\mathbf{W}_0^{-1,2}(\Omega))$ in the definition of a weak solution. We may identify $\mathbf{f}$ with a distribution in Q_T, acting on functions $\boldsymbol{\phi} \in \mathbf{C}_0^\infty(Q_T)$ through the formula

$$\langle\!\langle \mathbf{f}, \boldsymbol{\phi}\rangle\!\rangle_{Q_T} := \int_0^T \langle \mathbf{f}(t), \boldsymbol{\phi}(\,.\,,t)\rangle_\Omega \, dt. \tag{4.25}$$

(Here and further on, we denote by $\langle\!\langle\,.\,,\,.\,\rangle\!\rangle_{Q_T}$ the action of a distribution in Q_T on a function from $C_0^\infty(Q_T)$ or $\mathbf{C}_0^\infty(Q_T)$.)

We come to the definition:

Let $\mathbf{u}$ *be a weak solution to the IBVP (4.9)–(4.12). If there exists a distribution* p *in* Q_T *such that* $\mathbf{u}$ *and* p *satisfy the Navier-Stokes equation (4.9) in the sense of distributions in* Q_T *then* p *is called a pressure, associated with the weak solution* $\mathbf{u}$.

C) Existence of an Associated Pressure None of the three equivalent definitions of the weak solution to the IBVP (4.9)–(4.12) contains an explicit information on an associated pressure. This is why it is natural to ask whether to every weak solution of the IBVP (4.9)–(4.12) an associated pressure exists. In this paragraph, we show that the answer is positive.

Let $\mathbf{u}$ be a weak solution to the IBVP (4.9)–(4.12). Due to Lemma 4.3.1, item 1), equation (4.20) is equivalent to

$$\mathbf{u}(t) - \mathbf{u}(0) + \int_0^t \mathcal{P}_2\big[\nu\mathcal{A}\mathbf{u} + \mathcal{B}(\mathbf{u},\mathbf{u}) - \mathbf{f}\big]\,\mathrm{d}\tau \;=\; \mathbf{0}$$

for a.a. $t \in (0,T)$. Since $\mathbf{u}(t)$ and $\mathbf{u}(0)$ are in $\mathbf{L}^2_\sigma(\Omega)$, they coincide with $\mathcal{P}_2\mathbf{u}(t)$ and $\mathcal{P}_2\mathbf{u}(0)$, respectively. (See Sect. 4.2E.) Hence

$$\mathcal{P}_2\bigg(\mathbf{u}(t) - \mathbf{u}(0) + \int_0^t \big[\nu\mathcal{A}\mathbf{u} + \mathcal{B}(\mathbf{u},\mathbf{u}) - \mathbf{f}\big]\,\mathrm{d}\tau\bigg) \;=\; \mathbf{0}.$$

Define $\mathbf{F}(t) \in \mathbf{W}_0^{-1,2}(\Omega)$ by the formula

$$\mathbf{F}(t) \;:=\; \mathbf{u}(t) - \mathbf{u}(0) + \int_0^t \big[\nu\mathcal{A}\mathbf{u} + \mathcal{B}(\mathbf{u},\mathbf{u}) - \mathbf{f}\big]\,\mathrm{d}\tau. \tag{4.26}$$

Since $\langle \mathbf{F}(t), \boldsymbol{\psi}\rangle_\Omega = \langle \mathcal{P}_2\mathbf{F}(t), \boldsymbol{\psi}\rangle_{\Omega,\sigma} = 0$ for all $\boldsymbol{\psi} \in \mathbf{W}^{1,2}_{0,\sigma}(\Omega)$, $\mathbf{F}(t)$ is an element of $\mathbf{W}^{1,2}_{0,\sigma}(\Omega)^\perp$. Hence $E^{-1,2}\mathbf{F}(t) = \mathbf{F}(t)$ and $(I - E^{-1,2})\mathbf{F}(t) = \mathbf{0}$. Thus,

$$\begin{aligned}(I - E^{-1,2})\mathbf{u}(t) &- (I - E^{-1,2})\mathbf{u}(0) \\ &+ \int_0^t (I - E^{-1,2})\big[\nu\mathcal{A}\mathbf{u} + \mathcal{B}(\mathbf{u},\mathbf{u}) - \mathbf{f}\big]\,\mathrm{d}\tau \;=\; \mathbf{0}\end{aligned}$$

holds as an equation in $\mathbf{W}_0^{-1,2}(\Omega)$. Applying Lemma 4.3.1 (with $\mathbf{X} = \mathbf{W}_0^{-1,2}(\Omega)$), we observe that

$$\big[(I - E^{-1,2})\mathbf{u}\big]' + (I - E^{-1,2})\big[\nu\mathcal{A}\mathbf{u} + \mathcal{B}(\mathbf{u},\mathbf{u}) - \mathbf{f}\big] \;=\; \mathbf{0},$$

where $[(I - E^{-1,2})\mathbf{u}]'$ is the distributional derivative with respect to t of $(I - E^{-1,2})\mathbf{u}$, as a function from $(0,T)$ to $\mathbf{W}_0^{-1,2}(\Omega)$. This yields

$$\begin{aligned}\mathbf{u}' + \nu\mathcal{A}\mathbf{u} + \mathcal{B}(\mathbf{u},\mathbf{u}) \\ = \mathbf{f} + [E^{-1,2}\mathbf{u}]' + \nu E^{-1,2}\mathcal{A}\mathbf{u} + E^{-1,2}\mathcal{B}(\mathbf{u},\mathbf{u}) - E^{-1,2}\mathbf{f}.\end{aligned} \tag{4.27}$$

Let $\Omega_0 \subset\subset \Omega$ be a non-empty domain. By Lemma 4.2.5, there exist unique $p_0(t)$, $p_1(t)$, $p_2(t)$, $p_3(t)$ in $L^2_{pot}(\Omega)$ such that

$$\begin{aligned}\nabla p_0(t) &= -E^{-1,2}\mathbf{u}(t), & \nabla p_1(t) &= -\nu E^{-1,2}\mathcal{A}\mathbf{u}(t), \\ \nabla p_2(t) &= -E^{-1,2}\mathcal{B}\big(\mathbf{u}(t),\mathbf{u}(t)\big), & \nabla p_3(t) &= E^{-1,2}\mathbf{f}(t)\end{aligned} \tag{4.28}$$

and $\int_{\Omega_0} p_i(t)\,\mathrm{d}\mathbf{x} = 0$ ($i = 0, 1, 2, 3$) for a.a. $t \in (0,T)$. Using (4.16) and the boundedness of projection $E^{-1,2}$, we get

$$\begin{aligned} &\nabla p_0 \in L^\infty(0,T;\, \mathbf{W}_0^{-1,2}(\Omega)), &&\nabla p_1 \in L^2(0,T;\, \mathbf{W}_0^{-1,2}(\Omega)),\\ &\nabla p_2 \in L^{4/3}(0,T;\, \mathbf{W}_0^{-1,2}(\Omega)), &&\nabla p_3 \in L^2(0,T;\, \mathbf{W}_0^{-1,2}(\Omega)). \end{aligned} \tag{4.29}$$

Hence

$$\begin{aligned} &p_0 \in L^\infty(0,T;\, L^2_{loc}(\Omega)), &&p_1 \in L^2(0,T;\, L^2_{loc}(\Omega)),\\ &p_2 \in L^{4/3}(0,T;\, L^2_{loc}(\Omega)), &&p_3 \in L^2(0,T;\, L^2_{loc}(\Omega)). \end{aligned} \tag{4.30}$$

Equation (4.27) shows that

$$\begin{aligned} &\int_0^T \int_\Omega \big[-\mathbf{u}\cdot\boldsymbol{\psi}\,\eta'(t) + \nu\nabla\mathbf{u} : \nabla\boldsymbol{\psi}\,\eta(t) + \mathbf{u}\cdot\nabla\mathbf{u}\cdot\boldsymbol{\psi}\,\eta(t)\big]\,\mathrm{d}\mathbf{x}\,\mathrm{d}t\\ &\quad = \int_0^T \langle \mathbf{f}, \boldsymbol{\psi}\rangle_\Omega\,\eta(t)\,\mathrm{d}t - \int_0^T\int_\Omega p_0\,\mathrm{div}\,\boldsymbol{\psi}\,\eta'(t)\,\mathrm{d}\mathbf{x}\,\mathrm{d}t\\ &\qquad + \int_0^T\int_\Omega (p_1+p_2+p_3)\,\mathrm{div}\,\boldsymbol{\psi}\,\eta(t)\,\mathrm{d}t \end{aligned}$$

for all functions $\boldsymbol{\psi} \in \mathbf{W}_0^{1,2}(\Omega)$ and $\eta \in C_0^\infty((0,T))$. Since the set of all finite linear combinations of functions of the type $\boldsymbol{\psi}(\mathbf{x})\,\eta(t)$, where $\boldsymbol{\psi} \in \mathbf{W}_0^{1,2}(\Omega)$ and $\eta \in C_0^\infty((0,T))$, is dense in $\mathbf{C}_0^\infty(Q_T)$ in the topology of the space $W^{1,2}(0,T;\, \mathbf{W}_0^{1,2}(\Omega))$, we also have

$$\begin{aligned} &\int_0^T\int_\Omega \big[-\mathbf{u}\cdot\partial_t\boldsymbol{\varphi} + \nu\nabla\mathbf{u} : \nabla\boldsymbol{\varphi} + \mathbf{u}\cdot\nabla\mathbf{u}\cdot\boldsymbol{\varphi}\big]\,\mathrm{d}\mathbf{x}\,\mathrm{d}t\\ &= \int_0^T \langle\mathbf{f},\boldsymbol{\varphi}\rangle_\Omega\,\mathrm{d}t - \int_0^T\int_\Omega p_0\,\mathrm{div}\,\partial_t\boldsymbol{\varphi}\,\mathrm{d}\mathbf{x}\,\mathrm{d}t + \int_0^T\int_\Omega (p_1+p_2+p_3)\,\mathrm{div}\,\boldsymbol{\varphi}\,\mathrm{d}\mathbf{x}\,\mathrm{d}t \end{aligned}$$

for all $\boldsymbol{\varphi} \in \mathbf{C}_0^\infty(Q_T)$. This shows that if we put

$$p := \partial_t p_0 + p_1 + p_2 + p_3, \tag{4.31}$$

where $\partial_t p_0$ is the distributional derivative of p_0 with respect to t, then the pair $\mathbf{u}$, p satisfies the Navier-Stokes equation (4.9) in the sense of distributions in Q_T.

For a.a. $t \in (0,T)$, the functions $p_0(t)$ and $p_1(t)$ are harmonic in Ω. This follows from the identities

$$\begin{aligned} \int_\Omega p_0(t)\,\Delta\phi\,\mathrm{d}\mathbf{x} &= -\big\langle\nabla p_0(t), \nabla\phi\big\rangle_\Omega = \big\langle E^{-1,2}\mathbf{u}(t), \nabla\phi\big\rangle_\Omega = \big\langle\mathbf{u}(t), E^{1,2}\nabla\phi\big\rangle_\Omega\\ &= \big\langle\mathbf{u}(t), \nabla\phi\big\rangle_\Omega = \int_\Omega \mathbf{u}(t)\cdot\nabla\phi\,\mathrm{d}\mathbf{x} = 0 \quad (\text{for all } \phi \in C_0^\infty(\Omega)). \end{aligned}$$

(We have used (4.24).) Hence, by Weyl's lemma, $p_0(t)$ is a harmonic function in Ω. The fact that $p_1(t)$ is harmonic can be proven similarly.

D) Uniqueness of the Associated Pressure Up to an Additive Distribution of the Form (4.32) If G is a distribution in $(0, T)$ and $\psi \in C_0^\infty(Q_T)$, then we define a distribution g in Q_T by the formula

$$\langle\!\langle g, \psi \rangle\!\rangle_{Q_T} := \Big\langle G, \int_\Omega \psi \, \mathrm{d}\mathbf{x} \Big\rangle_{(0,T)}, \tag{4.32}$$

where $\langle G, \, . \rangle_{(0,T)}$ denotes the action of distribution G on a function from $C_0^\infty((0,T))$. Obviously, if $\boldsymbol{\phi} \in \mathbf{C}_0^\infty(Q_T)$, then

$$\langle\!\langle \nabla g, \boldsymbol{\phi} \rangle\!\rangle_{Q_T} = -\langle\!\langle g, \operatorname{div} \boldsymbol{\phi} \rangle\!\rangle_{Q_T} = -\Big\langle G, \int_\Omega \operatorname{div} \boldsymbol{\phi} \, \mathrm{d}\mathbf{x} \Big\rangle_{(0,T)} = 0, \tag{4.33}$$

because $\int_\Omega \operatorname{div} \boldsymbol{\phi}(\, . \,, t) \, \mathrm{d}\mathbf{x} = 0$ for all $t \in (0, T)$. Thus, $p + g$ (where p is given by (4.31)) is a pressure, associated with the weak solution $\mathbf{u}$ to the IBVP (4.9)–(4.12), too.

On the other hand, if $p + g$ is a pressure, associated with the weak solution $\mathbf{u}$, then g satisfies

$$0 = \langle\!\langle \nabla g, \boldsymbol{\phi} \rangle\!\rangle_{Q_T} = -\langle\!\langle g, \operatorname{div} \boldsymbol{\phi} \rangle\!\rangle_{Q_T} \tag{4.34}$$

for all $\boldsymbol{\phi} \in \mathbf{C}_0^\infty(Q_T)$. For $h \in C_0^\infty((0,T))$, define

$$\langle G, h \rangle_{(0,T)} := \langle\!\langle g, \psi \rangle\!\rangle_{Q_T}, \tag{4.35}$$

where $\psi \in C_0^\infty(Q_T)$ is chosen so that $h(t) = \int_\Omega \psi(\mathbf{x}, t) \, \mathrm{d}\mathbf{x}$ for all $t \in (0, T)$. The definition of the distribution G is independent of the concrete choice of function ψ due to these reasons: let ψ_1 and ψ_2 be two functions from $C_0^\infty(Q_T)$ such that $h(t) = \int_\Omega \psi_1(\mathbf{x}, t) \, \mathrm{d}\mathbf{x} = \int_\Omega \psi_2(\mathbf{x}, t) \, \mathrm{d}\mathbf{x}$ for $t \in (0, T)$. Denote by G_1, respectively, G_2, the distribution, defined by formula (4.35) with $\psi = \psi_1$, respectively, $\psi = \psi_2$. Since $\operatorname{supp}(\psi_1 - \psi_2)$ is a compact subset of Q_T and $\int_\Omega [\psi_1(\, . \,, t) - \psi_2(\, . \,, t)] \, \mathrm{d}\mathbf{x} = 0$ for all $t \in (0, T)$, there exists a function $\boldsymbol{\phi} \in \mathbf{C}_0^\infty(Q_T)$ such that $\operatorname{div} \boldsymbol{\phi} = \psi_1 - \psi_2$ in Q_T. (See, e.g., [29, Sec. III.3] or [10] for the construction of function $\boldsymbol{\phi}$.) Then

$$\langle G_1 - G_2, h \rangle_{(0,T)} := \langle\!\langle g, \psi_1 - \psi_2 \rangle\!\rangle_{Q_T} = \langle\!\langle g, \operatorname{div} \boldsymbol{\phi} \rangle\!\rangle_{Q_T},$$

which is equal to zero due to (4.34). Formula (4.35) and the identity $h(t) = \int_\Omega \psi(\mathbf{x}, t) \, \mathrm{d}\mathbf{x}$ show that the distribution g has the form (4.32).

The next theorem summarizes the results of Sects. 4.3.2C and 4.3.2D:

Theorem 4.3.2 *Let* **u** *be a weak solution to the Navier-Stokes IBVP (4.9)–(4.12). Then there exists an associated pressure* p *(as a distribution in* Q_T*) of the form (4.31), where* p_0, p_1, p_2, p_3 *satisfy (4.28)–(4.30). Moreover,*

1) *if* $\Omega_0 \subset\subset \Omega$, *then the functions* $p_0(t), \dots, p_3(t)$ *can be chosen uniquely so that they satisfy the additional conditions* $\int_{\Omega_0} p_i(t)\, \mathrm{d}\mathbf{x} = 0$ *for* $i = 0, 1, 2, 3$ *and a.a.* $t \in (0, T)$,
2) *the functions* $p_0(t)$ *and* $p_1(t)$ *are harmonic in* Ω *for a.a.* $t \in (0, T)$,
3) $p + g$ *is also a pressure, associated with the weak solution* **u**, *if and only if* g *is a distribution of the form (4.32).*

E) Remark If Ω is a bounded Lipschitz domain, then the statement of Theorem 4.3.2 can be improved so that $L^2_{loc}(\Omega)$ is replaced by $L^2(\Omega)$ in (4.30) and the choice $\Omega_0 = \Omega$ is also permitted in statement 2). This is enabled by Lemma 4.2.4, which shows that the range of projection $E^{-1,2}$ coincides with $\nabla(L^2(\Omega))$.

The construction of the associated pressure, presented in Sect. 4.3.2C, uses ideas from paper [83] by J. Wolf. An analogous construction has also been shown in [83] in the case that **u** is just the so called distributional solution of the system (4.9), (4.10) (where $\mathbf{f} = \operatorname{div} \mathbb{F}$ for some $\mathbb{F} \in L^2(Q_T)^{3\times 3}$) with bounded energy, which means that $\mathbf{u} \in C_w\big([0, T]; \mathbf{L}^2(\Omega)\big) \cap L^2(0, T;\ \mathbf{W}^{1,2}(\Omega))$, $\operatorname{div} \mathbf{u} = 0$ a.e. in Q_T and the distribution $\partial_t \mathbf{u} + \mathbf{u} \cdot \nabla \mathbf{u} - \nu \Delta \mathbf{u} - \operatorname{div} \mathbb{F}$ vanishes on all divergence-free functions from $\mathbf{C}_0^\infty(Q_T)$.

F) Example We give an example of a simple distributional solution to the system (4.9), (4.10), that is not smooth in dependence on t and the associated pressure cannot be identified with a function from $L^1_{loc}(Q_T)$. Although the solution does not satisfy the boundary condition (4.11), the example sheds light on the reasons why the pressure generally exists only as a distribution and not as a function. The example is taken from [30] and [68].

Let $\psi \in \mathbf{W}^{2,2}(\Omega)$ be a harmonic function in Ω and $a \in L^\infty(0, T)$. Put

$$\mathbf{u}(\mathbf{x}, t) := a(t)\, \nabla\psi(\mathbf{x}),$$

$$p(\mathbf{x}, t) := -a'(t)\, \psi(\mathbf{x}) - a^2(t)\, \frac{|\psi(\mathbf{x})|^2}{2}.$$

for $\mathbf{x} \in \Omega$ and $0 \le t < T$. Function **u** is divergence-free and the pair **u**, p satisfies the Navier-Stokes equation (4.9) (with $\mathbf{f} = \mathbf{0}$) in the sense of distributions in Q_T. If the derivative $a'(t)$ exists only as a distribution in $(0, T)$ that cannot be identified with a function from $L^1_{loc}((0, T))$, then p is a distribution in Q_T that cannot be identified with a function from $L^1_{loc}(Q_T)$. (Recall that such distributions are sometimes called *singular*, while the others are called *regular*.)

By analogy with Sect. 4.3.2D, one can show that the pressure p, associated with the velocity **u**, is unique up to an additive distribution g of the type (4.32). If $a' \notin L^1_{loc}((0, T))$, and p is therefore a singular distribution, then the distribution $p + g$ is also singular for any distribution g of the type (4.32). Thus, any pressure, associated with the velocity **u**, is only a singular distribution.

G) On Other Related Results The first results, showing that the pressure associated with a weak solution to the Navier-Stokes IBVP (4.9)–(4.12) generally exists only as a distribution, come (to our knowledge) from J. L. Lions (see [50, Remarque I.6.7] and R. Temam (see [79, pp. 307–308]). Of more recent works, we cite:

1) J. Simon [71] assumes that Ω is bounded and proves that there exists at least one weak solution $\mathbf{u}$ with an associated pressure $p \in W^{-1,\infty}(0,T;\, L^2_{loc}(\Omega))$. If Ω is a Lipschitz domain, then $p \in W^{-1,\infty}(0,T; L^2(\Omega))$.
2) G. P. Galdi [30] shows that the pressure exists as a distributional derivative with respect to t of a certain function P in Q_T, such that $P(\,.\,,t) \in L^2_{loc}(\Omega)$ for all $t \in (0,T)$ and if $\Omega' \subset\subset \Omega$, then function P can be modified by an appropriate additive function $C = C(t,\Omega')$ so that $P + C \in L^\infty(0,T;\, L^2(\Omega'))$.
3) H. Sohr [75, Sec. V.1.7] assumes that Ω is any domain in $\mathbb{R}^N$ (for $N = 2$ or $N = 3$) and $\mathbf{f}$ has the form $\mathbf{f}_0 + \operatorname{div}\mathbb{F}$, where $\mathbf{f}_0 \in L^1_{loc}\big([0,T);\, \mathbf{L}^2(\Omega)\big)$ and $\mathbb{F} \in L^s_{loc}\big([0,T);\, L^2(\Omega)^{3\times 3}\big)$ (for $s = 4/N$), and defines the weak solution to be an element of $L^\infty_{loc}\big([0,T);\, \mathbf{L}^2_\sigma(\Omega)\big) \cap L^2_{loc}\big([0,T);\, \mathbf{W}^{1,2}_{0,\sigma}(\Omega)\big)$ (for $0 < T \leq \infty$). Then he obtains $p = -\partial_t P$ for some $P \in L^s_{loc}\big([0,T);\, L^2_{loc}(\Omega)\big)$.

It is remarkable that if domain Ω is smooth then the pressure, associated with a weak solution $\mathbf{u}$ to the IBVP (4.9)–(4.12), exists as a regular distribution in Q_T, i.e. a distribution that can be identified with a function from $L^1_{loc}(Q_T)$. Of works on this theme, we cite H. Sohr and W. von Wahl [73], Y. Giga and H. Sohr [36] and Y. Taniuchi [78]. A brief description of some ideas and results from [36] is given in Sect. 4.3.3.

A pressure, associated with a more general system than (4.9), (4.10), has been constructed in paper [83] by J. Wolf. The author deals with a distribution of the type $\partial_t\mathbf{u} + \mathbf{F}$ in Q_T, where $\mathbf{u},\, \mathbf{F} \in L^1(0,T;\, \mathbf{W}^{-1,q}_0(\Omega))$ (for some $1 < q < \infty$), vanishing on all smooth solenoidal vector fields, and constructs a global associated pressure in the case that Ω is the so-called *∇_q-regular domain*, or a local associated pressure if Ω is an arbitrary domain. As the approach, used in [83], is also applicable to more general systems than the Navier-Stokes equations, e.g. to equations of motion of some classes of non-Newtonian incompressible fluids, we explain it in greater detail in Sect. 4.3.4.

H) Briefly on a Suitable Weak Solution If Ω is a smooth domain then a series of authors have shown that the problem (4.9)–(4.12) has the so-called *suitable weak solution,* which is a pair of functions $\mathbf{u},\ p$ such that $\mathbf{u}$ is a weak solution, p is an associated pressure, and $\mathbf{u},\ p$ satisfy the *generalized energy inequality*

$$
\begin{aligned}
2\nu \int_0^T \int_\Omega |\nabla\mathbf{u}|^2\phi \,\mathrm{d}\mathbf{x}\,\mathrm{d}t \;\leq\; & \int_0^T \int_\Omega \big[|\mathbf{u}|^2\,(\partial_t\phi + \nu\Delta\phi) + \big(|\mathbf{u}|^2 + 2p\big)\,\mathbf{u}\cdot\nabla\phi\big]\,\mathrm{d}\mathbf{x}\,\mathrm{d}t \\
& + 2\int_0^T \langle \mathbf{f}, \mathbf{u}\phi\rangle_\Omega \,\mathrm{d}t
\end{aligned}
\tag{4.36}
$$

for every infinitely differentiable non-negative scalar function ϕ with a compact support in Q_T. This inequality is also often called the *local (or localized) energy inequality*. It can be formally derived from (4.9)–(4.12) if one multiplies equation (4.9) by $\mathbf{u}\phi$, integrates over Q_T, and applies the integration by parts. In order to give a reasonable sense to the integral of $2p\,\mathbf{u}\cdot\nabla\phi$ in (4.36), it is necessary to include some assumptions on the integrability of p to the definition of a suitable weak solution: since $\mathbf{u}\in\mathbf{L}^{10/3}(Q_T)$, it is sufficient to assume that $p\in L^{10/7}(Q_T)$; most of the authors, however, consider $p\in L^{3/2}(Q_T)$ or $p\in L^{5/3}(Q_T)$. (The latter is consistent with the known integrability of an associated pressure in a smooth domain, see Sect. 4.3.3.) First results on the existence of a suitable weak solution come from V. Scheffer [66] and L. Caffarelli, R. Kohn, L. Nirenberg [12]. The authors of [12] assumed that Ω is either the whole space $\mathbb{R}^3$ or a smooth bounded domain in $\mathbb{R}^3$ and obtained $p\in L^{5/4}(Q_T)$, provided that $\mathbf{f}\in\mathbf{L}^2(Q_T)\cap\mathbf{L}^q_{loc}(Q_T)$ for some $q>\frac{5}{2}$ and $\operatorname{div}\mathbf{f}=0$. (If $p\in L^{5/4}(Q_T)$ then the integrability of $2p\,\mathbf{u}\cdot\nabla\phi$ in Q_T does not follow just from Hölder's inequality and one must use subtler arguments, see [12, subsection 2C].) Other related results, including a finer or more detailed information on the pressure that is a part of a suitable weak solution, can be found in the papers [47, 49, 78] and [80]. The main importance of a suitable weak solution, however, does not lie in the fact that it involves the pressure as a function (and not only as a distribution), but in the fact that the generalized energy inequality enables one to derive a series of local regularity criteria, e.g. criteria for regularity of the solution at a space-time point $(\mathbf{x}_0,t_0)\in Q_T$. (See, e.g., [9, 12, 23, 47, 49, 60, 80, 81], etc.) Recall that $(\mathbf{x}_0,t_0)$ is said to be a *regular point* of solution $\mathbf{u}$ if there exists a neighborhood $U(\mathbf{x}_0,t_0)$ in Q_T such that $\mathbf{u}$ is essentially bounded in $U(\mathbf{x}_0,t_0)$. One of the criteria, derived in [12], says that there exists $\epsilon^*>0$ such that if $(\mathbf{x}_0,t_0)\in Q_T$ and $\mathbf{u}$ is a suitable weak solution to the problem (4.9)–(4.12), satisfying

$$\limsup_{r\to 0+}\frac{1}{r}\int_{t_0-7r^2/8}^{t_0+r^2/8}\int_{B_r(\mathbf{x}_0)}|\nabla\mathbf{u}|^2\,\mathrm{d}\mathbf{x}\,\mathrm{d}t\;<\;\epsilon^* \tag{4.37}$$

then $(\mathbf{x}_0,t_0)$ is a regular point of solution $\mathbf{u}$. (Later on, it was shown that the same proposition also holds if the outside integral is considered over the interval (t_0-r^2,t_0) instead of $(t_0-\frac{7}{8}r^2,t_0+\frac{1}{8}r^2)$, see, e.g., [47], or it can be generalized so that $|\nabla\mathbf{u}|$ is replaced by $|\mathbf{curl}\,\mathbf{u}\times(\mathbf{u}/|\mathbf{u}|)|$, see [80].) The aforementioned regularity criterion (4.37) has been used in [12] to prove that the set of hypothetic *singular points* of solution $\mathbf{u}$ (i.e., points of Q_T that are not regular) has one-dimensional parabolic measure equal to zero (which implies that one-dimensional Hausdorff measure of the set of singular points is zero, too).

Note that if $\mathbf{f}$ is, e.g., in $\mathbf{L}^{10/7}(Q_T)$ then the product $\mathbf{f}\cdot\mathbf{u}$ is in $L^1(Q_T)$. Thus, the quantity

$$\mu\;:=\;-\partial_t|\mathbf{u}|^2+\nu\Delta|\mathbf{u}|^2-2\nu|\nabla\mathbf{u}|^2-\operatorname{div}(|\mathbf{u}|^2\mathbf{u})-2\operatorname{div}(p\mathbf{u})+2\mathbf{f}\cdot\mathbf{u} \tag{4.38}$$

is well defined as a distribution in Q_T. In this case, inequality (4.36) expresses the condition that the distribution μ is non-negative (we shortly write $\mu \geq 0$), which means that $\langle\!\langle \mu, \phi \rangle\!\rangle_{Q_T} \geq 0$ for all $\phi \in C_0^\infty(Q_T)$, $\phi \geq 0$.

Assuming that Ω is an arbitrary domain, J. Wolf [82] proved the existence of the so-called *local suitable weak solution*. It is an analogue of the suitable weak solution, required to satisfy the local energy inequality in each ball $B \subset \overline{B} \subset \Omega$ with a certain local pressure, constructed in the ball B. Using this inequality, the author derived Caccioppoli-type inequalities, which further enabled him to prove several local regularity criteria, analogous to the ones from [12] or [49], and to show that the set of hypothetic singular points of the local suitable weak solution in Q_T has one-dimensional parabolic measure equal to zero.

I) A Dissipative Weak Solution and the Role of Pressure in Its Partial Regularity Theory Recently, D. Chamorro, P.-G. Lemarié-Rieusset, and K. Mayoufi [15] introduced the notion of a *dissipative weak solution*. The crucial point in the definition of the dissipative weak solution is the fact that if $\mathbf{u}$, p is a distributional solution of the Navier-Stokes system (4.9), (4.10) in $Q := B_\rho(\mathbf{x}_0) \times (a, b)$ (where $\rho > 0$ and $-\infty < a < b < \infty$) such that $\mathbf{u} \in L^\infty(a, b;\, \mathbf{L}^2(B_\rho(\mathbf{x}_0))) \cap L^2(a, b;\, \mathbf{W}^{1,2}(B_\rho(\mathbf{x}_0)))$ then the product $p\mathbf{u}$ exists as a distribution in Q. More precisely: assume that γ is an infinitely differentiable function in $\mathbb{R}$, supported in $(-1, 1)$ and satisfying $\int_{-1}^{1} \gamma(t)\, \mathrm{d}t = 1$, and θ is an infinitely differentiable function in $\mathbb{R}^3$, supported in $B_1(\mathbf{0})$ and satisfying $\int_{\mathbb{R}^3} \theta(\mathbf{x})\, \mathrm{d}\mathbf{x} = 1$. Put $\gamma_\alpha(t) := \alpha^{-1}\gamma(t/\alpha)$, $\theta_\epsilon(\mathbf{x}) := \epsilon^{-3}\theta(\mathbf{x}/\epsilon)$ (for $\alpha, \epsilon > 0$) and $\varphi_{\alpha,\epsilon}(\mathbf{x}, t) := \gamma_\alpha(t)\,\theta_\epsilon(\mathbf{x})$. Suppose that $(\mathbf{x}_0, t_0) \in Q$ and r_0 is so small that $B_{r_0}(\mathbf{x}_0) \times (t_0 - r_0^2, t_0 + r_0^2) \subset Q$. Then, for $0 < \alpha < \frac{1}{2}r_0^2$ and $0 < \epsilon < \frac{1}{2}r_0$, the convolutions $p * \varphi_{\alpha,\epsilon}$ and $\mathbf{u} * \varphi_{\alpha,\epsilon}$ are well defined as functions of $\mathbf{x}$ and t for $t_0 - \frac{1}{2}r_0^2 < t < t_0 + \frac{1}{2}r_0^2$ and $|\mathbf{x} - \mathbf{x}_0| < \frac{1}{2}r_0$. The authors show that the limit

$$\lim_{\epsilon \to 0+}\ \lim_{\alpha \to 0+}\ \operatorname{div}\big[(p * \varphi_{\alpha,\epsilon})\,(\mathbf{u} * \varphi_{\alpha,\epsilon})\big] \tag{4.39}$$

exists as a distribution in Q and it does not depend on the concrete choice of the functions γ and θ. (See [15, Proposition 3.3].) This non-trivial result implies that, for $\mathbf{f} \in \mathbf{L}^{10/7}(Q)$, quantity μ defined by (4.38) is a distribution in Q. Now, we can define a *dissipative weak solution* $\mathbf{u}$, p to the system (4.9), (4.10) in Q to be a distributional solution of (4.9), (4.10) in Q, such that $\mathbf{u} \in L^\infty(a, b;\, \mathbf{L}^2(B_\rho(\mathbf{x}_0))) \cap L^2(a, b;\, \mathbf{W}^{1,2}(B_\rho(\mathbf{x}_0)))$ and $\mu \geq 0$.

It is important that the dissipative weak solution satisfies the same regularity criterion as the suitable weak solution—concretely: there exists $\epsilon^* > 0$ with the property that if $(\mathbf{x}_0, t_0) \in Q$ and $\mathbf{u}$, p is a dissipative weak solution to the system (4.9), (4.10) in Q, such that

$$\limsup_{r \to 0+} \frac{1}{r} \int_{t_0 - r^2}^{t_0} \int_{B_r(\mathbf{x}_0)} |\nabla \mathbf{u}|^2\, \mathrm{d}\mathbf{x}\, \mathrm{d}t\ <\ \epsilon^*, \tag{4.40}$$

then $(\mathbf{x}_0, t_0)$ is a regular point of the solution $\mathbf{u}$, p. (See [15, Theorem 4].) By analogy with a suitable weak solution, this regularity criterion enables one to show that the set of possible singular points of a dissipative weak solution has one-dimensional parabolic measure equal to zero.

4.3.3 The Case of a Smooth Domain Ω

In this subsection, we briefly explain some results from paper [36] by Y. Giga and H. Sohr. Although the authors deal with the N-dimensional Navier-Stokes equations (for $N \geq 2$), we focus on the case $N = 3$.

A) The Stokes Operator Suppose that domain Ω is either the whole space $\mathbb{R}^3$, or a half-space in $\mathbb{R}^3$, or a bounded or exterior domain in $\mathbb{R}^3$ with the boundary of the class $C^{2+(h)}$ for some $h > 0$. Denote by A_s the so-called *Stokes operator* in $\mathbf{L}^s_\sigma(\Omega)$ (for $1 < s < \infty$), i.e. $A_s := -P_s\Delta$ with the domain $D(A_s) = \mathbf{W}^{2,s}(\Omega) \cap \mathbf{W}^{1,s}_{0,\sigma}(\Omega)$. (Recall that P_s is the Helmholtz projection, see Sect. 4.2E.) Note that $A_s = A_r$ on $D(A_s) \cap D(A_r)$. It is well known that operator A_s is nonnegative, its spectrum is a subset of the interval $[0, \infty)$ on the real axis and $(-A_s)$ generates the analytic semigroup $\mathrm{e}^{-A_s t}$ in $\mathbf{L}^s_\sigma(\Omega)$. The domain of A_s, equipped with the graph norm, is a Banach space. The L^2-theory of the Stokes operator is explained, e.g., in [75]. The book [75] also contains a series of further references, where the readers can find the corresponding L^s-theory. One can deduce from [74] that if $1 < s \leq q < \infty$, $0 \leq \alpha, \beta \leq 1$, $2(\beta - \alpha) \geq 3(s^{-1} - q^{-1})$ then $D((A_s + I)^\beta) \hookrightarrow D((A_q + I)^\alpha)$ and

$$\big\|(A_q + I)^\alpha \boldsymbol{\psi}\|_q \leq c \,\big\|(A_s + I)^\beta \boldsymbol{\psi}\big\|_s \tag{4.41}$$

for $\boldsymbol{\psi} \in D((A_s + I)^\beta)$. Applying the duality argument, one can also derive from (4.41) that

$$\big\|(A_s + I)^{-\gamma} \boldsymbol{\psi}\big\|_q \leq c \,\|\boldsymbol{\psi}\|_s \tag{4.42}$$

for $\boldsymbol{\psi} \in \mathbf{L}^s_\sigma(\Omega)$ and $2\gamma \geq 3(s^{-1} - q^{-1})$.

B) The Linear (Stokes) Problem For $0 < \alpha < 1$ and $1 < r < \infty$, denote

$$\mathbf{D}^{\alpha,r}_s(\Omega) := \Big\{\boldsymbol{\psi} \in \mathbf{L}^s_\sigma(\Omega);\ \|\boldsymbol{\psi}\|_{\mathbf{D}^{\alpha,r}_s(\Omega)} := \|\boldsymbol{\psi}\|_s + \int_0^\infty \big\|t^{1-\alpha} A_s \mathrm{e}^{-A_s t} \boldsymbol{\psi}\big\|_s^r \,\frac{\mathrm{d}t}{t} < \infty\Big\}.$$

This space coincides with the real interpolation space $[D(A_s), \mathbf{L}^s_\sigma(\Omega)]_{1-\alpha,r}$, see [11].

Let $0 < T \leq \infty$. Let us at first consider the linear IVP

$$\mathbf{v}' + \nu A_s \mathbf{v} = \mathbf{F}, \qquad \mathbf{v} = \mathbf{v}_0 \tag{4.43}$$

in $\mathbf{L}^s_\sigma(\Omega)$. It is shown in [36] that if $\mathbf{F} \in L^r(0,T;\ \mathbf{L}^s_\sigma(\Omega))$ (where $1 < r < \infty$ and $1 < s < \infty$ if Ω is bounded or a half-space or the whole space $\mathbb{R}^3$, or $1 < s < \frac{3}{2}$ if Ω is an exterior domain) and $\mathbf{v}_0 \in \mathbf{D}_s^{1-r^{-1},r}(\Omega)$ then there exists a unique solution $\mathbf{v}$ of (4.43), such that $\mathbf{v} \in L^r(0,T_0;\ \mathbf{W}^{2,s}(\Omega))$ for each $0 < T_0 \le T$, $T_0 < \infty$. Moreover, there exists an associated pressure π such that the pair $\mathbf{v}$, π satisfies the evolutionary Stokes equation

$$\partial_t \mathbf{v} + \nabla\pi \;=\; \nu\Delta\mathbf{v} + \mathbf{F} \tag{4.44}$$

a.e. in Q_T and

$$\int_0^T \left(\|\partial_t\mathbf{v}\|_s^r + \|\nabla^2\mathbf{v}\|_s^r + \|\nabla\pi\|_s^r\right) \mathrm{d}t \le c_1\, \|\mathbf{v}_0\|^r_{\mathbf{D}_s^{1-r^{-1},r}(\Omega)} + c_2 \int_0^T \|\mathbf{F}\|_s^r \,\mathrm{d}t. \tag{4.45}$$

The proof is based on analysis of an abstract parabolic equation in a Banach space and former results of Giga, Sohr, Solonnikov, Miyakawa and others, see [36] for the complete list of references.

C) The Nonlinear (Navier-Stokes) Problem Suppose that $\mathbf{u}$ is a weak solution to the Navier-Stokes IBVP (4.9)–(4.12), with $\mathbf{f} \in L^r(0,T;\ \mathbf{L}^s(\Omega))$ and $\mathbf{u}_0 \in \mathbf{D}_s^{1-r^{-1},r}(\Omega)$. Applying Hölder's, Sobolev's and interpolation inequalities, one can verify that if $1 < s < \frac{3}{2}$, $2/r + 3/s = 4$ then $\mathbf{u}\cdot\nabla\mathbf{u} \in L^r(0,T;\ \mathbf{L}^s(\Omega))$. Thus, denoting $\mathbf{F} := P_s\mathbf{f} - P_s(\mathbf{u}\cdot\nabla\mathbf{u})$, we observe that $\mathbf{F} \in L^r(0,T;\ \mathbf{L}^s_\sigma(\Omega))$. Consequently, the problem (4.43) (with $\mathbf{v}_0 = \mathbf{u}_0$) has a unique solution $\mathbf{v}$, such that $\mathbf{v}$ and the associated pressure π satisfy (4.44) and (4.45). This, however, does not directly imply that $\mathbf{v} = \mathbf{u}$ and π is a pressure associated with solution $\mathbf{u}$. The reason is that the weak solution $\mathbf{u}$ is only known to satisfy the integral equation (4.13), respectively, also equations (4.19) and (4.20) (where the latter is an equation in $\mathbf{W}^{-1,2}_{0,\sigma}(\Omega)$), but not automatically also the first equation in (4.43) (which is an equation in $\mathbf{L}^s_\sigma(\Omega)$). Thus, the authors of [36] apply subtler arguments, which we explain in the case of a bounded domain Ω: we use (4.13) with $\boldsymbol{\phi}(\mathbf{x},t) = J_k(\mathbf{w}(\mathbf{x}))\,\vartheta(t)$, where $k \in \mathbb{N}$, $\mathbf{w} \in \mathbf{C}^\infty_{0,\sigma}(\Omega)$, $\vartheta \in C^1\big([0,T]\big)$, $\vartheta(T) = 0$, and $J_k := (I + k^{-1}A_s)^{-1}$. ($J_k(\mathbf{w})$ is the so-called *Yosida approximation* of function $\mathbf{w}$. The properties of Yosida's approximations are described in greater detail, e.g., in [75].) Thus, (4.13) yields

$$\int_0^T \int_\Omega \big[-\mathbf{u}\cdot J_k(\mathbf{w})\,\vartheta' + \nu\nabla\mathbf{u} : \nabla J_k(\mathbf{w})\,\vartheta + (\mathbf{u}\cdot\nabla\mathbf{u})\cdot J_k(\mathbf{w})\,\vartheta\big]\,\mathrm{d}\mathbf{x}\,\mathrm{d}t = \int_\Omega \mathbf{u}_0\cdot J_k(\mathbf{w})\,\vartheta(0)\,\mathrm{d}\mathbf{x} + \int_0^T \int_\Omega \mathbf{f}\cdot J_k(\mathbf{w})\,\vartheta\,\mathrm{d}\mathbf{x}\,\mathrm{d}t,$$

$$\int_0^T \int_\Omega \big[- J_k(\mathbf{u}) \cdot \mathbf{w}\, \vartheta' + \nu \nabla J_k(\mathbf{u}) : \nabla \mathbf{w}\, \vartheta + J_k(P_s(\mathbf{u} \cdot \nabla \mathbf{u})) \cdot \mathbf{w}\, \vartheta \big] \, \mathrm{d}\mathbf{x}\, \mathrm{d}t$$
$$= \int_\Omega J_k(\mathbf{u}_0) \cdot \mathbf{w}\, \vartheta(0)\, \mathrm{d}\mathbf{x} + \int_0^T \int_\Omega J_k(\mathbf{f}) \cdot \mathbf{w}\, \vartheta \, \mathrm{d}\mathbf{x}\, \mathrm{d}t. \tag{4.46}$$

(The integral of $\nu \nabla \mathbf{u} : \nabla J_k(\mathbf{w})\, \vartheta$ on the first line has been modified by means of the identities

$$\int_\Omega \nabla \mathbf{u} : \nabla J_k(\mathbf{w})\, \mathrm{d}\mathbf{x} = - \int_\Omega \mathbf{u} \cdot P_s \Delta J_k(\mathbf{w})\, \mathrm{d}\mathbf{x} = - \int_\Omega \mathbf{u} \cdot J_k(P_s \Delta \mathbf{w})\, \mathrm{d}\mathbf{x}$$
$$= - \int_\Omega J_k(\mathbf{u}) \cdot \Delta \mathbf{w}\, \mathrm{d}\mathbf{x} = \int_\Omega \nabla J_k(\mathbf{u}) : \nabla \mathbf{w}\, \mathrm{d}\mathbf{x}.)$$

Since $\|A_s J_k(\mathbf{u})\|_s \le c\, \|A_2 J_k(\mathbf{u})\|_2 \le c\, \|\mathbf{u}\|_2$, which is in $L^\infty(0,T)$, we observe from (4.46) that $J_k(\mathbf{u})$ satisfies the equation

$$J_k(\mathbf{u})' + \nu A_s J_k(\mathbf{u}) = \mathbf{F}_k := J_k(\mathbf{f}) - J_k[P_s(\mathbf{u} \cdot \nabla \mathbf{u})], \tag{4.47}$$

(in space $\mathbf{L}^s_\sigma(\Omega)$) and the initial condition $J_k(\mathbf{u})(0) = J_k(\mathbf{u}_0)$. Note that the solution $J_k(\mathbf{u})$ is unique, see Sect. 4.3.3B. Thus, estimate (4.45) yields

$$\int_0^T \big(\|J_k(\mathbf{u})'\|_s^r + \|\nabla^2 J_k(\mathbf{u})\|_s^r + \|\nabla p_k\|_s^r \big)\, \mathrm{d}t$$
$$\le c_1\, \|J_k(\mathbf{u}_0)\|^r_{\mathbf{D}_s^{1-r^{-1},r}(\Omega)} + c_2 \int_0^T \|\mathbf{F}_k\|_s^r\, \mathrm{d}t$$
$$\le c_3\, \|\mathbf{u}_0\|^r_{\mathbf{D}_s^{1-r^{-1},r}(\Omega)} + c_4 \int_0^T \big(\|\mathbf{f}\|_s^r + \|\mathbf{u} \cdot \nabla \mathbf{u}\|_s^r \big)\, \mathrm{d}t,$$

where p_k is a pressure, associated with the solution $J_k\mathbf{u}$ of equation (4.47). Constants c_3 and c_4 are independent of k. Hence, passing to the limit for $k \to \infty$, we deduce that there exists a scalar function p with $\nabla p \in L^r(0,T;\, \mathbf{L}^s(\Omega))$ such that $\mathbf{u}$, p satisfy the Navier-Stokes equation (4.9) a.e. in Q_T and

$$\int_0^T \big(\|\partial_t \mathbf{u}\|_s^r + \|\nabla^2 \mathbf{u}\|_s^r + \|\nabla p\|_s^r \big)\, \mathrm{d}t$$
$$\le c_3\, \|\mathbf{u}_0\|^r_{\mathbf{D}_s^{1-r^{-1},r}(\Omega)} + c_4 \int_0^T \big(\|\mathbf{f}\|_s^r + \|\mathbf{u} \cdot \nabla \mathbf{u}\|_s^r \big)\, \mathrm{d}t. \tag{4.48}$$

Function p can be chosen so that $p \in L^r(0,T;\, L^{3s/(3-s)}(\Omega))$. (This basically follows from the inclusion $\nabla p \in L^r(0,T;\, \mathbf{L}^s(\Omega))$ and the Poincaré-Sobolev inequality, see [36] for further references.)

Combining this procedure with the method of localization to a bounded domain $\Omega_R := \Omega \cap B_R(\mathbf{0})$ (described, e.g., in [61]) for sufficiently large R and considering the limit passage $R \to \infty$, we can also derive analogous results if $\Omega = \mathbb{R}^3$ or Ω is an exterior domain in $\mathbb{R}^3$. The case when Ω is a half-space can be reduced to the whole space $\mathbb{R}^3$ by an appropriate symmetric extension of $\mathbf{u}$ and p. Thus, we obtain the theorem (see Theorem 3.1 in [36]):

Theorem 4.3.3 *Let Ω be a bounded or exterior domain in $\mathbb{R}^3$ with the boundary of the class $C^{2+(h)}$ for some $h > 0$, or a half-space in $\mathbb{R}^3$ or the whole space $\mathbb{R}^3$. Let $T > 0$, $1 < s < \frac{3}{2}$, $1 < r < 2$, $2/r + 3/s = 4$, $\mathbf{f} \in L^r(0,T;\, \mathbf{L}^s(\Omega)) \cap L^2(0,T;\, \mathbf{L}^2(\Omega))$ and $\mathbf{u}_0 \in \mathbf{D}_s^{1-r^{-1},r}(\Omega) \cap \mathbf{L}^2_\sigma(\Omega)$. Let $\mathbf{u}$ be a weak solution to the Navier-Stokes IBVP (4.9)–(4.12) and p be an associated pressure. Then $\mathbf{u}$ and p satisfy estimate (4.48). Consequently, $\mathbf{u} \in L^r(0,T_0;\, \mathbf{W}^{2,s}(\Omega))$ for each $0 < T_0 \le T$, $T_0 < \infty$ and function p can be chosen so that it belongs to $L^r(0,T;\, L^{3s/(3-s)}(\Omega))$. Functions $\mathbf{u}$, p satisfy the equations (4.9), (4.10) a.e. in Q_T.*

D) Remark The assumptions $\mathbf{f} \in L^2(0,T;\, \mathbf{L}^2(\Omega))$ and $\mathbf{u}_0 \in \mathbf{L}^2_\sigma(\Omega)$ guarantee the consistency with the definition of a weak solution, see Sect. 4.3.1B. The pressure p is determined uniquely up to an additive function $g \in L^r(0,T)$.

The choice $r = \frac{5}{3}$, $s = \frac{15}{14}$ in Theorem 4.3.3 yields $p \in L^{5/3}(Q_T)$. This inclusion has already been obtained by Caffarelli, Kohn, and Nirenberg in [12] in the case $\Omega = \mathbb{R}^3$. (The authors come from the equation $\Delta p = \operatorname{div} \mathbf{f} - \nabla^2 : (\mathbf{u} \otimes \mathbf{u})$ and use the fact that $|\mathbf{u} \otimes \mathbf{u}|$ is in $L^{5/3}(Q_T)$. The same method as in [12], however, fails if $\Omega \ne \mathbb{R}^3$ due to the lack of appropriate boundary conditions for p.)

4.3.4 A Pressure Associated with a Weak Solution to More General Equations

We have shown in Sect. 4.3.1D that the Navier-Stokes system is equivalent to equation (4.20) in the space $\mathbf{W}^{-1,2}_{0,\sigma}(\Omega)$. In this subsection, we show that a certain pressure can also be assigned to function $\mathbf{u}$, satisfying a more general equation of the type

$$(\mathbf{u}')_\sigma + \mathcal{P}_q \mathbf{F} = \mathbf{0} \tag{4.49}$$

in the space $\mathbf{W}^{-1,q}_{0,\sigma}(\Omega)$, where $1 < q < \infty$. The content of this subsection is based on paper [83] by J. Wolf.

A) q-Suitable and ∇_q-Regular Domains, Projections $E^{1,q'}$ and $E^{-1,q}$, Mapping $\mathcal{F}_q^{(U)}$ Before we present the first main theorem of this subsection, we introduce several new notions:

Let $1 < q < \infty$. A sub-domain U of Ω is said to be *q-suitable* if the restriction $p|_U$ is in $L^q(U)$ for all $p \in L^q_{pot}(\Omega)$. (See Sect. 4.2D for the definition of $L^q_{pot}(U)$.) Obviously, as $p \equiv 1 \in L^q_{pot}(\Omega)$, $\text{meas}(U) < \infty$. Furthermore, if $U \subset\subset \Omega$, then U is q-suitable and if $L^q_{pot}(\Omega) = L^q(\Omega)$ then every sub-domain U of Ω is q-suitable.

Further on, we denote by $L^q_0(U)$ the space of those functions p from $L^q(U)$ that satisfy $\int_U p \, d\mathbf{x} = 0$.

We have introduced the projections $E^{1,2}$ (in $\mathbf{W}^{1,2}_0(\Omega)$) and $E^{-1,2}$ (in $\mathbf{W}^{-1,2}_0(\Omega)$) in Sect. 4.3.2A. If $1 < q < \infty$ then, however, the existence of analogous projections $E^{1,q'}$ (in $\mathbf{W}^{1,q'}_0(\Omega)$) and $E^{-1,q}$ (in $\mathbf{W}^{-1,q}_0(\Omega)$) with similar properties is not automatic. The properties, we have in mind, are:

$$\ker E^{1,q'} = \mathbf{W}^{1,q'}_{0,\sigma}(\Omega), \tag{4.50}$$

$$E^{1,q'}(\boldsymbol{\psi}) = E^{1,2}\boldsymbol{\psi} \quad \text{for all } \boldsymbol{\psi} \in \mathbf{C}^\infty_0(\Omega). \tag{4.51}$$

Domain Ω is said to be *∇_q-regular* if a projection $E^{1,q'} : \mathbf{W}^{1,q'}_0(\Omega) \to \mathbf{W}^{1,q'}_0(\Omega)$, satisfying (4.50) and (4.51), exists. In [83], the author proves sufficient conditions for Ω being ∇_q-regular (see [83, Lemma 5.3]) and shows that these conditions are satisfied, e.g., if Ω is a bounded or exterior domain in $\mathbb{R}^3$ with $\partial\Omega \in C^1$ or $\Omega = \mathbb{R}^3$. Obviously, every domain Ω in $\mathbb{R}^3$ is ∇_2-regular. The equalities (4.24) and (4.51) imply that $E^{1,q'}(\nabla\phi) = \nabla\phi$ for all $\phi \in C^\infty_0(\Omega)$.

We denote by $E^{-1,q}$ the adjoint projection to $E^{1,q'}$. Note that projections $E^{1,q'}$ and $E^{-1,q}$ are not unique, because $E^{1,q'}$ depends on $E^{1,2}$ (through condition (4.51)) and the projection $E^{1,2}$ depends on the way we define the scalar products in $\mathbf{W}^{1,2}_0(\Omega)$ and $\mathbf{W}^{-1,2}_0(\Omega)$. (One can also use $\mathcal{A} + \lambda I$ for $\lambda > 0$ instead of $\mathcal{A} + I$ in (4.21).) If $\mathbf{f} \in \mathbf{W}^{-1,q}_0(\Omega) \cap \mathbf{W}^{-1,2}_0(\Omega)$, then

$$\langle E^{-1,q}\mathbf{f}, \boldsymbol{\psi}\rangle_\Omega = \langle \mathbf{f}, E^{1,q}\boldsymbol{\psi}\rangle_\Omega = \langle \mathbf{f}, E^{1,2}\boldsymbol{\psi}\rangle_\Omega = \langle E^{-1,2}\mathbf{f}, \boldsymbol{\psi}\rangle_\Omega$$

for all $\boldsymbol{\psi} \in \mathbf{C}^\infty_0(\Omega)$, which shows that $E^{-1,q}\mathbf{f} = E^{-1,2}\mathbf{f}$.

If Ω is a bounded smooth domain, then projection $E^{-1,q}$ can also be described in another way. For this purpose, we use the information on solution of the equation

$$-\nu\Delta\mathbf{v} + \lambda\mathbf{v} + \nabla p = \mathbf{f} \tag{4.52}$$

for $\lambda \geq 0$. (Equation (4.52) is called the steady Stokes equation if $\lambda = 0$ and the steady Stokes-like equation if $\lambda > 0$.) It is proven in paper [28] that if Ω is a bounded domain in $\mathbb{R}^3$ with $\partial\Omega \in C^1$, $\mathbf{f} \in \mathbf{W}^{-1,q}_0(\Omega)$ $(1 < q < \infty)$ and $\lambda = 0$ then there exists a unique pair $(\mathbf{v}, p) \in \mathbf{W}^{1,q}_{0,\sigma}(\Omega) \times L^q_0(\Omega)$, satisfying equation (4.52) (as an equation in $\mathbf{W}^{-1,q}_0(\Omega)$) and the inequality

$$\|\nabla\mathbf{v}\|_q + \|p\|_q \leq c\,\|\mathbf{f}\|_{-1,q}. \tag{4.53}$$

If, moreover, $\partial\Omega \in C^2$ and $\mathbf{f} \in \mathbf{L}^q(\Omega)$ then, applying the estimates from [29, Ch. IV], one can deduce that the solution of (4.52) satisfies

$$\|\mathbf{v}\|_{2,q} + \|p\|_{1,q} \;\leq\; c\,\|\mathbf{f}\|_q. \tag{4.54}$$

These results can be extended by means of standard arguments to the case $\lambda > 0$.

Assume at first that $q = 2$ and $\lambda = \nu$. Define $\widetilde{E}^{-1,2}\mathbf{f} := \nabla p$. If $\nabla\xi \in \nabla(L^2(\Omega)) \equiv \mathbf{W}^{1,2}_{0,\sigma}(\Omega)^\perp$, then

$$\big(\nu(-\Delta\mathbf{v}+\mathbf{v}),\nabla\xi\big)_{-1,2} \;=\; \big(\nu(\mathcal{A}+I)\mathbf{v},\nabla\xi\big)_{-1,2} \;=\; \nu\big\langle\mathbf{v},\nabla\xi\big\rangle_\Omega \;=\; 0.$$

This shows that $-\Delta\mathbf{v}+\mathbf{v}$ is in the orthogonal complement to $\mathbf{W}^{1,2}_{0,\sigma}(\Omega)^\perp$ in $\mathbf{W}^{-1,2}_0(\Omega)$. Thus, as $\nabla p \in \mathbf{W}^{1,2}_{0,\sigma}(\Omega)^\perp$, we observe that $\nabla p \equiv \widetilde{E}^{-1,2}\mathbf{f} = E^{-1,2}\mathbf{f}$. As this holds for all $\mathbf{f} \in \mathbf{W}^{-1,2}_0(\Omega)$, we have

$$E^{-1,2} \;=\; \widetilde{E}^{-1,2}.$$

By analogy, if $1 < q < \infty$ and $\lambda = \nu$, then the aforementioned result on the solvability of equation (4.52) shows that $\mathbf{W}^{-1,q}_0(\Omega)$ can be decomposed to the direct sum $\mathbf{W}^{-1,q}_{\mathrm{div}}(\Omega) \oplus \nabla(L^q(\Omega))$, where $\mathbf{W}^{-1,q}_{\mathrm{div}}(\Omega) := \{\Delta\mathbf{v};\ \mathbf{v} \in \mathbf{W}^{1,q}_{0,\sigma}(\Omega)\}$. Define $\widetilde{E}^{-1,q}\mathbf{f} := \nabla p$. Obviously, $\widetilde{E}^{-1,q}$ is a projection in $\mathbf{W}^{-1,q}_0(\Omega)$. If $\mathbf{f} \in \mathbf{W}^{-1,q}_0(\Omega) \cap \mathbf{W}^{-1,2}_0(\Omega)$, then

$$\widetilde{E}^{-1,q}\mathbf{f} \;=\; \widetilde{E}^{-1,2}\mathbf{f} \;=\; E^{-1,2}\mathbf{f} \;=\; E^{-1,q}\mathbf{f}.$$

Now, we obtain from the density of $\mathbf{W}^{-1,q}_0(\Omega) \cap \mathbf{W}^{-1,2}_0(\Omega)$ in $\mathbf{W}^{-1,q}_0(\Omega)$ that

$$E^{-1,q} \;=\; \widetilde{E}^{-1,q}. \tag{4.55}$$

Finally, let domain Ω be ∇_q-regular and U be a q-suitable sub-domain of Ω. If $\mathbf{f} \in \mathbf{W}^{-1,q}_0(\Omega)$, then $E^{-1,q}\mathbf{f} \in \mathbf{W}^{1,q'}_{0,\sigma}(\Omega)^\perp$. (This follows from (4.50).) Hence, as $\mathbf{W}^{1,q'}_{0,\sigma}(\Omega)^\perp = \nabla(L^q_{pot}(\Omega))$ (due to Lemma 4.2.5), there exists $p \in L^q_{pot}(\Omega)$ such that $E^{-1,q}\mathbf{f} = \nabla p$. Out of infinitely many functions p with this property, we choose the one satisfying $\int_U p\,\mathrm{d}\mathbf{x} = 0$ and put $\mathcal{F}^{(U)}_{q,\Omega}\mathbf{f} = p$.

B) A Global Pressure, Associated with the Distribution $\mathbf{u}'+\mathbf{F}$ The next theorem concerns a ∇_q-regular domain Ω. It comes from [83].

Theorem 4.3.4 *Let $1 < q < \infty$, $\Omega \subset \mathbb{R}^3$ be a ∇_q-regular domain and U be a q-suitable sub-domain of Ω. Let $\mathbf{u}$, $\mathbf{F} \in L^1(0,T;\ \mathbf{W}^{-1,q}_0(\Omega))$ so that*

$$\int_0^T \big[\langle -\mathbf{u},\boldsymbol{\psi}\rangle_\Omega\,\vartheta'(t) + \langle\mathbf{F},\boldsymbol{\psi}\rangle_\Omega\,\vartheta(t)\big]\,\mathrm{d}t \;=\; 0 \tag{4.56}$$

for all $\boldsymbol{\psi} \in \mathbf{C}^{\infty}_{0,\sigma}(\Omega)$ *and* $\vartheta \in C_0^{\infty}((0,T))$. *Then*

$$\int_0^T \big[- \langle \mathbf{u}, \partial_t \boldsymbol{\phi} \rangle_{\Omega} + \langle \mathbf{F}, \boldsymbol{\phi} \rangle_{\Omega} \big] \, \mathrm{d}t \; = \int_0^T \int_{\Omega} \big(- p_0 \operatorname{div} \partial_t \boldsymbol{\phi} + p_1 \operatorname{div} \boldsymbol{\phi} \big) \, \mathrm{d}\mathbf{x} \, \mathrm{d}t \tag{4.57}$$

for all $\boldsymbol{\phi} \in \mathbf{C}_0^{\infty}(Q_T)$, *where* $p_0(t) = -\mathfrak{F}^{(U)}_{q,\Omega}\mathbf{u}(t)$ *and* $p_1(t) = -\mathfrak{F}^{(U)}_{q,\Omega}\mathbf{F}(t)$ *for a.a.* $t \in (0,T)$. *If, moreover,* $\operatorname{div} \mathbf{u} = 0$ *in the sense of distributions in* Q_T *then* $p_0(t)$ *is harmonic for a.a.* $t \in (0,T)$. *Furthermore, if* $\mathbf{u} \in C_{w^*}\big([0,T];\ \mathbf{W}_0^{-1,q}(\Omega)\big)$, *then* p_0 *is continuous in* $\Omega \times [0,T]$.

Proof Since $\mathbf{C}^{\infty}_{0,\sigma}(\Omega)$ is dense in $\mathbf{W}^{1,q'}_{0,\sigma}(\Omega)$, (4.56) holds for all $\boldsymbol{\psi} \in \mathbf{W}^{1,q'}_{0,\sigma}(\Omega) \equiv \ker E^{1,q'}$. The operator $I - E^{1,q'}$ is a projection of $\mathbf{W}_0^{1,q'}(\Omega)$ onto $\ker E^{1,q'}$. Thus, (4.56) yields

$$\int_0^T \big[\big\langle - \mathbf{u}, (I - E^{1,q'})\boldsymbol{\psi} \big\rangle_{\Omega} \, \vartheta'(t) + \big\langle \mathbf{F}, (I - E^{1,q'})\boldsymbol{\psi} \big\rangle_{\Omega} \, \vartheta(t) \big] \, \mathrm{d}t \; = \; 0$$

for all $\boldsymbol{\psi} \in \mathbf{W}_0^{1,q'}(\Omega)$ and $\vartheta \in C_0^{\infty}((0,T))$. Hence

$$\int_0^T \big[\big\langle - (I - E^{-1,q})\mathbf{u}, \boldsymbol{\psi} \big\rangle_{\Omega} \, \vartheta'(t) + \big\langle (I - E^{-1,q})\mathbf{F}, \boldsymbol{\psi} \big\rangle_{\Omega} \, \vartheta(t) \big] \, \mathrm{d}t \; = \; 0.$$

This shows that $[(I - E^{-1,q})\mathbf{u}]' + (I - E^{-1,q})\mathbf{F} = \mathbf{0}$ in the sense of distributions in $(0,T)$. Due to the density of all finite linear combinations of functions of the type $\boldsymbol{\psi}(\mathbf{x})\,\vartheta(t)$ (where $\boldsymbol{\psi} \in \mathbf{W}_0^{1,q'}(\Omega)$ and $\vartheta \in C_0^{\infty}((0,T))$) in $C_0^1((0,T);\ \mathbf{W}_0^{1,q'}(\Omega))$, we obtain

$$\int_0^T \big[\big\langle - (I - E^{-1,q})\mathbf{u}, \boldsymbol{\phi}'(t) \big\rangle_{\Omega} + \big\langle (I - E^{-1,q})\mathbf{F}, \boldsymbol{\phi}(t) \big\rangle_{\Omega} \big] \, \mathrm{d}t \; = \; 0 \tag{4.58}$$

for all $\boldsymbol{\phi} \in C_0^1((0,T);\ \mathbf{W}_0^{1,q'}(\Omega))$. Equation (4.57) now follows from (4.58), if we take into account that $E^{-1,q}\mathbf{u}(t) = -\nabla p_0(t)$ and $E^{-1,q}\mathbf{F}(t) = -\nabla p_1(t)$.

The fact that p_0 is harmonic can be proven by means of the same arguments (based on Weyl's lemma) as in Sect. 4.3.2C.

For the proof of the last part of Theorem 4.3.4, i.e. the continuity of p_0 in $\Omega \times [0,T]$, provided that $\mathbf{u} \in C_{w^*}\big([0,T];\ \mathbf{W}_0^{-1,q}(\Omega)\big)$, we refer to paper [83], proof of Theorem 6.1. □

C) Remark Equation (4.56) says that $\mathbf{u}$ and $\mathbf{F}$ satisfy (4.49), where (by analogy with Sect. 4.3.1D), $(\mathbf{u}')_{\sigma}$ denotes the distributional derivative (with respect to t) of $\mathbf{u}$, as a function from $(0,T)$ to $\mathbf{W}^{-1,q}_{0,\sigma}(\Omega)$. If $\mathbf{u}'$ denotes the distributional derivative (with respect to t) of $\mathbf{u}$, as a function from $(0,T)$ to $\mathbf{W}_0^{-1,q}(\Omega)$, then both the equations (4.56) and (4.49) express the fact that the distribution $\mathbf{u}' + \mathbf{F}$ (acting on functions from $C_0^\infty((0,T);\ \mathbf{W}_0^{1,q'}(\Omega)))$ vanishes on functions from $C_0^\infty((0,T);$ $\mathbf{W}_{0,\sigma}^{1,q'}(\Omega))$. Due to the definition of mapping $\mathcal{F}_{r,\Omega}^{(U)}$, both ∇p_0 and ∇p_1 belong to $L^1(0,T;\ \mathbf{W}_0^{-1,q}(\Omega))$. Thus, equation (4.57) can also be written in the form

$$\mathbf{u}' + \mathbf{F} + (\nabla p_0)' + \nabla p_1 \;=\; \mathbf{0},$$

where $(\nabla p_0)'$ is the distributional derivative (with respect to t) of ∇p_0, as a function from $(0,T)$ to $\mathbf{W}_0^{-1,q}(\Omega)$. Moreover, it also says that $\mathbf{u}$, $\mathbf{F}$, p_0, and p_1, as distributions in Q_T, satisfy the equation

$$\partial_t\mathbf{u} + \mathbf{F} + \nabla\partial_t p_0 + \nabla p_1 \;=\; \mathbf{0}.$$

This justifies calling $\partial_t p_0 + p_1$ an "associated Pressure."

D) A Local Pressure, Associated with the Distribution u′ + F If domain Ω is ∇_q-regular and q-suitable, then we write only $\mathcal{F}_{q,\Omega}$ instead of $\mathcal{F}_{q,\Omega}^{(\Omega)}$. Furthermore, if $1 < q_1, q_2 < \infty$ and $\boldsymbol{\psi} \in \mathbf{C}_0^\infty(\Omega)$, then $E^{1,q_1'}\boldsymbol{\psi} = E^{1,2}\boldsymbol{\psi} = E^{1,q_2'}\boldsymbol{\psi}$. Hence $E^{1,q_1'}\boldsymbol{\psi} = E^{1,q_2'}\boldsymbol{\psi}$ for all $\boldsymbol{\psi} \in \mathbf{W}_0^{1,q_1'}(\Omega) \cap \mathbf{W}_0^{1,q_2'}(\Omega)$ due to the density of $\mathbf{C}_0^\infty(\Omega)$ in both $\mathbf{W}_0^{1,q_1'}(\Omega)$ and $\mathbf{W}_0^{1,q_2'}(\Omega)$. This implies that $E^{-1,q_1}\mathbf{f} = E^{-1,q_2}\mathbf{f}$ for all $\mathbf{f} \in \mathbf{W}_0^{-1,q_1}(\Omega) \cap \mathbf{W}_0^{-1,q_2}(\Omega)$. Consequently, $\mathcal{F}_{q_1,\Omega}\mathbf{f} = \mathcal{F}_{q_2,\Omega}\mathbf{f}$. It is therefore logical to omit q in $\mathcal{F}_{q,\Omega}$ and to write only $\mathcal{F}_\Omega$.

The next theorem provides an information on the existence and structure of the local associated pressure in an arbitrary open set Ω in $\mathbb{R}^3$. It coincides with Theorem 6.4 in [83].

Theorem 4.3.5 *Let $1 < q < \infty$, Ω be an open set in $\mathbb{R}^3$. Let $\mathbf{u} \in L^1(0,T;\ \mathbf{L}_{loc}^1(\Omega))$ and $\mathbf{F} \in L^1(0,T;\ \mathbf{W}_{loc}^{-1,q}(\Omega))$ be such that equation (4.56) holds for all $\boldsymbol{\psi} \in \mathbf{C}_{0,\sigma}^\infty(\Omega)$ and $\vartheta \in C_0^\infty((0,T))$. Let $D \subset\subset \Omega$ be a domain with $\partial D \in C^1$. Then the equation*

$$-\int_0^T\!\!\int_D \mathbf{u}\cdot\partial_t\boldsymbol{\phi}\,\mathrm{d}\mathbf{x}\,\mathrm{d}t + \int_0^T \langle\mathbf{F},\boldsymbol{\phi}\rangle_D\,\mathrm{d}t$$
$$= \int_0^T\!\!\int_D \big(-p_{0,D}\,\mathrm{div}\,\partial_t\boldsymbol{\phi} + p_{1,D}\,\mathrm{div}\,\boldsymbol{\phi}\big)\,\mathrm{d}\mathbf{x}\,\mathrm{d}t \tag{4.59}$$

holds for all $\boldsymbol{\phi} \in \mathbf{C}_0^\infty(D\times(0,T))$, where $p_{0,D}(t) = -\mathcal{F}_D\mathbf{u}(t)|_D$ and $p_{1,D}(t) = -\mathcal{F}_D\mathbf{F}(t)|_D$ for a.a. $t \in (0,T)$.

Note that since $\mathbf{u} \in L^1(0,T;\ \mathbf{L}_{loc}^1(\Omega))$, $\mathbf{u}$ also satisfies $\mathbf{u} \in L^1(0,T;\ \mathbf{L}^1(D)) \subset L^1(0,T;\ \mathbf{W}_0^{-1,s}(D))$ (for all $1 < s < \frac{3}{2}$). The assumption $\mathbf{F} \in L^1(0,T;\ \mathbf{W}_{loc}^{-1,q}(\Omega))$ implies that $\mathbf{F} \in L^1(0,T;\ \mathbf{W}_0^{-1,q}(D))$. Hence $p_{0,D} \in L^1(0,T;\ L_0^s(D))$ and $p_{1,D} \in L^1(0,T;\ L_0^q(D))$.

Equation (4.59) shows that $\partial_t \mathbf{u}|_D + \mathbf{F}|_D + \nabla(\partial_t p_{0,D} + p_{1,D})$ is the zero distribution in $D \times (0, T)$, which justifies calling $\partial_t p_{0,D} + p_{1,D}$ a "local associated pressure."

The proof of Theorem 4.3.5 can be found in [83].

E) Corollaries of Theorem 4.3.5 Suppose that all assumptions of Theorem 4.3.5 are fulfilled. Then the following assertions hold:

(i) *If* $\operatorname{div}\mathbf{u} = 0$ *in the sense of distributions then the function* $p_{0,D}(t)$ *is harmonic in* D *for a.a.* $t \in (0, T)$.

(ii) *If* $\mathbf{u} \in L^1(0, T;\ \mathbf{L}^s_{loc}(\Omega))$ *for some* $1 < s < \infty$ *and* D *is a* C^2*-domain then* $p_{0,D} \in L^1(0, T;\ W^{1,s}(D))$ *and satisfies the estimate*

$$\|\nabla p_{0,D}(t)\|_{s;\,D} \ \le\ c(s, D)\, \|\mathbf{u}(t)\|_{s;\,D} \qquad for\ a.a.\ t \in (0, T). \tag{4.60}$$

If D *is only a* C^1*-domain, then we only have* $p_{0,D} \in L^1(0, T;\ W^{1,s}_{loc}(D))$.

(iii) *If* $\mathbf{F} = \mathbf{F}_1 + \ldots + \mathbf{F}_N$ *for* $\mathbf{F}_i \in L^1(0, T;\ \mathbf{W}^{-1,q_i}_{loc}(\Omega))$ *(for* $1 < q_i < \infty$; $i = 1, \ldots, N$*), then*

$$p_{1,D} \ =\ p^1_{1,D} + \ldots + p^N_{1,D},$$

where $p^1_{1,D} = \mathcal{F}_D(\mathbf{F}_1|_D), \ldots,\ p^N_{1,D} = \mathcal{F}_D(\mathbf{F}_N|_D)$. *Moreover, the estimates*

$$\|p^i_{1,D}(t)\|_{q_i;\,D} \ \le\ c(q_i, D)\, \|\mathbf{F}_i(t)|_D\|_{-1,q_i;\,D} \qquad (i = 1, \ldots, N)$$

hold for a.a. $t \in (0, T)$.

(iv) *If* $\mathbf{u} \in C_w\big([0, T];\ \mathbf{L}^1_{loc}(\Omega)\big)$ *and* $\operatorname{div}\mathbf{u} = 0$ *in the sense of distributions, then* $p_{0,D}$ *is continuous in* $D \times [0, T]$.

Assertion (i) can be proven in this way: the identity $p_{0,D}(t) = -\mathcal{F}_D(\mathbf{u}(t))$ means that $\nabla p_{0,D}(t) = -E^{-1,s}\mathbf{u}(t)|_D$ for $1 < s < \frac{3}{2}$. Let $\psi \in C^\infty_0(D)$. Then

$$\begin{aligned} \big\langle \Delta p_{0,D}(t), \psi \big\rangle_D \ &=\ -\big\langle \nabla p_{0,D}(t), \nabla\psi \big\rangle_D \ =\ \big\langle E^{-1,s}\mathbf{u}(t)|_D, \nabla\psi \big\rangle_D \\ &=\ \big\langle \mathbf{u}(t)|_D, E^{1,s'}\nabla\psi \big\rangle_D \ =\ \big\langle \mathbf{u}(t)_D, \nabla\psi \big\rangle_D \ =\ \int_D \mathbf{u}(t)\cdot\nabla\psi\ \mathrm{d}\mathbf{x} \\ &=\ \int_D \operatorname{div}\mathbf{u}(t)\ \psi\ \mathrm{d}\mathbf{x} \ =\ 0. \end{aligned}$$

As this holds for all $\psi \in C^\infty_0(D)$, $p_{0,D}(t)$ is harmonic in D.

The inclusion $p_{0,D} \in L^1(0, T;\ W^{1,s}(D))$ and inequality (4.60) in assertion (ii) follow from (4.53) and (4.55). Assertion (iii) follows from the linearity of the assignment $\mathbf{F}|_D \mapsto p_{0,D}$. Finally, assertion (iv) follows from Theorem 4.3.4.

4.4 An Interior Regularity of Pressure Under Serrin's Condition

In this section, we assume for simplicity that the external body force $\mathbf{f}$ in equation (4.9) is zero.

The next lemma presents the classical result on interior regularity of a weak solution $\mathbf{u}$ to the Navier-Stokes system (4.9), (4.10). It says that if $\mathbf{u}$ is a weak solution to the equations (4.9), (4.10) in $\Omega' \times (t_1, t_2)$, where Ω' is a domain in $\mathbb{R}^3$ and $t_1 < t_2$, satisfying certain additional condition of integrability in $\Omega' \times (t_1, t_2)$, then $\mathbf{u}$ is in a some sense regular in $\Omega' \times (t_1, t_2)$. (See T. Ohyama [63] and J. Serrin [68].)

Lemma 4.4.1 *Let Ω' be a domain in $\mathbb{R}^3$ and let $\mathbf{u}$ be a weak solution to the Navier-Stokes system (4.9), (4.10) (with $\mathbf{f} = \mathbf{0}$) in $\Omega' \times (t_1, t_2)$. Suppose that $\mathbf{u}$ satisfies the condition*

(i) $\mathbf{u} \in L^r(t_1, t_2;\, \mathbf{L}^s(\Omega'))$ *for some $r, s \in \mathbb{R}$ such that* $\dfrac{2}{r} + \dfrac{3}{s} = 1,\ 3 < s < \infty.$

Then $\mathbf{u}$ and its spatial derivatives (of all orders) are essentially bounded on every compact subset of $\Omega' \times (t_1, t_2)$.

The assumption that $\mathbf{u}$ is a weak solution to the Navier-Stokes system (4.9), (4.10) in $\Omega' \times (t_1, t_2)$ means that $\mathbf{u} \in L^\infty(0, T;\, \mathbf{L}^2(\Omega')) \cap L^2(0, T;\, \mathbf{W}^{1,2}(\Omega'))$, $\operatorname{div} \mathbf{u} = 0$ a.e. in $\Omega' \times (t_1.t_2)$ and $\mathbf{u}$ satisfies the integral equation

$$\int_{t_1}^{t_2} \int_{\Omega'} \big[-\mathbf{u} \cdot \partial_t \boldsymbol{\phi} + \nu \nabla \mathbf{u} : \nabla \boldsymbol{\phi} + \mathbf{u} \cdot \nabla \mathbf{u} \cdot \boldsymbol{\phi} \big] \, \mathrm{d}\mathbf{x} \, \mathrm{d}t \;=\; 0$$

for all $\boldsymbol{\phi} \in C_0^\infty\big((t_1, t_2);\, \mathbf{W}_{0,\sigma}^{1,2}(\Omega')\big)$. (This equation follows from (4.13).) The assumptions of the lemma are satisfied, e.g., in some neighborhood of every regular point $(\mathbf{x}_0, t_0)$ of the weak solution $\mathbf{u}$, where the solution is (by definition) essentially bounded. We do not mention boundary or initial conditions, because the lemma holds independently of these conditions.

We observe that Lemma 4.4.1 says nothing on the regularity of p and $\partial_t \mathbf{u}$ in $\Omega' \times (t_1, t_2)$. In fact, the regularity of $\partial_t \mathbf{u}$ is connected with the regularity of ∇p, because all other terms in the Navier-Stokes equation (4.9) contain just $\mathbf{u}$ or spatial derivatives of $\mathbf{u}$ and they are therefore "covered" by Lemma 4.4.1. In the next two subsections, we present some results on the regularity of p and $\partial_t \mathbf{u}$ in $\Omega' \times (t_1, t_2)$ provided that condition (i) holds. The results show the importance of boundary conditions, satisfied by $\mathbf{u}$ on $\partial\Omega \times (t_1, t_2)$, for the regularity of p and $\partial_t \mathbf{u}$ in $\Omega' \times (t_1, t_2)$.

4.4.1 The Case $\Omega = \mathbb{R}^3$

A) The Main Theorem Recall that the notation $\Omega'' \subset\subset \Omega'$ means that Ω'' is a bounded sub-domain of Ω' such that $\overline{\Omega''} \subset \Omega'$. The next theorem is in various versions proven in [72] and [59].

Theorem 4.4.2 *Let $\Omega = \mathbb{R}^3$. Let Ω' be a domain in $\mathbb{R}^3$, $-\infty < t_1 < t_2 < \infty$ and let $\mathbf{u}$ be a weak solution to the Navier-Stokes system (4.9), (4.10), satisfying condition (i) of Lemma 4.4.1 in $\Omega' \times (t_1, t_2)$. Let p be an associated pressure. Then $\partial_t \mathbf{u}$, ∇p, and all their spatial derivatives (of all orders) are in $\mathbf{L}^\infty(\Omega'' \times (\tau_1, \tau_2))$ for any $t_1 < \tau_1 < \tau_2 < t_2$ and $\Omega'' \subset\subset \Omega'$.*

Proof We deduce by means of Theorem 4.3.3 that the gradient of the associated pressure p is, together with $\partial_t \mathbf{u}$ and $\Delta \mathbf{u}$, in $L^s(\tau_1, \tau_2;\ \mathbf{L}^{3s/(4s-2)}(\mathbb{R}^3))$ for $s \in (1, 2)$. It follows from equation (4.9) that

$$\Delta p \;=\; -\mathrm{div}\,(\mathbf{u} \cdot \nabla \mathbf{u}) \tag{4.61}$$

in $\mathbb{R}^3 \times \{t\}$ for a.a. $t \in (\tau_1, \tau_2)$. We further assume that t is one of the times in (τ_1, τ_2) for which (4.61) holds.

Let η be a function from $C_0^\infty(\mathbb{R}^3)$, with values in the interval [0, 1], such that $\mathrm{supp}\,\eta \subset \Omega'$, $\eta = 1$ in Ω'' and $\epsilon := \mathrm{dist}\{\mathrm{supp}\,(\nabla \eta);\ \Omega''\} > 0$. Let $\mathbf{e}$ be a unit vector in $\mathbb{R}^3$. Then

$$\begin{aligned} \eta(\mathbf{x})\,\mathbf{e} \cdot \nabla p(\mathbf{x}, t) \;&=\; -\frac{1}{4\pi} \int_{\mathbb{R}^3} \frac{1}{|\mathbf{x} - \mathbf{y}|}\, \Delta\big[\eta(\mathbf{y})\,\mathbf{e} \cdot \nabla p(\mathbf{y}, t)\big]\, \mathrm{d}\mathbf{y} \\ &=\; -\frac{1}{4\pi}\big[P^{(1)}(\mathbf{x}, t) + 2P^{(2)}(\mathbf{x}, t) + P^{(3)}(\mathbf{x}, t)\big], \end{aligned}$$

for $\mathbf{x} \in \mathbb{R}^3$, where

$$\begin{aligned} P^{(1)}(\mathbf{x}, t) \;&=\; \int_{\mathrm{supp}\,\nabla\eta} \frac{1}{|\mathbf{x} - \mathbf{y}|}\, \Delta\eta(\mathbf{y})\, [\mathbf{e} \cdot \nabla p(\mathbf{y}, t)]\, \mathrm{d}\mathbf{y}, \\ P^{(2)}(\mathbf{x}, t) \;&=\; \int_{\mathrm{supp}\,\nabla\eta} \frac{1}{|\mathbf{x} - \mathbf{y}|}\, \nabla\eta(\mathbf{y}) \cdot \nabla\big[\mathbf{e} \cdot \nabla p(\mathbf{y}, t)\big]\, \mathrm{d}\mathbf{y}, \\ P^{(3)}(\mathbf{x}, t) \;&=\; \int_{\mathrm{supp}\,\eta} \frac{1}{|\mathbf{x} - \mathbf{y}|}\, \eta(\mathbf{y})\, \Delta\big[\mathbf{e} \cdot \nabla p(\mathbf{y}, t)\big]\, \mathrm{d}\mathbf{y}. \end{aligned}$$

We denote by $D^{\boldsymbol{\alpha}}$ the spatial derivative of order $|\boldsymbol{\alpha}|$, where $\boldsymbol{\alpha} = (\alpha_1, \alpha_2, \alpha_3)$ (for $\alpha_1, \alpha_2, \alpha_3 \in \mathbb{N} \cup \{0\}$) is a multi-index and $|\boldsymbol{\alpha}| \equiv \alpha_1 + \alpha_2 + \alpha_3$. If we want to specify whether the differentiation is considered with respect to $\mathbf{x}$ or $\mathbf{y}$, we use the notation $D_{\mathbf{x}}^{\boldsymbol{\alpha}}$ or $D_{\mathbf{y}}^{\boldsymbol{\alpha}}$.

Let us at first estimate $D^{\boldsymbol{\alpha}} P^{(3)}$. Applying (4.61) and Lemma 4.4.1, we obtain

$$
\begin{aligned}
|D^{\boldsymbol{\alpha}} P^{(3)}(\mathbf{x},t)| &= \left| D^{\boldsymbol{\alpha}}_{\mathbf{x}} \int_{\operatorname{supp}\eta} \frac{1}{|\mathbf{x}-\mathbf{y}|}\, \eta(\mathbf{y}) \big(\mathbf{e}\cdot\nabla_{\mathbf{y}}\Delta_{\mathbf{y}} p(\mathbf{y},t)\big)\, \mathrm{d}\mathbf{y} \right| \\
&= \left| \int_{\operatorname{supp}\eta} \frac{1}{|\mathbf{x}-\mathbf{y}|}\, D^{\boldsymbol{\alpha}}_{\mathbf{y}} \big\{ \eta(\mathbf{y})\, \mathbf{e}\cdot\nabla_{\mathbf{y}} \operatorname{div}_{\mathbf{y}}\big[\mathbf{u}(\mathbf{y},t)\cdot\nabla_{\mathbf{y}}\mathbf{u}(\mathbf{y},t)\big]\big\}\, \mathrm{d}\mathbf{y} \right| \\
&\leq c(\boldsymbol{\alpha}) \int_{\operatorname{supp}\eta} \frac{\mathrm{d}\mathbf{y}}{|\mathbf{x}-\mathbf{y}|} \leq c(\boldsymbol{\alpha})
\end{aligned} \tag{4.62}
$$

for $\mathbf{x} \in \Omega''$. The constant $c(\boldsymbol{\alpha})$ on the right- hand side of (4.62) is independent of $\mathbf{e}$, $\mathbf{x}$, and t. The derivative $D^{\boldsymbol{\alpha}} P^{(1)}$ satisfies

$$
D^{\boldsymbol{\alpha}} P^{(1)}(\mathbf{x},t) = \int_{\operatorname{supp}\nabla\eta} D^{\boldsymbol{\alpha}}_{\mathbf{x}} \left(\frac{1}{|\mathbf{x}-\mathbf{y}|}\right) \Delta_{\mathbf{y}}\eta(\mathbf{y})\, \mathbf{e}\cdot\nabla_{\mathbf{y}} p(\mathbf{y},t)\, \mathrm{d}\mathbf{y}. \tag{4.63}
$$

Further, we use the Helmholtz decomposition

$$
D^{\boldsymbol{\alpha}}_{\mathbf{x}} \left(\frac{1}{|\mathbf{x}-\mathbf{y}|}\right) \Delta_{\mathbf{y}}\eta(\mathbf{y})\, \mathbf{e} = \nabla_{\mathbf{y}}\varphi_1^{\mathbf{x}}(\mathbf{y}) + \mathbf{w}_1^{\mathbf{x}}(\mathbf{y}) \qquad \text{in } \mathbb{R}^3, \tag{4.64}
$$

where $\varphi_1^{\mathbf{x}}(\mathbf{y}) = O(|\mathbf{y}|^{-2})$ for $|\mathbf{y}| \to \infty$ and

$$
\Delta_{\mathbf{y}}\varphi_1^{\mathbf{x}}(\mathbf{y}) = \operatorname{div}_{\mathbf{y}} \left[D^{\boldsymbol{\alpha}}_{\mathbf{x}} \left(\frac{1}{|\mathbf{x}-\mathbf{y}|}\right) \Delta_{\mathbf{y}}\eta(\mathbf{y})\, \mathbf{e} \right] \qquad \text{in } \mathbb{R}^3. \tag{4.65}
$$

Note that for each fixed $\mathbf{x} \in \Omega''$, the right-hand side is an infinitely differentiable function of variable $\mathbf{y}$, supported for $\mathbf{y} \in \operatorname{supp}\nabla\eta$, whose all derivatives are bounded independently of $\mathbf{e}$ and $\mathbf{x}$. Hence $\nabla_{\mathbf{y}}\varphi_1^{\mathbf{x}} \in \mathbf{L}^r(\mathbb{R}^3)$ and $\mathbf{w}_1^{\mathbf{x}} \in \mathbf{L}^r_\sigma(\mathbb{R}^3)$ for each $r \in (1,\infty)$. Moreover,

$$
\left|\nabla^2_{\mathbf{y}}\varphi_1^{\mathbf{x}}(\mathbf{y})\right| \leq c(\boldsymbol{\alpha}) \qquad \text{for all } \mathbf{y} \in \mathbb{R}^3,
$$

where $c(\boldsymbol{\alpha})$ is independent of $\mathbf{e}$ and of $\mathbf{x}$. Thus, we have

$$
\begin{aligned}
|D^{\boldsymbol{\alpha}} P^{(1)}(\mathbf{x},t)| &= \left| \int_{\mathbb{R}^3} \nabla_{\mathbf{y}}\varphi_1^{\mathbf{x}}(\mathbf{y}) \cdot \nabla_{\mathbf{y}} p(\mathbf{y},t)\, \mathrm{d}\mathbf{y} \right| \\
&= \left| \int_{\mathbb{R}^3} \nabla_{\mathbf{y}}\varphi_1^{\mathbf{x}}(\mathbf{y}) \cdot \big[\partial_t\mathbf{u}(\mathbf{y},t) + \mathbf{u}(\mathbf{y},t)\cdot\nabla_{\mathbf{y}}\mathbf{u}(\mathbf{y},t) - \nu\Delta_{\mathbf{y}}\mathbf{u}(\mathbf{y},t)\big]\, \mathrm{d}\mathbf{y} \right| \\
&= \left| \int_{\mathbb{R}^3} \nabla_{\mathbf{y}}\varphi_1^{\mathbf{x}}(\mathbf{y}) \cdot \big[\mathbf{u}(\mathbf{y},t)\cdot\nabla_{\mathbf{y}}\mathbf{u}(\mathbf{y},t)\big]\, \mathrm{d}\mathbf{y} \right| \\
&= \left| \int_{\mathbb{R}^3} \nabla^2_{\mathbf{y}}\varphi_1^{\mathbf{x}}(\mathbf{y}) : \big[\mathbf{u}(\mathbf{y},t)\otimes\mathbf{u}(\mathbf{y},t)\big]\, \mathrm{d}\mathbf{y} \right|
\end{aligned}
$$

$$\leq\ c(\boldsymbol{\alpha}) \int_{\mathbb{R}^3} |\mathbf{u}(\mathbf{y},t)|^2 \, \mathrm{d}\mathbf{y} \ \leq\ c(\boldsymbol{\alpha}), \tag{4.66}$$

where $c(\boldsymbol{\alpha})$ is independent of $\mathbf{e}$, t, and $\mathbf{x}$. The term $D^{\boldsymbol{\alpha}} P^{(2)}$ can be estimated similarly: applying the integration by parts, we obtain

$$D^{\boldsymbol{\alpha}} P^{(2)}(\mathbf{x},t) \ = \ - \int_{\operatorname{supp}\nabla\eta} D^{\boldsymbol{\alpha}}_{\mathbf{x}} \operatorname{div}_{\mathbf{y}} \Big(\frac{1}{|\mathbf{x}-\mathbf{y}|} \, \nabla_{\mathbf{y}}\eta(\mathbf{y}) \Big) \, \mathbf{e} \cdot \nabla_{\mathbf{y}} p(\mathbf{y},t) \, \mathrm{d}\mathbf{y}. \tag{4.67}$$

Then we use the Helmholtz decomposition

$$D^{\boldsymbol{\alpha}}_{\mathbf{x}} \operatorname{div}_{\mathbf{y}} \Big(\frac{1}{|\mathbf{x}-\mathbf{y}|} \, \nabla_{\mathbf{y}}\eta(\mathbf{y}) \Big) \, \mathbf{e} \ = \ \nabla_{\mathbf{y}}\varphi_2^{\mathbf{x}}(\mathbf{y}) + \mathbf{w}_2^{\mathbf{x}}(\mathbf{y}), \tag{4.68}$$

and proceed in the same way as in (4.66).

We observe from (4.62), (4.66) and the same estimate of $D^{\boldsymbol{\alpha}} P^{(2)}$ that $D^{\boldsymbol{\alpha}} \nabla p \in \mathbf{L}^\infty(\Omega'' \times (\tau_1, \tau_2))$. From this, the Navier-Stokes equation (4.9) and Lemma 4.4.1, we deduce that $D^{\boldsymbol{\alpha}} \partial_t \mathbf{u} \in L^\infty(\Omega'' \times (\tau_1, \tau_2))$, too. This completes the proof. □

B) Application to the Procedure of Localization In this paragraph, we describe a situation, in which the information provided by Theorem 4.4.2 plays an important role.

A suitable weak solution $\mathbf{u}$ to the Navier-Stokes IBVP is often known (or assumed), in addition to the properties that directly follow from its definition, to have some other properties (e.g., a better integrability) in a sub-domain Ω' of Ω. In order to exploit these additional properties, one usually needs to localize the equations (4.9) and (4.10) to Ω'. It means to formulate a new problem, whose solution is in the spatial variable supported only in Ω' and coincides with $\mathbf{u}$ in a chosen sub-domain $\Omega'' \subset\subset \Omega'$. The usual way one can do it is to multiply solution $\mathbf{u}$ by a cut-off function $\eta \in C_0^\infty(\mathbb{R}^3)$ with the same properties as in the proof of Theorem 4.4.2. Let $\mathcal{D} \subset \Omega' \setminus \overline{\Omega''}$ be a "smooth" bounded domain (or a disjoint union of a finite number of such domains), containing $\operatorname{supp}\nabla\eta$. The boundary of $\mathcal{D}$ can be expressed as a union of two disjoint parts: $\partial\mathcal{D} = (\partial\mathcal{D})_1 \cup (\partial\mathcal{D})_2$, where $\eta = 0$ on $(\partial\mathcal{D})_1$ and $\eta = 1$ on $(\partial\mathcal{D})_2$.

Recall that due to [12], the set of hypothetic singular points of this solution—let us denote this set by $\mathcal{S}(\mathbf{u}, p)$—has the 1-dimensional Hausdorff measure equal to zero. Using this result and the fact that the singular set $\mathcal{S}(\mathbf{u}, p)$ is closed in $\Omega \times (0,T)$, it is shown in [59] that function η and domain $\mathcal{D}$ can be chosen so that $\big(\overline{\mathcal{D}} \times [0,T]\big) \cap \mathcal{S}(\mathbf{u}, p) = \emptyset$ and solution $\mathbf{u}$ is essentially bounded in some neighborhood of $\overline{\mathcal{D}} \times (0,T)$. Consequently, applying Lemma 4.4.1, we deduce that $\mathbf{u}$ has all spatial derivatives (of all orders) essentially bounded in $\mathcal{D} \times (\delta, T-\delta)$ for any $0 < \delta < \frac{1}{2}T$.

Since the product $\eta\mathbf{u}$ is not divergence-free, we put $\mathbf{v} := \eta\mathbf{u} - \mathbf{U}$, where the "corrector" $\mathbf{U}$ satisfies $\operatorname{div}\mathbf{U} = \nabla\eta \cdot \mathbf{u}$. This equation implies that $\operatorname{div}\mathbf{v} = 0$. The existence of an appropriate function $\mathbf{U}$ is guaranteed, e.g., by [29, Theorem III.3.2] or [10, Theorem 2.4]). It follows from these theorems that there exists a linear mapping $\mathfrak{B}$ from $W_0^{m,2}(\mathcal{D})$ (for each $m \in \{0\} \cup \mathbb{N}$) to $\mathbf{W}_0^{m+1,2}(\mathcal{D})$ such that for all $f \in W_0^{m,2}(\mathcal{D})$, satisfying $\int_{\mathcal{D}} f \, \mathrm{d}\mathbf{x} = 0$,

1. $\operatorname{div} \mathfrak{B} f = f$ a.e. in $\mathcal{D}$,
2. $\|\nabla^{m+1} \mathfrak{B} f\|_{2;\, \mathcal{D}} \leq c \, \|\nabla^m f\|_{2;\, \mathcal{D}}$.

Mapping $\mathfrak{B}$ is often called the Bogovskij operator. Denoting by $\mathbf{n}$ the outer normal vector field on $\partial\mathcal{D}$, we calculate:

$$\begin{aligned} \int_{\mathcal{D}} \nabla\eta \cdot \mathbf{u} \, \mathrm{d}\mathbf{x} &= \int_{(\partial\mathcal{D})_1} \eta \, \mathbf{u} \cdot \mathbf{n} \, \mathrm{d}S + \int_{(\partial\mathcal{D})_2} \eta \, \mathbf{u} \cdot \mathbf{n} \, \mathrm{d}S - \int_{\mathcal{D}} \eta \operatorname{div} \mathbf{u} \, \mathrm{d}\mathbf{x} \\ &= \int_{(\partial\mathcal{D})_2} \mathbf{u} \cdot \mathbf{n} \, \mathrm{d}S = 0. \end{aligned}$$

(The last equality holds because $(\partial\mathcal{D})_2$ consists of one or more closed surfaces, and the flux of $\mathbf{u}$ through each of them equals zero, because $\mathbf{u}$ is divergence-free.) Hence we may put $\mathbf{U}(\,.\,,t) := \mathfrak{B}\big[\nabla\eta \cdot \mathbf{u}(\,.\,,t)\big]$ for $t \in (\delta, T-\delta)$. Since $\nabla\eta \cdot \mathbf{u}(\,.\,,t) \in W_0^{m,2}(\mathcal{D})$ for any $m \in \{0\} \cup \mathbb{N}$, we obtain $\mathbf{U}(\,.\,,t) \in \mathbf{W}_0^{m+1,2}(\mathcal{D})$ for any $m \in \{0\} \cup \mathbb{N}$. As all spatial derivatives of $\nabla\eta \cdot \mathbf{u}$ are essentially bounded in $\mathcal{D} \times (\delta, T-\delta)$, we deduce that $\mathbf{U} \in L^\infty(\delta, T-\delta;\, \mathbf{W}_0^{m+1,2}(\mathcal{D}))$ for any $m \in \{0\} \cup \mathbb{N}$.

Extending $\mathbf{U}$ by zero outside $\mathcal{D}$, we observe that the function $\mathbf{v} \equiv \eta\mathbf{u} - \mathbf{U}$ is divergence-free in $\mathbb{R}^3 \times (\delta, T-\delta)$, coincides with $\mathbf{u}$ in $\Omega'' \times (\delta, T-\delta)$, equals zero in $(\mathbb{R}^3 \smallsetminus \Omega') \times (\delta, T-\delta)$ and all its spatial derivatives are essentially bounded in $\mathcal{D} \times (\delta, T-\delta)$. One can deduce that for any $\vartheta \in L^{5/3}(0,T)$, the pair $\mathbf{v}$, $\eta(p+\vartheta)$ is a suitable weak solution to the Navier-Stokes system

$$\partial_t \mathbf{v} + \mathbf{v} \cdot \nabla\mathbf{v} + \nabla[\zeta\,(p+\vartheta)] = \nu\Delta\mathbf{v} + \mathbf{g}, \tag{4.69}$$

$$\operatorname{div} \mathbf{v} = 0 \tag{4.70}$$

in $\mathbb{R}^3 \times (\delta, T-\delta)$, where

$$\begin{aligned} \mathbf{g} = {} & -\partial_t \mathbf{U} - \mathbf{U} \cdot \nabla(\eta\mathbf{u}) - (\eta\mathbf{u}) \cdot \nabla\mathbf{U} + \mathbf{U} \cdot \nabla\mathbf{U} + (\eta\mathbf{u} \cdot \nabla\eta)\,\mathbf{u} \\ & - \eta\,(1-\eta)\,\mathbf{u} \cdot \nabla\mathbf{u} - 2\nu\,\nabla\eta \cdot \nabla\mathbf{u} - \nu\,\mathbf{u}\,\Delta\eta + \nu\Delta\mathbf{U} + (p+\vartheta)\,\nabla\eta. \end{aligned} \tag{4.71}$$

When studying the properties of solution $\mathbf{v}$ to the system (4.69), (4.70), it is important to know as much as possible about the right-hand side $\mathbf{g}$ in equation (4.69). It follows from the definition of the cut-off function η and the properties of function $\mathbf{U}$ that $\operatorname{supp} \mathbf{g} \subset \overline{\mathcal{D}} \times [0,T]$.

We already know that all spatial derivatives of $\mathbf{u}$ are in $\mathbf{L}^\infty(\mathcal{D} \times (\delta, T-\delta))$. Since $\mathbf{U} = \mathfrak{B}[\nabla\eta \cdot \mathbf{u}]$, the time derivative $\partial_t \mathbf{U}$ equals $\mathfrak{B}[\nabla\eta \cdot \partial_t \mathbf{u}]$ in $\mathcal{D} \times (\delta, T-\delta)$. If $\Omega = \mathbb{R}^3$ (which we assume in this subsection), then we can apply Theorem 4.4.2

and deduce that all spatial derivatives of $\partial_t \mathbf{u}$ are in $\mathbf{L}^\infty(\mathcal{D} \times (\delta, T-\delta))$, too. Thus, we observe that all terms on the right-hand side of (4.71), except for the last term $(p+\vartheta)\,\nabla\eta$, have all spatial derivatives in $\mathbf{L}^\infty(\mathbb{R}^3 \times (\delta, T-\delta))$ (supported in $\overline{\mathcal{D}} \times [0, T]$).

Since $p \in L^{5/3}(\mathbb{R}^3 \times (0, T))$ and ∇p has all spatial derivatives in $\mathbf{L}^\infty(\mathcal{D} \times (\delta, T-\delta))$ due to Theorem 4.4.2, the function $\vartheta \in L^{5/3}(0, T)$ can be chosen so that $p + \vartheta \in L^\infty(\mathcal{D} \times (\delta, T-\delta))$. (It is, e.g., sufficient to put $\vartheta(t) := \int_{\mathcal{D}} p(\mathbf{x}, t)\, \mathrm{d}\mathbf{x}$.) Then all spatial derivatives of $(p+\vartheta)\,\nabla\eta$ are in $\mathbf{L}^\infty(\mathbb{R}^3 \times (\delta, T-\delta))$, supported in $\overline{\mathcal{D}} \times [0, T]$. Consequently, the same statement also holds on the whole function $\mathbf{g}$. All these observations are summarized in the next theorem:

Theorem 4.4.3 *Let* $\mathbf{u}$, p *be a suitable weak solution to the problem (4.9), (4.10, (4.12) in* $\mathbb{R}^3 \times (0, T)$. *Let* Ω' *be a sub-domain of* $\mathbb{R}^3$ *and* $\Omega'' \subset\subset \Omega'$ *be a sub-domain of* Ω'. *Let* $\mathcal{D}$, η, *and* $\mathbf{U}$ *be the set, respectively, functions, described above in this paragraph, and* $\mathbf{v} := \eta\mathbf{u} - \mathbf{U}$. *Then function* $\mathbf{v}$ *satisfies* $\mathbf{v} = \mathbf{u}$ *in* $\Omega'' \times (0, T)$ *and* $\mathbf{v} = \mathbf{0}$ *in* $\mathbb{R}^3 \setminus \Omega'$. *If* $\vartheta \in L^{5/3}(0, T)$, *then the pair* $\mathbf{v}$, $\eta(p+\vartheta)$ *is a suitable weak solution to the system (4.69), (4.70) in* $\mathbb{R}^3 \times (0, T)$. *Function* $\mathbf{g}$ *on the right-hand side of (4.69) is supported in* $\overline{\mathcal{D}} \times [0, T]$ *and function* $\vartheta \in L^{5/3}(0, T)$ *can be chosen so that* $\mathbf{g}$ *has all spatial derivatives (of all orders) in* $\mathbf{L}^\infty(\mathbb{R}^3 \times (\delta, T-\delta))$ *for any* $0 < \delta < \frac{1}{2}T$.

Of many articles, where the procedure of localization has also been described and applied, we mention, e.g., [57].

4.4.2 *The Case of the No-Slip Boundary Condition*

The next theorem comes from the papers [57] and [72].

Theorem 4.4.4 *Let* Ω *be a bounded or exterior domain in* $\mathbb{R}^3$ *with the boundary at least of the class* $C^{2+(h)}$ *for some* $h > 0$ *or a half-space in* $\mathbb{R}^3$. *Let* Ω' *be a sub-domain of* Ω, $-\infty < t_1 < t_2 < \infty$ *and let* $\mathbf{u}$ *be a weak solution to the Navier-Stokes system (4.9), (4.10) in* $\Omega \times (t_1, t_2)$, *satisfying the no-slip boundary condition* $\mathbf{u} = \mathbf{0}$ *on* $\partial\Omega \times (t_1, t_2)$ *and condition (i) of Lemma 4.4.1 in* $\Omega' \times (t_1, t_2)$. *Let* p *be an associated pressure. Then* $\partial_t\mathbf{u}$, ∇p *and all their spatial derivatives (of all orders) are in* $L^s(\tau_1, \tau_2; \mathbf{L}^\infty(\Omega''))$ *for any* $s \in (1, 2)$, $t_1 < \tau_1 < \tau_2 < t_2$ *and* $\Omega'' \subset\subset \Omega'$.

Proof We proceed in the same way as in the proof of Theorem 4.4.2 up to formula (4.63). The Helmholtz decomposition analogous to (4.64) does not help us, because, unlike in (4.66), the integral over Ω of $\nabla\varphi_1^{\mathbf{x}} \cdot \Delta\mathbf{u}$ need not be equal to zero. (This is due to the fact that although $\Delta\mathbf{u}$ is divergence-free, the no-slip boundary condition (4.11) does not provide any information on the normal component of $\Delta\mathbf{u}$ on $\partial\Omega$.) However, using formula (4.63), the inequality $|\mathbf{x} - \mathbf{y}| \geq \epsilon > 0$ (for $\mathbf{x} \in \Omega''$ and $\mathbf{y} \in \operatorname{supp} \nabla\eta$) and the inclusion $\nabla p \in L^s(\tau_1, \tau_2; L^{3s/(4s-2)}(\Omega))$ for $s \in (1, 2)$ (following from Theorem 4.3.3), we obtain

$$\int_{\tau_1}^{\tau_2} \|D^{\alpha} P^{(1)}(\,.\,,t)\|_{\infty;\,\Omega''}^{s}\, dt$$

$$= \int_{\tau_1}^{\tau_2} \biggl(\operatorname*{ess\,sup}_{\mathbf{x}\in\Omega''} \biggl| \int_{\operatorname{supp} \nabla\eta} D_{\mathbf{x}}^{\alpha} \Bigl(\frac{1}{|\mathbf{x}-\mathbf{y}|} \Bigr) \Delta_{\mathbf{y}}\eta(\mathbf{y})\, \mathbf{e}\cdot \nabla p(\mathbf{y},t)\, d\mathbf{y} \biggr| \biggr)^{s} dt$$

$$\leq c(\boldsymbol{\alpha},s) \int_{\tau_1}^{\tau_2} \biggl(\int_{\operatorname{supp} \nabla\eta} |\nabla p(\mathbf{y},t|\, d\mathbf{y} \biggr)^{s} dt$$

$$\leq c(\boldsymbol{\alpha},s) \int_{\tau_1}^{\tau_2} \biggl(\int_{\operatorname{supp} \nabla\eta} |\nabla p(\mathbf{y},t)|^{\frac{3s}{4s-2}}\, d\mathbf{y} \biggr)^{\frac{4s-2}{3}} dt \,<\, \infty. \tag{4.72}$$

The term $D^{\alpha} P^{(2)}$ can be estimated similarly. The proof can now be completed by means of the same arguments as at the end of the proof of Theorem 4.4.2. □

Note that if Ω is a domain as in Theorem 4.4.4 then Theorem 4.4.3 also holds with the only change: function $\mathbf{g}$ has all spatial derivatives (of all orders) in $L^s(\delta, T-\delta;\ \mathbf{L}^{\infty}(\Omega))$ for any $s \in (1,2)$.

4.4.3 The Case of the Navier and Navier-Type Boundary Conditions

A) The Navier and Navier-Type Boundary Conditions We assume that Ω is a domain in $\mathbb{R}^3$ with the boundary of the class at least $C^{2+(h)}$ for some $h > 0$. Instead of the no-slip boundary condition (4.11), equations (4.9), (4.10) are also often considered with the so-called *Navier boundary conditions*

$$\text{a)}\ \mathbf{u}\cdot\mathbf{n} = 0, \qquad \text{b)}\ \bigl[\mathbb{T}_{\mathrm{d}}(\mathbf{u})\cdot\mathbf{n}\bigr]_{\tau} + \kappa\,\mathbf{u} = \mathbf{0} \tag{4.73}$$

on $\Gamma_T \equiv \partial\Omega \times (0,T)$. Condition (4.73 a) expresses the impermeability of the boundary of Ω. Symbol $\mathbb{T}_{\mathrm{d}}(\mathbf{u})$ in condition (4.73 b) denotes the dynamic stress tensor induced by the velocity field $\mathbf{u}$, subscript τ denotes the tangential component and κ is the coefficient of friction between the fluid and the boundary. (We suppose that κ is a nonnegative constant.) In a Newtonian fluid, tensor $\mathbb{T}_{\mathrm{d}}(\mathbf{u})$ equals $2\nu\mathbb{D}(\mathbf{u})$, where $\mathbb{D}(\mathbf{u}) := (\nabla\mathbf{u})_{\mathrm{sym}} := \frac{1}{2}\,[\nabla\mathbf{u} + (\nabla\mathbf{u})^T]$. Tensor $\mathbb{D}(\mathbf{u})$ is usually called the rate of deformation tensor. Condition (4.73 b), proposed by H. Navier in 1824, expresses the requirement that the tangential component of the stress, with which the fluid acts on the boundary, is proportional to the tangential velocity.

Under the no-flux condition (4.73 a)), $[\mathbb{T}_{\mathrm{d}}(\mathbf{u})\cdot\mathbf{n}]_{\tau}$ satisfies the formula

$$[\mathbb{T}_{\mathrm{d}}(\mathbf{u})\cdot\mathbf{n}]_{\tau} \;=\; \nu\,\mathbf{curl}\,\mathbf{u}\times\mathbf{n} - 2\nu\mathbf{u}\cdot\nabla\mathbf{n}$$

on $\partial\Omega$, see, e.g., [4] or [17]. Thus, condition (4.73 b) can also be written in the equivalent form

$$\nu\, \mathbf{curl}\ \mathbf{u} \times \mathbf{n} - 2\nu \mathbf{u} \cdot \nabla \mathbf{n} + \kappa \mathbf{u} \;=\; \mathbf{0}.$$

Assuming that $\kappa = 0$ and neglecting the term $2\nu\mathbf{u} \cdot \nabla\mathbf{n}$ (which is anyway equal to zero on flat parts of $\partial\Omega$), we observe that (4.73) reduces to

$$\text{a) } \mathbf{u} \cdot \mathbf{n} = 0, \qquad \text{b) } \mathbf{curl}\, \mathbf{u} \times \mathbf{n} = \mathbf{0}. \tag{4.74}$$

We call these conditions the *Navier-type boundary conditions* due to their relation to Navier's conditions (4.73).

B) A Note on the Theory of the Navier-Stokes Equations with the Boundary Conditions (4.73) or (4.74) As to Navier's boundary conditions (4.73), the readers can find the definition of a weak solution to the problem (4.9), (4.10), (4.12), (4.73) and the proof of the global in time existence of a weak solution in the papers [17] (with $\mathbf{f} = \mathbf{0}$), [58] (in a time-varying domain Ω) and [65] (in a half-space). Theorems on the local in time existence of a strong solution are proven, e.g., in [17] (for $\mathbf{f} = \mathbf{0}$) and [45] (in a bounded domain Ω). Steady problems are studied in [3] and [4].

Analogous results for the equations (4.9), (4.10) with the Navier-type boundary conditions (4.74) can be found in [84] (the definition of a weak solution and its global in time existence) and in [1] and [16] (the local in time existence of a strong solution). In all these papers, the authors assume for simplicity that $\mathbf{f} = \mathbf{0}$.

If $\mathbf{u}$ is a weak solution to the equations (4.9), (4.10) with the boundary conditions (4.73) or (4.74), then an associated pressure can be, in principle, constructed in a similar way as in Sect. 4.3.2C. We can also refer to paper [83], where the author constructed an associated pressure to the distributional solution of the system (4.9), (4.10), regardless of the boundary conditions. (See Sect. 4.3.2E for some other details.)

In spite of the aforementioned results, it is necessary to note that the theory of the Navier-Stokes equations (4.9), (4.10) with the boundary conditions (4.73) or (4.74) is far from being so elaborated as in the case of the no-slip boundary condition (4.11). The main reason is that the boundary conditions (4.73) and (4.74) have so far been given much less attention than the no-slip boundary condition (4.11), although especially conditions (4.73) have a good physical sense and the boundary conditions (4.73) and (4.74) often enable one to derive better properties of solutions than in the case of the no-slip condition (4.11). This state corresponds to the fact that an analogue of Theorem 4.3.3 for the Navier-Stokes equations with the boundary conditions (4.73) or (4.74) is, to the best of our knowledge, not available in literature. (Such an analogue is known to hold for the non-stationary Stokes problem in a smooth bounded domain, see [69], or in a large class of smooth unbounded domains, see [24].) This is why we further on in this subsection assume that Ω is a smooth bounded domain, $\mathbf{u}$ is a weak solution to the equations (4.9), (4.10) (with $\mathbf{f} = \mathbf{0}$) in $\Omega \times (t_1, t_2)$ satisfying the boundary conditions (4.73) or (4.74), p is an associated pressure and in order to justify operations we will perform in the proof of Theorem 4.4.5, we use the condition

(ii) $\partial_t \mathbf{u}(\,.\,,t)$, $\nabla p(\,.\,,t) \in \mathbf{L}^r(\Omega)$ and $\nabla^2\mathbf{u}(\,.\,,t) \in L^r(\Omega)^{3\times 3}$ for a.a. $t \in (t_1,t_2)$ and some $r > 1$.

C) An Interior Regularity of Pressure The next theorem is taken from the papers [46] (the case of boundary conditions (4.73)) and [61] (the case of boundary conditions (4.74)).

Theorem 4.4.5 *Let Ω be a bounded domain in $\mathbb{R}^3$ with the boundary at least of the class $C^{2+(h)}$ for some $h > 0$. Let Ω' be a sub-domain of Ω, $-\infty < t_1 < t_2 < \infty$ and let $\mathbf{u}$ be a weak solution to the Navier-Stokes system (4.9), (4.10) in $\Omega \times (t_1,t_2)$, satisfying the slip boundary conditions (4.73) or (4.74) on $\partial\Omega \times (t_1,t_2)$ and condition (i) of Lemma 4.4.1 in $\Omega' \times (t_1,t_2)$. Let solution $\mathbf{u}$ and the associated pressure p also satisfy condition (ii). Then ∇p and $\partial_t\mathbf{u}$ and all their spatial derivatives (of all orders) are in $L^s(\tau_1,\tau_2;\,\mathbf{L}^\infty(\Omega''))$ for any $t_1 < \tau_1 < \tau_2 < t_2$, $\Omega'' \subset\subset \Omega'$ and $s = 4$ (the case of boundary conditions (4.73)) or $s = \infty$ (the case of boundary conditions (4.74)).*

Proof Let us at first assume that $\mathbf{u}$ satisfies the boundary conditions (4.73). We can again proceed in the same way as in the proof of Theorem 4.4.2 up to formula (4.64). Function $\varphi_1^{\mathbf{x}}$ is now supposed to satisfy equation (4.65) in Ω and the Neumann boundary condition

$$\frac{\partial\varphi_1^{\mathbf{x}}}{\partial\mathbf{n}_{\mathbf{y}}}(\mathbf{y}) \;=\; D_{\mathbf{x}}^{\boldsymbol{\alpha}}\left(\frac{1}{|\mathbf{x}-\mathbf{y}|}\right)\Delta_{\mathbf{y}}\eta(\mathbf{y})\,\mathbf{e}\cdot\mathbf{n}_{\mathbf{y}} \;=\; 0 \tag{4.75}$$

for $\mathbf{y} \in \partial\Omega$. By analogy with (4.66), we have

$$\begin{aligned}|D^{\boldsymbol{\alpha}}P^{(1)}(\mathbf{x},t)| \;&=\; \left|\int_\Omega \nabla_{\mathbf{y}}\varphi_1^{\mathbf{x}}(\mathbf{y})\cdot\nabla_{\mathbf{y}}p(\mathbf{y},t)\,\mathrm{d}\mathbf{y}\right| \\ &=\; \left|\int_\Omega \nabla_{\mathbf{y}}\varphi_1^{\mathbf{x}}(\mathbf{y})\cdot\big[\partial_t\mathbf{u}(\mathbf{y},t)+\mathbf{u}(\mathbf{y},t)\cdot\nabla_{\mathbf{y}}\mathbf{u}(\mathbf{y},t)-\nu\Delta_{\mathbf{y}}\mathbf{u}(\mathbf{y},t)\big]\,\mathrm{d}\mathbf{y}\right|. \end{aligned} \tag{4.76}$$

The integral of $\nabla_{\mathbf{y}}\varphi_1^{\mathbf{x}}\cdot\partial_t\mathbf{u}$ is equal to zero, because $\partial_t\mathbf{u}(\,.\,,t) \in \mathbf{L}^r_\sigma(\Omega)$. The integral of $\nabla_{\mathbf{y}}\varphi_1^{\mathbf{x}}\cdot[\mathbf{u}\cdot\nabla_{\mathbf{y}}\mathbf{u}]$ equals

$$\int_{\partial\Omega}[\nabla_{\mathbf{y}}\varphi_1^{\mathbf{x}}(\mathbf{y})\cdot\mathbf{u}(\mathbf{y},t)]\,[\mathbf{u}(\mathbf{y},t)\cdot\mathbf{n}_{\mathbf{y}}]\,\mathrm{d}_{\mathbf{y}}S - \int_\Omega \nabla_{\mathbf{y}}^2\varphi_1^{\mathbf{x}}(\mathbf{y}) : \big[\mathbf{u}(\mathbf{y},t)\otimes\mathbf{u}(\mathbf{y},t)\big]\,\mathrm{d}\mathbf{y}.$$

The first integral is equal to zero, because $\mathbf{u}$ satisfies condition (4.73 a). In order to estimate the second integral, we use the inequality

$$\begin{aligned}|\nabla_{\mathbf{y}}\varphi_1^{\mathbf{x}}(\mathbf{y})|_{0+(h)}+|\nabla_{\mathbf{y}}^2\varphi_1^{\mathbf{x}}(\mathbf{y})|_{0+(h)} \;&\leq\; c\left|\mathrm{div}_{\mathbf{y}}\left[D_{\mathbf{x}}^{\boldsymbol{\alpha}}\left(\frac{1}{|\mathbf{x}-\mathbf{y}|}\right)\Delta_{\mathbf{y}}\eta(\mathbf{y})\,\mathbf{e}\right]\right|_{0+(h)} \\ &\leq\; c(\boldsymbol{\alpha}), \end{aligned} \tag{4.77}$$

where $|\,.\,|_{0+(h)}$ is the norm in Hölder's space $C^{0+(h)}(\overline{\Omega})$, see [54]. (Recall that $\mathbf{x} \in \Omega''$ and $\Delta_{\mathbf{y}}\eta(\mathbf{y})$ is supported for $\mathbf{y} \in \operatorname{supp}\nabla\eta$, which has a positive distance from Ω''.) The constant on the right-hand side is independent of $\mathbf{e}$ and $\mathbf{x}$. Thus, we get

$$\left| \int_{\Omega} \nabla^2_{\mathbf{y}} \varphi_1^{\mathbf{x}}(\mathbf{y}) : \big[\mathbf{u}(\mathbf{y},t) \otimes \mathbf{u}(\mathbf{y},t)\big] \, \mathrm{d}\mathbf{y} \right| \leq c(\boldsymbol{\alpha}) \int_{\Omega} |\mathbf{u}(\mathbf{y},t)|^2 \, \mathrm{d}\mathbf{y} \leq c(\boldsymbol{\alpha}).$$

The integral of $\nabla_{\mathbf{y}}\varphi_1^{\mathbf{x}} \cdot \nu\Delta_{\mathbf{y}}\mathbf{u}$ in (4.76) can be treated as follows:

$$\begin{aligned}
\int_{\Omega} \nabla_{\mathbf{y}}\varphi_1^{\mathbf{x}}(\mathbf{y}) \cdot \nu\Delta_{\mathbf{y}}\mathbf{u}(\mathbf{y},t) \, \mathrm{d}\mathbf{y} &= \int_{\Omega} \nabla_{\mathbf{y}}\varphi_1^{\mathbf{x}}(\mathbf{y}) \cdot \operatorname{div}_{\mathbf{y}} \mathbb{T}_{\mathrm{d}}(\mathbf{u}(\mathbf{y},t)) \, \mathrm{d}\mathbf{y} \\
&= \int_{\partial\Omega} \nabla_{\mathbf{y}}\varphi_1^{\mathbf{x}}(\mathbf{y}) \cdot \big[\mathbb{T}_{\mathrm{d}}(\mathbf{u}(\mathbf{y},t)) \cdot \mathbf{n}_{\mathbf{y}}\big] \, \mathrm{d}\mathbf{y} - \int_{\Omega} \nabla^2_{\mathbf{y}}\varphi_1^{\mathbf{x}}(\mathbf{y}) : \mathbb{T}_{\mathrm{d}}(\mathbf{u}(\mathbf{y},t)) \, \mathrm{d}\mathbf{y}.
\end{aligned}$$

Since $\nabla_{\mathbf{y}}\varphi_1^{\mathbf{x}}(\mathbf{y}) \cdot [\mathbb{T}_{\mathrm{d}}(\mathbf{u}(\mathbf{y},t)) \cdot \mathbf{n}_{\mathbf{y}}] = \nabla_{\mathbf{y}}\varphi_1^{\mathbf{x}}(\mathbf{y}) \cdot [\mathbb{T}_{\mathrm{d}}(\mathbf{u}(\mathbf{y},t)) \cdot \mathbf{n}_{\mathbf{y}}]_{\tau} = -\nabla_{\mathbf{y}}\varphi_1^{\mathbf{x}}(\mathbf{y}) \cdot \kappa\, \mathbf{u}(\mathbf{y},t)$ (because $\nabla_{\mathbf{y}}\varphi_1^{\mathbf{x}}(\mathbf{y})$ is tangent on $\partial\Omega$ and due to condition (4.73 b)), we have

$$\begin{aligned}
&\int_{\Omega} \nabla_{\mathbf{y}}\varphi_1^{\mathbf{x}}(\mathbf{y}) \cdot \nu\Delta_{\mathbf{y}}\mathbf{u}(\mathbf{y},t) \, \mathrm{d}\mathbf{y} \\
&\quad = -\kappa \int_{\partial\Omega} \nabla_{\mathbf{y}}\varphi_1^{\mathbf{x}}(\mathbf{y}) \cdot \mathbf{u}(\mathbf{y},t) \, \mathrm{d}\mathbf{y} - 2\nu \int_{\Omega} \nabla^2_{\mathbf{y}}\varphi_1^{\mathbf{x}}(\mathbf{y}) : \nabla_{\mathbf{y}}\mathbf{u}(\mathbf{y},t) \, \mathrm{d}\mathbf{y} \\
&\quad = -\kappa \int_{\partial\Omega} \nabla_{\mathbf{y}}\varphi_1^{\mathbf{x}}(\mathbf{y}) \cdot \mathbf{u}(\mathbf{y},t) \, \mathrm{d}\mathbf{y} - 2\nu \int_{\partial\Omega} \nabla^2_{\mathbf{y}}\varphi_1^{\mathbf{x}}(\mathbf{y}) : \big[\mathbf{u}(\mathbf{y},t) \otimes \mathbf{n}_{\mathbf{y}}\big] \, \mathrm{d}_{\mathbf{y}}S \\
&\qquad + 2\nu \int_{\Omega} \nabla_{\mathbf{y}}\Delta_{\mathbf{y}}\varphi_1^{\mathbf{x}}(\mathbf{y}) \cdot \mathbf{u}(\mathbf{y},t) \, \mathrm{d}\mathbf{y}.
\end{aligned}$$

A simple integration by parts shows that the third integral on the right-hand side is equal to zero. Hence

$$\begin{aligned}
&\int_{\Omega} \nabla_{\mathbf{y}}\varphi_1^{\mathbf{x}}(\mathbf{y}) \cdot \nu\Delta_{\mathbf{y}}\mathbf{u}(\mathbf{y},t) \, \mathrm{d}\mathbf{y} \\
&\quad = -\kappa \int_{\partial\Omega} \nabla_{\mathbf{y}}\varphi_1^{\mathbf{x}}(\mathbf{y}) \cdot \mathbf{u}(\mathbf{y},t) \, \mathrm{d}\mathbf{y} - 2\nu \int_{\partial\Omega} \nabla^2_{\mathbf{y}}\varphi_1^{\mathbf{x}}(\mathbf{y}) : \big[\mathbf{u}(\mathbf{y},t) \otimes \mathbf{n}_{\mathbf{y}}\big] \, \mathrm{d}_{\mathbf{y}}S.
\end{aligned}$$

Thus, applying inequality (4.77), we get

$$\begin{aligned}
&\left| \int_{\Omega} \nabla_{\mathbf{y}}\varphi_1^{\mathbf{x}}(\mathbf{y}) \cdot \nu\Delta_{\mathbf{y}}\mathbf{u}(\mathbf{y},t) \, \mathrm{d}\mathbf{y} \right| \leq c(\boldsymbol{\alpha}) \int_{\partial\Omega} |\mathbf{u}(\mathbf{y},t)| \, \mathrm{d}_{\mathbf{y}}S \leq c(\boldsymbol{\alpha}) \, \|\mathbf{u}(\,.\,,t)\|_{2;\,\partial\Omega} \\
&\quad \leq c(\boldsymbol{\alpha}) \left(\|\mathbf{u}(\,.\,,t)\|_2 + \|\mathbf{u}(\,.\,,t)\|_2^{1/2} \, \|\mathbf{u}(\,.\,,t)\|_{1,2}^{1/2} \right) \leq c(\boldsymbol{\alpha}) \left(1 + \|\mathbf{u}(\,.\,,t)\|_{1,2}^{1/2} \right)
\end{aligned}$$

(see Theorem II.4.1 in [29]). From this, we observe that $D^{\boldsymbol{\alpha}} P^{(1)} \in L^4(\tau_1, \tau_2; \mathbf{L}^\infty(\Omega''))$. The proof can now be finished by means of the same arguments as in the proof of Theorem 4.4.2, after estimate (4.66).

If $\mathbf{u}$ satisfies the Navier-type boundary conditions (4.74), then $\Delta\mathbf{u}(\,.\,,t)\cdot\mathbf{n}=0$ on $\partial\Omega$ in the sense of traces for a.a. $t \in (t_1, t_2)$ (see, e.g., [16]), which means that

$$\int_\Omega \nabla_{\mathbf{y}}\varphi_1^{\mathbf{x}}(\mathbf{y})\cdot\Delta_{\mathbf{y}}\mathbf{u}(\mathbf{y},t)\,\mathrm{d}\mathbf{y} \;=\; 0$$

for a.a. $t \in (t_1, t_2)$. Thus, we can estimate $D^{\boldsymbol{\alpha}} P^{(1)}$ and $D^{\boldsymbol{\alpha}} P^{(2)}$ in the same way as in the proof of Theorem 4.4.2 and we finally obtain $D^{\boldsymbol{\alpha}}\nabla p \in L^\infty(\tau_1, \tau_2; \mathbf{L}^\infty(\Omega''))$. Consequently, $D^{\boldsymbol{\alpha}}\partial_t\mathbf{u} \in L^\infty(\tau_1, \tau_2; \mathbf{L}^\infty(\Omega''))$, too. □

One can now deduce from (4.71) that if Ω is a domain as in Theorem 4.4.5 then Theorem 4.4.3 also holds with the change that function $\mathbf{g}$ has all spatial derivatives (of all orders) in $L^s(\delta, T-\delta;\, \mathbf{L}^\infty(\Omega))$ for $s=4$ (in the case of boundary conditions (4.73)) or $s=\infty$ (in the case of boundary conditions (4.74)).

4.5 An Influence of Pressure on Regularity of a Weak Solution

The notion of a regular point of a weak solution to the Navier-Stokes system (4.9), (4.10) is defined in Sect. 4.3.2H. This paragraph deals with suitable weak solutions, but the regular point can be defined in the same way for any weak solution. Thus, $(\mathbf{x}_0, t_0) \in \Omega\times(0,T)$ is a regular point of solution $\mathbf{u}$ if $\mathbf{u}$ is essentially bounded in some space-time neighborhood of $(\mathbf{x}_0, t_0)$. (Solution $\mathbf{u}$ is in fact better than just essentially bounded in the neighborhood of the regular point $(\mathbf{x}_0, t_0)$: if $\mathbf{f}$ is "sufficiently smooth" then one can show that $\mathbf{u}$ is at least Hölder-continuous in some neighborhood of $(\mathbf{x}_0, t_0)$, see, e.g., [49] or [47].) The definition of a regular point can also be naturally extended to points in $\partial\Omega\times(0,T)$. The question whether all points of $\overline{\Omega}\times(0,T)$ are regular, or whether there can appear a singularity in $\Omega\times(0,T)$ or $\partial\Omega\times(0,T)$, is open, independently of smoothness of domain Ω or the given functions $\mathbf{u}_0$ and $\mathbf{f}$. This is also true if $\mathbf{u}$ is the so-called Leray-Hopf weak solution, which is a weak solution satisfying the energy inequality

$$\begin{aligned}\|\mathbf{u}(\,.\,,t)\|_2^2 + 2\nu\int_0^t \|\nabla\mathbf{u}(\,.\,,\tau)\|_2^2\,\mathrm{d}\tau \\ \leq 2\int_0^t \big\langle\mathbf{f}(\,.\,,\tau),\mathbf{u}(\,.\,,\tau)\big\rangle_\Omega\,\mathrm{d}\tau + \|\mathbf{u}_0\|_2^2 \qquad (4.78)\end{aligned}$$

for all $t \in [0,T)$ (see [30]), or a suitable weak solution.

There exist many additional conditions (the so-called *criteria of regularity*) which, imposed on a weak solution $\mathbf{u}$, guarantee that solution $\mathbf{u}$ has no singular points. The next lemma uses the so-called Serrin's condition, see (4.79). It is taken from [30] (Lemmas 5.4–5.6, Theorem 5.2) and [75] (Theorem V.1.8.2).

Lemma 4.5.1 *Let $\Omega \subset \mathbb{R}^3$ be an arbitrary domain with a uniformly C^2-boundary. Let $\mathbf{f} \in \mathbf{C}_0^\infty(\overline{Q_T})$ and let $\mathbf{u}$ be a weak solution to the Navier-Stokes IBVP (4.9)–(4.12). Suppose that*

$$\mathbf{u} \in L^r(0,T;\, \mathbf{L}^s(\Omega)) \quad \textit{for some } 2 \le r < \infty \textit{ and } 3 < s \le \infty,$$

$$\textit{satisfying } \frac{2}{r} + \frac{3}{s} = 1. \tag{4.79}$$

Then $\mathbf{u}$ and an appropriate associated pressure p (after a possible redefinition on a set of measure zero) satisfy

$$\mathbf{u} \in C^0\big((\epsilon,T];\, \mathbf{W}^{1,2}(\Omega)\big) \cap L^2(\epsilon,T;\, \mathbf{W}^{2,2}(\Omega)), \quad \partial_t^l \mathbf{u} \in L^2(\epsilon,T;\, \mathbf{L}^2_\sigma(\Omega)),$$
$$\nabla p \in L^2(\epsilon,T;\, \mathbf{L}^2(\Omega))$$

for all $l \in \mathbb{N}$, where ϵ is an arbitrary number from $(0,T)$. If, moreover, $\partial\Omega$ is uniformly of the class C^m (for $m \in \mathbb{N}$, $m \ge 2$), then

$$\partial_t^l \mathbf{u} \in L^2(\epsilon,T;\, \mathbf{W}^{k,2}(\Omega)), \quad \partial_t^l \nabla p \in L^2(\epsilon,T;\, \mathbf{W}^{k-2,2}(\Omega))$$

for all $l \in \{0\} \cup \mathbb{N}$ and $k = 2, \dots, m$. If $\partial\Omega$ is uniformly of the class C^∞, then $\mathbf{u} \in \mathbf{C}^\infty\big(\overline{\Omega} \times (0,T]\big)$ and $p \in C^\infty\big(\overline{\Omega} \times (0,T]\big)$.

Note that while the statements of Lemma 4.5.1 basically come (or are deduced) from Serrin's works in the 60s of the last century, analogous results for $r = \infty$ and $s = 3$ are known only since 2003, see [19].

A series of further criteria of regularity have been formulated for other quantities, e.g., two components or just one component of velocity, the gradient of velocity (or only some components of the gradient of velocity), vorticity (or only two components of vorticity), eigenvalues of the rate of deformation tensor, etc. In the next subsection, we give a survey of criteria that impose additional conditions on an associated pressure.

4.5.1 A Brief Survey of Regularity Criteria Based on Pressure

We restrict ourselves to the case $\mathbf{f} = \mathbf{0}$. We make this assumption 1) for simplicity and 2) because most of the results we will cite have also been obtained for $\mathbf{f} = \mathbf{0}$. The proofs could be, however, easily modified so that the results also hold for smooth and appropriately integrable $\mathbf{f} \neq \mathbf{0}$.

Many authors deal with the N-dimensional Navier-Stokes equations for general N in their works. Although we have so far considered only the case $N = 3$, we further cite the results with the N that appears in the paper.

1) S. Kaniel [39] (1969) assumed that Ω is a bounded smooth domain in $\mathbb{R}^3$, $\mathbf{u}_0 \in \mathbf{W}^{1,2}_{0,\sigma}(\Omega)$, $\mathbf{u}$ is a Leray–Hopf weak solution to the Navier-Stokes IBVP (4.9)–(4.12) in $\Omega \times (0, T)$ and p is an associated pressure. The author proved that the condition $p \in L^\infty(0, T;\ L^q(\Omega))$ (for some $q > \frac{12}{5}$) guarantees that solution $\mathbf{u}$ is regular.
2) H. Beirão da Veiga [5] (1998) proved that if $p \in L^\gamma_w(\Omega \times (0, T))$ for some $\gamma > \frac{5}{2}$ then solution $\mathbf{u}$ is regular. (Here, L^γ_w denotes the weak L^γ-space.)
3) Kaniel's result from [39] was later improved by L. Berselli [7] (1999), who assumed that Ω is a bounded smooth domain in $\mathbb{R}^N$, $N \geq 3$, and showed that if $\mathbf{u}_0 \in \mathbf{L}^q_\sigma(\Omega)$ and $p \in L^q(0, T;\ L^{Nq/(q+N-2)}(\Omega))$ for some $q > N$ then $\mathbf{u}$ is regular. More precisely, $\mathbf{u} \in C\big([0, T);\ \mathbf{L}^q_\sigma(\Omega)\big)$ and $|\mathbf{u}|^{q/2} \in L^2(0, T;\ W^{1,2}_0(\Omega))$.
4) H. Beirão da Veiga [6] (2000) considered Ω to be bounded and smooth domain in $\mathbb{R}^N$, $N \geq 3$, and assumed that $\mathbf{u}_0 \in \mathbf{L}^q_\sigma(\Omega)$ for some $q > N$ and $p/(1+|\mathbf{u}|) \in L^r(0, T;\ L^s(\Omega))$ for $2/r + N/s = 1$, $s > N$. Then he proved the regularity of $\mathbf{u}$, namely $\mathbf{u} \in C\big([0, T);\ \mathbf{L}^q_\sigma(\Omega)\big)$ and $|\mathbf{u}|^{q/2} \in L^2(0, T;\ W^{1,2}_0(\Omega))$.
5) D. Chae and J. Lee [14] (2001) considered the Navier-Stokes problem in $\mathbb{R}^3 \times (0, T)$. They showed that if $\mathbf{u}_0 \in \mathbf{L}^2_\sigma(\mathbb{R}^3) \cap \mathbf{L}^q(\mathbb{R}^3)$ for some $q > 3$, $\mathbf{u}$ is the Leray-Hopf weak solution in $(0, T)$ and p is in $L^r(0, T;\ L^s(\mathbb{R}^3))$ for some $1 < r \leq \infty$, $\frac{3}{2} < s < \infty$, satisfying $2/r+3/s < 2$, or $p \in L^1(0, T;\ L^\infty(\mathbb{R}^3))$, or if $p \in L^\infty(0, T;\ L^{3/2}(\mathbb{R}^3))$ with the corresponding norm sufficiently small, then $\mathbf{u}$ is a regular solution in $\Omega \times [0, T)$.
6) L. Berselli and G. P. Galdi [8] (2002) extended the result from [14] to the case of N spatial dimensions (for $N \geq 3$) for $2/r + N/s = 2$ (with $N/2 < s \leq \infty$). If $N^2/2(N-1) \leq s \leq 3$, then, in addition to the case $\Omega = \mathbb{R}^N$, domain Ω can also be a half-space or a bounded or exterior domain in $\mathbb{R}^N$ with C^∞ boundary. Moreover, the authors also considered (for Ω being one of these types of domains) the condition $\nabla p \in L^r(0, T;\ \mathbf{L}^s(\Omega))$ with $2/r + N/s = 3$, $N^2/(3N-2) \leq s \leq N$. The results of [8] are explained in greater detail in Sect. 4.5.2.

It is well known that if $\Omega = \mathbb{R}^N$ then solutions to the Navier-Stokes equations are scale-invariant, which means that if $\mathbf{u}(\mathbf{x}, t)$, $p(\mathbf{x}, t)$ satisfy equations (4.9), (4.10) in $\mathbb{R}^N \times (0, T)$ and $\lambda > 0$ then $\lambda\mathbf{u}(\lambda\mathbf{x}, \lambda^2 t)$, $\lambda^2 p(\lambda\mathbf{x}, \lambda^2 t)$ satisfy the same equations in $\mathbb{R}^N \times (0, T/\lambda^2)$. Moreover, the norm of $\lambda\mathbf{u}(\lambda\mathbf{x}, \lambda^2 t)$ in $L^r(0, T/\lambda^2;\ \mathbf{L}^s(\mathbb{R}^N))$ (where $2 \leq r \leq \infty$, $N \leq s \leq \infty$, $2/r + N/s = 1$) and the norm of $\lambda^2 p(\lambda\mathbf{x}, \lambda^2 t)$ in $L^r(0, T/\lambda^2;\ \mathbf{L}^s(\mathbb{R}^N))$ (where $1 \leq r \leq \infty$, $N/2 \leq s \leq \infty$, $2/r+N/s = 2$) are both independent of λ. From this point of view, the condition $p \in L^r(0, T;\ L^s(\mathbb{R}^N))$ with $2/r + N/s = 2$ is natural.

7) G. Seregin and V. Šverák [67] (2002) considered a suitable weak solution $\mathbf{u}$, p in $\mathbb{R}^3 \times (0, \infty)$. The authors say that a scalar function $g : \mathbb{R}^3 \times (0, \infty) \to [0, \infty)$

satisfies condition (C) if to any $t_0 > 0$ there exists $R_0 > 0$ such that

$$A(t_0) := \sup_{\mathbf{x}_0 \in \mathbb{R}^3} \sup_{t_0 - R_0^2 \le t \le t_0} \int_{|\mathbf{x}-\mathbf{x}_0|<R_0} \frac{g(\mathbf{x}, t)}{|\mathbf{x} - \mathbf{x}_0|} \, \mathrm{d}\mathbf{x} < \infty$$

and for each fixed $\mathbf{x}_0 \in \mathbb{R}^3$ and each fixed $R \in (0, R_0]$, the function

$$t \longmapsto \int_{|\mathbf{x}-\mathbf{x}_0|<R} \frac{g(\mathbf{x}, t)}{|\mathbf{x} - \mathbf{x}_0|} \, \mathrm{d}\mathbf{x}$$

is continuous at t_0 from the left. The main result of [67] says that if there exists function g satisfying condition (C) so that the normalized pressure

$$p(\mathbf{x}, t) := \frac{1}{4\pi} \int_{\mathbb{R}^3} \frac{1}{|\mathbf{x} - \mathbf{y}|} \, [\partial_i u_j \, \partial_j u_i](\mathbf{y}, t) \, \mathrm{d}\mathbf{y}$$

satisfies

$$|\mathbf{u}(\mathbf{x}, t)|^2 + 2p(\mathbf{x}, t) \le g(\mathbf{x}, t) \qquad \text{for all } \mathbf{x} \in \mathbb{R}^3, \ 0 < t < \infty \tag{4.80}$$

or

$$p(\mathbf{x}, t) \ge -g(\mathbf{x}, t) \qquad \text{for all } \mathbf{x} \in \mathbb{R}^3, \ 0 < t < \infty \tag{4.81}$$

then $\mathbf{u}$ is Hölder-continuous in $\mathbb{R}^3 \times (0, \infty)$, i.e. $\mathbf{u}$ is regular. This result is explained in greater detail, together with the principle of the proof, in Sect. 4.5.4.

8) J. Nečas and J. Neustupa [55] (2002, for $\Omega = \mathbb{R}^3$) proved the regularity of a suitable weak solution $\mathbf{u}$, p at the space-time point $(\mathbf{x}_0, t_0)$ under the conditions that there exist $r > 0$ (arbitrarily small) and $\rho > 0$ (arbitrarily large) such that

- the negative part p_- of the pressure is integrable with exponents $\alpha \in [\frac{3}{2}, \infty)$ (in time) and $\beta \in (\frac{3}{2}, \infty)$ (in space), such that $2/\alpha + 3/\beta = 2$, over

$$V_r^\rho := \big\{(\mathbf{x}, t) \in Q_T;\ t_0 - r^2/\rho^2 < t < t_0,\ |\mathbf{x} - \mathbf{x}_0| < \epsilon(t)\rho\big\}$$

  where $\epsilon(t) := \sqrt{t_0 - t}$,
- the velocity is integrable with the exponents a (in time) and b (in space) such that $a \in [3, \infty)$, $b \in (3, \infty)$, such that $2/a + 3/b = 1$, over

$$U_r^\rho := \big\{(\mathbf{x}, t) \in Q_T;\ t_0 - r^2/\rho^2 < t < t_0,\ \epsilon(t)\rho < |\mathbf{x} - \mathbf{x}_0| < r\big\}.$$

We observe that the assumptions impose the condition of integrability on p_- in a backward neighborhood $B_r(\mathbf{x}_0) \times (t_0 - r^2/\rho^2, t_0)$ of $(\mathbf{x}_0, t_0)$, intersected with the interior of the space-time paraboloid $|\mathbf{x} - \mathbf{x}_0|^2 = (t_0 - t)\rho$, and on $\mathbf{u}$

in $B_r(\mathbf{x}_0) \times (t_0 - r^2/\rho^2, t_0)$, intersected with the exterior of the paraboloid $|\mathbf{x} - \mathbf{x}_0|^2 = (t_0 - t)\rho$. As ρ can be arbitrarily large, the paraboloid can be arbitrarily wide. The results of [55] have been generalized in paper [62], where, instead of p_-, the author imposes conditions on the positive part of a certain linear combination of u_1^2, u_2^2, u_3^2, and p. An appropriate choice of the coefficients in the linear combination leads, e.g., to conditions on the positive part of the so-called Bernoulli pressure $p + \frac{1}{2}|\mathbf{u}|^2$ or on the negative part of p.

9) Y. Zhou [85] (2006) considered $\Omega = \mathbb{R}^3$ and $\mathbf{u}_0 \in \mathbf{L}^2_\sigma(\mathbb{R}^3) \cap \mathbf{L}^q(\mathbb{R}^3)$ for some $q \geq 4$. Applying a different method than in [8], the author proved the regularity of $\mathbf{u}$ under the assumption that the associated pressure p satisfies $\nabla p \in L^r(0, T;\, \mathbf{L}^s(\mathbb{R}^3))$, $2/r + 3/s \leq 3$, $\frac{2}{3} \leq r < \infty$ or $\nabla p \in L^\infty(0, T;\, \mathbf{L}^1(\mathbb{R}^3))$ with the corresponding norm sufficiently small. This result has been further extended by the same author to the case of N spatial dimensions (for $N = 3$ or $N = 4$) and the initial velocity in $\mathbf{L}^2_\sigma(\mathbb{R}^N) \cap \mathbf{L}^q(\mathbb{R}^N)$ for some $q > N$, see [86] (2006).

10) In papers [37] (2006) and [38] (2010), K. Kang and J. Lee considered a smooth bounded domain $\Omega \subset \mathbb{R}^N$ (for $N \geq 3$) and the initial velocity $\mathbf{u}_0$ in $\mathbf{L}^2_\sigma(\Omega) \cap \mathbf{L}^N(\Omega)$. They assumed that $\mathbf{u}$ is a Leray-Hopf weak solution and p is an associated pressure, and they proved the regularity of $\mathbf{u}$ provided that

- $p \in L^r(0, T;\, L^s(\Omega))$ for $\dfrac{2}{r} + \dfrac{N}{s} \leq 2$, $\dfrac{N}{2} < s \leq \infty$ if $N = 3, 4$,
  or $\dfrac{N^2}{2(N-1)} \leq s \leq \infty$ if $N \geq 5$,
- or $\nabla p \in L^r(0, T;\, \mathbf{L}^s(\Omega))$ for $\dfrac{2}{r} + \dfrac{N}{s} \leq 3$, $1 < s \leq \infty$.

We describe the results of [37] and [38] in greater detail in Sect. 4.5.3.

11) Q. Chen, Z. Zhang [18] (2007) assumed that $\mathbf{u}_0 \in \mathbf{L}^2_\sigma(\mathbb{R}^3) \cap \mathbf{L}^q(\mathbb{R}^3)$ for some $q > 3$, $\mathbf{u}$ is a Leray-Hopf weak solution of the problem (4.9), (4.10), (4.12) and p is an associated pressure. They proved the regularity of $\mathbf{u}$, provided that $p \in L^1(0, T;\, \dot{B}^0_{\infty,\infty}(\mathbb{R}^3))$. (Here $\dot{B}^0_{\infty,\infty}(\mathbb{R}^3)$ is the homogeneous Besov space.)

12) M. Struwe [76] (2007) extended the result from [86] to N spatial dimensions for $N \geq 3$, considering either $\Omega = \mathbb{R}^N$ or the spatially periodic problem in $\mathbb{R}^N$.

13) J. Fan and T. Ozawa [20] (2008) refined the result from [85] so that, instead of assuming $\nabla p \in L^{3/2}(0, T;\, \mathbf{L}^\infty(\mathbb{R}^3))$, they use the condition $\nabla p \in L^{3/2}(0, T;\, BMO(\mathbb{R}^3))$, where $BMO(\mathbb{R}^3)$ denotes the space of vector functions with bounded mean oscillation in $\mathbb{R}^3$.

14) J. Fan, S. Jiang, and G. Ni [21] (2008) considered $\Omega = \mathbb{R}^N$ (for $N \geq 3$) and obtained sufficient conditions for regularity of a weak solution $\mathbf{u}$, formulated by means of norms of an associated pressure in Morrey and Besov spaces. Their results generalize some theorems from [8, 86] and [13].

15) Z. Cai, J. Fan, and J. Zhai [13] (2010) considered $\Omega = \mathbb{R}^3$ and proved the regularity of a Leray–Hopf weak solution $\mathbf{u}$ to the Navier-Stokes IBVP (4.9), (4.10), (4.12) (with an associated pressure p), assuming that

$\mathbf{u}_0 \in \mathbf{L}^2_\sigma(\mathbb{R}^3) \cap \mathbf{L}^q(\mathbb{R}^3)$ for some $q \geq 4$ and at least one of the next conditions holds:

- $p \in L^r(0, T;\ L^s_w(\mathbb{R}^3))$ for $\dfrac{2}{r} + \dfrac{3}{s} = 2,\ \ 1 \leq r < \infty$,
- $p \in L^\infty(0, T;\ L^{3/2}_w(\mathbb{R}^3))$ with the corresponding norm "sufficiently small,"
- $\nabla p \in L^r(0, T;\ \mathbf{L}^s_w(\mathbb{R}^3))$ for $\dfrac{2}{r} + \dfrac{3}{s} = 3,\ \ \dfrac{2}{3} < s < \infty,\ \ 1 < r < \infty$,
- $\nabla p \in L^{2/3}(0, T;\ \mathbf{L}^\infty_w(\mathbb{R}^3))$ with the corresponding norm "sufficiently small."

16) T. Suzuki [77] (2012) considered a bounded smooth domain Ω in $\mathbb{R}^3$, the initial velocity in $\mathbf{W}^{1,2}_{0,\sigma}(\Omega) \cap \mathbf{L}^\infty(\Omega)$ and proved the regularity of an existing weak solution $\mathbf{u}$ if an associated pressure p is in the Lorentz space $L^r_w(0, T;\ L^q_w(\Omega))$ with the corresponding norm "sufficiently Small," where

$$\frac{2}{r} + \frac{3}{q} = 2, \quad \frac{5}{2} \leq q \leq 3, \qquad 2 \leq r \leq \frac{5}{2},$$

or, alternatively, if ∇p is in $L^r_w(0, T;\ \mathbf{L}^q_w(\Omega))$ with the corresponding norm "sufficiently Small," where

$$\frac{2}{r} + \frac{3}{q} = 3, \quad \frac{5}{2} \leq q < 3, \qquad 1 \leq r \leq \frac{5}{3}.$$

17) S. Bosia, M. Conti, and V. Pata [9] (2014) considered either $\Omega = \mathbb{R}^3$ or Ω being a smooth bounded domain in $\mathbb{R}^3$, and a Leray–Hopf weak solution $\mathbf{u}$. They proved the regularity criterion, saying that the inequality

$$\liminf_{\epsilon \to 0+} \epsilon^{(r-1)/2} \int_0^{T_\epsilon} \|\nabla p(\,.\,, t)\|_s^{r(1-\epsilon)}\, \mathrm{d}t \; < \; K,$$

where $T_\epsilon := T - \mathrm{e}^{-1/\epsilon}$, $\epsilon > 0$, $K = K(\Omega, r, s) > 0$ (K can be explicitly evaluated) and $2/r + 3/s = 3$, $1 < r \leq 3$, implies that solution $\mathbf{u}$ is regular in $\Omega \times (0, T)$.

4.5.2 More from the Paper [8] by Berselli and Galdi

The authors consider $\Omega \subset \mathbb{R}^N$ for $N \geq 3$. However, we restrict ourselves only to $N = 3$ in this subsection.

A) The Serrin-Type Condition Imposed on p, the Case $\Omega = \mathbb{R}^3$ Assume at first that $\Omega = \mathbb{R}^3$. Let $\mathbf{u}$ be a Leray-Hopf weak solution to the Navier-Stokes problem (4.9) (where $\mathbf{f} = \mathbf{0}$), (4.10), (4.12) with an associated pressure $p \in L^r(0, T;\ L^s(\Omega))$ for some $1 \leq r < \infty$, $\frac{3}{2} < s \leq \infty$, such that $2/r + 3/s = 2$.

Let the initial velocity $\mathbf{u}_0$ be in $\mathbf{L}^2_\sigma(\mathbb{R}^3) \cap \mathbf{L}^3(\Omega)$. This assumption guarantees the existence of a strong solution $\widetilde{\mathbf{u}}$ to the problem (4.9), (4.10), (4.12) on some time interval $(0, T_0) \subset (0, T)$, see [35]. (Here, we use the notion "strong solution" in the sense that $\widetilde{\mathbf{u}}$ is in $BC\big([0, T_0);\, \mathbf{L}^3(\mathbb{R}^3)\big) \cap L^l(0, T_0;\, \mathbf{L}^q(\mathbb{R}^3))$ for $l \geq 2$, $q > 3$, satisfying $2/l + 3/q = 1$.) Moreover, $\widetilde{\mathbf{u}}$ is infinitely differentiable in $\mathbb{R}^3 \times (0, T_0]$, see [35]. As the weak solution $\mathbf{u}$ satisfies the energy inequality (because it is the Leray-Hopf solution), it coincides with $\widetilde{\mathbf{u}}$ on $(0, T_0)$. Denote by T^* the supremum of $T_0 \in (0, T]$ with the aforementioned properties. Assume that $t \in (0, T^*)$.

The advantage of the whole space-case $\Omega = \mathbb{R}^3$ is that one can use the inequality

$$\|p(\,.\,,t)\|_q \;\leq\; c\,\|\mathbf{u}(\,.\,,t)\|_{2q}^2 \qquad \text{for } 1 < q < \infty, \tag{4.82}$$

where $c = c(q)$. Inequality (4.82) can be derived from the equation $\Delta p = -\mathrm{div}\,(\mathbf{u}\cdot\nabla\mathbf{u}) \equiv \nabla^2 : (\mathbf{u} \otimes \mathbf{u})$ by means of the Calderón-Zygmund theorem.

Multiplying the Navier-Stokes equation (4.9) (with $\mathbf{f} = \mathbf{0}$) by $|\mathbf{u}|\mathbf{u}$, integrating in $\mathbb{R}^3$ and applying the integration by parts, we obtain

$$\frac{1}{3}\,\frac{\mathrm{d}}{\mathrm{d}t}\,\|\mathbf{u}\|_3^3 + \nu \int_{\mathbb{R}^3} |\mathbf{u}|\,|\nabla\mathbf{u}|^2\,\mathrm{d}\mathbf{x} + \frac{4\nu}{9}\,\big\|\nabla|\mathbf{u}|^{3/2}\big\|_2^2 \;\leq\; \frac{2}{3}\int_{\mathbb{R}^3} |p|\,|\mathbf{u}|^{1/2}\,\big|\nabla|\mathbf{u}|^{3/2}\big|\,\mathrm{d}\mathbf{x}. \tag{4.83}$$

1) The case $3 \leq s < \infty$. Applying Hölder's inequality, interpolation inequalities, the Calderón-Zygmund inequality (4.82), and Young's inequality, we get

$$\begin{aligned}
\int_{\mathbb{R}^3} |p|\,|\mathbf{u}|^{1/2}\,\big|\nabla|\mathbf{u}|^{3/2}\big|\,\mathrm{d}\mathbf{x} \;&\leq\; \|p\|_3\,\|\mathbf{u}\|_3^{1/2}\,\big\|\nabla|\mathbf{u}|^{3/2}\big\|_2 \\
&\leq\; \|p\|_s^{\frac{s}{2s-3}}\,\|p\|_{3/2}^{\frac{s-3}{2s-3}}\,\|\mathbf{u}\|_3^{1/2}\,\big\|\nabla|\mathbf{u}|^{3/2}\big\|_2 \\
&\leq\; c\,\|p\|_s^{\frac{s}{2s-3}}\,\|\mathbf{u}\|_3^{\frac{2(s-3)}{2s-3}}\,\|\mathbf{u}\|_3^{1/2}\,\big\|\nabla|\mathbf{u}|^{3/2}\big\|_2 \\
&\leq\; c\,\|p\|_s^r\,\|\mathbf{u}\|_3^{\frac{6s-15}{2s-3}} + \frac{2\nu}{9}\,\big\|\nabla|\mathbf{u}|^{3/2}\big\|_2^2.
\end{aligned} \tag{4.84}$$

(Recall that $r = 2s/(2s-3)$.) Observe that, since $3 \leq s < \infty$, we have $1 \leq \frac{6s-15}{2s-3} = 3 - \frac{6}{2s-3} < 3$. When we estimate the right-hand side of (4.83) by means of (4.84) then the second term on the right-hand side can be absorbed by the third term on the left-hand side. Moreover, as $\|p\|_s^r$ is integrable in $(0, T)$, we obtain (applying Gronwall's inequality to (4.83)) the estimate $\|\mathbf{u}(\,.\,,t)\|_3 \leq c < \infty$ for $t \in (0, T^*)$, where c is independent of T^*. Hence the right-hand side of (4.83) is integrable in $(0, T^*)$. Consequently, $\nabla|\mathbf{u}|^{3/2} \in L^2(0, T^*;\, \mathbf{L}^2(\mathbb{R}^3))$, which yields (by Sobolev's inequality) $\mathbf{u} \in L^3(0, T^*;\, \mathbf{L}^9(\mathbb{R}^3))$. If T^* is the so-called epoch of irregularity of solution $\mathbf{u}$, which means that $\mathbf{u}(\,.\,,t)$ "blows-up" for $t \to T^*-$, then

$$\|\mathbf{u}(\,.\,,t)\|_9 \;\geq\; \frac{c}{(T^* - t)^{1/3}}$$

for t in some left neighborhood of T_*, see [35] or [30, Theorem 7.3]. This, however, implies that $\int_0^{T^*} \|\mathbf{u}(\,.\,,t)\|_9^3\, dt = \infty$, which is in contradiction with the inclusion $\mathbf{u} \in L^3(0,T^*;\, \mathbf{L}^9(\mathbb{R}^3))$. Hence $T^* = T$ and it cannot be an epoch of irregularity. (The fact that T^* is not an epoch of irregularity can also be deduced from the inclusion $\mathbf{u} \in L^3(0,T^*;\, \mathbf{L}^9(\mathbb{R}^3))$ and Lemma 4.5.1.)

2) *The case* $\frac{9}{4} \le s < 3$. Instead of (4.84), we estimate the integral on the right-hand side of (4.83) as follows:

$$
\begin{aligned}
\int_{\mathbb{R}^3} |p|\,|\mathbf{u}|^{1/2}\,\big|\nabla|\mathbf{u}|^{3/2}\big|\,d\mathbf{x} &\le \|p\|_s\,\|\mathbf{u}\|_{s/(s-2)}^{1/2}\,\big\|\nabla|\mathbf{u}|^{3/2}\big\|_2 \\
&\le \|p\|_s\,\|\mathbf{u}\|_3^{\frac{4s-9}{2s}}\,\|\mathbf{u}\|_9^{\frac{3(3-s)}{2s}}\,\big\|\nabla|\mathbf{u}|^{3/2}\big\|_2 \le c\,\|p\|_s\,\|\mathbf{u}\|_3^{\frac{4s-9}{2s}}\,\big\|\nabla|\mathbf{u}|^{3/2}\big\|_2^{\frac{3}{s}} \\
&\le c\,\|p\|_s^r\,\|\mathbf{u}\|_3^{\frac{(4s-9)r}{2s}} + \frac{2\nu}{9}\,\big\|\nabla|\mathbf{u}|^{3/2}\big\|_2^{\frac{3}{s}\frac{r-1}{r}} \\
&= c\,\|p\|_s^r\,\|\mathbf{u}\|_3^{\frac{4s-9}{2s-3}} + \frac{2\nu}{9}\,\big\|\nabla|\mathbf{u}|^{3/2}\big\|_2^2.
\end{aligned}
$$

(We have used Hölder's inequality, the interpolation Inequality, and Sobolev's inequality.) Since $\frac{9}{4} \le s < 3$, we have $\frac{4s-9}{2s-3} = 3 - \frac{6s}{3(2s-3)} < 3$ and we can proceed as in case 1).

3) *The case* $\frac{3}{2} < s < \frac{9}{4}$. Applying Hölder's inequality, the interpolation inequality, the Calderón-Zygmund inequality (4.82) and Sobolev's inequality, we estimate the integral on the right-hand side of (4.83) in this way:

$$
\begin{aligned}
\int_{\mathbb{R}^3} |p|\,|\mathbf{u}|^{1/2}\,\big|\nabla|\mathbf{u}|^{3/2}\big|\,d\mathbf{x} &\le \|p\|_{9/4}\,\|\mathbf{u}\|_9^{1/2}\,\big\|\nabla|\mathbf{u}|^{3/2}\big\|_2 \\
&\le \|p\|_s^{\frac{2s}{9-2s}}\,\|p\|_{9/2}^{\frac{9-4s}{9-2s}}\,\|\mathbf{u}\|_9^{1/2}\,\big\|\nabla|\mathbf{u}|^{3/2}\big\|_2 \\
&\le c\,\|p\|_s^{\frac{2s}{9-2s}}\,\|\mathbf{u}\|_9^{\frac{2(9-4s)}{9-2s}}\,\|\mathbf{u}\|_9^{1/2}\,\big\|\nabla|\mathbf{u}|^{3/2}\big\|_2 \\
&\le c\,\|p\|_s^{\frac{2s}{9-2s}}\,\big\|\nabla|\mathbf{u}|^{3/2}\big\|_2^{\frac{24-8s}{9-2s}} \le c\,\|p\|_s^{\frac{2s}{2s-3}} + \frac{2\nu}{9}\,\big\|\nabla|\mathbf{u}|^{3/2}\big\|_2^2 \\
&= c\,\|p\|_s^r + \frac{2\nu}{9}\,\big\|\nabla|\mathbf{u}|^{3/2}\big\|_2^2.
\end{aligned}
$$

4) *The case* $s = \infty$. We estimate the integral on the right-hand side of (4.83) by means of Hölder's inequality, the interpolation inequality, the Calderón-Zygmund inequality (4.82), and Young's inequality:

$$
\begin{aligned}
\int_{\mathbb{R}^3} |p|\,|\mathbf{u}|^{1/2}\,\big|\nabla|\mathbf{u}|^{3/2}\big|\,d\mathbf{x} &\le \|p\|_\infty^{1/2}\,\|p\|_{3/2}^{1/2}\,\|\mathbf{u}\|_3^{1/2}\,\big\|\nabla|\mathbf{u}|^{3/2}\big\|_2 \\
&\le \|p\|_\infty^{1/2}\,\|\mathbf{u}\|_3^{3/2}\,\big\|\nabla|\mathbf{u}|^{3/2}\big\|_2 \le c\,\|p\|_\infty\,\|\mathbf{u}\|_3^3 + \frac{2\nu}{9}\,\big\|\nabla|\mathbf{u}|^{3/2}\big\|_2^2.
\end{aligned}
$$

In both cases 3) and 4), one can again further proceed as in case 1). We have proven the theorem (see [8, Theorem 1.1]):

Theorem 4.5.2 *Let* $\mathbf{u}_0 \in \mathbf{L}^2_\sigma(\mathbb{R}^3) \cap \mathbf{L}^3(\mathbb{R}^3)$. *Let* $\mathbf{u}$ *be a Leray-Hopf weak solution to the Navier-Stokes problem (4.9), (4.10), (4.12) (with* $\Omega = \mathbb{R}^3$ *and* $\mathbf{f} = \mathbf{0}$*) such that the associated pressure* p *satisfies* $p \in L^r(0, T;\ L^s(\Omega))$ *for some* $1 \le r < \infty$, $\frac{3}{2} < s \le \infty$, *such that* $2/r + 3/s = 2$. *Then solution* $\mathbf{u}$ *is regular. (More precisely,* $\mathbf{u}$ *belongs to the class* $C^\infty\big(\mathbb{R}^3 \times (0, T]\big)$.*)*

B) Remark (on the Assumption $\mathbf{u}_0 \in \mathbf{L}^3(\mathbb{R}^3)$) The assumption $\mathbf{u}_0 \in \mathbf{L}^3(\mathbb{R}^3)$ can be omitted if $\mathbf{u}$ is a weak solution, satisfying the so-called strong energy inequality

$$\|\mathbf{u}(\,.\,, t_2)\|_2^2 + 2\nu \int_{t_1}^{t_2} \|\nabla \mathbf{u}(\,.\,, \tau)\|_2^2 \, \mathrm{d}\tau \;\le\; \|\mathbf{u}(\,.\,, t_1)\|_2^2 \tag{4.85}$$

for a.a. $t_1 \in (0, T)$ and all $t_2 \in [t_1, T)$ (see [30]). In this case, we use the fact that to any $\epsilon > 0$ there exists $t_1 \in (0, \epsilon)$ such that $\mathbf{u}_1 := \mathbf{u}(\,.\,, t_1) \in \mathbf{W}^{1,2}_{0,\sigma}(\mathbb{R}^3)$ and inequality (4.85) holds for all $t_2 \in [t_1, T)$. There exists $t_1 < T_0 \le T$ and a strong solution $\widetilde{\mathbf{u}}$ of the equations (4.9), (4.10) on the time interval (t_1, T_0), satisfying the initial condition $\widetilde{\mathbf{u}}(\,.\,, t_1) = \mathbf{u}_1$. Since the weak solution $\mathbf{u}$ satisfies inequality (4.85), it can be identified with $\widetilde{\mathbf{u}}$ on the interval (t_1, T_0). Further on, we use the same arguments as in Sect. 4.5.2A and show that $\mathbf{u}$ is in fact a strong solution on (t_1, T) and $\mathbf{u} \in C^\infty\big(\mathbb{R}^3 \times (t_1, T)\big)$. This implies that $\mathbf{u}$ is infinitely differentiable in $\mathbb{R}^3 \times (0, T)$.

C) Remark (Concerning the Case $\Omega \neq \mathbb{R}^3$) If Ω is a half-space of $\mathbb{R}^3$ or a bounded or exterior domain in $\mathbb{R}^3$ with the boundary of the class C^∞, then Theorem 4.5.2 still holds under the restriction $\frac{9}{4} \le s \le 3$. The reason is that a) the cited results from [35] still hold in these cases and b) we do not need the Calderón-Zygmund inequality (4.82) for $\frac{9}{4} \le s < 3$ (case 2). Moreover, if $s = 3$, then (4.84) reduces to the estimates

$$\begin{aligned} \int_\Omega |p|\,|\mathbf{u}|^{1/2}\,\big|\nabla|\mathbf{u}|^{3/2}\big|\,\mathrm{d}\mathbf{x} \;&\le\; \|p\|_3\,\|\mathbf{u}\|_3^{1/2}\,\big\|\nabla|\mathbf{u}|^{3/2}\big\|_2 \\ &\le\; c\,\|p\|_3^2\,\|\mathbf{u}\|_3 + \frac{2\nu}{9}\,\big\|\nabla|\mathbf{u}|^{3/2}\big\|_2^2, \end{aligned}$$

where we also do not need inequality (4.82).

D) The Serrin-Type Condition Imposed on ∇p Assume that $\Omega = \mathbb{R}^3$ or Ω is one of the domains named in Sect. 4.5.2C and $\mathbf{u}$, T^* and $t \in (0, T^*)$ are the same as in Sect. 4.5.2A. We assume that the associated pressure p satisfies the condition $\nabla p \in L^r(0, T;\ \mathbf{L}^s(\Omega))$ with $\frac{9}{7} \le s \le 3$, $2/r + 3/s = 3$. Multiplying again the Navier-Stokes equation (4.9) (with $\mathbf{f} = \mathbf{0}$) by $\mathbf{u}\,|\mathbf{u}|$ and integrating in Ω, we obtain the same inequality as (4.83), where, however, we do not apply integration by parts to the integral on the right-hand side and instead of it, we estimate the integral of

$\nabla p \cdot \mathbf{u}\,|\mathbf{u}|$ as follows:

$$\left| \int_\Omega \nabla p \cdot \mathbf{u}\,|\mathbf{u}|\; d\mathbf{x} \right| \le \int_\Omega |\nabla p|\,|\mathbf{u}|^2\; d\mathbf{x} \le \|\nabla p\|_s \, \big\| |\mathbf{u}|^{3/2} \big\|_{4s'/3}^{4/3}$$

$$\le c\, \|\nabla p\|_s \, \|\mathbf{u}\|_3^{\frac{9-2s'}{2s'}} \, \big\| |\mathbf{u}|^{3/2} \big\|_6^{\frac{2s'-3}{s'}} \le c\, \|\nabla p\|_s \, \|\mathbf{u}\|_3^{\frac{9-2s'}{2s'}} \, \big\| \nabla |\mathbf{u}|^{3/2} \big\|_2^{\frac{2s'-3}{s'}}$$

$$\le c\, \|\nabla p\|_s^{2s'/3} \, \|\mathbf{u}\|_3^{\frac{9-2s'}{3}} + \frac{2\nu}{9} \, \big\| \nabla |\mathbf{u}|^{3/2} \big\|_2^2$$

$$= c\, \|\nabla p\|_s^{r} \, \|\mathbf{u}\|_3^{\frac{9-2s'}{3}} + \frac{2\nu}{9} \, \big\| \nabla |\mathbf{u}|^{3/2} \big\|_2^2,$$

where s' is the conjugate exponent to s. (Recall that $r = 2s/(3s-3)$.) The exponent $(9-2s')/3$ is less than 3, hence we can again complete the proof of regularity of $\mathbf{u}$ as in case 1 in Sect. 4.5.2A. Note that the restriction $\frac{9}{7} \le s \le 3$ enables us to interpolate the norm $\big\| |\mathbf{u}|^{3/2} \big\|_{4s'/3}$ between $\|\mathbf{u}\|_3$ and $\big\| |\mathbf{u}|^{3/2} \big\|_6$. We obtain the theorem (see [8, Theorem 3.3]):

Theorem 4.5.3 *Let Ω be either the whole space $\mathbb{R}^3$ or a half-space in $\mathbb{R}^3$ or a bounded or exterior domain in $\mathbb{R}^3$ with the boundary of the class C^∞. Let $\mathbf{u}_0 \in \mathbf{L}^2_\sigma(\Omega) \cap \mathbf{L}^3(\Omega)$, $\mathbf{u}$ be a Leray-Hopf weak solution to the problem (4.9)–(4.12) (with $\mathbf{f} = \mathbf{0}$) and p be an associated pressure. If $\nabla p \in L^r(0,T;\, \mathbf{L}^s(\Omega))$ for some $\frac{9}{7} \le s \le 3$, $2/r + 3/s = 3$, then $\mathbf{u}$ belongs to the class $C^\infty\big(\mathbb{R}^3 \times (0,T]\big)$.*

Note that if $\mathbf{u}_0$ is only known to belong to $\mathbf{L}^2_\sigma(\Omega)$ and $\mathbf{u}$ is a weak solution satisfying the strong energy inequality then the statement on regularity of $\mathbf{u}$ can be proven in the version modified in the same way as in Remark 4.2.2.

4.5.3 More from the Papers [37] and [38] by Kang and Lee

Assume that Ω is a smooth bounded domain in $\mathbb{R}^3$. (As in paper [8], the authors consider $\Omega \subset \mathbb{R}^N$ for $N \ge 3$ in [37] and [38], but we restrict ourselves only to the case $N = 3$ in this subsection.) Let $\mathbf{u}$, p, and T^* have the same meaning as in Sect. 4.5.2A. Without loss of generality, we can assume that the pressure p is normalized so that $\int_\Omega p(\,.\,,t)\; d\mathbf{x} = 0$ for a.a. $t \in (0,T)$. Then, due to Poincaré's inequality, we have

$$\big\| p(\,.\,,t) \big\|_{\frac{3l}{3-l}} \le c\, \big\| \nabla p(\,.\,,t) \big\|_l \tag{4.86}$$

for a.a. $t \in (0,T^*)$ and $1 \le l < 3$.

A) The Serrin-Type Condition Imposed on p Assume that $p \in L^r(0,T;\, L^s(\Omega))$, where the exponents r and s satisfy the restrictions named in item 10) of Sect. 4.5.1, i.e. $1 \le r < \infty$, $\frac{3}{2} < s \le \infty$, such that $2/r + 3/s \le 2$. By analogy with [8], Kang and Lee also at first multiply equation (4.9) (at time $t \in (0, T^*)$) by $|\mathbf{u}|\,\mathbf{u}$ and integrate over Ω in paper [37]. They obtain the same inequality as (4.83), where $\mathbb{R}^3$ is only replaced by Ω. The method from paper [8], which works well if $\Omega = \mathbb{R}^3$, is now applicable only if $\frac{9}{4} \le s \le 3$, because in this case the Calderón-Zygmund inequality (4.82) is not needed. (See Remark 4.5.2.C.) This is why Kang and Lee use another approach for $\frac{3}{2} < s < \frac{9}{4}$ or $3 < s \le \infty$.

1) The case $\frac{3}{2} < s < \frac{9}{4}$. The authors estimate the integral on the right-hand side of (4.83) in this way:

$$\begin{aligned}\int_\Omega |p|\,|\mathbf{u}|^{1/2}\,\big|\nabla|\mathbf{u}|^{3/2}\big|\,\mathrm{d}\mathbf{x} &\le c\,\|p\|_{9/4}\,\big\||\mathbf{u}|^{1/2}\big\|_{18}\,\big\|\nabla|\mathbf{u}|^{3/2}\big\|_2\\ &\le c\,\|p\|_{9/4}\,\big\|\nabla|\mathbf{u}|^{3/2}\big\|_2^{4/3} \le c\,\|p\|_s^{\theta}\,\|p\|_{9/2}^{1-\theta}\,\big\|\nabla|\mathbf{u}|^{3/2}\big\|_2^{4/3}\\ &\le \|p\|_s^{\theta}\,\|\nabla p\|_{9/5}^{1-\theta}\,\big\|\nabla|\mathbf{u}|^{3/2}\big\|_2^{4/3} \le \frac{c}{\epsilon}\,\|p\|_s^{\frac{3\theta}{2\theta-1}} + c\epsilon\,\|\nabla p\|_{9/5}^{\frac{3(1-\theta)}{2(2-\theta)}}\,\big\|\nabla|\mathbf{u}|^{3/2}\big\|_2^{\frac{2}{2-\theta}},\end{aligned}$$

where $\theta = 2s/(9-2s)$ and $\epsilon > 0$ is sufficiently small. Substituting this to (4.83) and integrating with respect to time from t_1 to t_2 (where $0 < t_1 < t_2 \le T^*$), one obtains

$$\begin{aligned}\|\mathbf{u}(\,.\,,t_2)\|_3^3 &- \|\mathbf{u}(\,.\,,t_1)\|_3^3 + \frac{4\nu}{3}\int_{t_1}^{t_2}\big\|\nabla|\mathbf{u}|^{3/2}\big\|_2^2\,\mathrm{d}t\\ &\le \frac{c}{\epsilon}\int_{t_1}^{t_2}\|p\|_s^{\frac{3\theta}{2\theta-1}}\,\mathrm{d}t + c\epsilon\int_{t_1}^{t_2}\|\nabla p\|_{9/5}^{\frac{3(1-\theta)}{2(2-\theta)}}\,\big\|\nabla|\mathbf{u}|^{3/2}\big\|_2^{\frac{2}{2-\theta}}\,\mathrm{d}t\\ &\le \frac{c}{\epsilon}\int_{t_1}^{t_2}\|p\|_s^{\frac{3\theta}{2\theta-1}}\,\mathrm{d}t + c\epsilon\left(\int_{t_1}^{t_2}\|\nabla p\|_{9/5}^{3/2}\,\mathrm{d}t\right)^{\frac{1-\theta}{2-\theta}}\left(\int_{t_1}^{t_2}\big\|\nabla|\mathbf{u}|^{3/2}\big\|_2^2\,\mathrm{d}t\right)^{\frac{1}{2-\theta}}.\end{aligned}$$

As $\mathbf{u}$, p can be treated as a solution of the Stokes equation with the right-hand side $-\mathbf{u}\cdot\nabla\mathbf{u}$, one can apply inequality (4.45) and we estimate the integral of $\|\nabla p\|_{9/5}^{3/2}$ as follows:

$$\begin{aligned}\int_{t_1}^{t_2}\|p\|_{9/5}^{3/2}\,\mathrm{d}t &\le c\left[\|\mathbf{u}(\,.\,,t_1)\|_{\mathbf{D}^{1/3,3/2}_{9/5}(\Omega)}^{3/2} + \int_{t_1}^{t_2}\|\mathbf{u}\cdot\nabla\mathbf{u}\|_{9/5}^{3/2}\,\mathrm{d}t\right]\\ &\le c\left[\|\mathbf{u}(\,.\,,t_1)\|_{2,\,9/5}^{3/2} + \int_{t_1}^{t_2}\big\||\mathbf{u}|^{1/2}\nabla|\mathbf{u}|^{3/2}\big\|_{9/5}^{3/2}\,\mathrm{d}t\right]\\ &\le c\left[\|\mathbf{u}(\,.\,,t_1)\|_{2,\,9/5}^{3/2} + \int_{t_1}^{t_2}\big\|\nabla|\mathbf{u}|^{3/2}\big\|_2^2\,\mathrm{d}t\right]. \qquad (4.87)\end{aligned}$$

Hence

$$\|\mathbf{u}(\,.\,,t_2)\|_3^3 - \|\mathbf{u}(\,.\,,t_1)\|_3^3 + \frac{4\nu}{3}\int_{t_1}^{t_2} \big\|\nabla|\mathbf{u}|^{3/2}\big\|_2^2\, \mathrm{d}t$$

$$\leq \frac{c}{\epsilon}\int_{t_1}^{t_2} \|p\|_s^{\frac{3\theta}{2\theta-1}}\, \mathrm{d}t$$

$$+\, c\epsilon\bigg[\big\|\mathbf{u}(\,.\,,t_1)\big\|_{2,9/5}^{3/2} + \int_{t_1}^{t_2} \big\|\nabla|\mathbf{u}|^{3/2}\big\|_2^2\, \mathrm{d}t\bigg]^{\frac{1-\theta}{2-\theta}} \bigg(\int_{t_1}^{t_2} \big\|\nabla|\mathbf{u}|^{3/2}\big\|_2^2\, \mathrm{d}t\bigg)^{\frac{1}{2-\theta}}$$

$$\leq \frac{c}{\epsilon}\int_{t_1}^{t_2} \|p\|_s^q\, \mathrm{d}t + c\,\big\|\mathbf{u}(\,.\,,t_1)\big\|_{2,9/5}^{3/2} + c\epsilon\int_{t_1}^{t_2} \big\|\nabla|\mathbf{u}|^{3/2}\big\|_2^2\, \mathrm{d}t,$$

where $q := (3\theta)/(2\theta-1)$. If ϵ is small enough, then the last term on the right-hand side can be absorbed by the left-hand side so that at least $\frac{2}{3}\nu\int_{t_1}^{t_2}\big\|\nabla|\mathbf{u}|^{3/2}\big\|_2^2\,\mathrm{d}t$ still remains on the left-hand side. Since $\theta = 2s/(9-2s)$, one can verify that $2/q+3/s = 2$, which means that the first integral on the right-hand side is bounded above by a constant, independent of t_1 and t_2. Thus, the integral $\int_{t_1}^{t_2}\big\|\nabla|\mathbf{u}|^{3/2}\big\|_2^2\,\mathrm{d}t$ is bounded above by a constant, independent of t_2. Considering $t_2 \to T^*-$ and applying Sobolev's inequality, we observe that $\mathbf{u} \in L^3(t_1, T^*;\, \mathbf{L}^9(\Omega))$. The proof of regularity of the solution $\mathbf{u}$, p on the whole time interval $(0, T)$ can now be completed by means of the same arguments as in Sect. 4.5.2A.

2) The case $3 < s < \infty$. In paper [38]) (which is an erratum and addendum to [37]), the authors multiply equation (4.9) (at time $t \in (0, T^*)$) by $|\mathbf{u}|^2\mathbf{u}$ and integrate over Ω. By analogy with (4.83), they obtain the inequalities

$$\frac{1}{4}\frac{\mathrm{d}}{\mathrm{d}t}\|\mathbf{u}\|_4^4 + \nu\int_\Omega |\mathbf{u}|^2|\nabla\mathbf{u}|^2\,\mathrm{d}\mathbf{u} + \frac{\nu}{4}\int_\Omega \big|\nabla|\mathbf{u}|^2\big|^2\,\mathrm{d}\mathbf{x} \leq c\int_\Omega |p|\,|\nabla\mathbf{u}|\,|\mathbf{u}|^2\,\mathrm{d}\mathbf{x}$$

$$\leq c\,\|\nabla\mathbf{u}\|_2\,\|p\|_s\,\big\||\mathbf{u}|^2\big\|_{2s/(s-2)} \leq c\,\|\nabla\mathbf{u}\|_2\,\|p\|_s\,\|\mathbf{u}\|_{4s/(s-2)}^2$$

$$\leq c\,\|\nabla\mathbf{u}\|_2\,\|p\|_s\,\|\mathbf{u}\|_4^{\frac{2(s-3)}{s}}\,\big\||\mathbf{u}|^2\big\|_6^{\frac{3}{s}} \leq c\,\|\nabla\mathbf{u}\|_2\,\|p\|_s\,\|\mathbf{u}\|_4^{\frac{2(s-3)}{s}}\,\big\|\nabla|\mathbf{u}|^2\big\|_2^{\frac{3}{s}}$$

$$\leq c\,\|\nabla\mathbf{u}\|_2^2\,\|p\|_s^r + c\,\|p\|_s^{2-r}\,\|\mathbf{u}\|_4^{\frac{4(s-3)}{s}}\,\big\|\nabla|\mathbf{u}|^2\big\|_2^{\frac{6}{s}}$$

$$\leq c\,\|\nabla\mathbf{u}\|_2^2\,\|p\|_s^r + \frac{\nu}{8}\,\big\|\nabla|\mathbf{u}|^2\big\|_2^2 + c\,\|p\|_s^{\frac{s(2-r)}{s-3}}\,\|\mathbf{u}\|_4^4$$

$$= c\,\|\nabla\mathbf{u}\|_2^2\,\|p\|_s^r + \frac{\nu}{8}\,\big\|\nabla|\mathbf{u}|^2\big\|_2^2 + c\,\|p\|_s^r\,\|\mathbf{u}\|_4^4. \tag{4.88}$$

The second term on the right-hand side can be absorbed by the left-hand side. In order to estimate the first term on the right-hand side, we multiply equation (4.9) by $\Delta\mathbf{u}$, integrate in Ω, and apply integration by parts. We obtain the inequalities

$$\frac{1}{2}\,\frac{\mathrm{d}}{\mathrm{d}t}\,\|\nabla\mathbf{u}\|_2^2+\nu\,\|\Delta\mathbf{u}\|_2^2\;\le\;\left|\int_\Omega \mathbf{u}\cdot\nabla\mathbf{u}\cdot\Delta\mathbf{u}\,\mathrm{d}\mathbf{x}\right|\;\le\;\frac{\nu}{2}\,\|\Delta\mathbf{u}\|_2^2+c\,\big\|\mathbf{u}\,|\nabla\mathbf{u}|\,\big\|_2^2.$$

Hence

$$\big\|\nabla\mathbf{u}(\,.\,,t)\big\|_2^2+\nu\int_{t_1}^{t}\|\Delta\mathbf{u}\|_2^2\,\mathrm{d}\tau\;\le\;\big\|\nabla\mathbf{u}(\,.\,,t_1)\big\|_2^2+c\int_{t_1}^{t}\big\|\mathbf{u}\,|\nabla\mathbf{u}|\,\big\|_2^2\,\mathrm{d}\tau$$

for a.a. $t_1 < t < T^*$. Thus, inequality (4.88) yields

$$\frac{1}{4}\,\frac{\mathrm{d}}{\mathrm{d}t}\,\|\mathbf{u}\|_4^4+\nu\int_\Omega|\mathbf{u}|^2|\nabla\mathbf{u}|^2\,\mathrm{d}\mathbf{x}+\frac{\nu}{8}\int_\Omega\big|\nabla|\mathbf{u}|^2\big|^2\,\mathrm{d}\mathbf{x}$$
$$\le c\left(\big\|\nabla\mathbf{u}(\,.\,,t_1)\big\|_2^2+\int_{t_1}^{t_2}\big\|\mathbf{u}\,|\nabla\mathbf{u}|\,\big\|_2^2\,\mathrm{d}\tau\right)\|p\|_s^r+c\,\|p\|_s^r\,\|\mathbf{u}\|_4^4$$

at a.a. times $t\in(t_1,t_2]$, where $t_2 < T^*$. Integrating with respect to t from t_1 to t_2, we get

$$\big\|\mathbf{u}(\,.\,,t_2)\big\|_4^4+4\nu\int_{t_1}^{t_2}\big\|\mathbf{u}\,|\nabla\mathbf{u}|\,\big\|_2^2\,\mathrm{d}t+\frac{\nu}{2}\int_{t_1}^{t_2}\big\|\nabla|\mathbf{u}|^2\big\|_2^2\,\mathrm{d}t$$
$$\le c\left(\big\|\nabla\mathbf{u}(\,.\,,t_1)\big\|_2^2+\int_{t_1}^{t_2}\big\|\mathbf{u}\,|\nabla\mathbf{u}|\,\big\|_2^2\,\mathrm{d}\tau\right)\int_{t_1}^{t_2}\|p\|_s^r\,\mathrm{d}t+c\int_{t_1}^{t_2}\|p\|_s^r\,\|\mathbf{u}\|_4^4\,\mathrm{d}t,$$

$$\big\|\mathbf{u}(\,.\,,t_2)\big\|_4^4+\left(4\nu-\int_{t_1}^{t_2}\|p\|_s^r\,\mathrm{d}t\right)\int_{t_1}^{t_2}\big\|\mathbf{u}\,|\nabla\mathbf{u}|\,\big\|_2^2\,\mathrm{d}t+\frac{\nu}{2}\int_{t_1}^{t_2}\big\|\nabla|\mathbf{u}|^2\big\|_2^2\,\mathrm{d}t$$
$$\le c\,\big\|\nabla\mathbf{u}(\,.\,,t_1)\big\|_2^2\int_{t_1}^{t_2}\|p\|_s^r\,\mathrm{d}t+c\int_{t_1}^{t_2}\|p\|_s^r\,\|\mathbf{u}\|_4^4\,\mathrm{d}t.$$

Since $\|p\|_s^r$ is integrable in $(0,T)$, we may assume that $t_1 < T^*$ is so close to T^* that

$$4\nu-\int_{t_1}^{T^*}\|p\|_s^r\,\mathrm{d}t\;\ge\;0.$$

Then we have

$$\big\|\mathbf{u}(\,.\,,t_2)\big\|_4^4+\frac{\nu}{2}\int_{t_1}^{t_2}\big\|\nabla|\mathbf{u}|^2\big\|_2^2\,\mathrm{d}t$$
$$\le c\,\big\|\nabla\mathbf{u}(\,.\,,t_1)\big\|_2^2\int_{t_1}^{T}\|p\|_s^r\,\mathrm{d}t+c\int_{t_1}^{t_2}\|p\|_s^r\,\|\mathbf{u}\|_4^4\,\mathrm{d}t$$
$$\le c_1+c_2\int_{t_1}^{t_2}\|p\|_s^r\,\|\mathbf{u}\|_4^4\,\mathrm{d}t,$$

where $c_1 = c_1(t_1)$ and c_2 is independent of t_1. Applying Gronwall's inequality, we derive an estimate of $\|\mathbf{u}(\,.\,, t_2)\|_4$, uniform for t_2 in a left neighborhood of T^*. Consequently T^* cannot be an epoch of irregularity.

If $s = \infty$, then one can proceed practically in the same way as in the case $3 < s < \infty$.

Summarizing these results, we can formulate the theorem (see Theorem 1.1 in [38]):

Theorem 4.5.4 *Let Ω be a bounded domain in $\mathbb{R}^3$ with the C^∞ boundary and $\mathbf{u}_0 \in \mathbf{L}^2_\sigma(\Omega) \cap \mathbf{L}^3(\Omega)$. Let $\mathbf{u}$ be a Leray-Hopf weak solution to the Navier-Stokes problem (4.9)–(4.12) (with $\mathbf{f} = \mathbf{0}$) such that the associated pressure p satisfies $p \in L^r(0, T;\, L^s(\Omega))$ for some $1 \le r < \infty$, $\frac{3}{2} < s \le \infty$, such that $2/r + 3/s \le 2$. Then solution $\mathbf{u}$ is smooth in $\overline{\Omega} \times (0, T]$.*

B) The Serrin-Type Condition Imposed on ∇p The next theorem uses a condition on ∇p instead of p. It extends the statement of Theorem 4.5.3. (See [37, Theorem 1.2].)

Theorem 4.5.5 *Let Ω be a bounded domain in $\mathbb{R}^3$ with the C^∞ boundary and $\mathbf{u}_0 \in \mathbf{L}^2_\sigma(\Omega) \cap \mathbf{L}^3(\Omega)$. Let $\mathbf{u}$ be a Leray-Hopf weak solution to the Navier-Stokes problem (4.9)–(4.12) (with $\mathbf{f} = \mathbf{0}$) such that the associated pressure p satisfies $\nabla p \in L^r(0, T;\, \mathbf{L}^s(\Omega))$ for some $1 \le r < \infty$, $1 < s \le \infty$, such that $2/r + 3/s \le 3$. Then solution $\mathbf{u}$ is smooth in $\overline{\Omega} \times (0, T]$.*

Proof The assertion is straightforward if $s < 3$ because of Sobolev's inequality and Theorem 4.5.4. Thus, let $s \ge 3$. Assume, by contradiction, that there exists $T^* \in (0, T]$ such that $\int_0^t \|\nabla|\mathbf{u}|^{3/2}\|_2^2\, \mathrm{d}\tau < \infty$ for all $t \in (0, T^*)$ and $\lim_{t \to T^*-} \int_0^t \|\nabla|\mathbf{u}|^{3/2}\|_2^2\, \mathrm{d}\tau = \infty$. As in Sect. 4.5.2D, we need to estimate the integral $\int_\Omega \nabla p \cdot \mathbf{u}\, |\mathbf{u}|\, \mathrm{d}\mathbf{x}$. We proceed as follows:

$$\begin{aligned} \int_\Omega \nabla p \cdot \mathbf{u}\, |\mathbf{u}|\, \mathrm{d}\mathbf{x} &\le c\, \|\nabla p\|_s^{1/2}\, \|\nabla p\|_{3/2}^{1/2}\, \|\mathbf{u}\|_{12s/(4s-3)}^2 \\ &\le c\, \|\nabla p\|_s^{1/2}\, \|\nabla p\|_{3/2}^{1/2}\, \|\mathbf{u}\|_3^{\frac{8s-9}{4s}}\, \|\mathbf{u}\|_9^{\frac{9}{4s}} \\ &\le c\, \|\nabla p\|_s^{1/2}\, \|\nabla p\|_{3/2}^{1/2}\, \|\mathbf{u}\|_3^{\frac{8s-9}{4s}}\, \big\|\nabla|\mathbf{v}|^{3/2}\big\|_2^{\frac{3}{2s}}. \end{aligned} \tag{4.89}$$

Let $0 < t_1 < t_2 < T^*$. Integrating inequality (4.83), with the right-hand side replaced by the right-hand side of (4.89), from t_1 to t_2, we obtain

$$\begin{aligned} &\|\mathbf{u}(\,.\,, t_2)\|_3^3 - \|\mathbf{u}(\,.\,, t_1)\|_3^3 + \frac{4\nu}{3} \int_{t_1}^{t_2} \big\|\nabla|\mathbf{u}|^{3/2}\big\|_2^2\, \mathrm{d}t \\ &\quad \le c \int_{t_1}^{t_2} \|\nabla p\|_s^{1/2}\, \|\nabla p\|_{3/2}^{1/2}\, \|\mathbf{u}\|_3^{\frac{8s-9}{4s}}\, \big\|\nabla|\mathbf{v}|^{3/2}\big\|_2^{\frac{3}{2s}}\, \mathrm{d}t \end{aligned}$$

$$\leq c\left(\int_{t_1}^{t_2} \|\nabla p\|_s^{\frac{2s}{3(s-1)}} \|\mathbf{u}\|_3^{\frac{8s-9}{3(s-1)}} \, \mathrm{d}t\right)^{\frac{3(s-1)}{4s}} \left(\int_{t_1}^{t_2} \|\nabla p\|_{3/2}^2 \, \mathrm{d}t\right)^{\frac{1}{4}}$$

$$\cdot \left(\int_{t_1}^{t_2} \big\|\nabla|\mathbf{u}|^{3/2}\big\|_2^2 \, \mathrm{d}t\right)^{\frac{3}{4s}}$$

$$\leq c\left(\operatorname*{ess\,sup}_{t_1<t<t_2} \|\mathbf{u}(\,.\,,t)\|_3\right)^{\frac{8s-9}{4s}} \left(\int_{t_1}^{t_2} \|\nabla p\|_s^r \, \mathrm{d}t\right)^{\frac{1}{2r}} \left(\int_{t_1}^{t_2} \|\nabla p\|_{3/2}^2 \, \mathrm{d}t\right)^{\frac{1}{4}}$$

$$\cdot \left(\int_{t_1}^{t_2} \big\|\nabla|\mathbf{u}|^{3/2}\big\|_2^2 \, \mathrm{d}t\right)^{\frac{3}{4s}} =: RHS. \tag{4.90}$$

The integral of $\|\nabla p\|_{3/2}^2$ in (t_1, t_2) can be estimated by analogy with (4.87):

$$\int_{t_1}^{t_2} \|\nabla p\|_{3/2}^2 \, \mathrm{d}t \leq c\left[\big\|\mathbf{u}(\,.\,,t_1)\big\|_{\mathbf{D}_{3/2}^{1/2,2}(\Omega)}^2 + \int_{t_1}^{t_2} \|\mathbf{u}\cdot\nabla\mathbf{u}\|_{3/2}^2 \, \mathrm{d}t\right]$$

$$\leq c\,\big\|\mathbf{u}(\,.\,,t_1)\big\|_{2,\,3/2}^2 + c\int_{t_1}^{t_2} \|\mathbf{u}\|_3 \,\big\|\nabla|\mathbf{u}|^{3/2}\big\|_2^2 \, \mathrm{d}t$$

$$\leq c\,\big\|\mathbf{u}(\,.\,,t_1)\big\|_{2,\,3/2}^2 + c\left(\operatorname*{ess\,sup}_{t_1<t<t_2} \|\mathbf{u}(\,.\,,t)\|_3\right) \int_{t_1}^{t_2} \big\|\nabla|\mathbf{u}|^{3/2}\big\|_2^2 \, \mathrm{d}t.$$

Hence

$$RHS \leq c\left(\operatorname*{ess\,sup}_{t_1<t<t_2} \|\mathbf{u}(\,.\,,t)\|_3\right)^{\frac{9(s-1)}{4s}} \left(\int_{t_1}^{t_2} \|\nabla p\|_s^r \, \mathrm{d}t\right)^{\frac{1}{2r}} \left(\int_{t_1}^{t_2} \big\|\nabla|\mathbf{u}|^{3/2}\big\|_2^2 \, \mathrm{d}t\right)^{\frac{s+3}{4s}}$$

$$+ c\left(\operatorname*{ess\,sup}_{t_1<t<t_2} \|\mathbf{u}(\,.\,,t)\|_3\right)^{\frac{8s-9}{4s}} \left(\int_{t_1}^{t_2} \|\nabla p\|_s^r \, \mathrm{d}t\right)^{\frac{1}{2r}} \big\|\mathbf{u}(\,.\,,t_1)\big\|_{2,\,3/2}^{\frac{1}{2}}$$

$$\cdot \left(\int_{t_1}^{t_2} \big\|\nabla|\mathbf{u}|^{3/2}\big\|_2^2 \, \mathrm{d}t\right)^{\frac{3}{4s}}$$

$$\leq c\left(\operatorname*{ess\,sup}_{t_1<t<t_2} \|\mathbf{u}(\,.\,,t)\|_3\right)^3 \left[\int_{t_1}^{t_2} \|\nabla p\|_s^r \, \mathrm{d}t + \left(\int_{t_1}^{t_2} \|\nabla p\|_s^r \, \mathrm{d}t\right)^{\frac{6}{r(8s-9)}}\right]$$

$$+ c\,\big\|\mathbf{u}(\,.\,,t_1)\big\|_{2,\,3/2}^{\frac{3}{2}} + c\epsilon \int_{t_1}^{t_2} \big\|\nabla|\mathbf{u}|^{3/2}\big\|_2^2 \, \mathrm{d}t.$$

Substituting to (4.90) and choosing ϵ sufficiently small, we get

$$\big\|\mathbf{u}(\,.\,,t_2)\big\|_3^3 - \big\|\mathbf{u}(\,.\,,t_1)\big\|_3^3 + \frac{2\nu}{3}\int_{t_1}^{t_2} \big\|\nabla|\mathbf{u}|^{3/2}\big\|_2^2 \, \mathrm{d}t$$

$$
\leq c \left(\operatorname*{ess\,sup}_{t_1<t<t_2} \|\mathbf{u}(\,.\,,t)\|_3 \right)^3 \left[\int_{t_1}^{t_2} \|\nabla p\|_s^r \, \mathrm{d}t + \left(\int_{t_1}^{t_2} \|\nabla p\|_s^r \, \mathrm{d}t \right)^{\frac{6}{r(8s-9)}} \right]
$$

$$
+ c \left\| \mathbf{u}(\,.\,,t_1) \right\|_{2,\,3/2}^{\frac{3}{2}},
$$

$$
\left(\operatorname*{ess\,sup}_{t_1<t<t_2} \|\mathbf{u}(\,.\,,t)\|_3 \right)^3 + \frac{2\nu}{3} \int_{t_1}^{t_2} \left\| \nabla |\mathbf{u}|^{3/2} \right\|_2^2 \mathrm{d}t \;\leq\; \left\| \mathbf{u}(\,.\,,t_1) \right\|_3^3
$$

$$
+ c \left(\operatorname*{ess\,sup}_{t_1<t<t_2} \|\mathbf{u}(\,.\,,t)\|_3 \right)^3 \left[\int_{t_1}^{t_2} \|\nabla p\|_s^r \, \mathrm{d}t + \left(\int_{t_1}^{t_2} \|\nabla p\|_s^r \, \mathrm{d}t \right)^{\frac{6}{r(8s-9)}} \right]
$$

$$
+ c \left\| \mathbf{u}(\,.\,,t_1) \right\|_{2,\,3/2}^{\frac{3}{2}}.
$$

If t_1 is chosen sufficiently close to T^*, then the term on the last but one line can be absorbed by the first term on the left-hand side and we get the estimate

$$
\frac{1}{2} \left(\operatorname*{ess\,sup}_{t_1<t<t_2} \|\mathbf{u}(\,.\,,t)\|_3 \right)^3 + \frac{2\nu}{3} \int_{t_1}^{t_2} \left\| \nabla |\mathbf{u}|^{3/2} \right\|_2^2 \mathrm{d}t \;\leq\; \left\| \mathbf{u}(\,.\,,t_1) \right\|_3^3 + c \left\| \mathbf{u}(\,.\,,t_1) \right\|_{2,\,3/2}^{\frac{3}{2}}.
$$

Constant c on the right-hand side is independent of t_1 and t_2. Applying the same arguments as in Sect. 4.5.3A (the end of case 1)) and 4.5.2 A, we can now show that T^* in fact cannot be an epoch of irregularity and the solution $\mathbf{u}$, p is regular on the whole time interval $(0, T)$. □

4.5.4 *The Role of the Negative Part of Pressure*

In this subsection, we explain main ideas and results from paper [67] by G. Seregin and V. Šverák. The authors consider $\Omega = \mathbb{R}^3$, $\mathbf{f} = \mathbf{0}$, $T = \infty$, they deal with a Leray-Hopf weak solution $\mathbf{u}$ and they fix the associated pressure p so that it satisfies the identity

$$
p(\mathbf{x}, t) \;=\; \frac{1}{4\pi} \int_{\mathbb{R}^3} \frac{1}{|\mathbf{x}-\mathbf{y}|} \operatorname{div}\operatorname{div} \left[\mathbf{u}(\mathbf{y}, t) \otimes \mathbf{u}(\mathbf{y}, t) \right] \mathrm{d}\mathbf{y}. \tag{4.91}
$$

This pressure satisfies $\nabla p \in L^r(\delta, T';\, \mathbf{L}^s(\mathbb{R}^3))$ for $0 < \delta < T' < \infty$ and $1 < r < 2$, $1 < s < \frac{3}{2}$, such that $2/r + 3/s = 4$ (in accordance with Theorem 4.3.3). The pair $\mathbf{u}$, p satisfies the Navier-Stokes equation (4.9) in the sense of distributions in Q_T. A major part of arguments, used in [67], concerns suitable weak solutions, because the Leray-Hopf weak solution is suitable up to the first epoch of irregularity. Here we further assume from the beginning, for simplicity, that $\mathbf{u}$, p is a suitable weak solution. Function $\mathbf{u}$ is supposed to have been modified on a set of measure zero so that

$$
\liminf_{t\to t_0} \; \|\mathbf{u}(\,.\,,t)\|_2 \;\geq\; \|\mathbf{u}(\,.\,,t_0)\|_2 \qquad \text{for all } 0 < t_0 < T \tag{4.92}
$$

and $\mathbf{u}$ is weakly continuous from $(0, T)$ to $\mathbf{L}^2_\sigma(\mathbb{R}^3)$. (The authors argue that (4.92) is possible, because as the set of hypothetic singular points of $\mathbf{u}$ on the time level $t = t_0$ has the 1-dimensional Hausdorff measure zero, one has $\mathbf{u}(\mathbf{x}, t) \to \mathbf{u}(\mathbf{x}, t_0)$ as $t \to t_0$ for a.a. $\mathbf{x} \in \mathbb{R}^3$. Then (4.92) follows from Fatou's lemma.) Then the authors prove the two lemmas (see Lemma 3.2 and Lemma 3.3 in [67]):

Lemma 4.5.6 *Given $\Omega_0 \subset\subset \mathbb{R}^3$, $t_0 \in (0, T)$ and $\delta_0 \in (0, \sqrt{t_0})$. If*

$$a(\Omega_0, t_0, \delta_0) := \sup_{\mathbf{x}_0,\, t,\, R} \frac{1}{R} \|\mathbf{u}(\,.\,, t)\|^2_{2;\, B_R(\mathbf{x}_0)} < \infty$$

(where the supremum is taken over $\mathbf{x}_0 \in \Omega_0$, $t \in [t_0 - \delta_0^2, t_0]$ and $R > 0$) then

$$\lim_{t \to t_0-} \|\mathbf{u}(\,.\,, t) - \mathbf{u}(\,.\,, t_0)\|_{2;\, \Omega_0} = 0. \tag{4.93}$$

Lemma 4.5.7 *Let $\mathbf{u}$ be the aforementioned suitable weak solution to the Navier-Stokes IVP (4.9), (4.10), (4.12) in $Q_T \equiv \mathbb{R}^3 \times (0, T)$. There exists $\epsilon_* > 0$ such that if $B_{R_*}(\mathbf{x}_0) \times (t_0 - R_*^2, t_0) \subset Q_T$ for some $R_* > 0$ and*

$$\sup_{0<R<R_*} \sup_{t_0 - R^2 \le t \le t_0} \frac{1}{R} \|\mathbf{u}(\,.\,, t)\|^2_{2;\, B_R(\mathbf{x}_0)} < \epsilon_* \tag{4.94}$$

then $(\mathbf{x}_0, t_0)$ is a regular point of solution $\mathbf{u}$.

Lemma 4.5.6 is formulated for a general domain $\Omega \subset \mathbb{R}^3$ in [67]; here, we present the version corresponding to $\Omega = \mathbb{R}^3$. Although the body force $\mathbf{f}$ in equation (4.9) is supposed to be zero in the main theorem (see Theorem 4.5.8), Lemma 4.5.7 admits $\mathbf{f} \neq \mathbf{0}$, $\mathbf{f} \in \mathbf{M}_{2,\gamma}(Q_T)$ (the Morrey space) for some $\gamma > 0$. In this case, $\epsilon_* > 0$ depends on γ.

Further, the authors formulate condition (C): *A function $g : Q_T \to [0, \infty)$ is said to satisfy condition (C) if to any $t_0 > 0$ there exists $R_0 > 0$ such that*

$$a) \quad A(t_0) := \sup_{\mathbf{x}_0 \in \mathbb{R}^3} \sup_{t_0 - R_0^2 \le t \le t_0} \int_{B_{R_0}(\mathbf{x}_0)} \frac{g(\mathbf{x}, t)}{|\mathbf{x} - \mathbf{x}_0|} \, d\mathbf{x} < \infty,$$

$$b) \quad foreachfixed \mathbf{x}_0 \in \mathbb{R}^3 andforeachfixed R \in (0, R_0], thefunction$$
$$t \mapsto \int_{B_R(\mathbf{x}_0)} g(\mathbf{x}, t)/|\mathbf{x} - \mathbf{x}_0| \, d\mathbf{x} \; iscontinuousfromtheleftatpoint t_0.$$

The main result is as follows:

Theorem 4.5.8 *Let $\mathbf{u}_0 \in \mathbf{W}^{1,2}_{0,\sigma}(\mathbb{R}^3)$ and let $\mathbf{u}$, p be the aforementioned suitable weak solution of the Navier-Stokes IVP (4.9), (4.10), (4.12) in $Q_T \equiv \mathbb{R}^3 \times (0, T)$. Let there exist a function g satisfying condition (C) such that at least one of the conditions*

$$1)\quad |\mathbf{u}(\mathbf{x},t)|^2+2p(\mathbf{x},t)\ \le\ g(\mathbf{x},t)\quad \textit{for } \mathbf{x}\in\mathbb{R}^3,\ t\in(0,T), \tag{4.95}$$

$$2)\quad p(\mathbf{x},t)\ \ge\ -g(\mathbf{x},t)\qquad\qquad \textit{for } \mathbf{x}\in\mathbb{R}^3,\ t\in(0,T) \tag{4.96}$$

holds. Then $\mathbf{u}$ *is Hölder continuous in* Q_T.

Note that $T=\infty$ is also admitted in [67]. Condition 1) of Theorem 4.5.8 is fulfilled, e.g., if the quantity $\frac{1}{2}|\mathbf{u}|^2+p$ (which is the so-called *Bernoulli pressure*) is bounded above. (Indeed, if the inequality $\frac{1}{2}\,|\mathbf{u}(\mathbf{x},t)|^2+p(\mathbf{x},t)\le c_5$ holds a.e. in $\mathbb{R}^3\times(0,T)$ then one can simply put $g(\mathbf{x},t):=c_5$. This function g satisfies condition (C).) Similarly, condition 2) of Theorem 4.5.8 is fulfilled if p is bounded below. Thus, we observe that the positive part of p plays no role and solution $\mathbf{u}$ is regular just if the negative part of p is bounded.

Principles of the Proof of Theorem 4.5.8

I. The authors at first show that for any $\mathbf{x}_0\in\mathbb{R}^3$, any $t\in(0,T)$ and any $R>0$, the following crucial identities hold:

$$\begin{aligned}\int_{B_R(\mathbf{x}_0)}\frac{1}{|\mathbf{y}-\mathbf{x}_0|}\,\big(2p(\mathbf{y},t)+|\widehat{\mathbf{u}}^{\mathbf{x}_0}(\mathbf{y},t)|^2\big)\,\mathrm{d}\mathbf{y}&\\ =\int_{B_R(\mathbf{x}_0)}\frac{1}{R}\,\big(3p(\mathbf{y},t)+|\mathbf{u}(\mathbf{y},t)|^2\big)\,\mathrm{d}\mathbf{y}&\\ =R^2\int_{\mathbb{R}^3\smallsetminus B_R(\mathbf{x}_0)}\nabla^2_{\mathbf{y}}\Big(\frac{1}{|\mathbf{y}-\mathbf{x}_0|}\Big):\big[\mathbf{u}(\mathbf{y},t)\otimes\mathbf{u}(\mathbf{y},t)\big]\,\mathrm{d}\mathbf{y},&\end{aligned}\tag{4.97}$$

where

$$\widehat{\mathbf{u}}^{\mathbf{x}_0}(\mathbf{y},t)\ :=\ \mathbf{u}(\mathbf{y},t)-\frac{[\mathbf{u}(\mathbf{y},t)\cdot(\mathbf{y}-\mathbf{x}_0)]\,(\mathbf{y}-\mathbf{x}_0)}{|\mathbf{y}-\mathbf{x}_0|^2}.$$

($\widehat{\mathbf{u}}^{\mathbf{x}_0}(\mathbf{y},t)$ is the orthogonal projection of $\mathbf{u}(\mathbf{y},t)$ into the plane, perpendicular to $\mathbf{y}-\mathbf{x}_0$.) In order to verify (4.97), consider a "smooth" function $h:(0,\infty)\to[0,\infty)$. Then, by (4.91), we have

$$\begin{aligned}\int_{B_R(\mathbf{x}_0)}h\big(|\mathbf{x}_0-\mathbf{y}|\big)\,p(\mathbf{y},t)\,\mathrm{d}\mathbf{y}&\\ =\frac{1}{4\pi}\int_{\mathbb{R}^3}\operatorname{div}\operatorname{div}\big[\mathbf{u}(\mathbf{x},t)\otimes\mathbf{u}(\mathbf{x},t)\big]\int_{B_R(\mathbf{x}_0)}\frac{h\big(|\mathbf{x}_0-\mathbf{y}|\big)}{|\mathbf{x}-\mathbf{y}|}\,\mathrm{d}\mathbf{y}\,\mathrm{d}\mathbf{x}.&\end{aligned}\tag{4.98}$$

Using the substitution to the spherical coordinates centered at point $\mathbf{x}_0$ in the integral $\int_{B_R(\mathbf{x}_0)}h\big(|\mathbf{x}_0-\mathbf{y}|\big)/|\mathbf{x}-\mathbf{y}|\,\mathrm{d}\mathbf{y}$, one can calculate that

$$\int_{B_R(\mathbf{x}_0)} \frac{h(|\mathbf{x}_0-\mathbf{y}|)}{|\mathbf{x}-\mathbf{y}|}\, \mathrm{d}\mathbf{y} = \frac{4\pi}{|\mathbf{x}-\mathbf{x}_0|} \int_0^{|\mathbf{x}-\mathbf{x}_0|} \rho^2\, h(\rho)\, \mathrm{d}\rho + 4\pi \int_{|\mathbf{x}-\mathbf{x}_0|}^R \rho\, h(\rho)\, \mathrm{d}\rho$$

holds for $|\mathbf{x}-\mathbf{x}_0| \le R$ and

$$\int_{B_R(\mathbf{x}_0)} \frac{h(|\mathbf{x}_0-\mathbf{y}|)}{|\mathbf{x}-\mathbf{y}|}\, \mathrm{d}\mathbf{y} = \frac{4\pi}{|\mathbf{x}-\mathbf{x}_0|} \int_0^R \rho^2\, h(\rho)\, \mathrm{d}\rho$$

holds for $|\mathbf{x}-\mathbf{x}_0| > R$. Thus, applying the integration by parts to the right-hand side of (4.98) and interchanging $\mathbf{x}$ with $\mathbf{y}$ on the right-hand side, we get

$$\begin{aligned} &\int_{B_R(\mathbf{x}_0)} h(|\mathbf{x}_0-\mathbf{y}|)\, p(\mathbf{y},t)\, \mathrm{d}\mathbf{y} \\ &\quad = \int_{B_R(\mathbf{x}_0)} \big[\mathbf{u}(\mathbf{y},t)\otimes\mathbf{u}(\mathbf{y},t)\big] \\ &\qquad : \nabla_{\mathbf{y}}^2 \Big(\frac{1}{|\mathbf{y}-\mathbf{x}_0|} \int_0^{|\mathbf{y}-\mathbf{x}_0|} \rho^2\, h(\rho)\, \mathrm{d}\rho + \int_{|\mathbf{y}-\mathbf{x}_0|}^R \rho\, h(\rho)\, \mathrm{d}\rho \Big)\, \mathrm{d}\mathbf{y} \\ &\quad + \Big(\int_0^R \rho^2\, h(\rho)\, \mathrm{d}\rho \Big) \int_{\mathbb{R}^3\smallsetminus B_R(\mathbf{x}_0)} \big[\mathbf{u}(\mathbf{y},t)\otimes\mathbf{u}(\mathbf{y},t)\big] : \nabla_{\mathbf{y}}^2 \Big(\frac{1}{|\mathbf{y}-\mathbf{x}_0|} \Big)\, \mathrm{d}\mathbf{y}. \end{aligned}$$

Taking at first $h(\rho) = \rho^{-1}$, we show that the first line in (4.97) equals the third line. Then, taking $h(\rho) = 1$, we show that the second line equals the third line.

II. Assume, by contradiction, that t_0 is the first instant of time when a singular point of solution $\mathbf{u}$ appears. As $\mathbf{u}_0 \in \mathbf{W}^{1,2}_{0,\sigma}(\mathbb{R}^3)$, t_0 is positive and solution $\mathbf{u}$ is regular in $(0, t']$ for each $t' \in (0, t_0)$. The next step is to show that

$$\lim_{R\to\infty} \sup_{0\le t\le t_0} \int_{\mathbb{R}^3\smallsetminus B_R(\mathbf{0})} |\mathbf{u}(\mathbf{x},t)|^2\, \mathrm{d}\mathbf{x} = 0. \tag{4.99}$$

For this purpose, we can use the energy equality

$$\int_{\mathbb{R}^3} |\mathbf{u}(\mathbf{x},t)|^2\, \mathrm{d}\mathbf{x} + 2\nu \int_0^t \int_{\mathbb{R}^3} |\nabla\mathbf{u}(\mathbf{x},\tau)|^2\, \mathrm{d}\mathbf{x}\, \mathrm{d}\tau = \int_{\mathbb{R}^3} |\mathbf{u}_0(\mathbf{x})|^2\, \mathrm{d}\mathbf{x}$$

and the generalized energy equality

$$\begin{aligned} \int_{\mathbb{R}^3} |\mathbf{u}(\mathbf{x},t)|^2\, \phi(\mathbf{x})\, \mathrm{d}\mathbf{x} + 2\nu \int_0^t \int_{\mathbb{R}^3} |\nabla\mathbf{u}(\mathbf{x},\tau)|^2\, \phi(\mathbf{x})\, \mathrm{d}\mathbf{x}\, \mathrm{d}\tau &= \int_{\mathbb{R}^3} |\mathbf{u}_0(\mathbf{x})|^2\, \phi(\mathbf{x})\, \mathrm{d}\mathbf{x} \\ &+ \nu \int_0^t \int_{\mathbb{R}^3} |\mathbf{u}(\mathbf{x},\tau)|^2\, \Delta\phi(\mathbf{x})\, \mathrm{d}\mathbf{x}\, \mathrm{d}\tau \end{aligned}$$

$$+\int_0^t \int_{\mathbb{R}^3} \big(|\mathbf{u}(\mathbf{x},\tau)|^2 + 2p(\mathbf{x},\tau)\big)\, \mathbf{u}(\mathbf{x},\tau)\cdot\nabla\phi(\mathbf{x})\, \mathrm{d}\mathbf{x}\,\mathrm{d}\tau$$

(for each $\phi \in C_0^\infty(\mathbb{R}^3)$), which hold for all $t \in [0, t_0)$ due to the regularity of $\mathbf{u}$. Subtracting these equalities, we get

$$\begin{aligned}
&\int_{\mathbb{R}^3} |\mathbf{u}(\mathbf{x},t)|^2\,[1-\phi(\mathbf{x})]\,\mathrm{d}\mathbf{x} + 2\nu\int_0^t\int_{\mathbb{R}^3} |\nabla\mathbf{u}(\mathbf{x},\tau)|^2\,[1-\phi(\mathbf{x})]\,\mathrm{d}\mathbf{x}\,\mathrm{d}\tau \\
&\quad = \int_{\mathbb{R}^3} |\mathbf{u}_0(\mathbf{x})|^2\,[1-\phi(\mathbf{x})]\,\mathrm{d}\mathbf{x} - \nu\int_0^t\int_{\mathbb{R}^3} |\mathbf{u}(\mathbf{x},\tau)|^2\,\Delta\phi(\mathbf{x})\,\mathrm{d}\mathbf{x}\,\mathrm{d}\tau \\
&\qquad - \int_0^t\int_{\mathbb{R}^3} \big(|\mathbf{u}(\mathbf{x},\tau)|^2 + 2p(\mathbf{x},\tau)\big)\,\mathbf{u}(\mathbf{x},\tau)\cdot\nabla\phi(\mathbf{x})\,\mathrm{d}\mathbf{x}\,\mathrm{d}\tau. \qquad (4.100)
\end{aligned}$$

If we choose function ψ so that $0 \le \psi \le 1$, $\psi = 1$ in $B_{1/2}(\mathbf{0})$, $\psi = 0$ in $\mathbb{R}^3 \setminus B_1(\mathbf{0})$ and put $\phi_R(\mathbf{x}) := \psi(\mathbf{x}/R)$ (for $R > 0$), then (4.100) (with ϕ_R instead of ϕ) yields

$$\begin{aligned}
\int_{\mathbb{R}^3\setminus B_R(\mathbf{0})} |\mathbf{u}(\mathbf{x},t)|^2\,\mathrm{d}\mathbf{x} &\le \int_{\mathbb{R}^3} |\mathbf{u}(\mathbf{x},t)|^2\,[1-\phi_R(\mathbf{x})]\,\mathrm{d}\mathbf{x} \\
&\le \int_{\mathbb{R}^3\setminus B_{R/2}(\mathbf{0})} |\mathbf{u}_0(\mathbf{x})|^2\,\mathrm{d}\mathbf{x} + \frac{c\nu}{R^2}\int_0^t\int_{B_R(\mathbf{0})\setminus B_{R/2}(\mathbf{0})} |\mathbf{u}(\mathbf{x},\tau)|^2\,\mathrm{d}\mathbf{x}\,\mathrm{d}\tau \\
&\quad + \frac{c}{R}\int_0^t\int_{B_R(\mathbf{0})\setminus B_{R/2}(\mathbf{0})} \big(|\mathbf{u}(\mathbf{x},\tau)|^3 + |p(\mathbf{x},\tau)|\,|\mathbf{u}(\mathbf{x},\tau)|\big)\,\mathrm{d}\mathbf{x}\,\mathrm{d}\tau,
\end{aligned}$$

where c is independent of R and t. Hence

$$\begin{aligned}
&\sup_{0\le t\le t_0} \int_{\mathbb{R}^3\setminus B_R(\mathbf{0})} |\mathbf{u}(\mathbf{x},t)|^2\,\mathrm{d}\mathbf{x} \\
&\quad \le \int_{\mathbb{R}^3\setminus B_{R/2}(\mathbf{0})} |\mathbf{u}_0(\mathbf{x})|^2\,\mathrm{d}\mathbf{x} + \frac{c\nu}{R^2}\int_0^{t_0}\int_{B_R(\mathbf{0})\setminus B_{R/2}(\mathbf{0})} |\mathbf{u}(\mathbf{x},\tau)|^2\,\mathrm{d}\mathbf{x}\,\mathrm{d}\tau \\
&\qquad + \frac{c}{R}\int_0^{t_0}\int_{B_R(\mathbf{0})\setminus B_{R/2}(\mathbf{0})} \big(|\mathbf{u}(\mathbf{x},\tau)|^3 + |p(\mathbf{x},\tau)|\,|\mathbf{u}(\mathbf{x},\tau)|\big)\,\mathrm{d}\mathbf{x}\,\mathrm{d}\tau. \qquad (4.101)
\end{aligned}$$

Obviously, the two terms on the second line tend to zero for $R \to \infty$. In order to show that the term on the third line also tends to zero for $R \to \infty$, it is sufficient to take into account that

$$\int_0^{t_0}\int_{\mathbb{R}^3} |\mathbf{u}(\mathbf{x},\tau)|^3\,\mathrm{d}\mathbf{x}\,\mathrm{d}\tau \;\le\; c\,\big(\operatorname*{ess\,sup}_{0\le\tau\le t_0} \|\mathbf{u}(\,.\,,\tau)\|_2\big)^{\frac{3}{2}} \left(\int_0^{t_0} \|\nabla\mathbf{u}(\,.\,,\tau)\|_2^2\,\mathrm{d}\tau\right)^{\frac{3}{2}}$$

$$\leq\ c(t_0)\ \|\mathbf{u}_0\|_2^3,$$

$$\int_0^{t_0}\int_{\mathbb{R}^3} |p(\mathbf{x},\tau)|^{3/2}\ \mathrm{d}\mathbf{x}\,\mathrm{d}\tau\ \leq\ c\int_0^{t_0}\int_{\mathbb{R}^3} |\mathbf{u}(\mathbf{x},\tau)|^3\ \mathrm{d}\mathbf{x}\,\mathrm{d}\tau.$$

(The inequalities on the first line hold due to Hölder's, Sobolev's, and energy inequalities, the inequality on the second line follows from Calderón-Zygmund theorem.) Thus, (4.99) follows from (4.101), considering $R \to \infty$.

III. Assume further that condition 1) of Theorem 4.5.8 holds. Our aim is to prove that

$$\lim_{t\to t_0-}\int_{\mathbb{R}^3} |\mathbf{u}(\mathbf{x},t)-\mathbf{u}(\mathbf{x},t_0)|^2\ \mathrm{d}\mathbf{x}\ =\ 0. \tag{4.102}$$

We begin with the identities (4.97), written in the form

$$\begin{aligned}
&\int_{B_R(\mathbf{x}_0)} \frac{1}{|\mathbf{x}-\mathbf{x}_0|}\ \big(2p(\mathbf{x},t)+|\mathbf{u}(\mathbf{x},t)|^2\big)\ \mathrm{d}\mathbf{x} - \int_{B_R(\mathbf{x}_0)} \frac{1}{|\mathbf{x}-\mathbf{x}_0|}\ |\widetilde{\mathbf{u}}^{\mathbf{x}_0}(\mathbf{x},t)|^2\ \mathrm{d}\mathbf{x}\\
&\quad = -\frac{1}{2R}\int_{B_R(\mathbf{x}_0)} |\mathbf{u}(\mathbf{x},t)|^2\ \mathrm{d}\mathbf{x} + \frac{3}{2R}\int_{B_R(\mathbf{x}_0)} \big(2p(\mathbf{x},t)+|\mathbf{u}(\mathbf{x},t)|^2\big)\ \mathrm{d}\mathbf{x}\\
&\quad = R^2\int_{\mathbb{R}^3\smallsetminus B_R(\mathbf{x}_0)} \nabla^2_{\mathbf{x}}\Big(\frac{1}{|\mathbf{x}-\mathbf{x}_0|}\Big) : \big[\mathbf{u}(\mathbf{x},t)\otimes\mathbf{u}(\mathbf{x},t)\big]\ \mathrm{d}\mathbf{x},
\end{aligned} \tag{4.103}$$

where

$$\widetilde{\mathbf{u}}^{\mathbf{x}_0}(\mathbf{x},t)\ :=\ \mathbf{u}(\mathbf{x},t)-\widehat{\mathbf{u}}^{\mathbf{x}_0}(\mathbf{x},t)\ =\ \frac{[\mathbf{u}(\mathbf{x},t)\cdot(\mathbf{x}-\mathbf{x}_0)]\,(\mathbf{x}-\mathbf{x}_0)}{|\mathbf{x}-\mathbf{x}_0|^2}$$

(= the orthogonal projection of $\mathbf{u}(\mathbf{x},t)$ into the direction of $\mathbf{x}-\mathbf{x}_0$). Hence, using at first the identity between the first two lines in (4.103) and at the very end also the identity between the first and the third line in (4.103), we get

$$\begin{aligned}
\frac{1}{2R}\int_{B_R(\mathbf{x}_0)} |\mathbf{u}(\mathbf{x},t)|^2\ \mathrm{d}\mathbf{x}\ &=\ \frac{3}{2R}\int_{B_R(\mathbf{x}_0)} \big(|\mathbf{u}(\mathbf{x},t)|^2+2p(\mathbf{x},t)\big)\ \mathrm{d}\mathbf{x}\\
&\quad + \int_{B_R(\mathbf{x}_0)} \frac{1}{|\mathbf{x}-\mathbf{x}_0|}\ |\widetilde{\mathbf{u}}^{\mathbf{x}_0}(\mathbf{x},t)|^2\ \mathrm{d}\mathbf{x} - \int_{B_R(\mathbf{x}_0)} \frac{1}{|\mathbf{x}-\mathbf{x}_0|}\big(|\mathbf{u}(\mathbf{x},t)|^2+2p(\mathbf{x},t)\big)\ \mathrm{d}\mathbf{x}\\
&\leq\ \frac{3}{2R}\int_{B_R(\mathbf{x}_0)} g(\mathbf{x},t)\ \mathrm{d}\mathbf{x} - \int_{B_R(\mathbf{x}_0)} \frac{1}{|\mathbf{x}-\mathbf{x}_0|}\ g(\mathbf{x},t)\ \mathrm{d}\mathbf{x}\\
&\quad + \int_{B_R(\mathbf{x}_0)} \frac{1}{|\mathbf{x}-\mathbf{x}_0|}\ |\widetilde{\mathbf{u}}^{\mathbf{x}_0}(\mathbf{x},t)|^2\ \mathrm{d}\mathbf{x}
\end{aligned}$$

$$
\begin{aligned}
&+\int_{B_R(\mathbf{x}_0)} \frac{1}{|\mathbf{x}-\mathbf{x}_0|}\, \big[g(\mathbf{x},t) - \big(|\mathbf{u}(\mathbf{x},t)|^2 + 2p(\mathbf{x},t)\big)\big]\, \mathrm{d}\mathbf{x} \\
&\leq \frac{1}{2}\int_{B_R(\mathbf{x}_0)} \frac{1}{|\mathbf{x}-\mathbf{x}_0|}\, g(\mathbf{x},t)\, \mathrm{d}\mathbf{x} + \int_{B_R(\mathbf{x}_0)} \frac{1}{|\mathbf{x}-\mathbf{x}_0|}\, |\widetilde{\mathbf{u}}^{\mathbf{x}_0}(\mathbf{x},t)|^2\, \mathrm{d}\mathbf{x} \\
&\quad + \int_{B_R(\mathbf{x}_0)} \frac{1}{|\mathbf{x}-\mathbf{x}_0|}\, \big[g(\mathbf{x},t) - \big(|\mathbf{u}(\mathbf{x},t)|^2 + 2p(\mathbf{x},t)\big)\big]\, \mathrm{d}\mathbf{x} \qquad (4.104) \\
&= \frac{3}{2}\int_{B_R(\mathbf{x}_0)} \frac{1}{|\mathbf{x}-\mathbf{x}_0|}\, g(\mathbf{x},t)\, \mathrm{d}\mathbf{x} + \int_{B_R(\mathbf{x}_0)} \frac{1}{|\mathbf{x}-\mathbf{x}_0|}\, |\widetilde{\mathbf{u}}^{\mathbf{x}_0}(\mathbf{x},t)|^2\, \mathrm{d}\mathbf{x} \\
&\quad - \int_{B_R(\mathbf{x}_0)} \frac{1}{|\mathbf{x}-\mathbf{x}_0|}\, \big(|\mathbf{u}(\mathbf{x},t)|^2 + 2p(\mathbf{x},t)\big)\, \mathrm{d}\mathbf{x} \\
&= \frac{3}{2}\int_{B_R(\mathbf{x}_0)} \frac{1}{|\mathbf{x}-\mathbf{x}_0|}\, g(\mathbf{x},t)\, \mathrm{d}\mathbf{x} \\
&\quad + \int_{B_R(\mathbf{x}_0)} \frac{1}{|\mathbf{x}-\mathbf{x}_0|}\, \big(|\widetilde{\mathbf{u}}^{\mathbf{x}_0}(\mathbf{x},t)|^2 - |\mathbf{u}(\mathbf{x},t)|^2 - 2p(\mathbf{x},t)\big)\, \mathrm{d}\mathbf{x} \\
&= \frac{3}{2}\int_{B_R(\mathbf{x}_0)} \frac{1}{|\mathbf{x}-\mathbf{x}_0|}\, g(\mathbf{x},t)\, \mathrm{d}\mathbf{x} \\
&\quad - R^2 \int_{\mathbb{R}^3 \smallsetminus B_R(\mathbf{x}_0)} \nabla_{\mathbf{x}}^2\Big(\frac{1}{|\mathbf{x}-\mathbf{x}_0|}\Big) : \big[\mathbf{u}(\mathbf{x},t) \otimes \mathbf{u}(\mathbf{x},t)\big]\, \mathrm{d}\mathbf{x}.
\end{aligned}
$$

Let $R_0 = R_0(t_0)$ be the number given by condition (C). Obviously, if $R > R_0$, then

$$
\frac{1}{R}\int_{B_R(\mathbf{x}_0)} |\mathbf{u}(\mathbf{x},t)|^2\, \mathrm{d}\mathbf{x} \;\leq\; \frac{1}{R_0}\int_{\mathbb{R}^3} |\mathbf{u}(\mathbf{x},t)|^2\, \mathrm{d}\mathbf{x} \;\leq\; \frac{c}{R_0}, \qquad (4.105)
$$

where c is independent of $\mathbf{x}_0$, t, and R. On the other hand, if $0 < R \leq R_0$, then (4.104) yields

$$
\begin{aligned}
&\frac{1}{2R}\int_{B_R(\mathbf{x}_0)} |\mathbf{u}(\mathbf{x},t)|^2\, \mathrm{d}\mathbf{x} \\
&\quad\leq \frac{1}{2}\int_{B_R(\mathbf{x}_0)} \frac{1}{|\mathbf{x}-\mathbf{x}_0|}\, g(\mathbf{x},t)\, \mathrm{d}\mathbf{x} + \int_{B_R(\mathbf{x}_0)} \frac{1}{|\mathbf{x}-\mathbf{x}_0|}\, |\widetilde{\mathbf{u}}^{\mathbf{x}_0}(\mathbf{x},t)|^2\, \mathrm{d}\mathbf{x} \\
&\qquad + \int_{B_R(\mathbf{x}_0)} \frac{1}{|\mathbf{x}-\mathbf{x}_0|}\, \big[g(\mathbf{x},t) - \big(|\mathbf{u}(\mathbf{x},t)|^2 + 2p(\mathbf{x},t_0)\big)\big]\, \mathrm{d}\mathbf{x} \\
&\quad\leq \frac{1}{2}\int_{B_{R_0}(\mathbf{x}_0)} \frac{1}{|\mathbf{x}-\mathbf{x}_0|}\, g(\mathbf{x},t)\, \mathrm{d}\mathbf{x} + \int_{B_{R_0}(\mathbf{x}_0)} \frac{1}{|\mathbf{x}-\mathbf{x}_0|}\, |\widetilde{\mathbf{u}}^{\mathbf{x}_0}(\mathbf{x},t)|^2\, \mathrm{d}\mathbf{x}
\end{aligned}
$$

$$+\int_{B_{R_0}(\mathbf{x}_0)} \frac{1}{|\mathbf{x}-\mathbf{x}_0|} \left[g(\mathbf{x},t) - \left(|\mathbf{u}(\mathbf{x},t)|^2 + 2p(\mathbf{x},t)\right)\right] \mathrm{d}\mathbf{x}$$

$$= \frac{3}{2}\int_{B_{R_0}(\mathbf{x}_0)} \frac{1}{|\mathbf{x}-\mathbf{x}_0|}\, g(\mathbf{x},t)\, \mathrm{d}\mathbf{x}$$

$$- R_0^2 \int_{\mathbb{R}^3 \smallsetminus B_{R_0}(\mathbf{x}_0)} \nabla_{\mathbf{x}}^2\Big(\frac{1}{|\mathbf{x}-\mathbf{x}_0|}\Big) : \big[\mathbf{u}(\mathbf{x},t)\otimes\mathbf{u}(\mathbf{x},t)\big]\, \mathrm{d}\mathbf{x}$$

$$\leq \frac{3}{2}A(t_0) + \frac{c}{R_0}\, \|\mathbf{u}(\,.\,,t)\|_2^2 \;\leq\; \frac{3}{2}A(t_0) + \frac{c}{R_0},$$

where c is independent of $\mathbf{x}_0$, t, and R. These estimates, (4.105) and Lemma 4.5.6 imply that (4.93) holds for any $\Omega_0 \subset\subset \mathbb{R}^3$. Combining this result with (4.99), we deduce that (4.102) holds.

IV. Let $\mathbf{x}_0$ be an arbitrary point in $\mathbb{R}^3$ and ϵ_* be the number from Lemma 4.5.7. There exists R_* so small that

$$\frac{\epsilon_*}{2} > \frac{1}{2}\int_{B_{R*}(\mathbf{x}_0)} \frac{1}{|\mathbf{x}-\mathbf{x}_0|}\, g(\mathbf{x},t_0)\, \mathrm{d}\mathbf{x} + \int_{B_{R*}(\mathbf{x}_0)} \frac{1}{|\mathbf{x}-\mathbf{x}_0|}\, |\widetilde{\mathbf{u}}^{\mathbf{x}_0}(\mathbf{x},t_0)|^2\, \mathrm{d}\mathbf{x}$$

$$+\int_{B_{R*}(\mathbf{x}_0)} \frac{1}{|\mathbf{x}-\mathbf{x}_0|} \left[g(\mathbf{x},t_0) - \left(|\mathbf{u}(\mathbf{x},t_0)|^2 + 2p(\mathbf{x},t_0)\right)\right] \mathrm{d}\mathbf{x},$$

which equals

$$\frac{3}{2}\int_{B_{R*}(\mathbf{x}_0)} \frac{1}{|\mathbf{x}-\mathbf{x}_0|}\, g(\mathbf{x},t_0)\, \mathrm{d}\mathbf{x}$$

$$- R_*^2 \int_{\mathbb{R}^3 \smallsetminus B_{R*}(\mathbf{x}_0)} \nabla_{\mathbf{x}}^2\Big(\frac{1}{|\mathbf{x}-\mathbf{x}_0|}\Big) : \big[\mathbf{u}(\mathbf{x},t_0)\otimes\mathbf{u}(\mathbf{x},t_0)\big]\, \mathrm{d}\mathbf{x}$$

due to the last identity in (4.104). The function

$$t \longmapsto \frac{3}{2}\int_{B_{R*}(\mathbf{x}_0)} \frac{1}{|\mathbf{x}-\mathbf{x}_0|}\, g(\mathbf{x},t)\, \mathrm{d}\mathbf{x}$$

$$- R_*^2 \int_{\mathbb{R}^3 \smallsetminus B_{R*}(\mathbf{x}_0)} \nabla_{\mathbf{x}}^2\Big(\frac{1}{|\mathbf{x}-\mathbf{x}_0|}\Big) : \big[\mathbf{u}(\mathbf{x},t)\otimes\mathbf{u}(\mathbf{x},t)\big]\, \mathrm{d}\mathbf{x}$$

is left-continuous at point t_0. (The left-continuity of the first term on the right-hand side follows from condition (C) and the left-continuity of the second term follows from (4.102) and the boundedness of $\nabla_{\mathbf{x}}^2|\mathbf{x}-\mathbf{x}_0|^{-1}$ in $\mathbb{R}^3 \smallsetminus B_{R_*}(\mathbf{x}_0)$.) Thus, there exists $\delta_* > 0$, $\delta_* \leq \sqrt{t_0/2}$ such that

$$\frac{\epsilon_*}{2} > \frac{1}{2}\int_{B_{R*}(\mathbf{x}_0)} \frac{1}{|\mathbf{x}-\mathbf{x}_0|}\, g(\mathbf{x},t)\, \mathrm{d}\mathbf{x} + \int_{B_{R*}(\mathbf{x}_0)} \frac{1}{|\mathbf{x}-\mathbf{x}_0|}\, |\widetilde{\mathbf{u}}^{\mathbf{x}_0}(\mathbf{x},t)|^2\, \mathrm{d}\mathbf{x}$$

$$+ \int_{B_{R_*}(\mathbf{x}_0)} \frac{1}{|\mathbf{x}-\mathbf{x}_0|} \left[g(\mathbf{x},t) - \left(|\mathbf{u}(\mathbf{x},t)|^2 + 2p(\mathbf{x},t) \right) \right] \mathrm{d}\mathbf{x}$$

for all $t \in [t_0 - \delta_*^2, t_0]$. Hence, by (4.104), we have

$$\begin{aligned} \frac{1}{2R} \int_{B_R(\mathbf{x}_0)} & |\mathbf{u}(\mathbf{x},t)|^2 \, \mathrm{d}\mathbf{x} \\ & \leq \frac{1}{2} \int_{B_R(\mathbf{x}_0)} \frac{1}{|\mathbf{x}-\mathbf{x}_0|} \, g(\mathbf{x},t) \, \mathrm{d}\mathbf{x} + \int_{B_R(\mathbf{x}_0)} \frac{1}{|\mathbf{x}-\mathbf{x}_0|} \, |\widetilde{\mathbf{u}}^{\mathbf{x}_0}(\mathbf{x},t)|^2 \, \mathrm{d}\mathbf{x} \\ & \quad + \int_{B_R(\mathbf{x}_0)} \frac{1}{|\mathbf{x}-\mathbf{x}_0|} \left[g(\mathbf{x},t) - \left(|\mathbf{u}(\mathbf{x},t)|^2 + 2p(\mathbf{x},t) \right) \right] \mathrm{d}\mathbf{x} \\ & \leq \frac{1}{2} \int_{B_{R_*}(\mathbf{x}_0)} \frac{1}{|\mathbf{x}-\mathbf{x}_0|} \, g(\mathbf{x},t) \, \mathrm{d}\mathbf{x} + \int_{B_{R_*}(\mathbf{x}_0)} \frac{1}{|\mathbf{x}-\mathbf{x}_0|} \, |\widetilde{\mathbf{u}}^{\mathbf{x}_0}(\mathbf{x},t)|^2 \, \mathrm{d}\mathbf{x} \\ & \quad + \int_{B_{R_*}(\mathbf{x}_0)} \frac{1}{|\mathbf{x}-\mathbf{x}_0|} \left[g(\mathbf{x},t) - \left(|\mathbf{u}(\mathbf{x},t)|^2 + 2p(\mathbf{x},t) \right) \right] \mathrm{d}\mathbf{x} \; < \; \frac{\epsilon_*}{2} \end{aligned}$$

for all $t \in [t_0 - \delta_*^2, t_0]$ and $R \in (0, R_*]$. Thus, by Lemma 4.5.7, $(\mathbf{x}_0, t_0)$ is a regular point of solution $\mathbf{u}$. Since the point $\mathbf{x}_0$ has been chosen arbitrarily in $\mathbb{R}^3$, solution $\mathbf{u}$ has no singular points on the time level $t = t_0$. This is a contradiction with the assumption that the first singular point appears at time $t = t_0$.

V. If condition 2) of Theorem 4.5.8 holds instead of condition 1), then we can proceed similarly. The principal idea is that we always modify (4.97) so that we obtain $g(\mathbf{x},t) + p(\mathbf{x},t)$ instead of $g(\mathbf{x},t) - \left(|\mathbf{u}(\mathbf{x},t)|^2 + 2p(\mathbf{x},t) \right)$. □

Acknowledgements The author has been supported by the Mathematical Institute of the Czech Academy of Sciences (RVO 67985840) and by the Czech Science Foundation (grant No. 16-03230S).

Bibliography

1. H. Al Baba, Ch. Amrouche, M. Escobedo: L^p–theory for the time dependent Navier-Stokes problem with Navier-type boundary conditions. *Proc. of the 13th International Conference Zaragoza-Pau on Mathematics and its Applications,* Monografías Matemáticas García de Galdeano **40**, Prensas Univ. Zaragoza, Zaragoza 2016, pp. 1–8.
2. Ch. Amrouche, N. Seloula: L^p–theory for the Navier–Stokes equations with pressure boundary conditions. *Discr. Contin. Dyn. Syst. Ser. S* **6** (2013), no. 5, 1113–1137.
3. Ch. Amrouche, A. Rejaiba: L^p-theory for Stokes and Navier–Stokes equations with Navier boundary condition. *J. Differential Equations* **256** (2014), 1515–1547.
4. Ch. Amrouche, A. Rejaiba: Navier-Stokes equations with Navier boundary condition. *Math. Meth. Appl. Sci.* **39** (2016), 5091–5112.

5. H. Beirão da Veiga: Concerning the regularity of the solutions to the Navier-Stokes equations via the truncation method, Part II. In *Équations aux dérivées partielles et applications,* Gauthier-Villars, Paris 1998, pp. 127–138.
6. H. Beirão da Veiga: A sufficient condition on the pressure for the regularity of weak solutions to the Navier-Stokes equations. *J. Math. Fluid Mech.* **2** (2000), 96–106.
7. L. Berselli: Sufficient conditions for the regularity of the solutions of the Navier-Stokes equations. *Math. Meth. Appl. Sci.* **22** (1999), 1079–1085.
8. L. Berselli, G. P. Galdi: Regularity criterions involving the pressure for the weak solutions to the Navier-Stokes equations. *Proc. Amer. Math. Soc.* **130** (2002), no. 12, 3585–3595.
9. S. Bosia, M. Conti, V. Pata: A regularity criterion for the Navier-Stokes equations in terms of the pressure gradient. *Cent. Eur. J. Math.* **12** (2014), 1015–1025.
10. W. Borchers, H. Sohr: On the equations $\operatorname{rot} v = g$ and $\operatorname{div} u = f$ with zero boundary conditions. *Hokkaido Math. J.* **19** (1990), 67–87.
11. P. Butzer, H. Berens: *Semi-Groups of Operators and Approximation.* Springer, Berlin-Heidelberg-New York 1967.
12. L. Caffarelli, R. Kohn, L. Nirenberg: Partial regularity of suitable weak solutions of the Navier-Stokes equations. *Commun. Pure Appl. Math.* **35** (1982), 771–831.
13. Z. Cai, J. Fan, J. Zhai: Regularity criteria in weak spaces for 3-dimensional Navier-Stokes equations in terms of the pressure. *Differential Integral Equations* **23** (2010), no. 11–12, 1023–1033.
14. D. Chae, J. Lee: Regularity criterion in terms of pressure for the Navier-Stokes equations. *Nonlinear Analysis* **46** (2001), 727–735.
15. D. Chamorro, P.-G. Lemarié-Rieusset, K. Mayoufi: The role of pressure in the partial regularity theory for weak solutions of the Navier-Stokes equations. *Arch. Rat. Mech. Anal.* **228** (2018), 237–277.
16. G. Q. Chen, D. Osborne, Z. Qian: The Navier-Stokes equations with the kinematic and vorticity boundary conditions on non–flat boundaries. *Acta Math. Sci.* **29**B (2009), no. 4, 919–948.
17. G. Q. Chen, Z. Qian: A study of the Navier-Stokes equations with the kinematic and Navier boundary conditions. *Indiana Univ. Math. J.* **59** (2010), no. 2, 721–760.
18. Q. Chen, Z. Zhang: Regularity criterion via the pressure on weak solutions to the 3D Navier-Stokes equations. *Proc. Amer. Math. Soc.* **136** (2007), no. 6, 1829–1837.
19. L. Escauriaza, G. Seregin, V. Šverák: $L_{3,\infty}$–solutions of the Navier–Stokes equations and backward uniqueness. *Russian Math. Surveys* **58** (2003), 2, 211–250.
20. J. Fan, T. Ozawa: Regularity criterion for weak solutions to the Navier-Stokes equations in terms of pressure. *J. Inequal. Appl.* 2008, article ID 412678, 6 pages, DOI: 10.1155/2008/412678.
21. J. Fan, S. Jiang, G. Ni: On regularity criteria for the n-dimensional Navier–Stokes equations in terms of the pressure. *J. Differential Equations* **244** (2008), 2963–2979.
22. R. Farwig, H. Kozono, H. Sohr: The Helmholtz decomposition in arbitrary unbounded domains - a theory beyond L^2. *Proc. of Equadiff 11,* ed. M. Fila et al, Comenius University Press, ISBN 978-80-227-2624-5, Bratislava 2005, pp. 77–85.
23. R. Farwig, H. Kozono, H. Sohr: Energy–based regularity criteria for the Navier-Stokes equations. *J. Math. Fluid Mech.* **8** (2008), no. 3, 428–442.
24. R. Farwig, V. Rosteck: Maximal regularity of the Stokes system with Navier boundary condition in general unbounded domains. To appear in *J. Math. Soc. Japan.*
25. M. Feistauer, T. Neustupa: On non-stationary viscous incompressible flow through a cascade of profiles. *Math. Meth. Appl. Sci.* **29** (2006), no. 16, 1907–1941.
26. M. Feistauer, T. Neustupa: On the existence of a weak solution of viscous incompressible flow past a cascade of profiles with an arbitrarily large inflow. *J. Math. Fluid Mech.* **15** (2013), no. 4, 701–715.
27. D. Fujiwara, H. Morimoto: An L_r-theorem of the Helmholtz decomposition of vector fields. *J. Fac. Sci. Univ. Tokyo Sec. IA Math.* **24** (1977), no. 3, 685–700.
28. G. P. Galdi, C. G. Simader, H. Sohr: On the Stokes problem in Lipschitz domains. *Ann. Mat. Pura Appl. (IV)* **167** (1994), 147–163.

29. G. P. Galdi: *An Introduction to the Mathematical Theory of the Navier-Stokes Equations.* 2nd edition, Springer 2011.
30. G. P. Galdi: An Introduction to the Navier-Stokes initial–boundary value problem. In *Fundamental Directions in Mathematical Fluid Mechanics,* ed. G. P. Galdi, J. Heywood, R. Rannacher, series "Advances in Mathematical Fluid Mechanics". Birkhäuser, Basel 2000, pp. 1–98.
31. G. P. Galdi, J. Neustupa: Steady-state Navier–Stokes flow around a moving body. In *Handbook of Mathematical Analysis in Mechanics of Viscous Fluids,* eds. Y. Giga and A. Novotný, Springer 2018, pp. 341–417.
32. F. Gazzola: A note on the evolution of Navier-Stokes equations with a pressure-dependent viscosity. *Z. Angew. Math. Phys.* **48** (1997), 760–773.
33. F. Gazzola, P. Secchi: Some results about stationary Navier-Stokes equations with a pressure-dependent viscosity. In *Navier-Stokes equations: theory and numerical methods (Varenna 1997),* ed. R. Salvi, *Pitman Res. Notes Math. Ser.*, Vol. 388, Longman, Harlow 1998, pp. 31–37.
34. J. Geng, Z. Shen: The Neumann problem and Helmholtz decomposition in convex domains. *J. Funct. Anal.* **259** (8), 2010, 2147–2164.
35. Y. Giga: Solutions for semilinear parabolic equations in L^p and regularity of weak solutions of the Navier-Stokes system. *J. Differential Equations* **62** (1986), no. 2, 186–212.
36. Y. Giga, H. Sohr: Abstract L^p-estimates for the Cauchy problem with applications to the Navier-Stokes equations in exterior domains. *J. Funct. Anal.* **102** (1991), 72–94.
37. K. Kang, J. Lee: On regularity criteria in conjunction with the pressure of the Navier–Stokes equations. *Internat. Math. Res. Notes* Vol. 2006, Issue 9, Article ID 80762, 25 pp.
38. K. Kang, J. Lee: Erratum: On regularity criteria in conjunction with the pressure of the Navier–Stokes equations. *Internat. Math. Res. Notes* Vol. 2010, Issue 9, 1772–1774.
39. S. Kaniel: A sufficient condition for smoothness of solutions of the Navier-Stokes equations. *Israel J. Math.* **6** (1969), 354–358.
40. H. Kozono, T. Yanagisawa: L^r-Helmholtz decomposition and its application to the Navier-Stokes equations. In *Lectures on the Analysis of Nonlinear Partial Differential Equations, Part 3,* Morningside Lect. Math. 3, Int. Press, Somerville, MA, 2013, pp. 237–290.
41. S. Kračmar, J. Neustupa: A weak solvability of a steady variational inequality of the Navier–Stokes type with mixed boundary conditions. *Nonlinear Analysis* **47** (2001), no. 6, 4169–4180.
42. S. Kračmar, J. Neustupa: Modeling of the unsteady flow through a channel with an artificial outflow condition by the Navier–Stokes variational inequality. *Math. Nachr.* **291** (2018), no. 11–12, 1–14.
43. P. Kučera, Z. Skalák: Solutions to the Navier?Stokes equations with mixed boundary conditions. *Acta Appl. Math.* **54** (1998), no. 3, 275–288.
44. P. Kučera: Basic properties of the non-steady Navier-Stokes equations with mixed boundary conditions in a bounded domain. *Ann. Univ. Ferrara Sez. VII Sci. Mat.* **55** (2009), 289–308.
45. P. Kučera, J. Neustupa: On robustness of a strong solution to the Navier–Stokes equations with Navier's boundary conditions in the L^3–norm. *Nonlinearity* **30** (2017), no. 4, 1564–1583.
46. P. Kučera, S. Nečasová, J. Neustupa: A pressure associated with a weak solution to the Navier-Stokes equations with Navier's boundary conditions.
47. O. A. Ladyzhenskaya, G. Seregin: On partial regularity of suitable weak solutions to the three-dimensional Navier-Stokes equations. *J. Math. Fluid Mech.* **1** (1999), 356–387.
48. M. Lanzendörfer, J. Stebel: On pressure boundary conditions for steady flows of incompressible fluids with pressure and shear rate dependent viscosities. *Appl. Math.* **56** (2011), no. 3, 265–285.
49. F. Lin: A new proof of the Caffarelli-Kohn-Nirenberg theorem. *Commun. Pure Appl. Math.* **51** (1998), 241–257.
50. J. L. Lions: *Quelques méthodes de résolution des problèmes âux limites non linéaire.* Dunod, Gauthier-Villars, Paris 1969.
51. J. Málek, J. Nečas, K. R. Rajagopal: Global analysis of the flows of fluids with pressure-dependent viscosities. *Arch. Rat. Mech. Anal.* **165** (2002), no. 3, 243–269.

52. J. Málek, K. R. Rajagopal: Mathematical properties of the solutions to the equations governing the flow of fluids with pressure and shear rate dependent viscosities. In *Handbook of Mathematical Fluid Dynamics,* Vol. 4, Chap. 7, Elsevier/North-Holland, Amsterdam 2007, pp. 407–444.
53. S. Marušić: On the Navier–Stokes system with pressure boundary condition. *Ann. Univ. Ferrara Sez. VII Sci. Mat.* **53** (2007), 319–331.
54. G. Nardi: Schauder estimate for solutions of Poisson's equation with Neumann boundary condition. *Enseign. Math.* **60** (2014), no. 2, 421–435.
55. J. Nečas, J. Neustupa: New conditions for local regularity of a suitable weak solution to the Navier-Stokes equations. *J. Math. Fluid Mech.* **4** (2002), 237–256.
56. Š. Nečasová, J. Wolf: On the existence of global strong solutions to the equations modelling a motion of a rigid body around a viscous fluid. *Discr. Contin. Dyn. Syst.* **38** (2016), no. 3, 1539–1562.
57. J. Neustupa, P. Penel: Anisotropic and geometric criteria for interior regularity of weak solutions to the 3D Navier-Stokes equations. In *Mathematical Fluid Mechanics, Recent Results and Open Questions,* ed. J. Neustupa and P. Penel, Birkhäuser, Basel 2001, pp. 237–268.
58. J. Neustupa, P. Penel: A weak solution to the Navier-Stokes system with Navier's boundary condition in a time-varying domain. In *Recent Developments of Mathematical Fluid Mechanics,* Series: *Advances in Mathematical Fluid Mechanics,* ed. H. Amann, Y. Giga, H. Kozono, H. Okamoto, M. Yamazaki, Birhäuser-Verlag, Springer, Basel 2016, pp. 375–400.
59. J. Neustupa: The boundary regularity of a weak solution of the Navier-Stokes equation and connection with the interior regularity of pressure. *Appl. Math.* **6** (2003), 547–558.
60. J. Neustupa: A geometric improvement of a velocity–pressure local regularity criterion for a suitable weak solution to the Navier-Stokes equations. *Mathematica Bohem.* **139** (2014), no. 4, 685–698.
61. J. Neustupa, H. Al Baba: The interior regularity of pressure associated with a weak solution to the Navier-Stokes equations with the Navier-type boundary conditions. *J. Math. Anal. Appl.* **463** (2018), no. 1, 222–234.
62. J. Neustupa: A contribution to the theory of regularity of a weak solution to the Navier-Stokes equations via one component of velocity and other related quantities. *J. Math. Fluid Mech.* **20** (2018), no. 3, 1249–1267.
63. T. Ohyama: Interior regularity of weak solutions of the time dependent Navier-Stokes equations. *Proc. Japan. Acad.* **36** (1960), 273–277.
64. W. Rudin: *Functional Analysis.* Mc Graw-Hill Inc., New York 1973.
65. J. Saal: Stokes and Navier-Stokes equations with Robin boundary conditions in a half-space. *J. Math. Fluid Mech.* **8** (2006), 211–241.
66. V. Scheffer: Partial regularity of solutions to the Navier-Stokes equations. *Pacific J. Math.* **66** (1976), 535–552.
67. G. Seregin, V. Šverák: Navier-Stokes equations with lower bounds on the pressure. *Arch. Rat. Mech. Anal.* **163** (2002), 65–86.
68. J. Serrin: On the interior regularity of weak solutions of the Navier-Stokes equations. *Arch. Rat. Mech. Anal.* **9** (1962) 187–195.
69. R. Shimada: On the $L_p - L_q$ maximal regularity for Stokes equations with Robin boundary condition in a bounded domain. *Math. Meth. Appl. Sci.* **30** (2007), 257–289.
70. Ch. Simader, H. Sohr: A new approach to the Helmholtz decomposition and the Neumann problem in L^q-spaces for bounded and exterior domains. In *Mathematical Problems Relating to the Navier-Stokes Equation,* Series: *Adv. Math. Appl. Sci.* 11, ed. G. P. Galdi, World Sci. Publ., River Edge, NJ, 1992, pp. 1–35.
71. J. Simon: On the existence of pressure for solutions of the variational Navier-Stokes equations. *J. Math. Fluid Mech.* **1** (1999), no. 3, 225–234.
72. Z. Skalák, P. Kučera: Regularity of pressure in the neighbourhood of regular points of weak solutions of the Navier-Stokes equations. *Appl. Math.* **48** (2003), no. 6, 573–586.
73. H. Sohr, W. von Wahl: On the regularity of the pressure of weak solutions of Navier-Stokes equations. *Arch. Math.* **46** (1986), 428–439.

74. H. Sohr: Zur Regularitätstheorie der instationären Gleichungen von Navier-Stokes. *Math. Z.* **184** (1983), 359–375.
75. H. Sohr: *The Navier-Stokes Equations. An Elementary Functional Analytic Approach.* Birkhäuser Advanced Texts, Basel-Boston-Berlin 2001.
76. M. Struwe: On a Serrin-type regularity criterion for the Navier-Stokes equations in terms of the pressure. *J. Math. Fluid Mech.* **9** (2007), 235–242.
77. T. Suzuki: Regularity criteria of weak solutions in terms of pressure in Lorentz spaces to the Navier-Stokes equations. *J. Math. Fluid Mech.* **14** (2012), no. 4, 653–660.
78. Y. Taniuchi: On generalized energy inequality of Navier-Stokes equations. *Manuscripta Math.* **94** (1997), 365–384.
79. R. Temam: *Navier-Stokes Equations.* North-Holland, Amsterdam-New York-Oxford 1977.
80. J. Wolf: A direct proof of the Caffarelli-Kohn-Nirenberg theorem. In *Parabolic and Navier-Stokes equations, Banach Center Publ.* **81**, Part 2, Polish Acad. of Sci., Inst. Math., Warsaw 2008, pp. 533–552.
81. J. Wolf: A new criterion for partial regularity of suitable weak solutions to the Navier-Stokes equations. In *Advances in Mathematical Fluid Mechanics,* ed. R. Rannacher, A. Sequeira, Springer, Berlin, 2010, pp. 613–630.
82. J. Wolf: On the local regularity of suitable weak solutions to the generalized Navier-Stokes equations. *Ann. Univ. Ferrara Sez. VII Sci. Mat.* **61** (2015), 149–171.
83. J. Wolf: On the local pressure of the Navier-Stokes equations and related systems. *Adv. Differential Equations* **22** (2017), no. 5–6, 305–338.
84. Y. L. Xiao, Z. P. Xin: On the vanishing viscosity limit for the 3D Navier-Stokes equations with a slip boundary condition. *Commun. Pure Appl. Math.* **60** (2007), 1027–1055.
85. Y. Zhou: On regularity criteria in terms of pressure for the Navier-Stokes equations in $\mathbb{R}^3$. *Proc. Amer. Math. Soc.* **134** (2006), 149–156.
86. Y. Zhou: On a regularity criterion in terms of the gradient of pressure for the Navier-Stokes equations in $\mathbb{R}^N$. *Z. Angew. Math. Phys.* **57** (2006), 384–392.

Chapter 5
Flows of Fluids with Pressure Dependent Material Coefficients

Miroslav Bulíček

Dedicated to the memory of Josef Žabenský

Abstract It has been well documented in many studies that the material parameters of a fluid may essentially depend on the pressure and that they can vary by several orders of magnitude. The material parameters, for which this dependence is observed, are mainly the viscosity (due to the internal forces in the fluid) and the friction (due to fluid–(rigid) solid interaction). In addition, these large variations with respect to the pressure in the material parameters occur although the variations of the density are almost negligible (in comparison with other parameters). Therefore it is still reasonable to describe the above mentioned phenomena in many fluids by incompressible models. Likewise, the viscosity and the drag of many fluids vary with the shear rate and such shear (rate)-dependent viscosity and friction are extensively used, ranging from geophysics, chemical engineering, and bio-material science up to the food industry, enhanced oil recovery, carbon dioxide sequestration, or extraction of unconventional oil deposits, etc.

The aim of this study is to present an overview of available results for models with very complicated rheological laws used in engineering praxes. As particular examples that fit into the class of models studied here, we refer to the Darcy model, to the Brinkman models, and to the Bingham models. Nevertheless, the aim of this study is much more ambitious and we go much beyond these standard models and present a kind of unifying theory, which is based on the use of the so-called maximal monotone graphs, which seems to be very appropriate from the point of view of mathematical analysis of the problem.

M. Bulíček (✉)
Charles University, Prague, Czech Republic
e-mail: mbu8060@karlin.mff.cuni.cz

T. Bodnár et al. (eds.), *Fluids Under Pressure*, Advances in Mathematical Fluid Mechanics, https://doi.org/10.1007/978-3-030-39639-8_5

5.1 Introduction

It has been well documented in many studies (see, e.g., [3, 8, 37, 48, 56, 62]) that the viscosity of a fluid can vary by several orders of magnitude with the pressure. Since the friction due to fluid–(rigid) solid interaction usually dominates the friction between layers of the fluid itself, the relation between the drag coefficient and the pressure is even more substantial. In addition, these large variations with respect to the pressure in the material parameters occur although the variations of the density are almost negligible (in comparison with other parameters). Therefore it is still reasonable to describe the above mentioned phenomena in many fluids by incompressible models. Likewise, the viscosity of many fluids varies with the shear rate. See, for example, [7] and [43] for illustrative lists of areas where incompressible fluids with shear (rate)-dependent viscosity are extensively used, ranging from geophysics, chemical engineering, and bio-material science up to the food industry. Both phenomena also play an important role in understanding the problems of enhanced oil recovery, carbon dioxide sequestration, or extraction of unconventional oil deposits, etc.

In this survey paper, we consider mainly the steady flow that is capable of describing the above mentioned properties and which is frequently used in the engineering praxis. For simplicity, we also frequently omit the inertial forces and rather focus on the dependence of the analysis on the constitutive equations. Nevertheless, at the end of the chapter we also briefly list the available results for the full system, i.e., the system with inertial forces (the convective term) and also the unsteady problem.

We shall consider that the flow takes place in an open, bounded, connected domain $\Omega \subset \mathbb{R}^d$, which is supposed to have at least Lipschitz boundary. We focus mainly on physically most relevant cases $d = 2, 3$ but usually state all results for general dimension. First, we briefly recall a system of nonlinear partial differential equations (PDEs) we have in mind that falls into/generalizes the classical fluid flow models of Stokes, Darcy, Forchheimer, and Brinkman

$$\begin{aligned} -\operatorname{div}[2\nu(p, |\boldsymbol{D}\boldsymbol{v}|^2)\boldsymbol{D}\boldsymbol{v}] + \beta(p, |\boldsymbol{v}|, |\boldsymbol{D}\boldsymbol{v}|^2)\boldsymbol{v} + \nabla p = \boldsymbol{f} \quad &\text{in } \Omega, \\ \operatorname{div} \boldsymbol{v} = 0 \quad &\text{in } \Omega. \end{aligned} \tag{5.1}$$

In the PDE problem (5.1) we look for the quantities $\boldsymbol{v} : \Omega \to \mathbb{R}^d$, the velocity field, and $p : \Omega \to \mathbb{R}$, the pressure. The symbol $\boldsymbol{D} = \frac{1}{2}(\nabla + \nabla^T)$ denotes the symmetric part of the gradient. The density of external body forces $\boldsymbol{f}$ is for the sake of convenience in some cases supposed to be of the form

$$\boldsymbol{f} = -\operatorname{div} \boldsymbol{F},$$

where $\boldsymbol{F} : \Omega \to \mathbb{R}^{d \times d}$ is a given tensor-valued function. In addition, we set the density of the fluid to be equal to a positive constant and re-scaled the problem (5.1) accordingly.

The system (5.1) already captures many classical models. Let us recall the basic classification of such models.

5.1.1 Linear Models

We start our list with the linear models, i.e., the models, where $\boldsymbol{v}$ and p appear linearly.

(L1) For $\beta \equiv 0$ and $\nu \equiv \nu_0$ being a positive constant called *viscosity* the problem (5.1) reduces to the classical Stokes equation [61], describing a steady **slow** flow of an incompressible fluid, where the pressure p is determined up to a constant.

(L2) Conversely, if $\nu \equiv 0$ and $\beta = \beta_0$ is a positive constant, the PDE system (5.1) simplifies to the Darcy equation [25]. This is virtually the simplest PDE system capable of describing the flow of a single fluid through a rigid porous solid due to the pressure gradient. The number β_0 is the **drag coefficient**.

(L3) Thirdly, if $\nu \equiv \nu_0$ and $\beta \equiv \beta_0$ are positive constants, (5.1) simplifies to the Brinkman equation [9, 10], representing another popular model capable of describing certain flows through porous media.

Each of the three aforementioned PDE systems is linear, which may be advantageous from the theoretical point of view, and is also of use in engineering praxes. However, there are also many nonlinear models belonging to the setting of (5.1) that are technologically more important as they exhibit experimentally confirmed features, which cannot be captured by linear models. For instance, taking pressure dependent viscosity and drag coefficient in (5.1) leads to a *ceiling flux* (a saturation phenomenon, see [59]), while approaches based on the Darcy and the Brinkman models result in a flux that is linearly increasing with the pressure.

Our principal interest is to investigate the flows in which the material moduli (the drag coefficient and the viscosity) depend nonlinearly on the pressure p and the shear rate $|\boldsymbol{D}\boldsymbol{v}|^2 = \boldsymbol{D}\boldsymbol{v} \cdot \boldsymbol{D}\boldsymbol{v}$.

5.1.2 Nonlinear Models

Next, we recall the further possible settings leading to several classical models. The list is meant just for illustrative purposes with no ambition to be exhaustive. Here, $\nu(p, |\boldsymbol{D}\boldsymbol{v}|^2)$ is generalized viscosity and $\beta(p, |\boldsymbol{v}|, |\boldsymbol{D}\boldsymbol{v}|^2)$ the generalized drag coefficient. The constants ν_0, ν_1, β_0, β_1, β_2 are always assumed to be greater than zero.

(N1) For zero viscosity $\nu(p, |\boldsymbol{D}\boldsymbol{v}|^2) \equiv 0$ and the drag coefficient of the form $\beta(p, |\boldsymbol{v}|, |\boldsymbol{D}\boldsymbol{v}|^2) = \beta_0 + \beta_1|\boldsymbol{v}|$ we get the so-called Darcy–Forchheimer model [28]. Variants can be obtained by considering $\beta(p, |\boldsymbol{v}|, |\boldsymbol{D}\boldsymbol{v}|^2) = \beta_0 + \beta_1|\boldsymbol{v}|^q$ with some $q > 0$.

(N2) Zero drag coefficients $\beta(p, |\boldsymbol{v}|, |\boldsymbol{D}\boldsymbol{v}|^2) \equiv 0$ and pressure dependent viscosity $\nu(p, |\boldsymbol{D}\boldsymbol{v}|^2) = \nu_0 \exp(\nu_1 p)$ lead to the Barus model [6].

(N3) Shear rate dependent viscosity $\nu(p, |\boldsymbol{D}\boldsymbol{v}|^2) = \nu_0(\varepsilon + |\boldsymbol{D}\boldsymbol{v}|^2)^{\frac{r-2}{2}}$ and zero drag $\beta(p, |\boldsymbol{v}|, |\boldsymbol{D}\boldsymbol{v}|^2) \equiv 0$ with $\varepsilon \geq 0$ produce the power-law fluid models, see, e.g., [57, 58] and further references in [14].

(N4) The models with the pressure and the shear rate dependent viscosity of the form $\nu(p, |\boldsymbol{D}\boldsymbol{v}|^2) = \nu_0(\varepsilon + (1 + \exp(\nu_1 p))^{-q} + |\boldsymbol{D}\boldsymbol{v}|^2)^{\frac{r-2}{2}}$ and zero drag $\beta(p, |\boldsymbol{v}|, |\boldsymbol{D}\boldsymbol{v}|^2) \equiv 0$ with the parameters $q \in (0, \frac{r-1}{2\nu_1(2-r)}\varepsilon^{\frac{2-r}{2}})$, $r \in (1, 2)$, and $\varepsilon > 0$ lead to a model for which the global-in-time existence of weak solutions was established in [41].

(N5) Finally, the model that resembles the Bingham fluid $\nu(p, |\boldsymbol{D}\boldsymbol{v}|^2) = \dfrac{2\nu_0 p}{|\boldsymbol{D}\boldsymbol{v}|}$ with zero drag $\beta(p, |\boldsymbol{v}|, |\boldsymbol{D}\boldsymbol{v}|^2) \equiv 0$ leads to the Schaeffer model [55], which was proposed to describe flowing granular materials.

5.1.3 General Model—The Setting of Constitutive Equations

In fact we want to treat a more general model than (5.1). For this purpose, we consider as a starting point for the further investigation of the following system:

$$
\begin{aligned}
-\operatorname{div} \boldsymbol{S} + \boldsymbol{m} &= -\nabla p + \boldsymbol{f}, \\
\boldsymbol{S} &= \boldsymbol{S}^T, \\
\operatorname{div} \boldsymbol{v} &= 0.
\end{aligned} \tag{5.2}
$$

Here $\boldsymbol{S}$ stands for the deviatoric part of the Cauchy stress tensor $\boldsymbol{T}$ and p for the mean normal stress, i.e., the pressure. This means

$$
\boldsymbol{S} = \boldsymbol{T} - \frac{1}{d}\operatorname{Tr}(\boldsymbol{T})\boldsymbol{I} \qquad \text{and} \qquad p = -\frac{1}{d}\operatorname{Tr}(\boldsymbol{T}),
$$

so that $\boldsymbol{T} = \boldsymbol{S} - p\boldsymbol{I}$ with $\boldsymbol{I}$ being the identity tensor. The symbol $\boldsymbol{m}$ signifies the force acting on the fluid due to its interaction with the rigid solid. The first equation in (5.2) is the balance of linear momentum for a steady slow, the second equation represents the balance of angular momentum, and the last one is the incompressibility constraint.

Note that the choice

$$
\boldsymbol{S} = 2\nu(p, |\boldsymbol{D}\boldsymbol{v}|^2)\boldsymbol{D}\boldsymbol{v}, \tag{5.3}
$$

$$
\boldsymbol{m} = \beta(p, |\boldsymbol{v}|, |\boldsymbol{D}\boldsymbol{v}|^2)\boldsymbol{v}, \tag{5.4}
$$

where $\boldsymbol{v}$ denotes the difference of the velocity of the fluid (which is $\boldsymbol{v}$) and the velocity of the rigid solid, which is supposed to be zero, directly leads to (5.1).

However, we want to consider more sophisticated models. Therefore we shall assume the following **implicit** constitutive laws:

$$\boldsymbol{h}(\boldsymbol{m}, \boldsymbol{v}, \boldsymbol{D}\boldsymbol{v}, p) = \boldsymbol{0} \qquad \text{in } \Omega, \tag{5.5}$$

$$\boldsymbol{H}(\boldsymbol{S}, \boldsymbol{D}\boldsymbol{v}, p) = \boldsymbol{0} \qquad \text{in } \Omega, \tag{5.6}$$

where $\boldsymbol{h} : \mathbb{R}^d \times \mathbb{R}^d \times \mathbb{R}^{d\times d}_{\text{sym}} \times \mathbb{R} \to \mathbb{R}^d$ and $\boldsymbol{H} : \mathbb{R}^{d\times d}_{\text{sym}} \times \mathbb{R}^{d\times d}_{\text{sym}} \times \mathbb{R} \to \mathbb{R}^{d\times d}_{\text{sym}}$ are continuous mappings. The purpose of such setting is twofold. First, we can capture more general models than those in (5.3)–(5.4). Second, the implicit setting (5.5)–(5.6) also resembles the way how the models with pressure dependent material coefficients can be rigorously obtained. Indeed, usually the pressure is understood as a Lagrange multiplier and there is absolutely no reason why it should appear in the constitutive equations. However, if one switches to the implicit setting from the very beginning, then one can rigorously derive the models of the form (5.3)–(5.4). This idea was developed by Rajagopal and we refer to [50] and references therein for the setting of incompressible fluids. A complete thermodynamic basis for obtaining the Darcy, Forchheimer, and Brinkman models and their generalizations falling within the class given by (5.5)–(5.6) completed with (5.2) was developed in a recent work by Srinivasan and Rajagopal [60] and is based the theory of interacting continua as developed in [52, 54, 67]. The model (5.2) can be understood as a system describing slow steady (all inertial forces are neglected) flows of a single liquid through a rigid porous solid. Moreover, following ideas presented earlier in [51], they arrived at a general reduced thermodynamical identity that takes form

$$\xi = \boldsymbol{S} \cdot \boldsymbol{D}\boldsymbol{v} + \boldsymbol{m} \cdot \boldsymbol{v},$$

where ξ stands for the rate of dissipation, which should be **non-negative** by the second law of thermodynamics. Consequently, the model (5.3)–(5.4) is thermodynamically compatible as long as $\nu, \beta \geq 0$. Notice that there are two mechanisms for entropy production, one coming from the viscous part of the Cauchy stress tensor and the second one coming from the drag term. The authors of [60] further discussed how one can derive (5.3) and (5.4) purely from the knowledge of appropriately chosen constitutive equations for ξ. It appeared that applying the criterion of maximal rate of entropy production one deduces (5.3)–(5.4) by a proper choice of ξ, see [60] for details.

In what follows, we will consider as general classes of (5.5)–(5.6) as possible, but we will have to distinguish two cases. The first case, when we assume the zero viscosity will lead to the model, where all the information about the velocity field must be observed from $\boldsymbol{m}$. This will lead to a certain structural assumptions on $\boldsymbol{h}$ in (5.5). Namely, we shall assume that it is independent of $\boldsymbol{D}\boldsymbol{v}$ and that it induces a p continuously parameterized **maximal monotone graph** with respect to the velocity field $\boldsymbol{v}$. In the second case, when (5.6) allows also nonzero $\boldsymbol{S}$, we will be able to treat more general drag parameters, or more precisely more general forms of (5.5). However, we shall require that $\boldsymbol{H}$ induces the maximal monotone graph.

Furthermore, in case that the graph is also p dependent, we shall require certain smoothness of the graph with respect to the pressure and the shear rate. We refer to precise assumptions stated in the forthcoming sections.

5.1.4 Boundary Conditions

To conclude this introductory part, we must complete the problem (5.2) by proper boundary conditions. We denote $\boldsymbol{n}$ to be an outward normal vector at $\partial\Omega$ and consider that $\Gamma_{1,2} \subset \partial\Omega$ are relatively open parts of the boundary such that $\Gamma_1 \cap \Gamma_2 = \emptyset$ and $\overline{\Gamma_1 \cup \Gamma_2} = \partial\Omega$. For us, Γ_1 is the **exterior boundary**, where we want to prescribe the velocity field, and Γ_2 is the **accessible boundary**, where we want to prescribe the stress, see [4]. This means

$$\begin{aligned} \boldsymbol{v} &= \boldsymbol{v}_0 \quad \text{on } \Gamma_1, \\ (\boldsymbol{S} - p\boldsymbol{I})\boldsymbol{n} &= \boldsymbol{s}_0 \quad \text{on } \Gamma_2, \end{aligned} \tag{5.7}$$

where $\boldsymbol{v}_0$ and $\boldsymbol{s}_0$ are given. One can also consider a more general boundary condition that decomposes the boundary value on the normal part and the tangential part that can lead to slip boundary conditions. For slip boundary conditions, we may obtain essentially the same results but we do not discuss it here. We refer the interested reader to [15, 21, 22] for very complex boundary conditions considered in the fluid flow.

However, in case that $\boldsymbol{S}$ is vanishing in the constitutive law (5.6), then we can control only the normal component of the above[1] quantities. This means we assume that

$$\begin{aligned} (\boldsymbol{v} - \boldsymbol{v}_0) \cdot \boldsymbol{n} &= 0 \quad \text{on } \Gamma_1, \\ p &= p_0 \quad \text{on } \Gamma_2, \end{aligned} \tag{5.8}$$

where $\boldsymbol{v}_0$ and p_0 are given.

In what follows we keep this splitting and for the Darcy-like models we will consider (5.8) and for Brinkman-like models we will consider (5.7).

5.1.5 Notation and Structure of the Paper

We shall use the standard notation $L^r(\Omega)$ for the Lebesgue space and $W^{1,r}(\Omega)$ for the Sobolev space, respectively, and use the simplified notations

[1] We can consider the normal component of $\boldsymbol{v}$ since we control the divergence of $\boldsymbol{v}$.

$$\|z\|_r := \|z\|_{L^r(\Omega)} = \left(\int_\Omega |z|^r \, dx\right)^{\frac{1}{r}},$$

$$\|z\|_{1,r} := \|z\|_{W^{1,r}(\Omega)} = \left(\int_\Omega |z|^r + |\nabla z|^r \, dx\right)^{\frac{1}{r}}.$$

For any of the above spaces, we also denote

$$\mathring{X}(\Omega) := \left\{ f \in X(\Omega) \,\Big|\, \int_\Omega f(x)\, dx = 0 \right\}.$$

No explicit distinction between spaces of scalar- and vector-valued functions will be made. Confusion should never come to pass as we employ small bolded letters to denote vectors and bolded capitals for tensors. Further, we use the following notation for the mean value of an integrable function f:

$$f_\Omega := \frac{1}{|\Omega|} \int_\Omega f(x)\, dx.$$

Next, to denote a space of function vanishing on a part of the boundary $\Gamma \subset \partial\Omega$, we use the abbreviation[2]

$$W^{1,r}_\Gamma(\Omega) := \left\{ f \in W^{1,r}(\Omega) \,\Big|\, f = 0 \text{ on } \Gamma \right\}.$$

Moreover, in case that the $(d-1)$ Hausdorff measure of Γ is positive ($|\Gamma|_{d-1} > 0$), we use the Poincaré inequality and equip the space $W^{1,r}_\Gamma(\Omega)$ with the equivalent norm

$$\|f\|_{1,r} := \|f\|_{W^{1,r}_\Gamma(\Omega)} = \|\nabla f\|_r.$$

Next, we focus on the function spaces that will be used for the theoretical investigation and which are directly related to our choice of Γ_1 and Γ_2. For any $r \in (1, \infty)$ we define the following function spaces[3] (related to the velocity field)

$$W^{1,r}_{\Gamma_1,\mathrm{div}}(\Omega) := \left\{ \boldsymbol{f} \in W^{1,r}(\Omega) \mid \operatorname{div} \boldsymbol{f} = 0 \text{ in } \Omega,\ \boldsymbol{f} = \boldsymbol{0} \text{ on } \Gamma_1 \right\},$$

$$W^{-1,r'}_{\Gamma_1}(\Omega) := \left(W^{1,r}_{\Gamma_1}(\Omega)\right)^*,$$

[2]Whenever we write f on Γ, we are automatically employing the trace operator.

[3]The space $L^r_{\mathrm{div},\Gamma_1}(\Omega)$ can be equivalently defined as

$$L^r_{\mathrm{div},\Gamma_1}(\Omega) := \left\{ \boldsymbol{\varphi} \in L^r(\Omega) \mid \text{ for all } u \in W^{1,r'}_{\Gamma_2}(\Omega) \quad \int_\Omega \nabla u \cdot \boldsymbol{\varphi}\, dx = 0 \right\}.$$

$$L^r_{\mathrm{div}}(\Omega) := \left\{ \varphi \in L^r(\Omega) \mid \forall u \in \mathcal{C}_0^\infty(\Omega) \int_\Omega \varphi \cdot \nabla u \, dx = 0 \right\},$$

$$L^r_{\mathrm{div},\Gamma_1}(\Omega) := \left\{ \varphi \in L^r_{\mathrm{div}}(\Omega) \mid \varphi \cdot n = 0 \text{ on } \Gamma_1 \right\}.$$

In case that $\Gamma_1 = \partial\Omega$, we simply use the standard notation $W^{-1,r'}(\Omega) := W^{-1,r'}_{\Gamma_1}(\Omega)$. In addition, to preserve the Poincaré inequality, if $\Gamma_2 = \emptyset$ then we define

$$W^{1,r}_{\Gamma_2}(\Omega) := \mathring{W}^{1,r}(\Omega).$$

Furthermore, if $|\Gamma_1|_{(d-1)} \neq 0$, then we will frequently use the Korn inequality

$$\|f\|_{W^{1,r}_{\Gamma_1,\mathrm{div}}(\Omega)} = \|f\|_{1,r} \leq C(r,\Omega)\|Df\|_r, \tag{5.9}$$

valid for all $r \in (1,\infty)$ and all $f \in W^{1,r}_{\Gamma_1,\mathrm{div}}(\Omega)$ provided that Ω is Lipschitz. All above mentioned spaces are for $r \in [1,\infty)$ separable and for $r \in (1,\infty)$ reflexive.

If $U, V \subset \mathbb{R}^d$, we say V is compactly contained in U, symbolically $V \Subset U$, if V is bounded and $\overline{V} \subset U$. The symbol $\cdot$ stands for the scalar product and we again do not distinguish whether the product is taken on $\mathbb{R}^d$ or $\mathbb{R}^{d\times d}$. For $r \in (1,\infty)$ we denote $r' = r/(r-1)$ the dual exponent and $r^* = dr/(d-r)$ the embedding exponent, provided further $r < d$. If $r = d$, then r^* is an arbitrary number from $[1,\infty)$ and for $r > d$ we set $r^* := \infty$. Further, in the rest of the paper we will use the notion C for the generic constant that can depend only on the given data. In case it is dependent on any parameter essentially, it will be clearly denoted in the text.

Finally, we introduce the notion of a cut-off that will be used later. For any $k > 0$ we define the function $T_k : \mathbb{R} \to \mathbb{R}$ as

$$T_k(x) := \begin{cases} x & \text{for } |x| < k, \\ k\dfrac{x}{|x|} & \text{for } |x| \geq k. \end{cases} \tag{5.10}$$

Completely analogously we define also the vector-valued cut-off function $\boldsymbol{T}_k : \mathbb{R}^d \to \mathbb{R}^d$.

The rest of the chapter is organized as follows. In Sect. 5.2 we focus on the Darcy model and its generalization. This means we will always neglect the viscous part. We will state two results, which are related to the maximal monotone graph setting for the drag term. In addition, in some cases we will obtain the minimum and/or the maximum principle for the pressure. Section 5.3 is devoted to the generalized Brinkman–Darcy system. We employ here also the viscous part and prove the main result for the case that all the material parameters depend on the pressure in some convenient way. Moreover, we give a sketchy proof for the existence result in case that the viscous part is independent of the pressure. This will allow us to consider more general and also implicit constitutive relationships (just with the proviso that $\boldsymbol{S}$ is independent of the pressure). Finally, in Sect. 5.4, we add to the system also the convective term and time derivative, i.e., we will switch from **slow** flow presented in Sects. 5.2–5.3 to **non-approximate** flow. We just briefly recall the available results and describe what are the key difficulties in treatment of such problems.

5.2 Generalized Darcy–Forchheimer System

Our aim in this section is to describe the most general theory for the Darcy–Forchheimer like systems and our primary motivation is to generalize the models presented in (L2) and (N1). This means we use the model (5.2) with (5.5)–(5.6) but we set $\boldsymbol{H}(\boldsymbol{S}, \boldsymbol{D}, p) := \boldsymbol{S}$, which leads to $\boldsymbol{S} \equiv \boldsymbol{0}$. Further, since we do not control the gradient of the velocity field, we assume that $\boldsymbol{h}$ in (5.5) is independent of $\boldsymbol{D}\boldsymbol{v}$. Finally, we prescribe the boundary conditions (5.8). Thus, the resulting problem reads as follows: For given Γ_1, Γ_2, $\boldsymbol{f} : \Omega \to \mathbb{R}^d$, $\boldsymbol{v}_0 : \Omega \to \mathbb{R}^d$, $p_0 : \Omega \to \mathbb{R}$, and $\boldsymbol{h} : \mathbb{R}^d \times \mathbb{R}^d \times \mathbb{R} \to \mathbb{R}^d$ find a triplet $(\boldsymbol{m}, \boldsymbol{v}, p) : \Omega \to \mathbb{R}^d \times \mathbb{R}^d \times \mathbb{R}$ solving

$$
\begin{aligned}
\nabla p + \boldsymbol{m} &= \boldsymbol{f} && \text{in } \Omega, \\
\operatorname{div} \boldsymbol{v} &= 0 && \text{in } \Omega, \\
\boldsymbol{h}(\boldsymbol{m}, \boldsymbol{v}, p) &= \boldsymbol{0} && \text{in } \Omega, \\
(\boldsymbol{v} - \boldsymbol{v}_0) \cdot \boldsymbol{n} &= 0 && \text{on } \Gamma_1, \\
p - p_0 &= 0 && \text{on } \Gamma_2.
\end{aligned}
\tag{5.11}
$$

We continue as follows. First, we state the assumption on the implicit law given by $\boldsymbol{h}$, then formulate the main existence theorem and finally give the proof. This part is heavily inspired by the recent results presented in [23]. Nevertheless, we present here certain extension that is not covered by the results in [23].

5.2.1 *Assumptions on h—Maximal Monotone Graph Setting*

We shall assume that the mapping $\boldsymbol{h} : \mathbb{R}^d \times \mathbb{R}^d \times \mathbb{R} \to \mathbb{R}^d$ in $(5.11)_3$ is continuous and we will identify its null points with the p **parameterized maximal monotone** r **graph** $\mathcal{A} \subset \mathbb{R}^d \times \mathbb{R}^d \times \mathbb{R}$ via the following:

$$
\boldsymbol{h}(\boldsymbol{m}, \boldsymbol{v}, p) = \boldsymbol{0} \quad \Longleftrightarrow \quad (\boldsymbol{m}, \boldsymbol{v}, p) \in \mathcal{A}\,. \tag{5.12}
$$

We recall the concept of maximal monotone graphs. Thus, in what follows, we assume that $r \in (1, \infty)$ and say that $\mathcal{A}$ is a p-parameterized maximal monotone r graph if $\mathcal{A} \subset \mathbb{R}^d \times \mathbb{R}^d \times \mathbb{R}$ satisfies each of the conditions listed below:

(A1) inclusion of the origin

$$\forall p \in \mathbb{R} : (\mathbf{0}, \mathbf{0}, p) \in \mathcal{A},$$

(A2) monotonicity

$$\forall (\boldsymbol{m}_1, \boldsymbol{v}_1, p), (\boldsymbol{m}_2, \boldsymbol{v}_2, p) \in \mathcal{A} : (\boldsymbol{m}_1 - \boldsymbol{m}_2) \cdot (\boldsymbol{v}_1 - \boldsymbol{v}_2) \geq 0,$$

(A3) maximality

$$(\boldsymbol{m}', \boldsymbol{v}', p) \in \mathbb{R}^d \times \mathbb{R}^d \times \mathbb{R},$$

$$\forall (\boldsymbol{m}, \boldsymbol{v}, p) \in \mathcal{A} : (\boldsymbol{m}' - \boldsymbol{m}) \cdot (\boldsymbol{v}' - \boldsymbol{v}) \geq 0 \implies (\boldsymbol{m}', \boldsymbol{v}', p) \in \mathcal{A},$$

(A4) (r, r')-coercivity for $\boldsymbol{v}$ and $\boldsymbol{m}$

$$\exists c_1 > 0, c_2 \geq 0 \, \forall (\boldsymbol{m}, \boldsymbol{v}, p) \in \mathcal{A} : \boldsymbol{m} \cdot \boldsymbol{v} \geq c_1(|\boldsymbol{v}|^r + |\boldsymbol{m}|^{r'}) - c_2,$$

(A5) existence of a Carathéodory selection, i.e., $\exists \boldsymbol{m}^* : \mathbb{R}^d \times \mathbb{R} \to \mathbb{R}^d$ such that

(i) $\boldsymbol{m}^*(\cdot, p) : \mathbb{R}^d \to \mathbb{R}^d$ is measurable for every $p \in \mathbb{R}$,
(ii) $\boldsymbol{m}^*(\boldsymbol{v}, \cdot) : \mathbb{R} \to \mathbb{R}^d$ is continuous for a.e. $\boldsymbol{v} \in \mathbb{R}^d$,
(iii) $\forall (\boldsymbol{v}, p) \in \mathbb{R}^d \times \mathbb{R} : (\boldsymbol{m}^*(\boldsymbol{v}, p), \boldsymbol{v}, p) \in \mathcal{A}$.

Although the growth conditions for the mapping $\boldsymbol{m}^*$ are not stated explicitly, we can use **(A5)**$_{(iii)}$, **(A4)**, and the Young inequality to obtain

$$c_1(|\boldsymbol{v}|^r + |\boldsymbol{m}^*(\boldsymbol{v}, p)|^{r'}) - c_2 \leq \boldsymbol{m}^*(\boldsymbol{v}, p) \cdot \boldsymbol{v} \leq \frac{c_1 |\boldsymbol{m}^*(\boldsymbol{v}, p)|^{r'}}{r'} + \frac{|\boldsymbol{v}|^r}{c_1^{r-1} r}.$$

Hence, from the above inequality, we get

$$\forall (\boldsymbol{v}, p) \in \mathbb{R}^d \times \mathbb{R} : |\boldsymbol{m}^*(\boldsymbol{v}, p)| \leq \frac{c_2}{r c_1} + \frac{|\boldsymbol{v}|^{r-1}}{c_1^r}. \tag{5.13}$$

In addition, in order to be able to cover also more general situation, we need to generalize the setting of the r graph to the setting on nonuniform r graph. It means that instead of **(A4)** with uniform constants c_1 and c_2 we allow certain non-uniformity, i.e.,

(A4*) there exists $c_2 \geq 0$ and strictly positive continuous function $\alpha : \mathbb{R} \to \mathbb{R}_+$ such that

$$\forall (\boldsymbol{m}, \boldsymbol{v}, p) \in \mathcal{A} : \boldsymbol{m} \cdot \boldsymbol{v} \geq \alpha(p)(|\boldsymbol{v}|^r + |\boldsymbol{m}|^{r'}) - c_2.$$

It is evident that if we replace **(A4)** by **(A4*)**, we also need to change the estimate (5.13) in the following way: There exists a continuous function $\beta : \mathbb{R} \to \mathbb{R}_+$ such that

$$\forall (\boldsymbol{v}, p) \in \mathbb{R}^d \times \mathbb{R} : |\boldsymbol{m}^*(\boldsymbol{v}, p)| \le \beta(p)(1 + |\boldsymbol{v}|^{r-1}). \tag{5.14}$$

Finally, to cover also some exponential growths and to obtain certain minimum or/and maximum principles, we shall require an additional strict monotonicity assumption at the origin

(A6) strict monotonicity with respect to $\boldsymbol{m}$ at the origin

$$\forall (\boldsymbol{m}, \boldsymbol{v}, p) \in \mathcal{A} : \boldsymbol{m} \cdot \boldsymbol{v} = 0 \implies \boldsymbol{m} = \boldsymbol{0}.$$

Note that if the constant $c_2 = 0$ in **(A4)**, then **(A6)** is met automatically.

5.2.2 Motivation and Examples

As was already discussed in Sect. 5.1.3, the system (5.11) describes steady (slow) flows of fluids through porous media (see, for example, Nield and Bejan [47]) and it can be also considered as a special case in the hierarchy of models used for interacting continua (as presented in Rajagopal [51]), where we ignore the viscous effects but take into account only the drag force. The $\boldsymbol{m}$, representing the interaction force (linear momentum) between a fluid and a rigid solid. The velocity $\boldsymbol{v}$ is the relative velocity between the solid and the liquid and therefore it is frame indifferent. So it is a natural candidate to dictate the structure of the drag force $\boldsymbol{m}$. We have already discussed the simplest Darcy law in (L2), i.e., $\boldsymbol{m} = \alpha \boldsymbol{v}$ with $\alpha > 0$. The drawback of this model is that it does not relate well to reality for other than sufficiently small velocities, see [47]. Therefore, the Darcy law is frequently generalized to $\boldsymbol{m} = \alpha(|\boldsymbol{v}|)\boldsymbol{v}$, known as the Forchheimer model with α an affine function, which falls into the class (N1). The next natural step is to consider $\boldsymbol{m} = \alpha(p, |\boldsymbol{v}|)\boldsymbol{v}$ (still within the class (N1)) as a means of capturing a pressure-related viscosity [37, 63] leading so to the generalized Darcy–Forchheimer model. However, it was argued in [50] that not even such setting is always satisfactory and one is led to the implicit models $(5.11)_3$.

Let us now introduce few prototypic examples of the choice $\boldsymbol{h}$ which lead to the graph $\mathcal{A}$ fulfilling **(A1)**–**(A5)**. The Darcy or the Darcy–Forchheimer model with the specific choice

$$\boldsymbol{m} = \boldsymbol{m}(\boldsymbol{v}, p) = \alpha(p)|\boldsymbol{v}|^{r-2}\boldsymbol{v}, \tag{5.15}$$

with $r > 1$ and continuous function α fulfilling for all $p \in \mathbb{R}$ $0 < \alpha_1 \le \alpha(p) \le \alpha_2 < \infty$ falls into the class of maximal monotone graphs. Another very classical example is

$$|\boldsymbol{m}| \le \sigma(p) \Leftrightarrow \boldsymbol{v} = \boldsymbol{0} \quad \text{and} \quad |\boldsymbol{m}| > \sigma(p) \Leftrightarrow \boldsymbol{m} = \sigma(p)\frac{\boldsymbol{v}}{|\boldsymbol{v}|} + \gamma(p)|\boldsymbol{v}|^{r-2}\boldsymbol{v}, \tag{5.16}$$

with $\sigma(p)$ and $\gamma(p)$ being continuous functions fulfilling $0 < \alpha_1 \le \sigma(p), \gamma(p) \le \alpha_2 < \infty$. Note that this model is very similar to the Herschel–Bulkley responses between the Cauchy stress and the velocity gradient used in non-Newtonian fluids. The number $\sigma(p)$ plays the role of certain critical value of the force so that there is no motion until this threshold is reached. The above law can be rewritten as

$$\gamma(p)|\boldsymbol{v}|^{r-2}\boldsymbol{v} = (|\boldsymbol{m}| - \sigma(p))_+ \frac{\boldsymbol{m}}{|\boldsymbol{m}|},$$

where the symbol $a_+ := \max\{0, a\}$. This setting corresponds to $\boldsymbol{h}(\boldsymbol{m}, \boldsymbol{v}, p) = \boldsymbol{0}$ with

$$\boldsymbol{h}(\boldsymbol{m}, \boldsymbol{v}, p) = \gamma(p)|\boldsymbol{v}|^{r-2}\boldsymbol{v} - (|\boldsymbol{m}| - \sigma(p))_+ \frac{\boldsymbol{m}}{|\boldsymbol{m}|}.$$

Clearly, $\boldsymbol{h}$ is continuous and in addition the corresponding graph $\mathcal{A}$ satisfies **(A1)**–**(A5)**, see, e.g., [50] for the proof.

The above examples share the important property that α, σ, and γ are strictly positive bounded functions and thus we have **(A4)**. This however rules out the following example:

$$\boldsymbol{m}(\boldsymbol{v}, p) = \alpha_1 \exp(\alpha_2 p)\boldsymbol{v}, \qquad \alpha_{1,2} > 0 \tag{5.17}$$

that is actually of outmost importance since this is the model where the drag force changes rapidly even for small fluctuation of pressure, see, e.g., the experimental study [6] or [46] for theoretical justification.

Although the coefficient $\alpha_2 \ll 1$, see [6] for experimental data, it is evident that (5.17) violates **(A4)** and consequently also (5.13). Nevertheless, it can be covered by **(A4*)** and so we have a nonuniform estimate (5.14). For this purpose we also include this into our analysis. One can also consider variants of (5.17) in which only large pressures play a significant role, e.g.,

$$\boldsymbol{m}(\boldsymbol{v}, p) = \max\{\alpha_1, \alpha_1 \exp(\alpha_2 p)\}\boldsymbol{v}, \tag{5.18}$$

that will require less restrictive assumption on data.

It is worth mentioning that we restricted ourselves only onto the setting of r-graphs. However, one can easily extend the whole theory also to the case when the r growth is replaced by a general ψ growth with ψ being the Young function. As far as the Young function satisfies the so-called Δ_2 and ∇_2 condition, the analysis can go step by step as in the present paper. We refer the interested reader to [14], where the monotone operator theory for maximal monotone ψ graphs is described in details and applied to non-Newtonian fluids problems, see also [2, 29, 44] for further properties of maximal monotone graph setting.

5.2.3 Results

In the theory for flows through porous media there is a huge amount of results when $\boldsymbol{m}$ is a continuous monotone function of $\boldsymbol{v}$; however, in the setting of maximal monotone graph (i.e., we allow discontinuity and dependence on the pressure), the existence result was proven quite recently in [23] for graphs fulfilling **(A1)–(A5)**, in particular for models (5.15) and (5.16). In [23] the model (5.17) was also treated under the condition that the external forces $\boldsymbol{f}$ are conservative, i.e., $\boldsymbol{f} = \nabla g$ for some function g. More precisely, it was proven that if one replaces **(A4)** by a weaker condition **(A4*)** and add **(A6)** then one can get the satisfactory theory for (5.17) in case that there is no outflow/inflow on Γ_1.

We mimic the proof here, but we still get a generalization of the result presented in [23]. Within the setting of **(A1)–(A5)** we are able to establish the existence of a solution to the problem (5.11) fulfilling the first three equations pointwise (almost everywhere) in Ω (Theorem 5.2.1). Further, in case that $\boldsymbol{f}$ is conservative and there is no inflow on Γ_1, i.e., $\boldsymbol{v}_0 \cdot \boldsymbol{n} \geq 0$ we obtain the existence of solution that fulfills the minimum principle for the pressure. Similarly, if there is no outflow, we obtain a minimum principle for the pressure (Theorem 5.2.2) and finally we extend the theory also to the case when only **(A4*)** is valid and under certain hypothesis on data we get the existence of a solution for models like (5.18) and (5.17) (Theorem 5.2.3). These results thus also extend the work [23], because we do not require the strict monotonicity assumption **(A6)** here.

Next, we list main theorems.

Theorem 5.2.1 (General Existence Result) *Let Ω be a Lipschitz domain and $r \in (1, \infty)$ be given. Assume $\boldsymbol{f} \in L^{r'}(\Omega)^d$, $\boldsymbol{v}_0 \in L^r_{\operatorname{div}}(\Omega)$, and $p_0 \in W^{1,r'}(\Omega)$. Moreover, assume that $\mathcal{A}$ is a maximal monotone r-graph in the sense of* **(A1)–(A5)**. *Then there exists a triplet $(\boldsymbol{m}, \boldsymbol{v}, p) \in L^{r'}(\Omega) \times L^r_{\operatorname{div}}(\Omega) \times W^{1,r'}(\Omega)$ solving* (5.11), *i.e.,* $(5.11)_1$–$(5.11)_3$ *are satisfied a.e. in Ω and*

$$\boldsymbol{v} - \boldsymbol{v}_0 \in L^r_{\operatorname{div},\Gamma_1}(\Omega),$$

$$p - p_0 \in W^{1,r'}_{\Gamma_2}(\Omega).$$

Theorem 5.2.2 (Minimum/Maximum Principle) *Let assumptions of Theorem 5.2.1 be satisfied and Ω be additionally connected. In addition assume that $\boldsymbol{f} = \nabla g$ for some $g \in W^{1,r'}(\Omega)$. Then there exists a solution $(\boldsymbol{m}, \boldsymbol{v}, p)$ satisfying*

(i) $\boldsymbol{v}_0 \cdot \boldsymbol{n} \geq 0$ *on* Γ_1 *implies* $p - g \leq \operatorname{ess\,sup}_{\Gamma_2}(p_0 - g)$ *a.e. in* Ω.

(ii) $\boldsymbol{v}_0 \cdot \boldsymbol{n} \leq 0$ *on* Γ_1 *implies* $p - g \geq \operatorname{ess\,inf}_{\Gamma_2}(p_0 - g)$ *a.e. in* Ω.

In particular, if $\boldsymbol{v}_0 \cdot \boldsymbol{n} = 0$ *on* Γ_1, Γ_2 *is non-trivial in the sense* $|\Gamma_2|_{d-1} > 0$, $p_0 \in L^\infty(\Gamma_2)$, *and* $g \in L^\infty(\Omega) \cap W^{1,r'}(\Omega)$, *then* $p \in L^\infty(\Omega)$.

Furthermore, there exists a constant K *such that every solution satisfies*

(iii) $\boldsymbol{v}_0 \cdot \boldsymbol{n} \geq 0$ *on* Γ_1 *implies* $p - g \leq \operatorname{ess\,sup}_{\Gamma_2}(p_0 - g) + K$ *a.e. in* Ω.

(iv) $\boldsymbol{v}_0 \cdot \boldsymbol{n} \leq 0$ *on* Γ_1 *implies* $p - g \geq \operatorname{ess\,inf}_{\Gamma_2}(p_0 - g) - K$ *a.e. in* Ω.

In addition, if $\mathcal{A}$ *satisfies* **(A6)**, *then every solution satisfies (i) and (ii).*

Theorem 5.2.3 (Extended Existence Theorem) *Let* Ω *be Lipschitz and connected. Assume that* $\boldsymbol{f} = \nabla g$ *such that* $g \in L^\infty(\Omega) \cap W^{1,r'}(\Omega)$ *and let* $p_0 \in W^{1,r'}(\Omega) \cap L^\infty(\Gamma_2)$ *and* $\boldsymbol{v}_0 \in L^r_{\mathrm{div}}(\Omega)$ *fulfilling* $\boldsymbol{v}_0 \cdot \boldsymbol{n} = 0$ *on* Γ_1 *be given. Assume that* $|\Gamma_2|_{d-1} > 0$. *Moreover, let the graph* $\mathcal{A}$ *satisfy* **(A1)–(A3)** *and* **(A4*)–(A5)**. *Then the existence result of Theorem 5.2.1 still holds and we have in addition that* $p \in L^\infty(\Omega)$.

The result in Theorem 5.2.3 could be further extended, for example, if the function α in **(A4)*** vanishes/explodes only for $p \to \infty$, we just need to have control on the pressure from above. This is the case of example (5.18). In this case we can just use Theorem 5.2.2 to prove only the maximum principle. In this situation the only requirement is that there is no inflow. We formulate such result in separate theorem but we will not provide the proof and leave it to the reader.

Theorem 5.2.4 (Extended Existence Theorem II) *Let* Ω *be Lipschitz and connected. Assume that* $\boldsymbol{f} = \nabla g$ *such that* $g \in L^\infty(\Omega) \cap W^{1,r'}(\Omega)$ *and let* $p_0 \in W^{1,r'}(\Omega) \cap L^\infty(\Gamma_2)$ *and* $\boldsymbol{v}_0 \in L^r_{\mathrm{div}}(\Omega)$. *Assume that* $|\Gamma_2|_{d-1} > 0$. *Moreover, let the graph* $\mathcal{A}$ *satisfy* **(A1)–(A3)** *and* **(A4*)–(A5)**. *Then the existence result of Theorem 5.2.1 still holds provided that one of the following is true*

i) $\boldsymbol{v}_0 \cdot \boldsymbol{n} \geq 0$ *on* Γ_1 *and for any* $\bar{p} < \infty$ *there exists* $c_p > 0$ *such that for all* $p \in (-\infty, \bar{p})$ *the function* $\alpha(p)$ *from* **(A4*)** *satisfies* $c_{\bar{p}}^{-1} \leq \alpha(p) \leq c_{\bar{p}}$,
ii) $\boldsymbol{v}_0 \cdot \boldsymbol{n} \leq 0$ *on* Γ_1 *and for any* $\bar{p} > -\infty$ *there exists* $c_{\bar{p}} > 0$ *such that for all* $p \in (\bar{p}, \infty)$ *the function* $\alpha(p)$ *from* **(A4*)** *satisfies* $c_{\bar{p}}^{-1} \leq \alpha(p) \leq c_{\bar{p}}$.

Concerning the uniqueness, we do not provide the proof here. It can be proven trivially that if the graph $\mathcal{A}$ is independent of p and strictly monotone with respect to $\boldsymbol{v}$, then the velocity $\boldsymbol{v}$ is unique. Similarly, if the graph is strictly monotone[4] with

[4]We say that the graph $\mathcal{A}$ is strictly monotone with respect to $\boldsymbol{v}$ if for all $(\boldsymbol{m}_i, \boldsymbol{v}_i) \in \mathcal{A}$, there holds

$$(\boldsymbol{m}_1 - \boldsymbol{m}_2) \cdot (\boldsymbol{v}_1 - \boldsymbol{v}_2) = 0 \quad \Longrightarrow \quad \boldsymbol{v}_1 = \boldsymbol{v}_2.$$

Similarly, we say that the graph is strictly monotone with respect to $\boldsymbol{m}$ if there holds

$$(\boldsymbol{m}_1 - \boldsymbol{m}_2) \cdot (\boldsymbol{v}_1 - \boldsymbol{v}_2) = 0 \quad \Longrightarrow \quad \boldsymbol{m}_1 = \boldsymbol{m}_2.$$

respect to $\boldsymbol{m}$, then $\boldsymbol{m}$ and also p are unique. Indeed, it follows from the fact that if $(\boldsymbol{m}_i, \boldsymbol{v}_i, p_i)$ are solutions, then they satisfy

$$\int_\Omega (\boldsymbol{m}_i - \boldsymbol{m}_j) \cdot (\boldsymbol{v}_i - \boldsymbol{v}_j)\, dx = 0$$

and the uniqueness can be claimed from the considered strict monotonicity.

In the rest of this part we provide proofs of the above results. The first result is based on the proper use of the maximal monotone graph setting, see also the survey [13]. The minimum/maximum principle is proved for the regularized graph but remains valid also for the limiting object—here the procedure is different to [23]. Finally based on the minimum/maximum principle we prove Theorem 5.2.3.

5.2.4 Graph Regularization

Since we want to apply the standard fix point argument, we shall mollify the graph $\mathcal{A}$ by mollifying the mapping $\boldsymbol{m}^*$. For this purpose we denote for arbitrary $0 < \delta \leq 1$

$$\omega_\delta(\boldsymbol{x}, t) := \delta^{-d} \omega \left(\frac{\boldsymbol{x}}{\delta}\right),$$

where ω is the usual mollification kernel on $\mathbb{R}^d$. With its help we define

$$\boldsymbol{m}_\delta(\boldsymbol{x}, t) := \int_{\mathbb{R}^d} \boldsymbol{m}^*(\boldsymbol{x} - \boldsymbol{y}, t)\, \omega_\delta(\boldsymbol{y})\, d\boldsymbol{y}\,.$$

Clearly, since $\boldsymbol{m}^*$ is Carathéodory, the mapping $\boldsymbol{m}_\delta$ is continuous. In next lemma, we recall and show the most important properties of such mollification.

Lemma 5.2.5 *Let $\mathcal{A}$ satisfy* **(A1)–(A5)**, *$\boldsymbol{m}^*$ be the selection, $\boldsymbol{m}_\delta$ the regularization, and $\Omega \subset \mathbb{R}^d$ measurable. Then there exists positive constants C_1 and C_2 such that the graph $\mathcal{A}_\delta$ given by $\mathcal{A}_\delta := (\boldsymbol{m}_\delta(\boldsymbol{v}, p), \boldsymbol{v})$ fulfills* **(A1)–(A5)**, *where we replace c_1 and c_2 in* **(A4)** *by C_1 and C_2. In addition, we have the following convergence result: Assume that for $\delta \to 0_+$ we have*

$$\begin{aligned} p^\delta &\to p && \textit{strongly in } L^1(\Omega), \\ \boldsymbol{v}^\delta &\rightharpoonup \boldsymbol{v} && \textit{weakly in } L^r(\Omega), \\ \boldsymbol{m}_\delta(\boldsymbol{v}^\delta, p^\delta) &\rightharpoonup \boldsymbol{m} && \textit{weakly in } L^{r'}(\Omega), \end{aligned}$$

$$\limsup_{\delta \to 0_+} \int_\Omega \boldsymbol{m}_\delta(\boldsymbol{v}^\delta, p^\delta) \cdot \boldsymbol{v}^\delta\, dx \leq \int_\Omega \boldsymbol{m} \cdot \boldsymbol{v}\, dx.$$

Then $(\boldsymbol{v}, \boldsymbol{m}, p) \in \mathcal{A}$ almost everywhere in Ω.

Proof The first part of the proof, i.e., that $\mathcal{A}_\delta$ is a maximal monotone (r, r') graph, is rather standard and we refer the reader, for example, to [14, 23]. Hence, we focus just on the convergence result. First of all, it follows from the assumptions that

$$\limsup_{\delta\to 0_+} \int_\Omega \boldsymbol{m}_\delta(\boldsymbol{v}^\delta, p^\delta)\cdot(\boldsymbol{v}^\delta - \boldsymbol{v})\,dx \le 0. \tag{5.19}$$

Next, using the assumption **(A4)**, we have that

$$|\boldsymbol{m}_\delta(\boldsymbol{v}, p^\delta)|^{r'} \le C(1 + |\boldsymbol{v}|^r).$$

Moreover, using the definition of $\boldsymbol{m}_\delta$, the properties of $\boldsymbol{m}^*$, and the strong convergence of p^δ, we can find $\bar{\boldsymbol{m}}$ such that

$$\boldsymbol{m}_\delta(\boldsymbol{v}, p^\delta) \to \bar{\boldsymbol{m}} \qquad \text{a.e. in } \Omega.$$

Consequently, combining two above properties, we can use the Lebesgue dominated converge theorem and conclude that

$$\boldsymbol{m}_\delta(\boldsymbol{v}, p^\delta) \to \bar{\boldsymbol{m}} \qquad \text{strongly in } L^{r'}(\Omega). \tag{5.20}$$

Consequently, using the monotonicity, we have that

$$\begin{aligned}
&\limsup_{\delta\to 0_+} \int_\Omega |(\boldsymbol{m}_\delta(\boldsymbol{v}^\delta, p^\delta) - \boldsymbol{m}_\delta(\boldsymbol{v}, p^\delta))\cdot(\boldsymbol{v}^\delta - \boldsymbol{v})|\,dx \\
&= \limsup_{\delta\to 0_+} \int_\Omega (\boldsymbol{m}_\delta(\boldsymbol{v}^\delta, p^\delta) - \boldsymbol{m}_\delta(\boldsymbol{v}, p^\delta))\cdot(\boldsymbol{v}^\delta - \boldsymbol{v})\,dx \le 0
\end{aligned}$$

and therefore for any $\varphi \in L^\infty(\Omega)$ we deduce that

$$\begin{aligned}
&\lim_{\delta\to 0_+} \int_\Omega \boldsymbol{m}_\delta(\boldsymbol{v}^\delta, p^\delta)\cdot \boldsymbol{v}^\delta \varphi\,dx \\
&= \lim_{\delta\to 0_+} \int_\Omega (\boldsymbol{m}_\delta(\boldsymbol{v}^\delta, p^\delta) - \boldsymbol{m}_\delta(\boldsymbol{v}, p^\delta))\cdot(\boldsymbol{v}^\delta - \boldsymbol{v})\varphi\,dx \\
&\quad + \lim_{\delta\to 0_+} \int_\Omega \boldsymbol{m}_\delta(\boldsymbol{v}, p^\delta))\cdot(\boldsymbol{v}^\delta - \boldsymbol{v})\varphi + \boldsymbol{m}_\delta(\boldsymbol{v}^\delta, p^\delta)\cdot \boldsymbol{v}\varphi\,dx \\
&= \int_\Omega \boldsymbol{m}\cdot \boldsymbol{v}\varphi\,dx,
\end{aligned} \tag{5.21}$$

where we used (5.20) and the assumptions of lemma.

Next, let $\boldsymbol{v}' \in \mathbb{R}^d$ be arbitrary such that $\boldsymbol{m}^*(\boldsymbol{v}', \cdot)$ is continuous. Using the monotonicity and (5.13), we get

$$\begin{aligned}&(\boldsymbol{m}_\delta(\boldsymbol{v}^\delta, p^\delta) - \boldsymbol{m}^*(\boldsymbol{v}', p^\delta)) \cdot (\boldsymbol{v}^\delta - \boldsymbol{v}')\\&= \int_{\mathbb{R}^d} (\boldsymbol{m}^*(\boldsymbol{v}^\delta - \boldsymbol{y}, p^\delta) - \boldsymbol{m}^*(\boldsymbol{v}', p^\delta)) \cdot (\boldsymbol{v}^\delta - \boldsymbol{v}')\, \omega_\delta(\boldsymbol{y})\, d\boldsymbol{y}\\&\geq \int_{\mathbb{R}^d} (\boldsymbol{m}^*(\boldsymbol{v}^\delta - \boldsymbol{y}, p^\delta) - \boldsymbol{m}^*(\boldsymbol{v}', p^\delta)) \cdot \boldsymbol{y}\, \omega_\delta(\boldsymbol{y})\, d\boldsymbol{y}\\&\geq -C\delta(1 + |\boldsymbol{v}^\delta|^{r-1} + |\boldsymbol{v}'|^{r-1}).\end{aligned}$$

Therefore, for arbitrary non-negative $\varphi \in L^\infty(\Omega)$ we obtain

$$\begin{aligned}&\lim_{\delta\to 0_+} \int_\Omega (\boldsymbol{m}_\delta(\boldsymbol{v}^\delta, p^\delta) - \boldsymbol{m}^*(\boldsymbol{v}', p^\delta)) \cdot (\boldsymbol{v}^\delta - \boldsymbol{v}')\varphi\, dx\\&\quad\geq -C \lim_{\delta\to 0_+} \delta \int_\Omega (1 + |\boldsymbol{v}^\delta|^{r-1} + |\boldsymbol{v}'|^{r-1})\varphi\, dx = 0.\end{aligned} \tag{5.22}$$

Finally, since $\boldsymbol{m}^*(\boldsymbol{v}', \cdot)$ is continuous and bounded and p^δ converges strongly, we can combine (5.22) and (5.21) to get

$$\begin{aligned}0 &\leq \lim_{\delta\to 0_+} \int_\Omega (\boldsymbol{m}_\delta(\boldsymbol{v}^\delta, p^\delta) - \boldsymbol{m}^*(\boldsymbol{v}', p^\delta)) \cdot (\boldsymbol{v}^\delta - \boldsymbol{v}')\varphi\, dx\\&= \int_\Omega (\boldsymbol{m} - \boldsymbol{m}^*(\boldsymbol{v}', p)) \cdot (\boldsymbol{v} - \boldsymbol{v}')\varphi\, dx.\end{aligned}$$

Since φ is arbitrary non-negative, we conclude that for almost all $x \in \Omega$ and almost all $\boldsymbol{v}' \in \mathbb{R}^d$ there holds

$$0 \leq (\boldsymbol{m}(x) - \boldsymbol{m}^*(\boldsymbol{v}', p(x))) \cdot (\boldsymbol{v}(x) - \boldsymbol{v}').$$

Applying now [23, Lemma 1], we deduce that for almost all $x \in \Omega$ we have $(\boldsymbol{m}, \boldsymbol{v}, p) \in \mathcal{A}$. □

5.2.5 *Quasi-Compressible δ-Approximation and δ-Mollification*

In this part, we show the existence of a solution to the so-called quasi-compressible approximation with regularized graph $\mathcal{A}_\delta$, which reads as

$$\begin{aligned} \nabla p + \boldsymbol{m}_\delta(\boldsymbol{v}, p) &= \boldsymbol{f} && \text{in } \Omega, \\ \operatorname{div} \boldsymbol{v} &= \delta \operatorname{div}(|\nabla(p - p_0)|^{r'-2}\nabla(p - p_0)) && \text{in } \Omega, \\ (\boldsymbol{v} - \boldsymbol{v}_0) \cdot \boldsymbol{n} &= \delta|\nabla(p - p_0)|^{r'-2}\nabla(p - p_0) \cdot \boldsymbol{n} && \text{on } \Gamma_1, \\ p - p_0 &= 0 && \text{on } \Gamma_2. \end{aligned} \tag{5.23}$$

Equation $(5.23)_3$ is the quasi-compressible approximation. Indeed, letting formally $\delta \to 0_+$ we obtain $\operatorname{div} \boldsymbol{v} = 0$. Further, we mollified the graph $\mathcal{A}$ by the parameter δ in order to use the standard fix point theorem for continuous mappings.

We focus now on the existence of a solution to the above problem. The proof is based on the Galerkin approximation. Let $\{\boldsymbol{w}_i\}_{i\in\mathbb{N}} \subset L^r(\Omega)$ and $\{q_i\}_{i\in\mathbb{N}} \subset W^{1,r'}_{\Gamma_2}(\Omega)$ be linearly independent, with linear spans dense in $L^r(\Omega)$ and $W^{1,r'}_{\Gamma_2}(\Omega)$, respectively. We introduce the Galerkin approximation, which takes the form: for $n \in \mathbb{N}$ to find

$$\boldsymbol{v}_n(x) = \boldsymbol{v}_0(x) + \sum_{i=1}^{n} a_n^i \boldsymbol{w}_i(x), \tag{5.24}$$

$$p_n(x) = p_0(x) + \sum_{i=1}^{n} b_n^i q_i(x), \tag{5.25}$$

satisfying for all $i = 1, \ldots, n$

$$\int_\Omega \nabla p_n \cdot \boldsymbol{w}_i + \int_\Omega \boldsymbol{m}_\delta(\boldsymbol{v}_n, p_n) \cdot \boldsymbol{w}_i = \int_\Omega \boldsymbol{f} \cdot \boldsymbol{w}_i, \tag{5.26}$$

$$\delta \int_\Omega |\nabla(p_n - p_0)|^{r'-2}\nabla(p_n - p_0) \cdot \nabla q_i = \int_\Omega (\boldsymbol{v}_n - \boldsymbol{v}_0) \cdot \nabla q_i. \tag{5.27}$$

The system (5.26)–(5.27) is the system of $2n$ algebraic equations for unknowns $\{(a_n^i, b_n^i)\}_{i=1}^n$.

The existence of a solution to the above approximative problem follows from the application of a corollary of the Brouwer fix point theorem [40, Lemma 4.3] and is a consequence of the a priori estimates deduced below. Multiplying $(5.26)_i$ by a_n^i and $(5.27)_i$ by b_n^i and summing the resultant $2n$ equalities over $i = 1, \ldots, n$, we obtain

$$\int_\Omega \nabla p_n \cdot (\boldsymbol{v}_n - \boldsymbol{v}_0) + \int_\Omega \boldsymbol{m}_\delta(\boldsymbol{v}_n, p_n) \cdot (\boldsymbol{v}_n - \boldsymbol{v}_0) = \int_\Omega \boldsymbol{f} \cdot (\boldsymbol{v}_n - \boldsymbol{v}_0), \tag{5.28}$$

$$\delta \int_\Omega |\nabla(p_n - p_0)|^{r'} = \int_\Omega (\boldsymbol{v}_n - \boldsymbol{v}_0) \cdot \nabla(p_n - p_0). \tag{5.29}$$

Thus, summing the resulting equation, we have

$$\delta\|\nabla(p_n - p_0)\|_{r'}^{r'} + \int_\Omega \boldsymbol{m}_\delta(\boldsymbol{v}_n, p_n)\cdot(\boldsymbol{v}_n - \boldsymbol{v}_0) = \int_\Omega (\boldsymbol{f} - \nabla p_0)\cdot(\boldsymbol{v}_n - \boldsymbol{v}_0). \tag{5.30}$$

If we consider $\delta \in (0, 1)$, we can recall Lemma 5.2.5 for (r, r')-coercivity of $\boldsymbol{m}_\delta$, the Hölder and the Young inequalities to deduce

$$\delta\|\nabla(p_n - p_0)\|_{r'}^{r'} + \|\boldsymbol{m}_\delta(\boldsymbol{v}_n, p_n)\|_{r'}^{r'} + \|\boldsymbol{v}_n\|_r^r \le C(\|\boldsymbol{f} - \nabla p_0\|_{r'}, \|\boldsymbol{v}_0\|_r). \tag{5.31}$$

In particular, the constant C is independent of δ and n and this allows us to let $n \to \infty$ in (5.26)–(5.27).

Due to the reflexivity of the underlying spaces and thanks to compact embedding, we can extract a subsequence that we do not relabel, such that

$$\begin{aligned}
\boldsymbol{v}_n &\rightharpoonup \boldsymbol{v} && \text{weakly in } L^r(\Omega),\\
p_n - p_0 &\rightharpoonup p - p_0 && \text{weakly in } W^{1,r'}_{\Gamma_2}(\Omega),\\
p_n - p_0 &\to p - p_0 && \text{strongly in } L^{r'}(\Omega),\\
|\nabla(p_n - p_0)|^{r'-2}\nabla(p_n - p_0) &\rightharpoonup \boldsymbol{\chi} && \text{weakly in } L^r(\Omega),\\
\boldsymbol{m}_\delta(\boldsymbol{v}_n, p_n) &\rightharpoonup \boldsymbol{m} && \text{weakly in } L^{r'}(\Omega).
\end{aligned} \tag{5.32}$$

Convergence results (5.32) allow us to pass to the limit in (5.26)–(5.27). Indeed, using the density property of $\{\boldsymbol{w}_i\}$ in $L^r(\Omega)^d$ and $\{q_i\}$ in $W^{1,r'}_{\Gamma_2}(\Omega)$, we obtain

$$\begin{aligned}
\int_\Omega \nabla p\cdot\boldsymbol{w} + \int_\Omega \boldsymbol{m}\cdot\boldsymbol{w} &= \int_\Omega \boldsymbol{f}\cdot\boldsymbol{w}, && \forall \boldsymbol{w} \in L^r(\Omega),\\
\delta\int_\Omega \boldsymbol{\chi}\cdot\nabla q &= \int_\Omega (\boldsymbol{v} - \boldsymbol{v}_0)\cdot\nabla q, && \forall q \in W^{1,r'}_{\Gamma_2}(\Omega).
\end{aligned} \tag{5.33}$$

Moreover, it follows from the weak lower semicontinuity and (5.31) that

$$\delta\|\nabla(p - p_0)\|_{r'}^{r'} + \|\boldsymbol{m}\|_{r'}^{r'} + \|\boldsymbol{v}\|_r^r \le C(\|\boldsymbol{f} - \nabla p_0\|_{r'}, \|\boldsymbol{v}_0\|_r). \tag{5.34}$$

Our goal is to show that $\boldsymbol{m} = \boldsymbol{m}_\delta(\boldsymbol{v}, p)$ and $\boldsymbol{\chi} = |\nabla(p - p_0)|^{r'-2}\nabla(p - p_0)$ almost everywhere in Ω. We will use the standard monotone operator theory, namely the Minty method [44].

First, from (5.33) we deduce

$$\delta \int_\Omega \chi \cdot \nabla(p - p_0) + \int_\Omega \boldsymbol{m} \cdot (\boldsymbol{v} - \boldsymbol{v}_0) = \int_\Omega (\boldsymbol{f} - \nabla p_0) \cdot (\boldsymbol{v} - \boldsymbol{v}_0). \tag{5.35}$$

Next, we let $n \to \infty$ in (5.30), use the convergence results (5.32), and compare the result with (5.35) to conclude

$$\lim_{n\to\infty} \delta \int_\Omega |\nabla(p_n - p_0)|^{r'} + \int_\Omega \boldsymbol{m}_\delta(\boldsymbol{v}_n, p_n) \cdot \boldsymbol{v}_n = \delta \int_\Omega \chi \cdot \nabla(p - p_0) + \int_\Omega \boldsymbol{m} \cdot \boldsymbol{v}. \tag{5.36}$$

This is the usual starting point for the application of the Minty method. Hence, let $\boldsymbol{w} \in L^r(\Omega)$ and $\boldsymbol{u} \in L^{r'}(\Omega)$ be arbitrary. Then it follows from the continuity of $\boldsymbol{m}_\delta$, **(A4)**, and the strong convergence of p_n that

$$\boldsymbol{m}_\delta(\boldsymbol{w}, p_n) \to \boldsymbol{m}_\delta(\boldsymbol{w}, p) \qquad \text{in } L^{r'}(\Omega). \tag{5.37}$$

Then, we use the monotonicity of the corresponding operators to observe

$$\begin{aligned}
0 \le\ & \delta \int_\Omega \left(|\nabla(p_n - p_0)|^{r'-2} \nabla(p_n - p_0) - |\boldsymbol{u}|^{r'-2} \boldsymbol{u} \right) \cdot (\nabla(p_n - p_0) - \boldsymbol{u}) \\
& + \int_\Omega \left(\boldsymbol{m}_\delta(\boldsymbol{v}_n, p_n) - \boldsymbol{m}_\delta(\boldsymbol{w}, p_n) \right) \cdot (\boldsymbol{v}_n - \boldsymbol{w}) \\
=\ & \delta \int_\Omega |\nabla(p_n - p_0)|^{r'} + \int_\Omega \boldsymbol{m}_\delta(\boldsymbol{v}_n, p_n) \cdot \boldsymbol{v}_n \\
& - \delta \int_\Omega |\nabla(p_n - p_0)|^{r'-2} \nabla(p_n - p_0) \cdot \boldsymbol{u} + |\boldsymbol{u}|^{r'-2} \boldsymbol{u} \cdot (\nabla(p_n - p_0) - \boldsymbol{u}) \\
& - \int_\Omega \boldsymbol{m}_\delta(\boldsymbol{v}_n, p_n) \cdot \boldsymbol{w} + \boldsymbol{m}_\delta(\boldsymbol{w}, p_n) \cdot (\boldsymbol{v}_n - \boldsymbol{w}).
\end{aligned} \tag{5.38}$$

Finally, we use (5.32), (5.36), and (5.37) and let $n \to \infty$ in (5.38) to get

$$0 \le \delta \int_\Omega \left(\chi - |\boldsymbol{u}|^{r'-2} \boldsymbol{u} \right) \cdot (\nabla(p - p_0) - \boldsymbol{u}) + \int_\Omega \left(\boldsymbol{m} - \boldsymbol{m}_\delta(\boldsymbol{w}, p) \right) \cdot (\boldsymbol{v} - \boldsymbol{w}).$$

Setting $\boldsymbol{w} := \boldsymbol{v} \pm \varepsilon \bar{\boldsymbol{w}}$ and $\boldsymbol{u} := \nabla(p - p_0) \pm \varepsilon \bar{\boldsymbol{u}}$ with arbitrary $\bar{\boldsymbol{w}} \in L^r(\Omega)$ and arbitrary $\bar{\boldsymbol{u}} \in L^{r'}(\Omega)$ in the above equation, dividing by ε, and letting $\varepsilon \to 0_+$ leads to

$$0 = \delta \int_\Omega \left(\chi - |\nabla(p - p_0)|^{r'-2} \nabla(p - p_0) \right) \cdot \bar{\boldsymbol{u}} + \int_\Omega \left(\boldsymbol{m} - \boldsymbol{m}_\delta(\boldsymbol{v}, p) \right) \cdot \bar{\boldsymbol{w}} \tag{5.39}$$

and consequently

$$\chi = |\nabla(p - p_0)|^{r'-2}\nabla(p - p_0) \qquad \text{and} \qquad \boldsymbol{m} = \boldsymbol{m}_\delta(\boldsymbol{v}, p) \tag{5.40}$$

almost everywhere in Ω, which shows the existence of solution to (5.23).

5.2.6 δ-Limit

The last step is to let $\delta \to 0+$ in (5.23) to obtain the solution of the original problem. We denote $(\boldsymbol{v}_\delta, p_\delta)$ a solution constructed in the previous section. Recalling (5.34) and (5.40) we have

$$\delta \|\nabla(p_\delta - p_0)\|_{r'}^{r'} + \|\boldsymbol{m}_\delta(\boldsymbol{v}_\delta, p_\delta)\|_{r'}^{r'} + \|\boldsymbol{v}_\delta\|_r^r \le C(\|\boldsymbol{f} - \nabla p_0\|_{r'}, \|\boldsymbol{v}_0\|_r). \tag{5.41}$$

Similarly as before, we have

$$\begin{aligned} \boldsymbol{v}_\delta &\rightharpoonup \boldsymbol{v} && \text{weakly in } L^r(\Omega), \\ \delta\, |\nabla(p_\delta - p_0)|^{r'-2}\nabla(p_\delta - p_0) &\to \boldsymbol{0} && \text{strongly in } L^r(\Omega), \\ \boldsymbol{m}_\delta(\boldsymbol{v}_\delta, p_\delta) &\rightharpoonup \boldsymbol{m} && \text{weakly in } L^{r'}(\Omega). \end{aligned} \tag{5.42}$$

Furthermore, it follows from $(5.33)_1$ that

$$\nabla p_\delta = \boldsymbol{f} - \boldsymbol{m}_\delta(\boldsymbol{v}_\delta, p_\delta) \text{ a.e. in } \Omega.$$

Consequently, using (5.42) we see that the right-hand side converges weakly in $L^{r'}(\Omega)$ and since $p_\delta = p_0$ on Γ_2 and by using the compact embedding we also obtain that

$$\begin{aligned} p_\delta - p_0 &\rightharpoonup p - p_0 && \text{weakly in } W^{1,r'}_{\Gamma_2}(\Omega), \\ p_\delta - p_0 &\to p - p_0 && \text{strongly in } L^{r'}(\Omega). \end{aligned} \tag{5.43}$$

Thus, we can let $\delta \to 0_+$ in (5.33) (recall here, we use (5.33) with $(\boldsymbol{v}, p)$ replaced by $(\boldsymbol{v}_\delta, p_\delta)$, to deduce

$$\begin{aligned} \nabla p + \boldsymbol{m} &= \boldsymbol{f} && \text{a.e. in } \Omega, \\ \int_\Omega (\boldsymbol{v} - \boldsymbol{v}_0) \cdot \nabla q &= 0, && \forall q \in W^{1,r'}_{\Gamma_2}(\Omega). \end{aligned} \tag{5.44}$$

Clearly, $(5.44)_2$ gives $(\boldsymbol{v} - \boldsymbol{v}_0) \in L^r_{\mathrm{div},\Gamma_1}(\Omega)$. Thus, it just remains to check that $(\boldsymbol{v}, \boldsymbol{m}, p) \in \mathcal{A}$ almost everywhere in Ω. For this purpose, we use Lemma 5.2.5 and we see that it is enough to check that

$$\limsup_{\delta \to 0_+} \int_\Omega \boldsymbol{m}_\delta(\boldsymbol{v}_\delta, p_\delta) \cdot \boldsymbol{v}_\delta \, dx \le \int_\Omega \boldsymbol{m} \cdot \boldsymbol{v} \, dx. \tag{5.45}$$

We use subsequently the weak convergence results (5.42) and (5.33) and (5.40). Hence,

$$\begin{aligned}
&\limsup_{\delta \to 0_+} \int_\Omega \boldsymbol{m}_\delta(\boldsymbol{v}_\delta, p_\delta) \cdot \boldsymbol{v}_\delta \, dx \\
&\overset{(5.42)}{=} \limsup_{\delta \to 0_+} \int_\Omega \boldsymbol{m}_\delta(\boldsymbol{v}_\delta, p_\delta) \cdot (\boldsymbol{v}_\delta - \boldsymbol{v}_0) \, dx + \int_\Omega \boldsymbol{m} \cdot \boldsymbol{v}_0 \, dx \\
&\overset{(5.33),(5.42)}{=} \limsup_{\delta \to 0_+} \int_\Omega -\nabla p_\delta \cdot (\boldsymbol{v}_\delta - \boldsymbol{v}_0) \, dx + \int_\Omega \boldsymbol{f} \cdot (\boldsymbol{v} - \boldsymbol{v}_0) + \boldsymbol{m} \cdot \boldsymbol{v}_0 \, dx \\
&\overset{(5.42)}{=} \limsup_{\delta \to 0_+} \int_\Omega -\nabla (p_\delta - p) \cdot (\boldsymbol{v}_\delta - \boldsymbol{v}_0) \, dx + \int_\Omega (\boldsymbol{f} - \nabla p) \cdot (\boldsymbol{v} - \boldsymbol{v}_0) + \boldsymbol{m} \cdot \boldsymbol{v}_0 \, dx \\
&\overset{(5.42),(5.44)_1}{=} \limsup_{\delta \to 0_+} \int_\Omega -\nabla (p_\delta - p_0) \cdot (\boldsymbol{v}_\delta - \boldsymbol{v}_0) \, dx \\
&\quad + \int_\Omega -\nabla (p_0 - p) \cdot (\boldsymbol{v} - \boldsymbol{v}_0) \, dx + \int_\Omega \boldsymbol{m} \cdot \boldsymbol{v} \, dx \\
&\overset{(5.33)_2,(5.44)_2}{=} \delta \limsup_{\delta \to 0_+} \int_\Omega -|\nabla (p_\delta - p_0)|^{r'} \, dx + \int_\Omega \boldsymbol{m} \cdot \boldsymbol{v} \, dx \le \int_\Omega \boldsymbol{m} \cdot \boldsymbol{v} \, dx.
\end{aligned}$$

Therefore, we can use Lemma 5.2.5 and conclude that the triple $(\boldsymbol{m}, \boldsymbol{v}, p)$ is a solution to (5.11).

5.2.7 Maximum and Minimum Principle

Here, we focus on the qualitative properties of solution when $\boldsymbol{f}$ is conservative. It means we want to study the problem

$$\begin{aligned}
\nabla p + \boldsymbol{m} &= \nabla g && \text{in } \Omega, \\
\operatorname{div} \boldsymbol{v} &= 0 && \text{in } \Omega, \\
\boldsymbol{h}(\boldsymbol{m}, \boldsymbol{v}, p) &= \boldsymbol{0} && \text{in } \Omega, \\
(\boldsymbol{v} - \boldsymbol{v}_0) \cdot \boldsymbol{n} &= 0 && \text{on } \Gamma_1, \\
p - p_0 &= 0 && \text{on } \Gamma_2
\end{aligned} \tag{5.46}$$

with $\boldsymbol{h}$ inducing the graph $\mathcal{A}$. Our goal is to prove Theorem 5.2.2. We denote

$$a := \operatorname*{ess\,sup}_{x\in\Gamma_2}(p_0(x) - g(x)) < \infty.$$

Then for arbitrary $k \geq 0$ we define

$$\boldsymbol{w} = T_k((p - g - a)_+)\boldsymbol{v},$$

where we denoted $z_+ := \max\{0, z\}$ for arbitrary $z \in \mathbb{R}$ and T_k is the standard cut-off function defined in (5.10). In addition, we introduce

$$\Theta_k(t) := \int_0^t T_k(s)\,ds.$$

It follows from the definition that $\Theta_k((p - g - a)_+) \in W^{1,r'}_{\Gamma_2}(\Omega)$ is non-negative and that $\nabla\Theta_k((p - g - a)_+) = T_k((p - g - a)_+)\nabla(p - g)$. In addition, since $T_k((p - g - a)_+)$ is bounded for every fix k we have that $\boldsymbol{w} \in L^r(\Omega)$ for every fix k. Next, since $(\mathbf{0}, \mathbf{0}, p) \in \mathcal{A}$ and the graph is monotone, we also know that $\boldsymbol{m} \cdot \boldsymbol{v} \geq 0$ a.e. in Ω.

Taking the scalar product of (5.46) with $\boldsymbol{w}$ and integrating over Ω, we obtain

$$\begin{aligned}\int_\Omega |\boldsymbol{m} \cdot \boldsymbol{v} T_k((p - g - a)_+)|\,dx &= \int_\Omega \boldsymbol{m} \cdot \boldsymbol{v} T_k((p - g - a)_+)\,dx\\
&= -\int_\Omega T_k((p - g - a)_+)\nabla(p - g) \cdot \boldsymbol{v}\,dx = -\int_\Omega \nabla\Theta_k((p - g - a)_+) \cdot \boldsymbol{v}\,dx\\
&= -\int_{\Gamma_1} \Theta_k((p - g - a)_+)\boldsymbol{v}_0 \cdot \boldsymbol{n}\,dS \leq 0,\end{aligned} \tag{5.47}$$

where we used the assumption $\boldsymbol{v}_0 \cdot \boldsymbol{n} \geq 0$ on Γ_1. Thus,

$$\boldsymbol{m} \cdot \boldsymbol{v}\chi_{\{p-g-a\geq 0\}} = 0 \qquad \text{a.e. in } \Omega. \tag{5.48}$$

Consequently, using **(A4)**, we get

$$|\boldsymbol{m}|\chi_{\{p-g-a\geq 0\}} \leq C.$$

Therefore, it follows from $(5.46)_1$ that

$$|\nabla(p - g - a)_+| = |\nabla(p - g)|\chi_{\{p-g-a\geq 0\}} = |\boldsymbol{m}|\chi_{\{p-g-a\geq 0\}} \leq C.$$

Consequently, $(p - g - a)_+$ is Lipschitz and therefore p must be bounded from above, which finishes the proof of (iii) in Theorem 5.2.2. The statement (iv) can be proved similarly.

Let us now consider that $\mathcal{A}$ satisfy **(A6)**. Then (5.48) implies that

$$|\boldsymbol{m}|\chi_{\{p-g-a\geq 0\}} = 0 \quad \text{a.e. in } \Omega.$$

Consequently, we deduce from $(5.46)_1$ that

$$\nabla(p - g - a)_+ = 0 \quad \text{a.e. in } \Omega$$

and since Ω is connected we have that $(p - g - a)_+$ is constant. Using also the fact that $(p-g-a)_+ \in W^{1,r'}_{\Gamma_2}(\Omega)$, we get that $(p-g-a)_+ = 0$ and (i) in Theorem 5.2.2 follows. The property (ii) is proven similarly.

Finally, we show the existence of solution satisfying (i) and (ii) even if $\mathcal{A}$ does not satisfy **(A6)**. To do so, we first modify graph $\mathcal{A}$ to the graph $\mathcal{A}_h$, which satisfies **(A6)**, for which we will prove (i) and then let $h \to 0+$ to get the existence of a solution to the original problem. Since the estimate in (i), (ii) depends only on p_0 and g, the limiting solution will satisfy (i), (ii) as well. In what follows, we focus on the existence of solution satisfying (i). The proof for (ii) or for both of them is done in the same way.

Let us first introduce

$$\mathcal{A}_h := \{(\boldsymbol{m}, \boldsymbol{v}, p) \subset \mathbb{R}^d \times \mathbb{R}^d \times \mathbb{R} \mid \exists \boldsymbol{v}' \in \mathbb{R}^d,\ (\boldsymbol{m}, \boldsymbol{v}', p) \in \mathcal{A},\ \boldsymbol{v} = \boldsymbol{v}' + h|\boldsymbol{m}|^{r'-2}\boldsymbol{m}\}.$$

This graph now satisfies **(A1)–(A6)**. Therefore we can use Theorem 5.2.1 to get the existence of solution $(\boldsymbol{m}_h, \boldsymbol{v}_h, p_h)$ to

$$\begin{aligned} \nabla p_h + \boldsymbol{m}_h &= \nabla g && \text{in } \Omega, \\ \operatorname{div} \boldsymbol{v} &= 0 && \text{in } \Omega, \\ (\boldsymbol{m}_h, \boldsymbol{v}_h, p_h) &\in \mathcal{A}_h && \text{in } \Omega, \\ (\boldsymbol{v}_h - \boldsymbol{v}_0) \cdot \boldsymbol{n} &= 0 && \text{on } \Gamma_1, \\ p_h - p_0 &= 0 && \text{on } \Gamma_2. \end{aligned} \tag{5.49}$$

Repeating the proof of Theorem 5.2.1, we also have

$$\|\boldsymbol{v}_h\|_r + \|\boldsymbol{m}_h\|_{r'} + \|p_h - p_0\|_{1,r'} \leq C.$$

In addition, having **(A6)** we also get

$$p_h - g \leq \operatorname*{ess\,sup}_{x\in\Gamma_2}(p_0(x) - g(x)) \qquad \text{a.e. in } \Omega \tag{5.50}$$

Next, we repeat the proof of Theorem 5.2.1 and obtain

$$\begin{aligned} \boldsymbol{v}_h &\rightharpoonup \boldsymbol{v} && \text{weakly in } L^r(\Omega), \\ \boldsymbol{m}_h &\rightharpoonup \boldsymbol{m} && \text{weakly in } L^{r'}(\Omega), \\ p_h - p_0 &\rightharpoonup p - p_0 && \text{weakly in } W^{1,r'}_{\Gamma_2}(\Omega), \\ p_h - p_0 &\to p - p_0 && \text{strongly in } L^{r'}(\Omega). \end{aligned} \tag{5.51}$$

Hence, we can let $h \to 0_+$ in (5.49) to get that $(\boldsymbol{m}, \boldsymbol{v})$ satisfies (5.46) provided we show that $(\boldsymbol{m}, \boldsymbol{v}) \in \mathcal{A}$ almost everywhere in Ω. To do so, we define $\boldsymbol{v}'_h := \boldsymbol{v}_h - h|\boldsymbol{m}_h|^{r-2}\boldsymbol{m}_h$. Then, it follows from the definition of $\mathcal{A}_h$ that $(\boldsymbol{m}_h, \boldsymbol{v}'_h, p_h) \in \mathcal{A}$. In addition, it follows from (5.51) that

$$\boldsymbol{v}'_h \rightharpoonup \boldsymbol{v} \quad \text{in } L^r(\Omega). \tag{5.52}$$

Next, we check that

$$\limsup_{h\to 0_+} \int_\Omega \boldsymbol{m}_h \cdot \boldsymbol{v}'_h \, dx \le \int_\Omega \boldsymbol{m} \cdot \boldsymbol{v} \, dx. \tag{5.53}$$

To do so, we use $(5.49)_1$, $(5.46)_1$, and (5.51) to deduce

$$\begin{aligned} &\limsup_{h\to 0_+} \int_\Omega \boldsymbol{m}_h \cdot \boldsymbol{v}'_h \, dx = \limsup_{h\to 0_+} \int_\Omega \boldsymbol{m}_h \cdot \boldsymbol{v}_h - h|\boldsymbol{m}_h|^{r'} \, dx \\ &\le \limsup_{h\to 0_+} \int_\Omega \boldsymbol{m}_h \cdot \boldsymbol{v}_h \, dx = \limsup_{h\to 0_+} \int_\Omega (\nabla g - \nabla p_h) \cdot \boldsymbol{v}_h \, dx \\ &= \int_\Omega \nabla(g - p_0) \cdot \boldsymbol{v} \, dx + \limsup_{h\to 0_+} \int_\Omega \nabla(p_0 - p_h) \cdot \boldsymbol{v}_h \, dx \\ &= \int_\Omega \nabla(g - p_0) \cdot \boldsymbol{v} \, dx + \limsup_{h\to 0_+} \int_\Omega \nabla(p_0 - p_h) \cdot \boldsymbol{v}_0 \, dx \\ &= \int_\Omega \nabla(g - p_0) \cdot \boldsymbol{v} \, dx + \int_\Omega \nabla(p_0 - p) \cdot \boldsymbol{v}_0 \, dx = \int_\Omega \boldsymbol{m} \cdot \boldsymbol{v} \, dx, \end{aligned}$$

where we used the fact that $(\boldsymbol{v}_h - \boldsymbol{v}_0)$ as well as $(\boldsymbol{v} - \boldsymbol{v}_0)$ belong to $L^r_{\mathrm{div},\Gamma_1}(\Omega)$.

Then we mimic the proof of Lemma 5.2.5. Thus let $\bar{\boldsymbol{v}} \in \mathbb{R}^d$ be arbitrary. Then using (5.51) and the assumption on $\boldsymbol{m}^*$ we have

$$m^*(\bar{\boldsymbol{v}}, p_h) \to \boldsymbol{m}^*(\bar{\boldsymbol{v}}, p) \quad \text{strongly in } L^r(\Omega).$$

Consequently, it follows from (5.51) and (5.53) that

$$\limsup_{h\to 0_+} \int_\Omega |(\boldsymbol{m} - \boldsymbol{m}^*(\bar{\boldsymbol{v}}, p_h)) \cdot (\boldsymbol{v}'_h - \bar{\boldsymbol{v}})|\, dx$$

$$= \limsup_{h\to 0_+} \int_\Omega (\boldsymbol{m} - \boldsymbol{m}^*(\bar{\boldsymbol{v}}, p_h)) \cdot (\boldsymbol{v}'_h - \bar{\boldsymbol{v}})\, dx \le 0.$$

Repeating now step by step the second part of the proof of Lemma 5.2.5, we deduce that $(\boldsymbol{m}, \boldsymbol{v}) \in \mathcal{A}$ almost everywhere in Ω. Thus, the proof is complete.

5.2.8 Proof of Theorems 5.2.3–5.2.4

Here, we first focus on the proof of Theorem 5.2.3. Let us denote

$$K := \|g\|_{\infty,\Omega} + \|p_0 - g\|_{\infty,\Gamma_2}$$

and recall the definition of T_K in (5.10). Then we define the problem

$$\begin{aligned} \nabla p + \boldsymbol{m} &= \nabla g && \text{in } \Omega,\\ \operatorname{div} \boldsymbol{v} &= 0 && \text{in } \Omega,\\ \boldsymbol{h}(\boldsymbol{m}, \boldsymbol{v}, T_K(p)) &= \boldsymbol{0} && \text{in } \Omega,\\ (\boldsymbol{v} - \boldsymbol{v}_0) \cdot \boldsymbol{n} &= 0 && \text{on } \Gamma_1,\\ p - p_0 &= 0 && \text{on } \Gamma_2. \end{aligned} \tag{5.54}$$

Using the assumptions imposed on the graph $\mathcal{A}$, we see that the graph $\mathcal{A}_K$ induced by the null points

$$\boldsymbol{h}(\boldsymbol{m}, \boldsymbol{v}, T_K(p)) = \boldsymbol{0}$$

satisfies **(A1)**–**(A5)** with one proviso, namely with
(A4) $\exists c_2 \ge 0\ \forall (\boldsymbol{m}, \boldsymbol{v}, p) \in \mathcal{A} : \boldsymbol{m} \cdot \boldsymbol{v} \ge c_1(K)(|\boldsymbol{v}|^r + |\boldsymbol{m}|^{r'}) - c_2(K)$.
Therefore, we can use Theorem 5.2.1 to establish the existence of a solution to (5.54). In addition, using (5.2.2) we can construct a solution that fulfills

$$\|p\|_\infty \le K.$$

Consequently, we have that $T_K(p) = p$ almost everywhere in Ω and it is evident that the constructed solution of (5.54) is also a solution to (5.11).

The proof of Theorem 5.2.4 is done similarly. Let us illustrate it quickly in the case that $p_0 - g \ge a$ and $\boldsymbol{v}_0 \cdot \boldsymbol{n} \le 0$ on Γ_1. Then we can replace $(5.54)_3$ by

$$\boldsymbol{h}(\boldsymbol{m}, \boldsymbol{v}, \max\{T_K(p), p\}) = 0,$$

where $K \leq |a - \|(g)_-\|_\infty|$. Consequently, we can use Theorem 5.2.1 to get the existence of a solution and then apply (i) from Theorem 5.2.2 to obtain that $p \geq -K$ almost everywhere in Ω, which finishes the proof.

5.3 Generalized Brinkman–Darcy–Forchheimer Model

We recall here the basic setting of the problem, we study in this part. It consists of the system (5.2), boundary conditions (5.7), and the implicit constitutive relations (5.5)–(5.6), which summarized takes the form

$$\begin{aligned} -\operatorname{div} \boldsymbol{S} + \boldsymbol{m} &= -\nabla p + \boldsymbol{f} && \text{in } \Omega, \\ \boldsymbol{S} &= \boldsymbol{S}^T && \text{in } \Omega, \\ \operatorname{div} \boldsymbol{v} &= 0 && \text{in } \Omega, \\ \boldsymbol{h}(\boldsymbol{m}, \boldsymbol{v}, \boldsymbol{D}\boldsymbol{v}, p) &= \boldsymbol{0} && \text{in } \Omega, \\ \boldsymbol{H}(\boldsymbol{S}, \boldsymbol{D}\boldsymbol{v}, p) &= \boldsymbol{0} && \text{in } \Omega, \\ \boldsymbol{v} &= \boldsymbol{v}_0 && \text{on } \Gamma_1, \\ (\boldsymbol{S} - p\boldsymbol{I})\boldsymbol{n} &= \boldsymbol{s}_0 && \text{on } \Gamma_2. \end{aligned} \tag{5.55}$$

Our aim is to establish the existence of a solution under suitable assumptions on the implicit constitutive relations for $\boldsymbol{h}$ and $\boldsymbol{H}$. In particular, we want to combine all models (N2)–(N5) and study them[5] simultaneously. To simplify the presentation we neglected the convective term in the equation and consider only the case when $\Gamma_1 \neq \emptyset$ to preserve the Korn inequality. However, in case $\Gamma_1 = \emptyset$ one could still solve the problem by assuming zero mean value for the velocity field. Similarly, we assume that $\Gamma_2 \neq \emptyset$ in order to have prescribed pressure. However, also the case when $\Gamma_2 = \emptyset$ can be treated easily but one has to prescribe a mean value[6] of the pressure over some set $\Omega_0 \subset \Omega$. To simplify the formulation of a weak solution and to avoid technical difficulties, we shall assume that $\boldsymbol{f}$ and $\boldsymbol{s}_0$ satisfy

$$\boldsymbol{f} = -\operatorname{div} \boldsymbol{F} \text{ in } \Omega, \qquad \boldsymbol{F}\boldsymbol{n} = \boldsymbol{s}_0 \text{ on } \Gamma_2, \tag{5.56}$$

for some given tensor-valued function $\boldsymbol{F} : \Omega \to \mathbb{R}^{d\times d}$.

---

[5]The generalization of (N1) was already studied in the preceding section.

[6]Ideally, one would like to fix the pressure at one point. But since we deal only with Lebesgue integrable functions, such procedure cannot be applied. Nevertheless, we can try to mimic the fixing the value of p at some x_0 just by fixing the integral

$$\int_{B(x_0)} p\,dx$$

over some very small ball B centered at x_0.

At the end of Sect. 5.3, we give a brief overview of available results with convective term. Note that the method presented here is sufficient for obtaining the existence of solution also with the convective term under one condition—we need to be able to derive the uniform a priori estimates. While this is rather standard in our case, in case that the convective term is presented we need to deal with an additional boundary integral, which usually makes difficulties and cannot be treated easily.

5.3.1 Assumptions on h and $\mathcal{H}$

Similarly as for the Darcy or the generalized Darcy problem, we need to specify the assumption on the implicit relationships $\boldsymbol{h}$ and $\boldsymbol{H}$. We again identify the null points of $\boldsymbol{h}$ and $\boldsymbol{H}$ with graphs, i.e.,

$$\boldsymbol{h}(\boldsymbol{m}, \boldsymbol{v}, p, \boldsymbol{D}) = \boldsymbol{0} \Leftrightarrow (\boldsymbol{m}, \boldsymbol{v}, p, \boldsymbol{D}) \in \mathcal{A} \qquad \boldsymbol{H}(\boldsymbol{S}, \boldsymbol{D}, p) = \boldsymbol{0} \Leftrightarrow (\boldsymbol{S}, \boldsymbol{D}, p) \in \mathcal{B}$$

and instead of $(5.55)_4$–$(5.55)_5$ we look for

$$(\boldsymbol{S}, \boldsymbol{D}\boldsymbol{v}, p) \in \mathcal{B}, \qquad (\boldsymbol{m}, \boldsymbol{v}, \boldsymbol{D}\boldsymbol{v}, p) \in \mathcal{A} \qquad \text{a.e. in } \Omega.$$

Next, we introduce the assumptions on the graphs $\mathcal{A}$, $\mathcal{B}$. We will consider two cases. Either we shall assume that both graphs are maximally monotone, but with $\mathcal{B}$ being independent of the pressure p, or that the graph $\mathcal{B}$ ensures certain uniform monotonicity and smoothness but the graph $\mathcal{A}$ can be "almost" arbitrary, just represented by continuous possibly non-monotone functions.

5.3.1.1 Monotonicity Assumptions

The set of assumptions introduced here is inspired by the setting of the previous section. Hence for $\mathcal{A}$, we shall assume that $q \in (1, \infty)$ and say that $\mathcal{A}$ is $p, \boldsymbol{D}$ parameterized maximal monotone q graph if $\mathcal{A} \subset \mathbb{R}^d \times \mathbb{R}^d \times \mathbb{R}^{d\times d}_{\text{sym}} \times \mathbb{R}$ satisfies each of the conditions listed below:

(A1) inclusion of the origin

$$\forall p \in \mathbb{R}, \forall \boldsymbol{D} \in \mathbb{R}^{d\times d}_{\text{sym}} : (\boldsymbol{0}, \boldsymbol{0}, p, \boldsymbol{D}) \in \mathcal{A},$$

(A2) monotonicity

$$\forall (\boldsymbol{m}_1, \boldsymbol{v}_1, p, \boldsymbol{D}), (\boldsymbol{m}_2, \boldsymbol{v}_2, p, \boldsymbol{D}) \in \mathcal{A} : (\boldsymbol{m}_1 - \boldsymbol{m}_2) \cdot (\boldsymbol{v}_1 - \boldsymbol{v}_2) \geq 0,$$

(A3) maximality

$$(\boldsymbol{m}', \boldsymbol{v}', p, \boldsymbol{D}) \in \mathbb{R}^d \times \mathbb{R}^d \times \mathbb{R} \times \mathbb{R}^{d\times d}_{\mathrm{sym}},$$

$$\forall (\boldsymbol{m}, \boldsymbol{v}, p, \boldsymbol{D}) \in \mathcal{A} : (\boldsymbol{m}' - \boldsymbol{m}) \cdot (\boldsymbol{v}' - \boldsymbol{v}) \geq 0 \implies (\boldsymbol{m}', \boldsymbol{v}', p, \boldsymbol{D}) \in \mathcal{A},$$

(A4) (q, q')-coercivity for $\boldsymbol{v}$ and $\boldsymbol{m}$

$$\exists c_1 > 0, c_2 \geq 0 \, \forall (\boldsymbol{m}, \boldsymbol{v}, p, \boldsymbol{D}) \in \mathcal{A} : \boldsymbol{m} \cdot \boldsymbol{v} \geq c_1(|\boldsymbol{v}|^q + |\boldsymbol{m}|^{q'}) - c_2,$$

(A5) existence of a Carathéodory selection, i.e., $\exists \boldsymbol{m}^* : \mathbb{R}^d \times \mathbb{R} \times \mathbb{R}^{d\times d}_{\mathrm{sym}} \to \mathbb{R}^d$ such that

(i) $\boldsymbol{m}^*(\cdot, p, \boldsymbol{D}) : \mathbb{R}^d \to \mathbb{R}^d$ is measurable for every $(p, \boldsymbol{D}) \in \mathbb{R} \times \mathbb{R}^{d\times d}_{\mathrm{sym}}$,
(ii) $\boldsymbol{m}^*(\boldsymbol{v}, \cdot) : \mathbb{R} \times \mathbb{R}^{d\times d}_{\mathrm{sym}} \to \mathbb{R}^d$ is continuous for a.e. $\boldsymbol{v} \in \mathbb{R}^d$,
(iii) $\forall (\boldsymbol{v}, p, \boldsymbol{D}) \in \mathbb{R}^d \times \mathbb{R} \times \mathbb{R}^{d\times d}_{\mathrm{sym}} : (\boldsymbol{m}^*(\boldsymbol{v}, p), \boldsymbol{v}, p, \boldsymbol{D}) \in \mathcal{A}$.

Notice that from **(A4)**, we again have

$$\forall (\boldsymbol{v}, p, \boldsymbol{D}) \in \mathbb{R}^d \times \mathbb{R} \times \mathbb{R}^{d\times d}_{\mathrm{sym}} : |\boldsymbol{m}^*(\boldsymbol{v}, p, \boldsymbol{D})| \leq \frac{c_2}{r c_1} + \frac{|\boldsymbol{v}|^{r-1}}{c_1^r}. \tag{5.57}$$

For the graph $\mathcal{B}$ we shall assume the very similar assumptions, but require that it is independent of p (which is the major drawback) and strictly monotone with respect to $\boldsymbol{S}$. Furthermore, in case that $\mathcal{A}$ depends on $\boldsymbol{D}\boldsymbol{v}$ we shall also require that it is strictly monotone with respect to $\boldsymbol{D}$.

Thus, for $\mathcal{B}$ we assume that $r \in (1, \infty)$ and say that $\mathcal{B}$ is maximal monotone r graph if $\mathcal{B} \subset \mathbb{R}^{d\times d}_{\mathrm{sym}} \times \mathbb{R}^{d\times d}_{\mathrm{sym}}$ satisfies each of the conditions listed below:

(B1) inclusion of the origin

$$(\boldsymbol{0}, \boldsymbol{0}) \in \mathcal{B},$$

(B2) monotonicity

$$\forall (\boldsymbol{S}_1, \boldsymbol{D}_1), (\boldsymbol{S}_2, \boldsymbol{D}_2) \in \mathcal{B} : (\boldsymbol{S}_1 - \boldsymbol{S}_2) \cdot (\boldsymbol{D}_1 - \boldsymbol{D}_2) \geq 0,$$

(B3) maximality

$$(\boldsymbol{S}', \boldsymbol{D}') \in \mathbb{R}^{d\times d}_{\mathrm{sym}} \times \mathbb{R}^{d\times d}_{\mathrm{sym}},$$

$$\forall (\boldsymbol{S}, \boldsymbol{D}) \in \mathcal{B} : (\boldsymbol{S}' - \boldsymbol{S}) \cdot (\boldsymbol{D}' - \boldsymbol{D}) \geq 0 \implies (\boldsymbol{S}', \boldsymbol{D}') \in \mathcal{B},$$

(B4) (r, r')-coercivity for $\boldsymbol{D}$ and $\boldsymbol{S}$

$$\exists c_1 > 0, c_2 \geq 0 \, \forall (\boldsymbol{S}, \boldsymbol{D}) \in \mathcal{B} : \boldsymbol{S} \cdot \boldsymbol{D} \geq c_1(|\boldsymbol{D}|^r + |\boldsymbol{S}|^{r'}) - c_2,$$

(B5) existence of a Carathéodory selection, i.e., $\exists S^* : \mathbb{R}^{d\times d}_{\text{sym}} \to \mathbb{R}^{d\times d}_{\text{sym}}$ such that

(i) $S^*(\cdot) : \mathbb{R}^{d\times d}_{\text{sym}}$ is measurable,
(ii) $\forall D \in \mathbb{R}^{d\times d}_{\text{sym}} : (S^*(D), D) \in \mathcal{B}$.

Furthermore, in case that $\mathcal{A}$ depends on p we need that $\mathcal{B}$ is strictly monotone with respect to S (note that then the selection S^* is uniquely defined)

(B6)$_S$ strict monotonicity with respect to S

$$\forall (S_1, D_1), (S_2, D_2) \in \mathcal{B} : (S_1 - S_2)\cdot(D_1 - D_2) = 0 \quad \Longrightarrow \quad S_1 = S_2.$$

Moreover, if $\mathcal{A}$ depends on D then we need that $\mathcal{B}$ is strictly monotone with respect to D, i.e.,

(B6)$_D$ strict monotonicity with respect to D

$$\forall (S_1, D_1), (S_2, D_2) \in \mathcal{B} : (S_1 - S_2)\cdot(D_1 - D_2) = 0 \quad \Longrightarrow \quad D_1 = D_2.$$

We end this set of assumption by the consequence of **(B4)**, namely

$$\forall D \in \mathbb{R}^{d\times d} : |S^*(D)| \le \frac{c_2}{rc_1} + \frac{|D|^{r-1}}{c_1^r}. \tag{5.58}$$

5.3.1.2 Uniform Monotonicity Assumptions on $\mathcal{B}$—Continuity of $\mathcal{A}$

For the class of models here, we introduce much more restrictive assumptions on $\mathcal{B}$ but the key advantage will be that the graph $\mathcal{B}$ may depend on the pressure, which is not allowed for the assumptions introduced above. Furthermore, we do not require any monotonicity of m, just continuity is enough. Here, we restrict ourselves only to the explicit models given by (5.3)–(5.4), i.e.,

$$S = 2\nu(p, |Dv|^2)Dv, \tag{5.59}$$

$$m = \beta(p, |v|, |Dv|^2)v. \tag{5.60}$$

Finally, we state our assumption on the graphs $\mathcal{A}$, $\mathcal{B}$ but now given by the formulae (5.59)–(5.60). Here, the assumptions on S are inspired by the pioneering work [42], where they appeared for the first time. Moreover, the assumptions on $\mathcal{A}$ and $\mathcal{B}$ (or more precisely on S and β, respectively) will be somehow related, which was not the case in the preceding subsection, where r and q were arbitrary. Thus, we assume that $S : \mathbb{R}^{d\times d}_{\text{sym}} \times \mathbb{R} \to \mathbb{R}^{d\times d}_{\text{sym}}$ and $\beta : \mathbb{R} \times \mathbb{R}_+ \times \mathbb{R}_+ \longrightarrow \mathbb{R}_+$ are continuous functions fulfilling for some $r \in (1, 2]$ the following set of assumptions:

(C1) the uniform r-monotonicity and continuity: There exist positive constants C_1 and C_2 such that for all $B, D \in \mathbb{R}^{d\times d}_{\text{sym}}$ and all $p \in \mathbb{R}$

$$C_1(1+|\boldsymbol{D}|^2)^{(r-2)/2}|\boldsymbol{B}|^2 \leq \frac{\partial \boldsymbol{S}(p,\boldsymbol{D})}{\partial \boldsymbol{D}} \cdot (\boldsymbol{B}\otimes\boldsymbol{B}) \leq C_2(1+|\boldsymbol{D}|^2)^{(r-2)/2}|\boldsymbol{B}|^2. \tag{5.61}$$

(C2) small Lipschitz continuity with respect to p: For all $\boldsymbol{D} \in \mathbb{R}^{d\times d}_{\mathrm{sym}}$ and $p \in \mathbb{R}$

$$\left|\frac{\partial \boldsymbol{S}(p,\boldsymbol{D})}{\partial p}\right| \leq \gamma_0(1+|\boldsymbol{D}|^2)^{(r-2)/4}, \quad \text{with } 0<\gamma_0<\frac{C_1}{C_1+C_2}. \tag{5.62}$$

(C3) There exist $c>0$, $q_0 \in (0,1)$, $q_1 \in [0, \min\{r^*, \frac{(2-q_0)r}{q_0}\})$ and $q_2 \in [1,r)$ such that for all $(p,\boldsymbol{v},\boldsymbol{D}) \in \mathbb{R}\times\mathbb{R}^d\times\mathbb{R}^{d\times d}_{\mathrm{sym}}$

$$0 \leq \beta(p,|\boldsymbol{v}|,|\boldsymbol{D}|^2) \leq c(1+|p|^{q_0}+|\boldsymbol{v}|^{q_1}+|\boldsymbol{D}|^{q_2}). \tag{5.63}$$

Let us mention here that the above assumptions are, e.g., satisfied for

$$\boldsymbol{S}(p,\boldsymbol{D}) = (1+\gamma(p)+|\boldsymbol{D}|^2)^{\frac{r-2}{2}}\boldsymbol{D}$$

with a proper choice of γ. Unfortunately, and even without the drag force term $\boldsymbol{m}$, there is no universal large data theory for the model of the form

$$\boldsymbol{S} = 2\nu(p)\boldsymbol{D}.$$

There are only available some partial results concerning flows in special geometries (see [34, 35, 49, 64, 65, 68]), we are aware of merely a few, rather preliminary studies concerning flows in general domains and related mainly to small data like results (see [32, 33] and [53]).

Furthermore, the above assumptions even do not allow $\boldsymbol{S}$ being unbounded function of the pressure p. Nevertheless, we want to mention here that for spatially periodic boundary conditions and without the presence of the drag term $\boldsymbol{m}$, there was shown the existence of solution even for models, where $\boldsymbol{S}$ can grow sublinearly with respect to the pressure, see [17]. Since the method in [17] is rather technical, we do not provide the theory for such models here, but encourage the interested reader to compare the methods presented herewith those ones in [17].

5.3.2 Result

We formulate here the two key theorems about the existence of a weak solution. The first deals with the maximal monotone graph setting but we do not allow the graph $\mathcal{A}$ (the graph related the viscous stress to the velocity gradient) to depend on the pressure.

Theorem 5.3.1 (Maximal Monotone Graphs) *Let $d \geq 2$ and $\Omega \subset \mathbb{R}^d$ be an open, bounded, connected set with a Lipschitz boundary. Consider $r \in (1, \infty)$, $q \in (1, \infty)$ $\boldsymbol{F} \in L^{r'}(\Omega)$, $\boldsymbol{v}_0 \in W^{1,r}(\Omega)$ with $\operatorname{div} \boldsymbol{v}_0 = 0$. Assume the graphs $\mathcal{A}$ and $\mathcal{B}$ satisfy Assumptions* **(A1)–(A5)** *and* **(B1)–(B5)**. *Furthermore, assume that if $\mathcal{A}$ is parameterized by p, then* **(B6)**$_S$ *holds and if $\mathcal{A}$ is parameterized by $\boldsymbol{D}$ then* **(B6)**$_D$ *holds. Then there exists a weak solution to the equation* (5.1)*, i.e., a quadruple $(\boldsymbol{S}, \boldsymbol{m}, \boldsymbol{v}, p)$ such that $(\boldsymbol{v} - \boldsymbol{v}_0) \in W^{1,r}_{\Gamma_1,\operatorname{div}}(\Omega) \cap L^q(\Omega)$, $p \in L^1(\Omega)$, $\boldsymbol{S} \in L^{r'}(\Omega)$, $\boldsymbol{m} \in L^{q'}(\Omega)$ such that for all $\boldsymbol{\varphi} \in W^{1,\infty}_{\Gamma_1}(\Omega)$*

$$\int_\Omega [\boldsymbol{S} \cdot \boldsymbol{D}\boldsymbol{\varphi} + \boldsymbol{m} \cdot \boldsymbol{\varphi} - p \operatorname{div} \boldsymbol{\varphi}]\, dx = \int_\Omega \boldsymbol{F} \cdot \nabla \boldsymbol{\varphi}\, dx \tag{5.64}$$

and

$$(\boldsymbol{m}, \boldsymbol{v}, p, \boldsymbol{D}\boldsymbol{v}) \in \mathcal{A}, \quad (\boldsymbol{S}, \boldsymbol{D}\boldsymbol{v}) \in \mathcal{B} \qquad \textit{almost everywhere in } \Omega.$$

The second result is stated without the graph setting and we rather use here the explicit constitutive laws. On the one hand, we certainly lose some generality in the considered models. On the other hand, contrary to the result of Theorem 5.3.1,we are able to cover the case when $\boldsymbol{S}$ depends on the pressure. Moreover, we allow the graph $\mathcal{A}$ (now described by $\beta\boldsymbol{v}$) to be unbounded function of the pressure, which was also not allowed in Theorem 5.3.1.

Theorem 5.3.2 (Uniformly Monotone Graphs) *Let $d \geq 2$ and $\Omega \subset \mathbb{R}^d$ be an open, bounded, connected set with a Lipschitz boundary. Consider $\boldsymbol{F} \in L^{r'}(\Omega)$, $\boldsymbol{v}_0 \in W^{1,r}(\Omega) \cap L^\infty(\Omega)$, $r \in (1, 2]$ and suppose that Assumptions* **(C1)–(C3)** *hold. Then there exists a weak solution to the equation* (5.1)*, i.e., a pair $(\boldsymbol{v}, p)$ such that $(\boldsymbol{v} - \boldsymbol{v}_0) W^{1,r}_{\Gamma_1,\operatorname{div}}(\Omega)$, $p \in L^{d'}(\Omega)$, and $\beta(p, \boldsymbol{v}, |\boldsymbol{D}\boldsymbol{v}|^2)\boldsymbol{v} \in L^1(\Omega)$ fulfilling for all $\boldsymbol{\varphi} \in W^{1,\infty}_{\Gamma_1}(\Omega) \cap L^\infty(\Omega)$*

$$\int_\Omega \Big[\boldsymbol{S}(p, \boldsymbol{D}\boldsymbol{v}) \cdot \boldsymbol{D}\boldsymbol{\varphi} + \beta(p, |\boldsymbol{v}|, |\boldsymbol{D}\boldsymbol{v}|^2)\boldsymbol{v} \cdot \boldsymbol{\varphi} - p \operatorname{div} \boldsymbol{\varphi}\Big]\, dx = \int_\Omega \boldsymbol{F} \cdot \nabla \boldsymbol{\varphi}\, dx. \tag{5.65}$$

We will provide the proof of Theorem 5.3.2 in what follows. At the end, we just briefly comment the proof of Theorem 5.3.1 because it is almost identical but easier than for Theorem 5.3.2.

The results presented here are based on the methods and tools developed mainly in the works [12, 13, 30]. There were investigated the very similar models but without the presence of the drag force. On the other hand, in this paper there is included the convective term, which makes the problem of the same difficulty as presented in this chapter. Also the presence of the convective term gives the natural bound on the parameter $r > 2d/(d+2)$ in the above mentioned works. In addition, we consider here more natural boundary conditions and do not require $\boldsymbol{v}_0 = \boldsymbol{0}$

as is the usual assumption in the theory of fluid flow. Furthermore, in the above mentioned papers, the pressure is fixed by prescribing its mean value, which does not seem to be so realistic. On the other hand, in this chapter, we are able to fix the pressure by boundary conditions, which is certainly of physical interest, see also [39]. Finally, and it is also worth mentioning, in the above works, the restriction on the value of γ_0 in (5.62) heavily depends on the shape of the domain and is much worse than (5.62). Here, we followed the scheme developed in [18] and obtained up-to-date the least restrictive condition improving thus the results in [12, 30]. In other words, our viscosity ν allows a faster growth rate in the pressure variable, albeit still a sublinear one. Aside from C_1 and C_2, the bound γ_0 also used to detrimentally depend on geometry of the set Ω through the Bogovskiĭ operator on Ω (for more information about the constant, see [31, Lemma III.3.1]). The idea behind the enhancement in our work is to replace the Bogovskiĭ operator with the Newtonian potential at some point. We recall the key properties of the Newtonian potential in Lemma 5.3.7.

To finish this part, we recall a model, for which the existence analysis perfectly works, and which is in addition used in the praxis. The prototypic example for the viscous part is

$$\boldsymbol{S} = \left(\nu_0 + \frac{\alpha(p)}{1 + |\boldsymbol{D}|} \right) \boldsymbol{D},$$

where $\nu_0 > 0$ is a constant and $\alpha(\cdot)$ is a smooth function satisfying

$$0 \le \alpha(\cdot) \le \alpha_0 \quad \text{for some } \alpha_0 > 0 \text{ and } \quad |\alpha'(\cdot)| \le \frac{\nu_0}{2\nu_0 + \alpha_0}.$$

Note that this is a very good approximation of the Schaeffer model [55] mentioned in the introductory part. For the drag term, we may consider

$$\boldsymbol{m} = \left(\beta_0 + \beta_1 |\boldsymbol{v}|^{q-2} \right) \boldsymbol{v}.$$

The number β_0 is then called the Darcy coefficient and β_1 the Forchheimer coefficient (see [36]). Note that two above cases satisfy the assumption of Theorem 5.3.2 provided that $q < r^*$. Furthermore, since $\boldsymbol{m}$ is independent of the pressure, the restriction on the size of q would not be needed and we could get the existence of a solution for all values of $q \in (1, \infty)$. Since such a generalization is only minimal and also straightforward, we do not present the proof here.

5.3.3 Auxiliary Tools

We recall here important tools and results that will be used to prove Theorem 5.3.1–5.3.2.

5.3.3.1 Algebraic Tools

The first one is a set of algebraic inequalities that follows from (5.61)–(5.62). Note that from the lemma stated below, it also follows that $\mathcal{B}$ is maximal monotone r-graph.

Lemma 5.3.3 ([30], Lemmas 3.3, 3.4) *Let* $\boldsymbol{S}$ *satisfy* (5.61)–(5.62). *Then there exists positive constants* c_1, c_2 *such that for all* $p \in \mathbb{R}$ *and* $\boldsymbol{D} \in \mathbb{R}^{d\times d}_{\mathrm{sym}}$

$$\boldsymbol{S}(p, \boldsymbol{D}) \cdot \boldsymbol{D} \geq c_1 |\boldsymbol{D}|^r - c_2 \tag{5.66}$$

and

$$|\boldsymbol{S}(p, \boldsymbol{D})| \leq c_2(1 + |\boldsymbol{D}|)^{r-1}. \tag{5.67}$$

Furthermore, if we define

$$I^{1,2} := \int_0^1 (1 + |\overline{\boldsymbol{D}}(s)|^2)^{(r-2)/2} |\boldsymbol{D}^1 - \boldsymbol{D}^2|^2 \, ds,$$

where $\overline{\boldsymbol{D}}(s) = \boldsymbol{D}^2 + s(\boldsymbol{D}^1 - \boldsymbol{D}^2)$. *Then*

$$\frac{C_1}{2} I^{1,2} \leq (\boldsymbol{S}(p^1, \boldsymbol{D}^1) - \boldsymbol{S}(p^2, \boldsymbol{D}^2)) \cdot (\boldsymbol{D}^1 - \boldsymbol{D}^2) + \frac{\gamma_0^2}{2C_1} |p^1 - p^2|^2 \tag{5.68}$$

and

$$\begin{aligned} |(\boldsymbol{S}(p^1, \boldsymbol{D}^1) - \boldsymbol{S}(p^2, \boldsymbol{D}^2))| &\leq \gamma_0 |p^1 - p^2| \\ &\quad + C_2 \int_0^1 (1 + |\overline{\boldsymbol{D}}(s)|^2)^{(r-2)/2} |\boldsymbol{D}^1 - \boldsymbol{D}^2| \, ds. \end{aligned} \tag{5.69}$$

5.3.3.2 Linear PDE Operators

Next, we state several results about the classical operators used in the theory of fluid flow.

Lemma 5.3.4 (Stokes Operator: Theorems IV.1.1, IV.4.1, IV.4.4 in [31]) *Let* $\Omega \subset \mathbb{R}^d$ *be a Lipschitz domain,* $d \geq 2$. *There exists a continuous linear operator*

$$\mathcal{H} : W^{-1,2}(\Omega) \longrightarrow W^{1,2}_{0,\mathrm{div}}(\Omega) \times \mathring{L}^2(\Omega)$$

assigning to any $\boldsymbol{f} \in W^{-1,2}(\Omega)$ *the unique weak solution* $(\boldsymbol{v}, p)$ *of the Stokes problem*

$$
\begin{aligned}
-\Delta \boldsymbol{v} + \nabla p &= \boldsymbol{f} && \textit{in } \Omega, \\
\operatorname{div} \boldsymbol{v} &= 0 && \textit{in } \Omega, \\
\boldsymbol{v} &= \boldsymbol{0} && \textit{on } \partial\Omega, \\
p_\Omega &= 0.
\end{aligned}
$$

Moreover, if $\boldsymbol{f} \in W^{-1,2}(\Omega) \cap W^{k,q}_{loc}(\Omega)$ *for certain* $1 < q < \infty$ *and* $k \geq -1$, *then* $\mathcal{H}(\boldsymbol{f}) \in W^{k+2,q}_{loc}(\Omega) \times W^{k+1,q}_{loc}(\Omega)$ *and one has the estimate*

$$
\left\|\nabla^{k+2}\boldsymbol{v}\right\|_{q;\Omega''} + \left\|\nabla^{k+1} p\right\|_{q;\Omega''} \leq c\big(\|\boldsymbol{f}\|_{k,q;\Omega'} + \|\boldsymbol{v}\|_{k+1,q;\Omega'} + \|p\|_{k,q;\Omega'}\big) \tag{5.70}
$$

for any $\Omega'' \Subset \Omega' \Subset \Omega$, *where* $c = c(d, q, k, \Omega', \Omega'')$.

Lemma 5.3.5 (Bogovskiĭ Operator: [31], Theorem III.3.3) *Let* $\Omega \subset \mathbb{R}^d$ *be a Lipschitz domain,* $d \geq 2$ *and* $1 < q < \infty$. *There is a continuous linear operator*

$$
\mathcal{B} : \mathring{L}^q(\Omega) \longrightarrow W^{1,q}_0(\Omega)
$$

assigning to any $f \in \mathring{L}^q(\Omega)$ *a weak solution* $\boldsymbol{v}$ *of the divergence equation*

$$
\begin{aligned}
\operatorname{div} \boldsymbol{v} &= f && \textit{in } \Omega, \\
\boldsymbol{v} &= \boldsymbol{0} && \textit{on } \partial\Omega.
\end{aligned}
$$

The above lemma naturally requires that f has zero mean value. However, since in our case we are not able to preserve the mean value of the pressure (since it is given by boundary conditions) we have the following result (see also [39])

Lemma 5.3.6 (Bogovskiĭ Operator II) *Let* $d \geq 2$, *and* $1 < q < \infty$ *and* $\Omega \subset \mathbb{R}^d$ *be a Lipschitz domain with boundary* $\partial\Omega = \overline{\Gamma_2 \cup \Gamma_1}$, *where* Γ_1 *and* Γ_2 *are relatively open disjoint subsets of* $\partial\Omega$. *Assume that* $|\Gamma_2|_{d-1} \neq 0$. *There is a continuous linear operator*

$$
\mathcal{B}_{\Gamma_1} : L^q(\Omega) \longrightarrow W^{1,q}_{\Gamma_1}(\Omega)
$$

assigning to any $f \in L^q(\Omega)$ *a weak solution* $\boldsymbol{v}$ *of the divergence equation*

$$
\begin{aligned}
\operatorname{div} \boldsymbol{v} &= f && \textit{in } \Omega, \\
\boldsymbol{v} &= \boldsymbol{0} && \textit{on } \Gamma_1.
\end{aligned}
$$

Proof Since Γ_1 and Γ_2 are relatively open and disjoint, we can find a ball $B_0 \subset \mathbb{R}^d$ such that $B_0 \cap \Gamma_1 = \emptyset$, $|B_0 \cap \Omega| < |B_0|$, and $|(\mathbb{R}^d \setminus \Omega) \cap B_0| < |B_0|$. Next, we denote by $\mathcal{B}^{\Omega}$ the operator from Lemma 5.3.5 related to Ω and $\mathcal{B}^{B_0}$ the operator related to B_0. Finally, we define

$$\begin{aligned}\mathcal{B}_{\Gamma_1}(f) &:= \mathcal{B}^{\Omega}\left(f - \frac{\chi_{\{B_0\cap\Omega\}}}{|B_0 \cap \Omega|}|\Omega| f_{\Omega}\right)\\ &\quad + \mathcal{B}^{B_0}\left(\frac{\chi_{\{B_0\cap\Omega\}}}{|B_0 \cap \Omega|}|\Omega| f_{\Omega} - \frac{\chi_{\{B_0\cap(\mathbb{R}^d\setminus\Omega)\}}}{|B_0 \cap (\mathbb{R}^d \setminus \Omega)|}|\Omega| f_{\Omega}\right)\Bigg|_{\Omega},\end{aligned}$$

where the result of the second operator is extended by $\mathbf{0}$ outside B_0 and then restricted to Ω. Note that the definition is meaningful since

$$\begin{aligned}&\int_{\Omega} f - \frac{\chi_{\{B_0\cap\Omega\}}}{|B_0 \cap \Omega|}|\Omega| f_{\Omega} = 0,\\ &\int_{B_0} \frac{\chi_{\{B_0\cap\Omega\}}}{|B_0 \cap \Omega|}|\Omega| f_{\Omega} - \frac{\chi_{\{B_0\cap(\mathbb{R}^d\setminus\Omega)\}}}{|B_0 \cap (\mathbb{R}^d \setminus \Omega)|}|\Omega| f_{\Omega} = 0.\end{aligned}$$

Since both operators are linear and continuous, we directly obtain the same statement also for $\mathcal{B}_{\Gamma_1}$. Finally, using the properties of $\mathcal{B}$, we can compute in Ω

$$\begin{aligned}\operatorname{div} \mathcal{B}_{\Gamma_1}(f) &:= \operatorname{div} \mathcal{B}^{\Omega}\left(f - \frac{\chi_{\{B_0\cap\Omega\}}}{|B_0 \cap \Omega|}|\Omega| f_{\Omega}\right)\\ &\quad + \operatorname{div} \mathcal{B}^{B_0}\left(\frac{\chi_{\{B_0\cap\Omega\}}}{|B_0 \cap \Omega|}|\Omega| f_{\Omega} - \frac{\chi_{\{B_0\cap(\mathbb{R}^d\setminus\Omega)\}}}{|B_0 \cap (\mathbb{R}^d \setminus \Omega)|}|\Omega| f_{\Omega}\right)\Bigg|_{\Omega}\\ &= f - \frac{\chi_{\{B_0\cap\Omega\}}}{|B_0 \cap \Omega|}|\Omega| f_{\Omega} + \frac{\chi_{\{B_0\cap\Omega\}}}{|B_0 \cap \Omega|}|\Omega| f_{\Omega} = f.\end{aligned}$$

The proof is complete. □

Lemma 5.3.7 (Laplace Operator: [18], Lemma 4.4) *Let $\Omega \subset \mathbb{R}^d$ be open bounded set, $d \geq 2$ and $1 < q < \infty$. There is a continuous linear operator*

$$\mathcal{N} : L^q(\Omega) \longrightarrow W^{2,q}(\Omega)$$

fulfilling for any $f \in L^q(\Omega)$

$$-\Delta \mathcal{N}(f) = f \qquad \text{in } \Omega.$$

Moreover, we have $\left\|\nabla^2 \mathcal{N}(f)\right\|_2 \leq \|f\|_2$.

5.3.3.3 Convergence Tools

Finally, we introduce three tools used for the identification of weak limits. They are celebrated Biting lemma, invented by Chacon [11], Div–Curl lemma, found independently by Murat and Tartar [45, 66], and the Lipschitz approximation lemma for Sobolev functions, established by Acerbi and Fusco [1].

Lemma 5.3.8 (Biting Lemma, [5]) *Let $\Omega \subset \mathbb{R}^d$ have a finite Lebesgue measure and $\{f^k\}_{k=1}^\infty$ be a bounded sequence in $L^1(\Omega)$. Then there exist a function $f \in L^1(\Omega)$, a subsequence $\{f^j\}_{j=1}^\infty$ of $\{f^k\}_{k=1}^\infty$, and a non-increasing sequence of measurable sets $E_n \subset \Omega$ with $\lim_{n\to\infty} |E^n| = 0$, such that $f^j \rightharpoonup f$ in $L^1(\Omega \setminus E_n)$ for every fixed n.*

Lemma 5.3.9 (Div–Curl Lemma, [27], Theorem 10.21) *Let $\Omega \subset \mathbb{R}^d$ be open. Assume $\boldsymbol{u}^n \rightharpoonup \boldsymbol{u}$ in $L^p(\Omega)$ and $\boldsymbol{v}^n \rightharpoonup \boldsymbol{v}$ in $L^q(\Omega)$, where $1/p + 1/q = 1/r < 1$. In addition, let $\{\operatorname{div} \boldsymbol{u}^n\}_{n=1}^\infty$ be relatively compact in $W^{-1,s}(\Omega)$ and $\{\operatorname{curl} \boldsymbol{v}^n\}_{n=1}^\infty$ be[7] relatively compact in $W^{-1,s}(\Omega)$ for a certain $s > 1$. Then $\boldsymbol{u}^n \cdot \boldsymbol{v}^n \rightharpoonup \boldsymbol{u} \cdot \boldsymbol{v}$ in $L^r(\Omega)$.*

Lemma 5.3.10 (Lipschitz Approximation of Sobolev Functions, [1]) *Let $\Omega \subset \mathbb{R}^d$ be a Lipschitz open set and $p \geq 1$. There exists a constant c such that, for every $u \in W^{1,p}(\Omega)$ and every $\lambda > 0$ there exists $u_\lambda \in W^{1,\infty}(\Omega)$ satisfying*

$$\|u_\lambda\|_{1,\infty} \leq \lambda, \tag{5.71}$$

$$|\{u \neq u_\lambda\}| \leq c \frac{\|u\|_{1,p}^p}{\lambda^p}, \tag{5.72}$$

$$\|u_\lambda\|_{1,p} \leq c \, \|u\|_{1,p} \,. \tag{5.73}$$

5.3.4 *Quasi-Compressible and L^∞ Approximation*

Similarly as in Sect. 5.2.5, we introduce a cascade of approximations. First, we allow certain quasi-compressibility, which will be parameterized by ε, and we cut the drag term $\boldsymbol{m}$ with the help of parameter k. This means that (5.55) and (5.56), completed by the constitutive relations (5.59)–(5.60) will be approximated by

$$\begin{aligned} -\operatorname{div} \boldsymbol{S}(p, \boldsymbol{D}\boldsymbol{v}) + \beta_k \boldsymbol{v}_k &= -\nabla p - \operatorname{div} \boldsymbol{F} && \text{in } \Omega, \\ \varepsilon \Delta p &= \operatorname{div} \boldsymbol{v} && \text{in } \Omega, \\ \boldsymbol{v} = \boldsymbol{v}_0, \ \nabla p \cdot \boldsymbol{n} &= 0 && \text{on } \Gamma_1, \\ \boldsymbol{S}\boldsymbol{n} = \boldsymbol{F}\boldsymbol{n}, \ p &= 0 && \text{on } \Gamma_2, \end{aligned} \tag{5.74}$$

[7] $\operatorname{curl} = \frac{1}{2}(\nabla - \nabla^T)$.

where $\varepsilon > 0$ and $k \in \mathbb{N}$. The approximative quantities β_k and $\boldsymbol{v}_k$ are defined as

$$\beta_k(p, \boldsymbol{v}, |\boldsymbol{D}\boldsymbol{v}|^2) := \frac{\beta(p, \boldsymbol{v}, |\boldsymbol{D}\boldsymbol{v}|^2)}{1 + k^{-1}\beta(p, \boldsymbol{v}, |\boldsymbol{D}\boldsymbol{v}|^2)}, \qquad \boldsymbol{v}_k := \frac{\boldsymbol{v}}{1 + k^{-1}|\boldsymbol{v}|}. \tag{5.75}$$

Since $\beta \geq 0$ we have that β_k as well as $\boldsymbol{v}_k$ are bounded. We will proceed as follows. First, we let $\varepsilon \to 0_+$ to recover incompressibility and second we let $k \to \infty$ to recover the drag term.

The existence of a solution to (5.74) can be proven by standard monotone operator theory and we do not provide the proof here. Note that due to the presence of Δp, we have the compactness of the pressure for free, second due to the truncations β_k and $\boldsymbol{v}_k$, the drag term is just compact perturbation and therefore we can use the standard Minty method [44] to obtain the existence of solution. To summarize, we have the following result (see also [18, Lemma 5.1] for detailed proof, for $\Gamma_1 = \partial\Omega$ and $\boldsymbol{v}_0 = \boldsymbol{0}$, but the proof can be straightforwardly extended also to our setting):

Lemma 5.3.11 *Let all assumptions of Theorem 5.3.2 be satisfied. Then for every $\varepsilon > 0$ and $k \in \mathbb{N}$ there exist $(\boldsymbol{v}^{\varepsilon,k}, p^{\varepsilon,k}) \in W^{1,r}(\Omega) \times \big(W^{1,2}_{\Gamma_1}(\Omega) \cap L^{r'}(\Omega)\big)$ satisfying $(\boldsymbol{v}^{\varepsilon,k} - \boldsymbol{v}_0) \in W^{1,r}_{\Gamma_1}(\Omega)$ such that for all $\varphi \in W^{1,2}_{\Gamma_1}(\Omega) \cap L^{r'}(\Omega)$*

$$\varepsilon \int_\Omega \nabla p^{\varepsilon,k} \cdot \nabla \varphi \, dx + \int_\Omega \varphi \operatorname{div} \boldsymbol{v}^{\varepsilon,k} \, dx = 0 \tag{5.76}$$

and for all $\boldsymbol{\varphi} \in W^{1,r}_{\Gamma_1}(\Omega)$

$$\begin{aligned} \int_\Omega \boldsymbol{S}(p^{\varepsilon,k}, \boldsymbol{D}\boldsymbol{v}^{\varepsilon,k}) \cdot \boldsymbol{D}\boldsymbol{\varphi} + \beta_k(p^{\varepsilon,k}, |\boldsymbol{v}^{\varepsilon,k}|, |\boldsymbol{D}\boldsymbol{v}^{\varepsilon,k}|^2)\boldsymbol{v}_k^{\varepsilon,k} \cdot \boldsymbol{\varphi} \, dx \\ = \int_\Omega p^{\varepsilon,k} \operatorname{div} \boldsymbol{\varphi} + \boldsymbol{F} \cdot \nabla \boldsymbol{\varphi} \, dx. \end{aligned} \tag{5.77}$$

5.3.5 Incompressible Limit $\varepsilon \to 0_+$

In this section, we fix k and denote $(\boldsymbol{v}^\varepsilon, p^\varepsilon)$ a solution constructed in Lemma 5.3.11 and our goal is to let $\varepsilon \to 0_+$.

5.3.5.1 Uniform Estimates

We start with the uniform ε-independent estimates. We set $\varphi = p^\varepsilon$ in (5.76) and $\boldsymbol{\varphi} = \boldsymbol{v}^\varepsilon - \boldsymbol{v}_0$ in (5.77). Using the fact that $\operatorname{div} \boldsymbol{v}_0 = 0$ and summing up the resultant identities, we obtain

$$\varepsilon \|\nabla p^\varepsilon\|_2^2 + \int_\Omega \boldsymbol{S}(p^\varepsilon, \boldsymbol{D}\boldsymbol{v}^\varepsilon) \cdot \boldsymbol{D}\boldsymbol{v}^\varepsilon \, dx + \int_\Omega \beta_k \boldsymbol{v}_k^\varepsilon \cdot \boldsymbol{v}^\varepsilon \, dx$$
$$= \int_\Omega \boldsymbol{F} \cdot \nabla(\boldsymbol{v}^\varepsilon - \boldsymbol{v}_0) \, dx + \int_\Omega \boldsymbol{S}(p^\varepsilon, \boldsymbol{D}\boldsymbol{v}^\varepsilon) \cdot \boldsymbol{D}\boldsymbol{v}_0 \, dx + \int_\Omega \beta_k \boldsymbol{v}_k^\varepsilon \cdot \boldsymbol{v}_0 \, dx,$$

where $\beta_k^\varepsilon := \beta_k(p^\varepsilon, |\boldsymbol{v}^\varepsilon|, |\boldsymbol{D}\boldsymbol{v}^\varepsilon|^2)$.

It follows from (5.63) and the definition of β_k^ε and $\boldsymbol{v}_k^\varepsilon$ we have $\beta_k^\varepsilon \boldsymbol{v}_k^\varepsilon \cdot \boldsymbol{v}^\varepsilon \geq 0$. Thus, the properties (5.66)–(5.67), the Poincaré, the Young and the Korn inequalities, and the fact that $\boldsymbol{F} \in L^{r'}(\Omega)$ and $\boldsymbol{v}_0 \in W^{1,r}(\Omega)$ lead to (here C denotes a generic constant depending on data but not on $\varepsilon > 0$)

$$\sqrt{\varepsilon}\|\nabla p^\varepsilon\|_2 + \|\boldsymbol{v}^\varepsilon\|_{1,r} + \|\boldsymbol{S}(p^\varepsilon, \boldsymbol{D}\boldsymbol{v}^\varepsilon)\|_{r'} \leq C(k). \tag{5.78}$$

We also need to bound the pressure. Therefore we set

$$\boldsymbol{\varphi} := \mathcal{B}_{\Gamma_1}\big(|p^\varepsilon|^{r'-2} p^\varepsilon\big)$$

and observe that $\boldsymbol{\varphi} \in W^{1,r}_{\Gamma_1}(\Omega)$. In addition, it follows from Lemma 5.3.6 that $\|\boldsymbol{\varphi}\|_{1,r} \leq C \||p^\varepsilon|^{r'-1}\|_r = C\|p^\varepsilon\|_{r'}^{r'-1}$ with C independent of ε. Then insertion of $\boldsymbol{\varphi}$ into (5.77), use of the Hölder inequality, and the a priori estimate (5.78) give

$$\|p^\varepsilon\|_{r'}^{r'} = \int_\Omega \boldsymbol{S}(p^\varepsilon, \boldsymbol{D}\boldsymbol{v}^\varepsilon) \cdot \boldsymbol{D}\boldsymbol{\varphi} \, dx + \int_\Omega \beta_k^\varepsilon \boldsymbol{v}_\varepsilon^\varepsilon \cdot \boldsymbol{\varphi} \, dx - \int_\Omega \boldsymbol{F} \cdot \nabla \boldsymbol{\varphi} \, dx$$
$$\leq C(k) \, \|\boldsymbol{\varphi}\|_{1,r} \leq C(k) \, \|p^\varepsilon\|_{r'}^{r'-1}.$$

Consequently,

$$\|p^\varepsilon\|_{r'} \leq C(k). \tag{5.79}$$

The estimates (5.78) and (5.79) imply that we may extract a subsequence that we do not relabel such that (as $\varepsilon \to 0_+$):

$$\begin{aligned}
\boldsymbol{v}^\varepsilon &\rightharpoonup \boldsymbol{v} && \text{weakly in } W^{1,r}(\Omega),\\
\boldsymbol{v}^\varepsilon &\to \boldsymbol{v} && \text{a.e. in } \Omega,\\
p^\varepsilon &\rightharpoonup p && \text{weakly in } L^{r'}(\Omega),\\
\varepsilon \nabla p^\varepsilon &\to \boldsymbol{0} && \text{strongly in } L^2(\Omega),\\
\boldsymbol{S}(p^\varepsilon, \boldsymbol{D}\boldsymbol{v}^\varepsilon) &\rightharpoonup \overline{\boldsymbol{S}} && \text{weakly in } L^{r'}(\Omega),\\
\beta_k^\varepsilon \boldsymbol{v}_k^\varepsilon &\rightharpoonup \overline{\beta_k \boldsymbol{v}_k} && \text{weakly in } L^q(\Omega) \text{ for any } q \in [1, \infty),
\end{aligned} \tag{5.80}$$

where by the bar, we denoted the weak limits. Then, the limit $\varepsilon \to 0_+$ applied to (5.76) and the convergence result $(5.80)_4$ guarantee $\operatorname{div} \boldsymbol{v} = 0$ in Ω and consequently, we have $(\boldsymbol{v} - \boldsymbol{v}_0) \in W^{1,r}_{\Gamma_1,\operatorname{div}}(\Omega)$. Similarly, (5.77) combined with (5.80) yields

$$\int_\Omega \overline{\boldsymbol{S}} \cdot \boldsymbol{D}\boldsymbol{\varphi}\, dx + \int_\Omega \overline{\beta_k \boldsymbol{v}_k} \cdot \boldsymbol{\varphi}\, dx - \int_\Omega p \operatorname{div} \boldsymbol{\varphi}\, dx = \int_\Omega \boldsymbol{F} \cdot \nabla \boldsymbol{\varphi}\, dx \tag{5.81}$$

for all $\boldsymbol{\varphi} \in W^{1,r}_{\Gamma_1}(\Omega)$.

We need to identify $\overline{\boldsymbol{S}}$ and $\overline{\beta_k \boldsymbol{v}_k}$. This will be done by showing the pointwise convergence of p^ε and $\boldsymbol{D}\boldsymbol{v}^\varepsilon$. Then $\overline{\boldsymbol{S}} = \boldsymbol{S}(p, \boldsymbol{D}\boldsymbol{v})$ and $\overline{\beta_k \boldsymbol{v}_k} = \beta_k\big(p, |\boldsymbol{v}|, |\boldsymbol{D}\boldsymbol{v}|^2\big)\boldsymbol{v}_k$, which can be observed from $(5.80)_5$, $(5.80)_6$, and the Vitali theorem. We show the pointwise convergence on any compactly contained subdomain $\Omega' \Subset \Omega$.

5.3.5.2 Pointwise Convergence of p^ε

Hence, let Ω' be fix and $\eta \in \mathcal{C}^\infty_0(\Omega)$ be such that $0 \le \eta \le 1$ and $\eta \equiv 1$ in Ω'. Next, we define $u^\varepsilon = \mathcal{N}((p^\varepsilon - p)\eta)$ (this also means that $\Delta u^\varepsilon = (p - p^\varepsilon)\eta$ in Ω), where the operator $\mathcal{N}$ was introduced in Lemma 5.3.7. Due to the linearity of the operator $\mathcal{N}$ and also thanks to the weak convergence result $(5.80)_3$, we have

$$u^\varepsilon \rightharpoonup 0 \quad \text{weakly in } W^{2,r'}(\Omega), \tag{5.82}$$

$$u^\varepsilon \to 0 \quad \text{strongly in } W^{1,r'}(\Omega), \tag{5.83}$$

where for (5.83) we used the compact embedding. Using the definition of u^ε, we obtain

$$\|(p^\varepsilon - p)\eta\|_2^2 = -\int_\Omega p^\varepsilon \eta \Delta u^\varepsilon\, dx + \underbrace{\int_\Omega p\eta\Delta u^\varepsilon\, dx}_{\to 0}. \tag{5.84}$$

Since the second term vanishes if we let $\varepsilon \to 0_+$ (it is consequence of $(5.80)_3$, (5.82), and the fact that $r' \ge 2$), we focus on the first term, which can be rewritten as

$$-\int_\Omega p^\varepsilon \eta \Delta u^\varepsilon\, dx = -\int_\Omega p^\varepsilon \operatorname{div}(\eta \nabla u^\varepsilon)\, dx + \underbrace{\int_\Omega p^\varepsilon \nabla\eta \cdot \nabla u^\varepsilon\, dx}_{\to 0}.$$

Similarly as above, the second term also vanishes as $\varepsilon \to 0_+$, which follows from (5.83). Finally, since η is compactly supported, we can use (5.77) with $\boldsymbol{\varphi} := \eta \nabla u^\varepsilon$ to identify the first term as

$$-\int_\Omega p^\varepsilon \operatorname{div}(\eta\nabla u^\varepsilon)\,dx = -\int_\Omega \boldsymbol{S}(p^\varepsilon, \boldsymbol{D}\boldsymbol{v}^\varepsilon)\cdot\nabla(\eta\nabla u^\varepsilon)\,dx - \underbrace{\int_\Omega \beta_k^\varepsilon \boldsymbol{v}_k^\varepsilon\cdot\eta\nabla u^\varepsilon\,dx}_{\to 0}$$

$$+\underbrace{\int_\Omega \boldsymbol{F}\cdot\nabla(\eta\nabla u^\varepsilon)\,dx.}_{\to 0}$$

The latter two terms tend to zero, which follows from (5.82), (5.83), and the fact that $\|\beta_k^\varepsilon \boldsymbol{v}_k^\varepsilon\|_\infty \le C(k)$. Further, we rewrite the first term as

$$-\int_\Omega \boldsymbol{S}(p^\varepsilon, \boldsymbol{D}\boldsymbol{v}^\varepsilon)\cdot\nabla(\eta\nabla u^\varepsilon)\,dx = -\int_\Omega \boldsymbol{S}(p^\varepsilon, \boldsymbol{D}\boldsymbol{v}^\varepsilon)\cdot\eta\nabla^2 u^\varepsilon\,dx$$

$$-\underbrace{\int_\Omega \boldsymbol{S}(p^\varepsilon, \boldsymbol{D}\boldsymbol{v}^\varepsilon)\cdot(\nabla u^\varepsilon\otimes\nabla\eta)\,dx}_{\to 0}$$

and the last term converges to zero by (5.83). Lastly, we want to use the structural assumptions on $\boldsymbol{S}$ and therefore we split the integral as

$$-\int_\Omega \boldsymbol{S}(p^\varepsilon, \boldsymbol{D}\boldsymbol{v}^\varepsilon)\cdot\eta\nabla^2 u^\varepsilon\,dx = -\underbrace{\int_\Omega \boldsymbol{S}(p, \boldsymbol{D}\boldsymbol{v})\cdot\eta\nabla^2 u^\varepsilon\,dx}_{\to 0}$$

$$+\int_\Omega (\boldsymbol{S}(p, \boldsymbol{D}\boldsymbol{v}) - \boldsymbol{S}(p^\varepsilon, \boldsymbol{D}\boldsymbol{v}^\varepsilon))\cdot\eta\nabla^2 u^\varepsilon\,dx.$$

The first integral on the right-hand side vanishes for $\varepsilon \to 0_+$ by (5.82). To summarize, up to now we obtained that

$$\limsup_{\varepsilon\to 0_+} \left\|(p^\varepsilon - p)\eta\right\|_2^2 = \limsup_{\varepsilon\to 0_+}\int_\Omega (\boldsymbol{S}(p, \boldsymbol{D}\boldsymbol{v}) - \boldsymbol{S}(p^\varepsilon, \boldsymbol{D}\boldsymbol{v}^\varepsilon))\cdot\eta\nabla^2 u^\varepsilon\,dx. \tag{5.85}$$

The right-hand side will be estimated by using the assumptions on $\boldsymbol{S}$. Indeed, we have from (5.69) that

$$\int_\Omega (\boldsymbol{S}(p, \boldsymbol{D}\boldsymbol{v}) - \boldsymbol{S}(p^\varepsilon, \boldsymbol{D}\boldsymbol{v}^\varepsilon))\cdot\eta\nabla^2 u^\varepsilon\,dx \le \gamma_0\int_\Omega |(p-p^\varepsilon)\eta||\nabla^2 u^\varepsilon|\,dx$$

$$+ C_2\int_\Omega\int_0^1 (1+|\overline{\boldsymbol{D}}(s)|^2)^{(r-2)/2}|\boldsymbol{D}(\boldsymbol{v}-\boldsymbol{v}^\varepsilon)||\nabla^2 u^\varepsilon|\eta\,ds\,dx, \tag{5.86}$$

where $\overline{D}(s) = Dv^\varepsilon + s(Dv - Dv^\varepsilon)$. To shorten the notation, we also denote

$$I^\varepsilon = \int_0^1 (1 + |\overline{D}(s)|^2)^{(r-2)/2} |D(v - v^\varepsilon)|^2 \, ds.$$

Then we use the fact that $1 \le r \le 2$ and $0 \le \eta \le 1$, and the Hölder inequality to conclude from (5.86)

$$\begin{aligned}
&\int_\Omega (S(p, Dv) - S(p^\varepsilon, Dv^\varepsilon)) \cdot \eta \nabla^2 u^\varepsilon \, dx \\
&\quad \le \gamma_0 \left\| (p^\varepsilon - p)\eta \right\|_2^2 + C_2 \Big(\int_\Omega I^\varepsilon \eta \, dx \Big)^{1/2} \left\| (p^\varepsilon - p)\eta \right\|_2 \qquad (5.87) \\
&\quad \le \frac{1 + \gamma_0}{2} \left\| (p^\varepsilon - p)\eta \right\|_2^2 + \frac{C_2^2}{2(1 - \gamma_0)} \left\| I^\varepsilon \eta \right\|_1 .
\end{aligned}$$

Notice that in the above estimate we used the fact that $\|\nabla^2 u^\varepsilon\|_2 \le \|(p^\varepsilon - p)\eta\|_2$, see Lemma 5.3.7. Combination of (5.85) and (5.87) leads to

$$\limsup_{\varepsilon \to 0_+} \left\| (p^\varepsilon - p)\eta \right\|_2^2 \le \frac{C_2^2}{(1 - \gamma_0)^2} \limsup_{\varepsilon \to 0_+} \left\| I^\varepsilon \eta \right\|_1 . \qquad (5.88)$$

It remains to estimate $\left\| I^\varepsilon \eta \right\|_1$. Using (5.68), we have

$$\begin{aligned}
\frac{C_1}{2} \left\| I^\varepsilon \eta \right\|_1 \le &\int_\Omega (S(p, Dv) - S(p^\varepsilon, Dv^\varepsilon)) \cdot D(v - v^\varepsilon)\eta \, dx \\
&+ \frac{\gamma_0^2}{2C_1} \left\| (p^\varepsilon - p)\eta \right\|_2^2
\end{aligned} \qquad (5.89)$$

and substituting it into (5.88) and using the fact that $\gamma_0 < C_1/(C_1 + C_2)$ (see the assumption (5.62)), we arrive at

$$\begin{aligned}
&\limsup_{\varepsilon \to 0_+} \left\| (p^\varepsilon - p)\eta \right\|_2^2 \\
&\quad \le C(\gamma_0, C_1, C_2) \limsup_{\varepsilon \to 0_+} \int_\Omega (S(p^\varepsilon, Dv^\varepsilon) - S(p, Dv)) \cdot D(v^\varepsilon - v)\eta \, dx.
\end{aligned} \qquad (5.90)$$

The property (5.67) and the convergence $(5.80)_1$ yield

$$\lim_{\varepsilon \to 0_+} \int_\Omega S(p, Dv) \cdot D(v - v^\varepsilon)\eta \, dx = 0.$$

Towards handling the other integral in (5.90), we set $\boldsymbol{\varphi}_\varepsilon = (\boldsymbol{v}^\varepsilon - \boldsymbol{v})\eta$ and write

$$\int_\Omega \boldsymbol{S}(p^\varepsilon, \boldsymbol{D}\boldsymbol{v}^\varepsilon) \cdot \boldsymbol{D}(\boldsymbol{v}^\varepsilon - \boldsymbol{v})\eta \, dx = \int_\Omega \boldsymbol{S}(p^\varepsilon, \boldsymbol{D}\boldsymbol{v}^\varepsilon) \cdot \boldsymbol{D}\boldsymbol{\varphi}_\varepsilon \, dx + \underbrace{\int_\Omega \boldsymbol{S}(p^\varepsilon, \boldsymbol{D}\boldsymbol{v}^\varepsilon) \cdot (\nabla\eta \otimes (\boldsymbol{v} - \boldsymbol{v}^\varepsilon)) \, dx}_{\to 0}.$$

The latter integral vanishes for $\varepsilon \to 0_+$ by (5.80). For the remaining part, we employ the weak formulation (5.77) tested with $\boldsymbol{\varphi}_\varepsilon = (\boldsymbol{v}^\varepsilon - \boldsymbol{v})\eta$ to get

$$\int_\Omega \boldsymbol{S}(p^\varepsilon, \boldsymbol{D}\boldsymbol{v}^\varepsilon) \cdot \boldsymbol{D}\boldsymbol{\varphi}_\varepsilon \, dx = \int_\Omega p^\varepsilon \operatorname{div} \boldsymbol{\varphi}_\varepsilon \, dx - \underbrace{\int_\Omega \beta_k^\varepsilon \boldsymbol{v}_k^\varepsilon \cdot \boldsymbol{\varphi}_\varepsilon \, dx}_{\to 0} + \underbrace{\int_\Omega \boldsymbol{F} \cdot \nabla \boldsymbol{\varphi}_\varepsilon \, dx}_{\to 0}. \tag{5.91}$$

The last two terms vanish for $\varepsilon \to 0_+$ by using (5.80). Finally, the first term in (5.91) can be rewritten by using (5.76) and the fact that $\operatorname{div} \boldsymbol{v} = 0$ as

$$\begin{aligned} \int_\Omega p^\varepsilon \operatorname{div} \boldsymbol{\varphi}_\varepsilon \, dx &= \int_\Omega p^\varepsilon \eta \operatorname{div} \boldsymbol{v}^\varepsilon \, dx + \int_\Omega p^\varepsilon (\boldsymbol{v}^\varepsilon - \boldsymbol{v}) \cdot \nabla\eta \, dx \\ &= -\varepsilon \int_\Omega \nabla(p^\varepsilon \eta) \cdot \nabla p^\varepsilon \, dx + \int_\Omega p^\varepsilon (\boldsymbol{v}^\varepsilon - \boldsymbol{v}) \cdot \nabla\eta \, dx \\ &\le \underbrace{\int_\Omega p^\varepsilon (\boldsymbol{v}^\varepsilon - \boldsymbol{v} - \varepsilon \nabla p^\varepsilon) \cdot \nabla\eta \, dx}_{\to 0}, \end{aligned}$$

where the last integral tends to zero thanks to (5.80). Plugging this result into (5.91), we get that the integral on the right-hand side of (5.90) satisfies

$$\limsup_{\varepsilon \to 0_+} \int_\Omega (\boldsymbol{S}(p, \boldsymbol{D}\boldsymbol{v}) - \boldsymbol{S}(p^\varepsilon, \boldsymbol{D}\boldsymbol{v}^\varepsilon)) \cdot \boldsymbol{D}(\boldsymbol{v} - \boldsymbol{v}^\varepsilon)\eta \, dx \le 0. \tag{5.92}$$

Inserting this information back into (5.90), we conclude

$$\lim_{\varepsilon \to 0_+} \left\| (p^\varepsilon - p)\eta \right\|_2 = 0. \tag{5.93}$$

Since η and Ω' are arbitrary, it follows that $p^\varepsilon \to p$ almost everywhere in Ω (at least for a subsequence).

5.3.5.3 Pointwise Convergence of Dv^ε

Here, we show the strong convergence of $\boldsymbol{D}\boldsymbol{v}^\varepsilon$. Using (5.92)–(5.93) in (5.89), we obtain

$$\lim_{\varepsilon\to 0_+}\int_{\Omega'} I^\varepsilon\,dx \le \lim_{\varepsilon\to 0_+}\int_{\Omega} I^\varepsilon \eta\,dx = 0. \tag{5.94}$$

Next, using the Hölder inequality and the fact that $r \in (1,2]$ and recalling the definition of $\overline{\boldsymbol{D}}(s)$, we can calculate

$$\begin{aligned}
&\int_{\Omega'} |\boldsymbol{D}(\boldsymbol{v}^\varepsilon - \boldsymbol{v})|^r\,dx \\
&= \int_{\Omega'} \left(\int_0^1 (1+|\overline{\boldsymbol{D}}(s)|^2)^{\frac{r-2}{2}} |\boldsymbol{D}(\boldsymbol{v}^\varepsilon - \boldsymbol{v})|^2 (1+|\overline{\boldsymbol{D}}(s)|^2)^{\frac{2-r}{2}}\,ds\right)^{r/2} dx \\
&\le \int_{\Omega'} (I^\varepsilon)^{r/2} (1+|\boldsymbol{D}\boldsymbol{v}^\varepsilon|^2 + |\boldsymbol{D}\boldsymbol{v}|^2)^{\frac{r(2-r)}{4}}\,dx \\
&\le \left(\int_{\Omega'} I^\varepsilon\,dx\right)^{r/2} \left(\int_{\Omega'} (1+|\boldsymbol{D}\boldsymbol{v}^\varepsilon|^2 + |\boldsymbol{D}\boldsymbol{v}|^2)^{r/2}\,dx\right)^{\frac{2-r}{2}}.
\end{aligned}$$

Hence, (5.94) and (5.80) directly imply the required convergence

$$\lim_{\varepsilon\to 0_+}\int_{\Omega'} |\boldsymbol{D}(\boldsymbol{v}^\varepsilon - \boldsymbol{v})|^r\,dx = 0.$$

This together with (5.93) allows us to identify the weak limits in (5.81) and denoting $(\boldsymbol{v}^k, p^k) := (\boldsymbol{v}, p)$, where $(\boldsymbol{v}^k - \boldsymbol{v}_0) \in W^{1,r}_{\Gamma_1,\mathrm{div}}(\Omega)$ and $p^k \in L^{r'}(\Omega)$, we see that this couple satisfies for all $\boldsymbol{\varphi} \in W^{1,r}_{\Gamma_1}(\Omega)$

$$\begin{aligned}
\int_\Omega \boldsymbol{S}(p^k, \boldsymbol{D}\boldsymbol{v}^k)\cdot \boldsymbol{D}\boldsymbol{\varphi}\,dx &+ \int_\Omega \beta_k(p^k, |\boldsymbol{v}^k|, |\boldsymbol{D}\boldsymbol{v}^k|^2)\boldsymbol{v}^k_k \cdot \boldsymbol{\varphi}\,dx \\
&- \int_\Omega p^k \operatorname{div}\boldsymbol{\varphi}\,dx = \int_\Omega \boldsymbol{F}\cdot\nabla\boldsymbol{\varphi}\,dx,
\end{aligned} \tag{5.95}$$

where we employed the notation (5.75) for β_k and $\boldsymbol{v}_k$.

5.3.6 Removing the L^∞ Truncation ($k \to \infty$)

This final part concerns the limit $k \to \infty$. There are two essential differences from the previous part. First, the a priori estimates are not so straightforward due to the possible growth of β and also due to the presence of the boundary condition $\boldsymbol{v}_0$. In case $\boldsymbol{v}_0 = 0$, the estimates would be much easier. Second, due to the poor estimates on the pressure and on the drag term, we cannot use the standard monotone theory approach but we will have to use a different technique. This consists of the pressure decomposition (on the part which is compact but has only poor integrability and the second part, which has a natural integrability but is possibly noncompact), the use of the Div–Curl lemma, and also the Lipschitz approximation of Sobolev functions.

5.3.6.1 Uniform k-Independent Estimates

We set $\boldsymbol{\varphi} = \boldsymbol{v}^k - \boldsymbol{v}_0$ in the relation (5.95). Using (5.66), (5.67), the Korn and the Young inequality, we find that

$$\|\boldsymbol{v}^k\|_{1,r}^r + \|\boldsymbol{S}(p^k, \boldsymbol{D}\boldsymbol{v}^k)\|_{r'}^{r'} + \int_\Omega \beta_k \boldsymbol{v}_k^k \cdot (\boldsymbol{v}^k - \boldsymbol{v}_0)\, dx \le C(1 + \|\boldsymbol{v}_0\|_{1,r}^r + \|\boldsymbol{F}\|_{r'}^{r'}) \tag{5.96}$$

Using the definition of $\boldsymbol{v}_k^k$, and denoting $M := \|\boldsymbol{v}_0\|_\infty$, we also have that

$$\begin{aligned}\int_\Omega \beta_k \boldsymbol{v}_k^k \cdot (\boldsymbol{v}^k - \boldsymbol{v}_0)\, dx &\ge \int_\Omega \beta_k |\boldsymbol{v}_k^k|(|\boldsymbol{v}^k| - |\boldsymbol{v}_0|)\, dx \\ &\ge \frac{1}{2}\int_\Omega \beta_k |\boldsymbol{v}_k^k|^2\, dx - 4M^2 \int_{\{|\boldsymbol{v}^k| \le 2M\}} \beta_k\, dx.\end{aligned}$$

Thus, substituting this estimate into (5.96) and using the assumptions on $\boldsymbol{F}$ and $\boldsymbol{v}_0$, and the assumption on β (5.63) and the definition of β_k, we deduce that

$$\begin{aligned}&\|\boldsymbol{v}^k\|_{1,r}^r + \|\boldsymbol{S}(p^k, \boldsymbol{D}\boldsymbol{v}^k)\|_{r'}^{r'} + \int_\Omega \beta_k |\boldsymbol{v}_k^k|^2\, dx \le C\left(1 + \int_{\{|\boldsymbol{v}^k| \le 2M\}} \beta_k\, dx\right) \\ &\le C\int_\Omega 1 + |p^k|^{q_0} + |\boldsymbol{D}\boldsymbol{v}^k|^{q_2}\, dx \le C(1 + \|p^k\|_1^{q_0} + \|\boldsymbol{v}^k\|_{1,r}^{q_2}),\end{aligned}$$

where we used the facts that $q_0 < 1$ and $q_2 < r$. Consequently, applying the Young inequality and the fact that $q_2 < r$ we have

$$\|\boldsymbol{v}^k\|_{1,r}^r + \|\boldsymbol{S}(p^k, \boldsymbol{D}\boldsymbol{v}^k)\|_{r'}^{r'} + \int_\Omega \beta_k |\boldsymbol{v}_k^k|^2\, dx \le C\left(1 + \|p^k\|_1^{q_0}\right) \tag{5.97}$$

Next, to estimate the pressure term, we use the modification of the Bogovskiĭ operator in Lemma 5.3.6 and set for arbitrary $z < d'$

$$\boldsymbol{\varphi} = \mathcal{B}_{\Gamma_1}\left(|p^k|^{z-2} p^k\right).$$

Note that $\varphi \in W_0^{1,z'}(\Omega) \hookrightarrow L^\infty(\Omega)$ and $\|\varphi\|_{1,z'} \le C(z)\|p^k\|_z^{z-1}$ due to the continuity of $\mathcal{B}_{\Gamma_1}$. Using φ as a test function in (5.95) yields

$$\|p^k\|_z^z = \int_\Omega \boldsymbol{S}(p^k, \boldsymbol{D}\boldsymbol{v}^k) : \boldsymbol{D}\varphi\, dx + \int_\Omega \beta_k \boldsymbol{v}_k^k \cdot \varphi\, dx - \int_\Omega \boldsymbol{F} \cdot \nabla\varphi\, dx.$$

Next, since $r \le 2$, we have that $z \ge r$ and we can use the Hölder inequality, the estimate (5.97), and the assumption (5.63) to obtain (note that $q_0 < 1$ in (5.63))

$$\begin{aligned}
\|p^k\|_z &\le C(\|\boldsymbol{S}(p^k, \boldsymbol{D}\boldsymbol{v}^k)\|_{r'} + \|\beta_k \boldsymbol{v}_k^k\|_1 + \|\boldsymbol{F}\|_{r'}) \\
&\le C\left(1 + \left(\int_{\{|\boldsymbol{v}^k| \le 2M\}} \beta_k\, dx\right)^{\frac{1}{r'}} + \int_\Omega \beta_k |\boldsymbol{v}_k^k|\, dx\right) \\
&\le C\left(1 + \left(\int_{\{|\boldsymbol{v}^k| \le 2M\}} \beta_k\, dx\right)^{\frac{1}{r'}} + \left(\int_\Omega \beta_k |\boldsymbol{v}_k^k|^2\, dx\right)^{\frac12} \left(\int_\Omega \beta_k\, dx\right)^{\frac12}\right) \\
&\le C\left(1 + \|p^k\|_z^{\frac{q_0}{r'}} + \|\boldsymbol{v}^k\|_{1,r}^{\frac{q_2}{r'}} + \|p^k\|_z^{\frac{q_0}{2}} \left(\|p^k\|_z^{\frac{q_0}{2}} + \|\boldsymbol{v}^k\|_{1,r}^{\frac{q_1}{2}} + \|\boldsymbol{v}^k\|_{1,r}^{\frac{q_2}{2}}\right)\right).
\end{aligned}$$

Thus, the Young inequality implies that ($r' \ge 2 - q_0$)

$$\|p^k\|_z \le C\left(1 + \|\boldsymbol{v}^k\|_{1,r}^{\frac{q_2}{2-q_0}} + \|\boldsymbol{v}^k\|_{1,r}^{\frac{q_1}{2-q_0}}\right). \tag{5.98}$$

Substituting this into (5.97), we gain

$$\|\boldsymbol{v}^k\|_{1,r}^r + \|\boldsymbol{S}(p^k, \boldsymbol{D}\boldsymbol{v}^k)\|_{r'}^{r'} + \int_\Omega \beta_k |\boldsymbol{v}_k^k|^2\, dx \le C\left(1 + \|\boldsymbol{v}^k\|_{1,r}^{\frac{q_2 q_0}{2-q_0}} + \|\boldsymbol{v}^k\|_{1,r}^{\frac{q_1 q_0}{2-q_0}}\right).$$

Finally, we recall the assumptions on q_0, q_1, and q_2 stated in (5.63). Since $q_2 < r$ and $q_0 < 1$ we have that $q_2 q_0/(2 - q_0) < r$ and similarly, since $q_1 < (2 - q_0)r/q_0$, we deduce that

$$\|p^k\|_z + \|\boldsymbol{v}^k\|_{1,r}^r + \|\boldsymbol{S}(p^k, \boldsymbol{D}\boldsymbol{v}^k)\|_{r'}^{r'} + \int_\Omega \beta_k |\boldsymbol{v}_k^k|^2\, dx \le C. \tag{5.99}$$

In addition, while estimating the pressure, we also deduced on the way that $\|\beta_k \boldsymbol{v}_k^k\|_1 \le C$. But it would work with $L^{1+\delta}(\Omega)$ norm for some small $\delta > 0$ as well, and instead of considering $z < d'$ in estimating the pressure, we can improve our estimates to

$$\|p^k\|_{d'} + \|\beta_k \boldsymbol{v}_k^k\|_{1+\delta} < C. \tag{5.100}$$

Notice that the right-open intervals for q_0, q_1, and q_2 from assumption (5.63) are indispensable for such a claim. Therefore, it is possible by the estimates (5.99) and (5.100), to let $k \to \infty$ and presuppose (after a relabeling of the sequence) that

$$
\begin{aligned}
\boldsymbol{v}^k &\rightharpoonup \boldsymbol{v} && \text{weakly in } W^{1,r}(\Omega),\\
\boldsymbol{v}^k &\to \boldsymbol{v} && \text{a.e. in } \Omega,\\
p^k &\rightharpoonup p && \text{weakly in } L^{d'}(\Omega),\\
\boldsymbol{S}(p^k, \boldsymbol{D}(\boldsymbol{v}^k)) &\rightharpoonup \overline{\boldsymbol{S}} && \text{weakly in } L^{r'}(\Omega),\\
\beta_k \boldsymbol{v}_k^k &\rightharpoonup \overline{\beta \boldsymbol{v}} && \text{weakly in } L^{1+\delta}(\Omega),
\end{aligned}
\tag{5.101}
$$

where $(\boldsymbol{v} - \boldsymbol{v}_0) \in W^{1,r}_{\Gamma_1,\mathrm{div}}(\Omega)$. Using the above result, we moved (5.95) on to

$$\int_\Omega \overline{\boldsymbol{S}} \cdot \boldsymbol{D}\boldsymbol{\varphi}\, dx + \int_\Omega \overline{\beta \boldsymbol{v}} \cdot \boldsymbol{\varphi}\, dx - \int_\Omega p \operatorname{div} \boldsymbol{\varphi}\, dx = \int_\Omega \boldsymbol{F} \cdot \nabla \boldsymbol{\varphi}$$

valid for any $\boldsymbol{\varphi} \in W^{1,r}_{\Gamma_1}(\Omega) \cap L^\infty(\Omega)$ with $\operatorname{div} \boldsymbol{\varphi} \in L^d(\Omega)$. Not unlike the limit $\varepsilon \to 0_+$, the identification of the weak limits $\overline{\boldsymbol{S}}$ and $\overline{\beta \boldsymbol{v}}$ can and will be performed via the pointwise convergence of p^k and $\boldsymbol{D}\boldsymbol{v}^k$.

5.3.6.2 Decomposition of p^k

Contrary to the preceding limit procedure, we do not know whether p^k is a bounded sequence in $L^{r'}$ and consequently also in L^2, which is essential in the monotonicity relation (5.68). This is why we decompose p^k into two parts: one being converging almost everywhere (but strongly only in $L^1(\Omega)$) and the other still converging only weakly, though now in $L^{r'}(\Omega)$, whence the monotonicity property may be used. We will prove the convergence results in an arbitrary compactly contained subdomain $\Omega' \Subset \Omega$, which is from now considered fix.

Referring to Lemma 5.3.4 and noticing that both $\operatorname{div} \boldsymbol{S}(p^k, \boldsymbol{D}\boldsymbol{v}^k) - \operatorname{div} \boldsymbol{F}$ and $\beta_k \boldsymbol{v}_k^k$ belong to $W^{-1,2}(\Omega)$ (for each k), we may define

$$
\begin{aligned}
(\boldsymbol{v}_1^k, p_1^k) &:= \mathcal{H}(\operatorname{div} \boldsymbol{S}(p^k, \boldsymbol{D}\boldsymbol{v}^k) - \operatorname{div} \boldsymbol{F}),\\
(\boldsymbol{v}_2^k, p_2^k) &:= \mathcal{H}(-\beta^K \boldsymbol{T}_K \boldsymbol{v}^k).
\end{aligned}
\tag{5.102}
$$

The uniqueness of solutions to the Stokes problem and (5.95) imply

$$
\begin{aligned}
\boldsymbol{v}_1^k + \boldsymbol{v}_2^k &= 0,\\
p_1^k + p_2^k + p_\Omega^k &= p^k.
\end{aligned}
\tag{5.103}
$$

Note that we must add the mean value p^k_Ω since both p^k_1 and p^k_2 have zero mean value. From (5.99) and the continuity of $\mathcal{H}$ we observe

$$\|\boldsymbol{v}^k_1\|_{1,2} + \|p^k_1\|_2 \le C. \tag{5.104}$$

Further, we may apply (5.70) to $(5.102)_2$ with $k = 0$ and with the help of (5.100) deduce (for $\delta \ll 1$)

$$\begin{aligned}
\left\|\nabla^2 \boldsymbol{v}^k_2\right\|_{1+\delta;\Omega'} + \left\|\nabla p^k_2\right\|_{1+\delta;\Omega'} &\le c\left(\left\|\beta_k \boldsymbol{v}^k_k\right\|_{1+\delta} + \left\|\boldsymbol{v}^k_2\right\|_{1,1+\delta} + \left\|p^k_2\right\|_{1+\delta}\right)\\
&\le c\left(\left\|\beta_k \boldsymbol{v}^k_k\right\|_{1+\delta} + \left\|\boldsymbol{v}^k_1\right\|_{1,2} + \left\|p^k - p^k_1 - p^k_\Omega\right\|_{d'}\right)\\
&\le C.
\end{aligned}$$

Next step is to show the optimal estimate on p^k_1, i.e., to show that p^k_1 is in $L^{r'}$ locally in Ω. Note that if $r = 2$, we already got such a result and therefore we focus on the case $r < 2$. To do so, we use iteratively the following scheme. We assume that Ω'', Ω''' are such that $\Omega' \Subset \Omega'' \Subset \Omega''' \Subset \Omega$. Since $r' > 2$ we elicit the existence of a $r' - 2 \ge \sigma > 0$ such that (5.70) may be employed again, this time with $k = -1$, leading to

$$\begin{aligned}
&\left\|\boldsymbol{v}^k_1\right\|_{1,2+\sigma;\Omega''} + \left\|p^k_1\right\|_{2+\sigma;\Omega''}\\
&\le c\left(\left\|\operatorname{div} \boldsymbol{S}(p^k, \boldsymbol{D}\boldsymbol{v}^k) - \operatorname{div} \boldsymbol{F}\right\|_{-1,2+\sigma} + \left\|\boldsymbol{v}^k_1\right\|_{2+\sigma;\Omega'''} + \left\|p^k_1\right\|_{-1,2+\sigma;\Omega'''}\right)\\
&\le c\left(\left\|\boldsymbol{S}(p^k, \boldsymbol{D}\boldsymbol{v}^k)\right\|_{r'} + \left\|\boldsymbol{F}\right\|_{r'} + \left\|\boldsymbol{v}^k_1\right\|_{1,2;\Omega'''} + \left\|p^k_1\right\|_{2;\Omega'''}\right) \le C,
\end{aligned}$$

where for the last estimate we used (5.104). Repeating the same argument, we can finally bootstrap the above estimate and it yields

$$\left\|\boldsymbol{v}^k_1\right\|_{1,r';\Omega'} + \left\|p^k_1\right\|_{r';\Omega'} \le c\left(\left\|\boldsymbol{S}(p^k, \boldsymbol{D}\boldsymbol{v}^k)\right\|_{r'} + \left\|\boldsymbol{F}\right\|_{r'} + \left\|\boldsymbol{v}^k_1\right\|_{1,2} + \left\|p^k_1\right\|_{2}\right) \le C,$$

where C depends just on data and also on Ω'. Consequently, using (5.103), we deduce the following

$$\left(\left\|\boldsymbol{v}^k_1\right\|_{1,r';\Omega'} + \left\|\boldsymbol{v}^k_1\right\|_{2,1;\Omega'} + \left\|p^k_1\right\|_{r';\Omega'} + \left\|p^k_2\right\|_{1,1;\Omega'}\right) \le C(\Omega'). \tag{5.105}$$

Hence, using the reflexivity and the compact embedding, we may assume that for a subsequence that is not relabeled there holds

$$\begin{aligned}
p^k_1 &\rightharpoonup p_1 && \text{weakly in } L^{r'}(\Omega'),\\
p^k_2 &\to p_2 && \text{a.e. in } \Omega',\\
p^k_\Omega &\to p_\Omega && \text{in } \mathbb{R}
\end{aligned} \tag{5.106}$$

and obviously, (5.103) yields trivially $p_1+p_2+p_\Omega = p$. Thus, to show the pointwise convergence of p^k, it is enough to prove the pointwise convergence of p_1^k.

5.3.6.3 Convergence of p_1^k

We start the proof with an application of the Div–Curl lemma. We first notice that (5.102) and (5.105) imply

$$\left\| \operatorname{div}\left(\boldsymbol{S}(p^k, \boldsymbol{D}\boldsymbol{v}^k) - p_1^k \boldsymbol{I} - \boldsymbol{F}\right)\right\|_{1;\Omega'} = \left\|\nabla p_2^k + \beta_k \boldsymbol{v}_k^k\right\|_{1;\Omega'} \leq C(\Omega').$$

As $L^1(\Omega') \hookrightarrow\hookrightarrow W^{-1,q'}(\Omega')$ for $q > d$, this estimate together with (5.101) and (5.106) is crucial for using Div–Curl lemma 5.3.9. Indeed, let $s > r$ and a sequence $\{\boldsymbol{\varphi}^k\}_{k=1}^\infty$ fulfill

$$\boldsymbol{\varphi}^k \rightharpoonup \boldsymbol{\varphi} \quad \text{weakly in } W^{1,s}(\Omega'). \tag{5.107}$$

Then $1/r' + 1/s < 1$, $\operatorname{curl} \nabla \boldsymbol{\varphi}^k = 0$, and Div–Curl lemma 5.3.9 imply

$$\left(\boldsymbol{S}(p^k, \boldsymbol{D}\boldsymbol{v}^k) - p_1^k \boldsymbol{I}\right) \cdot \nabla \boldsymbol{\varphi}^k \rightharpoonup \left(\overline{\boldsymbol{S}} - p_1 \boldsymbol{I}\right) \cdot \nabla \boldsymbol{\varphi} \quad \text{weakly in } L^1(\Omega') \tag{5.108}$$

because, we evidently have $\boldsymbol{F} \cdot \nabla \boldsymbol{\varphi}^k \rightharpoonup \boldsymbol{F} \cdot \nabla \boldsymbol{\varphi}$ weakly in $L^1(\Omega')$. The convergence result (5.108) is the starting point for further analysis.

First, we apply (5.108) to obtain a result for p_1^k. To do so, we consider arbitrary $L > 0$ and consider (5.108) with $\boldsymbol{\varphi}^k = \nabla \psi_L^k$, where[8] (see Lemma 5.3.7 for notation)

$$\psi_L^k = \mathcal{N}(T_L(p_1^k - p_1)). \tag{5.109}$$

Because of truncation, we have (for a subsequence if needed)

$$T_L(p_1^k - p_1) \rightharpoonup \overline{T}_L \quad \text{weakly in } L^q(\Omega) \text{ for all } q \in [1, \infty),$$

and hence by the continuity of $\mathcal{N}$ (see Lemma 5.3.7) also

$$\psi_L^k \rightharpoonup \psi_L = \mathcal{N}(\overline{T}_L) \quad \text{weakly in } W^{2,q}(\Omega) \text{ for all } q \in [1, \infty). \tag{5.110}$$

[8] This definition means that $-\Delta \psi_L^k = T_L(p_1^k - p_1)$.

Thus, it follows from (5.108) that

$$\left(S(p^k, Dv^k) - p_1^k I\right) \cdot \nabla^2 \psi_L^k \rightharpoonup \left(\overline{S} - p_1 I\right) \cdot \nabla^2 \psi_L \quad \text{weakly in } L^1(\Omega').$$

Using the pointwise convergence of p_2^k (see (5.106)), the above result can be rewritten as

$$\begin{aligned}&\left(S(p^k, Dv^k) - S(p_1 + p_2^k + p_\Omega^k, Dv) - (p_1^k - p_1) I\right) \cdot \nabla^2 \psi_L^k \\ &\quad \rightharpoonup \left(\overline{S} - S(p, Dv)\right) \cdot \nabla^2 \psi_L \quad \text{weakly in } L^1(\Omega').\end{aligned}$$

Therefore, using the definition of ψ_L^k, we find that for any measurable $\Omega'' \subset \Omega'$

$$\begin{aligned}\limsup_{k\to\infty} \int_{\Omega''} |p_1^k - p_1||T_L(p_1^k - p_1)|\, dx &= \limsup_{k\to\infty} \int_{\Omega''} -(p_1^k - p_1)\Delta\psi_L^k\, dx \\ &= \limsup_{k\to\infty} \int_{\Omega''} -(p_1^k - p_1)\, I \cdot \nabla^2 \psi_L^k\, dx \\ &\le \limsup_{k\to\infty} \int_{\Omega''} |S(p^k, Dv^k) - S(p_1 + p_2^k + p_\Omega^k, Dv)||\nabla^2 \psi_L^k|\, dx \\ &\quad + \limsup_{k\to\infty} \left| \int_{\Omega''} \left(\overline{S} - S(p, Dv)\right) \cdot \nabla^2 \psi_L\, dx \right|. \end{aligned} \tag{5.111}$$

We continue by estimating the right-hand side. For the first term, the relation (5.69) implies that

$$\begin{aligned}&\int_{\Omega''} |S(p^k, Dv^k) - S(p_1 + p_2^k + p_\Omega^k, Dv)||\nabla^2 \psi_L^k|\, dx \\ &\le \gamma_0 \int_{\Omega''} |p_1^k - p_1||\nabla^2 \psi_L^k|\, dx \\ &\quad + C_2 \int_{\Omega''} \int_0^1 (1 + |\overline{D}(s)|^2)^{(r-2)/2} |D(v^k - v)||\nabla^2 \psi_L^k|\, ds\, dx,\end{aligned} \tag{5.112}$$

where we denoted $\overline{D}(s) = Dv^k + s(Dv - Dv^k)$. Defining

$$I^K := \int_0^1 (1 + |\overline{D}(s)|^2)^{(r-2)/2} |D(v^k - v)|^2 ds,$$

using the Hölder inequality and the fact that $r \le 2$, we see that (5.112) leads to

$$\begin{aligned}&\int_{\Omega''} |S(p^k, Dv^k) - S(p_1 + p_2^k + p_\Omega^k, Dv)||\nabla^2 \psi_L^k|\, dx \\ &\quad \le \gamma_0 \left\| p_1^k - p_1 \right\|_{2;\Omega''} \left\| \nabla^2 \psi_L^k \right\|_{2;\Omega''} + C_2 \left\| I^k \right\|_{1;\Omega''}^{1/2} \left\| \nabla^2 \psi_L^k \right\|_{2;\Omega''},\end{aligned}$$

which substituted into (5.111) gives

$$\begin{aligned}
&\limsup_{k\to\infty}\int_{\Omega''}|p_1^k-p_1||T_L(p_1^k-p_1)|\,dx\\
&\quad\le\limsup_{k\to\infty}\left(\gamma_0\left\|p_1^k-p_1\right\|_{2;\Omega''}\left\|\nabla^2\psi_L^k\right\|_{2;\Omega''}+C_2\left\|I^k\right\|_{1;\Omega''}^{1/2}\left\|\nabla^2\psi_L^k\right\|_{2;\Omega''}\right)\\
&\qquad+\left|\int_{\Omega''}\left(\overline{\boldsymbol{S}}-\boldsymbol{S}(p,\boldsymbol{Dv})\right)\cdot\nabla^2\psi_L\,dx\right|.
\end{aligned}\tag{5.113}$$

Next, we need to "replace" the term $|\nabla^2\psi_L^k|$ by the one containing $(p_1^k-p_1)$. To do so, we compare the weak limits of $|\nabla^2\psi_L^k|^2$ and $|\Delta\psi_L^k|^2$. Hence, thanks (5.110) we may assume without loss of generality that both $|\nabla^2\psi_L^k|^2$ and $|\Delta\psi_L^k|^2$ converge weakly in $L^q(\Omega')$ for any $q\in[1,\infty)$ as $k\to\infty$. To compare these weak limits, it suffices to investigate

$$\lim_{k\to\infty}\int_{\Omega'}\left(|\nabla^2\psi_L^k|^2-|\Delta\psi_L^k|^2\right)\varphi\,dx$$

for arbitrary $\varphi\in\mathcal{C}_c^\infty(\Omega')$. Using the integration by parts, we find that

$$\begin{aligned}
&\lim_{k\to\infty}\int_{\Omega'}(|\nabla^2\psi_L^k|^2-|\Delta\psi_L^k|^2)\varphi\,dx\\
&\quad=\lim_{k\to\infty}\int_{\Omega'}\left(\nabla^2\psi_L^k\cdot\nabla^2\psi_L^k\varphi-|\Delta\psi_L^k|^2\varphi\right)dx\\
&\quad=\lim_{k\to\infty}\int_{\Omega'}\left(-\nabla\psi_L^k\cdot\nabla\Delta\psi_L^k\varphi-(\nabla\psi_L^k\otimes\nabla\varphi)\cdot\nabla^2\psi_L^k-|\Delta\psi_L^k|^2\varphi\right)dx\\
&\quad=\lim_{k\to\infty}\int_{\Omega'}\left(\nabla\psi_L^k\cdot\nabla\varphi\Delta\psi_L^k-(\nabla\psi_L^k\otimes\nabla\varphi)\cdot\nabla^2\psi_L^k\right)dx\\
&\quad=\int_{\Omega'}\left(\nabla\psi_L\cdot\nabla\varphi\Delta\psi_L-(\nabla\psi_L\otimes\nabla\varphi)\cdot\nabla^2\psi_L\right)dx\\
&\quad=\int_{\Omega'}(|\nabla^2\psi_L|^2-|\Delta\psi_L|^2)\varphi\,dx.
\end{aligned}$$

Therefore, we can use the density argument and the fact that (5.110) holds for any $q<\infty$, to get for all measurable $\Omega''\subset\Omega'$

$$\lim_{k\to\infty}\int_{\Omega''}\left(|\nabla^2\psi_L^k|^2-|\Delta\psi_L^k|^2\right)dx=\int_{\Omega''}\left(|\nabla^2\psi_L|^2-|\Delta\psi_L|^2\right)dx,$$

which can be turned to

$$\limsup_{k\to\infty}\int_{\Omega''}|\nabla^2\psi_L^k|^2\,dx \le \limsup_{k\to\infty}\int_{\Omega''}|\Delta\psi_L^k|^2\,dx + \int_{\Omega''}\left(|\nabla^2\psi_L|^2-|\Delta\psi_L|^2\right)dx.$$

Hence, substituting this relation into (5.113), employing the pointwise estimate

$$|\Delta\psi_L^k|^2 = |T_L(p_1^k-p_1)|^2 \le |p_1^k-p_1|^2$$

and the a priori estimates (5.101) and (5.104), we get

$$\begin{aligned}
&\limsup_{k\to\infty}\int_{\Omega''}|p_1^k-p_1||T_L(p_1^k-p_1)|\,dx\\
&\le \limsup_{k\to\infty}\left(\gamma_0\left\|p_1^k-p_1\right\|_{2;\Omega''}+C_2\left\|I^k\right\|_{1;\Omega''}^{1/2}\right)\\
&\qquad\times\left(\left\|p_1^k-p_1\right\|_{2;\Omega''}^2+\int_{\Omega''}\left(|\nabla^2\psi_L|^2-|\Delta\psi_L|^2\right)dx\right)^{1/2}\\
&\qquad+\left|\int_{\Omega''}\left(\overline{S}-S(p,Dv)\right)\cdot\nabla^2\psi_L\,dx\right| \qquad (5.114)\\
&\le \limsup_{k\to\infty}\left(\gamma_0\left\|p_1^k-p_1\right\|_{2;\Omega''}^2+C_2\left\|I^k\right\|_{1;\Omega''}^{1/2}\left\|p_1^k-p_1\right\|_{2;\Omega''}\right)\\
&\qquad+C\left|\int_{\Omega''}\left(|\nabla^2\psi_L|^2-|\Delta\psi_L|^2\right)dx\right|^{1/2}\\
&\qquad+\left|\int_{\Omega''}\left(\overline{S}-S(p,Dv)\right)\cdot\nabla^2\psi_L\,dx\right|.
\end{aligned}$$

Finally, we make a special selection of Ω'' so that we can remove the truncation T_L from the above relation. We recall Biting Lemma 5.3.8, which we apply on the sequence

$$f^k := |p_1^k|^{r'} + |S(p^k, Dv^k)|^{r'}. \qquad (5.115)$$

Since $\{f^k\}_{k=1}^\infty$ is a bounded sequence in $L^1(\Omega')$, which can be observed (5.101) and (5.105), the Biting lemma guarantees the existence of a non-increasing sequence of measurable sets $E_n \subset \Omega'$ fulfilling $\lim_{n\to\infty}|E_n| = 0$ such that (modulo a subsequence) for each $n \in \mathbb{N}$ the sequence $\{f^k\}_{k=1}^\infty$ is uniformly equi-integrable in $\Omega_n := \Omega' \setminus E_n$. On these sets we now consider the limit $L \to \infty$. First of all, we show that the left-hand side of (5.114) gives us the full information about $\|p_1^k - p_1\|_2$. We denote $\Omega_L^k = \{|p_1^k - p_1| > L\}$ and it follows from (5.104) that $|\Omega_L^k| \le C/L^2$. Hence, using the uniform equi-integrability of $|p_1^k|^{r'}$ in Ω_n (note that $r' \ge 2$), we have

$$\limsup_{L\to\infty}\limsup_{k\to\infty}\int_{\Omega_n}|p_1^k-p_1||p_1^k-p_1-T_L(p_1^k-p_1)|\,dx$$
$$\leq \limsup_{L\to\infty}\limsup_{k\to\infty}\int_{\Omega_n\cap\Omega_L^k}|p_1^k-p_1|^2\,dx = 0.$$

Consequently,

$$\limsup_{k\to\infty}\int_{\Omega_n}|p_1^k-p_1|^2\,dx = \limsup_{L\to\infty}\limsup_{k\to\infty}\int_{\Omega_n}|p_1^k-p_1||T_L(p_1^k-p_1)|\,dx. \tag{5.116}$$

The remaining L-dependent terms in (5.114) tend to zero as $L\to\infty$, provided that

$$\psi_L \to 0 \quad \text{strongly in } W^{2,2}(\Omega). \tag{5.117}$$

Since $\mathcal{N}$ (see Lemma 5.3.7) is continuous, the problem (5.109) implies that to check (5.117), it is enough (see (5.110)) to deduce

$$\overline{T}_L \to 0 \quad \text{strongly in } L^2(\Omega). \tag{5.118}$$

To achieve this, we first draw from (5.104) that

$$T_L(p_1^k-p_1)-(p_1^k-p_1) \rightharpoonup \overline{T}_L \quad \text{weakly in } L^2(\Omega)$$

and therefore from the weak lower semicontinuity of the L^1-norm we find that

$$\left\|\overline{T}_L\right\|_1 \leq \liminf_{k\to\infty}\left\|T_L(p_1^k-p_1)-(p_1^k-p_1)\right\|_1 \leq 2\limsup_{k\to\infty}\int_{\Omega_L^k}|p_1^k-p_1|\,dx \leq C/L$$

and therefore (as $L\to\infty$)

$$\overline{T}_L \to 0 \quad \text{strongly in } L^1(\Omega). \tag{5.119}$$

Furthermore, we can strengthen the above convergence result from $L^1(\Omega)$ into $L^2(\Omega)$. To do so, we use the Lebesgue dominated convergence theorem. Denoting $\nu\in L^2(\Omega)$ the weak limit (recall that p_1^k is a bounded sequence in $L^2(\Omega)$)

$$|p_1^k-p_1| \rightharpoonup \nu \quad \text{weakly in } L^2(\Omega),$$

a simple estimate $|T_L(p_1^k-p_1)|\leq|p_1^k-p_1|$ implies $|\overline{T}_L|^2\leq\nu^2\in L^1(\Omega)$. Using the Lebesgue dominated convergence and (5.119), we have (5.118) and consequently also (5.117). Thus, going back to (5.114) and setting there $\Omega'':=\Omega_n$ and using (5.116), we have that

$$\limsup_{k\to\infty}\left\|p_1^k-p_1\right\|_{2;\Omega_n}^2 \leq \limsup_{k\to\infty}\left(\gamma_0\left\|p_1^k-p_1\right\|_{2;\Omega_n}^2+C_2\left\|I^k\right\|_{1;\Omega_n}^{1/2}\left\|p_1^k-p_1\right\|_{2;\Omega_n}\right).$$

The Young inequality then easily implies that for each n (note that $\gamma_0 < 1$)

$$\limsup_{k\to\infty} \left\| p_1^k - p_1 \right\|_{2;\Omega_n} \le \frac{C_2}{1-\gamma_0} \limsup_{k\to\infty} \left\| I^k \right\|_{1;\Omega_n}^{1/2}. \tag{5.120}$$

The final step is to provide the estimate for the right-hand side of (5.120). We again use the observation (5.108) based on the Div–Curl lemma 5.3.9. We fix $\lambda > 0$, recall Lemma 5.3.10 about Lipschitz approximations of Sobolev functions, and set $\boldsymbol{\varphi}^k := \boldsymbol{v}_\lambda^k$ in (5.108), where $\boldsymbol{v}_\lambda^k$ denotes the Lipschitz approximation of $\boldsymbol{v}^k$. Note that due to (5.71), fulfillment of the condition (5.107) may be taken for granted. Hence

$$\left(\boldsymbol{S}(p^k, \boldsymbol{D}\boldsymbol{v}^k) - p_1^k \boldsymbol{I}\right) \cdot \nabla \boldsymbol{v}_\lambda^k \rightharpoonup \left(\overline{\boldsymbol{S}} - p_1 \boldsymbol{I}\right) \cdot \nabla \overline{\boldsymbol{v}}_\lambda \quad \text{weakly in } L^1(\Omega'), \tag{5.121}$$

where

$$\boldsymbol{v}_\lambda^k \rightharpoonup \overline{\boldsymbol{v}}_\lambda \quad \text{weakly in } W^{1,q}(\Omega) \text{ for all } q \in [1, \infty).$$

Evidently, (5.121) implies

$$\lim_{k\to\infty} \int_{\Omega_n} \left(\boldsymbol{S}(p^k, \boldsymbol{D}\boldsymbol{v}^k) - p_1^k \boldsymbol{I}\right) \cdot \nabla \boldsymbol{v}_\lambda^k \, dx = \int_{\Omega_n} \left(\overline{\boldsymbol{S}} - p_1 \boldsymbol{I}\right) \cdot \nabla \overline{\boldsymbol{v}}_\lambda \, dx \tag{5.122}$$

for each n, where Ω_n are still the subsets specified above when we applied the Biting lemma to the sequence given in (5.115). Reordering the terms in (5.122), we have

$$\begin{aligned}
&\lim_{k\to\infty} \int_{\Omega_n} \left(\boldsymbol{S}(p^k, \boldsymbol{D}\boldsymbol{v}^k) - p_1^k \boldsymbol{I}\right) \cdot \nabla \boldsymbol{v}^k \, dx \\
&= \lim_{k\to\infty} \int_{\Omega_n} \left(\boldsymbol{S}(p^k, \boldsymbol{D}\boldsymbol{v}^k) - p_1^k \boldsymbol{I}\right) \cdot \nabla (\boldsymbol{v}^k - \boldsymbol{v}_\lambda^k) \, dx \\
&\quad + \lim_{k\to\infty} \int_{\Omega_n} \left(\boldsymbol{S}(p^k, \boldsymbol{D}\boldsymbol{v}^k) - p_1^k \boldsymbol{I}\right) \cdot \nabla \boldsymbol{v}_\lambda^k \, dx \\
&= \lim_{k\to\infty} \int_{\Omega_n} \left(\boldsymbol{S}(p^k, \boldsymbol{D}\boldsymbol{v}^k) - p_1^k \boldsymbol{I}\right) \cdot \nabla (\boldsymbol{v}^k - \boldsymbol{v}_\lambda^k) \, dx + \int_{\Omega_n} \left(\overline{\boldsymbol{S}} - p_1 \boldsymbol{I}\right) \cdot \nabla \overline{\boldsymbol{v}}_\lambda \, dx.
\end{aligned}$$

Since the left-hand side is independent of λ, we obtain

$$\begin{aligned}
&\lim_{k\to\infty} \int_{\Omega_n} \left(\boldsymbol{S}(p^k, \boldsymbol{D}\boldsymbol{v}^k) - p_1^k \boldsymbol{I}\right) \cdot \nabla \boldsymbol{v}^k \, dx \\
&= \lim_{\lambda\to\infty} \lim_{k\to\infty} \int_{\Omega_n} \left(\boldsymbol{S}(p^k, \boldsymbol{D}\boldsymbol{v}^k) - p_1^k \boldsymbol{I}\right) \cdot \nabla (\boldsymbol{v}^k - \boldsymbol{v}_\lambda^k) \, dx \\
&\quad + \lim_{\lambda\to\infty} \int_{\Omega_n} \left(\overline{\boldsymbol{S}} - p_1 \boldsymbol{I}\right) \cdot \nabla \overline{\boldsymbol{v}}_\lambda \, dx.
\end{aligned} \tag{5.123}$$

First term on the right vanishes. Indeed, thanks to (5.73) we have

$$\int_{\Omega_n} \left| \left(\boldsymbol{S}(p^k, \boldsymbol{D}\boldsymbol{v}^k) - p_1^k \boldsymbol{I} \right) \cdot \nabla(\boldsymbol{v}^k - \boldsymbol{v}_\lambda^k) \right| \, dx$$
$$\leq C \left\| \boldsymbol{v}^k \right\|_{1,r} \left\| \boldsymbol{S}(p^k, \boldsymbol{D}\boldsymbol{v}^k) - p_1^k \boldsymbol{I} \right\|_{r'; \Omega_n \cap \{\boldsymbol{v}^k \neq \boldsymbol{v}_\lambda^k\}}$$

Therefore, the uniform equi-integrability of

$$|\boldsymbol{S}(p^k, \boldsymbol{D}\boldsymbol{v}^k)|^{r'} + |p_1^k|^{r'}$$

in Ω_n (see (5.115)), the estimate (5.99), and the smallness property (5.72), i.e., $|\{\boldsymbol{v}^k \neq \boldsymbol{v}_\lambda^k\}| \leq C/\lambda^r$, lead to

$$\lim_{\lambda \to \infty} \lim_{k \to \infty} \int_{\Omega_n} \left(\boldsymbol{S}(p^k, \boldsymbol{D}\boldsymbol{v}^k) - p_1^k \boldsymbol{I} \right) \cdot \nabla(\boldsymbol{v}^k - \boldsymbol{v}_\lambda^k) \, dx = 0. \tag{5.124}$$

Next, we focus on the identification of the second term on the right-hand side of (5.123). The weak lower semicontinuity of a norm and (5.73) give

$$\|\overline{\boldsymbol{v}}_\lambda\|_{1,r} \leq \liminf_{k \to \infty} \left\| \boldsymbol{v}_\lambda^k \right\|_{1,r} \leq C \limsup_{k \to \infty} \left\| \boldsymbol{v}^k \right\|_{1,r} \leq C.$$

Accordingly, we may assume that there exists $\overline{\boldsymbol{v}}$ such that (as $\lambda \to \infty$)

$$\overline{\boldsymbol{v}}_\lambda \rightharpoonup \overline{\boldsymbol{v}} \quad \text{weakly in } W^{1,r}(\Omega).$$

To link $\boldsymbol{v}$ and $\overline{\boldsymbol{v}}$, we use the compact embedding, (5.72), and (5.73) to get

$$\|\boldsymbol{v} - \overline{\boldsymbol{v}}_\lambda\|_1 = \lim_{k \to \infty} \int_\Omega |\boldsymbol{v}^k - \boldsymbol{v}_\lambda^k| \, dx = \lim_{k \to \infty} \int_{\{\boldsymbol{v}^k \neq \boldsymbol{v}_\lambda^k\}} |\boldsymbol{v}^k - \boldsymbol{v}_\lambda^k| \, dx \leq C/\lambda^{r-1},$$

which directly lead to

$$\overline{\boldsymbol{v}}_\lambda \to \boldsymbol{v} \quad \text{strongly in } L^1(\Omega),$$

and due to uniqueness of weak limits

$$\overline{\boldsymbol{v}}_\lambda \rightharpoonup \boldsymbol{v} \quad \text{weakly in } W^{1,r}(\Omega).$$

Thus, using (5.124) and the above relation, we can let $\lambda \to \infty$ on the right-hand side of (5.123) to gain

$$\lim_{k\to\infty} \int_{\Omega_n} \left(\boldsymbol{S}(p^k, \boldsymbol{D}\boldsymbol{v}^k) - p_1^k \boldsymbol{I} \right) \cdot \nabla \boldsymbol{v}^k dx = \int_{\Omega_n} \left(\overline{\boldsymbol{S}} - p_1 \boldsymbol{I} \right) \cdot \nabla \boldsymbol{v} \, dx.$$

Since $\boldsymbol{v}^k$ as well as $\boldsymbol{v}$ are divergence-free, this is actually equivalent to

$$\lim_{k\to\infty} \int_{\Omega_n} \boldsymbol{S}(p^k, \boldsymbol{D}\boldsymbol{v}^k) \cdot \boldsymbol{D}\boldsymbol{v}^k dx = \int_{\Omega_n} \overline{\boldsymbol{S}} \cdot \boldsymbol{D}\boldsymbol{v} \, dx.$$

Finally, the strong convergence of $p_2^k + p_\Omega^k$ (see (5.106)) implies

$$\lim_{k\to\infty} \int_{\Omega_n} (\boldsymbol{S}(p^k, \boldsymbol{D}\boldsymbol{v}^k) - \boldsymbol{S}(p_1 + p_2^k + p_\Omega^k, \boldsymbol{D}\boldsymbol{v})) \cdot \boldsymbol{D}(\boldsymbol{v}^k - \boldsymbol{v}) \, dx = 0. \tag{5.125}$$

Hence, recalling (5.68) we see that (5.125) gives the desired estimate on the behavior of I^k

$$\limsup_{k\to\infty} \left\| I^k \right\|_{1;\Omega_n} \le \frac{\gamma_0^2}{C_1^2} \limsup_{k\to\infty} \left\| p_1^k - p_1 \right\|_{2;\Omega_n}^2. \tag{5.126}$$

Substituting (5.126) into (5.120), we have

$$\limsup_{k\to\infty} \left\| p_1^k - p_1 \right\|_{2;\Omega_n} \le \frac{C_2 \gamma_0}{C_1(1-\gamma_0)} \limsup_{k\to\infty} \left\| p_1^k - p_1 \right\|_{2;\Omega_n}.$$

Using the assumption (5.62) on γ_0, i.e.,

$$\gamma_0 < \frac{C_1}{C_1 + C_2} \Longleftrightarrow \frac{C_2 \gamma_0}{C_1(1-\gamma_0)} < 1,$$

we have

$$\lim_{k\to\infty} \left\| p_1^k - p_1 \right\|_{2;\Omega_n} = 0. \tag{5.127}$$

Since $\lim_{n\to\infty} \left| \Omega' \setminus \Omega_n \right| = 0$, we may suppose $p_1^k \to p_1$ a.e. in Ω' (at least for a subsequence).

5.3.6.4 Convergence of $\boldsymbol{D}\boldsymbol{v}^k$

We have yet to affirm the strong convergence of $\boldsymbol{D}\boldsymbol{v}^k$. But this is now done exactly as in the ε-passage. Following the computation below (5.94), we see that

$$\|\boldsymbol{D}(\boldsymbol{v}^k - \boldsymbol{v})\|_{r;\Omega_n}^2 \leq C \int_{\Omega_n} I^k dx$$

and by (5.68) we further deduce

$$C\|\boldsymbol{D}(\boldsymbol{v}^k - \boldsymbol{v})\|_{r;\Omega_n}^2 \leq \int_{\Omega_n} (\boldsymbol{S}(p^k, \boldsymbol{D}\boldsymbol{v}^k) - \boldsymbol{S}(p_1 + p_2^k + p_\Omega^k, \boldsymbol{D}\boldsymbol{v})) \cdot \boldsymbol{D}(\boldsymbol{v}^k - \boldsymbol{v})\, dx$$
$$+ \frac{\gamma_0^2}{2C_1}\|p_1^k - p_1\|_{2;\Omega_n}^2.$$

The right-hand side tends to zero as $k \to \infty$ by (5.125) and (5.127). Thus, we may assume $\boldsymbol{D}\boldsymbol{v}^k \to \boldsymbol{D}\boldsymbol{v}$ a.e. in Ω'.

5.3.7 Notes to the Proof of Theorem 5.3.1

We sketch here the proof of Theorem 5.3.1. The proof is again based on the following quasi-compressible approximation

$$\begin{aligned}
(\boldsymbol{S}, \boldsymbol{D}\boldsymbol{v}) \in \mathcal{B},\ (\boldsymbol{m}, \boldsymbol{v}, p, \boldsymbol{D}\boldsymbol{v}) \in \mathcal{A} \qquad & \text{in } \Omega,\\
-\operatorname{div} \boldsymbol{S} + \boldsymbol{m} = -\nabla p - \operatorname{div} \boldsymbol{F} \qquad & \text{in } \Omega,\\
\varepsilon\Delta p = \operatorname{div} \boldsymbol{v} \qquad & \text{in } \Omega,\\
\boldsymbol{v} = \boldsymbol{v}_0,\ \nabla p \cdot \boldsymbol{n} = 0 \qquad & \text{on } \Gamma_1,\\
\boldsymbol{S}\boldsymbol{n} = \boldsymbol{F}\boldsymbol{n},\ p = 0 \qquad & \text{on } \Gamma_2.
\end{aligned} \tag{5.128}$$

The existence of a solution $(\boldsymbol{S}^\varepsilon, \boldsymbol{m}^\varepsilon, \boldsymbol{v}^\varepsilon, p^\varepsilon)$ to the above system can be proved by using the monotone operator theory, combined with the compactness of the pressure. Moreover, one can obtain the uniform estimate

$$\sqrt{\varepsilon}\|p^\varepsilon\|_{1,2} + \|\boldsymbol{S}^\varepsilon\|_{r'} + \|\boldsymbol{m}^\varepsilon\|_{q'} + \|\boldsymbol{v}^\varepsilon\|_q \leq C(\|\boldsymbol{F}\|_{r'}, \|\boldsymbol{v}_0\|_{1,r}, \|\boldsymbol{v}_0\|_q).$$

The uniform estimate for the pressure can be obtained similarly as in preceding sections. But the essential part is to identify the nonlinearities. If we test the second equation in (5.128) by $(\boldsymbol{v}^\varepsilon - \boldsymbol{v})$, where $\boldsymbol{v}$ is the weak limit, we get

$$\limsup_{\varepsilon \to 0_+} \int_\Omega \boldsymbol{S}^\varepsilon \cdot \nabla(\boldsymbol{v}^\varepsilon - \boldsymbol{v}) + \boldsymbol{m}^\varepsilon \cdot (\boldsymbol{v}^\varepsilon - \boldsymbol{v})\, dx = 0.$$

Thus, thanks to the monotonicity of $\mathcal{A}$ we can deduce

$$\limsup_{\varepsilon \to 0_+} \int_\Omega \boldsymbol{S}^\varepsilon \cdot \nabla(\boldsymbol{v}^\varepsilon - \boldsymbol{v})\, dx \le \limsup_{\varepsilon \to 0_+} \int_\Omega \boldsymbol{m}^*(\boldsymbol{v}, \boldsymbol{D}\boldsymbol{v}^\varepsilon, p^\varepsilon) \cdot (\boldsymbol{v} - \boldsymbol{v}^\varepsilon)\, dx.$$

Due to the compact embedding, we also have

$$\boldsymbol{v}^\varepsilon \to \boldsymbol{v} \quad \text{a.e. in } \Omega, \tag{5.129}$$

which implies

$$\boldsymbol{m}^*(\boldsymbol{v}, \boldsymbol{D}\boldsymbol{v}^\varepsilon, p^\varepsilon) \cdot (\boldsymbol{v} - \boldsymbol{v}^\varepsilon) \to 0 \text{ a.e. in } \Omega.$$

Thus, thanks to the growth assumptions **(A4)**, we can use the Vitali theorem to observe

$$\limsup_{\varepsilon \to 0_+} \int_\Omega \boldsymbol{m}^*(\boldsymbol{v}, \boldsymbol{D}\boldsymbol{v}^\varepsilon, p^\varepsilon) \cdot (\boldsymbol{v} - \boldsymbol{v}^\varepsilon)\, dx = 0$$

and therefore also

$$\limsup_{\varepsilon \to 0_+} \int_\Omega \boldsymbol{S}^\varepsilon \cdot \nabla(\boldsymbol{v}^\varepsilon - \boldsymbol{v})\, dx \le 0. \tag{5.130}$$

The above relation directly leads to $(\boldsymbol{S}, \delta \boldsymbol{v}) \in \mathcal{B}$. Next, if $\mathcal{A}$ is independent of $\boldsymbol{D}\boldsymbol{v}$ and p, then (5.129) can be used to show that $(\boldsymbol{m}, \boldsymbol{v}, \boldsymbol{D}\boldsymbol{v}, p) \in \mathcal{A}$. Further if the graph $\mathcal{B}$ is strictly monotone with respect to $\boldsymbol{D}$, i.e., if **(A6)$_{\boldsymbol{D}}$** holds, then it follows from (5.130) that

$$\boldsymbol{D}\boldsymbol{v}^\varepsilon \to \boldsymbol{D}\boldsymbol{v} \quad \text{a.e. in } \Omega.$$

Consequently, if the graph $\mathcal{A}$ is independent of p but depending on $\boldsymbol{D}\boldsymbol{v}$, then we again obtain $(\boldsymbol{m}, \boldsymbol{v}, \boldsymbol{D}\boldsymbol{v}, p) \in \mathcal{A}$. Finally, if the graph $\mathcal{A}$ is also p-dependent, then we need to prove the pointwise convergence of the pressure p^ε. To do so, we use the strict monotonicity of $\mathcal{B}$ with respect to $\boldsymbol{S}$, i.e., **(A6)$_{\boldsymbol{D}}$** and conclude from (5.130) that

$$\boldsymbol{S}^\varepsilon \to \boldsymbol{S}\boldsymbol{v} \quad \text{a.e. in } \Omega.$$

Then we can decompose the pressure again onto two parts $p^\varepsilon = p_1^\varepsilon + p_2^\varepsilon$, where[9]

[9]Here the symbol $\sim$ denotes just that the sequences on both sides behave in the same way, i.e., if one converges weakly, the second one converges also weakly.

$$p_1^\varepsilon \rightharpoonup \boldsymbol{S}^\varepsilon \qquad \Longrightarrow \qquad p_1^\varepsilon \to p_1 \text{ a.e. in } \Omega,$$

and with p_2^ε being uniformly bounded in $W^{1,1}(\Omega)$ and consequently from the compact embedding we have

$$\nabla p_2^\varepsilon \rightharpoonup \boldsymbol{m}^\varepsilon \qquad \Longrightarrow \qquad p_2^\varepsilon \to p_2 \text{ a.e. in } \Omega.$$

Therefore, $p^\varepsilon \to p$ almost everywhere in Ω and we can show that $(\boldsymbol{m}, \boldsymbol{v}, \boldsymbol{D}\boldsymbol{v}, p) \in \mathcal{A}$.

5.4 Generalized Navier–Stokes–Brinkman Model

This section is meant just as a short overview of available results for systems where we do not neglect the inertial terms. It is split into two parts, first, we consider the steady case and then the unsteady case.

5.4.1 *Steady Model*

In the steady case, the full model reads as

$$\begin{aligned}
-\operatorname{div} \boldsymbol{S} + \operatorname{div}(\boldsymbol{v} \otimes \boldsymbol{v}) + \boldsymbol{m} &= -\nabla p + \boldsymbol{f} && \text{in } \Omega,\\
\boldsymbol{S} &= \boldsymbol{S}^T && \text{in } \Omega,\\
\operatorname{div} \boldsymbol{v} &= 0 && \text{in } \Omega,\\
\boldsymbol{h}(\boldsymbol{m}, \boldsymbol{v}, \boldsymbol{D}\boldsymbol{v}, p) &= \boldsymbol{0} && \text{in } \Omega,\\
\boldsymbol{H}(\boldsymbol{S}, \boldsymbol{D}\boldsymbol{v}, p) &= \boldsymbol{0} && \text{in } \Omega,\\
\boldsymbol{v} &= \boldsymbol{v}_0 && \text{on } \Gamma_1,\\
(\boldsymbol{S} - p\boldsymbol{I})\boldsymbol{n} &= \boldsymbol{s}_0 && \text{on } \Gamma_2.
\end{aligned} \tag{5.131}$$

Hence, the "only" difference from the model (5.55) is the presence of the term

$$\operatorname{div}(\boldsymbol{v} \otimes \boldsymbol{v}),$$

which seems to be only a minor difficulty. In certain sense it is true, since the methods presented in the previous sections are powerful enough to handle the convective term only as a compact perturbation. Therefore, we can formulate the following "theorem," which we do not prove but rather make several comments.

Meta-Theorem I *Let either the assumptions* **(A1)–(A5)** *and* **(B1)–(B6)** *or the assumptions* **(C1)–(C3)** *be satisfied with* $r > 2d/(d+2)$. *Assume that the boundary conditions are chosen such that we can get formally the uniform estimate for* $\boldsymbol{v} \in W^{1,r}(\Omega)$ *and* $\boldsymbol{m} \in L^{1+\delta}(\Omega)$. *Then there exists a weak solution to the problem* (5.131).

Clearly, the assumption $r > 2d/(d+2)$ is needed to have[10] integrable convective term $\boldsymbol{v} \otimes \boldsymbol{v}$. But what is much more difficult in (5.131) is to find a way, how to get the a priori estimates. Clearly, in case $\Gamma_1 = \partial\Omega$ and $\boldsymbol{v}_0 = \mathbf{0}$ there is no difference between the cases with or without the convective term. However, in case that $\boldsymbol{v}_0 \neq \mathbf{0}$ then the only available way how to derive the uniform estimate is presented in [38], where the author consider only the case when $|\boldsymbol{v}_0 \cdot \boldsymbol{n}| \ll 1$ and certain range of r $r \in (r_{\min}, 2)$ with some r_0 much bigger than $2d/(d+2)$. The second possibility is when $\Gamma_2 = \partial\Omega$. However, when one tries to derive an estimate in this case, then it is natural to multiply the balance of linear momentum by $\boldsymbol{v}$ and integrate over Ω. Unfortunately, the convective term does not vanish and there will be a remainder

$$\frac{1}{2}\int_{\partial\Omega} (\boldsymbol{v} \cdot \boldsymbol{n})|\boldsymbol{v}|^2.$$

This term does not have any sign in principle and cannot be handled by the viscous part (Of course, if $r > 3$ this term can be estimated by the viscous part). Therefore, it is sometimes assumed that s_0 in the boundary condition is of the form (see, e.g., [39])

$$s_0 = \bar{s}_0 - \frac{1}{2}\int_{\partial\Omega} (\boldsymbol{v} \cdot \boldsymbol{n})|\boldsymbol{v}|^2, \tag{5.132}$$

which then lead to the a priori estimates. Further, the condition (5.132) can be much more general, all we need is that it is able to handle the term $\int_{\partial\Omega}(\boldsymbol{v} \cdot \boldsymbol{n})|\boldsymbol{v}|^2$. Unfortunately, such artificial boundary conditions of the form (5.132) have almost no physical meaning.

To finish this part, we just recall here [12], where the most general existence theory is available and also [18], where was proven the up-to-date optimal constraint on γ in (5.62). Furthermore, the most general available theory for $\boldsymbol{S}$ being independent of pressure can be obtained with the help of [19], where the authors considered the setting of maximal monotone graphs but without the drag terms.

[10] It follows from the embedding $W^{1,r}(\Omega) \hookrightarrow L^{2+\delta}(\Omega)$ for some $\delta > 0$ provided that $r > 2d/(d+2)$.

5.4.2 Unsteady Model

In the unsteady case the only formal difference is due to the presence of the time derivative of the velocity field. The resulting systems thus may read as

$$
\begin{aligned}
\partial_t \boldsymbol{v} - \operatorname{div} \boldsymbol{S} + \operatorname{div}(\boldsymbol{v} \otimes \boldsymbol{v}) + \boldsymbol{m} &= -\nabla p + \boldsymbol{f} && \text{in } \Omega \times (0, T), \\
\boldsymbol{S} &= \boldsymbol{S}^T && \text{in } \Omega \times (0, T), \\
\operatorname{div} \boldsymbol{v} &= 0 && \text{in } \Omega \times (0, T), \\
\boldsymbol{h}(\boldsymbol{m}, \boldsymbol{v}, \boldsymbol{D}\boldsymbol{v}, p) &= \boldsymbol{0} && \text{in } \Omega \times (0, T), \\
\boldsymbol{H}(\boldsymbol{S}, \boldsymbol{D}\boldsymbol{v}, p) &= \boldsymbol{0} && \text{in } \Omega \times (0, T), \\
\boldsymbol{v} &= \boldsymbol{v}_0 && \text{on } \Gamma_1 \times (0, T), \\
(\boldsymbol{S} - p\boldsymbol{I})\boldsymbol{n} &= \boldsymbol{s}_0 && \text{on } \Gamma_2 \times (0, T), \\
\boldsymbol{v}(0) &= \boldsymbol{v}_1 && \text{in } \Omega,
\end{aligned}
\tag{5.133}
$$

where we added the time interval $(0, T)$ and also the initial condition $\boldsymbol{v}_1$. Also for this setting, we have available methods to handle the convective term as a perturbation, see [26] and [14]. However, in the evolutionary case we are in serious troubles with the pressure. Indeed, if we prescribe the Dirichlet boundary condition, it is not known even in simplest cases, whether the pressure exists as an integrable function and all we know is that it is only distribution with respect to time variable. Therefore, it is difficult to define the pressure pointwisely and so to include it into the constitutive equations. Fortunately, in case that the constitutive equations are independent of the pressure, we may still define a weak solution by using only divergence-free test functions $\boldsymbol{\varphi}$ in (5.64) and thus the pressure disappears in the weak formulation. Hence, we again formulate the "result" without any ambition to prove it.

Meta-Theorem II *Let either the assumptions* **(A1)–(A5)** *and* **(B1)–(B6)** *or the assumptions* **(C1)–(C3)** *be satisfied with* $r > 2d/(d+2)$ *and assume that the graphs* $\mathcal{A}$ *and* $\mathcal{B}$ *do not depend on the pressure. Assume that the boundary and initial conditions are chosen such that we can get formally the uniform estimate for* $\boldsymbol{v} \in L^r(0, T; W^{1,r}(\Omega))$ *and* $\boldsymbol{m} \in L^{1+\delta}(\Omega \times (0, T))$. *Then there exists a weak solution to the problem* (5.133).

To add the final comments to the theorem, we would like to point out that for homogeneous Dirichlet boundary conditions or for the Navier slip condition, the a priori estimates are available and consequently the existence theory can be built. For more general boundary conditions we must deal with the remainder from the convective term and we must again modify the boundary conditions accordingly, see [26] for the Dirichlet boundary conditions, without the drag term and with $\boldsymbol{S}$ being a continuous function of $\boldsymbol{D}$, and [14] for the setting of maximal monotone graphs.

Before we formulate the theorem also for $\mathcal{A}$ and $\mathcal{B}$ being dependent on the pressure, we recall here the boundary conditions for which we are able to obtain the existence of the integrable pressure. Thus, instead of conditions on Γ_1 and Γ_2, we just prescribe the Navier boundary conditions

$$\begin{aligned} \boldsymbol{v} \cdot \boldsymbol{n} &= 0 \quad \text{on } (0,T) \times \partial\Omega, \\ (\boldsymbol{S}\boldsymbol{n})_{\tau} + \alpha \boldsymbol{v} &= \boldsymbol{0} \quad \text{on } (0,T) \times \partial\Omega, \end{aligned} \tag{5.134}$$

where for any vector $\boldsymbol{f}$ the symbol $\boldsymbol{f}_{\tau}$ denotes its projection to the tangent plane on the boundary Ω, i.e., $\boldsymbol{f}_{\tau} := \boldsymbol{f} - (\boldsymbol{f} \cdot \boldsymbol{n})\boldsymbol{n}$. For these boundary conditions, we can now formulate the following "theorem."

Meta-Theorem III *Let assumptions* **(C1)–(C3)** *be satisfied with* $r > 2d/(d+2)$ *and* Ω *is a domain of class* $\mathcal{C}^{1,1}$. *Then there exists a weak solution to the problem* (5.133) *with boundary conditions given by* (5.134).

This result can be further extended to the more general and more realistic stick-slip boundary conditions, see [20, 21]. Although this result is not formulated anywhere (due to the presence of the drag term), it can be obtained with the help of the methods developed mainly in [14, 16, 24]. Furthermore, Meta-Theorem III could be also reformulated in the way: If we can get a priori estimates for $\boldsymbol{v}$ and p, then the weak solution exists.

We end this chapter just by recalling/summarizing the essential open problems for models describing the flow of incompressible fluids with pressure dependent material coefficients.

(1) Is it possible to get a priori estimate for steady model with Dirichlet boundary conditions on Γ_1 when $|\boldsymbol{v}_0 \cdot \boldsymbol{n}| \gg 1$ and with Neumann boundary condition on Γ_2? And if yes, for which range of r's?

(2) Is it possible to obtain an integrable pressure for unsteady model with Dirichlet boundary conditions?

Acknowledgements The author thanks to the project No. 16-03230S and the project No. 18-12719S financed by Czech Science Foundation.

Bibliography

1. E. Acerbi and N. Fusco. An approximation lemma for $W^{1,p}$ functions. In *Material instabilities in continuum mechanics (Edinburgh, 1985–1986)*, Oxford Sci. Publ., pages 1–5. Oxford Univ. Press, New York, 1988.
2. G. Alberti and L. Ambrosio. A geometrical approach to monotone functions in $\mathbb{R}^n$. *Math. Z.*, 230(2):259–316, 1999.
3. E.C. Andrade. Viscosity of liquids. *Nature*, 125(3148):309–310, 1930.
4. E. Aulisa, L. Bloshanskaya, L. Hoang, and A. Ibragimov. Analysis of generalized Forchheimer flows of compressible fluids in porous media. *J. Math. Phys.*, 50(10):103102, 44, 2009.

5. J.M. Ball and F. Murat. Remarks on Chacon's biting lemma. *Proc. Amer. Math. Soc.*, 107(3):655–663, 1989.
6. C. Barus. Isotherms, isopiestics and isometrics relative to viscosity. *Am. J. Sci.*, 45(3):87–96, 1893.
7. R.B. Bird, R.C. Armstrong, O. Hassager, and C.F. Curtiss. *Dynamics of Polymeric Liquids*. Number v. 1–2 in Board of advisors, engineering. Wiley, 1977.
8. P.W. Bridgman. *The Physics of High Pressure*. Dover books on physics and mathematical physics. Dover Publ., 1970.
9. H.C. Brinkman. A calculation of the viscous force exerted by a flowing fluid on a dense swarm of particles. *Appl. Sci. Res.*, A1:27–34, 1947a.
10. H.C. Brinkman. On the permeability of the media consisting of closely packed porous particles. *Appl. Sci. Res.*, A1:81–86, 1947b.
11. J.K. Brooks and R.V. Chacon. Continuity and compactness of measures. *Adv. in Math.*, 37(1):16–26, 1980.
12. M. Bulíček and V. Fišerová. Existence theory for steady flows of fluids with pressure and shear rate dependent viscosity, for low values of the power-law index. *Z. Anal. Anwend.*, 28(3):349–371, 2009.
13. M. Bulíček, P. Gwiazda, J. Málek, K.R. Rajagopal, and A. Świerczewska-Gwiazda. On flows of fluids described by an implicit constitutive equation characterized by a maximal monotone graph. In J. C. Robinson, J. L. Rodrigo, and W. Sadowski, editors, *Mathematical Aspects of Fluid Mechanics*, London Mathematical Society Lecture Note Series (No. 402), pages 26–54. Cambridge University Press, 2012.
14. M. Bulíček, P. Gwiazda, J. Málek, and A. Świerczewska-Gwiazda. On unsteady flows of implicitly constituted incompressible fluids. *SIAM J. Math. Anal.*, 44(4):2756–2801, 2012.
15. M. Bulíček and J. Málek. On unsteady internal flows of Bingham fluids subject to threshold slip on the impermeable boundary. In H. Amann, Y. Giga, H. Kozono, H. Okamoto, and M. Yamazaki, editors, *Recent Developments of Mathematical Fluid Mechanics*, Advances in Mathematical Fluid Mechanics, pages 135–156. Birkhäuser, 2016.
16. M. Bulíček, J. Málek, and K.R. Rajagopal. Navier's slip and evolutionary Navier-Stokes-like systems with pressure and shear-rate dependent viscosity. *Indiana Univ. Math. J.*, 56(1):51–85, 2007.
17. M. Bulíček, J. Málek, and K.R. Rajagopal. Analysis of the flows of incompressible fluids with pressure dependent viscosity fulfilling $\nu(p,\cdot) \to +\infty$ as $p \to +\infty$. *Czechoslovak Math. J.*, 59(2):503–528, 2009.
18. M. Bulíček, J. Málek, and J. Žabenský. On generalized Stokes' and Brinkman's equations with a pressure- and shear-dependent viscosity and drag coefficient. *Nonlinear Anal. Real World Appl.*, 26(0):109–132, 2015.
19. M. Bulíček, P. Gwiazda, J. Málek, and A. Świerczewska Gwiazda. On steady flows of incompressible fluids with implicit power-law-like rheology. *Adv. Calc. Var.*, 2(2):109–136, 2009.
20. M. Bulíček and J. Málek. On unsteady internal flows of Bingham fluids subject to threshold slip on the impermeable boundary. In *Recent developments of mathematical fluid mechanics*, Adv. Math. Fluid Mech., pages 135–156. Birkhäuser/Springer, Basel, 2016.
21. M. Bulíček and J. Málek. Internal flows of incompressible fluids subject to stick-slip boundary conditions. *Vietnam J. Math.*, 45(1–2):207–220, 2017.
22. M. Bulíček and J. Málek. Large data analysis for Kolmogorov's two-equation model of turbulence. *Nonlinear Analysis: Real World Applications*, 50:104–143, 2019.
23. M. Bulíček, J. Málek, and J. Žabenský. A generalization of the Darcy-Forchheimer equation involving an implicit, pressure-dependent relation between the drag force and the velocity. *J. Math. Anal. Appl.*, 424(1):785–801, 2015.
24. M. Bulíček and J. Žabenský. Large data existence theory for unsteady flows of fluids with pressure- and shear-dependent viscosities. *Nonlinear Anal.*, 127:94–127, 2015.
25. H. Darcy. *Les Fontaines publiques de la ville de Dijon*. V. Dalmont, 1856.

26. L. Diening, M. Růžička, and J. Wolf. Existence of weak solutions for unsteady motions of generalized Newtonian fluids. *Ann. Sc. Norm. Super. Pisa Cl. Sci. (5)*, 9(1):1–46, 2010.
27. E. Feireisl and A. Novotný. *Singular limits in thermodynamics of viscous fluids*. Advances in Mathematical Fluid Mechanics. Birkhäuser Verlag, Basel, 2009.
28. P. Forchheimer. Wasserbewegung durch Boden. *Zeitschrift des Vereines Deutscher Ingenieuer*, 45:1782–1788, 1901.
29. G. Francfort, F. Murat, and L. Tartar. Monotone operators in divergence form with x-dependent multivalued graphs. *Boll. Unione Mat. Ital. Sez. B Artic. Ric. Mat. (8)*, 7(1):23–59, 2004.
30. M. Franta, J. Málek, and K.R. Rajagopal. On steady flows of fluids with pressure- and shear-dependent viscosities. *Proc. R. Soc. Lond. Ser. A Math. Phys. Eng. Sci.*, 461(2055):651–670, 2005.
31. G.P. Galdi. *An introduction to the mathematical theory of the Navier-Stokes equations*. Springer Monographs in Mathematics. Springer, New York, second edition, 2011. Steady-state problems.
32. F. Gazzola. A note on the evolution Navier-Stokes equations with a pressure-dependent viscosity. *Z. Angew. Math. Phys.*, 48(5):760–773, 1997.
33. F. Gazzola and P. Secchi. Some results about stationary Navier-Stokes equations with a pressure-dependent viscosity. In *Navier-Stokes equations: theory and numerical methods (Varenna, 1997)*, volume 388 of *Pitman Res. Notes Math. Ser.*, pages 31–37. Longman, Harlow, 1998.
34. J. Hron, J. Málek, J. Nečas, and K.R. Rajagopal. Numerical simulations and global existence of solutions of two-dimensional flows of fluids with pressure- and shear-dependent viscosities. *Math. Comput. Simulation*, 61(3–6):297–315, 2003. MODELLING 2001 (Pilsen).
35. J. Hron, J. Málek, V. Průša, and K.R. Rajagopal. Further remarks on simple flows of fluids with pressure-dependent viscosities. *Nonlinear Anal. Real World Appl.*, 12(1):394–402, 2011.
36. V. Kalantarov and S. Zelik. Smooth attractors for the Brinkman-Forchheimer equations with fast growing nonlinearities. *Commun. Pure Appl. Anal.*, 11(5):2037–2054, 2012.
37. K. Kannan and K.R. Rajagopal. Flow through porous media due to high pressure gradients. *Appl. Math. Comput.*, 199(2):748–759, 2008.
38. M. Lanzendörfer. On steady inner flows of an incompressible fluid with the viscosity depending on the pressure and the shear rate. *Nonlinear Anal. Real World Appl.*, 10(4):1943–1954, 2009.
39. M. Lanzendörfer and J. Stebel. On pressure boundary conditions for steady flows of incompressible fluids with pressure and shear rate dependent viscosities. *Appl. Math.*, 56(3):265–285, 2011.
40. J.-L. Lions. *Quelques méthodes de résolution des problèmes aux limites non linéaires*. Dunod, 1969.
41. J. Málek, J. Nečas, and K.R. Rajagopal. Global analysis of the flows of fluids with pressure-dependent viscosities. *Arch. Ration. Mech. Anal.*, 165(3):243–269, 2002.
42. J. Málek, J. Nečas, and K.R. Rajagopal. Global existence of solutions for flows of fluids with pressure and shear dependent viscosities. *Appl. Math. Lett.*, 15(8):961–967, 2002.
43. J. Málek, K.R. Rajagopal, and M. Růžička. Existence and regularity of solutions and the stability of the rest state for fluids with shear dependent viscosity. *Math. Models Methods Appl. Sci.*, 5(6):789–812, 1995.
44. G.J. Minty. Monotone (nonlinear) operators in Hilbert space. *Duke Math. J.*, 29:341–346, 1962.
45. F. Murat. Compacité par compensation. *Ann. Scuola Norm. Sup. Pisa Cl. Sci. (4)*, 5(3):489–507, 1978.
46. K.B. Nakshatrala and K.R. Rajagopal. A numerical study of fluids with pressure-dependent viscosity flowing through a rigid porous medium. *Internat. J. Numer. Methods Fluids*, 67(3):342–368, 2011.
47. D.A. Nield and A. Bejan. *Convection in porous media*. Springer-Verlag, New York, second edition, 1999.
48. J.M. Paton and C.J. Schaschke. Viscosity measurement of biodiesel at high pressure with a falling sinker viscometer. *Chemical Engineering Research and Design*, 87(11A):1520–1526, 2009.

49. V. Průša. Revisiting Stokes first and second problems for fluids with pressure-dependent viscosities. *International Journal of Engineering Science*, 48(12):2054–2065, 2010.
50. K.R. Rajagopal. On implicit constitutive theories for fluids. *J. Fluid Mech.*, 550:243–249, 2006.
51. K.R. Rajagopal. On a hierarchy of approximate models for flows of incompressible fluids through porous solids. *Math. Models Methods Appl. Sci.*, 17(2):215–252, 2007.
52. K.R. Rajagopal and L. Tao. *Mechanics of mixtures*, volume 35 of *Series on Advances in Mathematics for Applied Sciences*. World Scientific Publishing Co., Inc., River Edge, NJ, 1995.
53. M. Renardy. Some remarks on the Navier-Stokes equations with a pressure-dependent viscosity. *Comm. Partial Differential Equations*, 11(7):779–793, 1986.
54. I. Samohýl. *Thermodynamics of irreversible processes in fluid mixtures*, volume 12 of *Teubner-Texte zur Physik [Teubner Texts in Physics]*. BSB B. G. Teubner Verlagsgesellschaft, Leipzig, 1987. Approached by rational thermodynamics, With German, French, Spanish and Russian summaries.
55. D.G. Schaeffer. Instability in the evolution equations describing incompressible granular flow. *J. Differential Equations*, 66(1):19–50, 1987.
56. C.J. Schaschke, S. Allio, and E. Holmberg. Viscosity measurement of vegetable oil at high pressure. *Food and Bioproducts Processing*, 84(C3):173–178, 2006.
57. T. Schwedoff. Recherches expérimentales sur la cohésion des liquides. I. rigidité des liquides. *J. Phys. Theor. Appl.*, 8:341–359, 1889.
58. T. Schwedoff. Recherches expérimentales sur la cohésion des liquides. II. viscosité des liquides. *J. Phys. Theor. Appl.*, 9:34–46, 1890.
59. S. Srinivasan, A. Bonito, and K.R. Rajagopal. Flow of a fluid through porous solid due to high pressure gradients. *J. Porous Media*, 16(3):193–203, 2013.
60. S. Srinivasan and K.R. Rajagopal. A thermodynamic basis for the derivation of the Darcy, Forchheimer and Brinkman models for flows through porous media and their generalizations. *International Journal of Non-Linear Mechanics*, 58:162–166, 2014.
61. G.G. Stokes. On the Effect of the Internal Friction of Fluids on the Motion of Pendulums. *Transactions of the Cambridge Philosophical Society*, 9:8–106, 1851.
62. S.C. Subramanian and K.R. Rajagopal. A note on the flow through porous solids at high pressures. *Comput. Math. Appl.*, 53(2):260–275, 2007.
63. S.C. Subramanian and K.R. Rajagopal. A note on the flow through porous solids at high pressures. *Comput. Math. Appl.*, 53(2):260–275, 2007.
64. S.A. Suslov and T.D. Tran. Revisiting plane Couette-Poiseuille flows of a piezo-viscous fluid. *J. Non-Newton. Fluid Mech*, 154:170–178, 2008.
65. S.A. Suslov and T.D. Tran. Stability of plane Poiseuille-Couette flows of a piezo-viscous fluid. *J. Non-Newton. Fluid Mech*, 156:139–149, 2009.
66. L. Tartar. Compensated compactness and applications to partial differential equations. In *Nonlinear analysis and mechanics: Heriot-Watt Symposium, Vol. IV*, volume 39 of *Res. Notes in Math.*, pages 136–212. Pitman, Boston, Mass.-London, 1979.
67. C. Truesdell. Sulle basi della termomeccanica. I, II. *Atti Accad. Naz. Lincei. Rend. Cl. Sci. Fis. Mat. Nat. (8)*, 22:33–38, 158–166, 1957.
68. M. Vasudevaiah and K.R. Rajagopal. On fully developed flows of fluids with a pressure dependent viscosity in a pipe. *Appl. Math.*, 50(4):341–353, 2005.

Chapter 6
Finite Element Pressure Stabilizations for Incompressible Flow Problems

Volker John, Petr Knobloch, and Ulrich Wilbrandt

Abstract The simulation of incompressible flow problems with pairs of velocity-pressure finite element spaces that do not satisfy the discrete inf-sup condition requires a so-called pressure stabilization. This chapter provides a survey of available methods which are presented for the Stokes problem to concentrate on the main ideas and to avoid additional difficulties originating from more complicated models. The methods can be divided into residual-based stabilizations and stabilizations that utilize only the pressure. For the first class, a comprehensive numerical analysis is presented, whereas for the second class, the presentation is more concise except for a detailed analysis of a local projection stabilization method. Connections of various pressure stabilizations to inf-sup stable discretizations with velocity spaces enriched by bubble functions are also discussed. Numerical studies compare several of the available pressure stabilizations.

Keywords Stokes equations · Discrete inf-sup condition · Pressure-stabilized Petrov–Galerkin (PSPG) method · Galerkin least squares (GLS) method · Douglas–Wang method · Local projection stabilization (LPS) method · Velocity finite element spaces with bubble functions

MSC 2010 65N30, 65N12, 65N15

6.1 Introduction

The behavior of incompressible flows is modeled by the incompressible Navier–Stokes equations, given here already in dimensionless form,

---

V. John (✉) · U. Wilbrandt
Weierstrass Institute, Berlin, Germany
e-mail: john@wias-berlin.de

P. Knobloch
Charles University, Prague, Czech Republic

T. Bodnár et al. (eds.), *Fluids Under Pressure*, Advances in Mathematical Fluid Mechanics, https://doi.org/10.1007/978-3-030-39639-8_6

$$\begin{aligned} \partial_t \mathbf{u} - \nu \Delta \mathbf{u} + (\mathbf{u} \cdot \nabla)\mathbf{u} + \nabla p &= \mathbf{f} \text{ in } (0, T] \times \Omega, \\ \nabla \cdot \mathbf{u} &= 0 \text{ in } (0, T] \times \Omega, \end{aligned} \tag{6.1}$$

where $\Omega \subset \mathbb{R}^d$, $d \in \{2, 3\}$, is the flow domain, T the final time, $\mathbf{u}$ the velocity field, p the pressure, ν the (dimensionless) kinematic viscosity, and $\mathbf{f}$ represents forces acting on the fluid. The first equation describes the conservation of linear momentum and the second equation, the so-called continuity equation, the conservation of mass. System (6.1) has to be equipped with an initial velocity condition and with boundary conditions on the boundary $\partial\Omega$.

There are three aspects that might lead to difficulties in the analysis and numerical simulation of the incompressible Navier–Stokes equations:

- It is a coupled system with two unknowns, where the pressure does not appear in the continuity equation. One obtains a so-called saddle point problem.
- The Navier–Stokes equations form a nonlinear system.
- In the case of (very) small viscosities, the first order term $(\mathbf{u} \cdot \nabla)\mathbf{u}$ dominates in the momentum equation. This situation corresponds to turbulent flows. System (6.1) is convection-dominated and its numerical simulation requires special approaches, so-called turbulence models.

This review will discuss numerical methods for treating the coupling of velocity and pressure. To concentrate on this issue, it suffices to consider the (scaled) stationary Stokes equations with homogeneous Dirichlet boundary conditions

$$\begin{aligned} -\nu \Delta \mathbf{u} + \nabla p &= \mathbf{f} \text{ in } \Omega, \\ \nabla \cdot \mathbf{u} &= 0 \text{ in } \Omega, \\ \mathbf{u} &= \mathbf{0} \text{ on } \partial\Omega. \end{aligned} \tag{6.2}$$

System (6.2) is a linear saddle point problem. The theory of linear saddle point problems was developed in the early 1970s in the seminal papers [6, 24]. In this theory, the weak or variational form of (6.2) is studied. It turns out that this form is well posed, i.e., there exists a unique solution that depends continuously on the right-hand side, if the spaces V for the velocity and Q for the pressure are chosen appropriately.

Applying a Galerkin finite element method to discretize the variational form of the Stokes equations, i.e., solely replacing the infinite-dimensional spaces V and Q with finite-dimensional spaces V^h and Q^h, leads to a finite-dimensional linear saddle point problem, whose algebraic form is

$$\begin{pmatrix} A & B^T \\ B & 0 \end{pmatrix} \begin{pmatrix} \underline{u} \\ \underline{p} \end{pmatrix} = \begin{pmatrix} \underline{f} \\ 0 \end{pmatrix}.$$

From the theory of linear saddle point problems, it follows that the Galerkin finite element method is only well posed for appropriate choices of the finite element spaces. Concretely, the spaces have to satisfy a discrete inf-sup condition

$$\inf_{q^h \in Q^h \setminus \{0\}} \sup_{\mathbf{v}^h \in V^h \setminus \{\mathbf{0}\}} \frac{(\nabla \cdot \mathbf{v}^h, q^h)}{\|\nabla \mathbf{v}^h\|_{L^2(\Omega)} \|q^h\|_{L^2(\Omega)}} \geq \beta_{\mathrm{is}}^h > 0. \tag{6.3}$$

For obtaining optimal order convergence, β_{is}^h has to be independent of the mesh width h.

In practice, it turns out that the inequality (6.3) requires the use of different finite element spaces for velocity and pressure. However, it was proved that the lowest order spaces, using continuous linear or d-linear functions for the finite element velocity and piecewise constant functions for the discrete pressure, do not satisfy (6.3). Thus, implementing finite element methods that respect (6.3) requires some effort. Another issue in practice is that many standard preconditioners for iterative solvers of linear systems of equations cannot be applied to linear saddle point problems due to the zeros in the main diagonal of the system matrix.

In view of these drawbacks, numerical methods were developed in order to circumvent the discrete inf-sup condition (6.3). The main idea of these so-called pressure stabilizations consists in introducing a pressure term in the finite element continuity equation to remove the saddle point character of the discrete problem, leading to an algebraic system of the form

$$\begin{pmatrix} A & D \\ B & -C \end{pmatrix} \begin{pmatrix} \underline{u} \\ \underline{p} \end{pmatrix} = \begin{pmatrix} \underline{f} \\ \underline{g} \end{pmatrix}. \tag{6.4}$$

Several methods were proposed in the 1980s, the first one by Brezzi and Pitkäranta in [26] and a number of residual-based pressure stabilizations in [39, 54, 55]. At the end of the 1990s and during the 2000s, new approaches were developed, which often contain terms where only the pressure appears, e.g., in [13, 29, 34, 38]. In recent years, variants of stabilized methods were proposed that allow an easier implementation as previous variants, e.g., in [7, 12, 31], or a finite element error analysis was presented with less regularity assumptions on the solution of the continuous problem in [83].

Altogether, there are many different proposals for pressure stabilizations. However, to the best of our knowledge, there is no up-to-date comprehensive survey of this topic in the literature available. In addition, it was pointed out as an open problem in [57] that *Systematic assessments of the proposed stabilized methods are missing that clarify their advantages and drawbacks and give finally proposals which ones should be preferred in simulations.* The present paper aims to close these gaps to some extent. However, there will be also some limitations of this survey. It is restricted to conforming finite element methods and to the discussion of the a priori error analysis.

Throughout the paper, standard notation for Lebesgue and Sobolev spaces is used. The inner product of $L^2(G)^d$, $d \in \{1, 2, 3\}$, will be denoted by $(\cdot, \cdot)_G$ and the notation $(\cdot, \cdot)$ is used if $G = \Omega$. All constants C, C_1, etc., do neither depend on the viscosity coefficient ν nor on the mesh width h. The notation C indicates a generic constant that can have different values at different places.

The paper is organized as follows. Section 6.2 introduces the considered finite element spaces and provides some properties which are used in the numerical analysis. Available convergence results for inf-sup stable discretizations are summarized in Sect. 6.3, to allow an easy comparison with the results for pressure-stabilized discretizations. The topic of Sect. 6.4 is the class of residual-based stabilizations. For some of them, the finite element analysis is presented in detail. Stabilizations that use only the pressure are described in Sect. 6.5. A detailed presentation of the analysis is provided for a local projection stabilization (LPS) scheme. Section 6.6 describes the connection of some stabilized discretizations to inf-sup stable methods that are enriched with bubble functions. Finally, numerical studies involving three residual-based stabilizations and one LPS method are presented in Sect. 6.7.

6.2 Weak Form of the Stokes Equations, Finite Element Spaces

In this section we summarize assumptions on the data of the Stokes problem (6.2) and on the finite element spaces V^h and Q^h which will be used throughout the remaining part of this chapter. It will be assumed that Ω is a bounded domain with a polygonal resp. polyhedral Lipschitz-continuous boundary, the viscosity ν is a positive constant, and $\mathbf{f} \in L^2(\Omega)^d$.

A weak form of the Stokes equations (6.2) reads: Given $\mathbf{f} \in L^2(\Omega)^d$, find $(\mathbf{u}, p) \in H_0^1(\Omega)^d \times L_0^2(\Omega)$ such that

$$\nu(\nabla\mathbf{u}, \nabla\mathbf{v}) - (\nabla\cdot\mathbf{v}, p) + (\nabla\cdot\mathbf{u}, q) = (\mathbf{f}, \mathbf{v}) \quad \forall\, (\mathbf{v}, q) \in H_0^1(\Omega)^d \times L_0^2(\Omega). \tag{6.5}$$

We shall use the notation $V = H_0^1(\Omega)^d$ and $Q = L_0^2(\Omega)$. The unique solvability of (6.5) is closely connected with the fact that the spaces V and Q satisfy the inf-sup condition

$$\inf_{q\in Q\setminus\{0\}} \sup_{\mathbf{v}\in V\setminus\{\mathbf{0}\}} \frac{(\nabla\cdot\mathbf{v}, q)}{\|\nabla\mathbf{v}\|_{L^2(\Omega)}\, \|q\|_{L^2(\Omega)}} \geq \beta_{\mathrm{is}} > 0. \tag{6.6}$$

The inequality

$$\|\nabla\cdot\mathbf{v}\|_{L^2(\Omega)} \leq \|\nabla\mathbf{v}\|_{L^2(\Omega)} \quad \forall\, \mathbf{v} \in V \tag{6.7}$$

will be used in the analysis. The space of weakly divergence-free functions is given by

$$V_{\mathrm{div}} = \{\mathbf{v} \in V \; : \; (\nabla\cdot\mathbf{v}, q) = 0 \;\forall\, q \in Q\}.$$

We assume that we are given a family $\{\mathcal{T}^h\}$ of triangulations of Ω consisting of simplices, quadrilaterals, or hexahedra and possessing the usual compatibility properties, which are parametrized by a positive parameter h. The set of interior faces (edges for $d = 2$) will be denoted by $\mathcal{E}^h$. We denote $h_K := \operatorname{diam}(K)$ and $h_E := \operatorname{diam}(E)$ for any $K \in \mathcal{T}^h$ and $E \in \mathcal{E}^h$ and assume that $h_K \leq h$ for all $K \in \mathcal{T}^h$. For each face $E \in \mathcal{E}^h$, we denote by $\mathbf{n}_E$ a fixed unit normal vector to E and by $[|q|]_E$ the jump of the function q across the face E such that $[|q|]_E > 0$ if q decreases in the direction of $\mathbf{n}_E$.

For each $\mathcal{T}^h$, we introduce finite element spaces $V^h \subset V$ and $Q^h \subset Q$ containing piecewise (mapped) polynomials of degree $k \geq 1$ and $l \geq 0$, respectively. We assume that the finite element spaces V^h and Q^h possess standard interpolation properties. More precisely, we denote by $I^h : V \cap H^{k+1}(\Omega)^d \to V_h$ and $J^h : Q \cap H^{l+1}(\Omega) \to Q_h$ interpolation operators satisfying

$$\left(\sum_{K\in\mathcal{T}^h} h_K^{-2} \left\|\mathbf{v} - I^h\mathbf{v}\right\|_{L^2(K)}^2\right)^{1/2} + \left\|\nabla(\mathbf{v} - I^h\mathbf{v})\right\|_{L^2(\Omega)}$$
$$+ \left(\sum_{K\in\mathcal{T}^h} h_K^2 \left\|\Delta(\mathbf{v} - I^h\mathbf{v})\right\|_{L^2(K)}^2\right)^{1/2}$$
$$+ \left(\sum_{E\in\mathcal{E}^h} h_E^{-1} \left\|\mathbf{v} - I^h\mathbf{v}\right\|_{L^2(E)}^2\right)^{1/2} \leq C\,h^k \,\|\mathbf{v}\|_{H^{k+1}(\Omega)}\,, \tag{6.8}$$

$$\left(\sum_{K\in\mathcal{T}^h} h_K^{-2} \left\|q - J^h q\right\|_{L^2(K)}^2\right)^{1/2} + \left(\sum_{K\in\mathcal{T}^h} \left\|\nabla(q - J^h q)\right\|_{L^2(K)}^2\right)^{1/2}$$
$$+ \left(\sum_{K\in\mathcal{T}^h} \sum_{E\subset\partial K} h_E^{-1} \left\|q - (J^h q)|_K\right\|_{L^2(E)}^2\right)^{1/2} \leq C\,h^l \,\|q\|_{H^{l+1}(\Omega)}\,, \tag{6.9}$$

for $\mathbf{v} \in V \cap H^{k+1}(\Omega)^d$ and $q \in Q \cap H^{l+1}(\Omega)$. The operator I^h may be the standard Lagrange interpolation. The definition of J^h depends on the construction of Q^h. For example, if $Q^h \subset H^1(\Omega)$, and hence $l \geq 1$, the operator J^h may be defined as the Lagrange interpolation projected into Q. If the functions in Q^h are discontinuous across faces, the operator J^h may be defined as the projection into a polynomial space on each element of the triangulation.

In addition, for $\mathbf{v} \in V$, we shall use a piecewise (multi)linear interpolant $\mathcal{I}^h\mathbf{v} \in V^h$ (e.g., the Clément or Scott–Zhang interpolant) satisfying

$$\left(\sum_{K\in\mathcal{T}^h} h_K^{-2}\left\|\mathbf{v}-\mathcal{I}^h\mathbf{v}\right\|_{L^2(K)}^2\right)^{1/2}+\left\|\nabla\mathcal{I}^h\mathbf{v}\right\|_{L^2(\Omega)}$$
$$+\left(\sum_{E\in\mathcal{E}^h} h_E^{-1}\left\|\mathbf{v}-\mathcal{I}^h\mathbf{v}\right\|_{L^2(E)}^2\right)^{1/2}\le C\left\|\nabla\mathbf{v}\right\|_{L^2(\Omega)}. \qquad (6.10)$$

Similarly, for $q \in Q \cap H^1(\Omega)$, we introduce an interpolant $\mathcal{J}^h q \in Q_h$ satisfying

$$\left(\sum_{K\in\mathcal{T}^h} h_K^{-2}\left\|q-\mathcal{J}^h q\right\|_{L^2(K)}^2\right)^{1/2}+\left(\sum_{K\in\mathcal{T}^h}\left\|\nabla\mathcal{J}^h q\right\|_{L^2(K)}^2\right)^{1/2}$$
$$+\left(\sum_{K\in\mathcal{T}^h}\sum_{E\subset\partial K} h_E^{-1}\left\|q-(\mathcal{J}^h q)|_K\right\|_{L^2(E)}^2\right)^{1/2}\le C\left\|\nabla q\right\|_{L^2(\Omega)}. \qquad (6.11)$$

Finally, for each $\mathcal{T}^h$, it is assumed that the following inverse inequality holds:

$$\left\|\Delta\mathbf{v}^h\right\|_{L^2(K)}\le C_{\text{inv}} h_K^{-1}\left\|\nabla\mathbf{v}^h\right\|_{L^2(K)} \qquad \forall\ \mathbf{v}^h\in V^h,\ K\in\mathcal{T}^h. \qquad (6.12)$$

Note that C_{inv} depends on the polynomial degree. It was shown in [53] for some types of mesh cells that it increases with increasing polynomial degree. For example, it has the value 0, 48, 149.1 for $P_1(K)$, $P_2(K)$, and $P_3(K)$, respectively, in the case that K is a right isosceles triangle.

6.3 Inf-Sup Stable Finite Element Discretizations

Inf-sup stable pairs of finite element spaces satisfy the discrete inf-sup condition (6.3). For the well-posedness of the discrete problem, the introduction of a pressure stabilization is not necessary. This section provides a survey on the most important results from the finite element convergence theory for inf-sup stable finite element discretizations to facilitate the comparison with the convergence results for stabilized discretizations presented in the subsequent sections.

Let the spaces V^h and Q^h satisfy the discrete inf-sup condition (6.3). Then, the conforming discretization of the Stokes problem reads as follows: Find $\left(\mathbf{u}^h, p^h\right) \in V^h \times Q^h$ such that

$$\nu\left(\nabla\mathbf{u}^h,\nabla\mathbf{v}^h\right)-\left(\nabla\cdot\mathbf{v}^h,p^h\right)+\left(\nabla\cdot\mathbf{u}^h,q^h\right)=\left(\mathbf{f},\mathbf{v}^h\right)\quad\forall\ \left(\mathbf{v}^h,q^h\right)\in V^h\times Q^h. \qquad (6.13)$$

The natural norms for the analysis of the Stokes problem are the $L^2(\Omega)$ norm of the velocity gradient and the $L^2(\Omega)$ norm of the pressure. Since the error analysis for these norms utilizes typical tools and it is rather short, the proofs will be presented in detail. The presentation follows [56, Section 4.2.1].

A crucial role in the analysis plays the subspace of discretely divergence-free functions

$$V_{\mathrm{div}}^h = \left\{ \mathbf{v}^h \in V^h \, : \, \left(\nabla \cdot \mathbf{v}^h, q^h \right) = 0 \, \forall \, q^h \in Q^h \right\}.$$

The solution of (6.13) belongs to this subspace. Note that in general functions from this subspace are not weakly divergence-free, i.e., it holds $V_{\mathrm{div}}^h \not\subset V_{\mathrm{div}}$.

Theorem 6.3.1 (Error Estimate for the $L^2(\Omega)$ Norm of the Velocity Gradient) *Let $(\mathbf{u}, p) \in V \times Q$ be the unique solution of the Stokes problem* (6.5) *and assume that the spaces V^h and Q^h satisfy* (6.3). *Then, the solution of the conforming discretization* (6.13) *satisfies the error estimate*

$$\left\| \nabla(\mathbf{u} - \mathbf{u}^h) \right\|_{L^2(\Omega)} \leq 2 \inf_{\mathbf{v}^h \in V_{\mathrm{div}}^h} \left\| \nabla(\mathbf{u} - \mathbf{v}^h) \right\|_{L^2(\Omega)} + \frac{1}{\nu} \inf_{q^h \in Q^h} \left\| p - q^h \right\|_{L^2(\Omega)}. \tag{6.14}$$

Proof The proof starts by formulating the error equation. Since $V^h \subset V$, functions from V^h can be used as test functions in (6.5). Subtracting (6.13) from (6.5) and setting $q = q^h = 0$ yields the so-called error equation

$$\nu \left(\nabla(\mathbf{u} - \mathbf{u}^h), \nabla \mathbf{v}^h \right) - \left(\nabla \cdot \mathbf{v}^h, p - p^h \right) = 0 \qquad \forall \, \mathbf{v}^h \in V^h. \tag{6.15}$$

Now, restricting the test functions to the space V_{div}^h, the second term on the left-hand side is modified such that an approximation term with respect to the pressure is obtained. One observes that $\left(\nabla \cdot \mathbf{v}^h, q^h \right) = 0$ for all $\mathbf{v}^h \in V_{\mathrm{div}}^h$ and $q^h \in Q^h$, which leads to

$$\nu \left(\nabla(\mathbf{u} - \mathbf{u}^h), \nabla \mathbf{v}^h \right) - \left(\nabla \cdot \mathbf{v}^h, p - q^h \right) = 0 \qquad \forall \, \mathbf{v}^h \in V_{\mathrm{div}}^h, \, q^h \in Q^h. \tag{6.16}$$

Next, an approximation error for the velocity is introduced. To this end, the error is decomposed into

$$\mathbf{u} - \mathbf{u}^h = \left(\mathbf{u} - \mathbf{w}^h \right) - \left(\mathbf{u}^h - \mathbf{w}^h \right) = \boldsymbol{\eta} - \boldsymbol{\phi}^h,$$

where $\mathbf{w}^h$ denotes an arbitrary interpolant of $\mathbf{u}$ in V_{div}^h. Hence, $\boldsymbol{\eta}$ is an approximation error which depends only on the finite element space V_{div}^h. The goal consists in estimating $\boldsymbol{\phi}^h \in V_{\mathrm{div}}^h$ by approximation errors as well. Therefore, this decomposition is inserted in (6.16) and the test function $\mathbf{v}^h = \boldsymbol{\phi}^h$ is chosen. It follows that

$$\nu \left\| \nabla \boldsymbol{\phi}^h \right\|_{L^2(\Omega)}^2 = \nu \left(\nabla \boldsymbol{\phi}^h, \nabla \boldsymbol{\phi}^h \right) = \nu \left(\nabla \boldsymbol{\eta}, \nabla \boldsymbol{\phi}^h \right) - \left(\nabla \cdot \boldsymbol{\phi}^h, p - q^h \right) \quad \forall\, q^h \in Q^h. \tag{6.17}$$

The first term on the right-hand side is estimated with the Cauchy–Schwarz inequality

$$\nu \left| \left(\nabla \boldsymbol{\eta}, \nabla \boldsymbol{\phi}^h \right) \right| \le \nu \left\| \nabla \boldsymbol{\eta} \right\|_{L^2(\Omega)} \left\| \nabla \boldsymbol{\phi}^h \right\|_{L^2(\Omega)}.$$

For the second term, one uses in addition (6.7), which gives

$$\begin{aligned} \left| - \left(\nabla \cdot \boldsymbol{\phi}^h, p - q^h \right) \right| &\le \left\| p - q^h \right\|_{L^2(\Omega)} \left\| \nabla \cdot \boldsymbol{\phi}^h \right\|_{L^2(\Omega)} \\ &\le \left\| p - q^h \right\|_{L^2(\Omega)} \left\| \nabla \boldsymbol{\phi}^h \right\|_{L^2(\Omega)}. \end{aligned}$$

Inserting these estimates in (6.17) and dividing by $\nu \left\| \nabla \boldsymbol{\phi}^h \right\|_{L^2(\Omega)} \neq 0$ yields

$$\left\| \nabla \boldsymbol{\phi}^h \right\|_{L^2(\Omega)} \le \left\| \nabla \boldsymbol{\eta} \right\|_{L^2(\Omega)} + \frac{1}{\nu} \left\| p - q^h \right\|_{L^2(\Omega)}.$$

This estimate is trivially true if $\left\| \nabla \boldsymbol{\phi}^h \right\|_{L^2(\Omega)} = 0$.

With the triangle inequality, it follows that

$$\begin{aligned} \left\| \nabla (\mathbf{u} - \mathbf{u}^h) \right\|_{L^2(\Omega)} &\le \left\| \nabla \boldsymbol{\phi}^h \right\|_{L^2(\Omega)} + \left\| \nabla \boldsymbol{\eta} \right\|_{L^2(\Omega)} \\ &\le 2 \left\| \nabla \boldsymbol{\eta} \right\|_{L^2(\Omega)} + \frac{1}{\nu} \left\| p - q^h \right\|_{L^2(\Omega)} \end{aligned}$$

for all $\mathbf{w}^h \in V_{\text{div}}^h$ and for all $q^h \in Q^h$, such that (6.14) follows. □

Theorem 6.3.2 (Error Estimate for the $L^2(\Omega)$ Norm of the Pressure) *Let the assumption of Theorem 6.3.1 be satisfied. Then the following error estimate holds:*

$$\begin{aligned} \left\| p - p^h \right\|_{L^2(\Omega)} &\le \frac{2\nu}{\beta_{\text{is}}^h} \inf_{\mathbf{v}^h \in V_{\text{div}}^h} \left\| \nabla (\mathbf{u} - \mathbf{v}^h) \right\|_{L^2(\Omega)} \\ &\quad + \left(1 + \frac{2}{\beta_{\text{is}}^h} \right) \inf_{q^h \in Q^h} \left\| p - q^h \right\|_{L^2(\Omega)}. \end{aligned} \tag{6.18}$$

Proof Let $q^h \in Q^h$ be arbitrary, then the triangle inequality implies

$$\left\| p - p^h \right\|_{L^2(\Omega)} \le \left\| p - q^h \right\|_{L^2(\Omega)} + \left\| p^h - q^h \right\|_{L^2(\Omega)}.$$

Replacing the right-hand side of the momentum equation of the finite element Stokes problem (6.13) by the left-hand side of the the momentum equation of the continuous Stokes problem (6.5) for $\mathbf{v}^h \in V^h$ yields

$$-\left(\nabla\cdot\mathbf{v}^h, p^h - q^h\right) = -\nu\left(\nabla\mathbf{u}^h, \nabla\mathbf{v}^h\right) + \left(\mathbf{f}, \mathbf{v}^h\right) + \left(\nabla\cdot\mathbf{v}^h, q^h\right)$$
$$= \nu\left(\nabla\left(\mathbf{u} - \mathbf{u}^h\right), \nabla\mathbf{v}^h\right) - \left(\nabla\cdot\mathbf{v}^h, p - q^h\right)$$

for all $\left(\mathbf{v}^h, q^h\right) \in V^h \times Q^h$. With the discrete inf-sup condition (6.3), the Cauchy–Schwarz inequality, and (6.7), it follows now that

$$\begin{aligned}
&\left\|p^h - q^h\right\|_{L^2(\Omega)} \\
&\le \frac{1}{\beta_{\text{is}}^h} \sup_{\mathbf{v}^h\in V^h\setminus\{\mathbf{0}\}} \frac{-\left(\nabla\cdot\mathbf{v}^h, p^h - q^h\right)}{\left\|\nabla\mathbf{v}^h\right\|_{L^2(\Omega)}} \\
&= \frac{1}{\beta_{\text{is}}^h} \sup_{\mathbf{v}^h\in V^h\setminus\{\mathbf{0}\}} \frac{\nu\left(\nabla\left(\mathbf{u} - \mathbf{u}^h\right), \nabla\mathbf{v}^h\right) - \left(\nabla\cdot\mathbf{v}^h, p - q^h\right)}{\left\|\nabla\mathbf{v}^h\right\|_{L^2(\Omega)}} \\
&\le \frac{1}{\beta_{\text{is}}^h} \sup_{\mathbf{v}^h\in V^h\setminus\{\mathbf{0}\}} \frac{\nu\left\|\nabla\left(\mathbf{u} - \mathbf{u}^h\right)\right\|_{L^2(\Omega)}\left\|\nabla\mathbf{v}^h\right\|_{L^2(\Omega)} + \left\|p - q^h\right\|_{L^2(\Omega)}\left\|\nabla\mathbf{v}^h\right\|_{L^2(\Omega)}}{\left\|\nabla\mathbf{v}^h\right\|_{L^2(\Omega)}} \\
&= \frac{1}{\beta_{\text{is}}^h}\left(\nu\left\|\nabla\left(\mathbf{u} - \mathbf{u}^h\right)\right\|_{L^2(\Omega)} + \left\|p - q^h\right\|_{L^2(\Omega)}\right) \quad \forall\, q^h \in Q^h.
\end{aligned}$$

Inserting the error bound (6.14) for the velocity yields the error estimate (6.18) for the pressure. □

The best approximation error in the subspace V_{div}^h can be estimated by the best approximation error in V^h

$$\inf_{\mathbf{v}^h\in V_{\text{div}}^h}\left\|\nabla\left(\mathbf{u} - \mathbf{v}^h\right)\right\|_{L^2(\Omega)} \le \left(1 + \frac{1}{\beta_{\text{is}}^h}\right)\inf_{\mathbf{w}^h\in V^h}\left\|\nabla\left(\mathbf{u} - \mathbf{w}^h\right)\right\|_{L^2(\Omega)}, \tag{6.19}$$

e.g., see [56, Lemma 3.60]. With respect to the dependency on the discrete inf-sup constant, estimate (6.19) is the worst case estimate. For many pairs of finite element spaces, an alternative estimate using a quasi-local Fortin projection is possible which does not depend on the inverse of β_{is}^h, see [48]. Applying (6.19) to the error bounds (6.14) and (6.18) gives the following estimate.

Corollary 6.3.3 (Error Estimate) *Let the spaces V^h and Q^h satisfy* (6.3) *with β_{is}^h bounded from below by $\beta_0 > 0$ independent of h. Assume that the solution of* (6.5) *satisfies $(\mathbf{u}, p) \in H^{k+1}(\Omega)^d \times H^{l+1}(\Omega)$, then one has the error estimate*

$$\nu \|\nabla(\mathbf{u} - \mathbf{u}^h)\|_{L^2(\Omega)} + \|p - p^h\|_{L^2(\Omega)}$$
$$\le C \left(\nu h^k \|\mathbf{u}\|_{H^{k+1}(\Omega)} + h^{l+1} \|p\|_{H^{l+1}(\Omega)} \right). \tag{6.20}$$

Another norm of interest is the $L^2(\Omega)$ norm of the velocity because its square is proportional to the kinetic energy of the flow. Applying the Poincaré–Friedrichs inequality, one observes that the estimate from Corollary 6.3.3 also holds for $\nu \left\|\mathbf{u} - \mathbf{u}^h\right\|_{L^2(\Omega)}$. However, such an error estimate is suboptimal with respect to h. In what follows, an optimal estimate of the velocity error in the $L^2(\Omega)$ norm will be derived using the usual Aubin–Nitsche technique. To this end, a regularity assumption on the Stokes problem in the following sense will be needed.

Definition 6.3.4 The Stokes problem (6.2) is regular if, for any $\mathbf{f} \in L^2(\Omega)^d$, the solution of the weak formulation (6.5) satisfies $(\mathbf{u}, p) \in H^2(\Omega)^d \times H^1(\Omega)$ and it holds

$$\nu \|\mathbf{u}\|_{H^2(\Omega)} + \|p\|_{H^1(\Omega)} \le C \|\mathbf{f}\|_{L^2(\Omega)}$$

with a constant C independent of $\mathbf{f}$ and ν.

Theorem 6.3.5 (**L^2 Estimate of the Velocity Error**) *Let the spaces V^h and Q^h satisfy* (6.3) *with β_{is}^h bounded from below by $\beta_0 > 0$ independent of h. Assume that the solution of* (6.5) *satisfies $(\mathbf{u}, p) \in H^{k+1}(\Omega)^d \times H^{l+1}(\Omega)$ and let the Stokes problem* (6.2) *be regular. Then there holds the error estimate*

$$\left\|\mathbf{u} - \mathbf{u}^h\right\|_{L^2(\Omega)} \le C \left(h^{k+1} \|\mathbf{u}\|_{H^{k+1}(\Omega)} + \frac{h^{l+2}}{\nu} \|p\|_{H^{l+1}(\Omega)} \right). \tag{6.21}$$

Proof Let $(\mathbf{z}, r) \in V \times Q$ be the solution of the problem

$$\nu(\nabla\mathbf{z}, \nabla\mathbf{v}) - (\nabla\cdot\mathbf{v}, r) + (\nabla\cdot\mathbf{z}, q) = \nu(\mathbf{u} - \mathbf{u}^h, \mathbf{v}) \qquad \forall\, (\mathbf{v}, q) \in V \times Q. \tag{6.22}$$

Then, according to the regularity assumption, one has $(\mathbf{z}, r) \in H^2(\Omega)^d \times H^1(\Omega)$ and

$$\nu \|\mathbf{z}\|_{H^2(\Omega)} + \|r\|_{H^1(\Omega)} \le C\nu \left\|\mathbf{u} - \mathbf{u}^h\right\|_{L^2(\Omega)}. \tag{6.23}$$

Since $\mathbf{u} - \mathbf{u}^h \in V$, one can set $\mathbf{v} = \mathbf{u} - \mathbf{u}^h$ and $q = 0$ in (6.22), which gives

$$\nu \left\|\mathbf{u} - \mathbf{u}^h\right\|_{L^2(\Omega)}^2 = \nu(\nabla\mathbf{z}, \nabla(\mathbf{u} - \mathbf{u}^h)) - (\nabla\cdot(\mathbf{u} - \mathbf{u}^h), r). \tag{6.24}$$

Let $\mathbf{z}^I \in V^h$ be the continuous piecewise (multi)linear Lagrange interpolant of $\mathbf{z}$ satisfying (6.8) with $k = 1$ and let $r^I = \mathfrak{J}^h r \in Q^h$ be an interpolant of r satisfying (6.11). Then

$$\left\| \nabla(\mathbf{z} - \mathbf{z}^I) \right\|_{L^2(\Omega)} \leq \; Ch \left\| \mathbf{z} \right\|_{H^2(\Omega)} \leq Ch \left\| \mathbf{u} - \mathbf{u}^h \right\|_{L^2(\Omega)}, \tag{6.25}$$

$$\left\| r - r^I \right\|_{L^2(\Omega)} \leq Ch \left\| \nabla r \right\|_{L^2(\Omega)} \leq C\nu h \left\| \mathbf{u} - \mathbf{u}^h \right\|_{L^2(\Omega)}. \tag{6.26}$$

It follows from (6.24) that

$$\begin{aligned} \nu \left\| \mathbf{u} - \mathbf{u}^h \right\|_{L^2(\Omega)}^2 &= \nu(\nabla(\mathbf{z} - \mathbf{z}^I), \nabla(\mathbf{u} - \mathbf{u}^h)) - (\nabla \cdot (\mathbf{u} - \mathbf{u}^h), r - r^I) \\ &\quad + \nu(\nabla \mathbf{z}^I, \nabla(\mathbf{u} - \mathbf{u}^h)) - (\nabla \cdot (\mathbf{u} - \mathbf{u}^h), r^I). \end{aligned} \tag{6.27}$$

Applying the Cauchy–Schwarz inequality and (6.25), (6.26), the first two terms in (6.27) can be estimated by

$$\begin{aligned} &\nu(\nabla(\mathbf{z} - \mathbf{z}^I), \nabla(\mathbf{u} - \mathbf{u}^h)) - (\nabla \cdot (\mathbf{u} - \mathbf{u}^h), r - r^I) \\ &\leq C\nu h \left\| \nabla(\mathbf{u} - \mathbf{u}^h) \right\|_{L^2(\Omega)} \left\| \mathbf{u} - \mathbf{u}^h \right\|_{L^2(\Omega)}. \end{aligned} \tag{6.28}$$

Setting $\mathbf{v}^h = \mathbf{z}^I$ in (6.15), using the fact that $\nabla \cdot \mathbf{z} = 0$ and applying the Cauchy–Schwarz inequality and (6.25), one derives

$$\begin{aligned} \nu(\nabla \mathbf{z}^I, \nabla(\mathbf{u} - \mathbf{u}^h)) &= \left(\nabla \cdot (\mathbf{z}^I - \mathbf{z}), p - p^h \right) \\ &\leq \left\| \nabla(\mathbf{z} - \mathbf{z}^I) \right\|_{L^2(\Omega)} \left\| p - p^h \right\|_{L^2(\Omega)} \\ &\leq Ch \left\| \mathbf{u} - \mathbf{u}^h \right\|_{L^2(\Omega)} \left\| p - p^h \right\|_{L^2(\Omega)}. \end{aligned}$$

Finally, the last term in (6.27) vanishes since, according to (6.5) and (6.13), $\mathbf{u}$ is weakly divergence-free and $\mathbf{u}^h$ is discretely divergence-free. Combining the above estimates gives

$$\nu \left\| \mathbf{u} - \mathbf{u}^h \right\|_{L^2(\Omega)} \leq Ch \left(\nu \left\| \nabla(\mathbf{u} - \mathbf{u}^h) \right\|_{L^2(\Omega)} + \left\| p - p^h \right\|_{L^2(\Omega)} \right)$$

and the statement of the theorem follows from Corollary 6.3.3. □

It should be noted that the velocity error bounds (6.20) and (6.21) improve substantially if an inf-sup stable pair of finite element spaces is used with $V_{\text{div}}^h \subset V_{\text{div}}$. Such pairs exist, e.g., the Scott–Vogelius pair P_2/P_1^{disc} applied on special meshes. Then, the pressure term in the error equation (6.16) vanishes and consequently the pressure terms vanish on the right-hand sides of the estimates (6.20) and (6.21). The consequences are that the velocity error bounds do not depend on the pressure and they do not depend explicitly on inverse powers of the viscosity. Even for spaces with $V_{\text{div}}^h \not\subset V_{\text{div}}$, an approach has been developed such that the velocity error bounds have these two properties, see [70, 71] or the recent survey paper [58]. To derive velocity error bounds with these two properties for pressure-stabilized methods, as presented in the following sections, is impossible.

For inf-sup stable pairs of finite element spaces, error estimates with respect to the norms of other Lebesgue spaces can be proved. In particular, estimates in $L^\infty(\Omega)$ were derived in [33, 46, 47, 51, 52] that are of the form

$$\nu \left\| \nabla \left(\mathbf{u} - \mathbf{u}^h \right) \right\|_{L^\infty(\Omega)} + \left\| p - p^h \right\|_{L^\infty(\Omega)} \\ \leq C \left(\nu \inf_{\mathbf{v}^h \in V^h} \left\| \nabla \left(\mathbf{u} - \mathbf{v}^h \right) \right\|_{L^\infty(\Omega)} + \inf_{q^h \in V^h} \left\| p - q^h \right\|_{L^\infty(\Omega)} \right). \tag{6.29}$$

In [46], even an estimate of the form

$$\nu \left\| \nabla \left(\mathbf{u} - \mathbf{u}^h \right) \right\|_{L^r(\Omega)} + \left\| p - p^h \right\|_{L^r(\Omega)} \tag{6.30}$$
$$\leq C \left(\nu \inf_{\mathbf{v}^h \in V^h} \left\| \nabla \left(\mathbf{u} - \mathbf{v}^h \right) \right\|_{L^r(\Omega)} + \inf_{q^h \in V^h} \left\| p - q^h \right\|_{L^r(\Omega)} \right), \quad 2 \leq r \leq \infty$$

was shown. The current state of the art is that estimates of form (6.29) and (6.30) can be proved for convex polyhedral domains.

6.4 Residual-Based Stabilizations

For another review of residual-based stabilizations, it is referred to [42].

6.4.1 A Framework

A framework for the derivation of residual-based stabilizations was presented in [19]. Starting point is the regularization of the Galerkin finite element method (6.13) with respect to the norm of Q^h

$$\nu \left(\nabla \mathbf{u}^h, \nabla \mathbf{v}^h \right) - \left(\nabla \cdot \mathbf{v}^h, p^h \right) + \left(\nabla \cdot \mathbf{u}^h, q^h \right) + \delta \left(p^h, q^h \right) = \left(\mathbf{f}, \mathbf{v}^h \right),$$

where $\delta > 0$ is a stabilization parameter. However, this stabilization acts like a penalty term which prevents the method from being optimally convergent for higher order finite element spaces. Thus, this stabilization should be replaced by a stabilization that is, on the one hand, similarly strong but, on the other hand, possesses a sufficiently small consistency error. Using [49, Cor. 2.1], it is known that there are positive constants C_1 and C_2 such that

$$C_1 \|q\|_{L^2(\Omega)} \leq \|\nabla q\|_{H^{-1}(\Omega)} \leq C_2 \|q\|_{L^2(\Omega)} \quad \forall\, q \in Q,$$

i.e., the $H^{-1}(\Omega)^d$ norm of ∇q is equivalent to the $L^2(\Omega)$ norm of q. Consequently, $\|\nabla q^h\|^2_{H^{-1}(\Omega)}$ has the same stabilization effect like $\|q^h\|^2_{L^2(\Omega)}$. The term $(\nabla p^h, \nabla q^h)_{-1}$ can be included in a stabilization term naturally by using the residual, where $(\cdot,\cdot)_{-1}$ is the inner product in $H^{-1}(\Omega)^d$, see [19] for a definition of this inner product.

For simplicity of presentation, only the case $Q^h \subset H^1(\Omega)$ is considered. The prototype of a residual-based stabilization from [19] has the form: Find $(\mathbf{u}^h, p^h) \in V^h \times Q^h$ such that

$$\begin{aligned} &\nu\left(\nabla \mathbf{u}^h, \nabla \mathbf{v}^h\right) - \left(\nabla\cdot \mathbf{v}^h, p^h\right) + \left(\nabla \cdot \mathbf{u}^h, q^h\right) \\ &\quad + \delta\left(-\nu\Delta \mathbf{u}^h + \nabla p^h, \kappa\nu\Delta \mathbf{v}^h + \nabla q^h\right)_{-1} \\ &= \left(\mathbf{f}, \mathbf{v}^h\right) + \delta\left(\mathbf{f}, \kappa\nu\Delta\mathbf{v}^h + \nabla q^h\right)_{-1} \quad \forall\, \left(\mathbf{v}^h, q^h\right) \in V^h \times Q^h, \end{aligned} \tag{6.31}$$

with $\kappa \in \{-1, 0, 1\}$ and $\delta > 0$. There are still two issues in (6.31). First, $(\cdot,\cdot)_{-1}$ is not computable and second, $\Delta\mathbf{u}^h$, $\Delta\mathbf{v}^h$ are not defined. Thanks to the regularity assumption on Q^h, the functions ∇p^h, ∇q^h are well defined.

A standard way to resolve these issues consists in approximating $(\cdot,\cdot)_{-1}$ by a weighted $L^2(\Omega)$ inner product, leading to the following problem: Find $(\mathbf{u}^h, p^h) \in V^h \times Q^h$ such that for all $(\mathbf{v}^h, q^h) \in V^h \times Q^h$

$$\begin{aligned} &\nu\left(\nabla \mathbf{u}^h, \nabla \mathbf{v}^h\right) - \left(\nabla\cdot \mathbf{v}^h, p^h\right) + \left(\nabla \cdot \mathbf{u}^h, q^h\right) \\ &\quad + \sum_{K\in\mathcal{T}^h} \delta h_K^2\left(-\nu\Delta \mathbf{u}^h + \nabla p^h, \kappa\nu\Delta \mathbf{v}^h + \nabla q^h\right)_K \\ &= \left(\mathbf{f}, \mathbf{v}^h\right) + \sum_{K\in\mathcal{T}^h} \delta h_K^2\left(\mathbf{f}, \kappa\nu\Delta\mathbf{v}^h + \nabla q^h\right)_K. \end{aligned} \tag{6.32}$$

For $\kappa = 0$, one obtains the PSPG method, which is discussed in Sect. 6.4.2, for $\kappa = 1$ the symmetric GLS method, see Sect. 6.4.3, and for $\kappa = -1$ the non-symmetric GLS method presented in Sect. 6.4.4.

In [19], a new proposal for approximating the inner product in $H^{-1}(\Omega)^d$ was presented. This proposal is discussed briefly in Sect. 6.4.5.

Definition 6.4.1 (Absolutely and Conditionally Stable Methods) A stabilized discrete method is called absolutely stable if it is stable for all $\delta > 0$. Otherwise, if it is stable only for a restricted set of parameters, it is called conditionally stable.

6.4.2 The PSPG Method

The Pressure Stabilizing Petrov–Galerkin (PSPG) method was proposed for finite element spaces with continuous discrete pressures in [55]. In the case of piecewise polynomial but discontinuous finite element pressure spaces, an additional term is necessary, which was introduced in [39, 54].

The PSPG method has the form: Find $\left(\mathbf{u}^h, p^h\right) \in V^h \times Q^h$ such that

$$A_{\mathrm{pspg}}\left(\left(\mathbf{u}^h, p^h\right), \left(\mathbf{v}^h, q^h\right)\right) = L_{\mathrm{pspg}}\left(\left(\mathbf{v}^h, q^h\right)\right) \quad \forall \left(\mathbf{v}^h, q^h\right) \in V^h \times Q^h, \tag{6.33}$$

where the bilinear form $A_{\mathrm{pspg}} : \left(\tilde{V} \times \tilde{Q}\right) \times \left(V \times \tilde{Q}\right) \to \mathbb{R}$ is given by

$$\begin{aligned} A_{\mathrm{pspg}}\left((\mathbf{u}, p), (\mathbf{v}, q)\right) &= \nu\left(\nabla \mathbf{u}, \nabla \mathbf{v}\right) - \left(\nabla \cdot \mathbf{v}, p\right) + \left(\nabla \cdot \mathbf{u}, q\right) \\ &\quad + \sum_{E \in \mathcal{E}^h} \gamma_E \left([|p|]_E, [|q|]_E\right)_E + \sum_{K \in \mathcal{T}^h} \left(-\nu \Delta \mathbf{u} + \nabla p, \delta_K \nabla q\right)_K \end{aligned} \tag{6.34}$$

and the linear form $L_{\mathrm{pspg}} : V \times \tilde{Q} \to \mathbb{R}$ by

$$L_{\mathrm{pspg}}\left((\mathbf{v}, q)\right) = (\mathbf{f}, \mathbf{v}) + \sum_{K \in \mathcal{T}^h} \left(\mathbf{f}, \delta_K \nabla q\right)_K, \tag{6.35}$$

with

$$\tilde{V} = \left\{\mathbf{v} \in V : \mathbf{v}|_K \in H^2(K)^d \text{ for all } K \in \mathcal{T}^h\right\}, \tag{6.36}$$

$$\tilde{Q} = \left\{q \in Q : q|_K \in H^1(K) \text{ for all } K \in \mathcal{T}^h\right\} \tag{6.37}$$

and nonnegative stabilization parameters γ_E and δ_K. Their appropriate choices will be based on the study of the existence and uniqueness of a solution of (6.33), see Lemma 6.4.3, and on finite element error estimates, see Theorem 6.4.6. The volume integrals in the stabilization terms contain the so-called strong residual of the Stokes equations.

The definition of $\tilde{Q}$ ensures that the jumps of the pressure across the faces of the mesh cells are well defined. If $Q^h \subset H^1(\Omega)$, then the jumps of the pressure vanish almost everywhere on the faces. From the practical point of view, the case of piecewise polynomial and continuous discrete pressure functions is very important such that then even $Q^h \subset C(\overline{\Omega})$.

Lemma 6.4.2 (A Norm in $V^h \times Q^h$ Containing the Stabilization Terms) *Let $\delta_K > 0$ for all $K \in \mathcal{T}^h$ and, in the case $Q^h \not\subset H^1(\Omega)$, let $\gamma_E > 0$ for all $E \in \mathcal{E}^h$. Then*

$$\left\| \left(\mathbf{v}^h, q^h \right) \right\|_{\text{pspg}} = \left(\nu \left\| \nabla \mathbf{v}^h \right\|_{L^2(\Omega)}^2 + \sum_{E \in \mathcal{E}^h} \gamma_E \left\| \left[\left| q^h \right| \right]_E \right\|_{L^2(E)}^2 + \sum_{K \in \mathcal{T}^h} \delta_K \left\| \nabla q^h \right\|_{L^2(K)}^2 \right)^{1/2} \tag{6.38}$$

defines a norm in $V^h \times Q^h$.

Proof Expression (6.38) is the square root of a sum of squares of seminorms. Thus, it is clearly a seminorm itself. It remains to prove that from $\left\| \left(\mathbf{v}^h, q^h \right) \right\|_{\text{pspg}} = 0$, it follows that $\mathbf{v}^h = \mathbf{0}$ and $q^h = 0$.

Let $\left\| \left(\mathbf{v}^h, q^h \right) \right\|_{\text{pspg}} = 0$, then all terms in (6.38) vanish. In particular, it holds $\left\| \nabla \mathbf{v}^h \right\|_{L^2(\Omega)} = 0$. Since this expression is a norm in V^h, it follows that $\mathbf{v}^h = \mathbf{0}$. With this result, one gets

$$0 = \sum_{E \in \mathcal{E}^h} \gamma_E \left\| \left[\left| q^h \right| \right]_E \right\|_{L^2(E)}^2 + \sum_{K \in \mathcal{T}^h} \delta_K \left\| \nabla q^h \right\|_{L^2(K)}^2 .$$

Because δ_K is assumed to be positive for all mesh cells, it follows that $\left\| \nabla q^h \right\|_{L^2(K)} = 0$ for all $K \in \mathcal{T}^h$. If $Q^h \subset H^1(\Omega)$, then $\left\| \left[\left| q^h \right| \right]_E \right\|_{L^2(E)} = 0$ for all faces. Otherwise, one gets this property from the assumption $\gamma_E > 0$ for all faces. Altogether, it follows that q^h is constant on Ω. The only globally constant function in Q^h is $q^h = 0$. Hence $\left\| \left(\mathbf{v}^h, q^h \right) \right\|_{\text{pspg}}$ defines a norm on $V^h \times Q^h$. □

Lemma 6.4.3 (Existence and Uniqueness of a Solution of (6.33)**)** *Let the assumptions of Lemma 6.4.2 be satisfied and let*

$$\delta_K \le \frac{h_K^2}{\nu C_{\text{inv}}^2}. \tag{6.39}$$

Then the PSPG problem (6.33) *possesses a unique solution.*

Proof First, the coercivity of the bilinear form $A_{\text{pspg}}(\cdot, \cdot)$ with respect to the norm $\|\cdot\|_{\text{pspg}}$ will be shown for any $(\mathbf{v}^h, q^h) \in V^h \times Q^h$. One obtains with the Cauchy–Schwarz inequality, the inverse inequality (6.12), the Young inequality, and the condition (6.39) on the stabilization parameters

$$\begin{aligned}
&A_{\text{pspg}}\left(\left(\mathbf{v}^h, q^h \right), \left(\mathbf{v}^h, q^h \right) \right) \\
&\ge \nu \left\| \nabla \mathbf{v}^h \right\|_{L^2(\Omega)}^2 + \sum_{E \in \mathcal{E}^h} \gamma_E \left\| \left[\left| q^h \right| \right]_E \right\|_{L^2(E)}^2 + \sum_{K \in \mathcal{T}^h} \delta_K \left\| \nabla q^h \right\|_{L^2(K)}^2 \\
&\quad - \sum_{K \in \mathcal{T}^h} \delta_K \nu \left\| \Delta \mathbf{v}^h \right\|_{L^2(K)} \left\| \nabla q^h \right\|_{L^2(K)}
\end{aligned}$$

$$\geq \left\| \left(\mathbf{v}^h, q^h \right) \right\|_{\text{pspg}}^2 - \sum_{K \in \mathcal{T}^h} \delta_K h_K^{-1} C_{\text{inv}} \nu \left\| \nabla \mathbf{v}^h \right\|_{L^2(K)} \left\| \nabla q^h \right\|_{L^2(K)}$$

$$\geq \left\| \left(\mathbf{v}^h, q^h \right) \right\|_{\text{pspg}}^2 - \frac{1}{2} \sum_{K \in \mathcal{T}^h} \frac{\delta_K C_{\text{inv}}^2 \nu^2}{h_K^2} \left\| \nabla \mathbf{v}^h \right\|_{L^2(K)}^2 - \frac{1}{2} \sum_{K \in \mathcal{T}^h} \delta_K \left\| \nabla q^h \right\|_{L^2(K)}^2$$

$$\geq \frac{1}{2} \left\| \left(\mathbf{v}^h, q^h \right) \right\|_{\text{pspg}}^2 . \tag{6.40}$$

The PSPG problem (6.33) is equivalent to a system of linear algebraic equations with a square matrix. The coercivity (6.40) implies that the homogeneous PSPG problem (for $\mathbf{f} = \mathbf{0}$) has only the trivial solution. Consequently, the matrix is non-singular, which proves the lemma. □

Since the stabilization parameters have to satisfy (6.39), they depend on the local mesh size. Hence, the norm $\|\cdot\|_{\text{pspg}}$ is a mesh-dependent norm. Note that in the case that $\Delta \mathbf{v}^h|_K = \mathbf{0}$ for all mesh cells K, as it is given, e.g., for P_1 finite elements, the restriction (6.39) on the stabilization parameter is not necessary.

Lemma 6.4.4 (Stability Estimate) *Let the assumptions of Lemmas 6.4.2 and 6.4.3 be satisfied. Then the solution of the PSPG problem* (6.33) *satisfies the stability estimate*

$$\left\| \left(\mathbf{u}^h, p^h \right) \right\|_{\text{pspg}} \leq \frac{C}{\nu^{1/2}} \|\mathbf{f}\|_{L^2(\Omega)} + 2 \left(\sum_{K \in \mathcal{T}^h} \delta_K \|\mathbf{f}\|_{L^2(K)}^2 \right)^{1/2} . \tag{6.41}$$

Proof Using the Cauchy–Schwarz inequality, the Poincaré–Friedrichs inequality, and the Cauchy–Schwarz inequality for sums, one obtains

$$\begin{aligned}
& L_{\text{pspg}} \left(\left(\mathbf{v}^h, q^h \right) \right) \\
& \leq \|\mathbf{f}\|_{L^2(\Omega)} \left\| \mathbf{v}^h \right\|_{L^2(\Omega)} + \sum_{K \in \mathcal{T}^h} \delta_K \|\mathbf{f}\|_{L^2(K)} \left\| \nabla q^h \right\|_{L^2(K)} \\
& \leq C \|\mathbf{f}\|_{L^2(\Omega)} \left\| \nabla \mathbf{v}^h \right\|_{L^2(\Omega)} \\
& \quad + \left(\sum_{K \in \mathcal{T}^h} \delta_K \|\mathbf{f}\|_{L^2(K)}^2 \right)^{1/2} \left(\sum_{K \in \mathcal{T}^h} \delta_K \left\| \nabla q^h \right\|_{L^2(K)}^2 \right)^{1/2} \\
& \leq \left(\frac{C}{\nu^{1/2}} \|\mathbf{f}\|_{L^2(\Omega)} + \left(\sum_{K \in \mathcal{T}^h} \delta_K \|\mathbf{f}\|_{L^2(K)}^2 \right)^{1/2} \right) \left\| \left(\mathbf{v}^h, q^h \right) \right\|_{\text{pspg}} ,
\end{aligned}$$

for all $\left(\mathbf{v}^h, q^h\right) \in V^h \times Q^h$. Inserting this estimate in (6.33) and setting $\left(\mathbf{v}^h, q^h\right) = \left(\mathbf{u}^h, p^h\right)$, the stability estimate follows using the coercivity (6.40). □

Lemma 6.4.5 (Consistency and Galerkin Orthogonality) *Let the solution of* (6.5) *satisfy* $(\mathbf{u}, p) \in H^2(\Omega)^d \times H^1(\Omega)$ *and let* $(\mathbf{u}^h, p^h) \in V^h \times Q^h$ *be the solution of the PSPG method* (6.33). *The PSPG method is consistent, i.e., it holds*

$$A_{\text{pspg}}\left((\mathbf{u}, p), \left(\mathbf{v}^h, q^h\right)\right) = L_{\text{pspg}}\left(\left(\mathbf{v}^h, q^h\right)\right) \quad \forall \left(\mathbf{v}^h, q^h\right) \in V^h \times Q^h \tag{6.42}$$

and it satisfies the Galerkin orthogonality

$$A_{\text{pspg}}\left(\left(\mathbf{u} - \mathbf{u}^h, p - p^h\right), \left(\mathbf{v}^h, q^h\right)\right) = 0 \quad \forall \left(\mathbf{v}^h, q^h\right) \in V^h \times Q^h. \tag{6.43}$$

Proof The residual vanishes for $(\mathbf{u}, p)$ and with that the residual-based stabilization terms in A_{pspg} and L_{pspg} are equal. Moreover, the stabilization term with pressure jumps vanishes since $p \in H^1(\Omega)$. Thus, only the terms from the weak formulation (6.5) remain and since the finite element spaces are conforming, (6.42) holds.

The Galerkin orthogonality is obtained by subtracting (6.33) from (6.42). □

To prove error estimates for the solution of (6.33), we shall need additional assumptions on the stabilization parameters. It will be assumed that there are positive constants δ_0, δ_1 and γ_0, γ_1 independent of ν and h such that

$$0 < \delta_0 \frac{h_K^2}{\nu} \le \delta_K \le \delta_1 \frac{h_K^2}{\nu} \qquad \forall K \in \mathcal{T}^h \tag{6.44}$$

and

$$0 < \gamma_0 \frac{h_E}{\nu} \le \gamma_E \le \gamma_1 \frac{h_E}{\nu} \qquad \forall E \in \mathcal{E}^h. \tag{6.45}$$

Theorem 6.4.6 (Error Estimate) *Let the solution of* (6.5) *satisfy* $(\mathbf{u}, p) \in H^{k+1}(\Omega)^d \times H^{l+1}(\Omega)$ *and let* $\left(\mathbf{u}^h, p^h\right) \in V^h \times Q^h$ *be the solution of the PSPG problem* (6.33). *Assume that the stabilization parameters satisfy* (6.44) *and* (6.45) *with* $\delta_1 \le 1/C_{\text{inv}}^2$. *Then the following error estimate holds:*

$$\left\| \left(\mathbf{u} - \mathbf{u}^h, p - p^h\right) \right\|_{\text{pspg}} \le C \left(\nu^{1/2} h^k \|\mathbf{u}\|_{H^{k+1}(\Omega)} + \frac{h^{l+1}}{\nu^{1/2}} \|p\|_{H^{l+1}(\Omega)} \right). \tag{6.46}$$

Proof The triangle inequality gives

$$\left\| \left(\mathbf{u} - \mathbf{u}^h, p - p^h\right) \right\|_{\text{pspg}}$$

$$\leq \left\| \left(\mathbf{u} - I^h \mathbf{u}, p - J^h p \right) \right\|_{\text{pspg}} + \left\| \left(\mathbf{u}^h - I^h \mathbf{u}, p^h - J^h p \right) \right\|_{\text{pspg}}, \tag{6.47}$$

where I^h and J^h are the interpolation operators satisfying (6.8) and (6.9). Both terms on the right-hand side of (6.47) are estimated separately.

One obtains with the interpolation estimates (6.8) and (6.9), and with the assumptions (6.44) and (6.45) on the stabilization parameters

$$\begin{aligned}
&\left\| \left(\mathbf{u} - I^h \mathbf{u}, p - J^h p \right) \right\|_{\text{pspg}}^2 \\
&\leq \nu \left\| \nabla \left(\mathbf{u} - I^h \mathbf{u} \right) \right\|_{L^2(\Omega)}^2 + \frac{\gamma_1 h}{\nu} \sum_{E \in \mathcal{E}^h} \left\| \left[\left| p - J^h p \right|\right]_E \right\|_{L^2(E)}^2 \\
&\quad + \frac{\delta_1 h^2}{\nu} \sum_{K \in \mathcal{T}^h} \left\| \nabla \left(p - J^h p \right) \right\|_{L^2(K)}^2 \\
&\leq C \left(\nu h^{2k} \|\mathbf{u}\|_{H^{k+1}(\Omega)}^2 + \frac{h^{2(l+1)}}{\nu} \|p\|_{H^{l+1}(\Omega)}^2 \right).
\end{aligned} \tag{6.48}$$

The estimate of the second term of (6.47) starts with the coercivity (6.40) and the Galerkin orthogonality (6.43)

$$\begin{aligned}
&\left\| \left(\mathbf{u}^h - I^h \mathbf{u}, p^h - J^h p \right) \right\|_{\text{pspg}}^2 \\
&\leq 2 A_{\text{pspg}} \left(\left(\mathbf{u}^h - I^h \mathbf{u}, p^h - J^h p \right), \left(\mathbf{u}^h - I^h \mathbf{u}, p^h - J^h p \right) \right) \\
&= 2 A_{\text{pspg}} \left(\left(\mathbf{u} - I^h \mathbf{u}, p - J^h p \right), \left(\mathbf{u}^h - I^h \mathbf{u}, p^h - J^h p \right) \right).
\end{aligned} \tag{6.49}$$

Now, splitting the right-hand side of (6.49) into the terms according to (6.34), each term will be estimated separately. The goal of these estimates is to obtain interpolation errors and to hide the other terms in the left-hand side of (6.49).

Using the Cauchy–Schwarz inequality, the Young inequality, and the interpolation estimate (6.8), one obtains for the viscous term

$$\begin{aligned}
&\nu \left(\nabla \left(\mathbf{u} - I^h \mathbf{u} \right), \nabla \left(\mathbf{u}^h - I^h \mathbf{u} \right) \right) \\
&\leq \nu \left\| \nabla \left(\mathbf{u} - I^h \mathbf{u} \right) \right\|_{L^2(\Omega)} \left\| \nabla \left(\mathbf{u}^h - I^h \mathbf{u} \right) \right\|_{L^2(\Omega)} \\
&\leq 4\nu \left\| \nabla \left(\mathbf{u} - I^h \mathbf{u} \right) \right\|_{L^2(\Omega)}^2 + \frac{\nu}{16} \left\| \nabla \left(\mathbf{u}^h - I^h \mathbf{u} \right) \right\|_{L^2(\Omega)}^2 \\
&\leq C \nu h^{2k} \|\mathbf{u}\|_{H^{k+1}(\Omega)}^2 + \frac{\nu}{16} \left\| \nabla \left(\mathbf{u}^h - I^h \mathbf{u} \right) \right\|_{L^2(\Omega)}^2.
\end{aligned}$$

The last term can be absorbed in the left-hand side of (6.49). In a similar way, using (6.9), one gets

$$\left(\nabla\cdot\left(\mathbf{u}^h - I^h\mathbf{u}\right), p - J^h p\right) \le C\frac{h^{2(l+1)}}{\nu}\,\|p\|^2_{H^{l+1}(\Omega)} + \frac{\nu}{16}\left\|\nabla\left(\mathbf{u}^h - I^h\mathbf{u}\right)\right\|^2_{L^2(\Omega)}.$$

The estimate of the next term requires an integration by parts

$$\begin{aligned}\left(\nabla\cdot\left(\mathbf{u} - I^h\mathbf{u}\right), p^h - J^h p\right) &= \sum_{E\in\mathcal{E}^h}\left(\left(\mathbf{u} - I^h\mathbf{u}\right)\cdot\mathbf{n}_E, \left[\left|p^h - J^h p\right|\right]_E\right)_E\\ &\quad - \sum_{K\in\mathcal{T}^h}\left(\mathbf{u} - I^h\mathbf{u}, \nabla\left(p^h - J^h p\right)\right)_K. \qquad (6.50)\end{aligned}$$

Both terms on the right-hand side of (6.50) are estimated more or less in the same way, e.g., one obtains for the last term with the Cauchy–Schwarz inequality, the Young inequality, the property (6.44) of the stabilization parameters, and the interpolation estimate (6.8)

$$\begin{aligned}&\sum_{K\in\mathcal{T}^h}\left(\mathbf{u} - I^h\mathbf{u}, \nabla\left(p^h - J^h p\right)\right)_K \le \sum_{K\in\mathcal{T}^h}\left\|\mathbf{u} - I^h\mathbf{u}\right\|_{L^2(K)}\left\|\nabla\left(p^h - J^h p\right)\right\|_{L^2(K)}\\
&\le 4\sum_{K\in\mathcal{T}^h}\frac{1}{\delta_K}\left\|\mathbf{u} - I^h\mathbf{u}\right\|^2_{L^2(K)} + \frac{1}{16}\sum_{K\in\mathcal{T}^h}\delta_K\left\|\nabla\left(p^h - J^h p\right)\right\|^2_{L^2(K)}\\
&\le \frac{4\nu}{\delta_0}\sum_{K\in\mathcal{T}^h}h_K^{-2}\left\|\mathbf{u} - I^h\mathbf{u}\right\|^2_{L^2(K)} + \frac{1}{16}\sum_{K\in\mathcal{T}^h}\delta_K\left\|\nabla\left(p^h - J^h p\right)\right\|^2_{L^2(K)}\\
&\le \frac{C\nu}{\delta_0}h^{2k}\|\mathbf{u}\|^2_{H^{k+1}(\Omega)} + \frac{1}{16}\sum_{K\in\mathcal{T}^h}\delta_K\left\|\nabla\left(p^h - J^h p\right)\right\|^2_{L^2(K)}.\end{aligned}$$

The estimate of the other term on the right-hand side of (6.50) uses (6.45). All stabilization terms are estimated with the same tools used so far. One gets

$$\begin{aligned}&\sum_{K\in\mathcal{T}^h}\left(-\nu\Delta\left(\mathbf{u} - I^h\mathbf{u}\right), \delta_K\nabla\left(p^h - J^h p\right)\right)_K\\
&\le C\nu h^{2k}\|\mathbf{u}\|^2_{H^{k+1}(\Omega)} + \frac{1}{16}\sum_{K\in\mathcal{T}^h}\delta_K\left\|\nabla\left(p^h - J^h p\right)\right\|^2_{L^2(K)},\end{aligned}$$

and

$$\begin{aligned}&\sum_{K\in\mathcal{T}^h}\left(\nabla\left(p - J^h p\right), \delta_K\nabla\left(p^h - J^h p\right)\right)_K\\
&\le C\frac{h^{2(l+1)}}{\nu}\|p\|^2_{H^{l+1}(\Omega)} + \frac{1}{16}\sum_{K\in\mathcal{T}^h}\delta_K\left\|\nabla\left(p^h - J^h p\right)\right\|^2_{L^2(K)}.\end{aligned}$$

Finally, for the term with the pressure jumps, one gets with (6.9)

$$\sum_{E\in\mathcal{E}^h} \gamma_E \left(\left[\left|p - J^h p\right|\right]_E, \left[\left|p^h - J^h p\right|\right]_E\right)_E$$

$$\leq C\frac{h^{2(l+1)}}{\nu} \|p\|^2_{H^{l+1}(\Omega)} + \frac{1}{16}\sum_{E\in\mathcal{E}^h} \gamma_E \left\|\left[\left|p^h - J^h p\right|\right]_E\right\|^2_{L^2(E)}.$$

Collecting all estimates proves the statement of the theorem. □

To derive an error estimate for the pressure in the L^2 norm, the following auxiliary problem (a kind of Stokes projection) will be considered: Find $(\mathbf{w}, r) \in V \times Q$ such that

$$\begin{aligned} (\nabla \mathbf{w}, \nabla \mathbf{v}) - (\nabla \cdot \mathbf{v}, r) &= 0 & \forall\, \mathbf{v} \in V, \\ -(\nabla \cdot \mathbf{w}, q) &= \left(p - p^h, q\right) & \forall\, q \in Q. \end{aligned} \tag{6.51}$$

It follows from the theory of linear saddle point problems that (6.51) possesses a unique solution.

Lemma 6.4.7 (Stability Estimate for (6.51)**)** *For the unique solution of* (6.51) *there holds the stability estimate*

$$\|\nabla \mathbf{w}\|_{L^2(\Omega)} + \|r\|_{L^2(\Omega)} \leq C \left\|p - p^h\right\|_{L^2(\Omega)}. \tag{6.52}$$

The constant depends on the inverse of β_{is} *from* (6.6).

Proof Using (6.6), (6.51), and the Cauchy–Schwarz inequality gives

$$\begin{aligned} \beta_{\text{is}} \|r\|_{L^2(\Omega)} &\leq \sup_{\mathbf{v}\in V\setminus\{\mathbf{0}\}} \frac{(\nabla \cdot \mathbf{v}, r)}{\|\nabla \mathbf{v}\|_{L^2(\Omega)}} = \sup_{\mathbf{v}\in V\setminus\{\mathbf{0}\}} \frac{(\nabla \mathbf{w}, \nabla \mathbf{v})}{\|\nabla \mathbf{v}\|_{L^2(\Omega)}} \\ &\leq \sup_{\mathbf{v}\in V\setminus\{\mathbf{0}\}} \frac{\|\nabla \mathbf{w}\|_{L^2(\Omega)} \|\nabla \mathbf{v}\|_{L^2(\Omega)}}{\|\nabla \mathbf{v}\|_{L^2(\Omega)}} = \|\nabla \mathbf{w}\|_{L^2(\Omega)}. \end{aligned} \tag{6.53}$$

Inserting $(\mathbf{v}, q) = (\mathbf{w}, r)$ in (6.51), subtracting both equations, and applying the Cauchy–Schwarz inequality and (6.53) yields

$$\begin{aligned} \|\nabla \mathbf{w}\|^2_{L^2(\Omega)} &= -\left(p - p^h, r\right) \leq \left\|p - p^h\right\|_{L^2(\Omega)} \|r\|_{L^2(\Omega)} \\ &\leq \frac{1}{\beta_{\text{is}}} \left\|p - p^h\right\|_{L^2(\Omega)} \|\nabla \mathbf{w}\|_{L^2(\Omega)}. \end{aligned} \tag{6.54}$$

Combining (6.53) and (6.54) leads to

$$\|\nabla \mathbf{w}\|_{L^2(\Omega)} + \|r\|_{L^2(\Omega)} \le \left(1 + \frac{1}{\beta_{\text{is}}}\right) \|\nabla \mathbf{w}\|_{L^2(\Omega)} \le \frac{1}{\beta_{\text{is}}}\left(1 + \frac{1}{\beta_{\text{is}}}\right) \left\|p - p^h\right\|_{L^2(\Omega)}.$$

□

Theorem 6.4.8 (**L^2 Estimate of the Pressure Error**) *Assume that the solution of* (6.5) *satisfies* $(\mathbf{u}, p) \in H^{k+1}(\Omega)^d \times H^{l+1}(\Omega)$ *and that the stabilization parameters satisfy* (6.44) *and* (6.45) *with* $\delta_1 \le 1/C_{\text{inv}}^2$. *Then there holds the error estimate*

$$\left\|p - p^h\right\|_{L^2(\Omega)} \le C\left(\nu h^k \|\mathbf{u}\|_{H^{k+1}(\Omega)} + h^{l+1}\|p\|_{H^{l+1}(\Omega)}\right).$$

Proof Let $(\mathbf{w}, r)$ be the solution of (6.51). Let $\mathcal{I}^h\mathbf{w} \in V^h$ be an interpolant of $\mathbf{w}$ satisfying (6.10). Inserting $q = p - p^h$ in (6.51) gives

$$\left\|p - p^h\right\|_{L^2(\Omega)}^2 = -\left(\nabla \cdot \mathbf{w}, p - p^h\right) = -\left(\nabla \cdot \left(\mathbf{w} - \mathcal{I}^h\mathbf{w}\right), p - p^h\right) - \left(\nabla \cdot \left(\mathcal{I}^h\mathbf{w}\right), p - p^h\right). \quad (6.55)$$

Consider now the second term on the right-hand side of (6.55). The Galerkin orthogonality (6.43) with $\mathbf{v}^h = \mathcal{I}^h\mathbf{w}$ and $q^h = 0$ leads to

$$0 = \nu\left(\nabla\left(\mathbf{u} - \mathbf{u}^h\right), \nabla\mathcal{I}^h\mathbf{w}\right) - \left(\nabla \cdot \left(\mathcal{I}^h\mathbf{w}\right), p - p^h\right).$$

Hence, one obtains with the Cauchy–Schwarz inequality, (6.10), and (6.52)

$$\left|\left(\nabla \cdot \left(\mathcal{I}^h\mathbf{w}\right), p - p^h\right)\right| \le C\nu \left\|\nabla(\mathbf{u} - \mathbf{u}^h)\right\|_{L^2(\Omega)} \left\|p - p^h\right\|_{L^2(\Omega)}. \quad (6.56)$$

The estimate of the first term on the right-hand side of (6.55) starts with integration by parts, followed by the Cauchy–Schwarz inequality and application of (6.44), (6.45), (6.10), and (6.52)

$$\begin{aligned} &-\left(\nabla \cdot \left(\mathbf{w} - \mathcal{I}^h\mathbf{w}\right), p - p^h\right) \\ &= \sum_{K\in\mathcal{T}^h} \left(\mathbf{w} - \mathcal{I}^h\mathbf{w}, \nabla\left(p - p^h\right)\right)_K - \sum_{E\in\mathcal{E}^h} \left(\left(\mathbf{w} - \mathcal{I}^h\mathbf{w}\right) \cdot \mathbf{n}_E, \left[\left|p - p^h\right|\right]_E\right)_E \\ &\le \left(\sum_{K\in\mathcal{T}^h} \delta_K^{-1} \left\|\mathbf{w} - \mathcal{I}^h\mathbf{w}\right\|_{L^2(K)}^2\right)^{1/2} \left(\sum_{K\in\mathcal{T}^h} \delta_K \left\|\nabla\left(p - p^h\right)\right\|_{L^2(K)}^2\right)^{1/2} \end{aligned}$$

$$+\left(\sum_{E\in\mathcal{E}^h}\gamma_E^{-1}\left\|\mathbf{w}-\mathcal{I}^h\mathbf{w}\right\|_{L^2(E)}^2\right)^{1/2}\left(\sum_{E\in\mathcal{E}^h}\gamma_E\left\|\left[\left|p-p^h\right|\right]_E\right\|_{L^2(E)}^2\right)^{1/2}$$

$$\le C\nu^{1/2}\left(\frac{1}{\delta_0^{1/2}}+\frac{1}{\gamma_0^{1/2}}\right)\|\nabla\mathbf{w}\|_{L^2(\Omega)}\left\|(\mathbf{u}-\mathbf{u}^h,p-p^h)\right\|_{\mathrm{pspg}}$$

$$\le C\nu^{1/2}\left(\frac{1}{\delta_0^{1/2}}+\frac{1}{\gamma_0^{1/2}}\right)\left\|p-p^h\right\|_{L^2(\Omega)}\left\|(\mathbf{u}-\mathbf{u}^h,p-p^h)\right\|_{\mathrm{pspg}}. \qquad (6.57)$$

Combining the estimates (6.55), (6.56), and (6.57) yields

$$\left\|p-p^h\right\|_{L^2(\Omega)}\le C\nu^{1/2}\left\|(\mathbf{u}-\mathbf{u}^h,p-p^h)\right\|_{\mathrm{pspg}},$$

where the constant C depends on $\delta_0^{-1/2}$ and $\gamma_0^{-1/2}$. Thus, the final estimate follows from Theorem 6.4.6. □

Theorem 6.4.9 (**L^2 Estimate of the Velocity Error**) *Let the stabilization parameters satisfy* (6.44) *and* (6.45) *with* $\delta_1 \le 1/C_{\mathrm{inv}}^2$ *and let the Stokes problem* (6.2) *be regular. Assume that the solution of* (6.5) *satisfies* $(\mathbf{u},p)\in H^{k+1}(\Omega)^d\times H^{l+1}(\Omega)$, *then there holds the error estimate*

$$\left\|\mathbf{u}-\mathbf{u}^h\right\|_{L^2(\Omega)}\le C\left(h^{k+1}\|\mathbf{u}\|_{H^{k+1}(\Omega)}+\frac{h^{l+2}}{\nu}\|p\|_{H^{l+1}(\Omega)}\right).$$

Proof We start as in the proof of Theorem 6.3.5 and introduce $(\mathbf{z},r)\in V\times Q$ defined by (6.22) and the interpolants $(\mathbf{z}^I,r^I)\in V^h\times Q^h$ satisfying (6.25) and (6.26). Then one again arrives at (6.27) since, up to (6.28), the proof is independent of the analyzed method. We shall use the fact that, in view of (6.11), (6.23), and (6.45), the interpolant r^I satisfies

$$\left(\sum_{K\in\mathcal{T}^h}\left\|\nabla r^I\right\|_{L^2(K)}^2\right)^{1/2}\le C\|\nabla r\|_{L^2(\Omega)}\le C\nu\left\|\mathbf{u}-\mathbf{u}^h\right\|_{L^2(\Omega)}, \qquad (6.58)$$

$$\left(\sum_{E\in\mathcal{E}^h}\gamma_E\left\|\left[\left|r-r^I\right|\right]_E\right\|_{L^2(E)}^2\right)^{1/2}\le C\nu^{1/2}h\left\|\mathbf{u}-\mathbf{u}^h\right\|_{L^2(\Omega)}. \qquad (6.59)$$

To estimate the last two terms in (6.27), we employ the Galerkin orthogonality (6.43). Since $\mathbf{z}^I\in V^h$, we may set $(\mathbf{v}^h,q^h)=(\mathbf{z}^I,0)$ in (6.43), which gives

$$\nu\left(\nabla(\mathbf{u}-\mathbf{u}^h),\nabla\mathbf{z}^I\right)-\left(\nabla\cdot\mathbf{z}^I,p-p^h\right)=0. \qquad (6.60)$$

Furthermore, for $\left(\mathbf{v}^h, q^h\right) = (\mathbf{0}, r^I)$, one deduces from (6.43) that

$$\left(\nabla\cdot(\mathbf{u}-\mathbf{u}^h), r^I\right) + \sum_{E\in\mathcal{E}^h} \gamma_E \left(\left[\left|p-p^h\right|\right]_E, \left[\left|r^I\right|\right]_E\right)_E$$
$$+ \sum_{K\in\mathcal{T}^h} \left(-\nu\Delta(\mathbf{u}-\mathbf{u}^h) + \nabla(p-p^h), \delta_K \nabla r^I\right)_K = 0. \quad (6.61)$$

Thus, using the property $\nabla\cdot\mathbf{z} = 0$ and the fact that $r \in H^1(\Omega)$, one has

$$\nu(\nabla\mathbf{z}^I, \nabla(\mathbf{u}-\mathbf{u}^h)) - (\nabla\cdot(\mathbf{u}-\mathbf{u}^h), r^I)$$
$$= \left(\nabla\cdot(\mathbf{z}^I-\mathbf{z}), p-p^h\right) + \sum_{E\in\mathcal{E}^h} \gamma_E \left(\left[\left|p-p^h\right|\right]_E, \left[\left|r^I-r\right|\right]_E\right)_E$$
$$+ \sum_{K\in\mathcal{T}^h} \left(-\nu\Delta(\mathbf{u}-\mathbf{u}^h) + \nabla(p-p^h), \delta_K \nabla r^I\right)_K. \quad (6.62)$$

Then, applying the Cauchy–Schwarz inequality, (6.44), (6.25), (6.58), and (6.59), one derives

$$\nu(\nabla\mathbf{z}^I, \nabla(\mathbf{u}-\mathbf{u}^h)) - (\nabla\cdot(\mathbf{u}-\mathbf{u}^h), r^I)$$
$$\le \left\|\nabla(\mathbf{z}-\mathbf{z}^I)\right\|_{L^2(\Omega)} \left\|p-p^h\right\|_{L^2(\Omega)}$$
$$+ \sum_{E\in\mathcal{E}^h} \gamma_E \left\|\left[\left|p-p^h\right|\right]_E\right\|_{L^2(E)} \left\|\left[\left|r-r^I\right|\right]_E\right\|_{L^2(E)}$$
$$+ \sum_{K\in\mathcal{T}^h} \nu\delta_K \left\|\Delta(\mathbf{u}-\mathbf{u}^h)\right\|_{L^2(K)} \left\|\nabla r^I\right\|_{L^2(K)}$$
$$+ \sum_{K\in\mathcal{T}^h} \delta_K \left\|\nabla(p-p^h)\right\|_{L^2(K)} \left\|\nabla r^I\right\|_{L^2(K)}$$
$$\le Ch\left(\left\|p-p^h\right\|^2_{L^2(\Omega)} + \nu\sum_{E\in\mathcal{E}^h} \gamma_E \left\|\left[\left|p-p^h\right|\right]_E\right\|^2_{L^2(E)}\right.$$
$$\left. + \nu\sum_{K\in\mathcal{T}^h} \delta_K \left\|\nabla(p-p^h)\right\|^2_{L^2(K)}\right)^{1/2} \left\|\mathbf{u}-\mathbf{u}^h\right\|_{L^2(\Omega)}$$
$$+ C\nu h\left(\sum_{K\in\mathcal{T}^h} h_K^2 \left\|\Delta(\mathbf{u}-\mathbf{u}^h)\right\|^2_{L^2(K)}\right)^{1/2} \left\|\mathbf{u}-\mathbf{u}^h\right\|_{L^2(\Omega)}. \quad (6.63)$$

To estimate the last term, we employ the triangle inequality and (6.12) to obtain

$$
\begin{aligned}
& h_K \left\| \Delta(\mathbf{u} - \mathbf{u}^h) \right\|_{L^2(K)} \\
& \le h_K \left\| \Delta(\mathbf{u} - I^h\mathbf{u}) \right\|_{L^2(K)} + h_K \left\| \Delta(I^h\mathbf{u} - \mathbf{u}^h) \right\|_{L^2(K)} \\
& \le h_K \left\| \Delta(\mathbf{u} - I^h\mathbf{u}) \right\|_{L^2(K)} + C_{\mathrm{inv}} \left\| \nabla(I^h\mathbf{u} - \mathbf{u}^h) \right\|_{L^2(K)} \\
& \le h_K \left\| \Delta(\mathbf{u} - I^h\mathbf{u}) \right\|_{L^2(K)} + C_{\mathrm{inv}} \left\| \nabla(I^h\mathbf{u} - \mathbf{u}) \right\|_{L^2(K)} + C_{\mathrm{inv}} \left\| \nabla(\mathbf{u} - \mathbf{u}^h) \right\|_{L^2(K)}.
\end{aligned}
$$

Then (6.8) implies that

$$
\left(\sum_{K \in \mathcal{T}^h} h_K^2 \left\| \Delta(\mathbf{u} - \mathbf{u}^h) \right\|_{L^2(K)}^2 \right)^{1/2} \le C h^k \|\mathbf{u}\|_{H^{k+1}(\Omega)} + C \left\| \nabla(\mathbf{u} - \mathbf{u}^h) \right\|_{L^2(\Omega)}. \tag{6.64}
$$

Combining (6.27), (6.28), (6.63), and (6.64) gives

$$
\begin{aligned}
\nu \left\| \mathbf{u} - \mathbf{u}^h \right\|_{L^2(\Omega)} \le\ & C \nu^{1/2} h \left\| \left(\mathbf{u} - \mathbf{u}^h, p - p^h \right) \right\|_{\mathrm{pspg}} \\
& + C h \left\| p - p^h \right\|_{L^2(\Omega)} + C \nu h^{k+1} \|\mathbf{u}\|_{H^{k+1}(\Omega)}
\end{aligned}
$$

and the statement of the theorem follows from Theorems 6.4.6 and 6.4.8. □

One observes the usual scaling properties of the error estimates with respect to ν: small values of ν lead to large bounds for velocity errors due to large weights of the pressure contributions in the error bounds whereas large values of ν lead to large bounds for pressure errors due to the scaling of the velocity terms in the error bounds.

For discontinuous pressure approximations, the jump term in (6.34) can be replaced by a so-called local jump term, as proposed in [60, 81]. In this approach, there is an outer sum over appropriate macro mesh cells and than an inner sum of jumps over edges that are strictly in the interior of the macro mesh cells. Numerical studies of this method can be found in [81] and a finite element error analysis for P_1/P_0 and Q_1/Q_0 in [60]. The analysis for the Q_1/Q_0 case was extended to special anisotropic meshes in [69].

If the PSPG method is used with the P_1/P_0 finite element, then it is possible to compute a divergence-free velocity field in $H_{\mathrm{div}}(\Omega)$, where

$$
\begin{aligned}
H_{\mathrm{div}}(\Omega) = \Big\{ \mathbf{v} \,:\, & \mathbf{v} \in L^2(\Omega), \nabla \cdot \mathbf{v} \in L^2(\Omega), \nabla \cdot \mathbf{v} = 0, \text{ and } \mathbf{v} \cdot \mathbf{n} = 0 \text{ on } \partial\Omega \\
& \text{in the sense of traces} \Big\}.
\end{aligned}
$$

with an inexpensive post-processing step, see [12]. The idea consists in adding to $\mathbf{u}^h$ a correction $\mathbf{u}^h_{\mathrm{RT}_0} \in \mathrm{RT}_0$, the Raviart–Thomas space of lowest order, such that

$\nabla \cdot \left(\mathbf{u}^h + \mathbf{u}^h_{\mathrm{RT}_0}\right) = 0$ in $L^2(\Omega)$. Details of this approach and some numerical results can be found also in [56, Remark 4.102, Example 4.103].

The paper [79] studies a stabilization of somewhat general form, which contains as special cases the PSPG method and the inf-sup stable MINI element from [5]. Error estimates are derived for both, the $H^1(\Omega)$ and the $L^2(\Omega)$ norm of the velocity and the pressure. A PSPG method with weak imposition of the boundary condition using a penalty-free Nitsche method was analyzed in [22]. It was shown in [9] that a PSPG-type method, with an appropriate stabilization parameter, can be used to stabilize discrete inf-sup conditions of the dual Darcy problem and of the curl formulation of Maxwell's problem.

Remark 6.4.10 (Anisotropic Meshes) The PSPG method for the Q_1/Q_1 pair of finite element spaces on anisotropic quadrilateral grids aligned with the Cartesian coordinate axes was studied in [16]. The definition of the stabilization parameter includes both edge lengths of the quadrilateral cells.

The PSPG method on anisotropic grids was studied for the P_1/P_1 pair of spaces in [73]. A finite element analysis is presented, where the stabilization parameter is of the form

$$\delta_K = \delta \frac{h_{K,\min}}{\nu},$$

with $h_{K,\min}$ being the smaller characteristic length of K obtained via the polar decomposition of the matrix from the affine map from a standard reference cell to K.

A PSPG method on anisotropic grids in boundary layers, in the context of the Oseen equations, was studied in [2]. For the Stokes equations, the stabilization parameter has the form

$$\delta_K = \delta \frac{h_{K,\min}}{C_{\mathrm{inv}}^2 \nu},$$

where $h_{K,\min}$ is some kind of minimal length of the mesh cell K, e.g., the shortest edge for mesh cells of brick form. △

A modification of the PSPG method for continuous discrete pressure that is stable for stabilization parameters $\delta = \delta_0 h^2/\nu$ with arbitrary $\delta_0 > 0$, in contrast to condition (6.39), will be discussed briefly in Sect. 6.4.5.

6.4.3 The (Symmetric) Galerkin Least Squares (GLS) Method

The (symmetric) Galerkin Least Squares (GLS) method uses, like the PSPG method (6.33)–(6.35), the residual of the strong form of the equation. In contrast to the PSPG method, the operator of the strong form of the equation is applied also to the test functions. Hence, the application of a GLS method is a little bit more expensive than the use of the PSPG method.

The symmetric GLS method was proposed in [54]. It has the following form: Find $\left(\mathbf{u}^h, p^h\right) \in V^h \times Q^h$ such that

$$A_{\text{sgls}}\left(\left(\mathbf{u}^h, p^h\right), \left(\mathbf{v}^h, q^h\right)\right) = L_{\text{sgls}}\left(\left(\mathbf{v}^h, q^h\right)\right) \quad \forall \left(\mathbf{v}^h, q^h\right) \in V^h \times Q^h, \tag{6.65}$$

with

$$\begin{aligned} A_{\text{sgls}}\left((\mathbf{u}, p), (\mathbf{v}, q)\right) &= \nu\left(\nabla \mathbf{u}, \nabla \mathbf{v}\right) - \left(\nabla \cdot \mathbf{v}, p\right) + \left(\nabla \cdot \mathbf{u}, q\right) \\ &\quad + \sum_{E \in \mathcal{E}^h} \gamma_E \left([|p|]_E, [|q|]_E\right)_E + \sum_{K \in \mathcal{T}^h} \left(-\nu \Delta \mathbf{u} + \nabla p, \delta_K \left(\nu \Delta \mathbf{v} + \nabla q\right)\right)_K, \end{aligned} \tag{6.66}$$

$$L_{\text{sgls}}\left((\mathbf{v}, q)\right) = (\mathbf{f}, \mathbf{v}) + \sum_{K \in \mathcal{T}^h} \left(\mathbf{f}, \delta_K \left(\nu \Delta \mathbf{v} + \nabla q\right)\right)_K. \tag{6.67}$$

Remark 6.4.11 The discretization (6.65) can be equivalently written in the form

$$\tilde{A}_{\text{sgls}}\left(\left(\mathbf{u}^h, p^h\right), \left(\mathbf{v}^h, q^h\right)\right) = \tilde{L}_{\text{sgls}}\left(\left(\mathbf{v}^h, q^h\right)\right) \quad \forall \left(\mathbf{v}^h, q^h\right) \in V^h \times Q^h, \tag{6.68}$$

where $\tilde{A}_{\text{sgls}}\left((\mathbf{u}, p), (\mathbf{v}, q)\right) = A_{\text{sgls}}\left((\mathbf{u}, p), (\mathbf{v}, -q)\right)$ and $\tilde{L}_{\text{sgls}}\left((\mathbf{v}, q)\right) = L_{\text{sgls}}\left((\mathbf{v}, -q)\right)$. It is easy to see that the bilinear form $\tilde{A}_{\text{sgls}}$ is symmetric, which is the reason for calling the discretization (6.65) *symmetric* GLS method. The form (6.68) is typically used in implementations. However, to unify the presentation of the various methods, we consider (6.65) for the analysis. △

To simplify the subsequent considerations, the analysis will be given only for the case of continuous pressure finite element spaces, i.e., $Q^h \subset H^1(\Omega)$. In this case, the pressure jumps across faces in (6.66) vanish. Discontinuous pressure approximations are discussed briefly in Remark 6.4.18.

As before, it will be assumed that the stabilization parameter δ_K satisfies (6.44). Defining an extended $L^2(\Omega)$ norm for the pressure

$$\|q\|_{\text{ext}} = \left(\frac{1}{\nu}\|q\|_{L^2(\Omega)}^2 + \sum_{K \in \mathcal{T}^h} \delta_K \|\nabla q\|_{L^2(K)}^2\right)^{1/2},$$

the norm for the analysis of the symmetric GLS method is given by

$$\|(\mathbf{v}, q)\|_{\text{sgls}} = \left(\nu \|\nabla \mathbf{v}\|_{L^2(\Omega)}^2 + \|q\|_{\text{ext}}^2\right)^{1/2}. \tag{6.69}$$

In contrast to the bilinear form of the PSPG method, the bilinear form A_{sgls} is not coercive. However, we shall show that it satisfies an inf-sup condition, which is sufficient for proving the unique solvability and error estimates for the symmetric GLS method. First, let us prove the following auxiliary result.

Lemma 6.4.12 (Weaker Estimate in the Spirit of the Discrete Inf-Sup Condition) *There are positive constants C_1 and C_2 independent of h such that for all $q \in Q \cap H^1(\Omega)$, it holds*

$$\sup_{\mathbf{v}^h \in V^h \setminus \{\mathbf{0}\}} \frac{(\nabla \cdot \mathbf{v}^h, q)}{\left\| \nabla \mathbf{v}^h \right\|_{L^2(\Omega)}} \geq C_1 \|q\|_{L^2(\Omega)} - C_2 \left(\sum_{K \in \mathcal{T}^h} h_K^2 \|\nabla q\|_{L^2(K)}^2 \right)^{1/2} . \tag{6.70}$$

Proof Choose $q \in Q \cap H^1(\Omega) \setminus \{0\}$ arbitrarily but fixed. The idea of the proof consists in constructing a function $\mathbf{w}^h \in V^h$ such that an inequality of form (6.70) is already satisfied with $\mathbf{w}^h$.

In view of the inf-sup condition (6.6), there exists $\mathbf{w} \in V$ such that

$$\nabla \cdot \mathbf{w} = q, \quad \|\nabla \mathbf{w}\|_{L^2(\Omega)} \leq \frac{1}{\beta_{\text{is}}} \|q\|_{L^2(\Omega)},$$

see [56, Cor. 3.44]. It follows that

$$\frac{(\nabla \cdot \mathbf{w}, q)}{\|\nabla \mathbf{w}\|_{L^2(\Omega)}} = \frac{(q, q)}{\|\nabla \mathbf{w}\|_{L^2(\Omega)}} \geq \beta_{\text{is}} \|q\|_{L^2(\Omega)} . \tag{6.71}$$

Let $\mathbf{w}^h = \mathcal{I}^h \mathbf{w} \in V^h$ be an interpolant of $\mathbf{w}$ satisfying (6.10). Then, using (6.71), integration by parts, the Cauchy–Schwarz inequality, the Cauchy–Schwarz inequality for sums, and (6.10) yields

$$\begin{aligned}
&\left(\nabla \cdot \mathbf{w}^h, q\right) \\
&= \left(\nabla \cdot \left(\mathbf{w}^h - \mathbf{w}\right), q\right) + (\nabla \cdot \mathbf{w}, q) \\
&\geq \left(\mathbf{w} - \mathbf{w}^h, \nabla q\right) + \beta_{\text{is}} \|q\|_{L^2(\Omega)} \|\nabla \mathbf{w}\|_{L^2(\Omega)} \\
&\geq - \left(\sum_{K \in \mathcal{T}^h} h_K^{-2} \left\| \mathbf{w} - \mathbf{w}^h \right\|_{L^2(K)}^2 \right)^{1/2} \left(\sum_{K \in \mathcal{T}^h} h_K^2 \|\nabla q\|_{L^2(K)}^2 \right)^{1/2} \\
&\quad + \beta_{\text{is}} \|q\|_{L^2(\Omega)} \|\nabla \mathbf{w}\|_{L^2(\Omega)} \\
&\geq -C \|\nabla \mathbf{w}\|_{L^2(\Omega)} \left(\sum_{K \in \mathcal{T}^h} h_K^2 \|\nabla q\|_{L^2(K)}^2 \right)^{1/2} + \beta_{\text{is}} \|q\|_{L^2(\Omega)} \|\nabla \mathbf{w}\|_{L^2(\Omega)} \\
&= \left[\beta_{\text{is}} \|q\|_{L^2(\Omega)} - C \left(\sum_{K \in \mathcal{T}^h} h_K^2 \|\nabla q\|_{L^2(K)}^2 \right)^{1/2} \right] \|\nabla \mathbf{w}\|_{L^2(\Omega)} .
\end{aligned} \tag{6.72}$$

If the expression in the square brackets in (6.72) is positive, it follows that $\mathbf{w}^h \neq \mathbf{0}$ and then using (6.10) and (6.72) yields

$$\frac{\left(\nabla \cdot \mathbf{w}^h, q\right)}{\left\|\nabla \mathbf{w}^h\right\|_{L^2(\Omega)}} \geq C \frac{\left(\nabla \cdot \mathbf{w}^h, q\right)}{\|\nabla \mathbf{w}\|_{L^2(\Omega)}} \geq C_1 \|q\|_{L^2(\Omega)} - C_2 \left(\sum_{K \in \mathcal{T}^h} h_K^2 \|\nabla q\|_{L^2(K)}^2 \right)^{1/2}. \tag{6.73}$$

If the right-hand side of (6.73) (which is a multiple of the expression in the square brackets in (6.72)) is nonpositive, one chooses an arbitrary $\mathbf{w}^h \in V^h \setminus \{\mathbf{0}\}$ for which the left-hand side of (6.73) is nonnegative, such that (6.73) holds also in this case. □

Lemma 6.4.13 (Inf-sup Condition for the Bilinear Form A_{sgls}**)** *Let* $Q^h \subset H^1(\Omega)$. *Let the conditions* (6.44) *on* $\{\delta_K\}$ *be satisfied and let*

$$\delta_1 < \frac{1}{C_{\text{inv}}^2}. \tag{6.74}$$

Then, there is a positive constant C *such that for all* $\left(\mathbf{v}^h, q^h\right) \in V^h \times Q^h$, *it holds*

$$\sup_{(\mathbf{w}^h, r^h) \in V^h \times Q^h \setminus \{(\mathbf{0},0)\}} \frac{A_{\text{sgls}}\left(\left(\mathbf{v}^h, q^h\right), \left(\mathbf{w}^h, r^h\right)\right)}{\left\|\left(\mathbf{w}^h, r^h\right)\right\|_{\text{sgls}}} \geq C \left\|\left(\mathbf{v}^h, q^h\right)\right\|_{\text{sgls}}. \tag{6.75}$$

Proof Consider an arbitrary pair $\left(\mathbf{v}^h, q^h\right) \in V^h \times Q^h$. The idea of the proof consists in constructing a pair $\left(\mathbf{w}^h, r^h\right) \in V^h \times Q^h \setminus \{(\mathbf{0}, 0)\}$ that satisfies inequality (6.75).

First, assume that $q^h \neq 0$. According to the proof of Lemma 6.4.12, there is $\mathbf{z}^h \in V^h \setminus \{\mathbf{0}\}$ such that (6.70) holds for $\mathbf{v}^h = \mathbf{z}^h$ and $q = q^h$ without the supremum. Note that (6.70) also holds when $\mathbf{z}^h$ is multiplied by any positive number. Accordingly, one chooses $\mathbf{z}^h$ such that

$$\left\|\nabla \mathbf{z}^h\right\|_{L^2(\Omega)} = \frac{1}{\nu} \left\|q^h\right\|_{L^2(\Omega)}. \tag{6.76}$$

Now, the pair for which the satisfaction of (6.75) will be shown is

$$\left(\mathbf{w}^h, r^h\right) = \left(\mathbf{v}^h - \kappa \mathbf{z}^h, q^h\right), \tag{6.77}$$

where κ will be chosen appropriately in the forthcoming analysis. It is

$$\begin{aligned} &A_{\text{sgls}}\left(\left(\mathbf{v}^h, q^h\right), \left(\mathbf{w}^h, r^h\right)\right) \\ &= A_{\text{sgls}}\left(\left(\mathbf{v}^h, q^h\right), \left(\mathbf{v}^h, q^h\right)\right) + \kappa A_{\text{sgls}}\left(\left(\mathbf{v}^h, q^h\right), \left(-\mathbf{z}^h, 0\right)\right). \end{aligned} \tag{6.78}$$

Both terms on the right-hand side of this identity will be studied separately.

With the Cauchy–Schwarz inequality, one obtains

$$
\begin{aligned}
& A_{\text{sgls}}\left(\left(\mathbf{v}^h, q^h\right), \left(-\mathbf{z}^h, 0\right)\right) \\
&= -\nu\left(\nabla \mathbf{v}^h, \nabla \mathbf{z}^h\right) + \left(\nabla \cdot \mathbf{z}^h, q^h\right) - \sum_{K\in\mathcal{T}^h} \delta_K \left(-\nu\Delta \mathbf{v}^h + \nabla q^h, \nu\Delta \mathbf{z}^h\right)_K \\
&\ge -\nu \left\|\nabla \mathbf{v}^h\right\|_{L^2(\Omega)} \left\|\nabla \mathbf{z}^h\right\|_{L^2(\Omega)} + \left(\nabla \cdot \mathbf{z}^h, q^h\right) \\
&\quad + \nu^2 \sum_{K\in\mathcal{T}^h} \delta_K \left(\Delta \mathbf{v}^h, \Delta \mathbf{z}^h\right)_K - \nu \sum_{K\in\mathcal{T}^h} \delta_K \left(\nabla q^h, \Delta \mathbf{z}^h\right)_K. \qquad (6.79)
\end{aligned}
$$

Each term on the right-hand side of (6.79) will be estimated from below.

Using (6.70) (without supremum) with $(\mathbf{v}^h, q) = (\mathbf{z}^h, q^h)$, (6.76), and (6.44) yields

$$
\begin{aligned}
\left(\nabla \cdot \mathbf{z}^h, q^h\right) &\ge \left(C_1 \left\|q^h\right\|_{L^2(\Omega)} - C_2 \left(\sum_{K\in\mathcal{T}^h} h_K^2 \left\|\nabla q^h\right\|_{L^2(K)}^2\right)^{1/2}\right) \left\|\nabla \mathbf{z}^h\right\|_{L^2(\Omega)} \\
&= \frac{C_1}{\nu} \left\|q^h\right\|_{L^2(\Omega)}^2 - \frac{C_2}{\nu} \left(\sum_{K\in\mathcal{T}^h} h_K^2 \left\|\nabla q^h\right\|_{L^2(K)}^2\right)^{1/2} \left\|q^h\right\|_{L^2(\Omega)} \\
&\ge \frac{C_1}{\nu} \left\|q^h\right\|_{L^2(\Omega)}^2 - \frac{C_2}{\delta_0^{1/2}} \left(\sum_{K\in\mathcal{T}^h} \delta_K \left\|\nabla q^h\right\|_{L^2(K)}^2\right)^{1/2} \frac{1}{\nu^{1/2}} \left\|q^h\right\|_{L^2(\Omega)}.
\end{aligned}
$$

Applying the Cauchy–Schwarz inequality, the inverse inequality (6.12), (6.44), the Cauchy–Schwarz inequality for sums, and (6.76) yields

$$
\begin{aligned}
\nu^2 \sum_{K\in\mathcal{T}^h} \delta_K \left(\Delta \mathbf{v}^h, \Delta \mathbf{z}^h\right)_K &\ge -\nu^2 \sum_{K\in\mathcal{T}^h} \delta_K \left\|\Delta \mathbf{v}^h\right\|_{L^2(K)} \left\|\Delta \mathbf{z}^h\right\|_{L^2(K)} \\
&\ge -\nu^2 C_{\text{inv}}^2 \sum_{K\in\mathcal{T}^h} \delta_K h_K^{-2} \left\|\nabla \mathbf{v}^h\right\|_{L^2(K)} \left\|\nabla \mathbf{z}^h\right\|_{L^2(K)} \\
&\ge -\nu C_{\text{inv}}^2 \delta_1 \sum_{K\in\mathcal{T}^h} \left\|\nabla \mathbf{v}^h\right\|_{L^2(K)} \left\|\nabla \mathbf{z}^h\right\|_{L^2(K)} \\
&\ge -\nu C_{\text{inv}}^2 \delta_1 \left\|\nabla \mathbf{v}^h\right\|_{L^2(\Omega)} \left\|\nabla \mathbf{z}^h\right\|_{L^2(\Omega)} \\
&= -C_{\text{inv}}^2 \delta_1 \nu^{1/2} \left\|\nabla \mathbf{v}^h\right\|_{L^2(\Omega)} \frac{1}{\nu^{1/2}} \left\|q^h\right\|_{L^2(\Omega)}.
\end{aligned}
$$

The estimate of the third term uses the same tools

$$
\begin{aligned}
&-\nu \sum_{K\in\mathcal{T}^h} \delta_K \left(\nabla q^h, \Delta \mathbf{z}^h\right)_K \\
&\ge -\nu \left(\sum_{K\in\mathcal{T}^h} \delta_K \left\|\nabla q^h\right\|_{L^2(K)}^2\right)^{1/2} \left(\sum_{K\in\mathcal{T}^h} \delta_K \left\|\Delta \mathbf{z}^h\right\|_{L^2(K)}^2\right)^{1/2} \\
&\ge -C_{\text{inv}}\delta_1^{1/2} \left(\sum_{K\in\mathcal{T}^h} \delta_K \left\|\nabla q^h\right\|_{L^2(K)}^2\right)^{1/2} \frac{1}{\nu^{1/2}} \left\|q^h\right\|_{L^2(\Omega)}.
\end{aligned}
$$

Inserting all estimates in (6.79) and applying (6.76) leads to

$$
\begin{aligned}
&A_{\text{sgls}}\left(\left(\mathbf{v}^h, q^h\right), \left(-\mathbf{z}^h, 0\right)\right) \\
&\ge -\left(1 + C_{\text{inv}}^2\delta_1\right) \nu^{1/2} \left\|\nabla \mathbf{v}^h\right\|_{L^2(\Omega)} \frac{1}{\nu^{1/2}} \left\|q^h\right\|_{L^2(\Omega)} + \frac{C_1}{\nu} \left\|q^h\right\|_{L^2(\Omega)}^2 \\
&\quad -\left(\frac{C_2}{\delta_0^{1/2}} + C_{\text{inv}}\delta_1^{1/2}\right) \left(\sum_{K\in\mathcal{T}^h} \delta_K \left\|\nabla q^h\right\|_{L^2(K)}^2\right)^{1/2} \frac{1}{\nu^{1/2}} \left\|q^h\right\|_{L^2(\Omega)} \\
&= -C_3\, \nu^{1/2} \left\|\nabla \mathbf{v}^h\right\|_{L^2(\Omega)} \frac{1}{\nu^{1/2}} \left\|q^h\right\|_{L^2(\Omega)} + \frac{C_1}{\nu} \left\|q^h\right\|_{L^2(\Omega)}^2 \\
&\quad -C_4 \left(\sum_{K\in\mathcal{T}^h} \delta_K \left\|\nabla q^h\right\|_{L^2(K)}^2\right)^{1/2} \frac{1}{\nu^{1/2}} \left\|q^h\right\|_{L^2(\Omega)},
\end{aligned}
$$

with positive constants C_3 and C_4 that do not depend on ν, but C_4 depends on $\delta_0^{-1/2}$. The application of the Young inequality with some $\varepsilon > 0$ gives

$$
\begin{aligned}
&A_{\text{sgls}}\left(\left(\mathbf{v}^h, q^h\right), \left(-\mathbf{z}^h, 0\right)\right) \\
&\ge \left(C_1 - \frac{\varepsilon}{2}\left(C_3 + C_4\right)\right) \frac{1}{\nu} \left\|q^h\right\|_{L^2(\Omega)}^2 - \frac{C_3}{2\varepsilon}\nu \left\|\nabla \mathbf{v}^h\right\|_{L^2(\Omega)}^2 \\
&\quad -\frac{C_4}{2\varepsilon} \sum_{K\in\mathcal{T}^h} \delta_K \left\|\nabla q^h\right\|_{L^2(K)}^2.
\end{aligned}
$$

Choosing now $0 < \varepsilon < 2C_1/(C_3 + C_4)$ leads to

$$A_{\text{sgls}}\left(\left(\mathbf{v}^h, q^h\right), \left(-\mathbf{z}^h, 0\right)\right)$$
$$\geq C_5 \frac{1}{\nu} \left\| q^h \right\|_{L^2(\Omega)}^2 - C_6 \nu \left\| \nabla \mathbf{v}^h \right\|_{L^2(\Omega)}^2 - C_7 \sum_{K \in \mathcal{T}^h} \delta_K \left\| \nabla q^h \right\|_{L^2(K)}^2, \quad (6.80)$$

with positive constants C_5, C_6, and C_7.

Now, the first term on the right-hand side of (6.78) will be estimated. Using the definition (6.66) gives

$$A_{\text{sgls}}\left(\left(\mathbf{v}^h, q^h\right), \left(\mathbf{v}^h, q^h\right)\right)$$
$$= \nu \left\| \nabla \mathbf{v}^h \right\|_{L^2(\Omega)}^2 + \sum_{K \in \mathcal{T}^h} \delta_K \left(-\nu \Delta \mathbf{v}^h + \nabla q^h, \nu \Delta \mathbf{v}^h + \nabla q^h\right)_K$$
$$= \nu \left\| \nabla \mathbf{v}^h \right\|_{L^2(\Omega)}^2 + \sum_{K \in \mathcal{T}^h} \delta_K \left\| \nabla q^h \right\|_{L^2(K)}^2 - \nu^2 \sum_{K \in \mathcal{T}^h} \delta_K \left\| \Delta \mathbf{v}^h \right\|_{L^2(K)}^2.$$

By using (6.12) and (6.44), one obtains

$$A_{\text{sgls}}\left(\left(\mathbf{v}^h, q^h\right), \left(\mathbf{v}^h, q^h\right)\right)$$
$$\geq \nu \left\| \nabla \mathbf{v}^h \right\|_{L^2(\Omega)}^2 + \sum_{K \in \mathcal{T}^h} \delta_K \left\| \nabla q^h \right\|_{L^2(K)}^2 - \nu C_{\text{inv}}^2 \delta_1 \left\| \nabla \mathbf{v}^h \right\|_{L^2(\Omega)}^2$$
$$= \left(1 - C_{\text{inv}}^2 \delta_1\right) \nu \left\| \nabla \mathbf{v}^h \right\|_{L^2(\Omega)}^2 + \sum_{K \in \mathcal{T}^h} \delta_K \left\| \nabla q^h \right\|_{L^2(K)}^2.$$

By the assumption (6.74) on δ_1, the term in the parentheses is positive. Hence, with a positive constant C_8, it is

$$A_{\text{sgls}}\left(\left(\mathbf{v}^h, q^h\right), \left(\mathbf{v}^h, q^h\right)\right) \geq C_8 \nu \left\| \nabla \mathbf{v}^h \right\|_{L^2(\Omega)}^2 + \sum_{K \in \mathcal{T}^h} \delta_K \left\| \nabla q^h \right\|_{L^2(K)}^2. \quad (6.81)$$

Inserting (6.80) and (6.81) in (6.78) yields

$$A_{\text{sgls}}\left(\left(\mathbf{v}^h, q^h\right), \left(\mathbf{w}^h, r^h\right)\right)$$
$$\geq \left(C_8 - \kappa C_6\right) \nu \left\| \nabla \mathbf{v}^h \right\|_{L^2(\Omega)}^2 + \kappa C_5 \frac{1}{\nu} \left\| q^h \right\|_{L^2(\Omega)}^2$$
$$+ \left(1 - \kappa C_7\right) \sum_{K \in \mathcal{T}^h} \delta_K \left\| \nabla q^h \right\|_{L^2(K)}^2.$$

Choosing now $0 < \kappa < \min\{C_8/C_6, 1/C_7\}$ leads to the existence of a positive constant C_9 such that

$$A_{\text{sgls}}\left(\left(\mathbf{v}^h, q^h\right), \left(\mathbf{w}^h, r^h\right)\right) \ge C_9 \left\|\left(\mathbf{v}^h, q^h\right)\right\|_{\text{sgls}}^2 . \tag{6.82}$$

Considering the denominator of (6.75), using the definition (6.77) of $\left(\mathbf{w}^h, r^h\right)$, the triangle inequality, and (6.76) yields

$$\begin{aligned}
\left\|\left(\mathbf{w}^h, r^h\right)\right\|_{\text{sgls}} &= \left(\nu \left\|\nabla\left(\mathbf{v}^h - \kappa \mathbf{z}^h\right)\right\|_{L^2(\Omega)}^2 + \left\|q^h\right\|_{\text{ext}}^2\right)^{1/2} \\
&\le \left(2\nu \left\|\nabla \mathbf{v}^h\right\|_{L^2(\Omega)}^2 + 2\kappa^2 \nu \left\|\nabla \mathbf{z}^h\right\|_{L^2(\Omega)}^2 + \left\|q^h\right\|_{\text{ext}}^2\right)^{1/2} \\
&= \left(2\nu \left\|\nabla \mathbf{v}^h\right\|_{L^2(\Omega)}^2 + 2\kappa^2 \frac{1}{\nu} \left\|q^h\right\|_{L^2(\Omega)}^2 + \left\|q^h\right\|_{\text{ext}}^2\right)^{1/2} \\
&\le \left(2 + 2\kappa^2\right)^{1/2} \left\|\left(\mathbf{v}^h, q^h\right)\right\|_{\text{sgls}} = C_{10} \left\|\left(\mathbf{v}^h, q^h\right)\right\|_{\text{sgls}}
\end{aligned} \tag{6.83}$$

with a positive constant C_{10} that is independent of ν.

Combining (6.82) and (6.83) gives the inf-sup condition (6.75).

Finally, if $q_h = 0$, the inf-sup condition (6.75) immediately follows from (6.81). □

The proof of the inf-sup condition for the bilinear form A_{sgls} requires an upper bound of the stabilization parameter, hence this method is not absolutely stable. Note that the bound (6.74) for δ_1 depends on the polynomial degree, compare the note after (6.12).

Lemma 6.4.14 (Existence and Uniqueness of a Solution of (6.65)**)** *Let the assumptions of Lemma 6.4.13 be satisfied, then the symmetric GLS problem* (6.65) *possesses a unique solution.*

Proof The existence and uniqueness of the solution follows analogously as in the proof of Lemma 6.4.3 since the inf-sup condition (6.75) implies that the homogeneous symmetric GLS problem has only the trivial solution. □

Lemma 6.4.15 (Consistency and Galerkin Orthogonality) *Let the solution of* (6.5) *satisfy* $(\mathbf{u}, p) \in H^2(\Omega)^d \times H^1(\Omega)$ *and let* $(\mathbf{u}^h, p^h) \in V^h \times Q^h$ *be the solution of the symmetric GLS method* (6.65). *This method is consistent, i.e., it holds*

$$A_{\text{sgls}}\left((\mathbf{u}, p), \left(\mathbf{v}^h, q^h\right)\right) = L_{\text{sgls}}\left(\left(\mathbf{v}^h, q^h\right)\right) \quad \forall \left(\mathbf{v}^h, q^h\right) \in V^h \times Q^h \tag{6.84}$$

and it satisfies the Galerkin orthogonality

$$A_{\mathrm{sgls}}\left(\left(\mathbf{u}-\mathbf{u}^h, p-p^h\right), \left(\mathbf{v}^h, q^h\right)\right) = 0 \quad \forall \left(\mathbf{v}^h, q^h\right) \in V^h \times Q^h. \tag{6.85}$$

Proof The lemma follows in the same way as Lemma 6.4.5. □

Theorem 6.4.16 (Error Estimate) *Let the assumptions of Lemma 6.4.13 be satisfied. Assume that the solution of* (6.5) *satisfies* $(\mathbf{u}, p) \in H^{k+1}(\Omega)^d \times H^{l+1}(\Omega)$, *then there holds the error estimate*

$$\left\|\left(\mathbf{u}-\mathbf{u}^h, p-p^h\right)\right\|_{\mathrm{sgls}} \le C\left(\nu^{1/2}h^k \|\mathbf{u}\|_{H^{k+1}(\Omega)} + \frac{h^{l+1}}{\nu^{1/2}} \|p\|_{H^{l+1}(\Omega)}\right).$$

Proof Let I^h and J^h be the interpolation operators satisfying (6.8) and (6.9). From the proof of Lemma 6.4.13, it is known that there is a pair $\left(\mathbf{v}^h, q^h\right) \in V^h \times Q^h$ such that

$$\left\|\left(\mathbf{u}^h - I^h\mathbf{u}, p^h - J^h p\right)\right\|_{\mathrm{sgls}} \le C\frac{A_{\mathrm{sgls}}\left(\left(\mathbf{u}^h - I^h\mathbf{u}, p^h - J^h p\right), \left(\mathbf{v}^h, q^h\right)\right)}{\left\|\left(\mathbf{v}^h, q^h\right)\right\|_{\mathrm{sgls}}}.$$

With the Galerkin orthogonality (6.85) of the symmetric GLS method, one obtains

$$\left\|\left(\mathbf{u}^h - I^h\mathbf{u}, p^h - J^h p\right)\right\|_{\mathrm{sgls}} \le C\frac{A_{\mathrm{sgls}}\left(\left(\mathbf{u} - I^h\mathbf{u}, p - J^h p\right), \left(\mathbf{v}^h, q^h\right)\right)}{\left\|\left(\mathbf{v}^h, q^h\right)\right\|_{\mathrm{sgls}}}. \tag{6.86}$$

Now, all terms of the numerator of the right-hand side of (6.86) will be estimated such that the contribution from $\left(\mathbf{v}^h, q^h\right)$ can be bounded by $\left\|\left(\mathbf{v}^h, q^h\right)\right\|_{\mathrm{sgls}}$. With the Cauchy–Schwarz inequality and (6.7), one obtains

$$\begin{aligned}
\nu\left(\nabla\left(\mathbf{u} - I^h\mathbf{u}\right), \nabla\mathbf{v}^h\right) &\le \nu\left\|\nabla\left(\mathbf{u} - I^h\mathbf{u}\right)\right\|_{L^2(\Omega)} \left\|\nabla\mathbf{v}^h\right\|_{L^2(\Omega)},\\
\left(\nabla\cdot\mathbf{v}^h, p - J^h p\right) &\le \left\|p - J^h p\right\|_{L^2(\Omega)} \left\|\nabla\mathbf{v}^h\right\|_{L^2(\Omega)},\\
\left(\nabla\cdot\left(\mathbf{u} - I^h\mathbf{u}\right), q^h\right) &\le \left\|\nabla\left(\mathbf{u} - I^h\mathbf{u}\right)\right\|_{L^2(\Omega)} \left\|q^h\right\|_{L^2(\Omega)}.
\end{aligned}$$

The terms coming from the stabilization are estimated individually, using also the inverse inequality (6.12) and the upper bound (6.44) of the parameter δ_K:

$$\begin{aligned}
&\sum_{K\in\mathcal{T}^h} \delta_K \left(-\nu\Delta\left(\mathbf{u} - I^h\mathbf{u}\right), \nu\Delta\mathbf{v}^h\right)_K\\
&\le C\nu\left(\sum_{K\in\mathcal{T}^h} h_K^2 \left\|\Delta\left(\mathbf{u} - I^h\mathbf{u}\right)\right\|_{L^2(K)}^2\right)^{1/2} \left\|\nabla\mathbf{v}^h\right\|_{L^2(\Omega)},
\end{aligned}$$

$$\sum_{K\in\mathcal{T}^h}\delta_K\left(-\nu\Delta\left(\mathbf{u}-I^h\mathbf{u}\right),\nabla q^h\right)_K$$
$$\le C\nu^{1/2}\left(\sum_{K\in\mathcal{T}^h}h_K^2\left\|\Delta\left(\mathbf{u}-I^h\mathbf{u}\right)\right\|_{L^2(K)}^2\right)^{1/2}\left(\sum_{K\in\mathcal{T}^h}\delta_K\left\|\nabla q^h\right\|_{L^2(K)}^2\right)^{1/2},$$
$$\sum_{K\in\mathcal{T}^h}\delta_K\left(\nabla\left(p-J^hp\right),\nu\Delta\mathbf{v}^h\right)_K$$
$$\le C\nu^{1/2}\left(\sum_{K\in\mathcal{T}^h}\delta_K\left\|\nabla\left(p-J^hp\right)\right\|_{L^2(K)}^2\right)^{1/2}\left\|\nabla\mathbf{v}^h\right\|_{L^2(\Omega)},$$
$$\sum_{K\in\mathcal{T}^h}\delta_K\left(\nabla\left(p-J^hp\right),\nabla q^h\right)_K$$
$$\le\left(\sum_{K\in\mathcal{T}^h}\delta_K\left\|\nabla\left(p-J^hp\right)\right\|_{L^2(K)}^2\right)^{1/2}\left(\sum_{K\in\mathcal{T}^h}\delta_K\left\|\nabla q^h\right\|_{L^2(K)}^2\right)^{1/2}.$$

Collecting terms and using the definition (6.69) of the symmetric GLS norm yields

$$\begin{aligned}&A_{\text{sgls}}\left(\left(\mathbf{u}-I^h\mathbf{u},p-J^hp\right),\left(\mathbf{v}^h,q^h\right)\right)\\&\le C\Big[\left\|\left(\mathbf{u}-I^h\mathbf{u},p-J^hp\right)\right\|_{\text{sgls}}\\&\quad+\left(\nu\sum_{K\in\mathcal{T}^h}h_K^2\left\|\Delta\left(\mathbf{u}-I^h\mathbf{u}\right)\right\|_{L^2(K)}^2\right)^{1/2}\Big]\left\|\left(\mathbf{v}^h,q^h\right)\right\|_{\text{sgls}}.\end{aligned}\tag{6.87}$$

The triangle inequality gives

$$\left\|\left(\mathbf{u}-\mathbf{u}^h,p-p^h\right)\right\|_{\text{sgls}}\le\left\|\left(\mathbf{u}-I^h\mathbf{u},p-J^hp\right)\right\|_{\text{sgls}}+\left\|\left(\mathbf{u}^h-I^h\mathbf{u},p^h-J^hp\right)\right\|_{\text{sgls}}$$

and hence, inserting (6.87) in (6.86), one obtains

$$\begin{aligned}\left\|\left(\mathbf{u}-\mathbf{u}^h,p-p^h\right)\right\|_{\text{sgls}}&\le C\left\|\left(\mathbf{u}-I^h\mathbf{u},p-J^hp\right)\right\|_{\text{sgls}}\\&\quad+C\left(\nu\sum_{K\in\mathcal{T}^h}h_K^2\left\|\Delta\left(\mathbf{u}-I^h\mathbf{u}\right)\right\|_{L^2(K)}^2\right)^{1/2}.\end{aligned}$$

The terms on the right-hand side of this estimate can be estimated using (6.8), (6.9), and (6.44), giving the statement of the theorem. □

Theorem 6.4.17 (**L^2 Estimate of the Velocity Error**) *Let the assumptions of Lemma 6.4.13 be satisfied and let the Stokes problem* (6.2) *be regular. Assume that the solution of* (6.5) *satisfies* $(\mathbf{u}, p) \in H^{k+1}(\Omega)^d \times H^{l+1}(\Omega)$, *then there holds the error estimate*

$$\left\|\mathbf{u}-\mathbf{u}^h\right\|_{L^2(\Omega)} \le C\left(h^{k+1}\,\|\mathbf{u}\|_{H^{k+1}(\Omega)} + \frac{h^{l+2}}{\nu}\,\|p\|_{H^{l+1}(\Omega)}\right).$$

Proof The proof is very similar to the proof of Theorem 6.4.9. First, we again repeat the part of the proof of Theorem 6.3.5 up to (6.28). Second, applying the Galerkin orthogonality (6.85) in an analogous way as in the proof of Theorem 6.4.9, one obtains

$$\begin{aligned}
&\nu(\nabla \mathbf{z}^I, \nabla(\mathbf{u}-\mathbf{u}^h)) - (\nabla\cdot(\mathbf{u}-\mathbf{u}^h), r^I)\\
&= \left(\nabla\cdot(\mathbf{z}^I-\mathbf{z}), p-p^h\right)\\
&\quad + \sum_{K\in\mathcal{T}^h}\left(-\nu\Delta(\mathbf{u}-\mathbf{u}^h)+\nabla(p-p^h), \delta_K(-\nu\Delta\mathbf{z}^I+\nabla r^I)\right)_K. \qquad (6.88)
\end{aligned}$$

To estimate the additional terms (in comparison to (6.62)), one may use the estimate

$$\left(\sum_{K\in\mathcal{T}^h} h_K^2\left\|\Delta\mathbf{z}^I\right\|_{L^2(K)}^2\right)^{1/2} \le Ch\left\|\mathbf{u}-\mathbf{u}^h\right\|_{L^2(\Omega)}, \qquad (6.89)$$

which follows from the triangle inequality, (6.8) and (6.23). Then, the right-hand side of (6.88) can be estimated by the right-hand side of (6.63) (the jump term now vanishes), which leads to the estimate

$$\begin{aligned}
&\nu(\nabla \mathbf{z}^I, \nabla(\mathbf{u}-\mathbf{u}^h)) - (\nabla\cdot(\mathbf{u}-\mathbf{u}^h), r^I)\\
&\le C\nu^{1/2}h\left\|p-p^h\right\|_{\text{ext}}\left\|\mathbf{u}-\mathbf{u}^h\right\|_{L^2(\Omega)}\\
&\quad + C\nu h\left(\sum_{K\in\mathcal{T}^h} h_K^2\left\|\Delta(\mathbf{u}-\mathbf{u}^h)\right\|_{L^2(K)}^2\right)^{1/2}\left\|\mathbf{u}-\mathbf{u}^h\right\|_{L^2(\Omega)}. \qquad (6.90)
\end{aligned}$$

Combining (6.27), (6.28), (6.90), and (6.64) gives

$$\nu\left\|\mathbf{u}-\mathbf{u}^h\right\|_{L^2(\Omega)} \le C\nu^{1/2}h\left\|\left(\mathbf{u}-\mathbf{u}^h, p-p^h\right)\right\|_{\text{sgls}} + C\nu h^{k+1}\,\|\mathbf{u}\|_{H^{k+1}(\Omega)}$$

and the theorem follows from Theorem 6.4.16. □

Remark 6.4.18 (Discontinuous Pressure Finite Element spaces) The proof of the inf-sup condition (6.75) relies on (6.70). It can be shown that an inequality of this form holds also for discontinuous pressure spaces, see [43]. Then, for low order spaces, one has to include pressure jumps in the method, as for the PSPG method. The optimal choice of the stabilization parameter for the pressure jumps in (6.66) is $\gamma_E \sim h_E/\nu$, see [54] for details. For high order spaces, the inclusion of such jumps is not necessary. High order means that $P_d \subset V^h$ for simplicial meshes and $Q_2 \subset V^h$ for quadrilateral or hexahedral meshes. With such spaces, the known discrete inf-sup stability of V^h/P_0 or V^h/Q_0 is utilized in the proof. △

In [3], a multiscale enrichment of the velocity finite element space is proposed that leads to a family of stabilized methods. The enrichment functions are defined locally, but the functions of the ansatz space do not vanish on the boundary of the mesh cells. After performing some manipulations and applying static condensation, the resulting method contains the symmetric GLS stabilization term and a jump term at the faces. The stabilization parameter of the jump term is known exactly. One member of the family uses the jump of the Cauchy stress tensor across the faces. This method is called algebraic subgrid scale method (ASGS) in [8], where it was analyzed for the Brinkman equations (Stokes equations plus a zeroth order velocity term in the momentum balance).

An a priori analysis and an a posteriori analysis for the symmetric GLS method with minimal regularity conditions on the solution of the weak problem are presented in [83]. It uses a technique developed in [50].

6.4.4 The Non-Symmetric Galerkin Least Squares Method (Douglas–Wang Method)

A method that looks similar to the symmetric GLS method (6.65)–(6.67) was proposed in [39]: Find $(\mathbf{u}^h, p^h) \in V^h \times Q^h$ such that

$$A_{\mathrm{nsgls}}\left(\left(\mathbf{u}^h, p^h\right), \left(\mathbf{v}^h, q^h\right)\right) = L_{\mathrm{nsgls}}\left(\left(\mathbf{v}^h, q^h\right)\right) \quad \forall \left(\mathbf{v}^h, q^h\right) \in V^h \times Q^h, \tag{6.91}$$

with

$$\begin{aligned} A_{\mathrm{nsgls}}\left((\mathbf{u}, p), (\mathbf{v}, q)\right) &= \nu\left(\nabla\mathbf{u}, \nabla\mathbf{v}\right) - \left(\nabla\cdot\mathbf{v}, p\right) + \left(\nabla\cdot\mathbf{u}, q\right) \\ &\quad + \sum_{E\in\mathcal{E}^h} \gamma_E\left([|p|]_E, [|q|]_E\right)_E + \sum_{K\in\mathcal{T}^h} \left(-\nu\Delta\mathbf{u} + \nabla p, \delta_K\left(-\nu\Delta\mathbf{v} + \nabla q\right)\right)_K, \end{aligned} \tag{6.92}$$

$$L_{\mathrm{nsgls}}\left((\mathbf{v}, q)\right) = (\mathbf{f}, \mathbf{v}) + \sum_{K\in\mathcal{T}^h} \left(\mathbf{f}, \delta_K\left(-\nu\Delta\mathbf{v} + \nabla q\right)\right)_K. \tag{6.93}$$

The difference between (6.65)–(6.67) and (6.91)–(6.93) is just the sign in front of $\nu\Delta\mathbf{v}$ in the residual-based stabilization terms. The method (6.91)–(6.93) is non-symmetric.

Again, the presentation of the analysis will be restricted to continuous pressure finite element spaces, i.e., $Q^h \subset H^1(\Omega)$. To prove error estimates, we shall again use the assumptions (6.44) on the stabilization parameters, to ensure a correct scaling with respect to h_K and ν. However, an important difference to the previous two methods is that the stability holds without any upper bound on the stabilization parameters, cf. Lemmas 6.4.3, 6.4.13, and 6.4.20.

The following norm will be used:

$$\|(\mathbf{v}, q)\|_{\text{nsgls}} = \left(\nu \|\nabla\mathbf{v}\|^2_{L^2(\Omega)} + \sum_{K\in\mathcal{T}^h} \delta_K \|-\nu\Delta\mathbf{v} + \nabla q\|^2_{L^2(K)}\right)^{1/2}. \tag{6.94}$$

Lemma 6.4.19 (**$\|(\cdot,\cdot)\|_{\text{nsgls}}$ Defines a Norm**) *If $Q^h \subset H^1(\Omega)$, then the expression defined in* (6.94) *is a norm in $V^h \times Q^h$ for any set of positive stabilization parameters $\{\delta_K\}$.*

Proof Clearly, $\|(\cdot,\cdot)\|_{\text{nsgls}}$ defines a seminorm as a sum of norms and seminorms. It remains to show that $\left\|(\mathbf{v}^h, q^h)\right\|_{\text{nsgls}} = 0$ implies $(\mathbf{v}^h, q^h) = (\mathbf{0}, 0)$.

From $\left\|(\mathbf{v}^h, q^h)\right\|_{\text{nsgls}} = 0$, it follows that $\left\|\nabla\mathbf{v}^h\right\|_{L^2(\Omega)} = 0$, hence that $\mathbf{v}^h = \mathbf{0}$. Now, one has

$$\sum_{K\in\mathcal{T}^h} \delta_K \left\|\nabla q^h\right\|^2_{L^2(K)} = 0.$$

Since all δ_K are positive, one finds that q^h is piecewise constant. The only piecewise constant function that belongs to $H^1(\Omega) \cap L^2_0(\Omega)$ is $q^h = 0$. □

Lemma 6.4.20 (Existence and Uniqueness of a Solution of (6.91)) *For any set of positive stabilization parameters $\{\delta_K\}$, the non-symmetric GLS problem* (6.91) *with $Q^h \subset H^1(\Omega)$ has a unique solution.*

Proof For $(\mathbf{v}^h, q^h) \in V^h \times Q^h$, one has

$$A_{\text{nsgls}}\left(\left(\mathbf{v}^h, q^h\right), \left(\mathbf{v}^h, q^h\right)\right)$$
$$= \nu \left\|\nabla\mathbf{v}^h\right\|^2_{L^2(\Omega)} + \sum_{K\in\mathcal{T}^h} \delta_K \left\|-\nu\Delta\mathbf{v}^h + \nabla q^h\right\|^2_{L^2(K)} = \left\|(\mathbf{v}^h, q^h)\right\|^2_{\text{nsgls}} \tag{6.95}$$

and hence the bilinear form given in (6.92) is coercive. Now, the existence and uniqueness of the solution follows in the same way as in the proof of Lemma 6.4.3.

□

Lemma 6.4.21 (Consistency and Galerkin Orthogonality) *Let the solution of* (6.5) *satisfy* $(\mathbf{u}, p) \in H^2(\Omega)^d \times H^1(\Omega)$ *and let* $(\mathbf{u}^h, p^h) \in V^h \times Q^h$ *be the solution of the non-symmetric GLS method* (6.91). *This method is consistent, i.e., it holds*

$$A_{\text{nsgls}}\left((\mathbf{u}, p), \left(\mathbf{v}^h, q^h\right)\right) = L_{\text{nsgls}}\left(\left(\mathbf{v}^h, q^h\right)\right) \quad \forall \left(\mathbf{v}^h, q^h\right) \in V^h \times Q^h \tag{6.96}$$

and it satisfies the Galerkin orthogonality

$$A_{\text{nsgls}}\left(\left(\mathbf{u} - \mathbf{u}^h, p - p^h\right), \left(\mathbf{v}^h, q^h\right)\right) = 0 \quad \forall \left(\mathbf{v}^h, q^h\right) \in V^h \times Q^h. \tag{6.97}$$

Proof The proof follows the lines of the proof of Lemma 6.4.5. □

Lemma 6.4.22 (Estimate of the Term with the Divergence) *Let* $Q^h \subset H^1(\Omega)$ *and let the stabilization parameters satisfy* (6.44). *Then, for any* $\mathbf{v} \in V$, *any* $(\mathbf{z}^h, q^h) \in V^h \times Q^h$ *and for any* $\varepsilon > 0$, *it holds*

$$\left|\left(\nabla \cdot \mathbf{v}, q^h\right)\right| \le \varepsilon \left\|(\mathbf{z}^h, q^h)\right\|_{\text{nsgls}}^2 + \frac{\nu}{4\varepsilon}\left(\frac{1}{\delta_0} + C_{\text{inv}}^2\right) \sum_{K \in \mathcal{T}^h} h_K^{-2} \|\mathbf{v}\|_{L^2(K)}^2. \tag{6.98}$$

Proof Applying integration by parts and using that $q^h \in H^1(\Omega)$ yields

$$\left(\nabla \cdot \mathbf{v}, q^h\right) = -\left(\mathbf{v}, \nabla q^h\right).$$

For each mesh cell K, it is for arbitrary $\mathbf{z}^h \in V^h$

$$-\left(\mathbf{v}, \nabla q^h\right)_K = -\left(\mathbf{v}, -\nu\Delta\mathbf{z}^h + \nabla q^h\right)_K + \left(\mathbf{v}, -\nu\Delta\mathbf{z}^h\right)_K.$$

Using the triangle inequality, the Cauchy–Schwarz inequality, the property (6.44), as well as the Young inequality gives for any $\varepsilon, \varepsilon_1 > 0$

$$\begin{aligned}
&\left|\left(\nabla \cdot \mathbf{v}, q^h\right)\right| \\
&\le \sum_{K \in \mathcal{T}^h} \left|\left(\mathbf{v}, -\nu\Delta\mathbf{z}^h + \nabla q^h\right)_K\right| + \sum_{K \in \mathcal{T}^h} \left|\left(\mathbf{v}, -\nu\Delta\mathbf{z}^h\right)_K\right| \\
&\le \frac{\nu}{4\delta_0\varepsilon} \sum_{K \in \mathcal{T}^h} h_K^{-2} \|\mathbf{v}\|_{L^2(K)}^2 + \varepsilon \sum_{K \in \mathcal{T}^h} \delta_K \left\|-\nu\Delta\mathbf{z}^h + \nabla q^h\right\|_{L^2(K)}^2 \\
&\quad + \varepsilon_1 \sum_{K \in \mathcal{T}^h} h_K^2 \left\|\nu\Delta\mathbf{z}^h\right\|_{L^2(K)}^2 + \frac{1}{4\varepsilon_1} \sum_{K \in \mathcal{T}^h} h_K^{-2} \|\mathbf{v}\|_{L^2(K)}^2.
\end{aligned}$$

Utilizing the inverse inequality (6.12) yields

$$\varepsilon_1 \sum_{K\in\mathcal{T}^h} h_K^2 \left\|\nu\Delta\mathbf{z}^h\right\|_{L^2(K)}^2 \leq \varepsilon_1 C_{\mathrm{inv}}^2 \nu^2 \sum_{K\in\mathcal{T}^h} \left\|\nabla\mathbf{z}^h\right\|_{L^2(K)}^2.$$

Choosing $\varepsilon_1 = \varepsilon C_{\mathrm{inv}}^{-2}\nu^{-1}$ and collecting terms gives (6.98). □

Theorem 6.4.23 (Error Estimate) *Assume that the solution of* (6.5) *satisfies* $(\mathbf{u}, p) \in H^{k+1}(\Omega)^d \times H^{l+1}(\Omega)$, *that* $Q^h \subset H^1(\Omega)$, *and that the stabilization parameters satisfy* (6.44), *then there holds the error estimate*

$$\left\|\left(\mathbf{u}-\mathbf{u}^h, p-p^h\right)\right\|_{\mathrm{nsgls}} \leq C\left(\nu^{1/2}h^k \|\mathbf{u}\|_{H^{k+1}(\Omega)} + \frac{h^{l+1}}{\nu^{1/2}}\|p\|_{H^{l+1}(\Omega)}\right).$$

Proof Let I^h and J^h be the interpolation operators satisfying (6.8) and (6.9). Using the coercivity (6.95) and the Galerkin orthogonality (6.97) yields

$$\begin{aligned}
&\left\|\left(\mathbf{u}^h - I^h\mathbf{u}, p^h - J^hp\right)\right\|_{\mathrm{nsgls}}^2 \\
&= A_{\mathrm{nsgls}}\left(\left(\mathbf{u}^h - I^h\mathbf{u}, p^h - J^hp\right), \left(\mathbf{u}^h - I^h\mathbf{u}, p^h - J^hp\right)\right) \\
&= A_{\mathrm{nsgls}}\left(\left(\mathbf{u} - I^h\mathbf{u}, p - J^hp\right), \left(\mathbf{u}^h - I^h\mathbf{u}, p^h - J^hp\right)\right).
\end{aligned}$$

Applying the Cauchy–Schwarz inequality, Lemma 6.4.22 with $\mathbf{z}^h = \mathbf{u}^h - I^h\mathbf{u}$ and $\varepsilon = 1/4$, and the Young inequality gives

$$\begin{aligned}
&\left\|\left(\mathbf{u}^h - I^h\mathbf{u}, p^h - J^hp\right)\right\|_{\mathrm{nsgls}}^2 \\
&\leq \nu\left\|\nabla\left(\mathbf{u} - I^h\mathbf{u}\right)\right\|_{L^2(\Omega)}\left\|\nabla\left(\mathbf{u}^h - I^h\mathbf{u}\right)\right\|_{L^2(\Omega)} \\
&\quad + \left\|p - J^hp\right\|_{L^2(\Omega)}\left\|\nabla\left(\mathbf{u}^h - I^h\mathbf{u}\right)\right\|_{L^2(\Omega)} \\
&\quad + \left|\left(\nabla\cdot\left(\mathbf{u} - I^h\mathbf{u}\right), p^h - J^hp\right)\right| \\
&\quad + \sum_{K\in\mathcal{T}^h} \delta_K \left\|-\nu\Delta\left(\mathbf{u} - I^h\mathbf{u}\right) + \nabla(p - J^hp)\right\|_{L^2(K)} \\
&\qquad \times \left\|-\nu\Delta\left(\mathbf{u}^h - I^h\mathbf{u}\right) + \nabla(p^h - J^hp)\right\|_{L^2(K)} \\
&\leq \frac{1}{2}\left\|\left(\mathbf{u}^h - I^h\mathbf{u}, p^h - J^hp\right)\right\|_{\mathrm{nsgls}}^2 + 2\left\|\left(\mathbf{u} - I^h\mathbf{u}, p - J^hp\right)\right\|_{\mathrm{nsgls}}^2 \\
&\quad + \frac{2}{\nu}\left\|p - J^hp\right\|_{L^2(\Omega)}^2 + \nu\left(\frac{1}{\delta_0} + C_{\mathrm{inv}}^2\right)\sum_{K\in\mathcal{T}^h} h_K^{-2}\left\|\mathbf{u} - I^h\mathbf{u}\right\|_{L^2(K)}^2.
\end{aligned}$$

The proof is finished by applying the triangle inequality

$$\left\|\left(\mathbf{u}-\mathbf{u}^h, p-p^h\right)\right\|_{\text{nsgls}} \le \left\|\left(\mathbf{u}-I^h\mathbf{u}, p-J^h p\right)\right\|_{\text{nsgls}} + \left\|\left(\mathbf{u}^h-I^h\mathbf{u}, p^h-J^h p\right)\right\|_{\text{nsgls}}$$

and using (6.44), (6.8), and (6.9). □

Theorem 6.4.24 (**L^2 Estimate of the Pressure Error**) *Assume that the solution of* (6.5) *satisfies* $(\mathbf{u}, p) \in H^{k+1}(\Omega)^d \times H^{l+1}(\Omega)$, *that* $Q^h \subset H^1(\Omega)$, *and that the stabilization parameters satisfy* (6.44), *then there holds the error estimate*

$$\left\|p - p^h\right\|_{L^2(\Omega)} \le C\left(\nu h^k \|\mathbf{u}\|_{H^{k+1}(\Omega)} + h^{l+1}\|p\|_{H^{l+1}(\Omega)}\right).$$

Proof Like in the proof of Theorem 6.4.8, we start with (6.55). The Galerkin orthogonality (6.97) with $(\mathbf{v}^h, q^h) = (\mathcal{J}^h\mathbf{w}, 0)$ gives

$$\begin{aligned} 0 &= \nu\left(\nabla\left(\mathbf{u}-\mathbf{u}^h\right), \nabla\mathcal{J}^h\mathbf{w}\right) - \left(\nabla\cdot\left(\mathcal{J}^h\mathbf{w}\right), p-p^h\right) \\ &\quad - \sum_{K\in\mathcal{T}^h}\left(-\nu\Delta\left(\mathbf{u}-\mathbf{u}^h\right) + \nabla\left(p-p^h\right), \delta_K\nu\Delta\mathcal{J}^h\mathbf{w}\right)_K. \end{aligned}$$

Hence, using the Cauchy–Schwarz inequality and applying (6.44), (6.12), (6.10), and (6.52), one obtains

$$\left|\left(\nabla\cdot\left(\mathcal{J}^h\mathbf{w}\right), p-p^h\right)\right| \le C\nu^{1/2}\left\|\left(\mathbf{u}-\mathbf{u}^h, p-p^h\right)\right\|_{\text{nsgls}}\left\|p-p^h\right\|_{L^2(\Omega)}. \tag{6.99}$$

The estimate (6.57) reduces to

$$\begin{aligned} &-\left(\nabla\cdot\left(\mathbf{w}-\mathcal{J}^h\mathbf{w}\right), p-p^h\right) \\ &\le \frac{C\nu^{1/2}}{\delta_0^{1/2}}\left\|p-p^h\right\|_{L^2(\Omega)}\left(\sum_{K\in\mathcal{T}^h}\delta_K\left\|\nabla\left(p-p^h\right)\right\|_{L^2(K)}^2\right)^{1/2}. \end{aligned} \tag{6.100}$$

To estimate the last term in (6.100), we apply the triangle inequality, (6.44), and (6.64), which gives

$$\begin{aligned} &\sum_{K\in\mathcal{T}^h}\delta_K\left\|\nabla\left(p-p^h\right)\right\|_{L^2(K)}^2 \\ &\le 2\left\|(\mathbf{u}-\mathbf{u}^h, p-p^h)\right\|_{\text{nsgls}}^2 + 2\delta_1\nu\sum_{K\in\mathcal{T}^h}h_K^2\left\|\Delta(\mathbf{u}-\mathbf{u}^h)\right\|_{L^2(K)}^2 \\ &\le C\left\|(\mathbf{u}-\mathbf{u}^h, p-p^h)\right\|_{\text{nsgls}}^2 + C\nu h^{2k}\|\mathbf{u}\|_{H^{k+1}(\Omega)}^2. \end{aligned}$$

Combining this estimate with (6.55), (6.99), and (6.100) yields

$$\left\| p - p^h \right\|_{L^2(\Omega)} \leq C\nu^{1/2} \left\| (\mathbf{u} - \mathbf{u}^h, p - p^h) \right\|_{\text{nsgls}} + C\nu h^k \left\| \mathbf{u} \right\|_{H^{k+1}(\Omega)},$$

where the constant C depends on $\delta_0^{-1/2}$. Applying Theorem 6.4.23 finishes the proof. □

Theorem 6.4.25 (**L^2 Estimate of the Velocity Error**) *Let the Stokes problem* (6.2) *be regular. Assume that the solution of* (6.5) *satisfies* $(\mathbf{u}, p) \in H^{k+1}(\Omega)^d \times H^{l+1}(\Omega)$, *that* $Q^h \subset H^1(\Omega)$, *and that the stabilization parameters satisfy* (6.44), *then there holds the error estimate*

$$\left\| \mathbf{u} - \mathbf{u}^h \right\|_{L^2(\Omega)} \leq C \left(h^{k+1} \left\| \mathbf{u} \right\|_{H^{k+1}(\Omega)} + \frac{h^{l+2}}{\nu} \left\| p \right\|_{H^{l+1}(\Omega)} \right).$$

Proof We proceed analogously as in the proofs of Theorems 6.4.9 and 6.4.17. Again, the starting point is the identity (6.27), where the first two terms on the right-hand side can be estimated by (6.28). From the Galerkin orthogonality (6.97), one obtains

$$\begin{aligned}
&\nu(\nabla \mathbf{z}^I, \nabla(\mathbf{u} - \mathbf{u}^h)) - (\nabla \cdot (\mathbf{u} - \mathbf{u}^h), r^I) \\
&= \left(\nabla \cdot (\mathbf{z}^I - \mathbf{z}), p - p^h \right) \\
&\quad + \sum_{K \in \mathcal{T}^h} \left(-\nu\Delta(\mathbf{u} - \mathbf{u}^h) + \nabla(p - p^h), \delta_K (\nu\Delta\mathbf{z}^I + \nabla r^I) \right)_K .
\end{aligned}$$

Thus, using (6.44), (6.25), (6.58), and (6.89), one derives

$$\begin{aligned}
&\nu(\nabla \mathbf{z}^I, \nabla(\mathbf{u} - \mathbf{u}^h)) - (\nabla \cdot (\mathbf{u} - \mathbf{u}^h), r^I) \\
&\leq \left\| \nabla(\mathbf{z} - \mathbf{z}^I) \right\|_{L^2(\Omega)} \left\| p - p^h \right\|_{L^2(\Omega)} \\
&+ \left\| \left(\mathbf{u} - \mathbf{u}^h, p - p^h \right) \right\|_{\text{nsgls}} \left(\sum_{K \in \mathcal{T}^h} \delta_K \left\| \nu\Delta\mathbf{z}^I + \nabla r^I \right\|_{L^2(K)}^2 \right)^{1/2} \\
&\leq Ch \left(\left\| p - p^h \right\|_{L^2(\Omega)} + \nu^{1/2} \left\| \left(\mathbf{u} - \mathbf{u}^h, p - p^h \right) \right\|_{\text{nsgls}} \right) \left\| \mathbf{u} - \mathbf{u}^h \right\|_{L^2(\Omega)} . \qquad (6.101)
\end{aligned}$$

Combining (6.27), (6.28), and (6.101) gives

$$\nu \left\| \mathbf{u} - \mathbf{u}^h \right\|_{L^2(\Omega)} \leq Ch \left(\left\| p - p^h \right\|_{L^2(\Omega)} + \nu^{1/2} \left\| \left(\mathbf{u} - \mathbf{u}^h, p - p^h \right) \right\|_{\text{nsgls}} \right)$$

and the theorem follows from Theorems 6.4.23 and 6.4.24. □

In the case of discontinuous pressure approximations, the optimal choice of the stabilization parameter for the pressure jumps is $\gamma_E = \mathcal{O}(h_E/\nu)$, see [39].

In [4], an extension of the non-symmetric GLS method is proposed. This method possesses jump terms that contain the residual of the stress tensor on the internal edges, i.e., the jump of the normal derivative of the finite element velocity and the jump of the finite element pressure. It is unconditionally stable for a norm where, in comparison with $\|(\cdot,\cdot)\|_{\text{nsgls}}$ defined in (6.94), the Laplacian of the velocity is absent but the residual of the stress tensor at the inner faces is present. An a priori analysis and an a posteriori analysis of this method are provided in [4].

For P_1/P_1 finite elements, the non-symmetric GLS method with a weak imposition of the boundary condition via a penalty-free Nitsche method was studied in [18].

6.4.5 An Absolutely Stable Modification of the PSPG Method

The PSPG method from Sect. 6.4.2 is only conditionally stable, see the upper bound (6.39) for the stabilization parameter used in Lemma 6.4.3 to prove the coercivity. In [19], an absolutely stable modification of the PSPG method was proposed which we now briefly describe.

The PSPG method (6.33) will be now considered with $Q_h \subset H^1(\Omega)$ and $\delta_K = \delta := \delta_0 h^2/\nu$, which can be used on a uniform grid. In (6.34), the operator Δ is applied elementwise. In [19], it was replaced by the discrete Laplacian $\Delta^h : V \to V^h$ defined by

$$\left(\Delta^h \mathbf{u}, \mathbf{v}^h\right) = -\left(\nabla \mathbf{u}, \nabla \mathbf{v}^h\right) \quad \forall\, \mathbf{u} \in V,\ \mathbf{v}^h \in V^h. \tag{6.102}$$

Then the modified PSPG method reads: Find $\left(\mathbf{u}^h, p^h\right) \in V^h \times Q^h$ such that

$$\begin{aligned} \nu\left(\nabla \mathbf{u}^h, \nabla \mathbf{v}^h\right) - \left(\nabla \cdot \mathbf{v}^h, p^h\right) + \left(\nabla \cdot \mathbf{u}^h, q^h\right) + \delta\left(-\nu \Delta^h \mathbf{u}^h + \nabla p^h, \nabla q^h\right) \\ = (\mathbf{f}, \mathbf{v}^h) + \delta\left(\mathbf{f}, \nabla q^h\right) \quad \forall\left(\mathbf{v}^h, q^h\right) \in V^h \times Q^h. \end{aligned} \tag{6.103}$$

Thus, the modified PSPG method requires the additional solution of problem (6.102), which is a linear system with the mass matrix as system matrix. In practical computations, the mass matrix can be replaced by a lumped mass matrix or local projection. Using (6.102), method (6.103) can be rewritten as

$$\left(-\nu \Delta^h \mathbf{u}^h + \nabla p^h, \mathbf{v}^h + \delta \nabla q^h\right) + \left(\nabla \cdot \mathbf{u}^h, q^h\right) = \left(\mathbf{f}, \mathbf{v}^h + \delta \nabla q^h\right), \tag{6.104}$$

such that it has the form of a Petrov–Galerkin method.

Recall that the conditional stability of the PSPG method stems from the fact that the coercivity is based on estimating the term $\sum_{K\in\mathcal{T}^h} \delta_K \left(-\nu\Delta\mathbf{v}^h, \nabla q^h\right)_K$ by $\left\|\left(\mathbf{v}^h, q^h\right)\right\|_{\text{pspg}}^2$. This step inevitably leads to a bound on δ_K. However, replacing Δ by Δ^h, it is possible to get rid of this term by a suitable choice of $\mathbf{v}^h$. Indeed, defining $\mathbf{v}^h$ as the L^2 projection of $-\delta\nabla q^h$ onto V^h, we see from (6.104) that the respective term disappears. This together with further tools enabled to prove in [19] that, for any $\delta_0 > 0$, the bilinear form corresponding to the modified PSPG method satisfies an inf-sup condition of the type (6.75) with respect to the norm $\nu^{1/2}\left\|\nabla\mathbf{v}^h\right\|_{L^2(\Omega)} + \nu^{-1/2}\left\|q^h\right\|_{L^2(\Omega)}$ with a constant dependent on δ_0.

The modified PSPG method (6.103) is obviously not consistent in general. However, this inconsistency is very weak so that the optimal order of convergence with respect to the mentioned norm could be proved in [19].

6.5 Stabilizations Using Only the Pressure

This section is dedicated to methods that use only the pressure in the stabilization term. Hence, there is no need to compute the residual and the use of second derivatives of the finite element functions is not necessary. However, many methods connect pressure degrees of freedom that do not belong to the same mesh cell. Consequently, the stencil of the matrix C in (6.4) is denser than for residual-based stabilizations.

After having introduced a framework in Sect. 6.5.1, a number of methods will be presented briefly. A detailed analysis is provided for a Local Projection Stabilization (LPS) method in Sect. 6.5.4.

6.5.1 A Framework

An abstract approach for the derivation and analysis of pressure-stabilized schemes was presented in [25], see also [21, Chapter 6.3]. For the Stokes equations, the considered scheme has the form: Find $\left(\mathbf{u}^h, p^h\right) \in V^h \times Q^h$ such that for all $\left(\mathbf{v}^h, q^h\right) \in V^h \times Q^h$

$$\nu\left(\nabla\mathbf{u}^h, \nabla\mathbf{v}^h\right) - \left(\nabla\cdot\mathbf{v}^h, p^h\right) + \left(\nabla\cdot\mathbf{u}^h, q^h\right) + \delta S\left(\left(\mathbf{u}^h, p^h\right), \left(\mathbf{v}^h, q^h\right)\right) = \left(\mathbf{f}, \mathbf{v}^h\right), \tag{6.105}$$

with $\delta > 0$ and $S : \left(V^h \times Q^h\right)\times\left(V^h \times Q^h\right) \to \mathbb{R}$ being a bilinear form that should be chosen such that (6.105) is a stable discrete scheme. There are two essential assumptions on S. The bilinear form should be bounded with a constant independent of h. Likewise, uniformly in h, there should exist a Hilbert space $\mathcal{H}$, some operator

$G^h \in \mathcal{L}\left(V^h \times Q^h, \mathcal{H}\right)$, and a constant $C > 0$ such that for all $\left(\mathbf{v}^h, q^h\right) \in V^h \times Q^h$

$$S\left(\left(\mathbf{v}^h, q^h\right), \left(\mathbf{v}^h, q^h\right)\right) \geq C \left\| G^h\left(\left(\mathbf{v}^h, q^h\right)\right)\right\|_{\mathcal{H}}^2 .$$

For the abstract problem considered in [25], more operators, assumptions, etc. were introduced. Then, stability and error estimates, e.g., with respect to the errors in the norms of V and Q were derived.

Let $Q^h \subset H^1(\Omega)$. The application of the abstract theory presented in [25] to the Stokes equations considers pressure stabilizations that use only the pressure. A first example consists in taking

$$S\left(\left(\mathbf{u}^h, p^h\right), \left(\mathbf{v}^h, q^h\right)\right) = \left(\nabla p^h, \nabla q^h\right),$$

$G^h\left(\mathbf{v}^h, q^h\right) = \nabla q^h$, $\mathcal{H} = L^2(\Omega)^d$, and $\delta = \mathcal{O}\left(h^2\right)$, which gives the method of Brezzi–Pitkäranta, see Sect. 6.5.2. A second example consists in choosing

$$S\left(\left(\mathbf{u}^h, p^h\right), \left(\mathbf{v}^h, q^h\right)\right) = \left(\left(I - P_{\overline{V^h}}\right) \nabla p^h, \nabla q^h\right),$$

with $P_{\overline{V^h}}$ being the $L^2(\Omega)$ projection operator onto $\overline{V^h}$, where $\overline{V^h}$ is defined with the same polynomials as V^h but without incorporating the boundary conditions in the definition. In this method, one has $\mathcal{H} = L^2(\Omega)^d$ and $G^h\left(\mathbf{v}^h, q^h\right) = \left(I - P_{\overline{V^h}}\right) \nabla q^h$. One obtains the method proposed in [34], see Sect. 6.5.3. Concerning the choice of δ, one finds in [34], where bounds for the pressure error in different norms than in [25] were proved, that one gets stability for $\delta \geq Ch^2$ and optimal convergence for $\delta = \mathcal{O}\left(h^2\right)$. In [25], see also [21, Chapter 8.13.3], it is shown that for $V^h \times Q^h = P_1 \times P_1$, stability and optimal convergence are obtained with $\delta = \mathcal{O}(1)$.

For a detailed investigation on how several methods introduced in this section fit into the framework of [25], it is referred to [27].

6.5.2 The Brezzi–Pitkäranta Method

The Brezzi–Pitkäranta method from [26] was the first stabilization method for circumventing the discrete inf-sup condition (6.3). This method was proposed for the P_1/P_1 pair of finite element spaces and it has the form: Find $(\mathbf{u}^h, p^h) \in V^h \times Q^h = P_1 \times P_1$ such that

$$\begin{aligned} \nu\left(\nabla \mathbf{u}^h, \nabla \mathbf{v}^h\right) - \left(\nabla \cdot \mathbf{v}^h, p^h\right) &= \left(\mathbf{f}, \mathbf{v}^h\right) \ \forall\, \mathbf{v}^h \in V^h, \\ -\left(\nabla \cdot \mathbf{u}^h, q^h\right) - \sum_{K \in \mathcal{T}^h} \left(\nabla p^h, \delta_K^p \nabla q^h\right)_K &= 0 \qquad \forall\, q^h \in Q^h. \end{aligned} \tag{6.106}$$

Considering a uniform family of triangulations, the optimal order convergence of the solution of (6.106) with respect to $\left\|\nabla\left(\mathbf{u}-\mathbf{u}^h\right)\right\|_{L^2(\Omega)}$ and $\left\|p-p^h\right\|_{L^2(\Omega)}$ was proved for the stabilization parameter $\delta_K^p = \mathcal{O}\left(h^2\right)$ (for $\nu = 1$). As discussed above, method (6.106) fits into the framework presented in Sect. 6.5.1.

As it is often noted in the literature, the Brezzi–Pitkäranta method imposes artificial boundary conditions for the finite element pressure. Considering for simplicity $\delta_K^p = \delta$, then the strong form of the continuity equation of (6.106) reads as

$$-\nabla \cdot \mathbf{u} + \delta \Delta p = 0.$$

Deriving in the usual way the corresponding weak form leads to

$$-\left(\nabla \cdot \mathbf{u}, q\right) - \delta\left(\nabla p, \nabla q\right) + \delta \int_{\partial\Omega} \left(\nabla p \cdot \mathbf{n}\right) q \, \mathrm{d}s = 0 \quad \forall\, q \in Q.$$

Since no boundary integral appears in (6.106), one finds that an artificial boundary condition of the form

$$\delta\left(\nabla p^h \cdot \mathbf{n}\right) = 0 \quad \text{on } \partial\Omega$$

for the discrete pressure is introduced with this method.

A stabilized method of Brezzi–Pitkäranta-type with a nonlinear stabilization parameter is presented in [77], the so-called pressure Laplacian stabilization (PLS) method. The stabilization parameter depends on the residuals of the finite element continuity and momentum equations.

6.5.3 Stabilization with Global Fluctuations of the Pressure Gradient

In [34], it was shown that for constructing a pressure-stable method, it is not necessary to use the full gradient of the discrete pressure, as in (6.106). Denoting by $\overline{V^h}$ the velocity finite element space with the same polynomials as V^h but without prescribed boundary conditions, then it is proposed in [34] to apply the following method: Find $\left(\mathbf{u}^h, p^h, \overline{\nabla p^h}\right) \in V^h \times Q^h \times \overline{V^h}$ such that

$$
\begin{aligned}
\nu\left(\nabla \mathbf{u}^h, \nabla \mathbf{v}^h\right) - \left(\nabla \cdot \mathbf{v}^h, p^h\right) &= \left(\mathbf{f}, \mathbf{v}^h\right) \ \forall\ \mathbf{v}^h \in V^h, \\
-\left(\nabla \cdot \mathbf{u}^h, q^h\right) - \sum_{K \in \mathcal{T}^h} \left(\nabla p^h - \overline{\nabla p^h}, \delta_K^p \nabla q^h\right)_K &= 0 \qquad \forall\ q^h \in Q^h, \\
\left(\nabla p^h - \overline{\nabla p^h}, \overline{\mathbf{v}^h}\right) &= 0 \qquad \forall\ \overline{\mathbf{v}^h} \in \overline{V^h}.
\end{aligned}
\tag{6.107}
$$

The third equation of (6.107) defines $\overline{\nabla p^h}$ to be the $L^2(\Omega)$ projection of ∇p^h onto $\overline{V^h}$. In this way, one can interpret $\overline{\nabla p^h}$ as being large scales of ∇p^h and the difference $\nabla p^h - \overline{\nabla p^h}$ as being fluctuations. Only the fluctuations appear in the stabilization term of the discrete continuity equation. It was already discussed above that this method fits into the framework described in Sect. 6.5.1.

A finite element analysis of the method can be found in [34]. This analysis considers a family of quasi-uniform triangulations and $\delta_K^p = \delta$. For $\delta = \mathcal{O}\left(h^2\right)$, the stability of the finite element solution and optimal error estimates for $\left\|\nabla\left(\mathbf{u} - \mathbf{u}^h\right)\right\|_{L^2(\Omega)}$, $\left\|\nabla\left(p - p^h\right)\right\|_{L^2(\Omega)}$, and $\left\|\nabla p - \overline{\nabla p^h}\right\|_{L^2(\Omega)}$ were proved. Extensions of the analysis that allow the choice of local stabilization parameters and to the steady-state Navier–Stokes equations can be found in [35].

Another analysis of method (6.107) can be found in [66].[1] The error estimate from [66] bounds $\left\|p - p^h\right\|_{L^2(\Omega)}$ whereas the estimate from [34] gives a bound for $h\left\|\nabla\left(p - p^h\right)\right\|_{L^2(\Omega)}$.

A method of type (6.107) was analyzed for the Brinkman equations in [8]. As additional terms, a grad-div stabilization, using fluctuations of the divergence, and jump terms across faces, which involve the Cauchy stress tensor, appear. The analysis covers both limit cases of the Brinkman equations, namely the Stokes and the Darcy equations.

6.5.4 *Local Projection Stabilization (LPS) Methods*

To assure the stability of the PSPG method (6.33), it would be sufficient to consider the term

$$
\sum_{K \in \mathcal{T}^h} \delta_K \left(\nabla p, \nabla q\right)_K \tag{6.108}
$$

instead of the residual-based terms in (6.34) and (6.35). This would provide several advantages (e.g., symmetry of the stabilization, simpler implementation, absolute stability) but it would not lead to optimal error estimates. A remedy preserving

[1]Reading [66], one is wondering that there is no reference to [34] for method (6.107). From the article's history, one finds that [66] was submitted shortly after [34] was published.

most of the advantages of (6.108) without compromising the convergence rates of the PSPG method is to apply locally suitable projection operators to ∇p and ∇q in (6.108) so that the consistency error can be estimated in the desired way.

It is convenient to define the mentioned local projections on macroelements. Precisely, one introduces a set $\mathcal{M}^h$ consisting of a finite number of open subsets M of Ω such that $\overline{\Omega} = \cup_{M \in \mathcal{M}^h} \overline{M}$. In contrast to $\mathcal{T}^h$, the sets in $\mathcal{M}^h$ are allowed to overlap. For any $K \in \mathcal{T}^h$, $E \in \mathcal{E}^h$, and $M \in \mathcal{M}^h$ it is assumed that either $K \subset \overline{M}$ or $K \subset \overline{\Omega} \setminus M$ and that either $E \subset \overline{M}$ or $E \subset \Omega \setminus \overline{M}$. Furthermore, for any $M \in \mathcal{M}^h$, one introduces a finite-dimensional space $D_M \subset L^2(M)^d$ and a continuous linear projection operator π_M which maps the space $L^2(M)^d$ onto the space D_M. Then one defines the so-called fluctuation operator $\kappa_M = id - \pi_M$, where id is the identity operator on $L^2(M)^d$. Finally, the term (6.108) is replaced by

$$\sum_{M \in \mathcal{M}^h} \delta_M (\kappa_M(\nabla^h p), \kappa_M(\nabla^h q))_M,$$

where $(\nabla^h q)|_K = \nabla(q|_K)$ for any $K \in \mathcal{T}^h$.

Thus, the local projection stabilization (LPS) method reads: Find $\left(\mathbf{u}^h, p^h\right) \in V^h \times Q^h$ such that

$$A_{\text{lps}}\left(\left(\mathbf{u}^h, p^h\right), \left(\mathbf{v}^h, q^h\right)\right) = (\mathbf{f}, \mathbf{v}^h) \quad \forall \left(\mathbf{v}^h, q^h\right) \in V^h \times Q^h, \tag{6.109}$$

where the bilinear form $A_{\text{lps}} : \left(V \times \tilde{Q}\right) \times \left(V \times \tilde{Q}\right) \to \mathbb{R}$ is given by

$$\begin{aligned} A_{\text{lps}}\left((\mathbf{u}, p), (\mathbf{v}, q)\right) &= \nu\left(\nabla \mathbf{u}, \nabla \mathbf{v}\right) - \left(\nabla \cdot \mathbf{v}, p\right) + \left(\nabla \cdot \mathbf{u}, q\right) \\ &\quad + \sum_{E \in \mathcal{E}^h} \gamma_E \left([|p|]_E, [|q|]_E\right)_E + \sum_{M \in \mathcal{M}^h} \delta_M (\kappa_M(\nabla^h p), \kappa_M(\nabla^h q))_M \end{aligned}$$

and the space $\tilde{Q}$ was defined in (6.37).

We make analogous assumptions on the stabilization parameters as for the residual-based methods, i.e., it is assumed that the parameters $\{\gamma_E\}$ satisfy (6.45) and that

$$0 < \delta_0 \frac{h_M^2}{\nu} \le \delta_M \le \delta_1 \frac{h_M^2}{\nu} \qquad \forall\, M \in \mathcal{M}^h \tag{6.110}$$

with some positive constants δ_0, δ_1 and $h_M := \text{diam}(M)$.

To perform an analysis of the method and prove optimal error estimates, a key assumption is the validity of the inf-sup conditions

$$\sup_{\mathbf{v} \in V_M \setminus \{\mathbf{0}\}} \frac{(\mathbf{v}, \boldsymbol{q})_M}{\|\mathbf{v}\|_{L^2(M)}} \ge \beta_{LP} \|\boldsymbol{q}\|_{L^2(M)} \qquad \forall\, \boldsymbol{q} \in D_M,\ M \in \mathcal{M}^h \tag{6.111}$$

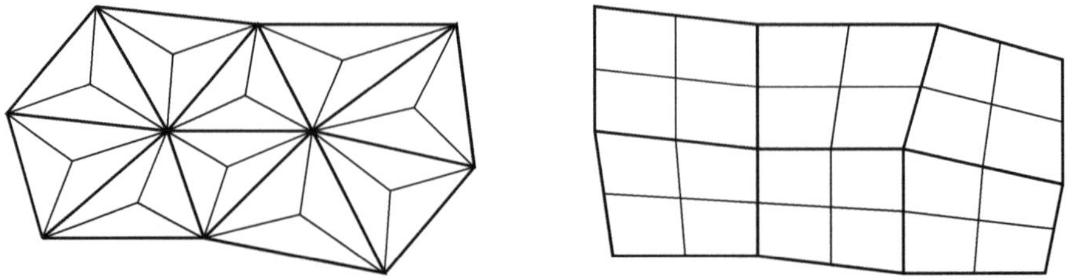

Fig. 6.1 Relation between the meshes $\mathcal{M}^h$ (bold lines) and $\mathcal{T}^h$ (bold and fine lines) in the two-level method

with $V_M = \{\mathbf{v}^h \in V^h : \mathbf{v}^h = \mathbf{0} \text{ in } \Omega \setminus M\}$ and a constant β_{LP} independent of h. This property limits possible combinations of spaces V^h and D_M.

Using local projections onto macro mesh cells for pressure stabilization was proposed for the Q_1/Q_0 pair of finite element spaces already in [82]. The original local projection stabilization [13, 14] was designed as a two-level method. Given a triangulation of Ω, the elements of this triangulation are considered as the set $\mathcal{M}^h$. Then this triangulation is refined as depicted in Fig. 6.1 for the two-dimensional case, i.e., each triangle is divided into three triangles by connecting its vertices with the barycenter and each quadrilateral is divided into four quadrilaterals by connecting midpoints of opposite edges. This gives the triangulation $\mathcal{T}^h$. If the space V^h is defined on $\mathcal{T}^h$ like before (i.e., it contains locally (mapped) polynomials of degree $k \geq 1$), then the inf-sup conditions (6.111) hold for $D_M = P_{k-1}(M)^d$.

Another choice of the spaces V^h and D_M (a one-level method) was proposed in [72]. In this case $\mathcal{M}^h = \mathcal{T}^h$ and to satisfy the inf-sup conditions (6.111) with $D_M = P_{k-1}(M)^d$ the space V^h is enriched elementwise by bubble functions.

Finally, let us describe a choice of the spaces V^h and D_M based on a set $\mathcal{M}^h$ consisting of overlapping sets M as proposed in [61]. Assuming that each element of $\mathcal{T}^h$ has at least one vertex in Ω, then for each interior vertex a macroelement consisting of elements of $\mathcal{T}^h$ sharing this vertex is defined. For this set $\mathcal{M}^h$, one can use our standard choice of V^h and local spaces $D_M = P_{k-1}(M)^d$ to satisfy the inf-sup conditions (6.111).

Note that the first two ways of constructing the spaces V^h and D_M lead to a significant increase of the number of degrees of freedom, either due to an enrichment by bubble functions (in the one-level method) or due to a refinement of the given triangulation (in the two-level method). On the other hand, in the variant with overlapping sets M, the number of degrees of freedom remains the same as if one would use, e.g., a residual-based stabilization.

We refer to [61, 72] for details on the definitions of the spaces and for proofs of the inf-sup conditions.

In view of the examples of the spaces D_M, it is reasonable to assume that there exist interpolation operators $j_M : L^2(M)^d \to D_M$ such that, for $m = 0, \ldots, k$, one has

$$\|\boldsymbol{q} - j_M \boldsymbol{q}\|_{L^2(M)} \leq C\, h_M^m\, \|\boldsymbol{q}\|_{H^m(M)} \qquad \forall\, \boldsymbol{q} \in H^m(M)^d,\ M \in \mathcal{M}^h. \tag{6.112}$$

Finally, let us state a few natural assumptions needed for the subsequent analysis. We assume that there are various positive constants independent of h such that

$$\mathrm{card}\{M' \in \mathcal{M}^h\,;\; M \cap M' \neq \emptyset\} \leq C_{\mathcal{M}} \qquad \forall\, M \in \mathcal{M}^h, \tag{6.113}$$

$$\mathrm{card}\{K \in \mathcal{T}^h\,;\; K \subset \overline{M}\} \leq C_{\mathcal{T}} \qquad \forall\, M \in \mathcal{M}^h, \tag{6.114}$$

$$\mathrm{card}\{M \in \mathcal{M}^h\,;\; K \subset \overline{M}\} \leq C_{\mathcal{T}} \qquad \forall\, K \in \mathcal{T}^h, \tag{6.115}$$

$$\mathrm{card}\{E \in \mathcal{E}^h\,;\; E \subset \overline{M}\} \leq C_{\mathcal{E}} \qquad \forall\, M \in \mathcal{M}^h, \tag{6.116}$$

$$\mathrm{card}\{M \in \mathcal{M}^h\,;\; E \subset \overline{M}\} \leq C_{\mathcal{E}} \qquad \forall\, E \in \mathcal{E}^h, \tag{6.117}$$

$$\|\kappa_M\|_{\mathcal{L}(L^2(M)^d, L^2(M)^d)} \leq C_\kappa \qquad \forall\, M \in \mathcal{M}^h, \tag{6.118}$$

$$h_M \leq C'_{\mathcal{M}}\, h_{M'} \qquad \forall\, M, M' \in \mathcal{M}^h,\; M \cap M' \neq \emptyset. \tag{6.119}$$

Furthermore, for any $E \in \mathcal{E}^h$ and $M \in \mathcal{M}^h$ with $E \subset \overline{M}$, we assume that

$$h_M \leq C'_{\mathcal{E}}\, h_E\,, \tag{6.120}$$

$$\|v\|_{L^2(E)} \leq C_e\,(h_M^{-1/2}\, \|v\|_{L^2(M)} + h_M^{1/2}\, \|\nabla v\|_{L^2(M)}) \qquad \forall\, v \in H^1(M). \tag{6.121}$$

Finally, we shall need the inverse inequality

$$\left\|\nabla \mathbf{v}^h\right\|_{L^2(M)} \leq \bar{C}_{\mathrm{inv}}\, h_M^{-1}\, \|\mathbf{v}^h\|_{L^2(M)} \qquad \forall\, \mathbf{v}^h \in V^h,\; M \in \mathcal{M}^h\,. \tag{6.122}$$

Let us now investigate the stability of the LPS method. One obviously has

$$A_{\mathrm{lps}}\left((\mathbf{v}, q)\,, (\mathbf{v}, q)\right) = |(\mathbf{v}, q)|^2_{\mathrm{lps}} \qquad \forall\; (\mathbf{v}, q) \in V \times \tilde{Q}, \tag{6.123}$$

where

$$|(\mathbf{v}, q)|_{\mathrm{lps}} = \Bigg(\nu\, \|\nabla \mathbf{v}\|^2_{L^2(\Omega)} + \sum_{E \in \mathcal{E}^h} \gamma_E\, \left\|[|q|]_E\right\|^2_{L^2(E)} + \sum_{M \in \mathcal{M}^h} \delta_M\, \left\|\kappa_M(\nabla^h q)\right\|^2_{L^2(M)}\Bigg)^{1/2}.$$

The functional $|\cdot|_{\mathrm{lps}}$ is only a seminorm on $V \times \tilde{Q}$. In what follows we shall prove that the bilinear form A_{lps} is stable on $V^h \times Q^h$ with respect to the norm

$$\|(\mathbf{v},q)\|_{\text{lps}} = \left(\nu \|\nabla \mathbf{v}\|^2_{L^2(\Omega)} + \sum_{E\in\mathcal{E}^h} \gamma_E \left\|[|q|]_E\right\|^2_{L^2(E)} + \sum_{M\in\mathcal{M}^h} \delta_M \left\|\nabla^h q\right\|^2_{L^2(M)}\right)^{1/2}$$

in the sense of an inf-sup condition. The norm $\|\cdot\|_{\text{lps}}$ is an analogue of the PSPG norm (6.38) and the proof that it is a norm on $V \times \tilde{Q}$ is the same as in Lemma 6.4.2. One even has the following result.

Lemma 6.5.1 (Relation to the PSPG Norm) *Given stabilization parameters* $\{\delta_M\}$ *and* $\{\gamma_E\}$ *satisfying* (6.110) *and* (6.45), *respectively, one has*

$$\|(\mathbf{v},q)\|_{\text{lps}} = \|(\mathbf{v},q)\|_{\text{pspg}} \qquad \forall\ (\mathbf{v},q) \in V \times \tilde{Q},$$

where the norm $\|\cdot\|_{\text{pspg}}$ *is defined using stabilization parameters* $\{\delta_K\}$ *satisfying*

$$0 < \delta_0 \frac{h_K^2}{\nu} \le \delta_K \le \delta_1' \frac{h_K^2}{\nu} \qquad \forall\ K \in \mathcal{T}^h \tag{6.124}$$

with a constant δ_1' *independent of h and* ν.

Proof One has

$$\begin{aligned}\sum_{M\in\mathcal{M}^h} \delta_M \left\|\nabla^h q\right\|^2_{L^2(M)} &= \sum_{M\in\mathcal{M}^h} \delta_M \sum_{\substack{K\in\mathcal{T}^h,\\ K\subset\overline{M}}} \|\nabla q\|^2_{L^2(K)} \\ &= \sum_{K\in\mathcal{T}^h} \sum_{\substack{M\in\mathcal{M}^h,\\ K\subset\overline{M}}} \delta_M \|\nabla q\|^2_{L^2(K)} \\ &= \sum_{K\in\mathcal{T}^h} \delta_K \|\nabla q\|^2_{L^2(K)} \end{aligned}\tag{6.125}$$

with

$$\delta_K := \sum_{\substack{M\in\mathcal{M}^h,\\ K\subset\overline{M}}} \delta_M. \tag{6.126}$$

For any $M \in \mathcal{M}^h$ such that $K \subset \overline{M}$ one gets $\delta_K \ge \delta_M \ge \delta_0 h_M^2/\nu \ge \delta_0 h_K^2/\nu$. On the other hand, using (6.120) and (6.115), it follows that

$$\delta_K \le \delta_1 \sum_{\substack{M\in\mathcal{M}^h,\\ K\subset\overline{M}}} \frac{h_M^2}{\nu} \le \delta_1 \left(C_{\mathcal{E}}'\right)^2 C_{\mathcal{T}} \frac{h_K^2}{\nu}.$$

□

Lemma 6.5.2 (Inf-sup Condition for the Bilinear Form A_{lps}) *Let the conditions* (6.110) *and* (6.45) *on the stabilization parameters* $\{\delta_M\}$ *and* $\{\gamma_E\}$ *be satisfied and let the inf-sup conditions* (6.111) *hold. Then, there is a positive constant C such that for all* $(\mathbf{v}^h, q^h) \in V^h \times Q^h$, *it holds*

$$\sup_{(\mathbf{w}^h, r^h)\in V^h\times Q^h\setminus\{(\mathbf{0},0)\}} \frac{A_{\mathrm{lps}}\left(\left(\mathbf{v}^h, q^h\right), \left(\mathbf{w}^h, r^h\right)\right)}{\left\|\left(\mathbf{w}^h, r^h\right)\right\|_{\mathrm{lps}}} \geq C \left\|\left(\mathbf{v}^h, q^h\right)\right\|_{\mathrm{lps}}.$$

Proof Consider any $(\mathbf{v}^h, q^h) \in V^h \times Q^h$ and set $\boldsymbol{s} = \nabla^h q^h$. Then $\boldsymbol{s} \in L^2(\Omega)^d$ and, using the identity

$$(\nabla \cdot \mathbf{w}, q) + (\mathbf{w}, \nabla^h q) = \sum_{E\in\mathcal{E}^h} (\mathbf{w} \cdot \mathbf{n}_E, [|q|]_E)_E \quad \forall\, \mathbf{w} \in V,\ q \in \tilde{Q}, \tag{6.127}$$

that follows from integration by parts, one obtains

$$\begin{aligned} A_{\mathrm{lps}}((\mathbf{v}^h, q^h), (\mathbf{z}^h, 0)) \geq (\mathbf{z}^h, \boldsymbol{s}) - \nu \left\|\nabla \mathbf{v}^h\right\|_{L^2(\Omega)} \left\|\nabla \mathbf{z}^h\right\|_{L^2(\Omega)} \\ - \sum_{E\in\mathcal{E}^h} \left(\mathbf{z}^h \cdot \mathbf{n}_E, \left[\left|q^h\right|\right]_E\right)_E \end{aligned} \tag{6.128}$$

for any $\mathbf{z}^h \in V^h$. Our aim is to choose the function $\mathbf{z}^h$ in such a way that the term $(\mathbf{z}^h, \boldsymbol{s})$ provides a control of

$$S := \sum_{M\in\mathcal{M}^h} \delta_M \|\boldsymbol{s}\|^2_{L^2(M)}.$$

For this one can employ the inf-sup conditions (6.111) which imply that, for any $M \in \mathcal{M}^h$, there exists $\mathbf{z}_M \in V_M$ such that (cf., e.g., [40])

$$(\mathbf{z}_M, \boldsymbol{q})_M = \delta_M (\boldsymbol{s}, \boldsymbol{q})_M \qquad \forall\, \boldsymbol{q} \in D_M, \tag{6.129}$$

$$\|\mathbf{z}_M\|_{L^2(M)} \leq \beta_{LP}^{-1} \delta_M \|\boldsymbol{s}\|_{L^2(M)}. \tag{6.130}$$

Consider any $M \in \mathcal{M}^h$. Since $\pi_M \boldsymbol{s} \in D_M$, one gets

$$\begin{aligned} (\mathbf{z}_M, \boldsymbol{s}) &= (\mathbf{z}_M, \pi_M \boldsymbol{s})_M + (\mathbf{z}_M, \kappa_M \boldsymbol{s})_M \\ &= \delta_M (\boldsymbol{s}, \pi_M \boldsymbol{s})_M + (\mathbf{z}_M, \kappa_M \boldsymbol{s})_M \\ &= \delta_M \|\boldsymbol{s}\|^2_{L^2(M)} - \delta_M (\boldsymbol{s}, \kappa_M \boldsymbol{s})_M + (\mathbf{z}_M, \kappa_M \boldsymbol{s})_M. \end{aligned}$$

Due to (6.130) and the Young inequality, one has

$$
\begin{aligned}
|\delta_M (\boldsymbol{s}, \kappa_M \boldsymbol{s})_M - (\mathbf{z}_M, \kappa_M \boldsymbol{s})_M| &\le (\delta_M \|\boldsymbol{s}\|_{L^2(M)} + \|\mathbf{z}_M\|_{L^2(M)}) \|\kappa_M \boldsymbol{s}\|_{L^2(M)} \\
&\le \delta_M (1 + \beta_{LP}^{-1}) \|\boldsymbol{s}\|_{L^2(M)} \|\kappa_M \boldsymbol{s}\|_{L^2(M)} \\
&\le \frac{\delta_M}{2} \|\boldsymbol{s}\|_{L^2(M)}^2 + (1 + \beta_{LP}^{-2}) \delta_M \|\kappa_M \boldsymbol{s}\|_{L^2(M)}^2
\end{aligned}
$$

and hence

$$
(\mathbf{z}_M, \boldsymbol{s}) \ge \frac{\delta_M}{2} \|\boldsymbol{s}\|_{L^2(M)}^2 - (1 + \beta_{LP}^{-2}) \delta_M \|\kappa_M \boldsymbol{s}\|_{L^2(M)}^2 .
$$

Thus, setting $\mathbf{z}^h = \sum_{M \in \mathcal{M}^h} \mathbf{z}_M$, one gets

$$
(\mathbf{z}^h, \boldsymbol{s}) \ge \frac{1}{2} S - (1 + \beta_{LP}^{-2}) \sum_{M \in \mathcal{M}^h} \delta_M \|\kappa_M \boldsymbol{s}\|_{L^2(M)}^2 .
$$

In view of (6.113), one has

$$
\begin{aligned}
\|\nabla \mathbf{z}^h\|_{L^2(\Omega)}^2 &\le \sum_{M' \in \mathcal{M}^h} \|\nabla \mathbf{z}^h\|_{L^2(M')}^2 \\
&\le \sum_{M' \in \mathcal{M}^h} \Bigg(\sum_{\substack{M \in \mathcal{M}^h, \\ M \cap M' \neq \emptyset}} \|\nabla \mathbf{z}_M\|_{L^2(M')} \Bigg)^2 \\
&\le C_{\mathcal{M}} \sum_{M' \in \mathcal{M}^h} \sum_{M \in \mathcal{M}^h} \|\nabla \mathbf{z}_M\|_{L^2(M')}^2 \\
&= C_{\mathcal{M}} \sum_{M \in \mathcal{M}^h} \sum_{\substack{M' \in \mathcal{M}^h, \\ M \cap M' \neq \emptyset}} \|\nabla \mathbf{z}_M\|_{L^2(M')}^2 \\
&\le C_{\mathcal{M}}^2 \sum_{M \in \mathcal{M}^h} \|\nabla \mathbf{z}_M\|_{L^2(M)}^2 .
\end{aligned}
$$

Using (6.122), (6.130), and (6.110), one derives

$$
\nu \, \|\nabla \mathbf{z}_M\|_{L^2(M)}^2 \le \bar{C}_{\text{inv}}^2 \nu h_M^{-2} \|\mathbf{z}_M\|_{L^2(M)}^2 \le \delta_1 \bar{C}_{\text{inv}}^2 \beta_{LP}^{-2} \delta_M \|\boldsymbol{s}\|_{L^2(M)}^2
$$

and hence

$$
\nu^{1/2} \left\| \nabla \mathbf{z}^h \right\|_{L^2(\Omega)} \le C_1 S^{1/2}, \tag{6.131}
$$

with $C_1 = \delta_1^{1/2} C_{\mathcal{M}} \bar{C}_{\text{inv}} \beta_{LP}^{-1}$. Finally, using the Cauchy–Schwarz inequality, (6.121), (6.122), (6.130), (6.110), (6.45), (6.120), (6.116), and (6.117), the last term in (6.128) can be estimated by

$$
\begin{aligned}
\left| \sum_{E \in \mathcal{E}^h} \left(\mathbf{z}^h \cdot \mathbf{n}_E, \left[\left| q^h \right|\right]_E \right)_E \right| &\le \sum_{\substack{E \in \mathcal{E}^h,\, M \in \mathcal{M}^h, \\ E \subset \overline{M}}} \|\mathbf{z}_M\|_{L^2(E)} \left\| \left[\left| q^h \right|\right]_E \right\|_{L^2(E)} \\
&\le C_e \left(1 + \bar{C}_{\text{inv}}\right) \beta_{LP}^{-1} \sum_{\substack{E \in \mathcal{E}^h,\, M \in \mathcal{M}^h, \\ E \subset \overline{M}}} h_M^{-1/2} \delta_M \|\mathbf{s}\|_{L^2(M)} \left\| \left[\left| q^h \right|\right]_E \right\|_{L^2(E)} \\
&\le C_{\mathcal{E}}\, C_e \left(1 + \bar{C}_{\text{inv}}\right) \beta_{LP}^{-1} \left(C'_{\mathcal{E}} \frac{\delta_1}{\gamma_0} \right)^{1/2} S^{1/2} \left(\sum_{E \in \mathcal{E}^h} \gamma_E \left\| \left[\left| q^h \right|\right]_E \right\|_{L^2(E)}^2 \right)^{1/2} .
\end{aligned}
$$

Thus, combining the above inequalities and applying the Young inequality, it follows that

$$
A_{\text{lps}}((\mathbf{v}^h, q^h), (\mathbf{z}^h, 0)) \ge \frac{1}{4} S - C_2 \left| (\mathbf{v}^h, q^h) \right|_{\text{lps}}^2 ,
$$

where C_2 depends only on $C_{\mathcal{M}}$, $C_{\mathcal{E}}$, $C'_{\mathcal{E}}$, C_e, $\bar{C}_{\text{inv}}$, δ_1, γ_0, and β_{LP}. Setting

$$
\mathbf{w}^h = 4\,\mathbf{z}^h + (1 + 4\,C_2)\,\mathbf{v}^h , \qquad r^h = (1 + 4\,C_2)\,q^h
$$

and using (6.123), one obtains

$$
A_{\text{lps}}((\mathbf{v}^h, q^h), (\mathbf{w}^h, r^h)) \ge S + \left| (\mathbf{v}^h, q^h) \right|_{\text{lps}}^2 \ge \left\| (\mathbf{v}^h, q^h) \right\|_{\text{lps}}^2 .
$$

From (6.131), it follows that

$$
\begin{aligned}
\left\| (\mathbf{w}^h, r^h) \right\|_{\text{lps}} &\le 4\,\nu^{1/2} \left\| \nabla \mathbf{z}^h \right\|_{L^2(\Omega)} + (1 + 4\,C_2) \left\| (v^h, q^h) \right\|_{\text{lps}} \\
&\le (1 + 4\,C_1 + 4\,C_2) \left\| (\mathbf{v}^h, q^h) \right\|_{\text{lps}} ,
\end{aligned}
$$

which proves the theorem. □

We now move on to error estimates. First, let us investigate the consistency of the method.

Lemma 6.5.3 (Consistency Error) *Let the solution of* (6.5) *satisfy* $(\mathbf{u}, p) \in H_0^1(\Omega)^d \times H^1(\Omega)$ *and let* $(\mathbf{u}^h, p^h) \in V^h \times Q^h$ *be the solution of the LPS method* (6.109). *The LPS method is not consistent and it holds*

$$A_{\mathrm{lps}}\left(\left(\mathbf{u}-\mathbf{u}^h, p-p^h\right), \left(\mathbf{v}^h, q^h\right)\right) = \sum_{M\in\mathcal{M}^h} \delta_M(\kappa_M(\nabla p), \kappa_M(\nabla^h q^h))_M \tag{6.132}$$

for all $\left(\mathbf{v}^h, q^h\right) \in V^h \times Q^h$.

Proof The lemma is a simple consequence of (6.5) and (6.109). □

The term on the right-hand side of (6.132) represents the consistency error and is estimated in the following lemma.

Lemma 6.5.4 (Estimate of the Consistency Error) *Let* $\{\delta_M\}$ *satisfy* (6.110) *and let* $p \in H^{m+1}(\Omega)$ *with* $0 \le m \le k$. *Then, for any* $q^h \in Q^h$, *one has*

$$\sum_{M\in\mathcal{M}^h} \delta_M(\kappa_M(\nabla p), \kappa_M(\nabla^h q^h))_M \le C\nu^{-1/2}h^{m+1}\left\|p\right\|_{H^{m+1}(\Omega)} \left\|\left(\mathbf{0}, q^h\right)\right\|_{\mathrm{lps}}.$$

Proof Applying the Cauchy–Schwarz inequality, (6.110), (6.118), and (6.112), one obtains for any $q^h \in Q^h$

$$\begin{aligned}
&\sum_{M\in\mathcal{M}^h} \delta_M\left(\kappa_M(\nabla p), \kappa_M(\nabla^h q)\right)_M \\
&\le \sum_{M\in\mathcal{M}^h} \delta_M \left\|\kappa_M(\nabla p - j_M\nabla p)\right\|_{L^2(M)} \left\|\kappa_M(\nabla^h q^h)\right\|_{L^2(M)} \\
&\le \frac{C_\kappa^2\delta_1^{1/2}}{\nu^{1/2}} \left(\sum_{M\in\mathcal{M}^h} h_M^2 \left\|\nabla p - j_M\nabla p\right\|_{L^2(M)}^2\right)^{1/2} \left(\sum_{M\in\mathcal{M}^h} \delta_M \left\|\nabla^h q^h\right\|_{L^2(M)}^2\right)^{1/2} \\
&\le C\nu^{-1/2} \left(\sum_{M\in\mathcal{M}^h} h_M^{2m+2} \left\|\nabla p\right\|_{H^m(M)}^2\right)^{1/2} \left(\sum_{M\in\mathcal{M}^h} \delta_M \left\|\nabla^h q^h\right\|_{L^2(M)}^2\right)^{1/2}
\end{aligned}$$

and the lemma follows using (6.114) and (6.115). □

Theorem 6.5.5 (Error Estimate) *Let the solution of* (6.5) *satisfy* $(\mathbf{u}, p) \in H^{k+1}(\Omega)^d \times H^{l+1}(\Omega)$ *and let* $\left(\mathbf{u}^h, p^h\right) \in V^h \times Q^h$ *be the solution of the LPS problem* (6.109). *Assume that the stabilization parameters satisfy* (6.110) *and* (6.45) *and that the inf-sup conditions* (6.111) *hold. Then the following error estimate holds:*

$$\left\|\left(\mathbf{u}-\mathbf{u}^h, p-p^h\right)\right\|_{\mathrm{lps}} \le C\left(\nu^{1/2}h^k\left\|\mathbf{u}\right\|_{H^{k+1}(\Omega)} + \frac{h^{\min\{k,l\}+1}}{\nu^{1/2}}\left\|p\right\|_{H^{l+1}(\Omega)}\right).$$

Proof Let I^h and J^h be the interpolation operators satisfying (6.8) and (6.9). From the proof of Lemma 6.5.2, it is known that there is a pair $\left(\mathbf{v}^h, q^h\right) \in V^h \times Q^h$ such that

$$\left\|\left(\mathbf{u}^h - I^h\mathbf{u}, p^h - J^h p\right)\right\|_{\mathrm{lps}} \le C \frac{A_{\mathrm{lps}}\left(\left(\mathbf{u}^h - I^h\mathbf{u}, p^h - J^h p\right), \left(\mathbf{v}^h, q^h\right)\right)}{\left\|\left(\mathbf{v}^h, q^h\right)\right\|_{\mathrm{lps}}}.$$

With Lemmas 6.5.3 and 6.5.4, one obtains

$$\begin{aligned}\left\|\left(\mathbf{u}^h - I^h\mathbf{u}, p^h - J^h p\right)\right\|_{\mathrm{lps}} &\le C \frac{A_{\mathrm{lps}}\left(\left(\mathbf{u} - I^h\mathbf{u}, p - J^h p\right), \left(\mathbf{v}^h, q^h\right)\right)}{\left\|\left(\mathbf{v}^h, q^h\right)\right\|_{\mathrm{lps}}} \\ &\quad + C\nu^{-1/2} h^{\min\{k,l\}+1} \|p\|_{H^{\min\{k,l\}+1}(\Omega)}.\end{aligned}$$

Applying the Cauchy–Schwarz inequality, one gets

$$\begin{aligned}A_{\mathrm{lps}}\left(\left(\mathbf{u} - I^h\mathbf{u}, p - J^h p\right), \left(\mathbf{v}^h, q^h\right)\right) &\le \left|(\mathbf{u} - I^h\mathbf{u}, p - J^h p)\right|_{\mathrm{lps}} \left|(\mathbf{v}^h, q^h)\right|_{\mathrm{lps}} \\ &\quad - (\nabla\cdot\mathbf{v}^h, p - J^h p) + (\nabla\cdot(\mathbf{u} - I^h\mathbf{u}), q^h)\end{aligned}$$

and

$$-(\nabla\cdot\mathbf{v}^h, p - J^h p) \le \nu^{-1/2} \left\|p - J^h p\right\|_{L^2(\Omega)} \left\|(\mathbf{v}^h, q^h)\right\|_{\mathrm{lps}}.$$

Using (6.127), the Cauchy–Schwarz inequality, (6.110), (6.45), and (6.8), one derives

$$\begin{aligned}(\nabla\cdot(\mathbf{u} - I^h\mathbf{u}), q^h) &= -(\mathbf{u} - I^h\mathbf{u}, \nabla^h q^h) + \sum_{E\in\mathcal{E}^h}\left((\mathbf{u} - I^h\mathbf{u})\cdot\mathbf{n}_E, \left[\left|q^h\right|\right]_E\right)_E \\ &\le \frac{\nu^{1/2}}{\delta_0^{1/2}}\left(\sum_{K\in\mathcal{T}^h} h_K^{-2}\left\|\mathbf{u} - I^h\mathbf{u}\right\|_{L^2(K)}^2\right)^{1/2}\left(\sum_{M\in\mathcal{M}^h}\delta_M\left\|\nabla^h q^h\right\|_{L^2(M)}^2\right)^{1/2} \\ &\quad + \frac{\nu^{1/2}}{\gamma_0^{1/2}}\left(\sum_{E\in\mathcal{E}^h} h_E^{-1}\left\|\mathbf{u} - I^h\mathbf{u}\right\|_{L^2(E)}^2\right)^{1/2}\left(\sum_{E\in\mathcal{E}^h}\gamma_E\left\|\left[\left|q^h\right|\right]_E\right\|_{L^2(E)}^2\right)^{1/2} \\ &\le C\nu^{1/2} h^k \|\mathbf{u}\|_{H^{k+1}(\Omega)}\left\|(\mathbf{v}^h, q^h)\right\|_{\mathrm{lps}}.\end{aligned}$$

Combining the above inequalities and using the triangle inequality, (6.9), (6.118) and Lemma 6.5.1, one obtains

$$\begin{aligned}\left\|\left(\mathbf{u} - \mathbf{u}^h, p - p^h\right)\right\|_{\mathrm{lps}} &\le C\left\|\left(\mathbf{u} - I^h\mathbf{u}, p - J^h p\right)\right\|_{\mathrm{pspg}} \\ &\quad + C\nu^{1/2} h^k \|\mathbf{u}\|_{H^{k+1}(\Omega)} + C\nu^{-1/2} h^{\min\{k,l\}+1}\|p\|_{H^{l+1}(\Omega)}\end{aligned}$$

and the statement of the theorem follows from (6.48). □

Theorem 6.5.6 (**L^2 Estimate of the Pressure Error**) *Assume that the solution of* (6.5) *satisfies* $(\mathbf{u}, p) \in H^{k+1}(\Omega)^d \times H^{l+1}(\Omega)$, *that the stabilization parameters satisfy* (6.110) *and* (6.45) *and that the inf-sup conditions* (6.111) *hold. Then there holds the error estimate*

$$\left\| p - p^h \right\|_{L^2(\Omega)} \le C \left(\nu h^k \left\| \mathbf{u} \right\|_{H^{k+1}(\Omega)} + h^{\min\{k,l\}+1} \left\| p \right\|_{H^{l+1}(\Omega)} \right).$$

Proof Using $\{\delta_K\}$ defined in (6.126), the proof of Theorem 6.4.8 can be repeated without any changes. Then the statement of the present theorem follows from Lemma 6.5.1 and Theorem 6.5.5. □

Theorem 6.5.7 (**L^2 Estimate of the Velocity Error**) *Let the stabilization parameters satisfy* (6.110) *and* (6.45), *let the inf-sup conditions* (6.111) *hold, and let the Stokes problem* (6.2) *be regular. Assume that the solution of* (6.5) *satisfies* $(\mathbf{u}, p) \in H^{k+1}(\Omega)^d \times H^{l+1}(\Omega)$, *then there holds the error estimate*

$$\left\| \mathbf{u} - \mathbf{u}^h \right\|_{L^2(\Omega)} \le C \left(h^{k+1} \left\| \mathbf{u} \right\|_{H^{k+1}(\Omega)} + \frac{h^{\min\{k,l\}+2}}{\nu} \left\| p \right\|_{H^{l+1}(\Omega)} \right).$$

Proof Up to (6.60), the proof of Theorem 6.4.9 remains valid also in this case. Then, instead of (6.61), one obtains from (6.132)

$$\begin{aligned}\left(\nabla \cdot (\mathbf{u} - \mathbf{u}^h), r^I\right) &+ \sum_{E \in \mathcal{E}^h} \gamma_E \left(\left[\left| p - p^h \right|\right]_E, \left[\left| r^I \right|\right]_E \right)_E \\ &+ \sum_{M \in \mathcal{M}^h} \delta_M (\kappa_M(\nabla^h (p - p^h)), \kappa_M(\nabla^h r^I))_M \\ &\qquad = \sum_{M \in \mathcal{M}^h} \delta_M (\kappa_M(\nabla p), \kappa_M(\nabla^h r^I))_M .\end{aligned}$$

Thus, instead of (6.62), one obtains the following expression for the last two terms in (6.27)

$$\begin{aligned}&\nu(\nabla \mathbf{z}^I, \nabla(\mathbf{u} - \mathbf{u}^h)) - (\nabla \cdot (\mathbf{u} - \mathbf{u}^h), r^I) \\ &= \left(\nabla \cdot (\mathbf{z}^I - \mathbf{z}), p - p^h\right) + \sum_{E \in \mathcal{E}^h} \gamma_E \left(\left[\left| p - p^h \right|\right]_E, \left[\left| r^I - r \right|\right]_E \right)_E \\ &\quad + \sum_{M \in \mathcal{M}^h} \delta_M (\kappa_M(\nabla^h (p - p^h)), \kappa_M(\nabla^h r^I))_M \\ &\quad - \sum_{M \in \mathcal{M}^h} \delta_M (\kappa_M(\nabla p), \kappa_M(\nabla^h r^I))_M .\end{aligned}$$

Analogously as in (6.63), but using also Lemma 6.5.4, (6.118), (6.125), and (6.124), one derives

$$\begin{aligned}
&\nu(\nabla \mathbf{z}^I, \nabla(\mathbf{u}-\mathbf{u}^h)) - (\nabla\cdot(\mathbf{u}-\mathbf{u}^h), r^I)\\
&\le Ch\left(\left\|p-p^h\right\|_{L^2(\Omega)} + \nu^{1/2}\left\|\left(\mathbf{0}, p-p^h\right)\right\|_{\mathrm{lps}}\right.\\
&\qquad\left. + h^{\min\{k,l\}+1}\left\|p\right\|_{H^{l+1}(\Omega)}\right)\left\|\mathbf{u}-\mathbf{u}^h\right\|_{L^2(\Omega)}.
\end{aligned}$$

Combining this estimate with (6.27) and (6.28), the theorem follows using Theorems 6.5.5 and 6.5.6. □

The LPS method for the Stokes problem was introduced in [13]. A generalization and unified analysis was presented in [72] where the stability with respect to a norm containing the $L^2(\Omega)$ norm of the pressure was established. The techniques presented here are a special case of the analysis published in [64]. As one can see, the LPS method leads to analogous stability and convergence results as residual-based approaches. However, in comparison with residual-based stabilizations, an important advantage of LPS methods is that they do not create additional couplings between various unknowns. A drawback is that the local projections couple pressure degrees of freedom that do not belong to the same mesh cell. Hence, the sparsity pattern of the pressure-pressure matrix C in (6.4) is denser as, e.g., for residual-based discretizations.

Remark 6.5.8 (LPS Method with Scott–Zhang-Type Projector) An LPS method that uses a particular Scott–Zhang-type projector, which is well defined for $L^1(\Omega)$ functions, is proposed in [7]. Like the LPS method with overlapping macroelements, it neither requires nested meshes nor an enrichment of spaces by bubble functions. However, similarly as for the other versions of the LPS method, the projector leads to a wider sparsity pattern of the pressure-pressure matrix. A finite element error analysis of this method and few numerical comparisons with the symmetric GLS method from Sect. 6.4.3 for P_1/P_1 finite elements, which is in this case equivalent to the PSPG method, are presented in [7]. The method is absolutely stable. There are no assumptions on upper bounds of the stabilization parameter in the analysis and the numerical studies show even a slight improvement of the accuracy for large stabilization parameters. △

A stabilizing term of the form

$$\sum_{K\in\mathcal{T}^h} \frac{\alpha}{\nu} h_K^2 \left(\left(I-P^h\right)\left(\nabla p^h\right), \left(I-P^h\right)\left(\nabla q^h\right)\right)_K$$

is proposed in [31] for P_k/P_k finite elements, where P^h is some stable approximation operator from $L^2(\Omega)^d$ into the space of continuous piecewise polynomial functions of degree $k-1$. This operator was chosen in the numerical studies from [31] as an extension of a nodal interpolation operator. The arising method is called term-by-term stabilized method. The differences to already existing methods are discussed in detail in [31]. Depending on the actual choice of P^h, it can be considered as an LPS method that is defined on a single mesh and with standard finite element spaces. If P^h is chosen to be a global $L^2(\Omega)$ projection, then the image of the projection operator is different than for the method from [34]. In [31], a finite element convergence analysis is presented that proves optimal orders for the $L^2(\Omega)$ norm of the velocity gradient and of the pressure.

A two-level LPS method was studied in [76]. Using this method, the pressure gradient from the LPS stabilization term can be locally eliminated, which facilitates the implementation of this LPS method.

In [11], the so-called residual local projection (RELP) method is proposed for low order pairs of finite element spaces. It contains an LPS term for the pressure. An additional pressure-pressure coupling is introduced by jump terms of the stress tensor across faces of the mesh cells. Special cases of the RELP method coincide with methods from [38] and [3]. The finite element error analysis presented in [11] shows optimal convergence for the $L^2(\Omega)$ norms of the velocity gradient and of the pressure. A similar method, where the jumps of the stress tensor are replaced by jumps of the pressure, is proposed and analyzed in [12]. The methods from [11, 12] do not need multiple levels or extra degrees of freedom for computing the local projection and all computations can be performed on the mesh cell level. However, the stencil of some matrix blocks gets enlarged due to the jump terms.

6.5.5 *Stabilization with Fluctuations of the Pressure*

A pressure-stabilized method that uses fluctuations of the pressure itself, instead of the gradient of the pressure as the methods discussed in Sects. 6.5.3 and 6.5.4, was proposed in [38].

Let $Q^h = P_k$ or $Q^h = Q_k$, $k \ge 1$, then the method utilizes the $L^2(\Omega)$ projection $P_{L^2}^{k-1} : Q^h \to P_{k-1}^{\text{disc}}$ onto the discontinuous piecewise polynomial space of degree $k-1$. Since the image space consists of discontinuous finite element functions, the projection operator $P_{L^2}^{k-1}$ can be computed locally, i.e., mesh cell by mesh cell. The discrete continuity equation of the method proposed in [38] reads as follows

$$-\left(\nabla \cdot \mathbf{u}^h, q^h\right) - \frac{1}{\nu}\left(p^h - P_{L^2}^{k-1} p^h, q^h - P_{L^2}^{k-1} q^h\right) = 0 \quad \forall\, q^h \in Q^h. \tag{6.133}$$

There is no user-chosen parameter in (6.133).

A finite element analysis of this method for the equal order pairs P_1/P_1 and Q_1/Q_1 of lowest order is performed in [20]. The inverse of the viscosity does not appear in the stabilization term in contrast to (6.133). An extension of the method to the pairs P_1/P_0 and Q_1/Q_0 is also proposed. The analysis shows that in all cases the method is unconditionally stable and optimal error bounds were derived, e.g., linear convergence for $\left\|\nabla\left(\mathbf{u} - \mathbf{u}^h\right)\right\|_{L^2(\Omega)}$ and $\left\|p - p^h\right\|_{L^2(\Omega)}$.

A similar method, which uses projections in a pressure space defined on a coarser grid, was developed in [65]. The derivation of this method used ideas from the variational multiscale framework.

The stabilization term proposed in [38] can be written in the form

$$c^h\left(p^h, q^h\right) = \frac{\alpha}{\nu}\underline{p}^T\left(\tilde{M} - M\right)\underline{q}, \tag{6.134}$$

where $\alpha \in \mathbb{R}$ ($\alpha = 1$ in [38]), $\underline{p}$ and $\underline{q}$ are the vector representations of p^h and q^h with respect to the standard basis of Q^h, M is the mass matrix with respect to this basis, and $\tilde{M}$ is the mass matrix from the functions arising in the $L^2(\Omega)$ projection.

Subsequently, further methods with stabilization terms of type (6.134) were proposed in [15] and [67]. The method of [15] uses as $\tilde{M}$ an under-integrated mass matrix. Concrete examples for the bilinear form from (6.134) are given for P_1 finite elements in two dimensions, where

$$c^h\left(p^h, q^h\right) = \frac{\alpha}{\nu}\int_\Omega\left(I_1^h(p^h q^h) - p^h q^h\right)\mathrm{d}\boldsymbol{x}$$

and for P_2 finite elements in 2d where

$$c^h\left(p^h, q^h\right) = \frac{\alpha}{\nu}\int_\Omega\left(I_3^h(p^h q^h) - p^h q^h\right)\mathrm{d}\boldsymbol{x}.$$

Here, I_k^h, $k \geq 1$, is the Lagrangian interpolation operator onto the space of continuous piecewise polynomial functions of degree k. Optimal estimates for the $L^2(\Omega)$ errors of the velocity gradient and the pressure are derived in [15].

The method from [67] uses two local Gauss integrations to define the matrices, where $\tilde{M}$ is defined by a first order Gaussian integration in each direction. This method is proposed in [67] for P_1/P_1 and Q_1/Q_1 finite elements in two dimensions. It is already observed in [15] that for these cases the method from [67] is equivalent to the method already proposed in [38]. However, the methods from [15] and [38] are not equivalent.

6.5.6 *Continuous Interior Penalty Methods*

Continuous Interior Penalty (CIP) methods use jumps of the pressure gradient or the normal derivative of the pressure across faces of mesh cells for stabilizing the inf-sup condition. The first method of this class was proposed in [28, 29]. However, the use of jumps across faces of the mesh cells for pressure stabilization dates back to a method proposed in [82]. For the Q_1/Q_0 pair of finite element spaces, this method uses jumps of the pressure itself.

In [29], the stabilization term, which defines the matrix $-C$ in (6.4), has the form

$$\frac{1}{2}\sum_{K\in\mathcal{T}^h}\left(\delta_0 h_K^{s+1}\sum_{E\subset\partial K}\left(\left[\left|\nabla p^h\cdot\mathbf{n}_E\right|\right]_E,\left[\left|\nabla q^h\cdot\mathbf{n}_E\right|\right]_E\right)_E\right), \qquad (6.135)$$

where

$$s=\begin{cases}2 & \text{if } \nu\ge h,\\ 1 & \text{if } \nu<h.\end{cases}$$

Additionally, a jump term containing the divergence of $\mathbf{u}^h$ is included in the method studied in [29]. A finite element analysis for the P_1/P_1 pair of spaces was presented. Assuming that $p\in H^2(\Omega)$, the estimate

$$\begin{aligned}&\left\|\nabla\left(\mathbf{u}-\mathbf{u}^h\right)\right\|_{L^2(\Omega)}+\frac{1}{\nu}\left\|p-p^h\right\|_{L^2(\Omega)}\\ &\le Ch\Bigg[\max\left\{\frac{1+\delta_0}{\nu^{1/2}},1\right\}\max\left\{h^{s/2},h^{(2-s)/2}\right\}\|\mathbf{u}\|_{H^2(\Omega)}\\ &\quad+\max\left\{\frac{1}{\nu^{1/2}},\frac{1}{\nu},\frac{\delta_0^{1/2}}{\nu^{1/2}}\right\}\max\left\{h^{s/2},h^{(2-s)/2}\right\}\|p\|_{H^2(\Omega)}\Bigg]\end{aligned}$$

was proved. It follows that in the case $\nu<h$, the error reduction of $\left\|p-p^h\right\|_{L^2(\Omega)}$ is of order 1.5. Also the case that only $p\in H^1(\Omega)$ holds was studied in [29].

The stabilization term of the method from [28] uses the jumps of the pressure gradient instead of the normal derivative.

Using classical CIP stabilizations, e.g., (6.135), connects pressure degrees of freedom that do not belong to a common mesh cell. Hence, the matrix stencil of C is denser than, e.g., for residual-based stabilizations.

A so-called local CIP method was introduced and analyzed in [30]. The advantage of this method is that it allows static condensation. As a result, the matrix stencil of the matrix C is substantially smaller than for the classical CIP methods. The local CIP method uses a so-called macro-mesh $\mathcal{M}^h$, where each mesh cell $M\in\mathcal{M}^h$ consists of a small number of simplicial cells $K\in\mathcal{T}^h$. Then, the stabilization term has the form

$$\sum_{M\in\mathcal{M}^h}\left[\sum_{K\subset M}\delta_K h_K\left(\sum_{E\subset\partial K,E\subset\mathrm{int}(M)}\left(\left[\left|\nabla p^h\right|\right]_E,\left[\left|\nabla q^h\right|\right]_E\right)_E\right)\right],$$

where $\mathrm{int}(M)$ is the interior of M and

$$\delta_K=\min\left\{\frac{h_K^2}{\nu},h_K\right\}.$$

As a particular case of the error analysis presented in [30], one obtains the estimates for P_k/P_k finite elements, $k \geq 1$,

$$\begin{aligned}&\left\|\nabla\left(\mathbf{u}-\mathbf{u}^h\right)\right\|_{L^2(\Omega)}\\ &\leq Ch^k\left(\left(1+\frac{h}{\nu}\right)^{1/2}|\mathbf{u}|_{H^{k+1}(\Omega)}+\min\left\{\frac{h^{1/2}}{\nu^{1/2}},\frac{h}{\nu}\right\}|p|_{H^{k+1}(\Omega)}\right)\end{aligned}$$

and

$$\begin{aligned}&\left\|p-p^h\right\|_{L^2(\Omega)}\\ &\leq Ch^k\left((\nu+h)^{1/2}|\mathbf{u}|_{H^{k+1}(\Omega)}+\left(h+\min\left\{h^{1/2},\frac{h}{\nu^{1/2}}\right\}\right)|p|_{H^{k+1}(\Omega)}\right).\end{aligned}$$

The error reduction for the pressure is of order $k+0.5$ as long as $\nu < h$. This higher order, even $k+1$ for $\nu \ll h$, was observed in the numerical studies of [30].

6.6 Connections to Inf-Sup Stable Methods with Bubble Functions

If a pair of finite element spaces for approximating the velocity and pressure does not satisfy the discrete inf-sup condition (6.3), one can construct a stable pair of spaces by adding suitable functions to the velocity space. The velocity space V^h then has the form $V^h = V_1^h \oplus V_2^h$, where V_1^h typically assures the approximation properties of the space V^h and V_2^h guarantees the fulfillment of the inf-sup condition (6.3) for the given pressure space Q^h. In this section we shall consider only spaces of this type.

The functions contained in the space V_2^h are often called bubble functions. It was realized very soon [78] that there is a close relationship between stabilized methods and Galerkin methods with bubble functions. Namely, if one drops the bubble part of the solution of a Galerkin method with bubble functions, then one sometimes gets functions which represent a solution of a stabilized method. For linear problems, this relationship was also established in an abstract framework via virtual bubbles in [10]. There are many more papers devoted to investigations of the mentioned relationship, see, e.g., [63] for references.

In this section we go a step further and consider also modifications of the conforming discretization for the spaces $V^h = V_1^h \oplus V_2^h$, Q^h to obtain equivalent representations for a wider class of stabilized methods based on the spaces V_1^h, Q^h. Such equivalences are helpful for a better understanding of the properties of stabilized methods and for their theoretical investigations. Moreover, the technique of modified discretizations can be used for designing new stabilized methods.

The theory available for the modified discretizations then automatically provides existence and convergence statements for the corresponding stabilized methods.

There are many examples of finite element spaces of the mentioned type. The simplest choice for the spaces V_1^h and Q^h are piecewise constant functions for Q^h and continuous piecewise (bi-, tri-)linear functions for V_1^h. To satisfy the inf-sup condition, it suffices to use a space V_2^h consisting of one vector-valued edge/face-bubble function per each inner edge/face, see [17, 41]. In the triangular/tetrahedral case, spaces Q^h, V_1^h consisting of continuous piecewise linear functions may be stabilized using V_2^h consisting of d vector-valued element bubble functions per each element. This pair of spaces is known as the MINI element, cf. [5]. In two dimensions, the same space V_2^h can be used if V_1^h consists of continuous piecewise quadratic functions and Q^h of discontinuous piecewise linear functions, cf. [36]. A generalization of [5] to the quadrilateral case is described in [74]. Further examples of spaces V_1^h, V_2^h and Q^h can be found, e.g., in [49].

Since $V^h = V_1^h \oplus V_2^h$ (which implies that $V_1^h \cap V_2^h = \{\mathbf{0}\}$), any function $\mathbf{v}^h \in V^h$ can be written in the form $\mathbf{v}^h = \mathbf{v}_1^h + \mathbf{v}_2^h$ where the functions $\mathbf{v}_1^h \in V_1^h$ and $\mathbf{v}_2^h \in V_2^h$ are uniquely determined. When there will be no danger of ambiguity, we shall also use the notations $\mathbf{v}_1^h$ and $\mathbf{v}_2^h$ for arbitrary functions belonging to V_1^h and V_2^h, respectively. The conforming discretization (6.13) can be equivalently written in the form: Find $\mathbf{u}_1^h \in V_1^h$, $\mathbf{u}_2^h \in V_2^h$, and $p^h \in Q^h$ such that

$$\nu(\nabla \mathbf{u}_1^h, \nabla \mathbf{v}_1^h) + \nu(\nabla \mathbf{u}_2^h, \nabla \mathbf{v}_1^h) - (\nabla \cdot \mathbf{v}_1^h, p^h) = (\mathbf{f}, \mathbf{v}_1^h) \quad \forall\, \mathbf{v}_1^h \in V_1^h, \tag{6.136}$$

$$\nu(\nabla \mathbf{u}_1^h, \nabla \mathbf{v}_2^h) + \nu(\nabla \mathbf{u}_2^h, \nabla \mathbf{v}_2^h) - (\nabla \cdot \mathbf{v}_2^h, p^h) = (\mathbf{f}, \mathbf{v}_2^h) \quad \forall\, \mathbf{v}_2^h \in V_2^h, \tag{6.137}$$

$$-(\nabla \cdot \mathbf{u}_1^h, q^h) - (\nabla \cdot \mathbf{u}_2^h, q^h) = 0 \quad \forall\, q^h \in Q^h. \tag{6.138}$$

It is assumed that the approximation properties of the space V^h are determined by the space V_1^h and hence the interpolation operator I^h may be assumed to map $V \cap H^{k+1}(\Omega)^d$ into V_1^h. Then it turns out (cf. Lemma 6.6.4 below) that the component $\mathbf{u}_1^h$ of $\mathbf{u}^h$ has the same asymptotic approximation properties as $\mathbf{u}^h$. Therefore, it makes sense to consider $\mathbf{u}_1^h$ as an approximation of the velocity $\mathbf{u}$ whereas $\mathbf{u}_2^h$ serves as a stabilization tool only. Note that one can compute $\mathbf{u}_2^h$ from (6.137) as a function of $\mathbf{u}_1^h$ and p^h. Substituting this $\mathbf{u}_2^h$ into (6.136) and (6.138), one obtains a discrete problem for $\mathbf{u}_1^h$ and p^h where the terms $\nu(\nabla \mathbf{u}_2^h, \nabla \mathbf{v}_1^h)$ and $(\nabla \cdot \mathbf{u}_2^h, q^h)$ give rise to stabilization terms.

Let us demonstrate the procedure just described for the MINI element proposed in [5]. In this case the spaces V_1^h and Q^h consist of continuous piecewise linear functions with respect to a simplicial triangulation $\mathcal{T}^h$. Furthermore,

$$V_2^h = \left[\operatorname{span}\{\varphi_K\}_{K \in \mathcal{T}^h}\right]^d, \tag{6.139}$$

where φ_K are scalar element bubble functions defined on K as the product of the barycentric coordinates on K and vanishing outside of K. Thus, $\varphi_K|_K \in$

$P_{d+1}(K) \cap H_0^1(K)$. The proof of the inf-sup stability relies on the construction of a Fortin operator, see [5] or [56, Section 3.6.1] for details. The component $\mathbf{u}_2^h$ of $\mathbf{u}^h$ can be expressed in the form

$$\mathbf{u}_2^h = \sum_{K \in \mathcal{T}^h} \mathbf{u}_K \, \varphi_K$$

with uniquely determined numbers $\mathbf{u}_K \in \mathbb{R}^d$. To eliminate $\mathbf{u}_2^h$ from (6.136)–(6.138), one can employ that

$$(\nabla \mathbf{u}_2^h, \nabla \mathbf{v}_1^h) = (\nabla \mathbf{u}_1^h, \nabla \mathbf{v}_2^h) = 0 \qquad \forall \, \mathbf{v}_1^h \in V_1^h, \, \mathbf{v}_2^h \in V_2^h \, . \tag{6.140}$$

Indeed, since the bubble functions vanish on ∂K and the Laplacian of a linear function vanishes, too, one finds by integration by parts

$$(\nabla \mathbf{u}_1^h, \nabla \mathbf{v}_2^h)_K = ((\mathbf{n}_{\partial K} \cdot \nabla)\mathbf{u}_1^h, \mathbf{v}_2^h)_{\partial K} - (\Delta \mathbf{u}_1^h, \mathbf{v}_2^h)_K = 0$$

for any $K \in \mathcal{T}^h$. Similarly, employing that the gradient of a linear function is constant, one gets

$$-(\nabla \cdot \mathbf{v}_2^h, p^h)_K = (\mathbf{v}_2^h, \nabla p^h)_K = \nabla p^h|_K \cdot \int_K \mathbf{v}_2^h \, \mathrm{d}\boldsymbol{x} \, . \tag{6.141}$$

Setting $\mathbf{v}_2^h = \boldsymbol{e}^i \, \varphi_K$, $i = 1, \ldots, d$, where $\boldsymbol{e}^i$ is the unit vector in the direction of the ith coordinate axis, and applying (6.140) and (6.141), one obtains from (6.137)

$$u_{K,i} \, \nu \, \|\nabla \varphi_K\|_{L^2(K)}^2 + \partial_{x_i} p^h|_K \int_K \varphi_K \, \mathrm{d}\boldsymbol{x} = (f_i, \varphi_K) = \bar{f}_i^h|_K \int_K \varphi_K \, \mathrm{d}\boldsymbol{x} \, ,$$

where $\bar{f}_i^h$ are components of the piecewise constant function $\bar{\mathbf{f}}^h$ defined by averaging of $\mathbf{f}$ with the weights φ_K, i.e.,

$$\bar{\mathbf{f}}^h|_K := \frac{\int_K \mathbf{f} \, \varphi_K \, \mathrm{d}\boldsymbol{x}}{\int_K \varphi_K \, \mathrm{d}\boldsymbol{x}} \, , \qquad K \in \mathcal{T}^h \, .$$

Thus, in view of (6.141), the second term in (6.138) becomes

$$\begin{aligned}
-(\nabla \cdot \mathbf{u}_2^h, q^h) &= \sum_{K \in \mathcal{T}^h} \nabla q^h|_K \cdot \mathbf{u}_K \int_K \varphi_K \, \mathrm{d}\boldsymbol{x} \\
&= \sum_{K \in \mathcal{T}^h} \nabla q^h|_K \cdot (\bar{\mathbf{f}}^h - \nabla p^h)|_K \, \frac{\left(\int_K \varphi_K \, \mathrm{d}\boldsymbol{x}\right)^2}{\nu \, \|\nabla \varphi_K\|_{L^2(K)}^2} \\
&= \sum_{K \in \mathcal{T}^h} (\bar{\mathbf{f}}^h - \nabla p^h, \delta_K \nabla q^h)_K \, ,
\end{aligned}$$

where

$$\delta_K = \frac{\left(\int_K \varphi_K \, \mathrm{d}\boldsymbol{x}\right)^2}{\nu \, \|\nabla \varphi_K\|^2_{L^2(K)} \, |K|} .$$

Therefore, inserting (6.140) in (6.136), one ends up with the following problem for the linear part of the approximate solution: Find $\mathbf{u}_1^h \in V_1^h$ and $p^h \in Q^h$ such that

$$\begin{aligned} \nu(\nabla \mathbf{u}_1^h, \nabla \mathbf{v}_1^h) - (\nabla \cdot \mathbf{v}_1^h, p^h) &= (\mathbf{f}, \mathbf{v}_1^h) & \forall\, \mathbf{v}_1^h \in V_1^h , \\ (\nabla \cdot \mathbf{u}_1^h, q^h) + \sum_{K \in \mathcal{T}^h} (\nabla p^h, \delta_K \nabla q^h)_K &= \sum_{K \in \mathcal{T}^h} (\bar{\mathbf{f}}^h, \delta_K \nabla q^h)_K & \forall\, q^h \in Q^h . \end{aligned}$$

It is known that $\int_K \varphi_K \, \mathrm{d}\boldsymbol{x} = \mathcal{O}(|K|) = \mathcal{O}(h_K^d)$ and $\|\nabla \varphi_K\|_{L^2(K)} = \mathcal{O}(h_K^{d/2-1})$, e.g., see [1, Lemma 3.2, Theorem 3.3], and hence δ_K satisfies (6.44). Thus, one finds that the MINI element leads for the linear part of the solution to the PSPG method for $V^h/Q^h = P_1/P_1$ (up to the averaging of the right-hand side), see (6.33).

To recover the PSPG method for other finite elements than the MINI element or to obtain other stabilized methods, it would be convenient to drop some of the terms from (6.136) and (6.137) representing a coupling between the spaces V_1^h and V_2^h. Such modifications of the discrete problem (6.136)–(6.138) were studied in [62] with the aim to reduce the size of the stiffness matrix which may be significantly increased by enriching the velocity space V_1^h by the space V_2^h. Surprisingly, it was shown that not all the terms in (6.136)–(6.138) are necessary for the solvability of the discrete problem and for optimal convergence properties of the approximate solutions. One can even proceed in a more general fashion and to multiply the terms $\nu(\nabla \mathbf{u}_2^h, \nabla \mathbf{v}_1^h)$, $\nu(\nabla \mathbf{u}_1^h, \nabla \mathbf{v}_2^h)$, and $\nu(\nabla \mathbf{u}_2^h, \nabla \mathbf{v}_2^h)$ by some real numbers α_1, α_2, and α_3, respectively. In other words, the bilinear form $\nu(\nabla \mathbf{u}^h, \nabla \mathbf{v}^h)$ in (6.13) is replaced by the bilinear form

$$\begin{aligned} a^h(\mathbf{u}^h, \mathbf{v}^h) = \nu(\nabla \mathbf{u}_1^h, \nabla \mathbf{v}_1^h) + \alpha_1 \nu(\nabla \mathbf{u}_2^h, \nabla \mathbf{v}_1^h) + \alpha_2 \nu(\nabla \mathbf{u}_1^h, \nabla \mathbf{v}_2^h) \\ + \alpha_3 \nu(\nabla \mathbf{u}_2^h, \nabla \mathbf{v}_2^h) . \end{aligned} \qquad (6.142)$$

The multiplication by α_3 is considered since numerical experiments suggest that it can reduce the velocity error for small ν. In addition, the right-hand side of (6.13) will be replaced by a functional $\mathbf{f}^h \in H^{-1}(\Omega)^d$. In particular, $\mathbf{f}^h$ defined by

$$\langle \mathbf{f}^h, \mathbf{v}^h \rangle = (\mathbf{f}, \mathbf{v}_1^h) \qquad \forall\, \mathbf{v}^h \in V^h \qquad (6.143)$$

represents replacing the right-hand side of (6.137) by zero. Note that the relation (6.143) defines a functional $\mathbf{f}^h \in [V^h]'$ which can be extended to $\mathbf{f}^h \in H^{-1}(\Omega)^d$ according to the Hahn–Banach theorem.

Thus, the following discretization of the Stokes problem will be considered in the following: Find $(\widetilde{\mathbf{u}}^h, \widetilde{p}^h) \in V^h \times Q^h$ such that

$$a^h(\widetilde{\mathbf{u}}^h, \mathbf{v}^h) - (\nabla \cdot \mathbf{v}^h, \widetilde{p}^h) + (\nabla \cdot \widetilde{\mathbf{u}}^h, q^h) = \langle \mathbf{f}^h, \mathbf{v}^h \rangle \quad \forall \left(\mathbf{v}^h, q^h\right) \in V^h \times Q^h, \tag{6.144}$$

with a^h defined in (6.142).

To analyze the problem (6.144), we shall make additional assumptions on the finite element spaces and the triangulations $\mathcal{T}^h$. We assume that there exists a reference element $\widehat{K}$ such that, for each element $K \in \mathcal{T}^h$, one can introduce a regular one-to-one mapping $F_K : \widehat{K} \to K$ with $F_K(\widehat{K}) = K$. Moreover, it will be assumed that the triangulations $\mathcal{T}^h$ are shape regular in the sense that

$$\|\widehat{\nabla} F_K\|_{L^\infty(\widehat{K})} \le C\, h_K\,, \qquad \|\nabla F_K^{-1}\|_{L^\infty(K)} \le C\, h_K^{-1} \qquad \forall\, K \in \mathcal{T}^h\,.$$

Thus, denoting for any element $K \in \mathcal{T}^h$ and any $v \in L^2(K)$

$$\widehat{v}_K = v \circ F_K\,,$$

one has, for any $K \in \mathcal{T}^h$,

$$C\, h_K^d\, \|\widehat{v}_K\|_{L^2(\widehat{K})}^2 \le \|v\|_{L^2(K)}^2 \le \widetilde{C}\, h_K^d\, \|\widehat{v}_K\|_{L^2(\widehat{K})}^2 \quad \forall\, v \in L^2(K)\,, \tag{6.145}$$

$$C\, h_K^{d-2}\, \|\widehat{\nabla}\widehat{v}_K\|_{L^2(\widehat{K})}^2 \le \|\nabla v\|_{L^2(K)}^2 \le \widetilde{C}\, h_K^{d-2}\, \|\widehat{\nabla}\widehat{v}_K\|_{L^2(\widehat{K})}^2\; \forall\, v \in H^1(K)\,. \tag{6.146}$$

It will be assumed that

$$V_1^h = \{\mathbf{v} \in H_0^1(\Omega)^d \;:\; \mathbf{v} \circ F_K \in \widehat{V}_1 \quad \forall\, K \in \mathcal{T}^h\}\,, \tag{6.147}$$

$$V_2^h \subset \{\mathbf{v} \in H_0^1(\Omega)^d \;:\; \mathbf{v} \circ F_K \in \widehat{V}_2 \quad \forall\, K \in \mathcal{T}^h\}\,, \tag{6.148}$$

where $\widehat{V}_1, \widehat{V}_2 \subset H^1(\widehat{K})^d$ are finite-dimensional spaces satisfying $\widehat{V}_1 \cap \widehat{V}_2 = \{\mathbf{0}\}$. The inclusion in (6.148) is considered to cover the case when the vector bubbles in V_2^h are defined using normal vectors to edges or faces of the triangulation. Then one can prove the following two important results.

Lemma 6.6.1 *The space* $V^h = V_1^h \oplus V_2^h$ *with* V_1^h *and* V_2^h *satisfying* (6.147) *and* (6.148)*, respectively, satisfies*

$$\left\|\nabla \mathbf{v}_1^h\right\|_{L^2(\Omega)} + \left\|\nabla \mathbf{v}_2^h\right\|_{L^2(\Omega)} \le C \left\|\nabla \mathbf{v}^h\right\|_{L^2(\Omega)} \qquad \forall\, \mathbf{v}^h \in V^h\,. \tag{6.149}$$

Proof Since, in a finite-dimensional space, any bounded sequence contains a convergent subsequence, it is easy to show by contradiction that

$$0 < \widehat{C}_1 := \inf_{\widehat{\mathbf{v}}_1 \in \widehat{V}_1,\, \|\widehat{\mathbf{v}}_1\|_{H^1(\widehat{K})} = 1}\; \inf_{\widehat{\mathbf{v}}_2 \in \widehat{V}_2} \|\widehat{\mathbf{v}}_1 + \widehat{\mathbf{v}}_2\|_{H^1(\widehat{K})}\,.$$

This implies that $\widehat{C}_1 \, \|\widehat{\mathbf{v}}_1\|_{H^1(\widehat{K})} \le \|\widehat{\mathbf{v}}_1 + \widehat{\mathbf{v}}_2\|_{H^1(\widehat{K})}$ for any $\widehat{\mathbf{v}}_1 \in \widehat{V}_1, \widehat{\mathbf{v}}_2 \in \widehat{V}_2$. Thus, it follows from the equivalence of norms in finite-dimensional spaces that $\widehat{C}_1 \, \|\widehat{\nabla}\widehat{\mathbf{v}}_1\|_{L^2(\widehat{K})} \le \widehat{C}_2 \, \|\widehat{\nabla}(\widehat{\mathbf{v}}_1 + \widehat{\mathbf{v}}_2)\|_{L^2(\widehat{K})}$ for any $\widehat{\mathbf{v}}_1 \in \widehat{V}_1 \cap L_0^2(\widehat{K})^d, \widehat{\mathbf{v}}_2 \in \widehat{V}_2 \cap L_0^2(\widehat{K})^d$ and hence for any $\widehat{\mathbf{v}}_1 \in \widehat{V}_1, \widehat{\mathbf{v}}_2 \in \widehat{V}_2$. Applying (6.146) and summing over all elements of the triangulation, one gets $\|\nabla \mathbf{v}_1^h\|_{L^2(\Omega)} \le C \, \|\nabla(\mathbf{v}_1^h + \mathbf{v}_2^h)\|_{L^2(\Omega)}$ for any $\mathbf{v}_1^h \in V_1^h, \mathbf{v}_2^h \in V_2^h$ and the lemma follows. □

Lemma 6.6.2 *Let* $\widehat{V}_2 \cap P_0(\widehat{K})^d = \{\mathbf{0}\}$. *Then the space* V_2^h *satisfying* (6.148) *satisfies*

$$\left\| \mathbf{v}_2^h \right\|_{L^2(\Omega)} \le C\, h \left\| \nabla \mathbf{v}_2^h \right\|_{L^2(\Omega)} \qquad \forall\ \mathbf{v}_2^h \in V_2^h . \tag{6.150}$$

Proof It follows from the equivalence of norms in finite-dimensional spaces that $\|\widehat{\mathbf{v}}_2\|_{L^2(\widehat{K})} \le C \, \|\widehat{\nabla}\widehat{\mathbf{v}}_2\|_{L^2(\widehat{K})}$ for any $\widehat{\mathbf{v}}_2 \in \widehat{V}_2$. Then (6.150) follows using (6.145) and (6.146). □

Remark 6.6.3 The assumption $\widehat{V}_2 \cap P_0(\widehat{K})^d = \{\mathbf{0}\}$ is satisfied for all common bubble spaces V_2^h. Thus, in particular, (6.149) and (6.150) hold for all the examples of spaces V_1^h and V_2^h presented at the beginning of this section. △

The following lemma shows that, for a finite element discretization of *any* problem, the V_2^h component of the approximate solution can be dropped without influencing the asymptotic convergence properties of the approximate solution.

Lemma 6.6.4 (Estimates for the Components of $\mathbf{v}^h \in V^h$) *Consider any* $\mathbf{v} \in V \cap H^{k+1}(\Omega)^d$ *and* $\mathbf{v}^h \in V^h$. *Then one has*

$$\begin{aligned} \left\| \nabla(\mathbf{v} - \mathbf{v}_1^h) \right\|_{L^2(\Omega)} + \left\| \nabla \mathbf{v}_2^h \right\|_{L^2(\Omega)} &\le C \left\| \nabla(\mathbf{v} - \mathbf{v}^h) \right\|_{L^2(\Omega)} \\ &\quad + C\, h^k \, \|\mathbf{v}\|_{H^{k+1}(\Omega)} , \end{aligned} \tag{6.151}$$

$$\begin{aligned} \left\| \mathbf{v} - \mathbf{v}_1^h \right\|_{L^2(\Omega)} + \left\| \mathbf{v}_2^h \right\|_{L^2(\Omega)} &\le \left\| \mathbf{v} - \mathbf{v}^h \right\|_{L^2(\Omega)} \\ &\quad + C \left\{ h \left\| \nabla(\mathbf{v} - \mathbf{v}^h) \right\|_{L^2(\Omega)} + h^{k+1} \, \|\mathbf{v}\|_{H^{k+1}(\Omega)} \right\} . \end{aligned} \tag{6.152}$$

Proof Due to (6.150) and (6.149), one has for $m = 0, 1$

$$\left| \mathbf{v}_2^h \right|_{H^m(\Omega)} = \left| (\mathbf{v}^h - I^h \mathbf{v})_2 \right|_{H^m(\Omega)} \le C\, h^{1-m} \left\| \nabla(\mathbf{v}^h - I^h \mathbf{v}) \right\|_{L^2(\Omega)}$$

and hence it follows using the triangle inequality that

$$\left|\mathbf{v}-\mathbf{v}_1^h\right|_{H^m(\Omega)}+\left|\mathbf{v}_2^h\right|_{H^m(\Omega)}\leq\left|\mathbf{v}-\mathbf{v}^h\right|_{H^m(\Omega)}$$
$$+C\,h^{1-m}\left\{\left\|\nabla(\mathbf{v}-\mathbf{v}^h)\right\|_{L^2(\Omega)}+\left\|\nabla(\mathbf{v}-I^h\,\mathbf{v})\right\|_{L^2(\Omega)}\right\}.$$

Now (6.151) and (6.152) follow using (6.8). □

Now let us investigate the properties of the discrete problem (6.144).

Theorem 6.6.5 (Existence and Uniqueness of a Solution of (6.144)**)** *Let the constants $\alpha_1, \alpha_2, \alpha_3$ used in the definition of a^h satisfy $\alpha_3 > 0$ and $|\alpha_1+\alpha_2| \leq 2\sqrt{\alpha_3}$ and let the spaces V^h and Q^h satisfy the discrete inf-sup condition* (6.3). *Then, for any $\mathbf{f}^h \in H^{-1}(\Omega)^d$, the problem* (6.144) *has a unique solution.*

Proof Denoting $\alpha = (\alpha_1 + \alpha_2)/2$, one has for any $\mathbf{v}^h \in V^h$

$$a^h(\mathbf{v}^h,\mathbf{v}^h)=\nu\,(\nabla(\mathbf{v}_1^h+\alpha\,\mathbf{v}_2^h),\nabla(\mathbf{v}_1^h+\alpha\,\mathbf{v}_2^h))+\nu\,(\alpha_3-\alpha^2)\,(\nabla\mathbf{v}_2^h,\nabla\mathbf{v}_2^h)$$
$$=\nu\left\|\nabla(\mathbf{v}_1^h+\alpha\,\mathbf{v}_2^h)\right\|_{L^2(\Omega)}^2+\nu\,(\alpha_3-\alpha^2)\left\|\nabla\mathbf{v}_2^h\right\|_{L^2(\Omega)}^2$$

and hence it follows from (6.149) and the triangle inequality that, for some $C > 0$,

$$C\,\nu\left\|\nabla\mathbf{v}^h\right\|_{L^2(\Omega)}^2\leq a^h(\mathbf{v}^h,\mathbf{v}^h)\qquad\forall\ \mathbf{v}^h\in V^h\,. \tag{6.153}$$

This and the discrete inf-sup condition (6.3) imply that the problem (6.144) has only the trivial solution if $\mathbf{f}^h = \mathbf{0}$. Consequently, since the problem (6.144) is equivalent to a linear algebraic system with a square matrix, it has a unique solution for any $\mathbf{f}^h \in H^{-1}(\Omega)^d$. □

Theorem 6.6.6 (Error Estimate) *Let the assumptions of Theorem 6.6.5 be satisfied and let β_{is}^h from* (6.3) *be bounded from below by $\beta_0 > 0$ independent of h. Assume that the solution of* (6.5) *satisfies $(\mathbf{u}, p) \in H^{k+1}(\Omega)^d \times H^{l+1}(\Omega)$, then one has the error estimate*

$$\nu\,\|\nabla(\mathbf{u}-\widetilde{\mathbf{u}}^h)\|_{L^2(\Omega)}+\|p-\widetilde{p}^h\|_{L^2(\Omega)}\leq C\,\|\mathbf{f}-\mathbf{f}^h\|_{[V^h]'}$$
$$+\,C\left(\nu h^k\,\|\mathbf{u}\|_{H^{k+1}(\Omega)}+h^{l+1}\,\|p\|_{H^{l+1}(\Omega)}\right)+\nu\,|1-\alpha_2|\,h\,\|\mathbf{u}\|_{H^2(\Omega)}\,. \tag{6.154}$$

Proof Subtracting (6.13) from (6.144), one obtains for $q^h = 0$ and any $\mathbf{v}^h \in V^h$

$$a^h(\widetilde{\mathbf{u}}^h-\mathbf{u}^h,\mathbf{v}^h)-(\nabla\cdot\mathbf{v}^h,\widetilde{p}^h-p^h)=\langle\mathbf{f}^h-\mathbf{f},\mathbf{v}^h\rangle+\nu\,(1-\alpha_1)\,(\nabla\mathbf{u}_2^h,\nabla\mathbf{v}_1^h)$$
$$+\,\nu\,(1-\alpha_2)\,(\nabla\mathbf{u}_1^h,\nabla\mathbf{v}_2^h)+\nu\,(1-\alpha_3)\,(\nabla\mathbf{u}_2^h,\nabla\mathbf{v}_2^h)\,. \tag{6.155}$$

One infers applying (6.150) that, for any $\mathbf{v}_2^h \in V_2^h$,

$$(\nabla \mathbf{u}, \nabla \mathbf{v}_2^h) \le \|\Delta \mathbf{u}\|_{L^2(\Omega)} \left\|\mathbf{v}_2^h\right\|_{L^2(\Omega)} \le Ch\,\|\mathbf{u}\|_{H^2(\Omega)} \left\|\nabla \mathbf{v}_2^h\right\|_{L^2(\Omega)}. \qquad (6.156)$$

Writing $\mathbf{u}_1^h = (\mathbf{u}_1^h - \mathbf{u}) + \mathbf{u}$, one gets

$$(\nabla \mathbf{u}_1^h, \nabla \mathbf{v}_2^h) \le C \left\{ \left\|\nabla(\mathbf{u}_1^h - \mathbf{u})\right\|_{L^2(\Omega)} + h\,\|\mathbf{u}\|_{H^2(\Omega)} \right\} \left\|\nabla \mathbf{v}_2^h\right\|_{L^2(\Omega)}.$$

Now, denoting

$$A_h = \|\mathbf{f} - \mathbf{f}^h\|_{[V^h]'} + \nu \left\|\nabla(\mathbf{u} - \mathbf{u}_1^h)\right\|_{L^2(\Omega)} + \nu \left\|\nabla \mathbf{u}_2^h\right\|_{L^2(\Omega)} + \nu\,|1 - \alpha_2|\,h\,\|\mathbf{u}\|_{H^2(\Omega)},$$

one derives from (6.155) applying (6.149) that

$$a^h(\widetilde{\mathbf{u}}^h - \mathbf{u}^h, \mathbf{v}^h) - (\nabla \cdot \mathbf{v}^h, \widetilde{p}^h - p^h) \le C A_h \left\|\nabla \mathbf{v}^h\right\|_{L^2(\Omega)} \quad \forall\, \mathbf{v}^h \in V^h. \qquad (6.157)$$

Using (6.151) and (6.20), one obtains

$$A_h \le \|\mathbf{f} - \mathbf{f}^h\|_{[V^h]'} + C\left(\nu h^k\,\|\mathbf{u}\|_{H^{k+1}(\Omega)} + h^{l+1}\,\|p\|_{H^{l+1}(\Omega)}\right) + \nu\,|1 - \alpha_2|\,h\,\|\mathbf{u}\|_{H^2(\Omega)}. \qquad (6.158)$$

Setting $\mathbf{v}^h = \widetilde{\mathbf{u}}^h - \mathbf{u}^h$ in (6.157) and using the fact that $\mathbf{v}^h$ is discretely divergence-free, one gets from (6.153)

$$\nu\,\|\nabla(\widetilde{\mathbf{u}}^h - \mathbf{u}^h)\|_{L^2(\Omega)} \le C A_h. \qquad (6.159)$$

Using the Cauchy–Schwarz inequality and (6.149) gives

$$a^h(\mathbf{w}^h, \mathbf{v}^h) \le C\,\nu\,\|\nabla \mathbf{w}^h\|_{L^2(\Omega)} \|\nabla \mathbf{v}^h\|_{L^2(\Omega)} \qquad \forall\, \mathbf{w}^h, \mathbf{v}^h \in V^h,$$

which together with (6.157) and (6.159) implies that

$$(\nabla \cdot \mathbf{v}^h, \widetilde{p}^h - p^h) \le C A_h\,\|\nabla \mathbf{v}^h\|_{L^2(\Omega)} \qquad \forall\, \mathbf{v}^h \in V^h.$$

Thus, applying (6.3), one gets

$$\|\widetilde{p}^h - p^h\|_{L^2(\Omega)} \le C A_h. \qquad (6.160)$$

Now, using the triangle inequality, (6.20), and (6.158)–(6.160), one obtains (6.154). □

Theorem 6.6.7 **(L^2 Estimate of the Velocity Error)** *Let the assumptions of Theorem 6.6.5 be satisfied and let $\beta_{\rm is}^h$ from* (6.3) *be bounded from below by $\beta_0 > 0$ independent of h. Let the solution of* (6.5) *satisfy $(\mathbf{u}, p) \in H^{k+1}(\Omega)^d \times H^{l+1}(\Omega)$ and let the Stokes problem* (6.2) *be regular. Then there holds the error estimate*

$$\begin{aligned}\|\mathbf{u} - \widetilde{\mathbf{u}}^h\|_{L^2(\Omega)} &\le \frac{C}{\nu}\|\mathbf{f} - \mathbf{f}^h\|_{[V_1^h]'} + \frac{Ch}{\nu}\|\mathbf{f} - \mathbf{f}^h\|_{[V^h]'} \\ &+ C\left(h^{k+1}\|\mathbf{u}\|_{H^{k+1}(\Omega)} + \frac{h^{l+2}}{\nu}\|p\|_{H^{l+1}(\Omega)}\right) + |1-\alpha_2|\, h^2\, \|\mathbf{u}\|_{H^2(\Omega)}.\end{aligned} \tag{6.161}$$

Proof Let $(\mathbf{z}, r) \in V \times Q$ be the solution of the problem (6.22) with $\mathbf{u} - \mathbf{u}^h$ replaced by $\widetilde{\mathbf{u}}^h - \mathbf{u}^h$. Then all the relations (6.23)–(6.28) also hold with $\mathbf{u} - \mathbf{u}^h$ replaced by $\widetilde{\mathbf{u}}^h - \mathbf{u}^h$. Thus, using the fact that $\widetilde{\mathbf{u}}^h - \mathbf{u}^h$ is discretely divergence-free, one obtains

$$\begin{aligned}\nu\|\widetilde{\mathbf{u}}^h - \mathbf{u}^h\|_{L^2(\Omega)}^2 &\le C\nu h\, \|\nabla(\widetilde{\mathbf{u}}^h - \mathbf{u}^h)\|_{L^2(\Omega)}\|\widetilde{\mathbf{u}}^h - \mathbf{u}^h\|_{L^2(\Omega)} \\ &+ \nu(\nabla\mathbf{z}^I, \nabla(\widetilde{\mathbf{u}}^h - \mathbf{u}^h)),\end{aligned} \tag{6.162}$$

where $\mathbf{z}^I$ satisfies

$$\left\|\nabla(\mathbf{z} - \mathbf{z}^I)\right\|_{L^2(\Omega)} \le Ch\,\|\mathbf{z}\|_{H^2(\Omega)} \le Ch\,\|\widetilde{\mathbf{u}}^h - \mathbf{u}^h\|_{L^2(\Omega)}. \tag{6.163}$$

Since $\mathbf{z}^I \in V_1^h$, one has

$$\nu(\nabla(\widetilde{\mathbf{u}}^h - \mathbf{u}^h), \nabla\mathbf{z}^I) = a^h(\widetilde{\mathbf{u}}^h - \mathbf{u}^h, \mathbf{z}^I) + (1-\alpha_1)\nu(\nabla(\widetilde{\mathbf{u}}_2^h - \mathbf{u}_2^h), \nabla\mathbf{z}^I)$$

and hence it follows from (6.155) that

$$\begin{aligned}\nu(\nabla(\widetilde{\mathbf{u}}^h - \mathbf{u}^h), \nabla\mathbf{z}^I) =& \langle\mathbf{f}^h - \mathbf{f}, \mathbf{z}^I\rangle + (\nabla\cdot\mathbf{z}^I, \widetilde{p}^h - p^h) + (1-\alpha_1)\nu(\nabla\widetilde{\mathbf{u}}_2^h, \nabla\mathbf{z}^I) \\ =& \langle\mathbf{f}^h - \mathbf{f}, \mathbf{z}^I\rangle - (\nabla\cdot(\mathbf{z} - \mathbf{z}^I), \widetilde{p}^h - p^h) \\ &- (1-\alpha_1)\nu(\nabla\widetilde{\mathbf{u}}_2^h, \nabla(\mathbf{z} - \mathbf{z}^I)) + (1-\alpha_1)\nu(\nabla\widetilde{\mathbf{u}}_2^h, \nabla\mathbf{z}).\end{aligned}$$

Applying (6.163) and (6.156) with $\mathbf{u}$ replaced by $\mathbf{z}$, one gets

$$\begin{aligned}\nu(\nabla(\widetilde{\mathbf{u}}^h - \mathbf{u}^h), \nabla\mathbf{z}^I) \le& C\,\|\mathbf{f} - \mathbf{f}^h\|_{[V_1^h]'}\|\widetilde{\mathbf{u}}^h - \mathbf{u}^h\|_{L^2(\Omega)} \\ &+ Ch\left(\nu\|\nabla\widetilde{\mathbf{u}}_2^h\|_{L^2(\Omega)} + \|\widetilde{p}^h - p^h\|_{L^2(\Omega)}\right)\|\widetilde{\mathbf{u}}^h - \mathbf{u}^h\|_{L^2(\Omega)}.\end{aligned}$$

Substituting this estimate into (6.162) and using the triangle inequality and (6.149), one obtains

$$\nu\|\widetilde{\mathbf{u}}^h - \mathbf{u}^h\|_{L^2(\Omega)} \le C\,\|\mathbf{f} - \mathbf{f}^h\|_{[V_1^h]'}$$
$$+ Ch\left(\nu\|\nabla \mathbf{u}_2^h\|_{L^2(\Omega)} + \nu\|\nabla(\widetilde{\mathbf{u}}^h - \mathbf{u}^h)\|_{L^2(\Omega)} + \|\widetilde{p}^h - p^h\|_{L^2(\Omega)}\right).$$

Then, (6.161) follows as a consequence of the triangle inequality, (6.151), (6.154), (6.20), and (6.21). □

Remark 6.6.8 If $\mathbf{f}^h$ is defined by (6.143), then, for any $\mathbf{v}^h \in V^h$, one has $\langle \mathbf{f} - \mathbf{f}^h, \mathbf{v}^h\rangle = (\mathbf{f}, \mathbf{v}_2^h) \le \|\mathbf{f}\|_{L^2(\Omega)}\,\|\mathbf{v}_2^h\|_{L^2(\Omega)}$ and, using (6.150) and (6.149), one deduces that $\mathbf{f}^h$ satisfies $\|\mathbf{f} - \mathbf{f}^h\|_{[V^h]'} \le Ch\,\|\mathbf{f}\|_{L^2(\Omega)}$. Moreover, one has $\|\mathbf{f} - \mathbf{f}^h\|_{[V_1^h]'} = 0$. Thus, if $k = 1$, the problem (6.144) leads to optimal error estimates with respect to h for any constants $\alpha_1, \alpha_2, \alpha_3$ satisfying $\alpha_3 > 0$ and $|\alpha_1 + \alpha_2| \le 2\sqrt{\alpha_3}$. If $k > 1$, optimal error estimates are obtained for $\mathbf{f}^h = \mathbf{f}$ and $\alpha_2 = 1$. △

Now let us discuss the relation of the modified discretization (6.144) to stabilized methods. For simplicity, we confine ourselves to the two-dimensional case. It is convenient to write the problem (6.144) in the equivalent form

$$\nu(\nabla\widetilde{\mathbf{u}}_1^h, \nabla\mathbf{v}_1^h) + \alpha_1\nu(\nabla\widetilde{\mathbf{u}}_2^h, \nabla\mathbf{v}_1^h) - (\nabla\cdot\mathbf{v}_1^h, \widetilde{p}^h) = \langle \mathbf{f}^h, \mathbf{v}_1^h\rangle \quad \forall\, \mathbf{v}_1^h \in V_1^h, \tag{6.164}$$

$$\alpha_2\nu(\nabla\widetilde{\mathbf{u}}_1^h, \nabla\mathbf{v}_2^h) + \alpha_3\nu(\nabla\widetilde{\mathbf{u}}_2^h, \nabla\mathbf{v}_2^h) - (\nabla\cdot\mathbf{v}_2^h, \widetilde{p}^h) = \langle \mathbf{f}^h, \mathbf{v}_2^h\rangle \quad \forall\, \mathbf{v}_2^h \in V_2^h, \tag{6.165}$$

$$-(\nabla\cdot\widetilde{\mathbf{u}}_1^h, q^h) - (\nabla\cdot\widetilde{\mathbf{u}}_2^h, q^h) = 0 \quad \forall\, q^h \in Q^h. \tag{6.166}$$

First, let us consider the case $\alpha_1 = \alpha_2 = 0$, $\alpha_3 = 1$, and $\mathbf{f}^h$ defined by (6.143). Let V_1^h consist of continuous piecewise (bi)linear functions. If Q^h consists of piecewise constant functions, then one can set $V_2^h = \operatorname{span}\{\varphi_E\,\mathbf{n}_E\}_{E\in\mathcal{E}^h}$, where $\varphi_E \in H_0^1(\Omega)$ are scalar finite element functions assigned to interior edges E of the triangulation $\mathcal{T}^h$ which have their supports in the two elements adjacent to E and satisfy $\int_E \varphi_E\,\mathrm{d}s \ne 0$, see, e.g., [17, 41] for particular examples of φ_E. The vectors $\mathbf{n}_E$ are again fixed normal vectors to the edges E. Defining φ_E in such a way that the interiors of the supports of any two functions $\varphi_E, \varphi_{E'}$ with $E \ne E'$ are disjoint, one can compute $\mathbf{u}_2^h$ from (6.165) and substitute it in (6.166), which gives

$$(\nabla\cdot\widetilde{\mathbf{u}}_1^h, q^h) + \sum_{E\in\mathcal{E}^h} \gamma_E\left(\left[\left|\widetilde{p}^h\right|\right]_E, \left[\left|q^h\right|\right]_E\right)_E = 0 \quad \forall\, q^h \in Q^h,$$

where

$$\gamma_E = \frac{|\int_E \varphi_E\,\mathrm{d}s|^2}{\nu\,\|\nabla\varphi_E\|_{L^2(\Omega)}^2\, h_E}.$$

The usual scaling argument shows that γ_E satisfies (6.45). Thus, for the considered spaces V^h and Q^h, the modified discretization is equivalent to the PSPG method (6.33) for the spaces V_1^h and Q^h.

If V_1^h is as above and Q^h consists of continuous piecewise (bi)linear functions, one can consider a general space $V_2^h = \mathrm{span}\{\varphi_i^h \boldsymbol{t}_i^h\}_{i=1}^{N^h}$ where $\boldsymbol{t}_i^h$ are unit vectors and $\varphi_i^h \in H_0^1(\Omega)$ are finite element functions having their supports in one element or in two elements possessing a common edge. To distinguish which element or elements a function φ_i^h belongs to, points A_i^h different from the vertices of $\mathcal{T}^h$ are introduced. If A_i^h lies in the interior of some element $K \in \mathcal{T}^h$, one requires that $\mathrm{supp}\,\varphi_i^h \subset K$ and if A_i^h lies on an edge E, one requires that $\mathrm{supp}\,\varphi_i^h$ lies in the two elements adjacent to E and that $\boldsymbol{t}_i^h$ is parallel to E. For triangular meshes, one assumes that there exist two points $A_i^h \in K$ for each element K. In the quadrilateral case, three points $A_i^h \in K$ are supposed for any K. In other words, one needs two functions φ_i^h per element in the triangular case and three functions φ_i^h per element in the quadrilateral case. In both cases, each function may be common to two elements. Under further assumptions, see [63] for details, which are satisfied for the spaces considered here, it can be shown that this space V_2^h assures the validity of the inf-sup condition (6.3). For example, in the triangular case, the space V_2^h defined in (6.139) and leading to the MINI element can be put into the above general framework. Then, for each element K, one has two bubble functions φ_i^h which coincide and are equal to φ_K. If $\mathcal{T}^h$ consists of quadrilaterals, the stability is assured by four bubble functions on each element, see [74].

In particular, as a special case of the general framework from the previous paragraph, one can use spaces of the type $V_2^h = \mathrm{span}\{\varphi_E \boldsymbol{t}_E\}_{E\in\overline{\mathcal{E}}^h}$, where $\boldsymbol{t}_E$ are tangential vectors to the edges E and $\overline{\mathcal{E}}^h$ denotes the set of all edges of the triangulation $\mathcal{T}^h$. The functions φ_E are constructed in such a way that the interiors of their supports are mutually disjoint and they vanish on $\partial\Omega$ also if $E \subset \partial\Omega$, see [63]. Using again the modified discretization (6.164)–(6.166) with $\alpha_1 = \alpha_2 = 0$, $\alpha_3 = 1$, and $\mathbf{f}^h$ defined by (6.143), one infers analogously as above that the piecewise (bi)linear part of the solution to (6.164)–(6.166) satisfies

$$(\nabla \cdot \tilde{\mathbf{u}}_1^h, q^h) + \sum_{E\in\overline{\mathcal{E}}^h} \gamma_E \left(\frac{\partial \tilde{p}^h}{\partial \boldsymbol{t}_E}, \frac{\partial q^h}{\partial \boldsymbol{t}_E} \right)_E = 0 \qquad \forall\, q^h \in Q^h$$

with

$$\gamma_E = \frac{|\int_\Omega \varphi_E \,\mathrm{d}\boldsymbol{x}|^2}{\nu\, \|\nabla \varphi_E\|_{L^2(\Omega)}^2 \, h_E},$$

which is a different type of stabilization than those discussed in the preceding sections.

The spaces V_1^h, Q^h consisting of continuous piecewise (bi)linear functions can be also used with a space V_2^h of the type $V_2^h = \text{span}\{\varphi_i^h \boldsymbol{t}_i^h\}_{i=1}^{N^h}$. In the triangular case, let V_2^h be the space of the MINI element defined by (6.139) and in the quadrilateral case, it will be assumed that the elements of $\mathcal{T}^h$ are rectangles and, for any element $K \in \mathcal{T}^h$, four functions φ_i^h with disjoint supports in K will be used. The corresponding vectors $\boldsymbol{t}_i^h$ are parallel to the edges of K, see [63] for details. Using the same modified discretization as above and eliminating $\widetilde{\mathbf{u}}_2^h$, one obtains

$$(\nabla \cdot \widetilde{\mathbf{u}}_1^h, q^h) + \sum_{K \in \mathcal{T}^h} \left(\nabla \widetilde{p}^h, \delta_K \nabla q^h\right)_K = 0 \qquad \forall\, q^h \in Q^h\,,$$

where δ_K satisfies (6.44), i.e., one recovers the Brezzi–Pitkäranta method (6.106) for the spaces V_1^h, Q^h. If $\mathbf{f}^h = \mathbf{f}$ and $\mathbf{f}$ is piecewise (bi)linear, one obtains the stabilization

$$(\nabla \cdot \widetilde{\mathbf{u}}_1^h, q^h) + \sum_{K \in \mathcal{T}^h} \left(\nabla \widetilde{p}^h - \mathbf{f}, \delta_K \nabla q^h\right)_K = 0 \qquad \forall\, q^h \in Q^h\,,$$

which corresponds to the PSPG method (6.33) for the spaces V_1^h, Q^h.

Finally, let the spaces V_1^h, Q^h consist of continuous piecewise quadratic functions on triangles. These spaces do not satisfy the inf-sup condition (6.3). Dividing any element $K \in \mathcal{T}^h$ into four equal triangles by connecting midpoints of edges and introducing two vector bubble functions from the MINI element on each subtriangle having a common vertex with K, one obtains a space V_2^h assuring the stability. If one eliminates this space V_2^h from the original conforming discretization (6.136)–(6.138) and assumes that $\mathbf{f}$ is piecewise linear, one obtains the symmetric GLS method (6.65) for the spaces V_1^h, Q^h. However, one can also use the modified discretization (6.164)–(6.166) with $\alpha_1 = -1$, $\alpha_2 = \alpha_3 = 1$ and $\mathbf{f}^h = \mathbf{f}$ which guarantees the same asymptotic convergence rates of the discrete solution as (6.136)–(6.138). Then, eliminating the space V_2^h, one obtains the non-symmetric GLS method (6.91) for the spaces V_1^h, Q^h.

6.7 Numerical Studies

In this section, numerical studies with some of the stabilized methods will be presented: the PSPG method (6.33)–(6.35), the symmetric GLS method (6.65)–(6.67), the non-symmetric GLS method (6.91)–(6.93), and the LPS method that utilizes a modified Scott–Zhang projector, see Remark 6.5.8. In addition, the Brezzi–Pitkäranta method (6.106) with P_1/P_1 finite elements was incorporated in our studies. For the sake of brevity, the results with this method are not presented here since, in our experience, they were not better than, e.g., the results obtained with the PSPG method.

Two examples were studied:

- an example for the Stokes equations (6.2) with prescribed solution, which studies standard errors, their order of convergence, and the dependency on the viscosity coefficient and on the stabilization parameter,
- an example for the stationary Navier–Stokes equations, a flow around a cylinder, which investigates the accuracy of computed quantities at the cylinder that are of physical relevance, the dependency of the results on the discretization of the nonlinear term, and which provides a comparison to results obtained with inf-sup stable pairs of finite elements.

All simulations were performed with the code PARMOON, [44, 84]. Linear systems of equations were solved with the sparse direct solver UMFPACK [37].

Remark 6.7.1 (Comparative Numerical Studies in the Literature) There are few numerical studies that compare several stabilized methods already available in the literature.

- Numerical studies at simple Stokes problems in [77] compare the PLS method, see Sect. 6.5.2, the symmetric GLS method from Sect. 6.4.3, and the method with orthogonal subscales from Sect. 6.5.3.
- A brief numerical comparison of the method based on two local Gauss integrations from [67], which is equivalent to the method from [38], and the symmetric GLS method from Sect. 6.4.3 for P_1/P_1 finite elements (in this situation the latter method is equivalent to the PSPG method from Sect. 6.4.2 and the non-symmetric GLS method from Sect. 6.4.4) can be found in [68]. For a driven cavity problem, it was observed that the pressure approximation close to the boundary is more accurate for the first method.
- A brief comparison of the LPS method mentioned in Remark 6.5.8 and the PSPG method for P_1/P_1 finite elements can be found in [7].

△

6.7.1 Stokes Problem with Prescribed Solution

This example studies some of the stabilized discretizations in the framework of the numerical analysis: the Stokes equations (6.2) possess a smooth solution with homogeneous Dirichlet boundary data. In addition, the solution does not depend on the viscosity coefficient ν. Errors in standard norms were monitored. The dependency of the errors and the order of convergence on the viscosity coefficient ν and on the stabilization parameter was investigated.

Consider the domain $\Omega = (0, 1)^2$ together with a polynomial solution

$$\mathbf{u} = \begin{pmatrix} u_1 \\ u_2 \end{pmatrix} = \begin{pmatrix} \partial_y \phi \\ -\partial_x \phi \end{pmatrix},$$

where ϕ is the stream function given by

$$\phi(x, y) = 1000\,x^2(1-x)^4y^3(1-y)^2.$$

Due to this construction, the solution is divergence-free; furthermore, it has homogeneous Dirichlet boundary values on $\partial\Omega$. The corresponding pressure p therefore should have zero mean value, $p \in L_0^2(\Omega)$. For this example, it is set to be

$$p = \pi^2\left(xy^3\cos(2\pi x^2 y) - x^2 y\sin(2\pi xy)\right) + \frac{1}{8}.$$

The right-hand side $\mathbf{f}$ in (6.2) is set accordingly. Figure 6.2 shows visualizations of the prescribed solution.

The stabilized methods involved in our studies are already mentioned at the beginning of this section. For all of them, P_k/P_k pairs of finite element spaces were considered with $k \in \{1, 2, 3\}$. Note that for the P_1/P_1 pair of finite element

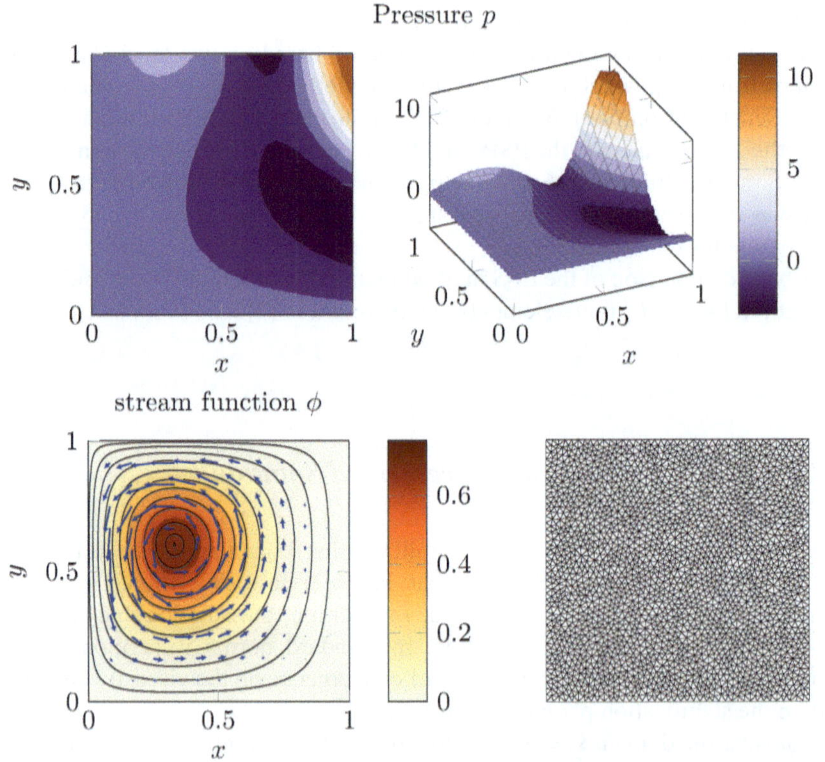

Fig. 6.2 Visualizations of the solution of the Stokes example. Top: pressure p. Bottom left: stream function ϕ together with streamlines and arrows of the velocity $\mathbf{u}$. Bottom right: for illustration, a generated grid which is coarser than the ones actually used

Table 6.1 Details on the generated grids used for the Stokes example

Grid level	Number of cells	Number of degrees of freedom		
		P_1/P_1	P_2/P_2	P_3/P_3
0	38728	58821	233823	525009
1	52464	79545	316479	710805
2	68628	103911	413703	929379
3	86398	130686	520563	1169634
4	106838	161466	643443	1445934

spaces, the PSPG method, the symmetric GLS method, and the non-symmetric GLS method coincide. The stabilization parameters of all methods have the form $\delta_K = \delta_0 h_K^2/\nu$ and the numerical studies considered for most methods $\delta_0 = 10^i$, $i \in \{-3, -2.5, \ldots, 0\}$. Only for the symmetric GLS method, we found that these stabilization parameters were too large, since an irregular behavior of the monitored errors could be observed, compare Fig. 6.4 below. For this method, results obtained with $\delta_0 = 10^i$, $i \in \{-5, -4.5, \ldots, -2\}$, will be presented. Simulations for $\nu = 10^j$, $j \in \{-6, -5, \ldots, 0\}$, were performed.

In our computational studies, unstructured grids of varying fineness have been employed, see Fig. 6.2 for an example and Table 6.1 for more details. The grids were generated with GMSH [45]. The convergence order has been computed via the formula $\log(e^H/e^h)/\log(H/h)$, where H and h are the characteristic grid lengths[2] while e^H and e^h are the respective errors on these grids.

In Figs. 6.3, 6.4, 6.5, 6.6, errors as well as convergence orders are shown for the studied methods. The errors are those obtained on the finest grid level 4 and the order of convergence was computed with the errors on the two finest levels. The PSPG method, the non-symmetric GLS method, and the LPS method only show a weak dependency on δ_0. This behavior can be expected from the analysis of the non-symmetric GLS method since it is absolutely stable. Furthermore, the errors for the velocity are larger for smaller ν while the errors for the pressure are larger for larger ν. This behavior reflects also the analytical results. These three methods also show a similar behavior in the estimated convergence orders. The orders of error reduction are often higher than expected for small ν, compare also [56, Fig. 4.14]. This effect was observed also for inf-sup stable discretizations, e.g., see [56, Fig. 4.9]. To the best of our knowledge, an explanation for this phenomenon is not known so far. The symmetric GLS method shows a more irregular behavior with respect to the dependency on δ_0, see Fig. 6.4 for larger values of δ_0 and higher order finite elements. For small values of δ_0, one can observe the same behavior as it is described above for the other methods.

Figure 6.7 presents a comparison of the methods among each other. For performing this comparison, for each value of the viscosity ν, the most accurate result with respect to the L^2 error of the velocity on the finest level was selected for each method, among all values of the stabilization parameter. It can be seen in

[2]In the case of uniform refinement, it is $H = 2h$.

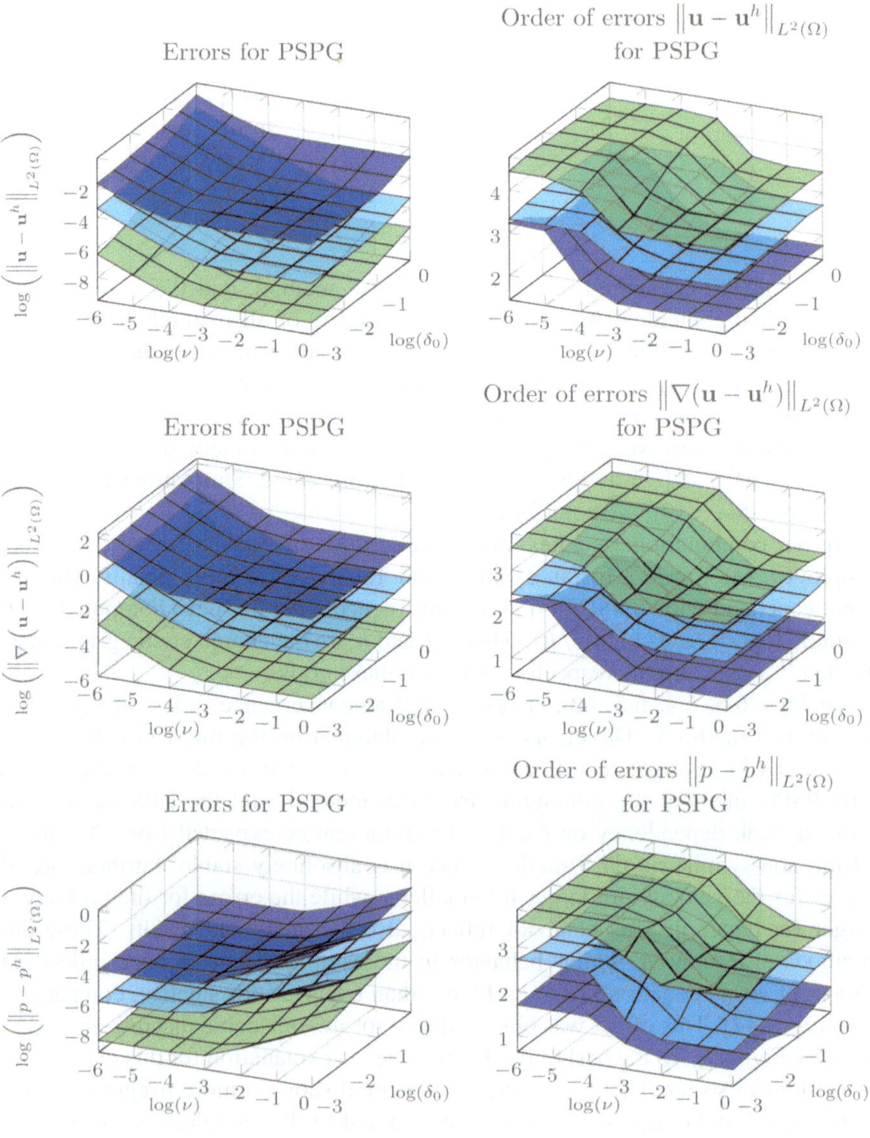

Fig. 6.3 The errors (left) and computed orders of convergence (right) with respect to the L^2 (top) and H^1 semi-norm (middle) of the velocity, as well as the L^2 norm of the pressure (bottom) for the PSPG method and P_1/P_1 (blue), P_2/P_2 (cyan), and P_3/P_3 (green)

Fig. 6.7 that in many cases, in particular for P_1/P_1 and P_3/P_3 finite elements, the curves are almost on top of each other, i.e., all methods gave very similar results. Only for the P_2/P_2 finite element and small viscosities, the non-symmetric GLS

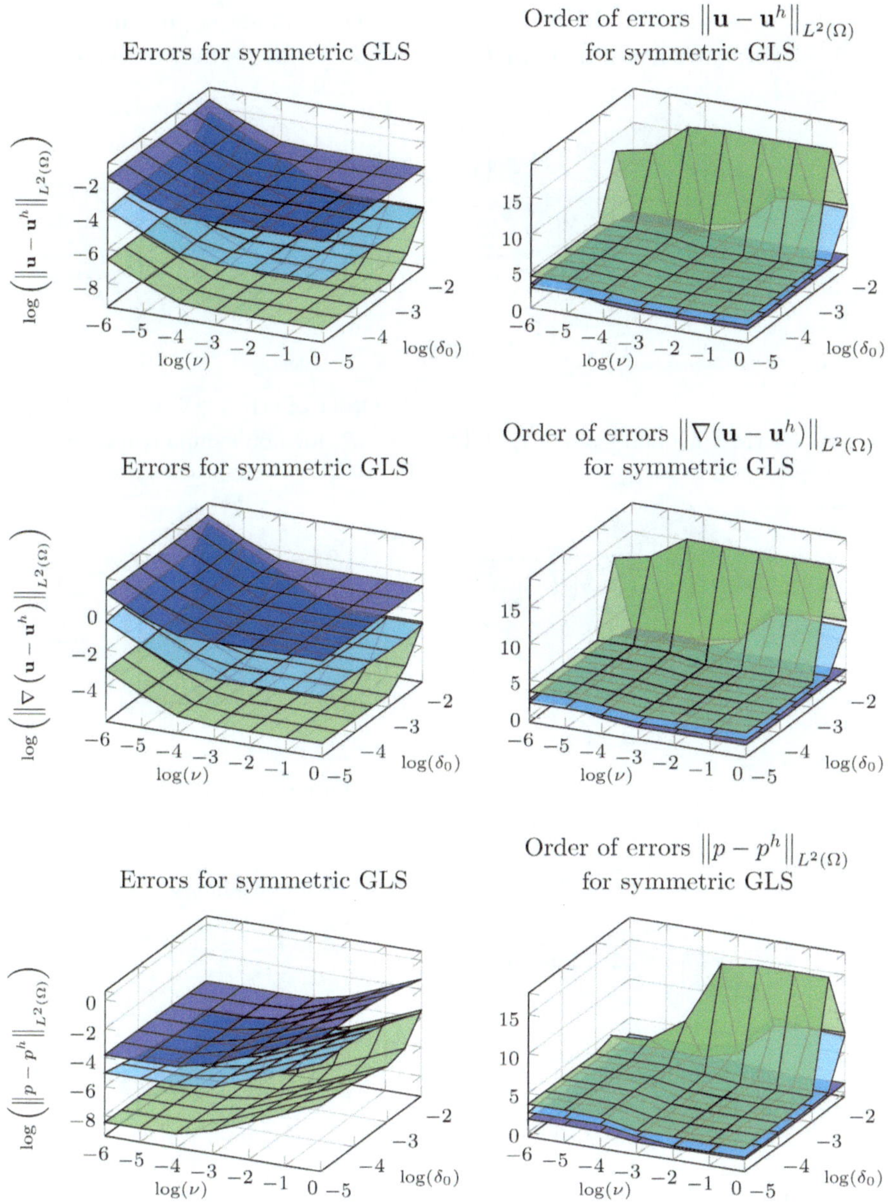

Fig. 6.4 The errors (left) and computed orders of convergence (right) with respect to the L^2 (top) and H^1 semi-norm (middle) of the velocity, as well as the L^2 norm of the pressure (bottom) for the symmetric GLS method and P_1/P_1 (blue), P_2/P_2 (cyan), and P_3/P_3 (green)

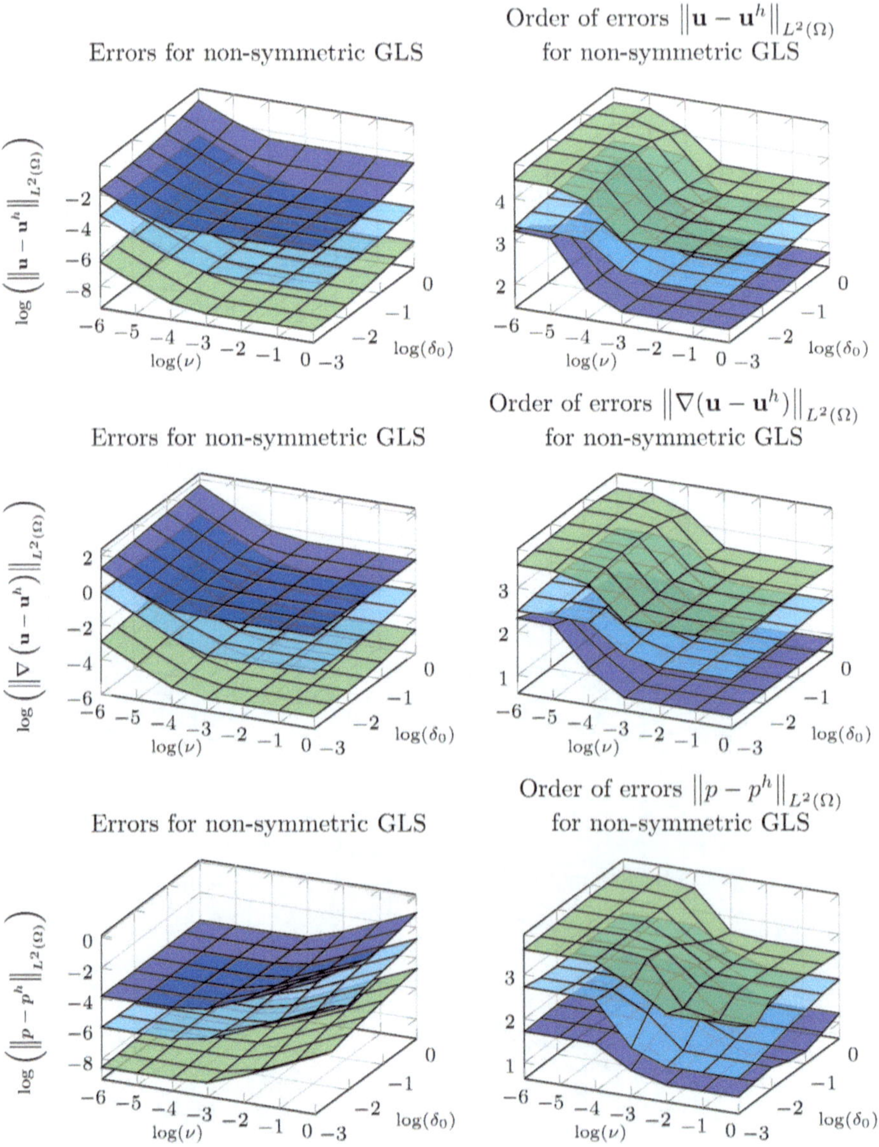

Fig. 6.5 The errors (left) and computed orders of convergence (right) with respect to the L^2 (top) and H^1 semi-norm (middle) of the velocity, as well as the L^2 norm of the pressure (bottom) for the non-symmetric GLS method and P_1/P_1 (blue), P_2/P_2 (cyan), and P_3/P_3 (green)

method led to slightly higher velocity errors and the LPS method to notably higher pressure errors than the other methods.

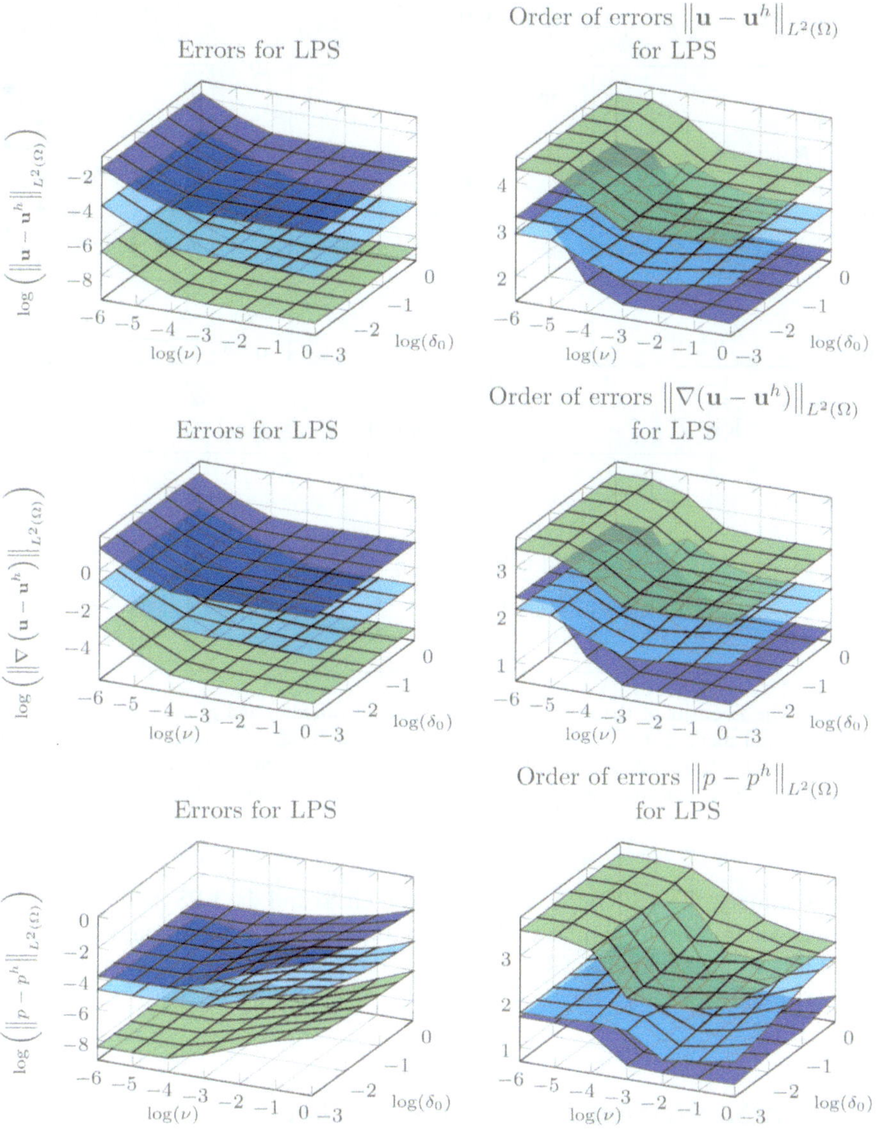

Fig. 6.6 The errors (left) and computed orders of convergence (right) with respect to the L^2 (top) and H^1 semi-norm (middle) of the velocity, as well as the L^2 norm of the pressure (bottom) for the LPS method and P_1/P_1 (blue), P_2/P_2 (cyan), and P_3/P_3 (green)

We like to note that we obtained similar results as presented in Figs. 6.3, 6.4, 6.5, 6.6, 6.7 on structured grids that were generated by refining a coarse grid consisting of two triangles regularly.

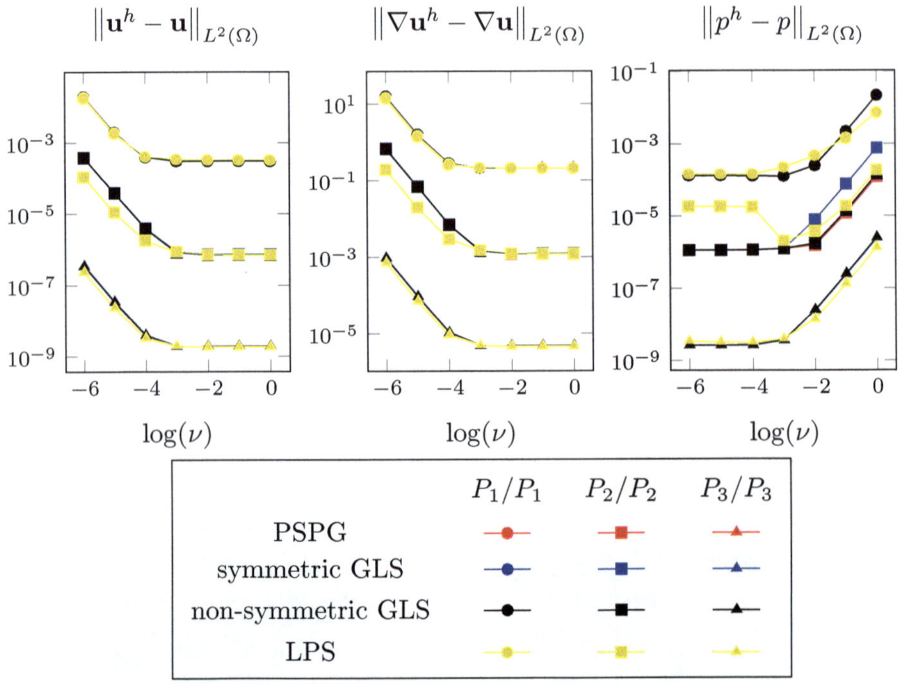

Fig. 6.7 The L^2errors of the velocity (left), its gradient (middle) and of the pressure (right) for the considered methods and different polynomial degrees on the finest level 4. The stabilization parameter δ_0 is always chosen such that the L^2error of the velocity is smallest among the used values for δ_0. Note that lines are often on top of each other, i. e., the methods led to very similar errors

In summary, the PSPG, the non-symmetric GLS, and the LPS methods behaved in this example quite similarly. The most remarkable observation was that the instability of the symmetric GLS method for large stabilization parameters became visible already for rather small values of δ_0.

6.7.2 A Steady-State Flow Around a Cylinder

The second example serves for assessing the stabilized discretizations mentioned at the beginning of this section at a more challenging example. It is given by the stationary Navier–Stokes equations

$$\begin{aligned} -\nu\Delta\mathbf{u} + (\mathbf{u}\cdot\nabla)\mathbf{u} + \nabla p &= \mathbf{f} \text{ in } \Omega, \\ \nabla\cdot\mathbf{u} &= 0 \text{ in } \Omega, \end{aligned} \tag{6.167}$$

and it requires the computation of coefficients which are of importance in applications. Furthermore, comparisons to some inf-sup stable discretizations are also provided.

A standard benchmark problem for (6.167) is the so-called *flow around a cylinder* problem defined in [80]. It is given by $\Omega = (0, 2.2) \times (0, 0.41) \setminus B_{0.1}(0.2, 0.2)$, where $B_r(x, y)$ is a (compact) two-dimensional cylinder (circle) with radius r centered at (x, y), $\nu = 10^{-3}$, and $\mathbf{f} = \mathbf{0}$. On the top and bottom boundary as well as at the cylinder homogeneous Dirichlet condition are prescribed. At the outflow boundary $\Gamma_{\text{out}} = \{2.2\} \times [0, 0.41]$, homogeneous Neumann (so-called do-nothing) conditions are imposed while the flow is driven entirely through a parabolic inflow on the left boundary,

$$\mathbf{u}(0, y) = \begin{pmatrix} 1.2y(1-y) \\ 0 \end{pmatrix}.$$

Benchmark parameters are the drag and lift coefficients at the cylinder and the pressure difference Δp between the front and the back of the cylinder, see [80] or [56, Ex. D5]. Reference values were computed in [59, 75], see also [56, Ex. D5]:

$$c_{\text{drag,ref}} = 5.57953523384, \quad c_{\text{lift,ref}} = 0.010618948146,$$
$$\Delta p_{\text{ref}} = 0.11752016697.$$

For discretizing the Navier–Stokes equations (6.167), one has to choose the discrete form of the nonlinear term. Several forms were proposed, which are equivalent only if the velocity is weakly divergence-free. However, finite element velocity fields usually do not possess this property. In our numerical studies, the so-called convective form

$$\left(\left(\mathbf{u}^h \cdot \nabla\right)\mathbf{u}^h, \mathbf{v}^h\right),$$

the skew-symmetric form

$$\frac{1}{2}\left[\left(\left(\mathbf{u}^h \cdot \nabla\right)\mathbf{u}^h, \mathbf{v}^h\right) - \left(\left(\mathbf{u}^h \cdot \nabla\right)\mathbf{v}^h, \mathbf{u}^h\right)\right],$$

and the energy momentum and angular momentum conserving (EMAC) form [32]

$$\left(2\mathbb{D}\left(\mathbf{u}^h\right)\mathbf{u}^h, \mathbf{v}^h\right) + \left(\left(\nabla \cdot \mathbf{u}^h\right)\mathbf{u}^h, \mathbf{v}^h\right), \quad \mathbb{D}(\mathbf{u}) = \frac{\nabla \mathbf{u} + (\nabla \mathbf{u})^T}{2},$$

were tested. Note that the EMAC form has on the one hand several favorable properties with respect to conservation of quantities, but on the other hand, it computes a modified pressure. For calculating the benchmark parameters, a reconstruction of the actual pressure is necessary.

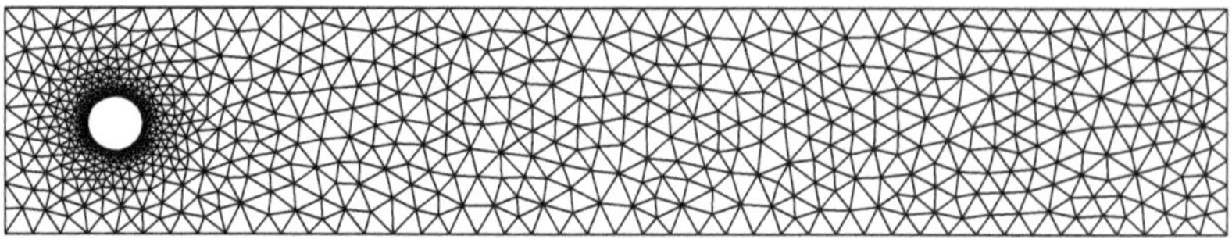

Fig. 6.8 Coarse grid, generated with GMSH, used for the flow around a cylinder example

Table 6.2 Information on the unstructured grids used in the simulations, the coarsest grid (level 0) is shown in Fig. 6.8

Grid level	Number of cells	Number of degrees of freedom					
		P_1/P_1	P_2/P_1	P_2/P_2	P_3/P_2	P_3/P_3	P_4/P_3
0	1629	2697	7753	10281	18595	22752	34324
1	5340	8475	24805	32970	59980	73485	111175
2	11202	17475	51529	68556	125014	153243	232105
3	19076	29493	87307	116214	212180	260163	394281
4	29193	44880	133186	177339	324031	397377	602455
5	41973	64260	191046	254439	465171	570537	865215

The nonlinear systems were solved with a Picard iteration. It was stopped if the Euclidean norm of the residual vector was smaller than 10^{-10} or after 10,000 iterations.

Results are presented for simulations conducted on unstructured grids, which were generated with GMSH, see Fig. 6.8 and Table 6.2. On each grid, the P_k/P_k, $k \in \{1, 2, 3\}$, finite element spaces were applied for the stabilized methods and the P_k/P_{k-1}, $k \in \{2, 3, 4\}$, inf-sup stable Taylor–Hood pairs of finite element spaces for the Galerkin method. Stabilization parameters of the form $\delta_K = \delta_0 h_K^2/\nu$ with $\delta_0 = 10^i$, $i \in \{-5, -4.5, \ldots, 0\}$, were considered. In all figures, the results for the stabilization parameter with the smallest error with respect to the drag coefficient are presented.

The accuracy for the computed benchmark parameters is illustrated in Figs. 6.9, 6.10, 6.11. For the drag coefficient, Fig. 6.9, it can be observed that the results obtained with the convective and skew-symmetric form are usually more accurate than those computed with the EMAC form. Using the inf-sup stable Taylor–Hood pairs of spaces gave often more accurate results than using the pressure-stabilized discretizations. For higher order pairs of spaces, the LPS method was a little bit more accurate than the other methods. For the lift coefficient, Fig. 6.10, again the EMAC form led to somewhat less accurate results than the other forms of the discrete convective term. Among the stabilized methods, no substantial differences of the accuracy can be observed. For higher order pairs of spaces, the Taylor–Hood discretization was sometimes somewhat more accurate than the stabilized methods. The results for the pressure difference are shown in Fig. 6.11. Again, the results computed with the Taylor–Hood pair of finite elements were usually among the most

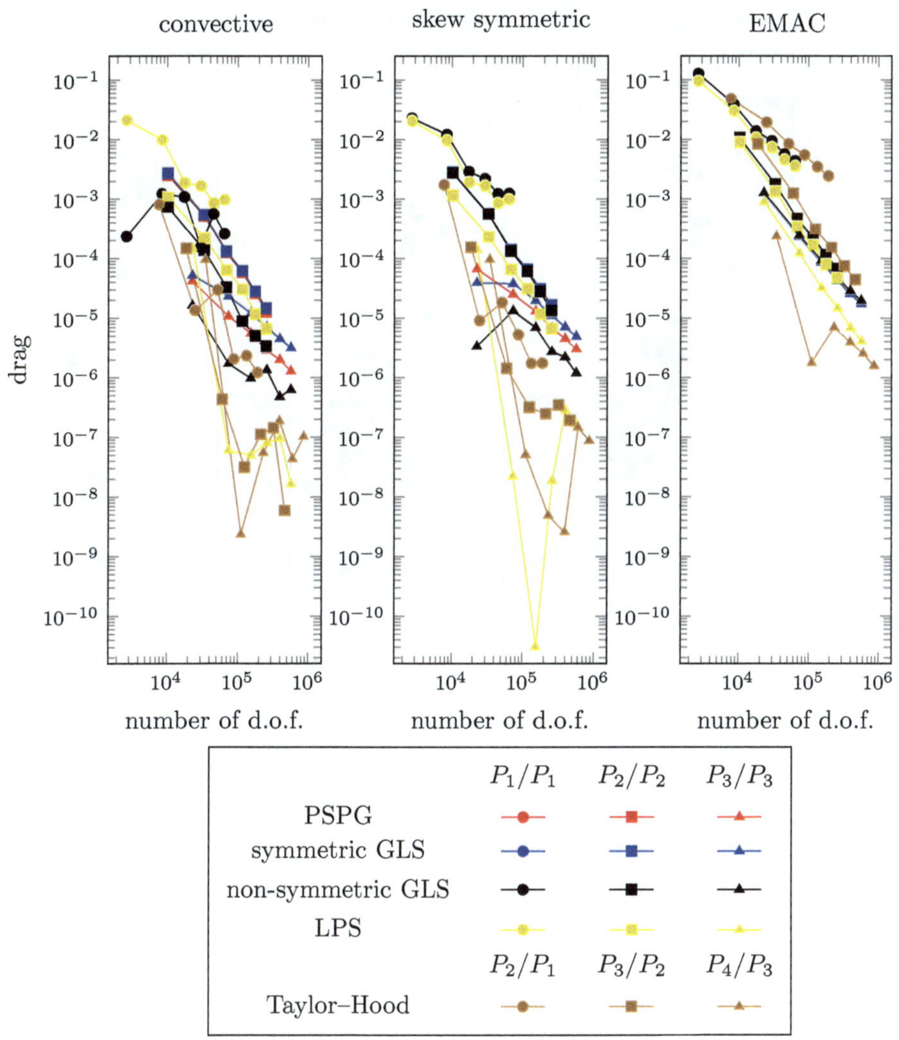

Fig. 6.9 Computed absolute differences to reference value for drag using the convective (left), skew symmetric (center), and EMAC (right) nonlinear form on the unstructured grids, see Table 6.2 and Fig. 6.8

accurate ones. For the stabilized discretizations, there is no clear picture. Often, the results from the LPS method belong to the better ones.

Information with respect to the number of nonlinear iterations for solving the Navier–Stokes equations is provided in Fig. 6.12. Apart from coarse grids, it can be seen that there are only minor differences between the discretization methods. The lowest number of iterations, usually below 20, was needed for the convective form of the convective term and the largest number, generally more than 50, for the

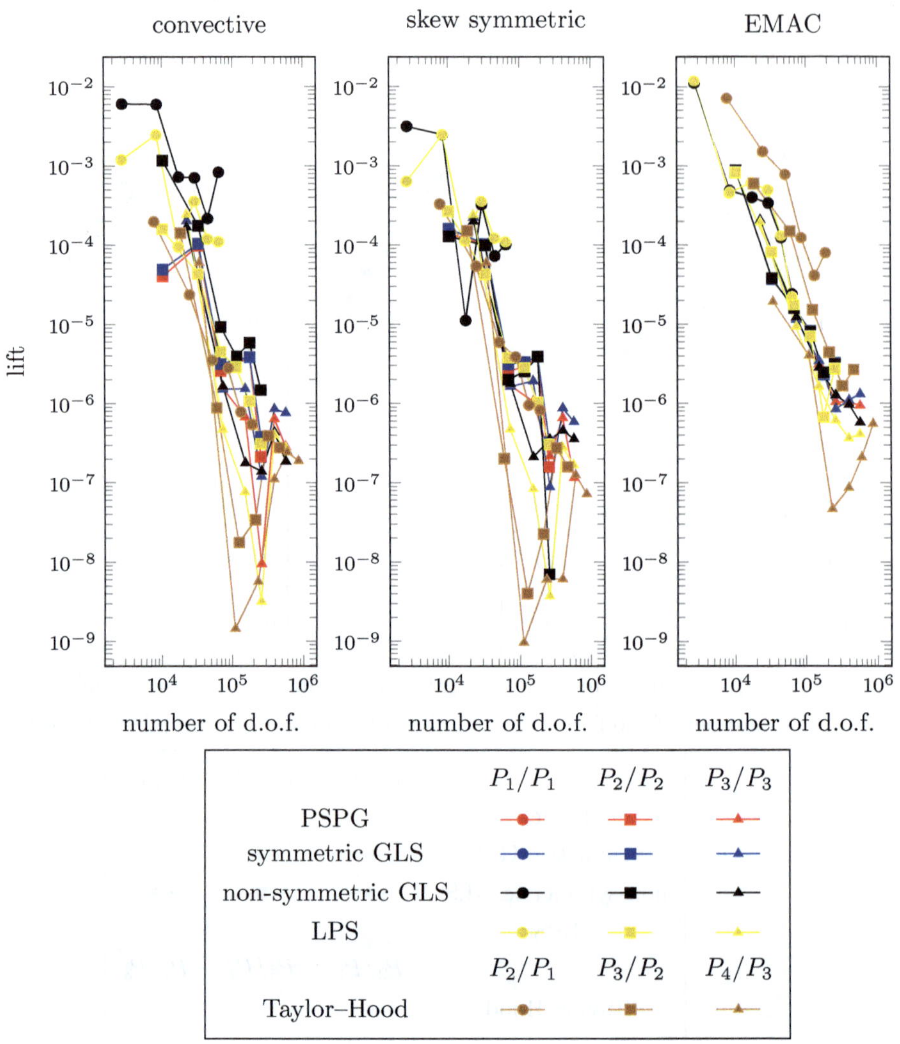

Fig. 6.10 Computed absolute differences to reference value for lift using the convective (left), skew symmetric (center), and EMAC (right) nonlinear form on the unstructured grids, see Table 6.2 and Fig. 6.8

EMAC form of the convective term. It should be noted that there are values of δ_0 for some of the pressure-stabilized discretizations where the nonlinear iteration took much more steps than presented in Fig. 6.12, even reaching the maximal prescribed number was observed.

Very similar results as presented in Figs. 6.9, 6.10, 6.11, 6.12 were obtained on the more structured triangular grid from [56, Figure 6.5].

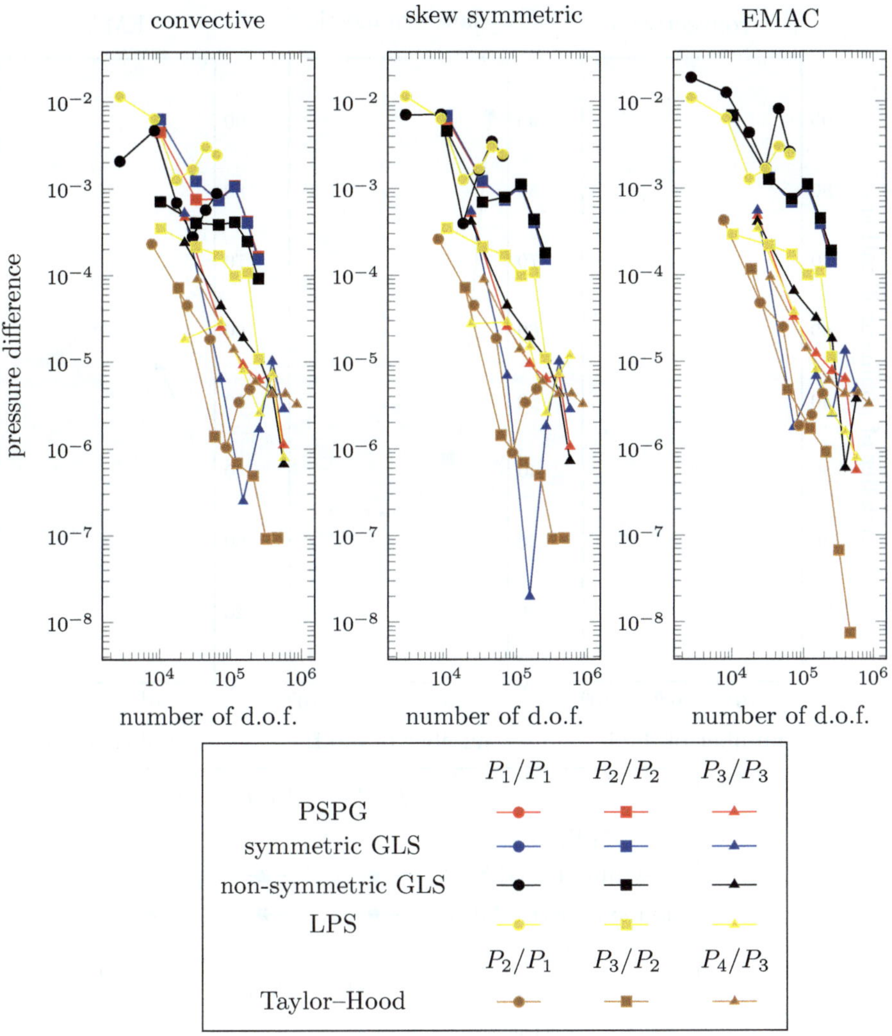

Fig. 6.11 Computed absolute differences to reference value for the pressure difference at the cylinder using the convective (left), skew symmetric (center), and EMAC (right) nonlinear form on the unstructured grids, see Table 6.2 and Fig. 6.8

To summarize, no fundamental differences between the pressure-stabilized discretizations could be observed in this example. However, it could be seen that the benchmark parameters computed with the inf-sup stable Taylor–Hood pair of finite element spaces were often more accurate.

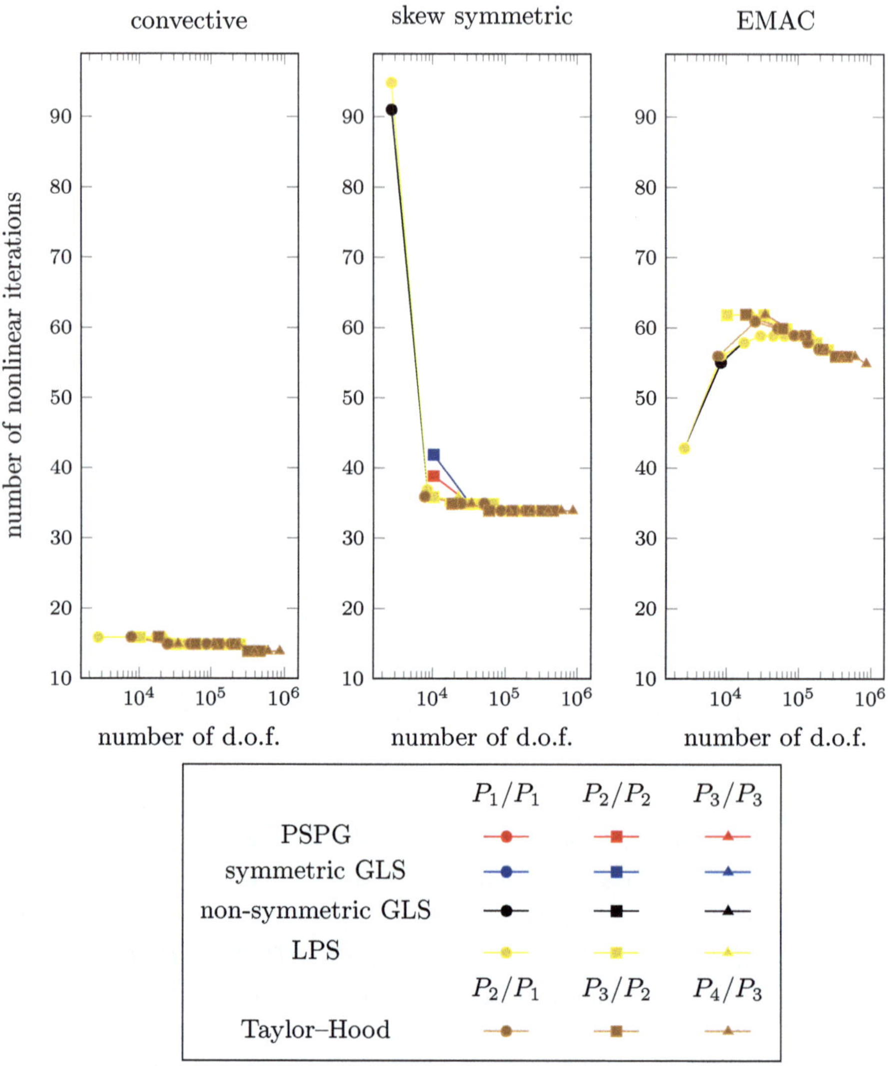

Fig. 6.12 The number of nonlinear iterations needed using the convective (left), skew symmetric (center), and EMAC (right) nonlinear form on the unstructured grids, see Table 6.2 and Fig. 6.8

6.8 Outlook

The Stokes equations (6.2) are the simplest equations for modeling flows with incompressible fluids. This section provides brief comments concerning the application of pressure-stabilized methods to more complicated equations, like the steady-state or time-dependent Navier–Stokes equations (6.1).

Stabilizations that use only the pressure are independent of the type of equation. In particular, for time-dependent problems, the matrix block C in (6.4) has to be assembled only in the initial time step, if the space Q^h does not change in the whole time interval. Later, only the matrix block A changes, due to the nonlinearity of the Navier–Stokes equations. The assembling procedure is more expensive for residual-based stabilizations, since there, the matrix blocks A, B, and D change whenever a new assembling is performed, because the convective field in the nonlinear term of the residual changes.

The matrix block C has for residual-based stabilizations the standard sparsity pattern that comes from the pressure finite element space Q^h. Pressure-based stabilizations require in general an extended sparsity pattern, since degrees of freedom of Q^h are coupled that do not belong to a common mesh cell.

Implementing residual-based stabilizations for certain temporal discretizations, like the Crank–Nicolson scheme, is somewhat involved, since the residual at former time steps is needed. In this respect the use of BDF schemes is easier.

In connection with optimization for flow problems, one finds in the literature, e.g. [23, Sec. 7.5], that symmetric stabilizations are of advantage, since then optimizing and discretizing commute. Stabilizations that use only the pressure possess this property, whereas the only symmetric residual-based stabilization is the symmetric GLS method. But this method has the drawback of being not absolutely stable.

Acknowledgements The work of P. Knobloch was supported through the grant No. 16-03230S of the Czech Science Foundation.

Bibliography

1. Mark Ainsworth and J. Tinsley Oden. A posteriori error estimation in finite element analysis. *Comput. Methods Appl. Mech. Engrg.*, 142(1–2):1–88, 1997.
2. Thomas Apel, Tobias Knopp, and Gert Lube. Stabilized finite element methods with anisotropic mesh refinement for the Oseen problem. *Appl. Numer. Math.*, 58(12):1830–1843, 2008.
3. Rodolfo Araya, Gabriel R. Barrenechea, and Frédéric Valentin. Stabilized finite element methods based on multiscaled enrichment for the Stokes problem. *SIAM J. Numer. Anal.*, 44(1):322–348, 2006.
4. Rodolfo Araya, Gabriel R. Barrenechea, and Frédéric Valentin. A stabilized finite-element method for the Stokes problem including element and edge residuals. *IMA J. Numer. Anal.*, 27(1):172–197, 2007.
5. D. N. Arnold, F. Brezzi, and M. Fortin. A stable finite element for the Stokes equations. *Calcolo*, 21(4):337–344, 1984.
6. Ivo Babuška. Error-bounds for finite element method. *Numer. Math.*, 16:322–333, 1970/1971.
7. Santiago Badia. On stabilized finite element methods based on the Scott-Zhang projector. Circumventing the inf-sup condition for the Stokes problem. *Comput. Methods Appl. Mech. Engrg.*, 247/248:65–72, 2012.
8. Santiago Badia and Ramon Codina. Unified stabilized finite element formulations for the Stokes and the Darcy problems. *SIAM J. Numer. Anal.*, 47(3):1971–2000, 2009.

9. Santiago Badia and Ramon Codina. Stokes, Maxwell and Darcy: a single finite element approximation for three model problems. *Appl. Numer. Math.*, 62(4):246–263, 2012.
10. Claudio Baiocchi, Franco Brezzi, and Leopoldo P. Franca. Virtual bubbles and Galerkin-least-squares type methods (Ga.L.S.). *Comput. Methods Appl. Mech. Engrg.*, 105(1):125–141, 1993.
11. Gabriel R. Barrenechea and Frédéric Valentin. Consistent local projection stabilized finite element methods. *SIAM J. Numer. Anal.*, 48(5):1801–1825, 2010.
12. Gabriel R. Barrenechea and Frédéric Valentin. Beyond pressure stabilization: a low-order local projection method for the Oseen equation. *Internat. J. Numer. Methods Engrg.*, 86(7):801–815, 2011.
13. R. Becker and M. Braack. A finite element pressure gradient stabilization for the Stokes equations based on local projections. *Calcolo*, 38(4):173–199, 2001.
14. R. Becker and M. Braack. A two–level stabilization scheme for the Navier–Stokes equations. In M. Feistauer, V. Dolejší, P. Knobloch, and K. Najzar, editors, *Numerical Mathematics and Advanced Applications*, pages 123–130. Springer-Verlag, Berlin, 2004.
15. Roland Becker and Peter Hansbo. A simple pressure stabilization method for the Stokes equation. *Comm. Numer. Methods Engrg.*, 24(11):1421–1430, 2008.
16. Roland Becker and Rolf Rannacher. Finite element solution of the incompressible Navier-Stokes equations on anisotropically refined meshes. In *Fast solvers for flow problems (Kiel, 1994)*, volume 49 of *Notes Numer. Fluid Mech.*, pages 52–62. Friedr. Vieweg, Braunschweig, 1995.
17. Christine Bernardi and Geneviève Raugel. Analysis of some finite elements for the Stokes problem. *Math. Comp.*, 44(169):71–79, 1985.
18. Laura Blank, Alfonso Caiazzo, Franz Chouly, Alexei Lozinski, and Joaquin Mura. Analysis of a stabilized penalty-free Nitsche method for the Brinkman, Stokes, and Darcy problems. *ESAIM Math. Model. Numer. Anal.*, 52(6):2149–2185, 2018.
19. Pavel Bochev and Max Gunzburger. An absolutely stable pressure-Poisson stabilized finite element method for the Stokes equations. *SIAM J. Numer. Anal.*, 42(3):1189–1207, 2004.
20. Pavel B. Bochev, Clark R. Dohrmann, and Max D. Gunzburger. Stabilization of low-order mixed finite elements for the Stokes equations. *SIAM J. Numer. Anal.*, 44(1):82–101, 2006.
21. Daniele Boffi, Franco Brezzi, and Michel Fortin. *Mixed finite element methods and applications*, volume 44 of *Springer Series in Computational Mathematics*. Springer, Heidelberg, 2013.
22. Thomas Boiveau and Erik Burman. A penalty-free Nitsche method for the weak imposition of boundary conditions in compressible and incompressible elasticity. *IMA J. Numer. Anal.*, 36(2):770–795, 2016.
23. M. Braack, E. Burman, V. John, and G. Lube. Stabilized finite element methods for the generalized Oseen problem. *Comput. Methods Appl. Mech. Engrg.*, 196(4–6):853–866, 2007.
24. F. Brezzi. On the existence, uniqueness and approximation of saddle-point problems arising from Lagrangian multipliers. *Rev. Française Automat. Informat. Recherche Opérationnelle Sér. Rouge*, 8(R-2):129–151, 1974.
25. F. Brezzi and M. Fortin. A minimal stabilisation procedure for mixed finite element methods. *Numer. Math.*, 89(3):457–491, 2001.
26. F. Brezzi and J. Pitkäranta. On the stabilization of finite element approximations of the Stokes equations. In *Efficient solutions of elliptic systems (Kiel, 1984)*, volume 10 of *Notes Numer. Fluid Mech.*, pages 11–19. Friedr. Vieweg, Braunschweig, 1984.
27. Erik Burman. Pressure projection stabilizations for Galerkin approximations of Stokes' and Darcy's problem. *Numer. Methods Partial Differential Equations*, 24(1):127–143, 2008.
28. Erik Burman, Miguel A. Fernández, and Peter Hansbo. Continuous interior penalty finite element method for Oseen's equations. *SIAM J. Numer. Anal.*, 44(3):1248–1274, 2006.
29. Erik Burman and Peter Hansbo. Edge stabilization for the generalized Stokes problem: a continuous interior penalty method. *Comput. Methods Appl. Mech. Engrg.*, 195(19–22):2393–2410, 2006.
30. Erik Burman and Friedhelm Schieweck. Local CIP stabilization for composite finite elements. *SIAM J. Numer. Anal.*, 54(3):1967–1992, 2016.

31. T. Chacón Rebollo, M. Gómez Mármol, V. Girault, and I. Sánchez Muñoz. A high order term-by-term stabilization solver for incompressible flow problems. *IMA J. Numer. Anal.*, 33(3):974–1007, 2013.
32. Sergey Charnyi, Timo Heister, Maxim A. Olshanskii, and Leo G. Rebholz. On conservation laws of Navier–Stokes Galerkin discretizations. *Journal of Computational Physics*, 337:289–308, 2017.
33. Hongsen Chen. Pointwise error estimates for finite element solutions of the Stokes problem. *SIAM J. Numer. Anal.*, 44(1):1–28, 2006.
34. Ramon Codina and Jordi Blasco. A finite element formulation for the Stokes problem allowing equal velocity-pressure interpolation. *Comput. Methods Appl. Mech. Engrg.*, 143(3–4):373–391, 1997.
35. Ramon Codina and Jordi Blasco. Analysis of a pressure-stabilized finite element approximation of the stationary Navier-Stokes equations. *Numer. Math.*, 87(1):59–81, 2000.
36. M. Crouzeix and P.-A. Raviart. Conforming and nonconforming finite element methods for solving the stationary Stokes equations. I. *Rev. Française Automat. Informat. Recherche Opérationnelle Sér. Rouge*, 7(R-3):33–76, 1973.
37. Timothy A. Davis. Algorithm 832: UMFPACK V4.3—an unsymmetric-pattern multifrontal method. *ACM Trans. Math. Software*, 30(2):196–199, 2004.
38. Clark R. Dohrmann and Pavel B. Bochev. A stabilized finite element method for the Stokes problem based on polynomial pressure projections. *Internat. J. Numer. Methods Fluids*, 46(2):183–201, 2004.
39. Jim Douglas, Jr. and Jun Ping Wang. An absolutely stabilized finite element method for the Stokes problem. *Math. Comp.*, 52(186):495–508, 1989.
40. A. Ern and J.-L. Guermond. *Theory and Practice of Finite Elements*. Springer-Verlag, New York, 2004.
41. Michel Fortin. Old and new finite elements for incompressible flows. *Internat. J. Numer. Methods Fluids*, 1(4):347–364, 1981.
42. Leopoldo P. Franca, Thomas J. R. Hughes, and Rolf Stenberg. Stabilized finite element methods. In *Incompressible computational fluid dynamics: trends and advances.*, pages 87–107. Cambridge: Cambridge University Press, 1993.
43. Leopoldo P. Franca and Rolf Stenberg. Error analysis of Galerkin least squares methods for the elasticity equations. *SIAM J. Numer. Anal.*, 28(6):1680–1697, 1991.
44. S. Ganesan, V. John, G. Matthies, R. Meesala, S. Abdus, and U. Wilbrandt. An object oriented parallel finite element scheme for computing PDEs: Design and implementation. In *IEEE 23rd International Conference on High Performance Computing Workshops (HiPCW) Hyderabad*, pages 106–115. IEEE, 2016.
45. Christophe Geuzaine and Jean-François Remacle. Gmsh: A 3-D finite element mesh generator with built-in pre- and post-processing facilities. *Internat. J. Numer. Methods Engrg.*, 79(11):1309–1331, 2009.
46. V. Girault, R. H. Nochetto, and L. R. Scott. Max-norm estimates for Stokes and Navier-Stokes approximations in convex polyhedra. *Numer. Math.*, 131(4):771–822, 2015.
47. V. Girault, R. H. Nochetto, and R. Scott. Maximum-norm stability of the finite element Stokes projection. *J. Math. Pures Appl.*, 84(3):279–330, 2005.
48. V. Girault and L. R. Scott. A quasi-local interpolation operator preserving the discrete divergence. *Calcolo*, 40(1):1–19, 2003.
49. Vivette Girault and Pierre-Arnaud Raviart. *Finite element methods for Navier-Stokes equations*, volume 5 of *Springer Series in Computational Mathematics*. Springer-Verlag, Berlin, 1986. Theory and algorithms.
50. Thirupathi Gudi. A new error analysis for discontinuous finite element methods for linear elliptic problems. *Math. Comp.*, 79(272):2169–2189, 2010.
51. J. Guzmán and D. Leykekhman. Pointwise error estimates of finite element approximations to the Stokes problem on convex polyhedra. *Math. Comp.*, 81(280):1879–1902, 2012.
52. Johnny Guzmán and Manuel A. Sánchez. Max-norm stability of low order Taylor-Hood elements in three dimensions. *J. Sci. Comput.*, 65(2):598–621, 2015.

53. Isaac Harari and Thomas J. R. Hughes. What are C and h?: inequalities for the analysis and design of finite element methods. *Comput. Methods Appl. Mech. Engrg.*, 97(2):157–192, 1992.
54. Thomas J. R. Hughes and Leopoldo P. Franca. A new finite element formulation for computational fluid dynamics. VII. The Stokes problem with various well-posed boundary conditions: symmetric formulations that converge for all velocity/pressure spaces. *Comput. Methods Appl. Mech. Engrg.*, 65(1):85–96, 1987.
55. Thomas J. R. Hughes, Leopoldo P. Franca, and Marc Balestra. A new finite element formulation for computational fluid dynamics. V. Circumventing the Babuška-Brezzi condition: a stable Petrov-Galerkin formulation of the Stokes problem accommodating equal-order interpolations. *Comput. Methods Appl. Mech. Engrg.*, 59(1):85–99, 1986.
56. Volker John. *Finite element methods for incompressible flow problems*, volume 51 of *Springer Series in Computational Mathematics*. Springer, Cham, 2016.
57. Volker John, Petr Knobloch, and Julia Novo. Finite elements for scalar convection-dominated equations and incompressible flow problems: a never ending story? *Comput. Vis. Sci.*, 19(5–6):47–63, 2018.
58. Volker John, Alexander Linke, Christian Merdon, Michael Neilan, and Leo G. Rebholz. On the divergence constraint in mixed finite element methods for incompressible flows. *SIAM Rev.*, 59(3):492–544, 2017.
59. Volker John and Gunar Matthies. Higher-order finite element discretizations in a benchmark problem for incompressible flows. *Int. J. Numer. Methods Fluids*, 37(8):885–903, 2001.
60. Nasserdine Kechkar and David Silvester. Analysis of locally stabilized mixed finite element methods for the Stokes problem. *Math. Comp.*, 58(197):1–10, 1992.
61. P. Knobloch. A generalization of the local projection stabilization for convection–diffusion–reaction equations. *SIAM J. Numer. Anal.*, 48:659–680, 2010.
62. Petr Knobloch. Reduced finite element discretizations of the Stokes and Navier-Stokes equations. *Numer. Funct. Anal. Optim.*, 27(2):161–187, 2006.
63. Petr Knobloch and Lutz Tobiska. Stabilization methods of bubble type for the Q_1/Q_1-element applied to the incompressible Navier-Stokes equations. *M2AN Math. Model. Numer. Anal.*, 34(1):85–107, 2000.
64. Petr Knobloch and Lutz Tobiska. Improved stability and error analysis for a class of local projection stabilizations applied to the Oseen problem. *Numer. Methods Partial Differential Equations*, 29(1):206–225, 2013.
65. W. Layton. Model reduction by constraints, discretization of flow problems and an induced pressure stabilization. *Numer. Linear Algebra Appl.*, 12(5–6):547–562, 2005.
66. G. Leborgne. An optimally consistent stabilization of the inf-sup condition. *Numer. Math.*, 91(1):35–56, 2002.
67. Jian Li and Yinnian He. A stabilized finite element method based on two local Gauss integrations for the Stokes equations. *J. Comput. Appl. Math.*, 214(1):58–65, 2008.
68. Jian Li, Yinnian He, and Zhangxin Chen. Performance of several stabilized finite element methods for the Stokes equations based on the lowest equal-order pairs. *Computing*, 86(1):37–51, 2009.
69. Qifeng Liao and David Silvester. Robust stabilized Stokes approximation methods for highly stretched grids. *IMA J. Numer. Anal.*, 33(2):413–431, 2013.
70. Alexander Linke. On the role of the Helmholtz decomposition in mixed methods for incompressible flows and a new variational crime. *Comput. Methods Appl. Mech. Engrg.*, 268:782–800, 2014.
71. Alexander Linke, Gunar Matthies, and Lutz Tobiska. Robust arbitrary order mixed finite element methods for the incompressible Stokes equations with pressure independent velocity errors. *ESAIM Math. Model. Numer. Anal.*, 50(1):289–309, 2016.
72. G. Matthies, P. Skrzypacz, and L. Tobiska. A unified convergence analysis for local projection stabilizations applied to the Oseen problem. M2AN Math. Model. Numer. Anal., 41(4):713–742, 2007.
73. Stefano Micheletti, Simona Perotto, and Marco Picasso. Stabilized finite elements on anisotropic meshes: a priori error estimates for the advection-diffusion and the Stokes problems. *SIAM J. Numer. Anal.*, 41(3):1131–1162, 2003.

74. P. Mons and G. Rogé. L'élément Q_1-bulle/Q_1. *RAIRO Modél. Math. Anal. Numér.*, 26(4):507–521, 1992.
75. Guido Nabh. *On higher order methods for the stationary incompressible Navier-Stokes equations.* Heidelberg: Univ. Heidelberg, Naturwissenschaftlich-Mathematische Gesamtfakultät, 1998. Ph.D. thesis.
76. Kamel Nafa and Andrew J. Wathen. Local projection stabilized Galerkin approximations for the generalized Stokes problem. *Comput. Methods Appl. Mech. Engrg.*, 198(5–8):877–883, 2009.
77. Eugenio Oñate, Prashanth Nadukandi, Sergio R. Idelsohn, Julio García, and Carlos Felippa. A family of residual-based stabilized finite element methods for Stokes flows. *Internat. J. Numer. Methods Fluids*, 65(1–3):106–134, 2011.
78. Roger Pierre. Simple C^0 approximations for the computation of incompressible flows. *Comput. Methods Appl. Mech. Engrg.*, 68(2):205–227, 1988.
79. Roger Pierre. Regularization procedures of mixed finite element approximations of the Stokes problem. *Numer. Methods Partial Differential Equations*, 5(3):241–258, 1989.
80. M. Schäfer and S. Turek. Benchmark computations of laminar flow around a cylinder. (With support by F. Durst, E. Krause and R. Rannacher). In *Flow simulation with high-performance computers II. DFG priority research programme results 1993–1995*, pages 547–566. Wiesbaden: Vieweg, 1996.
81. D. J. Silvester and N. Kechkar. Stabilised bilinear-constant velocity-pressure finite elements for the conjugate gradient solution of the Stokes problem. *Comput. Methods Appl. Mech. Engrg.*, 79(1):71–86, 1990.
82. David Silvester. Optimal low order finite element methods for incompressible flow. *Comput. Methods Appl. Mech. Engrg.*, 111(3–4):357–368, 1994.
83. Rolf Stenberg and Juha Videman. On the error analysis of stabilized finite element methods for the Stokes problem. *SIAM J. Numer. Anal.*, 53(6):2626–2633, 2015.
84. Ulrich Wilbrandt, Clemens Bartsch, Naveed Ahmed, Najib Alia, Felix Anker, Laura Blank, Alfonso Caiazzo, Sashikumaar Ganesan, Swetlana Giere, Gunar Matthies, Raviteja Meesala, Abdus Shamim, Jagannath Venkatesan, and Volker John. ParMooN—A modernized program package based on mapped finite elements. *Comput. Math. Appl.*, 74(1):74–88, 2017.

Chapter 7
Finite-Volume Methods for Navier-Stokes Equations

Milovan Perić

Abstract Flows of engineering interest can only be predicted by using numerical solution methods, because the governing equations cannot be solved analytically. Among many possible approaches, finite-volume methods have become popular in this field and are the basis for most commercial and public computer codes used to simulate fluid flow. One of the most widely used methods, which is applicable to complex geometries and arbitrary polyhedral computational grids, is described in this chapter. The solution method was originally developed for incompressible flows and later extended to compressible flows at any speed. It is also described how to deal with moving grids which follow the motion of solid bodies. The applicability of the method to various flow types is demonstrated by simulating flow around sphere at various Reynolds numbers, ranging from a creeping flow at Re = 5 to a supersonic flow at Re = 5,000,000. The final example involves a sphere oscillating in water at 6000 Hz with a very small amplitude, which requires that compressibility of water is taken into account. The fact that the same method can be applied to such a wide class of flows, including a wide range of turbulence-modeling approaches, is the main reason why it is used in general-purpose commercial codes.

7.1 Introduction

Pressure plays a major role in all flows of engineering interest. In many cases, it is the sole driving force causing the fluid to flow (e.g., internal flows driven by pumps, fans, turbines, etc.). Even when the driving force is gravity (or other body forces), pressure plays a prominent role because it reacts to any change in geometry and is responsible for changes in flow direction, secondary flow, and many other phenomena. Environmental flows—in rivers, lakes, oceans, and the atmosphere—are also strongly affected by pressure variation, as is obvious from weather forecast

M. Perić (✉)
University of Duisburg-Essen, Duisburg, Germany
e-mail: milovan.peric@uni-due.de

T. Bodnár et al. (eds.), *Fluids Under Pressure*, Advances in Mathematical Fluid Mechanics, https://doi.org/10.1007/978-3-030-39639-8_7

maps which always refer to pressure highs and laws (which may last for several days and are so important that they are always given a name).

There are many mathematical models which describe fluid flow; pressure is part of all of them. The most general model are the Navier–Stokes equations, which have been derived about 200 years ago. However, these equations cannot be solved analytically, except for a handful of special cases in which so many terms in the equations can be neglected that they eventually become solvable (e.g., a fully developed flow in a pipe or between infinite parallel plates). All other models are based on simplifications of the Navier–Stokes equations. Euler equations are obtained by neglecting the viscous effects (i.e., treating the fluid as inviscid). While flows away from solid walls can often be described reasonably well by the Euler equations, they cannot account for boundary-layer effects and lead to zero net forces (drag and lift) exerted by flow on submerged solid bodies. Stokes equations neglect the inertia and are thus linear; they can be used to study flows at very low speeds (Reynolds numbers below 1). The simplest model is the assumption that the fluid is inviscid and the velocity field is irrotational—this leads to a *potential flow* approximation. Reynolds-averaged Navier–Stokes equations are obtained by performing averaging of all variables and the Navier–Stokes equations themselves, either over a sufficiently long time period or over a sufficiently large number of flow realizations. These equations do not resolve the temporal fluctuations, whose effect has to be approximated by using the so-called *turbulence models*.

Because none of the above-mentioned equations can be solved analytically, the only option left for predicting fluid flows is to use numerical solution methods. These methods, by definition, can only provide *approximate solutions*. The accuracy of *numerical solutions* depends on the quality of discrete approximations (interpolation, differentiation, integration), and the fineness of the space and time discretization. However, even the solution with an arbitrarily small numerical error may not represent the real flow accurately if the equations solved are not an accurate *mathematical model*. Only the Navier–Stokes equations of all the models described above deliver an accurate representation of real flows if the numerical errors are small enough. However, to achieve this (what is called a *direct numerical simulation* or DNS) for engineering flows at high Reynolds numbers is impractical, if not impossible, with current computers. The reason is that the scales (spatial and temporal) of fluctuations in turbulent flows reduce over-proportionally with increasing Reynolds number; the computing effort for DNS is roughly proportional to $(\mathrm{Re})^3$. Therefore, for many engineering applications using simplified equations is the only practical possibility.

There are many numerical methods which can be used to obtain approximate solutions of the Navier–Stokes equations. We shall deal here only with finite-volume (FV) methods, which are most widely used for this purpose among engineers and are the basis of all major commercial and public CFD software (CFD stands for *computational fluid dynamics*). The reasons for the popularity of this approach are:

- FV methods are based on the integral form of conservation equations and are thus by definition fully conservative. This imposes some limits on numerical errors and ensures that even solutions obtained with relatively coarse discretizations are physically meaningful and practically useful.
- All terms in the equations which need to be approximated have a physical meaning with which engineers are familiar (mass flux, momentum flux, pressure force, shear force, heat flux, volumetric source or sink, etc.). Many approximations are derived by relying on the understanding of the physical nature of the term which is being approximated (e.g., upwind-biased approximations of convection terms).
- FV methods are mathematically extremely simple; most of the widely used FV methods have been developed by engineers for engineers (mathematicians usually use finite-element methods, which are more mathematical and, although in principle equally applicable to solving the Navier–Stokes equations, not widely used in engineering practice).

In the next section FV method for arbitrary polyhedral control volumes is described; it corresponds to a large degree to methods used in most commercial and public CFD codes. The following two sections describe the two most widely used approaches for computing incompressible flows: fractional-step methods and pressure-correction methods based on the so-called SIMPLE algorithm. The extension of these methods to compressible flows is also described, although there are many other approaches derived specifically for compressible flows. Many engineering flows are subject to moving boundaries; the following section describes the handling of moving grids. The final section is devoted to some illustrative examples of application of the described methods, covering the range of flows from creeping (Reynolds number around 1) to hypersonic (Mach number around 5).

7.2 Finite-Volume Methods for Arbitrary Polyhedral Control Volumes

Finite-Volume (FV) methods start with mass and momentum conservation equations written in integral form. If the heat transfer needs to be considered (which is always the case when the flow is compressible), the energy conservation equation in integral form is included in the set of equations to be solved. Other equations describing transport or conservation of scalar quantities (like species concentration, kinetic energy of turbulence, etc.) may also need to be solved. All these equations represent a coupled system of non-linear equations whose solution may require special attention.

The *integral forms* of the mass conservation (continuity) and momentum equations (collectively called the Navier–Stokes equations) for a fixed control volume (CV) read

$$\frac{\partial}{\partial t}\int_V \rho \, \mathrm{d}V + \int_S \rho \boldsymbol{v}\cdot\boldsymbol{n} \, \mathrm{d}S = 0 \, . \tag{7.1}$$

$$\frac{\partial}{\partial t}\int_V \rho\boldsymbol{v} \, \mathrm{d}V + \int_S \rho\boldsymbol{v}\boldsymbol{v}\cdot\boldsymbol{n} \, \mathrm{d}S = \int_S \operatorname{div} \boldsymbol{T}\cdot\boldsymbol{n} \, \mathrm{d}S + \int_V \rho\boldsymbol{b} \, \mathrm{d}V \, . \tag{7.2}$$

Here ρ stands for fluid density, $\boldsymbol{v}$ is the fluid velocity, S is the CV surface with $\boldsymbol{n}$ being the unit normal vector directed outwards, V is the CV volume, $\boldsymbol{T}$ is the tensor representing surface forces due to pressure and viscous stresses, and $\boldsymbol{b}$ represents volumetric forces like gravity.

Most fluids encountered in engineering—gases as well as liquids—belong to the class called *Newtonian*; for such fluids the stress tensor $\boldsymbol{T}$ can be expressed as a function of pressure, velocity and viscosity, as follows:

$$\boldsymbol{T} = -\left(p + \frac{2}{3}\mu \operatorname{div} \boldsymbol{v}\right)\boldsymbol{I} + 2\mu\boldsymbol{D} \, , \tag{7.3}$$

where μ is the dynamic viscosity, $\boldsymbol{I}$ is the unit tensor, p is the static pressure, and $\boldsymbol{D}$ is the rate of strain (deformation) tensor:

$$\boldsymbol{D} = \frac{1}{2}\left[\nabla\boldsymbol{v} + (\nabla\boldsymbol{v})^{\mathrm{T}}\right] \, . \tag{7.4}$$

Both density and dynamic viscosity are fluid properties which are usually determined experimentally; they are, in general, functions of temperature, pressure, and concentration of chemical species, but in many practical applications they are assumed constant.

The same relation between the stress tensor and the velocity applies also for some non-Newtonian fluids, but the viscosity is then a function of the strain rate and/or temperature. Other non-Newtonian fluids require special treatment and this will not be covered here.

A *coordinate-free vector form* of the Navier–Stokes equations is readily obtained by applying Gauss' divergence theorem to Eqs. (7.1) and (7.2):

$$\frac{\partial\rho}{\partial t} + \operatorname{div}(\rho\boldsymbol{v}) = 0 \, . \tag{7.5}$$

$$\frac{\partial(\rho\boldsymbol{v})}{\partial t} + \operatorname{div}(\rho\boldsymbol{v}\boldsymbol{v}) = \operatorname{div} \boldsymbol{T} + \rho\boldsymbol{b} \, . \tag{7.6}$$

While the discretization is based on the integral form of equations, the vector form is more suitable for some aspects of numerical analysis, e.g., for deriving an equation for pressure or pressure correction when studying incompressible flows. This aspect will be addressed in the next section.

All fluids are, in principle, compressible; however, the effects of compressibility can often be neglected. In most engineering applications, the flow can be considered incompressible if Mach number is below ca. 0.3 and density can be assumed constant. However, density variation may have to be accounted for even if the Mach number is very low if buoyancy effects are important. Compressibility effects in liquids are usually only important if the amplitudes of pressure variation are high.

There are many numerical solution methods for compressible flows, i.e., for flows in which density variations are significant and the Mach number is the relevant parameter. The development of such methods was mostly driven by military applications from the cold-war era. We shall not deal with such methods here but rather concentrate on methods originally developed for incompressible flows and later extended to handle compressible flows as well. The most widely used commercial CFD codes as well as the public toolkit OpenFOAM are based on such methods.

When studying flows around moving bodies, one can often reduce the complexity of the problem by using a moving (body-fixed) coordinate system. For example, flow around a moving body is unsteady when viewed from a fixed coordinate frame (i.e., for an observer who does not move), but it may be steady when viewed from a coordinate system attached to the body and moving with it.

If the coordinate system itself moves, the above equations need to be modified; the momentum equation obtains then on the left-hand side the following extra terms:

$$\int_V \rho \left[\boldsymbol{a}_0 + \left(\frac{\mathrm{d}\boldsymbol{\omega}}{\mathrm{d}t} \times \boldsymbol{r} \right) + [\boldsymbol{\omega} \times (\boldsymbol{\omega} \times \boldsymbol{r})] + (2\boldsymbol{\omega} \times \boldsymbol{v}) \right] \mathrm{d}V. \tag{7.7}$$

Here $\boldsymbol{a}_0$ is the acceleration of the coordinate-system origin and $\boldsymbol{\omega}$ is the angular velocity vector describing the rotation of the coordinate system. However, all these extra terms vanish if the body moves linearly with a constant velocity; in that case equations are the same for both fixed and moving coordinate system. The difference lies in the meaning of the velocity vector $\boldsymbol{v}$: in the case of a coordinate system moving with the body it represents the velocity relative to the body, which is equal to zero at body surface (no-slip boundary condition). One can in such a situation easily convert relative velocity into absolute velocity and vice versa by adding or subtracting the velocity of the coordinate system.

The analysis in relative terms is often used in experiments as well as in numerical studies. In a wind tunnel, the car (or another vehicle) model is stationary and the air (and, if one wants to be faithful to nature, the floor) is moving; this approach simplifies the analysis greatly. If the body motion is irregular, the flow is unsteady from any viewpoint and there is no benefit from changing the reference frame. In some situations, when studying flow around bodies moving relative to each other, we have to move the grid (i.e., the control volumes) as well; this will be addressed in Sect. 7.5.

Often fluid flow is accompanied by other transport phenomena, like heat and mass transfer. This requires solution of additional scalar transport equations. The integral form of the equation describing conservation of a generic scalar quantity, ϕ, is analogous to the momentum equation and reads

$$\frac{\partial}{\partial t}\int_V \rho\phi \, \mathrm{d}V + \int_S \rho\phi\boldsymbol{v}\cdot\boldsymbol{n}\, \mathrm{d}S = \int_S \Gamma_\phi \nabla\phi\cdot\boldsymbol{n}\, \mathrm{d}S + \int_V q_\phi \, \mathrm{d}V \, , \tag{7.8}$$

where Γ_ϕ is the diffusivity for the quantity ϕ and q_ϕ is the volumetric source or sink. An example is the energy equation which, for most engineering flows, has the following form:

$$\frac{\partial}{\partial t}\int_V \rho h \mathrm{d}V + \int_S \rho h\boldsymbol{v}\cdot\boldsymbol{n}\, \mathrm{d}S = \int_S k\nabla T\cdot\boldsymbol{n}, \mathrm{d}S + \int_V (\boldsymbol{v}\cdot\nabla p + \boldsymbol{S} : \nabla\boldsymbol{v})\mathrm{d}V + \frac{\partial}{\partial t}\int_V p\mathrm{d}V, \tag{7.9}$$

where h is the enthalpy, T is the temperature, k is the thermal conductivity, $k = \mu c_\mathrm{p}/\mathrm{Pr}$, $\boldsymbol{S}$ is the viscous part of the stress tensor, $\boldsymbol{S} = \boldsymbol{T} + p\boldsymbol{I}$, Pr is the Prandtl number, and c_p is the specific heat at constant pressure. The source term represents work done by pressure and viscous forces; it may be neglected in incompressible flows.

Species concentration equations have the same form, with T replaced by the concentration c and Pr replaced by Sc, the Schmidt number.

The discretization and the analysis of the numerical method is usually carried out for the generic equation (7.8); when necessary, terms peculiar to one equation are explained separately.

We shall assume here that the control volumes have an arbitrary polyhedral shape. There are many special methods designed for Cartesian or other control volumes of regular shape or topology; these are special cases of the general method described below. We also assume that the grid is fitted to the solution domain boundary, i.e., the boundary conditions are directly applied at the CV-faces coinciding with the boundaries. Special methods—like immersed-boundary or penalty methods—which use regular grids and apply special discretization for cells cut by physical boundaries will not be covered here. For more details on immersed-boundary methods, see the review article by Peskin [26] or detailed descriptions in dissertations, e.g., by Peller [23] and Hylla [17], among others.

When using Cartesian base vectors and CVs defined by a certain number of vertexes connected by straight lines, the shape of the cell face is not important; because it is bounded by straight line segments, its projections onto Cartesian coordinate surfaces (which represent surface vector components) are the same, whatever the shape. Cartesian base vectors are preferred because the momentum equations then have the simplest form (no so-called curvature terms or apparent forces, which are needed when base vectors are changing direction). For this reason, CVs are defined by vertexes connected by straight lines.

The grids can be of various types (structured, block-structured, or unstructured); however, if arbitrary polyhedral CVs are allowed, the grid has to be assumed unstructured. This means that the location of grid entities relative to each other is not implicitly assumed based on indexes, but the data is rather sorted in vertex, face, and volume lists. First, all vertexes are listed with their index and three Cartesian coordinates. Then, face list is created, in which each face of each cell (note that internal faces are common to two cells; they are listed only once, but they include pointers to cells on either side) is defined by its index and indexes of vertexes which define a closed polygon (vertexes are connected by straight line segments in the order they are listed, with the last vertex being connected with the first one to close the polygon). Finally, list of cells is created, with cell index and the list of faces which enclose it.

Information that is stored in the face list also includes:

- Surface vector components (face projections onto Cartesian coordinate surfaces);
- Coordinates of the face centroid;
- Indexes of the cells on either side of the face (the convention is usually that the surface vector points from the first to the second listed cell);
- Coefficients of the matrix A for cells on each side which multiply neighbor cell variable value.

Information that is stored in cell list includes, among others:

- Cell volume;
- Coordinates of the cell centroid;
- Variable values and fluid properties;
- The coefficient from the main diagonal, A_P, from the matrix A (see Eq. (7.32)).

Polyhedral CVs are usually obtained by creating control volumes around vertexes of a tetrahedral grid, which is created first (but there are other approaches as well). One such possibility is demonstrated in two dimensions (2D) in Fig. 7.1. Tetrahedra are split into four hexahedra by connecting midpoints on edges with centroids of faces and volume (in 2D, this corresponds to Voronoi diagram—a dual to Delaunay triangulation). The control volume around one tetrahedron vertex is defined by joining all hexahedra created by splitting tetrahedra which share that vertex. All sub-faces shared by the same two control volumes are merged together and the computational point is placed in the CV-centroid; tetrahedral grid is then discarded. This minimizes the number of faces of polyhedral CVs.

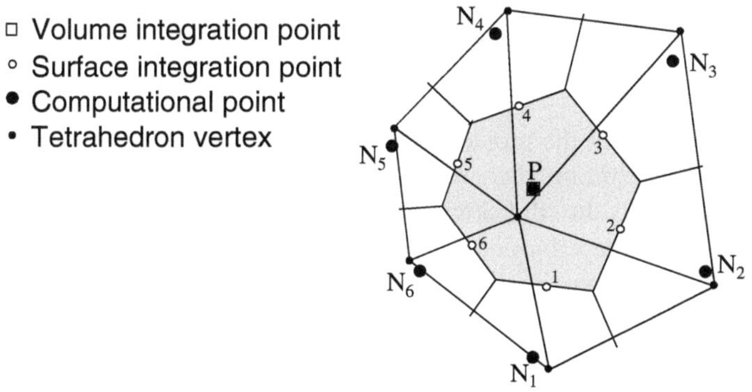

Fig. 7.1 On the construction of polyhedral control volumes from tetrahedral grids

Because the shape of a surface enclosed by straight edges is not important for the approximations of integrals described in the following subsections, one can choose a suitable, unique definition of the surface which makes computation of surface vectors and cell volume easier and ensures that CVs do not overlap. One possibility is to select an arbitrary auxiliary point in each face (e.g., one obtained by averaging the coordinates of all vertexes defining the face) and an auxiliary point inside cell (e.g., one obtained by averaging the coordinates of all vertexes defining the cell). Each face can then be subdivided into triangles by connecting each vertex with the auxiliary face point; from the centroid and surface vector of each triangle, one can easily obtain the centroid and surface vector of the face. Similarly, by connecting each vertex and each auxiliary face point with the auxiliary cell point, one can define tetrahedra and from their volume and centroid, one can obtain volume and centroid of the cell.

FV methods involve three levels of approximation: (i) integrals over surface, volume, or time; (ii) interpolation to obtain values at locations other than CV-center; (iii) differentiation, needed to compute gradients at CV-center or face center. The common approximations of surface and volume integrals from the conservation equation (7.8) are described next, followed by a description of most widely used interpolation and differentiation methods. More details—including an analysis of discretization errors of various combinations of approximations—can be found in a book by Ferziger et al. [13].

7.2.1 *Approximation of Surface and Volume Integrals for Polyhedral CVs*

The simplest approximation of surface integrals—applicable to any polygonal shape—is the midpoint-rule approximation. The flux of any vector $\boldsymbol{f}$ ($\boldsymbol{f} = \rho\phi\boldsymbol{v}$ for convection and $\boldsymbol{f} = \Gamma_\phi \nabla\phi$ for diffusion fluxes) through the face S_k (see Fig. 7.2) can be approximated as:

$$F_k = \int_{S_k} \boldsymbol{f} \cdot \boldsymbol{n} \, \mathrm{d}S \approx (\boldsymbol{f} \cdot \boldsymbol{n})_k S_k = \sum_i f_k^i S_k^i \, , \tag{7.10}$$

where S_k^i is the area of the projection of cell face k onto the Cartesian coordinate surface x_i = const. (or, in other words, the ith Cartesian component of the surface vector $S\boldsymbol{n}$), and f_k^i is the ith Cartesian component of the vector $\boldsymbol{f}$ at the center of the cell face. Thus, the values of f_k^i are taken to represent the mean value over the whole cell face, an approximation which is exact if the variation of the function is linear.

The truncation error of this integral approximation (when the *exact* values of f_k^i or approximations of second or higher order are used) is proportional to the square of the mesh spacing, i.e., it is a *second-order approximation*. This is the most widely

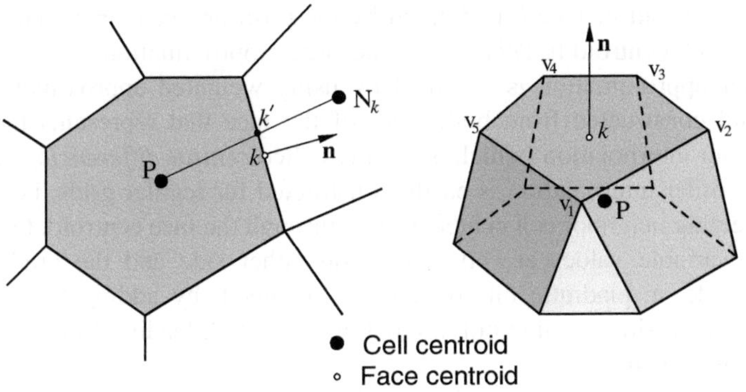

Fig. 7.2 A typical polyhedral CV and the notation used

used approximation; its second order makes it accurate enough for most engineering applications, and its simplicity makes the implementation in a computer code easy. Higher-order integral approximations are not so easy to develop for an arbitrary CV shape; this is why all major commercial codes use midpoint-rule approximation for integrals.

The simplest method to approximate a volume integral for arbitrary polyhedral cells is again the midpoint rule:

$$\int_V q_\phi \, \mathrm{d}V \approx (q_\phi)_\mathrm{P} \Delta V \ , \tag{7.11}$$

where $(q_\phi)_\mathrm{P}$ stands for the value of the specific source term at the CV-center (which corresponds to the computational node P). Thus, no interpolation is necessary to evaluate the integrand (the differentiation may, however, be necessary, depending on the particular expression for q_ϕ, which often does involve gradients of some quantities; for example, source terms in equations for most variables in widely used turbulence models involve velocity gradients).

This approximation is also of second order when the point P lies in the centroid of the volume ΔV.

7.2.2 *Interpolation Schemes*

There are many interpolation methods to choose from when variable values at face centroids are computed. For example, one-sided linear extrapolation (called "linear upwind scheme" if node P is upstream of the face centroid k) uses the variable value and the gradient at the cell centroid on one side of the face to approximate the value of the variable ϕ at the face centroid k:

$$\phi_k = \phi_\mathrm{P} + (\nabla \phi)_\mathrm{P} \cdot (\boldsymbol{r}_k - \boldsymbol{r}_\mathrm{P}) \ , \tag{7.12}$$

where the centroid of face k is defined by the position vector $\boldsymbol{r}_k$; analogously, $\boldsymbol{r}_{\mathrm{P}}$ defines the cell centroid P. This is a second-order approximation.

Another approximation is obtained by using weighted approximations of the above kind constructed from both sides of the face that represents the centered second-order interpolation (which is usually called *central-differencing scheme*).

Higher-order interpolation is easily constructed for regular grids, i.e., when the line connecting neighbor cell centers passes through the face centroid. For example, by using variable values at cell centers on either side and the gradient at the upstream side, a quadratic interpolation is obtained; by adding the gradient on the downstream side, a cubic polynomial can be fitted, leading to the fourth-order centered interpolation.

Note that higher-order interpolation when computing face centroid values does not increase the order of flux approximation, if midpoint rule is used to approximate surface integrals; its order is second even if the exact value of the variable at the face centroid is used. On the other hand, if a first-order upwind scheme is used to approximate the face centroid value (i.e., either ϕ_{P} or ϕ_{N_k}, depending on flow direction), then the overall approximation of the integral becomes first-order accurate, in spite of the second-order accuracy of the midpoint-rule approximation of the integral. A detailed explanation and demonstration of these issues is available in the book by Ferziger et al. [13].

7.2.3 *Differentiation Schemes*

The gradient of a variable at the CV-center is often required—for interpolation to face centroid (as in approximations of convection fluxes), to compute diffusion fluxes, and for source terms in turbulence model equations. A simple second-order approximation can be obtained using Gauss' theorem and face values of the variable. The derivative at CV-center can be evaluated using midpoint-rule approximation of the volume integral as follows:

$$\left(\frac{\partial \phi}{\partial x_i}\right)_{\mathrm{P}} \approx \frac{\int_V \frac{\partial \phi}{\partial x_i}\,\mathrm{d}V}{\Delta V} \,. \tag{7.13}$$

Because the derivative $\partial\phi/\partial x_i$ can be interpreted as divergence of the vector $\phi\,\boldsymbol{i}_i$, one can transform the volume integral in the above equation into a surface integral:

$$\int_V \frac{\partial \phi}{\partial x_i}\,\mathrm{d}V = \int_S \phi\,\boldsymbol{i}_i \cdot \boldsymbol{n}\,\mathrm{d}S \approx \sum_k \phi_k S_k^i \,. \tag{7.14}$$

Therefore, the components of the gradient vector at CV-center can be computed as follows:

$$\left(\frac{\partial \phi}{\partial x_i}\right)_{\mathrm{P}} \approx \frac{\sum_k \phi_k S_k^i}{\Delta V} \,. \tag{7.15}$$

In the above expression one can use the same values of ϕ_k which are used to calculate the convection flux; however, one can also use different approximations, as long as they are consistent. This approximation can be applied to arbitrary polyhedral grids.

Another second-order approximation of the gradient at cell center (node P) on any grid can be obtained by assuming linear variation of the variable in the vicinity of node P (linear shape function) and using central-difference approximations of derivatives along lines which connect the node P with neighbor nodes, e.g.:

$$(\nabla \phi)_{\mathrm{P}} \cdot (\boldsymbol{r}_{N_k} - \boldsymbol{r}_{\mathrm{P}}) \approx \phi_{N_k} - \phi_{\mathrm{P}} \,. \tag{7.16}$$

There are as many such expressions as there are neighbors of node P (the minimum is four for a tetrahedral CV). Thus, there are more equations than unknowns (three components of the gradient vector) so the least-squares method has to be used to compute the gradient; see Demirdžić and Muzaferija [9] for more details. Approximations of higher order are more difficult to construct for arbitrary polyhedral CVs; most commercial and public CFD codes employ one or both of the above methods.

In order to calculate the diffusion flux through a cell face using midpoint-rule approximation, one needs the gradient of ϕ at the cell-face center (or its component in the direction normal to cell face):

$$F_k^{\mathrm{d}} = \int_{S_k} \Gamma_\phi \nabla \phi \cdot \boldsymbol{n} \,\mathrm{d}S \approx (\Gamma_\phi \nabla \phi \cdot \boldsymbol{n})_k \, S_k \approx \left(\Gamma_\phi \frac{\partial \phi}{\partial n}\right)_k S_k \,. \tag{7.17}$$

The cell-center gradients can be interpolated to the cell-face centers; however, oscillatory solutions may develop in this case (see Ferziger et al. [13], for a detailed discussion). A large number of methods to deal with this issue has been developed over the past 20 years; Demirdžić [7] gives an overview and points out to one of the best approximations, which is given here without derivation:

$$F_k^{\mathrm{d}} \approx \Gamma_{\phi,k} S_k \frac{\phi_{N_k} - \phi_{\mathrm{P}}}{(\boldsymbol{r}_{N_k} - \boldsymbol{r}_{\mathrm{P}}) \cdot \boldsymbol{n}_k} + \underline{\Gamma_{\phi,k} S_k \left[(\nabla \phi)_k \cdot \boldsymbol{n}_k - \overline{(\nabla \phi)}_k \cdot \frac{\boldsymbol{r}_{N_k} - \boldsymbol{r}_{\mathrm{P}}}{(\boldsymbol{r}_{N_k} - \boldsymbol{r}_{\mathrm{P}}) \cdot \boldsymbol{n}_k} \right]} \,. \tag{7.18}$$

Because all solution methods for the Navier–Stokes equations are of iterative nature, only the first term on the right-hand side is treated implicitly (i.e., it contributes to the coefficient matrix A); the underlined term is calculated using prevailing values of the variables and treated as a deferred correction (i.e., it is added to the source term on the right-hand side). In this term, $(\nabla \phi)_k$ is obtained by interpolating gradients from the two CV-centers to the face centroid k, while $\overline{(\nabla \phi)}_k$ represents the average

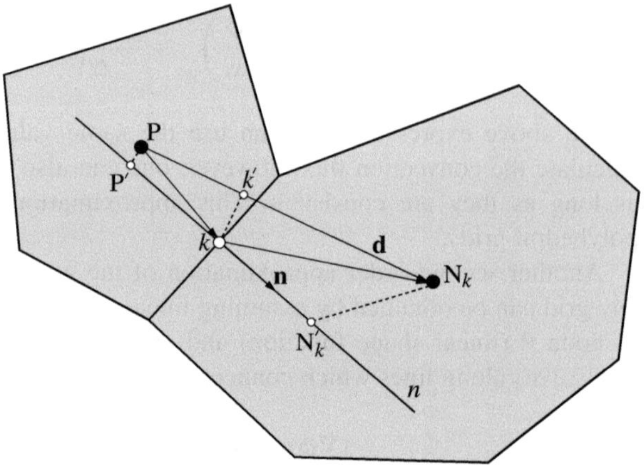

Fig. 7.3 On the method of computing normal component of the gradient at a cell-face

of the gradients at nodes on either side of the face k, P and N_k. The reason for this distinction is that the first term on the right-hand side, being the central-difference approximation, is second-order accurate at the midpoint between nodes P and N_k; the last term on the right-hand side should represent an approximation of the same derivative at the same location using interpolated cell-center gradients and cancel the first term out when the variation of ϕ is smooth, so gradients are interpolated to the midpoint between nodes P and N_k. The remaining term represents an approximation of the diffusion flux based on interpolated gradient. In the case of a non-smooth variable variation, the difference between the first and the last term tends to smooth the oscillations out.

Another second-order approximation, which is also applicable to arbitrary polyhedral grids, involves definition of auxiliary nodes P′ and N_k' on the normal passing through the cell-face center, see Fig. 7.3. The values of ϕ at these auxiliary nodes can be approximated using variable value and gradient at the nearby cell centers as follows:

$$\phi_{\mathrm{P}'} = \phi_{\mathrm{P}} + (\nabla\phi)_{\mathrm{P}} \cdot (\boldsymbol{r}_{\mathrm{P}'} - \boldsymbol{r}_{\mathrm{P}}) \; ; \quad \phi_{N_k'} = \phi_{N_k} + (\nabla\phi)_{N_k} \cdot (\boldsymbol{r}_{N_k'} - \boldsymbol{r}_{N_k}) \, . \tag{7.19}$$

The derivative with respect to n, which is the only one needed to calculate the diffusion flux, can now be approximated by a central difference using Eq. (7.19) as:

$$\left(\frac{\partial\phi}{\partial n}\right)_k \approx \frac{\phi_{N_k} - \phi_{\mathrm{P}}}{|\boldsymbol{r}_{N_k'} - \boldsymbol{r}_{\mathrm{P}'}|} + \left[\frac{(\nabla\phi)_{N_k} \cdot (\boldsymbol{r}_{N_k'} - \boldsymbol{r}_{N_k}) - (\nabla\phi)_{\mathrm{P}} \cdot (\boldsymbol{r}_{\mathrm{P}'} - \boldsymbol{r}_{\mathrm{P}})}{|\boldsymbol{r}_{N_k'} - \boldsymbol{r}_{\mathrm{P}'}|}\right]^{\mathrm{old}} . \tag{7.20}$$

The first term on the right-hand side can be treated implicitly, while the second term can be calculated using values from previous iteration (deferred correction). The auxiliary nodes can be selected to be equally away from face center k; there are many options.

Higher-order approximations of diffusion fluxes are difficult to develop for arbitrary polyhedral CVs. Most commercial and public CFD codes use one of the above two approaches to compute diffusion fluxes.

7.2.4 *Linearization*

The convection fluxes are non-linear. Other terms in equations may also be non-linear, especially when fluid properties vary substantially (e.g., if the flow is compressible or turbulence models are employed). The non-linear terms must be linearized upon discretization to render algebraic equation systems solvable. In most commercial and public CFD codes, the simple Picard-iteration is applied. For example, in the generic conservation equation for property ϕ, we have

$$F_k^{\mathrm{c}} = \int_{S_k} \rho\phi\boldsymbol{v}\cdot\boldsymbol{n}\,\mathrm{d}S \approx \phi_k \dot{m}_k \,. \tag{7.21}$$

Here $\dot{m}$ represents the mass flux which is calculated explicitly using velocity and density from previous iteration; it is treated in the current iteration as a known quantity. When the midpoint-rule approximation is used, the mass flux through the face k is expressed as $\dot{m}_k = (\rho v_n S)_k$, where v_n is the velocity component normal to the face and S_k is the face area. ϕ_k is treated as unknown and expressed via nodal values using one of the interpolation methods described above. Any other non-linear term can be linearized in a similar way.

7.2.5 *Boundary Conditions*

Some cell faces coincide with solution domain boundaries; they are attached to a single cell and are usually sorted separately in the list of faces according to the type of boundary: solid wall, symmetry plane, inlet, outlet, etc. The integrals over boundary faces must be either known or expressed via variable values at cell centers to make the algebraic equation systems solvable (the number of algebraic equations is equal to the number of cells; therefore, the number of unknowns must also be equal to the number of cells). This usually reduces—with respect to unknown variables—to applying either Dirichlet (specified variable values at the boundary) or Neumann boundary conditions (specified variable derivative). However, in FV methods one must always think in terms of fluxes through boundary faces and this leads to some special considerations which are described below. For the sake of simplicity, the implementation of boundary conditions is described for Cartesian cells only; the additional complexity for arbitrary non-orthogonal grids is described in detail in the book by Ferziger et al. [13].

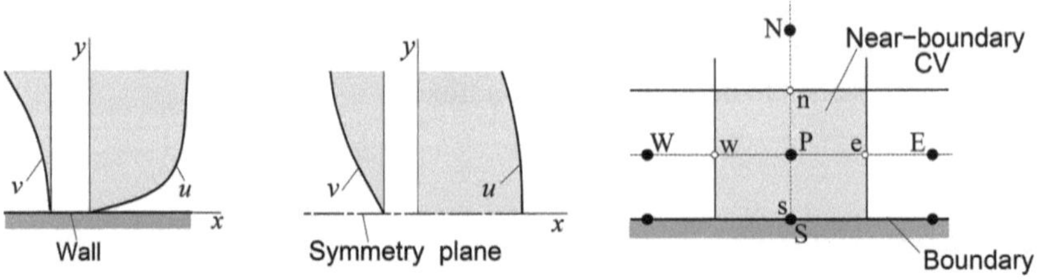

Fig. 7.4 On the boundary conditions for momentum equations at a wall and a symmetry plane

At a wall, the no-slip boundary condition applies, i.e., the velocity of the fluid is equal to the wall velocity. However, it is not enough to simply set the velocity at boundary nodes equal to wall velocity—one has to consider mass and momentum fluxes at wall faces. The velocity component normal to wall sets the mass flux to zero (for an impermeable wall); shear and normal forces at wall faces are momentum fluxes and need to be carefully evaluated.

The momentum equation for the wall-normal velocity component requires another condition to be enforced—the normal viscous stress is zero at a wall. This follows from the continuity equation, e.g., for a wall at $y = 0$ (see Fig. 7.4):

$$\left(\frac{\partial u}{\partial x}\right)_{\mathrm{wall}} = 0 \Rightarrow \left(\frac{\partial v}{\partial y}\right)_{\mathrm{wall}} = 0 \Rightarrow \tau_{yy} = 2\mu \left(\frac{\partial v}{\partial y}\right)_{\mathrm{wall}} = 0 \,. \tag{7.22}$$

Therefore, the diffusion flux in the v equation at the south boundary is

$$F_{\mathrm{s}}^{\mathrm{d}} = \int_{S_{\mathrm{s}}} \tau_{yy} \, \mathrm{d}S = 0 \,. \tag{7.23}$$

This should be implemented directly, rather than using only the condition that $v = 0$ at the wall. Because $v_{\mathrm{P}} \neq 0$, one would otherwise obtain a non-zero derivative in the discretized flux expression and thus a non-zero normal stress.

The shear stress at a wall boundary can be calculated by using a one-sided approximation of the derivative $\partial u / \partial y$; one possible approximation is (for the u equation and the situation from Fig. 7.4)

$$F_{\mathrm{s}}^{\mathrm{d}} = \int_{S_{\mathrm{s}}} \tau_{xy} \, \mathrm{d}S = \int_{S_{\mathrm{s}}} \mu \frac{\partial u}{\partial y} \, \mathrm{d}S \approx \mu_{\mathrm{s}} S_{\mathrm{s}} \frac{u_{\mathrm{P}} - u_{\mathrm{S}}}{y_{\mathrm{P}} - y_{\mathrm{S}}} \,. \tag{7.24}$$

In this approximation, linear variation of wall-parallel velocity component in wall-normal direction is assumed. This is correct in the viscous part of the turbulent boundary layer, but is not accurate for laminar flows, where a quadratic variation would be a better choice; this requires using the gradient at P as well to construct a parabola and take its derivative at the wall, which is easily done and will not be described in further detail. The difference between linear and quadratic approximation can be treated as a deferred correction, because it involves the gradient at P.

At a symmetry plane we have the opposite situation: the shear stress is zero, but the normal stress is not, because (for the situation from Fig. 7.4):

$$\left(\frac{\partial u}{\partial y}\right)_{\text{sym}} = 0\,; \quad \left(\frac{\partial v}{\partial y}\right)_{\text{sym}} \neq 0\,. \tag{7.25}$$

The diffusion flux in the u-equation and in all scalar equations is zero, but the diffusion flux in the v-equation is non-zero; it can be approximated as:

$$F_{\rm s}^{\rm d} = \int_{S_{\rm s}} \tau_{yy}\, \mathrm{d}S = \int_{S_{\rm s}} 2\mu \frac{\partial v}{\partial y}\, \mathrm{d}S \approx 2\mu_{\rm s} S_{\rm s} \frac{v_{\rm P} - v_{\rm S}}{y_{\rm P} - y_{\rm S}}\,, \tag{7.26}$$

where $v_{\rm S} = 0$ applies (which also sets the mass flux to zero).

The boundary pressure is needed to calculate the pressure forces in the momentum equations. We have to use extrapolation to obtain pressure at the boundaries where velocity is specified. In most cases, linear extrapolation is sufficiently accurate for a second-order method, but quadratic extrapolation is more accurate. There are cases in which a large pressure gradient near a wall in the wall-normal direction balances a body force (buoyancy, centrifugal force, etc.). If the pressure extrapolation is not accurate, this balance may not be satisfied and large velocity components directed towards wall or away from it may result. This can be avoided by calculating the normal velocity component for the first CV from the continuity equation, by adjusting the pressure extrapolation, or by local grid refinement.

At an inlet boundary, all variables (except pressure) are usually specified, which means that convection fluxes can be readily computed. The computation of diffusion fluxes requires derivatives in the direction normal to boundary; these are approximated using known variable value at the boundary and the variable value (and possibly its gradient) at the centroid of the next-to-boundary cell (one-sided approximation resulting from linear or quadratic extrapolation).

At the outlet, if the inlet mass fluxes are given, extrapolation of the velocity to the boundary (zero gradient, e.g., $u_{\rm E} = u_{\rm P}$) can usually be used for steady flows when the outflow boundary is far from the region of interest and the Reynolds number is large. The extrapolated velocity is then corrected to give exactly the same total mass flux as at inlet (this cannot be guaranteed by any extrapolation). The corrected velocities are then considered as prescribed for the following outer iteration.

7.2.6 Time-Integration Methods

When unsteady flows are simulated, one starts with an initial solution at time $t = t_0$ (which should satisfy all equations) and integrates the equations over a finite time interval Δt to obtain the solution at $t = t_0 + \Delta t$. One can then continue to find solutions at future time levels in the same way. Explicit methods, which compute the solution at the new time level using only information from previous time levels, are seldom used in general-purpose CFD codes, because they suffer from stability constraints: the time step must be smaller than a particular limit for the integration method to converge. Most commercial and public codes use therefore implicit methods, in which the new solution at any point depends not only on the old solutions at the same point, but also on the new solution at neighbor points. These methods are especially efficient when looking for a steady-state solution, or when computing time-dependent flows with slow transients.

The fully implicit Euler method is first-order accurate and therefore only useful for marching towards steady state. If one re-writes the conservation equations as

$$\frac{\partial \psi}{\partial t} = F \ , \tag{7.27}$$

where $\psi = \int_V \rho\phi \, \mathrm{d}V$ and F stands for the sum of convection fluxes, diffusion fluxes, and source terms, the implicit Euler method integrates this equation over a time interval Δt as follows:

$$\psi^{n+1} = \psi^n + F^{n+1}\Delta t \ . \tag{7.28}$$

This corresponds to the backward-difference approximation, i.e., it is assumed that F^{n+1} represents the mean value of F over the whole time step.

In order to simulate transient flows, one needs a method of at least second order; Crank–Nicolson scheme is widely used for this purpose. It is based on trapezoid rule, i.e., the mean value of F is approximated by averaging the values from two time steps:

$$\psi^{n+1} = \psi^n + \frac{1}{2}(F^n + F^{n+1})\Delta t \ . \tag{7.29}$$

This method requires saving both variable values and F (for each equation) from two time levels.

Another popular second-order time-integration method uses midpoint rule to integrate the equation from $t_{n+1} - \Delta t/2$ to $t_{n+1} + \Delta t/2$. Because F^{n+1} is determined at the center of the integration interval, the product $F^{n+1}\Delta t$ represents a second-order approximation of the integral over the integration period. The integral of Eq. (7.27) with this scheme reads

$$\psi^{n+3/2} = \psi^{n+1/2} + F^{n+1}\Delta t \ . \tag{7.30}$$

The problem is that variable values midway between time levels are not known and need to be expressed via interpolation in terms of the "nodal" values. By using quadratic interpolation through solution at three time levels (t_{n-1}, t_n and t_{n+1}), one obtains

$$\frac{3\,\psi^{n+1} - 4\,\psi^n + \psi^{n-1}}{2\,\Delta t} = F^{n+1} \ . \tag{7.31}$$

The left-hand side represents an approximation of the time derivative at t_{n+1} obtained by taking a derivative of a parabola fitted to the solution at three time levels. This scheme is called "fully implicit" because it only needs nodal variable values from previous time levels; fluxes and source terms are evaluated only at the new time level t_{n+1}. This allows for changes in grid topology (i.e., re-gridding) between two time steps; one only needs to interpolate the old solutions to the centroid location of new cells, and the simulation can be continued. Most commercial and public CFD codes include the above time-integration schemes.

7.2.7 The Algebraic Equation Systems and Their Solution

By summing all the flux approximations and source terms, an algebraic equation which relates the variable value at the CV-center to the values at neighbor CVs with which it has common faces is obtained. The numbers of equations and unknowns are both equal to the number of CVs so the system is in principle solvable. The algebraic equation for a particular CV has the form:

$$A_P \phi_P + \sum_k A_k \phi_{N_k} = Q_P \,, \tag{7.32}$$

where the index k runs over all CV-faces. The coefficients A_k typically contain contributions from convection and diffusion fluxes, while Q contains source terms and deferred corrections (i.e., all terms which are computed using prevailing values of variables).

For the solution domain as a whole, a matrix equation is obtained:

$$A\boldsymbol{\phi} = \boldsymbol{Q} \,, \tag{7.33}$$

where A is a square sparse coefficient matrix, $\boldsymbol{\phi}$ is a vector (or column matrix) containing the variable values at cell centers, and $\boldsymbol{Q}$ is the vector containing the terms on the right-hand side of Eq. (7.32). The number of non-zero elements in each row is equal to the number of cell faces of the corresponding CV. For unstructured grids, the ordering of cells in their list is usually optimized in order to reduce the bandwidth of the matrix A.

Although Eq. (7.33) appears linear, it is derived from discretized non-linear governing equations with the help of particular linearization techniques; one such equation is obtained for each velocity component and any other transport equation that may need to be solved (like turbulence model equations or energy equation). The system of coupled, non-linear equations must therefore be solved using an iterative solution method. Because of the required iterations to update the linearized terms, it is not necessary to solve the intermediate linear equations systems like Eq. (7.33) very accurately—both the coefficient matrix and the right-hand side will change in the next iteration anyway.

The iterative solution method involves two iteration loops:

- *Outer iteration* loop, used to update non-linear terms (which usually also couple equations for different variables, e.g., velocity components and temperature), and
- *Inner iteration* loop, within which linear equation systems are solved to a prescribed tolerance before the next outer iteration is started.

When solving the coupled system of equations involving mass, momentum, and energy conservation equations, there is another choice to be made:

- *Sequential solution* approach, i.e., solving for each variable (three velocity components, pressure, density, temperature, etc.) in turn, or
- *Coupled solution* approach, in which all variables are considered as a single vector of unknowns.

In the sequential solution approach, simpler linearization techniques can be used, less storage is required and each iteration is much cheaper than when the coupled solution approach is used. However, the number of iterations required to reach the converged steady-state solution is usually larger. When solving steady-state problems, coupled solvers are usually somewhat more efficient (depending on the problem being solved), but for transient problems, they are in most cases substantially more expensive. The reason is that, when marching in time, the solution at the next time step can be well initialized using solutions at previous time steps, meaning that usually only a few outer iterations (3 to 10) in the sequential solver are needed. Even the coupled solver needs a similar number of iterations due to non-linearity, so tighter coupling of unknowns often brings no benefits.

Linear equation systems (7.33) are usually solved using multigrid or conjugate-gradient type methods; details about most widely used methods are available in the book by Ferziger et al. [13].

7.3 Pressure-Velocity Coupling

When computing incompressible flows, the problem of determining the pressure requires special attention. Pressure appears in momentum equations, but it is not a dominant variable in any of them. On the other hand, all three velocity components appear in the continuity equation, but each of them has its own momentum equation in which it is the dominant variable. The continuity equation has thus to be used to compute pressure, but the way of doing so is not obvious.

One can obtain a pressure equation by taking the divergence of the vector form of momentum equation, but this approach is not widely used. Two classes of methods—which are more similar to each other than it appears—have established themselves over the past few decades in most CFD codes: fractional-step methods and SIMPLE-like algorithms. These methods are briefly described in the following two sections; for more details, see the book by Ferziger et al. [13].

7.3.1 Fractional-Step Methods for Incompressible Flows

The fractional-step concept is more a generic approach than a particular method; there are many variants of it. Two versions will be presented here: one non-iterative version, which is representative of many similar methods used to simulate unsteady flows (e.g., Choi and Moin [6]; Armfield et al. [3]), and one iterative version which is similar to the SIMPLE-class of methods. We refer here to the Navier–Stokes equations in vector form and use subsequently operator-notation, because the approach is not limited to any particular discretization and solution method:

$$\frac{\partial(\rho \boldsymbol{v})}{\partial t} + \mathrm{div}(\rho \boldsymbol{v}\boldsymbol{v}) = \mathrm{div}\, \boldsymbol{S} - \boldsymbol{\nabla} p \;, \tag{7.34}$$

$$\mathrm{div}(\rho \boldsymbol{v}) = 0 \;. \tag{7.35}$$

The first method is based on the Crank–Nicolson time-integration scheme:

$$\frac{(\rho \boldsymbol{v})^{n+1} - (\rho \boldsymbol{v})^n}{\Delta t} + C(\boldsymbol{v}^{n+1/2}) = \frac{L(\boldsymbol{v}^{n+1}) + L(\boldsymbol{v}^n)}{2} - \frac{G(p^n) + G(p^{n+1})}{2} \;, \tag{7.36}$$

$$D(\rho \boldsymbol{v})^{n+1} = 0 \;. \tag{7.37}$$

Here C stands for convection terms, L for viscous terms, G for pressure terms, and D is the discrete divergence operator for the continuity equation. Central differences of second order are usually used for all spatial derivatives. However, the fractional-step approach can be applied to any spatial and temporal discretization method. In this particular version, the convection fluxes needed at $t_{n+1/2}$ for the time integration with Crank–Nicolson scheme are obtained using explicit extrapolation from the two previous time steps, using linear extrapolation (which corresponds to the Adams-Bashforth 2nd-order explicit scheme):

$$C(\boldsymbol{v}^{n+1/2}) \approx \frac{3}{2} C(\boldsymbol{v}^n) - \frac{1}{2} C(\boldsymbol{v}^{n-1}) + O(\Delta t^2) \;, \tag{7.38}$$

where $C(\boldsymbol{v})$ describes the discretized net convection flux (the sum of convection fluxes over all CV-faces); $\boldsymbol{v}^n$ stand for the velocity field at time level t_n. This makes it possible to avoid outer iterations on momentum equations, because viscous and pressure terms are linear (when viscosity and density are constant). An implicit version would use:

$$C(\boldsymbol{v}^{n+1/2}) \approx \frac{C(\boldsymbol{v}^n) + C(\boldsymbol{v}^{n+1})}{2} \;, \tag{7.39}$$

in which case outer iterations cannot be avoided.

The pressure at t_{n+1} needs also to be first estimated; the usual approach is to use p^n in the first step (it will be corrected in the second step), but one can also apply explicit extrapolation to estimate p^{n+1} using an expression like (7.38).

The solution algorithm is as follows:

1. Solve the momentum equations for an estimate of the solution at t_{n+1}, $\boldsymbol{v}^*$, using p^n instead of p^{n+1}:

$$\frac{(\rho \boldsymbol{v})^* - (\rho \boldsymbol{v})^n}{\Delta t} + C(\boldsymbol{v}^{n+1/2}) = \frac{L(\boldsymbol{v}^*) + L(\boldsymbol{v}^n)}{2} - G(p^n) \, , \tag{7.40}$$

2. Define pressure correction p'

$$p^{n+1} = p^n + p' \tag{7.41}$$

and require that the corrected velocity and pressure satisfy the following form of the momentum equation:

$$\frac{(\rho \boldsymbol{v})^{n+1} - (\rho \boldsymbol{v})^n}{\Delta t} + C(\boldsymbol{v}^{n+1/2}) = \frac{L(\boldsymbol{v}^*) + L(\boldsymbol{v}^n)}{2} - \frac{G(p^n) + G(p^{n+1})}{2} \, . \tag{7.42}$$

3. Define corrected velocity as a function of the gradient of pressure correction by subtracting Eq. (7.40) from Eq. (7.42):

$$(\rho \boldsymbol{v})^{n+1} - (\rho \boldsymbol{v})^* = -\frac{\Delta t}{2} G(p') \, . \tag{7.43}$$

4. Define pressure-correction equation by requiring that the corrected velocity satisfies the continuity equation (7.37):

$$D(\rho \boldsymbol{v})^{n+1} = 0 \quad \Rightarrow \quad D(G(p')) = \frac{2D(\rho \boldsymbol{v})^*}{\Delta t} \, . \tag{7.44}$$

The left-hand side represents the Laplace-operator; the pressure-correction equation thus has the form of a Poisson equation.

5. Correct pressure according to Eq. (7.41) and velocity according to Eq. (7.43); if the pressure-correction equation (7.44) was solved accurately, the corrected velocity field will satisfy the discretized continuity equation (7.37).

Because the corrected velocities and pressure satisfy the simplified momentum equation (7.42)—in which $\boldsymbol{v}^*$ is used in viscous terms instead of corrected velocity $\boldsymbol{v}^{n+1}$—and not the proper one (7.36), this non-iterative solution method includes a so-called *splitting error*. It could be eliminated only by iterating momentum equation until the new velocity at t_{n+1} is included in all terms and not changing any more.

The splitting error can be estimated by realizing that the difference between $\boldsymbol{v}^{n+1}$ and $\boldsymbol{v}^*$ is (cf. Eq. (7.43))

$$\boldsymbol{v}^{n+1} - \boldsymbol{v}^* = \frac{\Delta t}{2\rho} G(p') = \frac{(\Delta t)^2}{2\rho} G\left(\frac{\delta p}{\delta t}\right) \tag{7.45}$$

because $p' = p^{n+1} - p^n \approx (\delta p/\delta t)\Delta t$. The difference between Eqs. (7.36) and (7.42) is

$$\frac{1}{2}\left[L(\boldsymbol{v}^{n+1}) - L(\boldsymbol{v}^*)\right] = \frac{(\Delta t)^2}{4\rho} L\left[G\left(\frac{\delta p}{\delta t}\right)\right] . \tag{7.46}$$

Thus, the splitting error is proportional to the time step squared and the time derivative of pressure. Because the Crank–Nicolson scheme is a second-order one, the splitting error is of the same order as the regular discretization error. However, the order of the method gives no information about the error magnitude—only about the rate of error reduction when time steps are small enough. Thus, if time step is too large for the given flow, or if pressure variation with time is significant, the error may be large.

This method is called non-iterative because it does not require the outer iteration loop: momentum equations are solved once, followed by the solution of the pressure-correction equation—and that concludes the time step. However, because convection is treated explicitly, there is a limit to how large the time step can be. The maximum allowable time step is problem-dependent; it depends on how far the fluid moves within one time step relative to cell size, on the ratio of viscous to convection terms and on pressure variation in space and time.

The splitting error can be eliminated by computing convection terms implicitly and bringing the pressure-correction equation into the iteration loop for non-linearity updates (outer iterations). By doing so, the computing effort is increased because the Poisson equation for pressure-correction is solved multiple times within each time step. However, the overhead is not as large as it may seem: when outer iterations are performed, the mass conservation does not have to be enforced to a very tight tolerance within each outer iteration, as is the case when it is solved only once per time step in the non-iterative version.

The alternative method is fully implicit and based on the three-time-level time-integration scheme presented earlier. With this scheme, the convection, diffusion, pressure, and any other source terms are computed only at the new time level t_{n+1}. The solution method is as follows:

1. At the mth outer iteration within the new time step, solve the momentum equation to estimate the new time level solution of the following form:

$$\frac{3(\rho\boldsymbol{v})^* - 4(\rho\boldsymbol{v})^n + (\rho\boldsymbol{v})^{n-1}}{2\Delta t} + C(\boldsymbol{v}^*) = L(\boldsymbol{v}^*) - G(p^{m-1}) , \tag{7.47}$$

   where $\boldsymbol{v}^*$ is the predictor value for $\boldsymbol{v}^m$; it needs to be corrected to enforce continuity.

2. Require that the corrected velocity and pressure satisfy this form of the momentum equation:

$$\frac{3(\rho \boldsymbol{v})^m - 4(\rho \boldsymbol{v})^n + (\rho \boldsymbol{v})^{n-1}}{2\Delta t} + C(\boldsymbol{v}^*) = L(\boldsymbol{v}^*) - G(p^m) \,. \tag{7.48}$$

By subtracting Eq. (7.47) from Eq. (7.48) we obtain the following relation between velocity and pressure correction:

$$\frac{3}{2\Delta t}\left[(\rho \boldsymbol{v})^m - (\rho \boldsymbol{v})^*\right] = -G(p') \quad \Rightarrow \quad (\rho \boldsymbol{v})' = -\frac{2\Delta t}{3} G(p') \,. \tag{7.49}$$

3. Impose the continuity requirement on velocity $\boldsymbol{v}^m$,

$$D(\rho \boldsymbol{v})^m = 0 \quad \Rightarrow \quad D(G(p')) = \frac{3}{2\Delta t} D(\rho \boldsymbol{v})^* \,, \tag{7.50}$$

and solve the resulting pressure-correction equation.
4. Increase the iteration counter and repeat steps 1 to 3 until residuals become sufficiently small. Then set $\boldsymbol{v}^{n+1} = \boldsymbol{v}^m$, $p^{n+1} = p^m$ and proceed to the next time level.

Note that the above pressure-correction equation looks exactly the same as that from the previous version, Eq. (7.44). However, the right-hand side is different, because $\boldsymbol{v}^*$ came from a different form of the momentum equation. In this case one does not have to solve a linear momentum equation or pressure-correction equation to a very tight tolerance, because the solution will be continued in the next outer iteration; usually, reducing residuals one order of magnitude in each outer iteration is sufficient if three or more outer iterations per time step are performed.

The above iterative implicit fractional-step method can be used to compute both steady and unsteady flows. In case of a steady-state flow, one can use larger time steps and does not have to iterate to a tight tolerance within each time step; one to two iterations are usually enough. For unsteady flows one has to select the time step so that the variation of variables in time is sufficiently resolved (e.g., order of 100 time steps per oscillation period in the case of periodic flows). Iterations within a time step should be continued until all equations are satisfied to a sufficient accuracy.

This version of fractional-step method is very similar to the SIMPLE algorithm described in the next section. More details and examples of application—including direct comparison with SIMPLE algorithm—is available in Ferziger et al. [13].

7.3.2 SIMPLE Algorithm for Incompressible Flows

SIMPLE and related algorithms are almost exclusively used in commercial and public general-purpose CFD codes. Only the SIMPLE method (Caretto et al. [5]) will be briefly described, because other methods represent only slight modifications and are described in literature (e.g., in Ferziger et al. [13]). Using the notation from the preceding section and the fully implicit three-time-level scheme, the equations to be solved are

$$\frac{3(\rho \boldsymbol{v})^{n+1} - 4(\rho \boldsymbol{v})^n + (\rho \boldsymbol{v})^{n-1}}{2\Delta t} + C(\boldsymbol{v}^{n+1}) = L(\boldsymbol{v}^{n+1}) - G(p^{n+1}) \,. \tag{7.51}$$

$$D(\rho \boldsymbol{v})^{n+1} = 0 \,. \tag{7.52}$$

It will be assumed here that density and viscosity are constant; the extension to handle also compressible flows will be described in Sect. 7.4. When all discretized terms are grouped together, we obtain for each velocity component a matrix equation of the form:

$$A^{m-1} u_i^m = Q_i^{m-1} - G_i(p^m) \,, \tag{7.53}$$

where m is the outer iteration counter. G_i is a shorthand notation for the i-component of the gradient operator. We use this notation although in FV methods pressure forces are computed at the faces of CV; the resulting net force can still be expressed as the product of pressure gradient at cell center and the cell volume. The source term Q contains all terms that are explicitly computed in terms of u_i^{m-1} (e.g., deferred correction due to higher-order convection discretization or non-orthogonality effects in viscous terms) as well as any body force or other terms that may depend on other variables at the new time level (e.g., on temperature). It also contains parts of the unsteady term that refer to solutions at previous time steps, see Eq. (7.51). For the sake of clarity, the superscript $^{m-1}$ for the matrix A and source term Q will be dropped from here on.

The matrix A can be split into a diagonal part A_D and an off-diagonal part A_OD, for reasons that will become obvious shortly. The superscript m denoting values from the current outer iteration is replaced by one (predictor stage) or two (corrector stage) asterisks.

At outer iteration m, the following simplified form of Eq. (7.53) is solved first, using the pressure from previous iteration:

$$(A_\mathrm{D} + A_\mathrm{OD}) u_i^* = Q_i - G_i(p^{m-1}) \,. \tag{7.54}$$

The velocity field obtained by solving this equation, $\boldsymbol{v}^*$, will in general not satisfy the continuity equation. In order to enforce the continuity constraint, both pressure and velocities are corrected as follows:

$$p^* = p^{m-1} + p' \quad , \quad u_i^{**} = u_i^* + u_i' \,. \tag{7.55}$$

The relation between velocity and pressure correction is obtained by requiring that the corrected velocity and pressure satisfy the following simplified version of Eq. (7.53):

$$A_\mathrm{D} u_i^{**} + A_\mathrm{OD} u_i^* = Q_i - G_i(p^*) \,. \tag{7.56}$$

Now by subtracting Eq. (7.54) from Eq. (7.56), one obtains the following relation between velocity and pressure corrections:

$$A_{\mathrm{D}} u_i' = -G_i(p') \quad \Rightarrow \quad u_i' = -(A_{\mathrm{D}})^{-1} G_i(p') \,. \tag{7.57}$$

This relation is simple because the diagonal matrix can be easily inverted; if u_i^{**} was applied to the off-diagonal matrix in Eq. (7.56) as well, the relation would be too complicated for the purpose of deriving a pressure-correction equation. The simplification is justified by the fact that, as outer iterations converge, all corrections tend to zero, so the final solution is not affected. However, this simplification does affect the convergence rate and under-relaxation needs to be introduced and optimized to assure convergence and computational efficiency (without under-relaxation the usable time step size would be limited; beyond limit, outer iterations would diverge).

By requesting that corrected velocities u_i^{**} satisfy the discretized continuity equation,

$$D(\rho \boldsymbol{v})^* + D(\rho \boldsymbol{v}') = 0 \,, \tag{7.58}$$

and expressing u_i' via p' with the help of expression (7.57), the pressure-correction equation is obtained:

$$D(\rho (A_{\mathrm{D}})^{-1} G(p')) = D(\rho \boldsymbol{v})^* \,. \tag{7.59}$$

Note that this equation has the same form as Eq. (7.50) from the previous section. The major difference is that $(A_{\mathrm{D}})^{-1}$ appears inside divergence operator on the left-hand side. In the case of fractional-step method, we had $2\Delta t/3$ at that place, but because this quantity does not vary in space, it was taken out of the divergence operator and appears now on the right-hand side of Eq. (7.50) as a reciprocal multiplier. Because A_{D} in general varies in space (it contains contributions from convection and diffusion terms), it has to be included inside divergence operator.

Once the pressure-correction equation is solved, the velocities and pressure are updated using Eqs. (7.57) and (7.55). These are taken to represent the solution at outer iteration m and the new iteration can start. This is known as the SIMPLE algorithm.

There are several modifications of the SIMPLE algorithm in literature; the SIMPLEC algorithm (van Doormal and Raithby [35]) and PISO algorithm (Issa [18]) are widely used.

As already noted, due to the neglect of the effect of velocity corrections in off-diagonal terms, the SIMPLE algorithm requires under-relaxation of changes in both velocities and pressure. Especially for steady flows computed using an infinite time step, its performance depends greatly on the value of the under-relaxation parameter used in the momentum equations. It has been first found by trial and error that convergence can be improved if one adds only a portion of p' to p^{m-1}, i.e., if one updates the pressure, contrary to the statement in expression (7.55), as follows:

$$p^m = p^{m-1} + \alpha_p p' \,, \tag{7.60}$$

where $0 \leq \alpha_p \leq 1$. The optimum value of α_p appears to be close to

$$\alpha_p = 1 - \alpha_u \,, \tag{7.61}$$

where α_u is the under-relaxation factor for velocities; see Raithby and Schneider [28] and Perić [24] for the derivation of this relation using different arguments. The under-relaxation of velocity changes is implicitly built into momentum equations using approach suggested by Patankar [22]:

$$\underbrace{\frac{A_\mathrm{P}}{\alpha_\phi}}_{A_\mathrm{P}^*} \phi_\mathrm{P}^m + \sum_k A_k \phi_{N_k}^m = \underbrace{Q_\mathrm{P} + \frac{1-\alpha_\phi}{\alpha_\phi} A_\mathrm{P} \phi_\mathrm{P}^{m-1}}_{Q_\mathrm{P}^*} \,, \tag{7.62}$$

where A_P^* and Q_P^* are modified main diagonal matrix elements and source vector components. This modified equation is solved within inner iterations. When the outer iterations converge, the terms involving α_ϕ cancel out and we obtain the solution of the original problem.

Methods of this kind are fairly efficient for solving steady-state problems; their convergence can be further improved by the multigrid strategy applied to outer iterations (see, e.g., Lilek et al. [21]). A nice feature of SIMPLE-like methods (including the implicit iterative fractional-step method described above) is that they are easily extended to solve additional transport equations (e.g., for chemical species, variables from turbulence models, etc.). Also, variable fluid properties are easily handled: they are simply assumed constant during one outer iteration and re-computed at the end of the loop, when all variables are updated. This is shown in Fig. 7.5; the outer-iterations loop is easily extended to update any non-linearity or deferred correction, or to solve additional equations.

Fig. 7.5 Flow chart of SIMPLE-like algorithms

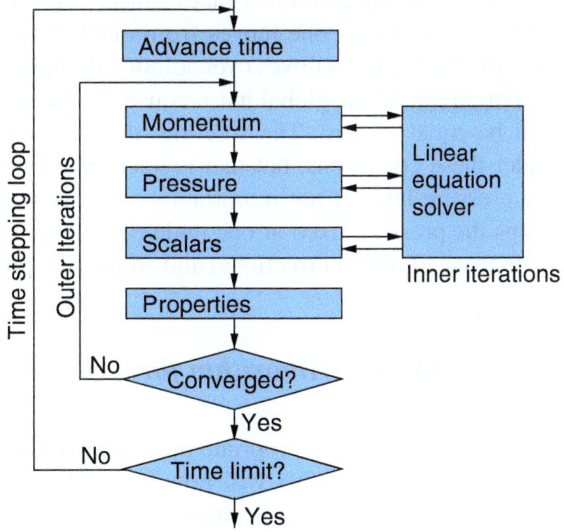

By comparing the SIMPLE method with fractional-step methods presented in the previous section, one can see that it is very similar to the implicit version of the latter. The only difference is that SIMPLE method, when deriving the pressure-correction equation, updates the velocity not only in the unsteady term (as fractional-step methods do), but also in the diagonal part of discretized convection and diffusion fluxes. This difference is not so important when unsteady flows are computed using small time steps, but it does become significant if large time steps are used to march towards a steady-state solution.

The pressure-correction equation derived from discretized momentum and continuity equation is easier to handle than a pressure equation which could be derived from differential equations (by applying divergence operator to the vector form of the momentum equation). The main reason is that the boundary conditions for the pressure-correction equation are easier to formulate. When the velocity component normal to solution domain boundary (i.e., the mass flux through the boundary cell-face) is known, it does not need to be corrected, i.e., $u' = 0$. This is equivalent to setting the gradient of pressure correction in the same direction to zero, see Eq. (7.57). Thus, the pressure-correction equation has Neumann boundary conditions on all boundaries where the velocity is prescribed (inlet, wall, symmetry, etc.). Even at boundaries where the velocity is not prescribed (e.g., outlet), one often first applies an extrapolation (or another kind of approximation) of the velocity and applies a global correction to make velocities at boundaries satisfy the global mass conservation. When this is done, the velocity is—for the sake of deriving the pressure-correction equation—considered as specified and the velocity correction is set to zero.

If an equation has zero-gradient boundary condition on all boundaries, the sum of all source terms in the interior has to vanish. This is indeed the case in the pressure-correction equation in FV methods, because its source term represents the sum of mass fluxes over cell faces of each CV. When source terms are summed over all CVs, mass fluxes through all inner faces cancel out, because the outward normal changes sign at each face as one moves from one CV to its neighbor. Thus, there remains only the sum of mass fluxes over solution domain boundaries, and when it is ensured that these satisfy the global mass-conservation equation, the sum of all source terms will be equal to zero. The solution of pressure-correction equation is not unique under these conditions, but that is not a problem because in incompressible flows the pressure level is not important—only pressure differences count. One usually keeps the pressure fixed at one node (by subtracting the pressure correction at that node from all other corrections) and adjusts the pressure at other nodes relative to it.

7.3.3 SIMPLE Algorithm for Arbitrary Polyhedral Grids

The derivation of an appropriate pressure-correction equation for a FV method and arbitrary polyhedral CVs will now be briefly described. The surface and volume integrals in the momentum equation are assumed to be approximated using second-order approximations (midpoint-rule integration, linear interpolation, and central differencing). The pressure term is also treated conservatively, i.e., in each equation the appropriate components of the pressure forces on the CV surface are calculated. However, one can express the pressure contribution to the source term also as a product of the mean pressure gradient and cell volume in a conservative way using the Gauss-theorem, see Eq. (7.15):

$$Q_{i,\mathrm{P}}^{p} = -\int_S p\,\boldsymbol{i}_i \cdot \boldsymbol{n}\,\mathrm{d}S = -\int_V \frac{\partial p}{\partial x_i}\,\mathrm{d}V$$

$$\sum_k p_k S_k^i \approx \left(\frac{\partial p}{\partial x_i}\right)_{\mathrm{P}} \Delta V \quad \Rightarrow \quad \left(\frac{\partial p}{\partial x_i}\right)_{\mathrm{P}} = \frac{\sum_k p_k S_k^i}{\Delta V} \,. \tag{7.63}$$

As was described earlier, the linearized momentum equations are solved first using pressure from previous iteration, thus producing the first estimate of the new outer iteration solution, u_i^{m*}. In order to enforce the continuity constraint, one needs to calculate mass fluxes through CV-faces using new velocity components u_i^{m*}. To this end one needs velocity vector at the cell face, which requires interpolation from neighbor CV-centers. Simple linear interpolation may lead to oscillations; the reason is that an oscillatory pressure distribution is not sensed by the pressure source term in Eq. (7.63) when the oscillation wavelength is two cells wide. Thus, interpolated velocity at the cell face—which contains interpolated pressure terms calculated at cell centers—will also not sense pressure gradient due to oscillatory pressure field. Rhie and Chow [29] presented an interpolation practice which avoids oscillations: the velocity is interpolated, but a correction to the interpolated value is applied, based on the difference between the interpolated pressure term and the one calculated at the cell face. The correction is negligible when the pressure variation is smooth, but becomes large if oscillations are present. The cell-face velocity is computed as:

$$v_{n,k}^{m*} = \overline{(v_n^{m*})}_k - (\Delta V)_k \overline{\left(\frac{1}{A_{\mathrm{P}}^{v_n}}\right)}_k \left[\left(\frac{\delta p^{m-1}}{\delta n}\right)_k - \overline{\overline{\left(\frac{\delta p^{m-1}}{\delta n}\right)}}_k\right] , \tag{7.64}$$

where $v_{n,k} = (\boldsymbol{v} \cdot \boldsymbol{n})_k$ is the velocity component normal to the cell face. The single overbar denotes linear interpolation and double overbar means arithmetic average of values at two neighboring CV-centers. The volume around cell-face k, $(\Delta V)_k$ is constructed by taking a dot product of surface vector $(\boldsymbol{n}S)_k$ and distance vector $\boldsymbol{r}_{\mathrm{N}_k} - \boldsymbol{r}_{\mathrm{P}}$, or simply by averaging the two neighbor volumes (this does not affect the final solution significantly because the term in brackets is small when the pressure field is smooth). Because the coefficient A_{P} is the same for all Cartesian velocity components, the same value is also used here as $A_{\mathrm{P}}^{v_n}$. The correction term is proportional to the third derivative of pressure multiplied by $(\Delta n)^2$ and vanishes when the variation of pressure is linear or quadratic; Δn is here the mesh spacing in the direction normal to the cell face. The term in brackets reduces with grid refinement as a second-order error term and is therefore consistent with other approximations (see Ferziger et al. [13], for a detailed discussion of this issue). The pressure derivative in the normal direction is approximated at the cell face using central-difference. Because that approximation is second-order accurate at a location midway between nodes P and N_k, see Fig. 7.2, the interpolated derivative should also be calculated at that location; thus, the double overbar in the above equation denotes arithmetic average rather than linear interpolation. This guarantees that the correction term is zero if pressure varies linearly or quadratically and is only significant for oscillatory variations.

The averaged derivative at a cell face can be obtained by averaging the components of the gradient vector from the two cell centers; these are calculated at each CV-center during solution of the momentum equations using one of the approximations for the gradient described above. Thus, one can write

$$\overline{\overline{\left(\frac{\delta p^{m-1}}{\delta n}\right)_k}} = \overline{\overline{(\nabla p)_k}} \cdot \boldsymbol{n}_k \, . \tag{7.65}$$

The pressure derivative in the direction normal to cell face can be computed using expression (7.20); sometimes, simplifications are introduced, as will be outlined shortly.

The mass flux through the face k can now be calculated as:

$$\dot{m}_k^{m*} = \int_{S_k} \rho \boldsymbol{v} \cdot \boldsymbol{n} \, \mathrm{d}S \approx (\rho v_n^{m*} S)_k \, . \tag{7.66}$$

The continuity equation is in general not satisfied by these mass fluxes, but results in a mass imbalance:

$$\sum_k \dot{m}_k^{m*} = \Delta \dot{m}_{\mathrm{P}} \, , \tag{7.67}$$

where the summation is over all cell faces (note that the same value of the mass flux is used for two CVs, but with opposite signs, i.e., $\dot{m}_k$ for a CV-centered around node P is equal to $-\dot{m}_k$ for a CV-centered around node N_k).

In order to enforce mass conservation, the mass fluxes need to be corrected by correcting the normal velocity component. By analogy with Eq. (7.57), the velocity correction can be expressed through the gradient of the pressure correction as follows (see also Eq. (7.64)):

$$(v_n')_k \approx -(\rho \, \Delta V \, S)_k \overline{\left(\frac{1}{A_{\mathrm{P}}^{v_n}}\right)_k} \left(\frac{\delta p'}{\delta n}\right)_k \, . \tag{7.68}$$

The derivative of p' with respect to n at the cell-face center can again be obtained from expression (7.20); however, the inclusion of the whole expression on the right-hand side of this equation would lead to a too complicated pressure-correction equation. Because the role of the pressure-correction equation is only to drive the solution process towards convergence, where all corrections must become negligible, and because the pressure-correction equation of the SIMPLE algorithm is not exact anyway, one can introduce here another approximation: neglect the contribution of non-orthogonality and express the derivative of p' with respect to n by only the first term on the right-hand side of Eq. (7.20):

$$\left(\frac{\delta p'}{\delta n}\right)_k \approx \frac{p'_{\mathrm{N}_k} - p'_{\mathrm{P}}}{(\boldsymbol{r}_{\mathrm{N}_k} - \boldsymbol{r}_{\mathrm{P}}) \cdot \boldsymbol{n}} \,. \tag{7.69}$$

When this simplified expression is introduced in Eq. (7.68) and the mass-flux corrections are expressed as in Eq. (7.66),

$$\dot{m}'_k \approx (\rho v'_n S)_k \,, \tag{7.70}$$

the requirement that the corrected fluxes satisfy the continuity equation

$$\sum_k \dot{m}_k^{m*} + \sum \dot{m}'_k = 0 \tag{7.71}$$

leads to the algebraic pressure-correction equation at each CV of the form:

$$A_{\mathrm{P}} p'_{\mathrm{P}} + \sum_k A_k p'_{N_k} = -\Delta \dot{m}_{\mathrm{P}} \,. \tag{7.72}$$

Once the pressure-correction equation is solved, one can correct the cell-face mass fluxes, cell-center velocities, and pressure. Only an α_p-portion of p' is added to p^{m-1}.

When the grid is not appreciably non-orthogonal, the neglected effects of non-orthogonality are hardly felt. However, when the grid is substantially non-orthogonal, one needs to further reduce α_p, typically to 0.1 or even below. One can take the effect of grid non-orthogonality into account through another corrector, in which the neglected contribution of gradients on the right-hand side of Eq. (7.69) is computed explicitly, i.e., it lags one correction step behind the principal term; see Ferziger et al. [13] for further details.

If mass fluxes are specified at boundaries, the pressure-correction equation has zero normal derivative there (Neumann boundary conditions, as described earlier). However, one may also specify pressure at inlet and outlet and let the mass flow rate be determined in the course of computation, such that the total pressure loss due to flow becomes equal to the specified pressure difference. In that case, $p' = 0$ at these boundaries (Dirichlet boundary condition), but the velocity is unknown. The velocity at the boundary is first computed using expression (7.64), in which interpolation between two nodes on either side of an internal cell-face is replaced by extrapolation towards boundary. Because no large gradients in flow direction are expected near inlet and outlet boundaries, one usually uses the simplest possible extrapolation—the values from the CV-center P. By setting the node N_k

at the boundary face center, expressions for velocity correction, (7.68), and the derivative of p' with respect to n, (7.69), remain the same. Because boundary pressure is prescribed, $p'_{\mathrm{N}_k} = 0$ but Eq. (7.68) produces a correction to the boundary velocity.

Because pressure is a scalar variable, one cannot compute three velocity vector components reliably from pressure boundary condition. One leaves the tangential components as extrapolated and corrects only the boundary-normal component as outlined above. However, that works well only if the flow is nearly orthogonal to the boundary, or at least everywhere directed inward or outward. Another possibility is to prescribe the flow direction in addition to pressure; in that case, the degrees of freedom for flow variation at the pressure boundary are reduced and the convergence is improved.

Pressure is often specified at outlet boundaries. Especially when fluid leaves the solution domain through more than one outlet, it is necessary to specify either the flow rate or the pressure at each outlet to make the solution unique. If pressure is specified at both inlet and outlet and the Reynolds number is very high, the convergence becomes slow because information travels slowly against the flow direction (inertial terms are much larger than viscous terms).

For more details on pressure boundary conditions, see Ferziger et al. [13] and Gresho and Sani [15], among others.

7.4 Pressure-Based Methods for Compressible Flows

A large number of methods specially designed to compute compressible flows have been developed in the past. These methods solve the continuity equation for density and obtain the pressure from an equation of state. Some special versions of discretization methods and special terminologies have been introduced in such methods. Very often the discretization is first done using a lower-order method and then the solution is *re-constructed* by adding a correction which leads to a higher-order approximation. Details on methods designed specifically for compressible flows can be found in Anderson et al. [2] and Hirsch [16], among other books.

There are also special coupled solution methods, which consider the density, velocity components, and temperature as one vector of unknowns; iterative solution methods with multigrid acceleration are usually used to solve the resulting algebraic equation systems. With special preconditioning (see Weiss and Smith [38]) such coupled solvers can be used for both compressible and incompressible flows; see Weiss et al. [37] for a description of one such method. These methods are somewhat more complicated to program and require more memory than the sequential solution methods of the SIMPLE-type, but due to the simultaneous solution for all variables they are potentially more robust. The advantages are usually visible when computing steady-state flows; when the flow is unsteady, segregated solution methods are usually more efficient.

Most of the methods designed specifically for compressible flows usually become inefficient if the Mach number is low in some regions, because in that case the density does not change appreciably, neither in space nor in time; they then need special preconditioning. On the other hand, the pressure-correction methods, originally developed for truly incompressible flows, can easily be extended to compressible flows, leading to efficient methods for all flow speeds. One such approach, based on the SIMPLE algorithm, is described below because it is simple and yet effective; it is also used in most commercial and public CFD codes.

To compute compressible gas flows, it is necessary to solve not only the continuity and momentum equations but also a conservation equation for the thermal energy and an equation of state. The energy equation—which is usually written as an equation for enthalpy—must retain terms which are neglected in incompressible flows. For example, viscous dissipation may be a significant heat source and conversion of internal energy to kinetic energy (and vice versa) by means of flow dilatation is also important. The integral form of energy equation was given in Eq. (7.9).

For a perfect gas with constant specific heats, c_p and c_v, the enthalpy becomes $h = c_\mathrm{p} T$, allowing the energy equation to be written in terms of the temperature. Furthermore, under these assumptions, the equation of state is

$$\rho = \frac{p}{RT} , \tag{7.73}$$

where R is the gas constant. In pressure-correction methods for all flow speeds, the equation of state is used to compute density once the pressure and temperature have been updated.

Liquids may also have to be considered as compressible if the pressure is high (either globally or locally, after some events like those leading to the so-called water-hammer problem) or if propagation of acoustic pressure waves is to be simulated. Equations of state for liquids can have many different forms, expressing density as a function of pressure and possibly also of temperature; often, temperature dependence can be neglected. For water, the Tait equation is often used:

$$\rho = \rho_0 + \left(\frac{p - p_0 + B}{B} \right)^{\frac{1}{m}} , \tag{7.74}$$

where, for pure water, p_0 and ρ_0 are the atmospheric pressure and the corresponding water density, $B = 0.3$ GPa and $m = 7.15$.

The major extension to the pressure-correction scheme described earlier for incompressible flows is the addition of the unsteady term in the continuity equation, and the correction of density in mass fluxes, in addition to velocity correction. The first step is the solution of momentum equations; here there are no changes compared to earlier described methods, because the density was always included in all terms as a variable. The changes happen in the construction of the pressure-correction equation. The discretized mass-conservation equation reads in the first step (implicit Euler method used, for the sake of simplicity):

$$\frac{\Delta V}{\Delta t}(\rho^{m-1} - \rho^{n})_{\mathrm{P}} + \sum_{k} \dot{m}_{k}^{m*} = \Delta \dot{m}_{\mathrm{P}} \, . \tag{7.75}$$

The new density is from the previous outer iteration $(m-1)$, and the new mass fluxes are based on the velocity field obtained by solving the momentum equation in the current outer iteration m using density and pressure from previous outer iteration.

In the correction step one has to take into account that by changing the pressure, both velocity and density will change. The velocity correction at cell-face center is expressed as a function of the normal gradient of pressure correction, see Eq. (7.68); the density correction at cell-face center can be expressed as a function of the pressure correction by using the equation of state:

$$\rho_{k}' \approx \left(\frac{\partial \rho}{\partial p}\right)_{T,k} p_{k}' = (C_{\rho} p')_{k}. \tag{7.76}$$

The same expression is used to correct the density at cell center in the unsteady term. For air (ideal gas) and for water (whose compressibility is modeled by the Tait equation), the following expressions for C_{ρ} are obtained:

$$C_{\rho}^{\mathrm{air}} = \frac{1}{RT} \quad \text{and} \quad C_{\rho}^{\mathrm{water}} = \frac{\rho}{m(p - p_{0} + B)} \, . \tag{7.77}$$

The corrected mass flux $\dot{m}^{m} = \dot{m}^{m*} + \dot{m}'$ can now be expressed as:

$$(\dot{m}^{m*} + \dot{m}')_{k} = (\rho^{m-1} + \rho')_{k} (v_{n}^{m*} + v_{n}')_{k} S_{k} \, . \tag{7.78}$$

If one neglects the product of density and velocity corrections (because it tends to zero faster than other terms), the mass-flux correction can be expressed as:

$$\dot{m}_{k}' \approx (\rho^{m-1} v_{n}' S)_{k} + (\rho' v_{n}^{m*} S)_{k} \, . \tag{7.79}$$

If one substitutes the velocity correction as given by Eq. (7.68) and the density correction as given by Eq. (7.76), after adopting an interpolation scheme for the cell-face center value of p' and upon introducing mass-flux corrections into the mass-conservation equation for corrected quantities,

$$\frac{\Delta V}{\Delta t} C_{\rho} p_{\mathrm{P}}' + \sum_{k} \dot{m}_{k}' + \Delta \dot{m}_{\mathrm{P}} = 0 \, , \tag{7.80}$$

a pressure-correction equation in discretized form is obtained. The expressions for the coefficients depend on the approximations of cell-face center values, which can be of any type described earlier for convection and diffusion fluxes. If shocks are present and one uses linear interpolation and central differences, oscillations around shocks will result. Some kind of damping is necessary to avoid them; one may blend first-order upwind schemes with central differences, or use some kind of non-oscillatory schemes or limiters (e.g., total-variation diminishing or TVD schemes, see Hirsch [16], or essentially non-oscillatory or ENO schemes, see e.g., Sonar [34], among others).

There are important differences between the pressure-correction equations for incompressible and for compressible flows. The former represents the discretized Poisson equation, i.e., an equation with diffusion (resulting from velocity correction at cell faces) and source (mass imbalance) terms only. In the compressible case, the pressure-correction equation contains in addition convection (resulting from density correction at cell faces) and unsteady terms, i.e., it resembles the generic transport equation. Also, the solution of the pressure-correction equation for incompressible flow—if the mass fluxes are prescribed at all boundaries—is indeterminate to within an additive constant; in the compressible case, the presence of convection terms makes the solution unique (the pressure must be prescribed somewhere on the boundary).

The nice feature of the pressure-correction equation is that it automatically adapts to the local state of the flow. The ratio of convection term and diffusion term is proportional to $(\mathrm{Ma})^2$; in regions of flow where the Mach number is low, the equation reduces to the form appropriate for incompressible flows. On the other hand, when and where the Mach number is large, the convection terms dominate and properly reflect the hyperbolic nature of the flow. Solving the pressure-correction equation is then equivalent to solving the continuity equation for density, as is traditionally done in methods designed for compressible flows.

This type of the method is especially attractive for flows in which both strongly compressible and almost incompressible regions exist. Examples are flows in valves and body wakes. Details of this kind of solution methods can be found in Demirdžić et al. [8], Karki and Patankar [19] and Van Doormal et al. [36], among others. In these references, as well as in Ferziger et al. [13], details on some specific boundary conditions in compressible flows are also given.

7.5 Computation of Flows With Moving Boundaries

In many engineering applications, the flow domain is changing in time due to the movement of boundaries. The movement may be determined by external effects (e.g., moving valves, pistons, propellers, etc.) or it may depend on the flow itself and have to be determined as part of the solution (fluid-structure interaction problems). In both cases the grid has to move with the boundary. If the base vectors remain fixed (as is the case when Cartesian components are used), the only change in the conservation equations is the appearance of the relative velocity in convection terms; however, to be conservative, one has to consider the so-called *Space Conservation*

Law (SCL). The conservation equations for space, mass, and momentum in integral form for a moving control volume read

$$\frac{\mathrm{d}}{\mathrm{d}t}\int_V \mathrm{d}V - \int_S \boldsymbol{v}_\mathrm{b} \cdot \boldsymbol{n}\,\mathrm{d}S = 0\,, \tag{7.81}$$

$$\frac{\mathrm{d}}{\mathrm{d}t}\int_V \rho\,\mathrm{d}V + \int_S \rho(\boldsymbol{v} - \boldsymbol{v}_\mathrm{b}) \cdot \boldsymbol{n}\,\mathrm{d}S = 0\,, \tag{7.82}$$

$$\frac{\mathrm{d}}{\mathrm{d}t}\int_V \rho u_i\,\mathrm{d}V + \int_S u_i\rho(\boldsymbol{v} - \boldsymbol{v}_\mathrm{b}) \cdot \boldsymbol{n}\,\mathrm{d}S = \int_S (\tau_{ij}\boldsymbol{i}_j - p\,\boldsymbol{i}_i) \cdot \boldsymbol{n}\,\mathrm{d}S + \int_V \rho b_i\,\mathrm{d}V\,. \tag{7.83}$$

Here $\boldsymbol{v}_\mathrm{b}$ is the velocity with which the CV-boundary is moving. Obviously, if the control volume boundary moves with the same velocity as the fluid, the convection fluxes are equal to zero. The control volume becomes then the *control mass* and we are dealing with the Lagrangian approach to the description of fluid motion.

Equation (7.81) describes the conservation of space when the CV changes its shape and position with time. If the grid is not moving, this equation becomes redundant and in all other equations $\boldsymbol{v}_\mathrm{b} = 0$ applies. Thus, the equations for non-moving CVs are only a special case of the equations for moving CVs. The solution method designed for moving grids can then be used for fixed grids as well.

It should be noted that the time derivative in the above equations has a different meaning in fixed and moving grids, although it is approximated in the same way. If the CV does not move, the time derivative represents the *local* change of the conserved quantity (i.e., variation in time at a fixed location in space), while in another extreme of a CV whose surface moves exactly with fluid velocity, it becomes the total (material) derivative because no fluid enters or leaves the CV. This change of meaning is accounted for by the convection fluxes, which become zero in the latter case. For an arbitrary CV motion, the time derivative represents the change over time in a CV whose location, shape, and size also change with time. The above equations can be derived from the Navier–Stokes equations for a fixed control volume by applying Leibniz rule when CV is allowed to vary in time; see Ferziger et al. [13] for more details.

Why is it important to obey the SCL can be seen by considering the mass-conservation equation (7.82) for a fluid of constant density; it can then be written as:

$$\frac{\mathrm{d}}{\mathrm{d}t}\int_V \mathrm{d}V - \int_S \boldsymbol{v}_\mathrm{b} \cdot \boldsymbol{n}\,\mathrm{d}S + \int_S \boldsymbol{v} \cdot \boldsymbol{n}\,\mathrm{d}S = 0\,. \tag{7.84}$$

The first two terms represent the SCL and add up to zero, cf. Eq. (7.81); thus, for fluids with constant density, the mass-conservation equation reduces to

$$\int_S \boldsymbol{v} \cdot \boldsymbol{n}\,\mathrm{d}S = 0 \quad \text{or} \quad \mathrm{div}\,\boldsymbol{v} = 0 \tag{7.85}$$

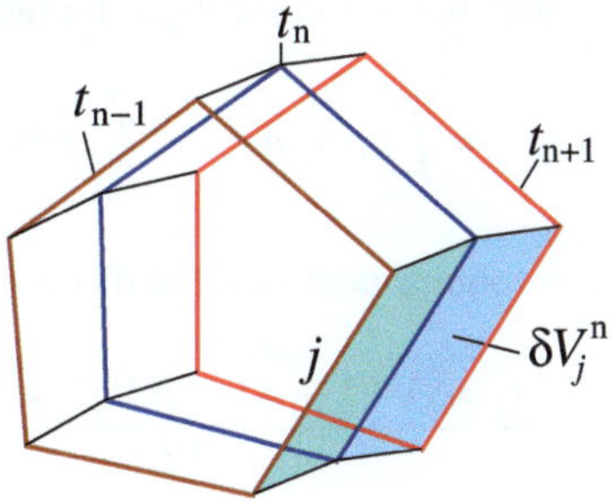

Fig. 7.6 A typical 2D CV at three time levels and the volumes swept by cell faces

irrespective of whether the CV is fixed or moving. It is therefore important to ensure that the above two terms add up to zero in the discretized equations as well (i.e., the sum of volume fluxes through CV-faces due to their movement must equal the rate of change of volume); otherwise, velocity field will not be divergence-free, which is equivalent to introduction of artificial mass sources. These artificial mass sources or sinks may accumulate in time and spoil the solution, as demonstrated by Demirdžić and Perić [10].

The implementation of SCL in a FV method using fully implicit time-integration scheme with three time levels will now be briefly outlined. The extension to other time-integration schemes is straightforward.

The discretized SCL equation can be cast into the following form (see Eq. (7.31)):

$$\frac{3\,(\Delta V)^{n+1} - 4\,(\Delta V)^{n} + (\Delta V)^{n-1}}{2\,\Delta t} = \sum_k (\boldsymbol{v}_\mathrm{b} \cdot \boldsymbol{n})_k^{n+1} S_k^{n+1} \,, \tag{7.86}$$

where the summation is over all faces of the CV. Note that the difference between CV volumes at consecutive time levels can be expressed as the sum of volumes δV_k swept by each CV face when moving from its position at one time step to the next one, see Fig. 7.6, i.e.:

$$(\Delta V)^{n+1} - (\Delta V)^{n} = \sum_k \delta V_k^n \,. \tag{7.87}$$

When this expression is introduced into Eq. (7.86), one finds out that the SCL is satisfied identically if the volume fluxes through cell faces are defined as:

$$\dot{V}_k^{n+1} = \left(\int_{S_k} \boldsymbol{v}_\mathrm{b} \cdot \boldsymbol{n}\,\mathrm{d}S\right)^{n+1} \approx [(\boldsymbol{v}_\mathrm{b} \cdot \boldsymbol{n})_k S_k]^{n+1} \approx \frac{3\,\delta V_k^n - \delta V_k^{n-1}}{2\,\Delta t} \,. \tag{7.88}$$

Therefore, the volumes swept by each face over one Δt, δV_k, are calculated from the grid positions at two time levels and used to define volume fluxes $\dot{V}_k^{n+1}$; there is then no need to define explicitly the velocity of the CV surface, $\boldsymbol{v}_\mathrm{b}$. The alternative is to calculate the grid velocity $\boldsymbol{v}_\mathrm{b}$ from the movement of the cell-face center and calculate the volume flux by multiplying it with the new surface vector; however, it is difficult then to make sure that SCL is satisfied.

The mass flux through the face k can now be calculated as (see Eq. (7.66)):

$$\dot{m}_k^{m*} = \int_{S_k} \rho \boldsymbol{v} \cdot \boldsymbol{n} \, \mathrm{d}S - \int_{S_k} \rho \boldsymbol{v}_\mathrm{b} \cdot \boldsymbol{n} \, \mathrm{d}S \approx (\rho^{m-1} v_n^{m*} S^{m-1})_k - (\rho \dot{V})_k^{m-1} . \qquad (7.89)$$

The requirement that the discretized mass-conservation equation be satisfied:

$$\frac{3\,(\rho_\mathrm{P} \Delta V)^m - 4\,(\rho_\mathrm{P} \Delta V)^n + (\rho_\mathrm{P} \Delta V)^{n-1}}{2\,\Delta t} + \sum_k \dot{m}_k^{m*} + \sum_k \dot{m}_k' = 0 \, , \qquad (7.90)$$

and the introduction of the above-defined expressions for $\dot{V}_k$ and $\dot{m}_k'$ lead finally to a pressure-correction equation of the same form as in the case of fixed grids (the same approach is valid for both compressible and incompressible flows). If the grid position at the new time level is known (e.g., in piston-driven flows, flows around rotating machinery, etc.), the volume fluxes through cell faces $\dot{V}_k^{n+1}$ are not dependent on the outer iteration counter m; otherwise, the volume fluxes need to be corrected during outer iterations together with other corrections. This is the case when a two-way fluid-structure interaction is considered, or flow around floating bodies is computed.

When the grid position is known at each time level, the implementation of the grid movement in the solution procedure is simple; see Demirdžić and Perić [11] and Ferziger et al. [13] for more details and examples of application. When the boundary movement is not known in advance, an iterative procedure within each time step (outer iterations) has to be used, as outlined above.

7.6 Examples of Solutions of the Navier–Stokes Equations

In this section, examples of application of the SIMPLE algorithm to compute incompressible and compressible flows are presented. The aim is to demonstrate the performance of the solution method under different flow regimes, from creeping flow to a supersonic flow with shocks. All computations were performed with the commercial CFD code Simcenter STAR-CCM+ from Siemens.

The flow around a sphere at various Reynolds numbers has been selected for demonstration purpose. The sphere has a diameter of 61.4 mm (corresponding to a billiard ball, which was used in experimental studies by Bakić [4]); at the downstream stagnation point, it is fixed to a cylindrical rod with a diameter of 8 mm. The solution domain is a box, extending 1 m away from sphere in all directions, except for flow at Re = 50,000; in this case the experimental setup from Bakić [4] is used, with wind tunnel walls being 150 mm away from sphere in y and z-direction. The computational grids consist predominantly of Cartesian hexahedral cells, locally refined near sphere and in the wake. Around sphere, there is a layer of prismatic cells with varying base (from a triangle to an arbitrary polygon). This

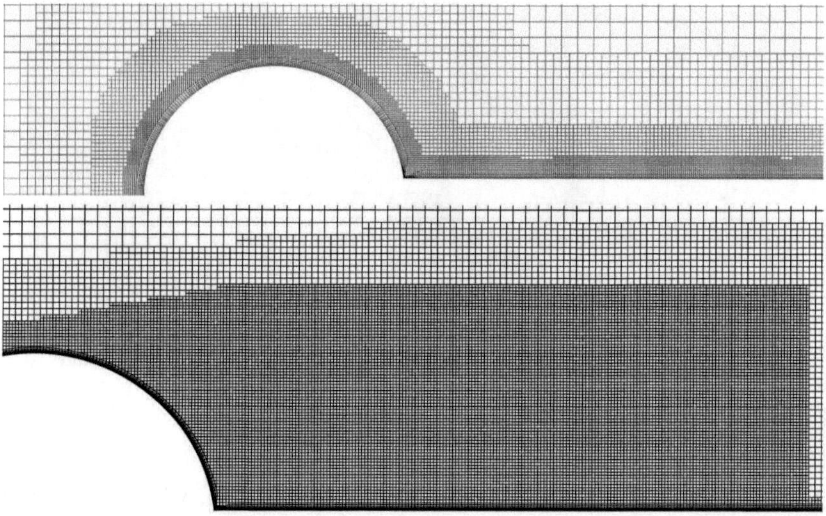

Fig. 7.7 Details of grids used to compute flows around sphere: Re = 50 (upper) and Re = 50,000 (lower)

layer is created by projecting faces resulting from cutting Cartesian cells by an offset surface at a prescribed distance from sphere surface; such grids are called *trimmed*. The refinement in the wake is different for different Reynolds numbers; Fig. 7.7 shows parts of the grid near sphere for Re = 50 and Re = 50,000. For these two Reynolds numbers, grid-dependence study was performed to estimate discretization errors; at other Reynolds numbers, computation was performed on a single grid whose fineness has been selected such that it is adequate to provide a reasonably accurate representation of the flow (based on experience).

7.6.1 Creeping Flow (Re = 5)

At very low Reynolds numbers, the flow is dominated by viscous effects and it stays attached to the sphere surface. A practical example is sedimentation of solid particles in a stagnant liquid: small particles are slowly falling down, creating a creeping flow around them. At very small Reynolds numbers (below 1) velocities are very low; the convection terms, which involve velocities squared, are negligible compared to viscous terms (which involve products of viscosity and velocity gradients). By neglecting convection terms in the momentum equations, one obtains Stokes equations, which are linear and thus easier to solve. Commercial CFD codes usually do not offer the option of deactivating convection terms, so they are computed even though their contribution to the solution is negligible.

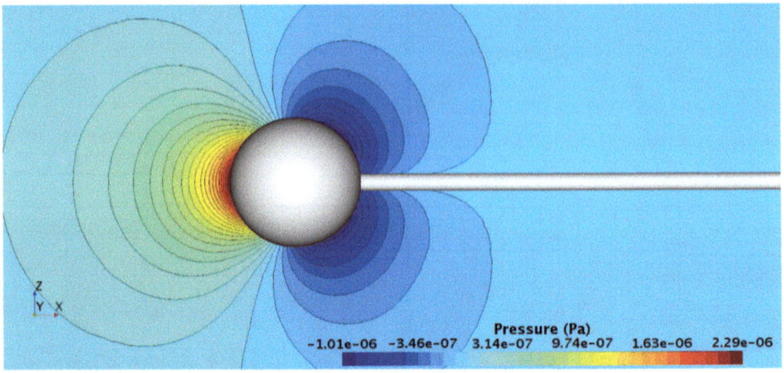

Fig. 7.8 Details of pressure distribution around sphere at Re = 5

Stokes derived an analytical expression for sphere drag, which is valid up to Re ≈ 1;

$$C_\mathrm{d} = \frac{F_x}{0.5\,\rho U^2 R^2 \pi} = \frac{24}{\mathrm{Re}}\,, \tag{7.91}$$

where F_x is the flow-induced force on the sphere in the flow direction. In the same way lift coefficients are determined by replacing F_x by F_y or F_z, where y and z are coordinates in the cross-flow direction. The Reynolds number is defined with free-stream velocity and sphere diameter, Re = $(\rho U_\infty D)/\mu$.

At Re = 5, the flow is steady and it was computed in a solution domain including only a quarter of the sphere, with symmetry conditions at the horizontal and vertical symmetry planes, $y = 0$ and $z = 0$. With air ($\rho = 1.2$ kg/m^3, $\mu = 0.000018$ Pa s) flowing around the sphere of the selected diameter, the velocity is very low—1.2215 mm/s. For this reason the pressure variation is also very small: the difference between maximum and minimum pressure on sphere surface is just 0.0000033 Pa. The computed pressure distribution around sphere is shown in Fig. 7.8. Note that the location of minimum pressure is on sphere surface ca. 135° downstream of front stagnation point. The drag coefficient of 6.719 corresponds well to experimental observations.

7.6.2 Steady Laminar Flow with Separation (Re = 50)

At Re = 50, the flow is still steady but it exhibits a recirculation zone: the flow separates from sphere surface at an angle of ca. $\theta = 135°$ from the front stagnation point and re-attaches at the supporting rod about $0.4D$ behind sphere, see Fig. 7.9. Because the flow is steady and axi-symmetric, it was also computed in a quarter of the flow domain with imposed symmetry conditions (the flow could have actually been computed as two-dimensional axi-symmetric).

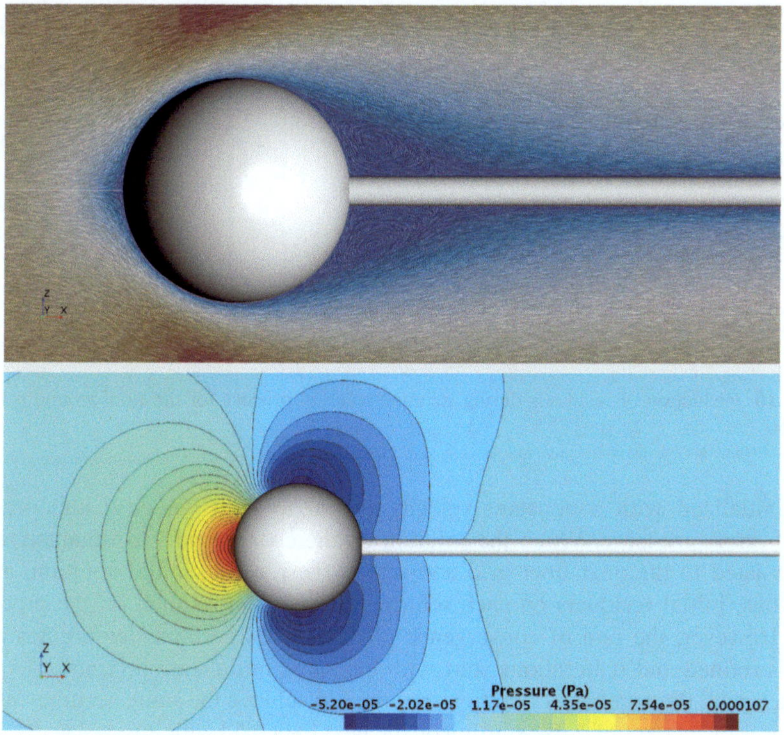

Fig. 7.9 Details of flow pattern (upper) and pressure distribution around sphere (lower) at Re = 50

Figure 7.9 shows the flow pattern around the sphere and the pressure distribution. The velocity is now 10 times higher than for Re = 5 and therefore the difference between maximum and minimum pressure is also higher—0.00018 Pa. Note that the location of minimum pressure has moved upstream with increasing Reynolds number: it is now at ca. 92° downstream of front stagnation point.

For this Reynolds number, the flow was computed on three systematically refined grids (grid spacing reduced by a factor 1.5), consisting of 270,617 CV, 477,661 CV, and 1,844,873 CV, respectively. The difference in computed drag between the medium and the fine grid was smaller than 0.01%, which indicates that the discretization errors are very low. For the second-order discretization method used, Richardson extrapolation predicts the error on the fine grid being equal to 1/1.25 of the difference between solutions on medium and fine grid. Thus, the grid-independent solution is expected to differ from the solution obtained on the fine grid by less than 0.01%. The computed drag coefficient on the fine grid of 1.550 corresponds well to experimental observations.

Figure 7.10 shows the residual norms (sum of absolute values over all CVs) as functions of the number of outer iterations for all three grids. The residuals are normalized so that the initial value of the norm is around 1. Because the solution

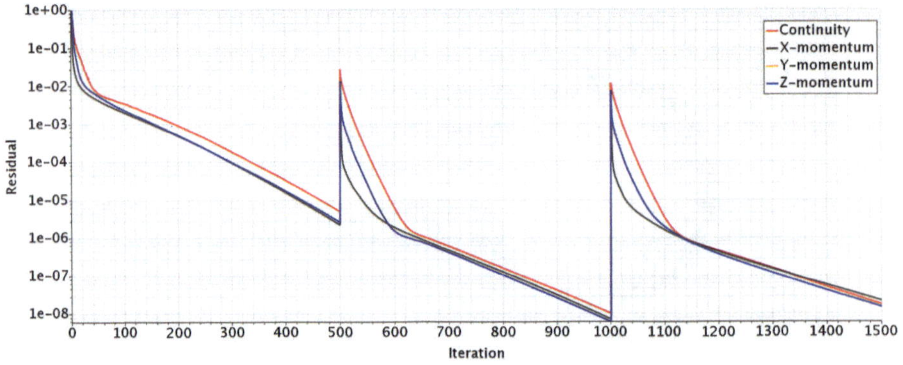

Fig. 7.10 Reduction of residuals during iterative solution process for the flow around sphere at Re = 50

was initialized with a constant velocity (equal to inlet velocity) and pressure, initial errors are large. When the solution on the coarse grid was obtained, it was interpolated to the next finer grid and served there as the initial solution, and so on. Thus, initial residuals on each refined grid are smaller than on the preceding grid. However, the rate of convergence of SIMPLE algorithm deteriorates as the grid is refined and if iterations start with the same initialization (constant values) on each grid, the number of iterations required to achieve the same accuracy usually increases by a factor of 2. By starting on the next finer grid with interpolated solution from preceding grid, the required number of iterations is significantly reduced. A further increase in efficiency can be achieved by applying multigrid method to outer iterations, but this feature is not available in commercial CFD software; see Lilek et al. [21] or Ferziger et al. [13] for a description of such a multigrid method. In the present example, residuals were reduced to a much lower level than is normally necessary in industrial applications; a reduction of 3 to 4 orders of magnitude relative to initial residual level is usually sufficient to guarantee that solution variables do not change at the 3 to 4 most significant digits.

7.6.3 Unsteady Laminar Flow with Separation (Re = 500)

By increasing the inflow velocity by a factor of 10, the Reynolds number of 500 is obtained. Now the flow is still laminar, but it is no longer steady—it exhibits a strong variation in time. This variation is not regularly periodic, as is the case with flow around circular cylinder. For a reasonably accurate simulation of this flow, the grid design needs to be modified and it also needs to be finer than for Re = 50. Now the flow in the wake is highly complex and one therefore needs a finer grid there; this is true for any time-accurate simulation of unsteady flow for larger Reynolds numbers. The grid used here was similar to the one shown in Fig. 7.7 for Re = 50,000; it had

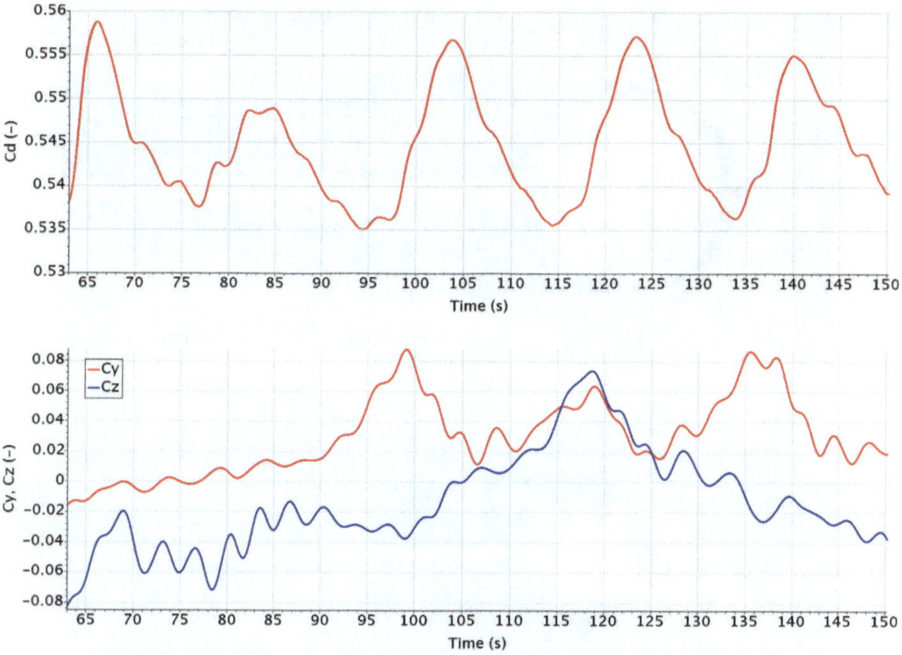

Fig. 7.11 Variation of drag (upper) and two lift coefficient (lower) over time in the flow around sphere at Re = 500

4,726,424 CV in total, the thickness of the cell layer next to sphere surface was 0.18 mm and the grid spacing in sphere wake was 0.9375 mm (half of that value in a 5 mm thicker layer around sphere surface).

As can be seen in Fig. 7.11, both the drag coefficient and the two lift coefficients vary irregularly in time; with a further increase in flow speed, the flow in sphere wake becomes turbulent. The drag coefficient oscillates around a mean value of 0.545 with an oscillation period of ca. 20 s, which corresponds well to experimental observations. The two lift coefficients are supposed to oscillate around zero value, but the period of oscillation is order of magnitude longer than for the drag; one would need to simulate the flow over a very long period of time in order to obtain zero mean values. The long-time average of the unsteady flow would be axi-symmetric, but this mean flow would not necessarily correspond to the steady-state solution that would be obtained if one computed the flow as two-dimensional axi-symmetric.

Figure 7.12 shows instantaneous flow pattern and pressure distribution in one axial section and in Fig. 7.13 the axial velocity and the axial component of the vorticity vector in a cross-section one diameter behind sphere are shown. These figures indicate the range of velocity and pressure variation and also confirm that the flow is not axi-symmetric and also not yet turbulent. Note that the location of

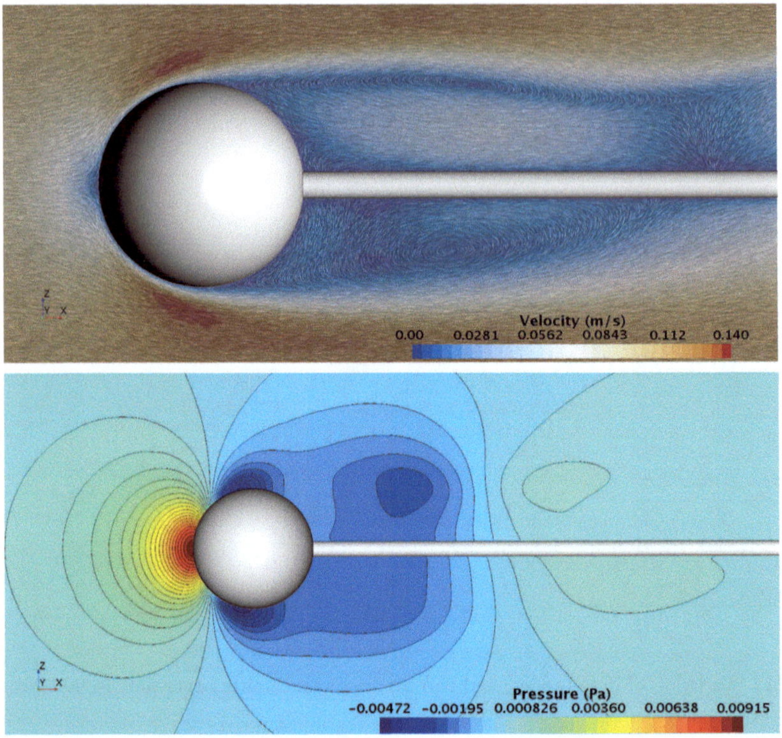

Fig. 7.12 Details of instantaneous flow pattern (upper) and pressure distribution (lower) around sphere at Re = 500

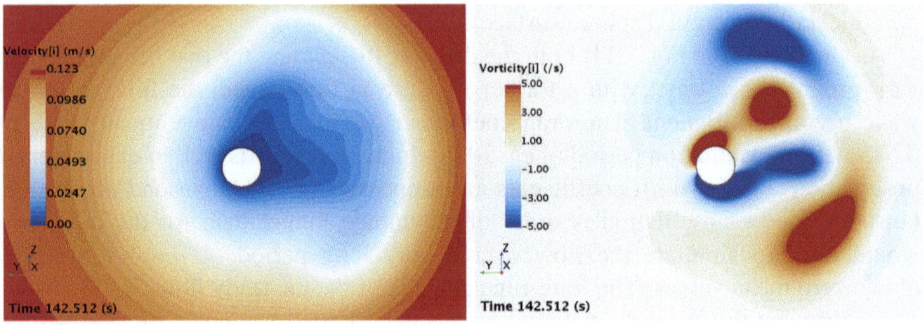

Fig. 7.13 Details of instantaneous contours of axial velocity component (left) and axial component of vorticity vector (right) in a cross-section one diameter behind sphere at Re = 500

minimum pressure on sphere surface now oscillates, but it is further upstream than at Re = 50 (always at $\theta < 90°$). Animation of the flow shows that a couple of vortex rings and tubes are generated in the sphere wake, which oscillate around the support rod, growing and collapsing alternately.

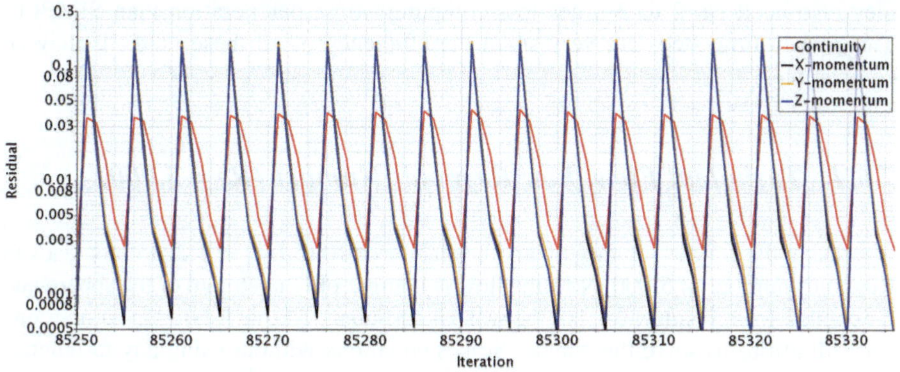

Fig. 7.14 Variation of residuals with outer iterations in the simulation of flow around sphere at Re = 500

In this simulation, the time step was 0.01 s; it is very short relative to the long period of drag oscillation, but one can see from Fig. 7.11 that all forces show oscillations at various frequencies; the shortest periods are of the order of 3 s, which means that there are about 300 time steps per such an oscillation period. However, forces are integral quantities and they oscillate less than the local velocity and pressure values at any location behind sphere. Visualization of vorticity variation in the wake shows strong fluctuation at higher frequencies, so the selected time step was appropriate in order to accurately resolve all relevant temporal variations in the flow.

Figure 7.14 shows variation of residuals as functions of outer iterations for several time steps. The residuals in momentum equations are reduced almost three orders of magnitude after 5 outer iterations. It appears that the residuals in the continuity equation are not reduced as well; however, one should take into account that the plotted mass residuals are recorded before solving the pressure-correction equation (they are computed using uncorrected velocity components after the momentum equations are solved). After solving the pressure-correction equation in the last outer iterations, the continuity equation is satisfied to within the convergence tolerance specified for the inner iterations in the pressure-correction equation. Thus, at the end of each time step, the mass residuals are at least one order of magnitude lower than the level shown in Fig. 7.14 and therefore at the same level as the residuals from momentum equations.

For unsteady flows, when time steps are small enough, one can specify a fixed number of outer iterations per time step (3 to 5) as is considered appropriate for the update of non-linear terms and deferred corrections. Experience shows that SIMPLE algorithm achieves higher accuracy for the same time step than PISO algorithm or non-iterative versions of fractional-step method. This is expected because the latter do not update all terms affected by linearization and deferred corrections, as they perform only a single outer iteration on momentum equations per time step. However, because non-iterative methods do not have the outer iteration loop, they also need ca. 2 to 3 times less computing time per iteration than SIMPLE. Thus, when time steps are very small the efficiency of all these methods may be comparable.

7.6.4 Turbulent Flow, Low Reynolds Number (Re = 5,000)

Increasing the inflow velocity by another factor of 10 to $U_\infty = 1.2215$ m/s leads to Reynolds number of 5,000. Now the flow is laminar only upstream of the sphere and far enough from it sideways, but the wake is turbulent. At this Reynolds number one can still afford to solve the Navier–Stokes equations without using any turbulence model; the flow is computed as laminar and unsteady—one only has to ensure that the grid is fine enough to resolve all scales of spatial variation in velocity and pressure, and that the time step is small enough to resolve all temporal variations of variables. If these conditions are met, one is performing a so-called *direct numerical simulation* (DNS) of a turbulent flow. Because the flow is turbulent only in the wake, the grid was locally refined to the required resolution only in this zone, as shown in Fig. 7.7 for Re = 50,000. The grid is additionally refined in the zone where the flow separates from the sphere (around 89° from the front stagnation point) and along the resulting shear layer, separating the high-speed flow around sphere from the recirculating flow in the wake. The first cell layer at sphere wall is 0.05 mm ($D/1230$) thick, while the Cartesian cells in the shear layer have the dimension of 0.235 mm ($D/260$); the major part of the wake is discretized by cubic cells with edge length 0.47 mm ($D/130$). The total number of cells in the grid was 21,068,524. The time step adequate for such a simulation has been estimated to be $\Delta t = 50\ \mu$s, meaning that the fluid flowing at the mean inflow velocity needs ca. 1,000 time steps to travel the distance equal to sphere diameter.

Figure 7.15 shows visualization of instantaneous flow pattern in simulation as well as in experiment performed by Seidl et al. [32]. In experiment, the flow is visualized by injecting ink into fluid (water) through two holes, one at $\theta = 60°$ and one at $\theta = 120°$, where the angle θ is measured from the front stagnation point. The ink traces seen in the picture represent the so-called streaklines—the lines created by particles which all passed though the same location in space at some time in the past. One can create the same kind of lines in a simulation by injecting massless particles and following their motion, but this was not done here. Instead, the flow features are visualized by plotting the component of vorticity vector normal to the section plane. The curling of lines of constant vorticity indicates the roll-up of the shear layer in much the same way as the experimental streaklines; the similarity of the two representations shown in Fig. 7.15 is obvious. One can clearly see from these pictures that the flow separates laminarly, and the first two roll-ups are also laminar; thereafter the regularity of fluid motion breaks down and the flow becomes turbulent. The end of the recirculation zone is approximately 2 diameters downstream of sphere.

Instantaneous pressure distribution shown in Fig. 7.15 confirms the existence of vortex rings in the shear layer, as indicated by nearly symmetric low pressure

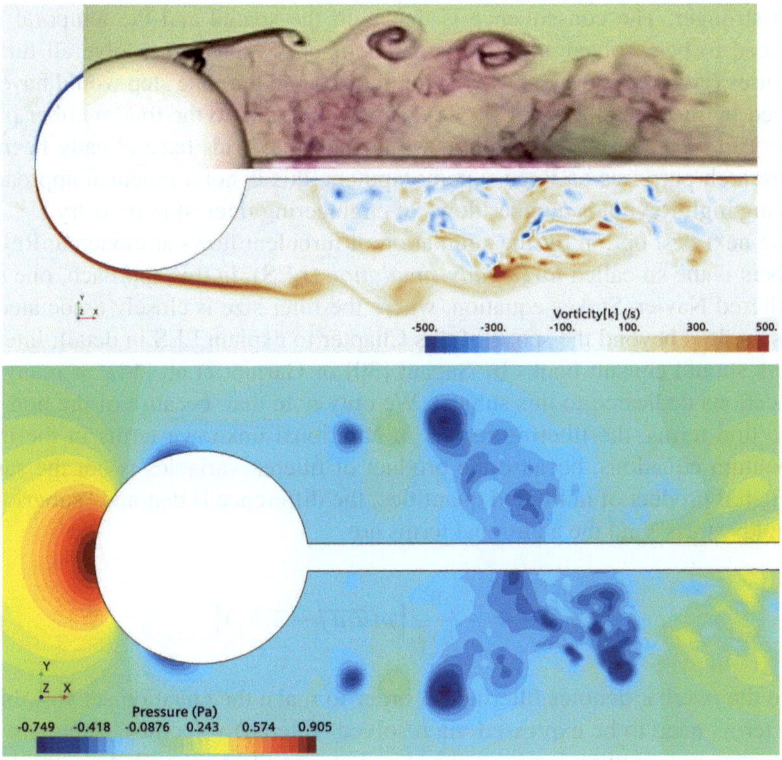

Fig. 7.15 Details of instantaneous flow pattern from experiment by Seidl et al. ([32]; upper part of upper figure) and from simulation (lower part of upper figure) and instantaneous pressure around sphere at Re = 5,000

zones. The vortex rings become stronger as they travel downstream and eventually turbulent, leading to a breakdown as they approach reattachment location $2.6D$ downstream of sphere center. The lowest pressure on sphere surface is found at about $\theta = 70°$, but pressure is much lower in vortex rings downstream of sphere.

The flow separation from sphere has been studied in detail by Achenbach [1]; more details about DNS of flow around sphere at Re = 5,000 can be found in Seidl et al. [32] and Seidl [31].

7.6.5 Turbulent Flow, Medium Reynolds Number (Re = 50,000)

With another order-of-magnitude increase of the inflow velocity to $U_\infty = 12.215$ m/s, the Reynolds number of 50,000 is obtained. The flow is still laminar upstream of the sphere and far enough from it sideways, but the turbulence in the wake is now much stronger. The consequence is that both the spatial and the temporal scales that need to be resolved are much smaller; if one wanted to resolve all turbulent structures (i.e., to perform a DNS), both the grid and the time step would have to be reduced by more than a factor of 100. This would lead to the total number of cells in the grid of the order of 5 billion; while such fine grids have already been used for research purposes on large supercomputers, this is not a practical approach for studying high Reynolds number flows of engineering interest in industry.

The next best option for the simulation of turbulent flows at moderate Reynolds numbers is the so-called *large-eddy simulation* (LES). In this approach, one solves the filtered Navier–Stokes equation, where the filter size is closely associated with grid size. It is beyond the scope of this Chapter to explain LES in detail; interested readers should consult books by Sagaut [30] or Garnier et al. [14], or many other publications dedicated to this subject. We only note that, because of the non-linear convection terms, the filtering results in additional unknown terms in the filtered momentum equations, because the product of filtered variables is not the same as the filtered product of unfiltered quantities; the difference is denoted "subgrid-scale Reynolds stress" and the additional terms are

$$\frac{\partial \tau_{ij}^{\mathrm{sgs}}}{\partial x_j} = -\frac{\partial}{\partial x_j}\left[\rho(\overline{u_i u_j} - \overline{u}_i\overline{u}_j)\right] \,, \tag{7.92}$$

where the overbar denotes filtering. In order to make the equation set solvable, the extra terms need to be expressed via resolved (filtered) velocities. There are many approaches in literature; here the oldest and the simplest subgrid-stress model, due to Smagorinsky [33], has been used. It is based on the concept of eddy-viscosity and expresses the subgrid-scale Reynolds stresses as:

$$\tau_{ij}^{\mathrm{sgs}} - \frac{1}{3}\tau_{kk}^{\mathrm{sgs}}\delta_{ij} \approx \mu_{\mathrm{t}}\left(\frac{\partial \overline{u}_i}{\partial x_j} + \frac{\partial \overline{u}_j}{\partial x_i}\right) = 2\mu_{\mathrm{t}}\overline{S}_{ij} \,, \tag{7.93}$$

where δ_{ij} is the Kronecker's delta and $\overline{S}_{ij}$ is the strain rate of the resolved velocity field. Thus, it is assumed that the effects of subgrid-scale Reynolds stresses correspond to locally increased viscosity of the fluid. The additional (eddy or turbulent) viscosity varies across the flow—it too has to be computed using the available information from the resolved (filtered) velocity field. The usual expression used in conjunction with the Smagorinsky model is

$$\mu_{\mathrm{t}} = C_{\mathrm{S}}^2\rho\Delta^2|\overline{S}| \,, \qquad \overline{S} = \sqrt{\overline{S}_{ij}\overline{S}_{ij}} \,, \tag{7.94}$$

where C_{S} is a model parameter (values vary between 0.05 and 0.2, depending on flow type and distance from walls) and Δ is the filter length scale (usually set equal to the grid size).

Now we are no longer solving the exact equations; therefore, in addition to discretization errors, the solution also contains modeling errors. Both error types are difficult to estimate in LES: when the grid and time step are refined, the larger portion of turbulence spectrum is resolved and a smaller portion is modeled, so both the discretization and the modeling errors reduce at the same time. The classical method of estimation of discretization errors by using a systematic grid refinement and applying Richardson extrapolation cannot be directly used. In order to separate discretization and modeling errors, one would need to use the same filter size on all grids; while this is in principle possible, it is not very practical and is seldom used. Because the estimation of modeling errors requires comparison with a more accurate reference solution (either from an accurate experiment or from a more accurate simulation), one simply refines both the grid and the time step and analyses the difference in resulting solution by comparing it with the reference solution.

Simulations were conducted using three grids: the coarse one had only 1,956,982 CVs, the medium one had 5,841,227 CVs and the fine grid had 33,356,370 CVs. The solution domain corresponds to the test section of the wind tunnel used in experiments of Bakić [4]: the tunnel cross-section is 300×300 mm and the sphere is positioned in its center. Inlet and outlet cross-sections are both 300 mm away from sphere center. Boundary layers at tunnel walls are not resolved by the grid, because it is assumed that their influence on the flow behind sphere is negligible; the main effects of tunnel walls is the flow confinement. The sphere with its diameter of 61.4 mm (cross-section area 0.00296 m^2) blocks about 3.3% of the tunnel cross-section area of 0.09 m^2. At this moderate Reynolds number, this blockage ratio is significantly affecting the flow—the results are not the same as would be obtained in an infinite environment. As the Reynolds number increases, this effect diminishes.

In LES one should resolve around 80% of turbulence spectrum and model around 20%; the reason is that large-scale turbulence is problem-dependent and difficult to model (and thus should be resolved), while the small-scale turbulence is more of a universal nature and thus easier to model (i.e., the modeling errors from the subgrid-scale model are smaller if this criterion is met). The time step was 20 μs for the coarse grid and 10 μs for the medium and fine grid. The two coarser grids used here certainly do not satisfy the above criterion and one can therefore expect larger errors in the solutions. However, the alternative approach for industrial flow simulation—the use of Reynolds-averaged Navier–Stokes (RANS) equations which model all of turbulence—produces very poor results for this kind of flow (with errors of the order of 50%) and the intention was to see whether LES-errors are smaller.

The reason why this kind of flow is difficult to predict using turbulence models lies in the fact that the flow separates as laminar from the sphere surface and then undergoes transition from laminar to turbulent in the separated shear layer, with only the wake being fully turbulent. The finest grid predicts the mean drag of $C_D = 0.492$, which corresponds closely to experimentally obtained values. The instantaneous values vary strongly in time, as can be seen from Fig. 7.16. The medium grid predicts 4.2% lower the mean drag ($C_D = 0.471$), which would still be acceptable in industrial applications. The result from the coarsest grid is 9% lower ($C_D = 0.448$), which is significantly lower than the experimental data. However, compared to the solutions with RANS equations this is still much closer to reality.

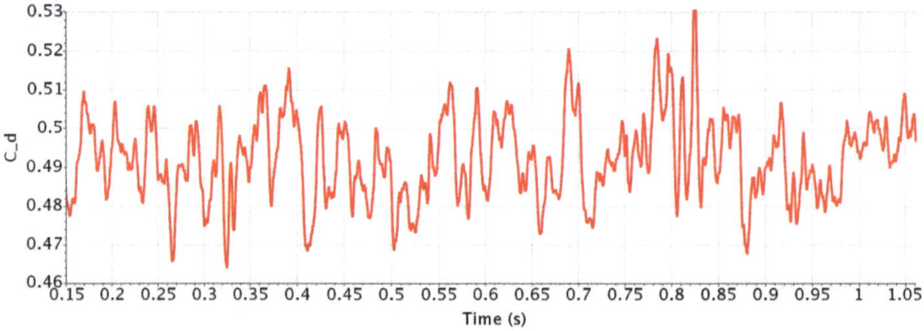

Fig. 7.16 Variation of sphere drag with time, computed on the fine grid at Re = 50,000

Figure 7.17 shows visualization of instantaneous flow pattern and pressure distribution around sphere, computed on the finest grid. One can see that the size of turbulent eddies has drastically reduced compared to the flow at 10 times lower Reynolds number. The requirement to use a very fine numerical grid is obvious. The simulation was started by using the solution from the coarser grid to initialize the flow field. After about 20,000 time steps, the collection of data for averaging was started. After 60,000 time steps, the average flow field has become almost axi-symmetric (one would need to run the simulation over a much longer period of time to obtain zero average lift and perfectly axi-symmetric mean flow field). Figure 7.18 shows visualization of time-averaged flow pattern and pressure distribution around sphere, computed on the finest grid. Note that the real time over which the solutions were averaged is only 0.6 s; during this time, fluid has flown over a distance equal to 120 sphere diameters.

Comparison of mean (Fig. 7.18) and instantaneous (Fig. 7.17) pressure distribution reveals that maximum pressure is almost the same (98.6 vs. 97.9 Pa), while minimum pressures differ drastically (mean: −57.5 Pa; instantaneous: −119 Pa). The lowest pressure in the mean field is found on sphere surface, at about $\theta = 75°$, while in the instantaneous field, the minimum values are found at centers of turbulent eddies in the shear layer. Although engineers are mostly interested in time-averaged values, it is often important to know the range in which instantaneous values vary. For example, in a flow through a pump the mean pressure may be everywhere above saturation level and one may think that cavitation does not occur, while instantaneous values may locally be well below saturation level, leading to cavitation and many engineering problems (noise, vibration, erosion, and performance loss).

Figure 7.19 shows variation of mean pressure coefficient over sphere surface; it is defined as follows:

$$C_\mathrm{p} = \frac{p - p_\infty}{\frac{1}{2}\rho U_\infty^2} \,. \tag{7.95}$$

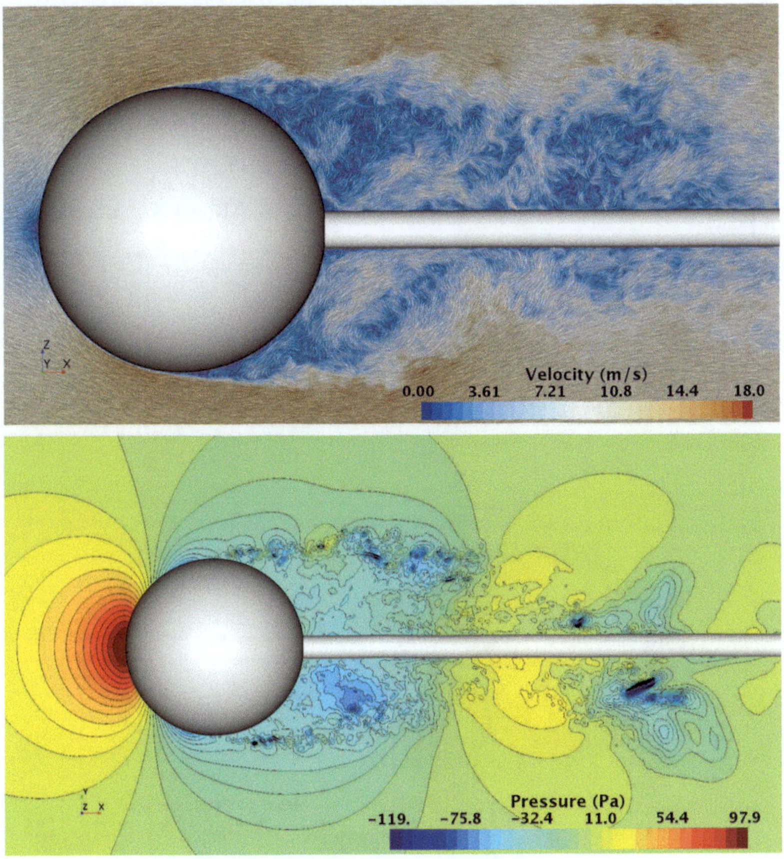

Fig. 7.17 Details of instantaneous flow pattern (upper) and pressure distribution (lower) in sphere wake at Re = 50,000

Here p stands for local pressure and p_∞ and U_∞ are the pressure and velocity in the far field (here: values at the inlet of solution domain). Engineers are used to expect that the pressure coefficient at a stagnation point equals unity (because at that location all kinetic energy is converted to pressure—the velocity is zero and the pressure has its maximum value). However, this expectation is justifiable only (i) at sufficiently high Reynolds numbers and (ii) in an infinite environment. In the present case, at a moderate Reynolds number Re = 50,000 and in a wind tunnel with a significant blockage, the value of pressure coefficient at the front stagnation point should be above 1. As one can see from Fig. 7.16, this is indeed the case. The computed mean pressure coefficient follows the ideal curve obtained analytically for inviscid flow (in which there is no drag—the pressure recovers at the downstream side to the same value found at the front stagnation point) with an offset tp to $\theta \approx 50°$; thereafter, the curves differ significantly. The lowest value of the pressure coefficient from the simulation is much higher than the theoretical value;

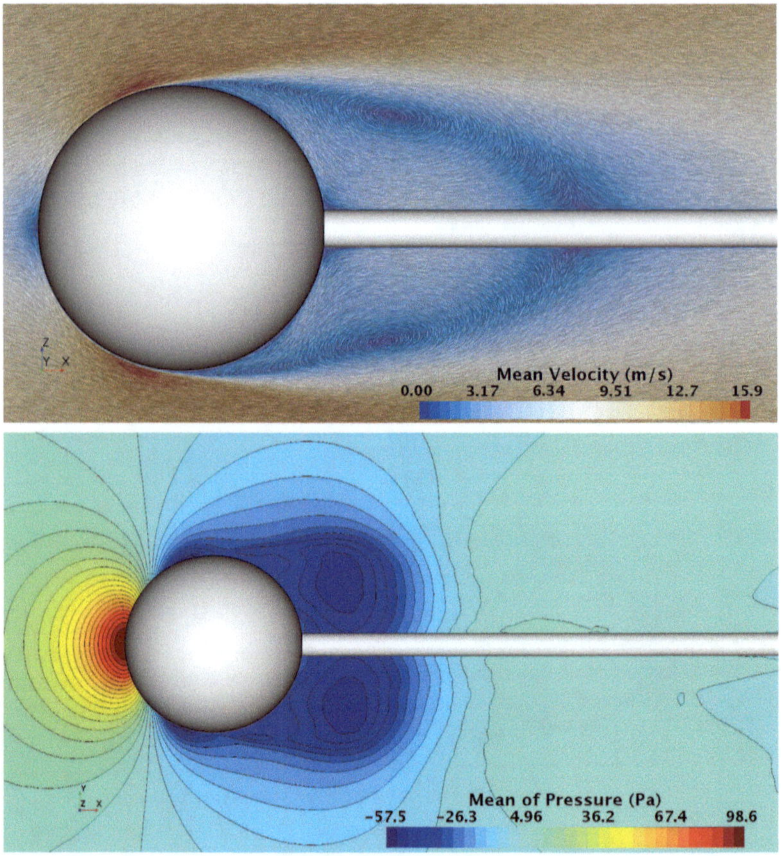

Fig. 7.18 Details of mean (averaged over 60,000 time steps) flow pattern (upper) and pressure distribution (lower) in sphere wake at Re = 50,000

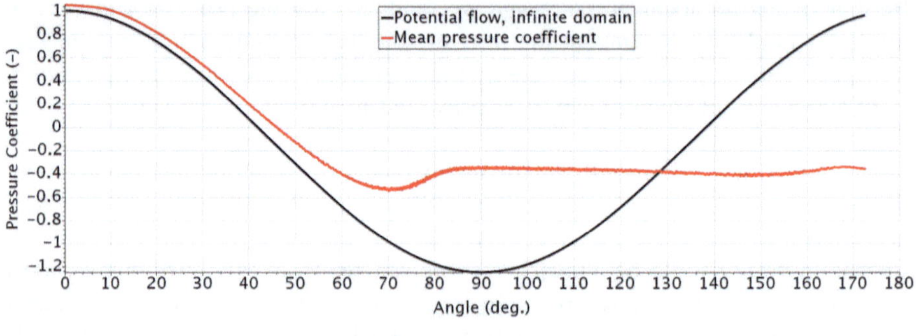

Fig. 7.19 Variation of pressure coefficient along sphere perimeter, computed on the fine grid at Re = 50,000

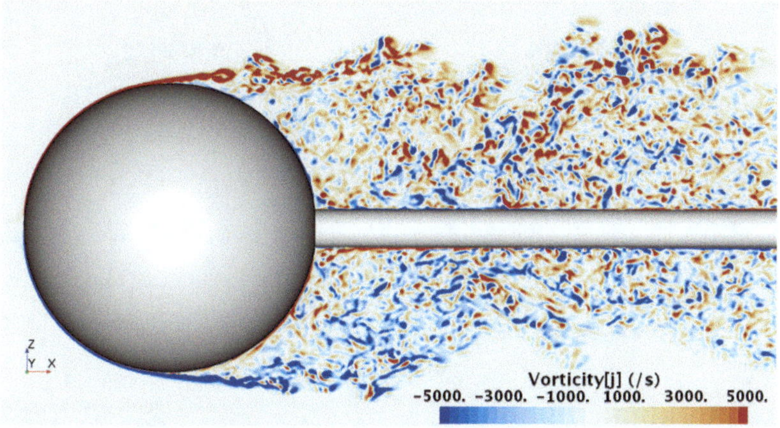

Fig. 7.20 Details of instantaneous distribution of the vorticity vector component normal to the plane $y = 0$ at Re = 50,000

the minimum is also found at a different location ($\theta \approx 70°$ rather than 90°). The pressure recovery is also only marginal: the pressure coefficient remains at the level of about -0.4 from $\theta \approx 80°$ until the rear stagnation point. This limited recovery is responsible for the high value of the drag coefficient.

As already indicated in Fig. 7.17, at the Reynolds number of 50,000 the turbulence structures (eddies) are much smaller compared to those at Re = 5,000. With a further increase in Reynolds number, these structures become smaller and smaller, thus requiring finer and finer grids. The size of resolved structures can be visualized by vorticity; Figs. 7.20 and 7.21 show contours of vorticity vector component normal to section planes.

By comparing vorticity contours in Fig. 7.20 with the grid shown in Fig. 7.7, one can see that the grid spacing imposes a limit to the size of structures that can be resolved. The effects of turbulent eddies which are smaller than the grid spacing on the resolved scales are provided by the subgrid-scale model. From Fig. 7.21 one can see that the intensity of turbulence is strongest at the edge of the wake (separated shear layer) until about one sphere diameter distance downstream of sphere center, where the flow approaches the mean reattachment location (see the mean flow pattern in Fig. 7.18). One can also see from these figures that the wake is initially nearly circular and centered around the stick (which is aligned with the mean flow direction), but with increasing distance from sphere center the wake becomes more asymmetric. The whole wake performs meandering motion around the stick holding the sphere; the stick itself does not have a large influence on this motion.

The combination of turbulent fluctuations and large-scale motion of sphere wake is what makes the RANS modeling of such flows difficult. In any cross-section downstream of sphere there is a small zone in which the flow is all the time turbulent, surrounded by a zone with intermittent turbulence, and at larger radius the flow

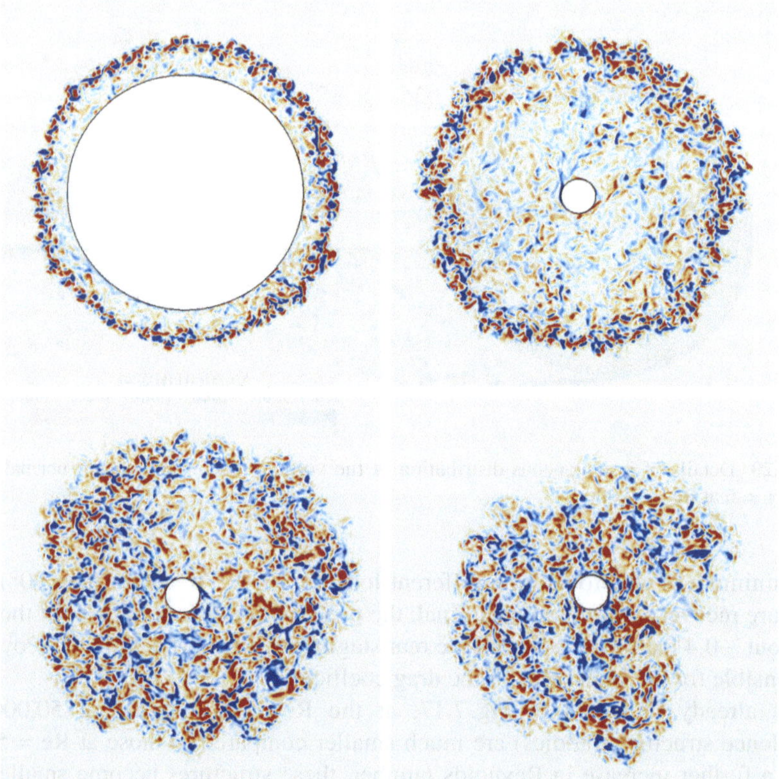

Fig. 7.21 Details of instantaneous contours of the axial component of vorticity vector in cross-sections $0.25D$ (upper left), $0.5D$ (upper right), $1D$ (lower left), and $1.5D$ (lower right) downstream of sphere center at Re = 50,000

remains laminar all the time (even though the flow is unsteady in the large part of the laminar zone, it is not turbulent). For the simulation of such flows, one should use approaches like LES which can account for intermittent turbulence zones. Simulation of turbulent flows using Reynolds-averaged Navier–Stokes equations is suitable only for flows which are statistically steady or turbulent throughout the whole solution domain; the large-scale unsteadiness can be handled as long as the flow is turbulent all the time.

Figure 7.22 shows comparison of computed mean axial velocity and rms-values of its fluctuation (time-averaged squares of axial velocity fluctuation, $\overline{u_x u_x}$) in a cross-section half a diameter behind sphere. In experiments by Bakić [4], velocity was measured along one line at constant x and variable y coordinate. In simulation, the number of samples used for time-averaging (60,000) is not sufficient to obtain perfect axi-symmetric mean values—for this the simulation would need to run much longer. Alternatively, one could perform averaging both in time and in circumferential spatial direction—this leads faster to stable mean values. However,

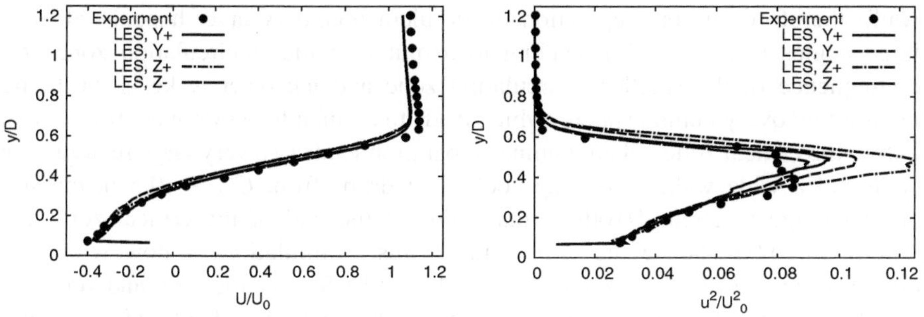

Fig. 7.22 Radial profiles of the mean axial velocity component (left) and the rms-value of its fluctuation (right) in a cross-section one diameter downstream of sphere center at Re = 50,000

because the computational grid was not axi-symmetric, such averaging is not easy to perform. Instead, time-averaged profiles are taken along 4 lines at $x = D$: increasing z coordinate at $y = 0$ (labeled Z+), decreasing z coordinate at $y = 0$ (labeled Z−), increasing y coordinate at $z = 0$ (labeled Y+), and decreasing y coordinate at $z = 0$ (labeled Y−). All 4 profiles are presented in Fig. 7.22; their vanishing discrepancy would indicate sufficiently long averaging period. In the present simulation, the mean velocity is reasonably close to the true average, while rms-values would need a significantly longer averaging time. Nevertheless, both profiles show a reasonably good agreement with experimental data. The inlet velocity with which the values are normalized were not exactly the same in experiment and simulation, as can be seen from the difference in velocities in the laminar zone away from sphere (above $y/D = 0.7$).

7.6.6 Turbulent Flow, High Reynolds Number (Re = 500,000)

With a 10 times higher Reynolds number (Re = 500,000), the recirculation zone behind sphere is much smaller and the wake is much narrower and steadier. This flow belongs to the so-called *post-critical* regime; between Re = 200,000 and 400,000 the so-called *drag crises* takes place. Up to this critical range, the flow is laminar prior to separation, which takes place at angles (measured from the front stagnation point) smaller than 90°. The shear layer becomes unstable (Kelvin-Helmholtz instability) and thus leads to a turbulent wake. Within the critical range of Reynolds numbers, the separation point moves further downstream and becomes unstable. When the separation point reaches angles around 110°, the instabilities in the shear layer after flow separation start interacting with sphere wall, leading to a turbulent boundary layer. The small recirculation zone is called "laminar separation bubble". The turbulent boundary layer can stay attached to the wall much longer than the laminar layer before separation, because of the higher kinetic energy (turbulence brings high-speed flow from more distant zones closer to the wall); therefore, the big separation of turbulent boundary layer happens at much large angles (around 135°), leading to a much smaller recirculation zone. The consequence of the smaller recirculation zone and narrower wake is the higher pressure recovery behind sphere, which leads to a much lower drag coefficient than before the critical zone, where laminar separation leads to a very large recirculation zone and a wide wake. The drag coefficient drops from $C_D \approx 0.5$ in the sub-critical range to about 1/10th of that value at the end of the critical zone, i.e., to $C_D \approx 0.05$—hence the term "drag crises". Simulation of flow within the critical zone is difficult; it requires methods like LES, very fine grid and very long simulation time, because of large-scale wake fluctuation and intermittent zones. After the critical range, RANS methods become applicable.

In this case a version of the so-called k-ε turbulence model was used to compute the flow around sphere at Re = 500,000. The solution domain is the same as for Re = 50,000 described in the previous section. In Reynolds-averaging, each variable is expressed as a sum of its mean (time-averaged) value and a fluctuation around the mean, e.g., $u_i = \overline{u}_i + u_i'$. When this decomposition is inserted into the Navier–Stokes equations and the time-averaging is applied to equations themselves, fluctuations drop out of each linear term, because by definition the time-average of a fluctuation is equal to zero. However, terms involving products of two variables lead to a new term involving the average of the product of fluctuations, which is—in a turbulent flow—not equal to zero:

$$\overline{u_i u_j} = \overline{(\overline{u}_i + u_i')(\overline{u}_j + u_j')} = \overline{u}_i \overline{u}_j + \overline{u_i' u_j'} \ . \tag{7.96}$$

The continuity equation looks the same in terms of time-averaged velocities, because all terms are linear (when the density is constant). In the momentum equations, the rate-of-change term, the pressure term, the viscous terms and the gravity term also remain unchanged, but the convection term leads to new unknowns involving products of velocity fluctuations. These unknowns are called *Reynolds-stress tensor*; because of symmetry, it has only 6 different components, see the last term on the right-hand side of Eq. (7.96). Because we only have 4 equations (continuity and 3 momentum component equations), we cannot compute the 10 unknowns (3 mean velocity vector components, mean pressure and 6 components of the Reynolds-stress tensor) without either adding six more equations or expressing the Reynolds-stress tensor components through the mean variables.

The k-ε turbulence model used here (lag EB k-ε model; see Lardeau [20], for details) introduces four additional equations that need to be solved: one for the kinetic energy of turbulence, k, one for its dissipation rate, ε, and two additional equations for variables whose meaning is less obvious. It is beyond the scope of this Chapter to explain turbulence models in detail; interested readers are directed to books by Pope [27], Wilcox [39] or Durbin and Pettersson-Reif [12], among others. It is here sufficient to note that the six components of the Reynolds-stress tensor are expressed as functions of time-averaged velocities and additional variables from turbulence model equations, which makes the problem solvable. In the so-called *eddy-viscosity* type of RANS turbulence models, to which the k-ε turbulence model used here belongs, the Reynolds stresses are expressed in terms of mean velocities and a turbulent viscosity in a similar way as was the case with subgrid-scale stresses in LES models, see Eq. (7.93). The turbulent viscosity is computed from k and ε and it can vary three or more orders of magnitude within the solution domain, thus making the equations even more non-linear and coupled and, consequently, more difficult to solve. There are many versions of eddy-viscosity RANS turbulence models, and as many models of a different kind, including those that solve transport equations for all Reynolds stresses; many produce poor results for flows around sphere, even at high Reynolds numbers. The method used here produced the best results among four models tried.

Figure 7.23 shows flow pattern and mean pressure distribution around the sphere at Re = 500,000. The flow separates from sphere surface at about 130° from front stagnation point and re-attaches about one diameter downstream from sphere center. The inlet velocity was left the same as for Re = 50,000 but the density was increased by a factor of 10 to 12 kg/m^3, because the compressibility effects would become significant if the velocity was increased by a factor of 10 compared to Re = 50,000. The grid had 6,592,424 CV; the grid spacing near sphere was 0.4375 mm and the

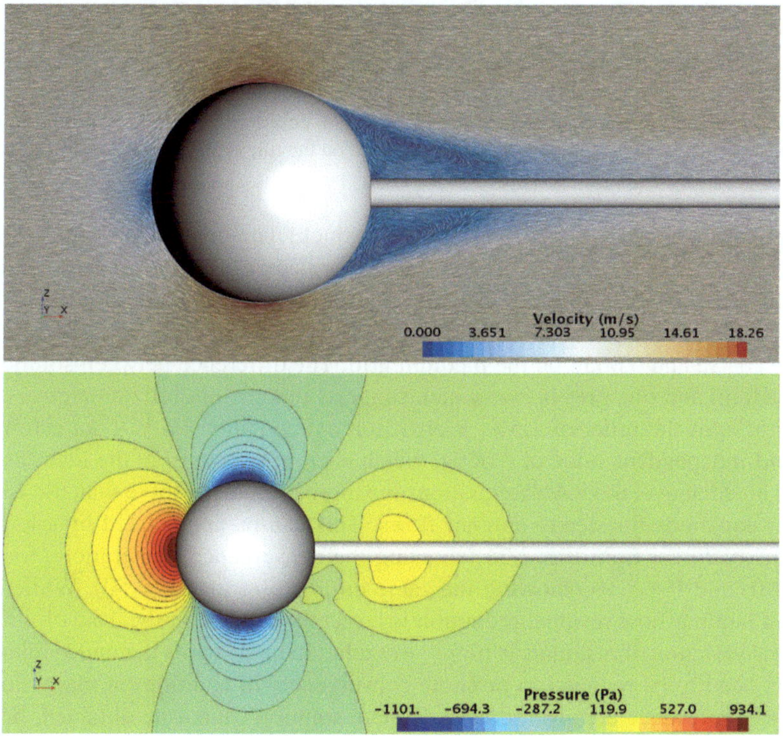

Fig. 7.23 Details of mean flow pattern (upper) and pressure distribution (lower) in sphere wake at Re = 500,000

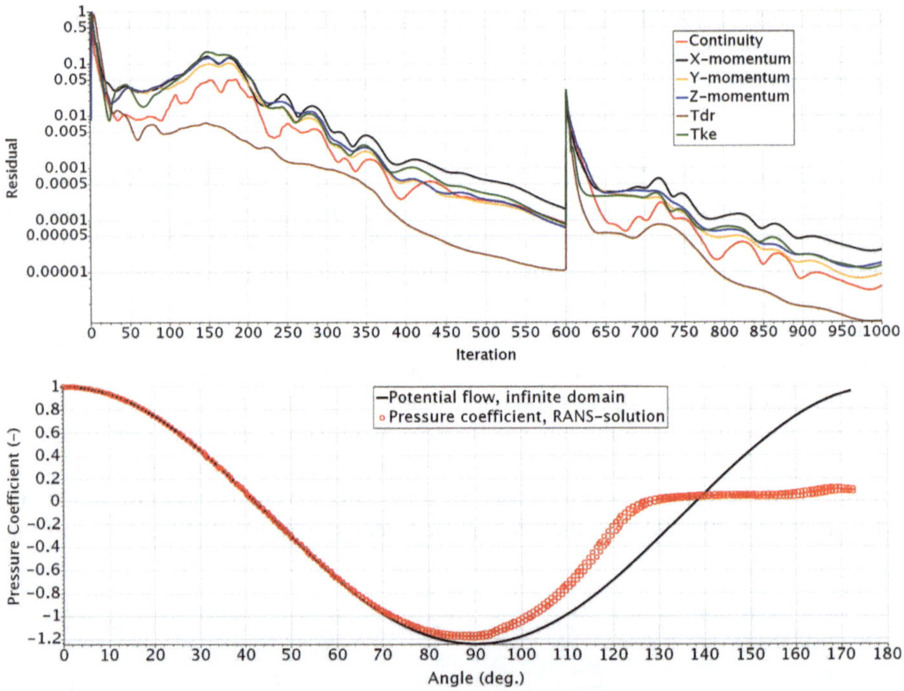

Fig. 7.24 Variation of residuals with outer iterations on coarse and medium grid (upper) and of pressure coefficient along sphere perimeter computed on the fine grid (lower), for Re = 500,000

first prism layer next to wall was 0.01 mm thick. The computed drag coefficient was $C_D = 0.0619$. Computation was also performed on two coarser grids, with grid spacing increased by a factor of 1.5 between two levels. The coarse grid had 728,923 CV and the medium one had 2,131,351 CV. The computed drag values were 0.0679 on the coarse and 0.0638 on the medium grid. The difference between solutions on the medium and fine grid is 3%, which suggests that the discretization error on the fine grid is of the order of 2.5%. Richardson extrapolation leads to an estimate of the grid-independent value of 0.0604, which is close to data from the literature.

As noted above, the RANS equations are more difficult to solve than the Navier–Stokes equations for steady laminar flows. This can be appreciated by comparing residual histories from Fig. 7.24 (showing the convergence of iterations for Re = 500,000) and Fig. 7.10 (showing the same information for Re = 50). While in the case of laminar flow residual reduction is perfectly smooth, in the case of turbulent flow the variation of residuals is highly irregular. In many cases the under-relaxation factors need to be reduced to promote convergence. In addition, in many flows of industrial relevance a converged steady-state solution cannot be obtained, because the flow in some zones exhibits inherent unsteadiness which cannot be characterized as turbulence (like vortex shedding). In such cases the computation has to be performed in unsteady mode, which leads to URANS (unsteady RANS).

It is also interesting to compare the distribution of pressure coefficient around sphere computed for Re = 50,000 (Fig. 7.16) and Re = 500,000 (Fig. 7.24). The solution domain—and thus the blockage ratio—is the same as for Re = 50,000. However, for the 10 times higher Re, the blockage does not play an important role; the pressure coefficient at the front stagnation point is now equal to 1, as expected. The computed values are shown for the upper and lower contour of sphere surface; they are not identical, because the flow is not absolutely symmetrical, although simulation on the finest grid was run for many iterations in order to reduce residuals to very low values. The difference between the two sides is not very large compared to the difference between the two Reynolds numbers. At a high Reynolds number, the pressure coefficient is almost identical to the analytical solution from potential theory up to $\theta = 70°$. The minimum value is also obtained at almost the same location, and the minimum value from simulation is only marginally higher than from theory. On the back side of the sphere, the distributions differ significantly. In turbulent flow the recovery is much faster than in potential theory, but the pressure coefficient does not increase above zero; it stays at around this level from $\theta = 70°$ onward. Compared to the flow at Re = 50,000 the recovery is much stronger, which is the reason why the drag coefficient is almost an order of magnitude lower.

7.6.7 *Supersonic Turbulent Flow (Re = 5,000,000)*

At higher Reynolds numbers compressibility effects need to be taken into account. Here the flow around sphere at a Reynolds number equal to 5 million and a free-stream Mach number equal to 1.78 is considered. In this case, sphere diameter was increased by a factor of 2 to 122.8 mm; fluid is air at the free-stream temperature of 300 K and is treated as an ideal gas. The solution domain is a box extending 2 m away from sphere in all directions. The computation was performed as two-dimensional axi-symmetric. The standard version of the k-ε turbulence model was used, and in addition to the continuity, momentum, and energy equations, two additional equations for turbulence model are solved (for k and ε).

Because the flow upstream of sphere is supersonic, no information about the presence of obstacle can propagate against the flow direction. Therefore, a bow shock has to form ahead of sphere, as shown in Fig. 7.25; its distance from sphere depends on the Mach number in the free stream (the higher the Mach number, the closer will the shock be to the sphere surface). At the shock, the flow changes direction because behind the shock the flow is subsonic and thus feels the presence of the body ahead of it. As the flow accelerates over the front part of sphere surface, the Mach number increases again to the supersonic range, with maximum values reaching almost 2.8 (see Fig. 7.25). At around $\theta \approx 120°$, the flow separates to create a recirculation zone which extends to about $1.35D$ behind sphere. Through another shock at the separation line, the Mach number at sphere surface drops from the maximum value around 2.8 to below 0.01.

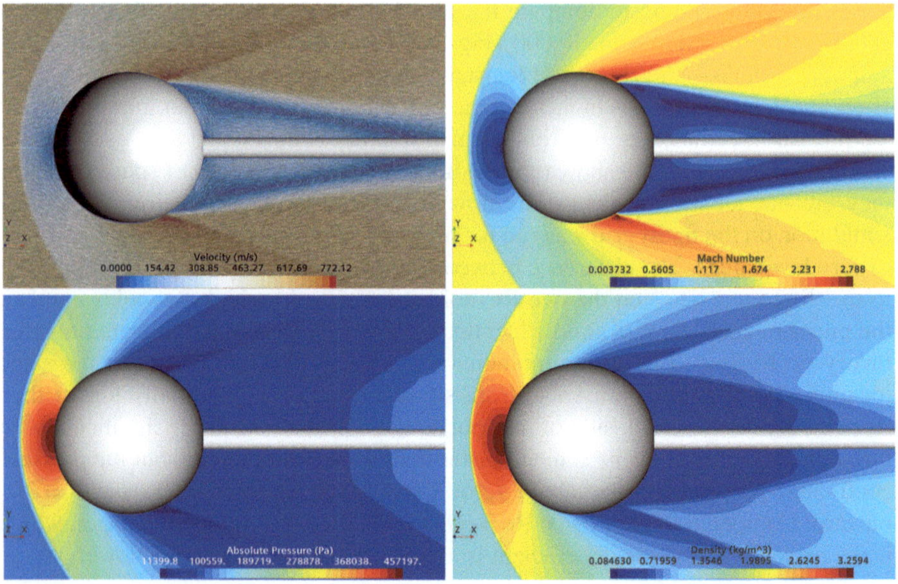

Fig. 7.25 Details of flow pattern (upper left), Mach number (upper right), absolute pressure (lower left) and density (lower right) around sphere and in its wake at Re = 5,000,000

The flow in the recirculation zone is practically incompressible, with minimum values of Mach number around 0.004. Thus, the solution method must cope with highly compressible flow zones, including shocks and Mach numbers close to 3, as well as with incompressible flow regions with Mach numbers below 0.01. The extension of SIMPLE algorithm to compressible flows copes with these situations without any problems (turbulence modeling is a much large challenge than compressibility effects).

Atmospheric pressure was specified at the outlet; at the front stagnation point, the maximum pressure of 3.57 bar above atmospheric is predicted, see Fig. 7.25. The minimum pressure values are found at the separation line ($\theta \approx 120°$). Within the largest part of the recirculation zone, pressure is almost constant; it increases again around reattachment point, as the flow converges from all sides towards the supporting rod. Density variation, which is also shown in Fig. 7.25, looks similar to pressure distribution; the maximum value of density is also obtained at the front stagnation point (ca. 3.26 kg/m^3—almost three times higher than the free-stream density), while the minimum values in the wake are about 14 times lower than in the free stream (ca. 0.085 kg/m^3).

As shown in Fig. 7.26, turbulence plays an important role only at the rear side of the sphere, especially within the recirculation zone. The maximum value of kinetic energy of turbulence is found in the shear layer close to the reattachment point. By assuming isotropic turbulence in the zone near reattachment location (as the LES at

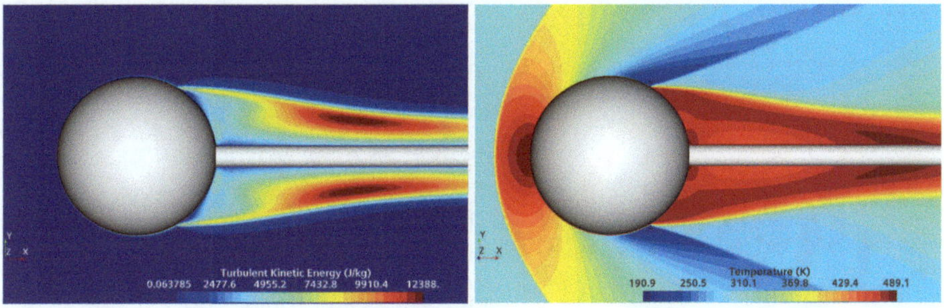

Fig. 7.26 Details of turbulent kinetic energy (left) and temperature (right) around sphere and in its wake at Re = 5,000,000

Re = 50,000 suggests), one can estimate that the turbulence intensity in this zone (defined as $\sqrt{\overline{u_i' u_i'}}/U_\infty$) is around 20%. The maximum values of turbulent viscosity are obtained further downstream of reattachment location, amounting to ca. 7,000 times the viscosity of air.

Temperature also varies over a wide range within the flow. From 300 K in the free stream ahead of bow shock, it increases to over 480 K at the front stagnation point, see Fig. 7.26. As the flow accelerates with increasing θ along sphere surface, temperature is decreasing and reaches the minimum value of around 190 K shortly before the separation line, where the Mach number is at its maximum. Within the recirculation zone temperature is again high, over 450 K.

7.6.8 Compressible Flow Around Sphere Oscillating in Water

The final test case involves sphere oscillating sinusoidally in x-direction with a frequency of 6,000 Hz and amplitude of 0.005 mm. In this case there was no support rod attached to the sphere, and the solution domain was a cube with edge length of 2 m, with sphere positioned in its center; sphere diameter was $D = 61.4$ mm. The whole grid moves with sphere. The fluid is water and it is considered as compressible, using the Tait equation to express density as a function of pressure. The fluid far away from sphere is assumed to be at rest. The time step was 0.4 μs, which means that there were 416 time steps per oscillation period. Although the amplitude of oscillation is very small (only 0.008% of sphere diameter), due to high frequency, large pressure oscillations occur. Thus, sphere motion results in pressure waves traveling away from sphere with the speed of sound, which is in water close to 1,500 m/s; the wavelength of these acoustic pressure waves is approximately 0.25 m. In order to avoid reflections from solution domain boundaries, the velocity components were forced to zero over a distance of 0.5 m from solution domain boundaries, with an exponential increase in forcing strength from 0 to the maximum

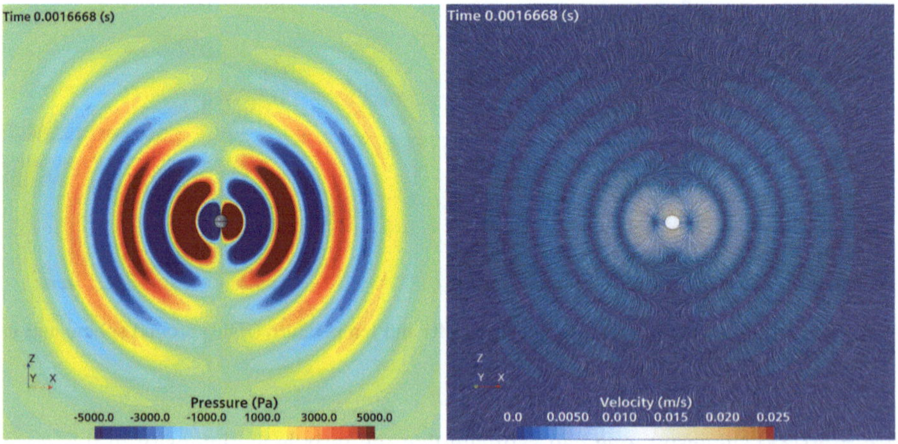

Fig. 7.27 Details of pressure (left) and velocity distribution (right) around sphere oscillating in water at 6,000 Hz with amplitude of 0.005 mm

value at the boundary. This is achieved by adding a source term to discretized momentum equations of the form:

$$q_i = -\gamma \left[\rho V (u_i - u_i^*)\right]_{\mathrm{P}} , \tag{7.97}$$

where γ is forcing strength (its dimension is s^{-1}) and $u_i^* = 0$ is the reference solution towards which the velocities are forced; V is the CV volume.

The optimum value of the maximum forcing strength at the boundary was determined using the theory presented by Perić [25]; for acoustic waves with frequency of 6,000 Hz, the optimum is found to be around 88,880 s^{-1}. This leads to pressure waves being damped as they propagate towards solution domain boundaries, without causing reflections which would travel back into the solution domain and interfere with waves emitted at the oscillating sphere. Figure 7.27 shows pressure and velocity fields after 10 oscillation periods; no disturbances resulting from reflection of pressure waves at boundaries are visible.

The maximum and minimum pressure on sphere surface during oscillation are 1.32 bar above and 1.32 bar below far-field pressure, respectively. Assuming that the far-field pressure is atmospheric (i.e., equal to 1 bar), the absolute pressure would be negative (with a minimum value of -0.32 bar) over some part of the oscillation period. During this time cavitation would take place, with water evaporating and leading to a vapor cavity attached to sphere. This has not been simulated here, but it is important to note that even such small-amplitude oscillation can lead to such effects if the frequency is high enough. The maximum and minimum pressure increase approximately linearly with increasing amplitude of oscillation, as does the force acting on the sphere. For the amplitude of 0.005 mm, the maximum force of 520 N is obtained. A 10 times larger amplitude of 0.05 mm would lead to pressures between 13 and -13 bar and the force on sphere up to 5,200 N.

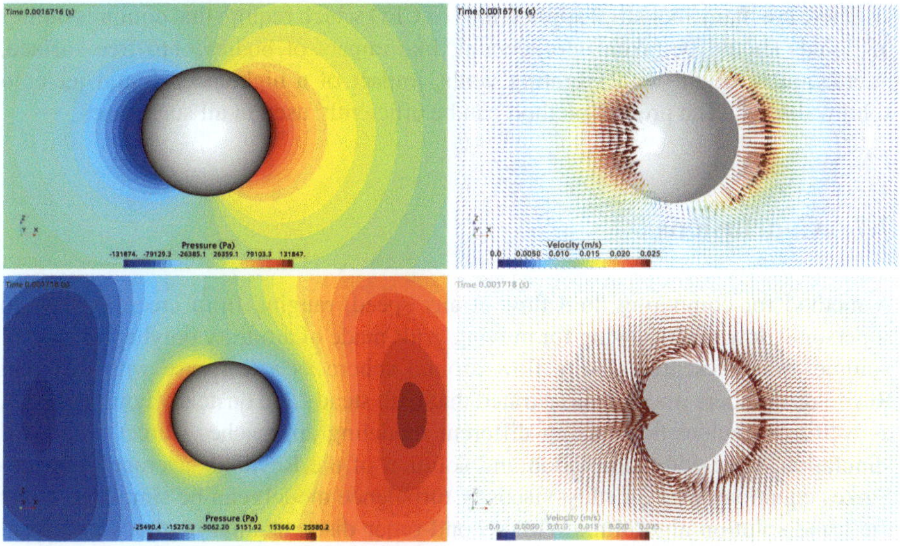

Fig. 7.28 Details of pressure (left) and velocity distribution (right) around sphere oscillating in water at 6,000 Hz with amplitude of 0.005 mm: $t = 0.0016716$ s (upper) and $t = 0.001718$ s (lower)

Figure 7.28 shows details of pressure and velocity field around sphere at two characteristic times which are quarter of oscillation period apart. The upper part of the figure shows the situation when the difference between maximum and minimum pressure on sphere surface is the largest—almost 2.64 bar. The sphere is accelerating to the right and displacing liquid ahead of it, thus creating high pressure: liquid is moving away because of resulting pressure gradient. On the other side, sphere wall is moving away from the surrounding liquid, thus creating under-pressure which is sucking liquid so that it can fill the space made free by departing wall. The lower part of the figure shows situation when the sphere starts decelerating: due to inertia, liquid is now on the right-hand side moving away faster than sphere wall, starting to create under-pressure on this side of sphere. On the other side, liquid approaches sphere surface faster that the wall moves away, thus leading to increasing pressure—the adverse pressure gradient should slow down the liquid motion. At around this time, the pressure difference between the right-hand side and left-hand side of sphere surface is the lowest—around 0.51 bar.

If the flow under the above conditions was computed as incompressible, there would be no pressure waves like those shown in Fig. 7.27: pressure contours around sphere would all the time look like those in the upper part of Fig. 7.28, only the pressure amplitudes would be varying and after half a period the blue and red would quickly swap the sides. Few diameters away from sphere pressure is almost constant all the time. The largest difference between maximum and minimum pressure on sphere surface is somewhat smaller than when water compressibility is taken into account (ca. 2.24 bar), but the lowest difference is nearly zero at the time of zero sphere acceleration.

There are many industrial applications of CFD which require that compressibility of liquids should be taken into account, like impact of bodies onto free surface, interaction of liquid with vibrating walls, impact of a high-speed liquid jet onto solid walls, etc. The present method can be efficiently used in all such cases.

7.7 Conclusions

A method for computing fluid flow at any speed, ranging from creeping flow to supersonic flow, and allowing for moving grids made of arbitrary polyhedral control volumes was presented. Using flow around sphere as a test case with different Reynolds numbers it was demonstrated that both steady and unsteady laminar flows, as well as turbulent flows using different modeling approaches (direct numerical simulation, large-eddy simulation and solution of the Reynolds-averaged Navier–Stokes equations with an eddy-viscosity turbulence model) can be simulated using the same solution algorithm. With second-order discretization methods and local grid refinement, solutions with accuracy sufficient for most engineering applications can be obtained with an acceptable computing effort. This is the main reason why discretization and solution methods presented here are implemented in most commercial and public CFD codes used in engineering applications.

Acknowledgements The author acknowledges contributions to the material presented here by J. H. Ferziger, R. L. Street, I. Demirdžić, S. Muzaferija, E. Schreck and many other present and past co-workers.

Bibliography

1. Elmar Achenbach. Vortex shedding from spheres. *J. Fluid Mech.*, **62**:209–221, 1974.
2. D. A. Anderson, J. C. Tannehill, and R. H. Pletcher. *Computational fluid mechanics and heat transfer*. Hemisphere, New York, 1984.
3. S. Armfield, N. Williamson, M. Kirkpatrick, and R. Street. A divergence free fractional-step method for the Navier-Stokes equations on non-staggered grids. *ANZIAM J.*, **51 (E)**:C654–C667, 2010.
4. V. Bakić. *Experimental investigation of turbulent flows around a sphere*. PhD dissertation, Technical University of Hamburg-Harburg, Germany, 2002.
5. L. S. Caretto, A. D. Gosman, S. V. Patankar, and D. B. Spalding. Two calculation procedures for steady three-dimensional flows with recirculation. In *Proc. Third Int. Conf. Numer. Methods Fluid Dyn.*, Paris, 1972.
6. H. Choi and P. Moin. Effects of the computational time step on numerical solutions of turbulent flow. *J. Comput. Phys.*, **113**:1–4, 1994.
7. I. Demirdžić. On the discretization of the diffusion term in finite-volume continuum mechanics. *Numer. Heat Transfer*, **68**:1–10, 2015.
8. I. Demirdžić, Ž. Lilek, and M. Perić. A collocated finite volume method for predicting flows at all speeds. *Int. J. Numer. Methods Fluids*, **16**:1029–1050, 1993.

9. I. Demirdžić and S. Muzaferija. Numerical method for coupled fluid flow, heat transfer and stress analysis using unstructured moving meshes with cells of arbitrary topology. *Comput. Methods Appl. MechĖngrg.*, **125**:235–255, 1995.
10. I. Demirdžić and M. Perić. Space conservation law in finite volume calculations of fluid flow. *Int. J. Numer. Methods Fluids*, **8**:1037–1050, 1988.
11. I. Demirdžić and M. Perić. Finite volume method for prediction of fluid flow in arbitrarily shaped domains with moving boundaries. *Int. J. Numer. Methods Fluids*, **10**:771–790, 1990.
12. P. A. Durbin and B. A. Pettersson Reif. *Statistical theory and modeling for turbulent flows*. Wiley, Chichester, England, 2nd edition, 2011.
13. J. H. Ferziger, M. Perić, and R. L. Street. *Computational methods for fluid dynamics*. Springer, Cham, Switzerland, 4th edition, 2019.
14. E. Garnier, N. Adams, and P. Sagaut. *Large-eddy simulation for compressible flows*. Springer, Berlin, 2009.
15. P. M. Gresho and R. L. Sani. On pressure boundary conditions for the incompressible Navier-Stokes equations. *Int. J. Numer. Methods Fluids*, **7**:11–46, 1990.
16. C. Hirsch. *Numerical computation of internal and external flows*, volume I. Butterworth-Heinemann (Elsevier), Burlington, MA, 2nd edition, 2007.
17. E. A. Hylla. *Eine Immersed Boundary Methode zur Simulation von Strömungen in komplexen und bewegten Geometrien*. PhD dissertation, Technische Universität Berlin, Berlin, Germany, 2013.
18. R. I. Issa. Solution of implicitly discretized fluid flow equations by operator-splitting. *J. Comput. Phys.*, **62**:40–65, 1986.
19. K. C. Karki and S. V. Patankar. Pressure based calculation procedure for viscous flows at all speeds in arbitrary configurations. *AIAA J.*, **27**:1167–1174, 1989.
20. S. Lardeau. Consistent strain/stress lag eddy-viscosity model for hybrid RANS/LES. In Y. Hoarau, S. H. Peng, D. Schwamborn, and A. Revell, editors, *Progress in Hybrid RANS-LES Modelling*, pages 39–51. Springer, Cham, Switzerland, 2018.
21. Ž. Lilek, S. Muzaferija, and M. Perić. Efficiency and accuracy aspects of a full-multigrid SIMPLE algorithm for three-dimensional flows. *Numer. Heat Transfer, Part B*, **31**:23–42, 1997a.
22. S. V. Patankar. *Numerical heat transfer and fluid flow*. McGraw-Hill, New York, 1980.
23. N. Peller. *Numerische Simulation turbulenter Strömungen mit Immersed Boundaries*. PhD dissertation, Technische Universität München, Fachgebiet Hydromechanik, Mitteilungen, 2010.
24. M. Perić. *A finite volume method for the prediction of three-dimensional fluid flow in complex ducts*. PhD dissertation, Imperial College, London, 1985.
25. R. Perić. *Minimizing undesired wave reflection at the domain boundaries in flow simulations with forcing zones*. PhD dissertation, Technische Universität Hamburg, Germany, 2019.
26. C. S. Peskin. The immersed boundary method. *Acta Numerica*, **11**:479–517, 2002.
27. S. B. Pope. *Turbulent flows*. Cambridge Univ. Press, Cambridge, 2000.
28. G. D. Raithby and G. E. Schneider. Numerical solution of problems in incompressible fluid flow: treatment of the velocity-pressure coupling. *Numer. Heat Transfer*, **2**:417–440, 1979.
29. C. M. Rhie and W. L. Chow. A numerical study of the turbulent flow past an isolated airfoil with trailing edge separation. *AIAA J.*, **21**:1525–1532, 1983.
30. P. Sagaut. *Large-eddy simulation for incompressible flows: An introduction*. Springer, Berlin, 3 edition, 2006.
31. V. Seidl. *Entwicklung und Anwendung eines parallelen Finite-Volumen-Verfahrens zur Strömungssimulation auf unstrukturierten Gittern mit lokaler Verfeinerung*. PhD dissertation, University of Hamburg, Germany, 1997.
32. V. Seidl, S. Muzaferija, and M. Perić. Parallel DNS with local grid refinement. *App. Sci. Res.*, **59**:379–394, 1998.
33. J. Smagorinsky. General circulation experiments with the primitive equations. Part I: The basic experiment. *Monthly Weather Rev.*, **91**:99–164, 1963.

34. Th. Sonar. On the construction of essentially non-oscillatory finite volume approximations to hyperbolic conservation laws on general triangulations: Polynomial recovery, accuracy and stencil selection. *Comput. Methods Appl. Mech. Engrg.*, **140**:157–181, 1997.
35. J. P. Van Doormal and G. D. Raithby. Enhancements of the SIMPLE method for predicting incompressible fluid flows. *Numer. Heat Transfer*, **7**:147–163, 1984.
36. J. P. Van Doormal, G. D. Raithby, and B. H. McDonald. The segregated approach to predicting viscous compressible fluid flows. *ASME J. Turbomachinery*, **109**:268–277, 1987.
37. J. M. Weiss, J. P. Maruszewski, and W. A. Smith. Implicit solution of preconditioned Navier-Stokes equations using algebraic multigrid. *AIAA J.*, **37**:29–36, 1999.
38. J. M. Weiss and W. A. Smith. Preconditioning applied to variable and constant density flows. *AIAA J.*, **33**:2050–2057, 1995.
39. D. C. Wilcox. *Turbulence modeling for CFD*. DCW Industries, Inc., La Cañada, CA, 3 edition, 2006.

MIX
Papier aus verantwortungsvollen Quellen
Paper from responsible sources
FSC® C105338

If you have any concerns about our products,
you can contact us on
ProductSafety@springernature.com

In case Publisher is established outside the EU,
the EU authorized representative is:
Springer Nature Customer Service Center GmbH
Europaplatz 3, 69115 Heidelberg, Germany

Printed by Libri Plureos GmbH
in Hamburg, Germany